Nutrients in Foods

GILBERT A. LEVEILLE • MARY ELLEN ZABIK • KAREN J. MORGAN

The Nutrition Guild • Cambridge, Massachusetts • 1983

Library of Congress Cataloging in Publication Data
Main entry under title:

Nutrients in foods.

1. Food—Composition—Tables. I. Leveille, Gilbert A., 1934- . [DNLM: 1. Food analysis—Tables. 2. Nutritive value—Tables. QU 16 N9755]
TX5551.N76 1983 641.1'0212 83-13448
ISBN 0-938550-00-4

Book and Cover Design: Richard C. Bartlett
Cover: "Autumn Fruits," Currier and Ives lithograph. Courtesy of the Print Collection, Heritage Plantation of Sandwich, MA. Photograph by William A. Newman.

Nutrient Table typesetting by Science Press, Inc.
Printed in the United States of America at Braun-Brumfield, Inc.

Published by: The Nutrition Guild
28 Hurlbut Street
Cambridge, MA 02138

ACKNOWLEDGEMENTS

The authors wish to thank the faculty, staff, and graduate students of the Department of Food Science and Human Nutrition, Michigan State University, and the staff and graduate students of the Department of Human Nutrition, Foods and Food Systems Management, University of Missouri, Columbia, for their help in preparing the materials used in this book. We further wish to thank David H. Agee, Ph.D., Harvard University, and Lawrence Kushi, D.Sc., Department of Nutrition, Harvard School of Public Health, for preparing the introduction to this book.

Finally, the authors want to thank Ronald M. Deutsch for his role in urging them to make these nutrient data available in book form.

The data come from the Michigan State University Nutrient Data Bank, updated as of 1982. The numerous sources of data in the MSU data base are discussed elsewhere in this introduction, however, we wish to acknowledge the food manufacturers who have supplied us with nutrient data for their products. These data are not generally available, but as a public service, many food companies responded to our desire to make available nutritional data on food as it is consumed by the general public. A list of these contributing food companies is found following the references.

Table of Contents

Nutrient Table Column Headings
Table of Contents

List of Tables

Introduction

ORGANIZATION

Nutrients in Foods

NUTRIENTS IN FOODS presents the nutrient composition of more than 2700 foods showing almost all dietary nutrient factors known to be essential for growth and maintenance of human beings. Omitted from the nutrient composition table are only those essential nutrient factors that are an integral part of some factor already listed, (e.g. cobalt in vitamin B-12), or factors contained in foods which are not known to be essential for humans, such as choline or silicon.

In addition to raw nutrient values, further information is listed in *NUTRIENTS IN FOODS* to facilitate determining nutrient availability or adequacy. Electrolytes are given both in milligrams (mg) and as milliequivalents (meq), vitamin E values are expressed in International Units (IU) and in milligrams of alpha and other tocopherol, and vitamin A is given in International Units and Retinol Equivalents (RE) of both preformed vitamin A and beta-carotene. Both crude and dietary fiber values are given, and the polyunsaturated to saturated fatty acid ratio (P/S ratio) is calculated. A factor to convert the nutrient values of the food from the given portion size into 100 gram (g) portions is given, allowing easy comparison of the nutrient composition of any two food items. As a guide to dietary planning, the percent of U.S.RDA is shown for the appropriate nutrients.

ENERGY-PRODUCING NUTRIENTS

Nutrients are loosely grouped in the table by type, starting with the energy-producing nutrients: carbohydrate, protein, and fat. These, along with water, provide the bulk in weight and volume of any diet and are the source for most of the energy humans require for maintenance and activity. Values for some components of these macronutrients are also included, such as total sugar and the non-digestible portion of carbohydrate given as both crude and dietary fiber. The essential amino acids, those that cannot be made by the body or cannot be made in amounts adequate for growth, are listed. In addition to total fat content, lipid values are given for saturated, monounsaturated and polyunsaturated fatty acids, cholesterol, and the major essential fatty acid, linoleic acid.

VITAMINS

The next major class of nutrients is the vitamins. These are organic substances, necessary in small amounts for proper functioning, but not synthesized by the body at all or not synthesized in adequate amounts. Vitamins act as coenzymes and in other ways in many metabolic reactions. Although not in themselves energy sources, several are required in biochemical reactions that produce energy. Vitamins may be divided into two major classes: those that are fat soluble and those that are water soluble.

The water-soluble vitamins—the B-complex vitamins and vitamin C—are listed first in the table. These vitamins vary greatly in chemical structure, are generally less stable than the fat-soluble vitamins, and are not easily stored in the body. They are more readily affected by processing, storage, and cooking methods; thus the nutrient value of these vitamins can be quite variable from sample to sample of a given food. These vitamins generally function as coenzymes in the metabolism and release of energy from carbohydrate, fat, and protein.

The fat-soluble vitamins, generally more stable than the water-soluble vitamins, consist solely of carbon, hydrogen, and oxygen, and they are absorbed, stored, and excreted in similar fashion. As they can be stored in liver and fatty tissue, daily intake is not required, and excess intake can lead to toxicity. These vitamins have traditionally been expressed in units of activity, International Units (IU), which represent the amount of the vitamin which will produce a specific increment of change in the nutritional health and growth of a laboratory rat. Vitamin A values are now also expressed in retinol equivalents (RE), retinol being the physiologically active form of vitamin A in humans and other animals.

MINERALS

The last category of nutrients in the table, minerals, can be classified into the major minerals, those requiring daily intakes of more than 100 mg, and the trace elements, those minerals requiring maximum amounts of at most a few milligrams per day.

Of the seven major minerals, sulphur is so universal in foods that it is generally not measured and it is not included in food composition tables.

MAJOR MINERALS

Calcium	Potassium
Chlorine	Sodium
Magnesium	Sulphur
Phosphorus	

TRACE MINERALS

Chromium	Cobalt
Copper	Fluorine
Iodine	Molybdenum
Iron	Nickel
Manganese	Silicon
Selenium	Tin
Zinc	Vanadium

Sodium, potassium, and chlorine function primarily as electrolytes in the body. Calcium, phosphorus, and magnesium are stored in bone and teeth while being involved in many important metabolic reactions.

Of the fourteen beneficial trace minerals normally found in the human body, nickel, silicon, tin, and vanadium, have not been definitely established as essential nutrients, and cobalt is a component of vitamin B-12. The dietary requirement and physiological function of iodine, iron, and zinc are well understood. The remaining essential trace minerals (chromium, copper, magnesium, manganese, and selenium) are known to be important components of various enzymes or other compounds, but their dietary requirement is unclear. Trace minerals have been compared to vitamins because they often function as cofactors that are a part of enzymes, similar to the coenzyme function of the B-complex vitamins, and their daily requirement, however important, is very small.

Contaminent trace minerals, such as lead, are not considered in this book.

HOW TO USE *NUTRIENTS IN FOODS*

Organization of the *NUTRIENTS IN FOODS* Table

Food composition handbooks typically list 15 to 30 nutrients for two to three thousand foods, with supplementary tables for a few of the less commonly referenced nutrients. *NUTRIENTS IN FOODS* displays 62 factors for more than 2,700 selected foods in a single, unified table which runs across two pages.

Food names are listed alphabetically in the first column of the two-page spread and numerical values are in two horizontal rows across the pages to the right. (see Table 1, Sample Pages 150-151.)

The names of nutrients and nutrient factors appear in slanted pairs at the top of each column and are arranged from left to right in traditional order, with energy-producing nutrients and their components first, followed by the water-soluble vitamins, the fat-soluble vitamins, the major minerals and electrolytes, and finally the trace minerals. Each food listing has 31 columns of nutrient data with double headings. The left of each pair of slanted headings identifies the upper of the two numbers, or number pairs in the case of nutrients with a % U.S.RDA (percent of United States Recommended Daily Allowances). The right heading identifies the lower of the numerical values.

To aid the eye in moving across the page, each food listing is separated from the one above and below by a hairline rule with the breaks in the rule dividing the numerical data visually into vertical "columns." There are eighteen food listings per page, labelled A through S, with the letters appearing in the far left and far right margins of each two-page spread.

As a further aid to visual comprehension, the first relevant defining words of a new food or food category are capitalized and the listing preceeded by a heavy rule in place of the hairline rule.

As this system of heavy lines is designed to be a visual assist rather than an universal scheme for categorizing foods, the "rule" for dividing a food category may differ from food type to food type. For example, the 70 BEEF listings (pp 32-40) are split into several categories, whereas the 18 VEAL listings (pp 276-280) are so few in number that the same degree of division by heavy lines might serve to confuse rather than help in using these tables.

At the top left and top right of each two-page spread are identical running heads which repeat the first and last food listing on that two-page spread. Use these as you would those in a dictionary.

Foods are arranged in a single alphabetical list starting with

ALLSPICE, ground

and ending with

ZUCCHINI, home recipe, sticks.

Foods are not organized by major category, such as MEATS, FRUITS, VEGETABLES, etc, although some lower-level categorizing is done for convenience. For example:

Apple crisp cookies are found under *COOKIES, apple crisp.*
Tomato soup is found under *SOUP, tomato.*

The user who does not find a food in strict alphabetical order based on a food type should consult the logical category. Examples of primary categories in *NUTRIENTS IN FOOD* are:

BABY FOOD
BEVERAGE
CAKE
CANDY
CANDY BAR
CEREAL
CHEESE
COOKIE
CRACKER
FLOUR
GRAVY
HORS D'OUEVRE
JAM/JELLY
JUICE
LUNCHEON MEAT
MEAT SUBSTITUTE
MILK
OIL
ORGAN MEATS
PIE
PUDDING
SAUCE
SAUSAGE
SCHOOL LUNCH
SNACK FOOD
SOUP
STEW
SYRUP
T.V.DINNER

Fast food chain offerings are found under the name of the company: ARTHUR TREACHER, MACDONALD'S, PIZZA HUT, TACO BELL.

Trademarked commercial foods have been denoted by capitalizing the initial letters of the trademark word or words rather than using the sign TM . Manufacturer names, where appropriate, are at the end of the food name, separated from it by a hyphen.

Sources of Data in *NUTRIENTS IN FOODS*

Nutrient information in this book is from the Michigan State University Nutrient Data Bank and was derived from a wide variety of sources. Much data is from United States Department of Agriculture (USDA) compilations, including the updates of *USDA Handbook #8* through *8-7, Composition of Foods; Sausages and Luncheon Meats, Raw, Processed, and Prepared* (Richardson, et al., 1980).

It should be noted that USDA data is not identified as to source. In the case of prepared foods, USDA values were frequently derived by analyzing a combination of similar preparations from a variety of manufacturers resulting in a putative "average" value for that particular item. Analyses of combination dishes from sources other than the USDA are generally identified as househould preparations ("home recipe"), or commercial preparations (giving the name of the manufacturer).

Home recipes come primarily from *The Better Homes and Gardens Cookbook* (Better Homes and Gardens, l976). In a few instances *The Joy of Cooking* (Rombauer, 1974) was the source.

Much proprietary food manufacturer data not available from the USDA has been supplied directly by commercial companies (see acknowledgements), with the balance coming from published reports and other research resources (see bibliography). The Michigan State University Nutrient Data Bank is continually updated so the nutrient data in this edition were current shortly before the book went to print.

How to Find Nutrient Data in *NUTRIENTS IN FOODS*

Use the sample two-page spread (pp 150-151) in Table 1 to familiarize yourself with the organization and use of the nutrient tables in this book.

1. Locate food items by referring to the running heads at the top of the pages.
2. Once you have located your desired food (for this example, 150-B, *HAZELNUTS, shelled)* it may help to place a straightedge or ruler just under or just above the two-page food listing. Use the margin letters A-S to position the ruler or to help the eye track across the page.

TABLE 1
Sample Pages 150–151

150 HAMBURGER HELPER, lasagne, prep / ICE CREAM bar, chocolate coated *Nutrients in Foods*

	FOOD	Portion	Weight in grams Conversion for 100 g	Kilocalories H_2O g	Total carbohydrate g Total fats g	Crude fiber g Dietary fiber g	Total protein g Total sugar g	% USRDA	Arginine mg Histidine mg	Isoleucine mg Leucine mg	Lysine mg Methionine mg	Phenylalanine mg Threonine mg	Valine mg Tryptophan mg	Cystine mg Tyrosine mg	Polyunsat. fatty acids g Monounsat. fatty acids g	Saturated fatty acids g P/S ratio	Linoleic acid g Cholesterol mg	Thiamin mg Ascorbic acid mg	% USRDA
A	Hamburger Helper,lasagne,prep	1 serving	174 *0.57*	336 100.3	32.0 14.2	unk. 0(a)	19.8 0.0	**35**	1016 537	875 1370	1462 415	688 738	928 196	201 568	1.82 5.93	6.19 0.29	0.39 61	0.15 4	**10** **7**
B	HAZELNUTS/filberts,shelled	1/4 C	34 *2.96*	214 2.0	5.6 21.1	1.01 unk.	4.3 0.0	**7**	737 97	288 317	141 47	182 140	316 71	56 147	2.33 16.83	1.56 1.50	2.23 0	0.16 1	**10** **2**
C	HERRING,canned,plain,solids & liquid	3-1/2x2x3/4" piece	75 *1.33*	156 47.2	0.0 10.2	0.00 0(a)	14.9 0.0	**33**	946 458	788 1316	1345 433	552 767	791 160	182 578	unk. unk.	1.93 unk.	1.93 73	unk. unk.	**unk.** **unk.**
D	herring,canned,w/tomato sauce	1.7" long + 1 Tbsp sauce	55 *1.82*	97 36.7	2.0 5.8	0.00 0(a)	8.7 0.0	**19**	551 267	443 652	756 252	322 373	460 87	106 336	unk. unk.	1.10 unk.	1.10 53	unk. unk.	**unk.** **unk.**
E	herring,pickled,pieces	1-3/4x7/8x 3/4"	15 *6.67*	33 8.9	0.0 2.3	0.00 0(a)	3.1 0.0	**7**	194 94	156 229	266 89	113 132	162 31	37 118	unk. unk.	0.43 unk.	0.43 13	unk. unk.	**unk.** **unk.**
F	herring,pickled,whole	1.7" long	50 *2.00*	112 29.7	0.0 7.5	0.00 0(a)	10.2 0.0	**23**	646 313	520 765	888 296	378 439	541 102	124 395	unk. unk.	1.43 unk.	1.43 43	unk. unk.	**unk.** **unk.**
G	herring,smoked kippered,filleted	1 sm 2-3/8x 1-3/8x1/4"	20 *5.00*	42 12.2	0.0 2.6	0.00 0(a)	4.4 0.0	**10**	281 136	234 391	400 129	164 228	235 48	54 172	unk. unk.	0.48 unk.	0.48 17	unk. unk.	**unk.** **unk.**
H	herring,smoked kippered,filleted	1 med 4-3/8x 1-3/8x1/4"	40 *2.50*	84 24.4	0.0 5.2	0.00 0(a)	8.9 0.0	**20**	562 273	469 783	800 258	329 456	470 95	108 344	unk. unk.	0.96 unk.	0.96 34	unk. unk.	**unk.** **unk.**
I	HICKORY NUTS	1/4 C	16 *6.25*	108 0.5	2.0 11.0	0.30 unk.	2.1 0.0	**3**	unk. unk.	unk. unk.	unk. unk.	unk. unk.	unk. unk.	unk. unk.	2.03 7.52	0.96 2.12	1.92 0	0.08 0	**6** **0**
J	HONEY	1 Tbsp	21 *4.76*	64 3.6	17.3 0.0	0.02 tr(a)	0.1 17.3	**0**	tr(a) tr(a)	tr(a) tr(a)	tr(a) tr(a)	tr(a) tr(a)	tr(a) tr(a)	tr(a) tr(a)	0.00 0.00	0.00 0.00	0.00 0	tr tr	**0** **0**
K	HONEYDEW MELON,raw,balls, edible portion	1/2 C	85 *1.18*	28 77.0	6.5 0.3	0.51 0.77	0.7 6.2	**1**	unk. unk.	unk. unk.	13 2	unk. unk.	unk. 1	unk. unk.	unk. unk.	tr(a) unk.	unk. 0	0.03 20	**2** **33**
L	honeydew melon,raw,fourth, edible portion	1/4 of 7" dia	373 *0.27*	123 337.8	28.7 1.1	2.24 3.36	3.0 27.2	**5**	unk. unk.	unk. unk.	56 7	unk. unk.	unk. 4	unk. unk.	unk. unk.	tr(a) unk.	unk. 0	0.15 86	**10** **143**
M	HORS D'OEUVRE,beef puff-Durkee	1 whole	14 *7.09*	47 unk.	3.1 4.5	unk. unk.	2.2 unk.	**3**	unk. unk.	unk. unk.	unk. unk.	unk. unk.	unk. unk.	unk. unk.	unk. unk.	unk. unk.	unk. unk.	unk. unk.	**unk.** **unk.**
N	hors d'oeuvre,cheese puff-Durkee	1 whole	14 *7.19*	58 unk.	2.9 5.7	unk. unk.	2.7 unk.	**4**	unk. unk.	unk. unk.	unk. unk.	unk. unk.	unk. unk.	unk. unk.	unk. unk.	unk. unk.	unk. unk.	unk. unk.	**unk.** **unk.**
O	hors d'oeuvre,franks-n-blanket-Durkee	1 whole	13 *7.87*	45 unk.	1.0 3.8	unk. unk.	1.8 unk.	**3**	unk. unk.	unk. unk.	unk. unk.	unk. unk.	unk. unk.	unk. unk.	unk. unk.	unk. unk.	unk. unk.	unk. unk.	**unk.** **unk.**
P	hors d'oeuvre,shrimp puff-Durkee	1 whole	14 *7.09*	44 unk.	3.0 4.3	unk. unk.	2.0 unk.	**3**	unk. unk.	unk. unk.	unk. unk.	unk. unk.	unk. unk.	unk. unk.	unk. unk.	unk. unk.	unk. unk.	unk. unk.	**unk.** **unk.**
Q	HORSERADISH,prep	1 Tbsp	15 *6.67*	6 13.1	1.4 tr	0.00 0(a)	0.2 1.1	**0**	unk. unk.	unk. unk.	unk. unk.	unk. unk.	unk. unk.	unk. unk.	tr(a) tr(a)	tr(a) 0.00	tr(a) 0(a)	unk. unk.	**unk.** **unk.**
R	HUSH PUPPIES,home recipe	1 average	56 *1.79*	147 25.8	17.6 7.0	0.26 unk.	3.6 1.2	**7**	100 90	176 387	181 84	165 155	205 36	53 190	2.83 1.68	1.38 2.05	2.78 45	0.11 1	**7** **2**
S	ICE CREAM bar,chocolate coated	1 average	47 *2.13*	149 22.1	12.1 10.5	tr(a) tr(a)	1.6 8.5	**4**	unk. unk.	unk. unk.	unk. unk.	unk. unk.	unk. unk.	unk. unk.	unk. unk.	unk. unk.	unk. unk.	0.01 0	**1** **0**

Source: Nutrients in Foods, 1983.

6

Riboflavin mg / Niacin mg	% USRDA	Vitamin B_6 mg / Folacin mcg	% USRDA	Vitamin B_{12} mcg / Pantothenic acid mg	% USRDA	Biotin mg / Vitamin A IU	% USRDA	Preformed A RE / Beta carotene RE	Vitamin D IU / Vitamin E IU	% USRDA	Total tocopherol mg / Alpha tocopherol mg	Other tocopherol mg / Total ash g	Calcium mg / Phosphorus mg	% USRDA	Sodium mg / Sodium meq	Potassium mg / Potassium meq	Chlorine mg / Chlorine meq	Iron mg / Magnesium mg	% USRDA	Zinc mg / Copper mg	% USRDA	Iodine mcg / Selenium mcg	% USRDA	Manganese mcg / Chromium mcg	
0.22	13	0.13	7	0.82	14	unk.	unk.	7	0	0	0.3	tr(a)	16	2	31	293	unk.	2.6	14	2.9	19	unk.	unk.	unk.	A
3.6	18	3	1	0.15	2	245	5	tr	0.3	1	0.3	0.85	127	13	1.3	7.5	unk.	14	3	0.05	3	tr(a)		tr(a)	
0.18	11	0.18	9	0.00	0	unk.	unk.	0	0	0	9.5	unk.	71	7	1	238	23	1.1	6	1.0	7	unk.	unk.	1420	B
0.3	2	24	6	0.39	4	36	1	4	11.4	38	7.1	0.84	114	11	0.0	6.1	0.6	59	15	0.43	22	1		unk.	
0.13	8	0.12	6	6.00	100	unk.	unk.	unk.	unk.	unk.	tr(a)	tr(a)	110	11	unk.	unk.	unk.	1.3	8	unk.	unk.	unk.	unk.	unk.	C
unk.	unk.	tr(a)	0	0.52	5	unk.	unk.	unk.	0.0	0	tr(a)	2.77	223	22	unk.	unk.	unk.	13	3	unk.	unk.	tr(a)		tr(a)	
0.06	4	unk.	unk.	unk.	unk.	unk.	unk.	unk.	unk.	unk.	unk.	unk.	unk.	unk.	unk.	unk.	unk.	unk.	unk.	unk.	unk.	unk.	unk.	unk.	D
1.9	10	unk.	unk.	unk.	unk.	unk.	unk.	unk.	unk.	unk.	unk.	1.81	134	13	unk.	unk.	unk.	unk.	unk.	unk.	unk.	unk.		unk.	
unk.	unk.	0.02	1	unk.	unk.	unk.	unk.	unk.	unk.	unk.	tr(a)	tr(a)	unk.	unk.	unk.	unk.	unk.	unk.	unk.	unk.	unk.	unk.	unk.	unk.	E
unk.	unk.	tr(a)	0	tr(a)	0	unk.	unk.	unk.	0.0	0	tr(a)	0.60	unk.	unk.	unk.	unk.	unk.	unk.	unk.	unk.	unk.	tr(a)		tr(a)	
unk.	unk.	0.06	3	unk.	unk.	unk.	unk.	unk.	unk.	unk.	tr(a)	unk.	unk.	unk.	unk.	unk.	unk.	unk.	unk.	unk.	unk.	unk.	unk.	unk.	F
unk.	unk.	tr(a)	0	tr(a)	0	unk.	unk.	unk.	0.0	0	tr(a)	2.00	unk.	unk.	unk.	unk.	unk.	unk.	unk.	unk.	unk.	tr(a)		tr(a)	
0.06	3	0.05	3	0.30	5	unk.	unk.	2	unk.	unk.	tr(a)	tr(a)	13	1	unk.	unk.	unk.	0.3	2	unk.	unk.	unk.	unk.	1	G
0.7	3	tr(a)	0	0.21	2	6	0	tr	0.0	0	tr(a)	0.80	51	5	unk.	unk.	unk.	unk.	unk.	unk.	unk.	tr(a)		tr(a)	
0.11	7	0.10	5	0.60	10	unk.	unk.	3	unk.	unk.	tr(a)	tr(a)	26	3	unk.	unk.	unk.	0.6	3	unk.	unk.	unk.	unk.	2	H
1.3	7	tr(a)	0	0.42	4	12	0	tr	0.0	0	tr(a)	1.60	102	10	unk.	unk.	unk.	unk.	unk.	unk.	unk.	tr(a)		tr(a)	
unk.	unk.	unk.	unk.	unk.	unk.	unk.	unk.	0(a)	0	0	unk.	unk.	tr	0	unk.	unk.	unk.	0.4	2	unk.	unk.	unk.	unk.	unk.	I
unk.	unk.	unk.	unk.	unk.	unk.	unk.	unk.	unk.	unk.	unk.	unk.	0.32	58	6	unk.	unk.	unk.	26	6	unk.	unk.	unk.		unk.	
0.01	1	tr	0	0.00	0	tr(a)	0	0	0	0	0.0	0.0	1	0	1	11	6	0.1	1	tr	0	unk.	unk.	6	J
0.1	0	1	0	0.04	0	0	0	0	0.0	0	0.0	0.04	1	0	0.0	0.3	0.2	1	0	0.04	2	unk.		unk.	
0.03	2	0.05	2	0.00	0	unk.	unk.	0	0(a)	0	0.1	unk.	3	0	10	213	35	0.1	1	0.1	1	unk.	unk.	15	K
0.5	3	unk.	unk.	0.18	2	34	1	3	0.1	0	unk.	0.51	9	1	0.4	5.5	1.0	6	1	0.03	2	0		0	
0.11	7	0.21	10	0.00	0	unk.	unk.	0	0(a)	0	0.4	unk.	11	1	45	936	153	0.4	2	0.4	3	unk.	unk.	67	L
2.2	11	unk.	unk.	0.77	8	149	3	15	0.4	2	unk.	2.24	37	4	1.9	23.9	4.3	25	6	0.15	8	0		0	
unk.	unk.	unk.	unk.	unk.	unk.	unk.	unk.	unk.	unk.	unk.	unk.	unk.	unk.	unk.	unk.	unk.	unk.	unk.	unk.	unk.	unk.	unk.	unk.	unk.	M
unk.	unk.	unk.	unk.	unk.	unk.	unk.	unk.	unk.	unk.	unk.	unk.	unk.	unk.	unk.	unk.	unk.	unk.	unk.	unk.	unk.	unk.	unk.		unk.	
unk.	unk.	unk.	unk.	unk.	unk.	unk.	unk.	unk.	unk.	unk.	unk.	unk.	unk.	unk.	unk.	unk.	unk.	unk.	unk.	unk.	unk.	unk.	unk.	unk.	N
unk.	unk.	unk.	unk.	unk.	unk.	unk.	unk.	unk.	unk.	unk.	unk.	unk.	unk.	unk.	unk.	unk.	unk.	unk.	unk.	unk.	unk.	unk.		unk.	
unk.	unk.	unk.	unk.	unk.	unk.	unk.	unk.	unk.	0(a)	0	unk.	unk.	unk.	unk.	unk.	unk.	unk.	unk.	unk.	unk.	unk.	unk.	unk.	unk.	O
unk.	unk.	unk.	unk.	unk.	unk.	unk.	unk.	unk.	unk.	unk.	unk.	unk.	unk.	unk.	unk.	unk.	unk.	unk.	unk.	unk.	unk.	unk.		unk.	
unk.	unk.	unk.	unk.	unk.	unk.	unk.	unk.	unk.	0(a)	0	unk.	unk.	unk.	unk.	unk.	unk.	unk.	unk.	unk.	unk.	unk.	unk.	unk.	unk.	P
unk.	unk.	unk.	unk.	unk.	unk.	unk.	unk.	unk.	unk.	unk.	unk.	unk.	unk.	unk.	unk.	unk.	unk.	unk.	unk.	unk.	unk.	unk.		unk.	
unk.	unk.	0.02	1	0(a)	0	unk.	unk.	0(a)	0(a)	0	unk.	unk.	9	1	14	43	unk.	0.1	1	0.2	1	unk.	unk.	unk.	Q
unk.	unk.	unk.	unk.	unk.	unk.	unk.	unk.	unk.	unk.	unk.	unk.	0.27	5	1	0.6	1.1	unk.	unk.	unk.	tr(a)	0	unk.		unk.	
0.12	7	0.15	7	0.16	3	0.003	1	0	11	3	4.1	tr	129	13	545	106	16	0.9	5	0.7	5	54.3	36	12	R
0.8	4	18	4	0.41	4	47	1	tr	1.1	4	1.1	2.29	223	22	23.7	2.7	0.4	44	11	0.09	4	tr		1	
0.07	4	0.02	1	0.26	4	unk.	unk.	unk.	2	1	unk.	unk.	55	6	24	84	unk.	0.0	0	unk.	unk.	unk.	unk.	unk.	S
0.4	2	unk.	unk.	unk.	unk.	145	3	unk.	unk.	unk.	0.1	0.42	42	4	1.0	2.1	unk.	9	2	unk.	unk.	unk.		unk.	

3. Next, locate the desired nutrient(s) (for this example, phenylalanine and threonine) in the column heads and follow that column down to the particular food listing.
4. The upper number (182) is the value for the left of the two column heads (phenylalanine), while the lower number (140) is the value for the right heading (threonine). The headings indicate that these numbers are given in milligrams (mg).
5. To relocate foods, jot down an "index." using page number and margin letter, e.g., to relocate *HONEY,* search for 150-J.
6. Where it is given, the percent of U.S.RDA for each nutrient is just to the right of the raw values. In the case of riboflavin and niacin, *HAZELNUTS* (1/4C) have 0.18 mg of riboflavin which represents approximately 11% of the U.S.RDA for that nutrient and 0.3 mg of niacin, approximately 2% of the U.S.RDA for that nutrient.
7. If you wish to compare the relative amounts of a given nutrient in different foods, use the conversion factor in column 3. For example, if you wish to compare hazelnuts with plain, canned herring as a source for threonine, you would multiply the threonine value for hazelnuts by its conversion factor and the same value for the herring serving by its conversion factor and then compare the two as shown below. Clearly, plain herring has roughly 2 1/2 (2.46) times as much threonine per gram as do hazelnuts.

Food Serving As Listed	Amount of Threonine in that Serving		Conversion Factor		Calculated Amount of Threonine/100 g
75 gms Herring	767 mg	x	1.3	=	1020 mg
34 gms Hazelnuts	140 mg	x	2.96	=	414 mg

Numerical Data

Nutrient data and the P/S ratio are rounded to the nearest appropriate decimal point and shown in regular arabic numerals. Percents of U.S.RDA are rounded to whole numbers and shown in boldface type. In column 3, the factor which converts nutrient data to 100 g portions is given in italics.

The source for all numerical values is the original portion size. Both the *nutrient value* and the *percent U.S.RDA* (where appropriate) are derived from the original portion value, not the nutrient value from the portion size and the percent U.S.RDA from the nutrient values. Values, both nutrient and U.S.RDA, are rounded off accordingly. This results in occasional apparent inconsistencies in percent U.S.RDA for similar items.

In order to promote accuracy, every nutrient-value position in the tables either has a numerical value for that nutrient or indicates what is known by a series of codes.

If it is known that a nutrient does not occur in a food, the value is given as *0,* (zero). Some values are listed as assumed zero, *0(a),* or assumed trace, *tr(a).* These are used to differentiate between analyzed data giving zero or trace from an estimated value. In both cases the percent U.S.RDA is shown as 0. The designation "unknown" *(unk)* is used for nutrient values for which there are currently no analyzed data , but it is assumed that a measurable amount of the nutrient is present in the food.

Also, in the first development of the Michigan State University Nutrient Data Bank,imputed values for folacin, B-6, and B-12 in cooked fruits, vegetables, and meats were from maximum loss for these nutrients reported by Harris and Karmas (1975). The percentages of loss used were 10% for cobalamin (B-12), 40% for pyroxidine (B-6), and 80% for folacin. As more accurate values have been and are being published, these imputed values are being changed.

In the case of Total Sugar and Selenium which do not have a designated percent U.S.RDA but which, for other reasons, occur in the same column with nutrients that do, the percent U.S.RDA space is left blank. This is the only instance in which a blank space occurs in the tables.

The U.S. Recommended Daily Allowance values used in *NUTRIENTS IN FOODS* are for adults and children over 4 years of age for all items except baby, junior, and infant foods. For strained and infant foods, U.S.RDA values are for infants under 13 months, while the percents U.S.RDA for toddler foods were calculated from values for children under 4 years of age. Table 5 summarizes U.S.RDA values for these age groups.

The percent U.S.RDA for protein was calculated based on the FDA regulations which use an RDA of 45 g if the protein efficiency ratio (PER) is equal to or greater than casein, or an RDA of 65 g if the PER is less. Respective values for children are 20 g or 28 g, and for infants 18 g or 25 g. For mixed dishes, those containing both high and low quality protein, the percent U.S.RDA is based on the value of the mixture of the two protein types or on the basis of the low quality protein if the level of each protein is not known.

Food companies were asked for analytic data rather than label declaration information. Although all companies provided information per 100 g, and most companies provided analytical data, there are some instances in which the data were obviously back-calculated from label information. Cereals are an excellent example of this—with numerous values contributing exactly 25% of the U.S.RDA.

Label data are less accurate for analytic purposes because the manufacturer may deliberately report somewhat lower levels of nutrients than actually present in the food to take into account the losses which may occur due to storage time and other conditions intervening between the manufacture date and the point at which the food is actually eaten. By the same token, the calories may be overstated to make sure the consumer is not getting more calories than reported on the label.

IMPORTANT CONSIDERATIONS IN USING *NUTRIENTS IN FOODS*

Factors Affecting Nutrient Availability and the Reliability of Nutrient Composition Tables

Variation of nutrient levels from published values is possible at many stages, with some nutrients being more vulnerable to change than others.

The initial nutrient content of foods depends in part on the soil in which food plants are grown or the feed which is consumed by livestock. Once an animal is slaughtered or a plant harvested, changes in nutrient composition begin, many of which are ongoing. Anyone who has eaten fresh corn on the cob, plunged in boiling water minutes after it was picked, and then compared the taste with corn from the grocery where it has been for several hours or days can readily appreciate the changes in sugar/starch balance that occur. Similarly, spontaneous Vitamin C loss may start after green vegetables are picked.

Beyond time and storage conditions, methods used in food processing and cooking, as well as environmental factors such as light, heat, and humidity, can alter the nutrient composition of foods.

Some nutrients are more subject to loss than others. Estimates of total protein intake in nutrient tables, such as *NUTRIENTS IN FOODS*, are probably close to that which is actually consumed. Values for total carbohydrate and fat consumption are similarly reliable, with the exception that the fat content of meat can be quite variable, particularly due to cooking and eating practices. Figures for kilocalories provided by the diet should also be reliable, as they are related to the consumption of protein, carbohydrate, and fat.

Vitamins and minerals are subject to much greater environmentally-induced variation than the macronutrients. The trace element content of foods in particular is much more variable, and there may be enormous variation in the mineral content of different samples of the same food. The selenium content of wheat grown in New England, for example, may be orders of magnitude different from that grown in Iowa.

The use of nutrient composition tables for determining the nutritional adequacy of diets is further affected because nutrients may not be available for use by the human system in the same concentration as they are present in foods. During digestion and absorption, some nutrients may be more available, i.e., more readily absorbed, than others, and this bioavailability can be affected by many things. Other foods or drugs present in the digestive tract, the form in which the nutrient is present in the food, current body stores and needs for the nutrient, disease states, and other conditions can all influence the ability to absorb given nutrients.

Once in the system, utilization of nutrients depends on the status and health of the individual, including such factors as age, sex, pregnancy, and lactation. Again, disease states or the presence or absence of other nutrients, drugs, etc. can influence the need for and utilization of specific absorbed nutrients.

The carrot and stick of this complex nutrition obstacle course are the often serious effects of too little (deficiency) or too much (toxicity) of specific nutrients in the body. The descriptive materials on the individual nutrients in this book will touch upon the major influences on nutrient availability and uptake.

Standards For Judging Nutritional Adequacy

A major goal of nutrition and dietetics is to provide guidelines for the adequate intake of essential nutrients: "How much is enough?" and "How much is too much?" are important questions. When dealing with these questions about the nutritional adequacy of diets, there are two standards commonly used and confused in the United States. These are the Recommended Dietary Allowances (RDA's) formulated by the Food and Nutrition Board of the National Academy of Sciences, and the United States Recommended Daily Allowances (U.S.RDA's) formulated by the U.S. Food and Drug Administration (FDA).

These two standards are related to but different from each other. In *NUTRIENTS IN FOODS*, the RDA values for nutrients are given in Table 2 and the RDA's for energy in Table 5. The U.S.RDA's (Table 4) are also expressed as the % U.S.RDA (percent of United States Recommended Daily Allowance) and in this form are given in the food tables where appropriate. To understand better the differences between these two standards, a brief description of each follows.

RECOMMENDED DIETARY ALLOWANCES (RDA) (FOOD AND NUTRITION BOARD)

Given the importance of having enough but not too much of many nutrients, much thought and effort has been put into establishing recommended levels of intake for the essential nutrients.

Periodically, since 1943, the Food and Nutrition Board of the U.S. National Academy of Sciences has been reformulating recommended dietary allowances for energy, protein, 3 fat-soluble and 7 water-soluble vitamins, and 6 minerals. The most recent reformulated RDA's, 1980, are shown in Table 2. These RDA's are defined as: "the levels of intake of essential nutrients considered, in the judgement of the Committee on Dietary Allowances of the Food and Nutrition Board on the basis of available scientific knowledge, to be adequate to meet the known nutritional needs of practically all healthy persons " (National Academy of Sciences, 1980).

The RDA's are meant to apply only to healthy populations, and should be met from the consumption of a wide variety of readily available foods. They should not be confused with the nutrient requirements of individuals, because these are too variable. Rather, the RDA's represent an average level of daily intake for populations, under the assumption that if the average intake of a nutrient over time approximates the RDA, then nutritional inadequacy for that nutrient will be rare in that population. Many individuals could actually consume much lower amounts of some nutrients than the RDA's specify and still have an adequate intake. This is because the RDA's are calculated to be higher than the actual requirement for most humans in order to take into account individual variability in requirements, as well as poor utilization of nutrients.

Establishing recommended intakes is a complicated matter that must take into account factors affecting absorption rates, the effect of precursor forms of some nutrients, sources of the nutrients, etc. For example, most RDA's for infants under six months old are based on the concentrations of these nutrients found in breast milk of healthy mothers, because it has been repeatedly observed that a diet of breast milk alone is adequate for all the nutrient needs of infants up until six months of age.

Unlike the allowances for nutrients, the RDA allowances for energy are not calculated to be greater than the minimum requirement consonant with good health but rather are calculated to approximate actual energy needs, since excess amounts, however small, will be stored as fat and thus contribute to the development of obesity. Table 5 gives energy allowances separate from nutrient allowances.

Nutrient needs may vary considerably depending on the age, sex, and reproductive status of the individual. In order to address this variation, the RDA's are calculated for four basic groups: *Infants,* divided into two age categories; *Children,* in three age categories; *adolescent and older males,* in five age categories; and *adolescent and older females,* in five age categories. Some RDA's are also altered for *pregnant and lactating women* so that a total of seventeen age-sex categories exist, each with its own RDA's.

The RDA's are for normal, healthy populations, so elevated physical activity, the presence of chronic diseases, and acute health problems (such as burns, wounds, parasite infections, and recovery from surgery), as well as other conditions may alter nutrient needs.

TABLE 2
Food and Nutrition Board, National Academy of Sciences-National Research Council
Recommended Daily Dietary Allowances,[a] Revised 1980
Designed for the maintenance of good nutrition of practically all healthy people in the U.S.A.

		Weight		Height			Fat-Soluble Vitamins			Water-Soluble Vitamins							Minerals					
	Age (years)	(kg)	(lb)	(cm)	(in)	Protein (g)	Vitamin A (μg RE)[b]	Vitamin D (μg)[c]	Vitamin E (mg α-TE)[d]	Vitamin C (mg)	Thiamin (mg)	Riboflavin (mg)	Niacin (mg NE)[e]	Vitamin B-6 (mg)	Folacin[f] (μg)	Vitamin B-12 (μg)	Calcium (mg)	Phosphorus (mg)	Magnesium (mg)	Iron (mg)	Zinc (mg)	Iodine (μg)
Infants	0.0–0.5	6	13	60	24	kg × 2.2	420	10	3	35	0.3	0.4	6	0.3	30	0.5[g]	360	240	50	10	3	40
	0.5–1.0	9	20	71	28	kg × 2.0	400	10	4	35	0.5	0.6	8	0.6	45	1.5	540	360	70	15	5	50
Children	1–3	13	29	90	35	23	400	10	5	45	0.7	0.8	9	0.9	100	2.0	800	800	150	15	10	70
	4–6	20	44	112	44	30	500	10	6	45	0.9	1.0	11	1.3	200	2.5	800	800	200	10	10	90
	7–10	28	62	132	52	34	700	10	7	45	1.2	1.4	16	1.6	300	3.0	800	800	250	10	10	120
Males	11–14	45	99	157	62	45	1000	10	8	50	1.4	1.6	18	1.8	400	3.0	1200	1200	350	18	15	150
	15–18	66	145	176	69	56	1000	10	10	60	1.4	1.7	18	2.0	400	3.0	1200	1200	400	18	15	150
	19–22	70	154	177	70	56	1000	7.5	10	60	1.5	1.7	19	2.2	400	3.0	800	800	350	10	15	150
	23–50	70	154	178	70	56	1000	5	10	60	1.4	1.6	18	2.2	400	3.0	800	800	350	10	15	150
	51+	70	154	178	70	56	1000	5	10	60	1.2	1.4	16	2.2	400	3.0	800	800	350	10	15	150
Females	11–14	46	101	157	62	46	800	10	8	50	1.1	1.3	15	1.8	400	3.0	1200	1200	300	18	15	150
	15–18	55	120	163	64	46	800	10	8	60	1.1	1.3	14	2.0	400	3.0	1200	1200	300	18	15	150
	19–22	55	120	163	64	44	800	7.5	8	60	1.1	1.3	14	2.0	400	3.0	800	800	300	18	15	150
	23–50	55	120	163	64	44	800	5	8	60	1.0	1.2	13	2.0	400	3.0	800	800	300	18	15	150
	51+	55	120	163	64	44	800	5	8	60	1.0	1.2	13	2.0	400	3.0	800	800	300	10	15	150
Pregnant						+30	+200	+5	+2	+20	+0.4	+0.3	+2	+0.6	+400	+1.0	+400	+400	+150	*h*	+5	+25
Lactating						+20	+400	+5	+3	+40	+0.5	+0.5	+5	+0.5	+100	+1.0	+400	+400	+150	*h*	+10	+50

[a] The allowances are intended to provide for individual variations among most normal persons as they live in the United States under usual environmental stresses. Diets should be based on a variety of common foods in order to provide other nutrients for which human requirements have been less well defined. See text for detailed discussion of allowances and of nutrients not tabulated. See Table 1 (p. 20) for weights and heights by individual year of age. See Table 3 (p. 23) for suggested average energy intakes.

[b] Retinol equivalents. 1 retinol equivalent = 1 μg retinol or 6 μg β carotene. See text for calculation of vitamin A activity of diets as retinol equivalents.

[c] As cholecalciferol. 10 μg cholecalciferol = 400 IU of vitamin D.

[d] α-tocopherol equivalents. 1 mg *d*-α tocopherol = 1 α-TE. See text for variation in allowances and calculation of vitamin E activity of the diet as α-tocopherol equivalents.

[e] 1 NE (niacin equivalent) is equal to 1 mg of niacin or 60 mg of dietary tryptophan.

[f] The folacin allowances refer to dietary sources as determined by *Lactobacillus casei* assay after treatment with enzymes (conjugases) to make polyglutamyl forms of the vitamin available to the test organism.

[g] The recommended dietary allowance for vitamin B-12 in infants is based on average concentration of the vitamin in human milk. The allowances after weaning are based on energy intake (as recommended by the American Academy of Pediatrics) and consideration of other factors, such as intestinal absorption; see text.

[h] The increased requirement during pregnancy cannot be met by the iron content of habitual American diets nor by the existing iron stores of many women; therefore the use of 30–60 mg of supplemental iron is recommended. Iron needs during lactation are not substantially different from those of nonpregnant women, but continued supplementation of the mother for 2–3 months after parturition is advisable in order to replenish stores depleted by pregnancy.

Source: National Academy of Sciences, 1980.

ESTIMATED SAFE AND ADEQUATE INTAKE (FOOD AND NUTRITION BOARD)

For nutrients whose requirements are less well understood, the Food and Nutrition Board has established suggested safe and adequate levels of intake (Table 3). These are also provided for several age and sex categories, though in less detail than the RDA's. The suggested levels of intake are not considered RDA's, since little is known about the actual requirements for these nutrients, and even less is known about the content of these nutrients in usual diets. However, three of these nutrients (biotin, pantothenic acid, and copper) have U.S.RDA values described in the next section.

U.S. RECOMMENDED DAILY ALLOWANCES (U.S.RDA) (FOOD AND DRUG ADMINISTRATION)

The RDA's developed by the Food and Nutrition Board are considered the best estimates of the amount of specific nutrients that should be ingested on average by a *population* to ensure adequate intake of the essential nutrients. However, they are not easy to use in deciding how best to obtain good nutrition when purchasing food.

In the early 1970s, as part of its efforts to promote nutritional awareness and health, the Food and Drug Administration (FDA) promulgated a simplified version of the RDA's, which was designed to appear on the limited space of food labels. The United States Recommended Daily Allowances (U.S.RDA's) thus created were designed to aid the public in making intelligent choices based on nutritional considerations when shopping for food. The U.S.RDA's simplified the seventeen age and sex categories of the RDA's into just four: *Infants under 13 months, children under 4 years, pregnant and lactating women, and all other individuals.*

To make the food labels more useful in constructing diets or single meals, the number given for each nutrient on the label is not the U.S.RDA, but rather is the percentage of the U.S.RDA provided by a single serving of the food contained in the package. The serving or portion size is based on the food packager's estimate of a reasonable amount of food that might be included in one meal by an average person. Since increasing or decreasing the serving size would also change the percent of U.S.RDA of the nutrients provided by one serving, Federal guidelines determine what can be considered a serving size, including a definition of an average person as being represented by an "average" U.S. adult male, middle-aged and in good health doing light labor.

The label must not only give the weight of the serving, but also indicate the number of servings contained in the package so that the consumer will have a quick estimate of how much of the package will provide the percentage of the nutrient indicated on the label. For example, if the number for vitamin C is 33%, three servings of the food would provide almost 100% of the U.S.RDA.

There are two basic reasons why nutrients are chosen to be included in the U.S.RDA list. The first is that these selected nutrients are the most completely understood, so that there is a reasonable basis for deriving estimates of daily allowances. The second is the assumption that, if one is consuming adequate amounts of most of these nutrients, the other essential nutrients not included in the U.S.RDA list will be supplied in the diet in adequate amounts.

U.S.RDA values have been established for protein, twelve vitamins, and seven minerals (Lecos, 1982), shown in Table 4. Of these, percentages for protein and five vitamins (vitamin A, vitamin C, thiamin, riboflavin, and niacin) must be shown on the food label. The remaining fourteen vitamins and minerals may be shown on the label at the manufacturer's discretion, unless they have added any one of the nineteen U.S.RDA nutrients to the food or made some nutritional claim in which case all nineteen U.S.RDA percentages must be shown.

An additional part of the nutritional labeling of foods is that they include the number of kilocalories per serving, as well as the number of grams of carbohydrate, fat,and protein. In some instances, the grams of saturated and polyunsaturated fatty acids, the P/S ratio, and the milligrams of cholesterol are also included. It should be noted that there are U.S.RDA values for two vitamins (biotin and pantothenic acid) and one mineral (copper) for which no RDA exists.

The U.S.RDA's are generally based upon the RDA that is highest among the seventeen age and sex categories for each of the nutrients. This means that each U.S.RDA is generally derived from the adult male RDA, although there are some exceptions. These include the the following: The U.S.RDA for riboflavin is set at the average RDA for males 15 to 22 years of age; the U.S.RDA's for calcium and phosphorus are set to the average RDA for adult males and females; and the U.S.RDA's for vitamin E and vitamin B-12 are set at twice the adult RDA.

It is clear that the U.S.RDA is higher than required by many individuals. Many adults need only 75% of the U.S.RDA for most nutrients, children only about 50%.

In addition to these quantitative differences between the RDA's and the U.S.RDA's, there exists a major qualitative difference in the U.S.RDA for protein. The U.S.RDA for protein, unlike the RDA, attempts to take into account the relative quality of the protein in terms of its amino acid

TABLE 3

Estimated Safe and Adequate Daily Dietary Intakes of Selected Vitamins and Minerals[a]

		Vitamins		
	Age (years)	Vitamin K (μg)	Biotin (μg)	Pantothenic Acid (mg)
Infants	0–0.5	12	35	2
	0.5–1	10–20	50	3
Children	1–3	15–30	65	3
and	4–6	20–40	85	3–4
Adolescents	7–10	30–60	120	4–5
	11+	50–100	100–200	4–7
Adults		70–140	100–200	4–7

		Trace Elements[b]					
	Age (years)	Copper (mg)	Manganese (mg)	Fluoride (mg)	Chromium (mg)	Selenium (mg)	Molybdenum (mg)
Infants	0–0.5	0.5–0.7	0.5–0.7	0.1–0.5	0.01–0.04	0.01–0.04	0.03–0.06
	0.5–1	0.7–1.0	0.7–1.0	0.2–1.0	0.02–0.06	0.02–0.06	0.04–0.08
Children	1–3	1.0–1.5	1.0–1.5	0.5–1.5	0.02–0.08	0.02–0.08	0.05–0.1
and	4–6	1.5–2.0	1.5–2.0	1.0–2.5	0.03–0.12	0.03–0.12	0.06–0.15
Adolescents	7–10	2.0–2.5	2.0–3.0	1.5–2.5	0.05–0.2	0.05–0.2	0.10–0.3
	11+	2.0–3.0	2.5–5.0	1.5–2.5	0.05–0.2	0.05–0.2	0.15–0.5
Adults		2.0–3.0	2.5–5.0	1.5–4.0	0.05–0.2	0.05–0.2	0.15–0.5

		Electrolytes		
	Age (years)	Sodium (mg)	Potassium (mg)	Chloride (mg)
Infants	0–0.5	115–350	350–925	275–700
	0.5–1	250–750	425–1275	400–1200
Children	1–3	325–975	550–1650	500–1500
and	4–6	450–1350	775–2325	700–2100
Adolescents	7–10	600–1800	1000–3000	925–2775
	11+	900–2700	1525–4575	1400–4200
Adults		1100–3300	1875–5625	1700–5100

[a] Because there is less information on which to base allowances, these figures are not given in the main table of RDA and are provided here in the form of ranges of recommended intakes.

[b] Since the toxic levels for many trace elements may be only several times usual intakes, the upper levels for the trace elements given in this table should not be habitually exceeded.

Source: National Academy of Sciences, 1980.

TABLE 4

U.S. Recommended Daily Allowances (U.S.RDA)

Vitamins, Minerals, and Protein	Unit of Measurement	Infants	Adults and Children 4 or More Years of Age	Children Under 4 Years of Age	Pregnant or Lactating Women
Fat-Soluble Vitamins					
Vitamin A[c]	IU	1,500.0	5,000.0	2,500.00	8,000.0
Vitamin D	IU	400.0	400.0	400.00	400.0
Vitamin E	IU	5.0	30.0	10.00	30.0
Water-Soluble Vitamins					
Vitamin C[c]	Mg	35.0	60.0	40.00	60.0
Folic Acid	Mg	0.1	0.4	0.20	0.8
Thiamine	Mg	0.5	1.5	0.70	1.7
Riboflavin	Mg	0.6	0.7	0.80	2.0
Niacin	Mg	8.0	20.0	9.00	20.0
Vitamin B_6	Mg	0.4	2.0	0.70	2.5
Vitamin B_{12}	Mcg	2.0	6.0	3.00	8.0
Biotin	Mg	0.5	0.3	0.15	0.3
Pantothenic Acid	Mg	3.0	10.0	5.00	10.0
Major Minerals					
Calcium	G	0.6	1.0	0.80	1.3
Phosphorus	G	0.5	1.0	0.80	1.3
Trace Minerals					
Iodine	Mcg	45.0	150.0	70.00	150.0
Iron	Mg	15.0	18.0	10.00	18.0
Magnesium	Mg	70.0	400.0	200.00	450.0
Copper	Mg	0.6	2.0	1.00	2.0
Zinc	Mg	5.0	15.0	8.00	15.0
Macronutrients					
Protein	G	18.0	45.0	20.00	

[a] Presence optional for adults and children 4 or more years of age in vitamin and mineral supplements.

[b] If protein efficiency ratio of protein is equal to or better than that of casein, U.S. RDA is 45 g for adults, 18 g for infants, and 20 g for children under 4.

[c] These nutrients must appear on the food label. Others are optional.

Source: FDA, 1981.

composition, since the U.S.RDA pertains to the specific food carrying the label rather than with a complete diet of mixed foods, as is the case with the RDA's. In determining the relative quality of the protein content of different foods, casein, the principal protein in milk, is taken as the standard. The unit of measurement for comparison is the Protein Efficiency Ratio (PER), which is based upon the increase in body weight of young animals fed different sources of protein. Thus, if a standard quantity of protein results in a relatively small amount of weight gain, that protein has a low PER.

If the quality of a protein in terms of PER is equal to or greater than that of casein, the U.S.RDA for that protein is set at 45 g per day, and the percent of U.S.RDA is calculated on that amount. If the PER for the protein is less than that of casein, then the U.S.RDA is set at 65 g per day to compensate for its lower quality, meaning that the individual would have to consume a relatively larger amount of that food to get the same protein percent of U.S.RDA as from a food with the higher PER. At the lower limits, if the PER of the protein is 20% or less than that of casein, the food label must read "not a significant source of protein."

In determining the U.S.RDA for other nutrients, any other differences in utilization not considered in the formulation of the RDA's were not taken into account. In the case of niacin, it should be noted that, as in *NUTRIENTS IN FOODS*, the U.S.RDA does not consider the presence of tryptophan in foods in calculating niacin values, so the proportion of U.S.RDA for total niacin activity will be underestimated.

It should be stressed that the U.S.RDA's and the rules that accompany them are strictly for food labeling purposes. At the present time, there is no legal requirement to provide nutrient information on a food label, unless specific nutrients have been added to a food as a supplement or some health claim is made for the nutritional value of the food. Thus, many of the food labels contain nutrient information that is voluntarily provided by the food companies as a public service. A recent FDA survey showed that 44% of packaged food sales in the United States carried nutrition labeling, half of those on a voluntary basis (Lecos, 82).

In *NUTRIENTS IN FOODS*, the percent U.S.RDA has been calculated for *all* foods where the nutrient value is known, not just for those foods which might carry nutrition labels in a store. U.S.RDA's are provided in this book for many foods for which no labeling provision has yet been made by the FDA. For some of these foods, such as fresh fruits, vegetables, and meats, the FDA and U.S. Department of Agriculture are working to develop rules for providing nutrient information.

TABLE 5

Mean Heights and Weights and Recommended Energy Intake[a]

	Age	Weight		Height		Energy Needs (with range)		
Category	**(years)**	**(kg)**	**(lb)**	**(cm)**	**(in.)**	**(kcal)**		**(MJ)**
Infants	0.0–0.5	6	13	60	24	kg × 115	(95–145)	kg × 0.48
	0.5–1.0	9	20	71	28	kg × 105	(80–135)	kg × 0.44
Children	1–3	13	29	90	35	1300	(900–1800)	5.5
	4–6	20	44	112	44	1700	(1300–2300)	7.1
	7–10	28	62	132	52	2400	(1650–3300)	10.1
Males	11–14	45	99	157	62	2700	(2000–3700)	11.3
	15–18	66	145	176	69	2800	(2100–3900)	11.8
	19–22	70	154	177	70	2900	(2500–3300)	12.2
	23–50	70	154	178	70	2700	(2300–3100)	11.3
	51–75	70	154	178	70	2400	(2000–2800)	10.1
	76+	70	154	178	70	2050	(1650–2450)	8.6
Females	11–14	46	101	157	62	2200	(1500–3000)	9.2
	15–18	55	120	163	64	2100	(1200–3000)	8.8
	19–22	55	120	163	64	2100	(1700–2500)	8.8
	23–50	55	120	163	64	2000	(1600–2400)	8.4
	51–75	55	120	163	64	1800	(1400–2200)	7.6
	76+	55	120	163	64	1600	(1200–2000)	6.7
Pregnancy						+300		
Lactation						+500		

[a] The data in this table have been assembled from the observed median heights and weights of children shown in Table 1, together with desirable weights for adults given in Table 2 for the mean heights of men (70 in.) and women (64 in.) between the ages of 18 and 34 years as surveyed in the U.S. population (HEW/NCHS data).

The energy allowances for the young adults are for men and women doing light work. The allowances for the two older age groups represent mean energy needs over these age spans, allowing for a 2-percent decrease in basal (resting) metabolic rate per decade and a reduction in activity of 200 kcal/day for men and women between 51 and 75 years, 500 kcal for men over 75 years, and 400 kcal for women over 75 years (see text). The customary range of daily energy output is shown in parentheses for adults and is based on a variation in energy needs of ±400 kcal at any one age (see text and Garrow, 1978), emphasizing the wide range of energy intakes appropriate for any group of people.

Energy allowances for children through age 18 are based on median energy intakes of children of these ages followed in longitudinal growth studies. The values in parentheses are 10th and 90th percentiles of energy intake, to indicate the range of energy consumption among children of these ages (see text).

Source: National Academy of Sciences, 1980.

COLUMN HEADING TOPICS

Units of Measure

PORTION SIZE

Foods are given in common U.S. household portions. These represent either the amount that might normally be consumed, or an amount that can be readily converted into portions usually consumed. Servings of meat and some other solid foods are often given in terms of physical measurement in inches (e.g. Table 1, Herring, items C-H). When the serving size is ambiguous—large, small, average—reference to the weight will help estimate the size. Pie portions given as 1/7th of a pie come from USDA data based upon food service round metal pie cutters which divide a pie into even sevenths. Abbreviations used in *NUTRIENTS IN FOODS* are shown in Table 9 at the back of the book. Please note that *C* stands for an 8-ounce measuring cup while 1 *cup* represents a standard 6-ounce teacup. Tablespoon and teaspoon refer to U.S. kitchen measuring spoons, and prepared instant coffee portions are in terms of teaspoons per 6 ounce cup.

WEIGHT

Food weights are expressed in metric terms (grams), most nutrients in grams (g), milligrams (mg), or micrograms (mcg), having the following ratio:

1 g = 1,000 mg = 1,000,000 mcg

A table of weights and measures (Table 10) is provided at the back of the book.

CONVERSION FACTOR

In column 3, listed under the weight in grams is an italicized number which can used to convert all values (except the P/S ratio) for specific foods into 100 g equivalents for comparative purposes. If you wish to compare, for example, the relative proportion of vitamin C in two or more foods, simply multiply the value of vitamin C for each food by the conversion factor, and you will have all values in terms of grams of vitamin C per 100 g of that food item. In the case of vitamin B-12, you would have all values in terms of micrograms of vitamin B-12 per 100 g of the food. See the section HOW TO USE *NUTRIENTS IN FOODS* for a more detailed example.

KILOCALORIES

Kilocalories (kcal) are a measure of the energy contained in food determined by the amount of heat released in complete oxidation of the foodstuff. By definition, a kilocalorie is the amount of heat needed to raise the temperature of one kilogram of water by one degree centrigrade. The kilojoule (kJ), the accepted metric measure of energy, is the amount of energy needed to move one kilogram a distance of one meter with a force of one newton. This is a measure of the effective value of stored energy rather than heat, although the two can be used similarly. To convert kilocalories to kilojoules, multiply kilocalories by 4.184. Since human requirements exceed 1,000 kJ, these are expressed in megajoules (mJ).

1 kcal	=	4,184 kJ
1 kJ	=	0.239 kcal
1,000 kcal	=	4.184 mJ
1 mJ	=	239 kcal

In determining the caloric value of a given food to the human body, the Atwater coefficients of physiological fuel value are used. These approximate the caloric value of protein (4 kcal per gram), carbohydrate (4 kcal per gram), and fat (9 kcal per gram). Ethanol, the alcohol contained in beer, wine, and spirits, also contributes calories (7 kcal per gram), and must be considered if present. None of the other nutrients provide energy, and hence are not considered in this equation. The Atwater coefficients do take into account the fact that not all of the energy as measured by complete oxidation in a bomb calorimeter is actually physiologically available, due to differences in the efficiency of absorption (coefficient of digestibility) of protein (92%), carbohydrates (99%), fats (95%), and ethanol (100%).

Energy is the activating force in all life processes. All activity, from metabolic process and maintenance of body temperature to physical action, are fueled by energy release. The amount of energy needed to support the processes required for growth and maintenance of the human body varies widely depending on age, body size, sex, and reproductive status, as well as on lifestyle, activity levels, and the environment. These factors have been taken into account in arriving at reasonable estimates of desired body weights and recommended energy intake for specific age/sex categories of human beings (see Table 3).

Energy malnutrition is generally considered to be one of the greatest nutrition health problems in the world today. In most of the Third World, energy deficiency, otherwise thought of as inadequate food or hunger, is a major contributor to death, especially in infants and children. Resistance to infection and disease is often lowered by inadequate energy intake, compromising health and promoting disabilities. Underweight women may cease menstruation or have problems with pregnancy, including a higher risk of premature or low birth weight infants.

By contrast, excess energy consumption above energy needs leads to obesity, a major nutrition-related health problem in the United States. Obesity, usually defined as more than 20% above recommended weight, is implicated in cardiovascular disease, liver and kidney disease, arthritis, and diabetes. Excess energy is stored as fat, and even seemingly small excess amounts consumed on a daily basis can eventually lead to serious excess storage problems.

The Macronutrients and Their Constituents

WATER

Water is a vital nutrient so critical to human life that a depletion of more than 20% is invariably fatal. Water composes 50-70% of the adult body weight, approximately 60% of which is intracellular. The balance of the water is in the cardiovascular system (32%) or in the interstitial areas (8%). Adipose tissue averages 30% water, bones 10%, and teeth just 5%, but muscle tissue is 72% water.

Water has four major physiological functions: as a regulator of internal body temperatures, as a lubricant for digestion and other body functions, as a solvent for body chemicals, and as an agent in digestive hydrolysis.

Most of daily water intake is in fluid form, but solid food provides as much as one-third of the water ingested. The amount of water contained in foods is also a major source of the variation in nutrient composition of different samples of the same food. Water gain or loss during storage, processing, or cooking of foods can alter the proportion of other nutrients provided by a given portion size of food.

Water deficiency (dehydration) can lead to serious complications. Low intake, however, is rarely the only cause. Diarrhea, hemmorhage, excessive sweating, loss through wounds or burns, and abuse of diuretics all can deplete water stores. Diabetes insipidus produces excessive water loss, and cholera fatality is often due to massive water and electrolyte loss. Failure to provide adequate water intake in the presence of these conditions can lead to severe complications related to electrolyte imbalance, such as cardiac failure and eventual death.

Excess retention of water in the tissues produces a swollen condition known as edema. This sign of serious systemic problems can in some instances lead to congestive heart failure.

Though rare, water toxicity can also occur, with diluted sodium levels in the body leading to giddiness, nausea, convulsions, coma, and perhaps death.

TOTAL CARBOHYDRATE

The chief role of carbohydrate for humans is to furnish energy to the body in its converted form of glucose. As glucose, it serves several physiological functions: It provides the energy for normal cell functioning; it forms part of the nucleic acids DNA and RNA, of the cell membranes, and of connective tissue; it is a substrate for the synthesis of non-essential amino acids; and it can be converted to fat for storage, or stored as glycogen in the liver and muscles.

Of dietary carbohydrates, three major forms are of primary nutrition interest. These are the simple sugars, or mono- and disaccharides, which are sweet to the taste; the complex carbohydrates, polysaccharides, or starches, which are the predominant digestible form of carbohydrate in cereal grains and legumes and are tasteless; and fiber, which is not broken down by human digestive enzymes but nonetheless has important physiological functions.

Glycogen, the principal carbohydrate in animal tissue, largely disappears upon death of the organism and so is not readily available in the diet. Dairy food contains lactose, a disaccharide. Table sugar (sucrose) is a disaccharide. The non-water portion of all vegetable foods is composed in large part of carbohydrate. Simple carbohydrates predominate in fruits and vegetables, complex carbohydrates dominating in cereals and legumes.

Regardless of the form in which carbohydrates exist in the diet, all carbohydrates must be converted to simple sugars through digestion before they can be absorbed and then converted by the liver to glucose for use by the human body. Even so, the form in which carbohydrate is ingested may have significant metabolic effects.

Absorption of lactose may be affected in those individuals who have a genetic inability to produce sufficient lactase to digest the sugar found in milk. In affected individuals, this lactose intolerance is usually absent in childhood, but becomes progressively more serious with age. Evidence indicates that most non-Caucasian populations are largely lactose intolerant. As an example, Chinese and Japanese cooking are noted for being devoid of milk and milk products.

There is no RDA for carbohydrates, although the Food and Nutrition Board recommends a minimum intake of 50 to 100 g per day. It is not uncommon for human diets to contain over four times this amount. Ingesting less than 50 g of carbohydrate per day may force the body to meet energy needs by metabolizing protein for energy, leading to wasting of body muscle mass.

CRUDE FIBER/DIETARY FIBER

Fiber is actually a number of undigestible carbohydrate and carbohydrate-like substances, largely cellulose, hemicellulose, pectin, lignin, and gums, that cannot be processed by human digestive enzymes and thus pass intact into the colon. By absorbing water and adding bulk to the fecal matter, fiber assists in effective elimination of wastes.

Fiber was generally ignored by nutritionists as the non-nutritive component of food until it was noted that the high intake of fiber in some traditional African diets was associated with a lower incidence of a wide variety of diseases relative to the West (Burkitt and Trowell, 1975). In particular those diseases related to the digestive tract, such as diverticulitis or hiatus hernia were of lower incidence. Cardiovascular disease, colon and breast cancer, and diabetes may also be related to low fiber consumption.

Despite the increased interest, no RDA has been established for fiber.

Original estimates of fiber content were based on the material undissolved by concentrated acid as well as concentrated base. This is just a minor portion of the total fiber present in foods, being primarily a measure of the cellulose and some of the lignin. However, research indicates that components of fiber other than cellulose may play physiologically important roles in the prevention of many diseases. Therefore, a more complete and useful measure of fiber is dietary fiber. Estimates of both the crude and dietary fiber content of foods are included in *NUTRIENTS IN FOODS*.

There is an indication that some components of fiber may inhibit the absorption of some minerals and trace elements. Thus, excess fiber intake may decrease the bioavailability of these nutrients in the diet. Excess intake of dry bran has been shown to cause intestinal blockage as well.

TOTAL SUGAR

In response to the growing concern about the high proportion of simple sugars (mono- and dissacharides) in the U.S. diet, *NUTRIENTS IN FOODS* provides nutrient data for total sugar content. In the average U.S. diet, it is estimated that simple sugars are responsible for over 20% of the calories, and three quarters of this is from sugar added during food processing, cooking, or at the table.

The total sugar content data in *NUTRIENTS IN FOODS* are the sum total of all simple sugars that occur in the foods, including sucrose (table sugar), fructose, lactose (milk sugar), galactose, maltose, and glucose. A single value is given because the data are most readily available in combined form. This is particularly true in the case of processed foods, baked goods, and confectionary goods, where shifts in ingredients may alter the relative percentage of constituent sugars, but not the total amount.

TOTAL FATS

Fats provide approximately 40% of the calories in the U.S. diet. Fat is the principal and most efficient storage form of energy, although it has many other functions as well. Cell membranes are predominantly made of fats as are the myelin sheaths surrounding nerve fibers. Additionally, some fats are important in the formation of steroid hormones, and in the absorption of the fat-soluble vitamins (A, D, E, and K). A subcutaneous layer of fat also serves to insulate the body, helping to maintain body temperature.

Beyond the influence of the total amount of dietary fats (lipids), the chemical structures of the different lipids impart very different physiological functions. In particular, those lipids whose chemical structure includes an acid group—the so-called saturated, monounsaturated and polyunsaturated fatty acids—have important physiological differences. It should be noted that not all lipids are fatty acids.

Two other fat components, cholesterol and linoleic acid, are of nutritional interest. The former is important in the level of blood lipids, and the latter is the principal essential fatty acid.

The minimum dietary requirement for total fat for an average U.S. adult, assuming other sources of energy (carbohydrate & protein), is met by a diet containing 15-25 g of the proper fats.

Fat-deficient diets disrupt tissue metabolism and the regulation of cholesterol metabolism as well as deny the body essential fatty acids which are the precursors for prostagladins and other critical metabolic factors.

SATURATED, MONOUNSATURATED, AND POLYUNSATURATED FATTY ACIDS, AND THE P/S RATIO

Depending on their chemical structure, fatty acids are classified as saturated (with no double bonds), monounsaturated (with one double bond), or polyunsaturated (with two or more double bonds). Saturated fatty acids tend to be solid at room temperature and are commonly known as fats, while polyunsaturated fatty acids tend to be liquid (oils).

In general, saturated fatty acids are more prevalent in meat from birds and mammals, where fat is the major storage form of energy and serves as a subcutaneous layer of insulation. Polyunsaturated fatty acids are more common in fish and plants, exceptions being coconut and palm oil. Polyunsaturated fatty acids predominate in cell membranes where fluidity to facilitate transport of nutrients and wastes into and out of cells is important.

Fatty acids with many double bonds are easily oxidized, and the food industry has utilized this property to make these polyunsaturated fatty acids more stable through the process of hydrogenization (indicated by (H) in *NUTRIENTS IN FOODS*). Polyunsaturated fatty acids reacted with hydrogen become more saturated and less susceptible to spoilage through oxidation. Another result may be increased solidity, as in margarine.

Monounsaturated fatty acids, which may be considered relatively neutral in most dietary recommendations, account for the bulk of fat content in most diets.

Diets are rarely limited to one form of fat or another, and the relative amount of the different fatty acids is an important factor in influencing the level of blood lipids. In this regard, an important measure of the fatty acid content of specific foods or meals is the ratio of the polyunsaturated fatty acids to the saturated fatty acids, or the P/S ratio. Generally, in terms of the elevation of blood cholesterol, the lower the P/S ratio, the greater the blood cholesterol level will probably be.

LINOLEIC ACID

It is important to realize that some of the fatty acids are also essential nutrients. Principal among these essential fatty acids is linoleic acid. Linolenic acid, and sometimes arachidonic acid, are also included among the essential fatty acids in the diet, but if linoleic acid is present in adequate amounts, then the other essential fatty acids will be present in adequate amounts as well. In addition, arachidonic acid can be synthesized from linoleic acid and is thus not a dietary essential in the presence of adequate dietary linoleic acid.

The essential fatty acids have many functions in the body, including being the major fatty component of cell membranes and playing an integral role in cholesterol metabolism. In addition, the essential fatty acids are converted into hormone-like substances known as prostaglandins and thromboxanes. These are crucial in the clotting of blood, the control of smooth muscle tension, and many other physiological processes.

Although there is no established RDA for linoleic acid, recommendations for the ideal amount of essential fatty acids to be consumed suggest an intake of approximately 2% of calories. This amount is met by virtually all diets, because linoleic acid is contained in most foods. Especially rich sources of linoleic acid are the vegetable oils.

CHOLESTEROL

Dietary cholesterol is of major interest because of its potential effect on the level of blood cholesterol. Among dietary factors, cholesterol was probably the first found to influence atherosclerosis, when rabbits were fed cholesterol by investigators early in the twentieth century. However, relative to the effect of saturated and polyunsaturated fatty acids on blood cholesterol, the effect of dietary cholesterol appears to be minor. Regardless, dietary cholesterol is one among many factors that can influence blood cholesterol levels and possibly the development of atherosclerosis and coronary heart disease.

It must be stressed that cholesterol also plays important positive physiological roles. It is a precursor for all the steroid hormones, including aldosterone and the sex hormones, estrogen and testosterone. In the presence of ultraviolet light, it is converted into a precursor of vitamin D, and it is a necessary component of cell membranes, helping to stabilize them and prevent their destruction.

Although it is important physiologically, cholesterol is not a dietary essential. Unlike most of the vitamins, minerals, and trace elements, it need not be consumed in the diet, because in the absence of dietary cholesterol it is produced by the body. Indeed, there is no recommended minimal intake as in the case of carbohydrate, which, strictly speaking, is also not a dietary essential.

Foods of animal origin are virtually the only source of dietary cholesterol. Egg yolks and organ meats tend to have the highest concentration of cholesterol, although all animal foods contain some cholesterol. Diets that consist predominantly of vegetable foods, such as are encountered in most of the Third World, have little dietary cholesterol.

TOTAL PROTEIN AND THE AMINO ACIDS

Proteins are composed of chains of amino acids and protein requirements are related to amino acid requirements. Only about half of the twenty-two amino acids used by the human body are usually considered to be essential, i.e., they cannot be synthesized by the body, or are synthesized in less than adequate amounts for body needs and thus must be provided in the diet. The other amino acids are normally synthesized by the body provided that adequate nitrogen is available.

Values for the following amino acids are given in *NUTRIENTS IN FOODS* (Arginine is usually not considered an essential amino acid; cystine and tyrosine only under certain circumstances).

Arginine	Methionine
Cystine	Phenylalanine
Histidine	Threonine
Isoleucine	Tryptophan
Leucine	Tyrosine
Lysine	Valine

Table 6 gives the estimated human amino acid requirements.

Except during the first few weeks of life, intact protein is not absorbed directly into the system. Instead, it must be broken down into peptides and amino acids which are then absorbed and used by the body. Therefore, in terms of the nutritional adequacy of protein, the protein source is unimportant as long as it can be converted into adequate quantities of the essential amino acids for human body needs. This latter is very important in assessing the quality of protein, which varies widely.

Scoring

The Food and Agricultural Organization and World Health Organization (FAO/WHO) have established a reference protein having the proper proportion of essential amino acids for human needs. Table 7 shows this reference protein with the amount of each amino acid in milligrams per one gram of protein. This table can be used for "scoring" the amino acid content of a food, meal, or day's intake of food to determine its nutritional adequacy. Since the content of individual essential amino acids is given in *NUTRIENTS IN FOODS*, foods or meals can be "scored" for adequacy of amino acid content using the FAO/WHO reference protein.

To "score" proteins using these tables, take the number of milligrams of each essential amino acid from the main nutrient tables and divide it by the number of grams of protein given for that food using the formula in Table 7.

It is important to understand that the overall score of a given protein is based upon the 'limiting' amino acid (the one that is lowest relative to the FAO/WHO reference protein). Generally, the limiting amino acids in cereal grains are lysine, tryptophan, and threonine; while in legumes, the limiting amino acids are the sulphur-containing amino acids, cystine and methionine.

TABLE 6
Estimated Amino Acid Requirements of Man[a]

Amino Acid	Requirement, mg/kg body weight/day			Amino Acid Pattern for High-Quality Proteins, mg/g of protein
	Infant (4–6 months)	Child (10–12 years)	Adult	
Histidine	33	?	?	17
Isoleucine	83	28	12	42
Leucine	135	42	16	70
Lysine	99	44	12	51
Total *S*-containing amino acids (methionine and cystine)	49	22	10	26
Total aromatic amino acids (phenylalanine and tyrosine)	141	22	16	73
Threonine	68	28	8	35
Tryptophan	21	4	3	11
Valine	92	25	14	48

SOURCE: FNB, 1975.

[a]Two grams per kilogram of body weight per day of protein of the quality listed in column 4 would meet the amino acid needs of the infant.

Availability

While the relative amounts of the essential amino acids in a protein determine its adequacy, other factors further affect the availability of dietary protein for digestion. Heating foods during normal cooking may aid in later digestion of protein, but high and excessive heat can increase the bonding between some amino acids and hinder digestion. Some foods, particularly legumes, contain substances that may interfere with the digestion of protein. Heat destroys some of these substances, so cooking is an important process in increasing the bioavailability of the protein in food. Protein digestion and absorption may also be affected by the presence of pancreatic or intestinal disease.

The RDA's for protein vary with age and sex, ranging from 2 g per kg body weight per day at infancy to 56 g per day for a 70 kg adult male and 44 g per day for a 55 kg adult female. During pregnancy and lactation, protein requirements rise.

Not only do foods of animal origin provide the greatest amount of protein, but meat provides the most complete proteins. Legumes and nuts have the highest levels among vegetable sources of protein, but many vegetable foods are incomplete sources of amino acids.

Protein deficiency is manifested in general weakness, diarrhea, collection of body fluids (edema), skin sores, and some hair loss. Fat accumulates in the liver, and serum albumin drops, as do blood amino acid levels. Pancreatic enzyme output also slows.

Severe protein deficiency, kwashiorkor, is self-reinforcing. Protein inadequacy lowers body resistance, and the resulting infections make further demands on available proteins. In addition, the drop in pancreatic enzymes reduces the system's ability to absorb protein. This downward spiral can lead to death.

TABLE 7
Amino Acid Scoring

Amino Acid	Level in mg/gm of Protein
Isoleucine	40
Leucine	70
Lysine	55
Methionine & cystine	35
Phenylalanine & tyrosine	60
Threonine	40
Valine	50
Tryptophan	10
Total	360

$$\text{Amino Acid Score} = \frac{\text{Amino Acid in mg per gm of test protein}}{\text{Amino Acid in mg per gm of reference protein}} \times 100$$

Source: Kreutler, 1980 (from FAO, 1973).

The Water-Soluble Vitamins

ASCORBIC ACID (VITAMIN C)

The primary function of vitamin C is the maintenance of collagen, a protein important in the formation of connective tissue in skin, bone, ligaments, and scar tissue. Vitamin C is also involved in phenylalanine and tyrosine metabolism, and may play an important role in the function of the adrenal glands and white blood cells. Many of these apparent functions are related to the antioxidant properties of ascorbic acid, which is also the reason vitamin C is popular as a food additive.

Vitamin C is recognized as the vitamin most subject to change by environmental factors, and thus published values of the vitamin C content of foods can differ markedly from that actually consumed. As a water-soluble vitamin, vitamin C can be leached into cooking water when boiling fruits or vegetables. High temperatures or prolonged cooking, the presence of baking soda, or the use of copper cooking vessels further destroys much of the vitamin C in foods. Exposure to light also reduces vitamin C content. Thus, the fresher an uncooked fruit or vegetable, the greater the vitamin C content of that food will be.

In addition, the absorption and utilization of vitamin C by the body is altered by a number of factors. Smoking, stress, fever, long-term use of cortisone or antibiotics, and the use of aspirin or oral contraceptives appear to reduce absorption of or increase requirements for vitamin C. Vitamin C requirements also apparently increase during wound healing.

Fruits and vegetables are good sources of vitamin C. Citrus fruits and juices are popularly known as vitamin C-rich foods. In addition, in the U.S. many canned juices are fortified with vitamin C. Other fruits and vegetables, such as strawberries, melons, green peppers, and the cruciferous vegetables from the cabbage family are good sources as well. For example, raw potatoes are a sufficiently good source to have figured in the successful reversal of the symptoms of scurvy (though French fries would not be a very good source for this vitamin).

The RDA for vitamin C is set at 60 mg per day for adults. This is purposely well above the requirements for the prevention of scurvy, which is around 10 mg per day.

The adult body stores about 1500-5000 mg of vitamin C. In deprivation diets when body stores fall below 600 mg, psychological changes are observed. Below 300 mg, symptoms of scurvy appear. Scurvy manifests itself in the breakdown of connective tissue and capillary hemorrhage, resulting in swollen joints, bleeding gums and other hemorrhaging, tooth loss, muscle cramps, aching bones, lowered resistance to infection, poor wound healing, lassitude, weakness, and ultimately death.

Many claims have been made about the preventive or curative effects of doses of vitamin C well above the RDA. Viral and bacterial infections such as the common cold, arthritis, drug addiction, and cancer are some of the diseases which have been claimed to be related to deficiency of vitamin C intake. These claims have generally not been substantiated by scientific investigation. Indeed, populations around the world and throughout history have been apparently healthy without taking megadoses of vitamin C. This observation and the possibility that, although it is readily excreted, megadoses of vitamin C may promote the formation of kidney stones or gout in some individuals suggest that it is prudent to obtain vitamin C from foods rather than from supplements.

THIAMIN (VITAMIN B-1)

Thiamin was the first vitamin whose physiological role in the body was described biochemically. As the coenzyme thiamin pyrophosphate it is particularly important in reactions central to energy metabolism.

The milling of cereal grains removes the bran and, with it, much of the thiamin that is naturally present. It is this fact that led to the discovery of this vitamin, because populations which consumed refined rice as a staple food often developed beriberi, which was subsequently discovered to be caused by thiamin deficiency. Since refining removes much of the vitamin, it is necessary to ensure that dietary amounts of thiamin will be adequate. This is usually done during food processing with the enrichment of commercial cereals, white flour, rice, etc.

Factors other than refining influence the amount of thiamin in food or in the diet. Thiamin can be lost in cooking water due to its water-solubility, and heat can easily destroy the activity of thiamin in an alkaline environment. Thiamin absorption can be decreased by alcohol, barbituates, or tea in large quantities, and there is a thiaminase present in raw fish that can also decrease the content of thiamin in the diet. This latter may be a contributing factor to the observation that populations in the Far East have historically been particularly prone to the development of thiamin deficiency.

Because thiamin is important in energy metabolism, increased caloric intake increases the requirement for this vitamin. In this regard, the RDA for thiamin is usually expressed in terms of milligrams per 1000 kcal. The symptoms of beriberi, which include nervous and cardiovascular system problems such as muscle weakness, anorexia, and mental confusion, may be related to decreased energy availability due to the absence of thiamin.

Although many foods contain significant amounts of thiamin, the only rich dietary sources are whole cereal grains, legumes, seeds, and yeast.

Toxicity from dietary intake is highly unlikely because thiamin is very efficiently excreted.

RIBOFLAVIN

Riboflavin, or vitamin B-2, participates in the metabolism of energy from fats, carbohydrates, and proteins. It is also involved in the production of red blood cells and the synthesis of corticosteroids. Due to its role in energy metabolism, daily recommended intakes of riboflavin are sometimes related to energy intake and expressed in milligrams per 1000 kcal.

Riboflavin readily decomposes if exposed to heat in an alkaline solution, although it is heat-stable in acidic solutions. Thus, the presence of baking soda, which is alkaline, reduces the amount of riboflavin present in foods. Riboflavin can also be leached into cooking water. Because it is destroyed by light, the dairy industry attempts to conserve the riboflavin present in milk by using opaque bottles and cartons. Riboflavin is relatively unaffected by oxidation.

In the digestive tract, the gout medication Probenecid decreases riboflavin absorption, sulfa drugs decrease utilization of riboflavin, and some oral contraceptives lower serum levels.

Dairy products are the major source of riboflavin, providing over half the riboflavin in the U.S. diet. Other good sources include green leafy vegetables, winter squash, whole grains, and organ meats. Beyond this, enriched bread and cereals provide a significant amount of this vitamin.

Riboflavin deficiency is rare. Reported cases have resulted largely from diet restrictions due to alcoholism, food fadism, etc. Signs of deficiency include angular stomatitis (an inflamed mouth with cracks at the corners), dry, scaly facial skin, and other abnormal conditions. However, these symptoms are non-specific, also being caused by other conditions, and the effects of riboflavin deficiency are minor.

No adverse effects of excess riboflavin intake are known.

NIACIN

Like riboflavin, niacin (also known as nicotinic acid or nicotinamide) is an important coenzyme involved in the release of energy from carbohydrate, fat, and protein. It is also involved in the synthesis of sex hormones and the metabolism of cholesterol as well as the proper functioning of the nervous system. Due to its involvement with energy metabolism, niacin requirements are given in terms of milligrams per 1000 kcal.

The amount of niacin present in foods is generally unaffected by storage, heat, and cooking, though some may be lost in cooking water. However, the niacin in foods, particularly in maize (corn) and perhaps potatoes, exists in a form that is very poorly absorbed by the human body unless treated by alkali during food preparation.

Niacin deficiency disease, or pellagra, reached endemic levels in the Southern U.S. at the beginning of this century. Diets which contained large amounts of maize were seen as partly responsible. Some traditional corn staples, such as Mexican tortillas, were made from maize treated with lime water, a practice that liberated the niacin in the maize so that it was available for absorption. Other foods, such as millet and sorghum, widely eaten in some Third World countries, contain high concentrations of niacin as well as tryptophan, but the presence of large amounts of leucine drastically reduces the amount of niacin available to the system. In the U.S. at present, dietary changes, in particular the enrichment of bread and commercial cereal with niacin, have virtually eliminated pellagra.

Dietary niacin is not the only source of niacin for the human body. It may also be synthesized from the amino acid tryptophan. Approximately 60 mg of tryptophan can replace or contribute 1 mg of niacin. Therefore, the niacin available from a diet is best measured in niacin equivalents, calculated as the total niacin in the diet plus one-sixtieth of the tryptophan. *In NUTRIENTS IN FOODS, niacin values reflect preformed dietary niacin only, so the tryptophan content of the diet should also be taken into account to derive the best estimate of the available niacin.*

It should be noted that the RDA for niacin is based on niacin equivalents, while the U.S.RDA is not.

Niacin occurs naturally in many foods, but is best available from legumes, fish, and organ meats. Whole grains contain relatively high levels of niacin, but it may be in largely unabsorbable form unless processed properly.

Niacin deficiency leads eventually to pellagra, characterized by dermatitis, inflammation of the mouth and digestive tract, and mental disorders, and results in death if not treated. Pellagra may be the result of deficiencies of not only niacin but also thiamin, riboflavin, and vitamin B-6; the latter being necessary for the synthesis of niacin from tryptophan.

Large oral doses of niacin in the form of nicotinic acid cause tingling, flushing, and a burning sensation. Long term use may result in liver toxicity and other serious problems. Even so, nicotinic acid is sometimes prescribed as a vasodilator for a number of vascular disorders, and pharmacological doses have been used in the treatment of hyperlipidemia.

VITAMIN B-6

Vitamin B-6 is involved in amino acid metabolism, including the synthesis of nonessential amino acids. The synthesis of hemoglobin and the conversion of tryptophan to niacin are also dependent on vitamin B-6.

The content of vitamin B-6 in foods may be reduced by exposure to ultraviolet light, and as a water-soluble vitamin, B-6 may be leached into cooking water. Otherwise vitamin B-6 is relatively stable and is widely distributed in foods. Whole grain cereals, legumes, bananas, and meats are particularly good sources.

Individual requirements for vitamin B-6 may increase if drugs which are pyridoxine antagonists, such as penicillamine, isoniazid, or oral contraceptives, are taken, and diuretics may deplete vitamin B-6 in the system. Absorption may be affected if the individual is an alcoholic or has celiac disease. High protein diets also appear to increase the requirement for vitamin B-6, and pregnancy is associated with a decrease in blood vitamin B-6 levels, causing most nutritional recommendations for pregnancy to include increasing vitamin B-6 intake.

Vitamin B-6 dietary deficiency is not considered a major health problem. However, induced deficiency effects are sufficiently serious to establish an RDA. These include disorders of the skin, nerves, tongue and mouth sores, nausea, insulin sensitivity, and in extreme cases severe convulsions. Since vitamin B-6 is involved in tryptophan conversion to niacin, it may play a minor role in pellagra.

Dietary toxicity of vitamin B-6 is not an issue.

FOLACIN

Along with vitamin B-12, folacin is commonly thought of as one of the anemia-preventing vitamins. Folacin is primarily involved in the synthesis of nucleotide bases needed for DNA formation, and is thus necessary for cell division. This is the basis for the classic megaloblastic anemia seen with folacin deficiency. As red blood cells are among the most rapidly dividing cells in the body, impairment of DNA synthesis would most likely affect these cells first.

The importance of folacin in DNA synthesis and cell division provides the basis for some of the chemotherapeutics used in the treatment of cancer. In an attempt to prevent rapidly multiplying cancer cells from proliferating, folacin antagonists such as methotrexate or aminopterin are administered, inhibiting the synthesis of DNA and preventing cell division.

Food preparation can adversely affect the folacin content of foods. Prolonged heating, discarding cooking water, and microwave cooking reduce the folacin content in food. Folacin absorption may also be inhibited by alcohol, anticonvulsants, and antitubercular drugs. Pregnancy and the use of oral contraceptives decrease the amount of folacin present in the blood, thus increasing requirements for the vitamin.

Folacin occurs widely in foodstuffs, particularly in liver, green leafy vegetables, and legumes. Whole grains, oranges, and bananas are also reasonable sources. However, other fruits, meats, and dairy food are relatively poor sources of folacin.

The adult RDA for folacin (400 mcg) is based on the assumed requirement of 100-200 mcg per day to maintain tissue levels and an average availability of 25-50%. The infant RDA for folacin is based on average levels in mothers' milk.

Folacin deficiency leading to megaloblastic anemia is rare in industrialized countries, but is not uncommon in parts of the third world.

Dietary folicin toxicity is not known, although some of its analogues such as the chemotherapeutic agent, methotrexate, are extremely toxic.

VITAMIN B-12

Vitamin B-12 was the last vitamin to be identified, being isolated almost simultaneously in the United States and Britain in 1948. One of the physiological functions of vitamin B-12 is closely linked with the action of folacin, and thus deficiency of vitamin B-12 also leads to impairment of DNA synthesis and megaloblastic anemia. However, vitamin B-12 also plays a role in the synthesis and metabolism of the fatty myelin sheath that surrounds nerve cells and is therefore necessary for proper neurological function. This second aspect distinguishes the signs of vitamin B-12 deficiency from those of folacin deficiency, because the megaloblastic anemia produced by the deficiency of either is eventually accompanied by nervous degeneration if the cause is a lack of vitamin B-12.

Vitamin B-12 is usually quite stable during food preparation although some losses can occur at boiling temperatures in alkaline solution. Age, iron and vitamin B-6 deficiencies, and gastritis decrease the absorption of vitamin B-12. The per cent absorption of vitamin B-12 varies inversely with current body levels.

Vitamin B-12 is found almost exclusively in foods of animal origin. Although plants do not need or manufacture vitamin B-12, it is possible for vegetable foods to contain some vitamin B-12, due to the presence of micro-organisms such as yeasts, molds, or bacteria that are often an intrinsic part of fermented foods. However, most commonly consumed vegetable foods are devoid of this vitamin, so they do not contribute vitamin B-12 to the diet.

Due to the lack of vitamin B-12 in most vegetable foods, dietary deficiency may occur in strict vegetarians if fermented foods are not included in the diet. However, this deficiency is extremely rare, even among strict vegetarians, suggesting that incidental contamination of foods with micro-organisms provides enough vitamin B-12 for physiological needs.

Unlike most of the other water-soluble vitamins, vitamin B-12 is efficiently stored in the liver so that daily consumption is not necessary. This efficient storage may contribute to the rarity of dietary vitamin B-12 deficiency.

An RDA has been established for vitamin B-12 even though dietary deficiency from inadequate intake is rare.

The more common cause of vitamin B-12 deficiency is the absence of Castle's intrinsic factor, a protein synthesized by the stomach and necessary for the efficient absorption of vitamin B-12 from the diet. This commonly occurs with aging, but may also result from an inborn error of metabolism. Lack of Castle's intrinsic factor results in pernicious anemia, a progressive and ultimately fatal disease characterized by megaloblastic red blood cells where these cells are larger but fewer in number. While megaloblastic anemia can also result from folic acid deficiency, in pernicious anemia nervous system problems eventually develop.

No toxic effects of vitamin B-12 are known.

PANTOTHENIC ACID

Pantothenic acid is an integral part of coenzyme A and acyl carrier protein. As such it participates in a wide variety of biochemical reactions in the body, including the breakdown of carbohydrate, protein, and fat as well as the synthesis of fatty acids, cholesterol, steroid hormones, and acetylcholine.

Pantothenic acid is unstable in alkaline (e.g., baking soda) or acid (e.g., vinegar) environments, and may also be lost at temperatures above boiling. Thawing of frozen foods may lead to significant loss of pantothenic acid in the dripping that occurs, and milling of cereal grains can remove half the pantothenic acid present in the whole grain.

The best sources of pantothenic acid include whole grains, legumes, some vegetables, yeast, egg yolk, and organ meats.

Dietary deficiency of panthothenic acid is not known to occur in humans. However, induced deficiencies can affect cells in many tissues, producing weakness, muscle cramps, impaired antibody production, duodenal ulcers, and lowered adrenal function. Therefore, there is some concern over potential pantothenic acid losses in food processing. Deficiency may also result if the intestinal flora are insufficient to synthesize large amounts of pantothenic acid.

Excess doses, 10-20 grams of pantothenic acid, may result in diarrhea.

Although there is no established RDA for pantothenic acid, the Food and Nutrition Board has estimated safe and adequate intake levels for this vitamin.

BIOTIN

Biotin is important in a wide variety of metabolic processes. As a carrier for carbon dioxide, it is important in the construction of carbon chains in fatty acid synthesis. It is also involved in amino acid metabolism, synthesis of pancreatic amylase, and the formation of antibodies, as well as in the utilization of folacin, vitamin B-12, and pantothenic acid.

Largely unaffected by heat and only partly water-soluble, biotin is relatively stable during cooking and processing. Biotin is also contained in a large variety of foods, the richest sources being organ meats and yeast, although legumes, nuts, and some vegetables also contain fair amounts. Thus, human diets are rarely if ever devoid of biotin.

It is questioned whether biotin needs to be provided in the diet at all since it is synthesized in large quantities by bacteria in the digestive tract. This may provide more than enough of the vitamin for physiological needs.

There is no RDA for biotin, although the Food and Nutrition Board does recommend a daily intake of 100-200 mcg.

The only reported cases of biotin deficiency in humans are from individuals who eat very large quantities of raw eggs (six dozen raw eggs weekly). Raw egg whites contain the protein avidin which binds biotin, preventing its absorption. When biotin deficiency is experimentally induced, symptoms include skin abnormalities, muscle pain, and nausea.

No toxic effects are known.

The Fat-Soluble Vitamins

VITAMIN A, RETINOL, AND BETA-CAROTENE

The role of vitamin A in vision is well-recognized. Its role in rhodopsin (visual purple) regeneration in the cones of the retina affects the rapidity with which the eye regains its ability to see in dim light after it has been exposed to bright light. The gradual loss of this night vision is the first symptom of vitamin A deficiency.

However, vitamin A has many other physiological roles, most of which are poorly understood. It is important in the growth and maintenance of body tissues, in particular the epithelial tissues and mucous membranes of the skin and the respiratory, reproductive and digestive systems. It may be through this mechanism that vitamin A is critical in protecting the body from infections. Vitamin A is also important in the proper development of bone, and in steroid hormone synthesis.

The physiologically active forms of vitamin A are retinol, retinal, and retinoic acid. The principal dietary forms of the vitamin are preformed retinol (given as 'preformed A' in *NUTRIENTS IN FOODS*), found in foods of animal origin, and beta-carotene, or pro-vitamin A, found primarily in foods of vegetable origin. Beta-carotene is known as pro-vitamin A because it can be converted into the physiologically-active form of vitamin A by the body. There are other carotenoids that have some vitamin A activity, but beta-carotene is the most important. Beta-carotene is responsible for the yellow-orange color found naturally in carrots and winter squash, and is sometimes used as a food coloring.

The various forms of vitamin A have different physiological activity, traditionally expressed in International Units (IU).

1 IU vitamin A = 0.3 mcg retinol = 0.6 mcg beta-carotene

More recently, vitamin A requirements have been expressed in micrograms of retinol equivalents (RE), since retinol is the physiologically active form of the vitamin.

1 retinol equivalent = 1 mcg retinol
= 6 mcg beta-carotene
= 12 mcg other provitamin A carotenids
= 3.33 IU vitamin activity from retinol
= 10 IU vitamin A activity from beta-carotene

NUTRIENTS IN FOODS lists retinol and beta-carotene in retinol equivalents, and the total vitamin A activity in International Units. To determine the retinol equivalents for a food or for a diet from sources other than *NUTRIENTS IN FOODS*, use the following formulae.

1. If retinol and β-carotene are given in micrograms, then:

$$\mu g \text{ retinol} + \frac{\mu g\ \beta\text{-carotene}}{6} = \text{retinol equivalents.}$$

Example: A diet contains 500 μg retinol and 1800 μg β-carotene.

$$500 + \frac{1800}{6} = 800 \text{ retinol equivalents.}$$

2. If both are given in IU, then:

$$\frac{\text{IU of retinol}}{3.33} + \frac{\text{IU of } \beta\text{-carotene}}{10} = \text{retinol equivalents.}$$

Example: A diet contains 1666 IU of retinol and 3000 IU of β-carotene.

$$\frac{1666}{3.33} + \frac{3000}{10} = 800 \text{ retinol equivalents.}$$

3. If β-carotene and other provitamin A carotenoids are given in micrograms, then:

$$\frac{\mu g\ \beta\text{-carotene}}{6} + \frac{\mu g \text{ other carotenoids}}{12} = \text{retinol equivalents.}$$

Example: A 100-g sample of sweet potatoes contains 2400 μg β-carotene and 480 μg of other provitamin A carotenoids.

$$\frac{2400}{6} + \frac{480}{12} = 440 \text{ retinol equivalents.}$$

Source: National Academy of Sciences, 1980.

Vitamin A is widely available in diets, with liver, yellow-orange and green leafy vegetables having the highest concentrations.

Neither retinol nor beta-carotene is particularly affected by cooking and processing, although cutting and chopping food may release vitamin A by rupturing cell membranes. However, oxidation reactions in the processing of dehydrated fruits and vegetables will lead to some loss of beta-carotene. In addition to the differing physiological activity of retinol and beta-carotene, the presence of excess alcohol, iron, mineral oil, or cortisone can reduce absorption of vitamin A.

Once in the system, zinc deficiency, protein deficiency, and hepatitis lower vitamin A serum levels, while oral contraceptives or kidney disease may increase serum levels, the latter reducing the amount excreted.

As a fat-soluble vitamin, there is little loss of retinol or beta-carotene through excretion in the urine, so vitamin A can be toxic. This toxicity of dietary vitamin A has been observed in arctic explorers who dined on polar bear liver, which contains extremely large concentrations of retinol. There have also been cases of retinol toxicity in people using supplemental retinol. Symptoms include skin problems, loss of appetite, headaches, central nervous system aberrations, and enlarged livers. On the other hand, beta-carotene does not seem to be toxic in large doses, the only symptom of excess intake being a benign yellow-orange coloring of the skin due to the excess yellow-orange pigment being stored in fat deposits throughout the body.

Vitamin A deficiency is also potentially a major problem and is one of the major causes of blindness in several parts of the Third World. With inadequate intakes of vitamin A during periods of rapid growth, the eyes can become keratinized and brittle and eventually lose their ability to function, a condition known as xeropthalmia.

The RDA for Vitamin A is very generally related to body weight, with the adult female allowances about 80% of the adult male, and RDAs for children and adolescents interpolated from infant and adult RDAs.

There has recently been increased interest in vitamin A, especially beta-carotene, because of the potential role it may play in the prevention of cancer. Of the many dietary factors that have been hypothesized as being related to cancer, the evidence for the preventive capabilities of beta-carotene is among the strongest (Ames, 1983). Therefore, beta-carotene is presently the subject of intense research. However, this potential role for beta-carotene is at present very poorly understood, and excess intakes of beta-carotene are not warranted.

VITAMIN D

Vitamin D plays an integral role in the metabolism of calcium and phosphorus. It is a critical agent in the mineralization of bone and in the mobilization of minerals from bone. Vitamin D also mediates the absorption of dietary calcium and phosphorus, and infuences the excretion and reabsorption of these minerals via the kidney.

Some people point out that vitamin D is not a vitamin in the strict sense, in that it can be synthesized by the body, and that, as calciferol, it more closely resembles a hormone. Indeed, many people throughout the world receive little or no vitamin D in their diet, obtaining all their needs for the vitamin from the synthesis of vitamin D through the action of ultraviolet light on 7-dehydrocholesterol in the skin. However, in many environments, in particular northern urban settings, adequate amounts of vitamin D cannot be provided through synthesis in the presence of sunlight and thus must be provided in the diet.

Naturally occurring dietary sources of vitamin D are uncommon. The only rich dietary sources known are fish liver oils, and these were used as traditional folk remedies in Scotland since at least the early eighteenth century. Other seafoods, such as herring, sardines, and tuna, contain vitamin D in small amounts. Eggs and dairy food contain even smaller amounts. The only known vegetable sources of the vitamin are fungi and yeasts. Fortification of foods with vitamin D is a common practice in some countries, and therefore vitamin D deficiency is now extremely rare. In the United States and Canada, milk has been fortified with vitamin D since the turn of the century, while in Great Britain and some other European countries, margarine is fortified with this vitamin. More recently, some cereal manufacturers include vitamin D as one of the many nutrients added to their product.

As a fat-soluble vitamin, vitamin D is absorbed from the diet in much the same way as other fatty substances. Bile is required for absorption, and the presence of mineral oil and some drugs can interfere with absorption.

Vitamin D is largely unaffected by heat and light during processing and storage, and since it is a fat-soluble vitamin, losses during cooking occur only if the fatty portion of fish is discarded prior to eating. Several drugs may affect vitamin D metabolism, notably cortisone, barbiturates, anti-cholesterol drugs, and some anticonvulsants.

Vitamin D deficiency produces rickets, a developmental disorder of bones most dramatically seen in growing children. Classical symptoms include malformed ribs, described as a 'rachitic rosary,' bowed legs and other bone deformation, and increased bone breakage. Exposure to sunlight will alleviate these symptoms.

In northern Europe, sunning children in winter is a widespread folk practice. It has been calculated that exposing the largely unpigmented cheeks of an otherwise bundled-up Scandinavian baby to a few hours a day of weak winter sun will stave off the onset of rickets. By contrast, urban children in Industrial Revolution England suffered severe rickets in an environment in which people bundled up against the cold, lived and worked inside, and the sky had a continual pall of smoke and haze which screened out much ultraviolet light. Now with widespread dietary vitamin D fortification, rickets seldom occurs. However, infants and children who rarely consume foods fortified with vitamin D and who are infrequently exposed to sunlight are at risk of developing rickets.

Vitamin D is the only nutrient for which RDA intake levels are set higher for infants and young children than for adults. This reflects the greater growth needs at these ages and the assumption that adults will get part of their vitamin D through exposure to sunlight.

Although rickets is most destructive in small children, adults may develop rickets after many pregnancies or in old age. Rickets can also be caused by genetic factors and by disease and disorders which affect the two metabolic forms of vitamin D, a condition known as vitamin D-resistant rickets.

As with other fat-soluble vitamins, vitamin D is not readily excreted in the urine, instead being stored in fatty tissue and in the liver. Because of this, toxicity can occur if too much vitamin D is ingested. This leads to abnormal calcification of tissue as the body attempts to maintain normal blood calcium levels in the presence of increased absorption of calcium from the diet. Vitamin D toxicity due to overexposure to the sun is not known to occur.

VITAMIN E AND THE TOCOPHEROLS

Like vitamin A, vitamin E is actually a family of molecules that have vitamin E activity. These molecules are generally called tocopherols and tocotrienols, with alpha-tocopherol the most potent physiologically. Beta-tocopherol has about 40% of the activity of alpha-tocopherol, gamma-tocopherol has about 8% of the activity, and alpha-tocotrienol about 20%. These four molecules are the principal forms of vitamin E found in the diet.

Physiologically, the principal role of vitamin E appears to be as an antioxidant. Vitamin E is a fat-soluble vitamin found throughout the body in fatty tissue, especially in cell membranes. In this location, vitamin E appears to inhibit the oxidation of the polyunsaturated fatty acids that compose much of the cell membrane structure.

Oxidation of polyunsaturated fatty acids in cell membranes is associated with cell damage, so vitamin E has been thought to play a preventive role in many degenerative diseases, including cardiovascular disease and cancer (Ames, 1983). Other properties ascribed to vitamin E include improvement of sexual potency and reproductive capabilities, and inhibition of the aging process. However, none of these roles for vitamin E has been clearly delineated to date, and it is not clear that people who take large doses of vitamin E derive any physiological benefit. Fortunately, although vitamin E is a fat-soluble vitamin and is thus not easily excreted from the body, there appears to be little or no harmful effect from large doses of this vitamin.

The amount of vitamin E obtained from foods may vary due to its high sensitivity to cooking and processing loss. Although it is largely unaffected by boiling, high frying heat will oxidize it. Exposure to light or air will degrade this vitamin, so storage must be monitored to avoid oxidation. Chlorine dioxide bleaching of flour destroys what vitamin E has not been removed by milling. Similarly, the refining of vegetable oil removes much of its vitamin E. Contact with copper and iron can also decrease the amount of vitamin E present in food. Mineral oil will reduce the absorption of Vitamin E, while increased amounts of dietary polyunsaturated fatty acids can increase requirements for the vitamin.

Indeed, the RDA for Vitamin E is directly related to body levels of polyunsaturated fatty acids (PUFA), and the published RDA values represent average adequate intakes in balanced U.S. diets. The adequacy of these intakes will vary if PUFA intake is large or small.

Despite the potential losses of vitamin E from foods during processing and cooking, primary dietary deficiency of vitamin E in humans is virtually unknown. Only in abnormal circumstances, such as in premature babies with low body stores of the vitamin or in patients with malabsorption of fat, is there a danger of precipitating symptoms associated with vitamin E deficiency such as hemolytic anemia.

The richest dietary sources of vitamin E are unrefined vegetable oils, including wheat germ, sunflower seed oil, and palm oil. Whole grains, eggs, and some vegetables are also good sources.

In *NUTRIENTS IN FOODS*, vitamin E content of foods is given in International Units, which are roughly equivalent to milligrams of alpha-tocopherol. The tocopherol content is also given as total tocopherol, alpha-tocopherol, and other tocopherol. Although the other tocopherols in the diet have differing degrees of vitamin E activity, they can be approximated overall as having 20% of the activity of alpha-tocopherol. Thus, the milligrams of the total tocopherols contained in the foods can be converted into milligrams of alpha-tocopherol equivalents by adding to the milligrams of alpha-tocopherol one-fifth of the milligrams of other tocopherol.

In some instances in *NUTRIENTS IN FOODS*, alpha and other tocoperol values may not exactly equal the total calculated tocopherols, because data are scattered for these values and may have come from more than one reliable source.

It should be noted that tocopherols, like retinol units, are sometimes expressed in terms of tocopherol equivalents (as, for example, in the RDA tables, Table 2).

1 alpha tocopherol equivalent = 1 mg alpha tocopherol
= 2 mg beta-carotene
= 10 mg gamma tocopherol

The numbers given in NUTRIENTS IN FOODS are NOT tocopherol equivalents.

The Major Minerals

TOTAL ASH

Ash is not a nutrient, strictly speaking. It is a measure of the residue of a food after total oxidation in a bomb calorimiter. As such it contains some, but not all, of the mineral content of organic matter, since some minerals volatize under extreme heat. *NUTRIENTS IN FOODS* gives ash as a basis for comparison with older food composition tables which listed Total Ash as one of the bits of data that could be derived using earlier analytical techniques.

CALCIUM

Calcium, the principal component of bone and teeth is also important in a wide variety of other functions, including the control of blood clotting, the activity of muscle tissue, and the function of a number of enzymes. Calcium is also involved in the synthesis of acetylcholine and the absorption of vitamin B-12.

Because of the integral role of calcium in a large number of physiological processes, blood calcium levels are closely regulated through the interaction of a number of hormones, including vitamin D. In the presence of a calcium deficiency, the body will draw upon the reserves of calcium located in bone.

Vitamin D is required for the absorption and metabolism of calcium. The presence of lactose and the digestive enzyme, lactase, enhances the absorption of calcium as well. Dietary protein increases urinary excretion of calcium so that relatively high levels of dietary protein, common in the United States, probably produce a net loss of calcium. Indeed, this is given as one justification for the relatively high RDA's for calcium given in the U.S. (800 mg for adults) versus that recommended by the FAO/WHO (400-500 mg).

The RDA for calcium is highest for both sexes between 11 and 18 years of age, just when the adolescent growth spurt is promoting great increase in body size and stature.

The calcium content of foods varies from relatively large concentrations in dairy foods, green leafy vegetables, nuts, and legumes, to only small amounts in refined cereal grains. Whole grains, meats, and other foods appear to have intermediate concentrations of calcium. Drinking water may also contribute varying amounts of calcium. Thus, the calcium content of diets worldwide varies considerably, from only 200 to 300 mg per day, to well over 1 g per day.

Phytic acid, a component of fiber, may bind calcium and inhibit its absorption. However, the presence of a phytase in cereal grains may counteract this effect in dietary contexts. In addition, there appears to be a systemic adaptation to high fiber diets that allows the body to maintain calcium balance.

Although there is a wide variation in intake of calcium, it does not appear that osteoporosis, a disease of aging involving decalcification and weakening of bones, is related to the amount of calcium available in the diet. In fact, other aspects of lifestyle may be more important in the development and maintenance of healthy bones. Perhaps the most important of these is physical activity. It is well known that bone, like other tissues, will atrophy during periods of disuse as when people are bedridden, or a limb is placed in a cast for extended periods. It is sometimes taught that regular physical activity, more than increased dietary intake of calcium, is the best way to prevent osteoporosis.

Although high levels of calcium in the body may have undesirable effects, this may be due not to excess dietary intake but rather to other problems, such as vitamin D imbalances.

PHOSPHORUS

After calcium, phosphorus is the most prevalent mineral in the body, where it is widely distributed, though 80-90% is contained in bone and teeth.

In addition to its major role in bone formation, phosphorus serves as a part of DNA and RNA, in lipid transport and glucose absorption, and in acid/base regulation. It is also involved in energy metabolism, in the function of many hormones, and is essential to the proper functioning of several of the B vitamins.

Since phosphorus is widely available in all plant and animal foods, dietary deficiency is unknown. Abnormal or impaired phosphorus metabolism is instead a result of disease, especially those involving the bones or kidneys. The only other cases of dietary phosphorus deficiency have been secondary to the consumption of large quantities of aluminum hydroxide antacids which can bind phosphorus, preventing its absorption.

The amount of dietary phosphorus is of interest because it influences the utilization of other nutrients. Because of the interrelationships of calcium and phosphorus absorption and metabolism, the amount of phosphorus in the diet can affect the amount of calcium absorbed. It is recommended that the relative proportion of these two minerals in the diet be nearly equal (1:1) in adults to optimize the availability of calcium.

As part of phytic acid, the amount of phosphorus in the diet may influence the absorption of other minerals, including iron, zinc, and copper. At the same time, phytic acid phosphorus is not available for absorption, being a part of fiber. Leavening of bread can liberate this phosphorus and inactivate the inhibitory effects of phytic acid on the absorption of other minerals.

The RDA for phosphorus is identical to the RDA for calcium at all ages except for the first year of life when calcium requirements are set somewhat higher.

No toxic effects of phosphorus have been described.

The Electrolytes

SODIUM, POTASSIUM, AND CHLORINE

These three minerals, particularly sodium and potassium, play crucial roles in maintaining the osmotic equilibrium of the body fluids. Sodium is the principal cation (a positively charged ion) in extracellular fluid, and potassium is the principal cation in intracellular fluid. Sodium is critical in acid/base balance and extracellular fluid volume, carbohydrate and protein metabolism, and the generation of nerve impulses. Like sodium, potassium functions to maintain fluid balance and volume. It further assists in carbohydrate metabolism, protein synthesis, muscle contraction, and nerve impulse conduction.

Chlorine, as the anion (a negatively-charged ion) chloride, is primarily present in extracellular fluid, but in red blood cells can easily cross the cell membrane to minimize fluctuations in fluid balance. This chloride shift is one reason why red blood cells are able to carry carbon dioxide to the lungs to be excreted. Chlorine is also important in maintaining acid/base equilibrium, and is a major component of gastric juice.

In *NUTRIENTS IN FOODS* the electrolytes are presented both in milligrams and in milliequivalents (meq). The important roles of the electrolytes are related to their ionic charge and not to their concentration by weight. Thus the milliequivalents become a more meaningful quantification. One meq of sodium equals 23 mg, 1 meq of potassium equals 39 mg, and 1 meq of chloride equals 35.5 mg.

Sodium, potassium, and chlorine are ubiquitous in all diets. Sodium is found in most meats, dairy products, and other foods. Potassium is also found in most meats and dairy products, as well as in fruits, vegetables, and grains. Both sodium and chloride exist in most diets as salt, and much dietary salt is added during cooking or at the table.

The normal sodium content of drinking water can vary, and in the United States the use of salt on roads in winter can, through water runoff, add significant amounts of sodium to drinking water supplies. Food processing is also a major source of added sodium in the diets of most Americans. The sodium content of canned and some frozen vegetables is usually greater than in the fresh food. Also, the potassium content of processed foods is usually less than that of the unprocessed food. Similarly, many popular snack items are salted and therefore can be significant sources of sodium in the diet.

No RDA has been established for any of the electrolytes, though the Food and Nutrition Board has published recommended safe and adequate levels of consumption for all three.

It has been suggested that long-term intake of large amounts of sodium may lead to elevated blood pressure, or hypertension. Because hypertension is a major risk factor for a number of diseases, including coronary heart disease, stroke, and kidney failure, elevated sodium intakes are considered a major public health concern. In addition, it has been suggested that the ratio of sodium to potassium may be important in determining the risk of hypertension, with a relative deficiency of potassium elevating blood pressure.

Physiological electrolyte deficiency is rarely a result of dietary deficiency but may be caused by other factors, including sweating, diarrhea, or vomiting, or the excessive use of diuretics. Extreme sodium loss may lead to circulatory failure. Potassium deficiency is rarely found, but kidney failure or serum dehydration can precipitate potassium toxicity. Both conditions tend to produce similar symptoms, potentially culminating in cardiac failure. Chlorine deficiency is generally associated with disturbances in sodium metabolism.

The potential effect of chronic excess sodium is hypertension, while excess potassium may lead to heart failure. No toxic effects of dietary chlorine are known, although the chemical chlorine is very toxic.

The Trace Minerals

IRON

Iron deficiency anemia is generally recognized as a major nutritional deficiency health problem in the U.S. today. Because iron is a component of hemoglobin, too little of this trace element will lead to lower circulating levels of hemoglobin and red blood cells, and thus to anemia. As a part of hemoglobin, iron plays a crucial role in oxygen transport from the lungs to the cells and in carbon dioxide transport from the cells back to the lungs. Iron is also a component of muscle myoglobin, where it supplies oxygen for use in muscle contraction. Additionally, iron is a cofactor of several enzymes and is involved in cell respiration and protein synthesis.

The body stringently regulates the absorption of iron from the small intestine because body stores of iron are efficiently recycled, and iron is not easily eliminated from the system. Excess iron can lead to a toxic condition known as hemosiderosis. While iron toxicity due to excess dietary intake is rare, it does occur, especially in chronic alcoholics who consume large quantities of cheap wines, which can be rich in iron. The Bantu, who cook in iron pots and drink a home brew rich in iron known as kaffir beer, also develop hemosiderosis. Overingestion of iron supplement pills is a common cause of poisoning in young children, often with fatal results.

However, iron deficiency, not toxicity, remains a more common concern, especially in women during the reproductive period of their lives when iron is lost from the body during menstruation and transferred to the fetus during pregnancy or to the infant during lactation. The amount of iron available in the diet is influenced by a number of factors. Two of the most important are the form of the iron, heme or non-heme, and the amount of vitamin C in the diet, which can increase the absorption of iron. If these factors are known, the relative amount of absorbable iron in a given meal can be estimated.

Heme iron is the form of iron contained in hemoglobin and myoglobin, and thus found only in meat. Approximately 40% of the iron in dietary meats is heme iron. Under normal conditions, approximately 23% of heme iron is absorbed, although this can increase to 35% when body stores are depleted. Non-heme iron, the other major form of dietary iron, is generally much more poorly absorbed than heme iron. However, the absorption of non-heme iron can be positively influenced by the presence of vitamin C in the diet, as well as by some as yet unknown factor present in meat.

To calculate the amount of absorbable iron in any given meal:

A. Use the Nutrient Tables to derive the following information.
   1. Total amount of iron in the meal (g)
   2. Total amount of heme iron (g)
      (= total iron from meat x 0.4)
   3. Amount of non-heme iron (g)
      (= total iron - heme iron)
   4. Ascorbic acid content of the meal (mg)
   5. Amount of meat, poultry, or fish in the meal (g)

B. Multiply the total mg of heme iron (item 2) by 23% (its standard absorption percentage).

C. Referring to Table 8, if a meal is of low iron availability (as defined by items 4 and 5), multiply the total mg of non-heme iron (item 3) by 3%; if of medium availability, by 5%; if of high availability, by 8%.

D. Add together the values thus derived for heme and non-heme iron to get the total mg of absorbable iron in that meal.

If you merely want an estimate of how much available iron will be found in any given meal, in general, if a meal contains less than 30 g (1 oz) of meat, poultry, or fish AND less than 25 mg of ascorbic acid, it will be of low iron availability because of the small amount of heme iron and the low absorption of non-heme iron in the absence of sufficient meat and vitamin C.

If a meal contains more than 90 g (3 oz) of meat, poultry, or fish OR more than 75 mg of ascorbic acid, iron availability will be high due to the high percentage of heme iron and/or the high vitamin C intake. A meal of high iron availability also results from 30-90 g of meat AND 25-75 mg of ascorbic acid.

The intermediate state, with 30-90 g of meat BUT LESS THAN 25 mg of ascorbic acid, OR with 25-75 mg of ascorbic acid BUT LESS THAN 30 g of meat, yields only medium iron availability to the system (see Table 8).

Good sources of dietary iron are poultry, fish, meat (especially liver), seafood, and legumes. Milk and eggs are poor sources, and fruits and vegetables have only non-heme iron. Vegetarians must increase ascorbic acid intake to compensate for the absence of the meat factor and the absence of heme iron. In addition, the presence of antacids, tannic acid, fiber, EDTA, egg yolks, milk, cheese, and phosphate salts can all lower iron absorption. Cooking in iron vessels can increase the iron content of the diet.

The established RDA for iron is based on the assumption that 10% of total iron is absorbed by the system.

MAGNESIUM

Magnesium is distributed throughout most human tissue, being the second most abundant cation within cells after potassium. It is also stored in bone, and is a cofactor for several enzymes. Magnesium is involved in muscle contraction, and is important in the utilization of several of the vitamins, including some of the B-vitamins, vitamin C, and vitamin E. The metabolism of magnesium is interrelated with that of calcium,

TABLE 8
Availability of Iron in Different Meals[a]

Type of Meal	Absorption of Iron Present in Meal (%) Nonheme Iron	Heme Iron
Low-Availability Meal	3	23
<30 g meat, poultry, fish <25 mg ascorbic acid		
Medium-Availability Meal	5	23
30–90 g meat, poultry, fish or 25–75 mg ascorbic acid		
High-Availability Meal	8	23
>90 g meat, poultry, fish or >75 mg ascorbic acid; or 30–90 g meat, poultry, fish plus 25–75 mg ascorbic acid		

Source: National Academy of Sciences, 1980 (adapted from Monsen et al., 1978).

phosphorus, and the major electrolytes as well. Calcium and vitamin D appear to increase the absorption of magnesium.

Dietary magnesium deficiency is considered rare, but in view of the importance of magnesium in the human system, an RDA was established for this mineral. Now, recent dietary surveys show intakes of magnesium to be marginal for many segments of the U.S. population.

Good sources of magnesium are dairy foods and green leafy vegetables, as magnesium is an integral part of chlorophyll.

ZINC

Zinc is widely distributed in the human body, being found in virtually all tissues. It is a critical component of many enzymes and is therefore involved in many metabolic pathways. These include the transport of vitamin A (retinol) in the bloodstream, and the synthesis of DNA and RNA. Zinc is also involved in the wound healing process.

Milling cereal grains reduces the amount of zinc present, but, unlike iron, zinc enrichment of refined flour is not a regular practice in the United States. However, the presence of phytates in grain may inhibit the absorption of dietary zinc so that the zinc in whole grains may actually be less available than the zinc in leavened or refined flour products. Regardless, whole grains remain good sources of dietary zinc, along with legumes and meats. Oysters are also particularly rich in zinc. Fruits, vegetables, and egg whites are poor sources.

Zinc deficiency has been seen in several populations around the world, such as in poorer communities in Iran, where the staple food is unleavened whole grain bread. It is of rapid onset as body stores are minimal. Many parts of the United States have zinc-deficient soils, and marginal zinc deficiency is a factor in some segments of the U.S. population. Studies of some areas have revealed an average dietary intake of 8.6 mg per day, well below the RDA of 15 mg per day.

Symptoms of zinc deficiency include poor growth and hypogonadism, decreased taste acuity, and poor appetite.

Zinc toxicity is extremely rare, occuring only under abnormal circumstances, such as home hemodialysis with water having extremely high zinc content, or in deliberate ingestion of zinc sulfate to induce vomiting.

COPPER

Like zinc, copper is a cofactor of many enzymes. Copper is critical in the absorption of iron and the formation of hemoglobin, and is important in the transport of iron as well. As a result, it is difficult to differentiate copper deficiency anemia in animals from that due to iron deficiency. Copper is further required for fatty acid metabolism, the production of the myelin sheaths that surround nerve fibers, bone formation, and the synthesis of RNA. It also helps oxidize vitamin C.

Copper deficiency has been observed in animals such as cattle raised on pastures with copper-poor soil. However, dietary copper deficiency in humans is not known to exist, except in abnormal conditions such as prolonged intravenous feeding with solutions lacking copper. Symptoms include anemia and bone diseases, skin sores, impaired respiration, and loss of taste. Since humans eat foods from a variety of sources, relying strictly on foods grown or raised on copper-poor soil is unlikely.

Dietary copper toxicity is virtually unknown, although excess can cause nausea and vomiting.

No RDA has been established for copper, although the Food and Nutrition Board has published safe and adequate intake values which have been used to formulate an U.S.RDA for this trace mineral.

IODINE

Iodine is unique among the essential minerals in that it forms an important part of hormones, in this case, thyroxin and triiodothyronine, thyroid substances which control the rate of metabolism in most cells in the body.

The iodide content of foods is determined principally by the environment in which they were grown or raised. Most soils have little iodine, so most foods tend to be poor sources of this mineral. The major exception is seafood, with ocean fish and sea vegetables particularly rich sources of iodine. It should be noted that some foods such as cabbages and turnips contain so-called goitrogenic substances that inhibit the normal metabolism of the thyroid gland and iodine.

Diets containing little seafood are naturally low in iodine which explains why iodine deficiency, or goiter, is endemic in some parts of the world, especially areas removed from the sea coasts. Goiter is characterized by an enlarged thyroid gland, which increases its mass in order to enhance its efficiency in the presence of lowered iodine levels. Cretinism, a serious mental retardation, is the result of iodine deprivation in utero. As such it became the first treatable serious developmental disability.

The value of iodine in treating goiter was first noticed in 1820 by Coindet, a Swiss physician, but it was not until a century later that iodine supplementation for goiter was well accepted.

The RDA for iodine (150 mcg for all individuals above ten years of age) is set at a level two to three times that required to prevent goiter (50-75 mcg), with slightly higher amounts recommended for pregnant or lactating women.

Table salt is now regularly fortified with iodine in the U.S. However, since not all salt is iodized, the consumer must always check the label to ensure adequate intake. As a result of iodized salt use, goiter is now largely unknown in the U.S., even among people who eat little or no seafood. Unfortunately, it is still endemic in other areas of the world.

Although extreme intakes of iodine may produce serious toxic effects in animals, few instances of toxic response are known in human populations, although iodine-induced goiter has been reported in some Japanese populations that consume large amounts of sea vegetables.

SELENIUM

Interest in selenium has increased in recent years as its role in the human body has become better understood. This mineral is a cofactor of the enzyme glutathione peroxidase and is thus an antioxidant within the cells of the body. It appears to complement the antioxidant properties of vitamin E, helping to maintain the integrity of cell membranes. As with vitamin E, selenium has been thought to be a potential preventive for cancer (Ames, 1983). While theoretically possible, this role for selenium has not yet been substantiated.

The selenium content of foods varies according to the selenium content of the soil on which it was grown and can be highly variable. Generally, cereal grains, fish, and meats contribute the greatest amounts of selenium to the diet, while fruits and vegetables contain very little selenium.

Selenium deficiency, usually coupled with vitamin E deficiency, is known to produce a number of diseases in non-human primates. Its dietary importance in man is assumed, though no direct evidence of selenium deficiency has been described.

In parts of North Dakota and Wyoming the concentration of selenium in the soil is particularly rich, and animals grazing there have been known to develop symptoms of selenium toxicity. Selenium toxicity is rare in humans, but may affect nails, hair, and teeth, accompanied with irritability and extreme fatigue. The observed toxicity of selenium in animals cautions against ingesting large quantities of the mineral.

No RDA has been established for this trace mineral, however, in light of the effects of selenium deprivation or toxicity in animals, the Food and Nutrition Board has estimated safe and adequate levels of intake.

MANGANESE

Manganese is a cofactor of enzymes involved in the transfer and metabolism of phosphate groups, and is thus important in energy metabolism. It is also involved in the synthesis of mucopolysaccharides in the formation of cartilage.

Dietary manganese deficiency and toxicity have not been observed in humans, and it is relatively non-toxic in animals as well. Thus, the manganese content of human diets is presumably adequate. Whole grains, legumes, and vegetables have the highest concentrations of manganese, while meats, milk, and refined cereals have the lowest amounts.

No RDA has been established. However, in view of its importance in the human system, the Food and Nutrition Board has published estimated safe and adequate intake values for this trace mineral.

CHROMIUM

Normal glucose metabolism in animals requires trivalent chromium, which is also important for its effect on insulin production. A few cases of impaired glucose tolerance responsive to chromium have been reported. The most likely causes of such marginal deficiency states relate to old age, pregnancy, or undernutrition. Diabetes may also impair chromium metabolism. However, cases of deficiency are rare, and similarly chromium toxicity has not been observed. This suggests that the chromium found in most human diets is probably safe and adequate.

The availability of chromium in the diet is in part dependent on the form in which it is ingested. The organically bound trivalent chromium is more readily available, and yeast is a particularly rich source. Other good sources of dietary chromium include whole grains, cheese, and meats. Leafy greens, on the other hand, are poor sources.

There is no established RDA for chromium, though, based on its role in critical metabolic functions, the Food and Nutrition Board has estimated safe and adequate intake levels for this trace mineral.

REFERENCES

Adams, C.F., 1975. *Nutritive Value of American Foods in Common Units*, Agriculture Handbook #456, USDA, Washington, D.C.

American Home Economics Association, 1975. *Handbook of Food Preparation*, 7th Edition. Washington, D.C.

Ames, B.N., 1983. Dietary carcinogens and anticarcinogens, Science, 221,(4617)(Sept.23,l983):125.

Ahmad, I. and A. Ali, 1980. The effect of oil processing operations on the total sulphur content in rapeseed oil. J. Am. Oil Chem. Soc., 57:7.

Anderson, B.A., 1976. VII. Pork Products: Comprehensive evaluation of fatty acids in foods. J. Am. Dietet. Assoc., 69:44.

Anderson, B.A., G.A. Fristrom and J.L. Weihrauch, 1977. X. Lamb and veal: Comprehensive evaluation of fatty acids in foods. J. Am. Dietet. Assoc., 70:53.

Anderson, B.A., J.A. Kinsella and B.K. Watt, 1975. II. Beef products: Comprehensive evaluation of fatty acids in foods. J. Am. Dietet. Assoc., 67:35.

Anon., l978. Perspective on fast food. Dietetic Currents, 5, Sept-Oct.

Anon., l979. USDA analyzes cereals for fine food sugars. CNI Weekly Report. 9(27):7.

Appledorf, H., 1974. Nutritional analyses of food served from fast-food chains. Food Technol., 28(4):50.

Aulek. D.J., 1977. *Nutritional Analyses of Food Served at MacDonald's Restaurants*. Warf Institute, Inc., Madison, Wisc.

Baker,D. and J.M.Holden, 1981. Fiber in breakfast cereals. J. Food Sci., 46:396.

Bauernfeind, J.C. and I.D.Desai, 1977. The tocopherol content of food and influencing factors. Fd. Sci. and Nutr., 337-382.

Bedford, C.L. and L.R. Baker, 1977. *Improving Nutrient Values of Certain Fruit and Vegetable Crops by Breeding, Plant Nutrition and Growth Regulators.* Final Report to Rakum Foundation from Michigan State University, E. Lansing.

Better Homes and Gardens, 1976. *New Cookbook*, Meredith Corporation, Des Moines, IA.

Block, R.J. and R.H. Mandl, 1958. Amino acid composition of bread proteins, J. Am. Dietet. Assoc., 34:72.

Brignoli, C.A., J.E. Kinsella and J.L. Weihrauch, 1976. V. Unhydrogenated fats and oils: Comprehensive evaluation of fatty acids in foods. J. Am. Dietet. Assoc., 68:224.

Bunnell, R.H., J. Keating, A. Quaresino, and G.K. Parman, 1965. Alpha-tocopherol content of foods. Am. J. Clin. Nutr., 17:1.

Burkett, D.P. and H.C.Trowell, 1975. *Refined Carbohydrates and Disease*. Academic Press, New York.

Butterfield, S. and D.H. Calloway, 1972. Folacin in wheat and selected foods, J. Am. Dietet. Assoc., 60:310.

Church, C.F. and H.N. Church, 1975. *Food Values of Portions Commonly Used*. 12th. ed. J.P.Lippencott, Philadelphia.

Cotterill, O.J., W.W. Marion and E.C.Naber, 1977. A nutrient re-evaluation of shell eggs. Poultry Sci., 56:1927.

Davidson, S., R. Passmore, J.F. Brock, and A.S. Truswell, 1979. *Human Nutrition and Dietetics*. Churchill Livingston, London.

Davis, K.C., 1973. Vitamin E content of selected baby foods. J. Food Sci., 38:422.

Davis, K.R., N.L. Heneker, D. LeTourneau, R.F. Cain, L.J. Peters and J. Mcginnis, 1980. Evaluation of the nutrient composition of wheat. I. Lipid constituents. Cereal Chem., 57:178.

Davis, K.R., R.F. Cain, L.J. Peters, D. LeTourneau and J. McGinnis, 1981. Evaluation of the nutrient composition of wheat. II. Proximate analysis. thiamin, rinoflavin, niacin, and pyridoxine. Cereal Chem., 58(2):116.

Dicks, M.W., 1965. *Vitamin E Content of Foods and Foods for Human and Animal Consumption*. Bulletin 435. Agricultural Experimental Station, University of Wyoming, Laramie.

Eheart, J.F. and B.S. Mason, 1967. Sugar and acid in the edible portion of fruits. J. Am. Dietet. Assoc., 50:130.

Exler , J.L. and J.L. Weihrauch, 1976. VIII. Finfish: Comprehensive evaluation of fatty acids in foods. J. Am. Dietet. Assoc., 71:518.

Exler, J.L. and J.L. Weihrauch, 1977. XII. Shellfish: Comprehensive evaluation of fatty acids in foods. J. Am. Dietet. Assoc., 71:518.

Feeley, R.M. and, P.E. Criner and B.K. Watt, 1972. Cholesterol content of foods. J. Am. Dietet. Assoc., 61-74.

Feeley, R.M. and B.K. Watt, 1970. Nutritive values of foods distributed under USDA food assistance programs. J. Am. Dietet. Assoc., 57:528.

FAO, 1973. *Energy and Protein Requirements*, FAO Nutrition Report Series #52, Food and Agriculture Association, Rome.

FAO, 1979. *Amino-Acid Content of Foods and Biological Data from Proteins.* Food and Agriculture Association, Rome.

FDA, 1981 (rev.). HHS Publication No (FDA) 81-246. U.S. Department of Health and Human Services, Rockville, MD.

FNB, 1975. *Improvement of protein nutriture.* Food and Nutrition Board, National Research Council, National Academy of Sciences, Washington, D.C.

Fowler, S., B. West and G. Sugart, 1971. *Food for Fifty*, 5th. ed. John Wiley and Sons, New York.

Freeland, J.H. and R.J. Cousins, 1976. Zinc content of selected foods. J. Am. Dietet. Assoc., 67:351.

Fristrom, G.A., B.C. Stewart, J.L. Wiehrauch and L.P. Posati, 1975. Comprehensive evaluation of fatty acids in foods. J. Am. Dietet. Assoc., 67:371.

Fristrom, G.A., B.C. Stewart, J.L. Weihrauch., and L.P. Posati, 1975. IV. Nuts, peanuts and soups: Comprehensive evaluation of fatty acids in foods. J. Am. Dietet. Assoc., 67:351.

Fristrom, G.A. and J.L. Weihrauch, 1976. IX. Comprehensive evaluation of fatty acids in foods. J. Am. Dietet. Assoc., 69:51.

Giebhart, S.E., R. Cutrufelli and R.H. Matthews, 1978. *Composition of Foods: Baby Foods: Raw, Processed, Prepared*. Consumer Foods Economics Institute. Agriculture Handbook No. 8-3. USDA, Washington, D.C.

Gormican, A., 1970. Inorganic elements in foods used in hospital menus. J. Am. Dietet. Assoc., 56:397.

Greger, J.L., S. Marhefka and A.H. Geisler, 1978. Magnesium content of selected foods. J.Food Sci. 43::1610.

Haeflein, K.A. and A.I. Rasmussen, 1977. Zinc content of selected foods. J. Am. Dietet. Assoc., 70:610.

Hardinge, M.G. and H. Crooks, 1958. Fatty acid composition of food fats. J. Am. Dietet. Assoc., 34:106.

Hardinge, M.G. and H. Crooks, 1961. Lesser known vitamins in foods. J. Am. Dietet. Assoc., 38:240.

Harris, R.S. and E. Karmas (eds), 1975. *Nutritional Evaluation of Food Processing.* The AVI Publishing Co, Westport, CT.

Hook, I. and I.K. Brandt, 1966. Copper content of some low-copper foods. J. Am. Dietet. Assoc., 49:20.

Inglett and Charalanbous (eds). *Tropical Foods: Chemistry & Nutrition,* Vol. 1, pp.33. Academic Press, New Yor.

Klein, B.P., C.H.Y. Kuo and G. Boyd, 1981. Folacin and ascorbic acid retention in fresh raw, microwave and conventionally cooked spinach. J. Food Sci. 46:64.

Kirkpatrick, D.C. and D.E. Coffin, 1975. Trace metal content of various cured meats. J. Sci. Fd. Agric. 26:4.

Koehler,H.H., H.C. Lee and M. Jacobson, 1977. Tocopherols in canned entrees and vended sandwiches. J. Am. Dietet. Assoc., 70:61.

Kraus, B., 1974. *The Dictionary of Sodium, Fats, and Cholesterol.* Grosset and Dunlap, New York.

Kreutler, P., A., 1980. *Nutrition in Perspective.* Prentice-Hall, Englewood Cliffs, NJ.

Kylen, A.M., and M. Rolland, 1975. Nutrients in seeds and sprouts of alfalfa, lentils, mung beans and soybeans. J. Food Sci. 40:100.

Lantz, E.M., H.W. Gaugh and A.M. Campbell, 1962. *Effects of planting date on the composition and cooking quality of pinto beans.* New Mexico Ag. Expt. Sta. Bull. 467 (June).

Lecos, C., 1982. RDAs: Key to nutrition. FDA Consumer, HHS Publication No. (FDA)82-21681. U.S.GPO, Washington, D.C.

Lee, C.Y., R.S. Schalleberger and M.T. Vittum, 1970. Free sugars in fruits and vegetables. Food Sciences N.Y. State Agr. Exp. Sta., Geneva, NY. 1:1.

Leverton, R.M. and G.V. Odell, 1959. *The Nutritive Value of Cooked Meat.* Agr. Exp. Sta., Oklahoma University. Publication MP-49.

Marsh, A.C., M.K. Moss and E.W.S. Murphy, 1977. *Composition of Foods: Spices and Herbs: Raw, Processed, Prepared* Consumer and Food Economics Institute, Agriculture Handbook No. 8-2. USDA, Washington, D.C.

Marsh, A.C., 1980. *Composition of Foods: Sauces and Gravies: Raw, Processed, and Prepared.* Consumer and Food Economics Institute. Agriculture Handbook No. 8-6, USDA, Washington D.C.

Marusich, W.L., E. DeRitter, E.F. Ogrinz, J. Keating, M. Mitrovic and R.H. Bannell, 1975. Effect of supplemental Vitamin E in control of rancidity in poultry meat. Poultry Sci. 54:83.

Matthews, R.H. and Y.J. Garrison, 1975. *Food Yields Summarized by Different Stages of Preparation.* Consumer and Food Economics Institutes, Northeastern Region, Agricultural Research Service. USDA Handbook No. 102.

McCarthy, M.A., B.W. Murphy, B.J. Ritchey and P.C. Washburn, 1977. Mineral content of legumes as related to nutrient labeling. Food Technol. 31(2):8.

McCarthy, M.A., M.L. Orr and B.K. Watt, 1968. Phenylalanine and tyrosine in vegetables and fruits. J. Am. Dietet. Assoc., 52:13.

Miljanich, P. and R. Ostwald, 1970. Fatty acids in newer brands of margarine. J. Am. Dietet. Assoc., 56:2.

Miller, G.T., V.R. Williams and D.S. Moschette, 1965. Phenylalanine content of fruit. J. Am. Dietet. Assoc., 46:4.

Monsen, E.R., L. Hallberg, M. Layrisse, D.M. Hegsted, J.D. Cook, W. Mertz, and C.A. Finch, 1978. Estimation of available dietary iron. Am. J. Clin. Nutr. 31: 134-141.

Murphy, E.W., B.W. Willes and B.K. Watt, 1975. Provisional tables on the zinc content of foods. J. Am. Dietet. Assoc., 66:34.

National Academy of Sciences, 1980. *Recommended Dietary Allowances.* 9th. ed. Food and Nutrition Board, Washington, D.C.

Nelson, G.Y. and M.R. Gram, 1961. Magnesium content of accessory foods. J. Am. Dietet. Assoc., 38:43.

Orr, M.L., 1969. *Pantothenic Acid, Vitamin B-6 and Vitamin B-12 in Foods.* Home Economics Research Report No. 36. USDA, Washington, D.C.

Orr, M.L., and B.K. Watt, 1957. *Amino Acid Content of Foods.* Home Economics Research Report No. 4, USDA, Washington, D.C.

Palansky, M.N. and E.W. Murphy, 1966. Vitamin B-6 components in fruits and nuts. J. Am. Dietet. Assoc., 48:108.

Passmore, R., B.N. Nichol, and M.N. Rao, 1974. *Handbook on Human Nutritional Requirements.* World Health Organization, Geneva.

Paul, A.A. and D.A.T. Southgate, 1978. *McCance and Widdowson's The Composition of Foods.* 4th. ed. Elsevier/North-Holland Biomedical Press, New York.

Pennington, J.A., 1975. *Dietary Nutrient Guide.* The AVI Publishing Co., Westport, CT.

Perloff, B.P. and R.R. Butram, 1977. Folacin in selected foods. J. Am. Dietet. Assoc., 70:161.

Posati, L.P., J.K. Kinsella, and B.K. Watt, 1975. I. Dairy Product products: Comprehensive evaluation of fatty acids in foods. J. Am. Dietet. Assoc., 66:48.

Posati, L.P., J.E. Kinsella and B.K. Watt, 1976. III. Eggs and egg products: Comprehensive evaluation of fatty acids in foods. J. Am. Dietet. Assoc., 67:111.

Posati, J.L. and M.L. Orr, 1976. *Composition of Foods: Dairy and Egg Products: Raw, Processed, and Prepared.* Consumer and Food Economics Institute, Agriculture Handbook No. 8-1. USDA, Washington, D.C.

Posati, L.P., 1979. *Composition of Foods: Poultry Products: Raw, Processed, Prepared.* Consumer and Food Economics Institute. Agriculture Handbook No. 8-5. USDA, Washington, D.C.

Reeves, J.B. III and J.L Weihrauch, 1979. *Composition of Foods: Fats and Oils: Raw Processed and Prepared.* Consumer and Food Economics Institute. Agriculture Handbook No. 8-4. USDA, Washington, D.C.

Richardson, M., L.P. Posati and B.A. Anderson, 1980. *Composition of Foods: Sausages and Luncheon Meats: Raw, Processed, Prepared.* Consumer and Food Economics Institute. Agriculture Handbook No. 8-7. USDA, Washington, D.C.

Rombauer, I.S., 1974. *The Joy of Cooking.* Signet Publishing Co., New York.

Shannan, I.L., 1980. Ice cream: potential dental hazard? J. of Dentistry for Children, July-August, 1980.

Shibasaki, K.. and E.W. Hesseltine, 1962. Miso fermentation. Economic Botany 16:18.

Shurtleff, W. and A. Aoyagi, 1976. *The Book of Miso.* Autumn Press.

Shurtleff, W. and A. Aoyagi, 1975. *The Book of Tofu.* Autumn Press.

Slover, H.T. and R.T. Thompson, 1971. Tocopherols in foods and fats. Lipids 6:291.

Slover, H.T. and R.H. Thompson, I980. Lipids in fast foods. J. Food Sci. 45:1583.

Smith, W.E., 1972. *Effects of Three Cooking Methods on Pesticide Residue in Chinook and Coho Salmon.* Master's thesis, Michigan State University, E. Lansing.

Southgate, D.A.T, A.A. Paul, A.C. Dean and A.A. Christie, 1978. Free sugars in foods. J. Human Nutr. 32:335.

Standel, B.R., D.R. Bassett, P.B. Policar and T. Thom, 1970. Fatty acid, cholesterol and proximate analyses of some ready-to-eat foods. J. Am. Dietet. Assoc., 56:392.

Staroscik, J.A., F.U. Gregorio and S.K. Reeder, 1980. Nutrients in fresh peeled oranges and grapefruit from California and Arizona. J. Am. Dietet. Assoc., 77(11):567.

Walsh, J.H., B.W.Wyse and R.G. Hansen, 1981. Pantothenic acid content of 75 processed and cooked foods. J. Am. Dietet. Assoc., 78:140.

Watt, B.K. and L.A. Merrill, 1963. *Composition of Foods - Raw, Processed, Prepared.* Agriculture Handbook No. 8, Revised. USDA, Washington, D.C.

Webb, B.H., A.H. Johnson and J.A. Alford, 1974. *Fundamentals of Dairy Chemistry*. 2nd. ed. The AVI Publishing Co., Westport, CT.

Weihrauch, J.L., J.E. Kinsella and B.K. Watt, 1976. VI. Cereal Products: Comprehensive evaluation of fatty acids in foods. J. Am. Dietet. Assoc., 68:335.

Zook, E.G., F.W. Greene and E.R. Morris, 1970. Nutrient composition of selected wheats and wheat products. VI. Distribution of manganese, copper, nickel, zinc, magnesium, lead, tin, cadmium, chromium and selenium as determined by atomic absorption spectroscopy and colorimetry. J. Am. Oil Chem. Soc., 47:120.

Zook, E.G. and J. Lehman, 1968. Mineral composition of fruits. J. Am. Dietet. Assoc., 52:225.

CONTRIBUTING FOOD COMPANIES

The authors thank the following companies for providing nutrient information for their products and for giving their permission to include this nutrient information in our book:

Abbott Laboratories
Alberto-Culver Co.
Anderson Clayton Foods
Armour and Co.
Avoset Food Corp.
Best Foods
Borden, Inc.
Campbell's Soup
Carnation
Central Soya
Chicago Dietetic Supply
Cumberland Packing Co.
Delmark Foodservice Co.
DelMonte Corp.
Diamond Crystal Salt Co.
Dr. Pepper Co.
Durkee Foods
Fisher Cheese Co.
Frito-Lay
General Foods Corporation
General Mills, Inc.
Golden Grain Macaroni Co.
Henri's Food Products Co.
Hershey Foods Corp.
Heublin, Inc., Grocery Products Group
Interstate Brands Corp.
ITT Continental Baking Co.,Inc.
J.M. Smucker Co.
Keebler Co.
Kellogg Co.
Kitchens of Sara Lee
LaChoy Food Products
Lawry's Foods, Inc.
Lever Brothers Co.
Life Savers, Inc.
Loma Linda Foods
MacDonald's Corp.
Mars, Inc.
Mead Johnson
Mrs. Paul's Kitchens, Inc.
Nabisco, Inc.
Nestle Co., Inc.
Oscar Meyer & Co.
Ovaltine Products, Inc.
Pet, Inc.
Pizza Hut, Inc.
Ralston Purina Co.
R.J.R. Foods, Inc.
Standard Brands, Inc.
Standard Milling Co.
Stokeley-Van Camp, Inc.
Stouffer Foods Corp.
The Coca-Cola Co.
The B. Manischewitz Co.
The Pillsbury Co.
The Quaker Oats Co.
The Seven-Up Co.
The Stroh Brewing Co.
Thomas J. Lipton, Inc.
Tootsie Roll Industries, Inc.,
V.M. Underwood Co.
Weight Watchers International, Inc.
C.G. Whitlock Process Co.
Worthington Foods

Nutrient Table

	FOOD	Portion	Weight in grams / *Conversion for 100 g*	Kilocalories / H_2O g	Total carbohydrate g / Total fats g	Crude fiber g / Dietary fiber g	Total protein g / Total sugar g	% USRDA	Arginine mg / Histidine mg	Isoleucine mg / Leucine mg	Lysine mg / Methionine mg	Phenylalanine mg / Threonine mg	Valine mg / Tryptophan mg	Cystine mg / Tyrosine mg	Polyunsat. fatty acids g / Monounsat. fatty acids g	Saturated fatty acids g / P/S ratio	Linoleic acid g / Cholesterol mg	Thiamin mg / Ascorbic acid mg	**% USRDA**
A	ALLSPICE,ground	1 tsp	2.0	5	1.4	0.43	0.1	**0**	unk.	unk.	unk.	unk.	unk.	unk.	0.05	0.05	0.05	tr	**0**
			50.00	0.2	0.2	unk.	0.0		unk.	unk.	unk.	unk.	unk.	unk.	0.01	0.93	0	1	**2**
B	ALMOND extract	1 tsp	5.0	16	0.0	0(a)	0.0	**0**	unk.	0(a)	0(a)	0(a)	0(a)	0(a)	0.00	0.00	0.00	unk.	**unk.**
			20.00	unk.	0.0	unk.	0.0		0(a)	0(a)	0(a)	0(a)	0(a)	0(a)	0.00	0.00	0	unk.	**unk.**
C	almonds,shelled,chopped	1/4 C	33	194	6.3	0.84	6.0	**9**	712	284	189	372	365	123	3.41	1.40	3.20	0.08	**5**
			3.08	1.5	17.6	1.72	1.4		168	473	84	198	57	201	11.86	2.44	0	tr	**0**
D	almonds,unshelled	1/2 C	39	120	3.9	0.56	3.7	**6**	436	174	117	229	225	75	2.14	0.86	1.97	0.05	**3**
			2.56	1.8	10.8	2.07	0.9		103	291	52	122	35	124	7.33	2.49	0	tr	**0**
E	ANCHOVY paste	1 Tbsp	21	42	0.8	0.00	4.2	**9**	265	213	366	159	234	51	1.47	unk.	unk.	unk.	**unk.**
			4.76	unk.	2.3	0(a)	unk.		129	300	126	189	42	163	unk.	unk.	unk.	unk.	**unk.**
F	anchovy,pickled,lightly salted,canned	4 anchovies	16	28	0.0	0.00	3.1	**7**	195	171	294	124	177	37	unk.	unk.	unk.	unk.	**unk.**
			6.25	9.4	1.6	0(a)	0.0		94	254	97	144	33	119	unk.	unk.	unk.	unk.	**unk.**
G	ANISE seed	1 tsp	2.2	7	1.1	0.32	0.4	**1**	unk.	unk.	unk.	unk.	unk.	unk.	0.07	0.07	0.07	unk.	**unk.**
			45.45	0.2	0.3	unk.	0.0		unk.	unk.	unk.	unk.	unk.	unk.	0.22	1.05	0	unk.	**unk.**
H	APPLE,fresh,any,w/skin,edible portion	1 average	138	80	20.0	1.38	0.3	**0**	8	10	16	7	11	4	unk.	tr(a)	unk.	0.04	**3**
			0.72	116.5	0.8	3.03	15.7		5	17	2	10	3	4	unk.	unk.	0	6	**9**
I	apple,fresh,any,w/skin, sliced or cubed	1/2 C	55	32	8.0	0.55	0.1	**0**	3	4	6	3	4	1	unk.	tr(a)	unk.	0.02	**1**
			1.82	46.4	0.3	1.21	6.3		2	7	1	4	1	2	unk.	unk.	0	2	**4**
J	apple,frozen,escalloped-Stouffer	1/2 pkg	170	208	41.7	unk.	0.3	**1**	unk.	unk.	unk.	unk.	unk.	unk.	unk.	unk.	unk.	tr	**0**
			0.59	119.3	4.4	unk.	unk.		unk.	unk.	unk.	unk.	unk.	unk.	unk.	unk.	unk.	tr	**0**
K	apple,home recipe,apple brown betty made w/enriched bread	1/2 C	108	162	31.9	0.54	1.7	**3**	unk.	unk.	unk.	unk.	unk.	unk.	unk.	1.61	0.32	0.06	**4**
			0.93	69.3	3.8	unk.	21.0		unk.	unk.	unk.	unk.	unk.	unk.	1.29	unk.	unk.	1	**2**
L	apple,home recipe,baked w/skin, w/2 Tbsp sugar	1 average	115	144	35.0	1.03	0.2	**0**	6	8	14	6	9	3	unk.	tr(a)	unk.	0.03	**2**
			0.87	unk.	0.7	2.19	31.5		4	14	2	9	2	3	unk.	unk.	0	3	**6**
M	apple,home recipe,dry,stewed	1/2 C	139	124	31.9	tr	0.1	**0**	3	4	6	3	4	1	0.00	0.00	0.00	0.01	**1**
			0.72	106.5	0.3	1.21	30.1		2	7	1	4	1	2	0.00	0.00	0	2	**4**
N	apple,home recipe,spiced	1/2 C	151	129	32.8	0.27	0.3	**1**	9	15	11	8	10	0	0.00	0.00	0.00	0.05	**3**
			0.66	116.6	0.8	3.03	27.7		4	14	4	8	0	4	0.00	0.00	0	6	**10**
O	APPLESAUCE,canned,sweetened	1/2 C	128	116	30.3	0.64	0.3	**0**	7	9	15	7	10	3	unk.	tr(a)	unk.	0.03	**2**
			0.78	96.5	0.1	2.42	29.1		5	16	2	9	3	4	unk.	unk.	0	1	**2**
P	applesauce,canned,unsweetened	1/2 C	122	50	13.2	0.73	0.2	**0**	7	9	14	6	10	3	unk.	tr(a)	unk.	0.02	**2**
			0.82	108.0	0.2	2.93	12.4		5	15	2	9	2	4	unk.	unk.	0	1	**2**
Q	APRICOT,canned,halves,heavy syrup, solids & liquid	1/2 C	129	111	28.4	0.52	0.8	**1**	10	14	22	12	19	40	unk.	unk.	unk.	0.03	**2**
			0.78	99.2	0.1	1.37	23.3		12	22	4	16	6	10	unk.	unk.	0	5	**9**
R	apricot,canned,halves,light syrup, solids & liquid	1/2 C	123	81	20.7	0.49	0.9	**1**	11	15	25	14	21	45	unk.	unk.	unk.	0.02	**2**
			0.81	100.7	0.1	1.51	13.9		14	25	4	18	7	11	unk.	unk.	0	5	**8**
S	apricot,canned,halves,water pack, solids & liquid	1/2 C	123	47	11.8	0.49	0.9	**1**	11	15	25	14	21	45	unk.	tr(a)	unk.	0.02	**2**
			0.81	109.6	0.1	1.85	7.5		14	25	4	18	7	11	unk.	unk.	0	5	**8**

Row	Riboflavin mg / Niacin mg	% USRDA	Vitamin B_6 mg / Folacin mcg	% USRDA	Vitamin B_{12} mcg / Pantothenic acid mg	% USRDA	Biotin mg / Vitamin A IU	% USRDA	Preformed A RE / Beta carotene RE	Vitamin D IU / Vitamin E IU	% USRDA	Total tocopherol mg / Alpha tocopherol mg	Other tocopherol mg / Total ash g	Calcium mg / Phosphorus mg	% USRDA	Sodium mg / Sodium meq	Potassium mg / Potassium meq	Chlorine mg / Chlorine meq	Iron mg / Magnesium mg	% USRDA	Zinc mg / Copper mg	% USRDA	Iodine mcg / Selenium mcg	% USRDA	Manganese mcg / Chromium mcg
A	tr	**0**	unk.	**unk.**	0.00	**0**	unk.	**unk.**	0(a)	0(a)	**0**	unk.	unk.	13	**1**	2	21	unk.	0.1	**1**	tr	**0**	unk.	**unk.**	unk.
	0.1	**0**	unk.	**unk.**	unk.	**unk.**	11	**0**	unk.	unk.	**unk.**	unk.	0.09	2	**0**	0.1	0.5	unk.	3	**1**	unk.	**unk.**	unk.		unk.
B	unk.	**unk.**	unk.	**unk.**	0(a)	**0**	unk.	**unk.**	0(a)	0(a)	**0**	unk.	unk.	unk.	**unk.**	unk.	unk.	unk.	0(a)	**0**	unk.	**unk.**	unk.	**unk.**	unk.
	unk.	**unk.**	unk.	**unk.**	unk.	**unk.**	unk.	**unk.**	unk.	unk.	**unk.**	unk.	unk.	unk.	**unk.**	unk.	unk.	unk.	unk.	**unk.**	unk.	**unk.**	unk.		unk.
C	0.30	**18**	0.03	**2**	0.00	**0**	0.006	**2**	0	0	**0**	unk.	unk.	76	**8**	1	251	1	1.5	**9**	0.8	**6**	unk.	**unk.**	783
	1.1	**6**	31	**8**	0.15	**2**	0	**0**	0	4.9	**16**	4.9	0.97	164	**16**	0.1	6.4	0.0	95	**24**	0.46	**23**	1		unk.
D	0.18	**11**	0.02	**1**	0.00	**0**	0.036	**12**	0	0	**0**	unk.	unk.	51	**5**	1	301	tr	0.9	**5**	0.5	**3**	unk.	**unk.**	0
	0.9	**5**	21	**5**	0.09	**1**	0	**0**	0	3.0	**10**	3.0	0.62	95	**10**	0.0	7.7	0.0	54	**13**	0.28	**14**	tr		unk.
E	unk.	**unk.**	unk.	**unk.**	unk.	**unk.**	unk.	**unk.**	unk.	unk.	**unk.**	unk.	unk.	3	**0**	2058	42	unk.	unk.	**unk.**	unk.	**unk.**	unk.	**unk.**	unk.
	unk.	**unk.**	unk.	**unk.**	unk.	**unk.**	unk.	**unk.**	unk.	unk.	**unk.**	unk.	unk.	45	**5**	89.5	1.1	unk.	unk.	**unk.**	unk.	**unk.**	unk.		unk.
F	unk.	**unk.**	0.02	**1**	unk.	**unk.**	unk.	**unk.**	unk.	unk.	**unk.**	unk.	unk.	27	**3**	unk.	unk.	unk.	unk.	**unk.**	unk.	**unk.**	unk.	**unk.**	2
	unk.	**unk.**	2	**0**	unk.	**unk.**	unk.	**unk.**	unk.	unk.	**unk.**	unk.	1.86	34	**3**	unk.	unk.	unk.	unk.	**unk.**	0.01	**1**	unk.		unk.
G	unk.	**unk.**	unk.	**unk.**	0.00	**0**	unk.	**unk.**	0(a)	0(a)	**0**	unk.	unk.	14	**1**	tr	32	unk.	0.8	**5**	0.1	**1**	unk.	**unk.**	unk.
	unk.	**unk.**	unk.	**unk.**	unk.	**unk.**	unk.	**unk.**	unk.	unk.	**unk.**	unk.	0.15	10	**1**	0.0	0.8	unk.	4	**1**	unk.	**unk.**	unk.		unk.
H	0.03	**2**	0.04	**2**	0.00	**0**	0.001	**0**	0	0(a)	**0**	1.0	tr	10	**1**	1	152	1	0.1	**1**	0.1	**1**	unk.	**unk.**	40
	0.1	**1**	11	**3**	0.14	**1**	124	**3**	12	1.2	**4**	1.0	0.41	4	**1**	0.1	3.9	0.0	4	**1**	0.05	**3**	0		0
I	0.01	**1**	0.02	**1**	0.00	**0**	0.000	**0**	0	0(a)	**0**	0.4	tr	4	**0**	1	60	1	0.1	**0**	tr	**0**	unk.	**unk.**	16
	0.1	**0**	4	**1**	0.06	**1**	50	**1**	5	0.5	**2**	0.4	0.16	6	**0**	0.0	1.5	0.0	2	**0**	0.02	**1**	0		0
J	0.37	**22**	unk.	**unk.**	unk.	**unk.**	unk.	**unk.**	unk.	unk.	**unk.**	unk.	unk.	tr	**0**	75	157	unk.	2.7	**15**	unk.	**unk.**	unk.	**unk.**	unk.
	tr	**0**	unk.	**unk.**	unk.	**unk.**	tr	**0**	unk.	unk.	**unk.**	unk.	unk.	unk.	**unk.**	3.3	4.0	unk.	unk.	**unk.**	unk.	**unk.**	unk.		unk.
K	0.04	**3**	unk.	**unk.**	unk.	**unk.**	unk.	**unk.**	unk.	unk.	**unk.**	unk.	unk.	19	**2**	164	107	unk.	0.6	**4**	unk.	**unk.**	unk.	**unk.**	unk.
	0.4	**2**	unk.	**unk.**	unk.	**unk.**	107	**2**	unk.	unk.	**unk.**	unk.	0.75	24	**2**	7.1	2.8	unk.	5	**1**	unk.	**unk.**	unk.		unk.
L	0.02	**1**	unk.	**unk.**	0(a)	**0**	unk.	**unk.**	0	0(a)	**0**	unk.	unk.	8	**1**	1	126	6	0.3	**2**	unk.	**unk.**	unk.	**unk.**	unk.
	0.1	**1**	unk.	**unk.**	unk.	**unk.**	107	**2**	11	unk.	**unk.**	unk.	unk.	12	**1**	0.0	3.2	0.2	tr	**0**	tr(a)	**0**	tr(a)		unk.
M	0.01	**1**	0.02	**1**	0.00	**0**	0.001	**0**	0	0	**0**	0.4	tr	2	**0**	1	61	1	0.1	**1**	0.2	**1**	unk.	**unk.**	26
	0.1	**0**	4	**1**	0.06	**1**	50	**1**	5	0.5	**2**	0.4	0.17	3	**0**	0.0	1.6	0.0	2	**1**	0.03	**1**	0		14
N	0.02	**1**	0.04	**2**	0.00	**0**	0.001	**0**	0	0	**0**	1.0	tr	18	**2**	2	158	0	0.5	**3**	0.2	**1**	unk.	**unk.**	45
	0.1	**1**	11	**3**	0.14	**1**	127	**3**	12	1.2	**4**	1.0	0.44	9	**1**	0.1	4.0	0.0	5	**1**	0.05	3	0		7
O	0.01	**1**	0.04	**2**	0.00	**0**	unk.	**unk.**	0	0(a)	**0**	unk.	unk.	5	**1**	3	83	6	0.6	**4**	unk.	**unk.**	unk.	**unk.**	13
	tr	**0**	1	**0**	0.11	**1**	51	**1**	5	unk.	**unk.**	unk.	0.25	6	**1**	0.1	2.1	0.2	6	**2**	0.01	**1**	tr(a)		19
P	0.01	**1**	unk.	**unk.**	unk.	**unk.**	unk.	**unk.**	0	0(a)	**0**	unk.	unk.	5	**1**	2	95	5	0.6	**3**	0.1	**1**	unk.	**unk.**	12
	tr	**0**	1	**0**	0.11	**1**	49	**1**	5	unk.	**unk.**	unk.	0.37	6	**1**	0.1	2.4	0.1	2	**1**	0.01	**1**	tr(a)		18
Q	0.03	**2**	0.07	**4**	0.00	**0**	unk.	**unk.**	0	0(a)	**0**	unk.	unk.	14	**1**	1	302	tr	0.4	**2**	0.1	**1**	unk.	**unk.**	13
	0.5	**3**	unk.	**unk.**	0.12	**1**	2245	**45**	224	unk.	**unk.**	unk.	0.52	19	**2**	0.1	7.7	0.0	9	**2**	0.05	**3**	tr(a)		19
R	0.02	**1**	unk.	**unk.**	0.00	**0**	unk.	**unk.**	0	0(a)	**0**	unk.	unk.	14	**1**	1	294	tr	0.4	**2**	0.1	**1**	unk.	**unk.**	unk.
	0.5	**3**	unk.	**unk.**	unk.	**unk.**	2189	**44**	219	unk.	**unk.**	unk.	0.61	18	**2**	0.0	7.5	0.0	unk.	**unk.**	unk.	**unk.**	tr(a)		unk.
S	0.02	**1**	0.07	**3**	0(a)	**0**	unk.	**unk.**	0	0(a)	**0**	unk.	unk.	15	**2**	1	303	tr	0.4	**2**	0.1	**1**	unk.	**unk.**	12
	0.5	**3**	1	**0**	0.12	**1**	2251	**45**	225	unk.	**unk.**	unk.	0.61	20	**2**	0.0	7.7	0.0	9	**2**	0.05	**3**	tr(a)		18

	FOOD	Portion	Weight in grams / Conversion for 100 g	Kilocalories / H_2O g	Total carbohydrate g / Total fats g	Crude fiber g / Dietary fiber g	Total protein g / Total sugar g	**% USRDA**	Arginine mg / Histidine mg	Isoleucine mg / Leucine mg	Lysine mg / Methionine mg	Phenylalanine mg / Threonine mg	Valine mg / Tryptophan mg	Cystine mg / Tyrosine mg	Polyunsat. fatty acids g / Monounsat. fatty acids g	Saturated fatty acids g / P/S ratio	Linoleic acid g / Cholesterol mg	Thiamin mg / Ascorbic acid mg	**% USRDA**
A	apricot,dry,halves,uncooked	1/2 C	65	169	43.2	1.95	3.3	**5**	42	unk.	unk.	unk.	unk.	unk.	unk.	tr(a)	unk.	0.01	**0**
			1.54	16.2	0.3	7.80	29.9		unk.	unk.	unk.	unk.	unk.	unk.	unk.	unk.	0	8	**13**
B	apricot,fresh,any,edible portion	1 average	36	18	4.6	0.21	0.4	**1**	5	6	10	6	9	19	unk.	tr(a)	unk.	0.01	**1**
			2.80	30.5	0.1	0.75	2.2		6	10	2	7	3	5	unk.	unk.	0	4	**6**
C	ARTHUR TREACHER'S cole slaw	1 serving	97	121	11.2	0.00	1.1	**2**	unk.	unk.	unk.	unk.	unk.	unk.	unk.	unk.	unk.	0.02	**1**
			1.03	75.4	8.0	unk.	unk.		unk.	unk.	unk.	unk.	unk.	unk.	unk.	unk.	unk.	unk.	**unk.**
D	Arthur Treacher's fish	1 piece	63	171	9.9	0.19	9.5	**21**	unk.	unk.	unk.	unk.	unk.	unk.	unk.	unk.	unk.	unk.	**unk.**
			1.59	32.2	10.4	unk.	unk.		unk.	unk.	unk.	unk.	unk.	unk.	unk.	unk.	unk.	unk.	**unk.**
E	Arthur Treacher's french fries	1 serving	102	274	37.0	0.69	3.6	**6**	unk.	unk.	unk.	unk.	unk.	unk.	unk.	unk.	unk.	0.14	**10**
			0.98	46.9	12.4	unk.	unk.		unk.	unk.	unk.	unk.	unk.	unk.	unk.	unk.	unk.	unk.	**unk.**
F	ARTICHOKE (globe/French) cooked, edible portion	1 average	120	31	11.9	2.88	3.4	**5**	unk.	unk.	unk.	unk.	unk.	unk.	unk.	unk.	unk.	0.08	**6**
			0.83	103.8	0.2	unk.	unk.		unk.	unk.	unk.	unk.	unk.	unk.	unk.	unk.	0	10	**16**
G	artichoke heart,cooked-Birds Eye	1 average	16	6	1.2	0.16	0.4	**1**	unk.	unk.	unk.	unk.	unk.	unk.	unk.	tr(a)	unk.	0.01	**1**
			6.45	13.7	0.1	unk.	0.9		unk.	unk.	unk.	unk.	unk.	unk.	unk.	unk.	0	1	**2**
H	ASPARAGUS,green,canned,cut, drained solids	1/2 C	118	25	4.0	0.94	2.8	**4**	108	80	105	69	108	27	unk.	tr(a)	unk.	0.07	**5**
			0.85	108.7	0.5	1.76	2.9		unk.	99	33	67	26	59	unk.	unk.	0	18	**29**
I	asparagus,green,canned,low sodium,cut, drained solids	1/2 C	118	23	3.6	0.82	3.0	**5**	117	110	143	95	147	29	unk.	tr(a)	unk.	0.07	**5**
			0.85	110.0	0.4	1.76	2.9		unk.	134	46	92	36	64	unk.	unk.	0	18	**29**
J	asparagus,green,fresh,whole,cooked, drained solids	1/2 C	90	18	3.2	0.63	2.0	**3**	76	71	93	61	95	19	unk.	tr(a)	unk.	0.14	**10**
			1.11	84.2	0.2	1.35	2.2		unk.	87	30	59	23	57	unk.	unk.	0	23	**39**
K	asparagus,green,frozen,whole,cooked, drained solids	1/2 C	95	22	3.6	0.76	3.0	**5**	117	75	98	65	101	29	unk.	tr(a)	unk.	0.15	**10**
			1.05	87.6	0.2	1.43	2.4		unk.	92	31	63	25	63	unk.	unk.	0	25	**41**
L	asparagus,white,canned,cut, drained solids	1/2 C	121	27	4.4	0.97	2.5	**4**	unk.	68	88	58	91	24	unk.	tr(a)	unk.	0.06	**4**
			0.83	111.7	0.6	1.82	3.0		unk.	82	17	56	23	53	unk.	unk.	0	18	**30**
M	asparagus,white,canned,low sodium,cut, drained solids	1/2 C	118	22	4.1	0.82	2.2	**3**	unk.	80	105	69	108	21	unk.	tr(a)	unk.	0.06	**4**
			0.85	110.4	0.2	1.76	2.9		unk.	99	33	67	27	46	unk.	unk.	0	18	**29**
N	AVOCADO,raw,cubed,edible portion	1/2 C	75	125	4.7	1.20	1.6	**2**	unk.	15	22	16	21	unk.	unk.	2.25	1.50	0.08	**6**
			1.33	55.5	12.3	1.50	0.4		8	24	7	13	5	10	5.25	unk.	0	11	**18**
O	avocado,raw,edible portion	1 average	227	378	14.3	3.62	4.8	**7**	unk.	46	67	48	63	unk.	unk.	6.79	4.53	0.25	**17**
			0.44	167.6	37.1	4.53	1.4		24	74	22	39	15	30	15.85	unk.	0	32	**53**
P	avocado,raw,mashed,edible portion	1/2 C	165	276	10.4	2.64	3.5	**5**	unk.	33	49	35	46	unk.	unk.	4.95	3.30	0.18	**12**
			0.61	122.1	27.1	3.30	1.0		17	54	16	29	11	22	11.55	unk.	0	23	**39**
Q	BABY FOOD,BAKED PRODUCT, arrowroot cookie	1 cookie	6.0	27	4.3	unk.	0.5	**2**	unk.	unk.	unk.	unk.	unk.	unk.	0.03	0.20	0.05	0.03	**4**
			16.67	0.3	0.9	unk.	unk.		unk.	unk.	unk.	unk.	unk.	unk.	0.52	0.13	tr	tr	**1**
R	baby food,baked product,cookie	1 cookie	6.5	28	4.4	0.03	0.8	**3**	unk.	unk.	unk.	unk.	unk.	unk.	0.04	0.24	0.06	0.09	**14**
			15.38	0.4	0.9	unk.	unk.		unk.	unk.	unk.	unk.	unk.	unk.	0.46	0.15	unk.	0	**0**
S	baby food,baked product,pretzel	1 pretzel	6.0	24	4.9	0.02	0.6	**3**	unk.	unk.	unk.	unk.	unk.	unk.	unk.	unk.	unk.	0.03	**4**
			16.67	0.2	0.1	unk.	unk.		unk.	unk.	unk.	unk.	unk.	unk.	unk.	unk.	unk.	tr	**1**

Riboflavin mg Niacin mg	% USRDA	Vitamin B_6 mg Folacin mcg	% USRDA	Vitamin B_{12} mcg Pantothenic acid mg	% USRDA	Biotin mg Vitamin A IU	% USRDA	Preformed A RE Beta carotene RE	Vitamin D IU Vitamin E IU	% USRDA	Total tocopherol mg Alpha tocopherol mg	Other tocopherol mg Total ash g	Calcium mg Phosphorus mg	% USRDA	Sodium mg Sodium meq	Potassium mg Potassium meq	Chlorine mg Chlorine meq	Iron mg Magnesium mg	% USRDA	Zinc mg Copper mg	% USRDA	Iodine mcg Selenium mcg	% USRDA	Manganese mcg Chromium mcg	
0.10	**6**	0.11	**6**	0.00	**0**	unk.	**unk.**	0	0(a)	**0**	unk.	unk.	44	**4**	17	636	23	3.6	**20**	0.1	**1**	unk.	**unk.**	unk.	A
2.1	**11**	3	**1**	0.49	**5**	7085	**142**	708	unk.	**unk.**	unk.	1.95	70	**7**	0.7	16.3	0.6	40	**10**	0.26	**13**	tr(a)		unk.	
0.01	**1**	0.02	**1**	0.00	**0**	unk.	**unk.**	0	0(a)	**0**	unk.	unk.	5	**0**	tr	100	tr	0.2	**1**	tr	**0**	unk.	**unk.**	17	B
0.2	**1**	1	**0**	0.09	**1**	964	**19**	96	unk.	**unk.**	unk.	0.25	8	**1**	0.0	2.6	0.0	3	**1**	0.03	**1**	0		0	
0.04	**2**	0.06	**3**	0.06	**1**	unk.	**unk.**	26	unk.	**unk.**	unk.	unk.	unk.	**unk.**	unk.	unk.	unk.	unk.	**unk.**	0.1	**1**	unk.	**unk.**	unk.	C
0.2	**1**	unk.	**unk.**	unk.	**unk.**	264	**5**	18	unk.	**unk.**	unk.	0.96	unk.	**unk.**	unk.	unk.	unk.	12	**3**	0.01	**1**	unk.		unk.	
unk.	**unk.**	unk.	**unk.**	unk.	**unk.**	0.002	**1**	unk.	unk.	**unk.**	unk.	unk.	13	**1**	312	352	unk.	0.2	**1**	0.3	**2**	1.7	**1**	unk.	D
unk.	**unk.**	1	**0**	unk.	**unk.**	13	**0**	unk.	unk.	**unk.**	unk.	1.01	125	**13**	13.6	9.0	unk.	20	**5**	0.04	**2**	unk.		unk.	
0.05	**3**	0.09	**5**	0.02	**0**	0.001	**0**	0	unk.	**unk.**	unk.	unk.	25	**3**	295	106.9	unk.	0.6	**3**	3.8	**26**	2.7	**2**	unk.	E
2.1	**11**	1	**0**	unk	**unk.**	0	**0**	0	unk.	**unk.**	unk.	1.81	125	**13**	12.8	27.3	unk.	39	**10**	0.14	**7**	unk.		unk.	
0.05	**3**	0.09	**4**	0(a)	**0**	unk.	**unk.**	0	0(a)	**0**	unk.	unk.	61	**6**	36	361	unk.	1.3	**7**	0.4	**3**	unk.	**unk.**	unk.	F
0.8	**4**	unk.	**unk.**	unk.	**unk.**	180	**4**	18	unk.	**unk.**	unk.	0.72	83	**8**	1.6	9.2	unk.	unk.	**unk.**	tr(a)	**0**	tr(a)		unk.	
0.02	**1**	0.01	**1**	0.00	**0**	unk.	**unk.**	0	0	**0**	unk.	unk.	3	**0**	7	38	unk.	0.1	**1**	0.0	**0**	unk.	**unk.**	unk.	G
0.1	**1**	15	**4**	0.03	**0**	24	**1**	2	unk.	**unk.**	unk.	0.09	9	**1**	0.3	1.0	unk.	5	**1**	0.01	**1**	tr(a)		unk.	
0.12	**7**	0.06	**3**	0.00	**0**	0.000	**0**	0	0(a)	**0**	unk.	unk.	22	**2**	277	195	36	2.2	**12**	1.1	**8**	unk.	**unk.**	unk.	H
0.9	**5**	unk.	**unk.**	0.23	**2**	940	**19**	94	unk.	**unk.**	2.9	1.53	62	**6**	12.1	5.0	1.0	12	**3**	tr(a)	**0**	tr(a)		unk.	
0.12	**7**	0.06	**3**	0.00	**0**	0.000	**0**	0	0(a)	**0**	unk.	unk.	22	**2**	4	195	36	2.2	**12**	1.1	**8**	unk.	**unk.**	unk.	I
0.9	**5**	unk.	**unk.**	0.23	**2**	940	**19**	94	unk.	**unk.**	2.9	0.47	62	**6**	0.1	5.0	1.0	12	**3**	tr(a)	**0**	tr(a)		unk.	
0.16	**10**	0.08	**4**	0.00	**0**	tr	**0**	0	0(a)	**0**	unk.	unk.	19	**2**	1	165	28	0.5	**3**	0.9	**6**	unk.	**unk.**	unk.	J
1.3	**6**	58	**14**	0.28	**3**	810	**16**	81	unk.	**unk.**	2.2	0.36	45	**5**	0.0	4.2	0.8	9	**2**	tr(a)	**0**	tr(a)		unk.	
0.13	**8**	0.09	**4**	0.00	**0**	tr	**0**	0	0(a)	**0**	unk.	unk.	21	**2**	1	226	29	1.0	**6**	0.9	**6**	unk.	**unk.**	unk.	K
1.0	**5**	61	**15**	0.29	**3**	741	**15**	74	unk.	**unk.**	2.4	0.57	64	**6**	0.0	5.8	0.8	9	**2**	tr(a)	**0**	tr(a)		unk.	
0.07	**4**	0.07	**3**	0.00	**0**	0.000	**0**	0	0(a)	**0**	unk.	unk.	19	**2**	286	169	38	1.2	**7**	1.2	**8**	unk.	**unk.**	unk.	L
0.8	**4**	unk.	**unk.**	0.24	**2**	97	**2**	10	unk.	**unk.**	3.0	1.81	50	**5**	12.4	4.3	1.1	12	**3**	tr(a)	**0**	tr(a)		unk.	
0.07	**4**	0.06	**3**	0.00	**0**	0.000	**0**	0	0(a)	**0**	unk.	unk.	19	**2**	5	164	36	1.2	**7**	1.1	**8**	unk.	**unk.**	unk.	M
0.8	**4**	unk.	**unk.**	0.23	**2**	94	**2**	9	unk.	**unk.**	2.9	0.47	48	**5**	0.2	4.2	1.0	12	**3**	tr(a)	**0**	tr(a)		unk.	
0.15	**9**	0.31	**16**	0.00	**0**	0.004	**1**	0	0(a)	**0**	0.9	unk.	8	**1**	3	453	8	0.4	**3**	0.3	**2**	unk.	**unk.**	101	N
1.2	**6**	38	**10**	0.80	**8**	218	**4**	22	1.1	**4**	unk.	0.90	32	**3**	0.1	11.6	0.2	34	**8**	0.30	**15**	0		0	
0.45	**27**	0.95	**48**	0.00	**0**	0.012	**4**	0	0(a)	**0**	2.7	unk.	23	**2**	9	1368	23	1.4	**8**	1.0	**7**	unk.	**unk.**	304	O
3.6	**18**	116	**29**	2.42	**24**	657	**13**	66	3.3	**11**	unk.	2.72	95	**10**	0.4	35.0	0.6	102	**26**	0.91	**45**	0		0	
0.33	**19**	0.69	**35**	0.00	**0**	0.009	**3**	0	0(a)	**0**	2.0	unk.	16	**2**	7	997	16	1.0	**6**	0.7	**5**	unk.	**unk.**	221	P
2.6	**13**	84	**21**	1.77	**18**	479	**10**	48	2.4	**8**	unk.	1.98	69	**7**	0.3	25.5	0.5	74	**19**	0.66	**33**	0		0	
0.03	**3**	tr	**0**	tr	**0**	unk.	**unk.**	unk.	unk.	**unk.**	unk.	unk.	2	**0**	21	9	unk.	0.2	**2**	tr	**1**	unk.	**unk.**	1338	Q
0.3	**4**	unk.	**unk.**	unk.	**unk.**	unk.	**unk.**	unk.	unk.	**unk.**	unk.	0.08	8	**1**	0.9	0.2	unk.	1	**1**	0.01	**1**	unk.		unk.	
0.21	**26**	0.38	**55**	0.30	**10**	unk.	**unk.**	unk.	unk.	**unk.**	unk.	unk.	7	**1**	12	33	unk.	0.3	**3**	0.1	**1**	unk.	**unk.**	tr	R
1.0	**12**	unk.	**unk.**	unk.	**unk.**	1	**0**	unk.	unk.	**unk.**	unk.	0.13	12	**1**	0.5	0.8	unk.	4	**2**	0.02	**2**	unk.		unk.	
0.02	**3**	0.00	**0**	unk.	**unk.**	unk.	**unk.**	unk.	unk.	**unk.**	unk.	unk.	1	**0**	16	8	unk.	0.2	**2**	0.1	**1**	unk.	**unk.**	12	S
0.2	**2**	unk.	**unk.**	unk.	**unk.**	0	**0**	unk.	unk.	**unk.**	unk.	0.06	7	**1**	0.7	0.2	unk.	3	**1**	0.03	**3**	unk.		unk.	

	FOOD	Portion	Weight in grams / *Conversion for 100 g*	Kilocalories / H_2O g	Total carbohydrate g / Total fats g	Crude fiber g / Dietary fiber g	Total protein g / Total sugar g	% USRDA	Arginine mg / Histidine mg	Isoleucine mg / Leucine mg	Lysine mg / Methionine mg	Phenylalanine mg / Threonine mg	Valine mg / Tryptophan mg	Cystine mg / Tyrosine mg	Polyunsat. fatty acids g / Monounsat. fatty acids g	Saturated fatty acids g / P/S ratio	Linoleic acid g / Cholesterol mg	Thiamin mg / Ascorbic acid mg	% USRDA
A	baby food,baked product,	1 biscuit	11	43	8.4	0.05	1.2	**5**	66	93	39	58	97	17	unk.	unk.	unk.	0.03	**4**
	teething biscuit		*9.09*	0.7	0.5	unk.	unk.		44	16	30	62	21	73	unk.	unk.	unk.	1	**3**
B	baby food,baked product,zwieback	1 piece	7.0	30	5.2	0.01	0.7	**3**	unk.	unk.	unk.	unk.	unk.	unk.	0.08	0.28	0.01	0.01	**2**
			14.29	0.3	0.7	unk.	unk.		unk.	unk.	unk.	unk.	unk.	unk.	0.23	0.29	1	tr	**1**
C	BABY FOOD,CEREAL & egg yolks	4-1/2 oz jar	128	101	7.9	unk.	3.2	**13**	unk.	unk.	unk	unk.	unk.	unk.	unk.	unk.	unk.	0.06	**9**
	w/bacon,strnd		*0.78*	110.0	6.4	unk.	unk.		unk.	unk.	unk.	unk.	unk.	unk.	unk.	unk.	unk.	1	**3**
D	baby food,cereal & egg yolks,strnd	4-1/2 oz jar	128	65	9.0	0.13	2.4	**10**	133	128	163	118	156	36	0.19	0.78	0.33	0.01	**2**
			0.78	113.7	2.3	unk.	unk.		58	220	65	105	35	108	0.93	0.25	81	1	**2**
E	baby food,cereal & eggs,strnd	4-1/2 oz jar	128	unk.	unk.	unk.	unk.	**unk.**	unk.	unk.	unk.	unk.	unk.	unk.	0.42	0.63	0.37	unk.	**unk.**
			0.78	111.6	1.9	unk.	unk.		unk.	unk.	unk.	unk.	unk.	unk.	0.73	0.67	66	unk.	**unk.**
F	baby food,cereal,barley,dry	1 Tbsp	2.4	9	1.8	0.03	0.3	**1**	14	10	9	16	14	6	unk.	unk.	unk.	0.07	**9**
			41.67	0.2	0.1	unk.	unk.		6	19	5	9	3	10	unk.	unk.	0(a)	tr	**0**
G	baby food,cereal,barley,	1 oz	28	32	4.6	0.06	1.3	**5**	55	67	79	68	80	19	unk.	unk.	unk.	0.14	**20**
	prep w/whole milk		*3.52*	21.2	0.9	unk.	unk.		33	114	29	53	17	57	unk.	unk.	unk.	tr	**1**
H	baby food,cereal,high protein	1 Tbsp	2.4	9	1.4	0.03	0.6	**2**	unk.	unk.	unk.	unk.	unk.	unk.	unk.	unk.	unk.	0.09	**13**
	w/apple & orange,dry		*41.67*	0.1	0.2	unk.	unk.		unk.	unk.	unk.	unk.	unk.	unk.	unk.	unk.	0(a)	tr	**0**
I	baby food,cereal,high protein	1 oz	28	32	3.8	0.06	2.0	**8**	unk.	unk.	unk.	unk.	unk.	unk.	unk.	unk.	unk.	0.19	**27**
	w/apple & orange,prep w/whole milk		*3.52*	21.1	1.1	unk.	unk.		unk.	unk.	unk.	unk.	unk.	unk.	unk.	unk.	unk.	tr	**1**
J	baby food,cereal,high protein,dry	1 Tbsp	2.4	9	1.1	0.06	0.9	**4**	67	42	57	45	45	17	unk.	unk.	unk.	0.06	**9**
			41.67	0.1	0.1	unk.	unk.		67	71	15	35	13	34	unk.	unk.	0(a)	tr	**0**
K	baby food,cereal,high protein,	1 oz	28	32	3.3	0.11	2.5	**10**	158	129	172	126	139	41	unk.	unk.	unk.	0.13	**19**
	prep w/whole milk		*3.52*	21.2	1.1	unk.	unk.		68	214	48	103	37	103	unk.	unk.	unk.	tr	**1**
L	baby food,cereal,mixed	4-3/4 oz jar	135	111	24.2	0.40	1.6	**7**	90	59	49	85	80	36	tr(a)	tr(a)	tr(a)	0.38	**54**
	w/applesauce & bananas,strnd		*0.74*	108.0	0.7	unk.	unk.		38	119	34	47	18	59	tr(a)	0.00	unk.	34	**86**
M	baby food,cereal,mixed w/bananas,dry	1 Tbsp	2.4	9	1.9	0.02	0.3	**1**	12	10	11	13	14	5	unk.	unk.	unk.	0.09	**13**
			41.67	0.1	0.1	unk.	unk.		8	23	5	9	3	11	unk.	unk.	0(a)	tr	**0**
N	baby food,cereal,mixed w/bananas,	1 oz	28	33	4.7	0.03	1.3	**5**	52	67	83	63	80	16	unk.	unk.	unk.	0.18	**26**
	prep w/whole milk		*3.52*	21.1	1.0	unk.	unk.		36	121	30	52	17	58	unk.	unk.	unk.	tr	**1**
O	baby food,cereal,mixed w/honey,dry	1 Tbsp	2.4	9	1.8	unk.	0.3	**1**	unk.	unk.	unk.	unk.	unk.	unk.	unk.	unk.	unk.	0.06	**9**
			41.67	0.1	0.1	unk.	unk.		unk.	unk.	unk.	unk.	unk.	unk.	unk.	unk.	0(a)	0	**0**
P	baby food,cereal,mixed w/honey,	1 oz	28	33	4.5	unk.	1.4	**6**	unk.	unk.	unk.	unk.	unk.	unk.	unk.	unk.	unk.	0.13	**18**
	prep w/whole milk		*3.52*	21.1	1.0	unk.	unk.		unk.	unk.	unk.	unk.	unk.	unk.	unk.	unk.	unk.	tr	**1**
Q	baby food,cereal,mixed,dry	1 Tbsp	2.4	9	1.8	0.02	0.3	**1**	17	11	10	16	16	8	unk.	unk.	unk.	0.06	**8**
			41.67	0.2	0.1	unk.	unk.		7	25	6	10	4	12	unk.	unk.	0(a)	tr	**0**
R	baby food,cereal,mixed,	1 oz	28	32	4.5	0.06	1.4	**6**	61	69	81	69	83	23	unk.	unk.	unk.	0.12	**17**
	prep w/whole milk		*3.52*	21.2	1.0	unk.	unk.		34	125	31	54	18	60	unk.	unk.	unk.	tr	**1**
S	baby food,cereal,oatmeal	4-3/4 oz jar	135	99	20.8	0.94	1.8	**7**	126	66	72	89	92	39	tr(a)	tr(a)	tr(a)	0.57	**81**
	w/applesauce & bananas,strnd		*0.74*	111.0	0.9	unk.	unk.		46	134	40	62	23	70	tr(a)	0.00	unk.	29	**74**

Riboflavin mg / Niacin mg	% USRDA	Vitamin B6 mg / Folacin mcg	% USRDA	Vitamin B12 mcg / Pantothenic acid mg	% USRDA	Biotin mg / Vitamin A IU	% USRDA	Preformed A RE / Beta carotene RE	Vitamin D IU / Vitamin E IU	% USRDA	Total tocopherol mg / Alpha tocopherol mg	Other tocopherol mg / Total ash g	Calcium mg / Phosphorus mg	% USRDA	Sodium mg / Sodium meq	Potassium mg / Potassium meq	Chlorine mg / Chlorine meq	Iron mg / Magnesium mg	% USRDA	Zinc mg / Copper mg	% USRDA	Iodine mcg / Selenium mcg	% USRDA	Manganese mcg / Chromium mcg	
0.06	7	0.01	2	0.01	0	unk.	unk.	unk.	unk.	unk.	unk.	unk.	29	4	40	36	unk.	0.4	4	0.1	1	unk.	unk.	10	A
0.5	5	unk.	unk.	unk.	unk.	13	1	unk.	unk.	unk.	unk.	0.25	18	2	1.7	0.9	unk.	6	3	0.04	4	unk.		unk.	
0.02	2	0.01	1	unk.	unk.	unk.	unk.	unk.	unk.	unk.	unk.	unk.	1	0	16	21	unk.	tr	0	0.1	1	unk.	unk.	19	B
0.9	10	unk.	unk.	unk.	unk.	4	0	unk.	unk.	unk.	unk.	0.10	4	1	0.7	0.5	unk.	1	1	0.02	2	unk.		unk.	
0.10	13	unk.	unk.	unk.	unk.	unk.	unk.	unk.	unk.	unk.	0.3	0.0	36	5	61	45	unk.	unk.	unk.	0.3	4	unk.	unk.	unk.	C
0.3	4	5	26	unk.	unk.	120	5	unk.	0.3	3	0.3	0.51	unk.	unk.	2.7	1.1	unk.	6	3	0.03	3	unk.		unk.	
0.05	6	0.03	4	90.88	3029	unk.	unk.	unk.	unk.	unk.	unk.	unk.	31	4	42	50	unk.	0.6	6	0.4	5	unk.	unk.	unk.	D
0.1	1	4	21	unk.	unk.	180	7	unk.	unk.	unk.	unk.	0.64	51	6	1.8	1.3	unk.	4	2	0.03	3	unk.		unk.	
unk.	unk.	unk.	unk.	unk.	unk.	unk.	unk.	unk.	unk.	unk.	unk.	unk.	unk.	unk.	unk.	unk.	unk.	unk.	unk.	unk.	unk.	unk.	unk.	unk.	E
unk.	unk.	12	60	unk.	unk.	unk.	unk.	unk.	unk.	unk.	unk.	unk.	unk.	unk.	unk.	unk.	unk.	unk.	unk.	unk.	unk.	unk.		unk.	
0.06	8	0.01	1	unk.	unk.	unk.	unk.	unk.	unk.	unk.	unk.	unk.	19	2	1	9	unk.	1.8	18	0.1	1	unk.	unk.	46	F
0.9	10	1	4	unk.	unk.	unk.	unk.	unk.	unk.	unk.	unk.	0.08	11	1	0.0	0.2	unk.	3	1	0.01	1	unk.		unk.	
0.16	21	0.03	4	unk.	unk.	unk.	unk.	unk.	unk.	unk.	unk.	unk.	65	8	14	55	unk.	3.5	35	0.2	3	unk.	unk.	unk.	G
1.7	19	3	13	unk.	unk.	30	1	unk.	unk.	unk.	unk.	0.34	43	5	0.6	1.4	unk.	9	4	unk.	unk.	unk.		unk.	
0.10	13	0.01	1	0.01	0	unk.	unk.	unk.	unk.	unk.	unk.	unk.	18	2	2	32	unk.	2.1	21	0.1	1	unk.	unk.	0	H
0.6	6	unk.	unk.	unk.	unk.	1	0	unk.	unk.	unk.	unk.	0.12	13	2	0.1	0.8	unk.	3	2	0.02	2	unk.		unk.	
0.24	30	0.03	4	0.10	3	unk.	unk.	unk.	unk.	unk.	unk.	unk.	63	8	16	98	unk.	4.1	41	0.2	3	unk.	unk.	unk.	I
1.1	13	unk.	unk.	unk.	unk.	28	1	unk.	unk.	unk.	unk.	0.43	47	6	0.7	2.5	unk.	11	5	unk.	unk.	unk.		unk.	
0.06	8	0.01	2	0(a)	0	unk.	unk.	unk.	unk.	unk.	unk.	unk.	17	2	1	3.2	unk.	1.8	18	0.1	1	unk.	unk.	86	J
0.8	9	5	23	unk.	unk.	unk.	unk.	unk.	unk.	unk.	unk.	0.13	15	2	0.0	0.8	unk.	5	3	0.03	3	unk.		unk.	
0.16	21	0.03	5	unk.	unk.	unk.	unk.	unk.	unk.	unk.	unk.	unk.	62	8	14	99	unk.	3.4	35	0.3	4	unk.	unk.	unk.	K
1.6	18	10	50	unk.	unk.	30	1	unk.	unk.	unk.	unk.	0.43	50	6	0.6	2.5	unk.	14	7	unk.	unk.	unk.		unk.	
0.47	59	0.18	26	unk.	unk.	unk.	unk.	unk.	unk.	unk.	0.7	0.4	8	1	3	55	unk.	8.9	89	0.3	3	unk.	unk.	unk.	L
5.3	59	5	24	unk.	unk.	24	1	unk.	0.9	9	0.3	0.40	31	4	0.1	1.4	unk.	11	5	unk.	unk.	unk.		unk.	
0.09	11	0.01	1	0.01	0	unk.	unk.	unk.	unk.	unk.	unk.	unk.	17	2	3	16	unk.	1.6	16	tr	0	unk.	unk.	36	M
0.5	6	unk.	unk.	unk.	unk.	3	0	unk.	unk.	unk.	unk.	0.07	9	1	0.1	0.4	unk.	2	1	0.01	1	unk.		unk.	
0.20	26	0.03	4	0.10	3	unk.	unk.	unk.	unk.	unk.	unk.	unk.	61	8	17	67	unk.	3.2	32	0.2	2	unk.	unk.	unk.	N
1.0	11	unk.	unk.	unk.	unk.	31	1	unk.	unk.	unk.	unk.	0.31	39	5	0.7	1.7	unk.	7	4	unk.	unk.	unk.		unk.	
0.07	8	unk.	unk.	unk.	unk.	unk.	unk.	unk.	unk.	unk.	unk.	unk.	28	4	1	6	unk.	1.6	16	unk.	unk.	unk.	unk.	unk.	O
0.9	10	unk.	unk.	unk.	unk.	1	0	unk.	unk.	unk.	unk.	0.09	16	2	0.0	0.2	unk.	unk.	unk.	unk.	unk.	unk.		unk.	
0.16	21	unk.	unk.	unk.	unk.	unk.	unk.	unk.	unk.	unk.	unk.	unk.	83	10	14	49	unk.	3.2	32	unk.	unk.	unk.	unk.	unk.	P
1.8	20	unk.	unk.	unk.	unk.	26	1	unk.	unk.	unk.	unk.	0.34	52	7	0.6	1.2	unk.	unk.	unk.	unk.	unk.	unk.		unk.	
0.07	8	0.00	0	unk.	unk.	unk.	unk.	unk.	unk.	unk.	unk.	unk.	18	2	1	10	unk.	1.5	15	0.1	1	unk.	unk.	62	Q
0.8	9	1	5	unk.	unk.	unk.	unk.	unk.	unk.	unk.	unk.	0.08	9	1	0.0	0.3	unk.	2	1	0.01	1	unk.		unk.	
0.16	21	0.02	3	unk.	unk.	unk.	unk.	unk.	unk.	unk.	unk.	unk.	62	8	13	57	unk.	3.0	30	0.2	3	unk.	unk.	unk.	R
1.6	18	3	16	unk.	unk.	30	1	unk.	unk.	unk.	unk.	0.31	40	5	0.6	1.4	unk.	8	4	unk.	unk.	unk.		unk.	
0.50	62	0.27	39	unk.	unk.	unk.	unk.	unk.	unk.	unk.	0.7	0.3	12	2	3	63	unk.	7.6	76	0.5	7	unk.	unk.	405	S
6.8	76	5	24	unk.	unk.	40	2	unk.	0.9	9	0.5	0.54	55	7	0.1	1.6	unk.	21	11	0.23	23	unk.		unk.	

	FOOD	Portion	Weight in grams / *Conversion for 100 g*	Kilocalories / H_2O g	Total carbohydrate g / Total fats g	Crude fiber g / Dietary fiber g	Total protein g / Total sugar g	% USRDA	Arginine mg / Histidine mg	Isoleucine mg / Leucine mg	Lysine mg / Methionine mg	Phenylalanine mg / Threonine mg	Valine mg / Tryptophan mg	Cystine mg / Tyrosine mg	Polyunsat. fatty acids g / Monounsat. fatty acids g	Saturated fatty acids g / P/S ratio	Linoleic acid g / Cholesterol mg	Thiamin mg / Ascorbic acid mg	% USRDA
A	baby food,cereal,oatmeal	1 Tbsp	2.4	9	1.8	0.02	0.3	**1**	17	12	13	15	17	7	unk.	unk.	unk.	0.09	**12**
	w/bananas,dry		*41.67*	0.1	0.1	unk.	unk.		9	23	6	12	4	12	unk.	unk.	0(a)	tr	**0**
B	baby food,cereal,oatmeal	1 oz	28	33	4.5	0.06	1.3	**5**	61	70	88	67	85	20	unk.	unk.	unk.	0.18	**25**
	w/bananas,prep w/whole milk		*3.52*	21.1	1.1	unk.	unk.		38	122	32	55	18	61	unk.	unk.	unk.	0	**0**
C	baby food,cereal,oatmeal w/honey,dry	1 Tbsp	2.4	9	1.7	unk.	0.3	**1**	unk.	unk.	unk.	unk.	unk.	unk.	unk.	unk.	unk.	0.07	**10**
			41.67	0.1	0.2	unk.	unk.		unk.	unk.	unk.	unk.	unk.	unk.	unk.	unk.	0(a)	0	**0**
D	baby food,cereal,oatmeal w/honey,	1 oz	28	33	4.3	unk.	1.4	**6**	unk.	unk.	unk.	unk.	unk.	unk.	unk.	unk.	unk.	0.14	**20**
	prep w/whole milk		*3.52*	21.2	1.1	unk.	unk.		unk.	unk.	unk.	unk.	unk.	unk.	unk.	unk.	unk.	tr	**1**
E	baby food,cereal,oatmeal,dry	1 Tbsp	2.4	10	1.7	0.03	0.3	**1**	25	13	14	13	19	12	unk.	unk.	unk.	0.07	**10**
			41.67	0.1	0.2	unk.	unk.		6	26	6	11	4	13	unk.	unk.	0(a)	tr	**0**
F	baby food,cereal,oatmeal,	1 oz	28	33	4.3	0.06	1.4	**6**	76	73	88	63	88	30	unk.	unk.	unk.	0.14	**21**
	prep w/whole milk		*3.52*	21.2	1.2	unk.	unk.		32	126	32	57	19	63	unk.	unk.	unk.	tr	**1**
G	baby food,cereal,rice	4-3/4 oz jar	135	107	23.1	0.27	1.6	**7**	54	65	92	78	92	19	tr(a)	tr(a)	tr(a)	0.35	**50**
	w/applesauce & bananas,strnd		*0.74*	109.3	0.5	unk.	unk.		43	157	27	54	16	76	tr(a)	0.00	unk.	43	**107**
H	baby food,cereal,rice	1 Tbsp	2.4	10	1.9	0.01	0.2	**1**	14	11	11	10	13	4	unk.	unk.	unk.	0.10	**14**
	w/bananas,dry		*41.67*	0.1	0.1	unk.	unk.		7	20	5	11	4	9	unk.	unk.	0(a)	0	**0**
I	baby food,cereal,rice w/bananas,prep	1 oz	28	33	4.8	0.03	1.2	**5**	56	69	83	56	78	16	unk.	unk.	unk.	0.19	**28**
	w/whole milk		*3.52*	21.1	1.0	unk.	unk.		34	115	30	56	18	55	unk.	unk.	unk.	tr	**1**
J	baby food,cereal,rice w/honey,dry	1 Tbsp	2.4	9	1.9	unk.	0.2	**1**	unk.	unk.	unk.	unk.	unk.	unk.	unk.	unk.	unk.	0.07	**9**
			41.67	0.1	0.1	unk.	unk.		unk.	unk.	unk.	unk.	unk.	unk.	unk.	unk.	0(a)	0	**0**
K	baby food,cereal,rice w/honey,	1 oz	28	33	4.9	unk.	1.1	**4**	unk.	unk.	unk.	unk.	unk.	unk.	unk.	unk.	unk.	0.14	**20**
	prep w/whole milk		*3.52*	21.1	0.9	unk.	unk.		unk.	unk.	unk.	unk.	unk.	unk.	unk.	unk.	unk.	tr	**1**
L	baby food,cereal,rice,dry	1 Tbsp	2.4	9	1.9	0.02	0.2	**1**	16	7	7	9	11	4	unk.	unk.	unk.	0.06	**9**
			41.67	0.2	0.1	unk.	unk.		5	13	4	8	2	8	unk.	unk.	0(a)	tr	**0**
M	baby food,cereal,rice,	1 oz	28	33	4.7	0.03	1.1	**4**	58	60	76	55	73	15	unk.	unk.	unk.	0.13	**19**
	prep w/whole milk		*3.52*	21.2	1.0	unk.	unk.		30	102	28	50	15	53	unk.	unk.	unk.	tr	**1**
N	BABY FOOD,DESSERT,apple betty,strnd	4-3/4 oz jar	135	97	26.5	unk.	0.5	**2**	unk.	unk.	unk.	unk.	unk.	unk.	0.00	0.00	0.00	0.01	**2**
			0.74	107.9	0.0	unk.	unk.		unk.	unk.	unk.	unk.	unk.	unk.	0.00	0.00	0(a)	47	**117**
O	baby food,dessert,caramel	4-3/4 oz jar	135	104	23.2	unk.	1.8	**7**	unk.	unk.	unk.	unk.	unk.	unk.	tr(a)	tr(a)	tr(a)	0.01	**2**
	pudding,strnd		*0.74*	108.5	0.9	unk.	unk.		unk.	unk.	unk.	unk.	unk.	unk.	tr(a)	0.00	unk.	3	**7**
P	baby food,dessert,cherry vanilla	4-3/4 oz jar	135	92	24.0	0.27	0.3	**1**	unk.	unk.	unk.	unk.	unk.	unk.	tr(a)	tr(a)	tr(a)	0.01	**2**
	pudding,strnd		*0.74*	110.0	0.4	unk.	unk.		unk.	unk.	unk.	unk.	unk.	unk.	tr(a)	0.00	unk.	1	**4**
Q	baby food,dessert,chocolate	4-1/2 oz jar	128	108	20.6	unk.	2.4	**10**	105	129	173	109	152	unk.	unk.	unk.	unk.	0.01	**2**
	custard pudding,strnd		*0.78*	102.3	2.2	unk.	unk.		60	233	59	111	unk.	88	unk.	unk.	unk.	2	**5**
R	baby food,dessert,cottage cheese	4-3/4 oz jar	135	93	17.8	1.75	3.9	**16**	81	132	190	143	159	unk.	tr(a)	tr(a)	tr(a)	0.03	**4**
	w/pineapple,strnd		*0.74*	111.5	1.1	unk.	unk.		72	244	35	99	unk.	109	tr(a)	0.00	unk.	31	**79**
S	baby food,dessert,Dutch apple,strnd	4-3/4 oz jar	135	92	22.5	0.40	0.0	**0**	unk.	unk.	unk.	unk.	unk.	unk.	0.04	0.78	0.03	0.01	**2**
			0.74	111.0	1.2	unk.	unk.		unk.	unk.	unk.	unk.	unk.	unk.	0.28	0.05	unk.	29	**72**

Riboflavin mg / Niacin mg	% USRDA	Vitamin B_6 mg / Folacin mcg	% USRDA	Vitamin B_{12} mcg / Pantothenic acid mg	% USRDA	Biotin mg / Vitamin A IU	% USRDA	Preformed A RE / Beta carotene RE	Vitamin D IU / Vitamin E IU	% USRDA	Total tocopherol mg / Alpha tocopherol mg	Other tocopherol mg / Total ash g	Calcium mg / Phosphorus mg	% USRDA	Sodium mg / Sodium meq	Potassium mg / Potassium meq	Chlorine mg / Chlorine meq	Iron mg / Magnesium mg	% USRDA	Zinc mg / Copper mg	% USRDA	Iodine mcg / Selenium mcg	% USRDA	Manganese mcg / Chromium mcg	
0.09	**11**	0.01	**2**	0.01	**0**	unk.	**unk.**	unk.	unk.	**unk.**	unk.	unk.	16	**2**	3	18	unk.	1.6	**16**	0.1	**1**	unk.	**unk.**	70	A
0.5	**5**	unk.	**unk.**	unk.	**unk.**	2	**0**	unk.	unk.	**unk.**	unk.	0.09	11	**1**	0.1	0.4	unk.	3	**1**	0.01	**1**	unk.		unk.	
0.21	**27**	0.03	**4**	0.09	**3**	unk.	**unk.**	unk.	unk.	**unk.**	unk.	unk.	59	**7**	17	70	unk.	3.2	**32**	0.2	**2**	unk.	**unk.**	unk.	B
1.0	**11**	unk.	**unk.**	unk.	**unk.**	28	**1**	unk.	unk.	**unk.**	unk.	0.34	43	**5**	0.8	1.8	unk.	9	**4**	unk.	**unk.**	unk.		unk	
0.07	**9**	unk.	**unk.**	unk.	**unk.**	unk.	**unk.**	unk.	unk.	**unk.**	unk.	unk.	28	**4**	1	6	unk.	1.6	**16**	unk.	**unk.**	unk.	**unk.**	unk.	C
0.9	**10**	unk.	**unk.**	unk.	**unk.**	1	**0**	unk.	unk.	**unk.**	unk.	0.11	18	**2**	0.0	0.2	unk.	unk.	**unk.**	unk.	**unk.**	unk.		unk.	
0.17	**21**	unk.	**unk.**	unk.	**unk.**	unk.	**unk.**	unk.	unk.	**unk.**	unk.	unk.	82	**10**	14	48	unk.	3.1	**32**	unk.	**unk.**	unk.	**unk.**	unk.	D
1.7	**19**	unk.	**unk.**	unk.	**unk.**	26	**1**	unk.	unk.	**unk.**	unk.	0.37	56	**7**	0.6	1.2	unk.	unk.	**unk.**	unk.	**unk.**	unk.		unk.	
0.06	**8**	tr	**1**	unk.	**unk.**	unk.	**unk.**	unk.	unk.	**unk.**	unk.	unk.	18	**2**	1	11	unk.	1.8	**18**	0.1	**1**	unk.	**unk.**	106	E
0.9	**10**	1	**4**	unk.	**unk.**	unk.	**unk.**	unk.	unk.	**unk.**	unk.	0.08	12	**2**	0.0	0.3	unk.	3	**1**	0.01	**1**	unk.		unk.	
0.16	**20**	0.02	**2**	unk.	**unk.**	unk.	**unk.**	unk.	unk.	**unk.**	unk.	unk.	62	**8**	13	58	unk.	3.4	**35**	0.3	**3**	unk.	**unk.**	unk.	F
1.7	**19**	3	**14**	unk.	**unk.**	30	**1**	unk.	unk.	**unk.**	unk.	0.31	45	**6**	0.6	1.5	unk.	10	**5**	unk.	**unk.**	unk.		unk.	
0.57	**71**	0.32	**45**	unk.	**unk.**	unk.	**unk.**	unk.	unk.	**unk.**	0.4	0.1	23	**3**	38	38	unk.	9.1	**91**	0.4	**4**	unk.	**unk.**	tr	G
5.4	**60**	3	**17**	unk.	**unk.**	28	**1**	unk.	0.4	**4**	0.3	0.40	16	**2**	1.6	1.0	unk.	4	**2**	0.04	**4**	unk.		unk.	
0.09	**11**	0.02	**2**	0.01	**0**	unk.	**unk.**	unk.	unk.	**unk.**	unk.	unk.	17	**2**	2	18	unk.	1.6	**16**	tr	**0**	unk.	**unk.**	tr	H
0.6	**6**	unk.	**unk.**	unk.	**unk.**	1	**0**	unk.	unk.	**unk.**	unk.	0.06	10	**1**	0.1	0.5	unk.	12	**6**	tr	**0**	unk.		unk.	
0.22	**27**	0.04	**6**	0.09	**3**	unk.	**unk.**	unk.	unk.	**unk.**	unk.	unk.	60	**8**	16	72	unk.	3.1	**31**	0.2	**2**	unk.	**unk.**	unk.	I
1.1	**12**	unk.	**unk.**	unk.	**unk.**	26	**1**	unk.	unk.	**unk.**	unk.	0.28	41	**5**	0.7	1.8	unk.	10	**5**	unk.	**unk.**	unk.		unk.	
0.07	**9**	unk.	**unk.**	unk.	**unk.**	unk.	**unk.**	unk.	unk.	**unk.**	unk.	unk.	28	**4**	1	2	unk.	1.6	**16**	unk.	**unk.**	unk.	**unk.**	unk.	J
0.9	**10**	unk.	**unk.**	unk.	**unk.**	1	**0**	unk.	unk.	**unk.**	unk.	0.10	15	**2**	0.0	0.0	unk.	unk.	**unk.**	unk.	**unk.**	unk.		unk.	
0.17	**22**	unk.	**unk.**	unk.	**unk.**	unk.	**unk.**	unk.	unk.	**unk.**	unk.	unk.	83	**10**	14	40	unk.	3.1	**31**	unk.	**unk.**	unk.	**unk.**	unk.	K
1.7	**19**	unk.	**unk.**	unk.	**unk.**	26	**1**	unk.	unk.	**unk.**	unk.	0.37	51	**6**	0.6	1.0	unk.	unk.	**unk.**	unk.	**unk.**	unk.		unk.	
0.05	**7**	0.01	**2**	0(a)	**0**	unk.	**unk.**	unk.	unk.	**unk.**	unk.	unk.	20	**3**	1	9	unk.	1.8	**18**	tr	**1**	unk.	**unk.**	142	L
0.7	**8**	1	**3**	unk.	**unk.**	unk.	**unk.**	unk.	unk.	**unk.**	unk.	0.09	14	**2**	0.0	0.2	unk.	5	**2**	0.01	**1**	unk.		unk.	
0.14	**18**	0.03	**5**	unk.	**unk.**	unk.	**unk.**	unk.	unk.	**unk.**	unk.	unk.	68	**9**	13	54	unk.	3.5	**35**	0.2	**2**	unk.	**unk.**	unk.	M
1.5	**16**	2	**12**	unk.	**unk.**	30	**1**	unk.	unk.	**unk.**	unk.	0.34	50	**6**	0.6	1.4	unk.	13	**6**	unk.	**unk.**	unk.		unk.	
0.05	**7**	unk.	**unk.**	0(a)	**0**	unk.	**unk.**	unk.	unk.	**unk.**	unk.	unk.	26	**3**	13	67	unk.	0.2	**2**	unk.	**unk.**	unk.	**unk.**	unk.	N
0.1	**1**	1	**3**	unk.	**unk.**	23	**1**	unk.	unk.	**unk.**	unk.	0.13	unk.	**unk.**	0.6	1.7	unk.	unk.	**unk.**	unk.	**unk.**	unk.		unk.	
0.11	**14**	unk.	**unk.**	unk.	**unk.**	unk.	**unk.**	unk.	unk.	**unk.**	unk.	unk.	59	**7**	36	70	unk.	0.2	**2**	unk.	**unk.**	unk.	**unk.**	unk.	O
0.1	**1**	1	**6**	unk.	**unk.**	49	**2**	unk.	unk.	**unk.**	unk.	0.54	unk.	**unk.**	1.6	1.8	unk.	unk.	**unk.**	unk.	**unk.**	unk.		unk.	
0.01	**2**	0.01	**2**	unk.	**unk.**	unk.	**unk.**	unk.	unk.	**unk.**	unk.	unk.	7	**1**	22	46	unk.	0.3	**3**	0.1	**1**	unk.	**unk.**	tr	P
0.1	**1**	tr	**2**	unk.	**unk.**	270	**11**	unk.	unk.	**unk.**	unk.	0.27	9	**1**	0.9	1.2	unk.	3	**1**	0.03	**3**	unk.		unk.	
0.13	**16**	0.02	**2**	unk.	**unk.**	unk.	**unk.**	unk.	unk.	**unk.**	unk.	unk.	78	**10**	29	110	unk.	0.5	**5**	0.4	**5**	unk.	**unk.**	tr	Q
0.1	**1**	6	**29**	unk.	**unk.**	59	**2**	unk.	unk.	**unk.**	unk.	0.77	63	**8**	1.3	2.8	unk.	15	**7**	0.10	**10**	unk.		unk.	
0.07	**8**	0.02	**2**	0.11	**4**	unk.	**unk.**	unk.	unk.	**unk.**	unk.	unk.	35	**4**	70	58	unk.	0.1	**2**	0.2	**3**	unk.	**unk.**	unk.	R
0.1	**1**	6	**30**	unk.	**unk.**	39	**2**	unk.	unk.	**unk.**	unk.	0.54	51	**6**	3.0	1.5	unk.	5	**3**	unk.	**unk.**	unk.		unk.	
0.01	**2**	0.01	**2**	unk.	**unk.**	unk.	**unk.**	unk.	unk.	**unk.**	unk.	unk.	7	**1**	22	45	unk.	0.3	**3**	tr	**0**	unk.	**unk.**	tr	S
0.1	**1**	1	**5**	unk.	**unk.**	66	**3**	unk.	unk.	**unk.**	unk.	0.13	4	**1**	0.9	1.1	unk.	tr	**0**	0.01	**1**	unk.		unk.	

	FOOD	Portion	Weight in grams / *Conversion for 100 g*	Kilocalories / H_2O g	Total carbohydrate g / Total fats g	Crude fiber g / Dietary fiber g	Total protein g / Total sugar g	**% USRDA**	Arginine mg / Histidine mg	Isoleucine mg / Leucine mg	Lysine mg / Methionine mg	Phenylalanine mg / Threonine mg	Valine mg / Tryptophan mg	Cystine mg / Tyrosine mg	Polyunsat. fatty acids g / Monounsat. fatty acids g	Saturated fatty acids g / P/S ratio	Linoleic acid g / Cholesterol mg	Thiamin mg / Ascorbic acid mg	**% USRDA**
A	baby food,dessert,fruit dessert,strnd	4-3/4 oz jar	135	80	21.6	0.27	0.4	**2**	unk.	unk.	unk.	unk.	unk.	unk.	0(a)	0(a)	0(a)	0.03	**4**
			0.74	112.6	0.0	unk.	unk.		unk.	unk.	unk.	unk.	unk.	unk.	0(a)	0.00	0(a)	3	**8**
B	baby food,dessert,orange pudding,strnd	4-3/4 oz jar	135	108	23.9	0.54	1.5	**6**	115	77	121	57	94	unk.	tr(a)	tr(a)	tr(a)	0.05	**8**
			0.74	107.7	1.2	unk.	unk.		42	153	20	61	unk.	61	tr(a)	0.00	unk.	12	**32**
C	baby food,dessert,peach cobbler,strnd	4-3/4 oz jar	135	88	24.0	unk.	0.4	**2**	unk.	unk.	unk.	unk.	unk.	unk.	0(a)	0(a)	0(a)	0.01	**2**
			0.74	110.4	0.0	unk.	unk.		unk.	unk.	unk.	unk.	unk.	unk.	0(a)	0.00	0(a)	28	**69**
D	baby food,dessert,peach melba,strnd	4-3/4 oz jar	135	81	22.3	unk.	0.3	**1**	unk.	unk.	unk.	unk.	unk.	unk.	0(a)	0(a)	0(a)	0.01	**2**
			0.74	111.9	0.0	unk.	unk.		unk.	unk.	unk.	unk.	unk.	unk.	0(a)	0.00	0(a)	42	**106**
E	baby food,dessert,pineapple orange,strnd	4-1/2 oz jar	128	90	24.4	unk.	0.3	**1**	unk.	unk.	unk.	unk.	unk.	unk.	0(a)	0(a)	0(a)	0.03	**4**
			0.78	103.0	0.0	unk.	unk.		unk.	unk.	unk.	unk.	unk.	unk.	0(a)	0.00	0(a)	18	**46**
F	baby food,dessert,pineapple	4-1/2 oz jar	128	104	26.0	0.90	1.7	**7**	unk.	unk.	unk.	unk.	unk.	unk.	tr(a)	tr(a)	tr(a)	0.05	**7**
	pudding,strnd		*0.78*	99.3	0.4	unk.	unk.		unk.	unk.	unk.	unk.	unk.	unk.	tr(a)	0.00	unk.	35	**87**
G	baby food,dessert,vanilla	4-1/2 oz jar	128	109	20.6	unk.	2.0	**8**	79	96	146	93	113	unk.	0.20	1.29	0.17	0.01	**2**
	custard pudding,strnd		*0.78*	102.3	2.6	unk.	unk.		52	76	59	76	unk.	76	0.79	0.16	unk.	1	**3**
H	BABY FOOD,DINNER,beef & egg	4-1/2 oz jar	128	68	9.0	0.38	2.9	**12**	180	159	212	134	151	33	unk.	unk.	unk.	0.05	**7**
	noodles,strnd		*0.78*	113.4	2.2	unk.	unk.		84	238	60	114	32	106	unk.	unk.	unk.	2	**4**
I	baby food,dinner,chicken &	4-1/2 oz jar	128	67	9.6	0.38	2.7	**11**	175	141	198	125	160	35	unk.	unk.	unk.	0.04	**6**
	noodles,strnd		*0.78*	113.3	1.9	unk.	unk.		67	230	54	113	31	101	unk.	unk.	unk.	1	**3**
J	baby food,dinner,chicken soup,	4-1/2 oz jar	128	74	10.8	0.38	3.2	**13**	197	156	236	137	183	35	unk.	unk.	unk.	0.01	**2**
	cream of,strnd		*0.78*	111.5	2.0	unk.	unk.		76	264	65	136	38	119	unk.	unk.	unk.	2	**4**
K	baby food,dinner,chicken soup,strnd	4-1/2 oz jar	128	64	9.2	unk.	2.0	**8**	unk.	unk.	unk.	unk.	unk.	unk.	unk.	unk.	unk.	0.03	**4**
			0.78	114.0	2.2	unk.	unk.		unk.	unk.	unk.	unk.	unk.	unk.	unk.	unk.	unk.	1	**3**
L	baby food,dinner,egg yolks,strnd	3.3 oz jar	94	191	0.9	unk.	9.4	**52**	658	532	737	389	603	161	unk.	4.88	0.04	0.07	**9**
			1.06	66.4	16.3	unk.	unk.		199	795	255	433	101	389	5.96	unk.	739	1	**3**
M	baby food,dinner,grits & egg	4-1/2 oz jar	128	unk.	unk.	unk.	2.3	**9**	108	111	142	125	136	35	unk.	unk.	unk.	0.04	**6**
	yolks,strnd		*0.78*	112.6	2.9	unk.	unk.		72	234	69	92	28	114	unk.	unk.	unk.	1	**2**
N	baby food,dinner,high meat/cheese,	4-1/2 oz jar	128	96	5.4	0.38	7.3	**29**	449	307	539	275	361	78	unk.	unk.	unk.	0.04	**6**
	beef w/vegetables,strnd		*0.78*	109.3	5.4	unk.	unk.		219	544	204	279	59	209	unk.	unk.	unk.	2	**6**
O	baby food,dinner,high meat/cheese,	4-1/2 oz jar	128	100	7.6	0.26	7.9	**32**	520	370	614	308	392	69	unk.	unk.	unk.	0.04	**6**
	chicken w/vegetables,strnd		*0.78*	107.1	4.6	unk.	unk.		230	625	209	332	83	230	unk.	unk.	unk.	1	**2**
P	baby food,dinner,high meat/cheese,	4-3/4 oz jar	135	157	25.5	1.21	8.5	**34**	288	374	585	394	474	61	unk.	unk.	unk.	0.05	**8**
	cottage cheese w/pineapple,strnd		*0.74*	97.2	3.0	unk.	unk.		238	752	267	293	128	4	unk.	unk.	unk.	2	**5**
Q	baby food,dinner,high meat/cheese,	4-1/2 oz jar	128	97	7.0	0.26	8.1	**32**	494	353	626	284	371	92	0.64	1.51	0.58	0.13	**18**
	ham w/vegetables,strnd		*0.78*	107.6	4.5	unk.	unk.		287	599	198	312	78	224	1.87	0.42	unk.	2	**6**
R	baby food,dinner,high meat/cheese,	4-1/2 oz jar	128	111	7.7	0.13	7.2	**29**	472	352	567	294	364	74	unk.	unk.	unk.	0.01	**2**
	turkey w/vegetables,strnd		*0.78*	106.2	6.1	unk.	unk.		247	571	187	303	77	230	unk.	unk.	unk.	2	**5**
S	baby food,dinner,high meat/cheese,	4-1/2 oz jar	128	88	7.8	0.38	7.5	**30**	520	343	608	294	378	91	unk.	unk.	unk.	0.03	**4**
	veal w/vegetables,strnd		*0.78*	108.3	3.5	unk.	unk.		244	596	174	307	78	221	unk.	unk.	unk.	2	**6**

Row	Riboflavin mg / Niacin mg	% USRDA	Vitamin B_6 mg / Folacin mcg	% USRDA	Vitamin B_{12} mcg / Pantothenic acid mg	% USRDA	Biotin mg / Vitamin A IU	% USRDA	Preformed A RE / Beta carotene RE	Vitamin D IU / Vitamin E IU	% USRDA	Total tocopherol mg / Alpha tocopherol mg	Other tocopherol mg / Total ash g	Calcium mg / Phosphorus mg	% USRDA	Sodium mg / Sodium meq	Potassium mg / Potassium meq	Chlorine mg / Chlorine meq	Iron mg / Magnesium mg	% USRDA	Zinc mg / Copper mg	% USRDA	Iodine mcg / Selenium mcg	% USRDA	Manganese mcg / Chromium mcg
A	0.01	**2**	0.05	**7**	0(a)	**0**	unk.	**unk.**	unk.	unk.	**unk.**	unk.	unk.	12	**2**	19	127	unk.	0.3	**3**	tr	**1**	unk.	**unk.**	135
	0.2	**2**	4	**22**	unk.	**unk.**	339	**14**	unk.	unk.	**unk.**	unk.	0.40	9	**1**	0.8	3.3	unk.	5	**3**	0.03	**3**	unk.		unk.
B	0.08	**10**	0.04	**5**	unk.	**unk.**	unk.	**unk.**	unk.	unk.	**unk.**	unk.	unk.	43	**5**	unk.	116	unk.	0.1	**1**	0.2	**2**	unk.	**unk.**	tr
	0.2	**2**	11	**53**	unk.	**unk.**	155	**6**	unk.	unk.	**unk.**	unk.	0.81	38	**5**	unk.	3.0	unk.	6	**3**	0.03	**3**	unk.		unk.
C	0.01	**2**	0.01	**1**	0(a)	**0**	unk.	**unk.**	unk.	unk.	**unk.**	unk.	unk.	5	**1**	9	73	unk.	unk.	**unk.**	0.2	**2**	unk.	**unk.**	tr
	0.4	**4**	1	**7**	unk.	**unk.**	192	**8**	unk.	unk.	**unk.**	unk.	0.27	7	**1**	0.4	1.9	unk.	3	**2**	0.07	**7**	unk		unk.
D	0.05	**7**	unk.	**unk.**	0(a)	**0**	unk.	**unk.**	unk.	unk.	**unk.**	unk.	unk.	13	**2**	12	112	unk.	0.4	**5**	unk.	**unk.**	unk.	**unk.**	unk.
	0.5	**5**	3	**13**	unk.	**unk.**	248	**10**	unk.	unk.	**unk.**	unk.	0.40	unk.	**unk.**	0.5	2.9	unk.	unk.	**unk.**	unk.	**unk.**	unk.		unk.
E	0.03	**3**	unk.	**unk.**	0(a)	**0**	unk.	**unk.**	unk.	unk.	**unk.**	unk.	unk.	14	**2**	13	60	unk.	0.2	**2**	unk.	**unk.**	unk.	**unk.**	unk.
	0.1	**1**	3	**17**	unk.	**unk.**	73	**3**	unk.	unk.	**unk.**	unk.	0.26	5	**1**	0.6	1.5	unk.	5	**3**	0.03	**3**	unk.		unk.
F	0.06	**8**	0.05	**7**	0.08	**3**	unk.	**unk.**	unk.	unk.	**unk.**	unk.	unk.	40	**5**	24	104	unk.	0.2	**2**	0.3	**3**	unk.	**unk.**	unk.
	0.1	**2**	7	**33**	unk.	**unk.**	51	**2**	unk.	unk.	**unk.**	unk.	0.64	40	**5**	1.1	2.6	unk.	12	**6**	unk.	**unk.**	unk.		unk.
G	0.10	**13**	0.03	**4**	unk.	**unk.**	unk.	**unk.**	unk.	unk.	**unk.**	unk.	unk.	70	**9**	36	84	unk.	0.3	**3**	0.3	**4**	unk.	**unk.**	tr
	0.1	**1**	8	**38**	0.32	**6**	82	**3**	unk.	unk.	**unk.**	unk.	0.64	58	**7**	1.6	2.2	unk.	6	**3**	0.04	**4**	unk.		unk.
H	0.05	**6**	0.27	**39**	0.12	**4**	unk.	**unk.**	unk.	unk.	**unk.**	unk.	unk.	12	**1**	37	60	unk.	0.5	**5**	0.4	**5**	unk.	**unk.**	unk.
	0.9	**10**	7	**33**	0.27	**6**	1053	**42**	unk.	unk.	**unk.**	unk.	0.51	37	**5**	1.6	1.5	unk.	10	**5**	0.01	**1**	unk.		unk.
I	0.06	**8**	0.04	**6**	unk.	**unk.**	unk.	**unk.**	unk.	unk.	**unk.**	0.3	0.2	28	**4**	20	50	unk.	0.6	**6**	0.4	**5**	unk.	**unk.**	unk.
	0.6	**7**	7	**35**	unk.	**unk.**	1158	**46**	unk.	0.4	**4**	0.2	0.51	31	**4**	0.9	1.3	unk.	12	**6**	0.05	**5**	unk.		unk.
J	0.05	**6**	0.06	**8**	0.08	**3**	unk.	**unk.**	unk.	unk.	**unk.**	unk.	unk.	45	**6**	24	100	unk.	0.4	**4**	0.3	**4**	unk.	**unk.**	tr
	0.5	**5**	unk.	**unk.**	unk.	**unk.**	929	**37**	unk.	unk.	**unk.**	unk.	0.51	37	**5**	1.1	2.5	unk.	10	**5**	0.01	**1**	unk.		unk.
K	0.04	**5**	unk.	**unk.**	unk.	**unk.**	unk.	**unk.**	unk.	unk.	**unk.**	unk.	unk.	47	**6**	20	unk.	unk.	0.3	**4**	unk.	**unk.**	unk.	**unk.**	unk.
	0.4	**4**	7	**33**	unk.	**unk.**	1772	**71**	unk.	unk.	**unk.**	unk.	0.51	unk.	**unk.**	0.9	unk.	unk.	unk.	**unk.**	unk.	**unk.**	unk.		unk.
L	0.25	**32**	0.15	**22**	1.45	**48**	unk.	**unk.**	unk.	unk.	**unk.**	1.6	1.0	71	**9**	37	72	unk.	2.6	**26**	1.0	**13**	unk.	**unk.**	tr
	tr	**0**	87	**433**	2.01	**40**	1176	**47**	unk.	1.9	**19**	0.6	1.13	270	**34**	1.6	1.8	unk.	8	**4**	0.03	**3**	unk.		unk.
M	0.09	**11**	0.03	**5**	0.05	**2**	unk.	**unk.**	unk.	unk.	**unk.**	unk.	unk.	36	**5**	unk.	72	unk.	0.6	**6**	0.3	**4**	unk.	**unk.**	unk.
	0.4	**4**	4	**19**	unk.	**unk.**	156	**6**	unk.	unk.	**unk.**	unk.	unk.	46	**6**	unk.	1.8	unk.	6	**3**	unk.	**unk.**	unk.		unk.
N	0.09	**11**	0.11	**16**	0.65	**22**	unk.	**unk.**	unk.	unk.	**unk.**	0.7	0.1	15	**2**	46	179	unk.	0.9	**9**	1.4	**17**	unk.	**unk.**	tr
	1.7	**19**	7	**37**	0.31	**6**	1004	**40**	unk.	0.9	**9**	0.7	0.77	61	**8**	2.0	4.6	unk.	11	**5**	0.05	**5**	unk.		unk.
O	0.09	**11**	0.08	**11**	0.18	**6**	unk.	**unk.**	unk.	unk.	**unk.**	0.6	0.3	67	**8**	35	76	unk.	1.3	**13**	0.9	**12**	unk.	**unk.**	tr
	1.3	**15**	1	**7**	0.41	**8**	737	**30**	unk.	0.7	**7**	0.3	0.64	69	**9**	1.5	1.9	unk.	10	**5**	0.04	**4**	unk.		unk.
P	0.19	**24**	0.06	**9**	0.31	**10**	unk.	**unk.**	unk.	unk.	**unk.**	unk.	unk.	88	**11**	201	128	unk.	0.1	**1**	0.3	**4**	unk.	**unk.**	27
	0.1	**2**	unk.	**unk.**	unk.	**unk.**	104	**4**	unk.	unk.	**unk.**	unk.	0.94	99	**12**	8.8	3.3	unk.	6	**3**	tr	**0**	unk.		unk.
Q	0.10	**13**	0.14	**19**	0.35	**12**	unk.	**unk.**	unk.	unk.	**unk.**	0.4	0.1	14	**2**	28	198	unk.	0.8	**8**	1.2	**15**	unk.	**unk.**	tr
	1.8	**20**	8	**40**	0.50	**10**	214	**9**	unk.	0.5	**5**	0.3	0.77	73	**9**	1.2	5.1	unk.	14	**7**	0.36	**36**	unk.		unk.
R	0.09	**11**	0.05	**7**	0.56	**19**	unk.	**unk.**	unk.	unk.	**unk.**	0.2	0.0	79	**10**	38	154	unk.	0.9	**9**	1.1	**14**	unk.	**unk.**	tr
	1.3	**14**	12	**61**	unk.	**unk.**	420	**17**	unk.	0.2	**2**	0.2	0.77	90	**11**	1.7	3.9	unk.	8	**4**	0.09	**9**	unk.		unk.
S	0.09	**11**	0.11	**15**	0.58	**19**	unk.	**unk.**	unk.	unk.	**unk.**	0.2	0.0	12	**1**	31	196	unk.	0.8	**8**	1.3	**16**	unk.	**unk.**	tr
	2.1	**23**	unk.	**unk.**	0.32	**6**	349	**14**	unk.	0.2	**2**	0.2	0.90	69	**9**	1.3	5.0	unk.	12	**6**	0.05	**5**	unk.		unk.

	FOOD	Portion	Weight in grams / *Conversion for 100 g*	Kilocalories / H_2O g	Total carbohydrate g / Total fats g	Crude fiber g / Dietary fiber g	Total protein g / Total sugar g	**% USRDA**	Arginine mg / Histidine mg	Isoleucine mg / Leucine mg	Lysine mg / Methionine mg	Phenylalanine mg / Threonine mg	Valine mg / Tryptophan mg	Cystine mg / Tyrosine mg	Polyunsat. fatty acids g / Monounsat. fatty acids g	Saturated fatty acids g / P/S ratio	Linoleic acid g / Cholesterol mg	Thiamin mg / Ascorbic acid mg	**% USRDA**
A	baby food,dinner,macaroni & cheese,strnd	4-1/2 oz jar	128	76	9.6	0.13	3.3	**13**	131	166	189	169	188	41	unk.	unk.	unk.	0.06	**9**
			0.78	111.5	2.7	unk.	unk.		79	300	115	100	41	157	unk.	unk.	unk.	2	**4**
B	baby food,dinner,macaroni,tomato & beef,strnd	4-1/2 oz jar	128	70	11.3	0.51	2.8	**11**	154	137	173	129	147	40	unk.	unk.	unk.	0.09	**13**
			0.78	111.7	1.4	unk.	unk.		74	229	46	104	32	95	unk.	unk.	unk.	2	**5**
C	baby food,dinner,mixed vegetables,strnd	4-1/2 oz jar	128	52	12.2	unk.	1.5	**6**	unk.	unk.	unk.	unk.	unk.	unk.	tr(a)	tr(a)	tr(a)	0.03	**4**
			0.78	113.5	0.1	unk.	unk.		unk.	unk.	unk.	unk.	unk.	unk.	tr(a)	0.00	0(a)	4	**9**
D	baby food,dinner,turkey & rice,strnd	4-1/2 oz jar	128	63	9.3	0.13	2.4	**10**	178	122	196	102	137	26	0.38	0.54	0.36	0.01	**2**
			0.78	114.0	1.7	unk.	unk.		58	195	67	100	26	87	0.59	0.71	13	2	**4**
E	baby food,dinner,vegetables & bacon,strnd	4-1/2 oz jar	128	88	11.0	0.51	2.0	**8**	141	84	109	79	109	29	0.46	1.52	0.42	0.04	**6**
			0.78	110.0	4.2	unk.	unk.		44	133	41	69	19	70	1.88	0.30	4	2	**4**
F	baby food,dinner,vegetables & beef,strnd	4-1/2 oz jar	128	68	9.0	0.26	2.6	**10**	151	106	179	87	128	24	unk.	unk.	unk.	0.03	**4**
			0.78	113.3	2.6	unk.	unk.		56	173	41	91	26	65	unk.	unk.	unk.	2	**4**
G	baby food,dinner,vegetables & chicken,strnd	4-1/2 oz jar	128	55	8.4	0.26	2.4	**10**	26	116	155	100	133	35	unk.	unk.	unk.	0.01	**2**
			0.78	115.2	1.4	unk.	unk.		49	174	46	88	27	76	unk.	unk.	unk.	1	**4**
H	baby food,dinner,vegetables & ham,strnd	4-1/2 oz jar	128	61	8.8	0.26	2.3	**9**	148	100	152	84	118	23	unk.	unk.	unk.	0.04	**6**
			0.78	114.2	2.2	unk.	unk.		60	165	47	84	23	68	unk.	unk.	unk.	2	**5**
I	baby food,dinner,vegetables & lamb,strnd	4-1/2 oz jar	128	67	8.8	0.38	2.6	**10**	170	115	187	101	124	22	unk.	unk.	unk.	0.03	**4**
			0.78	113.4	2.6	unk.	unk.		59	188	38	93	29	79	unk.	unk.	unk.	2	**4**
J	baby food,dinner,vegetables & liver,strnd	4-1/2 oz jar	128	50	8.8	0.26	2.8	**11**	168	133	209	133	175	37	tr(a)	tr(a)	tr(a)	0.03	**4**
			0.78	115.2	0.5	unk.	unk.		68	244	64	114	41	97	tr(a)	0.00	unk.	2	**6**
K	baby food,dinner,vegetables & turkey,strnd	4-1/2 oz jar	128	54	8.4	0.26	2.2	**9**	134	102	156	79	120	23	unk.	unk.	unk.	0.01	**2**
			0.78	115.3	1.5	unk.	unk.		45	165	47	84	23	73	unk.	unk.	unk.	1	**4**
L	baby food,dinner,vegetables,dumplings & beef,strnd	4-1/2 oz jar	128	61	9.9	unk.	2.6	**10**	unk.	unk.	unk.	unk.	unk.	unk.	tr(a)	tr(a)	tr(a)	0.06	**9**
			0.78	113.8	1.2	unk.	unk.		unk.	unk.	unk.	unk.	unk.	unk.	tr(a)	0.00	unk.	1	**3**
M	baby food,dinner,vegetables,noodles & chicken,strnd	4-1/2 oz jar	128	81	10.1	0.26	2.6	**10**	unk.	unk.	unk.	unk.	unk.	unk.	unk.	unk.	unk.	0.04	**6**
			0.78	111.6	3.2	unk.	unk.		unk.	unk.	unk.	unk.	unk.	unk.	unk.	unk.	unk.	1	**2**
N	baby food,dinner,vegetables,noodles & turkey,strnd	4-1/2 oz jar	128	56	8.7	0.26	1.5	**6**	unk.	unk.	unk.	unk.	unk.	unk.	unk.	unk.	unk.	0.03	**4**
			0.78	115.6	1.5	unk.	unk.		unk.	unk.	unk.	unk.	unk.	unk.	unk.	unk.	unk.	1	**3**
O	BABY FOOD,FRUIT JUICE,apple	4-3/5 oz can	130	61	15.2	tr	0.0	**0**	0(a)	0(a)	0(a)	0(a)	0(a)	0(a)	tr(a)	tr(a)	tr(a)	0.01	**2**
			0.77	114.4	0.1	unk.	unk.		0(a)	0(a)	0(a)	0(a)	0(a)	0(a)	tr(a)	0.00	0(a)	75	**188**
P	baby food,fruit juice,apple-cherry	4-3/5 oz can	130	53	12.9	tr	0.1	**1**	tr(a)	tr(a)	tr(a)	tr(a)	tr(a)	tr(a)	tr(a)	tr(a)	tr(a)	0.01	**2**
			0.77	116.3	0.3	unk.	unk.		tr(a)	tr(a)	tr(a)	tr(a)	tr(a)	tr(a)	tr(a)	0.00	0(a)	76	**190**
Q	baby food,fruit juice,apple-grape	4-3/5 oz can	130	60	14.8	tr(a)	0.1	**1**	tr(a)	tr(a)	tr(a)	tr(a)	tr(a)	tr(a)	tr(a)	tr(a)	tr(a)	0.01	**2**
			0.77	114.5	0.3	tr(a)	unk.		tr(a)	tr(a)	tr(a)	tr(a)	tr(a)	tr(a)	tr(a)	0.00	0(a)	70	**174**
R	baby food,fruit juice,apple-peach	4-3/5 oz can	130	55	13.6	tr(a)	0.3	**1**	tr(a)	tr(a)	tr(a)	tr(a)	tr(a)	tr(a)	tr(a)	tr(a)	tr(a)	0.01	**2**
			0.77	115.7	0.1	unk.	unk.		tr(a)	tr(a)	tr(a)	tr(a)	tr(a)	tr(a)	tr(a)	0.00	0(a)	76	**190**
S	baby food,fruit juice,apple-plum	4-3/5 oz can	130	64	16.0	tr(a)	0.1	**1**	tr(a)	tr(a)	tr(a)	tr(a)	tr(a)	tr(a)	0.00	0.00	0.00	0.03	**4**
			0.77	113.5	0.0	tr(a)	unk.		tr(a)	tr(a)	tr(a)	tr(a)	tr(a)	tr(a)	0.00	0.00	0(a)	76	**189**

Riboflavin mg / Niacin mg	% USRDA	Vitamin B_6 mg / Folacin mcg	% USRDA	Vitamin B_{12} mcg / Pantothenic acid mg	% USRDA	Biotin mg / Vitamin A IU	% USRDA	Preformed A RE / Beta carotene RE	Vitamin D IU / Vitamin E IU	% USRDA	Total tocopherol mg / Alpha tocopherol mg	Other tocopherol mg / Total ash g	Calcium mg / Phosphorus mg	% USRDA	Sodium mg / Sodium meq	Potassium mg / Potassium meq	Chlorine mg / Chlorine meq	Iron mg / Magnesium mg	% USRDA	Zinc mg / Copper mg	% USRDA	Iodine mcg / Selenium mcg	% USRDA	Manganese mcg / Chromium mcg	
0.09	11	0.02	3	0.04	1	unk.	unk.	unk.	unk.	unk.	unk.	unk.	69	9	93	59	unk.	0.4	4	0.5	6	unk.	unk.	64	A
0.6	6	2	9	unk.	unk.	35	1	unk.	unk.	unk.	unk.	0.90	73	9	4.1	1.5	unk.	9	4	0.09	9	unk.		unk.	
0.08	10	0.06	8	0.29	10	unk.	unk.	unk.	unk.	unk.	unk.	unk.	20	3	22	123	unk.	0.6	6	0.3	4	unk.	unk.	tr	B
1.0	12	26	129	unk.	unk.	680	27	unk.	unk.	unk.	unk.	0.64	54	7	0.9	3.1	unk.	11	5	0.01	1	unk.		unk.	
0.04	5	unk.	unk.	0(a)	0	unk.	unk.	unk.	unk.	unk.	unk.	unk.	28	4	10	155	unk.	0.4	4	unk.	unk.	unk.	unk.	unk.	C
0.6	7	10	51	unk.	unk.	3492	140	unk.	unk.	unk.	unk.	0.64	unk.	unk.	0.4	4.0	unk.	unk.	unk.	unk.	unk.	unk.		unk.	
0.03	3	0.04	6	unk.	unk.	unk.	unk.	unk.	unk.	unk.	unk.	unk.	27	3	22	52	unk.	0.3	3	0.4	5	unk.	unk.	tr	D
0.4	4	4	21	unk.	unk.	782	31	unk.	unk.	unk.	unk.	0.51	26	3	0.9	1.3	unk.	6	3	0.01	1	unk.		unk.	
0.04	5	0.10	15	0.13	4	unk.	unk.	unk.	unk.	unk.	unk	unk.	18	2	55	114	unk.	0.5	5	0.3	4	unk.	unk.	tr	E
0.7	8	12	59	unk.	unk.	3409	136	unk.	unk.	unk.	unk.	0.77	40	5	2.4	2.9	unk.	15	8	0.03	3	unk.		unk.	
0.04	5	0.07	10	0.32	11	unk.	unk.	unk.	unk.	unk.	unk.	unk.	15	2	27	129	unk.	0.5	5	0.4	5	unk.	unk.	tr	F
0.7	7	6	30	0.15	3	1521	61	unk.	unk.	unk.	unk.	0.64	53	7	1.2	3.3	unk.	7	3	0.01	1	unk.		unk.	
0.03	3	0.03	4	unk.	unk.	unk.	unk.	unk.	unk.	unk.	0.2	0.1	18	2	14	38	unk.	0.3	4	unk.	unk.	unk.	unk.	unk.	G
0.2	3	4	21	0.26	5	1416	57	unk.	0.3	3	0.2	0.38	32	4	0.6	1.0	unk.	unk.	unk.	unk.	unk.	unk.		unk.	
0.04	5	0.04	6	unk.	unk.	unk.	unk.	unk.	unk.	unk.	unk.	unk.	10	1	15	109	unk.	0.4	4	0.2	3	unk.	unk.	tr	H
0.5	6	6	31	unk.	unk.	873	35	unk.	unk.	unk.	unk.	0.64	29	4	0.7	2.8	unk.	6	3	0.01	1	unk.		unk.	
0.04	5	0.06	8	0.20	7	unk.	unk.	unk.	unk.	unk.	0.6	0.2	15	2	26	120	unk.	0.4	5	0.3	3	unk.	unk.	tr	I
0.7	8	5	23	0.20	4	2554	102	unk.	0.7	7	0.3	0.64	63	8	1.1	3.1	unk.	9	5	0.01	1	unk.		unk.	
0.35	43	0.10	15	unk.	unk.	unk.	unk.	unk.	unk.	unk.	0.4	0.1	9	1	23	120	unk.	2.8	28	0.6	8	unk.	unk.	tr	J
1.5	17	37	184	unk.	unk.	3981	159	unk.	0.5	5	0.3	0.64	51	6	1.0	3.1	unk.	8	4	0.05	5	unk.		unk.	
0.03	3	0.04	6	unk.	unk.	unk.	unk.	unk.	unk.	unk.	0.2	0.0	20	3	17	56	unk.	0.3	4	0.3	4	unk.	unk.	tr	K
0.4	4	3	17	0.24	5	1068	43	unk.	0.2	2	0.2	0.51	24	3	0.7	1.4	unk.	7	3	0.03	3	unk.		unk.	
0.05	6	unk.	unk.	unk.	unk.	unk.	unk.	unk.	unk.	unk.	unk.	unk.	18	2	63	unk.	unk.	0.5	5	0.5	6	unk.	unk.	unk.	L
0.7	8	9	46	unk.	unk.	532	21	unk.	unk.	unk.	unk.	0.64	unk.	unk.	2.7	unk.	unk.	8	4	unk.	unk.	unk.		unk.	
0.06	8	0.03	4	0.10	3	unk.	unk.	unk.	unk.	unk.	unk.	unk.	36	5	26	70	unk.	0.4	5	0.3	4	unk.	unk.	unk.	M
0.5	6	4	21	0.24	5	1814	73	unk.	unk.	unk.	unk.	0.51	40	5	1.1	1.8	unk.	unk.	unk.	0.07	7	unk.		unk.	
0.05	6	0.02	3	0.13	4	unk.	unk.	unk.	unk.	unk.	unk.	unk.	41	5	27	81	unk.	0.2	2	0.3	4	unk.	unk.	unk.	N
0.3	4	3	15	unk.	unk.	1268	51	unk.	unk.	unk.	unk.	0.51	32	4	1.2	2.1	unk.	10	5	unk.	unk.	unk.		unk.	
0.03	3	0.04	5	0(a)	0	unk.	unk.	0	unk.	unk.	unk.	unk.	5	1	4	118	unk.	0.7	7	tr	1	unk.	unk.	unk.	O
0.1	1	0	0	unk.	unk.	23	1	tr	unk.	unk.	unk.	0.26	7	1	0.2	3.0	unk.	4	2	unk.	unk.	unk.		unk.	
0.03	3	0.04	6	0(a)	0	unk.	unk.	0	0(a)	0	unk.	unk.	6	1	4	127	unk.	0.9	9	tr	0	unk.	unk.	52	P
0.1	1	tr	2	unk.	unk.	6	0	tr	unk.	unk.	unk.	0.39	8	1	0.2	3.3	unk.	5	3	0.01	1	unk.		unk.	
0.03	3	0.04	6	0(a)	0	unk.	unk.	unk.	0(a)	0	unk.	unk.	8	1	4	117	unk.	0.5	5	tr	0	unk.	unk.	52	Q
0.1	2	tr	2	unk.	unk.	8	0	unk.	unk.	unk.	unk.	0.26	7	1	0.2	3.0	unk.	5	3	0.01	1	unk.		unk.	
0.01	2	0.03	4	0(a)	0	unk.	unk.	unk.	0(a)	0	unk.	unk.	4	1	0	126	unk.	0.7	7	tr	1	unk.	unk.	130	R
0.3	3	2	8	unk.	unk.	82	3	unk.	unk.	unk.	unk.	0.39	4	1	0.0	3.2	unk.	3	2	0.01	1	unk.		unk.	
0.03	3	0.04	5	0(a)	0	unk.	unk.	unk.	0(a)	0	unk.	unk.	6	1	unk.	131	unk.	0.8	8	tr	1	unk.	unk.	156	S
0.3	3	tr	1	unk.	unk.	56	2	unk.	unk.	unk.	unk.	0.39	4	1	unk.	3.4	unk.	4	2	0.01	1	unk.		unk.	

	FOOD	Portion	Weight in grams / *Conversion for 100 g*	Kilocalories / H_2O g	Total carbohydrate g / Total fats g	Crude fiber g / Dietary fiber g	Total protein g / Total sugar g	% USRDA	Arginine mg / Histidine mg	Isoleucine mg / Leucine mg	Lysine mg / Methionine mg	Phenylalanine mg / Threonine mg	Valine mg / Tryptophan mg	Cystine mg / Tyrosine mg	Polyunsat. fatty acids g / Monounsat. fatty acids g	Saturated fatty acids g / P/S ratio	Linoleic acid g / Cholesterol mg	Thiamin mg / Ascorbic acid mg	% USRDA
A	baby food,fruit juice,apple-prune	4-3/5 oz can	130	95	23.4	tr(a)	0.3	1	tr(a)	tr(a)	tr(a)	tr(a)	tr(a)	tr(a)	tr(a)	tr(a)	tr(a)	0.01	2
			0.77	105.7	0.1	unk.	unk.		tr(a)	tr(a)	tr(a)	tr(a)	tr(a)	tr(a)	tr(a)	0.00	0(a)	88	219
B	baby food,fruit juice,mixed fruit	4-3/5 oz can	130	61	15.1	tr	0.1	1	tr(a)	tr(a)	tr(a)	tr(a)	tr(a)	tr(a)	tr(a)	tr(a)	tr(a)	0.03	4
			0.77	114.3	0.1	unk.	unk.		tr(a)	tr(a)	tr(a)	tr(a)	tr(a)	tr(a)	tr(a)	0.00	0(a)	83	207
C	baby food,fruit juice,orange	4-3/5 oz can	130	242	13.3	0.13	0.8	3	tr(a)	tr(a)	tr(a)	tr(a)	tr(a)	tr(a)	tr(a)	tr(a)	tr(a)	0.06	9
			0.77	115.0	0.4	unk.	unk.		tr(a)	tr(a)	tr(a)	tr(a)	tr(a)	tr(a)	tr(a)	0.00	0(a)	81	203
D	baby food,fruit juice,orange-apple	4-3/5 oz can	130	56	13.1	tr(a)	0.5	2	tr(a)	tr(a)	tr(a)	tr(a)	tr(a)	tr(a)	tr(a)	tr(a)	tr(a)	0.05	7
			0.77	115.6	0.3	unk.	unk.		tr(a)	tr(a)	tr(a)	tr(a)	tr(a)	tr(a)	tr(a)	0.00	0(a)	100	250
E	baby food,fruit juice,	4-3/5 oz can	130	61	14.9	tr(a)	0.5	2	tr(a)	tr(a)	tr(a)	tr(a)	tr(a)	tr(a)	tr(a)	tr(a)	tr(a)	0.05	7
	orange-apple-banana		*0.77*	113.9	0.1	unk.	unk.		tr(a)	tr(a)	tr(a)	tr(a)	tr(a)	tr(a)	tr(a)	0.00	0(a)	42	104
F	baby food,fruit juice,orange-apricot	4-3/5 oz can	130	60	14.2	tr(a)	1.0	0	tr(a)	tr(a)	tr(a)	tr(a)	tr(a)	tr(a)	tr(a)	tr(a)	tr(a)	0.08	11
			0.77	114.1	0.1	tr(a)	unk.		tr(a)	tr(a)	tr(a)	tr(a)	tr(a)	tr(a)	tr(a)	0.00	0(a)	112	279
G	baby food,fruit juice,orange-banana	4-3/5 oz can	130	65	15.5	0.13	0.9	4	tr(a)	tr(a)	tr(a)	tr(a)	tr(a)	tr(a)	tr(a)	tr(a)	tr(a)	0.06	9
			0.77	113.0	0.1	unk.	unk.		tr(a)	tr(a)	tr(a)	tr(a)	tr(a)	tr(a)	tr(a)	0.00	0(a)	44	111
H	baby food,fruit juice,	4-3/5 oz can	130	62	15.2	tr	0.6	3	tr(a)	tr(a)	tr(a)	tr(a)	tr(a)	tr(a)	tr(a)	tr(a)	tr(a)	0.06	9
	orange-pineapple		*0.77*	113.5	0.1	unk.	unk.		tr(a)	tr(a)	tr(a)	tr(a)	tr(a)	tr(a)	tr(a)	0.00	0(a)	69	173
I	baby food,fruit juice,prune-orange	4-3/5 oz can	130	91	21.8	tr(a)	0.8	3	tr(a)	tr(a)	tr(a)	tr(a)	tr(a)	tr(a)	tr(a)	tr(a)	tr(a)	0.05	7
			0.77	106.5	0.4	tr(a)	unk.		tr(a)	tr(a)	tr(a)	tr(a)	tr(a)	tr(a)	tr(a)	0.00	0(a)	83	207
J	BABY FOOD,FRUIT,apple blueberry,strnd	4-3/4 oz jar	135	82	22.0	0.27	0.3	1	tr(a)	tr(a)	tr(a)	tr(a)	tr(a)	tr(a)	tr(a)	tr(a)	tr(a)	0.03	4
			0.74	112.2	0.3	unk.	unk.		tr(a)	tr(a)	tr(a)	tr(a)	tr(a)	tr(a)	tr(a)	0.00	0(a)	38	94
K	baby food,fruit,apple raspberry,strnd	4-3/4 oz jar	135	78	21.2	unk.	0.3	1	tr(a)	tr(a)	tr(a)	tr(a)	tr(a)	tr(a)	tr(a)	tr(a)	tr(a)	0.01	2
			0.74	113.1	0.3	unk.	unk.		tr(a)	tr(a)	tr(a)	tr(a)	tr(a)	tr(a)	tr(a)	0.00	0(a)	36	90
L	baby food,fruit,applesauce	4-3/4 oz jar	135	61	15.7	0.94	0.3	1	tr(a)	tr(a)	tr(a)	tr(a)	tr(a)	tr(a)	tr(a)	tr(a)	tr(a)	0.01	2
	& apricots,strnd		*0.74*	118.4	0.3	unk.	unk.		tr(a)	tr(a)	tr(a)	tr(a)	tr(a)	tr(a)	tr(a)	0.00	0(a)	26	64
M	baby food,fruit,applesauce	4-3/4 oz jar	135	65	17.7	unk.	0.4	2	tr(a)	tr(a)	tr(a)	tr(a)	tr(a)	tr(a)	0.00	0.00	0.00	0.03	4
	& cherries,strnd		*0.74*	116.6	0.0	unk.	unk.		tr(a)	tr(a)	tr(a)	tr(a)	tr(a)	tr(a)	0.00	0.00	0(a)	45	113
N	baby food,fruit,applesauce	4-1/2 oz jar	128	47	12.9	unk.	0.1	1	tr(a)	tr(a)	tr(a)	tr(a)	tr(a)	tr(a)	tr(a)	tr(a)	tr(a)	0.03	4
	& pineapple,strnd		*0.78*	114.6	0.1	unk.	unk.		tr(a)	tr(a)	tr(a)	tr(a)	tr(a)	tr(a)	tr(a)	0.00	0(a)	36	90
O	baby food,fruit,applesauce,strnd	4-1/2 oz jar	128	52	14.0	0.64	0.3	1	7	8	14	6	10	tr(a)	tr(a)	tr(a)	tr(a)	0.01	2
			0.78	113.4	0.3	unk.	unk.		5	15	2	9	2	3	tr(a)	0.00	0(a)	49	123
P	baby food,fruit,apricots	4-3/4 oz jar	135	81	22.0	0.40	0.4	2	tr(a)	tr(a)	tr(a)	tr(a)	tr(a)	tr(a)	0.00	0.00	0.00	0.01	2
	w/tapioca,strnd		*0.74*	112.2	0.0	unk.	unk.		tr(a)	tr(a)	tr(a)	tr(a)	tr(a)	tr(a)	0.00	0.00	0(a)	29	73
Q	baby food,fruit,bananas & pineapple	4-3/4 oz jar	135	92	24.8	unk.	0.3	1	tr(a)	tr(a)	tr(a)	tr(a)	tr(a)	tr(a)	tr(a)	tr(a)	tr(a)	0.03	4
	w/tapioca,strnd		*0.74*	109.5	0.1	unk.	unk.		tr(a)	tr(a)	tr(a)	tr(a)	tr(a)	tr(a)	tr(a)	0.00	0(a)	29	72
R	baby food,fruit,bananas	4-3/4 oz jar	135	77	20.7	0.27	0.5	2	tr(a)	tr(a)	tr(a)	tr(a)	tr(a)	tr(a)	tr(a)	tr(a)	tr(a)	0.01	2
	w/tapioca,strnd		*0.74*	113.4	0.1	unk.	unk.		tr(a)	tr(a)	tr(a)	tr(a)	tr(a)	tr(a)	tr(a)	0.00	0(a)	23	56
S	baby food,fruit,guava & papaya	4-1/2 oz jar	128	81	21.8	0.64	0.3	1	tr(a)	tr(a)	tr(a)	tr(a)	tr(a)	tr(a)	tr(a)	tr(a)	tr(a)	0.01	2
	w/tapioca,strnd		*0.78*	105.6	0.1	unk.	unk.		tr(a)	tr(a)	tr(a)	tr(a)	tr(a)	tr(a)	tr(a)	0.00	0(a)	104	259

Riboflavin mg / Niacin mg	% USRDA	Vitamin B_6 mg / Folacin mcg	% USRDA	Vitamin B_{12} mcg / Pantothenic acid mg	% USRDA	Biotin mg / Vitamin A IU	% USRDA	Preformed A RE / Beta carotene RE	Vitamin D IU / Vitamin E IU	% USRDA	Total tocopherol mg / Alpha tocopherol mg	Other tocopherol mg / Total ash g	Calcium mg / Phosphorus mg	% USRDA	Sodium mg / Sodium meq	Potassium mg / Potassium meq	Chlorine mg / Chlorine meq	Iron mg / Magnesium mg	% USRDA	Zinc mg / Copper mg	% USRDA	Iodine mcg / Selenium mcg	% USRDA	Manganese mcg / Chromium mcg	
tr	**0**	0.05	**7**	0(a)	**0**	unk.	**unk.**	unk.	0(a)	**0**	unk.	unk.	12	**2**	6	192	unk.	1.2	**12**	tr	**1**	unk.	**unk.**	130	A
0.4	**4**	tr	**1**	unk.	**unk.**	tr	**0**	unk.	unk.	**unk.**	unk.	0.39	20	**2**	0.3	4.9	unk.	4	**2**	0.01	**1**	unk.		unk.	
0.01	**2**	0.06	**8**	0(a)	**0**	unk.	**unk.**	0	0(a)	**0**	unk.	unk	10	**1**	5	131	unk.	0.4	**4**	tr	**0**	unk.	**unk.**	416	B
0.2	**2**	9	**44**	unk.	**unk.**	55	**2**	5	unk.	**unk.**	unk.	0.26	7	**1**	0.2	3.4	unk.	9	**4**	0.03	**3**	unk.		unk.	
0.04	**5**	0.07	**10**	0(a)	**0**	unk.	**unk.**	0	0(a)	**0**	unk.	unk.	16	**2**	1	239	unk.	0.2	**2**	0.1	**1**	unk.	**unk.**	tr	C
0.3	**4**	34	**172**	unk.	**unk.**	71	**3**	7	unk.	**unk.**	unk.	0.52	14	**2**	0.1	6.1	unk.	13	**7**	0.03	**3**	unk.		unk.	
0.04	**5**	0.05	**7**	0(a)	**0**	unk.	**unk.**	unk.	0(a)	**0**	unk.	unk.	13	**2**	4	179	unk.	0.3	**3**	tr	**0**	unk.	**unk.**	tr	D
0.2	**3**	16	**79**	unk.	**unk.**	95	**4**	unk.	unk.	**unk.**	unk.	0.52	9	**1**	0.2	4.6	unk.	9	**4**	0.01	**1**	unk.		unk.	
0.04	**5**	0.08	**12**	0(a)	**0**	unk.	**unk.**	unk.	0(a)	**0**	unk.	unk.	6	**1**	5	174	unk.	0.5	**5**	tr	**1**	unk.	**unk.**	unk.	E
0.3	**4**	13	**63**	unk.	**unk.**	35	**1**	unk.	unk.	**unk.**	unk.	0.52	10	**1**	0.2	4.5	unk.	8	**4**	unk.	**unk.**	unk.		unk.	
0.04	**5**	0.07	**10**	0(a)	**0**	unk.	**unk.**	unk.	0(a)	**0**	unk.	unk.	8	**1**	8	259	unk.	0.5	**5**	0.2	**2**	unk.	**unk.**	tr	F
0.4	**4**	26	**129**	unk.	**unk.**	281	**11**	unk.	unk.	**unk.**	unk.	0.52	16	**2**	0.3	6.6	unk.	9	**4**	0.04	**4**	unk.		unk.	
0.05	**7**	unk.	**unk.**	0(a)	**0**	unk.	**unk.**	0	0(a)	**0**	unk.	unk.	22	**3**	4	260	unk.	0.1	**1**	unk.	**unk.**	unk.	**unk.**	unk.	G
0.2	**3**	32	**159**	unk.	**unk.**	60	**2**	6	unk.	**unk.**	unk.	0.52	unk.	**unk.**	0.2	6.6	unk.	unk.	**unk.**	unk.	**unk.**	unk.		unk.	
0.03	**3**	0.08	**12**	0(a)	**0**	unk.	**unk.**	0	0(a)	**0**	unk.	unk.	10	**1**	3	183	unk.	0.5	**6**	0.1	**1**	unk.	**unk.**	610	H
0.2	**3**	24	**121**	unk.	**unk.**	40	**2**	4	unk.	**unk.**	unk.	0.52	12	**2**	0.1	4.7	unk.	14	**7**	0.04	**4**	unk.		unk.	
0.16	**20**	0.08	**11**	0(a)	**0**	unk.	**unk.**	unk.	0(a)	**0**	unk.	unk.	16	**2**	3	235	unk.	1.1	**11**	0.1	**1**	unk.	**unk.**	tr	I
0.5	**6**	17	**85**	unk.	**unk.**	170	**7**	unk.	unk.	**unk.**	unk.	0.52	13	**2**	0.1	6.0	unk.	14	**7**	0.01	**1**	unk.		unk.	
0.04	**5**	0.05	**7**	0(a)	**0**	unk.	**unk.**	unk.	0(a)	**0**	unk.	unk.	5	**1**	3	93	unk.	0.3	**3**	0.2	**8**	unk.	**unk.**	tr	J
0.2	**2**	5	**24**	unk.	**unk.**	27	**1**	unk.	unk.	**unk.**	unk.	0.27	11	**1**	0.1	2.4	unk.	3	**1**	0.28	**28**	unk.		unk.	
0.04	**5**	0.05	**7**	0(a)	**0**	unk.	**unk.**	unk.	0(a)	**0**	unk.	unk.	7	**1**	3	108	unk.	0.3	**3**	0.2	**3**	unk.	**unk.**	tr	K
0.1	**2**	5	**23**	0.12	**2**	30	**1**	unk.	unk.	**unk.**	unk.	0.27	11	**1**	0.1	2.8	unk.	3	**1**	0.12	**12**	unk.		unk.	
0.04	**5**	0.04	**6**	0(a)	**0**	unk.	**unk.**	unk.	0(a)	**0**	unk.	unk.	8	**1**	4	162	unk.	0.3	**3**	0.1	**1**	unk.	**unk.**	tr	L
0.2	**2**	2	**9**	unk.	**unk.**	522	**21**	unk.	unk.	**unk.**	unk.	0.40	12	**2**	0.2	4.1	unk.	5	**3**	0.07	**7**	unk.		unk.	
0.05	**7**	unk.	**unk.**	0(a)	**0**	unk.	**unk.**	unk.	0(a)	**0**	unk.	unk.	13	**2**	3	130	unk.	0.5	**5**	unk.	**unk.**	unk.	**unk.**	unk.	M
0.1	**2**	1	**3**	unk.	**unk.**	53	**2**	unk.	unk.	**unk.**	unk.	0.40	unk.	**unk.**	0.1	3.3	unk.	unk.	**unk.**	unk.	**unk.**	unk.		unk.	
0.04	**5**	0.05	**7**	0(a)	**0**	unk.	**unk.**	unk.	0(a)	**0**	unk.	unk.	5	**1**	3	100	0	0.1	**1**	0.1	**1**	unk.	**unk.**	88	N
0.1	**1**	2	**12**	unk.	**unk.**	26	**1**	unk.	unk.	**unk.**	unk.	0.26	8	**1**	0.1	2.5	0.0	5	**2**	0.12	**12**	unk.		unk.	
0.04	**5**	0.04	**6**	0(a)	**0**	unk.	**unk.**	unk.	0(a)	**0**	0.7	tr	5	**1**	3	91	unk.	0.3	**3**	tr	**0**	unk.	**unk.**	tr	O
0.1	**1**	2	**12**	0.14	**3**	22	**1**	unk.	0.8	**8**	0.7	0.26	9	**1**	0.1	2.3	unk.	3	**1**	0.04	**4**	unk.		unk.	
0.01	**2**	0.04	**6**	0(a)	**0**	unk.	**unk.**	unk.	0(a)	**0**	1.1	0.2	12	**2**	11	163	unk.	0.4	**4**	tr	**1**	unk.	**unk.**	tr	P
0.3	**3**	2	**10**	0.18	**4**	97	**4**	unk.	1.3	**13**	0.9	0.40	14	**2**	0.5	4.2	unk.	4	**2**	0.01	**1**	unk.		unk.	
0.03	**3**	0.12	**18**	0(a)	**0**	unk.	**unk.**	unk.	0(a)	**0**	unk.	unk.	9	**1**	11	105	unk.	0.2	**2**	0.1	**1**	unk.	**unk.**	tr	Q
0.2	**3**	7	**37**	unk.	**unk.**	54	**2**	unk.	unk.	**unk.**	unk.	0.27	7	**1**	0.5	2.7	unk.	10	**5**	0.04	**4**	unk.		unk.	
0.04	**5**	0.16	**22**	0(a)	**0**	unk.	**unk.**	unk.	0(a)	**0**	0.3	0.0	7	**1**	12	119	unk.	0.3	**3**	0.1	**1**	unk.	**unk.**	tr	R
0.2	**3**	7	**37**	0.20	**4**	58	**2**	unk.	0.4	**4**	0.3	0.40	9	**1**	0.5	3.0	unk.	14	**7**	0.04	**5**	unk.		unk.	
0.03	**3**	0.02	**3**	0(a)	**0**	unk.	**unk.**	unk.	0(a)	**0**	unk.	unk.	9	**1**	5	95	unk.	0.3	**3**	0.1	**1**	unk.	**unk.**	unk.	S
0.3	**4**	unk.	**unk.**	unk.	**unk.**	236	**9**	unk.	unk.	**unk.**	unk.	0.26	8	**1**	0.2	2.4	unk.	6	**3**	unk.	**unk.**	unk.		unk.	

	FOOD	Portion	Weight in grams / Conversion for 100 g	Kilocalories / H_2O g	Total carbohydrate g / Total fats g	Crude fiber g / Dietary fiber g	Total protein g / Total sugar g	% USRDA	Arginine mg / Histidine mg	Isoleucine mg / Leucine mg	Lysine mg / Methionine mg	Phenylalanine mg / Threonine mg	Valine mg / Tryptophan mg	Cystine mg / Tyrosine mg	Polyunsat. fatty acids g / Monounsat. fatty acids g	Saturated fatty acids g / P/S ratio	Linoleic acid g / Cholesterol mg	Thiamin mg / Ascorbic acid mg	% USRDA
A	baby food,fruit,guava w/tapioca,strnd	4-1/2 oz jar	128	86	23.4	1.28	0.4	**2**	tr(a)	tr(a)	tr(a)	tr(a)	tr(a)	tr(a)	0.00	0.00	0.00	0.01	**2**
			0.78	103.9	0.0	unk.	unk.		tr(a)	tr(a)	tr(a)	tr(a)	tr(a)	tr(a)	0.00	0.00	0(a)	97	**241**
B	baby food,fruit,mango w/tapioca,strnd	4-3/4 oz jar	135	108	29.2	0.27	0.4	**2**	tr(a)	tr(a)	tr(a)	tr(a)	tr(a)	tr(a)	tr(a)	tr(a)	tr(a)	0.03	**4**
			0.74	104.9	0.3	unk.	unk.		tr(a)	tr(a)	tr(a)	tr(a)	tr(a)	tr(a)	tr(a)	0.00	0(a)	168	**420**
C	baby food,fruit,papaya & applesauce w/tapioca,strnd	4-1/2 oz jar	128	90	24.2	0.51	0.3	**1**	tr(a)	tr(a)	tr(a)	tr(a)	tr(a)	tr(a)	tr(a)	tr(a)	tr(a)	0.01	**2**
			0.78	103.2	0.1	unk.	unk.		tr(a)	tr(a)	tr(a)	tr(a)	tr(a)	tr(a)	tr(a)	0.00	0(a)	145	**362**
D	baby food,fruit,peaches,strnd	4-3/4 oz jar	135	96	25.5	0.94	0.7	**3**	14	11	25	15	33	unk.	tr(a)	tr(a)	tr(a)	0.01	**2**
			0.74	95.8	0.3	unk.	unk.		14	24	26	23	3	17	tr(a)	0.00	0(a)	42	**106**
E	baby food,fruit,pears & pineapple,strnd	4-1/2 oz jar	128	52	14.0	0.38	0.4	**2**	tr(a)	tr(a)	tr(a)	tr(a)	tr(a)	tr(a)	tr(a)	tr(a)	tr(a)	0.03	**4**
			0.78	113.3	0.3	0.18	unk.		tr(a)	tr(a)	tr(a)	tr(a)	tr(a)	tr(a)	tr(a)	0.00	0(a)	35	**88**
F	baby food,fruit,pears,strnd	4-1/2 oz jar	128	52	13.8	unk.	0.4	**2**	unk.	tr(a)	tr(a)	10	tr(a)	tr(a)	tr(a)	tr(a)	tr(a)	0.01	**2**
			0.78	113.2	0.3	unk.	unk.		tr(a)	tr(a)	tr(a)	tr(a)	tr(a)	15	tr(a)	0.00	0(a)	31	**78**
G	baby food,fruit,plums w/tapioca,strnd	4-3/4 oz jar	135	96	26.6	unk.	0.1	**1**	tr(a)	tr(a)	tr(a)	tr(a)	tr(a)	tr(a)	0.00	0.00	0.00	0.01	**2**
			0.74	108.0	0.0	unk.	unk.		tr(a)	tr(a)	tr(a)	tr(a)	tr(a)	tr(a)	0.00	0.00	0(a)	1	**4**
H	baby food,fruit,prunes w/tapioca,strnd	4-3/4 oz jar	135	94	25.0	0.40	0.8	**3**	tr(a)	tr(a)	tr(a)	tr(a)	tr(a)	tr(a)	tr(a)	tr(a)	tr(a)	0.03	**4**
			0.74	108.4	0.1	unk.	unk.		tr(a)	tr(a)	tr(a)	tr(a)	tr(a)	tr(a)	tr(a)	0.00	0(a)	1	**3**
I	BABY FOOD,MEAT,beef w/beef heart,strnd	3-1/2 oz jar	99	93	0.0	0.10	12.6	**70**	unk.	613	1040	518	683	129	0.21	2.06	0.16	0.02	**3**
			1.01	81.7	4.4	unk.	unk.		336	1000	326	506	125	361	1.59	0.10	unk.	2	**5**
J	baby food,meat,beef,strnd	3-1/2 oz jar	99	96	tr	tr	13.4	**74**	unk.	633	1005	516	741	107	unk.	unk.	unk.	0.01	**1**
			1.01	unk.	4.7	unk.	unk.		348	987	431	581	135	417	unk.	unk.	unk.	2	**5**
K	baby food,meat,chicken,strnd	3-1/2 oz jar	99	129	0.1	unk.	13.6	**75**	947	639	1133	552	682	178	1.90	2.01	1.83	0.01	**1**
			1.01	76.7	7.8	unk.	unk.		411	1047	363	609	154	435	3.22	0.95	unk.	2	**4**
L	baby food,meat,ham,strnd	3-1/2 oz jar	99	110	0.0	unk.	13.8	**76**	931	654	1168	525	709	169	0.77	1.92	0.70	0.14	**20**
			1.01	78.6	5.7	unk.	0(a)		467	1100	351	597	137	461	2.49	0.40	unk.	2	**5**
M	baby food,meat,lamb,strnd	3-1/2 oz jar	99	102	0.1	unk.	14.0	**78**	914	655	1237	548	707	192	0.20	2.29	0.13	0.02	**3**
			1.01	79.5	4.7	unk.	unk.		352	1101	437	636	139	488	1.76	0.09	unk.	1	**3**
N	baby food,meat,liver,strnd	3-1/2 oz jar	99	100	1.4	unk.	14.2	**79**	831	694	926	678	898	213	0.06	1.36	0.22	0.05	**7**
			1.01	78.5	3.8	unk.	unk.		363	1307	372	676	220	569	0.68	0.04	181	19	**48**
O	baby food,meat,pork,strnd	3-1/2 oz jar	99	123	0.0	unk.	13.9	**77**	939	673	1146	564	695	152	0.77	2.38	0.68	0.15	**21**
			1.01	77.6	7.0	unk.	0(a)		443	1116	394	611	135	505	3.24	0.32	unk.	2	**5**
P	baby food,meat,turkey,strnd	3-1/2 oz jar	99	113	0.1	unk.	14.2	**79**	900	710	1172	584	720	172	1.44	1.89	1.32	0.02	**3**
			1.01	78.1	5.7	unk.	unk.		362	1127	439	627	148	498	1.94	0.76	unk.	2	**5**
Q	baby food,meat,veal,strnd	3-1/2 oz jar	99	100	0.0	0.10	13.4	**74**	888	604	1074	518	650	173	0.16	2.27	0.10	0.02	**3**
			1.01	80.1	4.8	unk.	0(a)		411	1034	295	558	154	428	1.77	0.07	unk.	2	**6**
R	BABY FOOD,VEGETABLE,beans,green, buttered,strnd	4-1/2 oz jar	128	42	8.4	unk.	1.5	**6**	84	69	76	63	83	13	tr(a)	tr(a)	tr(a)	0.03	**4**
			0.78	116.2	1.0	unk.	unk.		42	100	23	65	18	54	tr(a)	0.00	tr(a)	11	**27**
S	baby food,vegetable,beans,green,strnd	4-1/2 oz jar	128	32	7.6	1.28	1.7	**7**	92	74	83	68	91	14	tr(a)	tr(a)	tr(a)	0.04	**6**
			0.78	117.8	0.1	unk.	unk.		46	109	24	70	19	59	tr(a)	0.00	0(a)	7	**17**

Riboflavin mg / Niacin mg	% USRDA	Vitamin B_6 mg / Folacin mcg	% USRDA	Vitamin B_{12} mcg / Pantothenic acid mg	% USRDA	Biotin mg / Vitamin A IU	% USRDA	Preformed A RE / Beta carotene RE	Vitamin D IU / Vitamin E IU	% USRDA	Total tocopherol mg / Alpha tocopherol mg	Other tocopherol mg / Total ash g	Calcium mg / Phosphorus mg	% USRDA	Sodium mg / Sodium meq	Potassium mg / Potassium meq	Chlorine mg / Chlorine meq	Iron mg / Magnesium mg	% USRDA	Zinc mg / Copper mg	% USRDA	Iodine mcg / Selenium mcg	% USRDA	Manganese mcg / Chromium mcg	
0.09	**11**	0.05	**8**	0(a)	**0**	unk.	**unk.**	unk.	0(a)	**0**	unk.	unk.	9	**1**	3	93	unk.	0.3	**3**	0.1	**1**	unk.	**unk.**	unk.	A
0.5	**6**	unk.	**unk.**	unk.	**unk.**	383	**15**	unk.	unk.	**unk.**	unk.	0.26	6	**1**	0.1	2.4	unk.	3	**1**	unk.	**unk.**	unk.		unk.	
0.04	**5**	0.16	**23**	0(a)	**0**	unk.	**unk.**	unk.	0(a)	**0**	unk.	unk.	5	**1**	5	80	unk.	0.1	**1**	0.1	**1**	unk.	**unk.**	unk.	B
0.3	**4**	unk.	**unk.**	unk.	**unk.**	898	**36**	unk.	unk.	**unk.**	unk.	0.27	8	**1**	0.2	2.0	unk.	5	**3**	unk.	**unk.**	unk.		unk.	
0.04	**5**	0.03	**4**	0(a)	**0**	unk.	**unk.**	unk.	0(a)	**0**	unk.	unk.	9	**1**	6	101	unk.	unk.	**unk.**	tr	**1**	unk.	**unk.**	unk.	C
0.1	**2**	unk.	**unk.**	unk.	**unk.**	97	**4**	unk.	unk.	**unk.**	unk.	0.26	6	**1**	0.3	2.6	unk.	6	**3**	unk.	**unk.**	unk.		unk.	
0.04	**5**	0.02	**3**	0(a)	**0**	unk.	**unk.**	unk.	0(a)	**0**	1.9	0.1	8	**1**	8	219	unk.	0.3	**3**	0.1	**1**	unk.	**unk.**	tr	D
0.8	**9**	5	**26**	0.18	**4**	217	**9**	unk.	2.3	**23**	1.8	0.40	16	**2**	0.3	5.6	unk.	5	**3**	0.01	**1**	unk.		unk.	
0.04	**5**	0.02	**3**	0(a)	**0**	unk.	**unk.**	unk.	0(a)	**0**	unk.	unk.	13	**2**	5	148	unk.	0.3	**3**	0.1	**1**	unk.	**unk.**	110	E
0.3	**3**	4	**18**	unk.	**unk.**	37	**2**	unk.	unk.	**unk.**	unk.	0.38	12	**1**	0.2	3.8	unk.	11	**6**	0.01	**1**	unk.		unk.	
0.04	**5**	0.01	**2**	0(a)	**0**	unk.	**unk.**	unk.	0(a)	**0**	0.9	0.2	10	**1**	3	166	unk.	0.3	**3**	0.1	**1**	unk.	**unk.**	tr	F
0.2	**3**	5	**23**	0.12	**2**	42	**2**	unk.	1.1	**11**	0.7	0.38	15	**2**	0.1	4.3	unk.	10	**5**	0.03	**3**	unk.		unk.	
0.04	**5**	0.03	**5**	0(a)	**0**	unk.	**unk.**	unk.	0(a)	**0**	0.9	0.3	8	**1**	8	115	unk.	0.3	**3**	0.1	**1**	unk.	**unk.**	tr	G
0.3	**3**	1	**6**	0.15	**3**	13	**1**	unk.	1.0	**10**	0.5	0.27	8	**1**	0.3	2.9	unk.	3	**2**	0.04	**4**	unk.		unk.	
0.11	**14**	0.11	**16**	0(a)	**0**	unk.	**unk.**	unk.	0(a)	**0**	0.6	0.1	20	**3**	7	239	unk.	0.5	**5**	0.1	**2**	unk.	**unk.**	tr	H
0.7	**8**	tr	**1**	0.19	**4**	612	**25**	unk.	0.7	**7**	0.5	0.67	20	**3**	0.3	6.1	unk.	11	**6**	0.01	**1**	unk.		unk.	
0.36	**45**	0.12	**17**	unk.	**unk.**	unk.	**unk.**	unk.	unk.	**unk.**	unk.	unk.	4	**1**	62	198	unk.	2.0	**20**	3.0	**37**	unk.	**unk.**	unk.	I
3.9	**43**	5	**25**	unk.	**unk.**	125	**5**	unk.	unk.	**unk.**	unk.	0.89	93	**12**	2.7	5.1	unk.	13	**6**	1.97	**197**	unk.		unk.	
0.13	**16**	0.13	**18**	unk.	**unk.**	unk.	**unk.**	unk.	unk.	**unk.**	unk.	unk.	7	**1**	180	193	unk.	1.3	**13**	3.0	**38**	unk.	**unk.**	tr	J
3.0	**34**	unk.	**unk.**	unk.	**unk.**	288	**12**	unk.	unk.	**unk.**	unk.	unk.	106	**13**	7.8	4.9	unk.	15	**7**	0.05	**5**	unk.		unk.	
0.15	**19**	0.20	**28**	unk.	**unk.**	unk.	**unk.**	unk.	unk.	**unk.**	0.5	0.2	63	**8**	47	140	unk.	1.4	**14**	1.2	**15**	unk.	**unk.**	tr	K
3.2	**36**	10	**52**	0.67	**14**	134	**5**	unk.	0.6	**6**	0.3	0.79	96	**12**	2.0	3.6	unk.	16	**8**	0.07	**7**	unk.		unk.	
0.15	**19**	0.25	**36**	unk.	**unk.**	unk.	**unk.**	unk.	unk.	**unk.**	0.4	tr	6	**1**	41	202	unk.	1.0	**10**	2.3	**29**	unk.	**unk.**	tr	L
2.6	**29**	2	**10**	0.50	**10**	38	**2**	unk.	0.5	**5**	0.4	0.99	80	**10**	1.8	5.2	unk.	17	**9**	0.07	**7**	unk.		unk.	
0.20	**25**	0.15	**22**	2.17	**72**	unk.	**unk.**	unk.	unk.	**unk.**	unk.	unk.	7	**1**	61	203	unk.	1.5	**15**	3.4	**43**	unk.	**unk.**	tr	M
2.9	**32**	2	**11**	0.41	**8**	85	**3**	unk.	unk.	**unk.**	unk.	0.79	96	**12**	2.7	5.2	unk.	16	**8**	0.07	**7**	unk.		unk.	
1.79	**224**	0.34	**49**	2.14	**71**	unk.	**unk.**	unk.	unk.	**unk.**	0.3	0.0	4	**1**	73	225	unk.	5.2	**52**	2.7	**33**	unk.	**unk.**	68	N
8.2	**92**	334	**1670**	unk.	**unk.**	37754	**1510**	unk.	0.4	**4**	0.3	1.19	201	**25**	3.2	5.8	unk.	15	**8**	1.68	**168**	unk.		unk.	
0.20	**25**	0.20	**29**	0.98	**33**	unk.	**unk.**	unk.	unk.	**unk.**	unk.	unk.	5	**1**	42	221	unk.	1.0	**10**	2.0	**25**	unk.	**unk.**	tr	O
2.2	**25**	2	**9**	unk.	**unk.**	38	**2**	unk.	unk.	**unk.**	unk.	1.09	93	**12**	1.8	5.6	unk.	16	**8**	0.07	**7**	unk.		unk.	
0.21	**26**	0.18	**26**	0.99	**33**	unk.	**unk.**	unk.	unk.	**unk.**	0.3	0.6	23	**3**	54	229	unk.	1.2	**12**	2.2	**28**	unk.	**unk.**	tr	P
3.6	**40**	11	**56**	0.56	**11**	554	**22**	unk.	0.4	**4**	0.5	0.89	125	**16**	2.4	5.8	unk.	16	**8**	0.07	**7**	unk.		unk.	
0.16	**20**	0.15	**21**	unk.	**unk.**	unk.	**unk.**	unk.	unk.	**unk.**	0.2	tr	7	**1**	63	214	unk.	1.3	**13**	2.7	**34**	unk.	**unk.**	tr	Q
3.5	**39**	6	**29**	0.43	**9**	46	**2**	unk.	0.3	**3**	0.2	0.89	97	**12**	2.8	5.5	unk.	16	**8**	0.05	**5**	unk.		unk.	
0.14	**18**	unk.	**unk.**	unk.	**unk.**	unk.	**unk.**	unk.	unk.	**unk.**	0.3	0.1	82	**10**	4	205	unk.	1.6	**16**	unk.	**unk.**	unk.	**unk.**	unk.	R
0.4	**5**	36	**182**	unk.	**unk.**	584	**23**	unk.	0.4	**4**	0.2	0.77	unk.	**unk.**	0.2	5.2	unk.	unk.	**unk.**	unk.	**unk.**	unk.		unk.	
0.12	**14**	0.05	**7**	0(a)	**0**	unk.	**unk.**	unk.	unk.	**unk.**	0.3	0.1	50	**6**	3	202	unk.	1.0	**10**	0.2	**3**	unk.	**unk.**	268	S
0.4	**5**	44	**221**	0.20	**4**	573	**23**	unk.	0.4	**4**	0.2	0.77	26	**3**	0.1	5.2	unk.	26	**13**	0.03	**3**	unk.		unk.	

	FOOD	Portion	Weight in grams / *Conversion for 100 g*	Kilocalories / H_2O g	Total carbohydrate g / Total fats g	Crude fiber g / Dietary fiber g	Total protein g / Total sugar g	**% USRDA**	Arginine mg / Histidine mg	Isoleucine mg / Leucine mg	Lysine mg / Methionine mg	Phenylalanine mg / Threonine mg	Valine mg / Tryptophan mg	Cystine mg / Tyrosine mg	Polyunsat. fatty acids g / Monounsat. fatty acids g	Saturated fatty acids g / P/S ratio	Linoleic acid g / Cholesterol mg	Thiamin mg / Ascorbic acid mg	**% USRDA**
A	baby food,vegetable,beets,strnd	4-1/2 oz jar	128	44	9.9	1.02	1.7	**7**	38	49	44	23	58	9	tr(a)	tr(a)	tr(a)	0.01	**2**
			0.78	115.3	0.1	unk.	unk.		27	59	13	41	15	45	tr(a)	0.00	0(a)	3	**8**
B	baby food,vegetable,carrots,	4-1/2 oz jar	128	46	9.5	unk.	1.0	**4**	68	32	27	33	41	8	tr(a)	tr(a)	tr(a)	0.03	**4**
	buttered,strnd		*0.78*	116.0	0.8	unk.	unk.		17	44	12	31	15	26	tr(a)	0.00	unk.	12	**29**
C	baby food,vegetable,carrots,strnd	4-1/2 oz jar	128	35	7.7	1.02	1.0	**4**	64	31	26	31	38	8	tr(a)	tr(a)	tr(a)	0.03	**4**
			0.78	118.1	0.1	unk.	unk.		15	41	12	28	14	24	tr(a)	0.00	0(a)	7	**18**
D	baby food,vegetable,corn,creamed,strnd	4-1/2 oz jar	128	73	18.0	0.38	1.8	**7**	77	83	104	63	99	23	tr(a)	tr(a)	tr(a)	0.01	**2**
			0.78	107.0	0.5	unk.	unk.		59	179	51	67	19	86	tr(a)	0.00	0(a)	3	**7**
E	baby food,vegetable,garden	4-1/2 oz jar	128	47	8.7	1.15	2.9	**12**	tr	110	148	113	127	26	tr(a)	tr(a)	tr(a)	0.08	**11**
	vegetables,strnd		*0.78*	115.2	0.3	unk.	unk.		58	183	55	93	35	122	tr(a)	tr(a)	0(a)	7	**18**
F	baby food,vegetable,mixed	4-1/2 oz jar	128	52	10.2	0.51	1.5	**6**	101	60	51	60	74	31	tr(a)	tr(a)	tr(a)	0.03	**4**
	vegetables,strnd		*0.78*	114.9	0.6	unk.	unk.		33	99	23	49	17	56	tr(a)	0.00	0(a)	2	**5**
G	baby food,vegetable,peas,buttered,strnd	4-1/2 oz jar	128	72	13.6	unk.	4.7	**19**	525	204	315	192	228	36	tr(a)	tr(a)	tr(a)	0.10	**15**
			0.78	107.8	1.4	unk.	unk.		101	316	54	183	46	160	tr(a)	0.00	unk.	15	**38**
H	baby food,vegetable,peas,creamed,strnd	4-1/2 oz jar	128	68	11.4	0.51	2.8	**11**	unk.	unk.	unk.	unk.	unk.	unk.	tr(a)	tr(a)	tr(a)	0.12	**17**
			0.78	110.7	2.4	unk.	unk.		unk.	unk.	unk.	unk.	unk.	unk.	tr(a)	0.00	0(a)	2	**5**
I	baby food,vegetable,peas,strnd	4-1/2 oz jar	128	51	10.4	0.15	4.5	**18**	498	193	300	192	227	33	tr(a)	tr(a)	tr(a)	0.10	**15**
			0.78	112.0	0.4	unk.	unk.		96	301	51	174	44	152	tr(a)	0.00	0(a)	9	**22**
J	baby food,vegetable,spinach,	4-1/2 oz jar	128	47	7.3	0.64	3.2	**13**	195	143	189	123	193	40	tr(a)	tr(a)	tr(a)	0.03	**4**
	creamed,strnd		*0.78*	114.7	1.7	unk.	unk.		82	283	70	129	46	147	tr(a)	0.00	0(a)	11	**28**
K	baby food,vegetable,squash,	4-1/2 oz jar	128	37	8.8	unk.	0.8	**3**	45	32	29	28	36	8	tr(a)	tr(a)	tr(a)	0.01	**2**
	buttered,strnd		*0.78*	117.4	0.4	unk.	unk.		15	46	10	24	12	28	tr(a)	0.00	unk.	10	**24**
L	baby food,vegetable,squash,strnd	4-1/2 oz jar	128	31	7.2	0.90	1.0	**4**	59	42	40	37	46	9	tr(a)	tr(a)	tr(a)	0.01	**2**
			0.78	118.7	0.3	unk.	unk.		20	60	13	32	15	36	tr(a)	0.00	0(a)	10	**25**
M	baby food,vegetable,sweet potatoes,	4-3/4 oz jar	135	76	15.9	unk.	1.2	**5**	61	57	49	70	82	18	tr(a)	tr(a)	tr(a)	0.03	**4**
	buttered,strnd		*0.74*	116.0	0.9	unk.	unk.		28	86	27	61	24	46	tr(a)	0.00	unk.	12	**31**
N	baby food,vegetable,sweet	4-3/4 oz jar	135	77	17.8	0.81	1.5	**6**	73	69	59	84	99	20	tr(a)	tr(a)	tr(a)	0.04	**6**
	potatoes,strnd		*0.74*	114.5	0.1	unk.	unk.		35	105	32	73	30	55	tr(a)	0.00	unk.	13	**33**
O	BABY FOOD-JUNIOR,CEREAL & egg yolks	7-1/2 oz jar	213	111	15.1	0.21	4.0	**16**	222	213	271	196	260	60	0.32	1.30	0.55	0.02	**3**
			0.47	188.9	3.8	unk.	unk.		96	366	109	175	58	179	1.58	0.25	unk.	1	**4**
P	baby food-junior,cereal,egg yolks	7-1/2 oz jar	213	179	15.1	unk.	5.3	**21**	unk.	unk.	unk.	unk.	unk.	unk.	unk.	unk.	unk.	0.04	**6**
	& bacon		*0.47*	180.8	11.1	unk.	unk.		unk.	unk.	unk.	unk.	unk.	unk.	unk.	unk.	unk.	3	**6**
Q	baby food-junior,cereal,mixed cereal	7-3/4 oz jar	220	183	40.5	0.44	2.6	**11**	143	95	77	136	128	57	tr(a)	tr(a)	tr(a)	0.64	**91**
	w/applesauce & bananas		*0.45*	175.1	0.9	unk.	unk.		59	189	53	75	29	95	tr(a)	0.00	unk.	20	**50**
R	baby food-junior,cereal,oatmeal	7-3/4 oz jar	220	165	34.5	0.88	2.9	**11**	207	108	119	147	152	66	unk.	unk.	unk.	0.53	**75**
	w/applesauce & bananas		*0.45*	180.0	1.5	unk.	unk.		75	220	68	101	37	114	unk.	unk.	unk.	42	**105**
S	baby food-junior,cereal,rice	7-3/4 oz jar	220	185	41.1	0.44	2.2	**9**	143	110	139	119	139	35	tr(a)	tr(a)	tr(a)	0.55	**79**
	w/mixed fruit		*0.45*	175.1	0.4	unk.	unk.		64	213	42	81	24	108	tr(a)	0.00	0(a)	44	**111**

Riboflavin mg / Niacin mg	% USRDA	Vitamin B_6 mg / Folacin mcg	% USRDA	Vitamin B_{12} mcg / Pantothenic acid mg	% USRDA	Biotin mg / Vitamin A IU	% USRDA	Preformed A RE / Beta carotene RE	Vitamin D IU / Vitamin E IU	% USRDA	Total tocopherol mg / Alpha tocopherol mg	Other tocopherol mg / Total ash g	Calcium mg / Phosphorus mg	% USRDA	Sodium mg / Sodium meq	Potassium mg / Potassium meq	Chlorine mg / Chlorine meq	Iron mg / Magnesium mg	% USRDA	Zinc mg / Copper mg	% USRDA	Iodine mcg / Selenium mcg	% USRDA	Manganese mcg / Chromium mcg	
0.05	**6**	0.03	**4**	0(a)	**0**	unk.	**unk.**	unk.	unk.	**unk.**	0.2	tr	18	**2**	106	233	unk.	0.4	**4**	0.1	**2**	unk.	**unk.**	230	A
0.2	**2**	39	**197**	unk.	**unk.**	42	**2**	unk.	0.2	**2**	0.2	1.15	18	**2**	4.6	6.0	unk.	27	**13**	0.01	**1**	unk.		unk.	
0.08	**10**	unk.	**unk.**	unk.	**unk.**	unk.	**unk.**	unk.	unk.	**unk.**	0.6	0.1	45	**6**	23	4	unk.	0.4	**4**	unk.	**unk.**	unk.	**unk.**	unk.	B
0.8	**9**	12	**60**	unk.	**unk.**	13860	**554**	unk.	0.8	**8**	0.5	0.77	unk.	**unk.**	1.0	0.1	unk.	unk.	**unk.**	unk.	**unk.**	unk.		unk.	
0.05	**6**	0.09	**13**	0(a)	**0**	unk.	**unk.**	unk.	unk.	**unk.**	0.6	0.1	28	**4**	47	251	unk.	0.5	**5**	0.2	**2**	unk.	**unk.**	51	C
0.6	**7**	19	**95**	0.31	**6**	14670	**587**	unk.	0.8	**8**	0.5	1.02	26	**3**	2.1	6.4	unk.	13	**7**	tr	**0**	unk.		unk.	
0.06	**8**	0.05	**8**	0(a)	**0**	unk.	**unk.**	unk.	unk.	**unk.**	0.5	0.3	26	**3**	55	115	unk.	0.4	**4**	0.4	**5**	unk.	**unk.**	tr	D
0.7	**7**	14	**72**	0.37	**7**	96	**4**	unk.	0.6	**6**	0.2	0.64	42	**5**	2.4	2.9	unk.	10	**5**	0.26	**26**	unk.		unk.	
0.09	**11**	0.13	**18**	0(a)	**0**	unk.	**unk.**	unk.	unk.	**unk.**	1.5	0.8	36	**5**	45	215	unk.	1.1	**11**	0.4	**6**	unk.	**unk.**	243	E
1.0	**11**	51	**257**	unk.	**unk.**	7766	**311**	unk.	1.8	**18**	0.8	0.90	36	**5**	1.9	5.5	unk.	35	**17**	0.04	**4**	unk.		unk.	
0.04	**5**	0.07	**10**	0(a)	**0**	unk.	**unk.**	unk.	unk.	**unk.**	0.5	0.2	17	**2**	17	163	unk.	0.4	**4**	0.2	**3**	unk.	**unk.**	128	F
0.4	**5**	5	**25**	0.03	**1**	5108	**204**	unk.	0.6	**6**	0.3	0.64	28	**4**	0.7	4.2	unk.	16	**8**	0.04	**4**	unk.		unk.	
0.09	**11**	unk.	**unk.**	unk.	**unk.**	unk.	**unk.**	unk.	unk.	**unk.**	1.1	0.9	50	**6**	10	124	unk.	1.4	**14**	unk.	**unk.**	unk.	**unk.**	unk.	G
1.8	**20**	44	**222**	unk.	**unk.**	422	**17**	unk.	1.3	**13**	0.2	0.51	unk.	**unk.**	0.4	3.2	unk.	unk.	**unk.**	unk.	**unk.**	unk.		unk.	
0.08	**10**	0.06	**9**	0(a)	**0**	unk.	**unk.**	unk.	unk.	**unk.**	0.9	0.9	17	**2**	18	113	unk.	0.7	**7**	0.5	**6**	unk.	**unk.**	unk.	H
1.0	**12**	29	**145**	unk.	**unk.**	110	**4**	unk.	1.1	**11**	tr	0.64	40	**5**	0.8	2.9	unk.	unk.	**unk.**	0.07	**7**	unk.		unk.	
0.08	**10**	0.09	**13**	0(a)	**0**	unk.	**unk.**	unk.	unk.	**unk.**	1.1	0.9	26	**3**	5	143	unk.	1.2	**12**	0.6	**8**	unk.	**unk.**	64	I
1.3	**15**	33	**166**	0.36	**7**	723	**29**	unk.	1.3	**13**	0.2	0.64	55	**7**	0.2	3.7	unk.	26	**13**	0.09	**9**	unk.		unk.	
0.13	**16**	0.10	**14**	0(a)	**0**	unk.	**unk.**	unk.	unk.	**unk.**	2.1	0.1	114	**14**	61	244	unk.	0.8	**8**	0.5	**6**	unk.	**unk.**	422	J
0.3	**3**	78	**389**	unk.	**unk.**	5338	**214**	unk.	2.5	**25**	2.0	1.28	69	**9**	2.7	6.3	unk.	54	**27**	0.08	**8**	unk.		unk.	
0.09	**11**	unk.	**unk.**	unk.	**unk.**	unk.	**unk.**	unk.	unk.	**unk.**	0.6	0.2	42	**5**	3	163	unk.	0.5	**5**	unk.	**unk.**	unk.	**unk.**	unk.	K
0.5	**5**	15	**76**	unk.	**unk.**	2122	**85**	unk.	0.7	**7**	0.4	0.64	unk.	**unk.**	0.1	4.2	unk.	unk.	**unk.**	unk.	**unk.**	unk.		unk.	
0.08	**10**	0.08	**12**	0(a)	**0**	unk.	**unk.**	unk.	unk.	**unk.**	0.6	0.2	31	**4**	3	229	unk.	0.4	**4**	0.1	**2**	unk.	**unk.**	26	L
0.4	**5**	2	**10**	0.28	**6**	2589	**104**	unk.	0.7	**7**	0.4	0.77	19	**2**	0.1	5.9	unk.	18	**9**	0.01	**1**	unk.		unk.	
0.05	**7**	unk.	**unk.**	unk.	**unk.**	unk.	**unk.**	unk.	unk.	**unk.**	0.7	0.0	28	**4**	11	281	unk.	0.6	**6**	unk.	**unk.**	unk.	**unk.**	unk.	M
0.4	**4**	18	**88**	unk.	**unk.**	9199	**368**	unk.	0.8	**8**	0.7	0.94	unk.	**unk.**	0.5	7.2	unk.	unk.	**unk.**	unk.	**unk.**	unk.		unk.	
0.04	**5**	0.13	**18**	unk.	**unk.**	unk.	**unk.**	unk.	unk.	**unk.**	0.7	0.0	22	**3**	27	355	unk.	0.5	**5**	0.1	**2**	unk.	**unk.**	181	N
0.5	**5**	13	**66**	0.53	**11**	8691	**348**	unk.	0.8	**8**	0.7	1.08	32	**4**	1.2	9.1	unk.	30	**15**	0.01	**1**	unk.		unk.	
0.11	**13**	0.04	**6**	0.13	**4**	unk.	**unk.**	unk.	unk.	**unk.**	unk.	unk.	51	**6**	70	75	unk.	1.1	**11**	0.6	**8**	unk.	**unk.**	tr	O
0.1	**1**	7	**35**	unk.	**unk.**	307	**12**	unk.	unk.	**unk.**	unk.	0.85	85	**11**	3.1	1.9	unk.	12	**6**	0.03	**3**	unk.		unk.	
0.11	**13**	unk.	**unk.**	unk.	**unk.**	unk.	**unk.**	unk.	unk.	**unk.**	unk.	unk.	53	**7**	98	79	unk.	1.1	**11**	0.6	**8**	unk.	**unk.**	unk.	P
0.3	**4**	9	**47**	unk.	**unk.**	143	**6**	unk.	unk.	**unk.**	unk.	0.64	unk.	**unk.**	4.3	2.0	unk.	11	**5**	0.04	**5**	unk.		unk.	
0.79	**99**	0.31	**44**	unk.	**unk.**	unk.	**unk.**	unk.	unk.	**unk.**	unk.	unk.	9	**1**	79	70	unk.	12.3	**123**	0.5	**7**	unk.	**unk.**	308	Q
8.9	**99**	8	**41**	unk.	**unk.**	42	**2**	unk.	unk.	**unk.**	unk.	0.88	64	**8**	3.4	1.8	unk.	24	**12**	0.11	**11**	unk.		unk.	
1.06	**132**	0.53	**75**	unk.	**unk.**	unk.	**unk.**	unk.	unk.	**unk.**	unk.	unk.	13	**2**	68	106	unk.	12.1	**121**	0.7	**9**	unk.	**unk.**	748	R
7.4	**82**	8	**39**	unk.	**unk.**	62	**3**	unk.	unk.	**unk.**	unk.	0.88	90	**11**	3.0	2.7	unk.	37	**18**	0.11	**11**	unk.		unk.	
1.30	**162**	0.54	**78**	unk.	**unk.**	unk.	**unk.**	unk.	unk.	**unk.**	unk.	unk.	44	**6**	24	73	unk.	10.4	**104**	0.4	**5**	unk.	**unk.**	374	S
6.0	**67**	unk.	**unk.**	unk.	**unk.**	33	**1**	unk.	unk.	**unk.**	unk.	0.88	51	**6**	1.0	1.9	unk.	13	**7**	0.04	**4**	unk.		unk.	

	FOOD	Portion	Weight in grams / Conversion for 100 g	Kilocalories / H_2O g	Total carbohydrate g / Total fats g	Crude fiber g / Dietary fiber g	Total protein g / Total sugar g	% USRDA	Arginine mg / Histidine mg	Isoleucine mg / Leucine mg	Lysine mg / Methionine mg	Phenylalanine mg / Threonine mg	Valine mg / Tryptophan mg	Cystine mg / Tyrosine mg	Polyunsat. fatty acids g / Monounsat. fatty acids g	Saturated fatty acids g / P/S ratio	Linoleic acid g / Cholesterol mg	Thiamin mg / Ascorbic acid mg	% USRDA
A	BABY FOOD-JUNIOR,DESSERT,apple	7-3/4 oz jar	220	154	41.8	unk.	0.9	**4**	unk.	unk.	unk.	unk.	unk.	unk.	0.00	0.00	0.00	0.02	**3**
	betty		*0.45*	177.1	0.0	unk.	unk.		unk.	unk.	unk.	unk.	unk.	unk.	0.00	0.00	0(a)	60	**149**
B	baby food-junior,dessert,caramel	7-1/2 oz jar	213	168	36.2	unk.	3.0	**12**	unk.	unk.	unk.	unk.	unk.	unk.	tr(a)	tr(a)	tr(a)	0.02	**3**
	pudding		*0.47*	171.3	1.9	unk.	unk.		unk.	unk.	unk.	unk.	unk.	unk.	tr(a)	0.00	unk.	5	**12**
C	baby food-junior,dessert,cherry	7-3/4 oz jar	220	152	40.5	0.22	0.4	**2**	unk.	unk.	unk.	unk.	unk.	unk.	tr(a)	tr(a)	tr(a)	0.02	**3**
	vanilla pudding		*0.45*	178.2	0.4	unk.	unk.		unk.	unk.	unk.	unk.	unk.	unk.	tr(a)	0.00	unk.	2	**6**
D	baby food-junior,dessert,chocolate	7-3/4 oz jar	220	196	38.3	unk.	4.2	**17**	185	229	304	191	268	unk.	unk.	unk.	unk.	0.02	**3**
	custard pudding		*0.45*	172.7	3.5	unk.	unk.		106	411	103	196	unk.	154	unk.	unk.	unk.	2	**6**
E	baby food-junior,dessert,cottage	7-3/4 oz jar	220	172	35.0	2.20	6.6	**26**	132	216	312	233	262	unk.	tr(a)	tr(a)	tr(a)	0.04	**6**
	cheese w/pineapple		*0.45*	176.4	1.5	unk.	unk.		117	398	57	163	unk.	178	tr(a)	0.00	unk.	52	**131**
F	baby food-junior,dessert,Dutch apple	7-3/4 oz jar	220	152	37.0	0.66	0.0	**0**	unk.	unk.	unk.	unk.	unk.	unk.	0.07	1.36	0.04	0.02	**3**
			0.45	180.6	2.2	unk.	unk.		unk.	unk.	unk.	unk.	unk.	unk.	0.51	0.05	unk.	47	**118**
G	baby food-junior,dessert,fruit	7-3/4 oz jar	220	139	37.8	0.66	0.7	**3**	unk.	unk.	unk.	unk.	unk.	unk.	0(a)	0(a)	0(a)	0.04	**6**
	dessert		*0.45*	180.8	0.0	unk.	unk.		unk.	unk.	unk.	unk.	unk.	unk.	0(a)	0.00	0(a)	7	**17**
H	baby food-junior,dessert,peach cobbler	7-3/4 oz jar	220	147	40.3	unk.	0.7	**3**	unk.	unk.	unk.	unk.	unk.	unk.	0(a)	0(a)	0(a)	0.02	**3**
			0.45	178.6	0.0	unk.	unk.		unk.	unk.	unk.	unk.	unk.	unk.	0(a)	0.00	0(a)	45	**113**
I	baby food-junior,dessert,peach melba	7-3/4 oz jar	220	132	36.1	unk.	0.7	**3**	unk.	unk.	unk.	unk.	unk.	unk.	0(a)	0(a)	0(a)	0.02	**3**
			0.45	182.6	0.0	unk.	unk.		unk.	unk.	unk.	unk.	unk.	unk.	0(a)	0.00	0(a)	57	**143**
J	baby food-junior,dessert,pineapple	7-3/4 oz jar	220	191	47.5	1.76	3.1	**12**	unk.	unk.	unk.	unk.	unk.	unk.	tr(a)	tr(a)	tr(a)	0.09	**13**
	pudding		*0.45*	167.4	0.9	unk.	unk.		unk.	unk.	unk.	unk.	unk.	unk.	tr(a)	0.00	unk.	59	**147**
K	baby food-junior,dessert,tropical	7-3/4 oz jar	220	132	36.1	unk.	0.4	**2**	unk.	unk.	unk.	unk.	unk.	unk.	0(a)	0(a)	0(a)	0.02	**3**
	fruit		*0.45*	183.0	0.0	unk.	unk.		unk.	unk.	unk.	unk.	unk.	unk.	0(a)	0.00	0(a)	41	**103**
L	baby food-junior,dessert,vanilla	7-3/4 oz jar	220	196	35.6	unk.	3.5	**14**	139	167	251	163	196	unk.	0.42	2.60	0.35	0.02	**3**
	custard pudding		*0.45*	174.7	5.1	unk.	unk.		90	130	101	130	unk.	130	1.58	0.16	unk.	2	**4**
M	BABY FOOD-JUNIOR,DINNER,beef &	7-1/2 oz jar	213	121	15.8	0.43	5.3	**21**	334	294	394	249	281	62	unk.	unk.	unk.	0.06	**9**
	egg noodles		*0.47*	187.0	4.0	unk.	unk.		158	443	113	211	58	198	unk.	unk.	unk.	3	**8**
N	baby food-junior,dinner,chicken &	7-1/2 oz jar	213	109	16.0	1.28	4.0	**16**	266	213	300	190	243	51	unk.	unk.	unk.	0.06	**9**
	noodles		*0.47*	188.9	3.0	unk.	unk.		100	351	83	173	47	153	unk.	unk.	unk.	3	**6**
O	baby food-junior,dinner,high	4-1/2 oz jar	128	109	6.8	0.38	8.1	**32**	495	338	594	303	398	87	unk.	unk.	unk.	0.05	**7**
	meat/cheese,beef w/vegetables		*0.78*	106.5	5.9	unk.	unk.		242	599	225	307	64	230	unk.	unk.	unk.	2	**6**
P	baby food-junior,dinner,high	4-1/2 oz jar	128	118	5.4	0.26	9.0	**36**	588	417	694	348	442	78	unk.	unk.	unk.	0.04	**6**
	meat/cheese,chicken w/vegetables		*0.78*	105.9	7.0	unk.	unk.		261	705	237	375	93	261	unk.	unk.	unk.	1	**4**
Q	baby food-junior,dinner,high	4-1/2 oz jar	128	99	7.8	0.26	8.2	**33**	502	360	636	288	378	93	0.60	1.43	0.54	0.14	**20**
	meat/cheese,ham w/vegetables		*0.78*	107.0	4.2	unk.	unk.		291	608	201	317	79	227	1.78	0.42	23	2	**6**
R	baby food-junior,dinner,high	4-1/2 oz jar	128	115	7.6	0.26	7.5	**30**	499	371	599	311	385	78	unk.	unk.	unk.	0.01	**2**
	meat/cheese,turkey w/vegetables		*0.78*	105.6	6.4	unk.	unk.		261	604	197	320	82	243	unk.	unk.	unk.	2	**4**
S	baby food-junior,dinner,high	4-1/2 oz jar	128	104	9.3	0.38	8.2	**33**	unk.	342	620	332	385	65	unk.	unk.	unk.	0.03	**4**
	meat/cheese,veal w/vegetables		*0.78*	unk.	3.6	unk.	unk.		269	591	207	320	96	242	unk.	unk.	unk.	2	**6**

Riboflavin mg / Niacin mg	% USRDA	Vitamin B_6 mg / Folacin mcg	% USRDA	Vitamin B_{12} mcg / Pantothenic acid mg	% USRDA	Biotin mg / Vitamin A IU	% USRDA	Preformed A RE / Beta carotene RE	Vitamin D IU / Vitamin E IU	% USRDA	Total tocopherol mg / Alpha tocopherol mg	Other tocopherol mg / Total ash g	Calcium mg / Phosphorus mg	% USRDA	Sodium mg / Sodium meq	Potassium mg / Potassium meq	Chlorine mg / Chlorine meq	Iron mg / Magnesium mg	% USRDA	Zinc mg / Copper mg	% USRDA	Iodine mcg / Selenium mcg	% USRDA	Manganese mcg / Chromium mcg	
0.11	**14**	unk.	**unk.**	0(a)	**0**	unk.	**unk.**	unk.	unk.	**unk.**	unk.	unk.	35	**4**	20	117	unk.	0.4	**4**	unk.	**unk.**	unk.	**unk.**	unk.	A
0.1	**1**	1	**4**	unk.	**unk.**	35	**1**	unk.	unk.	**unk.**	unk.	0.44	unk.	**unk.**	0.9	3.0	unk.	unk.	**unk.**	unk.	**unk.**	unk.		unk.	
0.15	**19**	unk.	**unk.**	unk.	**unk.**	unk.	**unk.**	unk.	unk.	**unk.**	unk.	unk.	117	**15**	60	124	unk.	0.3	**3**	unk.	**unk.**	unk.	**unk.**	unk.	B
0.1	**1**	2	**10**	unk.	**unk.**	70	**3**	unk.	unk.	**unk.**	unk.	0.85	unk.	**unk.**	2.6	3.2	unk.	unk.	**unk.**	unk.	**unk.**	unk.		unk.	
0.02	**3**	0.03	**4**	unk.	**unk.**	unk.	**unk.**	unk.	unk.	**unk.**	unk.	unk.	11	**1**	33	73	unk.	0.4	**4**	0.1	**2**	unk.	**unk.**	110	C
0.1	**1**	1	**3**	unk.	**unk.**	440	**18**	unk.	unk.	**unk.**	unk.	0.44	15	**2**	1.4	1.9	unk.	9	**5**	0.04	**4**	unk.		unk.	
0.24	**30**	0.03	**4**	unk.	**unk.**	unk.	**unk.**	unk.	unk.	**unk.**	unk.	unk.	134	**17**	55	196	unk.	0.9	**9**	0.7	**9**	unk.	**unk.**	unk.	D
0.2	**3**	11	**53**	unk.	**unk.**	101	**4**	unk.	unk.	**unk.**	unk.	1.32	112	**14**	2.4	5.0	unk.	22	**11**	unk.	**unk.**	unk.		unk.	
0.11	**14**	0.02	**3**	0.15	**5**	unk.	**unk.**	unk.	unk.	**unk.**	unk.	unk.	68	**9**	112	92	unk.	0.3	**3**	0.5	**7**	unk.	**unk.**	44	E
0.1	**1**	11	**56**	unk.	**unk.**	35	**1**	unk.	unk.	**unk.**	unk.	0.44	86	**11**	4.9	2.4	unk.	10	**5**	tr	**0**	unk.		unk.	
0.02	**3**	0.03	**4**	unk.	**unk.**	unk.	**unk.**	unk.	unk.	**unk.**	unk.	unk.	9	**1**	35	81	unk.	0.4	**4**	tr	**0**	unk.	**unk.**	tr	F
0.1	**1**	2	**8**	unk.	**unk.**	110	**4**	unk.	unk.	**unk.**	unk.	0.22	9	**1**	1.5	2.1	unk.	6	**3**	0.02	**2**	unk.		unk.	
0.02	**3**	0.07	**10**	0(a)	**0**	unk.	**unk.**	unk.	unk.	**unk.**	unk.	unk.	20	**3**	29	209	unk.	0.5	**5**	0.1	**1**	unk.	**unk.**	264	G
0.3	**3**	8	**39**	unk.	**unk.**	526	**21**	unk.	unk.	**unk.**	unk.	0.66	18	**2**	1.2	5.3	unk.	18	**9**	0.02	**2**	unk.		unk.	
0.04	**6**	0.02	**2**	0(a)	**0**	unk.	**unk.**	unk.	unk.	**unk.**	unk.	unk.	9	**1**	20	123	unk.	0.2	**2**	0.1	**2**	unk.	**unk.**	tr	H
0.6	**6**	2	**12**	unk.	**unk.**	312	**13**	unk.	unk.	**unk.**	unk.	0.44	13	**2**	0.9	3.1	unk.	9	**4**	0.02	**2**	unk.		unk.	
0.07	**8**	unk.	**unk.**	0(a)	**0**	unk.	**unk.**	unk.	unk.	**unk.**	unk.	unk.	24	**3**	20	205	unk.	0.7	**7**	unk.	**unk.**	unk.	**unk.**	unk.	I
0.6	**7**	4	**21**	unk.	**unk.**	431	**17**	unk.	unk.	**unk.**	unk.	0.88	unk.	**unk.**	0.9	5.2	unk.	unk.	**unk.**	unk.	**unk.**	unk.		unk.	
0.11	**14**	0.09	**13**	0.13	**4**	unk.	**unk.**	unk.	unk.	**unk.**	unk.	unk.	75	**9**	48	198	unk.	0.4	**4**	0.4	**5**	unk.	**unk.**	unk.	J
0.3	**3**	12	**62**	unk.	**unk.**	81	**3**	unk.	unk.	**unk.**	unk.	1.10	66	**8**	2.1	5.1	unk.	20	**10**	unk.	**unk.**	unk.		unk.	
0.07	**8**	unk.	**unk.**	0(a)	**0**	unk.	**unk.**	unk.	unk.	**unk.**	unk.	unk.	22	**3**	15	128	unk.	0.6	**6**	unk.	**unk.**	unk.	**unk.**	unk.	K
0.2	**2**	unk.	**unk.**	unk.	**unk.**	44	**2**	unk.	unk.	**unk.**	unk.	0.44	unk.	**unk.**	0.7	3.3	unk.	unk.	**unk.**	unk.	**unk.**	unk.		unk.	
0.18	**22**	0.04	**6**	unk.	**unk.**	unk.	**unk.**	unk.	unk.	**unk.**	unk.	unk.	123	**15**	64	136	unk.	0.6	**6**	0.5	**6**	unk.	**unk.**	tr	L
0.1	**1**	14	**68**	0.56	**11**	79	**3**	unk.	unk.	**unk.**	unk.	1.10	99	**12**	2.8	3.5	unk.	14	**7**	0.04	**4**	unk.		unk.	
0.09	**11**	0.07	**9**	0.21	**7**	unk.	**unk.**	unk.	unk.	**unk.**	unk.	unk.	17	**2**	36	98	unk.	0.9	**9**	0.7	**9**	unk.	**unk.**	tr	M
1.2	**14**	12	**59**	0.49	**10**	1397	**56**	unk.	unk.	**unk.**	unk.	1.06	64	**8**	1.6	2.5	unk.	16	**8**	0.02	**2**	unk.		unk.	
0.06	**8**	0.06	**8**	unk.	**unk.**	unk.	**unk.**	unk.	unk.	**unk.**	unk.	unk.	36	**5**	36	75	unk.	0.8	**8**	0.5	**6**	unk.	**unk.**	tr	N
1.1	**12**	11	**56**	unk.	**unk.**	1906	**76**	unk.	unk.	**unk.**	unk.	0.85	51	**6**	1.6	1.9	unk.	12	**6**	0.02	**2**	unk.		unk.	
0.10	**13**	0.11	**16**	0.74	**25**	unk.	**unk.**	unk.	unk.	**unk.**	unk.	unk.	15	**2**	42	191	unk.	1.0	**10**	1.4	**17**	unk.	**unk.**	tr	O
1.8	**20**	8	**42**	0.35	**7**	1014	**41**	unk.	unk.	**unk.**	unk.	0.77	67	**8**	1.8	4.9	unk.	12	**6**	0.05	**5**	unk.		unk.	
0.09	**11**	0.05	**7**	0.20	**7**	unk.	**unk.**	unk.	unk.	**unk.**	unk.	unk.	55	**7**	33	79	unk.	0.9	**10**	0.9	**12**	unk.	**unk.**	tr	P
1.3	**14**	1	**7**	0.44	**9**	1071	**43**	unk.	unk.	**unk.**	unk.	0.64	69	**9**	1.4	2.0	unk.	9	**5**	0.05	**5**	unk.		unk.	
0.12	**14**	0.12	**17**	0.36	**12**	unk.	**unk.**	unk.	unk.	**unk.**	unk.	unk.	13	**2**	28	207	unk.	0.8	**8**	1.1	**14**	unk.	**unk.**	tr	Q
1.5	**17**	8	**42**	0.51	**10**	333	**13**	unk.	unk.	**unk.**	unk.	0.77	72	**9**	1.2	5.3	unk.	14	**7**	0.08	**8**	unk.		unk.	
0.09	**11**	0.05	**8**	0.58	**19**	unk.	**unk.**	unk.	unk.	**unk.**	unk.	unk.	91	**11**	55	137	unk.	1.0	**10**	1.2	**15**	unk.	**unk.**	tr	R
1.0	**12**	13	**63**	unk.	**unk.**	810	**32**	unk.	unk.	**unk.**	unk.	0.77	81	**10**	2.4	3.5	unk.	7	**4**	0.09	**9**	unk.		unk.	
0.09	**11**	0.09	**13**	unk.	**unk.**	unk.	**unk.**	unk.	unk.	**unk.**	unk.	unk.	6	**1**	289	164	unk.	0.6	**6**	1.4	**17**	unk.	**unk.**	tr	S
2.0	**22**	unk.	**unk.**	unk.	**unk.**	470	**19**	unk.	unk.	**unk.**	unk.	unk.	87	**11**	12.6	4.2	unk.	14	**7**	0.14	**14**	unk.		unk.	

	FOOD	Portion	Weight in grams / *Conversion for 100 g*	Kilocalories / H_2O g	Total carbohydrate g / Total fats g	Crude fiber g / Dietary fiber g	Total protein g / Total sugar g	% USRDA	Arginine mg / Histidine mg	Isoleucine mg / Leucine mg	Lysine mg / Methionine mg	Phenylalanine mg / Threonine mg	Valine mg / Tryptophan mg	Cystine mg / Tyrosine mg	Polyunsat. fatty acids g / Monounsat. fatty acids g	Saturated fatty acids g / P/S ratio	Linoleic acid g / Cholesterol mg	Thiamin mg / Ascorbic acid mg	% USRDA
A	baby food-junior,dinner,lamb &	7-1/2 oz jar	213	138	18.5	unk.	4.9	**20**	unk.	unk.	unk.	unk.	unk.	unk.	unk.	unk.	unk.	0.09	**12**
	noodles		*0.47*	184.2	4.7	unk.	unk.		unk.	unk.	unk.	unk.	unk.	unk.	unk.	unk.	unk.	4	**10**
B	baby food-junior,dinner,macaroni &	7-1/2 oz jar	213	160	18.3	unk.	5.3	**21**	unk.	unk.	unk.	unk.	unk.	unk.	unk.	unk.	unk.	0.11	**15**
	bacon		*0.47*	181.3	7.0	unk.	unk.		unk.	unk.	unk.	unk.	unk.	unk.	unk.	unk.	unk.	4	**11**
C	baby food-junior,dinner,macaroni &	7-1/2 oz jar	213	130	17.5	0.21	5.5	**22**	217	277	315	281	313	68	unk.	unk.	unk.	0.13	**18**
	cheese		*0.47*	184.2	4.3	unk.	unk.		132	498	192	166	68	262	unk.	unk.	unk.	3	**7**
D	baby food-junior,dinner,macaroni	7-1/2 oz jar	213	128	18.1	unk.	6.8	**27**	unk.	unk.	unk.	unk.	unk.	unk.	unk.	unk.	unk.	0.13	**18**
	& ham		*0.47*	184.2	3.0	unk.	unk.		unk.	unk.	unk.	unk.	unk.	unk.	unk.	unk.	unk.	5	**12**
E	baby food-junior,dinner,macaroni,	7-1/2 oz jar	213	126	20.0	0.64	5.3	**21**	283	251	317	239	271	72	unk.	unk.	unk.	0.11	**15**
	tomato & beef		*0.47*	184.7	2.3	unk.	unk.		136	422	83	192	60	175	unk.	unk.	unk.	3	**8**
F	baby food-junior,dinner,mixed	7-1/2 oz jar	213	70	16.8	unk.	2.1	**9**	unk.	unk.	unk.	unk.	unk.	unk.	0(a)	0(a)	0(a)	0.02	**3**
	vegetables		*0.47*	193.0	0.0	unk.	unk.		unk.	unk.	unk.	unk.	unk.	unk.	0(a)	0.00	0(a)	7	**18**
G	baby food-junior,dinner,spaghetti,	7-1/2 oz jar	213	134	21.5	0.85	5.3	**21**	unk.	unk.	unk.	unk.	unk.	unk.	unk.	unk.	unk.	0.15	**21**
	tomato & meat		*0.47*	182.1	2.8	unk.	unk.		unk.	unk.	unk.	unk.	unk.	unk.	unk.	unk.	unk.	5	**12**
H	baby food-junior,dinner,split peas &	7-1/2 oz jar	213	151	24.1	0.64	7.0	**28**	624	285	479	326	328	60	unk.	unk.	unk.	0.11	**15**
	ham		*0.47*	177.9	2.8	unk.	unk.		192	515	115	262	72	245	unk.	unk.	unk.	4	**10**
I	baby food-junior,dinner,turkey & rice	7-1/2 oz jar	213	104	15.3	0.43	3.8	**15**	281	192	309	162	217	40	0.68	0.92	0.64	0.02	**3**
			0.47	190.2	3.0	unk.	unk.		92	307	104	158	40	138	1.02	0.74	unk.	3	**6**
J	baby food-junior,dinner,vegetables &	7-1/2 oz jar	213	151	16.2	0.43	3.8	**15**	258	175	232	164	207	55	0.89	2.98	0.83	0.11	**15**
	bacon		*0.47*	183.6	8.3	unk.	unk.		85	273	79	130	36	121	3.66	0.30	unk.	2	**6**
K	baby food-junior,dinner,vegetables &	7-1/2 oz jar	213	113	15.8	0.43	5.1	**21**	302	213	358	175	256	49	unk.	unk.	unk.	0.06	**9**
	beef		*0.47*	187.2	3.6	unk.	unk.		113	345	81	181	49	130	unk.	unk.	unk.	3	**8**
L	baby food-junior,dinner,vegetables &	7-1/2 oz jar	213	106	18.1	0.43	4.0	**16**	245	187	251	162	215	58	unk.	unk.	unk.	0.02	**3**
	chicken		*0.47*	187.9	2.3	unk.	unk.		79	283	75	143	43	124	unk.	unk.	unk.	3	**8**
M	baby food-junior,dinner,vegetables &	7-1/2 oz jar	213	111	14.9	0.43	5.1	**20**	343	232	354	194	271	55	unk.	unk.	unk.	0.09	**12**
	ham		*0.47*	188.3	3.6	unk.	unk.		141	381	109	194	55	158	unk.	unk.	unk.	3	**7**
N	baby food-junior,dinner,vegetables &	7-1/2 oz jar	213	109	15.1	0.43	4.5	**18**	300	202	328	177	217	38	unk.	unk.	unk.	0.04	**6**
	lamb		*0.47*	188.7	3.6	unk.	unk.		102	330	66	164	51	141	unk.	unk.	unk.	4	**9**
O	baby food-junior,dinner,vegetables &	7-1/2 oz jar	213	94	17.5	0.64	3.8	**15**	232	183	290	183	243	51	tr(a)	tr(a)	tr(a)	0.04	**6**
	liver		*0.47*	189.4	1.3	unk.	unk.		94	339	89	158	58	134	tr(a)	0.00	unk.	4	**10**
P	baby food-junior,dinner,vegetables &	7-1/2 oz jar	213	100	16.4	0.43	3.6	**15**	228	175	266	136	204	38	unk.	unk.	unk.	0.04	**6**
	turkey		*0.47*	189.6	2.6	unk.	unk.		77	281	81	145	38	124	unk.	unk.	unk.	2	**6**
Q	baby food-junior,dinner,vegetables,	7-1/2 oz jar	213	102	17.0	unk.	4.5	**18**	unk.	unk.	unk.	unk.	unk.	unk.	tr(a)	tr(a)	tr(a)	0.09	**12**
	dumplings & beef		*0.47*	188.7	1.7	unk.	unk.		unk.	unk.	unk.	unk.	unk.	unk.	tr(a)	0.00	unk.	2	**4**
R	baby food-junior,dinner,vegetables,	7-1/2 oz jar	213	136	19.4	0.43	3.6	**15**	unk.	unk.	unk.	unk.	unk.	unk.	unk.	unk.	unk.	0.09	**12**
	noodles & chicken		*0.47*	183.6	4.7	unk.	unk.		unk.	unk.	unk.	unk.	unk.	unk.	unk.	unk.	unk.	2	**4**
S	baby food-junior,dinner,vegetables,	7-1/2 oz jar	213	111	16.2	0.43	3.8	**15**	unk.	unk.	unk.	unk.	unk.	unk.	unk.	unk.	unk.	0.04	**6**
	noodles & turkey		*0.47*	188.9	3.2	unk.	unk.		unk.	unk.	unk.	unk.	unk.	unk.	unk.	unk.	unk.	2	**4**

Riboflavin mg / Niacin mg	% USRDA	Vitamin B_6 mg / Folacin mcg	% USRDA	Vitamin B_{12} mcg / Pantothenic acid mg	% USRDA	Biotin mg / Vitamin A IU	% USRDA	Preformed A RE / Beta carotene RE	Vitamin D IU / Vitamin E IU	% USRDA	Total tocopherol mg / Alpha tocopherol mg	Other tocopherol mg / Total ash g	Calcium mg / Phosphorus mg	% USRDA	Sodium mg / Sodium meq	Potassium mg / Potassium meq	Chlorine mg / Chlorine meq	Iron mg / Magnesium mg	% USRDA	Zinc mg / Copper mg	% USRDA	Iodine mcg / Selenium mcg	% USRDA	Manganese mcg / Chromium mcg	
0.15	**19**	unk.	**unk.**	unk.	**unk.**	unk.	**unk.**	unk.	unk.	**unk.**	unk.	unk.	38	**5**	38	164	unk.	0.8	**8**	unk.	**unk.**	unk.	**unk.**	unk.	A
1.4	**16**	unk.	**unk.**	unk.	**unk.**	1668	**67**	unk.	unk.	**unk.**	unk.	0.64	unk.	**unk.**	1.7	4.2	unk.	unk.	**unk.**	unk.	**unk.**	unk.		unk.	
0.17	**21**	unk.	**unk.**	unk.	**unk.**	unk.	**unk.**	unk.	unk.	**unk.**	unk.	unk.	151	**19**	166	179	unk.	0.8	**8**	unk.	**unk.**	unk.	**unk.**	unk.	B
1.3	**15**	unk.	**unk.**	unk.	**unk.**	2471	**99**	unk.	unk.	**unk.**	unk.	1.06	unk.	**unk.**	7.2	4.6	unk.	unk.	**unk.**	unk.	**unk.**	unk.		unk.	
0.13	**16**	0.03	**5**	0.06	**2**	unk.	**unk.**	unk.	unk.	**unk.**	unk.	unk.	109	**14**	162	94	unk.	0.6	**6**	0.8	**10**	unk.	**unk.**	106	C
1.2	**13**	3	**16**	unk.	**unk.**	28	**1**	unk.	unk.	**unk.**	unk.	1.49	126	**16**	7.0	2.4	unk.	15	**7**	0.07	**7**	unk.		unk.	
0.21	**27**	unk.	**unk.**	unk.	**unk.**	unk.	**unk.**	unk.	unk.	**unk.**	unk.	unk.	160	**20**	100	226	unk.	0.8	**8**	unk.	**unk.**	unk.	**unk.**	unk.	D
1.7	**19**	unk.	**unk.**	unk.	**unk.**	1120	**45**	unk.	unk.	**unk.**	unk.	1.06	unk.	**unk.**	4.3	5.8	unk.	unk.	**unk.**	unk.	**unk.**	unk.		unk.	
0.13	**16**	0.10	**14**	0.51	**17**	unk.	**unk.**	unk.	unk.	**unk.**	unk.	unk.	30	**4**	36	153	unk.	0.8	**8**	0.5	**7**	unk.	**unk.**	tr	E
1.6	**18**	unk.	**unk.**	unk.	**unk.**	1472	**59**	unk.	unk.	**unk.**	unk.	0.64	94	**12**	1.6	3.9	unk.	17	**8**	0.02	**2**	unk		unk.	
0.04	**5**	unk.	**unk.**	0(a)	**0**	unk.	**unk.**	unk.	unk.	**unk.**	unk.	unk.	36	**5**	19	239	unk.	0.7	**7**	unk.	**unk.**	unk.	**unk.**	unk.	F
0.9	**10**	14	**71**	unk.	**unk.**	5206	**208**	unk.	unk.	**unk.**	unk.	0.85	unk.	**unk.**	0.8	6.1	unk.	unk.	**unk.**	unk.	**unk.**	unk.		unk.	
0.15	**19**	0.13	**18**	unk.	**unk.**	unk.	**unk.**	unk.	unk.	**unk.**	unk.	unk.	38	**5**	43	230	unk.	1.2	**12**	0.7	**9**	unk.	**unk.**	tr	G
2.3	**26**	unk.	**unk.**	0.38	**8**	1476	**59**	unk.	unk.	**unk.**	unk.	1.28	79	**10**	1.8	5.9	unk.	28	**14**	0.06	**6**	unk.		unk.	
0.11	**13**	0.09	**13**	unk.	**unk.**	unk.	**unk.**	unk.	unk.	**unk.**	unk.	unk.	49	**6**	30	290	unk.	1.1	**11**	0.6	**8**	unk.	**unk.**	tr	H
1.0	**11**	unk.	**unk.**	unk.	**unk.**	1284	**51**	unk.	unk.	**unk.**	unk.	1.28	104	**13**	1.3	7.4	unk.	31	**16**	0.09	**9**	unk.		unk.	
0.06	**8**	0.06	**8**	unk.	**unk.**	unk.	**unk.**	unk.	unk.	**unk.**	unk.	unk.	49	**6**	32	72	unk.	0.6	**6**	0.6	**7**	unk.	**unk.**	tr	I
0.6	**7**	7	**33**	unk.	**unk.**	2256	**90**	unk.	unk.	**unk.**	unk.	0.85	36	**5**	1.4	1.8	unk.	9	**5**	0.04	**4**	unk.		unk.	
0.06	**8**	0.14	**20**	unk.	**unk.**	unk.	**unk.**	unk.	unk.	**unk.**	unk.	unk.	23	**3**	96	183	unk.	0.9	**9**	0.4	**5**	unk.	**unk.**	43	J
1.2	**13**	19	**96**	unk.	**unk.**	3357	**134**	unk.	unk.	**unk.**	unk.	1.28	81	**10**	4.2	4.7	unk.	24	**12**	0.04	**4**	unk.		unk.	
0.06	**8**	0.14	**20**	0.55	**19**	unk.	**unk.**	unk.	unk.	**unk.**	unk.	unk.	21	**3**	51	224	unk.	1.0	**10**	0.8	**10**	unk.	**unk.**	tr	K
1.4	**16**	10	**52**	0.27	**5**	3012	**121**	unk.	unk.	**unk.**	unk.	1.06	92	**11**	2.2	5.7	unk.	18	**9**	0.02	**2**	unk.		unk.	
0.04	**5**	0.08	**11**	unk.	**unk.**	unk.	**unk.**	unk.	unk.	**unk.**	unk.	unk.	30	**4**	19	55	unk.	0.6	**6**	3.2	**40**	unk.	**unk.**	64	L
0.7	**8**	8	**41**	0.50	**10**	2545	**102**	unk.	unk.	**unk.**	unk.	0.85	55	**7**	0.8	1.4	unk.	17	**8**	0.83	**83**	unk.		unk.	
0.04	**5**	0.07	**10**	unk.	**unk.**	unk.	**unk.**	unk.	unk.	**unk.**	unk.	unk.	17	**2**	38	196	unk.	0.5	**5**	0.4	**5**	unk.	**unk.**	tr	M
0.7	**8**	11	**56**	unk.	**unk.**	1310	**52**	unk.	unk.	**unk.**	unk.	1.06	55	**7**	1.7	5.0	unk.	12	**6**	0.04	**4**	unk.		unk.	
0.06	**8**	0.09	**13**	0.34	**11**	unk.	**unk.**	unk.	unk.	**unk.**	unk.	unk.	28	**4**	28	202	unk.	0.7	**7**	0.4	**6**	unk.	**unk.**	tr	N
1.2	**13**	8	**36**	0.34	**7**	3159	**126**	unk.	unk.	**unk.**	unk.	1.06	104	**13**	1.2	5.2	unk.	16	**8**	0.02	**2**	unk.		unk.	
0.49	**61**	0.21	**30**	unk.	**unk.**	unk.	**unk.**	unk.	unk.	**unk.**	unk.	unk.	21	**3**	28	190	unk.	3.9	**39**	1.0	**13**	unk.	**unk.**	85	O
2.5	**28**	68	**341**	unk.	**unk.**	8369	**335**	unk.	unk.	**unk.**	unk.	1.06	81	**10**	1.2	4.8	unk.	19	**10**	0.17	**17**	unk.		unk.	
0.04	**5**	0.05	**8**	unk.	**unk.**	unk.	**unk.**	unk.	unk.	**unk.**	unk.	unk.	28	**4**	36	53	unk.	0.7	**7**	0.6	**7**	unk.	**unk.**	43	P
0.5	**6**	6	**31**	0.45	**9**	1908	**76**	unk.	unk.	**unk.**	unk.	0.85	41	**5**	1.6	1.4	unk.	12	**6**	0.04	**4**	unk.		unk.	
0.09	**11**	unk.	**unk.**	unk.	**unk.**	unk.	**unk.**	unk.	unk.	**unk.**	unk.	unk.	30	**4**	111	unk.	unk.	1.0	**10**	0.7	**9**	unk.	**unk.**	unk.	Q
1.0	**12**	16	**79**	unk.	**unk.**	1406	**56**	unk.	unk.	**unk.**	unk.	1.28	unk.	**unk.**	4.8	unk.	unk.	unk.	**unk.**	unk.	**unk.**	unk.		unk.	
0.09	**11**	0.05	**7**	0.19	**6**	unk.	**unk.**	unk.	unk.	**unk.**	unk.	unk.	55	**7**	55	126	unk.	1.0	**10**	0.7	**9**	unk.	**unk.**	unk.	R
1.4	**16**	7	**36**	0.43	**9**	2239	**90**	unk.	unk.	**unk.**	unk.	1.49	70	**9**	2.4	3.2	unk.	unk.	**unk.**	0.12	**12**	unk.		unk.	
0.09	**11**	0.04	**6**	0.26	**9**	unk.	**unk.**	unk.	unk.	**unk.**	unk.	unk.	68	**9**	36	155	unk.	0.6	**6**	0.6	**8**	unk.	**unk.**	unk.	S
0.6	**7**	6	**30**	unk.	**unk.**	2117	**85**	unk.	unk.	**unk.**	unk.	0.85	62	**8**	1.6	4.0	unk.	19	**10**	unk.	**unk.**	unk.		unk.	

	FOOD	Portion	Weight in grams / Conversion for 100 g	Kilocalories / H_2O g	Total carbohydrate g / Total fats g	Crude fiber g / Dietary fiber g	Total protein g / Total sugar g	% USRDA	Arginine mg / Histidine mg	Isoleucine mg / Leucine mg	Lysine mg / Methionine mg	Phenylalanine mg / Threonine mg	Valine mg / Tryptophan mg	Cystine mg / Tyrosine mg	Polyunsat. fatty acids g / Monounsat. fatty acids g	Saturated fatty acids g / P/S ratio	Linoleic acid g / Cholesterol mg	Thiamin mg / Ascorbic acid mg	% USRDA
A	BABY FOOD-JUNIOR,FRUIT,apple	7-3/4 oz jar	220	136	36.5	0.44	0.0	**0**	tr(a)	tr(a)	tr(a)	tr(a)	tr(a)	tr(a)	tr(a)	tr(a)	tr(a)	0.04	**6**
	blueberry		*0.45*	182.2	tr	unk.	unk.		tr(a)	tr(a)	tr(a)	tr(a)	tr(a)	tr(a)	tr(a)	0.00	0(a)	31	**76**
B	baby food-junior,fruit,apple	7-3/4 oz jar	220	128	34.1	unk.	0.4	**2**	tr(a)	tr(a)	tr(a)	tr(a)	tr(a)	tr(a)	tr(a)	tr(a)	tr(a)	0.02	**3**
	raspberry		*0.45*	184.8	0.4	unk.	unk.		tr(a)	tr(a)	tr(a)	tr(a)	tr(a)	tr(a)	tr(a)	0.00	0(a)	64	**159**
C	baby food-junior,fruit,applesauce	7-1/2 oz jar	213	79	21.9	1.28	0.0	**0**	0	0	0	0	0	0	0.00	0.00	0.00	0.02	**3**
			0.47	190.6	0.0	unk.	unk.		0	0	0	0	0	0	0.00	0.00	0(a)	81	**201**
D	baby food-junior,fruit,applesauce &	7-3/4 oz jar	220	103	27.3	1.54	0.4	**2**	tr(a)	tr(a)	tr(a)	tr(a)	tr(a)	tr(a)	tr(a)	tr(a)	tr(a)	0.02	**3**
	apricots		*0.45*	191.2	0.4	unk.	unk.		tr(a)	tr(a)	tr(a)	tr(a)	tr(a)	tr(a)	tr(a)	0.00	0(a)	39	**98**
E	baby food-junior,fruit,applesauce &	7-3/4 oz jar	220	106	29.0	unk.	0.7	**3**	tr(a)	tr(a)	tr(a)	tr(a)	tr(a)	tr(a)	0.00	0.00	0.00	0.04	**6**
	cherries		*0.45*	189.9	0.0	unk.	unk.		tr(a)	tr(a)	tr(a)	tr(a)	tr(a)	tr(a)	0.00	0.00	0(a)	51	**128**
F	baby food-junior,fruit,applesauce &	7-1/2 oz jar	213	83	22.4	unk.	0.2	**1**	tr(a)	tr(a)	tr(a)	tr(a)	tr(a)	tr(a)	tr(a)	tr(a)	tr(a)	0.04	**6**
	pineapple		*0.47*	189.8	0.2	unk.	unk.		tr(a)	tr(a)	tr(a)	tr(a)	tr(a)	tr(a)	tr(a)	0.00	0(a)	57	**143**
G	baby food-junior,fruit,apricots	7-3/4 oz jar	220	139	38.1	1.10	0.7	**3**	tr(a)	tr(a)	tr(a)	tr(a)	tr(a)	tr(a)	0.00	0.00	0.00	0.02	**3**
	w/tapioca		*0.45*	180.6	0.0	unk.	unk.		tr(a)	tr(a)	tr(a)	tr(a)	tr(a)	tr(a)	0.00	0.00	0(a)	39	**98**
H	baby food-junior,fruit,bananas &	7-3/4 oz jar	220	143	39.2	unk.	0.4	**2**	tr(a)	tr(a)	tr(a)	tr(a)	tr(a)	tr(a)	0.00	0.00	0.00	0.04	**6**
	pineapple w/tapioca		*0.45*	179.7	0.0	0.09	unk.		tr(a)	tr(a)	tr(a)	tr(a)	tr(a)	tr(a)	0.00	0.00	0(a)	42	**106**
I	baby food-junior,fruit,bananas	7-3/4 oz jar	220	147	39.2	0.44	0.9	**4**	tr(a)	tr(a)	tr(a)	tr(a)	tr(a)	tr(a)	tr(a)	tr(a)	tr(a)	0.04	**6**
	w/tapioca		*0.45*	179.3	0.4	unk.	unk.		tr(a)	tr(a)	tr(a)	tr(a)	tr(a)	tr(a)	tr(a)	0.00	0(a)	57	**141**
J	baby food-junior,fruit,peaches	7-3/4 oz jar	220	156	41.6	1.54	1.1	**4**	23	18	40	25	55	unk.	tr(a)	tr(a)	tr(a)	0.02	**3**
			0.45	176.2	0.4	unk.	unk.		23	39	42	37	5	28	tr(a)	0.00	0(a)	42	**104**
K	baby food-junior,fruit,pears	7-1/2 oz jar	213	92	24.7	unk.	0.6	**3**	unk.	tr(a)	tr(a)	16	tr(a)	tr(a)	tr(a)	tr(a)	tr(a)	0.02	**3**
			0.47	187.0	0.2	unk.	unk.		tr(a)	tr(a)	tr(a)	tr(a)	tr(a)	26	tr(a)	0.00	0(a)	47	**117**
L	baby food-junior,fruit,pears &	7-1/2 oz jar	213	94	24.3	0.64	0.6	**3**	tr(a)	tr(a)	tr(a)	tr(a)	tr(a)	tr(a)	tr(a)	tr(a)	tr(a)	0.04	**6**
	pineapple		*0.47*	187.0	0.4	0.22	unk.		tr(a)	tr(a)	tr(a)	tr(a)	tr(a)	tr(a)	tr(a)	0.00	0(a)	36	**90**
M	baby food-junior,fruit,plums	7-3/4 oz jar	220	163	44.9	unk.	0.2	**1**	tr(a)	tr(a)	tr(a)	tr(a)	tr(a)	tr(a)	0.00	0.00	0.00	0.02	**3**
	w/tapioca		*0.45*	174.2	0.0	unk.	unk.		tr(a)	tr(a)	tr(a)	tr(a)	tr(a)	tr(a)	0.00	0.00	0(a)	2	**4**
N	baby food-junior,fruit,prunes	7-3/4 oz jar	220	154	41.1	0.66	1.3	**5**	tr(a)	tr(a)	tr(a)	tr(a)	tr(a)	tr(a)	tr(a)	tr(a)	tr(a)	0.04	**6**
	w/tapioca		*0.45*	176.2	0.2	unk.	unk.		tr(a)	tr(a)	tr(a)	tr(a)	tr(a)	tr(a)	tr(a)	0.00	0(a)	2	**4**
O	BABY FOOD-JUNIOR,MEAT,beef	3-1/2 oz jar	99	105	0.0	unk.	14.3	**80**	unk.	652	1194	555	726	167	0.16	2.56	0.11	0.01	**1**
			1.01	79.1	4.9	unk.	unk.		487	1150	441	629	145	477	1.64	0.06	unk.	2	**5**
P	baby food-junior,meat,chicken	3-1/2 oz jar	99	148	0.0	unk.	14.5	**81**	947	686	1216	593	733	191	2.31	2.45	2.23	0.01	**1**
			1.01	75.2	9.5	unk.	0(a)		441	1125	390	653	165	466	3.91	0.94	unk.	1	**4**
Q	baby food-junior,meat,chicken sticks	1 stick	10	79	0.1	0.02	1.5	**8**	100	73	116	67	76	13	unk.	unk.	unk.	tr	**0**
			10.00	6.8	1.4	unk.	unk.		46	114	32	57	12	50	unk.	unk.	unk.	tr	**0**
R	baby food-junior,meat,ham	3-1/2 oz jar	99	124	0.0	unk.	14.9	**83**	1012	711	1270	570	771	184	0.90	2.22	0.81	0.14	**20**
			1.01	77.7	6.6	unk.	0(a)		509	1196	382	649	148	501	2.88	0.41	unk.	2	**5**
S	baby food-junior,meat,lamb	3-1/2 oz jar	99	111	0.0	unk.	15.0	**84**	986	707	1336	592	762	207	0.22	2.53	0.15	0.02	**3**
			1.01	78.8	5.1	unk.	0(a)		380	1188	471	686	149	527	1.94	0.09	unk.	2	**4**

Riboflavin mg / Niacin mg	% USRDA	Vitamin B_6 mg / Folacin mcg	% USRDA	Vitamin B_{12} mcg / Pantothenic acid mg	% USRDA	Biotin mg / Vitamin A IU	% USRDA	Preformed A RE / Beta carotene RE	Vitamin D IU / Vitamin E IU	% USRDA	Total tocopherol mg / Alpha tocopherol mg	Other tocopherol mg / Total ash g	Calcium mg / Phosphorus mg	% USRDA	Sodium mg / Sodium meq	Potassium mg / Potassium meq	Chlorine mg / Chlorine meq	Iron mg / Magnesium mg	% USRDA	Zinc mg / Copper mg	% USRDA	Iodine mcg / Selenium mcg	% USRDA	Manganese mcg / Chromium mcg	
0.09	11	0.09	13	0(a)	0	unk.	unk.	unk.	0(a)	0	unk.	unk.	11	1	29	143	unk.	0.9	9	unk.	unk.	unk.	unk.	unk.	A
0.2	2	8	39	unk.	unk.	92	4	unk.	unk.	unk.	unk.	0.44	15	2	1.2	3.7	unk.	unk.	unk.	unk.	unk.	unk.		unk.	
0.07	8	0.07	11	0(a)	0	unk.	unk.	unk.	0(a)	0	unk.	unk.	11	1	4	158	unk.	0.5	5	0.3	4	unk.	unk.	tr	B
0.2	2	7	36	0.19	4	66	3	unk.	unk.	unk.	unk.	0.44	18	2	0.2	4.0	unk.	7	4	0.18	18	unk.		unk.	
0.06	8	0.06	9	0(a)	0	unk.	unk.	unk.	0(a)	0	1.1	tr	11	1	4	164	unk.	0.5	5	127.8	0.1	unk.	unk.	tr	C
0.1	1	4	18	0.21	4	19	1	unk.	1.3	13	1.1	0.43	13	2	0.2	4.2	unk.	7	3	0.06	1	unk.		unk.	
0.07	8	0.07	9	0(a)	0	unk.	unk.	unk.	0(a)	0	unk.	unk.	13	2	7	240	unk.	0.6	6	0.1	1	unk.	unk.	tr	D
0.3	4	3	15	unk.	unk.	746	30	unk.	unk.	unk.	unk.	0.66	22	3	0.3	6.1	unk.	11	5	0.07	7	unk.		unk.	
0.11	14	unk.	unk.	0(a)	0	unk.	unk.	unk.	0(a)	0	unk.	unk.	20	3	7	213	unk.	0.9	9	unk.	unk.	unk.	unk.	unk.	E
0.2	2	1	4	unk.	unk.	90	4	unk.	unk.	unk.	unk.	0.66	unk.	unk.	0.3	5.5	unk.	unk.	unk.	unk.	unk.	unk.		unk.	
0.06	8	0.08	12	0(a)	0	unk.	unk.	unk.	0(a)	0	unk.	unk.	9	1	4	162	unk.	0.2	2	0.1	2	unk.	unk.	256	F
0.2	2	4	21	unk.	unk.	45	2	unk.	unk.	unk.	unk.	0.43	13	2	0.2	4.1	unk.	14	7	0.15	15	unk.		unk.	
0.02	3	0.06	9	0(a)	0	unk.	unk.	unk.	0(a)	0	1.8	0.3	18	2	13	275	unk.	0.6	6	0.1	1	unk.	unk.	tr	G
0.4	5	4	18	0.30	6	1591	64	unk.	2.2	22	1.5	0.88	22	3	0.6	7.0	unk.	13	7	0.02	2	unk.		unk.	
0.04	6	0.18	26	0(a)	0	unk.	unk.	unk.	0(a)	0	unk.	unk.	15	2	13	150	unk.	0.5	5	0.1	1	unk.	unk.	tr	H
0.4	4	12	59	unk.	unk.	90	4	unk.	unk.	unk.	unk.	0.44	11	1	0.6	3.8	unk.	21	11	0.04	4	unk.		unk.	
0.04	6	0.31	44	0(a)	0	unk.	unk.	unk.	0(a)	0	0.5	0(a)	18	2	20	238	unk.	0.7	7	0.1	1	unk.	unk.	tr	I
0.5	5	14	70	0.38	8	97	4	unk.	0.6	6	0.5	0.44	20	3	0.9	6.1	unk.	21	11	0.09	9	unk.		unk.	
0.07	8	0.04	6	0(a)	0	unk.	unk.	unk.	0(a)	0	3.1	0.2	11	1	11	341	unk.	0.6	6	0.1	1	unk.	unk.	tr	J
1.4	16	9	43	0.29	6	392	16	unk.	3.7	37	2.9	0.66	24	3	0.5	8.7	unk.	14	7	0.02	2	unk.		unk.	
0.06	8	0.02	3	0(a)	0	unk.	unk.	unk.	0(a)	0	1.5	0.3	17	2	4	245	unk.	0.5	5	0.1	2	unk.	unk.	tr	K
0.4	5	8	41	0.20	4	72	3	unk.	1.8	18	1.2	0.43	26	3	0.2	6.3	unk.	22	11	0.02	2	unk.		unk.	
0.04	5	0.03	4	0(a)	0	unk.	unk.	unk.	0(a)	0	unk.	unk.	21	3	2	251	15	0.4	5	0.1	1	unk.	unk.	341	L
0.4	4	6	31	unk.	unk.	68	3	unk.	unk.	unk.	unk.	0.64	21	3	0.1	6.4	0.4	25	13	0.02	2	unk.		unk.	
0.07	8	0.06	9	0(a)	0	unk.	unk.	unk.	0(a)	0	1.4	0.5	13	2	18	183	unk.	0.5	5	0.1	1	unk.	unk.	tr	M
0.5	5	2	10	0.25	5	207	8	unk.	1.7	17	0.8	0.44	13	2	0.8	4.7	unk.	12	6	0.07	7	unk.		unk.	
0.18	22	0.19	27	0(a)	0	unk.	unk.	unk.	0(a)	0	1.0	0.2	33	4	4	356	unk.	0.7	7	0.2	3	unk.	unk.	tr	N
1.2	13	tr	2	0.31	6	895	36	unk.	1.2	12	0.8	1.10	33	4	0.2	9.1	unk.	32	16	0.02	2	unk.		unk.	
0.16	20	0.12	17	1.46	49	unk.	unk.	unk.	0(a)	0	0.7	0.0	8	1	65	188	unk.	1.6	16	3.0	38	unk.	unk.	17	O
3.2	36	6	28	0.35	7	102	4	unk.	0.9	9	0.7	0.79	71	9	2.8	4.8	unk.	15	8	0.05	5	unk.		unk.	
0.16	20	0.19	27	unk.	unk.	unk.	unk.	unk.	unk.	unk.	unk.	unk.	54	7	50	121	unk.	1.0	10	1.2	15	unk.	unk.	20	P
3.4	38	11	55	0.72	14	184	7	unk.	unk.	unk.	unk.	0.79	89	11	2.2	3.1	unk.	16	8	0.07	7	unk.		unk.	
0.02	3	0.01	2	unk.	unk.	unk.	unk.	unk.	unk.	unk.	unk.	unk.	7	1	48	11	unk.	0.2	2	0.1	1	unk.	unk.	tr	Q
0.2	2	unk.	unk.	unk.	unk.	318	13	unk.	unk.	unk.	unk.	0.12	12	2	2.1	0.3	unk.	2	1	0.01	1	unk.		unk.	
0.19	24	0.20	28	unk.	unk.	unk.	unk.	unk.	unk.	unk.	unk.	unk.	5	1	66	208	unk.	1.0	10	2.3	29	unk.	unk.	15	R
2.8	31	2	10	0.53	11	32	1	unk.	unk.	unk.	unk.	0.99	88	11	2.9	5.3	unk.	18	9	0.07	7	unk.		unk.	
0.19	24	0.18	26	2.25	75	unk.	unk.	unk.	unk.	unk.	unk.	unk.	7	1	72	209	unk.	1.6	16	3.4	43	unk.	unk.	19	S
3.2	35	2	10	0.42	8	27	1	unk.	unk.	unk.	unk.	0.89	90	11	3.1	5.3	unk.	18	9	0.07	7	unk.		unk.	

	FOOD	Portion	Weight in grams / *Conversion for 100 g*	Kilocalories / H_2O g	Total carbohydrate g / Total fats g	Crude fiber g / Dietary fiber g	Total protein g / Total sugar g	**% USRDA**	Arginine mg / Histidine mg	Isoleucine mg / Leucine mg	Lysine mg / Methionine mg	Phenylalanine mg / Threonine mg	Valine mg / Tryptophan mg	Cystine mg / Tyrosine mg	Polyunsat. fatty acids g / Monounsat. fatty acids g	Saturated fatty acids g / P/S ratio	Linoleic acid g / Cholesterol mg	Thiamin mg / Ascorbic acid mg	**% USRDA**
A	baby food-junior,meat,meat sticks	1 stick	10	18	0.1	0.02	1.3	**7**	87	67	103	61	69	7	0.16	0.58	0.15	0.01	**1**
			10.00	6.9	1.5	unk.	unk.		46	104	31	58	9	52	0.59	0.27	unk.	tr	**1**
B	baby food-junior,meat,turkey	3-1/2 oz jar	99	128	0.0	unk.	15.3	**85**	969	764	1261	628	774	185	1.74	2.29	1.59	0.02	**3**
			1.01	76.7	7.0	unk.	0(a)		389	1213	472	674	158	536	2.35	0.76	unk.	2	**6**
C	baby food-junior,meat,turkey sticks	1 stick	10	18	0.1	0.05	1.4	**8**	88	63	118	61	65	12	unk.	unk.	unk.	tr	**0**
			10.00	7.0	1.4	unk.	unk.		37	107	30	55	10	48	unk.	unk.	unk.	tr	**0**
D	baby food-junior,meat,veal	3-1/2 oz jar	99	109	0.0	0.20	15.1	**84**	1004	682	1215	585	736	196	0.17	2.37	0.10	0.02	**3**
			1.01	79.0	4.9	unk.	0(a)		464	1168	334	632	174	484	1.86	0.07	unk.	2	**5**
E	BABY FOOD-JUNIOR,VEGETABLE,beans,	7-1/4 oz jar	206	51	11.7	1.85	2.5	**10**	136	111	122	101	134	21	tr(a)	tr(a)	tr(a)	0.04	**6**
	green		*0.49*	190.5	0.2	unk.	unk.		68	161	37	105	29	87	tr(a)	0.00	0(a)	17	**43**
F	baby food-junior,vegetable,beans,	7-1/4 oz jar	206	66	12.6	unk.	2.7	**11**	146	119	132	109	144	23	tr(a)	tr(a)	tr(a)	0.02	**3**
	green,buttered		*0.49*	187.9	1.9	unk.	unk.		74	175	39	113	31	95	tr(a)	0.00	unk.	18	**44**
G	baby food-junior,vegetable,beans,	7-1/2 oz jar	213	68	15.3	unk.	2.1	**9**	102	102	92	96	124	23	tr(a)	tr(a)	tr(a)	0.04	**6**
	green,creamed		*0.47*	193.6	0.9	unk.	unk.		49	166	47	81	32	92	tr(a)	0.00	0(a)	6	**14**
H	baby food-junior,vegetable,carrots	7-1/2 oz jar	213	68	15.3	1.70	1.7	**7**	111	51	45	53	66	13	tr(a)	tr(a)	tr(a)	0.04	**6**
			0.47	193.8	0.4	unk.	unk.		28	70	19	49	23	43	tr(a)	0.00	0(a)	12	**29**
I	baby food-junior,vegetable,carrots,	7-1/2 oz jar	213	70	14.3	unk.	1.7	**7**	111	51	45	53	66	13	tr(a)	tr(a)	tr(a)	0.04	**6**
	buttered		*0.47*	194.7	1.3	unk.	unk.		28	70	19	49	23	43	tr(a)	0.00	unk.	16	**41**
J	baby food-junior,vegetable,corn,	7-1/2 oz jar	213	138	34.7	0.21	3.0	**12**	128	138	175	104	166	38	tr(a)	tr(a)	tr(a)	0.02	**3**
	creamed		*0.47*	173.4	0.9	unk.	unk.		98	300	85	111	32	145	tr(a)	0.00	0(a)	5	**12**
K	baby food-junior,vegetable,mixed	7-1/2 oz jar	213	87	17.5	1.06	3.0	**12**	198	119	100	117	147	60	tr(a)	tr(a)	tr(a)	0.06	**9**
	vegetables		*0.47*	190.4	0.9	unk.	unk.		66	194	45	96	32	111	tr(a)	0.00	0(a)	5	**13**
L	baby food-junior,vegetable,peas,	7-1/4 oz jar	206	124	23.3	unk.	7.2	**29**	814	315	488	299	352	56	tr(a)	tr(a)	tr(a)	0.14	**21**
	buttered		*0.49*	172.0	2.7	unk.	unk.		157	490	84	284	70	249	tr(a)	0.00	unk.	26	**65**
M	baby food-junior,vegetable,spinach,	7-1/2 oz jar	213	89	13.6	1.06	6.4	**26**	388	288	379	247	386	79	tr(a)	tr(a)	tr(a)	0.04	**6**
	creamed		*0.47*	187.9	3.0	unk.	unk.		162	545	141	258	94	294	tr(a)	0.00	0(a)	8	**19**
N	baby food-junior,vegetable,squash	7-1/2 oz jar	213	51	11.9	1.49	1.7	**7**	100	70	66	62	77	15	tr(a)	tr(a)	tr(a)	0.02	**3**
			0.47	197.7	0.4	unk.	unk.		34	102	23	53	26	60	tr(a)	0.00	0(a)	17	**42**
O	baby food-junior,vegetable,squash,	7-1/2 oz jar	213	64	13.6	unk.	1.5	**6**	85	60	55	53	66	13	tr(a)	tr(a)	tr(a)	0.02	**3**
	buttered		*0.47*	195.5	1.3	unk.	unk.		28	87	19	45	21	51	tr(a)	0.00	unk.	16	**41**
P	baby food-junior,vegetable,sweet	7-3/4 oz jar	220	132	30.6	1.32	2.4	**10**	117	110	95	134	156	33	tr(a)	tr(a)	tr(a)	0.07	**9**
	potatoes		*0.45*	185.0	0.2	unk.	unk.		55	167	51	117	46	88	tr(a)	0.00	0(a)	21	**53**
Q	baby food-junior,vegetable,sweet	7-3/4 oz jar	220	125	26.8	unk.	1.8	**7**	84	79	68	97	114	24	tr(a)	tr(a)	tr(a)	0.04	**6**
	potatoes,buttered		*0.45*	188.3	1.5	unk.	unk.		40	121	37	84	33	64	tr(a)	0.00	unk.	20	**51**
R	BABY FOOD-TODDLER, dinner, beef & rice	6-1/4 oz jar	177	145	15.6	0.53	8.8	**32**	598	434	666	349	487	96	unk.	unk.	unk.	0.04	**5**
	rice		*0.56*	145.0	5.1	unk.	unk.		227	694	218	354	81	285	unk.	unk.	unk.	7	**17**
S	baby food-toddler,dinner,beef lasagna	6-1/4 oz jar	177	136	17.7	0.35	7.4	**27**	448	365	526	322	409	85	unk.	unk.	unk.	0.12	**18**
			0.56	145.7	3.7	unk.	unk.		181	573	145	289	83	234	unk.	unk.	unk.	3	**8**

Riboflavin mg / Niacin mg	% USRDA	Vitamin B$_6$ mg / Folacin mcg	% USRDA	Vitamin B$_{12}$ mcg / Pantothenic acid mg	% USRDA	Biotin mg / Vitamin A IU	% USRDA	Preformed A RE / Beta carotene RE	Vitamin D IU / Vitamin E IU	% USRDA	Total tocopherol mg / Alpha tocopherol mg	Other tocopherol mg / Total ash g	Calcium mg / Phosphorus mg	% USRDA	Sodium mg / Sodium meq	Potassium mg / Potassium meq	Chlorine mg / Chlorine meq	Iron mg / Magnesium mg	% USRDA	Zinc mg / Copper mg	% USRDA	Iodine mcg / Selenium mcg	% USRDA	Manganese mcg / Chromium mcg	
0.02	2	0.01	1	unk.	unk.	unk.	unk.	unk.	unk.	unk.	unk.	unk.	3	0	55	11	unk.	0.1	1	0.3	3	unk.	unk.	tr	A
0.1	2	unk.	unk.	unk.	unk.	7	0	unk.	unk.	unk.	unk.	0.14	10	1	2.4	0.3	unk.	1	1	0.01	1	unk.		unk.	
0.25	31	0.16	24	1.06	35	unk.	unk.	unk.	unk.	unk.	unk.	unk.	28	4	71	178	unk.	1.3	13	2.2	28	unk.	unk.	21	B
3.4	38	12	60	0.60	12	562	23	unk.	unk.	unk.	unk.	0.79	94	12	3.1	4.6	unk.	16	8	0.07	7	unk.		unk.	
0.02	2	0.01	1	unk.	unk.	unk.	unk.	unk.	unk.	unk.	unk.	unk.	7	1	48	9	unk.	0.1	1	0.2	3	unk.	unk.	6	C
0.2	2	unk.	unk.	unk.	unk.	23	1	unk.	unk.	unk.	unk.	0.09	10	1	2.1	0.2	unk.	1	1	0.01	1	unk.		unk.	
0.18	22	0.12	17	unk.	unk.	unk.	unk.	unk.	unk.	unk.	unk.	unk.	6	1	68	234	unk.	1.2	12	2.7	34	unk.	unk.	13	D
3.8	42	7	33	0.45	9	49	2	unk.	unk.	unk.	unk.	0.89	97	12	3.0	6.0	unk.	18	9	0.05	5	unk.		unk.	
0.21	26	0.07	10	0(a)	0	unk.	unk.	unk.	unk.	unk.	unk.	unk.	134	17	4	264	unk.	2.2	22	0.4	5	unk.	unk.	unk.	E
0.7	7	67	337	0.31	6	892	36	unk.	unk.	unk.	unk.	0.82	39	5	0.2	6.7	unk.	45	23	0.10	10	unk.		unk.	
0.23	28	unk.	unk.	unk.	unk.	unk.	unk.	unk.	unk.	unk.	unk.	unk.	142	18	4	352	unk.	2.4	24	unk.	unk.	unk.	unk.	unk.	F
0.7	7	56	281	unk.	unk.	787	32	unk.	unk.	unk.	unk.	1.03	unk.	unk.	0.2	9.0	unk.	unk.	unk.	unk.	unk.	unk.		unk.	
0.11	13	0.03	4	tr(a)	0	unk.	unk.	unk.	unk.	unk.	unk.	unk.	68	9	26	138	unk.	0.6	6	0.3	4	unk.	unk.	277	G
0.5	5	unk.	unk.	unk.	unk.	319	13	unk.	unk.	unk.	unk.	1.06	41	5	1.1	3.5	unk.	26	13	0.04	4	unk.		unk.	
0.09	11	0.17	24	0(a)	0	unk.	unk.	unk.	unk.	unk.	unk.	unk.	49	6	104	430	unk.	0.8	8	0.3	4	unk.	unk.	64	H
1.1	12	37	184	0.59	12	25155	1006	unk.	unk.	unk.	unk.	1.49	43	5	4.5	11.0	unk.	22	11	0.02	2	unk.		unk.	
0.13	16	unk.	unk.	unk.	unk.	unk.	unk.	unk.	unk.	unk.	unk.	unk.	75	9	34	309	unk.	0.7	7	unk.	unk.	unk.	unk.	unk.	I
1.1	12	18	92	unk.	unk.	20959	838	unk.	unk.	unk.	unk.	1.28	unk.	unk.	1.5	7.9	unk.	unk.	unk.	unk.	unk.	unk.		unk.	
0.11	13	0.09	13	0(a)	0	unk.	unk.	unk.	unk.	unk.	unk.	unk.	38	5	111	173	unk.	0.6	6	0.4	6	unk.	unk.	tr	J
1.1	12	27	135	tr	0	164	7	unk.	unk.	unk.	unk.	1.06	70	9	4.8	4.4	unk.	21	11	0.03	3	unk.		unk.	
0.06	8	0.17	25	0(a)	0	unk.	unk.	unk.	unk.	unk.	unk.	unk.	23	3	77	362	unk.	0.9	9	0.5	6	unk.	unk.	128	K
1.4	16	9	44	0.55	11	8935	357	unk.	unk.	unk.	unk.	1.28	53	7	3.3	9.3	unk.	27	14	0.11	11	unk.		unk.	
0.16	21	unk.	unk.	unk.	unk.	unk.	unk.	unk.	unk.	unk.	unk.	unk.	93	12	10	241	unk.	2.1	21	unk.	unk.	unk.	unk.	unk.	L
2.8	32	75	373	unk.	unk.	845	34	unk.	unk.	unk.	unk.	0.82	unk.	unk.	0.4	6.2	unk.	unk.	unk.	unk.	unk.	unk.		unk.	
0.19	24	0.12	18	0(a)	0	unk.	unk.	unk.	unk.	unk.	unk.	unk.	241	30	117	471	unk.	3.0	30	0.8	10	unk.	unk.	809	M
0.6	6	147	733	unk.	unk.	7830	313	unk.	unk.	unk.	unk.	2.13	104	13	5.1	12.0	unk.	92	46	0.17	17	unk.		unk.	
0.15	19	0.15	21	0(a)	0	unk.	unk.	unk.	unk.	unk.	unk.	unk.	51	6	2	394	unk.	0.7	8	0.2	2	unk.	unk.	unk.	N
0.8	9	33	164	0.47	9	4290	172	unk.	unk.	unk.	unk.	1.28	34	4	0.1	10.1	unk.	26	13	0.12	12	unk.		unk.	
0.15	19	unk.	unk.	unk.	unk.	unk.	unk.	unk.	unk.	unk.	unk.	unk.	66	8	4	288	unk.	0.9	9	0.2	3	unk.	unk.	21	O
0.7	8	25	125	unk.	unk.	3259	130	unk.	unk.	unk.	unk.	1.06	unk.	unk.	0.2	7.3	unk.	27	13	0.02	2	unk.		unk.	
0.07	8	0.25	36	0(a)	0	unk.	unk.	unk.	unk.	unk.	unk.	unk.	35	4	48	535	unk.	0.9	9	0.2	3	unk.	unk.	1100	P
0.8	9	2	11	0.90	18	14599	584	unk.	unk.	unk.	unk.	1.54	53	7	2.1	13.7	unk.	49	25	0.02	2	unk.		unk.	
0.11	14	unk.	unk.	unk.	unk.	unk.	unk.	unk.	unk.	unk.	unk.	unk.	62	8	18	475	unk.	0.9	9	unk.	unk.	unk.	unk.	unk.	Q
0.7	7	29	147	unk.	unk.	13295	532	unk.	unk.	unk.	unk.	1.54	unk.	unk.	0.8	12.1	unk.	unk.	unk.	unk.	unk.	unk.		unk.	
0.12	16	0.25	35	unk.	unk.	unk.	unk.	unk.	unk.	unk.	unk.	unk.	19	2	632	212	unk.	1.2	12	1.3	17	unk.	unk.	0	R
2.4	26	unk.	unk.	unk.	unk.	889	36	unk.	unk.	unk.	unk.	2.48	62	8	27.5	5.4	unk.	25	12	0.09	9	unk.		unk.	
0.16	20	0.13	18	unk.	unk.	unk.	unk.	unk.	unk.	unk.	unk.	unk.	32	4	804	216	unk.	1.5	15	1.2	15	unk.	unk.	142	S
2.4	27	unk.	unk.	unk.	unk.	2059	82	unk.	unk.	unk.	unk.	2.48	71	9	34.9	5.5	unk.	17	9	0.09	9	unk.		unk.	

	FOOD	Portion	Weight in grams / Conversion for 100 g	Kilocalories / H_2O g	Total carbohydrate g / Total fats g	Crude fiber g / Dietary fiber g	Total protein g / Total sugar g	% USRDA	Arginine mg / Histidine mg	Isoleucine mg / Leucine mg	Lysine mg / Methionine mg	Phenylalanine mg / Threonine mg	Valine mg / Tryptophan mg	Cystine mg / Tyrosine mg	Polyunsat. fatty acids g / Monounsat. fatty acids g	Saturated fatty acids g / P/S ratio	Linoleic acid g / Cholesterol mg	Thiamin mg / Ascorbic acid mg	% USRDA
A	baby food-toddler,dinner,beef stew	6-1/4 oz jar	177	90	9.7	0.53	9.0	**32**	625	435	697	356	481	94	0.19	1.03	0.16	0.02	**3**
			0.56	153.8	2.1	unk.	unk.		227	687	251	372	94	281	0.71	0.19	22	5	**13**
B	baby food-toddler,dinner,chicken stew	6 oz jar	170	133	10.9	0.51	8.8	**32**	563	437	697	362	496	88	1.33	1.87	1.24	0.05	**7**
			0.59	141.6	6.3	unk.	unk.		218	688	190	376	97	292	2.62	0.71	49	3	**8**
C	baby food-toddler,dinner,spaghetti,	6-1/4 oz jar	177	133	19.1	0.71	9.4	**34**	503	474	554	427	506	117	unk.	unk.	unk.	0.11	**15**
	tomato & meat		*0.56*	144.4	1.8	unk.	unk.		250	731	184	349	115	326	unk.	unk.	unk.	7	**18**
D	baby food-toddler,dinner,vegetables	6-1/4 oz jar	177	127	14.0	unk.	7.4	**27**	471	370	540	320	405	83	0.30	1.86	0.27	0.07	**10**
	& ham		*0.56*	148.0	5.3	unk.	unk.		202	575	143	301	101	251	2.32	0.16	14	7	**16**
E	baby food-toddler,dinner,vegetables	6-1/4 oz jar	177	142	14.2	0.88	8.5	**30**	522	441	586	354	492	92	unk.	unk.	unk.	0.04	**5**
	& turkey		*0.56*	145.7	6.0	unk.	unk.		202	689	156	329	106	308	unk.	unk.	unk.	6	**15**
F	BACON FAT	1 Tbsp	15	135	0.0	tr(a)	0.0	**0**	0(a)	0(a)	0(a)	0(a)	0(a)	0(a)	1.68	5.88	1.50	0.00	**0**
			6.67	unk.	15.0	0.00	0.0		0(a)	0(a)	0(a)	0(a)	0(a)	0(a)	6.18	0.29	14	0	**0**
G	BACON,any,cooked	2 slices	15	92	0.5	0.00	4.6	**10**	292	169	249	184	184	53	unk.	2.55	0.75	0.08	**5**
			6.67	1.2	7.8	0(a)	0.0		123	309	60	131	41	117	3.75	unk.	unk.	0	**0**
H	bacon,Canadian,broiled or fried,drained	2 slices	42	116	0.1	0.00	11.6	**26**	736	618	1042	474	646	129	1.13	2.49	0.60	0.39	**26**
			2.38	21.0	7.3	0(a)	0.1		381	958	302	508	120	415	3.06	0.45	37	unk.	**unk.**
I	bacon,Canadian,fried-Oscar Mayer	2 slices	42	59	0.0	0(a)	8.4	**19**	560	451	712	368	457	99	0.21	0.84	0.17	0.29	**20**
			2.38	29.0	2.9	0(a)	0.3		290	665	238	432	120	316	1.18	0.25	18	11	**19**
J	BACOS-General Mills	1 Tbsp	9.0	36	2.3	unk.	3.7	**6**	unk.	unk.	unk.	unk.	unk.	unk.	unk.	unk.	unk.	1.14	**76**
			11.11	unk.	1.5	unk.	0.0		unk.	unk.	unk.	unk.	unk.	unk.	unk.	unk.	0(a)	tr	**0**
K	BAGEL,egg,home recipe	1 average	51	126	21.3	0.08	3.5	**6**	34	169	122	191	173	67	0.30	1.45	0.26	0.13	**9**
			1.95	21.8	2.7	0.79	1.6		84	284	59	121	45	118	0.75	0.21	26	0	**0**
L	bagel,water,home recipe	1 average	52	119	21.3	0.08	3.0	**5**	22	144	86	167	140	56	0.23	1.25	0.19	0.13	**9**
			1.92	23.6	2.2	0.79	1.6		67	242	42	96	37	96	0.51	0.18	5	0	**0**
M	BAKING POWDER (sodium aluminum	1 tsp	3.6	5	1.1	0.00	tr	**0**	tr(a)	tr(a)	tr(a)	tr(a)	tr(a)	tr(a)	tr(a)	tr(a)	tr(a)	0.00	**0**
	sulfate)		*27.78*	0.1	tr	0(a)	0.0		tr(a)	tr(a)	tr(a)	tr(a)	tr(a)	tr(a)	tr(a)	0.00	0	0	**0**
N	baking powder,low sodium	1 tsp	3.6	6	1.5	0.00	tr	**0**	tr(a)	tr(a)	tr(a)	tr(a)	tr(a)	tr(a)	0(a)	0(a)	0(a)	0.00	**0**
			27.78	0.1	tr	0(a)	0.0		tr(a)	tr(a)	tr(a)	tr(a)	tr(a)	tr(a)	0(a)	0.00	0	0	**0**
O	BAKING SODA	1 tsp	4.1	0	0.0	0.00	0.0	**0**	0(a)	0(a)	0(a)	0(a)	0(a)	0(a)	0.00	0.00	0.00	0(a)	**0**
			24.40	unk.	0.0	0(a)	0.0		0(a)	0(a)	0(a)	0(a)	0(a)	0(a)	0.00	0.00	0	0(a)	**0**
P	BAKLAVA,home recipe	1-1/4x1-1/4x	32	126	9.8	0.19	2.0	**4**	126	97	74	115	114	39	0.89	3.89	0.77	0.04	**3**
		1-3/4"	*3.13*	10.6	9.2	0.57	4.9		52	161	33	69	22	72	3.67	0.23	23	1	**2**
Q	BANANA,yellow,raw,edible portion	1 average	119	101	26.4	0.59	1.3	**2**	75	52	57	55	55	36	unk.	tr(a)	unk.	0.06	**4**
			0.84	90.1	0.2	4.05	22.4		80	67	17	40	15	34	unk.	unk.	0	12	**20**
R	banana,yellow,raw,sliced,	1/2 C	75	64	16.6	0.38	0.8	**1**	48	33	36	34	34	22	unk.	tr(a)	unk.	0.04	**3**
	edible portion		*1.33*	56.8	0.1	2.55	14.1		50	42	11	25	9	21	unk.	unk.	0	8	**13**
S	BASIL,ground	1 tsp	1.5	4	0.9	0.27	0.2	**0**	10	9	tr(a)	tr(a)	11	2	unk.	unk.	unk.	tr	**0**
			66.67	0.1	0.1	unk.	0.0		tr(a)	tr(a)	tr(a)	9	3	6	unk.	unk.	0	1	**2**

Riboflavin mg / Niacin mg	% USRDA	Vitamin B6 mg / Folacin mcg	% USRDA	Vitamin B12 mcg / Pantothenic acid mg	% USRDA	Biotin mg / Vitamin A IU	% USRDA	Preformed A RE / Beta carotene RE	Vitamin D IU / Vitamin E IU	% USRDA	Total tocopherol mg / Alpha tocopherol mg	Other tocopherol mg / Total ash g	Calcium mg / Phosphorus mg	% USRDA	Sodium mg / Sodium meq	Potassium mg / Potassium meq	Chlorine mg / Chlorine meq	Iron mg / Magnesium mg	% USRDA	Zinc mg / Copper mg	% USRDA	Iodine mcg / Selenium mcg	% USRDA	Manganese mcg / Chromium mcg	
0.12	16	0.13	19	unk.	unk.	unk.	unk.	unk.	unk.	unk.	unk.	unk.	16	2	611	251	unk.	1.3	13	1.8	22	unk.	unk.	71	A
2.3	26	unk.	unk.	unk.	unk.	2919	117	unk.	unk.	unk.	unk.	2.48	78	10	26.6	6.4	unk.	16	8	0.07	7	unk.		unk.	
0.12	15	0.08	11	unk.	unk.	unk.	unk.	unk.	unk.	unk.	unk.	unk.	61	8	683	156	unk.	1.1	11	0.8	10	unk.	unk.	99	B
2.0	22	unk.	unk.	unk.	unk.	1717	69	unk.	unk.	unk.	unk.	2.38	87	11	29.7	4.0	unk.	16	8	0.07	7	unk.		unk.	
0.18	22	0.15	21	unk.	unk.	unk.	unk.	unk.	unk.	unk.	unk.	unk.	39	5	634	289	unk.	1.6	16	1.1	14	unk.	unk.	177	C
2.8	31	unk.	unk.	unk.	unk.	784	31	unk.	unk.	unk.	unk.	2.30	80	10	27.6	7.4	unk.	24	12	0.12	12	unk.		unk.	
0.11	13	0.15	22	unk.	unk.	unk.	unk.	unk.	unk.	unk.	unk.	unk.	41	5	531	271	unk.	1.2	12	0.8	10	unk.	unk.	99	D
1.2	14	unk.	unk.	unk.	unk.	628	25	unk.	unk.	unk.	unk.	2.48	71	9	23.1	6.9	unk.	30	15	0.07	7	unk.		unk.	
0.16	20	0.11	15	unk.	unk.	unk.	unk.	unk.	unk.	unk.	unk.	unk.	81	10	591	294	unk.	1.1	11	0.8	10	unk.	unk.	159	E
1.0	12	unk.	unk.	unk.	unk.	3715	149	unk.	unk.	unk.	unk.	2.48	101	13	25.7	7.5	unk.	20	10	0.11	11	unk.		unk.	
0.00	0	unk.	unk.	tr(a)	0	tr(a)	0	0	0(a)	0	0.2	tr	0	0	0	0	tr(a)	0.0	0	tr	0	unk.	unk.	tr(a)	F
0.0	0	tr(a)	0	tr(a)	0	0	0	0	0.2	1	0.2	tr(a)	0	0	0.0	0.0	0.0	0	0	tr(a)	0	tr(a)		tr(a)	
0.05	3	unk.	unk.	unk.	unk.	tr(a)	0	0	0(a)	0	0.1	tr	2	0	153	35	unk.	0.5	3	unk.	unk.	unk.	unk.	unk.	G
0.8	4	tr(a)	0	tr(a)	0	0	0	0	0.1	0	0.1	0.94	34	3	6.7	0.9	unk.	4	1	unk.	unk.	tr(a)		tr(a)	
0.07	4	0.10	5	0.23	4	unk.	unk.	0	0	0	tr(a)	tr(a)	8	1	1073	181	unk.	1.7	10	1.7	11	unk.	unk.	6	H
2.1	11	5	1	0.14	1	0	0	0	0.0	0	tr(a)	1.97	92	9	46.7	4.6	unk.	10	3	0.03	2	tr(a)		5	
0.06	4	0.15	8	0.29	5	tr(a)	0	tr(a)	14	4	tr(a)	tr(a)	3	0	590	126	781	0.2	1	0.5	4	unk.	unk.	3	I
2.4	12	tr(a)	0	tr(a)	0	tr(a)	0	0(a)	0.0	0	tr(a)	1.81	86	9	25.7	3.2	22.0	7	2	0.02	1	tr(a)		1	
0.02	1	unk.	unk.	0(a)	0	unk.	unk.	0(a)	0(a)	0	unk.	unk.	23	2	305	134	unk.	0.9	5	unk.	unk.	unk.	unk.	unk.	J
0.2	1	unk.	unk.	unk.	unk.	tr	0	tr	unk.	unk.	unk.	unk.	23	2	13.3	3.4	unk.	unk.	unk.	unk.	unk.	unk.		unk.	
0.12	7	0.04	2	0.05	1	0.001	0	0	2	0	0.1	0.0	12	1	666	45	26	1.0	5	0.3	2	168.1	112	108	K
1.2	6	40	10	0.25	3	82	2	0	0.1	0	tr	1.90	46	5	29.0	1.1	0.7	21	5	0.06	3	14		16	
0.10	6	0.03	2	0.00	0	tr	0	0	0	0	tr	0.0	9	1	660	39	19	0.9	5	0.2	1	164.9	110	106	L
1.2	6	38	10	0.20	2	72	1	0	tr	0	tr	1.86	35	4	28.7	1.0	0.5	11	3	0.05	3	14		16	
0.00	0	0(a)	0	0(a)	0	0(a)	0	0	0(a)	0	0(a)	0(a)	70	7	394	5	unk.	0(a)	0	unk.	unk.	unk.	unk.	unk.	M
0.0	0	0(a)	0	0(a)	0	0	0	0	0.0	0	0(a)	2.26	105	10	17.1	0.1	unk.	unk.	unk.	unk.	unk.	unk.		unk.	
0.00	0	0(a)	0	0(a)	0	0(a)	0	0	0(a)	0	0(a)	0(a)	173	17	tr	394	unk.	0(a)	0	unk.	unk.	unk.	unk.	unk.	N
0.0	0	0(a)	0	0(a)	0	0	0	0	0.0	0	0(a)	unk.	263	26	0.0	10.1	unk.	unk.	unk.	unk.	unk.	unk.		unk.	
0(a)	0	0(a)	0	0(a)	0	0(a)	0	0(a)	0(a)	0	0(a)	0(a)	unk.	unk.	1123	unk.	unk.	unk.	unk.	unk.	unk.	unk.	unk.	unk.	O
0(a)	0	0(a)	0	0(a)	0	0(a)	0	0(a)	0(a)	0	0(a)	unk.	unk.	unk.	48.8	unk.	unk.	unk.	unk.	unk.	unk.	unk.		unk.	
0.08	5	0.01	1	0.03	1	0.001	1	0	2	1	0.1	0.0	23	2	78	62	7	0.5	3	0.3	2	5.6	4	166	P
0.4	2	8	2	0.09	1	219	4	0	0.2	1	1.0	0.43	43	4	3.4	1.6	0.2	23	6	0.10	5	3		6	
0.07	4	0.61	30	0.00	0	0.005	2	0	0(a)	0	0.5	0.2	4	0	1	474	149	0.1	1	0.2	2	unk.	unk.	155	Q
0.8	4	33	8	0.31	3	226	5	23	0.6	2	0.3	0.95	21	2	0.0	12.1	4.2	26	7	0.10	5	0		18	
0.04	3	0.38	19	0.00	0	0.003	1	0	0(a)	0	0.3	0.1	2	0	1	278	94	0.1	0	0.1	1	unk.	unk.	98	R
0.5	3	21	5	0.19	2	143	3	14	0.4	1	0.2	0.60	14	1	0.0	7.1	2.6	16	4	0.06	3	0		11	
0.00	0	unk.	unk.	0.00	0	unk.	unk.	0(a)	0(a)	0	unk.	unk.	32	3	1	51	unk.	0.6	4	0.1	1	unk.	unk.	unk.	S
0.1	1	unk.	unk.	unk.	unk.	141	3	unk.	unk.	unk.	unk.	0.21	7	1	0.0	1.3	unk.	6	2	unk.	unk.	unk.		unk.	

	FOOD	Portion	Weight in grams / Conversion for 100 g	Kilocalories / H_2O g	Total carbohydrate g / Total fats g	Crude fiber g / Dietary fiber g	Total protein g / Total sugar g	% USRDA	Arginine mg / Histidine mg	Isoleucine mg / Leucine mg	Lysine mg / Methionine mg	Phenylalanine mg / Threonine mg	Valine mg / Tryptophan mg	Cystine mg / Tyrosine mg	Polyunsat. fatty acids g / Monounsat. fatty acids g	Saturated fatty acids g / P/S ratio	Linoleic acid g / Cholesterol mg	Thiamin mg / Ascorbic acid mg	% USRDA
A	BAY LEAF	1 average	0.2	1	0.2	0.05	0.0	**0**	unk.	unk.	unk.	unk.	unk.	unk.	unk.	unk.	unk.	0.00	**0**
			500.00	tr	tr	unk.	0.0		unk.	unk.	unk.	unk.	unk.	unk.	unk.	unk.	0	tr	**0**
B	bay leaf,crumbled	1 tsp	0.6	2	0.4	0.16	0.0	**0**	unk.	unk.	unk.	unk.	unk.	unk.	0.01	0.01	0.01	0.00	**0**
			166.67	tr	0.1	unk.	0.0		unk.	unk.	unk.	unk.	unk.	unk.	0.01	1.00	0	tr	**1**
C	BEAN SPROUTS,cooked ,drained solids	1/2 C	63	18	3.2	0.44	2.0	**3**	unk.	112	136	96	118	unk.	unk.	tr(a)	unk.	0.06	**4**
			1.60	56.9	0.1	2.00	0.9		unk.	182	22	62	14	unk.	unk.	unk.	0	4	**6**
D	BEANS,COMMON,red (kidney),canned,	1/2 C	128	115	20.9	0.00	7.3	**11**	419	413	539	402	441	73	unk.	unk.	unk.	0.06	**4**
	solids & liquid		*0.78*	96.9	0.5	unk.	8.9		207	625	73	315	68	280	unk.	unk.	unk.	0(a)	**0**
E	beans,common,red (kidney),cooked,	1/2 C	93	109	19.8	1.39	7.2	**11**	416	404	534	397	433	58	unk.	unk.	unk.	0.10	**7**
	unsalted		*1.08*	63.8	0.5	unk.	6.5		208	621	72	310	65	185	unk.	unk.	unk.	0(a)	**0**
F	beans,common,white (great northern),	1/2 C	90	106	19.1	1.35	7.0	**11**	unk.	400	519	386	428	unk.	unk.	unk.	unk.	0.13	**8**
	cooked,unsalted		*1.11*	62.1	0.5	unk.	unk.		unk.	604	70	301	63	unk.	unk.	unk.	unk.	0	**0**
G	beans,common,white pea,dry (navy),	1/2 C	95	112	20.1	1.42	7.4	**11**	unk.	423	548	408	452	unk.	unk.	unk.	unk.	0.13	**9**
	cooked,unsalted		*1.05*	65.5	0.6	unk.	unk.		unk.	637	74	318	66	unk.	unk.	unk.	unk.	0	**0**
H	beans,common,white,canned w/pork &	3/4 C	178	266	37.6	3.02	11.0	**17**	unk.	630	820	609	676	32	unk.	3.55	0.71	0.11	**7**
	sweet sauce		*0.56*	117.9	8.3	15.32	unk.		unk.	951	110	477	99	293	3.55	unk.	unk.	0	**0**
I	beans,common,white,canned w/o pork	1/2 C	128	153	29.3	1.78	8.0	**12**	unk.	459	597	442	492	unk.	unk.	unk.	unk.	0.09	**6**
			0.78	87.3	0.6	unk.	unk.		unk.	692	80	347	73	unk.	unk.	unk.	unk.	3	**4**
J	beans,common,white,canned w/pork &	3/4 C	178	217	33.7	2.48	10.8	**17**	unk.	614	799	593	659	unk.	unk.	1.77	0.35	0.14	**10**
	tomato sauce		*0.56*	125.5	4.6	15.32	unk.		unk.	928	108	465	96	unk.	1.77	unk.	unk.	4	**6**
K	BEANS,FAVA,Italian style-Birds Eye	1/2 C	85	30	6.0	unk.	2.0	**3**	unk.	83	133	90	93	17	0(a)	0(a)	0(a)	0.09	**6**
			1.18	unk.	0.0	unk.	3.2		50	147	13	70	20	67	0(a)	0.00	0(a)	21	**35**
L	BEANS,GREEN,canned,cut & French	1/2 C	68	16	3.5	0.67	0.9	**2**	unk.	43	51	23	46	11	unk.	tr(a)	unk.	0.02	**1**
	style,drained solids		*1.48*	62.0	0.1	2.30	1.5		unk.	56	14	36	13	unk.	unk.	unk.	0	3	**5**
M	beans,green,canned,low sodium,cut	1/2 C	68	15	3.2	0.61	1.0	**2**	unk.	45	53	24	49	11	unk.	tr(a)	unk.	0.02	**1**
	style,drained solids		*1.48*	62.9	0.1	2.30	1.5		unk.	59	15	38	14	unk.	unk.	unk.	0	3	**5**
N	beans,green,casserole,home recipe	1/2 C	100	51	7.4	0.92	1.6	**3**	65	55	48	40	46	11	0.07	1.21	0.07	0.05	**3**
			1.00	86.5	2.1	1.39	2.7		6	64	16	49	20	24	0.54	0.06	5	31	**51**
O	beans,green,fresh,cut & French style,	1/2 C	63	16	3.4	0.63	1.0	**2**	43	45	52	24	48	11	unk.	tr(a)	unk.	0.04	**3**
	cooked,drained solids		*1.60*	57.7	0.1	2.13	1.4		unk.	58	15	38	13	unk.	unk.	unk.	0	8	**13**
P	beans,green,frozen,cut,cooked,	1/2 C	68	17	3.8	0.67	1.1	**2**	unk.	49	56	26	52	12	unk.	tr(a)	unk.	0.05	**3**
	drained solids		*1.48*	62.2	0.1	2.30	1.5		unk.	63	16	41	15	unk.	unk.	unk.	0	3	**6**
Q	beans,green,raw,edible portion	1/2 C	55	18	3.9	0.55	1.0	**2**	unk.	47	54	38	50	12	unk.	tr(a)	unk.	0.04	**3**
			1.82	49.6	0.1	1.60	2.1		unk.	60	15	40	15	35	unk.	unk.	0	10	**17**
R	BEANS,LIMA baby,frozen,cooked,drained	1/2 C	90	106	20.1	1.71	6.7	**10**	unk.	313	362	319	340	unk.	unk.	tr(a)	unk.	0.08	**5**
	solids		*1.11*	61.9	0.2	unk.	1.3		unk.	448	86	254	49	unk.	unk.	unk.	0	11	**18**
S	beans,lima (butter),mature dry beans,	1/2 C	95	131	24.3	1.61	7.8	**12**	unk.	449	520	458	488	unk.	unk.	tr(a)	unk.	0.12	**8**
	cooked		*1.05*	60.9	0.6	4.85	unk.		unk.	644	123	365	69	unk.	unk.	unk.	0	unk.	**unk.**

Row	Riboflavin mg / Niacin mg	% USRDA	Vitamin B_6 mg / Folacin mcg	% USRDA	Vitamin B_{12} mcg / Pantothenic acid mg	% USRDA	Biotin mg / Vitamin A IU	% USRDA	Preformed A RE / Beta carotene RE	Vitamin D IU / Vitamin E IU	% USRDA	Total tocopherol mg / Alpha tocopherol mg	Other tocopherol mg / Total ash g	Calcium mg / Phosphorus mg	% USRDA	Sodium mg / Sodium meq	Potassium mg / Potassium meq	Chlorine mg / Chlorine meq	Iron mg / Magnesium mg	% USRDA	Zinc mg / Copper mg	% USRDA	Iodine mcg / Selenium mcg	% USRDA	Manganese mcg / Chromium mcg
A	tr	**0**	unk.	**unk.**	unk.	**unk.**	unk.	**unk.**	0(a)	0(a)	**0**	unk.	unk.	2	**0**	0	1	unk.	0.1	**1**	unk.	**unk.**	unk.	**unk.**	unk.
	0.0	**0**	unk.	**unk.**	unk.	**unk.**	1	**0**	tr	unk.	**unk.**	unk.	0.01	0	**0**	0.0	0.0	unk.	unk.	**unk.**	unk.	**unk.**	unk.		unk.
B	tr	**0**	unk.	**unk.**	0.00	**0**	unk.	**unk.**	0(a)	0(a)	**0**	unk.	unk.	5	**1**	tr	3	unk.	0.3	**1**	tr	**0**	unk.	**unk.**	unk.
	tr	**0**	unk.	**unk.**	unk.	**unk.**	37	**1**	unk.	unk.	**unk.**	unk.	0.02	1	**0**	0.0	0.1	unk.	1	**0**	unk.	**unk.**	unk.		unk.
C	0.06	**4**	unk.	**unk.**	0(a)	**0**	unk.	**unk.**	0	0(a)	**0**	unk.	unk.	11	**1**	3	98	unk.	0.6	**3**	unk.	**unk.**	unk.	**unk.**	unk.
	0.4	**2**	unk.	**unk.**	unk.	**unk.**	13	**0**	1	unk.	**unk.**	unk.	0.25	30	**3**	0.1	2.5	unk.	unk.	**unk.**	tr(a)	**0**	tr(a)		unk.
D	0.05	**3**	0.42	**21**	0(a)	**0**	unk.	**unk.**	0	0(a)	**0**	unk.	unk.	37	**4**	4	337	unk.	2.3	**13**	unk.	**unk.**	unk.	**unk.**	unk.
	0.8	**4**	unk.	**unk.**	0.13	**1**	tr	**0**	tr	unk.	**unk.**	unk.	1.91	139	**14**	0.2	8.6	unk.	47	**12**	0.13	**6**	unk.		unk.
E	0.06	**3**	unk.	**unk.**	0(a)	**0**	unk.	**unk.**	0	0(a)	**0**	3.2	3.2	35	**4**	3	314	unk.	2.2	**12**	0.9	**6**	unk.	**unk.**	unk.
	0.6	**3**	34	**9**	0.19	**2**	tr	**0**	tr	3.8	**13**	tr	1.20	130	**13**	0.1	8.0	unk.	unk.	**unk.**	unk.	**unk.**	unk.		unk.
F	0.06	**4**	0.13	**6**	0(a)	**0**	unk.	**unk.**	0	0(a)	**0**	1.5	1.1	45	**5**	6	374	unk.	2.4	**14**	0.9	**6**	unk.	**unk.**	unk.
	0.6	**3**	unk.	**unk.**	0.00	**0**	0	**0**	0	1.8	**6**	0.4	1.26	133	**13**	0.3	9.6	unk.	33	**8**	unk.	**unk.**	unk.		unk.
G	0.07	**4**	unk.	**unk.**	tr(a)	**0**	unk.	**unk.**	0	0(a)	**0**	1.6	1.1	47	**5**	7	395	unk.	2.6	**14**	0.9	**6**	unk.	**unk.**	unk.
	0.7	**3**	38	**10**	0.36	**4**	0	**0**	0	1.9	**6**	0.4	1.33	141	**14**	0.3	10.1	unk.	unk.	**unk.**	unk.	**unk.**	unk.		unk.
H	0.07	**4**	unk.	**unk.**	0(a)	**0**	unk.	**unk.**	unk.	0(a)	**0**	unk.	unk.	112	**11**	674	unk.	unk.	4.1	**23**	3.0	**20**	unk.	**unk.**	355
	0.9	**4**	unk.	**unk.**	0.11	**1**	unk.	**unk.**	unk.	unk.	**unk.**	unk.	2.84	202	**20**	29.3	unk.	unk.	82	**21**	0.30	**15**	unk.		tr
I	0.05	**3**	unk.	**unk.**	tr(a)	**0**	unk.	**unk.**	0	0(a)	**0**	unk.	unk.	87	**9**	431	342	unk.	2.5	**14**	unk.	**unk.**	unk.	**unk.**	unk.
	0.8	**4**	31	**8**	unk.	**unk.**	76	**2**	8	unk.	**unk.**	unk.	2.17	154	**15**	18.8	8.7	unk.	47	**12**	unk.	**unk.**	unk.		unk.
J	0.05	**3**	unk.	**unk.**	tr(a)	**0**	unk.	**unk.**	0	0(a)	**0**	2.1	1.8	96	**10**	822	373	unk.	3.2	**18**	3.0	**20**	unk.	**unk.**	355
	1.1	**5**	unk.	**unk.**	0.16	**2**	231	**5**	23	2.5	**8**	0.2	2.84	163	**16**	35.8	9.5	unk.	89	**22**	0.30	**15**	unk.		106
K	0.10	**6**	0.08	**4**	0.00	**0**	unk.	**unk.**	unk.	0(a)	**0**	tr(a)	tr(a)	20	**2**	4	215	unk.	0.4	**2**	unk.	**unk.**	unk.	**unk.**	unk.
	0.8	**4**	unk.	**unk.**	0.20	**2**	500	**10**	unk.	0.0	**0**	tr(a)	unk.	20	**2**	0.2	5.5	unk.	16	**4**	unk.	**unk.**	tr(a)		unk.
L	0.03	**2**	0.03	**1**	0.00	**0**	tr	**0**	0	0(a)	**0**	0.3	0.1	30	**3**	159	64	unk.	1.0	**6**	0.2	**1**	unk.	**unk.**	unk.
	0.2	**1**	9	**2**	0.07	**1**	317	**6**	32	0.4	**1**	0.1	0.88	17	**2**	6.9	1.6	unk.	9	**2**	0.06	**3**	tr(a)		unk.
M	0.03	**2**	0.03	**1**	0.00	**0**	tr	**0**	0	0(a)	**0**	0.3	0.1	30	**3**	1	64	unk.	1.0	**6**	0.2	**1**	unk.	**unk.**	unk.
	0.2	**1**	9	**2**	0.07	**1**	317	**6**	32	0.4	**1**	0.1	0.27	17	**2**	0.1	1.6	unk.	9	**2**	0.06	**3**	tr(a)		unk.
N	0.07	**4**	0.08	**4**	0.00	**0**	0.001	**0**	0	tr(a)	**0**	0.3	0.0	29	**3**	131	148	4	0.6	**3**	0.1	**1**	23.3	**16**	34
	0.4	**2**	25	**6**	0.14	**1**	395	**8**	31	0.3	**1**	0.1	0.79	31	**3**	5.7	3.8	0.1	15	**4**	0.05	**3**	0		0
O	0.06	**3**	0.04	**2**	0.00	**0**	tr	**0**	0	0(a)	**0**	unk.	unk.	31	**3**	3	94	unk.	0.4	**2**	0.2	**1**	unk.	**unk.**	unk.
	0.3	**2**	25	**6**	0.12	**1**	338	**7**	34	unk.	**unk.**	0.1	0.25	23	**2**	0.1	2.4	unk.	9	**2**	0.06	**3**	tr(a)		unk.
P	0.06	**4**	0.04	**2**	0.00	**0**	tr	**0**	0	0(a)	**0**	0.2	0.1	27	**3**	1	103	unk.	0.5	**3**	0.2	**1**	unk.	**unk.**	unk.
	0.3	**1**	22	**6**	0.09	**1**	391	**8**	39	0.2	**1**	0.1	0.34	22	**2**	0.0	2.6	unk.	13	**3**	0.06	**3**	tr(a)		unk.
Q	0.06	**4**	0.04	**2**	0.00	**0**	tr	**0**	0	0(a)	**0**	unk.	unk.	31	**3**	4	134	18	0.4	**2**	0.2	**2**	unk.	**unk.**	247
	0.3	**1**	24	**6**	0.10	**1**	330	**7**	33	unk.	**unk.**	0.1	0.38	24	**2**	0.2	3.4	0.5	18	**4**	0.04	**2**	1		unk.
R	0.04	**3**	0.08	**4**	0.00	**0**	unk.	**unk.**	0	0(a)	**0**	tr	unk.	31	**3**	116	355	unk.	2.3	**13**	0.8	**5**	unk.	**unk.**	unk.
	1.1	**5**	28	**7**	0.16	**2**	198	**4**	20	0.0	**0**	tr	1.17	113	**11**	5.0	9.1	unk.	43	**11**	tr(a)	**0**	tr(a)		unk.
S	0.06	**3**	unk.	**unk.**	0.00	**0**	unk.	**unk.**	0	0(a)	**0**	unk.	unk.	28	**3**	2	581	2	2.9	**16**	0.9	**6**	unk.	**unk.**	unk.
	0.7	**3**	unk.	**unk.**	unk.	**unk.**	tr	**0**	tr	unk.	**unk.**	unk.	1.42	146	**15**	0.1	14.9	0.0	31	**8**	0.15	**8**	tr(a)		unk.

	FOOD	Portion	Weight in grams / Conversion for 100 g	Kilocalories / H_2O g	Total carbohydrate g / Total fats g	Crude fiber g / Dietary fiber g	Total protein g / Total sugar g	% USRDA	Arginine mg / Histidine mg	Isoleucine mg / Leucine mg	Lysine mg / Methionine mg	Phenylalanine mg / Threonine mg	Valine mg / Tryptophan mg	Cystine mg / Tyrosine mg	Polyunsat. fatty acids g / Monounsat. fatty acids g	Saturated fatty acids g / P/S ratio	Linoleic acid g / Cholesterol mg	Thiamin mg / Ascorbic acid mg	% USRDA
A	beans,lima,canned,drained solids	1/2 C	85	82	15.6	1.53	4.6	**7**	unk.	266	306	270	289	unk.	unk.	tr(a)	unk.	0.03	**2**
			1.18	63.5	0.3	unk.	1.2		unk.	380	73	215	41	unk.	unk.	unk.	0	5	**9**
B	beans,lima,canned,low sodium, drained solids	1/2 C	85	81	15.0	1.53	4.9	**8**	unk.	286	331	291	310	unk.	unk.	tr(a)	unk.	0.03	**2**
			1.18	64.3	0.3	unk.	1.2		unk.	409	79	232	44	unk.	unk.	unk.	0	5	**9**
C	beans,lima,fordhook,frozen,cooked, drained solids	1/2 C	85	84	16.2	1.36	5.1	**8**	unk.	365	422	371	396	unk.	unk.	tr(a)	unk.	0.06	**4**
			1.18	62.5	0.1	unk.	1.2		unk.	522	100	296	57	unk.	unk.	unk.	0	14	**24**
D	BEANS,PINTO,prep,home recipe	1/2 C	58	203	37.1	2.51	13.3	**21**	854	841	954	741	754	unk.	unk.	unk.	unk.	0.49	**33**
			1.72	4.8	0.7	unk.	unk.		427	1128	367	634	unk.	unk.	unk.	unk.	0(a)	0(a)	**0**
E	BEANS,REFRIED,home recipe	1/2 C	141	43	5.0	0.35	1.8	**3**	105	101	134	99	108	tr	0.27	0.69	0.17	0.03	**2**
			0.71	60.8	1.9	unk.	0.0		52	156	18	78	16	tr	0.71	0.38	2	0	**0**
F	BEANS,WAX,canned,drained solids	1/2 C	68	16	3.5	0.61	0.9	**2**	unk.	43	49	23	45	7	unk.	tr(a)	unk.	0.02	**1**
			1.48	62.2	0.2	unk.	1.5		22	55	14	36	13	31	unk.	unk.	0	3	**6**
G	beans,wax,canned,low sodium, drained solids	1/2 C	68	14	3.2	0.61	0.8	**1**	unk.	36	42	20	39	7	unk.	tr(a)	unk.	0.02	**1**
			1.48	63.2	0.1	unk.	1.5		19	47	12	31	11	26	unk.	unk.	0	3	**6**
H	beans,wax,fresh,cooked,drained solids	1/2 C	63	14	2.9	0.63	0.9	**1**	unk.	39	46	21	42	7	unk.	tr(a)	unk.	0.04	**3**
			1.60	58.4	0.1	unk.	1.4		20	51	13	33	13	29	unk.	unk.	0	8	**14**
I	beans,wax,frozen,cooked, drained solids	1/2 C	68	18	4.2	0.74	1.1	**2**	unk.	51	59	28	55	9	unk.	tr(a)	unk.	0.05	**3**
			1.48	61.8	0.1	unk.	1.5		26	67	18	44	16	37	unk.	unk.	0	4	**7**
J	BEEF DRIPPINGS	1 Tbsp	10	90	0.0	tr(a)	0.0	**0**	0	0	0	0	0	0	0.40	4.91	0.31	tr(a)	**0**
			10.00	unk.	10.0	0.00	0.0		0	0	0	0	0	0	3.60	0.08	11	tr(a)	**0**
K	BEEF JUICE	1 C	240	60	0.0	tr(a)	11.8	**18**	unk.	unk.	unk.	unk.	unk.	unk.	unk.	unk.	unk.	unk.	**unk.**
			0.42	unk.	1.4	0(a)	unk.		unk.	unk.	unk.	unk.	unk.	unk.	unk.	unk.	unk.	0(a)	**0**
L	BEEF MARROW	1 Tbsp	15	131	0.0	0(a)	0.3	**1**	unk.	unk.	unk.	unk.	unk.	unk.	unk.	unk.	unk.	unk.	**unk.**
			6.67	unk.	13.9	0(a)	unk.		unk.	unk.	unk.	unk.	unk.	unk.	unk.	unk.	unk.	unk.	**unk.**
M	BEEF,BRISKET,w/o bone,cooked,sliced, no visible fat	3 slices 5x 1x1/4" ea	75	167	0.0	0.00	22.3	**50**	1407	1073	1979	979	1115	284	0.90	3.38	0.30	0.04	**3**
			1.33	44.3	7.9	0(a)	0.0		759	1806	602	1022	249	802	2.92	0.27	68	unk.	**unk.**
N	beef,brisket,w/o bone,cooked,sliced, w/fat	3 slices 5x 1x1/4" ea	75	356	0.0	0.00	15.4	**34**	976	744	1373	679	773	197	3.15	10.95	0.60	0.03	**2**
			1.33	27.0	32.1	0(a)	0.0		527	1253	417	709	173	556	10.57	0.29	71	unk.	**unk.**
O	BEEF,CHUCK STEW MEAT,w/o bone, braised,cubed,no visible fat	five 1" cubes	85	182	0.0	0.00	25.5	**57**	1611	1227	2266	1121	1277	326	0.80	2.46	0.23	0.04	**3**
			1.18	50.7	8.1	0(a)	0.0		869	2067	688	1170	286	918	2.18	0.32	77	unk.	**unk.**
P	beef,chuck stew meat,w/o bone, braised,cubed,no visible fat	1/2 C	76	162	0.0	0.00	22.6	**50**	1431	1090	2013	996	1134	289	0.71	2.19	0.20	0.04	**3**
			1.32	45.1	7.2	0(a)	0.0		772	1836	612	1039	254	815	1.93	0.32	69	unk.	**unk.**
Q	beef,chuck stew meat,w/o bone, braised,cubed,w/fat	1/2 C	76	247	0.0	0.00	19.6	**44**	1240	945	1745	863	982	251	1.81	6.04	0.38	0.04	**3**
			1.32	37.3	18.0	0(a)	0.0		669	1592	530	901	220	707	5.81	0.30	71	unk.	**unk.**
R	BEEF,CORNED,canned	3x2x1/4" slice	28	60	0.0	0.00	7.1	**16**	405	332	555	291	392	90	unk.	1.68	tr	0.01	**0**
			3.57	16.6	3.4	0(a)	0.0		241	520	176	280	74	255	1.40	unk.	26	0	**0**
S	beef,corned,canned,hash,w/potatoes	1 C	227	411	24.3	1.13	20.0	**44**	unk.	unk.	unk.	817	1112	unk.	unk.	11.35	tr	0.02	**2**
			0.44	153.0	25.7	unk.	0.0		unk.	unk.	499	unk.	unk.	unk.	11.35	unk.	unk.	unk.	**unk.**

Riboflavin mg / Niacin mg	% USRDA	Vitamin B6 mg / Folacin mcg	% USRDA	Vitamin B12 mcg / Pantothenic acid mg	% USRDA	Biotin mg / Vitamin A IU	% USRDA	Preformed A RE / Beta carotene RE	Vitamin D IU / Vitamin E IU	% USRDA	Total tocopherol mg / Alpha tocopherol mg	Other tocopherol mg / Total ash g	Calcium mg / Phosphorus mg	% USRDA	Sodium mg / Sodium meq	Potassium mg / Potassium meq	Chlorine mg / Chlorine meq	Iron mg / Magnesium mg	% USRDA	Zinc mg / Copper mg	% USRDA	Iodine mcg / Selenium mcg	% USRDA	Manganese mcg / Chromium mcg	
0.04	3	unk.	unk.	0.00	0	unk.	unk.	0	0(a)	0	tr	unk.	24	2	201	189	unk.	2.0	11	0.8	5	unk.	unk.	unk.	A
0.4	2	unk.	unk.	unk.	unk.	161	3	16	0.0	0	tr	1.10	60	6	8.7	4.8	unk.	unk.	unk.	tr(a)	0	tr(a)		unk.	
0.04	3	unk.	unk.	0.00	0	unk.	unk.	0	0(a)	0	tr	unk.	24	2	3	189	unk.	2.0	11	0.8	5	unk.	unk.	unk.	B
0.4	2	unk.	unk.	unk.	unk.	161	3	16	0.0	0	tr	0.51	60	6	0.1	4.8	unk.	unk.	unk.	tr(a)	0	tr(a)		unk.	
0.04	3	0.08	4	0.00	0	unk.	unk.	0	0(a)	0	tr	unk.	17	2	86	362	unk.	1.4	8	0.8	5	unk.	unk.	unk.	C
0.8	4	26	7	0.15	2	196	4	20	0.0	0	tr	1.10	77	8	3.7	9.3	unk.	41	10	tr(a)	0	tr(a)		unk.	
0.12	7	0.31	16	0.00	0	unk.	unk.	unk.	0(a)	0	tr(a)	tr(a)	79	8	6	574	unk.	3.7	21	1.3	9	unk.	unk.	874	D
1.3	6	unk.	unk.	0.38	4	unk.	unk.	unk.	0.0	0	tr(a)	2.27	266	27	0.3	14.7	unk.	100	25	0.41	20	unk.		unk.	
0.01	1	0(a)	0	0.00	0	tr(a)	0	0	0(a)	0	0.5	0(a)	9	1	104	79	tr(a)	0.6	3	0.2	2	26.5	18	0	E
0.2	1	9	2	0.05	1	0	0	0	0.6	2	0(a)	0.57	67	7	4.5	2.0	0.0	0	0	tr	0	0		0	
0.03	2	0.03	1	0.00	0	unk.	unk.	0	0(a)	0	unk.	unk.	30	3	159	64	unk.	1.0	6	unk.	unk.	unk.	unk.	unk.	F
0.2	1	unk.	unk.	unk.	unk.	67	1	7	unk.	unk.	unk.	0.61	17	2	6.9	1.6	unk.	unk.	unk.	tr(a)	0	tr(a)		unk.	
0.03	2	0.03	1	0.00	0	unk.	unk.	0	0(a)	0	unk.	unk.	30	3	1	64	unk.	1.0	6	unk.	1	unk.	unk.	unk.	G
0.2	1	unk.	unk.	unk.	unk.	67	1	7	unk.	unk.	unk.	0.27	17	2	0.1	1.6	unk.	7	2	tr(a)	0	tr(a)		40	
0.06	3	0.03	1	0.00	0	unk.	unk.	0	0(a)	0	unk.	unk.	31	3	2	94	unk.	0.4	2	unk.	unk.	unk.	unk.	unk.	H
0.3	2	21	5	0.08	1	144	3	14	unk.	unk.	unk.	0.25	23	2	0.1	2.4	unk.	unk.	unk.	tr(a)	0	tr(a)		unk.	
0.05	3	0.03	1	0.00	0	unk.	unk.	0	0(a)	0	unk.	unk.	24	2	1	111	unk.	0.5	3	unk.	unk.	unk.	unk.	unk.	I
0.3	1	23	6	0.08	1	67	1	7	unk.	unk.	unk.	0.34	21	2	0.0	2.8	unk.	14	4	tr(a)	0	tr(a)		unk.	
tr(a)	0	tr(a)	0	tr(a)	0	tr(a)	0	0(a)	unk.	unk.	unk.	unk.	tr	0	0	0(a)	tr(a)	tr(a)	0	tr(a)	0	tr(a)	0	tr(a)	J
tr(a)	0	tr(a)	0	tr(a)	0	0(a)	0	0(a)	0.3	1	0.3	tr(a)	1	0	0.0	0.0	0	tr(a)	0	unk.	unk.	tr(a)		unk.	
unk.	unk.	unk.	unk.	unk.	unk.	unk.	unk.	0(a)	0(a)	0	unk.	unk.	19	2	unk.	unk.	unk.	106.6	592	unk.	unk.	unk.	unk.	unk.	K
unk.	unk.	unk.	unk.	unk.	unk.	0(a)	0	0(a)	unk.	unk.	unk.	unk.	74	7	unk.	unk.	unk.	unk.	unk.	unk.	unk.	unk.		unk.	
unk.	unk.	unk.	unk.	unk.	unk.	unk.	unk.	unk.	0(a)	0	unk.	unk.	unk.	unk.	unk.	unk.	unk.	0.1	1	unk.	unk.	unk.	unk.	unk.	L
unk.	unk.	unk.	unk.	tr(a)	0	unk.	unk.	unk.	unk.	unk.	unk.	unk.	unk.	unk.	unk.	unk.	unk.	unk.	unk.	unk.	unk.	tr(a)		tr(a)	
0.16	10	0.20	10	2.12	35	unk.	unk.	4	0	0	0.1	tr(a)	10	1	45	278	unk.	2.8	16	4.6	31	unk.	unk.	unk.	M
3.4	17	3	1	0.24	2	15	0	tr	0.1	0	0.1	0.60	110	11	2.0	7.1	unk.	14	3	0.16	8	tr(a)		tr(a)	
0.12	7	0.15	7	1.47	25	unk.	unk.	16	0	0	0.1	tr(a)	7	1	45	278	unk.	2.0	11	2.9	19	unk.	unk.	unk.	N
2.4	12	3	1	0.18	2	60	1	1	0.1	0	0.1	0.45	76	8	2.0	7.1	unk.	11	3	0.16	8	tr(a)		tr(a)	
0.20	12	0.22	11	2.24	37	unk.	unk.	5	0	0	0.1	tr(a)	11	1	51	314	unk.	3.2	18	5.3	35	unk.	unk.	unk.	O
3.9	20	3	1	0.27	3	17	0	tr	0.1	1	0.1	0.68	136	14	2.2	8.0	unk.	15	4	0.07	3	tr(a)		tr(a)	
0.17	10	0.20	10	1.99	33	unk.	unk.	4	0	0	0.1	tr(a)	10	1	45	279	unk.	2.9	16	4.7	31	unk.	unk.	unk.	P
3.5	17	3	1	0.24	2	15	0	tr	0.1	0	0.1	0.60	121	12	2.0	7.1	unk.	14	3	0.06	3	tr(a)		tr(a)	
0.15	9	0.20	10	1.73	29	unk.	unk.	8	0	0	0.1	tr(a)	8	1	45	279	unk.	2.5	14	3.6	24	unk.	unk.	unk.	Q
3.0	15	3	1	0.24	2	30	1	tr	0.1	0	0.1	0.53	106	11	2.0	7.1	unk.	11	3	0.06	3	tr(a)		tr(a)	
0.07	4	0.03	1	0.52	9	unk.	unk.	unk.	unk.	unk.	tr	tr(a)	6	1	268	17	unk.	1.2	7	0.9	6	unk.	unk.	unk.	R
1.0	5	1	0	0.17	2	unk.	unk.	unk.	tr	0	tr(a)	0.95	30	3	11.7	0.4	unk.	unk.	unk.	0.06	3	tr(a)		tr(a)	
0.20	12	0.17	9	unk.	unk.	unk.	unk.	unk.	unk.	unk.	unk.	unk.	30	3	1226	454	unk.	4.5	25	unk.	unk.	unk.	unk.	unk.	S
4.8	24	unk.	unk.	unk.	unk.	unk.	unk.	unk.	unk.	unk.	unk.	4.09	152	15	53.3	11.6	unk.	unk.	unk.	unk.	unk.	unk.		unk.	

	FOOD	Portion	Weight in grams / Conversion for 100 g	Kilocalories / H_2O g	Total carbohydrate g / Total fats g	Crude fiber g / Dietary fiber g	Total protein g / Total sugar g	% USRDA	Arginine mg / Histidine mg	Isoleucine mg / Leucine mg	Lysine mg / Methionine mg	Phenylalanine mg / Threonine mg	Valine mg / Tryptophan mg	Cystine mg / Tyrosine mg	Polyunsat. fatty acids g / Monounsat. fatty acids g	Saturated fatty acids g / P/S ratio	Linoleic acid g / Cholesterol mg	Thiamin mg / Ascorbic acid mg	% USRDA
A	beef,corned,cooked	5x1x1/4"	28	104	0.0	0.00	6.4	14	405	309	570	263	355	82	unk.	4.20	0.28	0.01	0
		slice	*3.57*	12.3	8.5	0(a)	0.0		218	520	160	294	72	231	3.64	unk.	26	0	0
B	beef,corned,home recipe,corn beef & egg casserole	3/4 C	168	278	18.6	0.30	18.0	38	871	895	1267	821	995	251	0.36	5.35	1.66	0.08	6
			0.60	111.7	14.7	1.94	4.2		526	1436	444	730	205	670	6.15	0.07	159	2	3
C	BEEF,DRY,chipped,uncooked	1 oz (approx	28	58	0.0	0.00	9.7	22	615	510	851	400	540	124	unk.	0.85	tr	0.02	1
		1/4 C)	*3.52*	13.5	1.8	0(a)	0.0		332	798	241	430	114	350	0.85	unk.	unk.	0	0
D	beef,dry,home recipe,creamed	2/3 C	133	211	7.4	0.01	16.2	36	916	866	1358	692	923	200	0.41	7.61	0.27	0.07	4
			0.75	91.6	12.8	0.14	3.9		525	1366	396	710	196	615	3.74	0.05	32	1	2
E	BEEF,GROUND,hamburger,patty,cooked	3" dia x 5/8"	85	243	0.0	0.00	20.6	46	1300	1141	1906	897	1210	263	2.38	8.07	0.51	0.08	5
			1.18	46.1	17.3	0(a)	0.0		700	1786	541	962	255	740	7.73	0.51	80	0	0
F	beef,ground,round,cooked, no visible fat	1/2 C	55	104	0.0	0.00	17.2	38	1087	829	1530	757	862	220	0.47	1.48	0.14	0.04	3
			1.82	33.7	3.4	0(a)	0.0		586	1396	465	790	192	619	1.29	0.32	50	0	0
G	beef,ground,round,cooked,w/fat	1/2 C	55	144	0.0	0.00	15.7	35	993	757	1398	691	788	201	0.99	3.46	0.22	0.04	3
			1.82	30.1	8.5	0(a)	0.0		536	1275	425	722	176	566	3.24	0.29	52	0	0
H	beef,ground,round,patty,cooked, no visible fat	3" dia x 5/8"	85	161	0.0	0.00	26.6	59	1680	1281	2365	1170	1332	340	0.73	2.29	0.21	0.07	5
			1.18	52.0	5.2	0(a)	0.0		906	2157	718	1221	297	957	1.99	0.32	77	0	0
I	beef,ground,round,patty,cooked,w/fat	3" dia x 5/8"	87	227	0.0	0.00	24.9	55	1571	1198	2212	1094	1246	318	1.57	5.48	0.35	0.07	5
			1.15	47.6	13.4	0(a)	0.0		847	2018	672	1141	278	895	5.13	0.29	82	0	0
J	beef,ground,sirloin,broiled, no visible fat	1/2 C	55	114	0.0	0.00	17.7	39	1119	852	1574	778	887	226	0.54	1.81	0.16	0.05	3
			1.82	32.3	4.2	0(a)	0.0		603	1436	478	812	198	637	1.55	0.30	50	0	0
K	beef,ground,sirloin,cooked,w/fat	1/2 C	55	213	0.0	0.00	12.6	28	799	609	1124	556	633	162	2.14	7.31	0.38	0.03	2
			1.82	24.1	17.6	0(a)	0.0		431	1026	342	580	141	455	7.20	0.29	52	0	0
L	BEEF,HASH,canned,w/potatoes, no added salt	1 C	225	290	28.7	0(a)	21.7	47	unk.	1111	1795	902	1199	unk.	unk.	unk.	unk.	0.13	9
			0.44	unk.	9.9	0(a)	0.0		unk.	1681	502	941	251	unk.	unk.	unk.	unk.	21	35
M	BEEF,HOME RECIPE,hamburger casserole	3/4 C	187	205	16.1	0.77	12.4	26	695	622	1014	518	687	126	1.31	4.79	0.52	0.11	8
			0.53	145.4	10.3	1.61	2.3		351	968	290	524	150	445	3.85	0.27	63	15	26
N	beef,home recipe,Italian loaf	3x2-3/4x3/4"	110	234	10.9	0.18	17.4	38	967	996	1457	831	1049	235	1.48	5.63	0.45	0.07	5
		slice	*0.91*	65.8	13.0	0.40	1.4		573	1548	472	792	217	706	4.95	0.26	94	5	9
O	beef,home recipe,meatloaf,sliced	4x2-1/2x1/2"	75	150	2.5	0.00	11.9	26	753	574	1060	524	597	152	unk.	unk.	unk.	0.10	7
		slice	*1.33*	48.1	9.9	0(a)	0.0		407	967	322	547	134	429	unk.	unk.	unk.	unk.	unk.
P	beef,home recipe,potpie,baked	4-1/4" dia	227	558	42.7	0.91	22.9	35	unk.	unk.	unk.	unk.	unk.	unk.	unk.	9.08	2.27	0.25	17
			0.44	125.1	32.9	unk.	unk.		unk.	unk.	unk.	unk.	unk.	unk.	20.43	unk.	48	7	11
Q	beef,home recipe,stroganoff	3/4 C	166	260	9.2	0.12	13.6	30	738	629	1020	530	635	156	0.96	10.93	0.48	0.07	5
			0.60	118.6	18.5	1.09	3.5		397	995	330	537	137	432	4.91	0.09	69	3	5
R	beef,home recipe,Swiss steak	3x3x1/2"	130	214	9.6	0.39	22.6	49	1342	1068	1913	982	1103	288	0.63	2.90	0.54	0.13	9
		piece	*0.77*	86.9	9.0	1.78	3.2		741	1789	581	1011	253	800	4.61	0.22	61	14	23
S	beef,home recipe,Yorkshire pudding	3" square	84	171	15.5	0.06	4.7	9	155	235	244	240	265	93	0.44	4.08	0.53	0.10	7
			1.19	52.9	9.6	0.58	1.4		125	391	109	190	64	191	3.84	0.11	86	tr	1

Riboflavin mg / Niacin mg	% USRDA	Vitamin B_6 mg / Folacin mcg	% USRDA	Vitamin B_{12} mcg / Pantothenic acid mg	% USRDA	Biotin mg / Vitamin A IU	% USRDA	Preformed A RE / Beta carotene RE	Vitamin D IU / Vitamin E IU	% USRDA	Total tocopherol mg / Alpha tocopherol mg	Other tocopherol mg / Total ash g	Calcium mg / Phosphorus mg	% USRDA	Sodium mg / Sodium meq	Potassium mg / Potassium meq	Chlorine mg / Chlorine meq	Iron mg / Magnesium mg	% USRDA	Zinc mg / Copper mg	% USRDA	Iodine mcg / Selenium mcg	% USRDA	Manganese mcg / Chromium mcg	
0.05	3	0.03	1	0.52	9	unk.	unk.	unk.	unk.	unk.	tr	unk.	3	0	487	42	unk.	0.8	5	0.9	6	unk.	unk.	unk.	A
0.4	2	1	0	0.17	2	unk.	unk.	unk.	tr	0	tr(a)	0.81	26	3	21.2	1.1	unk.	8	2	0.06	3	tr(a)		tr(a)	
0.30	18	0.16	8	1.26	21	0.000	0	0	20	5	3.2	1.5	94	9	694	179	2	3.0	17	2.1	14	3.4	2	12	B
2.3	12	20	5	0.92	9	469	9	12	3.9	13	1.7	2.82	163	16	30.2	4.6	0.1	19	5	0.23	11	2		2	
0.09	5	0.14	7	0.52	9	unk.	unk.	unk.	0	0	0.1	tr(a)	6	1	1221	57	unk.	1.4	8	1.0	7	unk.	unk.	unk.	C
1.1	5	1	0	0.17	2	unk.	unk.	unk.	0.1	0	0.1	3.29	115	12	53.1	1.4	unk.	11	3	0.03	2	tr(a)		tr(a)	
0.27	16	0.23	11	0.97	16	tr	0	tr	33	8	0.3	tr(a)	107	11	1737	206	3	2.1	12	1.7	11	5.7	4	19	D
1.7	8	7	2	0.50	5	392	8	tr	0.3	1	0.3	5.17	234	23	75.6	5.3	0.1	12	3	0.17	9	3		3	
0.18	11	0.17	8	1.95	33	unk.	unk.	9	0	0	0.1	tr(a)	9	1	40	382	unk.	2.7	15	3.7	25	unk.	unk.	unk.	E
4.6	23	3	1	0.20	2	34	1	tr	0.1	0	0.1	1.10	165	17	1.7	9.8	unk.	18	5	0.07	3	tr(a)		tr(a)	
0.13	8	0.14	7	1.64	27	unk.	unk.	1	0	0	0.1	tr(a)	7	1	33	203	unk.	2.0	11	3.2	21	unk.	unk.	unk.	F
3.3	17	2	1	0.18	2	6	0	tr	0.1	0	0.1	0.77	147	15	1.4	5.2	unk.	16	4	0.04	2	tr(a)		tr(a)	
0.12	7	0.14	7	1.50	25	unk.	unk.	4	0	0	0.1	tr(a)	7	1	33	203	unk.	1.9	11	2.6	18	unk.	unk.	unk.	G
3.1	15	2	1	0.18	2	16	0	tr	0.1	0	0.1	0.71	138	14	1.4	5.2	unk.	15	4	0.04	2	tr(a)		tr(a)	
0.20	12	0.22	11	2.53	42	unk.	unk.	2	0	0	0.1	tr(a)	11	1	51	314	unk.	3.1	18	4.9	33	unk.	unk.	unk.	H
5.1	26	3	1	0.27	3	9	0	tr	0.2	1	0.1	1.19	228	23	2.2	8.0	unk.	25	6	0.07	3	tr(a)		tr(a)	
0.19	11	0.23	11	2.37	39	unk.	unk.	7	0	0	0.1	tr(a)	10	1	52	322	unk.	3.0	17	4.2	28	unk.	unk.	unk.	I
4.9	24	3	1	0.28	3	26	1	tr	0.1	1	0.1	1.13	218	22	2.3	8.2	unk.	24	6	0.07	4	tr(a)		tr(a)	
0.14	8	0.14	7	1.69	28	unk.	unk.	1	0	0	0.1	tr(a)	7	1	33	203	unk.	2.1	12	3.2	21	unk.	unk.	unk.	J
3.5	18	2	1	0.18	2	5	0	tr	0.1	0	0.1	0.82	144	14	1.4	5.2	unk.	16	4	0.04	2	tr(a)		tr(a)	
0.10	6	0.11	5	1.20	20	unk.	unk.	7	0	0	0.1	tr(a)	5	1	33	203	unk.	1.6	9	2.2	15	unk.	unk.	unk.	K
2.6	13	2	1	0.13	1	27	1	tr	0.1	0	0.1	0.60	105	11	1.4	5.2	unk.	12	3	0.04	2	tr(a)		tr(a)	
0.20	12	0.17	8	unk.	unk.	tr(a)	0	unk.	0(a)	0	tr(a)	tr(a)	28	3	40	unk.	unk.	2.9	16	unk.	unk.	unk.	unk.	unk.	L
4.6	23	11	3	tr(a)	0	30	1	0(a)	unk.	unk.	tr(a)	unk.	171	17	1.7	unk.	unk.	unk.	unk.	unk.	unk.	tr(a)		tr(a)	
0.17	10	0.22	11	0.54	9	0.002	1	4	7	2	0.4	0.1	79	8	550	460	9	2.1	12	2.0	14	32.5	22	2	M
2.9	15	17	4	0.50	5	507	10	13	0.4	2	0.2	2.88	175	17	23.9	11.8	0.3	41	10	0.17	8	0		0	
0.21	12	0.15	8	0.86	14	0.003	1	5	3	1	0.3	tr	36	4	312	266	19	2.0	11	2.3	16	47.2	32	15	N
2.9	15	11	3	0.32	3	116	2	3	0.4	1	0.2	1.56	177	18	13.6	6.8	0.5	37	9	0.05	3	tr		1	
0.16	10	unk.	unk.	unk.	unk.	unk.	unk.	unk.	unk.	unk.	unk.	unk.	7	1	unk.	unk.	unk.	1.3	8	unk.	unk.	unk.	unk.	unk.	O
1.9	9	unk.	unk.	unk.	unk.	unk.	unk.	unk.	unk.	unk.	unk.	2.63	134	13	unk.	unk.	unk.	unk.	unk.	unk.	unk.	unk.		unk.	
0.27	16	unk.	unk.	unk.	unk.	unk.	unk.	unk.	unk.	unk.	unk.	unk.	32	3	645	361	unk.	4.1	23	unk.	unk.	unk.	unk.	unk.	P
4.5	23	unk.	unk.	unk.	unk.	1861	37	unk.	unk.	unk.	unk.	3.40	161	16	28.0	9.2	unk.	unk.	unk.	unk.	unk.	unk.		unk.	
0.22	13	0.14	7	0.66	11	0.002	1	1	0	0	0.2	tr	60	6	503	244	8	1.9	11	2.2	15	46.7	31	36	Q
5.0	25	12	3	0.31	3	692	14	10	0.2	1	0.2	2.19	159	16	21.9	6.2	0.2	22	5	0.06	3	4		4	
0.20	12	0.25	13	1.09	18	0.002	1	2	0	0	unk.	tr(a)	17	2	139	435	5	3.1	17	4.1	28	tr(a)	0	51	R
4.8	24	9	2	0.43	4	684	14	68	unk.	unk.	0.1	1.63	204	20	6.1	11.1	0.1	32	8	0.16	8	4		4	
0.13	8	0.04	2	0.25	4	0.003	1	0	6	2	0.2	tr(a)	38	4	215	unk.	43	1.0	5	0.5	3	60.1	40	84	S
0.7	4	10	3	0.39	4	66	1	0	0.2	1	tr	0.85	78	8	9.4	unk.	1.2	50	12	0.07	3	10		12	

	FOOD	Portion	Weight in grams Conversion for 100 g	Kilocalories H_2O g	Total carbohydrate g Total fats g	Crude fiber g Dietary fiber g	Total protein g Total sugar g	% USRDA	Arginine mg Histidine mg	Isoleucine mg Leucine mg	Lysine mg Methionine mg	Phenylalanine mg Threonine mg	Valine mg Tryptophan mg	Cystine mg Tyrosine mg	Polyunsat. fatty acids g Monounsat. fatty acids g	Saturated fatty acids g P/S ratio	Linoleic acid g Cholesterol mg	Thiamin mg Ascorbic acid mg	% USRDA
A	BEEF,PLATE short ribs,cooked,w/fat	2 med ribs	136 *0.74*	645 49.0	0.0 58.2	0.00 0(a)	28.0 0.0	**62**	1769 955	1349 2271	2490 756	1231 1285	1402 313	358 1008	7.21 23.80	24.21 0.30	1.22 128	0.05 unk.	**4** **unk.**
B	BEEF,ROAST,chuck,w/o bone,sliced, braised,no visible fat	3x2x3/4" slice	85 *1.18*	164 52.4	0.0 5.9	0.00 0(a)	25.9 0.0	**58**	1637 883	1248 2102	2304 700	1140 1189	1298 290	331 933	0.80 2.18	2.46 0.32	0.23 77	0.05 0(a)	**3** **0**
C	beef,roast,chuck,w/o bone,sliced, braised,w/fat	3x2x3/4" slice	85 *1.18*	246 45.0	0.0 16.3	0.00 0(a)	23.0 0.0	**51**	1506 785	1108 1867	2047 622	1012 1057	1153 258	294 829	2.04 6.54	6.80 0.30	0.42 80	0.04 0(a)	**3** **0**
D	beef,roast,rib,w/o bone,roasted, chopped,no visible fat	1/2 C	70 *1.43*	169 40.0	0.0 9.4	0.00 0(a)	19.7 0.0	**44**	1247 673	950 1600	1754 533	868 906	988 220	252 710	1.05 3.43	4.13 0.25	0.35 64	0.05 0	**3** **0**
E	beef,roast,rib,w/o bone,roasted, sliced,no visible fat	2 slices 4-1/8x 2-1/4x1/4" ea	83 *1.20*	200 47.5	0.0 11.1	0.00 0(a)	23.4 0.0	**52**	1478 798	1126 1897	2080 632	1029 1074	1172 261	299 842	1.24 4.07	4.90 0.25	0.41 76	0.06 0	**4** **0**
F	beef,roast,rib,w/o bone,roasted, sliced,w/fat	2 slices 4-1/8x 2-1/4x1/4" ea	83 *1.20*	365 33.2	0.0 32.7	0.00 0(a)	16.5 0.0	**37**	1043 563	795 1339	1468 446	726 758	827 185	211 594	3.89 13.28	13.69 0.28	0.75 78	0.04 0	**3** **0**
G	beef,roast,round rump,w/o bone, roasted,chopped,no visible fat	1/2 C	70 *1.43*	146 42.3	0.0 6.5	0.00 0(a)	20.4 0.0	**45**	1287 694	981 1651	1810 550	895 934	1020 228	260 733	0.84 2.38	2.73 0.31	0.25 64	0.05 0	**3** **0**
H	beef,roast,round rump,w/o bone, roasted,sliced,no visible fat	2 slices 4-1/8x 2-1/4x1/4" ea	83 *1.20*	173 50.1	0.0 7.7	0.00 0(a)	24.1 0.0	**54**	1526 823	1163 1958	2146 652	1062 1108	1209 271	309 869	1.00 2.82	3.24 0.31	0.30 76	0.06 0	**4** **0**
I	beef,roast,round rump,w/o bone, roasted,sliced,w/fat	2 slices 4-1/8x 2-1/4x1/8" ea	83 *1.20*	288 39.9	0.0 22.7	0.00 0(a)	19.6 0.0	**44**	1238 667	943 1588	1741 529	861 899	980 219	250 705	2.82 9.13	9.46 0.30	0.58 78	0.05 0	**3** **0**
J	beef,roast,sirloin tip,roasted, chopped,no visible fat	1/2 C	70 *1.43*	168 39.4	0.0 8.7	0.00 0(a)	20.9 0.0	**46**	1317 710	1004 1691	1854 563	917 957	1044 233	266 750	0.69 1.97	2.31 0.30	0.21 64	0.06 0	**4** **0**
K	beef,roast,sirloin tip,roasted, sliced,no visible fat	2 slices 2-1/2x 1-1/2x3/4" ea	85 *1.18*	204 47.9	0.0 10.6	0.00 0(a)	25.3 0.0	**56**	1600 863	1219 2054	2251 684	1113 1162	1268 283	323 911	0.83 2.40	2.80 0.30	0.25 77	0.07 0	**5** **0**
L	beef,roast,sirloin tip,roasted, sliced,w/fat	2 slices 2-1/2x 1-1/2x3/4" ea	85 *1.18*	414 29.8	0.0 38.2	0.00 0(a)	16.2 0.0	**36**	1025 553	781 1317	1442 439	714 745	813 182	207 584	3.31 11.13	11.30 0.29	0.59 80	0.05 0	**3** **0**
M	beef,roast,sliced,canned	3 slices 3x2x1/4" ea	84 *1.19*	188 50.4	0.0 10.9	0.00 0(a)	21.0 0.0	**47**	1326 729	1099 1720	1835 521	864 927	1166 245	265 712	1.51 4.96	5.29 0.29	0.34 76	0.02 0	**1** **0**
N	BEEF,STEAK,chuck blade,w/bone, braised,no visible fat	1/3 of 1.4 lb steak	107 *0.93*	266 60.5	0.0 14.9	0.00 0(a)	30.9 0.0	**69**	1953 1054	1488 2507	2748 835	1359 1419	1548 346	395 1113	1.71 5.46	6.42 0.27	0.53 97	0.05 0(a)	**4** **0**
O	beef,steak,chuck blade,w/bone, braised,w/fat	1/3 of 1.4 lb steak	125 *0.80*	534 50.4	0.0 45.9	0.00 0(a)	28.0 0.0	**62**	1769 954	1348 2270	2489 756	1231 1285	1401 314	358 1008	5.63 18.50	19.12 0.29	1.12 118	0.05 0(a)	**3** **0**
P	beef,steak,chuck blade,w/o bone, braised,no visible fat	4-1/8x2-3/4x 1/2" piece	85 *1.18*	212 48.0	0.0 11.8	0.00 0(a)	24.6 0.0	**55**	1551 837	1182 1992	2183 663	1079 1127	1230 275	314 884	1.36 4.33	5.10 0.27	0.42 unk.	0.04 0(a)	**3** **0**
Q	beef,steak,chuck blade,w/o bone, braised,w/fat	4-1/8x2-3/4x 1/2" piece	85 *1.18*	363 34.3	0.0 31.2	0.00 0(a)	19.0 0.0	**42**	1203 649	916 1544	1692 514	837 874	953 213	243 685	3.82 12.58	13.00 0.29	0.76 unk.	0.03 0(a)	**2** **0**
R	beef,steak,cubed,cooked, no visible fat	3-3/4x2-3/4x 1/2" (cubed)	85 *1.18*	222 46.5	0.0 13.1	0.00 0(a)	24.3 0.0	**54**	1535 828	1170 1971	2161 656	1068 1115	1217 272	310 875	1.53 5.01	5.35 0.29	0.34 80	0.07 0	**5** **0**
S	beef,steak,flank,w/o bone,braised, w/fat	**2-1/2x2-1/2x 3/4" piece**	85 *1.18*	167 52.2	0.0 6.2	0.00 0(a)	25.9 0.0	**58**	1637 883	1524 2387	2549 723	1198 1286	1618 341	331 933	0.79 2.27	2.63 0.30	0.24 77	0.50 0(a)	**3** **0**

Riboflavin mg / Niacin mg	% USRDA	Vitamin B_6 mg / Folacin mcg	% USRDA	Vitamin B_{12} mcg / Pantothenic acid mg	% USRDA	Biotin mg / Vitamin A IU	% USRDA	Preformed A RE / Beta carotene RE	Vitamin D IU / Vitamin E IU	% USRDA	Total tocopherol mg / Alpha tocopherol mg	Other tocopherol mg / Total ash g	Calcium mg / Phosphorus mg	% USRDA	Sodium mg / Sodium meq	Potassium mg / Potassium meq	Chlorine mg / Chlorine meq	Iron mg / Magnesium mg	% USRDA	Zinc mg / Copper mg	% USRDA	Iodine mcg / Selenium mcg	% USRDA	Manganese mcg / Chromium mcg	
0.22	**13**	0.27	**14**	2.67	**44**	unk.	**unk.**	29	0	**0**	0.1	tr(a)	12	**1**	49	224	unk.	3.7	**20**	5.2	**35**	unk.	**unk.**	0(a)	A
4.4	**22**	5	**1**	0.32	**3**	109	**2**	1	0.2	**1**	0.1	0.82	137	**14**	2.2	5.7	unk.	20	**5**	0.11	**5**	tr(a)		tr(a)	
0.20	**12**	0.17	**8**	2.28	**38**	unk.	**unk.**	2	0	**0**	0.1	tr(a)	12	**1**	51	314	unk.	3.2	**18**	5.3	**35**	unk.	**unk.**	unk.	B
3.9	**20**	3	**1**	0.20	**2**	8	**0**	tr	0.2	**1**	0.1	0.68	128	**13**	2.2	8.0	unk.	15	**4**	0.07	**3**	tr(a)		tr(a)	
0.18	**11**	0.22	**11**	2.02	**34**	unk.	**unk.**	7	0	**0**	0.1	tr(a)	10	**1**	51	314	unk.	2.9	**16**	4.1	**27**	unk.	**unk.**	unk.	C
3.6	**18**	3	**1**	0.27	**3**	25	**1**	tr	0.1	**0**	0.1	0.59	114	**11**	2.2	8.0	unk.	13	**3**	0.07	**3**	tr(a)		tr(a)	
0.15	**9**	0.18	**9**	1.88	**31**	unk.	**unk.**	4	0	**0**	0.1	tr(a)	8	**1**	42	259	unk.	2.5	**14**	4.1	**27**	unk.	**unk.**	unk.	D
3.6	**18**	3	**1**	0.22	**2**	14	**0**	tr	0.1	**0**	0.1	0.77	179	**18**	1.8	6.6	unk.	20	**5**	0.06	**3**	tr(a)		tr(a)	
0.17	**10**	0.22	**11**	2.22	**37**	unk.	**unk.**	4	0	**0**	0.1	tr(a)	10	**1**	50	307	unk.	3.0	**17**	4.8	**32**	unk.	**unk.**	unk.	E
4.2	**21**	3	**1**	0.27	**3**	17	**0**	tr	0.1	**1**	0.1	0.91	213	**21**	2.2	7.8	unk.	23	**6**	0.07	**3**	tr(a)		tr(a)	
0.12	**7**	0.16	**8**	1.57	**26**	unk.	**unk.**	18	0	**0**	0.1	tr(a)	7	**1**	50	307	unk.	2.2	**12**	3.2	**22**	unk.	**unk.**	unk.	F
3.0	**15**	3	**1**	0.20	**2**	67	**1**	1	0.1	**0**	0.1	0.58	154	**15**	2.2	7.8	unk.	17	**4**	0.07	**3**	tr(a)		tr(a)	
0.15	**9**	0.18	**9**	1.94	**32**	unk.	**unk.**	4	0	**0**	0.1	tr(a)	8	**1**	42	259	unk.	2.6	**14**	4.1	**27**	unk.	**unk.**	unk.	G
3.6	**18**	3	**1**	0.22	**2**	14	**0**	tr	0.1	**0**	0.1	0.91	170	**17**	1.8	6.6	unk.	20	**5**	0.06	**3**	tr(a)		tr(a)	
0.18	**11**	0.22	**11**	2.30	**38**	unk.	**unk.**	4	0	**0**	0.1	tr(a)	10	**1**	50	307	unk.	3.1	**17**	4.8	**32**	unk.	**unk.**	unk.	H
4.3	**22**	3	**1**	0.27	**3**	17	**0**	tr	0.1	**1**	0.1	1.08	202	**20**	2.2	7.8	unk.	24	**6**	0.07	**3**	tr(a)		tr(a)	
0.15	**9**	0.16	**8**	1.87	**31**	unk.	**unk.**	11	0	**0**	0.1	tr(a)	8	**1**	50	307	unk.	2.6	**14**	3.7	**25**	unk.	**unk.**	unk.	I
3.6	**18**	3	**1**	0.20	**2**	41	**1**	tr	0.1	**0**	0.1	0.83	164	**16**	2.2	7.8	unk.	23	**6**	0.07	**3**	tr(a)		tr(a)	
0.16	**10**	0.18	**9**	1.99	**33**	unk.	**unk.**	4	0	**0**	0.1	tr(a)	8	**1**	42	259	unk.	2.5	**14**	4.1	**27**	unk.	**unk.**	unk.	J
4.1	**20**	3	**1**	0.22	**2**	14	**0**	tr	0.1	**0**	0.1	0.98	167	**17**	1.8	6.6	unk.	20	**5**	0.06	**3**	tr(a)		tr(a)	
0.20	**12**	0.22	**11**	2.41	**40**	unk.	**unk.**	5	0	**0**	0.1	tr(a)	10	**1**	51	314	unk.	3.1	**17**	4.9	**33**	unk.	**unk.**	unk.	K
4.9	**25**	3	**1**	0.27	**3**	17	**0**	tr	0.2	**1**	0.1	1.19	203	**20**	2.2	8.0	unk.	25	**6**	0.07	**3**	tr(a)		tr(a)	
0.14	**8**	0.17	**8**	1.55	**26**	unk.	**unk.**	18	0	**0**	0.1	tr(a)	8	**1**	51	314	unk.	2.1	**12**	2.9	**19**	unk.	**unk.**	unk.	L
3.4	**17**	3	**1**	0.20	**2**	68	**1**	1	0.1	**0**	0.1	0.76	139	**14**	2.2	8.0	unk.	18	**5**	0.07	**3**	tr(a)		tr(a)	
0.19	**11**	0.17	**8**	2.00	**33**	unk.	**unk.**	unk.	0	**0**	0.1	tr(a)	13	**1**	50	218	unk.	2.0	**11**	5.2	**35**	unk.	**unk.**	unk.	M
3.5	**18**	3	**1**	0.20	**2**	34	**1**	unk.	0.1	**0**	0.1	1.68	97	**10**	2.2	5.6	unk.	unk.	**unk.**	0.07	**3**	tr(a)		tr(a)	
0.24	**14**	0.28	**14**	2.72	**45**	unk.	**unk.**	6	0	**0**	0.1	tr(a)	14	**1**	64	396	unk.	4.0	**22**	6.2	**41**	unk.	**unk.**	unk.	N
4.8	**24**	4	**1**	0.34	**3**	21	**0**	tr	0.2	**1**	0.1	0.75	153	**15**	2.8	10.1	unk.	19	**5**	0.09	**4**	tr(a)		tr(a)	
0.21	**13**	0.25	**12**	2.46	**41**	unk.	**unk.**	24	0	**0**	0.1	tr(a)	13	**1**	49	224	unk.	3.6	**20**	6.3	**42**	unk.	**unk.**	unk.	O
4.4	**22**	5	**1**	0.29	**3**	88	**2**	1	0.2	**1**	0.1	0.75	138	**14**	2.2	5.7	unk.	19	**5**	0.10	**5**	tr(a)		tr(a)	
0.19	**11**	0.22	**11**	2.16	**36**	unk.	**unk.**	5	0	**0**	0.1	tr(a)	11	**1**	51	314	unk.	3.1	**18**	4.9	**33**	unk.	**unk.**	unk.	P
3.8	**19**	3	**1**	0.27	**3**	17	**0**	tr	0.1	**1**	0.1	0.59	122	**12**	2.2	8.0	unk.	unk.	**unk.**	0.07	**3**	tr(a)		tr(a)	
0.14	**9**	0.17	**8**	1.67	**28**	unk.	**unk.**	16	0	**0**	0.1	tr(a)	8	**1**	33	152	unk.	2.5	**14**	4.3	**28**	unk.	**unk.**	unk.	Q
3.0	**15**	3	**7**	0.20	**2**	59	**1**	1	0.1	**0**	0.1	0.51	94	**9**	1.4	3.9	unk.	unk.	**unk.**	0.07	**3**	tr(a)		tr(a)	
0.19	**11**	0.22	**11**	2.14	**36**	unk.	**unk.**	7	0	**0**	0.1	tr(a)	10	**1**	51	314	unk.	3.0	**17**	4.1	**27**	unk.	**unk.**	unk.	R
4.8	**24**	3	**1**	0.27	**3**	26	**1**	tr	0.1	**1**	0.1	1.10	213	**21**	2.2	8.0	unk.	24	**6**	0.07	**3**	tr(a)		tr(a)	
0.20	**12**	0.22	**11**	2.28	**38**	unk.	**unk.**	2	0	**0**	0.1	tr(a)	12	**1**	51	314	unk.	9.2	**18**	5.3	**35**	unk.	**unk.**	unk.	S
3.9	**20**	3	**1**	0.27	**3**	8	**0**	tr	0.2	**1**	0.1	0.68	125	**13**	2.2	8.0	unk.	16	**4**	0.07	**3**	tr(a)		tr(a)	

	FOOD	Portion	Weight in grams Conversion for 100 g	Kilocalories H₂O g	Total carbohydrate g Total fats g	Crude fiber g Dietary fiber g	Total protein g Total sugar g	% USRDA	Arginine mg Histidine mg	Isoleucine mg Leucine mg	Lysine mg Methionine mg	Phenylalanine mg Threonine mg	Valine mg Tryptophan mg	Cystine mg Tyrosine mg	Polyunsat. fatty acids g Monounsat. fatty acids g	Saturated fatty acids g P/S ratio	Linoleic acid g Cholesterol mg	Thiamin mg Ascorbic acid mg	% USRDA
A	beef,steak,minute,cooked	3-3/4x2-3/4x	85	222	0.0	0.00	24.3	**54**	1535	1170	2161	1068	1217	310	1.53	5.35	0.34	0.07	**5**
		1/2"	*1.18*	46.5	13.1	0(a)	0.0		828	1971	656	1115	272	875	5.01	0.29	80	0	**0**
B	beef,steak,porterhouse,w/bone,trimmed,	1 med	172	385	0.0	0.00	51.9	**115**	3282	2501	4616	2282	2601	664	2.24	7.74	0.69	0.14	**9**
	broiled,no visible fat	(1 lb raw)	*0.58*	99.6	18.1	0(a)	0.0		1770	4211	1404	2382	581	1870	6.54	0.29	157	0(a)	**0**
C	beef,steak,porterhouse,w/bone,	1 med	301	1400	0.0	0.00	59.3	**132**	3744	2853	5271	2607	2968	759	15.35	52.98	2.71	0.18	**12**
	broiled,w/fat	(1 lb raw)	*0.33*	112.0	127.0	0(a)	0.0		2020	4807	1601	2721	662	2134	52.07	0.29	283	0(a)	**0**
D	beef,steak,porterhouse,w/o bone,trimmed,	4-1/8x2-3/4x	85	190	0.0	0.00	25.7	**57**	1622	1236	2281	1128	1285	328	1.10	3.82	0.34	0.07	**5**
	broiled,no visible fat	1/2"	*1.18*	49.2	8.9	0(a)	0.0		875	2081	694	1177	287	924	3.23	0.29	77	0(a)	**0**
E	beef,steak,porterhouse,w/o bone,	4-1/8x2-3/4x	85	395	0.0	0.00	16.7	**37**	1057	806	1488	736	838	214	4.33	14.96	0.76	0.05	**3**
	broiled,w/fat	1/2"	*1.18*	31.6	35.9	0(a)	0.0		570	1357	452	768	187	603	14.70	0.29	80	0(a)	**0**
F	beef,steak,rib,w/bone,broiled,trimmed,	1 med	121	292	0.0	0.00	34.1	**76**	2155	1642	3032	1500	1709	436	1.81	7.14	0.60	0.08	**6**
	no visible fat	(1 lb raw)	*0.83*	69.2	16.2	0(a)	0.0		1163	2766	922	1566	381	1228	5.93	0.25	110	0(a)	**0**
G	beef,steak,rib,w/bone,broiled,w/fat	1 med	229	1008	0.0	0.00	45.6	**101**	2879	2194	4051	2004	2281	582	10.74	37.78	2.06	0.11	**8**
		(1 lb raw)	*0.44*	91.6	90.2	0(a)	0.0		1553	3694	1230	2091	511	1640	36.64	0.28	215	0	**0**
H	beef,steak,rib,w/o bone,broiled,	4-1/8x2-3/4x	85	205	0.0	0.00	24.0	**53**	1514	1153	2130	1054	1200	306	1.27	5.01	0.42	0.06	**4**
	no visible fat	1/2"	*1.18*	48.6	11.4	0(a)	0.0		817	1943	648	1100	268	863	4.16	0.25	77	0(a)	**0**
I	beef,steak,rib,w/o bone,broiled,	4-1/8x2-3/4x	85	374	0.0	0.00	16.9	**38**	1068	814	1504	744	847	216	3.99	14.02	0.76	0.04	**3**
	w/fat	1/2"	*1.18*	34.0	33.5	0(a)	0.0		576	1371	456	776	190	609	13.60	0.28	80	0	**0**
J	beef,steak,round,w/o bone,braised,	4-1/8x2-3/4x	85	161	0.0	0.00	26.6	**59**	1680	1281	2365	1170	1332	340	0.73	2.29	0.21	0.07	**5**
	no visible fat	1/2"	*1.18*	52.0	5.2	0(a)	0.0		906	2157	718	1221	297	957	1.99	0.32	77	0	**0**
K	beef,steak,round,w/o bone,braised,	4-1/8x2-3/4x	85	222	0.0	0.00	24.3	**54**	1535	1170	2161	1068	1217	310	1.53	5.35	0.34	0.07	**5**
	w/fat	1/2"	*1.18*	46.5	13.1	0(a)	0.0		828	1971	656	1115	272	875	5.01	0.29	80	0	**0**
L	beef,steak,sirloin,w/bone,broiled,trimmed,	1/4 med (1	102	211	0.0	0.00	32.8	**73**	2075	1581	2919	1443	1644	419	1.00	3.37	0.31	0.09	**6**
	no visible fat	med=2 lb raw)	*0.98*	59.9	7.9	0(a)	0.0		1119	2663	887	1507	367	1182	2.88	0.30	93	0	**0**
M	beef,steak,sirloin,w/bone,broiled,	1/4 med (1	154	596	0.0	0.00	35.4	**79**	2238	1705	3148	1557	1773	453	6.01	20.48	1.08	0.09	**6**
	w/fat	med=2 lb raw)	*0.65*	67.6	49.3	0(a)	0.0		1207	2872	956	1625	396	1275	20.17	0.29	145	0	**0**
N	beef,steak,sirloin,w/o bone,broiled,	4-1/8x2-3/4x	85	176	0.0	0.00	27.4	**61**	1729	1317	2433	1203	1370	349	0.83	2.80	0.25	0.08	**5**
	no visible fat	1/2"	*1.18*	49.9	6.5	0(a)	0.0		932	2219	739	1255	306	985	2.40	0.30	77	0	**0**
O	beef,steak,sirloin,w/o bone,broiled,	4-1/8x2-3/4x	85	329	0.0	0.00	19.5	**43**	1235	941	1737	859	978	250	3.31	11.30	0.59	0.05	**3**
	w/fat	1/2"	*1.18*	37.3	27.2	0(a)	0.0		666	1585	528	897	218	704	11.13	0.29	80	0	**0**
P	beef,steak,t-bone,w/bone,broiled,trimmed,	1 med	111	248	0.0	0.00	33.9	**75**	2139	1630	3010	1488	1696	432	1.34	4.90	0.45	0.09	**6**
	no visible fat	(10.8 oz raw)	*0.90*	64.5	11.5	0(a)	0.0		1154	2746	915	1554	379	1219	4.23	0.27	101	0(a)	**0**
Q	beef,steak,t-bone,w/bone,broiled,	1 med	199	941	0.0	0.00	38.8	**86**	2450	1868	3447	1705	1941	495	10.54	35.80	1.79	0.12	**8**
	w/fat	(10.8 oz raw)	*0.50*	72.4	85.9	0(a)	0.0		1321	3145	1048	1780	434	1396	35.21	0.29	187	0(a)	**0**
R	beef,steak,t-bone,w/o bone,broiled,	4-1/8x2-3/4	85	190	0.0	0.00	25.8	**57**	1632	1244	2297	1136	1294	330	1.02	3.74	0.34	0.07	**5**
	no visible fat	x1/2"	*1.18*	49.2	8.8	0(a)	0.0		881	2095	698	1186	289	930	3.23	0.27	77	0(a)	**0**
S	beef,steak,t-bone,w/o bone,broiled,	4-1/8x2-3/4x	85	402	0.0	0.00	16.6	**37**	1047	798	1473	728	830	212	4.50	15.30	0.76	0.05	**3**
	w/fat	1/2"	*1.18*	30.9	36.7	0(a)	0.0		564	1344	448	761	185	597	15.04	0.29	80	0(a)	**0**

	Riboflavin mg / Niacin mg	% USRDA	Vitamin B6 mg / Folacin mcg	% USRDA	Vitamin B12 mcg / Pantothenic acid mg	% USRDA	Biotin mg / Vitamin A IU	% USRDA	Preformed A RE / Beta carotene RE	Vitamin D IU / Vitamin E IU	% USRDA	Total tocopherol mg / Alpha tocopherol mg	Other tocopherol mg / Total ash g	Calcium mg / Phosphorus mg	% USRDA	Sodium mg / Sodium meq	Potassium mg / Potassium meq	Chlorine mg / Chlorine meq	Iron mg / Magnesium mg	% USRDA	Zinc mg / Copper mg	% USRDA	Iodine mcg / Selenium mcg	% USRDA	Manganese mcg / Chromium mcg
A	0.19	**11**	0.22	**11**	2.14	**36**	unk.	**unk.**	7	0	**0**	0.1	tr(a)	10	**1**	51	3.14	unk.	3.0	**17**	4.1	**27**	unk.	**unk.**	unk.
	4.8	**24**	3	**1**	0.27	**3**	25	**1**	tr	0.01	**1**	0.1	1.10	213	**21**	2.2	8.0	unk.	24	**6**	0.07	**3**	tr(a)		tr(a)
B	0.40	**23**	0.45	**22**	4.85	**81**	unk.	**unk.**	9	0	**0**	0.2	tr(a)	21	**2**	103	636	unk.	6.4	**35**	10.0	**67**	unk.	**unk.**	unk.
	10.1	**51**	7	**2**	0.55	**6**	34	**1**	tr	0.2	**1**	0.2	2.41	416	**42**	4.5	16.3	unk.	50	**13**	0.14	**7**	tr(a)		tr(a)
C	0.48	**28**	0.60	**30**	5.54	**92**	unk.	**unk.**	57	0	**0**	0.2	tr(a)	27	**3**	181	1114	unk.	7.8	**44**	10.6	**71**	unk.	**unk.**	unk.
	12.6	**63**	12	**3**	0.71	**7**	211	**4**	2	0.3	**1**	0.2	2.71	506	**51**	7.9	28.5	unk.	63	**16**	0.24	**12**	tr(a)		tr(a)
D	0.20	**12**	0.22	**11**	2.40	**40**	unk.	**unk.**	5	0	**0**	0.1	0.2	10	**1**	51	314	unk.	3.1	**18**	4.9	**33**	unk.	**unk.**	unk.
	5.0	**25**	3	**1**	0.27	**3**	17	**0**	tr	0.1	**0**	0.1	1.19	206	**21**	2.2	8.0	unk.	25	**6**	0.07	**3**	tr(a)		tr(a)
E	0.14	**8**	0.17	**unk.**	1.56	**26**	unk.	**unk.**	16	0	**0**	0.1	tr(a)	8	**1**	51	314	unk.	2.2	**12**	3.0	**20**	unk.	**unk.**	unk.
	3.6	**18**	3	**1**	0.20	**2**	59	**1**	1	0.1	**0**	0.1	0.76	143	**14**	2.2	8.0	unk.	18	**5**	0.07	**3**	tr(a)		tr(a)
F	0.25	**15**	0.32	**16**	3.19	**53**	unk.	**unk.**	7	0	**0**	0.1	tr(a)	15	**2**	73	448	unk.	4.4	**24**	7.0	**47**	unk.	**unk.**	unk.
	6.2	**31**	5	**1**	0.39	**4**	24	**1**	tr	0.2	**1**	0.1	1.33	310	**31**	3.2	11.4	unk.	34	**9**	0.10	**5**	tr(a)		tr(a)
G	0.34	**20**	0.45	**23**	4.26	**71**	unk.	**unk.**	50	0	**0**	0.2	tr(a)	21	**2**	137	847	unk.	6.0	**33**	8.9	**59**	unk.	**unk.**	unk.
	8.2	**41**	9	**2**	0.54	**5**	183	**4**	2	0.2	**1**	0.2	1.60	426	**43**	6.0	21.7	unk.	46	**11**	0.18	**9**	tr(a)		tr(a)
H	0.18	**11**	0.22	**11**	2.24	**37**	unk	**unk**	5	0	**0**	0.1	tr(a)	10	**1**	51	314	unk.	3.1	**17**	4.9	**33**	unk.	**unk.**	unk.
	4.3	**22**	3	**1**	0.27	**3**	17	**0**	tr	0.1	**0**	0.1	0.93	218	**22**	2.2	8.0	unk.	24	**6**	0.07	**3**	tr(a)		tr(a)
I	0.13	**8**	0.17	**8**	1.58	**26**	unk.	**unk.**	18	0	**0**	0.1	tr(a)	8	**1**	51	314	unk.	2.2	**12**	3.3	**22**	unk.	**unk.**	unk.
	3.1	**15**	3	**1**	0.20	**2**	68	**1**	1	0.1	**0**	0.1	0.59	158	**16**	2.2	8.0	unk.	17	**4**	0.07	**3**	tr(a)		tr(a)
J	0.20	**12**	0.22	**11**	2.34	**39**	unk.	**unk.**	2	0	**0**	0.1	tr(a)	11	**1**	51	314	unk.	3.1	**18**	4.9	**33**	unk.	**unk.**	unk.
	5.1	**26**	3	**1**	0.27	**3**	9	**0**	tr	0.2	**1**	0.1	1.19	228	**23**	2.2	8.0	unk.	25	**6**	0.07	**3**	tr(a)		tr(a)
K	0.19	**11**	0.22	**11**	2.31	**39**	unk.	**unk.**	7	0	**0**	0.1	tr(a)	10	**1**	51	314	unk.	3.0	**17**	4.1	**27**	unk.	**unk.**	unk.
	4.8	**24**	3	**1**	0.27	**3**	25	**1**	tr	0.1	**1**	0.1	1.10	213	**21**	2.2	8.0	unk.	24	**6**	0.07	**3**	tr(a)		tr(a)
L	0.25	**15**	0.27	**13**	3.07	**51**	unk.	**unk.**	3	0	**0**	0.1	tr(a)	13	**1**	61	377	unk.	4.0	**22**	5.9	**39**	unk.	**unk.**	unk.
	6.5	**33**	4	**1**	0.33	**3**	10	**0**	tr	0.2	**1**	0.1	1.53	266	**27**	2.7	9.6	unk.	30	**7**	0.08	**4**	tr(a)		tr(a)
M	0.28	**16**	0.30	**15**	3.31	**55**	unk.	**unk.**	21	0	**0**	0.1	tr(a)	15	**2**	92	570	unk.	4.5	**25**	6.2	**41**	unk.	**unk.**	unk.
	7.2	**36**	6	**2**	0.36	**4**	77	**2**	1	0.2	**1**	0.1	1.69	294	**29**	4.0	14.6	unk.	32	**8**	0.12	**6**	tr(a)		tr(a)
N	0.21	**13**	tr	**0**	3.40	**57**	unk.	**unk.**	2	unk.	**unk.**	0.2	unk.	11	**1**	51	314	unk.	3.3	**18**	4.9	**33**	unk.	**unk.**	unk.
	5.4	**27**	unk.	**unk.**	unk.	**unk.**	8	**0**	tr	0.2	**1**	0.1	1.27	222	**22**	2.2	8.0	unk.	25	**6**	unk.	**unk.**	unk.		unk.
O	0.15	**9**	0.17	**8**	1.83	**31**	unk.	**unk.**	11	0	**0**	0.1	tr(a)	8	**1**	51	314	unk.	2.5	**14**	3.4	**23**	unk.	**unk.**	unk.
	4.0	**20**	3	**1**	0.20	**2**	42	**1**	tr	0.1	**0**	0.1	0.93	162	**16**	2.2	8.0	unk.	18	**5**	0.07	**3**	tr(a)		tr(a)
P	0.26	**15**	0.29	**15**	3.16	**53**	unk.	**unk.**	6	0	**0**	0.1	tr(a)	13	**1**	67	412	unk.	4.1	**23**	6.5	**43**	unk.	**unk.**	unk.
	6.6	**33**	4	**1**	0.36	**4**	23	**1**	tr	0.2	**1**	0.1	1.56	271	**27**	2.9	10.5	unk.	32	**8**	0.09	**5**	tr(a)		0
Q	0.32	**19**	0.39	**20**	3.62	**60**	unk.	**unk.**	43	0	**0**	0.2	tr(a)	16	**2**	119	736	unk.	5.2	**29**	6.9	**46**	unk.	**unk.**	unk.
	8.2	**41**	8	**2**	0.47	**5**	159	**3**	2	0.2	**1**	0.2	1.79	330	**33**	5.2	18.8	unk.	42	**10**	0.16	**8**	tr(a)		tr(a)
R	0.20	**12**	0.22	**11**	2.41	**40**	unk.	**unk.**	5	0	**0**	0.1	tr(a)	10	**1**	51	314	unk.	3.1	**18**	4.9	**33**	unk.	**unk.**	unk.
	5.0	**25**	3	**1**	0.27	**3**	17	**0**	tr	0.1	**0**	0.1	1.19	207	**21**	2.2	8.0	unk.	25	**6**	0.07	**3**	tr(a)		tr(a)
S	0.14	**8**	0.17	**8**	1.55	**26**	unk.	**unk.**	18	0	**0**	0.1	tr(a)	7	**1**	51	314	unk.	2.2	**12**	2.9	**20**	unk.	**unk.**	unk.
	3.5	**17**	3	**1**	0.20	**2**	68	**1**	1	0.1	**0**	0.1	0.76	141	**14**	2.2	8.0	unk.	18	**5**	0.07	**3**	tr(a)		tr(a)

	FOOD	Portion	Weight in grams *Conversion for 100 g*	Kilocalories H_2O g	Total carbohydrate g Total fats g	Crude fiber g Dietary fiber g	Total protein g Total sugar g	**% USRDA**	Arginine mg Histidine mg	Isoleucine mg Leucine mg	Lysine mg Methionine mg	Phenylalanine mg Threonine mg	Valine mg Tryptophan mg	Cystine mg Tyrosine mg	Polyunsat. fatty acids g Monounsat. fatty acids g	Saturated fatty acids g P/S ratio	Linoleic acid g Cholesterol mg	Thiamin mg Ascorbic acid mg	**% USRDA**
A	beef,steak,tenderloin,w/o bone, broiled,no visible fat	3-1/4x3-1/4x 3/4"	118 *0.85*	263 68.3	0.0 12.2	0.00 0(a)	35.9 0.0	**80**	2266 1222	1726 2909	3188 969	1576 1646	1796 401	458 1291	1.42 4.48	5.19 0.27	0.47 107	0.09 0(a)	**6** **0**
B	beef,steak,toploin (club),w/bone,trimmed, broiled,no visible fat	1 med (9.6 oz raw)	81 *1.23*	198 45.4	0.0 10.5	0.00 0(a)	24.0 0.0	**53**	1515 816	1154 1944	2131 647	1054 1100	1200 268	306 863	1.05 3.08	3.64 0.29	0.32 74	0.06 0(a)	**4** **0**
C	beef,steak,toploin (club),w/bone, broiled,w/fat	1 med (8 oz raw)	139 *0.72*	631 52.7	0.0 56.4	0.00 0(a)	28.6 0.0	**64**	1808 976	1379 2321	2545 773	1258 1314	1433 320	366 1030	7.09 24.05	24.46 0.29	1.25 131	0.08 0(a)	**6** **0**
D	beef,steak,toploin (strip),w/o bone, broiled,no visible fat	4-1/2x3x3/4" (8 oz raw)	150 *0.67*	335 86.8	0.0 15.4	0.00 0(a)	45.6 0.0	**101**	2880 1554	2195 3698	4053 1232	2004 2093	2283 510	582 1641	1.80 5.70	6.60 0.27	0.60 137	0.12 0(a)	**8** **0**
E	beef,steak,toploin (strip),w/o bone, broiled,w/fat	4-1/8x2-3/4x 1/2"	87 *1.15*	412 31.7	0.0 37.6	0.00 0(a)	17.0 0.0	**38**	1072 578	817 1375	1508 458	746 779	849 190	217 611	4.61 15.40	15.66 0.29	0.78 82	0.05 0(a)	**4** **0**
F	beef,steak,steakette,w/o bone,cooked,w/fat	2 oz raw	39 *2.54*	103 21.5	0.0 6.1	0.00 0(a)	11.2 0.0	**25**	710 383	541 911	999 303	494 516	563 126	143 404	0.71 2.32	2.48 0.29	0.16 37	0.03 0	**2** **0**
G	BEET GREENS,fresh,cooked, drained solids	1/2 C	73 *1.38*	13 67.9	2.4 0.1	0.80 unk.	1.2 unk.	**2**	unk. 24	51 79	67 21	72 47	62 14	14 36	unk. unk.	tr(a) unk.	unk. 0	0.05 11	**3** **18**
H	BEETS,canned,sliced or chopped, drained solids	1/2 C	85 *1.18*	31 75.9	7.5 0.1	0.68 2.30	0.8 6.7	**1**	56 11	19 24	23 8	12 15	24 7	9 14	0.00 unk.	0.00 0.00	unk. 0(a)	0.01 3	**1** **4**
I	beets,canned,sliced or chopped, low sodium,drained solids	1/2 C	85 *1.18*	31 76.3	7.4 0.1	0.68 2.30	0.8 6.7	**1**	50 18	28 52	42 8	12 29	24 10	6 25	0.00 unk.	0.00 0.00	unk. 0(a)	0.01 3	**1** **4**
J	beets,fresh,sliced or chopped,cooked, drained solids	1 med	50 *2.00*	16 45.4	3.6 0.0	0.40 1.35	0.5 3.2	**1**	36 7	17 19	30 2	9 11	17 5	6 8	unk. unk.	0.00 unk.	unk. 0(a)	0.01 3	**1** **5**
K	beets,fresh,whole,cooked, drained solids	1/2 C	85 *1.18*	27 77.3	6.1 0.1	0.68 2.30	0.9 5.5	**1**	61 11	29 31	50 3	15 19	28 8	9 14	unk. unk.	0.00 unk.	unk. 0(a)	0.03 5	**2** **9**
L	beets,pickled,sliced,drained solids- Del Monte	1/2 C	88 *1.14*	62 71.2	14.8 0.4	unk. unk.	0.6 5.7	**1**	unk. 8	19 21	32 2	10 13	19 5	unk. unk.	unk. unk.	tr(a) unk.	unk. 0	0.02 1	**1** **2**
M	BEVERAGE,ALCOHOLIC,ALE,mild	12 oz bottle /can	360 *0.28*	155 unk.	12.5 0.0	unk. unk.	1.7 2.8	**3**	unk. unk.	unk. unk.	unk. unk.	unk. unk.	unk. unk.	unk. unk.	0(a) 0(a)	0(a) 0.00	0(a) 0(a)	tr 0	**0** **0**
N	BEVERAGE,ALCOHOLIC,BEER,light, 4.2% alcohol by volume	12 oz bottle /can	360 *0.28*	112 348.1	7.1 0.0	0(a) 0(a)	1.2 unk.	**2**	unk. unk.	unk. unk.	unk. unk.	unk. unk.	unk. unk.	unk. unk.	0(a) 0(a)	0(a) 0.00	0(a) tr	0.00 tr	**0** **0**
O	beverage,alcoholic,beer, 4.5% alcohol by volume	12 oz bottle /can	360 *0.28*	151 331.6	13.7 0.0	0.00 0(a)	1.1 0.0	**2**	31 unk.	unk. unk.	unk. unk.	unk. unk.	unk. unk.	unk. unk.	0(a) 0(a)	0(a) 0.00	0(a) 0(a)	0.00 unk.	**0** **unk.**
P	BEVERAGE,ALCOHOLIC,BRANDY, California	1 oz glass	30 *3.33*	73 unk.	unk. unk.	unk. unk.	unk. unk.	**unk.**	unk. unk.	unk. unk.	unk. unk.	unk. unk.	unk. unk.	unk. unk.	unk. unk.	unk. unk.	unk. unk.	unk. unk.	**unk.** **unk.**
Q	BEVERAGE,ALCOHOLIC,COCKTAIL, daiquiri	2-1/2 oz glass	71 *1.41*	87 unk.	3.7 0.0	unk. 0(a)	0.1 3.7	**0**	unk. unk.	unk. unk.	unk. unk.	unk. unk.	unk. unk.	unk. unk.	unk. unk.	unk. unk.	unk. unk.	0.01 6	**1** **10**
R	beverage,alcoholic,cocktail,highball	8 oz glass	240 *0.42*	166 unk.	unk. unk.	unk. 0(a)	unk. unk.	**unk.**	unk. unk.	unk. unk.	unk. unk.	unk. unk.	unk. unk.	unk. unk.	unk. unk.	unk. unk.	unk. unk.	unk. unk.	**unk.** **unk.**
S	beverage,alcoholic,cocktail,Manhattan	2-1/2 oz glass	71 *1.41*	116 unk.	5.6 0.0	unk. 0(a)	tr 5.6	**0**	unk. unk.	unk. unk.	unk. unk.	unk. unk.	unk. unk.	unk. unk.	unk. unk.	unk. unk.	unk. unk.	0.00 0	**0** **0**

Riboflavin mg / Niacin mg	% USRDA	Vitamin B_6 mg / Folacin mcg	% USRDA	Vitamin B_{12} mcg / Pantothenic acid mg	% USRDA	Biotin mg / Vitamin A IU	% USRDA	Preformed A RE / Beta carotene RE	Vitamin D IU / Vitamin E IU	% USRDA	Total tocopherol mg / Alpha tocopherol mg	Other tocopherol mg / Total ash g	Calcium mg / Phosphorus mg	% USRDA	Sodium mg / Sodium meq	Potassium mg / Potassium meq	Chlorine mg / Chlorine meq	Iron mg / Magnesium mg	% USRDA	Zinc mg / Copper mg	% USRDA	Iodine mcg / Selenium mcg	% USRDA	Manganese mcg / Chromium mcg	
0.27	**16**	0.31	**15**	3.35	**56**	unk.	**unk.**	6	0	**0**	0.2	tr(a)	14	**1**	71	437	unk.	4.4	**24**	6.8	**46**	unk.	**unk.**	unk.	A
7.0	**35**	5	**1**	0.38	**4**	24	**1**	tr	0.2	**1**	0.2	1.65	287	**29**	3.1	11.2	unk.	34	**9**	0.09	**5**	tr(a)		tr(a)	
0.19	**11**	0.21	**11**	2.24	**37**	unk.	**unk.**	4	0	**0**	0.1	tr(a)	10	**1**	49	300	unk.	2.9	**16**	4.7	**31**	unk.	**unk.**	unk.	B
4.7	**24**	3	**1**	0.26	**3**	16	**0**	tr	0.1	**0**	0.1	1.13	193	**19**	2.1	7.7	unk.	23	**6**	0.06	**3**	tr(a)		tr(a)	
0.24	**14**	0.28	**14**	2.67	**45**	unk.	**unk.**	26	0	**0**	0.1	tr(a)	13	**1**	83	514	unk.	3.8	**21**	5.0	**33**	unk.	**unk.**	unk.	C
6.0	**30**	6	**1**	0.33	**3**	97	**2**	1	0.1	**0**	0.1	1.25	243	**24**	3.6	13.1	unk.	29	**7**	0.11	**6**	tr(a)		tr(a)	
0.34	**20**	0.39	**20**	4.26	**71**	unk.	**unk.**	8	0	**0**	0.2	tr(a)	18	**2**	90	555	unk.	5.5	**31**	8.7	**58**	unk.	**unk.**	unk.	D
8.8	**44**	6	**2**	0.48	**5**	30	**1**	tr	0.2	**1**	0.2	2.10	365	**36**	3.9	14.2	unk.	44	**11**	0.12	**6**	tr(a)		tr(a)	
0.14	**8**	0.17	**9**	1.58	**26**	unk.	**unk.**	19	0	**0**	0.1	tr(a)	7	**1**	52	322	unk.	2.3	**13**	3.0	**20**	unk.	**unk.**	unk.	E
3.6	**18**	3	**1**	0.20	**2**	70	**1**	1	0.1	**0**	0.1	0.78	144	**14**	2.3	8.2	unk.	18	**5**	0.07	**4**	tr(a)		tr(a)	
0.09	**5**	0.10	**5**	1.07	**18**	unk.	**unk.**	3	0	**0**	0.1	tr(a)	5	**1**	24	145	unk.	1.4	**8**	1.9	**13**	unk.	**unk.**	unk.	F
2.2	**11**	2	**0**	0.13	**1**	12	**0**	tr	0.1	**0**	0.1	0.51	98	**10**	1.0	3.7	unk.	11	**3**	0.03	**2**	tr(a)		tr(a)	
0.11	**6**	unk.	**unk.**	0.00	**0**	unk.	**unk.**	0	0(a)	**0**	unk.	unk.	72	**7**	55	241	unk.	1.4	**8**	unk.	**unk.**	unk.	**unk.**	unk.	G
0.2	**1**	unk.	**unk.**	unk.	**unk.**	3697	**74**	370	unk.	**unk.**	unk.	0.87	18	**2**	2.4	6.2	unk.	unk.	**unk.**	tr(a)	**0**	tr(a)		unk.	
0.03	**2**	0.04	**2**	0.00	**0**	unk.	**unk.**	0	0(a)	**0**	0.0	0(a)	16	**2**	201	142	unk.	0.6	**3**	0.3	**2**	unk.	**unk.**	unk.	H
0.1	**0**	unk.	**unk.**	0.08	**1**	17	**0**	2	0.0	**0**	0(a)	0.68	15	**2**	8.7	3.6	unk.	13	**3**	0.18	**9**	tr(a)		unk.	
0.03	**2**	0.04	**2**	0.00	**0**	unk.	**unk.**	0	0(a)	**0**	0.0	0(a)	16	**2**	39	142	unk.	0.6	**3**	0.3	**2**	unk.	**unk.**	unk.	I
0.1	**0**	unk.	**unk.**	0.08	**1**	17	**0**	2	0.0	**0**	0(a)	0.42	15	**2**	1.7	3.6	unk.	20	**5**	0.18	**9**	tr(a)		25	
0.02	**1**	0.02	**1**	0.00	**0**	unk.	**unk.**	0	0(a)	**0**	0.0	0(a)	7	**1**	22	104	unk.	0.3	**1**	0.2	**1**	unk.	**unk.**	unk.	J
0.1	**1**	39	**10**	0.04	**0**	10	**0**	1	0.0	**0**	0(a)	0.35	12	**1**	0.9	2.7	unk.	8	**2**	0.10	**5**	tr(a)		unk.	
0.03	**2**	0.03	**1**	0.00	**0**	unk.	**unk.**	0	0(a)	**0**	0.0	0(a)	12	**1**	37	177	unk.	0.4	**2**	0.3	**2**	unk.	**unk.**	unk.	K
0.3	**1**	66	**17**	0.06	**1**	17	**0**	2	0.0	**0**	0(a)	0.59	20	**2**	1.6	4.5	unk.	13	**3**	0.18	**9**	tr(a)		unk.	
0.04	**2**	0.04	**2**	0(a)	**0**	unk.	**unk.**	0	0(a)	**0**	unk.	unk.	16	**2**	291	144	unk.	0.4	**2**	0.3	**2**	unk.	**unk.**	tr	L
0.2	**1**	69	**17**	0.00	**0**	9	**0**	1	unk.	**unk.**	unk.	0.95	17	**2**	12.7	3.7	unk.	13	**3**	0.18	**9**	unk.		unk.	
0.11	**6**	unk.	**unk.**	unk.	**unk.**	unk.	**unk.**	0	0	**0**	unk.	unk.	47	**5**	unk.	unk.	unk.	0.4	**2**	unk.	**unk.**	unk.	**unk.**	unk.	M
0.8	**4**	unk.	**unk.**	unk.	**unk.**	0	**0**	0	unk.	**unk.**	unk.	unk.	65	**7**	unk.	unk.	unk.	unk.	**unk.**	unk.	**unk.**	unk.		unk.	
0.07	**4**	0.00	**0**	1.94	**32**	unk.	**unk.**	unk.	0	**0**	0.0	0.0	20	**2**	37	65	unk.	0.0	**0**	0.1	**1**	unk.	**unk.**	unk.	N
1.8	**9**	unk	**unk**	0.00	**0**	4	**0**	unk.	0.0	**0**	0.0	0.43	49	**5**	1.6	1.7	unk.	25	**6**	tr(a)	**0**	unk.		unk.	
0.11	**6**	0.22	**11**	0.00	**0**	unk.	**unk.**	unk.	0	**0**	0.0	0(a)	18	**2**	25	90	unk.	tr	**0**	0.4	**2**	unk.	**unk.**	unk.	O
2.2	**11**	49	**12**	0.29	**3**	unk.	**unk.**	unk.	0.0	**0**	0(a)	0.72	108	**11**	1.1	2.3	unk.	36	**9**	0.14	**7**	68		unk.	
unk.	**unk.**	unk.	**unk.**	unk.	**unk.**	unk.	**unk.**	unk.	unk.	**unk.**	unk.	unk.	unk.	**unk.**	unk.	unk.	unk.	unk.	**unk.**	unk.	**unk.**	unk.	**unk.**	unk.	P
unk.	**unk.**	unk.	**unk.**	unk.	**unk.**	unk.	**unk.**	unk.	unk.	**unk.**	unk.	unk.	unk.	**unk.**	unk.	unk.	unk.	unk.	**unk.**	0.01	**1**	unk.		unk.	
0.00	**0**	unk.	**unk.**	unk.	**unk.**	unk.	**unk.**	0	0	**0**	unk.	unk.	3	**0**	unk.	unk.	unk.	0.1	**0**	unk.	**unk.**	unk.	**unk.**	unk.	Q
tr	**0**	unk.	**unk.**	unk.	**unk.**	0	**0**	0	unk.	**unk.**	unk.	unk.	2	**0**	unk.	unk.	unk.	unk.	**unk.**	unk.	**unk.**	unk.		unk.	
unk.	**unk.**	unk.	**unk.**	unk.	**unk.**	unk.	**unk.**	unk.	unk.	**unk.**	unk.	unk.	unk.	**unk.**	unk.	unk.	unk.	unk.	**unk.**	unk.	**unk.**	unk.	**unk.**	unk.	R
unk.	**unk.**	unk.	**unk.**	unk.	**unk.**	unk.	**unk.**	unk.	unk.	**unk.**	unk.	unk.	unk.	**unk.**	unk.	unk.	unk.	unk.	**unk.**	unk.	**unk.**	unk.		unk.	
0.00	**0**	unk.	**unk.**	unk.	**unk.**	unk	**unk**	0	unk.	**unk.**	unk.	unk.	1	**0**	unk.	unk.	unk.	tr	**0**	unk.	**unk.**	unk.	**unk.**	unk.	S
tr	**0**	unk.	**unk.**	unk.	**unk.**	25	**1**	2	unk.	**unk.**	unk.	unk.	1	**0**	unk.	unk.	unk.	unk.	**unk.**	unk.	**unk.**	unk.		unk.	

	FOOD	Portion	Weight in grams / Conversion for 100 g	Kilocalories / H_2O g	Total carbohydrate g / Total fats g	Crude fiber g / Dietary fiber g	Total protein g / Total sugar g	**% USRDA**	Arginine mg / Histidine mg	Isoleucine mg / Leucine mg	Lysine mg / Methionine mg	Phenylalanine mg / Threonine mg	Valine mg / Tryptophan mg	Cystine mg / Tyrosine mg	Polyunsat. fatty acids g / Monounsat. fatty acids g	Saturated fatty acids g / P/S ratio	Linoleic acid g / Cholesterol mg	Thiamin mg / Ascorbic acid mg	**% USRDA**
A	beverage,alcoholic,cocktail,martini	2-1/2 oz	71	99	0.2	unk.	0.1	**0**	unk.	unk.	unk.	unk.	unk.	unk.	unk.	unk.	unk.	tr	**0**
		glass	*1.41*	unk.	0.0	0(a)	0.2		unk.	unk.	unk.	unk.	unk.	unk.	unk.	unk.	unk.	0	**0**
B	beverage,alcoholic,cocktail,	10 oz glass	300	210	2.7	unk.	unk.	**unk.**	unk.	unk.	unk.	unk.	unk.	unk.	unk.	unk.	unk.	unk.	**unk.**
	mint julep		*0.33*	unk.	unk.	0(a)	2.7		unk.	unk.	unk.	unk.	unk.	unk.	unk.	unk.	unk.	unk.	**unk.**
C	beverage,alcoholic,cocktail,	2-1/2 oz	71	127	2.5	unk.	unk.	**unk.**	unk.	unk.	unk.	unk.	unk.	unk.	unk.	unk.	unk.	unk.	**unk.**
	old-fashioned	glass	*1.41*	unk.	unk.	0(a)	2.5		unk.	unk.	unk.	unk.	unk.	unk.	unk.	unk.	unk.	unk.	**unk.**
D	beverage,alcoholic,cocktail,	2-1/2 oz	71	132	4.1	0.00	0.0	**4**	0(a)	0(a)	0(a)	0(a)	0(a)	0(a)	0.00	0.00	0.00	0.00	**0**
	sloe gin fizz	glass	*1.41*	49.8	0.0	0.00	2.5		0(a)	0(a)	tr(a)	0(a)	0(a)	0(a)	0.00	0.00	0	11	**19**
E	beverage,alcoholic,cocktail,	10 oz glass	300	180	9.0	unk.	0.3	**1**	unk.	unk.	unk.	unk.	unk.	unk.	unk.	unk.	unk.	0.03	**2**
	Tom Collins		*0.33*	unk.	0.0	0(a)	9.0		unk.	unk.	unk.	unk.	unk.	unk.	unk.	unk.	unk.	21	**35**
F	BEVERAGE,ALCOHOLIC,EGGNOG	4 oz glass	123	335	18.0	unk.	3.9	**6**	unk.	unk.	unk.	unk.	unk.	unk.	unk.	unk.	unk.	0.02	**2**
			0.81	unk.	15.9	0(a)	18.0		unk.	unk.	unk.	unk.	unk.	unk.	unk.	unk.	unk.	tr	**0**
G	BEVERAGE,ALCOHOLIC,GIN,100-proof	1-1/2 oz	42	124	tr	unk.	unk.	**unk.**	unk.	unk.	unk.	unk.	unk.	unk.	unk.	unk.	unk.	unk.	**unk.**
		jigger	*2.38*	24.1	unk.	0(a)	tr(a)		unk.	unk.	unk.	unk.	unk.	unk.	unk.	unk.	unk.	unk.	**unk.**
H	beverage,alcoholic,gin,94-proof	1-1/2 oz	42	115	tr	unk.	unk.	**unk.**	unk.	unk.	unk.	unk.	unk.	unk.	unk.	unk.	unk.	unk.	**unk.**
		jigger	*2.38*	25.3	unk.	0(a)	tr(a)		unk.	unk.	unk.	unk.	unk.	unk.	unk.	unk.	unk.	unk.	**unk.**
I	beverage,alcoholic,gin,90-proof	1-1/2 oz	42	110	tr	unk.	unk.	**unk.**	unk.	unk.	unk.	unk.	unk.	unk.	unk.	unk.	unk.	unk.	**unk.**
		jigger	*2.38*	26.1	unk.	0(a)	tr(a)		unk.	unk.	unk.	unk.	unk.	unk.	unk.	unk.	unk.	unk.	**unk.**
J	beverage,alcoholic,gin,86-proof	1-1/2 oz	42	105	tr	unk.	unk.	**unk.**	unk.	unk.	unk.	unk.	unk.	unk.	unk.	unk.	unk.	unk.	**unk.**
		jigger	*2.38*	26.9	unk.	unk.	tr(a)		unk.	unk.	unk.	unk.	unk.	unk.	unk.	unk.	unk.	unk.	**unk.**
K	beverage,alcoholic,gin,80-proof	1-1/2 oz	42	97	tr	unk.	unk.	**unk.**	unk.	unk.	unk.	unk.	unk.	unk.	unk.	unk.	unk.	unk.	**unk.**
		jigger	*2.38*	28.0	unk.	unk.	tr(a)		unk.	unk.	unk.	unk.	unk.	unk.	unk.	unk.	unk.	unk.	**unk.**
L	BEVERAGE,ALCOHOLIC,LIQUEUR,	2/3 oz glass	20	74	7.0	unk.	unk.	**unk.**	unk.	unk.	unk.	unk.	unk.	unk.	unk.	unk.	unk.	unk.	**unk.**
	anisette		*5.00*	unk.	unk.	0(a)	7.0		unk.	unk.	unk.	unk.	unk.	unk.	unk.	unk.	unk.	unk.	**unk.**
M	beverage,alcoholic,liqueur,	2/3 oz glass	20	64	6.0	unk.	unk.	**unk.**	unk.	unk.	unk.	unk.	unk.	unk.	unk.	unk.	unk.	unk.	**unk.**
	apricot brandy		*5.00*	unk.	unk.	0(a)	6.0		unk.	unk.	unk.	unk.	unk.	unk.	unk.	unk.	unk.	unk.	**unk.**
N	beverage,alcoholic,liqueur,benedictine	2/3 oz glass	20	69	6.6	unk.	unk.	**unk.**	unk.	unk.	unk.	unk.	unk.	unk.	unk.	unk.	unk.	unk.	**unk.**
			5.00	unk.	unk.	0(a)	6.6		unk.	unk.	unk.	unk.	unk.	unk.	unk.	unk.	unk.	unk.	**unk.**
O	beverage,alcoholic,liqueur,	2/3 oz glass	20	67	6.0	unk.	unk.	**unk.**	unk.	unk.	unk.	unk.	unk.	unk.	unk.	unk.	unk.	unk.	**unk.**
	creme de menthe		*5.00*	unk.	unk.	0(a)	6.0		unk.	unk.	unk.	unk.	unk.	unk.	unk.	unk.	unk.	unk.	**unk.**
P	beverage,alcoholic,liqueur,Curaçao	2/3 oz glass	20	54	5.7	unk.	unk.	**unk.**	unk.	unk.	unk.	unk.	unk.	unk.	unk.	unk.	unk.	unk.	**unk.**
			5.00	unk.	unk.	0(a)	5.7		unk.	unk.	unk.	unk.	unk.	unk.	unk.	unk.	unk.	unk.	**unk.**
Q	BEVERAGE,ALCOHOLIC,RUM,100-proof	1-1/2 oz	42	124	tr	unk.	unk.	**unk.**	unk.	unk.	unk.	unk.	unk.	unk.	unk.	unk.	unk.	unk.	**unk.**
		jigger	*2.38*	24.1	unk.	0(a)	tr(a)		unk.	unk.	unk.	unk.	unk.	unk.	unk.	unk.	unk.	unk.	**unk.**
R	beverage,alcoholic,rum,94-proof	1-1/2 oz	42	115	tr	unk.	unk.	**unk.**	unk.	unk.	unk.	unk.	unk.	unk.	unk.	unk.	unk.	unk.	**unk.**
		jigger	*2.38*	25.3	unk.	0(a)	tr(a)		unk.	unk.	unk.	unk.	unk.	unk.	unk.	unk.	unk.	unk.	**unk.**
S	beverage,alcoholic,rum,90-proof	1-1/2 oz	42	110	tr	unk.	unk.	**unk.**	unk.	unk.	unk.	unk.	unk.	unk.	unk.	unk.	unk.	unk.	**unk.**
		jigger	*2.38*	26.1	unk.	0(a)	tr(a)		unk.	unk.	unk.	unk.	unk.	unk.	unk.	unk.	unk.	unk.	**unk.**

Riboflavin mg / Niacin mg	% USRDA	Vitamin B_6 mg / Folacin mcg	% USRDA	Vitamin B_{12} mcg / Pantothenic acid mg	% USRDA	Biotin mg / Vitamin A IU	% USRDA	Preformed A RE / Beta carotene RE	Vitamin D IU / Vitamin E IU	% USRDA	Total tocopherol mg / Alpha tocopherol mg	Other tocopherol mg / Total ash g	Calcium mg / Phosphorus mg	% USRDA	Sodium mg / Sodium meq	Potassium mg / Potassium meq	Chlorine mg / Chlorine meq	Iron mg / Magnesium mg	% USRDA	Zinc mg / Copper mg	% USRDA	Iodine mcg / Selenium mcg	% USRDA	Manganese mcg / Chromium mcg	
tr	**0**	unk.	**unk.**	unk.	**unk.**	unk.	**unk.**	0	unk.	**unk.**	unk.	unk.	4	**0**	unk.	unk.	unk.	0.1	**0**	unk.	**unk.**	unk.	**unk.**	unk.	A
tr	**0**	unk.	**unk.**	unk.	**unk.**	3	**0**	tr	unk.	**unk.**	unk.	unk.	1	**0**	unk.	unk.	unk.	unk.	**unk.**	unk.	**unk.**	unk.		unk.	
unk.	**unk.**	unk.	**unk.**	unk.	**unk.**	unk.	**unk.**	unk.	unk.	**unk.**	unk.	unk.	unk.	**unk.**	unk.	unk.	unk.	unk.	**unk.**	unk.	**unk.**	unk.	**unk.**	unk.	B
unk.	**unk.**	unk.	**unk.**	unk.	**unk.**	unk.	**unk.**	unk.	unk.	**unk.**	unk.	unk.	unk.	**unk.**	unk.	unk.	unk.	unk.	**unk.**	unk.	**unk.**	unk.		unk	
unk.	**unk.**	unk.	**unk.**	unk.	**unk.**	unk.	**unk.**	unk.	unk.	**unk.**	unk.	unk.	unk.	**unk.**	unk.	unk.	unk.	unk.	**unk.**	unk.	**unk.**	unk.	**unk.**	unk.	C
unk.	**unk.**	unk.	**unk.**	unk.	**unk.**	unk.	**unk.**	unk.	unk.	**unk.**	unk.	unk.	unk.	**unk.**	unk.	unk.	unk.	unk.	**unk.**	unk.	**unk.**	unk.		unk.	
0.00	**0**	0.01	**1**	0.00	**0**	0(a)	**0**	0	0	**0**	0.0	0(a)	2	**0**	1	35	1	0.0	**0**	0.0	**0**	unk.	**unk.**	3	D
tr	**0**	tr	**0**	0.03	**0**	5	**0**	0	0.0	**0**	0(a)	0.07	2	**0**	0.0	0.9	0.0	2	**1**	0.04	**2**	0		1	
tr	**0**	unk.	**unk.**	unk.	**unk.**	unk.	**unk.**	0	0	**0**	unk.	unk.	9	**1**	unk.	unk.	unk.	tr	**0**	unk.	**unk.**	unk.	**unk.**	unk.	E
tr	**0**	unk.	**unk.**	unk.	**unk.**	0	**0**	0	unk.	**unk.**	unk.	unk.	9	**1**	unk.	unk.	unk.	unk.	**unk.**	unk.	**unk.**	unk.		unk.	
0.11	**7**	unk.	**unk.**	unk.	**unk.**	unk.	**unk.**	18	21	**5**	unk.	unk.	44	**4**	unk.	unk.	unk.	0.7	**4**	unk.	**unk.**	unk.	**unk.**	unk.	F
tr	**0**	unk.	**unk.**	unk.	**unk.**	84	**2**	2	unk.	**unk.**	unk.	unk.	74	**7**	unk.	unk.	unk.	unk.	**unk.**	unk.	**unk.**	unk.		unk.	
unk.	**unk.**	0.00	**0**	0.00	**0**	unk.	**unk.**	0(a)	0	**0**	0.0	0(a)	unk.	**unk.**	tr	1	unk.	unk.	**unk.**	tr	**0**	unk.	**unk.**	unk.	G
unk.	**unk.**	0	**0**	0.00	**0**	0	**0**	0(a)	0.0	**0**	0(a)	unk.	unk.	**unk.**	0.0	0.0	unk.	0	**0**	0.03	**2**	unk.		unk.	
unk.	**unk.**	0.00	**0**	0.00	**0**	unk.	**unk.**	0(a)	0	**0**	0.0	0(a)	unk.	**unk.**	tr	1	unk.	unk.	**unk.**	tr	**0**	unk.	**unk.**	unk.	H
unk.	**unk.**	0	**0**	0.00	**0**	0	**0**	0(a)	0.0	**0**	0(a)	unk.	unk.	**unk.**	0.0	0.0	unk.	0	**0**	0.03	**2**	unk.		unk.	
unk.	**unk.**	0.00	**0**	0.00	**0**	unk.	**unk.**	0(a)	0	**0**	0.0	0(a)	unk.	**unk.**	tr	1	unk.	unk.	**unk.**	tr	**0**	unk.	**unk.**	unk.	I
unk.	**unk.**	0	**0**	0.00	**0**	0	**0**	0(a)	0.0	**0**	0(a)	unk.	unk.	**unk.**	0.0	0.0	unk.	0	**0**	0.03	**2**	unk.		unk.	
unk.	**unk.**	0.00	**0**	0.00	**0**	unk.	**unk.**	0(a)	0	**0**	0.0	0(a)	unk.	**unk.**	tr	1	unk.	unk.	**unk.**	tr	**0**	unk.	**unk.**	unk.	J
unk.	**unk.**	0	**0**	0.00	**0**	0	**0**	0(a)	0.0	**0**	0(a)	unk.	unk.	**unk.**	0.0	0.0	unk.	0	**0**	0.03	**2**	unk.		unk.	
unk.	**unk.**	0.00	**0**	0.00	**0**	unk.	**unk.**	unk.	0	**0**	0.0	0(a)	unk.	**unk.**	tr	1	unk.	unk.	**unk.**	tr	**0**	unk.	**unk.**	unk.	K
unk.	**unk.**	0	**0**	0.00	**0**	unk.	**unk.**	unk.	0.0	**0**	0(a)	unk.	unk.	**unk.**	0.0	0.0	unk.	0	**0**	0.03	**2**	unk.		unk.	
unk.	**unk.**	unk.	**unk.**	unk.	**unk.**	unk.	**unk.**	unk.	unk.	**unk.**	unk.	unk.	unk.	**unk.**	unk.	unk.	unk.	unk.	**unk.**	unk.	**unk.**	unk.	**unk.**	unk.	L
unk.	**unk.**	unk.	**unk.**	unk.	**unk.**	unk.	**unk.**	unk.	unk.	**unk.**	unk.	unk.	unk.	**unk.**	unk.	unk.	unk.	unk.	**unk.**	unk.	**unk.**	unk.		unk.	
unk.	**unk.**	unk.	**unk.**	unk.	**unk.**	unk.	**unk.**	unk.	unk.	**unk.**	unk.	unk.	unk.	**unk.**	unk.	unk.	unk.	unk.	**unk.**	unk.	**unk.**	unk.	**unk.**	unk.	M
unk.	**unk.**	unk.	**unk.**	unk.	**unk.**	unk.	**unk.**	unk.	unk.	**unk.**	unk.	unk.	unk.	**unk.**	unk.	unk.	unk.	unk.	**unk.**	unk.	**unk.**	unk.		unk.	
unk.	**unk.**	unk.	**unk.**	unk.	**unk.**	unk.	**unk.**	unk.	unk.	**unk.**	unk.	unk.	unk.	**unk.**	unk.	unk.	unk.	unk.	**unk.**	unk.	**unk.**	unk.	**unk.**	unk.	N
unk.	**unk.**	unk.	**unk.**	unk.	**unk.**	unk.	**unk.**	unk.	unk.	**unk.**	unk.	unk.	unk.	**unk.**	unk.	unk.	unk.	unk.	**unk.**	unk.	**unk.**	unk.		unk.	
unk.	**unk.**	unk.	**unk.**	unk.	**unk.**	unk.	**unk.**	unk.	unk.	**unk.**	unk.	unk.	unk.	**unk.**	unk.	unk.	unk.	unk.	**unk.**	unk.	**unk.**	unk.	**unk.**	unk.	O
unk.	**unk.**	unk.	**unk.**	unk.	**unk.**	unk.	**unk.**	unk.	unk.	**unk.**	unk.	unk.	unk.	**unk.**	unk.	unk.	unk.	unk.	**unk.**	unk.	**unk.**	unk.		unk.	
unk.	**unk.**	unk.	**unk.**	unk.	**unk.**	unk.	**unk.**	unk.	unk.	**unk.**	unk.	unk.	unk.	**unk.**	unk.	unk.	unk.	unk.	**unk.**	unk.	**unk.**	unk.	**unk.**	unk.	P
unk.	**unk.**	unk.	**unk.**	unk.	**unk.**	unk.	**unk.**	unk.	unk.	**unk.**	unk.	unk.	unk.	**unk.**	unk.	unk.	unk.	unk.	**unk.**	unk.	**unk.**	unk.		unk.	
unk.	**unk.**	0.00	**0**	0.00	**0**	unk.	**unk.**	0(a)	0	**0**	0.0	0(a)	unk.	**unk.**	tr	1	unk.	unk.	**unk.**	tr	**0**	unk.	**unk.**	unk.	Q
unk.	**unk.**	0	**0**	0.00	**0**	0	**0**	0(a)	0.0	**0**	0(a)	unk.	unk.	**unk.**	0.0	0.0	unk.	0	**0**	0.03	**2**	unk.		unk.	
unk.	**unk.**	0.00	**0**	0.00	**0**	unk.	**unk.**	0(a)	0	**0**	0.0	0(a)	unk.	**unk.**	tr	1	unk.	unk.	**unk.**	tr	**0**	unk.	**unk**	unk.	R
unk.	**unk.**	0	**0**	0.00	**0**	0	**0**	0(a)	0.0	**0**	0(a)	unk.	unk.	**unk.**	0.0	0.0	unk.	0	**0**	0.03	**2**	unk.		unk.	
unk.	**unk.**	0.00	**0**	0.00	**0**	unk.	**unk.**	0(a)	0	**0**	0.0	0(a)	unk.	**unk.**	tr	1	unk.	unk.	**unk.**	tr	**0**	unk.	**unk.**	unk.	S
unk.	**unk.**	0	**0**	0.00	**0**	0	**0**	0(a)	0.0	**0**	0(a)	unk.	unk.	**unk.**	0.0	0.0	unk.	0	**0**	0.03	**2**	unk.		unk.	

	FOOD	Portion	Weight in grams / Conversion for 100 g	Kilocalories / H_2O g	Total carbohydrate g / Total fats g	Crude fiber g / Dietary fiber g	Total protein g / Total sugar g	% USRDA	Arginine mg / Histidine mg	Isoleucine mg / Leucine mg	Lysine mg / Methionine mg	Phenylalanine mg / Threonine mg	Valine mg / Tryptophan mg	Cystine mg / Tyrosine mg	Polyunsat. fatty acids g / Monounsat. fatty acids g	Saturated fatty acids g / P/S ratio	Linoleic acid g / Cholesterol mg	Thiamin mg / Ascorbic acid mg	% USRDA
A	beverage,alcoholic,rum,86-proof	1-1/2 oz	42	105	tr	unk.	unk.	**unk.**	unk.	unk.	unk.	unk.	unk.	unk.	unk.	unk.	unk.	unk.	**unk.**
		jigger	*2.38*	26.9	unk.	0(a)	tr(a)		unk.	unk.	unk.	unk.	unk.	unk.	unk.	unk.	unk.	unk.	**unk.**
B	beverage,alcoholic,rum,80-proof	1-1/2 oz	42	97	tr	unk.	unk.	**unk.**	unk.	unk.	unk.	unk.	unk.	unk.	unk.	unk.	unk.	unk.	**unk.**
		jigger	*2.38*	28.0	unk.	0(a)	tr(a)		unk.	unk.	unk.	unk.	unk.	unk.	unk.	unk.	unk.	unk.	**unk.**
C	BEVERAGE,ALCOHOLIC,SHERRY	2 oz glass	60	84	4.8	unk.	0.2	**0**	unk.	unk.	unk.	unk.	unk.	unk.	unk.	unk.	unk.	unk.	**unk.**
			1.67	unk.	unk.	unk.	2.2		unk.	unk.	unk.	unk.	unk.	unk.	unk.	unk.	unk.	unk.	**unk.**
D	BEVERAGE,ALCOHOLIC,VERMOUTH,	3-1/2 oz	100	105	1.0	unk.	unk.	**unk.**	unk.	unk.	unk.	unk.	unk.	unk.	unk.	unk.	unk.	unk.	**unk.**
	French-dry	glass	*1.00*	unk.	unk.	unk.	1.0		unk.	unk.	unk.	unk.	unk.	unk.	unk.	unk.	unk.	unk.	**unk.**
E	beverage,alcoholic,vermouth,	3-1/2 oz	100	167	12.0	unk.	unk.	**unk.**	unk.	unk.	unk.	unk.	unk.	unk.	unk.	unk.	unk.	unk.	**unk.**
	Italian-sweet	glass	*1.00*	unk.	unk.	unk.	12.0		unk.	unk.	unk.	unk.	unk.	unk.	unk.	unk.	unk.	unk.	**unk.**
F	BEVERAGE,ALCOHOLIC,VODKA,100-	1-1/2 oz	42	124	tr	unk.	unk.	**unk.**	unk.	unk.	unk.	unk.	unk.	unk.	unk.	unk.	unk.	unk.	**unk.**
	proof	jigger	*2.38*	24.1	unk.	0(a)	tr(a)		unk.	unk.	unk.	unk.	unk.	unk.	unk.	unk.	unk.	unk.	**unk.**
G	beverage,alcoholic,vodka,94-proof	1-1/2 oz	42	115	tr	unk.	unk.	**unk.**	unk.	unk.	unk.	unk.	unk.	unk.	unk.	unk.	unk.	unk.	**unk.**
		jigger	*2.38*	25.3	unk.	0(a)	tr(a)		unk.	unk.	unk.	unk.	unk.	unk.	unk.	unk.	unk.	unk.	**unk.**
H	beverage,alcoholic,vodka,90-proof	1-1/2 oz	42	110	tr	unk.	unk.	**unk.**	unk.	unk.	unk.	unk.	unk.	unk.	unk.	unk.	unk.	unk.	**unk.**
		jigger	*2.38*	26.1	unk.	0(a)	tr(a)		unk.	unk.	unk.	unk.	unk.	unk.	unk.	unk.	unk.	unk.	**unk.**
I	beverage,alcoholic,vodka,86-proof	1-1/2 oz	42	105	tr	unk.	unk.	**unk.**	unk.	unk.	unk.	unk.	unk.	unk.	unk.	unk.	unk.	unk.	**unk.**
		jigger	*2.38*	26.9	unk.	0(a)	tr(a)		unk.	unk.	unk.	unk.	unk.	unk.	unk.	unk.	unk.	unk.	**unk.**
J	beverage,alcoholic,vodka,80-proof	1-1/2 oz	42	97	tr	unk.	unk.	**unk.**	unk.	unk.	unk.	unk.	unk.	unk.	unk.	unk.	unk.	unk.	**unk.**
		jigger	*2.38*	28.0	unk.	0(a)	tr(a)		unk.	unk.	unk.	unk.	unk.	unk.	unk.	unk.	unk.	unk.	**unk.**
K	BEVERAGE,ALCOHOLIC,WHISKEY,100-	1-1/2 oz	42	124	tr	unk.	unk.	**unk.**	unk.	unk.	unk.	unk.	unk.	unk.	unk.	unk.	unk.	unk.	**unk.**
	proof	jigger	*2.38*	24.1	unk.	0(a)	tr(a)		unk.	unk.	unk.	unk.	unk.	unk.	unk.	unk.	unk.	unk.	**unk.**
L	beverage,alcoholic,whiskey,94-proof	1-1/2 oz	42	115	tr	unk.	unk.	**unk.**	unk.	unk.	unk.	unk.	unk.	unk.	unk.	unk.	unk.	unk.	**unk.**
		jigger	*2.38*	25.3	unk.	0(a)	tr(a)		unk.	unk.	unk.	unk.	unk.	unk.	unk.	unk.	unk.	unk.	**unk.**
M	beverage,alcoholic,whiskey,90-proof	1-1/2 oz	42	110	tr	unk.	unk.	**unk.**	unk.	unk.	unk.	unk.	unk.	unk.	unk.	unk.	unk.	unk.	**unk.**
		jigger	*2.38*	26.1	unk.	0(a)	tr(a)		unk.	unk.	unk.	unk.	unk.	unk.	unk.	unk.	unk.	unk.	**unk.**
N	beverage,alcoholic,whiskey,86-proof	1-1/2 oz	42	105	tr	unk.	unk.	**unk.**	unk.	unk.	unk.	unk.	unk.	unk.	unk.	unk.	unk.	unk.	**unk.**
		jigger	*2.38*	26.9	unk.	0(a)	tr(a)		unk.	unk.	unk.	unk.	unk.	unk.	unk.	unk.	unk.	unk.	**unk.**
O	beverage,alcoholic,whiskey,80-proof	1-1/2 oz	42	97	tr	unk.	unk.	**unk.**	unk.	unk.	unk.	unk.	unk.	unk.	unk.	unk.	unk.	unk.	**unk.**
		jigger	*2.38*	28.0	unk.	0(a)	tr(a)		unk.	unk.	unk.	unk.	unk.	unk.	unk.	unk.	unk.	unk.	**unk.**
P	BEVERAGE,ALCOHOLIC,WINE,dessert,	3-1/2 oz	105	144	8.1	unk.	0.1	**0**	unk.	unk.	unk.	unk.	unk.	unk.	unk.	unk.	unk.	0.01	**1**
	18.8% alcohol by volume	glass	*0.95*	80.5	0.0	unk.	8.1		unk.	unk.	unk.	unk.	unk.	unk.	unk.	unk.	unk.	unk.	**unk.**
Q	beverage,alcoholic,wine,table,	3-1/2 oz	104	88	4.4	unk.	0.1	**0**	unk.	unk.	unk.	unk.	unk.	unk.	unk.	unk.	unk.	tr	**0**
	12.2% alcohol by volume	glass	*0.96*	89.0	0.0	unk.	2.6		unk.	unk.	unk.	unk.	unk.	unk.	unk.	unk.	unk.	unk.	**unk.**
R	BEVERAGE,NONALC,CARB-Dr. Pepper	12 oz bottle	369	144	37.2	0(a)	0.0	**0**	0(a)	0(a)	0(a)	0(a)	0(a)	0(a)	0(a)	0(a)	0(a)	unk.	**unk.**
			0.27	unk.	0.0	0(a)	37.2		0(a)	0(a)	0(a)	0(a)	0(a)	0(a)	0(a)	0.00	0(a)	unk.	**unk.**
S	beverage,nonalc,carb-Dr. Pepper,	12 oz bottle	355	3	0.7	0.00	0.0	**0**	0(a)	0(a)	0(a)	0(a)	0(a)	0(a)	0(a)	0(a)	0(a)	unk.	**unk.**
	sugar-free		*0.28*	unk.	0.0	0(a)	0.0		0(a)	0(a)	0(a)	0(a)	0(a)	0(a)	0(a)	0.00	0(a)	unk.	**unk.**

Riboflavin mg / Niacin mg	% USRDA	Vitamin B_6 mg / Folacin mcg	% USRDA	Vitamin B_{12} mcg / Pantothenic acid mg	% USRDA	Biotin mg / Vitamin A IU	% USRDA	Preformed A RE / Beta carotene RE	Vitamin D IU / Vitamin E IU	% USRDA	Total tocopherol mg / Alpha tocopherol mg	Other tocopherol mg / Total ash g	Calcium mg / Phosphorus mg	% USRDA	Sodium mg / Sodium meq	Potassium mg / Potassium meq	Chlorine mg / Chlorine meq	Iron mg / Magnesium mg	% USRDA	Zinc mg / Copper mg	% USRDA	Iodine mcg / Selenium mcg	% USRDA	Manganese mcg / Chromium mcg	
unk.	unk.	0.00	0	0.00	0	unk.	unk.	0(a)	0	0	0.0	0(a)	unk.	unk.	tr	1	unk.	unk.	unk.	tr	0	unk.	unk.	unk.	A
unk.	unk.	0	0	0.00	0	0	0	0(a)	0.0	0	0(a)	unk.	unk.	unk.	0.0	0.0	unk.	0	0	0.03	2	unk.		unk.	
unk.	unk.	0.00	0	0.00	0	unk.	unk.	0(a)	0	0	0.0	0(a)	unk.	unk.	tr	1	unk.	unk.	unk.	tr	0	unk.	unk.	unk.	B
unk.	unk.	0	0	0.00	0	0	0	0(a)	0.0	0	0(a)	unk.	unk.	unk.	0.0	0.0	unk.	0	0	0.03	2	unk.		unk.	
unk.	unk.	unk.	unk.	unk.	unk.	unk.	unk.	unk.	unk.	unk.	unk.	unk.	unk.	unk.	unk.	unk.	unk.	unk.	unk.	unk.	unk.	unk.	unk.	unk.	C
unk.	unk.	unk.	unk.	unk.	unk.	unk.	unk.	unk.	unk.	unk.	unk.	unk.	unk.	unk.	unk.	unk.	unk.	unk.	unk.	unk.	unk.	unk.		unk.	
unk.	unk.	unk.	unk.	unk.	unk.	unk.	unk.	unk.	unk.	unk.	unk.	unk.	unk.	unk.	unk.	unk.	unk.	unk.	unk.	unk.	unk.	unk.	unk.	unk.	D
unk.	unk.	unk.	unk.	unk.	unk.	unk.	unk.	unk.	unk.	unk.	unk.	unk.	unk.	unk.	unk.	unk.	unk.	unk.	unk.	unk.	unk.	unk.		unk.	
unk.	unk.	unk.	unk.	unk.	unk.	unk.	unk.	unk.	unk.	unk.	unk.	unk.	unk.	unk.	unk.	unk.	unk.	unk.	unk.	unk.	unk.	unk.	unk.	unk.	E
unk.	unk.	unk.	unk.	unk.	unk.	unk.	unk.	unk.	unk.	unk.	unk.	unk.	unk.	unk.	unk.	unk.	unk.	unk.	unk.	unk.	unk.	unk.		unk.	
unk.	unk.	0.00	0	0.00	0	unk.	unk.	0(a)	0	0	0.0	0(a)	unk.	unk.	tr	1	unk.	unk.	unk.	tr	0	unk.	unk.	unk.	F
unk.	unk.	0	0	0.00	0	0	0	0(a)	0.0	0	0(a)	unk.	unk.	unk.	0.0	0.0	unk.	0	0	0.03	2	unk.		unk.	
unk.	unk.	0.00	0	0.00	0	unk.	unk.	unk.	0	0	0.0	0(a)	unk.	unk.	tr	1	unk.	unk.	unk.	tr	0	unk.	unk.	unk.	G
unk.	unk.	0	0	0.00	0	0	0	unk.	0.0	0	0(a)	unk.	unk.	unk.	0.0	0.0	unk.	0	0	0.03	2	unk.		unk.	
unk.	unk.	0.00	0	0.00	0	unk.	unk.	0(a)	0	0	0.0	0(a)	unk.	unk.	tr	1	unk.	unk.	unk.	tr	0	unk.	unk.	unk.	H
unk.	unk.	0	0	0.00	0	0	0	0(a)	0.0	0	0(a)	unk.	unk.	unk.	0.0	0.0	unk.	0	0	0.03	2	unk.		unk.	
unk.	unk.	0.00	0	0.00	0	unk.	unk.	0(a)	0	0	0.0	0(a)	unk.	unk.	tr	1	unk.	unk.	unk.	tr	0	unk.	unk.	unk.	I
unk.	unk.	0	0	0.00	0	0	0	0(a)	0.0	0	0(a)	unk.	unk.	unk.	0.0	0.0	unk.	0	0	0.03	2	unk.		unk.	
unk.	unk.	0.00	0	0.00	0	unk.	unk.	0(a)	0	0	0.0	0(a)	unk.	unk.	tr	1	unk.	unk.	unk.	tr	0	unk.	unk.	unk.	J
unk.	unk.	0	0	0.00	0	0	0	0(a)	0.0	0	0(a)	unk.	unk.	unk.	0.0	0.0	unk.	0	0	0.03	2	unk.		unk.	
unk.	unk.	0.00	0	0.00	0	unk.	unk.	0(a)	0	0	0.0	0(a)	unk.	unk.	tr	1	unk.	unk.	unk.	tr	0	unk.	unk.	unk.	K
unk.	unk.	0	0	0.00	0	0	0	0(a)	0.0	0	0(a)	unk.	unk.	unk.	0.0	0.0	unk.	0	0	0.03	2	unk.		unk.	
unk.	unk.	0.00	0	0.00	0	unk.	unk.	0(a)	0	0	0.0	0(a)	unk.	unk.	tr	1	unk.	unk.	unk.	tr	0	unk.	unk.	unk.	L
unk.	unk.	0	0	0.00	0	0	0	0(a)	0.0	0	0(a)	unk.	unk.	unk.	0.0	0.0	unk.	0	0	0.03	2	unk.		unk.	
unk.	unk.	0.00	0	0.00	0	unk.	unk.	0(a)	0	0	0.0	0(a)	unk.	unk.	tr	1	unk.	unk.	unk.	tr	0	unk.	unk.	unk.	M
unk.	unk.	0	0	0.00	0	0	0	0(a)	0.0	0	0(a)	unk.	unk.	unk.	0.0	0.0	unk.	0	0	0.03	2	unk.		unk.	
unk.	unk.	0.00	0	0.00	0	unk.	unk.	0(a)	0	0	0.0	0(a)	unk.	unk.	tr	1	unk.	unk.	unk.	tr	0	unk.	unk.	unk.	N
unk.	unk.	0	0	0.00	0	0	0	0(a)	0.0	0	0(a)	unk.	unk.	unk.	0.0	0.0	unk.	0	0	0.03	2	unk.		unk.	
unk.	unk.	0.00	0	0.00	0	unk.	unk.	0(a)	0	0	0.0	0(a)	unk.	unk.	tr	1	unk.	unk.	unk.	tr	0	unk.	unk.	unk.	O
unk.	unk.	0	0	0.00	0	0	0	0(a)	0.0	0	0(a)	unk.	unk.	unk.	0.0	0.0	unk.	0	0	0.03	2	unk.		unk.	
0.02	1	0.04	2	0.00	0	unk.	unk.	unk.	0	0	0.0	0(a)	8	1	4	79	unk.	unk.	unk.	0.1	1	unk.	unk.	unk.	P
0.2	1	tr	0	0.03	0	unk.	unk.	unk.	0.0	0	0(a)	0.21	unk.	unk.	0.2	2.0	unk.	9	2	0.08	4	unk.		unk.	
0.01	1	0.04	2	0.00	0	unk.	unk.	unk.	0	0	0.0	0(a)	9	1	5	96	unk.	0.4	2	0.1	1	unk.	unk.	unk.	Q
0.1	1	0	0	0.00	0	unk.	unk.	unk.	0.0	0	0(a)	0.21	10	1	0.2	2.4	unk.	9	2	0.11	6	unk.		unk.	
unk.	unk.	0.00	0	0.00	0	unk.	unk.	unk.	0	0	0.0	0(a)	10	1	28	2	unk.	unk.	unk.	0.4	3	unk.	unk.	unk.	R
unk.	unk.	0	0	0.00	0	unk.	unk.	unk.	0.0	0	0(a)	unk.	40	4	1.2	0.1	unk.	4	1	0.15	7	unk.		unk.	
unk.	unk.	0.00	0	0.00	0	unk.	unk.	unk.	0	0	0.0	0(a)	unk.	unk.	36	unk.	unk.	unk.	unk.	0.4	2	unk.	unk.	unk.	S
unk.	unk.	0	0	0.00	0	unk.	unk.	unk.	0.0	0	0(a)	unk.	42	4	1.6	unk.	unk.	unk.	unk.	0.14	7	unk.		unk.	

	FOOD	Portion	Weight in grams / *Conversion for 100 g*	Kilocalories / H_2O g	Total carbohydrate g / Total fats g	Crude fiber g / Dietary fiber g	Total protein g / Total sugar g	**% USRDA**	Arginine mg / Histidine mg	Isoleucine mg / Leucine mg	Lysine mg / Methionine mg	Phenylalanine mg / Threonine mg	Valine mg / Tryptophan mg	Cystine mg / Tyrosine mg	Polyunsat. fatty acids g / Monounsat. fatty acids g	Saturated fatty acids g / P/S ratio	Linoleic acid g / Cholesterol mg	Thiamin mg / Ascorbic acid mg	**% USRDA**
A	beverage,nonalc,carb-Fresca	12 oz bottle	355	4	0.0	0(a)	0.0	**0**	unk.	0	0	0	0	0	0.00	0.00	0.00	0(a)	**0**
			0.28	unk.	0.0	0(a)	0.0		0	0	0	0	0	0	0.00	0.00	0	0(a)	**0**
B	beverage,nonalc,carb-Seven-up	12 oz bottle	370	155	36.5	0(a)	0.0	**0**	0	0	0	0	0	0	0.00	0.00	0.00	0.00	**0**
			0.27	unk.	0.0	0(a)	36.5		0	0	0	0	0	0	0.00	0.00	0	0	**0**
C	beverage,nonalc,carb-Seven-up,diet	12 oz bottle	355	4	1.0	0.00	0.0	**0**	0	0	0	0	0	0	0.00	0.00	0.00	0.00	**0**
			0.28	unk.	0.0	0.00	0(a)		0	0	0	0	0	0	0.00	0.00	0	0	**0**
D	beverage,nonalc,carb-Tab	12 oz bottle	356	tr	tr	0(a)	0.0	**0**	unk.	0(a)	0(a)	0(a)	0(a)	0(a)	0(a)	0(a)	0(a)	0(a)	**0**
			0.28	unk.	0.0	0(a)	0(a)		0(a)	0(a)	0(a)	0(a)	0(a)	0(a)	0(a)	0.00	0(a)	0(a)	**0**
E	beverage,nonalc,carb,club soda	8 oz glass	237	0	tr	0.00	0.0	**0**	0(a)	0(a)	0(a)	0(a)	0(a)	0(a)	0(a)	0(a)	0(a)	0.00	**0**
			0.42	236.8	0.0	0(a)	0.0		0(a)	0(a)	0(a)	0(a)	0(a)	0(a)	0(a)	0.00	0	0	**0**
F	beverage,nonalc,carb,cola type	12 oz bottle	370	144	39.2	0.00	0.0	**0**	0(a)	0(a)	0(a)	0(a)	0(a)	0(a)	0(a)	0(a)	0(a)	0.00	**0**
			0.27	333.0	0.0	0(a)	37.4		0(a)	0(a)	0(a)	0(a)	0(a)	0(a)	0(a)	0.00	0	0	**0**
G	beverage,nonalc,carb,cream soda	12 oz bottle	371	160	40.8	0.00	0.0	**0**	0(a)	0(a)	0(a)	0(a)	0(a)	0(a)	0(a)	0(a)	0(a)	0.00	**0**
			0.27	330.2	0.0	0(a)	40.8		0(a)	0(a)	0(a)	0(a)	0(a)	0(a)	0(a)	0.00	0(a)	0	**0**
H	beverage,nonalc,carb,diet drink	12 oz bottle	355	4	3.5	0.00	0.0	**0**	0(a)	0(a)	0(a)	0(a)	0(a)	0(a)	unk.	unk.	unk.	0.00	**0**
			0.28	355.0	0.0	0(a)	0.0		0(a)	0(a)	0(a)	0(a)	0(a)	0(a)	unk.	unk.	0	0	**0**
I	beverage,nonalc,carb,fruit flavored	12 oz bottle	372	171	44.6	0.00	0.0	**0**	0(a)	0(a)	0(a)	0(a)	0(a)	0(a)	0(a)	0(a)	0(a)	0.00	**0**
			0.27	327.4	0.0	0(a)	44.6		0(a)	0(a)	0(a)	0(a)	0(a)	0(a)	0(a)	0.00	0	0	**0**
J	beverage,nonalc,carb,ginger ale,	12 oz bottle	366	113	29.3	0.00	0.0	**0**	0(a)	0(a)	0(a)	0(a)	0(a)	0(a)	0(a)	0(a)	0(a)	0.00	**0**
	pale dry / golden-Etener		*0.27*	336.7	0.0	0(a)	29.3		0(a)	0(a)	0(a)	0(a)	0(a)	0(a)	0(a)	0.00	0	0	**0**
K	beverage,nonalc,carb,quinine	12 oz bottle	366	113	29.3	0.00	0.0	**0**	0(a)	0(a)	0(a)	0(a)	0(a)	0(a)	0(a)	0(a)	0(a)	0.00	**0**
	soda (tonic)		*0.27*	336.7	0.0	0(a)	29.3		0(a)	0(a)	0(a)	0(a)	0(a)	0(a)	0(a)	0.00	0	0	**0**
L	beverage,nonalc,carb,root beer	12 oz bottle	370	152	40.0	0.00	0.0	**0**	0(a)	0(a)	0(a)	0(a)	0(a)	0(a)	0(a)	0(a)	0(a)	0.00	**0**
			0.27	331.1	0.0	0(a)	40.0		0(a)	0(a)	0(a)	0(a)	0(a)	0(a)	0(a)	0.00	0	0	**0**
M	BEVERAGE,NONALC,NONCARB,apple	8 oz glass	247	123	34.1	0.00	0.3	**0**	unk.	12	10	7	7	tr(a)	tr(a)	tr(a)	tr(a)	0.05	**3**
	cider		*0.40*	unk.	0.0	unk.	34.1		0	12	5	7	tr(a)	5	tr(a)	0.00	0	3	**5**
N	beverage,nonalc,noncarb,apple	8 oz glass	247	119	30.5	unk.	0.2	**0**	unk.	unk.	unk.	unk.	unk.	unk.	tr	tr(a)	unk.	0.00	**0**
	drink-Hi-C		*0.40*	unk.	unk.	unk.	30.5		unk.	unk.	unk.	unk.	unk.	unk.	unk.	0.00	0	133	**221**
O	beverage,nonalc,noncarb,citrus-	12 oz bottle	245	50	14.0	0(a)	0.0	**0**	0(a)	0(a)	0(a)	0(a)	0(a)	0(a)	0(a)	0(a)	0(a)	0.00	**0**
	Gatorade		*0.41*	unk.	0.0	0(a)	14.0		0(a)	0(a)	0(a)	0(a)	0(a)	0(a)	0(a)	0.00	0(a)	0	**0**
P	beverage,nonalc,noncarb,grape-Hi-C	8 oz glass	247	107	26.7	0(a)	0.0	**0**	unk.	0(a)	0(a)	0(a)	0(a)	0(a)	0(a)	0(a)	0(a)	tr	**0**
			0.40	unk.	0.0	unk.	unk.		0(a)	0(a)	0(a)	0(a)	0(a)	0(a)	0(a)	0.00	0(a)	80	**133**
Q	beverage,nonalc,noncarb,grape,	4 oz glass	125	66	16.6	tr	0.3	**0**	unk.	tr(a)	unk.	unk.	unk.	tr(a)	tr(a)	tr(a)	tr(a)	0.02	**2**
	frozen,diluted		*0.80*	108.0	tr	unk.	16.6		unk.	unk.	unk.	tr(a)	tr(a)	tr(a)	tr(a)	0.00	0	5	**8**
R	beverage,nonalc,noncarb,grape,	8 oz glass	265	118	30.1	0.05	0.0	**0**	unk.	unk.	unk.	unk.	unk.	unk.	0.00	0.00	0.00	0.00	**0**
	prep-Tang		*0.38*	233.6	0.0	unk.	29.8		unk.	unk.	unk.	unk.	unk.	unk.	0.00	0.00	0	118	**197**
S	beverage,nonalc,noncarb,grapefruit,	8 oz glass	265	114	28.2	0.01	0.0	**0**	unk.	unk.	unk.	unk.	unk.	unk.	0.00	0.00	0.00	0.00	**0**
	prep-Tang		*0.38*	233.6	0.0	unk.	27.5		unk.	unk.	unk.	unk.	unk.	unk.	0.00	0.00	0	118	**197**

Riboflavin mg / Niacin mg	% USRDA	Vitamin B_6 mg / Folacin mcg	% USRDA	Vitamin B_{12} mcg / Pantothenic acid mg	% USRDA	Biotin mg / Vitamin A IU	% USRDA	Preformed A RE / Beta carotene RE	Vitamin D IU / Vitamin E IU	% USRDA	Total tocopherol mg / Alpha tocopherol mg	Other tocopherol mg / Total ash g	Calcium mg / Phosphorus mg	% USRDA	Sodium mg / Sodium meq	Potassium mg / Potassium meq	Chlorine mg / Chlorine meq	Iron mg / Magnesium mg	% USRDA	Zinc mg / Copper mg	% USRDA	Iodine mcg / Selenium mcg	% USRDA	Manganese mcg / Chromium mcg	
0(a)	0	0(a)	0	0(a)	0	0(a)	0	0(a)	0(a)	0	0(a)	0(a)	unk.	unk.	57	unk.	unk.	unk.	unk.	unk.	unk.	unk.	unk.	unk.	A
0(a)	0	0(a)	0	0(a)	0	0(a)	0	0(a)	0.0	0	0(a)	0(a)	0	0	2.5	unk.	unk.	unk.	unk.	unk.	unk.	unk.		unk.	
0.00	0	0.00	0	0.00	0	0.000	0	0	0	0	0.0	0.0	unk.	unk.	33	unk.	0(a)	unk.	unk.	0(a)	0	0(a)	0	0(a)	B
0.0	0	0	0	0.00	0	0	0	0	0.0	0	0.0	0(a)	unk.	unk.	1.4	unk.	0.0	0(a)	0	0(a)	0	0(a)		0(a)	
0.00	0	0.00	0	0.00	0	0.000	0	0	0	0	0.0	0.0	tr(a)	0	48	tr(a)	tr(a)	tr(a)	0	0(a)	0	0(a)	0	0(a)	C
0.0	0	0	0	0.00	0	0	0	0	0.0	0	0.0	0(a)	tr(a)	0	2.1	0.0	0.0	tr(a)	0	0(a)	0	0(a)		0(a)	
0(a)	0	0(a)	0	0(a)	0	0(a)	0	0(a)	0(a)	0	0(a)	0(a)	unk.	unk.	45	unk.	unk.	unk.	unk.	unk.	unk.	unk.	unk.	unk.	D
0(a)	0	0(a)	0	0(a)	0	0(a)	0	0(a)	0.0	0	0(a)	tr(a)	45	5	2.0	unk.	unk.	unk.	unk.	unk.	unk.	unk.		unk.	
0.00	0	unk.	unk.	unk.	unk.	unk.	unk.	0	unk.	unk.	unk.	unk.	unk.	unk.	unk.	unk.	unk.	unk.	unk.	unk.	unk.	unk.	unk.	unk.	E
0.0	0	unk.	unk.	unk.	unk.	0	0	0	unk.	unk.	unk.	unk.	unk.	unk.	unk.	unk.	unk.	unk.	unk.	unk.	unk.	unk.		unk.	
0.00	0	0.00	0	0.00	0	unk.	unk.	0	0	0	0.0	0(a)	unk.	unk.	unk.	unk.	unk.	unk.	unk.	0.4	3	unk.	unk.	unk.	F
0.0	0	0	0	0.00	0	0	0	0	0.0	0	0(a)	unk.	unk.	unk.	unk.	unk.	unk.	unk.	unk.	0.15	7	unk.		6	
0.00	0	unk.	unk.	unk.	unk.	unk.	unk.	unk.	unk.	unk.	unk.	unk.	unk.	unk.	unk.	unk.	unk.	unk.	unk.	unk.	unk.	unk.	unk.	unk.	G
0.0	0	unk.	unk.	unk.	unk.	0	0	unk.	unk.	unk.	unk.	unk.	unk.	unk.	unk.	unk.	unk.	unk.	unk.	unk.	unk.	unk.		unk.	
0.00	0	unk.	unk.	unk.	unk.	unk.	unk.	0	unk.	unk.	unk.	unk.	unk.	unk.	unk.	unk.	unk.	unk.	unk.	unk.	unk.	unk.	unk.	unk.	H
0.0	0	unk.	unk.	unk.	unk.	0	0	0	unk.	unk.	unk.	unk.	unk.	unk.	unk.	unk.	unk.	unk.	unk.	unk.	unk.	unk.		unk.	
0.00	0	unk.	unk.	unk.	unk.	unk.	unk.	0	unk.	unk.	unk.	unk.	unk.	unk.	unk.	unk.	unk.	unk.	unk.	unk.	unk.	unk.	unk.	unk.	I
0.0	0	unk.	unk.	unk.	unk.	0	0	0	unk.	unk.	unk.	unk.	unk.	unk.	unk.	unk.	unk.	unk.	unk.	unk.	unk.	unk.		3	
0.00	0	0.00	0	0.00	0	unk.	unk.	0	0	0	0.0	0(a)	unk.	unk.	unk.	unk.	unk.	unk.	unk.	0.4	2	unk.	unk.	unk.	J
0.0	0	0	0	0.00	0	0	0	0	0.0	0	0(a)	unk.	unk.	unk.	unk.	unk.	unk.	4	1	0.11	6	unk.		unk.	
0.00	0	unk.	unk.	unk.	unk.	unk.	unk.	0	unk.	unk.	unk.	unk.	unk.	unk.	unk.	unk.	unk.	unk.	unk.	unk.	unk.	unk.	unk.	unk.	K
0.0	0	unk.	unk.	unk.	unk.	0	0	0	unk.	unk.	unk.	unk.	unk.	unk.	unk.	unk.	unk.	unk.	unk.	unk.	unk.	unk.		unk.	
0.00	0	unk.	unk.	unk.	unk.	unk.	unk.	0	unk.	unk.	unk.	unk.	unk.	unk.	unk.	unk.	unk.	unk.	unk.	unk.	unk.	unk.	unk.	unk.	L
0.0	0	unk.	unk.	unk.	unk.	0	0	0	unk.	unk.	unk.	unk.	unk.	unk.	unk.	unk.	unk.	unk.	unk.	unk.	unk.	unk.		6	
0.07	4	0.07	4	0(a)	0	0.001	0	0	0	0	1.3	0.5	15	2	10	247	unk.	1.2	7	0.1	1	unk.	unk.	52	M
0.0	0	4	1	0.05	1	99	2	1	1.5	5	0.8	0.49	27	3	0.4	6.3	unk.	10	3	tr(a)	0	tr(a)		unk.	
0.02	2	unk.	unk.	0(a)	0	unk.	unk.	unk.	0(a)	0	unk.	unk.	7	1	1	128	unk.	0.7	4	unk.	unk.	unk.	unk.	unk.	N
0.1	1	unk.	unk.	unk.	unk.	unk.	unk.	unk.	unk.	unk.	unk.	unk.	12	1	0.0	3.3	unk.	unk.	unk.	unk.	unk.	unk.		unk.	
0.00	0	0.00	0	0.00	0	0.000	0	0	0	0	0.0	0.0	0	0	130	24	94	unk.	unk.	unk.	unk.	unk.	unk.	unk.	O
0.0	0	0	0	0.00	0	0	0	0	0.0	0	0.0	unk.	24	2	5.6	0.6	2.7	unk.	unk.	unk.	unk.	unk.		unk.	
tr	0	tr(a)	0	0(a)	0	unk.	unk.	0	0(a)	0	0(a)	0(a)	tr	0	20	15	unk.	tr	0	unk.	unk.	unk.	unk.	unk.	P
tr	0	tr(a)	0	unk.	unk.	tr	0	tr	0.0	0	0(a)	unk.	0	0	0.9	0.4	unk.	1	0	tr(a)	0	tr(a)		unk.	
0.04	2	0.03	1	0.00	0	tr	0	0	0(a)	0	unk.	unk.	4	0	1	43	unk.	0.1	1	unk.	unk.	unk.	unk.	unk.	Q
0.2	1	3	1	0.05	1	tr	0	tr	unk.	unk.	unk.	0.12	5	1	0.0	1.1	unk.	5	1	tr(a)	0	tr(a)		unk.	
0.00	0	0.00	0	0.00	0	unk.	unk.	unk.	0	0	unk.	unk.	102	10	7	1	unk.	tr	0	tr	0	unk.	unk.	unk.	R
0.0	0	0	0	0.00	0	1974	40	unk.	unk.	unk.	unk.	0.45	108	11	0.3	0.0	unk.	tr	0	tr	0	tr(a)		unk.	
0.00	0	0.00	0	0.00	0	unk.	unk.	unk.	0	0	unk.	unk	129	13	2	1	unk.	tr	0	tr	0	unk.	unk.	unk.	S
0.0	0	0	0	0.00	0	1974	40	unk.	unk.	unk.	unk.	0.45	91	9	0.1	0.0	unk.	0	0	tr	0	tr(a)		unk.	

	FOOD	Portion	Weight in grams / Conversion for 100 g	Kilocalories / H_2O g	Total carbohydrate g / Total fats g	Crude fiber g / Dietary fiber g	Total protein g / Total sugar g	% USRDA	Arginine mg / Histidine mg	Isoleucine mg / Leucine mg	Lysine mg / Methionine mg	Phenylalanine mg / Threonine mg	Valine mg / Tryptophan mg	Cystine mg / Tyrosine mg	Polyunsat. fatty acids g / Monounsat. fatty acids g	Saturated fatty acids g / P/S ratio	Linoleic acid g / Cholesterol mg	Thiamin mg / Ascorbic acid mg	% USRDA
A	beverage,nonalc,noncarb,lemonade,	1 C	292	570	149.3	0.29	0.6	1	unk.	unk.	unk.	unk.	unk.	unk.	unk.	unk.	unk.	0.06	4
	frozen concentrate		*0.34*	141.7	0.3	0(a)	149.0		unk.	unk.	unk.	unk.	unk.	unk.	unk.	unk.	unk.	88	146
B	beverage,nonalc,noncarb,lemonade,	8 oz glass	248	109	28.3	tr	0.3	0	unk.	unk.	unk.	unk.	unk.	unk.	unk.	unk.	unk.	tr	0
	frozen,diluted		*0.40*	219.5	tr	0(a)	26.5		unk.	unk.	unk.	unk.	unk.	unk.	unk.	unk.	unk.	17	29
C	beverage,nonalc,noncarb,limeade,	1 C	292	546	144.5	tr	0.6	1	unk.	unk.	unk.	unk.	unk.	unk.	unk.	unk.	unk.	0.03	2
	frozen concentrate		*0.34*	146.0	0.3	0(a)	144.2		unk.	unk.	unk.	unk.	unk.	unk.	unk.	unk.	unk.	35	58
D	beverage,nonalc,noncarb,limeade,	8 oz glass	247	101	27.2	tr	tr	0	unk.	unk.	unk.	unk.	unk.	unk.	unk.	unk.	unk.	tr	0
	frozen,diluted		*0.40*	219.6	tr	0(a)	27.2		unk.	unk.	unk.	unk.	unk.	unk.	unk.	unk.	unk.	5	8
E	beverage,nonalc,noncarb,milk	8 oz glass	230	205	29.2	0.00	0.5	1	unk.	unk.	unk.	unk.	unk.	unk.	0.00	0.00	0.00	0.00	0
	drink,maltine		*0.43*	200.1	0.0	0(a)	29.2		unk.	unk.	unk.	unk.	unk.	unk.	0.00	0.00	0	0	0
F	beverage,nonalc,noncarb,mixed	8 oz glass	247	100	25.0	tr(a)	0.0	0	unk.	unk.	unk.	unk.	unk.	unk.	unk.	tr(a)	unk.	tr	0
	flavor-Hawaiian Punch		*0.40*	unk.	0.0	unk.	28.0		unk.	unk.	unk.	unk.	unk.	unk.	unk.	unk.	0(a)	30	50
G	beverage,nonalc,noncarb,mixed flavor,artif	8 oz glass	240	2	0.0	0(a)	0.0	0	unk.	0	0	0	0	0	0.00	0.00	0.00	0.00	0
	sweetened-Koolade		*0.42*	216.0	0.0	0(a)	0.0		0	0	0	0	0	0	0.00	0.00	0	10	17
H	beverage,nonalc,noncarb,mixed flavor,low	8 oz glass	247	44	11.1	tr(a)	0.1	0	unk.	tr(a)	unk.	unk.	unk.	tr(a)	tr(a)	tr(a)	tr(a)	unk.	unk.
	calorie-Hawaiian Punch		*0.40*	unk.	tr(a)	unk.	11.1		unk.	unk.	unk.	tr(a)	tr(a)	tr(a)	tr(a)	0.00	0(a)	0	0
I	beverage,nonalc,noncarb,mixed	8 oz glass	245	98	25.1	0.04	0.0	0	unk.	unk.	unk.	unk.	unk.	unk.	tr	tr	tr	0.00	0
	flavor,sweetened-Koolade		*0.41*	238.0	tr	unk.	25.0		unk.	unk.	unk.	unk.	unk.	unk.	tr	0.00	0	11	18
J	beverage,nonalc,noncarb,mixed	8 oz glass	240	2	0.0	0(a)	0.0	0	unk.	0	0	0	0	0	0.00	0.00	0.00	0(a)	0
	flavor,unsweetened-Koolade		*0.42*	216.0	0.0	0(a)	0.0		0	0	0	0	0	0	0.00	0.00	0	10	17
K	beverage,nonalc,noncarb,Mr. Misty Freeze	1 average	411	500	87.0	unk.	10.0	15	unk.	unk.	unk.	unk.	unk.	unk.	unk.	unk.	unk.	0.16	11
	-Dairy Queen		*0.24*	unk.	12.0	unk.	unk.		unk.	unk.	unk.	unk.	unk.	unk.	unk.	unk.	unk.	tr	0
L	beverage,nonalc,noncarb,orange &	8 oz glass	249	124	31.6	0.50	0.8	1	unk.	unk.	unk.	unk.	unk.	unk.	unk.	tr(a)	unk.	0.05	3
	apricot drink		*0.40*	215.9	0.2	unk.	31.1		unk.	unk.	unk.	unk.	unk.	unk.	unk.	unk.	0	40	66
M	beverage,nonalc,noncarb,orange	8 oz glass	247	107	26.7	0(a)	0.0	0	unk.	0(a)	0(a)	0(a)	0(a)	0(a)	0(a)	0(a)	0(a)	tr	0
	drink-Hi-C		*0.40*	unk.	0.0	0(a)	unk.		0(a)	0(a)	0(a)	0(a)	0(a)	0(a)	0(a)	0.00	0(a)	80	133
N	beverage,nonalc,noncarb,orange-Awake	8 oz glass	250	122	29.5	0.15	0.1	0	unk.	unk.	unk.	unk.	unk.	unk.	0.22	0.10	0.22	0.20	13
			0.40	219.3	0.4	unk.	27.3		unk.	unk.	unk.	unk.	unk.	unk.	0.07	2.25	0	121	202
O	beverage,nonalc,noncarb,orange-	8 oz glass	249	130	31.4	0.12	0.2	0	unk.	unk.	unk.	unk.	unk.	unk.	0.25	0.12	0.25	0.20	13
	Orange Plus		*0.40*	215.9	0.5	unk.	25.6		unk.	unk.	unk.	unk.	unk.	unk.	0.07	2.00	0	120	200
P	beverage,nonalc,noncarb,orange,	8 oz glass	265	117	29.3	0.01	0.0	0	unk.	unk.	unk.	unk.	unk.	unk.	tr	tr	tr	0.00	0
	prep-Tang		*0.38*	233.7	tr	unk.	28.4		unk.	unk.	unk.	unk.	unk.	unk.	tr	0.00	0	118	197
Q	beverage,nonalc,noncarb,	8 oz glass	250	135	34.0	tr(a)	0.5	1	unk.	tr(a)	unk.	unk.	unk.	tr(a)	unk.	unk.	unk.	0.05	3
	pineapple & grapefruit drink,canned		*0.40*	215.0	tr	unk.	34.0		unk.	unk.	unk.	tr(a)	tr(a)	tr(a)	unk.	unk.	unk.	40	67
R	beverage,nonalc,noncarb,	8 oz glass	250	135	33.8	tr(a)	0.5	1	unk.	tr(a)	unk.	unk.	unk.	tr(a)	unk.	tr(a)	unk.	0.05	3
	pineapple & orange drink,canned		*0.40*	215.0	0.2	unk.	33.8		unk.	unk.	unk.	tr(a)	tr(a)	tr(a)	unk.	unk.	0	40	67
S	BISCUIT,baking powder,home recipe	2-1/2" dia	40	155	18.7	0.07	3.0	5	unk.	148	103	162	145	54	0.16	1.95	0.61	0.11	7
			2.50	9.9	7.4	unk.	1.4		65	245	47	97	38	112	4.47	0.08	3	tr	0

Riboflavin mg / Niacin mg	% USRDA	Vitamin B₆ mg / Folacin mcg	% USRDA	Vitamin B₁₂ mcg / Pantothenic acid mg	% USRDA	Biotin mg / Vitamin A IU	% USRDA	Preformed A RE / Beta carotene RE	Vitamin D IU / Vitamin E IU	% USRDA	Total tocopherol mg / Alpha tocopherol mg	Other tocopherol mg / Total ash g	Calcium mg / Phosphorus mg	% USRDA	Sodium mg / Sodium meq	Potassium mg / Potassium meq	Chlorine mg / Chlorine meq	Iron mg / Magnesium mg	% USRDA	Zinc mg / Copper mg	% USRDA	Iodine mcg / Selenium mcg	% USRDA	Manganese mcg / Chromium mcg	
0.09	**5**	0.06	**3**	0.00	**0**	unk.	**unk.**	unk.	0(a)	**0**	0(a)	0(a)	12	**1**	6	205	unk.	0.6	**3**	unk.	**unk.**	unk.	**unk.**	unk.	A
0.9	**4**	unk.	**unk.**	0.14	**1**	58	**1**	unk.	0.0	**0**	0(a)	0.29	18	**2**	0.3	5.2	unk.	unk.	**unk.**	tr(a)	**0**	unk.		tr(a)	
0.02	**2**	0.01	**1**	0.00	**0**	unk.	**unk.**	unk.	0	**0**	0.0	0(a)	2	**0**	tr	40	unk.	tr	**0**	0.2	**2**	unk.	**unk.**	unk.	B
0.2	**1**	12	**3**	0.03	**0**	tr	**0**	unk.	0.0	**0**	0(a)	tr	3	**0**	0.0	1.0	unk.	248	**62**	0.02	**1**	unk.		tr(a)	
0.03	**2**	unk.	**unk.**	0.00	**0**	unk.	**unk.**	unk.	0(a)	**0**	0(a)	0(a)	15	**2**	tr	172	unk	0.3	**2**	unk.	**unk.**	unk.	**unk.**	unk.	C
0.3	**2**	unk.	**unk.**	unk.	**unk.**	tr	**0**	unk.	0.0	**0**	0(a)	0.58	18	**2**	0.0	4.4	unk.	unk.	**unk.**	unk.	**unk.**	unk.		unk.	
tr	**0**	unk.	**unk.**	0.00	**0**	unk.	**unk.**	unk.	0(a)	**0**	0(a)	0(a)	2	**0**	tr	32	unk.	tr	**0**	unk.	**unk.**	unk.	**unk.**	unk.	D
tr	**0**	unk.	**unk.**	unk.	**unk.**	tr	**0**	unk.	0.0	**0**	0(a)	tr	3	**0**	0.0	0.8	unk.	unk.	**unk.**	unk.	**unk.**	unk.		unk.	
0.09	**5**	unk.	**unk.**	unk.	**unk.**	unk.	**unk.**	0	unk.	**unk.**	unk.	unk.	0	**0**	unk.	unk.	unk.	0.9	**5**	unk.	**unk.**	unk.	**unk.**	unk.	E
0.5	**2**	unk.	**unk.**	unk.	**unk.**	0	**0**	0	unk.	**unk.**	unk.	0.23	35	**3**	unk.	unk.	unk.	unk.	**unk.**	unk.	**unk.**	unk.		unk.	
tr	**0**	unk.	**unk.**	0(a)	**0**	unk.	**unk.**	unk.	0(a)	**0**	unk.	unk.	20	**2**	49	62	unk.	tr	**0**	unk.	**unk.**	unk.	**unk.**	unk.	F
tr	**0**	unk.	**unk.**	unk.	**unk.**	tr	**0**	unk.	unk.	**unk.**	unk.	unk.	7	**1**	2.1	1.6	unk.	unk.	**unk.**	tr(a)	**0**	unk.		unk.	
0.00	**0**	0(a)	**0**	0.00	**0**	tr	**0**	0(a)	0(a)	**0**	0(a)	0(a)	56	**6**	tr	unk.	unk.	0.0	**0**	0.0	**0**	unk.	**unk.**	unk.	G
0.0	**0**	0	**0**	0.00	**0**	0	**0**	0(a)	0.0	**0**	0(a)	0.33	79	**8**	0.0	unk.	unk.	0	**0**	0.00	**0**	unk.		unk.	
unk.	**unk.**	unk.	**unk.**	0(a)	**0**	unk.	**unk.**	unk.	0(a)	**0**	unk.	unk.	unk.	**unk.**	unk.	unk.	unk.	tr(a)	**0**	unk.	**unk.**	unk.	**unk.**	unk.	H
unk.	**unk.**	unk.	**unk.**	unk.	**unk.**	unk.	**unk.**	unk.	unk.	**unk.**	unk.	unk.	unk.	**unk.**	unk.	unk.	unk.	unk.	**unk.**	tr(a)	**0**	tr(a)		unk.	
0.00	**0**	0.00	**0**	0.00	**0**	unk.	**unk.**	0	0	**0**	unk.	unk.	61	**6**	7	1	unk.	tr	**0**	0.0	**0**	unk.	**unk.**	unk.	I
0.0	**0**	0	**0**	0.00	**0**	0	**0**	15	unk.	**unk.**	unk.	0.37	82	**8**	0.3	0.0	unk.	0	**0**	tr	**0**	unk.		unk.	
0(a)	**0**	0(a)	**0**	0.00	**0**	0(a)	**0**	0(a)	0(a)	**0**	0(a)	0(a)	56	**6**	tr	unk.	unk.	0.0	**0**	0.0	**0**	unk.	**unk.**	unk.	J
0(a)	**0**	0	**0**	0.00	**0**	0	**0**	0(a)	0.0	**0**	unk.	0.33	79	**8**	0.0	unk.	unk.	0	**0**	0.00	**0**	unk.		unk.	
0.33	**19**	unk.	**unk.**	1.19	**20**	unk.	**unk.**	unk.	0(a)	**0**	unk.	unk.	300	**30**	200	unk.	unk.	tr	**0**	unk.	**unk.**	unk.	**unk.**	unk.	K
tr	**0**	unk.	**unk.**	unk.	**unk.**	200	**4**	unk.	unk.	**unk.**	unk.	unk.	unk.	**unk.**	8.7	unk.	unk.	unk.	**unk.**	unk.	**unk.**	unk.		unk.	
0.02	**2**	unk.	**unk.**	0(a)	**0**	unk.	**unk.**	0	0(a)	**0**	unk.	unk.	12	**1**	tr	234	unk.	0.2	**1**	unk.	**unk.**	unk.	**unk.**	unk.	L
0.5	**3**	unk.	**unk.**	unk.	**unk.**	1444	**29**	144	unk.	**unk.**	unk.	0.50	20	**2**	0.0	6.0	unk.	unk.	**unk.**	tr(a)	**0**	tr(a)		unk.	
tr	**0**	tr(a)	**0**	0(a)	**0**	unk.	**unk.**	0	0(a)	**0**	0(a)	0(a)	tr	**0**	23	17	unk.	tr	**0**	unk.	**unk.**	unk.	**unk.**	unk.	M
tr	**0**	tr(a)	**0**	unk.	**unk.**	tr	**0**	tr	0.0	**0**	0(a)	unk.	0	**0**	1.0	0.4	unk.	unk.	**unk.**	tr(a)	**0**	tr(a)		unk.	
0.02	**2**	0.00	**0**	0.00	**0**	unk.	**unk.**	0	0	**0**	unk.	unk.	80	**8**	20	319	unk.	0.9	**5**	tr	**0**	unk.	**unk.**	unk.	N
tr	**0**	unk.	**unk.**	unk.	**unk.**	0	**0**	0	unk.	**unk.**	unk.	0.92	44	**4**	0.9	8.1	unk.	tr	**0**	0.08	**4**	unk.		unk.	
1.02	**60**	0.11	**5**	0.00	**0**	unk.	**unk.**	0	0	**0**	unk.	unk.	54	**5**	12	374	unk.	0.8	**5**	tr	**0**	unk.	**unk.**	unk.	O
0.5	**3**	80	**20**	0.27	**3**	42	**1**	207	unk.	**unk.**	unk.	1.02	85	**9**	0.5	9.6	unk.	26	**7**	0.23	**11**	unk.		unk.	
0.00	**0**	0.00	**0**	0.00	**0**	unk.	**unk.**	unk.	0	**0**	unk.	unk.	64	**6**	2	65	unk.	tr	**0**	tr	**0**	unk.	**unk.**	unk.	P
0.0	**0**	0	**0**	0.00	**0**	1974	**40**	unk.	unk.	**unk.**	unk.	0.32	41	**4**	0.1	1.7	unk.	0	**0**	tr	**0**	unk.		unk.	
0.02	**2**	0.07	**4**	0.00	**0**	unk.	**unk.**	0	0(a)	**0**	unk.	unk.	13	**1**	tr	155	unk.	0.5	**3**	unk.	**unk.**	unk.	**unk.**	unk.	Q
0.2	**1**	unk.	**unk.**	unk.	**unk.**	25	**1**	3	unk.	**unk.**	unk.	0.50	13	**1**	0.0	4.0	unk.	25	**6**	tr(a)	**0**	tr(a)		unk.	
0.02	**2**	unk.	**unk.**	0(a)	**0**	unk.	**unk.**	0	0(a)	**0**	unk.	unk.	13	**1**	tr	175	unk.	0.5	**3**	unk.	**unk.**	unk.	**unk.**	unk.	R
0.2	**1**	unk.	**unk.**	unk.	**unk.**	125	**3**	13	unk.	**unk.**	unk.	0.50	15	**2**	0.0	4.5	unk.	25	**6**	tr(a)	**0**	tr(a)		unk.	
0.09	**5**	0.02	**1**	0.06	**1**	tr	**0**	0	8	**2**	tr	tr(a)	47	**5**	236	51	16	0.7	**4**	0.2	**2**	29.3	**20**	92	S
0.8	**4**	6	**1**	0.16	**2**	23	**1**	0	tr	**0**	tr	1.18	69	**7**	10.3	1.3	0.5	10	**2**	0.06	**3**	12		14	

	FOOD	Portion	Weight in grams / Conversion for 100 g	Kilocalories / H_2O g	Total carbohydrate g / Total fats g	Crude fiber g / Dietary fiber g	Total protein g / Total sugar g	% USRDA	Arginine mg / Histidine mg	Isoleucine mg / Leucine mg	Lysine mg / Methionine mg	Phenylalanine mg / Threonine mg	Valine mg / Tryptophan mg	Cystine mg / Tyrosine mg	Polyunsat. fatty acids g / Monounsat. fatty acids g	Saturated fatty acids g / P/S ratio	Linoleic acid g / Cholesterol mg	Thiamin mg / Ascorbic acid mg	% USRDA
A	BISQUICK-General Mills	1 C	113	480	76.0	tr(a)	8.0	**12**	unk.	unk.	unk.	unk.	unk.	unk.	unk.	unk.	unk.	0.45	**30**
			0.88	unk.	16.0	unk.	unk.		unk.	unk.	unk.	unk.	unk.	unk.	unk.	unk.	unk.	tr	**0**
B	BLACK-EYE PEAS,frozen,cooked,drained solids-Birds Eye	1/2 C	94	129	23.3	1.51	8.5	**13**	unk.	unk.	unk.	unk.	unk.	unk.	0.25	0.17	0.15	0.31	**21**
			1.09	61.0	0.6	unk.	8.1		unk.	unk.	unk.	unk.	unk.	unk.	0.03	1.47	0	4	**7**
C	BLACKBERRIES,canned,heavy syrup, solids & liquid	1/2 C	128	116	28.4	3.33	1.0	**2**	unk.	unk.	unk.	unk.	unk.	unk.	unk.	unk.	unk.	0.01	**1**
			0.78	97.4	0.8	7.30	25.9		unk.	unk.	unk.	unk.	unk.	unk.	unk.	unk.	0	9	**15**
D	blackberries,canned,water pack, solids & liquid	1/2 C	122	49	11.0	3.42	1.0	**2**	unk.	unk.	unk.	unk.	unk.	unk.	unk.	unk.	unk.	0.02	**2**
			0.82	108.9	0.7	7.69	8.5		unk.	unk.	unk.	unk.	unk.	unk.	unk.	unk.	0	9	**14**
E	blackberries,fresh	1/2 C	72	42	9.3	2.95	0.9	**1**	unk.	unk.	unk.	unk.	unk.	unk.	unk.	unk.	unk.	0.02	**1**
			1.39	60.8	0.6	5.26	5.0		unk.	unk.	unk.	unk.	unk.	unk.	unk.	unk.	0	15	**25**
F	BLUEBERRIES,canned,heavy syrup, solids & liquid	1/2 C	120	121	31.2	1.08	0.5	**1**	unk.	unk.	unk.	unk.	unk.	unk.	unk.	tr(a)	unk.	0.01	**1**
			0.83	87.8	0.2	unk.	30.1		unk.	unk.	unk.	unk.	unk.	unk.	unk.	unk.	0	7	**12**
G	blueberries,canned,water pack,solids & liquid	1/2 C	121	47	11.9	1.21	0.6	**1**	unk.	unk.	unk.	unk.	unk.	unk.	unk.	tr(a)	unk.	0.01	**1**
			0.83	108.1	0.2	unk.	10.6		unk.	unk.	unk.	unk.	unk.	unk.	unk.	unk.	0	8	**14**
H	blueberries,fresh	1/2 C	73	45	11.1	1.09	0.5	**1**	unk.	unk.	unk.	unk.	unk.	unk.	unk.	unk.	unk.	0.02	**1**
			1.38	60.3	0.4	unk.	7.7		unk.	unk.	unk.	unk.	unk.	unk.	unk.	unk.	0	10	**17**
I	blueberries,frozen,sweetened	1/2 C	115	121	30.5	1.03	0.7	**1**	unk.	unk.	unk.	unk.	unk.	unk.	unk.	unk.	unk.	0.05	**3**
			0.87	83.1	0.3	unk.	25.1		unk.	unk.	unk.	unk.	unk.	unk.	unk.	unk.	0	9	**15**
J	BLUEFISH,fillet,baked or broiled, w/butter or margarine	1 med	155	226	0.0	0.00	40.6	**90**	2303	2074	3577	1505	2154	476	unk.	unk.	unk.	0.17	**11**
			0.65	105.4	5.7	0(a)	0.0		1437	3091	1178	1748	406	1488	unk.	unk.	unk.	unk.	**unk.**
K	BOYSENBERRIES,frozen,unsweetened	1/2 C	63	30	7.2	1.70	0.8	**1**	unk.	unk.	unk.	unk.	unk.	unk.	unk.	tr(a)	unk.	0.01	**1**
			1.59	54.7	0.2	unk.	4.0		unk.	unk.	unk.	unk.	unk.	unk.	unk.	unk.	0	8	**14**
L	BRAZIL NUTS,shelled,salted	1/4 C	35	229	3.8	1.08	5.0	**8**	761	208	155	216	288	176	8.99	6.09	8.89	0.34	**22**
			2.86	1.6	23.4	1.44	0.6		955	395	329	148	65	169	7.77	1.48	0	35	**58**
M	brazil nuts,shelled,unsalted	1/4 C	35	229	3.8	1.08	5.0	**8**	761	208	155	216	288	176	8.99	6.09	8.89	0.34	**22**
			2.86	1.6	23.4	1.44	0.6		955	395	329	148	65	169	7.77	1.48	0	3	**6**
N	BREAD STICK	1 med	5.0	19	3.8	0.01	0.6	**1**	unk.	28	14	33	26	12	unk.	0.03	0.04	tr	**0**
			20.00	0.2	0.1	unk.	unk.		12	46	8	17	7	20	0.06	unk.	0(a)	tr	**0**
O	bread stick,cheese-Nabisco	1 stick	3.0	12	2.0	unk.	0.4	**1**	unk.	unk.	unk.	unk.	unk.	unk.	unk.	unk.	unk.	0.02	**1**
			33.33	0.2	0.3	unk.	unk.		unk.	unk.	unk.	unk.	unk.	unk.	unk.	unk.	unk.	unk.	**unk.**
P	bread stick,garlic-Nabisco	1 stick	3.0	12	2.1	unk.	0.4	**1**	unk.	unk.	unk.	unk.	unk.	unk.	unk.	unk.	unk.	0.01	**1**
			33.33	0.2	0.2	unk.	unk.		unk.	unk.	unk.	unk.	unk.	unk.	unk.	unk.	unk.	unk.	**unk.**
Q	bread stick,plain-Nabisco	1 stick	3.0	12	2.2	unk.	0.4	**1**	unk.	unk.	unk.	unk.	unk.	unk.	unk.	unk.	unk.	0.01	**1**
			33.33	0.1	0.2	unk.	unk.		unk.	unk.	unk.	unk.	unk.	unk.	unk.	unk.	unk.	unk.	**unk.**
R	bread stick,sesame-Nabisco	1 stick	3.0	13	2.0	unk.	0.4	**1**	unk.	unk.	unk.	unk.	unk.	unk.	unk.	unk.	unk.	0.01	**1**
			33.33	0.1	0.4	unk.	unk.		unk.	unk.	unk.	unk.	unk.	unk.	unk.	unk.	unk.	unk.	**unk.**
S	BREAD,Boston brown,canned	3-1/4" dia x 1/2"	45	95	20.5	0.31	2.5	**4**	unk.	116	101	119	136	unk.	unk.	unk.	unk.	0.05	**3**
			2.22	20.2	0.6	unk.	unk.		unk.	216	44	90	26	unk.	unk.	unk.	tr(a)	0	**0**

Riboflavin mg / Niacin mg	% USRDA	Vitamin B_6 mg / Folacin mcg	% USRDA	Vitamin B_{12} mcg / Pantothenic acid mg	% USRDA	Biotin mg / Vitamin A IU	% USRDA	Preformed A RE / Beta carotene RE	Vitamin D IU / Vitamin E IU	% USRDA	Total tocopherol mg / Alpha tocopherol mg	Other tocopherol mg / Total ash g	Calcium mg / Phosphorus mg	% USRDA	Sodium mg / Sodium meq	Potassium mg / Potassium meq	Chlorine mg / Chlorine meq	Iron mg / Magnesium mg	% USRDA	Zinc mg / Copper mg	% USRDA	Iodine mcg / Selenium mcg	% USRDA	Manganese mcg / Chromium mcg	
0.34	**20**	unk.	**unk.**	tr(a)	**0**	unk.	**unk.**	tr(a)	tr(a)	**0**	unk.	unk.	160	**16**	1400	198	unk.	2.2	**12**	unk.	**unk.**	unk.	**unk.**	unk.	A
4.0	**20**	unk.	**unk.**	unk.	**unk.**	tr	**0**	tr(a)	unk.	**unk.**	unk.	unk.	unk.	**unk.**	60.9	5.1	unk.	unk.	**unk.**	unk.	**unk.**	unk.		unk.	
0.07	**4**	0.10	**5**	0.00	**0**	unk.	**unk.**	unk.	0	**0**	unk.	unk.	26	**3**	6	414	unk.	2.1	**12**	1.5	**10**	unk.	**unk.**	unk.	B
0.8	**4**	122	**31**	0.22	**2**	76	**2**	unk.	unk.	**unk.**	unk.	0.98	135	**14**	0.3	10.6	unk.	52	**13**	0.19	**10**	unk.		unk.	
0.03	**2**	0.03	**2**	0.00	**0**	tr	**0**	0	0(a)	**0**	3.5	unk.	27	**3**	1	140	20	0.8	**4**	unk.	**unk.**	unk.	**unk.**	unk.	C
0.3	**1**	18	**5**	0.10	**1**	166	**3**	17	4.1	**14**	unk.	0.38	15	**2**	0.0	3.6	0.5	29	**7**	tr(a)	**0**	tr(a)		tr(a)	
0.02	**1**	0.03	**2**	unk.	**unk.**	0.001	**0**	0	0(a)	**0**	5.2	unk.	5	**1**	1	140	unk.	0.7	**4**	unk.	**unk.**	unk.	**unk.**	112	D
0.2	**1**	17	**4**	0.10	**1**	171	**3**	17	6.3	**21**	unk.	0.37	16	**2**	0.0	3.6	unk.	4	**1**	0.05	**2**	0		0	
0.03	**2**	0.04	**2**	0.00	**0**	tr	**0**	0	0(a)	**0**	1.9	1.5	10	**1**	1	122	11	0.3	**2**	unk.	**unk.**	unk.	**unk.**	1260	E
0.3	**1**	10	**3**	0.17	**2**	144	**3**	14	2.3	**8**	0.4	0.36	14	**1**	0.0	3.1	0.3	14	**4**	0.08	**4**	0		0	
0.01	**1**	0.05	**2**	0.00	**0**	unk.	**unk.**	0	0(a)	**0**	unk.	unk.	11	**1**	1	66	unk.	0.7	**4**	unk.	**unk.**	unk.	**unk.**	unk.	F
0.2	**1**	5	**1**	0.08	**1**	48	**1**	5	unk.	**unk.**	unk.	0.24	10	**1**	0.0	1.7	unk.	5	**1**	tr(a)	**0**	tr(a)		unk.	
0.01	**1**	0.05	**2**	0.00	**0**	unk.	**unk.**	0	0(a)	**0**	unk.	unk.	12	**1**	1	73	unk.	0.8	**5**	unk.	**unk.**	unk.	**unk.**	181	G
0.2	**1**	5	**1**	0.08	**1**	48	**1**	5	unk.	**unk.**	unk.	0.24	11	**1**	0.0	1.9	unk.	5	**1**	0.03	**2**	tr(a)		18	
0.04	**3**	0.05	**2**	0.00	**0**	unk.	**unk.**	0	0(a)	**0**	unk.	unk.	11	**1**	1	59	6	0.7	**4**	unk.	**unk.**	unk.	**unk.**	1667	H
0.4	**2**	6	**1**	0.11	**1**	72	**1**	7	unk.	**unk.**	unk.	0.22	9	**1**	0.0	1.5	0.2	4	**1**	0.08	**4**	tr(a)		unk.	
0.06	**3**	0.06	**3**	0.00	**0**	unk.	**unk.**	0	0(a)	**0**	unk.	unk.	7	**1**	1	76	unk.	0.5	**3**	unk.	**unk.**	unk.	**unk.**	unk.	I
0.5	**2**	5	**1**	0.14	**1**	34	**1**	3	unk.	**unk.**	unk.	0.34	13	**1**	0.0	1.9	unk.	5	**1**	tr(a)	**0**	tr(a)		unk.	
0.15	**9**	unk.	**unk.**	unk.	**unk.**	unk.	**unk.**	unk.	unk.	**unk.**	unk.	unk.	45	**5**	161	unk.	unk.	1.1	**6**	unk.	**unk.**	unk.	**unk.**	unk.	J
2.9	**15**	unk.	**unk.**	unk.	**unk.**	77	**2**	unk.	unk.	**unk.**	unk.	2.17	445	**45**	7.0	unk.	unk.	unk.	**unk.**	unk.	**unk.**	unk.		unk.	
0.08	**5**	0.04	**2**	0.00	**0**	unk.	**unk.**	0	0(a)	**0**	unk.	unk.	16	**2**	1	96	unk.	1.0	**6**	unk.	**unk.**	unk.	**unk.**	unk.	K
0.6	**3**	unk.	**unk.**	0.14	**1**	107	**2**	11	unk.	**unk.**	unk.	0.19	15	**2**	0.0	2.5	unk.	11	**3**	tr(a)	**0**	tr(a)		unk.	
0.04	**3**	0.06	**3**	0.00	**0**	unk.	**unk.**	0(a)	0	**0**	unk.	unk.	65	**7**	70	250	28	1.2	**7**	1.8	**12**	unk.	**unk.**	112	L
0.6	**3**	2	**0**	0.08	**1**	tr(a)	**0**	tr(a)	2.3	**8**	2.3	1.15	243	**24**	3.0	6.4	0.8	79	**20**	0.38	**19**	0		unk.	
0.04	**3**	0.06	**3**	0.00	**0**	unk.	**unk.**	0	0	**0**	unk.	unk.	65	**7**	tr	250	28	1.2	**7**	1.8	**12**	unk.	**unk.**	973	M
0.6	**3**	2	**0**	0.08	**1**	tr	**0**	tr	2.3	**8**	2.3	1.15	243	**24**	0.0	6.4	0.8	111	**28**	0.83	**42**	36		unk.	
tr	**0**	unk.	**unk.**	0(a)	**0**	unk.	**unk.**	0(a)	0(a)	**0**	unk.	unk.	1	**0**	84	5	unk.	tr	**0**	unk.	**unk.**	unk.	**unk.**	unk.	N
0.0	**0**	unk.	**unk.**	unk.	**unk.**	0(a)	**0**	0(a)	unk.	**unk.**	unk.	0.24	5	**1**	3.6	0.1	unk.	unk.	**unk.**	unk.	**unk.**	unk.		unk.	
0.02	**1**	unk.	**unk.**	unk.	**unk.**	unk.	**unk.**	unk.	0(a)	**0**	unk.	unk.	2	**0**	20	4	unk.	0.1	**1**	tr	**0**	unk.	**unk.**	unk.	O
0.2	**1**	unk.	**unk.**	unk.	**unk.**	unk.	**unk.**	unk.	unk.	**unk.**	unk.	0.07	6	**1**	0.9	0.1	unk.	1	**0**	0.01	**0**	unk.		unk.	
0.02	**1**	unk.	**unk.**	unk.	**unk.**	unk.	**unk.**	unk.	0(a)	**0**	unk.	unk.	1	**0**	16	4	unk.	0.1	**1**	tr	**0**	unk.	**unk.**	unk.	P
0.2	**1**	unk.	**unk.**	unk.	**unk.**	unk.	**unk.**	unk.	unk.	**unk.**	unk.	0.06	4	**0**	0.7	0.1	unk.	1	**0**	0.01	**0**	unk.		unk.	
0.02	**1**	unk.	**unk.**	unk.	**unk.**	unk.	**unk.**	unk.	0(a)	**0**	unk.	unk.	1	**0**	23	4	unk.	0.1	**1**	tr	**0**	unk.	**unk.**	unk.	Q
0.1	**1**	unk.	**unk.**	unk.	**unk.**	unk.	**unk.**	unk.	unk.	**unk.**	unk.	0.07	4	**0**	1.0	0.1	unk.	1	**0**	0.01	**0**	unk.		unk.	
0.02	**1**	unk.	**unk.**	unk.	**unk.**	unk.	**unk.**	unk.	0(a)	**0**	unk.	unk.	4	**0**	22	5	unk.	0.1	**1**	0.0	**0**	unk.	**unk.**	unk.	R
0.1	**1**	unk.	**unk.**	unk.	**unk.**	unk.	**unk.**	unk.	unk.	**unk.**	unk.	0.08	6	**1**	1.0	0.1	unk.	2	**1**	0.01	**1**	unk.		unk.	
0.03	**2**	unk.	**unk.**	tr(a)	**0**	unk.	**unk.**	0	tr(a)	**0**	unk.	unk.	40	**4**	113	131	unk.	0.9	**5**	unk.	**unk.**	unk.	**unk.**	unk.	S
0.5	**3**	unk.	**unk.**	unk.	**unk.**	0	**0**	0	unk.	**unk.**	unk.	1.17	72	**7**	4.9	3.4	unk.	unk.	**unk.**	unk.	**unk.**	unk.		unk.	

	FOOD	Portion	Weight in grams / Conversion for 100 g	Kilocalories / H_2O g	Total carbohydrate g / Total fats g	Crude fiber g / Dietary fiber g	Total protein g / Total sugar g	% USRDA	Arginine mg / Histidine mg	Isoleucine mg / Leucine mg	Lysine mg / Methionine mg	Phenylalanine mg / Threonine mg	Valine mg / Tryptophan mg	Cystine mg / Tyrosine mg	Polyunsat. fatty acids g / Monounsat. fatty acids g	Saturated fatty acids g / P/S ratio	Linoleic acid g / Cholesterol mg	Thiamin mg / Ascorbic acid mg	% USRDA
A	bread,bran raisin	1 lg slice	49	151	27.8	0.51	3.9	**6**	unk.	166	105	190	179	unk.	unk.	unk.	unk.	0.08	**5**
			2.04	unk.	4.3	unk.	6.6		unk.	260	58	112	47	unk.	unk.	unk.	0	0	**0**
B	bread,bran raisin,toasted	1 lg slice	48	151	27.2	0.5	3.8	**5**	unk.	163	100	186	175	unk.	unk.	unk.	unk.	0.08	**5**
			2.80	unk.	4.2	unk.	6.5		unk.	255	57	110	46	unk.	unk.	unk.	0	0	**0**
C	bread,cracked wheat	1 reg slice	25	66	13.0	0.13	2.2	**3**	unk.	93	59	107	100	55	unk.	0.11	0.16	0.03	**2**
			4.00	8.7	0.5	2.10	1.6		59	146	33	63	26	76	0.22	unk.	0(a)	tr	**0**
D	bread,cracked wheat,toasted	1 reg slice	21	66	13.0	0.13	2.2	**3**	unk.	78	49	89	84	55	unk.	0.12	0.17	0.02	**2**
			4.76	4.7	0.5	2.12	1.4		59	122	27	53	22	76	0.22	unk.	0(a)	tr	**0**
E	bread,French	4-3/4x4x 1/2" slice	25	73	13.8	0.05	2.3	**4**	unk.	64	51	124	96	46	unk.	0.25	0.20	0.07	**5**
			4.00	7.6	0.8	1.33	1.4		46	174	29	65	28	78	0.33	unk.	0	tr	**0**
F	bread,French,toasted	4-3/4x4x 1/2" slice	23	78	14.8	0.05	2.4	**4**	unk.	68	55	133	103	44	0.27	0.20	0.24	0.06	**4**
			4.35	4.4	0.8	unk.	1.3		49	186	31	70	29	83	0.40	1.35	0	tr	**0**
G	bread,home recipe,banana	3x2-1/2x 1/2" slice	34	116	14.6	0.12	2.1	**4**	100	99	83	109	110	43	1.20	1.19	1.24	0.06	**4**
			2.94	10.8	5.6	0.69	6.5		51	171	42	77	28	80	2.65	1.01	25	1	**1**
H	bread,home recipe,nut	1 slice	45	127	20.3	0.11	2.7	**5**	104	136	106	143	146	51	2.12	0.80	1.87	0.08	**5**
			2.22	16.9	4.0	0.59	9.1		65	225	50	97	35	105	0.83	2.63	13	tr	**0**
I	bread,home recipe,pumpkin	1 slice	45	127	20.1	0.27	2.0	**4**	86	96	82	100	103	37	1.10	2.00	0.90	0.05	**4**
			2.22	17.5	4.6	0.62	12.7		49	156	39	73	26	75	1.08	0.55	28	1	**1**
J	bread,home recipe,raisin	1 slice	45	140	27.4	0.17	3.2	**5**	unk.	139	101	151	139	50	0.21	1.28	0.18	0.11	**7**
			2.22	11.5	2.2	0.69	4.2		62	230	45	93	35	104	0.52	0.16	6	tr	**0**
K	bread,home recipe,sour dough	1 slice	28	76	15.9	0.06	2.1	**3**	unk.	101	56	118	97	42	0.12	0.05	0.11	0.09	**6**
			3.57	9.8	0.2	0.58	1.5		46	169	29	65	26	70	0.02	2.63	0	0	**0**
L	bread,Italian	4-1/2x3-1/4x 3/4" slice	30	83	16.9	0.06	2.7	**4**	unk.	76	61	148	115	55	unk.	tr	0.13	0.09	**6**
			3.33	9.5	0.2	unk.	1.7		55	208	34	78	33	93	0.07	unk.	0	0	**0**
M	bread,pita	1 average	64	182	39.8	unk.	8.6	**13**	unk.	unk.	unk.	unk.	unk.	unk.	unk.	unk.	unk.	unk.	**unk.**
			1.57	13.3	0.2	unk.	unk.		unk.	unk.	unk.	unk.	unk.	unk.	unk.	unk.	0	unk.	**unk.**
N	bread,raisin	1 reg slice	25	66	13.4	0.00	1.6	**3**	unk.	70	43	80	75	38	unk.	0.16	0.17	0.01	**1**
			4.00	8.8	0.7	0.40	3.4		63	109	24	33	20	48	0.32	unk.	0	tr	**0**
O	bread,raisin,toasted	1 reg slice	21	66	13.6	0.23	1.7	**3**	unk.	71	44	82	76	38	unk.	0.16	0.18	0.01	**1**
			4.76	4.6	0.7	0.40	2.8		64	110	24	33	20	48	0.34	unk.	0	tr	**0**
P	bread,Roman meal	1 slice	27	66	13.0	0.35	3.0	**5**	unk.	59	58	66	73	unk.	unk.	unk.	unk.	0.11	**7**
			3.70	unk.	1.1	unk.	1.8		unk.	94	22	52	15	unk.	unk.	unk.	tr	0	**0**
Q	bread,rye,dark (pumpernickel)	1 reg slice	25	62	13.3	0.27	2.3	**4**	unk.	98	93	106	118	unk.	unk.	unk.	unk.	0.06	**4**
			4.00	8.5	0.3	1.11	1.4		unk.	152	36	84	25	unk.	unk.	unk.	0(a)	0	**0**
R	bread,rye,dark (pumpernickel), snack loaf	1 sm slice	7.0	17	3.7	0.08	0.6	**1**	unk.	27	26	30	33	unk.	unk.	unk.	unk.	0.02	**1**
			14.29	2.4	0.1	unk.	0.4		unk.	42	10	23	7	unk.	unk.	unk.	0(a)	0	**0**
S	bread,rye,light	1 reg slice	25	61	13.0	0.10	2.3	**4**	unk.	98	73	110	118	50	unk.	unk.	unk.	0.04	**3**
			4.00	8.9	0.3	1.20	1.4		88	153	35	73	25	65	unk.	unk.	0(a)	0	**0**

Riboflavin mg / Niacin mg	% USRDA	Vitamin B_6 mg / Folacin mcg	% USRDA	Vitamin B_{12} mcg / Pantothenic acid mg	% USRDA	Biotin mg / Vitamin A IU	% USRDA	Preformed A RE / Beta carotene RE	Vitamin D IU / Vitamin E IU	% USRDA	Total tocopherol mg / Alpha tocopherol mg	Other tocopherol mg / Total ash g	Calcium mg / Phosphorus mg	% USRDA	Sodium mg / Sodium meq	Potassium mg / Potassium meq	Chlorine mg / Chlorine meq	Iron mg / Magnesium mg	% USRDA	Zinc mg / Copper mg	% USRDA	Iodine mcg / Selenium mcg	% USRDA	Manganese mcg / Chromium mcg	
0.09	**5**	unk.	**unk.**	0(a)	**0**	tr(a)	**0**	0(a)	0(a)	**0**	unk.	unk.	45	**5**	unk.	unk.	unk.	1.6	**9**	unk.	**unk.**	unk.	**unk.**	unk.	A
1.4	**7**	unk.	**unk.**	unk.	**unk.**	tr(a)	**0**	tr(a)	unk.	**unk.**	unk.	unk.	121	**12**	unk.	unk.	unk.	unk.	**unk.**	unk.	**unk.**	unk.		unk.	
0.09	**5**	unk.	**unk.**	0(a)	**0**	tr(a)	**0**	0(a)	0(a)	**0**	unk.	unk.	44	**4**	unk.	unk.	unk.	1.6	**9**	unk.	**unk.**	unk.	**unk.**	unk.	B
1.4	**7**	unk.	**unk.**	unk.	**unk.**	tr(a)	**0**	tr(a)	unk.	**unk.**	unk.	unk.	118	**12**	unk.	unk.	unk.	unk.	**unk.**	unk.	**unk.**	unk.		unk.	
0.02	**1**	0.02	**1**	0.00	**0**	tr	**0**	tr	0(a)	**0**	unk.	unk.	22	**2**	132	34	unk.	0.3	**2**	0.5	**3**	unk.	**unk.**	unk.	C
0.8	**4**	12	**3**	0.15	**2**	tr	**0**	0	unk.	**unk.**	unk.	0.52	32	**3**	5.8	0.9	unk.	9	**2**	0.05	**3**	unk.		unk.	
0.02	**1**	0.02	**1**	0.00	**0**	tr	**0**	tr	0(a)	**0**	unk.	unk.	22	**2**	132	34	unk.	0.3	**2**	0.5	**3**	unk.	**unk.**	unk.	D
0.8	**4**	12	**3**	0.15	**2**	tr	**0**	0	unk.	**unk.**	unk.	0.52	32	**3**	5.8	0.9	unk.	7	**2**	0.05	**3**	unk.		unk.	
0.05	**3**	0.01	**1**	0.00	**0**	tr	**0**	tr	0	**0**	tr	unk.	11	**1**	145	23	unk.	0.5	**3**	0.3	**2**	unk.	**unk.**	tr	E
0.6	**3**	10	**2**	0.09	**1**	tr	**0**	0	tr	**0**	unk.	0.47	21	**2**	6.3	0.6	unk.	6	**1**	0.01	**0**	unk.		unk.	
0.06	**3**	0.01	**1**	0.00	**0**	tr	**0**	tr	0	**0**	tr	unk.	11	**1**	155	24	unk.	0.6	**3**	0.4	**3**	unk.	**unk.**	tr	F
0.7	**3**	10	**3**	0.10	**1**	tr	**0**	0	tr	**0**	unk.	0.51	23	**2**	6.7	0.6	unk.	5	**1**	0.01	**0**	unk.		unk.	
0.05	**3**	0.07	**4**	0.05	**1**	0.002	**1**	0	2	**1**	0.1	tr	11	**1**	110	56	26	0.5	**3**	0.2	**2**	17.7	**12**	107	G
0.4	**2**	8	**2**	0.14	**1**	26	**1**	2	0.1	**0**	0.1	0.45	37	**4**	4.8	1.4	0.7	21	**5**	0.06	**3**	5		12	
0.07	**4**	0.04	**2**	0.08	**1**	0.002	**1**	0	7	**2**	1.0	0.0	37	**4**	214	55	15	0.6	**3**	0.3	**2**	38.0	**25**	128	H
0.5	**3**	7	**2**	0.14	**1**	25	**1**	tr	1.2	**4**	0.1	0.98	62	**6**	9.3	1.4	0.4	16	**4**	0.09	**5**	8		15	
0.05	**3**	0.04	**2**	0.06	**1**	0.002	**1**	0	3	**1**	0.1	0.0	17	**2**	177	70	21	0.6	**3**	0.3	**2**	27.0	**18**	98	I
0.4	**2**	7	**2**	0.17	**2**	764	**15**	66	0.1	**0**	0.1	0.58	35	**4**	7.7	1.8	0.6	18	**5**	0.06	**3**	4		13	
0.08	**5**	0.05	**3**	0.06	**1**	0.001	**0**	0	7	**2**	tr	0.0	38	**4**	90	146	26	1.0	**5**	0.3	**2**	17.9	**12**	2034	J
0.9	**4**	7	**2**	0.17	**2**	76	**2**	tr	0.1	**0**	tr	0.79	51	**5**	3.9	3.7	0.7	16	**4**	0.10	**5**	10		13	
0.07	**4**	0.02	**1**	0.00	**0**	tr	**0**	0	0	**0**	0.6	0.3	3	**0**	15	24	14	0.6	**3**	0.1	**1**	unk.	**unk.**	78	K
0.8	**4**	19	**5**	0.12	**1**	0	**0**	0	0.8	**3**	0.3	0.10	21	**2**	0.6	0.6	0.4	6	**2**	0.03	**2**	10		12	
0.06	**4**	0.02	**1**	0(a)	**0**	tr(a)	**0**	0	0(a)	**0**	tr	unk.	5	**1**	175	22	unk.	0.7	**4**	0.2	**1**	unk.	**unk.**	tr	L
0.8	**4**	12	**3**	0.11	**1**	0	**0**	0	tr	**0**	unk.	0.57	23	**2**	7.6	0.6	unk.	7	**2**	0.01	**0**	unk.		unk.	
unk.	**unk.**	unk.	**unk.**	unk.	**unk.**	unk.	**unk.**	unk.	unk.	**unk.**	unk.	unk.	unk.	**unk.**	305	unk.	unk.	unk.	**unk.**	unk.	**unk.**	unk.	**unk.**	unk.	M
unk.	**unk.**	unk.	**unk.**	unk.	**unk.**	unk.	**unk.**	unk.	unk.	**unk.**	unk.	1.65	unk.	**unk.**	13.3	unk.	unk.	unk.	**unk.**	unk.	**unk.**	unk.		unk.	
0.02	**1**	0.01	**1**	0.00	**0**	tr	**0**	tr	tr(a)	**0**	tr	unk.	18	**2**	91	58	unk.	0.3	**2**	0.3	**2**	unk.	**unk.**	unk.	N
0.6	**3**	7	**2**	0.10	**1**	tr	**0**	tr	tr	**0**	unk.	0.42	22	**2**	4.0	1.5	unk.	6	**2**	0.06	**3**	unk.		unk.	
0.02	**1**	0.01	**1**	0.00	**0**	tr	**0**	tr	tr(a)	**0**	tr	unk.	18	**2**	92	59	unk.	0.3	**2**	0.3	**2**	unk.	**unk.**	unk.	O
0.6	**3**	8	**2**	0.10	**1**	tr	**0**	tr	tr	**0**	unk.	0.42	22	**2**	4.0	1.5	unk.	7	**2**	0.06	**3**	unk.		unk.	
0.06	**4**	0.02	**1**	0(a)	**0**	tr(a)	**0**	0(a)	0(a)	**0**	unk.	unk.	1	**0**	162	unk.	unk.	19.2	**107**	unk.	**unk.**	unk.	**unk.**	unk.	P
0.9	**5**	16	**4**	0.16	**2**	0	**0**	tr	unk.	**unk.**	unk.	unk.	59	**6**	7.0	unk.	unk.	23	**6**	unk.	**unk.**	unk.		unk.	
0.03	**2**	0.04	**2**	0.00	**0**	tr(a)	**0**	0	0(a)	**0**	unk.	unk.	21	**2**	142	114	unk.	0.6	**3**	0.4	**3**	unk.	**unk.**	125	Q
0.3	**2**	6	**1**	0.13	**1**	0	**0**	0	0.3	**1**	0.3	0.65	57	**6**	6.2	2.9	unk.	18	**4**	0.04	**2**	unk.		15	
0.01	**1**	0.01	**1**	0.00	**0**	tr(a)	**0**	0	0(a)	**0**	unk.	unk.	6	**1**	40	32	unk.	0.2	**1**	0.1	**1**	unk.	**unk.**	35	R
0.1	**0**	2	**0**	0.03	**0**	0	**0**	0	0.1	**0**	0.1	0.18	16	**2**	1.7	0.8	unk.	5	**1**	0.01	**1**	unk.		4	
0.02	**1**	0.02	**1**	0.00	**0**	tr	**0**	0	0	**0**	0.1	tr	19	**2**	139	36	unk.	0.4	**2**	0.4	**3**	unk.	**unk.**	125	S
0.7	**4**	6	**1**	0.11	**1**	0	**0**	0	0.1	**0**	tr	0.55	37	**4**	6.1	0.9	unk.	11	**3**	0.05	**3**	unk.		15	

	FOOD	Portion	Weight in grams / Conversion for 100 g	Kilocalories / H_2O g	Total carbohydrate g / Total fats g	Crude fiber g / Dietary fiber g	Total protein g / Total sugar g	% USRDA	Arginine mg / Histidine mg	Isoleucine mg / Leucine mg	Lysine mg / Methionine mg	Phenylalanine mg / Threonine mg	Valine mg / Tryptophan mg	Cystine mg / Tyrosine mg	Polyunsat. fatty acids g / Monounsat. fatty acids g	Saturated fatty acids g / P/S ratio	Linoleic acid g / Cholesterol mg	Thiamin mg / Ascorbic acid mg	% USRDA
A	bread,rye,light,snack loaf	1 sm slice	7.0	17	3.6	0.03	0.6	**1**	unk.	27	20	31	33	14	unk.	unk.	unk.	0.01	**1**
			14.29	2.5	0.1	0.34	0.4		24	43	10	20	7	18	unk.	unk.	0(a)	0	**0**
B	bread,rye,light,toasted	1 reg slice	22	62	13.3	0.11	2.3	**4**	unk.	99	74	111	120	51	unk.	unk.	unk.	0.04	**3**
			4.55	5.5	0.3	1.21	1.3		90	155	35	74	27	67	unk.	unk.	0(a)	0	**0**
C	bread,Vienna	1 reg slice	25	73	13.8	0.05	2.3	**4**	unk.	64	51	124	96	50	unk.	0.16	0.20	0.07	**5**
			4.00	7.6	0.8	unk.	1.4		85	174	29	65	28	65	0.33	unk.	0	tr	**0**
D	bread,Vienna,toasted	1 reg slice	21	72	13.8	0.04	2.3	**4**	unk.	55	44	106	82	50	unk.	0.16	0.20	0.06	**4**
			4.67	4.1	0.7	unk.	1.2		85	149	25	56	24	65	0.33	unk.	0	tr	**0**
E	bread,white enriched	1 reg slice	25	68	12.6	0.05	2.2	**3**	unk.	104	65	117	100	65	unk.	0.18	0.12	0.06	**4**
			4.00	8.9	0.8	0.80	1.4		83	173	32	67	26	72	0.45	unk.	0	tr	**0**
F	bread,white enriched,thin sliced	1 thin slice	20	54	10.1	0.04	1.7	**3**	unk.	83	52	94	80	52	unk.	0.15	0.10	0.05	**3**
			5.00	7.1	0.6	0.64	1.2		66	138	25	54	21	57	0.36	unk.	0	tr	**0**
G	bread,white enriched,thin sliced -Weight Watchers	1 thin slice	14	35	6.5	unk.	1.0	**2**	unk.	unk.	unk.	unk.	unk.	unk.	unk.	unk.	unk.	0.06	**4**
			7.04	unk.	0.5	unk.	unk.		unk.	unk.	unk.	unk.	unk.	unk.	unk.	unk.	0(a)	0	**0**
H	bread,white enriched,thin sliced, toasted	1 thin slice	17	53	10.0	0.03	1.7	**3**	unk.	82	51	92	78	52	unk.	0.14	0.16	0.04	**3**
			5.88	4.3	0.6	0.62	1.0		65	135	25	53	20	57	0.29	unk.	0	tr	**0**
I	bread,white enriched,toasted	1 reg slice	22	69	12.9	0.04	2.2	**3**	unk.	106	66	119	101	67	unk.	0.18	0.21	0.05	**3**
			4.55	5.5	0.8	0.81	1.3		84	175	32	68	26	73	0.37	unk.	0	tr	**0**
J	bread,whole wheat	1 reg slice	23	56	11.0	0.37	2.4	**4**	unk.	106	71	117	113	72	unk.	0.14	0.22	0.06	**4**
			4.35	8.4	0.7	2.10	1.5		72	166	37	72	29	80	0.28	unk.	1	tr	**0**
K	bread,whole wheat,toasted	1 reg slice	19	55	10.8	0.36	2.4	**4**	unk.	104	71	117	113	71	unk.	0.13	0.21	0.05	**3**
			5.26	4.6	0.7	2.06	1.3		71	166	37	72	29	66	0.27	unk.	tr(a)	tr	**0**
L	BREADCRUMBS,dry,grated	1 C	100	392	73.4	0.30	12.6	**20**	unk.	593	334	687	561	252	unk.	1.00	1.00	0.22	**15**
			1.00	6.5	4.6	unk.	6.4		252	989	170	378	151	431	3.00	unk.	0	tr	**0**
M	breadcrumbs,salt free,dry,grated	1 C	100	392	73.4	0.30	12.6	**20**	unk.	593	334	687	561	252	unk.	1.00	1.00	0.22	**15**
			1.00	6.5	4.6	unk.	6.4		252	989	170	378	151	431	3.00	unk.	0	unk.	**unk.**
N	BREAKFAST SQUARES,all flavors -General Mills	1 square	43	190	22.5	unk.	6.0	**9**	unk.	unk.	unk.	unk.	unk.	unk.	unk.	unk.	unk.	0.19	**13**
			2.35	unk.	8.5	unk.	unk.		unk.	unk.	unk.	unk.	unk.	unk.	unk.	unk.	unk.	8	**13**
O	BROCCOLI,fresh,chopped,cooked, drained solids	1/2 C	78	20	3.5	1.16	2.4	**4**	138	91	105	87	122	27	unk.	tr(a)	unk.	0.07	**5**
			1.29	70.8	0.2	3.18	1.4		44	118	36	89	101	83	unk.	unk.	0	70	**116**
P	broccoli,fresh,stalk,cooked, drained solids	1 stalk	90	23	4.0	1.35	2.8	**4**	161	106	122	101	142	31	unk.	tr(a)	unk.	0.08	**5**
			1.11	82.2	0.3	3.69	1.6		51	137	41	103	117	96	unk.	unk.	0	81	**135**
Q	broccoli,frozen,chopped,cooked, drained solids	1/2 C	93	24	4.3	1.02	2.9	**4**	165	102	118	96	137	32	unk.	tr(a)	unk.	0.06	**4**
			1.08	84.5	0.2	3.79	1.7		52	131	41	99	120	99	unk.	unk.	0	68	**113**
R	broccoli,frozen,stalk,cooked, drained solids	1 sm stalk	30	8	1.4	0.33	0.9	**1**	54	33	38	31	44	10	unk.	tr(a)	unk.	0.02	**1**
			3.33	27.4	0.1	1.23	0.5		17	43	13	32	39	32	unk.	unk.	0	22	**37**
S	broccoli,home recipe,casserole	1 C	211	287	22.4	1.22	10.3	**19**	384	504	597	484	562	81	0.61	10.99	0.05	0.13	**8**
			0.48	157.1	18.3	4.07	5.2		206	790	197	367	122	322	4.74	0.06	52	70	**116**

Riboflavin mg / Niacin mg	% USRDA	Vitamin B_6 mg / Folacin mcg	% USRDA	Vitamin B_{12} mcg / Pantothenic acid mg	% USRDA	Biotin mg / Vitamin A IU	% USRDA	Preformed A RE / Beta carotene RE	Vitamin D IU / Vitamin E IU	% USRDA	Total tocopherol mg / Alpha tocopherol mg	Other tocopherol mg / Total ash g	Calcium mg / Phosphorus mg	% USRDA	Sodium mg / Sodium meq	Potassium mg / Potassium meq	Chlorine mg / Chlorine meq	Iron mg / Magnesium mg	% USRDA	Zinc mg / Copper mg	% USRDA	Iodine mcg / Selenium mcg	% USRDA	Manganese mcg / Chromium mcg	
0.00	**0**	0.01	**0**	0.00	**0**	tr	**0**	0	0	**0**	tr	tr	5	**1**	39	10	unk.	0.1	**1**	0.1	**1**	unk.	**unk.**	35	A
0.2	**1**	2	**0**	0.03	**0**	0	**0**	0	tr	**0**	tr	0.15	10	**1**	1.7	0.3	unk.	3	**1**	0.02	**1**	unk.		4	
0.02	**1**	0.03	**1**	0.00	**0**	tr	**0**	0	0	**0**	0.1	unk.	19	**2**	143	37	unk.	0.4	**2**	0.4	**3**	unk.	**unk.**	tr	B
0.6	**3**	6	**2**	0.11	**1**	0	**0**	0	0.1	**0**	unk.	0.57	38	**4**	6.2	0.9	unk.	11	**3**	0.05	**2**	unk.		unk.	
0.05	**3**	0.01	**1**	0.00	**0**	tr	**0**	tr	0(a)	**0**	tr	unk.	11	**1**	145	23	unk.	0.5	**3**	0.3	**2**	unk.	**unk.**	tr	C
0.6	**3**	10	**2**	0.09	**1**	tr	**0**	0	tr	**0**	unk.	0.47	21	**2**	6.3	0.6	unk.	6	**1**	0.06	**3**	unk.		unk.	
0.05	**3**	0.01	**1**	0.00	**0**	tr	**0**	tr	0(a)	**0**	tr	unk.	11	**1**	144	22	unk.	0.6	**3**	0.3	**2**	unk.	**unk.**	tr	D
0.6	**3**	10	**2**	0.09	**1**	tr	**0**	0	tr	**0**	unk.	0.47	21	**2**	6.3	0.6	unk.	4	**1**	0.06	**3**	unk.		unk.	
0.05	**3**	0.01	**1**	tr	**0**	tr	**0**	tr	tr(a)	**0**	0.1	tr	21	**2**	127	26	unk.	0.6	**4**	0.1	**1**	unk.	**unk.**	148	E
0.6	**3**	10	**2**	0.11	**1**	tr	**0**	0	0.1	**0**	tr	0.50	24	**2**	5.5	0.7	unk.	6	**2**	0.03	**1**	unk.		1	
0.04	**3**	0.01	**0**	tr	**0**	tr	**0**	tr	tr(a)	**0**	0.0	tr	17	**2**	101	21	unk.	0.5	**3**	0.1	**1**	unk.	**unk.**	118	F
0.5	**2**	8	**2**	0.09	**1**	tr	**0**	0	0.1	**0**	tr	0.40	19	**2**	4.4	0.5	unk.	5	**1**	0.02	**1**	unk.		1	
0.04	**2**	unk.	**unk.**	0(a)	**0**	unk.	**unk.**	0	0(a)	**0**	unk.	unk.	12	**1**	100	unk.	unk.	0.4	**2**	unk.	**unk.**	unk.	**unk.**	unk.	G
0.4	**2**	unk.	**unk.**	unk.	**unk.**	0	**0**	0	unk.	**unk.**	unk.	unk.	unk.	**unk.**	4.3	unk.	unk.	unk.	**unk.**	unk.	**unk.**	unk.		unk.	
0.04	**2**	0.01	**0**	tr	**0**	unk.	**unk.**	tr	tr(a)	**0**	tr	unk.	17	**2**	100	21	177	0.5	**3**	0.1	**1**	unk.	**unk.**	tr	H
0.6	**3**	8	**2**	0.07	**1**	tr	**0**	0	0.0	**0**	unk.	0.39	19	**2**	4.4	0.5	5.0	3	**1**	0.02	**1**	unk.		1	
0.05	**3**	0.01	**1**	tr	**0**	tr	**0**	tr	tr(a)	**0**	0.1	unk.	22	**2**	130	27	229	0.6	**4**	0.1	**1**	unk.	**unk.**	tr	I
0.6	**3**	10	**3**	0.11	**1**	tr	**0**	0	0.1	**0**	unk.	0.51	25	**3**	5.6	0.7	6.4	4	**1**	0.03	**1**	unk.		1	
0.03	**2**	0.04	**2**	0.00	**0**	tr	**0**	tr	tr(a)	**0**	0.5	0.4	23	**2**	121	63	unk.	0.5	**3**	0.4	**3**	unk.	**unk.**	943	J
0.6	**3**	13	**3**	0.17	**2**	tr	**0**	0	0.6	**2**	0.1	0.55	52	**5**	5.3	1.6	unk.	18	**5**	0.12	**6**	unk.		14	
0.03	**2**	0.04	**2**	0.00	**0**	tr	**0**	tr	tr(a)	**0**	0.5	unk.	22	**2**	119	62	unk.	0.5	**3**	0.4	**3**	unk.	**unk.**	unk.	K
0.6	**3**	13	**3**	0.17	**2**	tr	**0**	0	0.6	**2**	unk.	0.55	52	**5**	5.2	1.6	unk.	15	**4**	0.11	**6**	unk.		14	
0.30	**18**	0.06	**3**	tr(a)	**0**	tr(a)	**0**	tr	tr(a)	**0**	0.1	unk.	122	**12**	736	152	unk.	3.6	**20**	0.4	**3**	unk.	**unk.**	unk.	L
3.5	**18**	25	**6**	0.58	**6**	tr	**0**	0	0.2	**1**	unk.	2.90	141	**14**	32.0	3.9	unk.	34	**9**	0.33	**17**	unk.		unk.	
0.30	**18**	0.06	**3**	tr(a)	**0**	tr(a)	**0**	tr	tr(a)	**0**	0.1	unk.	122	**12**	40	152	unk.	3.6	**20**	0.4	**3**	unk.	**unk.**	unk.	M
3.5	**18**	25	**6**	0.58	**6**	tr	**0**	0	0.2	**1**	unk.	2.90	141	**14**	1.7	3.9	unk.	34	**9**	0.33	**17**	unk.		unk.	
0.21	**13**	0.25	**13**	0.75	**13**	tr(a)	**0**	unk.	50	**13**	tr(a)	tr(a)	125	**13**	255	unk.	unk.	2.2	**13**	unk.	**unk.**	18.8	**13**	tr(a)	N
2.5	**13**	50	**13**	tr(a)	**0**	625	**13**	unk.	0.0	**0**	tr(a)	unk.	125	**13**	11.1	unk.	unk.	50	**13**	unk.	**unk.**	tr(a)		unk.	
0.15	**9**	unk.	**unk.**	0.00	**0**	tr	**0**	0	0(a)	**0**	unk.	unk.	68	**7**	8	207	29	0.6	**3**	0.1	**1**	unk.	**unk.**	116	O
0.6	**3**	43	**11**	0.54	**5**	1937	**39**	194	0.9	**3**	0.9	0.62	48	**5**	0.3	5.3	0.8	9	**2**	1.08	**54**	tr(a)		unk.	
0.18	**11**	unk.	**unk.**	0.00	**0**	tr	**0**	0	0(a)	**0**	unk.	unk.	79	**8**	9	240	33	0.7	**4**	0.1	**1**	unk.	**unk.**	135	P
0.7	**4**	50	**13**	unk.	**unk.**	2250	**45**	225	1.0	**3**	1.0	0.72	56	**6**	0.4	6.1	0.9	11	**3**	1.26	**63**	tr(a)		unk.	
0.10	**6**	0.10	**5**	0.00	**0**	tr	**0**	0	0(a)	**0**	unk.	unk.	38	**4**	11	203	34	0.6	**4**	0.1	**1**	unk.	**unk.**	139	Q
0.5	**2**	52	**13**	0.48	**5**	1757	**35**	176	1.0	**3**	1.0	0.55	54	**5**	0.5	5.2	1.0	11	**3**	1.29	**65**	tr(a)		unk.	
0.03	**2**	0.03	**2**	0.00	**0**	tr	**0**	0	0(a)	**0**	unk.	unk.	12	**1**	4	66	11	0.2	**1**	0.1	**1**	unk.	**unk.**	45	R
0.1	**1**	17	**4**	0.16	**2**	570	**11**	57	0.3	**1**	0.3	0.18	17	**2**	0.2	1.7	0.3	4	**1**	0.42	**21**	tr(a)		unk.	
0.29	**17**	0.14	**7**	0.27	**5**	0.000	**0**	0	tr(a)	**0**	0.1	tr(a)	195	**20**	424	349	2	1.7	**9**	0.8	**5**	26.1	**17**	148	S
1.2	**6**	62	**16**	0.79	**8**	2415	**48**	177	0.2	**1**	0.1	2.40	190	**19**	18.5	8.9	0.0	35	**9**	1.42	**71**	1		1	

	FOOD	Portion	Weight in grams / *Conversion for 100 g*	Kilocalories / H_2O g	Total carbohydrate g / Total fats g	Crude fiber g / Dietary fiber g	Total protein g / Total sugar g	% USRDA	Arginine mg / Histidine mg	Isoleucine mg / Leucine mg	Lysine mg / Methionine mg	Phenylalanine mg / Threonine mg	Valine mg / Tryptophan mg	Cystine mg / Tyrosine mg	Polyunsat. fatty acids g / Monounsat. fatty acids g	Saturated fatty acids g / P/S ratio	Linoleic acid g / Cholesterol mg	Thiamin mg / Ascorbic acid mg	% USRDA
A	broccoli,raw,edible portion	1 stalk	76	24	4.5	1.13	2.7	**4**	157	103	119	99	139	30	unk.	tr(a)	unk.	0.08	**5**
			1.32	67.3	0.2	2.72	1.4		49	133	41	100	30	94	unk.	unk.	0	85	**142**
B	BRUSSELS SPROUTS,fresh,cooked, drained solids	1/2 C	78	28	5.0	1.24	3.3	**5**	203	136	146	117	143	19	unk.	tr(a)	unk.	0.06	**4**
			1.29	68.4	0.3	2.25	1.4		75	143	33	114	33	67	unk.	unk.	0	67	**112**
C	brussels sprouts,frozen,cooked, drained solids	1/2 C	78	26	5.0	0.93	2.5	**4**	154	104	112	82	109	15	unk.	tr(a)	unk.	0.06	**4**
			1.29	69.2	0.2	2.25	1.4		57	109	25	87	25	51	unk.	unk.	0	63	**105**
D	BULGUR WHEAT,DRY	1 C	171	605	129.4	2.91	19.1	**30**	unk.	748	559	1014	911	unk.	unk.	unk.	unk.	0.48	**32**
			0.58	17.1	2.6	unk.	unk.		unk.	1393	316	620	260	unk.	unk.	unk.	0(a)	0	**0**
E	BUN,frankfurter	1 average	40	119	21.2	0.08	3.3	**5**	unk.	165	109	179	48	unk.	unk.	0.40	0.40	0.11	**8**
			2.50	12.6	2.2	1.20	0.0		unk.	267	51	109	41	unk.	1.20	unk.	0	tr	**0**
F	bun,hamburger	1 average	40	119	21.2	0.08	3.3	**5**	unk.	165	109	179	48	unk.	0.26	0.40	0.40	0.11	**8**
			2.50	12.6	2.2	1.20	unk.		unk.	267	51	109	41	unk.	1.20	0.66	0	tr	**0**
G	BURRITO,home recipe	1 average	231	332	26.0	1.60	13.7	**26**	573	775	989	686	818	47	2.31	8.88	2.31	0.14	**10**
			0.43	171.5	20.1	unk.	1.6		264	1262	255	507	146	367	6.02	0.26	34	8	**13**
H	BUTTER regular	1 Tbsp	14	102	tr	0.00	0.1	**0**	4	7	10	6	8	1	0.43	7.17	0.26	0.00	**0**
			7.04	2.3	11.5	0.00	tr		3	12	3	5	2	6	2.90	0.06	31	0	**0**
I	butter unsalted	1 Tbsp	14	102	tr	0.00	0.1	**0**	4	7	10	6	8	1	0.43	7.17	0.26	0.00	**0**
			7.04	2.3	11.5	0.00	tr		3	12	3	5	2	6	2.90	0.06	31	0	**0**
J	CABBAGE,Chinese,raw,pieces	1/2 C	38	5	1.1	0.22	0.4	**1**	unk.	unk.	unk.	12	unk.	unk.	0(a)	0(a)	unk.	0.02	**1**
			2.67	35.6	tr	unk.	unk.		unk.	unk.	unk.	unk.	unk.	9	unk.	0.00	0(a)	9	**16**
K	cabbage,green chopped & cooked in small amount water,drained solids	1/2 C	73	14	3.1	0.58	0.8	**1**	66	31	38	17	24	17	0.00	0.00	unk.	0.03	**2**
			1.38	68.1	0.1	2.90	2.1		15	32	7	22	7	16	unk.	0.00	0	24	**40**
L	cabbage,green,leaf,cooked in large amount water	1 leaf	25	5	1.0	0.20	0.3	**0**	21	8	12	5	8	5	0.00	0.00	unk.	0.00	**0**
			4.00	23.6	0.0	1.00	0.7		5	10	2	7	2	5	unk.	0.00	0	6	**10**
M	cabbage,green,raw,coarsely chopped	1/2 C	35	8	1.9	0.28	0.4	**1**	38	14	23	11	15	10	0.00	0.00	unk.	0.02	**1**
			2.86	32.3	0.1	1.40	1.3		9	20	5	14	4	9	unk.	0.00	0	16	**27**
N	cabbage,red,raw,coarsely chopped	1/2 C	35	11	2.4	0.35	0.7	**1**	unk.	27	33	17	21	15	0.00	0.00	unk.	0.03	**2**
			2.86	31.6	0.1	unk.	1.6		13	28	6	20	6	14	unk.	0.00	0	21	**36**
O	cabbage,stuffed,home recipe	1 lg	174	373	17.6	0.30	24.7	**53**	1173	1290	1942	1143	1426	247	1.22	12.49	0.87	0.35	**23**
			0.57	104.7	22.4	1.07	4.3		813	2039	608	1021	332	1031	6.43	0.10	95	7	**11**
P	cabbage,stuffed,rolls,canned-Campbell	1 C	227	169	17.3	unk.	11.1	**17**	unk.	unk.	unk.	unk.	unk.	unk.	unk.	unk.	unk.	0.09	**6**
			0.44	188.6	6.1	unk.	unk.		unk.	unk.	unk.	unk.	unk.	unk.	unk.	unk.	unk.	0	**0**
Q	CAKE,angel food,mix,prep	1/12 of 9-3/4" cake	53	137	31.5	tr(a)	3.0	**6**	unk.	139	69	166	130	53	unk.	tr(a)	unk.	tr	**0**
			1.89	18.0	0.1	unk.	25.0		106	233	39	87	36	117	unk.	unk.	0	0	**0**
R	cake,apple,home recipe	2-1/8x2-1/8x 1-1/2" piece	44	145	20.9	0.02	1.2	**2**	27	59	50	64	63	24	0.17	2.20	1.18	0.04	**3**
			2.27	15.4	6.5	0.59	14.9		30	97	24	45	16	46	2.64	0.08	21	1	**1**
S	cake,banana,frozen-Sara Lee	1 piece	70	246	38.1	0.28	2.1	**3**	unk.	unk.	unk.	unk.	unk.	unk.	unk.	unk.	unk.	0.05	**3**
			1.43	19.0	9.7	unk.	18.8		unk.	unk.	unk.	unk.	unk.	unk.	unk.	unk.	unk.	0	**0**

Riboflavin mg / Niacin mg	% USRDA	Vitamin B_6 mg / Folacin mcg	% USRDA	Vitamin B_{12} mcg / Pantothenic acid mg	% USRDA	Biotin mg / Vitamin A IU	% USRDA	Preformed A RE / Beta carotene RE	Vitamin D IU / Vitamin E IU	% USRDA	Total tocopherol mg / Alpha tocopherol mg	Other tocopherol mg / Total ash g	Calcium mg / Phosphorus mg	% USRDA	Sodium mg / Sodium meq	Potassium mg / Potassium meq	Chlorine mg / Chlorine meq	Iron mg / Magnesium mg	% USRDA	Zinc mg / Copper mg	% USRDA	Iodine mcg / Selenium mcg	% USRDA	Manganese mcg / Chromium mcg	
0.17	10	0.15	7	0.00	0	tr	0	0	0(a)	0	unk.	unk.	78	8	11	288	57	0.8	5	unk.	unk.	unk.	unk.	113	A
0.7	3	52	13	0.88	9	1887	38	189	1.0	3	1.0	0.83	59	6	0.5	7.4	1.6	18	5	1.06	53	unk.		unk.	
0.11	6	0.09	5	0.00	0	tr	0	0	0(a)	0	unk.	unk.	25	3	8	212	12	0.9	5	0.3	2	unk.	unk.	unk.	B
0.6	3	28	7	0.28	3	403	8	40	0.7	2	0.7	0.62	56	6	0.3	5.4	0.3	10	3	0.04	2	tr(a)		unk.	
0.08	5	0.09	5	0.00	0	tr	0	0	0(a)	0	unk.	unk.	16	2	11	229	12	0.6	3	0.3	2	unk.	unk.	unk.	C
0.5	2	28	7	0.28	3	442	9	44	0.7	2	0.7	0.62	47	5	0.5	5.8	0.3	10	3	0.04	2	tr(a)		unk.	
0.24	14	0.55	27	0.00	0	unk.	unk.	0	0	0	unk.	unk.	50	5	0	392	unk.	6.3	35	7.5	50	23.9	16	unk.	D
7.7	39	65	16	1.42	14	0	0	0	3.3	11	3.3	2.74	578	58	0.0	10.0	unk.	274	68	unk.	unk.	unk.		unk.	
0.07	4	0.01	1	tr(a)	0	tr(a)	0	tr(a)	tr(a)	0	0.2	0.1	30	3	202	38	unk.	0.8	4	0.2	2	unk.	unk.	unk.	E
1.3	7	16	4	0.12	1	tr	0	tr(a)	0.2	1	tr	0.72	34	3	8.8	1.0	unk.	8	2	0.03	1	unk.		unk.	
0.07	4	0.01	1	tr(a)	0	tr(a)	0	tr(a)	tr(a)	0	0.2	0.1	30	3	202	38	unk.	0.8	4	0.2	2	unk.	unk.	unk.	F
1.3	7	16	4	0.12	1	tr	0	tr(a)	0.2	1	tr	0.72	34	3	8.8	1.0	unk.	8	2	0.03	1	unk.		unk.	
0.20	12	0.14	7	0.23	4	0.003	1	0	0	0	3.8	0.1	278	28	478	420	12	2.6	14	1.7	11	70.8	47	34	G
1.1	6	43	11	0.53	5	511	10	21	4.5	15	0.1	3.31	279	28	20.8	10.8	0.3	44	11	0.13	6	0		1	
tr	0	0.00	0	tr	0	unk.	unk.	unk.	unk.	unk.	0.2	0.0	3	0	117	4	unk.	tr	0	tr	0	unk.	unk.	unk.	H
tr	0	tr	0	unk.	unk.	434	9	unk.	0.3	1	0.2	0.30	3	0	5.1	0.1	unk.	tr	0	unk.	unk.	unk.		unk.	
tr	0	0.00	0	tr	0	unk.	unk.	unk.	unk.	unk.	0.2	0(a)	3	0	2	4	unk.	tr	0	tr	0	unk.	unk.	unk.	I
tr	0	tr	0	unk.	unk.	434	9	unk.	0.3	1	0.2	0.30	3	0	0.1	0.1	unk.	tr	0	unk.	unk.	unk.		unk.	
0.01	1	unk.	unk.	0(a)	0	unk.	unk.	0	0(a)	0	unk.	unk.	16	2	9	95	unk.	0.2	1	unk.	unk.	unk.	unk.	unk.	J
0.2	1	31	8	unk.	unk.	56	1	6	unk.	unk.	unk.	0.26	15	2	0.4	2.4	unk.	5	1	tr(a)	0	tr(a)		unk.	
0.03	2	0.07	4	0.00	0	unk.	unk.	0	0(a)	0	0.1	tr	32	3	10	118	unk.	0.2	1	0.3	2	unk.	unk.	unk.	K
0.2	1	13	3	0.07	1	94	2	9	0.1	0	tr	0.36	15	1	0.4	3.0	unk.	9	2	0.04	2	tr(a)		unk.	
0.00	0	unk.	unk.	0(a)	0	unk.	unk.	0	0(a)	0	unk.	unk.	11	1	3	38	unk.	0.1	0	0.1	1	unk.	unk.	unk.	L
tr	0	5	1	unk.	unk.	30	1	3	unk.	unk.	unk.	0.13	4	0	0.1	1.0	unk.	3	1	0.01	1	tr(a)		unk.	
0.02	1	0.06	3	0.00	0	0.001	0	0	0(a)	0	tr	tr	17	2	7	82	14	0.1	1	0.1	1	unk.	unk.	22	M
0.1	1	11	3	0.07	1	46	1	5	0.0	0	tr	0.24	10	1	0.3	2.1	0.4	5	1	0.02	1	tr(a)		1	
0.02	1	0.07	4	0.00	0	unk.	unk.	0	0(a)	0	tr	unk.	15	2	9	94	unk.	0.3	2	unk.	unk.	unk.	unk.	unk.	N
0.1	1	23	6	0.11	1	14	0	1	0.0	0	tr	0.24	12	1	0.4	2.4	unk.	6	2	tr(a)	0	tr(a)		unk.	
0.37	22	0.17	8	0.68	11	0.000	0	0	17	4	0.1	tr(a)	339	34	1007	335	2	2.5	14	3.4	23	2.8	2	10	O
2.9	15	22	6	0.56	6	776	16	3	0.1	0	0.1	4.74	401	40	43.8	8.6	0.0	34	9	0.13	7	1		1	
0.16	9	unk.	unk.	unk.	unk.	unk.	unk.	unk.	0(a)	0	tr(a)	tr(a)	50	5	799	291	unk.	1.4	8	unk.	unk.	unk.	unk.	unk.	P
2.7	14	unk.	unk.	tr(a)	0	681	14	unk.	0.0	0	tr(a)	3.90	unk.	unk.	34.7	7.5	unk.	unk.	unk.	unk.	unk.	tr(a)		unk.	
0.06	3	0.01	0	21.20	353	0.002	1	0	0	0	0.0	unk.	50	5	77	32	unk.	0.2	1	0.1	1	unk.	unk.	unk.	Q
0.1	0	6	2	0.11	1	0	0	0	0.0	0	unk.	0.37	63	6	3.4	0.8	unk.	6	2	0.02	1	unk.		unk.	
0.03	2	0.01	1	0.03	1	0.001	0	0	1	0	2.7	1.2	17	2	140	57	16	0.5	3	0.2	1	11.0	7	40	R
0.3	1	4	1	0.10	1	259	5	2	3.3	11	1.5	0.65	26	3	6.1	1.4	0.4	15	4	0.02	1	4		12	
0.10	6	unk.	unk.	tr(a)	0	unk.	unk.	unk.	0(a)	0	unk.	unk.	17	2	223	116	unk.	0.6	4	unk.	unk.	unk.	unk.	unk.	S
0.6	3	unk.	unk.	unk.	unk.	100	2	unk.	unk.	unk.	unk.	0.84	70	7	9.7	3.0	unk.	10	3	unk.	unk.	unk.		unk.	

	FOOD	Portion	Weight in grams / Conversion for 100 g	Kilocalories / H_2O g	Total carbohydrate g / Total fats g	Crude fiber g / Dietary fiber g	Total protein g / Total sugar g	% USRDA	Arginine mg / Histidine mg	Isoleucine mg / Leucine mg	Lysine mg / Methionine mg	Phenylalanine mg / Threonine mg	Valine mg / Tryptophan mg	Cystine mg / Tyrosine mg	Polyunsat. fatty acids g / Monounsat. fatty acids g	Saturated fatty acids g / P/S ratio	Linoleic acid g / Cholesterol mg	Thiamin mg / Ascorbic acid mg	% USRDA
A	cake,Boston cream,home recipe	1/8 of 9" dia	139	433	66.3	0.21	5.8	**11**	163	264	284	245	281	79	0.94	9.69	0.80	0.12	**8**
		cake	*0.72*	49.1	17.4	0.58	47.7		120	439	101	205	69	213	5.15	0.10	158	tr	**1**
B	cake,carrot,w/cream cheese icing,	2x2x1-3/4"	80	334	43.3	0.24	2.8	**5**	69	137	139	146	148	54	6.38	4.65	6.18	0.10	**6**
	home recipe		*1.25*	15.9	17.4	0.60	34.0		74	226	56	106	35	107	5.28	1.37	45	1	**2**
C	cake,cheese,home recipe	1/8 of 8" dia	125	414	35.2	0.22	7.1	**15**	276	361	535	371	405	91	1.26	16.90	0.87	0.07	**4**
		cake	*0.80*	53.8	27.9	0.62	27.8		226	642	165	298	76	318	7.04	0.07	159	9	**15**
D	cake,cherry French cream cheese,	1 piece	89	254	28.3	0.53	3.0	**5**	unk.	unk.	unk.	unk.	unk.	unk.	unk.	unk.	unk.	0.04	**2**
	frozen-Sara Lee		*1.12*	42.3	14.7	unk.	unk.		unk.	unk.	unk.	unk.	unk.	unk.	unk.	unk.	68	10	**17**
E	cake,chocolate or devil's food,	1/6 of 9" dia	92	287	39.0	0.20	4.4	**8**	116	188	197	188	209	68	0.30	4.75	0.91	0.08	**6**
	mix,prep	cake	*1.09*	33.7	13.7	0.43	24.4		97	311	84	148	50	152	7.21	0.06	57	tr	**0**
F	cake,chocolate or devil's food,	1/12 of 9" dia	92	350	51.2	0.28	4.0	**6**	unk.	313	248	221	322	110	unk.	8.28	1.29	0.03	**2**
	w/chocolate icing,home recipe	x 2-7/8"	*1.09*	21.7	16.2	0.83	39.6		304	488	110	193	74	230	4.78	unk	64	tr	**0**
G	cake,coffee,iced,home recipe	2-5/8x2-3/4x	108	341	59.8	0.06	2.5	**5**	unk.	124	113	130	134	46	0.31	4.12	0.68	0.07	**4**
		1-1/4"	*0.93*	33.9	10.9	0.40	49.5		61	205	49	92	31	95	5.39	0.07	35	tr	**0**
H	cake,coffee,nuts & icing,home recipe	2-5/8x2-3/4x	113	372	60.5	0.15	3.2	**6**	unk.	160	134	166	179	61	2.27	4.45	2.31	0.09	**6**
		1-1/4"	*0.88*	34.2	13.9	0.65	49.7		80	263	64	120	40	122	5.85	0.51	35	tr	**0**
I	cake,frozen,apple cake cobbler	1 piece	28	55	10.5	unk.	0.6	**1**	unk.	unk.	unk.	unk.	unk.	unk.	unk.	unk.	unk.	0.03	**2**
			3.53	unk.	1.4	unk.	unk.		unk.	unk.	unk.	unk.	unk.	unk.	unk.	unk.	unk.	0	**0**
J	cake,fruit,dark,home recipe	1/32 of 7" dia	43	163	25.7	0.26	2.1	**3**	unk.	146	116	112	150	52	unk.	1.38	1.29	0.06	**4**
		x 2-1/4"	*2.33*	7.8	6.6	unk.	11.5		142	228	52	99	39	107	3.48	unk.	19	tr	**0**
K	cake,fruit,dark,sliced,home recipe	2x1-1/2x1/4"	15	57	9.0	0.09	0.7	**1**	unk.	51	40	39	52	18	unk.	0.48	0.45	0.02	**1**
			6.67	2.7	2.3	unk.	4.0		49	79	18	34	13	37	1.21	unk.	7	tr	**0**
L	cake,gingerbread,home recipe	1/9 of 8-1/4"	76	276	41.5	1.54	2.9	**5**	unk.	127	94	143	132	56	0.25	2.90	1.01	0.10	**7**
		sq x 1-3/4"	*1.32*	17.4	12.2	0.58	22.4		63	212	49	92	35	99	7.55	0.08	28	2	**3**
M	cake,oatmeal,iced,home recipe	1/10 of 9" dia	120	356	58.5	0.13	3.2	**6**	71	122	116	130	135	52	0.64	7.59	0.46	0.09	**6**
		x 3"	*0.83*	44.2	12.9	0.38	46.4		65	204	55	97	33	99	3.47	0.08	74	tr	**0**
N	cake,peanut butter,home recipe	2x2x1-1/2"	52	211	24.0	0.31	5.8	**10**	unk.	277	247	323	323	102	2.37	2.30	2.23	0.07	**5**
			1.92	11.3	10.9	unk.	14.3		155	427	78	191	74	234	4.05	1.03	22	tr	**0**
O	cake,pineapple upsidedown-Campbell	1 piece	28	85	11.9	unk.	0.8	**1**	unk.	unk.	unk.	unk.	unk.	unk.	unk.	unk.	unk.	0.02	**1**
			3.52	unk.	3.7	unk.	unk.		unk.	unk.	unk.	unk.	unk.	unk.	unk.	unk.	unk.	1	**1**
P	cake,pound,all butter,frozen-Sara Lee	1 slice	30	122	14.2	0.15	1.5	**2**	unk.	unk.	unk.	unk.	unk.	unk.	unk.	unk.	unk.	0.06	**4**
			3.33	7.2	6.5	unk.	9.0		unk.	unk.	unk.	unk.	unk.	unk.	unk.	unk.	unk.	0	**0**
Q	cake,pound,sliced,home recipe	3-1/2x3x1/2"	29	119	15.9	0.03	1.9	**4**	unk.	151	336	101	159	52	unk.	1.45	0.99	0.01	**1**
			3.45	5.6	5.4	unk.	8.7		58	223	61	104	38	101	2.61	unk.	46	tr	**0**
R	cake,spice,w/caramel icing,	1/10 of 9" dia	130	411	65.9	0.12	4.2	**8**	unk.	219	222	212	245	78	0.75	5.23	1.42	0.10	**6**
	home recipe	x 3"	*0.77*	42.9	15.5	0.54	53.4		111	359	94	169	55	170	7.97	0.14	61	tr	**1**
S	cake,sponge,home recipe	1/12" of	44	135	23.9	0.03	2.4	**5**	78	119	123	121	134	47	0.28	1.81	0.24	0.05	**4**
		8-1/2" cake	*2.27*	13.8	3.4	0.29	16.6		63	197	55	96	32	96	0.99	0.16	48	tr	**0**

Riboflavin mg / Niacin mg	% USRDA	Vitamin B_6 mg / Folacin mcg	% USRDA	Vitamin B_{12} mcg / Pantothenic acid mg	% USRDA	Biotin mg / Vitamin A IU	% USRDA	Preformed A RE / Beta carotene RE	Vitamin D IU / Vitamin E IU	% USRDA	Total tocopherol mg / Alpha tocopherol mg	Other tocopherol mg / Total ash g	Calcium mg / Phosphorus mg	% USRDA	Sodium mg / Sodium meq	Potassium mg / Potassium meq	Chlorine mg / Chlorine meq	Iron mg / Magnesium mg	% USRDA	Zinc mg / Copper mg	% USRDA	Iodine mcg / Selenium mcg	% USRDA	Manganese mcg / Chromium mcg	
0.18	10	0.07	3	0.45	8	0.006	2	0	34	9	0.8	0.0	89	9	259	152	33	1.6	9	1.0	7	36.1	24	105	A
0.8	4	19	5	0.70	7	490	10	tr	1.0	3	0.2	1.54	158	16	11.3	3.9	0.9	31	8	0.12	6	10		39	
0.07	4	0.04	2	0.09	2	0.002	1	0	2	1	8.8	0.0	23	2	236	82	24	0.8	4	0.6	4	25.9	17	217	B
0.5	2	8	2	0.24	2	964	19	77	1.4	5	1.4	0.74	52	5	10.2	2.1	0.7	25	6	0.10	5	6		30	
0.18	11	0.06	3	0.44	7	0.003	1	0	10	3	0.3	tr	67	7	289	126	27	1.5	8	0.8	6	22.9	15	103	C
0.5	2	17	4	0.52	5	1085	22	1	0.4	1	0.1	1.21	114	11	12.6	3.2	0.8	30	7	0.04	2	5		21	
0.09	5	unk.	unk.	unk.	unk.	unk.	unk.	unk.	unk.	unk.	unk.	unk.	38	4	110	75	unk.	0.9	5	0.5	3	unk.	unk.	unk.	D
0.3	1	unk.	unk.	unk.	unk.	338	7	unk.	unk.	unk.	unk.	0.62	43	4	4.8	1.9	unk.	13	3	unk.	unk.	unk.		unk.	
0.13	8	0.04	2	0.21	4	0.004	1	0	15	4	0.5	0.0	45	5	187	120	33	1.1	6	0.5	3	29.8	20	71	E
0.6	3	14	3	0.31	3	61	1	tr	0.6	2	0.1	0.75	87	9	8.1	3.1	0.9	52	13	0.07	4	8		22	
0.07	4	0.05	2	tr(a)	0	0.006	2	unk.	8	2	0.1	unk.	54	5	386	120	unk.	0.7	4	0.6	4	unk.	unk.	unk.	F
0.3	1	6	1	0.18	2	396	8	unk.	0.2	1	0.1	1.29	85	9	16.8	3.1	unk.	14	3	0.29	14	unk.		unk.	
0.07	4	0.02	1	0.08	1	0.001	0	0	2	0	0.1	0(a)	47	5	174	115	34	1.3	7	0.4	3	16.6	11	76	G
0.5	3	4	1	0.17	2	157	3	0	0.1	0	tr	0.96	51	5	7.6	2.9	0.9	30	8	0.03	2	7		38	
0.09	5	0.05	3	0.08	1	0.003	1	0	2	0	0.1	0(a)	52	5	175	136	35	1.4	8	0.5	4	16.6	11	174	H
0.6	3	8	2	0.17	2	158	3	tr	0.1	0	0.1	1.05	69	7	7.6	3.5	1.0	37	9	0.10	5	7		41	
0.01	1	unk.	unk.	unk.	unk.	unk.	unk.	unk.	unk.	unk.	unk.	unk.	3	0	unk.	unk.	unk.	0.2	1	unk.	unk.	unk.	unk.	unk.	I
0.2	1	unk.	unk.	unk.	unk.	3	0	unk.	unk.	unk.	unk.	0.23	unk.	unk.	unk.	unk.	unk.	unk.	unk.	unk.	unk.	unk.		unk.	
0.06	4	0.03	2	0.06	1	0.003	1	unk.	6	1	0.3	unk.	31	3	68	213	unk.	1.1	6	0.3	2	unk.	unk.	unk.	J
0.3	2	1	0	0.17	2	52	1	unk.	0.4	1	unk.	0.90	49	5	3.0	5.4	unk.	7	2	0.04	2	unk.		unk.	
0.02	1	0.01	1	tr(a)	0	0.001	0	unk.	2	1	0.1	unk.	11	1	24	74	unk.	0.4	2	0.1	1	unk.	unk.	unk.	K
0.1	1	tr	0	0.06	1	18	0	unk.	0.1	0	unk.	0.31	17	2	1.0	1.9	unk.	2	1	0.01	1	unk.		unk.	
0.09	6	0.05	3	0.06	1	0.003	1	0	2	1	0.1	0.0	134	13	296	247	112	4.1	23	0.4	3	51.9	35	83	L
1.0	5	7	2	0.23	2	28	1	0	0.1	0	0.0	2.34	47	5	12.9	6.3	3.2	38	10	0.07	4	10		18	
0.08	5	0.03	1	0.09	2	0.002	1	0	3	1	0.3	0.0	49	5	302	152	47	1.9	10	0.5	3	42.0	28	73	M
0.6	3	7	2	0.21	2	454	9	0	0.3	1	0.2	1.51	68	7	13.2	3.9	1.3	50	13	0.05	3	7		36	
0.08	5	0.07	3	0.08	1	0.007	2	0	6	1	1.9	1.5	44	4	216	180	25	1.2	6	0.6	4	15.4	10	330	N
2.7	14	16	4	0.51	5	22	0	0	2.3	8	0.4	1.24	101	10	9.4	4.6	0.7	49	12	0.13	6	5		23	
0.02	1	unk.	unk.	unk.	unk.	unk.	unk.	unk.	unk.	unk.	unk.	unk.	9	1	unk.	unk.	unk.	0.4	2	unk.	unk.	unk.	unk.	unk.	O
0.4	2	unk.	unk.	unk.	unk.	69	1	unk.	unk.	unk.	unk.	0.14	unk.	unk.	unk.	unk.	unk.	unk.	unk.	unk.	unk.	unk.		unk.	
0.08	5	unk.	unk.	tr(a)	0	unk.	unk.	unk.	unk.	unk.	unk.	unk.	9	1	102	25	unk.	0.4	3	unk.	unk.	unk.	unk.	unk.	P
0.5	2	unk.	unk.	unk.	unk.	206	4	unk.	unk.	unk.	unk.	0.33	32	3	4.4	0.6	unk.	2	1	unk.	unk.	unk.		unk.	
0.03	2	0.01	1	tr(a)	0	0.001	0	unk.	6	1	2.1	unk.	12	1	52	23	unk.	0.2	1	0.2	1	unk.	unk.	unk.	Q
0.1	0	2	1	0.09	1	84	2	unk.	2.6	9	0.3	0.26	30	3	2.3	0.6	unk.	4	1	0.02	1	unk.		unk.	
0.13	8	0.04	2	0.15	3	0.003	1	0	9	2	0.1	0.0	66	7	350	128	40	1.4	8	0.6	4	52.1	35	107	R
0.7	3	8	2	0.27	3	125	3	0	0.2	1	0.0	1.37	85	9	15.2	3.3	1.1	43	11	0.06	3	9		37	
0.07	4	0.02	1	0.13	2	0.002	1	0	7	2	0.1	0.0	25	3	115	38	21	0.5	3	0.3	2	18.5	12	48	S
0.4	2	5	1	0.19	2	104	2	0	0.1	1	0.1	0.55	48	5	5.0	1.0	0.6	25	6	0.04	2	5		15	

	FOOD	Portion	Weight in grams / Conversion for 100 g	Kilocalories / H_2O g	Total carbohydrate g / Total fats g	Crude fiber g / Dietary fiber g	Total protein g / Total sugar g	% USRDA	Arginine mg / Histidine mg	Isoleucine mg / Leucine mg	Lysine mg / Methionine mg	Phenylalanine mg / Threonine mg	Valine mg / Tryptophan mg	Cystine mg / Tyrosine mg	Polyunsat. fatty acids g / Monounsat. fatty acids g	Saturated fatty acids g / P/S ratio	Linoleic acid g / Cholesterol mg	Thiamin mg / Ascorbic acid mg	% USRDA
A	cake,white,home recipe	1/6 of 9" dia x	75	273	38.0	0.05	3.6	**7**	96	183	171	184	201	79	0.13	3.01	0.90	0.08	**6**
		1-3/4"	*1.33*	20.9	12.0	0.52	unk.		79	301	87	132	49	142	7.48	0.04	3	tr	**0**
B	cake,white,w/uncooked white icing,	1/10 of 9" dia	140	481	79.7	0.05	3.6	**7**	98	184	176	184	202	76	0.34	6.45	0.99	0.08	**5**
	home recipe	x 3"	*0.71*	38.3	17.2	0.49	65.0		80	303	87	133	49	143	8.57	0.05	19	tr	**0**
C	cake,yellow,home recipe	1/6 of 9" dia	85	321	49.6	0.07	4.4	**8**	unk.	218	196	228	231	80	0.31	2.70	2.96	0.12	**8**
		x 1-3/4"	*1.18*	32.9	11.8	0.72	28.2		106	361	87	161	57	171	5.61	0.12	43	tr	**0**
D	cake,yellow,w/chocolate icing,	1/10 of 9" dia	125	463	78.2	0.15	4.0	**7**	unk.	181	167	187	193	65	0.72	9.69	0.55	0.09	**6**
	home recipe	x 3"	*0.80*	26.6	16.2	0.57	60.2		88	300	72	134	47	143	4.39	0.07	69	tr	**0**
E	CAKES,LUNCHBOX TYPE,devil's food	1 cake	43	136	25.5	unk.	1.7	**3**	unk.	unk.	unk.	unk.	unk.	unk.	unk.	unk.	unk.	0.00	**0**
	cupcake-Hostess		*2.35*	8.9	3.8	unk.	14.2		unk.	unk.	unk.	unk.	unk.	unk.	unk.	unk.	4	0	**0**
F	cakes,lunchbox type,Hoho-Hostess	1 cake	28	119	16.4	unk.	1.1	**2**	unk.	unk.	unk.	unk.	unk.	unk.	unk.	unk.	unk.	0.00	**0**
			3.52	2.8	5.9	unk.	unk.		unk.	unk.	unk.	unk.	unk.	unk.	unk.	unk.	13	0	**0**
G	cakes,lunchbox type,Snoball-Hostess	1 cake	43	136	25.1	unk.	1.3	**2**	unk.	unk.	unk.	unk.	unk.	unk.	unk.	unk.	unk.	0.00	**0**
			2.35	7.6	3.8	unk.	unk.		unk.	unk.	unk.	unk.	unk.	unk.	unk.	unk.	2	0	**0**
H	cakes,lunchbox type,Twinkie-Hostess	1 cake	43	144	25.9	unk.	1.3	**2**	unk.	unk.	unk.	unk.	unk.	unk.	unk.	unk.	unk.	0.06	**4**
			2.35	8.5	4.2	unk.	unk.		unk.	unk.	unk.	unk.	unk.	unk.	unk.	unk.	21	0	**0**
I	cakes,lunchbox type,Yodel's-Drake's	1 cake	25	115	16.0	unk.	1.0	**2**	unk.	unk.	unk.	unk.	unk.	unk.	unk.	unk.	unk.	0.06	**4**
			4.08	unk.	5.0	unk.	unk.		unk.	unk.	unk.	unk.	unk.	unk.	unk.	unk.	unk.	0	**0**
J	CANDIED FRUIT,cherry	1 cherry	3.5	12	3.0	0.02	0.0	**0**	unk.	unk.	unk.	unk.	unk.	unk.	unk.	tr(a)	unk.	tr	**0**
			28.57	0.4	tr	unk.	3.0		unk.	unk.	unk.	unk.	unk.	unk.	unk.	unk.	0(a)	0	**0**
K	candied fruit,citron	1 oz	28	89	22.8	0.40	0.1	**0**	unk.	tr(a)	tr(a)	tr(a)	tr(a)	tr(a)	unk.	unk.	unk.	unk.	**unk.**
			3.52	5.1	0.1	unk.	22.4		tr(a)	tr(a)	tr(a)	tr(a)	tr(a)	tr(a)	unk.	unk.	0(a)	unk.	**unk.**
L	candied fruit,lemon peel	1 oz	28	90	22.9	0.65	0.1	**0**	unk.	unk.	unk.	unk.	unk.	unk.	unk.	tr(a)	unk.	unk.	**unk.**
			3.52	4.9	0.1	unk.	22.2		unk.	unk.	unk.	unk.	unk.	unk.	unk.	unk.	0(a)	unk.	**unk.**
M	candied fruit,orange peel	1 oz	28	90	22.9	unk.	0.1	**0**	unk.	unk.	unk.	unk.	unk.	unk.	unk.	tr(a)	unk.	unk.	**unk.**
			3.52	4.9	0.1	unk.	22.2		unk.	unk.	unk.	unk.	unk.	unk.	unk.	unk.	0	unk.	**unk.**
N	candied fruit,pineapple slice	1 slice	57	179	45.2	0.45	0.4	**1**	unk.	unk.	unk.	unk.	unk.	unk.	unk.	tr(a)	unk.	0.07	**5**
			1.77	10.2	0.2	unk.	44.6		unk.	unk.	unk.	unk.	unk.	unk.	unk.	unk.	0(a)	13	**22**
O	CANDY BAR,chocolate coated coconut	1 oz bar	28	124	20.4	0.17	0.8	**1**	unk.	unk.	unk.	unk.	unk.	unk.	unk.	2.84	tr	0.01	**0**
	center		*3.52*	1.9	5.0	unk.	20.0		unk.	unk.	unk.	unk.	unk.	unk.	1.70	unk.	tr(a)	0	**0**
P	candy bar,chocolate coated honeycomb	1 oz bar	28	131	20.1	0.11	1.9	**3**	unk.	unk.	unk.	unk.	unk.	unk.	unk.	1.70	0.57	0.01	**1**
	w/peanut butter		*3.52*	0.5	5.5	unk.	19.9		unk.	unk.	unk.	unk.	unk.	unk.	3.12	unk.	tr(a)	tr	**0**
Q	candy bar,chocolate covered peppermint	1 oz bar	28	116	23.0	0.03	0.5	**1**	unk.	unk.	unk.	unk.	unk.	unk.	unk.	0.88	0.17	0.01	**1**
	pattie		*3.52*	1.6	3.0	unk.	23.0		unk.	unk.	unk.	unk.	unk.	unk.	1.96	unk.	tr(a)	tr	**0**
R	candy bar,chocolate plain	1 oz bar	28	148	16.2	0.11	2.2	**3**	unk.	34	37	45	40	23	3.41	5.17	0.20	0.02	**1**
			3.52	0.3	9.2	unk.	14.5		11	68	6	26	unk.	26	3.35	0.66	4	tr	**1**
S	candy bar,chocolate w/almonds	1 oz bar	28	151	14.6	0.20	2.6	**4**	tr(a)	unk.	unk.	unk.	unk.	unk.	unk.	4.57	0.60	0.02	**2**
			3.52	0.4	10.1	unk.	14.4		unk.	unk.	unk.	unk.	unk.	unk.	4.46	unk.	4	tr	**1**

	Riboflavin mg / Niacin mg	% USRDA	Vitamin B_6 mg / Folacin mcg	% USRDA	Vitamin B_{12} mcg / Pantothenic acid mg	% USRDA	Biotin mg / Vitamin A IU	% USRDA	Preformed A RE / Beta carotene RE	Vitamin D IU / Vitamin E IU	% USRDA	Total tocopherol mg / Alpha tocopherol mg	Other tocopherol mg / Total ash g	Calcium mg / Phosphorus mg	% USRDA	Sodium mg / Sodium meq	Potassium mg / Potassium meq	Chlorine mg / Chlorine meq	Iron mg / Magnesium mg	% USRDA	Zinc mg / Copper mg	% USRDA	Iodine mcg / Selenium mcg	% USRDA	Manganese mcg / Chromium mcg
A	0.10	**6**	0.02	**1**	0.07	**1**	0.001	**0**	0	8	**2**	tr	0.0	42	**4**	287	64	34	0.5	**3**	0.3	**2**	44.9	**30**	78
	0.6	**3**	5	**1**	0.14	**1**	23	**1**	0	tr	**0**	tr	1.22	58	**6**	12.5	1.6	1.0	10	**3**	0.06	**3**	9		23
B	0.11	**6**	0.02	**1**	0.07	**1**	0.001	**0**	0	7	**2**	0.1	0.0	46	**5**	325	69	32	0.6	**3**	0.6	**4**	42.8	**29**	92
	0.6	**3**	5	**1**	0.14	**1**	239	**5**	0	0.1	**0**	0.1	1.31	60	**6**	14.1	1.8	0.9	10	**3**	0.06	**3**	9		48
C	0.12	**7**	0.04	**2**	0.17	**3**	0.002	**1**	0	13	**3**	6.7	3.1	55	**6**	396	75	31	0.9	**5**	0.6	**4**	52.6	**35**	110
	0.9	**4**	9	**2**	0.30	**3**	446	**9**	0	8.0	**27**	3.4	1.54	86	**9**	17.2	1.9	0.9	30	**8**	0.08	**4**	13		30
D	0.12	**7**	0.03	**2**	0.14	**2**	0.002	**1**	0	10	**3**	0.4	0.0	51	**5**	343	96	27	1.0	**6**	0.7	**5**	41.6	**28**	102
	0.7	**4**	11	**3**	0.25	**3**	519	**10**	tr	0.5	**2**	0.2	1.43	87	**9**	14.9	2.5	0.8	35	**9**	0.07	**4**	10		47
E	0.06	**4**	unk.	**unk.**	tr(a)	**0**	unk.	**unk.**	0	tr(a)	**0**	unk.	unk.	19	**2**	212	unk.	unk.	0.6	**3**	unk.	**unk.**	unk.	**unk.**	unk.
	0.6	**3**	unk.	**unk.**	unk.	**unk.**	1	**0**	0	unk.	**unk.**	unk.	unk.	unk.	**unk.**	9.2	unk.	unk.	unk.	**unk.**	unk.	**unk.**	unk.		unk.
F	0.05	**3**	unk.	**unk.**	tr(a)	**0**	unk.	**unk.**	0	tr(a)	**0**	unk.	unk.	12	**1**	79	unk.	unk.	0.3	**2**	unk.	**unk.**	unk.	**unk.**	unk.
	0.1	**0**	unk.	**unk.**	unk.	**unk.**	23	**1**	0	unk.	**unk.**	unk.	unk.	unk.	**unk.**	3.4	unk.	unk.	unk.	**unk.**	unk.	**unk.**	unk.		unk.
G	0.04	**2**	unk.	**unk.**	tr(a)	**0**	unk.	**unk.**	0	tr(a)	**0**	unk.	unk.	12	**1**	164	unk.	unk.	0.5	**3**	unk.	**unk.**	unk.	**unk.**	unk.
	0.3	**2**	unk.	**unk.**	unk.	**unk.**	3	**0**	0	unk.	**unk.**	unk.	unk.	unk.	**unk.**	7.1	unk.	unk.	unk.	**unk.**	unk.	**unk.**	unk.		unk.
H	0.06	**4**	unk.	**unk.**	tr(a)	**0**	unk.	**unk.**	0	tr(a)	**0**	unk.	unk.	19	**2**	191	unk.	unk.	0.6	**3**	unk.	**unk.**	unk.	**unk.**	unk.
	0.5	**3**	unk.	**unk.**	0.10	**1**	0	**0**	0	unk.	**unk.**	unk.	unk.	unk.	**unk.**	8.3	unk.	unk.	3	**1**	unk.	**unk.**	unk.		unk.
I	0.07	**4**	unk.	**unk.**	tr(a)	**0**	unk.	**unk.**	0	tr(a)	**0**	unk.	unk.	0	**0**	unk.	unk.	unk.	0.9	**5**	unk.	**unk.**	unk.	**unk.**	unk.
	0.6	**3**	unk.	**unk.**	unk.	**unk.**	0	**0**	0	unk.	**unk.**	unk.	unk.	0	**0**	unk.	unk.	unk.	unk.	**unk.**	unk.	**unk.**	unk.		unk.
J	tr	**0**	unk.	**unk.**	0(a)	**0**	unk.	**unk.**	0	0(a)	**0**	unk.	unk.	tr	**0**	unk.	unk.	unk.	tr	**0**	unk.	**unk.**	unk.	**unk.**	unk.
	tr	**0**	unk.	**unk.**	unk.	**unk.**	7	**0**	1	unk.	**unk.**	unk.	0.02	tr	**0**	unk.	unk.	unk.	unk.	**unk.**	unk.	**unk.**	unk.		unk.
K	unk.	**unk.**	unk.	**unk.**	0(a)	**0**	unk.	**unk.**	0(a)	0(a)	**0**	unk.	unk.	24	**2**	82	34	unk.	0.2	**1**	unk.	**unk.**	unk.	**unk.**	unk.
	unk.	**unk.**	unk.	**unk.**	unk.	**unk.**	unk.	**unk.**	unk.	unk.	**unk.**	unk.	0.37	7	**1**	3.6	0.9	unk.	unk.	**unk.**	unk.	**unk.**	unk.		unk.
L	unk.	**unk.**	unk.	**unk.**	0(a)	**0**	unk.	**unk.**	0(a)	0(a)	**0**	unk.	unk.	unk.	**unk.**	14	3	unk.	unk.	**unk.**	unk.	**unk.**	unk.	**unk.**	unk.
	unk.	**unk.**	unk.	**unk.**	unk.	**unk.**	unk.	**unk.**	unk.	unk.	**unk.**	unk.	0.37	unk.	**unk.**	0.6	0.1	unk.	unk.	**unk.**	unk.	**unk.**	unk.		unk.
M	unk.	**unk.**	unk.	**unk.**	tr(a)	**0**	unk.	**unk.**	0(a)	0	**0**	unk.	unk.	unk.	**unk.**	14	3	unk.	unk.	**unk.**	unk.	**unk.**	unk.	**unk.**	unk.
	unk.	**unk.**	unk.	**unk.**	unk.	**unk.**	unk.	**unk.**	unk.	unk.	**unk.**	unk.	0.37	unk.	**unk.**	0.6	0.1	unk.	unk.	**unk.**	unk.	**unk.**	unk.		unk.
N	0.02	**1**	unk.	**unk.**	0(a)	**0**	unk.	**unk.**	0(a)	0	**0**	unk.	unk.	17	**2**	unk.	unk.	unk.	0.3	**2**	unk.	**unk.**	unk.	**unk.**	unk.
	0.2	**1**	unk.	**unk.**	unk.	**unk.**	59	**1**	6	unk.	**unk.**	unk.	0.45	14	**1**	unk.	unk.	unk.	unk.	**unk.**	unk.	**unk.**	unk.		unk.
O	0.02	**1**	unk.	**unk.**	tr(a)	**0**	unk.	**unk.**	0	tr(a)	**0**	unk.	unk.	14	**1**	56	47	unk.	0.3	**2**	unk.	**unk.**	unk.	**unk.**	unk.
	0.1	**0**	unk.	**unk.**	unk.	**unk.**	0	**0**	0	unk.	**unk.**	unk.	0.28	22	**2**	2.4	1.2	unk.	20	**5**	unk.	**unk.**	unk.		unk.
P	0.03	**2**	unk.	**unk.**	tr(a)	**0**	unk.	**unk.**	0	tr(a)	**0**	unk.	unk.	23	**2**	46	64	unk.	0.5	**3**	unk.	**unk.**	unk.	**unk.**	unk.
	0.8	**4**	unk.	**unk.**	unk.	**unk.**	tr	**0**	tr	unk.	**unk.**	unk.	0.43	38	**4**	2.0	1.6	unk.	unk.	**unk.**	unk.	**unk.**	unk.		unk.
Q	0.02	**1**	tr(a)	**0**	tr(a)	**0**	tr(a)	**0**	0	tr(a)	**0**	unk.	unk.	16	**2**	53	26	unk.	0.3	**2**	0.1	**1**	unk.	**unk.**	unk.
	tr	**0**	tr(a)	**0**	tr(a)	**0**	tr	**0**	tr	unk.	**unk.**	unk.	0.26	15	**2**	2.3	0.7	unk.	unk.	**unk.**	unk.	**unk.**	unk.		unk.
R	0.10	**6**	0.01	**0**	tr(a)	**0**	0.009	**3**	unk.	47	**12**	1.2	0.9	65	**7**	27	109	43	0.3	**2**	0.1	**1**	unk.	**unk.**	unk.
	0.1	**0**	2	**1**	0.03	**0**	77	**2**	unk.	1.4	**5**	0.3	0.54	66	**7**	1.2	2.8	1.2	16	**4**	0.02	**1**	unk.		unk.
S	0.12	**7**	0.01	**0**	tr(a)	**0**	0.008	**3**	unk.	20	**5**	unk.	unk.	65	**7**	23	126	unk.	0.5	**3**	0.1	**1**	unk.	**unk.**	unk.
	0.2	**1**	6	**1**	0.03	**0**	65	**1**	unk.	unk.	**unk.**	unk.	0.57	77	**8**	1.0	3.2	unk.	24	**6**	0.05	**2**	unk.		unk.

	FOOD	Portion	Weight in grams / Conversion for 100 g	Kilocalories / H_2O g	Total carbohydrate g / Total fats g	Crude fiber g / Dietary fiber g	Total protein g / Total sugar g	% USRDA	Arginine mg / Histidine mg	Isoleucine mg / Leucine mg	Lysine mg / Methionine mg	Phenylalanine mg / Threonine mg	Valine mg / Tryptophan mg	Cystine mg / Tyrosine mg	Polyunsat. fatty acids g / Monounsat. fatty acids g	Saturated fatty acids g / P/S ratio	Linoleic acid g / Cholesterol mg	Thiamin mg / Ascorbic acid mg	% USRDA
A	candy bar,chocolate w/peanuts	1 oz bar	28	154	12.7	0.00	4.0	6	unk.	unk.	unk.	unk.	unk.	unk.	unk.	4.57	1.53	0.07	5
			3.52	0.3	10.8	unk.	12.4		unk.	unk.	unk.	unk.	unk.	unk.	4.26	unk.	3	unk.	unk.
B	candy bar,Milky Way-M&M/Mars	1 oz bar	28	127	20.3	0.06	1.4	2	unk.	unk.	unk.	unk.	unk.	unk.	unk.	unk.	unk.	0.03	2
			3.52	unk.	4.5	unk.	17.7		unk.	unk.	unk.	unk.	unk.	unk.	unk.	unk.	unk.	tr	0
C	candy bar,peanut	1 oz bar	28	146	13.4	0.34	5.0	8	unk.	236	206	283	287	65	unk.	1.99	2.56	0.12	8
			3.52	0.4	9.1	unk.	13.3		140	350	51	155	64	206	3.98	unk.	0(a)	0	0
D	candy bar,Snicker-M&M/Mars	1 oz bar	28	138	16.8	0.20	2.8	4	unk.	unk.	unk.	unk.	unk.	unk.	unk.	unk.	unk.	0.01	1
			3.52	unk.	6.6	unk.	13.3		unk.	unk.	unk.	unk.	unk.	unk.	unk.	unk.	unk.	1	2
E	candy bar,Three Musketeers-M&M/Mars	1 oz bar	28	124	21.5	0.11	1.0	2	unk.	unk.	unk.	unk.	unk.	unk.	unk.	unk.	unk.	0.01	1
			3.52	unk.	3.9	unk.	19.8		unk.	unk.	unk.	unk.	unk.	unk.	unk.	unk.	unk.	1	2
F	CANDY,bridge mix-Nabisco	1 oz	28	126	19.5	unk.	1.5	2	unk.	unk.	unk.	unk.	unk.	unk.	unk.	unk.	unk.	unk.	unk.
			3.52	unk.	4.6	unk.	19.5		unk.	unk.	unk.	unk.	unk.	unk.	unk.	unk.	tr(a)	unk.	unk.
G	candy,butterscotch balls	1 average	5.0	20	4.7	0.00	tr	0	tr(a)	tr(a)	tr(a)	tr(a)	tr(a)	tr(a)	unk.	0.10	tr	0.00	0
		ball	*20.00*	0.1	0.2	unk.	4.7		tr(a)	tr(a)	tr(a)	tr(a)	tr(a)	tr(a)	0.05	unk.	unk.	0	0
H	candy,caramel,plain or chocolate	1 average	10	40	7.7	0.02	0.4	1	unk.	26	31	14	28	4	unk.	0.70	0.10	tr	0
		square	*10.00*	0.8	1.0	unk.	7.6		11	40	10	18	6	20	0.40	unk.	0	tr	0
I	candy,chocolate clusters w/peanuts	1 average	12	55	7.0	0.08	1.1	2	unk.	unk.	unk.	unk.	unk.	unk.	unk.	0.72	0.43	0.03	2
	& caramel		*8.33*	0.8	2.8	unk.	7.0		unk.	unk.	unk.	unk.	unk.	unk.	1.50	unk.	tr(a)	tr	0
J	candy,chocolate coated creams,	1 average	13	57	9.1	0.01	0.5	1	unk.	unk.	unk.	unk.	unk.	unk.	unk.	0.65	0.23	0.01	0
	assorted		*7.69*	1.0	2.2	unk.	9.1		unk.	unk.	unk.	unk.	unk.	unk.	1.30	unk.	tr(a)	tr	0
K	candy,chocolate coated malted	1 average	4.8	25	2.7	0.02	0.4	1	unk.	unk.	unk.	unk.	unk.	unk.	unk.	0.87	0.03	tr	0
	milk ball	ball	*20.83*	tr	1.6	unk.	2.1		unk.	unk.	unk.	unk.	unk.	unk.	0.57	unk.	1	tr	0
L	candy,chocolate coated nougat	1 oz bar	28	118	20.7	0.06	1.1	2	unk.	74	88	40	80	11	unk.	1.22	0.31	0.02	1
	caramel bar		*3.52*	2.2	3.9	unk.	20.0		31	114	28	51	17	57	2.44	unk.	tr(a)	tr	0
M	candy,chocolate covered almonds	1/4 C	41	235	16.4	0.62	5.1	8	tr(a)	unk.	unk.	unk.	unk.	unk.	unk.	3.06	2.27	0.05	3
			2.42	0.8	18.0	unk.	16.4		unk.	unk.	unk.	unk.	unk.	unk.	11.98	unk.	4	tr	1
N	candy,chocolate covered peanuts	1 oz	28	159	11.1	0.34	4.7	7	unk.	unk.	unk.	unk.	unk.	unk.	unk.	3.12	1.99	0.11	7
			3.52	0.3	11.7	unk.	10.8		unk.	unk.	unk.	unk.	unk.	unk.	6.25	unk.	3	unk.	unk.
O	candy,chocolate covered peppermint	1 average	15	64	12.5	tr(a)	0.2	0	unk.	unk.	unk.	unk.	unk.	unk.	unk.	unk.	unk.	unk.	unk.
	pattie-Nabisco		*6.80*	unk.	1.4	unk.	12.5		unk.	unk.	unk.	unk.	unk.	unk.	unk.	unk.	tr(a)	tr(a)	0
P	candy,chocolate covered raisins	1 oz	28	121	20.0	0.17	1.5	2	unk.	unk.	unk.	unk.	unk.	unk.	unk.	2.84	0.11	0.02	2
			3.52	1.4	4.9	unk.	13.0		unk.	unk.	unk.	unk.	unk.	unk.	1.70	unk.	2	tr	1
Q	candy,coated chocolate disks	1/4 C	42	197	30.8	0.13	2.2	3	unk.	unk.	unk.	unk.	unk.	unk.	unk.	4.65	0.17	0.03	2
			2.36	0.5	8.3	unk.	30.6		unk.	unk.	unk.	unk.	unk.	unk.	3.00	unk.	tr(a)	tr	0
R	candy,corn	1/4 C	50	182	44.8	tr(a)	0.0	0	tr(a)	tr(a)	tr(a)	tr(a)	tr(a)	tr(a)	tr(a)	0.25	0.22	tr	0
			2.00	3.8	1.0	unk.	44.8		tr(a)	tr(a)	tr(a)	tr(a)	tr(a)	tr(a)	0.50	0.00	0(a)	0	0
S	candy,fudge,chocolate,plain	1 inch cube	21	84	15.7	0.04	0.6	1	unk.	34	45	27	38	5	unk.	0.90	0.38	tr	0
			4.76	1.7	2.6	unk.	15.7		15	55	14	26	8	27	1.20	unk.	unk.	tr	0

Riboflavin mg / Niacin mg	% USRDA	Vitamin B_6 mg / Folacin mcg	% USRDA	Vitamin B_{12} mcg / Pantothenic acid mg	% USRDA	Biotin mg / Vitamin A IU	% USRDA	Preformed A RE / Beta carotene RE	Vitamin D IU / Vitamin E IU	% USRDA	Total tocopherol mg / Alpha tocopherol mg	Other tocopherol mg / Total ash g	Calcium mg / Phosphorus mg	% USRDA	Sodium mg / Sodium meq	Potassium mg / Potassium meq	Chlorine mg / Chlorine meq	Iron mg / Magnesium mg	% USRDA	Zinc mg / Copper mg	% USRDA	Iodine mcg / Selenium mcg	% USRDA	Manganese mcg / Chromium mcg	
0.07	4	0.04	2	tr(a)	0	0.009	3	unk.	33	8	unk.	unk.	49	5	19	138	unk.	0.4	2	0.3	2	unk.	unk.	unk.	A
1.4	7	11	3	0.20	2	51	1	unk.	unk.	unk.	unk.	0.57	84	8	0.8	3.5	unk.	26	7	0.07	4	unk.		unk.	
0.07	4	unk.	unk.	tr(a)	0	unk.	unk.	unk.	0(a)	0	unk.	unk.	29	3	59	61	unk.	0.2	1	unk.	unk.	unk.	unk.	unk.	B
0.1	0	unk.	unk.	unk.	unk.	unk.	unk.	unk.	unk.	unk.	unk.	unk.	41	4	25	1.6	unk.	9	2	unk.	unk.	unk.		unk.	
0.02	1	0.08	4	0(a)	0	0.006	2	0	0(a)	0	unk.	unk.	12	1	3	127	unk.	0.5	3	0.4	3	unk.	unk.	unk.	C
2.7	13	20	5	0.53	5	0	0	0	unk.	unk.	unk.	0.45	78	8	0.1	3.3	unk.	32	8	0.13	7	unk.		unk.	
0.05	3	unk.	unk.	tr(a)	0	unk.	unk.	unk.	tr(a)	0	unk.	unk.	32	3	77	93	unk.	0.2	1	unk.	unk.	unk.	unk.	unk.	D
0.8	4	unk.	unk.	unk.	unk.	34	1	unk.	unk.	unk.	unk.	unk.	54	5	3.4	2.4	unk.	18	5	unk.	unk.	unk.		unk.	
0.05	3	unk.	unk.	unk.	unk.	unk.	unk.	unk.	tr(a)	0	unk.	unk.	21	2	65	49	unk.	0.2	1	0.1	0	unk.	unk.	unk.	E
0.1	0	unk.	unk.	unk.	unk.	unk.	unk.	unk.	unk.	unk.	unk.	unk.	26	3	2.8	1.3	unk.	8	2	unk.	unk.	unk.		unk.	
unk.	unk.	unk.	unk.	unk.	unk.	unk.	unk.	unk.	unk.	unk.	unk.	unk.	unk.	unk.	unk.	unk.	unk	unk.	unk.	unk.	unk.	unk.	unk.	unk.	F
unk.	unk.	unk.	unk.	unk.	unk.	unk.	unk.	unk.	unk.	unk.	unk.	unk.	unk.	unk.	unk.	unk.	unk.	unk.	unk.	unk.	unk.	unk.		unk.	
tr	0	0(a)	0	0(a)	0	0(a)	0	unk.	0	0	unk.	unk.	1	0	3	tr	tr(a)	0.1	0	tr	0	tr(a)	0	0	G
tr	0	0(a)	0	0(a)	0	7	0	unk.	unk.	unk.	unk.	0.01	0	0	0.1	0.0	0.0	0	0	tr(a)	0	unk.		3	
0.02	1	tr	0	tr(a)	0	0.000	0	tr(a)	4	1	unk.	unk.	15	2	23	19	unk.	0.1	1	tr	0	unk.	unk.	unk.	H
tr	0	tr	0	0.00	0	1	0	unk.	unk.	unk.	unk.	0.15	12	1	1.0	0.5	unk.	tr	0	tr	0	unk.		unk.	
0.02	1	unk.	unk.	tr(a)	0	unk.	unk.	0	unk.	unk.	unk.	unk.	15	2	15	27	unk.	0.1	1	unk.	unk.	unk.	unk.	unk.	I
0.4	2	unk.	unk.	unk.	unk.	tr	0	tr	unk.	unk.	unk.	0.20	23	2	0.7	0.7	unk.	unk.	unk.	unk.	unk.	unk.		unk.	
0.01	1	unk.	unk.	tr(a)	0	unk.	unk.	0	tr(a)	0	unk.	unk.	17	2	24	23	unk.	0.1	0	unk.	unk.	unk.	unk.	unk.	J
tr	0	unk.	unk.	unk.	unk.	tr	0	tr	unk.	unk.	unk.	0.16	14	1	1.0	0.6	unk.	unk.	unk.	unk.	unk.	unk.		unk.	
0.02	1	tr	0	tr(a)	0	0.001	1	unk.	8	2	0.2	0.1	11	1	5	18	7	0.1	0	tr	0	unk.	unk.	unk.	K
tr	0	tr	0	0.00	0	13	0	unk.	0.2	1	0.1	0.09	11	1	0.2	0.5	0.2	3	1	tr	0	unk.		unk.	
0.05	3	0.01	0	tr(a)	0	0.001	1	unk.	unk.	unk.	unk.	unk.	36	4	49	60	unk.	0.5	3	0.1	1	unk.	unk.	unk.	L
0.1	0	1	0	0.00	0	11	0	unk.	unk.	unk.	unk.	0.40	35	4	2.1	1.5	unk.	9	2	unk.	unk.	unk.		unk.	
0.22	13	0.02	1	tr(a)	0	unk.	unk.	unk.	14	4	unk.	unk.	84	8	24	225	unk.	1.2	6	0.4	3	unk.	unk.	unk.	M
0.7	4	18	5	0.07	1	22	0	unk.	unk.	unk.	unk.	0.95	142	14	1.1	5.8	unk.	17	4	0.13	7	unk.		unk.	
0.05	3	0.05	2	tr(a)	0	0.009	3	0	29	7	unk.	unk.	33	3	17	143	unk.	0.4	2	0.3	2	unk.	unk.	unk.	N
2.1	11	13	3	0.25	3	tr	0	tr	unk.	unk.	unk.	0.57	85	9	0.7	3.7	unk.	30	7	0.09	4	unk.		unk	
unk.	unk.	tr(a)	0	tr(a)	0	tr(a)	0	unk.	tr(a)	0	unk.	unk.	unk.	unk.	unk.	unk.	unk.	unk.	unk.	0.1	0	unk.	unk.	unk.	O
unk.	unk.	tr(a)	0	tr(a)	0	unk.	unk.	unk.	unk.	unk.	unk.	unk.	unk.	unk.	unk.	unk.	unk.	unk.	unk.	unk.	unk.	unk.		unk.	
0.06	4	0.03	2	tr(a)	0	0.005	2	unk.	27	7	unk.	unk.	43	4	18	171	unk.	0.7	4	0.1	1	unk.	unk.	unk.	P
0.1	1	2	0	0.02	0	43	1	unk.	unk.	unk.	unk.	0.60	49	5	0.8	4.4	unk.	14	3	0.04	2	unk.		unk.	
0.08	5	unk.	unk.	tr(a)	0	unk.	unk.	0	unk.	unk.	unk.	unk.	57	6	30	106	unk.	0.5	3	unk.	unk.	unk.	unk.	unk.	Q
0.1	1	unk.	unk.	unk.	unk.	42	1	4	unk.	unk.	unk.	0.51	59	6	1.3	2.7	unk.	unk.	unk.	unk.	unk.	unk.		unk.	
tr	0	0(a)	0	0(a)	0	0(a)	0	0	0(a)	0	unk.	unk.	7	1	106	3	tr(a)	0.5	3	0.3	2	unk.	unk.	0	R
tr	0	0(a)	0	0(a)	0	0	0	0	unk.	unk.	unk.	0.35	3	0	4.6	0.1	0.0	tr	0	tr(a)	0	unk.		30	
0.02	1	unk.	unk.	tr(a)	0	unk.	unk.	0	unk.	unk.	unk.	unk.	16	2	40	31	unk.	0.2	1	unk.	unk.	unk.	unk.	unk.	S
tr	0	unk.	unk.	unk.	unk.	tr	0	tr	unk.	unk.	unk.	0.38	18	2	1.7	0.8	unk.	10	3	unk.	unk.	unk.		unk.	

	FOOD	Portion	Weight in grams / *Conversion for 100 g*	Kilocalories / H_2O g	Total carbohydrate g / Total fats g	Crude fiber g / Dietary fiber g	Total protein g / Total sugar g	% USRDA	Arginine mg / Histidine mg	Isoleucine mg / Leucine mg	Lysine mg / Methionine mg	Phenylalanine mg / Threonine mg	Valine mg / Tryptophan mg	Cystine mg / Tyrosine mg	Polyunsat. fatty acids g / Monounsat. fatty acids g	Saturated fatty acids g / P/S ratio	Linoleic acid g / Cholesterol mg	Thiamin mg / Ascorbic acid mg	% USRDA
A	candy,fudge,chocolate,w/nuts	1 inch cube	21	89	14.5	0.08	0.8	**1**	tr(a)	unk.	unk.	unk.	unk.	unk.	unk.	0.90	1.20	0.01	**1**
			4.76	1.6	3.7	unk.	14.4		unk.	unk.	unk.	unk.	unk.	unk.	1.28	unk.	unk.	tr	**0**
B	candy,fudge,vanilla,plain	1 inch cube	21	84	15.7	0.00	0.6	**1**	unk.	38	50	30	42	6	unk.	0.61	0.52	tr	**0**
			4.76	2.1	2.3	unk.	15.7		17	62	16	29	9	30	1.20	unk.	unk.	tr(a)	**0**
C	candy,fudge,vanilla,w/nuts	1 inch cube	21	89	14.4	0.04	0.9	**1**	unk.	unk.	unk.	unk.	unk.	unk.	unk.	0.61	1.28	0.01	**1**
			4.76	2.0	3.4	unk.	14.4		unk.	unk.	unk.	unk.	unk.	unk.	1.28	unk.	unk.	tr(a)	**0**
D	candy,gum drop	1 average	2.0	7	1.7	0.00	tr	**0**	tr(a)	tr(a)	tr(a)	tr(a)	tr(a)	tr(a)	unk.	unk.	unk.	0.00	**0**
			50.00	0.2	tr	unk.	1.7		tr(a)	tr(a)	tr(a)	tr(a)	tr(a)	tr(a)	unk.	unk.	0(a)	0	**0**
E	candy,gum drop leaf	1 average	10	35	8.7	0.00	0.0	**0**	tr(a)	tr(a)	tr(a)	tr(a)	tr(a)	tr(a)	unk.	unk.	unk.	0.00	**0**
			10.00	1.2	0.1	unk.	8.7		tr(a)	tr(a)	tr(a)	tr(a)	tr(a)	tr(a)	unk.	unk.	0(a)	0	**0**
F	candy,hard	1 average	5.0	19	4.9	0.00	0.0	**0**	0(a)	0(a)	0(a)	0(a)	0(a)	0(a)	unk.	unk.	unk.	0.00	**0**
		ball	*20.00*	0.1	0.1	unk.	4.9		0(a)	0(a)	0(a)	0(a)	0(a)	0(a)	unk.	unk.	0(a)	0	**0**
G	candy,jelly beans	1/4 C	55	202	51.2	0(a)	tr	**0**	tr(a)	tr(a)	tr(a)	tr(a)	tr(a)	tr(a)	unk.	unk.	unk.	0.00	**0**
			1.82	3.5	0.3	unk.	51.2		tr(a)	tr(a)	tr(a)	tr(a)	tr(a)	tr(a)	unk.	unk.	0(a)	0	**0**
H	candy,licorice	1 average	10	35	8.7	0.00	0.0	**0**	tr(a)	tr(a)	tr(a)	tr(a)	tr(a)	tr(a)	unk.	unk.	unk.	0.00	**0**
		stick	*10.00*	1.2	0.1	unk.	8.7		tr(a)	tr(a)	tr(a)	tr(a)	tr(a)	tr(a)	unk.	unk.	0(a)	0	**0**
I	candy,Life Saver	1 average	2.0	8	1.9	0.00	0.0	**0**	0(a)	0(a)	0(a)	0(a)	0(a)	0(a)	unk.	unk.	unk.	0.00	**0**
			50.00	tr	tr	unk.	1.9		0(a)	0(a)	0(a)	0(a)	0(a)	0(a)	unk.	unk.	0(a)	0	**0**
J	candy,lollipop or sucker	1 oz sucker	28	110	27.6	0.00	0.0	**0**	0(a)	0(a)	0(a)	0(a)	0(a)	0(a)	unk.	unk.	unk.	0.00	**0**
			3.52	0.4	0.3	unk.	27.6		0(a)	0(a)	0(a)	0(a)	0(a)	0(a)	unk.	unk.	tr(a)	0	**0**
K	candy,marshmallow w/circus peanut	1 average	8.0	26	6.4	0.00	0.2	**0**	unk.	unk.	unk.	unk.	unk.	unk.	tr(a)	tr(a)	tr(a)	0.00	**0**
			12.50	1.4	tr	unk.	6.4		unk.	unk.	unk.	unk.	unk.	unk.	tr(a)	0.00	0(a)	0	**0**
L	candy,mint,fondant pattie	1 average	8.8	32	7.9	tr(a)	0.0	**0**	unk.	tr(a)	tr(a)	tr(a)	tr(a)	tr(a)	unk.	0.04	0.04	tr	**0**
			11.36	0.7	0.2	unk.	7.9		tr(a)	tr(a)	tr(a)	tr(a)	tr(a)	tr(a)	0.09	unk.	0(a)	0	**0**
M	candy,peanut brittle	1 oz	28	120	23.0	0.14	1.6	**3**	unk.	77	67	92	93	27	unk.	0.60	0.91	0.05	**3**
			3.52	0.6	3.0	unk.	23.0		46	114	17	51	21	67	1.31	unk.	0(a)	0	**0**
N	candy,Reese's peanut butter cup -Hershey	1 average	17	86	9.3	0.00	1.9	**3**	unk.	unk.	unk.	unk.	unk.	unk.	unk.	unk.	unk.	0.02	**1**
			5.88	unk.	5.2	unk.	8.6		unk.	unk.	unk.	unk.	unk.	unk.	unk.	unk.	tr(a)	0(a)	**0**
O	candy,thin mint,chocolate-covered-Nabisco	1 average	10	42	8.1	tr(a)	0.2	**0**	unk.	unk.	unk.	unk.	unk.	unk.	unk.	unk.	unk.	unk.	**unk.**
			9.90	unk.	1.0	unk.	8.1		unk.	unk.	unk.	unk.	unk.	unk.	unk.	unk.	tr(a)	tr(a)	**0**
P	candy,Tootsie roll,miniature	1 sm	6.0	24	5.0	0.01	0.1	**0**	unk.	unk.	unk.	unk.	unk.	unk.	unk.	0.15	0.11	tr	**0**
			16.67	0.3	0.5	unk.	4.9		unk.	unk.	unk.	unk.	unk.	unk.	0.23	unk.	0(a)	unk.	**unk.**
Q	CANTALOUPE & honeydew melon balls, frozen w/syrup	1/2 C	115	71	18.1	0.34	0.7	**1**	unk.	unk.	18	21	unk.	unk.	unk.	unk.	unk.	0.03	**2**
			0.87	95.7	0.1	unk.	16.7		unk.	unk.	2	unk.	1	unk.	unk.	unk.	0	18	**31**
R	cantaloupe,raw,cubed,edible portion	1/2 C	80	24	6.0	0.24	0.6	**1**	unk.	unk.	14	17	unk.	unk.	unk.	unk.	unk.	0.03	**2**
			1.25	73.0	0.1	0.80	5.1		unk.	unk.	2	unk.	1	unk.	unk.	unk.	unk.	26	**44**
S	cantaloupe,raw,edible portion	1/4 of 5" dia	133	40	9.9	0.40	0.9	**1**	unk.	unk.	24	28	unk.	unk.	unk.	tr(a)	unk.	0.05	**4**
			0.75	120.8	0.1	1.33	8.5		unk.	unk.	3	unk.	1	unk.	unk.	unk.	0	44	**73**

Riboflavin mg / Niacin mg	% USRDA	Vitamin B_6 mg / Folacin mcg	% USRDA	Vitamin B_{12} mcg / Pantothenic acid mg	% USRDA	Biotin mg / Vitamin A IU	% USRDA	Preformed A RE / Beta carotene RE	Vitamin D IU / Vitamin E IU	% USRDA	Total tocopherol mg / Alpha tocopherol mg	Other tocopherol mg / Total ash g	Calcium mg / Phosphorus mg	% USRDA	Sodium mg / Sodium meq	Potassium mg / Potassium meq	Chlorine mg / Chlorine meq	Iron mg / Magnesium mg	% USRDA	Zinc mg / Copper mg	% USRDA	Iodine mcg / Selenium mcg	% USRDA	Manganese mcg / Chromium mcg	
0.02	**1**	unk.	**unk.**	tr(a)	**0**	unk.	**unk.**	0	unk.	**unk.**	unk.	unk.	17	**2**	36	37	unk.	0.3	**1**	unk.	**unk.**	unk.	**unk.**	unk.	A
0.1	**0**	unk.	**unk.**	unk.	**unk.**	tr	**0**	tr	unk.	**unk.**	unk.	0.38	24	**2**	1.6	0.9	unk.	unk.	**unk.**	unk.	**unk.**	unk.		unk.	
0.03	**2**	unk.	**unk.**	tr(a)	**0**	unk.	**unk.**	0	unk.	**unk.**	unk.	unk.	24	**2**	44	27	unk.	0.1	**1**	unk.	**unk.**	unk.	**unk.**	unk.	B
tr	**0**	unk.	**unk.**	unk.	**unk.**	tr	**0**	tr	unk.	**unk.**	unk.	0.23	17	**2**	1.9	0.7	unk.	10	**3**	unk.	**unk.**	unk.		unk.	
0.03	**2**	unk.	**unk.**	tr(a)	**0**	unk.	**unk.**	0	unk.	**unk.**	unk.	unk.	23	**2**	39	24	unk.	0.2	**1**	unk.	**unk.**	unk.	**unk.**	unk.	C
tr	**0**	unk.	**unk.**	unk.	**unk.**	tr	**0**	tr	unk.	**unk.**	unk.	0.25	24	**2**	1.7	0.6	unk.	unk.	**unk.**	unk.	**unk.**	unk.		unk.	
tr	**0**	tr(a)	**0**	0(a)	**0**	tr(a)	**0**	0	tr(a)	**0**	tr(a)	tr(a)	tr	**0**	1	tr	tr(a)	tr	**0**	tr	**0**	tr(a)	**0**	tr(a)	D
tr	**0**	tr(a)	**0**	tr(a)	**0**	0	**0**	0	0.0	**0**	tr(a)	tr	tr	**0**	0.0	0.0	0.0	tr(a)	**0**	tr(a)	**0**	tr(a)		tr(a)	
tr	**0**	tr(a)	**0**	0(a)	**0**	tr(a)	**0**	0	0(a)	**0**	tr(a)	tr(a)	1	**0**	3	0	tr(a)	0.0	**0**	0.1	**0**	tr(a)	**0**	tr(a)	E
tr	**0**	tr(a)	**0**	tr(a)	**0**	0	**0**	0	0.0	**0**	tr(a)	0.01	tr	**0**	0.1	0.0	0.0	tr(a)	**0**	tr(a)	**0**	tr(a)		6	
0.00	**0**	0(a)	**0**	0(a)	**0**	0(a)	**0**	0	0	**0**	tr(a)	tr(a)	1	**0**	2	tr	unk.	0.1	**1**	tr	**0**	tr(a)	**0**	tr(a)	F
0.0	**0**	0(a)	**0**	0(a)	**0**	0	**0**	0	0.0	**0**	tr(a)	0.01	0	**0**	0.1	0.0	unk.	tr(a)	**0**	tr(a)	**0**	tr(a)		3	
tr	**0**	0(a)	**0**	0(a)	**0**	0(a)	**0**	0	0(a)	**0**	tr(a)	tr(a)	7	**1**	7	1	tr(a)	0.6	**3**	0.3	**2**	unk.	**unk.**	tr(a)	G
tr	**0**	0(a)	**0**	0(a)	**0**	0	**0**	0	0.0	**0**	tr(a)	0.05	2	**0**	0.3	0.0	0.0	tr(a)	**0**	tr(a)	**0**	tr(a)		tr(a)	
tr	**0**	tr(a)	**0**	0(a)	**0**	tr(a)	**0**	0	0(a)	**0**	tr(a)	tr(a)	1	**0**	3	0	tr(a)	0.0	**0**	0.1	**0**	unk.	**unk.**	tr(a)	H
tr	**0**	tr(a)	**0**	tr(a)	**0**	0	**0**	0	0.0	**0**	tr(a)	0.01	tr	**0**	0.1	0.0	0.0	2	**1**	tr(a)	**0**	tr(a)		unk.	
0.00	**0**	0(a)	**0**	0(a)	**0**	0(a)	**0**	0	0	**0**	tr(a)	tr(a)	tr	**0**	1	tr	tr(a)	tr	**0**	tr(a)	**0**	tr(a)	**0**	tr(a)	I
0.0	**0**	0(a)	**0**	0(a)	**0**	0	**0**	0	0.0	**0**	tr(a)	0.01	0	**0**	0.0	0.0	0.0	tr(a)	**0**	tr(a)	**0**	tr(a)		1	
0.00	**0**	0(a)	**0**	0(a)	**0**	0(a)	**0**	0	0	**0**	tr(a)	unk.	6	**1**	9	1	tr(a)	0.5	**3**	0.2	**1**	unk.	**unk.**	tr(a)	J
0.0	**0**	0(a)	**0**	0(a)	**0**	0	**0**	0	0.0	**0**	tr(a)	0.09	2	**0**	0.4	0.0	0.0	tr(a)	**0**	tr(a)	**0**	tr(a)		17	
tr	**0**	tr(a)	**0**	0(a)	**0**	tr(a)	**0**	0	0(a)	**0**	tr(a)	tr(a)	1	**0**	3	0	tr(a)	0.1	**1**	0.0	**0**	unk.	**unk.**	tr(a)	K
tr	**0**	tr(a)	**0**	tr(a)	**0**	0	**0**	0	0.0	**0**	tr(a)	0.02	1	**0**	0.1	0.0	0.0	tr(a)	**0**	0.01	**1**	tr(a)		unk.	
tr	**0**	0(a)	**0**	0(a)	**0**	0(a)	**0**	0	0(a)	**0**	unk.	unk.	1	**0**	19	tr	tr(a)	0.1	**1**	0.1	**0**	unk.	**unk.**	0	L
tr	**0**	0(a)	**0**	0(a)	**0**	0	**0**	0	unk.	**unk.**	unk.	0.06	1	**0**	0.8	0.0	0.0	tr(a)	**0**	tr(a)	**0**	tr(a)		5	
0.01	**1**	0.02	**1**	0(a)	**0**	0.002	**1**	0	0(a)	**0**	unk.	unk.	10	**1**	9	43	unk.	0.7	**4**	0.3	**2**	unk.	**unk.**	unk.	M
1.0	**5**	7	**2**	0.17	**2**	0	**0**	0	unk.	**unk.**	unk.	0.26	27	**3**	0.4	1.1	unk.	10	**3**	0.05	**2**	unk.		unk.	
0.03	**2**	unk.	**unk.**	tr(a)	**0**	unk.	**unk.**	unk.	tr(a)	**0**	unk.	unk.	15	**2**	60	54	unk.	0.2	**1**	unk.	**unk.**	unk.	**unk.**	unk.	N
0.9	**5**	unk.	**unk.**	unk.	**unk.**	15	**0**	unk.	unk.	**unk.**	unk.	0.37	33	**3**	2.6	1.4	unk.	14	**4**	unk.	**unk.**	unk.		unk.	
unk.	**unk.**	unk.	**unk.**	tr(a)	**0**	unk.	**unk.**	unk.	tr(a)	**0**	unk.	unk.	unk.	**unk.**	unk.	unk.	unk.	unk.	**unk.**	tr	**0**	unk	**unk.**	unk.	O
unk.	**unk.**	unk.	**unk.**	unk.	**unk.**	unk.	**unk.**	unk.	unk.	**unk.**	unk.	unk.	unk.	**unk.**	unk.	unk.	unk.	unk.	**unk.**	unk.	**unk.**	unk.		unk.	
tr	**0**	unk.	**unk.**	tr(a)	**0**	unk.	**unk.**	0	tr(a)	**0**	unk.	unk.	4	**0**	12	7	unk.	0.1	**1**	unk.	**unk.**	unk.	**unk.**	unk.	P
tr	**0**	unk.	**unk.**	unk.	**unk.**	tr	**0**	tr	unk.	**unk.**	unk.	0.07	7	**1**	0.5	0.2	unk.	2	**1**	unk.	**unk.**	unk.		unk.	
0.02	**1**	unk.	**unk.**	0(a)	**0**	unk.	**unk.**	0	0(a)	**0**	unk.	unk.	11	**1**	10	216	unk.	0.3	**2**	unk.	**unk.**	unk.	**unk.**	unk.	Q
0.6	**3**	unk.	**unk.**	unk.	**unk.**	1771	**35**	177	unk.	**unk.**	unk.	0.46	14	**1**	0.4	5.5	unk.	unk.	**unk.**	tr(a)	**0**	tr(a)		unk.	
0.02	**1**	0.07	**3**	0.00	**0**	0.002	**1**	0	unk.	**unk.**	0.2	0.1	11	**1**	10	201	33	0.3	**2**	unk.	**unk.**	unk.	**unk.**	8	R
0.5	**2**	5	**1**	0.20	**2**	2720	**54**	272	0.3	**1**	0.1	0.40	13	**1**	0.4	5.1	0.9	7	**2**	0.01	**1**	0		12	
0.04	**2**	0.11	**6**	0.00	**0**	0.004	**1**	0	0(a)	**0**	0.4	0.2	19	**2**	16	333	54	0.5	**3**	tr	**0**	unk.	**unk.**	13	S
0.8	**4**	40	**10**	0.33	**3**	4505	**90**	450	0.5	**2**	0.2	0.66	21	**2**	0.7	8.5	1.5	11	**3**	0.02	**1**	0		20	

	FOOD	Portion	Weight in grams / *Conversion for 100 g*	Kilocalories / H_2O g	Total carbohydrate g / Total fats g	Crude fiber g / Dietary fiber g	Total protein g / Total sugar g	% USRDA	Arginine mg / Histidine mg	Isoleucine mg / Leucine mg	Lysine mg / Methionine mg	Phenylalanine mg / Threonine mg	Valine mg / Tryptophan mg	Cystine mg / Tyrosine mg	Polyunsat. fatty acids g / Monounsat. fatty acids g	Saturated fatty acids g / P/S ratio	Linoleic acid g / Cholesterol mg	Thiamin mg / Ascorbic acid mg	% USRDA
A	CARAWAY seed	1 tsp	2.2	7	1.1	0.28	0.4	1	unk.	unk.	unk.	unk.	unk.	unk.	0.07	0.01	0.07	0.01	1
			45.45	0.2	0.3	unk.	0.0		unk.	unk.	unk.	unk.	unk.	unk.	0.15	5.27	0	unk.	unk.
B	CARDAMON,ground	1 tsp	1.9	6	1.3	0.21	0.2	0	unk.	unk.	unk.	unk.	unk.	unk.	0.01	0.01	0.01	tr	0
			52.63	0.2	0.1	unk.	0.0		unk.	unk.	unk.	unk.	unk.	unk.	0.02	0.63	0	unk.	unk.
C	CARROT,canned,sliced,drained solids	1/2 C	78	23	5.2	0.62	0.6	1	28	23	26	22	29	8	unk.	tr(a)	unk.	0.02	1
			1.29	70.7	0.2	2.87	4.6		10	33	5	22	5	unk.	unk.	unk.	0	2	3
D	carrot,canned,sliced,low sodium, drained solids	1/2 C	78	19	4.3	0.62	0.6	1	28	23	26	22	29	8	unk.	tr(a)	unk.	0.02	1
			1.29	72.1	0.1	2.87	4.6		10	33	5	22	5	unk.	unk.	unk.	0	2	3
E	carrot,fresh,sliced,cooked, drained solids	1/2 C	78	24	5.5	0.77	0.7	1	31	26	30	24	32	8	unk.	tr(a)	unk.	0.04	3
			1.29	70.7	0.2	2.33	4.6		10	37	5	25	5	11	unk.	unk.	0	5	8
F	carrot,raw,edible portion	1 med	81	34	7.9	0.81	0.9	1	40	37	42	29	45	23	unk.	tr(a)	unk.	0.05	3
			1.23	71.4	0.2	2.35	4.8		14	53	8	35	8	14	unk.	unk.	0	6	11
G	carrot,raw,grated	1/2 C	55	23	5.3	0.55	0.6	1	27	25	29	20	31	16	unk.	tr(a)	unk.	0.03	2
			1.82	48.5	0.1	1.60	3.3		9	36	5	24	5	9	unk.	unk.	0	4	7
H	CASHEW nuts	1/4 C	35	196	10.3	0.49	6.0	9	unk.	397	259	307	518	171	2.69	3.22	2.54	0.15	10
			2.86	1.8	16.0	unk.	0.0		135	493	114	241	150	232	9.17	0.84	0	77	128
I	cashew nuts,salted	1/4 C	35	196	10.3	0.49	6.0	9	unk.	397	259	307	518	171	2.69	3.22	2.54	0.15	10
			2.86	1.8	16.0	unk.	0.0		135	493	114	241	150	232	9.17	0.84	0	unk.	unk.
J	CATSUP,any	1 Tbsp	15	16	3.8	0.07	0.3	1	unk.	9	12	7	9	3	unk.	tr(a)	unk.	0.01	1
			6.67	10.3	0.1	unk.	3.4		4	12	1	10	3	4	unk.	unk.	0(a)	2	4
K	catsup,low sodium	1 Tbsp	15	16	3.8	0.07	0.3	1	unk.	9	12	7	9	3	unk.	tr(a)	unk.	0.01	1
			6.67	10.3	0.1	unk.	3.4		4	12	1	10	3	4	unk.	unk.	0(a)	2	4
L	catsup,low sodium,low calorie-Tillie Lewis	1 Tbsp	17	8	1.7	0.08	0.3	1	unk.	10	14	8	10	3	unk.	tr(a)	unk.	0.02	1
			5.88	unk.	0.1	unk.	unk.		5	14	2	12	3	5	unk.	unk.	0(a)	unk.	unk.
M	CAULIFLOWER,fresh,cooked, drained solids	1/2 C	68	15	2.8	0.67	1.5	2	70	67	87	48	93	unk.	unk.	tr(a)	unk.	0.06	4
			1.48	62.6	0.1	1.49	2.1		30	104	31	65	20	22	unk.	unk.	0(a)	37	62
N	cauliflower,frozen,cooked, drained solids	1/2 C	90	16	3.0	0.72	1.7	3	77	74	95	53	103	unk.	unk.	tr(a)	unk.	0.04	2
			1.11	84.6	0.2	1.98	2.8		31	114	34	72	22	25	unk.	unk.	0(a)	37	62
O	cauliflower,raw,flowerettes	1 flowerette (10 per C)	10	3	0.5	0.10	0.3	0	12	12	15	8	16	unk.	unk.	tr(a)	unk.	0.01	1
			10.00	9.1	tr	0.22	0.3		5	18	5	11	3	4	unk.	unk.	0(a)	8	13
P	CAVIAR,sturgeon granular	1 Tbsp	16	42	0.5	0(a)	4.3	10	unk.	242	307	190	264	unk.	unk.	unk.	unk.	unk.	unk.
			6.25	7.4	2.4	0(a)	0.0		unk.	354	112	259	38	unk.	unk.	unk.	48	unk.	unk.
Q	CELERY seed	1 tsp	2.2	9	0.9	0.26	0.4	1	unk.	unk.	unk.	unk.	unk.	unk.	0.08	0.05	0.08	unk.	unk.
			45.45	0.1	0.6	unk.	0.0		unk.	unk.	unk.	unk.	unk.	unk.	0.34	1.70	0	tr	1
R	celery,chopped,cooked,drained solids	1/2 C	75	11	2.3	0.45	0.6	1	unk.	24	9	28	29	3	unk.	tr(a)	unk.	0.01	1
			1.33	71.5	0.1	1.31	0.9		9	41	8	20	6	8	unk.	unk.	0(a)	5	8
S	celery,raw,chopped	1/2 C	60	10	2.3	0.36	0.5	1	22	11	15	9	16	3	unk.	tr(a)	unk.	0.02	1
			1.67	56.5	0.1	1.05	0.7		1	13	1	10	unk.	4	unk.	unk.	0(a)	5	9

Riboflavin mg / Niacin mg	% USRDA	Vitamin B6 mg / Folacin mcg	% USRDA	Vitamin B12 mcg / Pantothenic acid mg	% USRDA	Biotin mg / Vitamin A IU	% USRDA	Preformed A RE / Beta carotene RE	Vitamin D IU / Vitamin E IU	% USRDA	Total tocopherol mg / Alpha tocopherol mg	Other tocopherol mg / Total ash g	Calcium mg / Phosphorus mg	% USRDA	Sodium mg / Sodium meq	Potassium mg / Potassium meq	Chlorine mg / Chlorine meq	Iron mg / Magnesium mg	% USRDA	Zinc mg / Copper mg	% USRDA	Iodine mcg / Selenium mcg	% USRDA	Manganese mcg / Chromium mcg	
0.01	**1**	unk.	**unk.**	0.00	**0**	unk.	**unk.**	0(a)	0(a)	**0**	unk.	unk.	15	**2**	tr	30	unk.	0.4	**2**	0.1	**1**	unk.	**unk.**	unk.	A
0.1	**0**	unk.	**unk.**	unk.	**unk.**	8	**0**	unk.	unk.	**unk.**	unk.	0.13	13	**1**	0.0	0.8	unk.	6	**1**	unk.	**unk.**	unk.		unk.	
tr	**0**	unk.	**unk.**	0.00	**0**	unk.	**unk.**	0(a)	0(a)	**0**	unk.	unk.	7	**1**	tr	21	unk.	0.3	**2**	0.1	**1**	unk.	**unk.**	unk.	B
tr	**0**	unk.	**unk.**	unk.	**unk.**	tr	**0**	unk.	unk.	**unk.**	unk.	0.11	3	**0**	0.0	0.5	unk.	4	**1**	unk.	**unk.**	unk.		unk.	
0.02	**1**	0.02	**1**	0.00	**0**	tr	**0**	0	0(a)	**0**	unk.	unk.	23	**2**	183	93	349	0.5	**3**	0.2	**2**	unk.	**unk.**	15	C
0.3	**2**	19	**5**	0.08	**1**	11625	**233**	1162	0.4	**1**	0.4	0.77	17	**2**	8.0	2.4	9.8	4	**1**	0.01	**0**	tr(a)		unk.	
0.02	**1**	0.02	**1**	0.00	**0**	tr	**0**	0	0(a)	**0**	unk.	unk.	23	**2**	30	93	349	0.5	**3**	0.2	**2**	unk.	**unk.**	unk.	D
0.3	**2**	19	**5**	0.08	**1**	11625	**233**	1162	0.4	**1**	0.4	0.39	17	**2**	1.3	2.4	9.8	4	**1**	0.01	**0**	tr(a)		unk.	
0.04	**2**	unk.	**unk.**	0.00	**0**	unk.	**unk.**	0	0(a)	**0**	0.2	0.1	26	**3**	26	172	22	0.5	**3**	0.2	**2**	unk.	**unk.**	15	E
0.4	**2**	19	**5**	unk.	**unk.**	8137	**163**	814	0.2	**1**	0.1	0.46	24	**2**	1.1	4.4	0.6	5	**1**	0.01	**0**	tr(a)		unk.	
0.04	**2**	0.12	**6**	0.00	**0**	0.002	**1**	0	0(a)	**0**	0.4	0.0	30	**3**	38	276	34	0.6	**3**	0.3	**2**	unk.	**unk.**	16	F
0.5	**2**	26	**7**	0.23	**2**	8910	**178**	891	0.4	**2**	0.4	0.65	29	**3**	1.7	7.1	1.0	15	**4**	0.01	**0**	tr(a)		1	
0.03	**2**	0.08	**4**	0.00	**0**	0.001	**1**	0	0(a)	**0**	0.2	0.0	20	**2**	26	188	23	0.4	**2**	0.2	**2**	unk.	**unk.**	11	G
0.3	**2**	18	**4**	0.15	**2**	6050	**121**	605	0.3	**1**	0.2	0.44	20	**2**	1.1	4.8	0.6	10	**3**	0.01	**0**	tr(a)		1	
0.09	**5**	unk.	**unk.**	0.00	**0**	unk.	**unk.**	0	0(a)	**0**	1.4	unk.	13	**1**	5	162	unk.	1.3	**7**	1.5	**10**	unk.	**unk.**	unk.	H
0.6	**3**	24	**6**	0.45	**5**	35	**1**	3	1.7	**6**	1.3	0.91	131	**13**	0.2	4.1	unk.	93	**23**	unk.	**unk.**	unk.		unk.	
0.09	**5**	unk.	**unk.**	0.00	**0**	unk.	**unk.**	0	0(a)	**0**	1.4	1.3	13	**1**	70	162	unk.	1.3	**7**	1.5	**10**	unk.	**unk.**	unk.	I
0.5	**2**	24	**6**	0.45	**5**	35	**1**	3	1.7	**6**	0.1	0.91	131	**13**	3.0	4.1	unk.	93	**23**	unk.	**unk.**	unk.		unk.	
0.01	**1**	0.02	**1**	0.00	**0**	0.001	**0**	0	0(a)	**0**	unk.	unk.	3	**0**	156	54	unk.	0.1	**1**	tr	**0**	unk.	**unk.**	unk.	J
0.2	**1**	1	**0**	0.03	**0**	210	**4**	21	unk.	**unk.**	unk.	0.54	8	**1**	6.8	1.4	unk.	3	**1**	0.08	**4**	unk.		unk.	
0.01	**1**	0.02	**1**	0.00	**0**	0.001	**0**	0	0(a)	**0**	unk.	unk.	3	**0**	3	54	unk.	0.1	**1**	0.2	**1**	unk.	**unk.**	unk.	K
0.2	**1**	1	**0**	0.03	**0**	210	**4**	21	unk.	**unk.**	unk.	0.54	8	**1**	0.1	1.4	unk.	3	**1**	0.08	**4**	unk.		unk.	
0.01	**1**	0.02	**1**	0(a)	**0**	0.001	**0**	0(a)	0(a)	**0**	tr	unk.	4	**0**	5	unk.	unk.	0.1	**1**	0.2	**1**	unk.	**unk.**	unk.	L
0.3	**1**	1	**0**	0.03	**0**	238	**5**	24	tr	**0**	unk.	unk.	9	**1**	0.2	unk.	unk.	4	**1**	0.09	**4**	unk.		unk.	
0.05	**3**	0.09	**4**	0.00	**0**	unk.	**unk.**	0	0(a)	**0**	unk.	unk.	14	**1**	6	139	unk.	0.5	**3**	0.2	**1**	unk.	**unk.**	unk.	M
0.4	**2**	23	**6**	0.34	**3**	40	**1**	4	unk.	**unk.**	unk.	0.40	28	**3**	0.3	3.6	unk.	unk.	**unk.**	tr(a)	**0**	tr(a)		unk.	
0.04	**3**	0.11	**6**	0.00	**0**	unk.	**unk.**	0	0(a)	**0**	0.1	tr	15	**2**	9	186	unk.	0.4	**3**	0.3	**2**	unk.	**unk.**	unk.	N
0.4	**2**	31	**8**	0.45	**5**	27	**1**	3	0.1	**0**	tr	0.54	34	**3**	0.4	4.8	unk.	12	**3**	tr(a)	**0**	tr(a)		unk.	
0.01	**1**	0.02	**1**	0.00	**0**	0.002	**1**	0	0(a)	**0**	unk.	unk.	2	**0**	1	29	3	0.1	**1**	tr	**0**	unk.	**unk.**	unk.	O
0.1	**0**	2	**1**	0.10	**1**	6	**0**	1	unk.	**unk.**	unk.	0.09	6	**1**	0.1	0.8	0.1	2	**1**	tr(a)	**0**	tr(a)		tr	
unk.	**unk.**	unk.	**unk.**	unk.	**unk.**	unk.	**unk.**	unk.	unk.	**unk.**	unk.	unk.	44	**4**	352	29	unk.	1.9	**11**	unk.	**unk.**	unk.	**unk.**	unk.	P
unk.	**unk.**	unk.	**unk.**	unk.	**unk.**	unk.	**unk.**	unk.	unk.	**unk.**	unk.	1.41	57	**6**	15.3	0.7	unk.	48	**12**	unk.	**unk.**	unk.		unk.	
unk.	**unk.**	unk.	**unk.**	0.00	**0**	unk.	**unk.**	0(a)	0(a)	**0**	unk.	unk.	39	**4**	4	31	unk.	1.0	**6**	0.2	**1**	unk.	**unk.**	unk.	Q
unk.	**unk.**	unk.	**unk.**	unk.	**unk.**	1	**0**	unk.	unk.	**unk.**	unk.	0.20	12	**1**	0.1	0.8	unk.	10	**2**	unk.	**unk.**	unk.		unk.	
0.02	**1**	0.03	**1**	0.00	**0**	unk.	**unk.**	0	0(a)	**0**	0.4	0.1	23	**2**	66	179	unk.	0.1	**1**	0.1	**1**	unk.	**unk.**	unk.	R
0.2	**1**	unk.	**unk.**	0.16	**2**	173	**4**	17	0.5	**2**	0.3	0.52	17	**2**	2.9	4.6	unk.	2	**1**	unk.	**unk.**	tr(a)		unk.	
0.02	**1**	0.04	**2**	0.00	**0**	unk.	**unk.**	0	0(a)	**0**	0.3	tr	23	**2**	76	205	82	0.2	**1**	0.1	**1**	unk.	**unk.**	12	S
0.2	**1**	4	**1**	0.26	**3**	144	**3**	14	0.3	**1**	0.3	0.60	17	**2**	3.3	5.2	2.3	5	**1**	0.01	**0**	tr(a)		18	

	FOOD	Portion	Weight in grams / Conversion for 100 g	Kilocalories / H_2O g	Total carbohydrate g / Total fats g	Crude fiber g / Dietary fiber g	Total protein g / Total sugar g	% USRDA	Arginine mg / Histidine mg	Isoleucine mg / Leucine mg	Lysine mg / Methionine mg	Phenylalanine mg / Threonine mg	Valine mg / Tryptophan mg	Cystine mg / Tyrosine mg	Polyunsat. fatty acids g / Monounsat. fatty acids g	Saturated fatty acids g / P/S ratio	Linoleic acid g / Cholesterol mg	Thiamin mg / Ascorbic acid mg	% USRDA
A	celery,raw,stalk	1 stalk	40	7	1.6	0.24	0.4	**1**	14	8	10	6	10	2	unk.	tr(a)	unk.	0.01	**1**
			2.50	37.6	tr	0.70	0.5		tr	9	1	6	unk.	3	unk.	unk.	0(a)	4	**6**
B	CEREAL,COOKED,bulgur (parboiled	1/4 C	38	63	13.1	0.30	2.3	**4**	unk.	unk.	81	unk.	unk.	unk.	0.12	0.03	0.11	0.02	**1**
	wheat) canned,unseasoned		*2.67*	21.0	0.3	unk.	0.0		unk.	unk.	56	unk.	130	unk.	0.03	3.67	0	0	**0**
C	cereal,cooked,Coco Wheats-	1/2 oz dry,	120	54	11.0	unk.	1.9	**3**	unk.	unk.	unk.	unk.	unk.	unk.	unk.	unk.	unk.	0.06	**4**
	Little Crow Foods	prep,(1/2 C)	*0.83*	105.5	0.2	unk.	unk.		unk.	unk.	unk.	unk.	unk.	unk.	unk.	unk.	unk.	unk.	**unk.**
D	cereal,cooked,corn grits,degermed,	1/2 oz dry,	123	62	13.5	0.12	1.5	**2**	unk.	67	42	unk.	unk.	unk.	0.06	0.01	0.06	0.05	**3**
	enriched	prep,(1/2 C)	*0.82*	106.7	0.1	unk.	0.0		unk.	190	unk.	58	9	unk.	0.02	5.00	0	0	**0**
E	cereal,cooked,corn meal white or	1/2 oz dry,	120	60	12.8	0.12	1.3	**2**	unk.	55	35	54	61	unk.	0.12	0.02	0.12	0.07	**5**
	yellow,degermed,enriched	prep,(1/2 C)	*0.83*	105.2	0.2	unk.	unk.		unk.	157	23	48	6	unk.	0.05	5.00	0	0	**0**
F	cereal,cooked,Cream Of Wheat,instant	1 oz dry,	200	110	22.0	tr	3.0	**5**	unk.	unk.	unk.	unk.	unk.	unk.	unk.	tr(a)	unk.	0.14	**9**
	-Nabisco	prep (3/4 C)	*0.50*	174.4	0.4	unk.	1.0		unk.	unk.	unk.	unk.	unk.	unk.	unk.	unk.	0(a)	tr	**0**
G	cereal,cooked,Cream Of Wheat,quick	1 oz dry,	200	110	22.0	tr	1.4	**2**	unk.	unk.	unk.	unk.	unk.	unk.	unk.	unk.	unk.	0.14	**9**
	-Nabisco	prep (3/4 C)	*0.50*	174.6	0.4	unk.	1.0		unk.	unk.	unk.	unk.	unk.	unk.	unk.	unk.	0(a)	0(a)	**0**
H	cereal,cooked,Cream Of Wheat,regular	1 oz dry,	200	114	22.0	tr	3.0	**5**	unk.	unk.	unk.	unk.	unk.	unk.	unk.	tr(a)	unk.	0.16	**11**
	-Nabisco	prep (1 C)	*0.50*	174.7	0.4	unk.	0.0		unk.	unk.	unk.	unk.	unk.	unk.	unk.	unk.	0(a)	0(a)	**0**
I	cereal,cooked,farina instant cooking,	1/2 oz dry,	123	67	14.0	0.12	2.1	**3**	unk.	unk.	unk.	unk.	unk.	unk.	unk.	unk.	unk.	0.09	**6**
	enriched	prep,(1/2 C)	*0.82*	105.2	0.1	unk.	0.0		unk.	unk.	unk.	unk.	unk.	unk.	unk.	unk.	0	0	**0**
J	cereal,cooked,farina quick cooking,	1/2 oz	123	53	10.9	unk.	1.6	**2**	unk.	599	245	unk.	unk.	unk.	unk.	unk.	unk.	0.06	**4**
	enriched	(1/2 C) dry	*0.82*	109.0	0.1	unk.	0.0		unk.	844	unk.	342	unk.	unk.	unk.	unk.	0	0	**0**
K	cereal,cooked,instant oatmeal	1-1/2 oz	43	153	29.1	0.92	5.2	**8**	unk.	unk.	unk.	unk.	unk.	unk.	unk.	tr(a)	unk.	0.56	**37**
	w/bran & raisins-Quaker	(1 pkg)	*2.35*	3.6	1.9	unk.	unk.		unk.	unk.	unk.	unk.	unk.	unk.	unk.	unk.	0(a)	0	**0**
L	cereal,cooked,Maltex-Standard Milling	3/4 oz dry,	128	86	15.4	0.38	2.8	**4**	unk.	unk.	unk.	unk.	unk.	unk.	unk.	tr(a)	unk.	0.12	**8**
		prep (1/2 C)	*0.78*	99.6	0.3	unk.	unk.		unk.	unk.	unk.	unk.	unk.	unk.	unk.	unk.	0	1	**2**
M	cereal,cooked,Maypo,Vermont Style	3/4 oz dry,	123	90	14.5	0.31	3.1	**5**	unk.	unk.	unk.	unk.	unk.	unk.	unk.	tr(a)	unk.	0.33	**22**
	-Standard Milling	prep (1/2 C)	*0.82*	94.6	0.8	unk.	1.0		unk.	unk.	unk.	unk.	unk.	unk.	unk.	unk.	0	13	**22**
N	cereal,cooked,Maypo,30 Second	7/8 oz dry,	110	110	17.0	0.24	4.2	**7**	unk.	unk.	unk.	unk.	unk.	unk.	unk.	tr(a)	unk.	0.40	**26**
	-Standard Milling	prep (1/2 C)	*0.91*	81.6	1.1	unk.	1.0		unk.	unk.	unk.	unk.	unk.	unk.	unk.	unk.	0	16	**27**
O	cereal,cooked,oatmeal or rolled oats,	1/2 oz dry,	123	67	11.9	0.24	2.4	**4**	unk.	143	115	143	165	unk.	0.50	0.21	0.48	0.10	**7**
	regular & instant cooking	prep (1/2 C)	*0.82*	106.0	1.2	unk.	tr		unk.	261	44	107	39	unk.	0.43	2.41	0	0	**0**
P	cereal,cooked,oats,quick & regular	1 oz dry,	131	110	20.7	0.49	4.6	**7**	unk.	unk.	unk.	unk.	unk.	unk.	unk.	unk.	unk.	0.21	**14**
	-Ralston	prep (2/3 C)	*0.77*	105.0	1.1	unk.	0.0		unk.	unk.	unk.	unk.	unk.	unk.	unk.	unk.	0(a)	1	**2**
Q	cereal,cooked,Pettijohns-Quaker	1 oz dry,	121	93	19.2	0.57	3.0	**5**	unk.	unk.	unk.	unk.	unk.	unk.	unk.	unk.	unk.	0.10	**7**
		prep (1/2 C)	*0.83*	unk.	0.5	unk.	0.0		unk.	unk.	unk.	unk.	unk.	unk.	unk.	unk.	0	0	**0**
R	cereal,cooked,Ralston,instant &	1 oz dry,	165	110	20.6	0.48	4.5	**7**	unk.	unk.	unk.	unk.	unk.	unk.	unk.	tr(a)	unk.	0.17	**11**
	regular-Ralston	prep (3/4 C)	*0.61*	139.7	1.1	unk.	0.0		unk.	unk.	unk.	unk.	unk.	unk.	unk.	unk.	0(a)	1	**2**
S	cereal,cooked,Wheatena	2/3 oz dry,	117	79	13.8	0.50	2.5	**4**	unk.	unk.	unk.	unk.	unk.	unk.	unk.	tr(a)	unk.	0.98	**66**
	-Standard Milling	prep (1/2 C)	*0.85*	94.3	0.3	unk.	0.0		unk.	unk.	unk.	unk.	unk.	unk.	unk.	unk.	0	tr	**0**

Riboflavin mg / Niacin mg	% USRDA	Vitamin B_6 mg / Folacin mcg	% USRDA	Vitamin B_{12} mcg / Pantothenic acid mg	% USRDA	Biotin mg / Vitamin A IU	% USRDA	Preformed A RE / Beta carotene RE	Vitamin D IU / Vitamin E IU	% USRDA	Total tocopherol mg / Alpha tocopherol mg	Other tocopherol mg / Total ash g	Calcium mg / Phosphorus mg	% USRDA	Sodium mg / Sodium meq	Potassium mg / Potassium meq	Chlorine mg / Chlorine meq	Iron mg / Magnesium mg	% USRDA	Zinc mg / Copper mg	% USRDA	Iodine mcg / Selenium mcg	% USRDA	Manganese mcg / Chromium mcg	
0.01	**1**	0.02	**1**	0.00	**0**	unk.	**unk.**	0	0(a)	**0**	0.2	tr	16	**2**	50	136	55	0.1	**1**	0.1	**0**	unk.	**unk.**	8	A
0.1	**1**	3	**1**	0.17	**2**	96	**2**	10	0.2	**1**	0.2	0.40	11	**1**	2.2	3.5	1.5	3	**1**	tr	**0**	tr(a)		12	
0.01	**1**	unk.	**unk.**	unk.	**unk.**	unk.	**unk.**	0	unk.	**unk.**	0.1	tr	8	**1**	225	33	unk.	0.5	**3**	unk.	**unk.**	unk.	**unk.**	unk.	B
0.9	**5**	unk.	**unk.**	unk.	**unk.**	0	**0**	0	unk.	**unk.**	tr	0.79	75	**8**	9.8	0.8	unk.	unk.	**unk.**	unk.	**unk.**	unk.		unk.	
0.04	**2**	unk.	**unk.**	unk.	**unk.**	unk.	**unk.**	unk.	unk.	**unk.**	unk.	unk.	unk.	**unk.**	unk.	unk.	unk.	0.4	**2**	unk.	**unk.**	unk.	**unk.**	unk.	C
0.5	**3**	unk.	**unk.**	unk.	**unk.**	unk.	**unk.**	unk.	unk.	**unk.**	unk.	unk.	unk.	**unk.**	unk.	unk.	unk.	unk.	**unk.**	unk.	**unk.**	unk.		unk.	
0.04	**2**	unk.	**unk.**	unk.	**unk.**	unk.	**unk.**	0	unk.	**unk.**	5.4	4.7	1	**0**	unk.	13	unk.	0.4	**2**	unk.	**unk.**	unk.	**unk.**	unk.	D
0.5	**2**	unk.	**unk.**	unk.	**unk.**	73	**2**	7	6.5	**22**	0.7	0.73	12	**1**	unk.	0.3	unk.	4	**1**	unk.	**unk.**	unk.		unk.	
0.05	**3**	unk.	**unk.**	unk.	**unk.**	unk.	**unk.**	0	unk.	**unk.**	2.9	2.1	1	**0**	unk.	19	unk.	0.5	**3**	0.1	**1**	unk.	**unk.**	unk.	E
0.6	**3**	11	**3**	0.83	**8**	72	**1**	7	0.9	**3**	0.8	0.36	17	**2**	unk.	0.5	unk.	8	**2**	unk.	**unk.**	unk.		unk.	
0.06	**4**	unk.	**unk.**	unk.	**unk.**	tr(a)	**0**	unk.	0(a)	**0**	tr(a)	tr(a)	150	**15**	4	33	unk.	8.1	**45**	0.3	**2**	unk.	**unk.**	tr(a)	F
1.2	**6**	unk.	**unk.**	unk.	**unk.**	unk.	**unk.**	unk.	unk.	**unk.**	tr(a)	0.60	150	**15**	0.2	0.8	unk.	10	**2**	0.06	**3**	tr(a)		unk.	
0.06	**4**	unk.	**unk.**	unk.	**unk.**	tr(a)	**0**	unk.	0(a)	**0**	tr(a)	tr(a)	150	**15**	131	37	unk.	8.1	**45**	0.3	**2**	unk.	**unk.**	tr(a)	G
1.2	**6**	unk.	**unk.**	tr(a)	**0**	unk.	**unk.**	unk.	unk.	**unk.**	tr(a)	0.92	150	**15**	5.7	0.9	unk.	9	**2**	0.06	**3**	tr(a)		unk.	
0.06	**4**	unk.	**unk.**	unk.	**unk.**	tr(a)	**0**	unk.	0(a)	**0**	tr(a)	tr(a)	6	**1**	2	34	unk.	8.1	**45**	0.3	**2**	unk.	**unk.**	tr(a)	H
1.2	**6**	unk.	**unk.**	unk.	**unk.**	unk.	**unk.**	unk.	unk.	**unk.**	tr(a)	0.18	40	**4**	0.1	0.9	unk.	8	**2**	0.06	**3**	tr(a)		unk.	
0.05	**3**	unk.	**unk.**	unk.	**unk.**	unk.	**unk.**	0	unk.	**unk.**	unk.	unk.	94	**9**	230	16	unk.	7.8	**44**	unk.	**unk.**	unk.	**unk.**	unk.	I
0.6	**3**	unk.	**unk.**	unk.	**unk.**	0	**0**	0	unk.	**unk.**	unk.	1.10	74	**7**	10.0	0.4	unk.	5	**1**	unk.	**unk.**	unk.		unk.	
0.04	**2**	unk.	**unk.**	unk.	**unk.**	unk.	**unk.**	0	0	**0**	unk.	unk.	73	**7**	202	12	unk.	6.1	**34**	unk.	**unk.**	unk.	**unk.**	unk.	J
0.5	**2**	unk.	**unk.**	unk.	**unk.**	0	**0**	0	unk.	**unk.**	unk.	0.86	81	**8**	8.8	0.3	unk.	4	**1**	unk.	**unk.**	unk.		unk.	
0.63	**37**	0.76	**38**	0.00	**0**	0.006	**2**	unk.	0	**0**	tr(a)	tr(a)	173	**17**	247	236	unk.	7.6	**42**	1.3	**9**	unk.	**unk.**	1587	K
8.1	**41**	155	**39**	0.45	**5**	1614	**32**	unk.	unk.	**unk.**	tr(a)	1.81	206	**21**	10.8	6.0	unk.	57	**14**	0.28	**14**	tr(a)		unk.	
0.10	**6**	0.06	**3**	0.04	**1**	tr	**0**	unk.	unk.	**unk.**	tr(a)	tr(a)	7	**1**	3	78	unk.	1.0	**6**	0.5	**4**	1.0	**1**	838	L
1.0	**5**	12	**3**	0.14	**1**	10	**0**	unk.	unk.	**unk.**	tr(a)	0.35	16	**2**	0.1	2.0	unk.	22	**6**	0.05	**3**	tr(a)		233	
0.37	**22**	0.44	**22**	1.32	**22**	tr	**0**	unk.	unk.	**unk.**	0.9	tr(a)	59	**6**	3	71	unk.	3.9	**22**	0.6	**4**	1.0	**1**	710	M
4.4	**22**	5	**1**	0.19	**2**	1102	**22**	unk.	1.1	**4**	tr(a)	0.48	10	**1**	0.1	1.8	unk.	26	**7**	0.05	**3**	tr(a)		12	
0.45	**27**	0.53	**27**	1.59	**27**	tr	**0**	unk.	unk.	**unk.**	0.9	tr(a)	79	**8**	2	82	unk.	6.1	**34**	0.7	**5**	2.0	**1**	9	N
5.3	**27**	8	**2**	0.16	**2**	1323	**27**	unk.	1.1	**4**	tr(a)	0.53	14	**1**	0.1	2.1	unk.	33	**8**	0.06	**3**	tr(a)		9	
0.02	**1**	unk.	**unk.**	unk.	**unk.**	unk.	**unk.**	0	0	**0**	4.0	1.2	11	**1**	267	75	unk.	0.7	**4**	0.6	**4**	unk.	**unk.**	unk.	O
0.1	**1**	unk.	**unk.**	unk.	**unk.**	0	**0**	0	4.7	**16**	2.8	0.98	70	**7**	11.6	1.9	unk.	26	**6**	unk.	**unk.**	unk.		unk.	
0.04	**2**	0.04	**2**	0.03	**0**	tr	**0**	unk.	unk.	**unk.**	tr(a)	tr(a)	14	**1**	2	105	unk.	1.2	**7**	1.1	**7**	unk.	**unk.**	tr(a)	P
0.3	**2**	13	**3**	0.21	**2**	unk.	**unk.**	unk.	0.2	**1**	0.2	0.63	135	**14**	0.1	2.7	unk.	42	**11**	0.14	**7**	tr(a)		unk.	
0.07	**4**	0.05	**2**	0.00	**0**	unk.	**unk.**	0	unk.	**unk.**	unk.	unk.	8	**1**	1	unk.	unk.	0.9	**5**	0.3	**2**	unk.	**unk.**	unk.	Q
1.2	**6**	unk.	**unk.**	unk.	**unk.**	0	**0**	0	unk.	**unk.**	unk.	unk.	105	**11**	0.0	unk.	unk.	unk.	**unk.**	unk.	**unk.**	unk.		unk.	
0.14	**8**	0.09	**5**	unk.	**unk.**	tr(a)	**0**	unk.	unk.	**unk.**	tr	tr	10	**1**	4	107	unk.	1.1	**7**	1.0	**7**	unk.	**unk.**	tr(a)	R
1.5	**8**	16	**4**	0.21	**2**	unk.	**unk.**	unk.	0.3	**1**	0.3	0.54	111	**11**	0.2	2.7	unk.	46	**12**	0.14	**7**	tr(a)		unk.	
0.05	**3**	0.01	**1**	tr	**0**	tr	**0**	unk.	unk.	**unk.**	2.0	0.7	7	**1**	1	86	unk.	0.9	**5**	0.6	**4**	1.0	**1**	9	S
0.1	**1**	11	**3**	0.05	**1**	9	**0**	unk.	1.4	**5**	1.4	0.37	29	**3**	0.0	2.2	unk.	25	**6**	0.05	**3**	tr(a)		3	

	FOOD	Portion	Weight in grams Conversion for 100 g	Kilocalories H_2O g	Total carbohydrate g Total fats g	Crude fiber g Dietary fiber g	Total protein g Total sugar g	% USRDA	Arginine mg Histidine mg	Isoleucine mg Leucine mg	Lysine mg Methionine mg	Phenylalanine mg Threonine mg	Valine mg Tryptophan mg	Cystine mg Tyrosine mg	Polyunsat. fatty acids g Monounsat. fatty acids g	Saturated fatty acids g P/S ratio	Linoleic acid g Cholesterol mg	Thiamin mg Ascorbic acid mg	% USRDA
A	cereal,cooked,whole meal wheat cereal -Ralston	1/2 oz dry, prep (1/2 C)	123 *0.82*	55 107.4	11.5 0.4	0.37 unk.	2.2 0.0	**3**	unk. unk.	unk. unk.	unk. unk.	unk. unk.	unk. unk.	unk. unk.	unk. unk.	unk. unk.	unk. 0	0.07 0	**5** **0**
B	CEREAL,DRY,instant grits,ham flavored -Quaker	0.8 oz pkg	23 *4.41*	79 1.4	17.0 0.3	0.13 unk.	2.4 unk.	**4**	unk. unk.	unk. unk.	unk. unk.	unk. unk.	unk. unk.	unk. unk.	unk. unk.	unk. unk.	unk. unk.	0.20 0	**13** **0**
C	cereal,dry,barley pearled light	1 oz (2 tsp)	28 *3.52*	99 3.2	22.4 0.3	0.14 unk.	2.3 unk.	**4**	unk. unk.	97 161	79 32	122 79	117 28	unk. unk.	unk. 0.28	tr unk.	0.28 0	0.03 0	**2** **0**
D	cereal,dry,barley,regular or quick -Quaker	1 oz (2 tsp)	28 *3.52*	101 2.8	21.4 0.3	0.17 unk.	3.0 unk.	**5**	unk. unk.	unk. unk.	unk. unk.	unk. unk.	unk. unk.	unk. unk.	unk. unk.	unk. unk.	unk. unk.	0.04 0	**3** **0**
E	cereal,dry,bolted white cornmeal mix -Quaker	1 oz (1/4 C)	28 *3.52*	99 3.0	20.8 0.7	0.20 unk.	2.4 0.0	**4**	unk. unk.	unk. unk.	unk. unk.	unk. unk.	unk. unk.	unk. unk.	unk. unk.	unk. unk.	unk. unk.	0.22 0	**15** **0**
F	cereal,dry,bolted white cornmeal -Quaker	1 oz (1/4 C)	28 *3.52*	103 3.3	21.5 1.2	0.30 unk.	2.3 0.0	**4**	unk. unk.	unk. unk.	unk. unk.	unk. unk.	unk. unk.	unk. unk.	unk. unk.	unk. unk.	unk. unk.	0.25 0	**17** **0**
G	cereal,dry,corn grits,regular or quick -Quaker	1 oz	28 *3.52*	102 3.0	22.5 0.2	0.14 unk.	2.4 0.0	**4**	unk. unk.	unk. unk.	unk. unk.	unk. unk.	unk. unk.	unk. unk.	unk. unk.	unk. unk.	unk. unk.	0.12 0	**8** **0**
H	cereal,dry,Cream Of Wheat,mix 'n eat, flavored-Nabisco	1-1/4 oz pkg	36 *2.82*	132 2.8	23.2 0.3	0.14 unk.	2.5 7.2	**4**	unk. unk.	unk. unk.	unk. unk.	unk. unk.	unk. unk.	unk. unk.	unk. unk.	tr(a) unk.	unk. 0(a)	0.38 unk.	**25** **unk.**
I	cereal,dry,Cream Of Wheat,mix 'n eat, plain-Nabisco	1 oz pkg	28 *3.52*	103 3.0	21.0 0.3	0.14 unk.	2.8 1.0	**4**	unk. unk.	unk. unk.	unk. unk.	unk. unk.	unk. unk.	unk. unk.	unk. unk.	tr(a) unk.	unk. 0(a)	0.37 unk.	**25** **unk.**
J	cereal,dry,degerminated white cornmeal-Quaker	1 oz (3 tsp)	28 *3.52*	102 3.0	22.1 0.5	0.17 unk.	2.3 0.0	**4**	unk. unk.	unk. unk.	unk. unk.	unk. unk.	unk. unk.	unk. unk.	unk. unk.	unk. unk.	unk. unk.	0.12 0	**8** **0**
K	cereal,dry,degerminated yellow cornmeal mix-Quaker	1 oz (3 tsp)	28 *3.52*	102 3.0	22.1 0.5	0.17 unk.	2.3 0.0	**4**	unk. unk.	unk. unk.	unk. unk.	unk. unk.	unk. unk.	unk. unk.	unk. unk.	unk. unk.	unk. unk.	0.12 0	**8** **0**
L	cereal,dry,farina/hot'n creamy-Quaker	1 oz	28 *3.52*	101 3.1	21.8 0.2	0.06 unk.	3.1 unk.	**5**	unk. unk.	unk. unk.	unk. unk.	unk. unk.	unk. unk.	unk. unk.	unk. unk.	unk. unk.	unk. 0	0.20 0	**13** **0**
M	cereal,dry,hot natural,Pettijohns /whole wheat-Quaker	1 oz (1/3 C)	28 *3.52*	100 3.0	20.6 0.5	0.57 unk.	3.2 0.0	**5**	unk. unk.	unk. unk.	unk. unk.	unk. unk.	unk. unk.	unk. unk.	unk. unk.	unk. unk.	unk. 0	0.11 0	**7** **0**
N	cereal,dry,instant grits-Quaker	0.8 oz pkg	23 *4.41*	79 1.7	17.7 0.2	0.11 unk.	2.1 unk.	**3**	unk. unk.	unk. unk.	unk. unk.	unk. unk.	unk. unk.	unk. unk.	unk. unk.	unk. unk.	unk. unk.	0.17 0	**12** **0**
O	cereal,dry,instant grits,bacon flavored-Quaker	0.8 oz pkg	23 *4.41*	81 1.6	17.3 0.3	0.11 unk.	2.1 unk.	**3**	unk. unk.	unk. unk.	unk. unk.	unk. unk.	unk. unk.	unk. unk.	unk. unk.	unk. unk.	unk. unk.	0.10 0	**7** **0**
P	cereal,dry,oatmeal	1 C	80 *1.25*	312 6.6	54.6 5.9	0.96 4.50	**11.4** tr	**18**	unk. 209	586 852	417 167	606 376	676 146	247 419	2.44 2.08	1.10 2.23	2.30 0	0.48 0	**32** **0**
Q	cereal,dry,oatmeal w/apple & cinnamon,instant-Quaker	1-1/4 oz pkg	35 *2.86*	133 2.3	25.7 1.6	0.42 unk.	4.1 unk.	**6**	unk. unk.	unk. unk.	unk. unk.	unk. unk.	unk. unk.	unk. unk.	unk. unk.	unk. unk.	unk. 0	0.47 0	**32** **0**
R	cereal,dry,oatmeal w/cinnamon & spices,instant-Quaker	1-5/8 oz pkg	46 *2.17*	176 2.8	34.8 1.9	0.37 unk.	5.1 unk.	**8**	unk. unk.	unk. unk.	unk. unk.	unk. unk.	unk. unk.	unk. unk.	unk. unk.	unk. unk.	unk. 0	0.56 0	**37** **0**
S	cereal,dry,oatmeal w/cinnamon spice, instant-Ralston	1-5/8 oz pkg	46 *2.17*	175 3.4	32.2 2.2	0.46 unk.	6.4 12.6	**10**	unk. unk.	unk. unk.	unk. unk.	unk. unk.	unk. unk.	unk. unk.	unk. unk.	tr(a) unk.	unk. 0(a)	0.19 unk.	**13** **unk.**

Riboflavin mg / Niacin mg	% USRDA	Vitamin B_6 mg / Folacin mcg	% USRDA	Vitamin B_{12} mcg / Pantothenic acid mg	% USRDA	Biotin mg / Vitamin A IU	% USRDA	Preformed A RE / Beta carotene RE	Vitamin D IU / Vitamin E IU	% USRDA	Total tocopherol mg / Alpha tocopherol mg	Other tocopherol mg / Total ash g	Calcium mg / Phosphorus mg	% USRDA	Sodium mg / Sodium meq	Potassium mg / Potassium meq	Chlorine mg / Chlorine meq	Iron mg / Magnesium mg	% USRDA	Zinc mg / Copper mg	% USRDA	Iodine mcg / Selenium mcg	% USRDA	Manganese mcg / Chromium mcg	
0.02	1	0.65	32	unk.	unk.	0.020	7	0	unk.	unk.	2.1	0.7	9	1	260	59	unk.	0.6	3	0.6	4	unk.	unk.	unk.	A
0.7	4	60	15	1.68	17	0	0	0	2.6	9	1.4	0.98	64	6	11.3	1.5	unk.	unk.	unk.	unk.	unk.	unk.		unk.	
0.12	7	0.05	3	0.00	0	0.001	0	0	0	0	unk.	unk.	6	1	533	42	unk.	1.2	7	0.1	1	18.3	12	0	B
1.4	7	8	2	0.59	6	0	0	0	unk.	unk.	unk.	1.49	26	3	22.9	1.1	unk.	7	2	0.02	1	unk.		unk.	
0.01	1	0.06	3	0.00	0	unk.	unk.	0	unk.	unk.	0.1	0.1	5	1	1	45	30	0.6	3	unk.	unk.	unk.	unk.	478	C
0.9	4	unk.	unk.	0.00	0	0	0	0	0.2	1	tr	0.26	54	5	0.0	1.2	0.8	11	3	0.21	11	unk.		unk.	
0.04	2	0.07	4	0.00	0	0.002	1	unk.	0	0	0.2	0.1	6	1	3	76	unk.	0.6	3	0.3	2	unk.	unk.	284	D
1.3	6	19	5	0.00	0	6	0	unk.	0.1	0	0.1	0.28	66	7	0.1	1.9	unk.	20	5	0.12	6	unk.		unk.	
0.14	8	0.14	7	0.00	0	0.002	1	0	0	0	unk.	unk.	60	6	328	181	unk.	1.1	6	0.3	2	unk.	unk.	0	E
1.6	8	19	5	0.00	0	0	0	0	unk.	unk.	0.6	1.42	181	18	14.3	4.6	unk.	21	5	0.04	2	unk.		unk.	
0.12	7	0.24	12	0.00	0	0.002	1	0	0	0	unk.	unk.	100	10	335	69	unk.	1.7	5	0.4	3	unk.	unk.	284	F
1.5	8	7	2	0.19	2	0	0	0	unk.	unk.	0.2	1.50	82	8	14.6	1.8	unk.	31	8	0.04	2	unk.		unk.	
0.07	4	0.03	2	0.00	0	0.001	0	0	0	0	unk.	unk.	1	0	tr	30	unk.	0.8	5	0.0	0	unk.	unk.	284	G
1.0	5	3	1	0.00	0	0	0	0	unk.	unk.	unk.	0.11	20	2	0.0	0.8	unk.	6	1	0.01	1	unk.		unk.	
0.26	15	unk.	unk.	unk.	unk.	tr(a)	0	unk.	unk.	unk.	tr(a)	tr(a)	20	2	284	55	tr(a)	8.1	45	0.2	2	unk.	unk.	tr(a)	H
5.0	25	99	25	unk.	unk.	1250	25	unk.	unk.	unk.	tr(a)	unk.	20	2	12.3	1.4	0.0	9	2	0.06	3	tr(a)		unk.	
0.26	15	unk.	unk.	unk.	unk.	tr(a)	0	unk.	unk.	unk.	tr(a)	tr(a)	20	2	264	38	unk.	8.1	45	0.2	2	unk.	unk.	tr(a)	I
5.0	25	100	25	unk.	unk.	1250	25	unk.	unk.	unk.	tr(a)	0.79	20	2	11.5	1.0	unk.	8	2	0.04	2	tr(a)		unk.	
0.07	4	0.10	5	0.00	0	0.001	0	0	0	0	unk.	unk.	1	0	1	53	unk.	0.8	5	0.3	2	unk.	unk.	0	J
1.0	5	15	4	0.00	0	0	0	0	unk.	unk.	0.4	0.17	37	4	0.0	1.4	unk.	12	3	0.02	1	unk.		unk.	
0.07	4	0.10	5	0.00	0	0.001	0	unk.	0	0	unk.	unk.	1	0	1	53	unk.	0.8	5	0.3	2	unk.	unk.	0	K
1.0	5	15	4	0.00	0	unk.	unk.	unk.	unk.	unk.	0.4	0.17	37	4	0.0	1.4	unk.	12	3	0.02	1	unk.		unk.	
0.13	8	0.02	1	0.00	0	0.001	0	0	0	0	unk.	unk.	6	1	0	27	unk.	1.4	8	0.2	1	12.8	9	0	L
1.5	8	19	5	0.12	1	0	0	0	unk.	unk.	0.0	0.12	26	3	0.0	0.7	unk.	6	1	0.06	3	unk.		unk.	
0.07	4	0.05	3	0.00	0	0.002	1	0	0	0	unk.	unk.	9	1	0	108	unk.	0.9	5	0.3	2	unk.	unk.	852	M
1.3	6	22	6	0.28	3	0	0	0	unk.	unk.	0.2	0.48	105	11	0.0	2.7	unk.	16	4	0.08	4	unk.		unk.	
0.9	5	0.05	2	0.00	0	0.001	0	0	0	0	unk.	unk.	7	1	349	29	unk.	1.0	6	0.1	1	1.1	1	0	N
1.3	7	2	1	0.43	4	0	0	0	unk.	unk.	unk.	1.03	17	2	15.0	0.8	unk.	5	1	0.01	1	unk.		unk.	
0.06	4	0.04	2	0.00	0	0.001	1	0	0	0	unk.	unk.	5	1	430	48	unk.	1.0	6	0.1	1	39.0	26	0	O
0.8	4	6	2	0.05	1	0	0	0	unk.	unk.	unk.	1.20	22	2	18.5	1.2	unk.	7	2	0.03	1	unk.		unk.	
0.11	7	0.11	6	0.00	0	unk.	unk.	0	0	0	unk.	unk.	42	4	2	282	unk.	3.6	20	2.7	18	unk.	unk.	2960	P
0.8	4	unk.	unk.	1.20	12	0	0	0	unk.	unk.	1.8	1.52	324	32	0.1	7.2	unk.	115	29	0.02	1	unk.		5	
0.28	17	0.69	35	0.00	0	0.004	1	unk.	0	0	unk.	unk.	156	16	220	106	unk.	6.0	33	0.7	5	unk.	unk.	1050	Q
5.1	25	135	34	0.35	4	1433	29	unk.	unk.	unk.	0.5	1.29	116	12	9.6	2.7	unk.	31	8	0.12	6	unk.		unk.	
0.34	20	0.77	38	0.00	0	0.007	2	unk.	0	0	unk.	unk.	172	17	280	104	unk.	6.6	37	1.0	6	unk.	unk.	1472	R
5.6	28	152	38	0.51	5	1578	32	unk.	unk.	unk.	unk.	1.48	146	15	12.2	2.7	unk.	51	13	0.12	6	unk.		unk.	
0.08	5	0.07	4	0.16	3	0.000	0	unk.	unk.	unk.	tr(a)	tr(a)	23	2	unk.	unk.	unk.	1.4	8	1.1	8	unk.	unk.	tr(a)	S
0.5	3	28	7	0.31	3	unk.	unk.	unk.	unk.	unk.	tr(a)	1.20	148	15	unk.	unk.	unk.	44	11	0.18	9	tr(a)		unk.	

	FOOD	Portion	Weight in grams / Conversion for 100 g	Kilocalories / H_2O g	Total carbohydrate g / Total fats g	Crude fiber g / Dietary fiber g	Total protein g / Total sugar g	% USRDA	Arginine mg / Histidine mg	Isoleucine mg / Leucine mg	Lysine mg / Methionine mg	Phenylalanine mg / Threonine mg	Valine mg / Tryptophan mg	Cystine mg / Tyrosine mg	Polyunsat. fatty acids g / Monounsat. fatty acids g	Saturated fatty acids g / P/S ratio	Linoleic acid g / Cholesterol mg	Thiamin mg / Ascorbic acid mg	% USRDA
A	cereal,dry,oatmeal w/maple & brown	1-1/2 oz pkg	42	161	31.5	0.31	4.9	**8**	unk.	unk.	unk.	unk.	unk.	unk.	unk.	unk.	unk.	0.52	**35**
	sugar,instant-Quaker		*2.38*	2.6	1.8	unk.	unk.		unk.	unk.	unk.	unk.	unk.	unk.	unk.	unk.	0	0	**0**
B	cereal,dry,oatmeal w/maple & brown	1-5/8 oz pkg	46	175	19.9	0.28	6.2	**10**	unk.	unk.	unk.	unk.	unk.	unk.	unk.	tr(a)	unk.	0.23	**15**
	sugar,instant-Ralston		*2.17*	3.4	1.4	unk.	7.8		unk.	unk.	unk.	unk.	unk.	unk.	unk.	unk.	0(a)	1	**2**
C	cereal,dry,oatmeal w/raisins &	1-1/2 oz pkg	42	157	31.0	0.36	4.6	**7**	unk.	unk.	unk.	unk.	unk.	unk.	unk.	unk.	unk.	0.50	**33**
	spices,instant-Quaker		*2.38*	3.1	1.8	unk.	unk.		unk.	unk.	unk.	unk.	unk.	unk.	unk.	unk.	unk.	0	**0**
D	cereal,dry,oatmeal,instant-Quaker	1/2 oz pkg	28	105	18.1	0.30	4.7	**7**	unk.	unk.	unk.	unk.	unk.	unk.	unk.	unk.	unk.	0.53	**35**
			3.52	2.6	1.7	unk.	tr		unk.	unk.	unk.	unk.	unk.	unk.	unk.	unk.	0	0	**0**
E	cereal,dry,oatmeal,instant-Ralston	1 oz pkg	28	110	20.6	0.48	4.5	**7**	unk.	unk.	unk.	unk.	unk.	unk.	unk.	tr(a)	unk.	0.22	**14**
		(1/4 C)	*3.52*	2.1	1.1	unk.	0.0		unk.	unk.	unk.	unk.	unk.	unk.	unk.	unk.	0(a)	1	**2**
F	cereal,dry,oatmeal,quick or old	1 C	80	307	52.0	0.90	13.6	**21**	unk.	unk.	unk.	unk.	unk.	unk.	unk.	unk.	unk.	0.59	**40**
	fashioned-Quaker		*1.25*	7.1	5.0	unk.	tr		unk.	unk.	unk.	unk.	unk.	unk.	unk.	unk.	0	0	**0**
G	cereal,dry,oats,quick & regular	1 C	114	442	72.2	1.61	19.7	**30**	unk.	unk.	unk.	unk.	unk.	unk.	unk.	tr(a)	unk.	0.84	**56**
	-Ralston		*0.88*	11.4	6.8	unk.	0.0		unk.	unk.	unk.	unk.	unk.	unk.	unk.	unk.	0(a)	5	**8**
H	cereal,dry,Ralston,instant &	1 oz (1/4 C)	28	110	20.7	0.49	4.6	**7**	unk.	unk.	unk.	unk.	unk.	unk.	unk.	tr(a)	unk.	0.17	**11**
	regular-Ralston		*3.52*	3.0	1.1	unk.	0.0		unk.	unk.	unk.	unk.	unk.	unk.	unk.	unk.	0(a)	unk.	**unk.**
I	CEREAL,RTE,All Bran-Kellogg	1 oz (1/3 C)	28	68	21.6	2.27	4.0	**6**	unk.	unk.	unk.	unk.	unk.	unk.	unk.	tr(a)	unk.	0.37	**25**
			3.52	unk.	0.6	8.90	14.0		unk.	unk.	unk.	unk.	unk.	unk.	unk.	unk.	0(a)	15	**25**
J	cereal,RTE,Alpha-Bits,sugar frosted	1 oz (1 C)	28	111	24.5	0.15	2.3	**4**	143	97	79	117	124	53	unk.	tr(a)	unk.	0.37	**25**
	oat cereal-Post		*3.52*	0.4	0.6	3.08	11.0		51	201	39	76	26	81	unk.	unk.	0	0(a)	**0**
K	cereal,RTE,Apple Jacks-Kellogg	1 oz (1 C)	28	111	26.0	tr	1.4	**2**	unk.	unk.	unk.	unk.	unk.	unk.	unk.	tr(a)	unk.	0.37	**25**
			3.52	unk.	0.0	0.31	15.0		unk.	unk.	unk.	unk.	unk.	unk.	unk.	unk.	0(a)	15	**25**
L	cereal,RTE,Boo Berry-General Mills	1 oz (1 C)	28	110	24.0	unk.	1.0	**2**	unk.	unk.	unk.	unk.	unk.	unk.	unk.	0(a)	unk.	0.38	**25**
			3.52	unk.	1.0	unk.	unk.		unk.	unk.	unk.	unk.	unk.	unk.	unk.	unk.	0(a)	15	**25**
M	cereal,RTE,Bran Buds-Kellogg	1 oz (1/3 C)	28	74	22.0	2.00	3.0	**5**	unk.	unk.	unk.	unk.	unk.	unk.	unk.	tr(a)	unk.	0.37	**25**
			3.52	unk.	1.0	8.00	8.0		unk.	unk.	unk.	unk.	unk.	unk.	unk.	unk.	0(a)	15	**25**
N	cereal,RTE,bran Chex-Ralston	1 oz (2/3 C)	28	90	23.0	1.30	2.9	**5**	unk.	unk.	unk.	unk.	unk.	unk.	unk.	unk.	unk.	0.37	**25**
			3.52	0.7	0.8	4.60	5.0		unk.	unk.	unk.	unk.	unk.	unk.	unk.	unk.	0(a)	15	**25**
O	cereal,RTE,bran flakes-Post	1 oz (2/3 C)	28	90	22.0	1.10	3.0	**5**	177	112	99	137	146	60	unk.	unk.	unk.	0.37	**25**
			3.52	0.9	0.5	3.80	4.9		76	199	47	96	52	89	unk.	unk.	0	0(a)	**0**
P	cereal,RTE,bran,unprocessed-Quaker	1 oz (1/4 C)	28	85	14.8	2.89	4.3	**7**	unk.	unk.	unk.	unk.	unk.	unk.	unk.	tr(a)	unk.	0.17	**12**
			3.52	2.6	0.8	unk.	0(a)		unk.	unk.	unk.	unk.	unk.	unk.	unk.	unk.	0(a)	0	**0**
Q	cereal,RTE,bran-Nabisco	1 oz (1/2 C)	28	75	21.0	2.13	3.2	**5**	unk.	unk.	unk.	unk.	unk.	unk.	unk.	tr(a)	unk.	0.68	**45**
			3.52	0.8	1.4	7.80	6.0		unk.	unk.	unk.	unk.	unk.	unk.	unk.	unk.	0(a)	27	**45**
R	cereal,RTE,Buc Wheats-General Mills	1 oz (3/4 C)	28	110	23.0	unk.	3.0	**5**	unk.	unk.	unk.	unk.	unk.	unk.	unk.	0(a)	unk.	0.68	**45**
			3.52	unk.	1.0	1.30	5.0		unk.	unk.	unk.	unk.	unk.	unk.	unk.	unk.	0(a)	27	**45**
S	cereal,RTE,C W Post family style	1 oz (1/4 C)	28	123	20.7	0.21	2.5	**4**	144	93	80	104	117	46	unk.	unk.	unk.	0.37	**25**
	cereal w/raisins-Post		*3.52*	1.0	5.0	0.05	8.0		48	154	37	74	28	76	unk.	unk.	0(a)	0(a)	**0**

Riboflavin mg / Niacin mg	% USRDA	Vitamin B_6 mg / Folacin mcg	% USRDA	Vitamin B_{12} mcg / Pantothenic acid mg	% USRDA	Biotin mg / Vitamin A IU	% USRDA	Preformed A RE / Beta carotene RE	Vitamin D IU / Vitamin E IU	% USRDA	Total tocopherol mg / Alpha tocopherol mg	Other tocopherol mg / Total ash g	Calcium mg / Phosphorus mg	% USRDA	Sodium mg / Sodium meq	Potassium mg / Potassium meq	Chlorine mg / Chlorine meq	Iron mg / Magnesium mg	% USRDA	Zinc mg / Copper mg	% USRDA	Iodine mcg / Selenium mcg	% USRDA	Manganese mcg / Chromium mcg	
0.31	**19**	0.73	**36**	0.00	**0**	0.007	**2**	unk.	0	**0**	unk.	unk.	160	**16**	278	101	unk.	6.3	**35**	0.8	**6**	unk.	**unk.**	1260	A
5.3	**26**	144	**36**	0.00	**0**	1485	**30**	unk.	unk.	**unk.**	0.3	1.44	141	**14**	12.0	2.6	unk.	84	**21**	0.20	**10**	unk.		unk.	
0.08	**5**	0.05	**3**	0.07	**1**	0.000	**0**	unk.	unk.	**unk.**	tr(a)	tr(a)	20	**2**	unk.	unk.	unk.	1.6	**9**	1.1	**7**	unk.	**unk.**	tr(a)	B
0.5	**2**	14	**3**	0.28	**3**	unk.	**unk.**	unk.	unk.	**unk.**	tr(a)	unk.	147	**15**	unk.	unk.	unk.	47	**12**	0.15	**8**	tr(a)		unk.	
0.35	**21**	0.74	**37**	0.00	**0**	0.006	**2**	unk.	0	**0**	unk.	unk.	163	**16**	223	148	unk.	6.5	**36**	0.8	**6**	unk.	**unk.**	1260	C
5.4	**27**	148	**37**	0.00	**0**	1449	**29**	unk.	unk.	**unk.**	0.3	1.40	131	**13**	9.7	3.8	unk.	56	**14**	0.16	**8**	unk.		unk.	
0.29	**17**	0.74	**37**	0.00	**0**	0.007	**2**	unk.	0	**0**	unk.	unk.	163	**16**	283	100	unk.	6.3	**35**	0.8	**6**	unk.	**unk.**	1132	D
5.5	**27**	149	**37**	0.28	**3**	1511	**30**	unk.	unk.	**unk.**	0.5	1.43	133	**13**	12.4	2.6	unk.	35	**9**	0.13	**6**	unk.		unk.	
0.04	**2**	0.04	**2**	tr	**0**	0.000	**0**	unk.	unk.	**unk.**	tr(a)	tr(a)	18	**2**	unk.	95	unk.	1.2	**7**	1.0	**6**	unk.	**unk.**	tr(a)	E
0.4	**2**	17	**4**	0.19	**2**	unk.	**unk.**	unk.	0.6	**2**	0.5	0.85	126	**13**	unk.	2.4	unk.	39	**10**	0.13	**7**	tr(a)		unk.	
0.09	**5**	0.17	**8**	0.00	**0**	0.000	**0**	0	0	**0**	unk.	unk.	40	**4**	3	288	unk.	3.4	**19**	2.5	**17**	47.6	**32**	3200	F
0.6	**3**	83	**21**	0.71	**7**	0	**0**	0	unk.	**unk.**	1.8	1.53	381	**38**	0.1	6.6	unk.	113	**28**	0.27	**14**	unk.		unk.	
0.16	**9**	0.17	**9**	0.08	**1**	tr	**0**	unk.	unk.	**unk.**	tr(a)	tr(a)	60	**6**	8	417	unk.	5.1	**28**	3.9	**26**	unk.	**unk.**	tr(a)	G
1.3	**7**	44	**11**	0.80	**8**	unk.	**unk.**	unk.	17.5	**58**	15.9	2.50	514	**51**	0.3	10.7	unk.	165	**41**	0.61	**30**	tr(a)		unk.	
0.14	**8**	0.10	**5**	0.08	**1**	tr	**0**	unk.	unk.	**unk.**	tr	tr	10	**1**	3	111	unk.	1.2	**7**	1.1	**8**	unk.	**unk.**	tr(a)	H
unk.	**unk.**	19	**5**	0.21	**2**	unk.	**unk.**	unk.	0.6	**2**	0.5	0.54	107	**11**	0.1	2.8	unk.	43	**11**	0.14	**7**	tr(a)		unk.	
0.43	**25**	0.50	**25**	unk.	**unk.**	tr(a)	**0**	unk.	50	**13**	tr(a)	tr(a)	22	**2**	320	349	unk.	4.5	**25**	3.8	**25**	unk.	**unk.**	tr(a)	I
5.0	**25**	100	**25**	tr(a)	**0**	1250	**25**	unk.	unk.	**unk.**	tr(a)	unk.	271	**27**	13.9	8.9	unk.	120	**30**	0.33	**17**	tr(a)		unk.	
0.43	**25**	0.50	**25**	1.50	**25**	tr(a)	**0**	unk.	50	**13**	tr(a)	tr(a)	8	**1**	195	110	unk.	1.8	**10**	1.5	**10**	unk.	**unk.**	tr(a)	J
5.0	**25**	100	**25**	tr(a)	**0**	1250	**25**	unk.	0.0	**0**	tr(a)	0.61	52	**5**	8.5	2.8	unk.	17	**4**	0.08	**4**	tr(a)		unk.	
0.43	**25**	0.50	**25**	unk.	**unk.**	tr(a)	**0**	unk.	50	**13**	tr(a)	tr(a)	3	**0**	115	28	unk.	4.5	**25**	3.8	**25**	unk.	**unk.**	tr(a)	K
5.0	**25**	100	**25**	tr(a)	**0**	1250	**25**	unk.	unk.	**unk.**	tr(a)	unk.	22	**2**	5.0	0.7	unk.	7	**2**	0.03	**2**	tr(a)		unk.	
0.43	**25**	0.50	**25**	1.50	**25**	tr(a)	**0**	unk.	40	**10**	tr(a)	unk.	20	**2**	215	50	unk.	4.5	**25**	0.3	**2**	unk.	**unk.**	tr(a)	L
5.0	**25**	unk.	**unk.**	tr(a)	**0**	1250	**25**	unk.	0.0	**0**	tr(a)	unk.	60	**6**	9.3	1.3	unk.	16	**4**	0.04	**2**	tr(a)		unk.	
0.43	**25**	0.50	**25**	unk.	**unk.**	tr(a)	**0**	unk.	50	**13**	tr(a)	tr(a)	19	**2**	185	303	unk.	4.5	**25**	3.8	**25**	unk.	**unk.**	tr(a)	M
5.0	**25**	100	**25**	tr(a)	**0**	1250	**25**	unk.	unk.	**unk.**	tr(a)	unk.	241	**24**	8.0	7.8	unk.	104	**25**	0.29	**15**	tr(a)		unk.	
0.15	**9**	0.50	**25**	1.50	**25**	0.002	**1**	unk.	unk.	**unk.**	unk.	unk.	17	**2**	263	228	unk.	4.5	**25**	1.2	**8**	unk.	**unk.**	tr(a)	N
5.0	**25**	100	**25**	0.29	**3**	62	**1**	unk.	unk.	**unk.**	unk.	1.48	189	**19**	11.4	5.8	unk.	73	**18**	0.22	**11**	tr(a)		unk.	
0.43	**25**	0.50	**25**	1.50	**25**	tr(a)	**0**	unk.	50	**13**	unk.	unk.	12	**1**	281	150	unk.	4.5	**25**	1.5	**10**	unk.	**unk.**	tr(a)	O
5.0	**25**	100	**25**	tr(a)	**0**	1250	**25**	unk.	unk.	**unk.**	unk.	1.30	179	**18**	12.2	3.8	unk.	61	**15**	0.22	**11**	tr(a)		unk.	
0.13	**8**	unk.	**unk.**	unk.	**unk.**	tr(a)	**0**	0	0	**0**	tr(a)	tr(a)	unk.	**unk.**	1	482	unk.	3.9	**22**	6.7	**45**	unk.	**unk.**	tr(a)	P
7.5	**37**	35	**9**	tr(a)	**0**	0	**0**	0	0.6	**2**	0.5	1.94	486	**49**	0.0	12.3	unk.	183	**46**	0.36	**18**	tr(a)		unk.	
0.76	**45**	0.90	**45**	2.70	**45**	tr(a)	**0**	unk.	unk.	**unk.**	tr(a)	tr(a)	20	**2**	221	375	unk.	3.5	**34**	2.5	**17**	unk.	**unk.**	tr(a)	Q
9.0	**45**	unk.	**unk.**	tr(a)	**0**	unk.	**unk.**	unk.	unk.	**unk.**	tr(a)	1.87	345	**35**	9.6	9.6	unk.	134	**19**	0.46	**23**	tr(a)		unk.	
0.77	**45**	0.90	**45**	2.70	**45**	0.000	**0**	unk.	unk.	**unk.**	4.5	unk.	20	**2**	255	115	unk.	8.1	**45**	0.6	**4**	unk.	**unk.**	tr(a)	R
9.0	**45**	0	**0**	2.50	**25**	1250	**25**	unk.	4.5	**15**	tr(a)	unk.	100	**10**	11.1	2.9	unk.	32	**8**	0.12	**6**	tr(a)		unk.	
0.43	**25**	0.50	**25**	1.50	**25**	tr(a)	**0**	unk.	50	**13**	0.6	unk.	13	**1**	51	71	unk.	4.5	**25**	0.4	**3**	unk.	**unk.**	unk.	S
5.0	**25**	100	**25**	tr(a)	**0**	1250	**25**	unk.	0.6	**2**	unk.	0.39	61	**6**	2.2	1.8	unk.	24	**6**	0.12	**6**	unk.		unk.	

	FOOD	Portion	Weight in grams / Conversion for 100 g	Kilocalories / H_2O g	Total carbohydrate g / Total fats g	Crude fiber g / Dietary fiber g	Total protein g / Total sugar g	% USRDA	Arginine mg / Histidine mg	Isoleucine mg / Leucine mg	Lysine mg / Methionine mg	Phenylalanine mg / Threonine mg	Valine mg / Tryptophan mg	Cystine mg / Tyrosine mg	Polyunsat. fatty acids g / Monounsat. fatty acids g	Saturated fatty acids g / P/S ratio	Linoleic acid g / Cholesterol mg	Thiamin mg / Ascorbic acid mg	% USRDA
A	cereal,RTE,C W Post family style	1 oz (1/4 C)	28	127	20.0	0.19	2.5	**4**	156	100	86	112	126	50	unk.	unk.	unk.	0.37	**25**
	cereal-Post		*3.52*	0.6	4.3	0.67	7.0		51	166	40	80	31	82	unk.	unk.	0(a)	0(a)	**0**
B	cereal,RTE,Cap'n Crunch-Quaker	1 oz (3/4 C)	28	120	22.9	0.17	1.5	**2**	unk.	unk.	unk.	unk.	unk.	unk.	unk.	unk.	unk.	0.51	**34**
			3.52	0.7	2.6	0.30	11.4		unk.	unk.	unk.	unk.	unk.	unk.	unk.	unk.	unk.	0	**0**
C	cereal,RTE,Cap'n Crunch's	1 oz (3/4 C)	28	120	22.9	0.17	1.5	**2**	unk.	unk.	unk.	unk.	unk.	unk.	unk.	unk.	unk.	0.48	**32**
	Crunchberries-Quaker		*3.52*	0.7	2.4	0.31	12.3		unk.	unk.	unk.	unk.	unk.	unk.	unk.	unk.	unk.	0	**0**
D	cereal,RTE,Cap'n Crunch's peanut	1 oz (3/4 C)	28	127	20.9	0.18	2.2	**3**	unk.	unk.	unk.	unk.	unk.	unk.	unk.	unk.	unk.	0.49	**33**
	butter-Quaker		*3.52*	0.5	3.7	0.30	9.1		unk.	unk.	unk.	unk.	unk.	unk.	unk.	unk.	unk.	0	**0**
E	cereal,RTE,Cheerios-General Mills	1 oz (1-1/4 C)	28	110	20.0	unk.	4.0	**6**	unk.	unk.	unk.	unk.	unk.	unk.	unk.	tr(a)	unk.	0.38	**25**
			3.52	unk.	2.0	1.03	1.0		unk.	unk.	unk.	unk.	unk.	unk.	unk.	unk.	0(a)	15	**25**
F	cereal,RTE,Cocoa Krispies-Kellogg	1 oz (3/4 C)	28	110	25.0	0.10	1.0	**2**	unk.	unk.	unk.	unk.	unk.	unk.	unk.	tr(a)	unk.	0.38	**25**
			3.52	unk.	0.0	0.08	12.0		unk.	unk.	unk.	unk.	unk.	unk.	unk.	unk.	0(a)	15	**25**
G	cereal,RTE,cocoa Pebbles-Post	1 oz (7/8 C)	28	110	24.0	0.20	1.4	**2**	296	107	110	108	169	22	unk.	tr(a)	unk.	0.37	**25**
			3.52	0.6	1.0	0.09	12.0		68	181	43	92	30	109	unk.	unk.	0(a)	0(a)	**0**
H	cereal,RTE,Cocoa Puffs-General Mills	1 oz (1 C)	28	110	25.0	unk.	1.0	**2**	unk.	unk.	unk.	unk.	unk.	unk.	unk.	tr(a)	unk.	0.37	**25**
			3.52	unk.	1.0	0.11	11.0		unk.	unk.	unk.	unk.	unk.	unk.	unk.	unk.	0(a)	15	**25**
I	cereal,RTE,Concentrate-Kellogg	1 oz (1/3 C)	28	110	15.0	0.10	12.0	**19**	unk.	909	852	1676	1022	unk.	unk.	tr(a)	unk.	0.75	**50**
			3.52	0.6	0.0	0.36	2.7		426	1392	568	710	199	unk.	unk.	unk.	0(a)	30	**50**
J	cereal,RTE,Cookie Crisp-Ralston	1 oz (1 C)	28	110	25.0	0.10	1.5	**2**	unk.	unk.	unk.	unk.	unk.	unk.	unk.	tr(a)	unk.	0.37	**25**
			3.52	0.6	1.1	0.11	11.0		unk.	unk.	unk.	unk.	unk.	unk.	unk.	unk.	0(a)	15	**25**
K	cereal,RTE,corn Chex-Ralston	1 oz (1 C)	28	110	25.0	0.10	2.0	**3**	unk.	unk.	unk.	unk.	unk.	unk.	unk.	tr(a)	unk.	0.38	**25**
			3.52	0.7	0.1	0.28	1.1		unk.	unk.	unk.	unk.	unk.	unk.	unk.	unk.	0(a)	15	**25**
L	cereal,RTE,corn flakes-Ralston	1 oz (1 C)	28	110	25.0	0.10	2.2	**3**	unk.	unk.	unk.	unk.	unk.	unk.	unk.	tr(a)	unk.	0.15	**10**
			3.52	0.7	0.1	unk.	2.1		unk.	unk.	unk.	unk.	unk.	unk.	unk.	unk.	0(a)	unk.	**unk.**
M	cereal,RTE,corn flakes-store brand	1 oz (1 C)	28	110	25.0	0.10	2.2	**3**	unk.	unk.	unk.	unk.	unk.	unk.	unk.	tr(a)	unk.	0.15	**10**
			3.52	0.7	0.1	0.31	2.1		unk.	unk.	unk.	unk.	unk.	unk.	unk.	unk.	0(a)	0	**0**
N	cereal,RTE,corn Total-General Mills	1 oz (1 C)	28	110	24.0	unk.	2.0	**3**	unk.	unk.	unk.	unk.	unk.	unk.	unk.	tr(a)	unk.	1.50	**100**
			3.52	unk.	1.0	unk.	3.0		unk.	unk.	unk.	unk.	unk.	unk.	unk.	unk.	0(a)	60	**100**
O	cereal,RTE,Corny-Snaps-Kellogg	1 oz (1 C)	28	120	24.0	0.10	2.0	**3**	unk.	unk.	unk.	unk.	unk.	unk.	unk.	tr(a)	unk.	0.37	**25**
			3.52	unk.	2.0	0.25	13.0		unk.	unk.	unk.	unk.	unk.	unk.	unk.	unk.	0(a)	15	**25**
P	cereal,RTE,Count Chocula-General Mills	1 oz (1 C)	28	110	24.0	unk.	2.0	**3**	unk.	unk.	unk.	unk.	unk.	unk.	unk.	tr(a)	unk.	0.37	**25**
			3.52	unk.	1.0	0.48	13.0		unk.	unk.	unk.	unk.	unk.	unk.	unk.	unk.	0(a)	15	**25**
Q	cereal,RTE,Country corn flakes	1 oz (1 C)	28	110	25.0	unk.	2.0	**3**	unk.	unk.	unk.	unk.	unk.	unk.	unk.	tr(a)	unk.	0.37	**25**
	-General Mills		*3.52*	unk.	1.0	unk.	3.0		unk.	unk.	unk.	unk.	unk.	unk.	unk.	unk.	0(a)	15	**25**
R	cereal,RTE,Country Morning	1 oz (1/3 C)	28	130	19.0	0.30	3.0	**5**	unk.	unk.	unk.	unk.	unk.	unk.	unk.	unk.	unk.	0.09	**6**
	w/raisins & dates-Kellogg		*3.52*	unk.	5.0	0.91	6.0		unk.	unk.	unk.	unk.	unk.	unk.	unk.	unk.	0(a)	tr	**0**
S	cereal,RTE,Country Morning-Kellogg	1 oz (1/3 C)	28	130	18.0	0.30	3.0	**5**	unk.	unk.	unk.	unk.	unk.	unk.	unk.	unk.	unk.	0.09	**6**
			3.52	unk.	5.0	0.91	7.0		unk.	unk.	unk.	unk.	unk.	unk.	unk.	unk.	0(a)	unk.	**unk.**

Riboflavin mg / Niacin mg	% USRDA	Vitamin B_6 mg / Folacin mcg	% USRDA	Vitamin B_{12} mcg / Pantothenic acid mg	% USRDA	Biotin mg / Vitamin A IU	% USRDA	Preformed A RE / Beta carotene RE	Vitamin D IU / Vitamin E IU	% USRDA	Total tocopherol mg / Alpha tocopherol mg	Other tocopherol mg / Total ash g	Calcium mg / Phosphorus mg	% USRDA	Sodium mg / Sodium meq	Potassium mg / Potassium meq	Chlorine mg / Chlorine meq	Iron mg / Magnesium mg	% USRDA	Zinc mg / Copper mg	% USRDA	Iodine mcg / Selenium mcg	% USRDA	Manganese mcg / Chromium mcg	
0.43	**25**	0.50	**25**	1.50	**25**	tr(a)	**0**	unk.	50	**13**	0.6	unk.	15	**2**	55	58	unk.	4.5	**25**	0.5	**3**	unk.	**unk.**	tr(a)	A
5.0	**25**	100	**25**	tr(a)	**0**	1250	**25**	unk.	2.5	**8**	unk.	0.38	69	**7**	2.4	1.5	unk.	24	**6**	0.12	**6**	tr(a)		unk.	
0.55	**32**	0.77	**39**	1.79	**30**	0.001	**1**	unk.	0	**0**	unk.	unk.	5	**1**	213	37	unk.	7.5	**42**	3.1	**21**	unk.	**unk.**	284	B
6.6	**33**	183	**46**	0.56	**6**	0	**0**	unk.	0.2	**1**	0.2	0.65	36	**4**	9.3	0.9	unk.	12	**3**	0.02	**1**	unk.		unk.	
0.55	**32**	0.50	**25**	2.03	**34**	0.001	**0**	unk.	0	**0**	unk.	unk.	9	**1**	198	40	unk.	7.3	**41**	2.9	**19**	unk.	**unk.**	284	C
6.6	**33**	90	**22**	2.43	**24**	0	**0**	unk.	0.2	**1**	0.2	0.64	38	**4**	8.6	1.0	unk.	11	**3**	0.01	**0**	unk.		unk.	
0.57	**33**	0.50	**25**	1.87	**31**	0.002	**1**	unk.	0	**0**	unk.	unk.	6	**1**	217	46	unk.	7.4	**41**	3.1	**21**	unk.	**unk.**	284	D
6.6	**33**	198	**49**	3.87	**39**	0	**0**	unk.	0.2	**1**	0.2	0.69	40	**4**	9.4	1.2	unk.	15	**4**	0.05	**2**	unk.		3	
0.42	**25**	0.50	**25**	1.50	**25**	tr(a)	**0**	unk.	40	**10**	tr(a)	tr(a)	40	**4**	328	112	unk.	4.5	**25**	0.9	**6**	unk.	**unk.**	tr(a)	E
5.0	**25**	unk.	**unk.**	tr(a)	**0**	1250	**25**	unk.	unk.	**unk.**	tr(a)	unk.	100	**10**	14.3	2.9	unk.	40	**10**	0.12	**8**	tr(a)		unk.	
0.43	**25**	0.50	**25**	unk.	**unk.**	tr(a)	**0**	unk.	50	**13**	tr(a)	tr(a)	6	**1**	200	47	unk.	1.8	**10**	1.5	**10**	unk.	**unk.**	tr(a)	F
5.0	**25**	100	**25**	tr(a)	**0**	1250	**25**	unk.	unk.	**unk.**	tr(a)	unk.	30	**3**	8.7	1.2	unk.	11	**2**	0.08	**4**	tr(a)		unk.	
0.43	**25**	0.50	**25**	1.50	**25**	tr(a)	**0**	unk.	50	**13**	tr(a)	tr(a)	5	**1**	165	47	unk.	0.7	**4**	1.5	**10**	unk.	**unk.**	tr(a)	G
5.0	**25**	100	**25**	tr(a)	**0**	1250	**25**	unk.	unk.	**unk.**	tr(a)	0.56	22	**2**	7.2	1.2	unk.	20	**5**	0.08	**4**	tr(a)		unk.	
0.43	**25**	0.50	**25**	1.50	**25**	tr(a)	**0**	unk.	0	**0**	tr(a)	tr(a)	tr	**0**	175	45	unk.	4.5	**25**	tr	**0**	unk.	**unk.**	tr(a)	H
5.0	**25**	unk.	**unk.**	tr(a)	**0**	tr(a)	**0**	unk.	unk.	**unk.**	tr(a)	unk.	20	**2**	7.6	1.1	unk.	8	**2**	0.04	**2**	tr(a)		unk.	
0.85	**50**	1.00	**50**	3.00	**50**	tr(a)	**0**	unk.	200	**50**	15.0	tr(a)	120	**12**	127	36	unk.	9.0	**50**	0.6	**4**	unk.	**unk.**	tr(a)	I
10.0	**50**	tr	**0**	unk.	**unk.**	2500	**50**	unk.	15.0	**50**	15.0	unk.	60	**6**	5.5	0.9	unk.	16	**4**	0.08	**4**	tr(a)		unk.	
0.43	**25**	0.50	**25**	1.50	**25**	tr(a)	**0**	unk.	100	**25**	tr(a)	tr(a)	6	**1**	200	27	unk.	4.5	**25**	0.3	**2**	unk.	**unk.**	tr(a)	J
5.0	**25**	3	**1**	0.06	**1**	1250	**25**	unk.	unk.	**unk.**	tr(a)	0.60	22	**2**	8.7	0.7	unk.	unk.	**unk.**	unk.	**unk.**	tr(a)		unk.	
0.07	**4**	0.50	**25**	1.50	**25**	tr(a)	**0**	unk.	unk.	**unk.**	tr(a)	tr(a)	3	**0**	271	23	unk.	1.8	**10**	0.1	**1**	unk.	**unk.**	tr(a)	K
5.0	**25**	100	**25**	0.03	**0**	143	**3**	unk.	unk.	**unk.**	tr(a)	0.85	11	**1**	11.8	0.6	unk.	4	**1**	0.02	**1**	tr(a)		unk.	
0.03	**2**	0.02	**1**	unk.	**unk.**	tr(a)	**0**	unk.	unk.	**unk.**	0.1	0.1	2	**0**	271	25	unk.	0.7	**4**	0.1	**1**	unk.	**unk.**	tr(a)	L
1.2	**6**	2	**1**	0.02	**0**	108	**2**	unk.	0.1	**1**	tr	0.82	11	**1**	11.8	0.6	unk.	3	**1**	0.03	**1**	tr(a)		unk.	
0.03	**2**	0.02	**1**	0.08	**1**	tr(a)	**0**	unk.	0	**0**	0.1	0.1	2	**0**	271	25	unk.	0.7	**4**	0.1	**1**	unk.	**unk.**	tr(a)	M
1.2	**6**	2	**1**	0.02	**0**	108	**2**	unk.	0.1	**0**	tr	0.82	10	**1**	11.8	0.6	unk.	3	**1**	0.03	**1**	tr(a)		unk.	
1.70	**100**	2.00	**100**	6.00	**100**	tr(a)	**0**	unk.	unk.	**unk.**	30.0	tr(a)	40	**4**	312	35	unk.	18.0	**100**	tr	**0**	unk.	**unk.**	tr(a)	N
20.0	**100**	400	**100**	tr(a)	**0**	5000	**100**	unk.	30.0	**100**	tr(a)	unk.	40	**4**	13.6	0.9	unk.	tr	**0**	tr	**0**	tr(a)		unk.	
0.43	**25**	0.50	**25**	unk.	**unk.**	tr(a)	**0**	unk.	40	**10**	tr(a)	tr(a)	10	**1**	207	72	unk.	1.8	**10**	1.5	**10**	unk.	**unk.**	tr(a)	O
5.0	**25**	100	**25**	unk.	**unk.**	1250	**25**	unk.	unk.	**unk.**	tr(a)	unk.	40	**4**	9.0	1.8	unk.	16	**4**	0.04	**2**	tr(a)		unk.	
0.43	**25**	0.50	**25**	1.50	**25**	tr(a)	**0**	unk.	40	**10**	tr(a)	tr(a)	20	**2**	200	65	unk.	4.5	**25**	0.3	**2**	unk.	**unk.**	tr(a)	P
5.0	**25**	unk.	**unk.**	tr(a)	**0**	1250	**25**	unk.	unk.	**unk.**	tr(a)	unk.	60	**6**	8.7	1.7	unk.	24	**6**	0.08	**4**	tr(a)		unk.	
0.43	**25**	0.50	**25**	1.50	**25**	tr(a)	**0**	unk.	40	**unk.**	tr(a)	tr(a)	tr	**0**	305	30	unk.	8.1	**45**	tr	**0**	unk.	**unk.**	tr(a)	Q
5.0	**25**	unk.	**unk.**	tr(a)	**0**	1250	**25**	unk.	unk.	**unk.**	tr(a)	unk.	tr	**0**	13.3	0.8	unk.	tr	**0**	tr	**0**	tr(a)		unk.	
0.07	**4**	tr	**0**	unk	**unk.**	tr(a)	**0**	tr(a)	0	**0**	tr(a)	tr(a)	20	**2**	72	165	unk.	0.7	**4**	0.6	**4**	unk.	**unk.**	tr(a)	R
0.4	**2**	tr	**0**	unk.	**unk.**	tr	**0**	tr(a)	unk.	**unk.**	tr(a)	unk.	80	**8**	3.1	4.2	unk.	32	**8**	0.12	**6**	tr(a)		unk.	
0.07	**4**	tr	**0**	unk.	**unk.**	tr	**0**	tr(a)	0	**0**	tr(a)	tr(a)	40	**4**	98	166	unk.	1.1	**6**	0.6	**4**	unk.	**unk.**	tr(a)	S
0.3	**2**	tr	**0**	tr(a)	**0**	1250	**25**	tr(a)	unk.	**unk.**	tr(a)	unk.	80	**8**	4.3	4.3	unk.	40	**10**	0.12	**6**	tr(a)		unk.	

	FOOD	Portion	Weight in grams / *Conversion for 100 g*	Kilocalories / H_2O g	Total carbohydrate g / Total fats g	Crude fiber g / Dietary fiber g	Total protein g / Total sugar g	**% USRDA**	Arginine mg / Histidine mg	Isoleucine mg / Leucine mg	Lysine mg / Methionine mg	Phenylalanine mg / Threonine mg	Valine mg / Tryptophan mg	Cystine mg / Tyrosine mg	Polyunsat. fatty acids g / Monounsat. fatty acids g	Saturated fatty acids g / P/S ratio	Linoleic acid g / Cholesterol mg	Thiamin mg / Ascorbic acid mg	**% USRDA**
A	cereal,RTE,Cracklin' Bran-Kellogg	1 oz (1/3 C)	28	120	20.0	1.14	2.0	**3**	unk.	unk.	unk.	unk.	unk.	unk.	unk.	tr(a)	unk.	0.37	**25**
			3.52	unk.	4.0	3.98	8.0		unk.	unk.	unk.	unk.	unk.	unk.	unk.	unk.	0(a)	15	**25**
B	cereal,RTE,Crazy Cow,chocolate -General Mills	1 oz (1 C)	28	110	24.0	unk.	2.0	**3**	unk.	unk.	unk.	unk.	unk.	unk.	unk.	tr(a)	unk.	0.38	**25**
			3.52	unk.	1.0	0.20	13.0		unk.	unk.	unk.	unk.	unk.	unk.	unk.	unk.	0(a)	15	**25**
C	cereal,RTE,Crazy Cow,strawberry -General Mills	1 oz (1 C)	28	110	25.0	unk.	1.0	**2**	unk.	unk.	unk.	unk.	unk.	unk.	unk.	tr(a)	unk.	0.38	**25**
			3.52	unk.	1.0	0.20	11.4		unk.	unk.	unk.	unk.	unk.	unk.	unk.	unk.	0(a)	15	**25**
D	cereal,RTE,crisp rice-store brand	1 oz (1 C)	28	110	24.6	0.09	1.8	**3**	unk.	unk.	unk.	unk.	unk.	unk.	unk.	tr(a)	unk.	0.12	**8**
			3.52	1.1	0.2	unk.	2.6		unk.	unk.	unk.	unk.	unk.	unk.	unk.	unk.	0(a)	2	**3**
E	cereal,RTE,crispy rice-Ralston	1 oz (1 C)	28	110	25.0	0.80	1.9	**3**	unk.	unk.	unk.	unk.	unk.	unk.	unk.	0(a)	unk.	0.14	**9**
			3.52	1.1	0.1	0.40	3.0		unk.	unk.	unk.	unk.	unk.	unk.	unk.	unk.	0(a)	1	**2**
F	cereal,RTE,fortified oat flakes-Post	1 oz (2/3 C)	28	104	19.7	0.36	6.2	**10**	315	272	317	256	293	120	unk.	tr(a)	unk.	0.37	**25**
			3.52	0.8	0.5	0.73	5.8		119	452	91	232	85	184	unk.	unk.	0	0(a)	**0**
G	cereal,RTE,Frankenberry-General Mills	1 oz (1 C)	28	110	30.0	unk.	1.0	**2**	unk.	unk.	unk.	unk.	unk.	unk.	unk.	tr(a)	unk.	0.38	**25**
			3.52	unk.	1.0	0.45	13.0		unk.	unk.	unk.	unk.	unk.	unk.	unk.	unk.	0(a)	15	**25**
H	cereal,RTE,Froot Loops-Kellogg	1 oz (1 C)	28	110	25.3	tr	1.4	**2**	unk.	unk.	unk.	unk.	unk.	unk.	unk.	tr(a)	unk.	0.37	**25**
			3.52	unk.	0.6	0.28	13.0		unk.	unk.	unk.	unk.	unk.	unk.	unk.	unk.	0(a)	15	**25**
I	cereal,RTE,frosted flakes-Kellogg	1 oz (2/3 C)	28	110	26.0	0.10	1.4	**2**	unk.	unk.	unk.	unk.	unk.	unk.	unk.	tr(a)	unk.	0.38	**25**
			3.52	unk.	0.0	0.30	10.8		unk.	unk.	unk.	unk.	unk.	unk.	unk.	unk.	0(a)	15	**25**
J	cereal,RTE,frosted Rice Krinkles-Post	1 oz (7/8 C)	28	110	26.0	0.04	1.2	**2**	91	59	47	48	72	19	unk.	tr(a)	unk.	0.37	**25**
			3.52	0.6	0.2	0.03	12.5		34	93	33	56	16	62	unk.	unk.	0	0(a)	**0**
K	cereal,RTE,frosted rice-Kellogg	1 oz (1 C)	28	100	25.6	0.10	1.4	**2**	unk.	unk.	unk.	unk.	unk.	unk.	unk.	tr(a)	unk.	0.38	**25**
			3.52	unk.	0.0	tr	11.0		unk.	unk.	unk.	unk.	unk.	unk.	unk.	unk.	0(a)	15	**25**
L	cereal,RTE,Frosty O's-General Mills	1 oz (1 C)	28	110	24.0	unk.	2.0	**3**	unk.	unk.	unk.	unk.	unk.	unk.	unk.	tr(a)	unk.	0.38	**25**
			3.52	unk.	1.0	unk.	11.0		unk.	unk.	unk.	unk.	unk.	unk.	unk.	unk.	0(a)	15	**25**
M	cereal,RTE,Fruit Brute-General Mills	1 oz (1 C)	28	110	24.0	unk.	2.0	**3**	unk.	unk.	unk.	unk.	unk.	unk.	unk.	tr(a)	unk.	0.38	**25**
			3.52	unk.	1.0	unk.	12.0		unk.	unk.	unk.	unk.	unk.	unk.	unk.	unk.	0(a)	15	**25**
N	cereal,RTE,fruity Pebbles-Post	1 oz (7/8 C)	28	110	24.4	0.05	1.2	**2**	93	66	48	49	74	20	unk.	tr(a)	unk.	0.37	**25**
			3.52	0.8	1.5	unk.	12.4		34	95	34	57	17	64	unk.	unk.	0	0(a)	**0**
O	cereal,RTE,Golden Grahams -General Mills	1 oz (1 C)	28	110	24.0	unk.	2.0	**3**	unk.	unk.	unk.	unk.	unk.	unk.	unk.	tr(a)	unk.	0.37	**25**
			3.52	unk.	1.0	0.48	8.5		unk.	unk.	unk.	unk.	unk.	unk.	unk.	unk.	0(a)	15	**25**
P	cereal,RTE,Grape-Nuts flakes-Post	1 oz (7/8 C)	28	100	23.0	0.44	3.0	**5**	150	120	90	149	147	59	unk.	tr(a)	unk.	0.37	**25**
			3.52	1.0	0.3	1.79	4.8		71	209	49	94	53	90	unk.	unk.	0(a)	0(a)	**0**
Q	cereal,RTE,Grape-Nuts-Post	1 oz (1/4 C)	28	100	23.3	0.46	3.3	**5**	173	139	104	173	170	68	unk.	tr(a)	unk.	0.37	**25**
			3.52	0.9	0.1	1.34	3.4		82	241	57	109	62	105	unk.	unk.	0(a)	0(a)	**0**
R	cereal,RTE,Heartland natural-Pet	1 oz (1/4 C)	28	120	18.0	unk.	3.0	**5**	unk.	unk.	unk.	unk.	unk.	unk.	unk.	unk.	unk.	0.12	**8**
			3.52	unk.	4.0	1.34	unk.		unk.	unk.	unk.	unk.	unk.	unk.	unk.	unk.	0(a)	unk.	**unk.**
S	cereal,RTE,Honeycomb sweet crisp corn -Post	1 oz (1-1/3 C)	28	112	25.0	0.11	1.7	**3**	74	66	43	85	86	37	unk.	tr(a)	unk.	0.37	**25**
			3.52	0.4	0.5	0.34	10.8		41	205	35	57	unk.	65	unk.	unk.	0	0(a)	**0**

Riboflavin mg / Niacin mg	% USRDA	Vitamin B_6 mg / Folacin mcg	% USRDA	Vitamin B_{12} mcg / Pantothenic acid mg	% USRDA	Biotin mg / Vitamin A IU	% USRDA	Preformed A RE / Beta carotene RE	Vitamin D IU / Vitamin E IU	% USRDA	Total tocopherol mg / Alpha tocopherol mg	Other tocopherol mg / Total ash g	Calcium mg / Phosphorus mg	% USRDA	Sodium mg / Sodium meq	Potassium mg / Potassium meq	Chlorine mg / Chlorine meq	Iron mg / Magnesium mg	% USRDA	Zinc mg / Copper mg	% USRDA	Iodine mcg / Selenium mcg	% USRDA	Manganese mcg / Chromium mcg	
0.43	**25**	0.50	**25**	unk.	**unk.**	tr(a)	**0**	unk.	50	**13**	tr(a)	tr(a)	17	**2**	170	163	unk.	1.8	**10**	1.5	**10**	unk.	**unk.**	tr(a)	A
5.0	**25**	100	**25**	tr(a)	**0**	1250	**25**	unk.	unk.	**unk.**	tr(a)	unk.	136	**14**	7.4	4.2	unk.	51	**13**	0.20	**10**	tr(a)		unk.	
0.43	**25**	0.50	**25**	1.50	**25**	tr(a)	**0**	unk.	40	**10**	tr(a)	tr	tr	**0**	203	72	unk.	4.5	**25**	tr	**0**	unk.	**unk.**	tr(a)	B
5.0	**25**	unk.	**unk.**	tr(a)	**0**	1250	**25**	unk.	unk.	**unk.**	tr(a)	unk.	20	**2**	8.8	1.8	unk.	8	**2**	0.08	**4**	tr(a)		unk.	
0.43	**25**	0.50	**25**	1.50	**25**	tr(a)	**0**	unk.	40	**10**	tr(a)	tr(a)	tr	**0**	170	25	unk.	4.5	**25**	tr	**0**	unk.	**unk.**	tr(a)	C
5.0	**25**	unk.	**unk.**	unk.	**unk.**	1250	**25**	unk.	unk.	**unk.**	tr(a)	unk.	20	**2**	7.4	0.6	unk.	tr	**0**	0.04	**2**	tr(a)		unk.	
0.03	**2**	0.03	**2**	0.10	**2**	tr(a)	**0**	unk.	0	**0**	tr(a)	tr(a)	4	**0**	204	24	unk.	0.7	**4**	0.4	**3**	unk.	**unk.**	tr(a)	D
2.0	**10**	4	**1**	0.02	**0**	0	**0**	unk.	unk.	**unk.**	tr(a)	0.62	30	**3**	8.9	0.6	unk.	9	**2**	0.06	**3**	tr(a)		unk.	
0.03	**2**	0.04	**2**	0.08	**1**	tr(a)	**0**	unk.	unk.	**unk.**	tr(a)	tr(a)	5	**1**	208	27	unk.	0.7	**4**	0.5	**3**	unk.	**unk.**	tr(a)	E
2.0	**10**	100	**25**	0.11	**1**	unk.	**unk.**	unk.	unk.	**unk.**	tr(a)	0.62	31	**3**	9.0	0.7	unk.	12	**3**	0.06	**3**	tr(a)		unk.	
0.43	**25**	0.50	**25**	1.50	**25**	tr(a)	**0**	unk.	50	**13**	tr(a)	tr(a)	38	**4**	253	138	unk.	8.1	**45**	0.8	**6**	unk.	**unk.**	tr(a)	F
5.0	**25**	100	**25**	tr(a)	**0**	1250	**25**	unk.	unk.	**unk.**	tr(a)	1.23	101	**10**	11.0	3.5	unk.	40	**10**	0.17	**8**	tr(a)		unk.	
0.43	**25**	0.50	**25**	1.50	**25**	tr(a)	**0**	unk.	40	**10**	tr(a)	tr(a)	20	**2**	145	45	unk.	4.5	**25**	tr	**2**	unk.	**unk.**	tr(a)	G
5.0	**25**	unk.	**unk.**	tr(a)	**0**	1250	**25**	unk.	unk.	**unk.**	tr(a)	unk.	60	**6**	6.3	1.1	unk.	16	**4**	0.04	**2**	tr(a)		unk.	
0.43	**25**	0.50	**25**	unk.	**unk.**	tr(a)	**0**	unk.	50	**13**	tr(a)	tr(a)	3	**0**	125	29	unk.	4.5	**25**	3.8	**25**	unk.	**unk.**	tr(a)	H
5.1	**26**	100	**25**	tr(a)	**0**	1250	**25**	unk.	0.0	**0**	tr(a)	unk.	25	**3**	5.4	0.7	unk.	8	**2**	0.03	**2**	tr(a)		unk.	
0.43	**25**	0.50	**25**	unk.	**unk.**	tr(a)	**0**	unk.	50	**13**	tr(a)	tr(a)	1	**0**	200	18	unk.	1.8	**10**	0.1	**0**	unk.	**unk.**	tr(a)	I
5.1	**25**	100	**25**	tr(a)	**0**	1250	**25**	unk.	unk.	**unk.**	tr(a)	unk.	8	**1**	8.7	0.5	unk.	2	**1**	0.01	**1**	tr(a)		unk.	
0.43	**25**	0.50	**25**	1.50	**25**	tr(a)	**0**	unk.	50	**13**	tr(a)	tr(a)	5	**1**	185	13	unk.	1.8	**10**	1.5	**10**	unk.	**unk.**	tr(a)	J
5.0	**25**	100	**25**	tr(a)	**0**	1250	**25**	unk.	unk.	**unk.**	tr(a)	0.52	21	**2**	8.0	0.3	unk.	9	**2**	0.05	**2**	tr(a)		unk.	
0.43	**25**	0.50	**25**	unk.	**unk.**	tr(a)	**0**	unk.	50	**13**	tr(a)	tr(a)	2	**0**	205	21	unk.	1.8	**10**	0.4	**3**	unk.	**unk.**	tr(a)	K
5.0	**25**	100	**25**	tr(a)	**0**	1250	**25**	unk.	unk.	**unk.**	tr(a)	unk.	21	**2**	8.9	0.5	unk.	7	**2**	0.03	**2**	tr(a)		unk.	
0.43	**25**	0.50	**25**	1.50	**25**	tr(a)	**0**	unk.	40	**10**	tr(a)	tr(a)	20	**2**	135	45	unk.	4.5	**25**	0.6	**4**	unk.	**unk.**	tr(a)	L
5.0	**25**	unk.	**unk.**	tr(a)	**0**	1250	**25**	unk.	unk.	**unk.**	tr(a)	unk.	80	**8**	5.9	1.1	unk.	24	**6**	0.08	**4**	tr(a)		unk.	
0.43	**25**	0.50	**25**	1.50	**25**	tr(a)	**0**	unk.	40	**10**	tr(a)	tr(a)	40	**4**	200	45	unk.	4.5	**25**	0.3	**2**	unk.	**unk.**	tr(a)	M
5.0	**25**	unk.	**unk.**	tr(a)	**0**	1250	**25**	unk.	unk.	**unk.**	tr(a)	unk.	60	**6**	8.7	1.1	unk.	16	**4**	0.04	**2**	tr(a)		unk.	
0.43	**25**	0.50	**25**	1.50	**25**	tr(a)	**0**	unk.	50	**13**	tr(a)	tr(a)	3	**0**	154	21	unk.	1.8	**10**	1.5	**10**	unk.	**unk.**	tr(a)	N
5.0	**25**	100	**25**	tr(a)	**0**	1250	**25**	unk.	unk.	**unk.**	tr(a)	0.50	17	**2**	6.7	0.5	unk.	8	**2**	0.04	**2**	tr(a)		unk.	
0.43	**25**	0.50	**25**	1.50	**25**	tr(a)	**0**	unk.	40	**10**	tr(a)	tr(a)	tr	**0**	355	60	unk.	4.5	**25**	tr	**0**	unk.	**unk.**	tr(a)	O
5.0	**25**	unk.	**unk.**	tr(a)	**0**	1250	**25**	unk.	unk.	**unk.**	tr(a)	unk.	40	**4**	15.4	1.5	unk.	8	**2**	0.04	**2**	tr(a)		unk.	
0.42	**25**	0.50	**25**	1.50	**25**	tr(a)	**0**	unk.	50	**13**	tr(a)	tr(a)	11	**1**	106	99	unk.	4.5	**25**	0.6	**4**	unk.	**unk.**	tr(a)	P
5.0	**25**	100	**25**	tr(a)	**0**	1250	**25**	unk.	unk.	**unk.**	tr(a)	0.85	85	**8**	4.6	2.5	unk.	31	**8**	0.14	**7**	tr(a)		unk.	
0.43	**25**	0.50	**25**	1.50	**25**	tr(a)	**0**	unk.	50	**13**	0.6	tr(a)	11	**1**	195	95	unk.	1.2	**7**	0.6	**4**	unk.	**unk.**	tr(a)	Q
5.0	**25**	100	**25**	tr(a)	**0**	1250	**25**	unk.	0.0	**0**	tr(a)	0.78	71	**7**	8.5	2.4	unk.	19	**5**	0.12	**6**	tr(a)		unk.	
0.04	**2**	unk.	**unk.**	unk.	**unk.**	tr(a)	**0**	unk.	0(a)	**0**	tr(a)	tr(a)	0	**0**	unk.	unk.	unk.	1.1	**6**	unk.	**unk.**	unk.	**unk.**	tr(a)	R
unk.	**unk.**	unk.	**unk.**	unk.	**unk.**	unk.	**unk.**	unk.	unk.	**unk.**	tr(a)	unk.	unk.	**unk.**	unk.	unk.	unk.	unk.	**unk.**	unk.	**unk.**	tr(a)		unk.	
0.43	**25**	0.50	**25**	1.50	**25**	tr(a)	**0**	unk.	50	**13**	tr(a)	tr(a)	5	**1**	211	91	unk.	1.8	**10**	1.5	**10**	unk.	**unk.**	tr(a)	S
5.0	**25**	100	**25**	tr(a)	**0**	1250	**25**	unk.	unk.	**unk.**	tr(a)	0.56	28	**3**	9.2	2.3	unk.	10	**3**	0.05	**3**	tr(a)		unk.	

	FOOD	Portion	Weight in grams / *Conversion for 100 g*	Kilocalories / H_2O g	Total carbohydrate g / Total fats g	Crude fiber g / Dietary fiber g	Total protein g / Total sugar g	**% USRDA**	Arginine mg / Histidine mg	Isoleucine mg / Leucine mg	Lysine mg / Methionine mg	Phenylalanine mg / Threonine mg	Valine mg / Tryptophan mg	Cystine mg / Tyrosine mg	Polyunsat. fatty acids g / Monounsat. fatty acids g	Saturated fatty acids g / P/S ratio	Linoleic acid g / Cholesterol mg	Thiamin mg / Ascorbic acid mg	**% USRDA**
A	cereal,RTE,Jean Lafoote's cinnamon	1 oz (3/4 C)	28	120	22.6	0.17	1.8	**3**	unk.	unk.	unk.	unk.	unk.	unk.	unk.	unk.	unk.	0.37	**25**
	crunch-Quaker		*3.52*	0.5	2.6	unk.	unk.		unk.	unk.	unk.	unk.	unk.	unk.	unk.	unk.	unk.	0	**0**
B	cereal,RTE,Kaboom-General Mills	1 oz (1 C)	28	110	24.0	unk.	2.0	**3**	unk.	unk.	unk.	unk.	unk.	unk.	unk.	unk.	unk.	0.68	**45**
			3.52	unk.	1.0	unk.	12.0		unk.	unk.	unk.	unk.	unk.	unk.	unk.	unk.	0(a)	27	**45**
C	cereal,RTE,Kellogg's Corn Flakes-Kellogg	1 oz (1 C)	28	108	24.7	tr	2.0	**3**	unk.	unk.	unk.	unk.	unk.	unk.	unk.	tr(a)	unk.	0.37	**25**
			3.52	unk.	0.0	0.14	2.0		unk.	unk.	unk.	unk.	unk.	unk.	unk.	unk.	0(a)	15	**25**
D	cereal,RTE,Kellogg's Sugar Frosted	1 oz (2/3 C)	28	110	26.0	0.10	1.0	**2**	unk.	unk.	unk.	unk.	unk.	unk.	unk.	tr(a)	unk.	0.37	**25**
	Flakes-Kellogg		*3.52*	unk.	0.0	0.20	11.6		unk.	unk.	unk.	unk.	unk.	unk.	unk.	unk.	0(a)	15	**25**
E	cereal,RTE,Kellogg's 40% Bran	1 oz (2/3 C)	28	88	22.7	1.14	3.0	**5**	unk.	unk.	unk.	unk.	unk.	unk.	unk.	tr(a)	unk.	0.37	**25**
	Flakes-Kellogg		*3.52*	unk.	0.6	3.98	4.9		unk.	unk.	unk.	unk.	unk.	unk.	unk.	unk.	0(a)	tr	**0**
F	cereal,RTE,King Vitamin cereal-Quaker	1 oz (3/4 C)	28	120	23.3	0.18	1.2	**2**	unk.	unk.	unk.	unk.	unk.	unk.	unk.	unk.	unk.	0.90	**60**
			3.52	0.5	2.4	unk.	unk.		unk.	unk.	unk.	unk.	unk.	unk.	unk.	unk.	unk.	36	**60**
G	cereal,RTE,Kix-General Mills	1 oz (1-1/2 C)	28	110	24.0	unk.	2.0	**3**	unk.	unk.	unk.	unk.	unk.	unk.	unk.	tr(a)	unk.	0.37	**25**
			3.52	unk.	1.0	0.39	1.4		unk.	unk.	unk.	unk.	unk.	unk.	unk.	unk.	0(a)	9	**15**
H	cereal,RTE,Life cereal-Quaker	1 oz (2/3 C)	28	105	19.7	0.38	5.6	**9**	unk.	unk.	166	unk.	unk.	unk.	unk.	unk.	unk.	0.58	**39**
			3.52	1.3	0.5	0.84	unk.		unk.	unk.	unk.	unk.	unk.	unk.	unk.	unk.	unk.	0	**0**
I	cereal,RTE,Lucky Charms-General Mills	1 oz (1 C)	28	110	24.0	unk.	2.0	**3**	unk.	unk.	unk.	unk.	unk.	unk.	unk.	tr(a)	unk.	0.38	**25**
			3.52	unk.	1.0	0.56	12.0		unk.	unk.	unk.	unk.	unk.	unk.	unk.	unk.	0(a)	15	**25**
J	cereal,RTE,natural cereal w/apples	1 oz (1/4 C)	28	134	17.9	0.40	3.1	**5**	unk.	unk.	unk.	unk.	unk.	unk.	unk.	unk.	unk.	0.09	**6**
	& cinnamon-Quaker		*3.52*	0.5	5.3	1.26	unk.		unk.	unk.	unk.	unk.	unk.	unk.	unk.	unk.	unk.	0	**0**
K	cereal,RTE,natural cereal w/raisins	1 oz (1/4 C)	28	134	17.8	0.36	3.1	**5**	unk.	unk.	unk.	unk.	unk.	unk.	unk.	unk.	unk.	0.08	**6**
	& dates-Quaker		*3.52*	1.0	5.2	unk.	unk.		unk.	unk.	unk.	unk.	unk.	unk.	unk.	unk.	unk.	0	**0**
L	cereal,RTE,natural cereal-Quaker	1 oz (1/4 C)	28	138	17.0	0.34	3.6	**6**	unk.	unk.	unk.	unk.	unk.	unk.	unk.	unk.	unk.	0.08	**6**
			3.52	0.6	6.1	1.04	unk.		unk.	unk.	unk.	unk.	unk.	unk.	unk.	unk.	unk.	0	**0**
M	cereal,RTE,Nature Valley granola	1 oz (1/3 C)	28	130	19.0	unk.	3.0	**5**	unk.	unk.	unk.	unk.	unk.	unk.	unk.	unk.	unk.	0.09	**6**
	cinnamon/raisin-General Mills		*3.52*	unk.	5.0	1.04	7.0		unk.	unk.	unk.	unk.	unk.	unk.	unk.	unk.	0(a)	tr	**0**
N	cereal,RTE,Nature Valley granola	1 oz (1/3 C)	28	130	18.0	unk.	3.0	**5**	unk.	unk.	unk.	unk.	unk.	unk.	unk.	unk.	unk.	0.09	**6**
	coconut/honey-General Mills		*3.52*	unk.	5.0	unk.	6.0		unk.	unk.	unk.	unk.	unk.	unk.	unk.	unk.	0(a)	tr	**0**
O	cereal,RTE,Nature Valley granola	1 oz (1/3 C)	28	130	20.0	unk.	3.0	**5**	unk.	unk.	unk.	unk.	unk.	unk.	unk.	unk.	unk.	0.09	**6**
	fruit/nut-General Mills		*3.52*	unk.	4.0	0.95	8.0		unk.	unk.	unk.	unk.	unk.	unk.	unk.	unk.	0(a)	tr	**0**
P	cereal,RTE,Nature Valley granola	1 oz (1/3 C)	28	130	19.0	unk.	3.0	**5**	unk.	unk.	unk.	unk.	unk.	unk.	unk.	unk.	unk.	0.09	**6**
	toasted oat-General Mills		*3.52*	unk.	5.0	unk.	6.0		unk.	unk.	unk.	unk.	unk.	unk.	unk.	unk.	0(a)	unk.	**unk.**
Q	cereal,RTE,Pep-Kellogg	1 oz (3/4 C)	28	100	24.0	0.30	2.0	**3**	unk.	unk.	unk.	unk.	unk.	unk.	unk.	tr(a)	unk.	1.50	**100**
			3.52	unk.	0.0	1.70	4.0		unk.	unk.	unk.	unk.	unk.	unk.	unk.	unk.	0(a)	15	**25**
R	cereal,RTE,Post Toasties-Post	1 oz (1-1/4 C)	28	109	24.5	0.11	2.0	**3**	71	84	43	111	111	49	unk.	tr(a)	unk.	0.37	**25**
			3.52	0.9	0.0	0.17	2.0		59	325	52	76	13	92	unk.	unk.	0(a)	0(a)	**0**
S	cereal,RTE,Product 19-Kellogg	1 oz (3/4 C)	28	108	24.0	tr	2.0	**3**	unk.	unk.	unk.	unk.	unk.	unk.	unk.	tr(a)	unk.	1.50	**100**
			3.52	unk.	0.3	0.23	3.0		unk.	unk.	unk.	unk.	unk.	unk.	unk.	unk.	0(a)	60	**100**

Riboflavin mg / Niacin mg	% USRDA	Vitamin B_6 mg / Folacin mcg	% USRDA	Vitamin B_{12} mcg / Pantothenic acid mg	% USRDA	Biotin mg / Vitamin A IU	% USRDA	Preformed A RE / Beta carotene RE	Vitamin D IU / Vitamin E IU	% USRDA	Total tocopherol mg / Alpha tocopherol mg	Other tocopherol mg / Total ash g	Calcium mg / Phosphorus mg	% USRDA	Sodium mg / Sodium meq	Potassium mg / Potassium meq	Chlorine mg / Chlorine meq	Iron mg / Magnesium mg	% USRDA	Zinc mg / Copper mg	% USRDA	Iodine mcg / Selenium mcg	% USRDA	Manganese mcg / Chromium mcg	
0.26	**15**	0.50	**25**	1.00	**17**	0.002	**1**	unk.	0	**0**	unk.	unk.	8	**1**	211	48	unk.	4.5	**25**	2.0	**13**	unk.	**unk.**	0	A
5.0	**25**	100	**25**	2.00	**20**	0	**0**	unk.	unk.	**unk.**	0.2	0.68	20	**2**	9.2	1.2	unk.	8	**2**	tr	**2**	unk.		unk.	
0.77	**45**	0.90	**45**	2.70	**45**	tr(a)	**0**	unk.	unk.	**unk.**	tr(a)	tr(a)	20	**2**	150	45	unk.	8.1	**45**	0.3	**2**	unk.	**unk.**	tr(a)	B
9.0	**45**	unk.	**unk.**	tr(a)	**0**	2254	**45**	unk.	unk.	**unk.**	tr(a)	unk.	60	**6**	6.5	1.1	unk.	16	**4**	0.04	**2**	tr(a)		unk.	
0.43	**25**	0.50	**25**	unk.	**unk.**	tr(a)	**0**	unk.	50	**13**	0.1	0.1	1	**0**	305	28	unk.	1.8	**10**	0.1	**1**	unk.	**unk.**	tr(a)	C
5.0	**25**	100	**25**	unk.	**unk.**	1250	**25**	unk.	0.2	**1**	0.2	unk.	12	**1**	13.3	0.7	unk.	3	**1**	0.02	**1**	tr(a)		unk.	
0.43	**25**	0.50	**25**	unk.	**unk.**	tr(a)	**0**	unk.	50	**13**	tr(a)	tr(a)	1	**0**	185	23	unk.	1.8	**10**	0.4	**3**	unk.	**unk.**	tr(a)	D
5.0	**25**	100	**25**	tr(a)	**0**	1250	**25**	unk.	unk.	**unk.**	tr(a)	unk.	36	**4**	8.1	0.6	unk.	2	**1**	0.03	**2**	tr(a)		unk.	
0.43	**25**	0.50	**25**	1.50	**25**	tr(a)	**0**	unk.	50	**13**	tr(a)	tr(a)	13	**1**	230	169	unk.	8.1	**45**	3.8	**25**	unk.	**unk.**	tr(a)	E
5.0	**25**	100	**25**	tr(a)	**0**	1252	**25**	unk.	unk.	**unk.**	tr(a)	unk.	131	**13**	10.0	4.3	unk.	49	**12**	0.17	**9**	tr(a)		unk.	
1.02	**60**	1.00	**50**	4.00	**67**	tr(a)	**0**	unk.	240	**60**	unk.	unk.	5	**1**	215	35	unk.	10.8	**60**	0.2	**2**	unk.	**unk.**	125	F
12.0	**60**	240	**60**	0.06	**1**	2502	**50**	unk.	15.0	**50**	15.0	0.72	23	**2**	9.5	0.9	unk.	unk.	**unk.**	0.04	**2**	unk.		unk.	
0.43	**25**	0.50	**25**	1.50	**25**	tr(a)	**0**	unk.	40	**10**	tr(a)	tr(a)	tr	**0**	315	45	unk.	8.1	**45**	tr	**0**	unk.	**unk.**	tr(a)	G
5.0	**25**	unk.	**unk.**	tr(a)	**0**	1252	**25**	unk.	unk.	**unk.**	tr(a)	unk.	40	**4**	13.7	1.2	unk.	8	**2**	0.04	**2**	tr(a)		unk.	
0.63	**37**	0.03	**2**	0.00	**0**	0.007	**2**	0	0	**0**	unk.	unk.	101	**11**	146	127	unk.	7.4	**41**	0.9	**6**	16.5	**11**	849	H
7.2	**36**	24	**6**	0.24	**2**	0	**0**	0	unk.	**unk.**	0.2	1.00	148	**15**	6.4	3.3	unk.	9	**2**	0.15	**8**	unk.		unk.	
0.43	**25**	0.50	**25**	1.50	**25**	tr(a)	**0**	unk.	40	**10**	tr(a)	tr(a)	20	**2**	205	60	unk.	4.5	**25**	0.3	**2**	unk.	**unk.**	tr(a)	I
5.0	**25**	unk.	**unk.**	tr(a)	**0**	1250	**25**	unk.	unk.	**unk.**	tr(a)	unk.	80	**8**	8.9	1.5	unk.	24	**6**	0.12	**6**	tr(a)		unk.	
0.16	**9**	0.03	**2**	0.08	**1**	unk.	**unk.**	0	0	**0**	unk.	unk.	43	**4**	14	140	unk.	0.8	**4**	0.5	**4**	unk.	**unk.**	507	J
0.5	**3**	5	**1**	0.30	**3**	0	**0**	0	0.2	**1**	unk.	0.55	95	**10**	0.6	3.6	unk.	19	**5**	0.12	**6**	unk.		unk.	
0.17	**10**	0.04	**2**	0.00	**0**	0.004	**1**	0	0	**0**	unk.	unk.	41	**4**	12	138	unk.	0.8	**5**	0.5	**4**	26.0	**17**	566	K
0.6	**3**	12	**3**	0.24	**2**	0	**0**	0	unk.	**unk.**	0.6	0.55	89	**9**	0.5	3.5	unk.	32	**8**	0.12	**6**	unk.		unk.	
0.15	**9**	0.05	**3**	0.28	**5**	0.000	**0**	0	0	**0**	unk.	unk.	49	**5**	12	140	unk.	0.8	**5**	0.6	**4**	9.1	**6**	566	L
0.6	**3**	14	**3**	0.23	**2**	0	**0**	0	unk.	**unk.**	0.8	0.56	104	**10**	0.5	3.6	unk.	34	**9**	0.13	**6**	unk.		unk.	
0.03	**2**	unk.	**unk.**	unk.	**unk.**	tr(a)	**0**	unk.	unk.	**unk.**	tr(a)	tr(a)	tr	**0**	45	104	unk.	1.1	**6**	0.6	**4**	unk.	**unk.**	tr	M
tr	**0**	unk.	**unk.**	tr(a)	**0**	tr	**0**	unk.	unk.	**unk.**	tr(a)	unk.	80	**8**	2.0	2.6	unk.	32	**8**	0.08	**4**	tr(a)		unk.	
0.03	**2**	unk.	**unk.**	unk.	**unk.**	tr(a)	**0**	unk.	unk.	**unk.**	tr(a)	tr(a)	20	**2**	55	105	unk.	1.1	**6**	0.6	**4**	unk.	**unk.**	32	N
tr	**0**	unk.	**unk.**	tr(a)	**0**	tr	**0**	unk.	unk.	**unk.**	tr(a)	unk.	80	**8**	2.4	2.7	unk.	32	**8**	0.12	**6**	tr(a)		80	
0.03	**2**	unk.	**unk.**	unk.	**unk.**	tr(a)	**0**	unk.	unk.	**unk.**	tr(a)	tr(a)	20	**2**	50	109	unk.	0.7	**4**	0.6	**4**	unk.	**unk.**	tr(a)	O
tr	**0**	unk.	**unk.**	tr(a)	**0**	tr	**0**	unk.	unk.	**unk.**	tr(a)	unk.	80	**8**	2.2	2.8	unk.	32	**8**	0.12	**6**	tr(a)		unk.	
0.03	**2**	unk.	**unk.**	unk.	**unk.**	tr(a)	**0**	unk.	unk.	**unk.**	tr(a)	tr(a)	20	**2**	48	97	unk.	1.1	**6**	0.6	**4**	unk.	**unk.**	tr(a)	P
tr	**0**	unk.	**unk.**	tr(a)	**0**	unk.	**unk.**	unk.	unk.	**unk.**	tr(a)	unk.	80	**8**	2.1	2.5	unk.	32	**8**	0.08	**4**	tr(a)		unk.	
0.43	**25**	0.50	**25**	unk.	**unk.**	tr(a)	**0**	unk.	40	**10**	30.0	30.0	10	**1**	356	134	unk.	1.8	**10**	1.5	**10**	unk.	**unk.**	tr(a)	Q
5.0	**25**	99	**25**	tr(a)	**0**	1250	**25**	unk.	30.0	**100**	tr(a)	tr	60	**6**	15.5	3.4	unk.	32	**8**	0.16	**8**	tr(a)		unk.	
0.43	**25**	0.50	**25**	1.50	**25**	tr(a)	**0**	unk.	50	**13**	tr(a)	tr(a)	2	**0**	130	36	unk.	0.7	**4**	0.1	**1**	unk.	**unk.**	tr(a)	R
0.50	**25**	100	**25**	tr(a)	**0**	1250	**25**	unk.	unk.	**unk.**	tr(a)	0.83	11	**1**	5.6	0.9	unk.	4	**1**	0.05	**3**	tr(a)		unk.	
1.70	**100**	2.00	**100**	6.00	**100**	tr(a)	**0**	unk.	400	**100**	30.0	30.0	4	**0**	325	44	unk.	18.0	**100**	0.3	**2**	unk.	**unk.**	tr(a)	S
20.2	**100**	400	**100**	tr(a)	**0**	1000	**100**	unk.	30.0	**100**	unk.	unk.	33	**3**	14.1	1.1	unk.	10	**3**	0.05	**3**	tr(a)		unk.	

	FOOD	Portion	Weight in grams / Conversion for 100 g	Kilocalories / H_2O g	Total carbohydrate g / Total fats g	Crude fiber g / Dietary fiber g	Total protein g / Total sugar g	% USRDA	Arginine mg / Histidine mg	Isoleucine mg / Leucine mg	Lysine mg / Methionine mg	Phenylalanine mg / Threonine mg	Valine mg / Tryptophan mg	Cystine mg / Tyrosine mg	Polyunsat. fatty acids g / Monounsat. fatty acids g	Saturated fatty acids g / P/S ratio	Linoleic acid g / Cholesterol mg	Thiamin mg / Ascorbic acid mg	% USRDA
A	cereal,RTE,puffed rice-Kellogg	3/8 oz (1 C)	11	40	9.0	0.03	1.0	**2**	unk.	unk.	unk.	unk.	unk.	unk.	unk.	unk.	unk.	unk.	**unk.**
			9.09	0.3	tr	0.10	unk.		unk.	unk.	unk.	unk.	unk.	unk.	unk.	unk.	unk.	unk.	**unk.**
B	cereal,RTE,puffed rice-Quaker	3/8 oz (1 C)	11	40	9.1	0.04	0.7	**1**	unk.	unk.	unk.	unk.	unk.	unk.	unk.	unk.	unk.	0.01	**1**
			9.09	0.4	tr	unk.	tr		unk.	unk.	unk.	unk.	unk.	unk.	unk.	unk.	0	0	**0**
C	cereal,RTE,puffed wheat-Kellogg	5/16 oz (1 C)	9.0	35	7.0	0.12	1.0	**2**	unk.	unk.	unk.	unk.	unk.	unk.	unk.	unk.	unk.	unk.	**unk.**
			11.11	0.2	0.2	0.36	0.2		unk.	unk.	unk.	unk.	unk.	unk.	unk.	unk.	unk.	unk.	**unk.**
D	cereal,RTE,puffed wheat-Quaker	5/16 oz (1 C)	9.0	35	7.0	0.20	1.7	**3**	unk.	unk.	unk.	unk.	unk.	unk.	unk.	unk.	unk.	0.02	**2**
			11.11	0.3	0.1	0.37	0.1		unk.	unk.	unk.	unk.	unk.	unk.	unk.	unk.	0	0	**0**
E	cereal,RTE,Quisp-Quaker	1 oz (1-1/8 C)	28	121	23.1	tr	1.5	**2**	unk.	unk.	unk.	unk.	unk.	unk.	unk.	2.50	0.05	0.51	**34**
			3.52	0.6	2.1	unk.	11.6		unk.	unk.	unk.	unk.	unk.	unk.	unk.	unk.	0	0	**0**
F	cereal,RTE,raisin bran w/cinnamon	1 oz (1/2 C)	28	100	23.0	0.94	2.6	**4**	unk.	unk.	unk.	unk.	unk.	unk.	unk.	tr(a)	unk.	0.38	**25**
	-Post		*3.52*	2.3	0.4	unk.	0.9		unk.	unk.	unk.	unk.	unk.	unk.	unk.	unk.	0(a)	0	**0**
G	cereal,RTE,raisin bran-Kellogg	1.3 oz (1 C)	37	110	27.0	1.10	4.0	**6**	unk.	unk.	unk.	unk.	unk.	unk.	unk.	tr(a)	unk.	0.38	**25**
			2.72	unk.	1.0	4.00	12.0		unk.	unk.	unk.	unk.	unk.	unk.	unk.	unk.	0(a)	tr	**0**
H	cereal,RTE,raisin bran-Post	1 oz (1/2 C)	28	86	21.6	0.92	2.5	**4**	130	84	74	102	109	45	unk.	tr(a)	unk.	0.37	**25**
			3.52	2.6	2.5	2.94	8.2		57	149	35	72	38	66	unk.	unk.	0(a)	0(a)	**0**
I	cereal,RTE,raisin bran-Ralston	1 oz (1/2 C)	28	90	22.6	0.83	0.9	**1**	unk.	unk.	unk.	unk.	unk.	unk.	unk.	tr(a)	unk.	0.28	**19**
			3.52	2.5	0.2	unk.	3.3		unk.	unk.	unk.	unk.	unk.	unk.	unk.	unk.	0(a)	1	**2**
J	cereal,RTE,raisin bran-store brand	1 oz (1/2 C)	28	100	21.9	0.68	2.0	**3**	unk.	unk.	unk.	unk.	unk.	unk.	unk.	tr(a)	unk.	0.05	**3**
			3.52	2.5	0.2	unk.	3.3		unk.	unk.	unk.	unk.	unk.	unk.	unk.	unk.	0(a)	0	**0**
K	cereal,RTE,rice Chex-Ralston	1 oz (1-1/8 C)	28	110	25.0	tr	1.6	**3**	unk.	unk.	unk.	unk.	unk.	unk.	unk.	tr(a)	unk.	0.38	**25**
			3.52	0.7	0.1	unk.	1.2		unk.	unk.	unk.	unk.	unk.	unk.	unk.	unk.	0(a)	15	**25**
L	cereal,RTE,Rice Krispies-Kellogg	1 oz (1 C)	28	108	24.7	unk.	2.0	**3**	unk.	unk.	unk.	unk.	unk.	unk.	unk.	tr(a)	unk.	0.37	**25**
			3.52	unk.	0.0	tr	3.0		unk.	unk.	unk.	unk.	unk.	unk.	unk.	unk.	0(a)	15	**25**
M	cereal,RTE,shredded wheat biscuit	5/6 oz	24	84	15.8	0.47	2.6	**4**	unk.	118	87	127	152	54	0.31	0.10	0.29	0.06	**4**
	-Nabisco	(1 biscuit)	*4.24*	1.4	0.5	2.40	0.1		62	180	37	107	22	62	0.09	3.17	0(a)	0(a)	**0**
N	cereal,RTE,shredded wheat biscuit	1.4 oz	40	147	31.0	0.64	3.3	**5**	unk.	unk.	unk.	unk.	unk.	unk.	unk.	unk.	unk.	0.11	**8**
	-Quaker	(2 biscuits)	*2.50*	2.5	0.4	unk.	unk.		unk.	unk.	unk.	unk.	unk.	unk.	unk.	unk.	0	0	**0**
O	cereal,RTE,shredded wheat spoon size	1 oz (2/3 C)	28	102	23.0	0.57	3.0	**5**	unk.	138	102	148	178	63	0.38	0.12	0.35	0.07	**25**
	-Nabisco		*3.52*	1.5	0.6	unk.	0(a)		73	211	43	125	26	73	0.11	3.17	0(a)	0	**0**
P	cereal,RTE,Special K-Kellogg	1 oz (1-1/4 C)	28	108	21.0	0.28	1.4	**2**	unk.	unk.	unk.	unk.	unk.	unk.	unk.	tr(a)	unk.	0.37	**25**
			3.52	unk.	0.3	0.14	2.0		unk.	unk.	unk.	unk.	unk.	unk.	unk.	unk.	0(a)	15	**25**
Q	cereal,RTE,Sugar Corn Pops-Kellogg	1 oz (1 C)	28	110	25.8	tr	1.4	**2**	unk.	unk.	unk.	unk.	unk.	unk.	unk.	tr(a)	unk.	0.37	**25**
			3.52	unk.	0.3	0.11	12.0		unk.	unk.	unk.	unk.	unk.	unk.	unk.	unk.	0(a)	15	**25**
R	cereal,RTE,Sugar Frosted Flakes	1 oz (3/4 C)	28	110	26.0	0.10	1.5	**2**	unk.	unk.	unk.	unk.	unk.	unk.	unk.	tr(a)	unk.	0.37	**25**
	-Ralston		*3.52*	0.6	0.4	unk.	9.4		unk.	unk.	unk.	unk.	unk.	unk.	unk.	unk.	0(a)	15	**25**
S	cereal,RTE,sugar frosted flakes	1 oz (3/4 C)	28	110	25.3	0.09	1.4	**2**	unk.	unk.	unk.	unk.	unk.	unk.	unk.	tr(a)	unk.	0.30	**20**
	-store brand		*3.52*	0.6	0.3	unk.	9.4		unk.	unk.	unk.	unk.	unk.	unk.	unk.	unk.	0(a)	9	**15**

Riboflavin *mg* / Niacin *mg*	% USRDA	Vitamin B_6 *mg* / Folacin *mcg*	% USRDA	Vitamin B_{12} *mcg* / Pantothenic acid *mg*	% USRDA	Biotin *mg* / Vitamin A *IU*	% USRDA	Preformed A *RE* / Beta carotene *RE*	Vitamin D *IU* / Vitamin E *IU*	% USRDA	Total tocopherol *mg* / Alpha tocopherol *mg*	Other tocopherol *mg* / Total ash *g*	Calcium *mg* / Phosphorus *mg*	% USRDA	Sodium *mg* / Sodium *meq*	Potassium *mg* / Potassium *meq*	Chlorine *mg* / Chlorine *meq*	Iron *mg* / Magnesium *mg*	% USRDA	Zinc *mg* / Copper *mg*	% USRDA	Iodine *mcg* / Selenium *mcg*	% USRDA	Manganese *mcg* / Chromium *mcg*	
unk.	unk.	0.04	2	unk.	unk.	unk.	unk.	unk.	unk.	unk.	tr	tr	1	0	tr	11	unk.	0.7	4	0.1	1	unk.	unk.	unk.	A
0.2	1	tr	0	unk.	unk.	unk.	unk.	unk.	tr	0	tr	unk.	10	1	0.0	0.3	unk.	2	1	0.03	2	unk.		unk.	
0.01	1	tr	0	0.00	0	tr	0	0	0	0	tr	tr	2	0	0	12	unk.	0.1	1	0.1	0	0.8	1	107	B
0.3	2	1	0	0.16	2	0	0	0	tr	0	tr	tr	9	1	0.0	0.3	unk.	3	1	0.01	1	unk.		unk.	
unk.	unk.	0.10	5	unk.	unk.	unk.	unk.	unk.	unk.	unk.	0.4	0.3	3	0	1	53	unk.	0.4	2	0.3	2	unk.	unk.	unk.	C
0.6	3	tr	0	unk.	unk.	unk.	unk.	unk.	0.4	1	0.1	unk.	32	3	0.1	1.3	unk.	8	2	0.04	2	unk.		unk.	
0.10	6	0.01	1	0.00	0	0.001	0	0	0	0	0.4	0.3	3	0	1	38	unk.	0.5	3	0.2	1	2.2	2	213	D
1.2	6	4	1	0.04	0	0	0	0	0.4	1	0.1	0.16	48	5	0.0	1.0	unk.	9	3	0.04	2	unk.		unk.	
0.72	42	0.86	43	2.45	41	0.001	0	0	0	0	unk.	unk.	9	1	228	42	unk.	6.0	33	0.2	1	60.8	41	284	E
5.5	28	8	2	0.04	0	0	0	0	unk.	unk.	0.2	0.68	24	2	9.9	1.1	unk.	10	3	0.01	1	unk.		unk.	
0.43	25	0.50	25	0.15	3	tr(a)	0	unk.	40	10	tr(a)	tr(a)	15	2	195	149	unk.	0.5	3	unk.	unk.	unk.	unk.	tr(a)	F
5.0	25	100	25	tr(a)	0	25	25	unk.	unk.	unk.	tr(a)	1.08	137	14	8.5	3.8	unk.	tr(a)	0	unk.	unk.	tr(a)		unk.	
0.41	25	0.50	25	1.52	25	tr(a)	0	unk.	50	13	tr(a)	tr(a)	14	1	215	178	unk.	8.1	45	3.5	25	unk.	unk.	tr(a)	G
5.0	25	100	25	tr(a)	0	1250	25	unk.	unk.	unk.	tr(a)	unk.	127	13	9.3	4.6	unk.	48	12	0.16	8	tr(a)		unk.	
0.43	25	0.50	25	1.50	25	tr(a)	0	unk.	50	13	tr(a)	tr(a)	13	1	237	175	unk.	4.5	25	1.5	10	unk.	unk.	tr(a)	H
5.0	25	100	25	tr(a)	0	1250	25	unk.	unk.	unk.	tr(a)	1.12	119	12	10.3	4.5	unk.	48	12	0.21	11	tr(a)		unk.	
0.32	19	0.38	19	1.13	19	tr(a)	0	unk.	75	19	tr(a)	tr(a)	14	1	247	146	unk.	3.4	19	0.9	6	unk.	unk.	tr(a)	I
3.8	19	75	19	0.21	2	942	19	unk.	unk.	unk.	tr(a)	1.11	126	13	10.8	3.7	unk.	43	11	0.16	8	tr(a)		unk.	
0.14	8	0.07	3	0.07	1	tr(a)	0	unk.	0	0	tr(a)	tr(a)	13	1	214	148	unk.	6.4	36	0.7	5	unk.	unk.	tr(a)	J
3.1	16	11	3	0.16	2	28	1	unk.	unk.	unk.	tr(a)	1.11	119	12	9.3	3.8	unk.	40	10	0.14	7	tr(a)		unk.	
unk.	unk.	0.50	25	1.50	25	tr	0	unk.	unk.	unk.	tr(a)	tr(a)	4	0	237	33	unk.	1.8	10	0.4	3	unk.	unk.	tr(a)	K
5.0	25	100	25	0.10	1	17	0	unk.	unk.	unk.	tr(a)	0.71	28	3	10.3	0.8	unk.	7	2	0.08	4	tr(a)		unk.	
0.43	25	0.50	25	unk.	unk.	tr(a)	0	unk.	50	13	tr(a)	tr(a)	3	0	305	30	unk.	1.8	10	0.5	4	unk.	unk.	tr(a)	L
5.1	25	100	25	tr(a)	0	1250	25	unk.	unk.	unk.	tr(a)	unk.	32	3	13.3	0.8	unk.	10	3	0.05	2	tr(a)		unk.	
0.03	2	0.06	3	0.00	0	tr(a)	0	0	0	0	0.2	0.1	10	1	3	81	unk.	0.9	5	0.7	5	unk.	unk.	684	M
1.2	6	14	4	0.17	2	0	0	0	0.2	1	0.1	0.35	81	8	0.1	2.1	unk.	31	8	0.17	8	tr(a)		14	
0.11	7	0.06	3	0.00	0	0.005	2	0	0	0	0.3	0.1	16	2	1	136	unk.	1.2	7	1.0	7	24.5	16	800	N
1.8	9	27	7	0.30	3	0	0	0	0.3	1	0.2	0.60	146	15	0.0	3.5	unk.	46	11	0.20	10	unk.		unk.	
0.03	2	0.07	4	0.00	0	tr(a)	0	0	0	0	0.2	0.1	11	1	4	105	unk.	1.2	7	1.0	6	unk.	unk.	821	O
1.5	25	17	4	0.20	2	0	0	0	0.1	0	0.1	0.48	99	10	0.2	2.7	unk.	39	10	0.20	10	tr(a)		17	
0.43	25	0.50	25	unk.	unk.	tr(a)	0	unk.	50	13	tr(a)	tr(a)	11	1	225	49	unk.	4.5	25	3.7	3	unk.	unk.	tr(a)	P
5.0	25	100	25	tr(a)	0	1250	25	unk.	unk.	unk.	tr(a)	unk.	54	5	9.8	1.2	unk.	17	4	0.12	6	tr(a)		unk.	
0.43	25	0.50	25	unk.	unk.	tr(a)	0	unk.	50	13	tr(a)	tr(a)	1	0	104	23	unk.	1.8	10	1.5	10	unk.	unk.	tr(a)	Q
5.0	25	100	25	tr(a)	0	1250	25	unk.	unk.	unk.	tr(a)	unk.	8	1	4.5	0.6	unk.	2	1	0.01	1	tr(a)		unk.	
0.43	25	0.50	25	1.50	25	tr(a)	0	unk.	100	25	tr(a)	tr(a)	3	0	184	18	unk.	0.7	4	0.6	4	unk.	unk.	tr(a)	R
5.0	25	2	1	0.01	0	1252	25	unk.	unk.	unk.	tr(a)	0.65	7	1	8.0	0.5	unk.	2	1	0.02	1	tr(a)		unk.	
0.34	20	0.60	30	1.50	25	tr(a)	0	unk.	120	30	tr(a)	tr(a)	3	0	184	18	unk.	0.7	4	0.1	0	unk.	unk.	tr(a)	S
3.0	15	1	0	0.01	0	1252	25	unk.	unk.	unk.	tr(a)	0.65	8	1	8.0	0.5	unk.	2	1	0.02	1	tr(a)		unk.	

	FOOD	Portion	Weight in grams / Conversion for 100 g	Kilocalories / H_2O g	Total carbohydrate g / Total fats g	Crude fiber g / Dietary fiber g	Total protein g / Total sugar g	**% USRDA**	Arginine mg / Histidine mg	Isoleucine mg / Leucine mg	Lysine mg / Methionine mg	Phenylalanine mg / Threonine mg	Valine mg / Tryptophan mg	Cystine mg / Tyrosine mg	Polyunsat. fatty acids g / Monounsat. fatty acids g	Saturated fatty acids g / P/S ratio	Linoleic acid g / Cholesterol mg	Thiamin mg / Ascorbic acid mg	**% USRDA**
A	cereal,RTE,Sugar Jets-General Mills	1 oz (1 C)	28	110	23.7	0.31	2.1	**3**	unk.	unk.	unk.	unk.	unk.	unk.	unk.	unk.	unk.	0.33	**22**
			3.52	1.0	1.1	unk.	unk.		unk.	unk.	unk.	unk.	unk.	unk.	unk.	unk.	unk.	0	**25**
B	cereal,RTE,Sugar Smacks-Kellogg	1 oz (3/4 C)	28	114	25.0	tr	1.7	**3**	unk.	unk.	unk.	unk.	unk.	unk.	unk.	tr(a)	unk.	0.37	**25**
			3.52	unk.	0.6	0.28	16.0		unk.	unk.	unk.	unk.	unk.	unk.	unk.	unk.	0(a)	15	**25**
C	cereal,RTE,Super Sugar Crisp-Post	1 oz (7/8 C)	28	110	25.7	0.27	1.8	**3**	84	68	47	84	82	34	unk.	tr(a)	unk.	0.37	**25**
			3.52	0.4	0.3	0.39	13.5		39	117	27	52	31	51	unk.	unk.	0(a)	0(a)	**0**
D	cereal,RTE,Super Sugar Crisp,orange honey-Post	1 oz (7/8 C)	28	110	25.0	0.34	2.0	**3**	unk.	unk.	unk.	unk.	unk.	unk.	unk.	tr(a)	unk.	0.38	**25**
			3.52	0.6	0.3	unk.	14.0		unk.	unk.	unk.	unk.	unk.	unk.	unk.	unk.	0(a)	0	**0**
E	cereal,RTE,Team flakes-Nabisco	1 oz (1 C)	28	110	24.0	0.28	1.9	**3**	unk.	unk.	unk.	unk.	unk.	unk.	unk.	tr(a)	unk.	0.37	**25**
			3.52	1.1	0.5	0.28	4.0		unk.	unk.	unk.	unk.	unk.	unk.	unk.	unk.	0(a)	0(a)	**0**
F	cereal,RTE,Toasted Miniwheats-Kellogg	1 oz (5 biscuits)	28	100	22.0	0.60	3.0	**5**	unk.	unk.	unk.	unk.	unk.	unk.	unk.	tr(a)	unk.	0.37	**25**
			3.52	unk.	0.0	2.81	1.0		unk.	unk.	unk.	unk.	unk.	unk.	unk.	unk.	0(a)	15	**25**
G	cereal,RTE,Total-General Mills	1 oz (1 C)	28	110	23.0	unk.	3.0	**5**	unk.	unk.	unk.	unk.	unk.	unk.	unk.	tr(a)	unk.	1.50	**100**
			3.52	unk.	1.0	1.90	2.4		unk.	unk.	unk.	unk.	unk.	unk.	unk.	unk.	0(a)	60	**100**
H	cereal,RTE,Trix-General Mills	1 oz (1 C)	28	110	25.0	unk.	1.0	**2**	unk.	unk.	unk.	unk.	unk.	unk.	unk.	tr(a)	unk.	0.38	**25**
			3.52	unk.	1.0	0.14	13.0		unk.	unk.	unk.	unk.	unk.	unk.	unk.	unk.	0(a)	15	**25**
I	cereal,RTE,wheat Chex-Ralston	1 oz (2/3 C)	28	100	23.0	0.60	3.0	**5**	unk.	unk.	unk.	unk.	unk.	unk.	unk.	tr(a)	unk.	0.37	**25**
			3.52	0.7	0.7	1.90	1.0		unk.	unk.	unk.	unk.	unk.	unk.	unk.	unk.	0(a)	15	**25**
J	cereal,RTE,Wheaties-General Mills	1 oz (1 C)	28	110	23.0	unk.	3.0	**5**	unk.	unk.	unk.	unk.	unk.	unk.	unk.	tr(a)	unk.	0.37	**25**
			3.52	unk.	1.0	1.76	2.3		unk.	unk.	unk.	unk.	unk.	unk.	unk.	unk.	0(a)	15	**25**
K	CHEESE,COTTAGE,creamed w/fruit added	1/2 C	113	140	15.0	tr	11.2	**25**	511	658	905	603	693	104	0.11	2.43	0.09	0.01	**1**
			0.89	81.5	3.8	0(a)	8.7		372	1150	337	496	124	597	0.90	0.05	12	tr	**0**
L	cheese,cottage,creamed,large curd	1/2 C	113	116	3.0	0.00	14.0	**31**	641	826	1136	757	870	131	0.16	3.21	0.11	0.02	**2**
			0.89	88.8	5.1	0(a)	3.0		467	1445	423	623	156	749	1.19	0.05	17	tr	**0**
M	cheese,cottage,creamed,small curd	1/2 C	105	108	2.8	0.00	13.1	**29**	598	771	1060	707	812	122	0.15	2.99	0.10	0.02	**1**
			0.95	82.9	4.7	0(a)	2.8		436	1348	395	582	146	699	1.11	0.05	16	tr	**0**
N	cheese,cottage,dry curd,not packed	1/2 C	73	62	1.3	0.00	12.5	**28**	571	736	1013	675	775	116	0.01	0.20	0.01	0.01	**1**
			1.38	57.8	0.3	0(a)	1.3		416	1288	377	555	139	667	0.07	0.04	5	0	**0**
O	cheese,cottage,lowfat,1% fat	1/2 C	113	81	3.1	0.00	14.0	**31**	638	823	1132	755	867	130	0.03	0.72	0.02	0.02	**2**
			0.88	93.2	1.2	0(a)	3.1		466	1440	421	621	156	746	0.27	0.05	5	tr	**0**
P	cheese,cottage,lowfat,2% fat	1/2 C	113	102	4.1	0.00	15.5	**35**	709	913	1255	837	962	144	0.07	1.38	0.05	0.02	**2**
			0.88	89.6	2.2	0(a)	4.1		516	1597	467	688	173	827	0.51	0.05	9	tr	**0**
Q	CHEESE,CREAM	1-3/4x1x1" piece	28	99	0.8	0.00	2.1	**5**	81	113	192	119	126	19	0.36	6.24	0.22	tr	**0**
			3.52	15.3	9.9	0(a)	0.8		77	208	51	91	19	102	2.38	0.06	31	0	**0**
R	CHEESE,FOOD,cold pack,American	2 Tbsp	28	94	2.4	0.00	5.6	**12**	unk.	unk.	unk.	unk.	unk.	unk.	0.20	4.36	0.13	0.01	**1**
			3.52	12.2	6.9	0(a)	2.4		unk.	unk.	unk.	unk.	unk.	unk.	1.71	0.05	18	0	**0**
S	cheese,food,pasteurized processed, American	3-1/2x3-3/8x 1/8" slice	28	93	2.1	0.00	5.6	**12**	unk.	unk.	unk.	unk.	unk.	unk.	0.20	4.38	0.13	0.01	**0**
			3.52	12.3	7.0	0(a)	2.1		unk.	unk.	unk.	unk.	unk.	unk.	1.72	0.05	18	0	**0**

Riboflavin mg / Niacin mg	% USRDA	Vitamin B_6 mg / Folacin mcg	% USRDA	Vitamin B_{12} mcg / Pantothenic acid mg	% USRDA	Biotin mg / Vitamin A IU	% USRDA	Preformed A RE / Beta carotene RE	Vitamin D IU / Vitamin E IU	% USRDA	Total tocopherol mg / Alpha tocopherol mg	Other tocopherol mg / Total ash g	Calcium mg / Phosphorus mg	% USRDA	Sodium mg / Sodium meq	Potassium mg / Potassium meq	Chlorine mg / Chlorine meq	Iron mg / Magnesium mg	% USRDA	Zinc mg / Copper mg	% USRDA	Iodine mcg / Selenium mcg	% USRDA	Manganese mcg / Chromium mcg	
0.40	24	0.60	30	1.60	27	unk.	unk.	unk.	unk.	unk.	unk.	unk.	50	5	220	unk.	unk.	3.3	18	unk.	unk.	unk.	unk.	unk.	A
3.3	25	6	2	0.00	0	unk.	unk.	unk.	unk.	unk.	unk.	1.14	116	12	9.6	unk.	unk.	32	8	unk.	unk.	unk.		unk.	
0.43	25	0.50	25	unk.	unk.	tr(a)	0	unk.	50	13	tr(a)	tr(a)	3	0	58	36	unk.	1.8	10	0.4	3	unk.	unk.	tr(a)	B
5.1	25	100	25	tr(a)	0	1250	25	unk.	unk.	unk.	tr(a)	unk.	30	3	2.5	0.9	unk.	13	3	0.05	2	tr(a)		unk.	
0.43	25	0.50	25	1.50	25	tr(a)	0	unk.	50	13	tr(a)	tr(a)	6	1	36	106	unk.	1.8	10	1.5	10	unk.	unk.	tr(a)	C
5.0	25	100	25	tr(a)	0	1250	25	unk.	unk.	unk.	tr(a)	0.26	52	5	1.6	2.7	unk.	17	4	0.09	5	tr(a)		unk.	
0.43	25	0.50	25	1.50	25	tr(a)	0	unk.	40	10	tr(a)	tr(a)	5	1	40	50	unk.	0.6	3	unk.	unk.	unk.	unk.	tr(a)	D
5.0	25	100	25	tr(a)	0	1250	25	unk.	unk.	unk.	tr(a)	0.31	37	4	1.7	1.3	unk.	tr(a)	0	unk.	unk.	tr(a)		unk.	
0.43	25	unk.	unk.	1.50	25	tr(a)	0	unk.	0(a)	0	tr(a)	tr(a)	4	0	175	48	unk.	1.7	10	0.4	3	unk.	unk.	tr(a)	E
5.0	25	unk.	unk.	unk.	unk.	1250	25	unk.	unk.	unk.	tr(a)	0.61	44	4	7.6	1.2	unk.	13	3	0.15	7	tr(a)		unk.	
0.43	25	0.50	25	unk.	unk.	tr(a)	0	unk.	40	10	tr(a)	tr(a)	8	1	14	163	unk.	1.8	10	1.5	10	unk.	unk.	tr(a)	F
5.0	25	100	25	tr(a)	0	1250	25	unk.	unk.	unk.	tr(a)	unk.	100	10	0.6	4.2	unk.	40	10	0.08	4	tr(a)		unk.	
1.70	100	2.00	100	6.00	100	0.000	0	unk.	unk.	unk.	30.0	tr(a)	40	4	373	122	unk.	18.0	100	0.6	4	unk.	unk.	tr(a)	G
20.0	100	400	100	unk.	unk.	5000	100	unk.	30.0	100	tr(a)	unk.	100	10	16.2	3.1	unk.	32	8	0.12	6	tr(a)		unk.	
0.43	25	0.50	25	1.50	25	tr(a)	0	unk.	40	10	tr(a)	tr(a)	tr	0	170	25	unk.	4.5	25	tr	0	unk.	unk.	tr(a)	H
5.0	25	unk.	unk.	tr(a)	0	1250	25	unk.	unk.	unk.	tr(a)	unk.	20	2	7.4	0.6	unk.	7	2	0.04	2	tr(a)		unk.	
0.10	6	0.50	25	1.50	25	tr	0	unk.	unk.	unk.	tr(a)	tr(a)	11	1	190	107	unk.	4.5	25	0.8	5	unk.	unk.	tr(a)	I
5.0	25	100	25	0.13	1	unk.	unk.	unk.	unk.	unk.	tr(a)	0.96	112	11	8.3	2.7	unk.	36	9	0.17	8	tr(a)		unk.	
0.43	25	0.50	25	1.50	25	tr(a)	0	unk.	40	10	tr(a)	tr(a)	unk.	unk.	371	111	unk.	4.5	25	0.6	4	unk.	unk.	tr(a)	J
5.0	25	unk.	unk.	tr(a)	0	1250	25	unk.	unk.	unk.	tr(a)	unk.	80	8	16.1	2.8	unk.	32	8	0.12	6	tr(a)		unk.	
0.14	8	0.06	3	0.55	9	unk.	unk.	unk.	unk.	unk.	unk.	unk.	54	5	458	76	unk.	0.1	1	0.3	2	unk.	unk.	unk.	K
0.1	1	11	3	0.19	2	139	3	unk.	unk.	unk.	unk.	1.47	118	12	19.9	1.9	unk.	5	1	unk.	unk.	unk.		unk.	
0.18	11	0.08	4	0.70	12	unk.	unk.	unk.	unk.	unk.	unk.	unk.	68	7	456	95	unk.	0.2	1	0.4	3	unk.	unk.	unk.	L
0.1	1	14	3	0.24	2	183	4	unk.	unk.	unk.	unk.	1.53	149	15	19.8	2.4	unk.	6	1	unk.	unk.	unk.		unk.	
0.17	10	0.07	4	0.65	11	unk.	unk.	unk.	unk.	unk.	unk.	unk.	63	6	425	88	unk.	0.1	1	0.4	3	unk.	unk.	unk.	M
0.1	1	13	3	0.22	2	171	3	unk.	unk.	unk.	unk.	1.43	139	14	18.5	2.3	unk.	5	1	unk.	unk.	unk.		unk.	
0.10	6	0.06	3	0.59	10	unk.	unk.	unk.	unk.	unk.	unk.	unk.	23	2	9	23	unk.	0.2	1	0.3	2	unk.	unk.	unk.	N
0.1	1	11	3	0.12	1	22	0	unk.	unk.	unk.	unk.	0.50	75	8	0.4	0.6	unk.	3	1	unk.	unk.	unk.		unk.	
0.18	11	0.08	4	0.71	12	unk.	unk.	unk.	unk.	unk.	unk.	unk.	69	7	459	97	unk.	0.2	1	0.4	3	unk.	unk.	unk.	O
0.1	1	14	3	0.24	2	42	1	unk.	unk.	unk.	unk.	1.57	151	15	20.0	2.5	unk.	6	1	unk.	unk.	unk.		unk.	
0.20	12	0.09	4	0.80	13	unk.	unk.	unk.	unk.	unk.	unk.	unk.	77	8	459	108	unk.	0.2	1	0.5	3	unk.	unk.	unk.	P
0.2	1	15	4	0.27	3	79	2	unk.	unk.	unk.	unk.	1.57	170	17	20.0	2.8	unk.	7	2	unk.	unk.	unk.		unk.	
0.05	3	0.01	1	0.12	2	unk.	unk.	unk.	unk.	unk.	unk.	unk.	23	2	84	34	unk.	0.3	2	0.2	1	unk.	unk.	unk.	Q
tr	0	4	1	0.08	1	405	8	unk.	unk.	unk.	unk.	0.33	30	3	3.7	0.9	unk.	2	0	unk.	unk.	unk.		unk.	
0.12	7	0.04	2	0.36	6	unk.	unk.	unk.	unk.	unk.	unk.	unk.	141	14	274	103	unk.	0.2	1	0.9	6	unk.	unk.	unk.	R
tr	0	1	0	0.00	0	200	4	unk.	unk.	unk.	unk.	1.26	114	11	11.9	2.6	unk.	9	2	unk.	unk.	unk.		unk.	
0.12	7	unk.	unk.	0.32	5	unk.	unk.	unk.	unk.	unk.	unk.	unk.	163	16	338	79	unk.	0.2	1	0.8	6	unk.	unk.	unk.	S
tr	0	unk.	unk.	0.16	2	259	5	unk.	unk.	unk.	unk.	1.52	130	13	14.7	2.0	unk.	9	2	unk.	unk.	unk.		unk.	

	FOOD	Portion	Weight in grams / Conversion for 100 g	Kilocalories / H_2O g	Total carbohydrate g / Total fats g	Crude fiber g / Dietary fiber g	Total protein g / Total sugar g	% USRDA	Arginine mg / Histidine mg	Isoleucine mg / Leucine mg	Lysine mg / Methionine mg	Phenylalanine mg / Threonine mg	Valine mg / Tryptophan mg	Cystine mg / Tyrosine mg	Polyunsat. fatty acids g / Monounsat. fatty acids g	Saturated fatty acids g / P/S ratio	Linoleic acid g / Cholesterol mg	Thiamin mg / Ascorbic acid mg	% USRDA
A	cheese,food,pasteurized processed,	2-3/4x2-1/4	27	87	1.2	0.00	5.9	**13**	unk.	unk.	unk.	unk.	unk.	unk.	unk.	unk.	unk.	tr	**0**
	Swiss	x1/4" slice	*3.70*	11.8	6.5	0(a)	1.2		unk.	unk.	unk.	unk.	unk.	unk.	unk.	unk.	22	0	**0**
B	CHEESE,HOME RECIPE,blintzes	2 average	132	186	18.2	0.05	12.6	**27**	543	696	892	653	759	170	0.66	2.97	0.58	0.11	**7**
			0.76	92.8	6.3	0.49	5.5		384	1193	347	548	154	608	2.05	0.22	149	tr	**1**
C	cheese,home recipe,fondue	1/2 C	123	303	4.9	0.02	17.0	**38**	552	916	1539	985	1274	173	0.98	10.74	0.74	0.04	**3**
			0.81	80.0	18.4	unk.	3.3		635	1762	467	619	239	1009	4.88	0.09	62	tr	**0**
D	cheese,home recipe,straws	5x2x5/8"	91	266	22.2	0.09	11.1	**23**	333	629	730	587	669	119	0.72	8.80	0.55	0.14	**9**
			1.10	41.0	14.5	0.86	1.0		342	996	258	390	143	490	3.65	0.08	82	0	**0**
E	CHEESE,NATURAL,blue	1-3/4x1x1"	28	100	0.7	0.00	6.1	**14**	202	320	527	309	443	31	0.23	5.31	0.15	0.01	**0**
		piece	*3.52*	12.0	8.2	0(a)	0.7		216	546	166	223	89	368	1.88	0.04	21	0	**0**
F	cheese,natural,blue,crumbled	1 Tbsp	8.4	30	0.2	0.00	1.8	**4**	60	95	156	91	131	9	0.07	1.57	0.05	tr	**0**
			11.90	3.6	2.4	0(a)	0.2		64	161	49	66	26	109	0.56	0.04	6	0	**0**
G	cheese,natural,brick	1-2/3x1x1"	28	105	0.8	0.00	6.6	**15**	248	323	603	350	418	37	0.22	5.33	0.14	tr	**0**
		piece	*3.52*	11.7	8.4	0(a)	0.8		234	637	160	250	92	317	2.10	0.04	27	0	**0**
H	cheese,natural,brie	1-2/3x1x1"	28	95	0.1	0.00	5.9	**13**	209	288	526	329	381	32	unk.	unk.	unk.	0.02	**1**
		piece	*3.52*	13.8	7.9	0(a)	0.1		203	548	168	213	91	341	unk.	unk.	28	0	**0**
I	cheese,natural,camembert	1-1/3 oz pkg	38	114	0.2	0.00	7.5	**17**	266	368	671	420	486	41	0.27	5.80	0.17	0.01	**1**
			2.63	19.7	9.2	0(a)	0.2		260	699	215	272	117	435	2.18	0.05	27	0	**0**
J	cheese,natural,caraway	1-2/3x1x1"	28	107	0.9	0.00	7.1	**16**	unk.	unk.	unk.	unk.	unk.	unk.	unk.	unk.	unk.	0.01	**1**
		piece	*3.52*	11.2	8.3	0(a)	0.9		unk.	unk.	unk.	unk.	unk.	unk.	unk.	unk.	unk.	0	**0**
K	cheese,natural,cheddar	1-2/3x1x1"	28	114	0.4	0.00	7.1	**16**	267	439	588	372	472	35	0.27	5.99	0.16	0.01	**0**
		piece	*3.52*	10.4	9.4	0(a)	0.4		248	677	185	252	91	341	5.99	0.04	30	0	**0**
L	cheese,natural,cheddar,chopped	1 C	113	455	1.4	0.00	28.1	**63**	1063	1747	2341	1481	1879	141	1.06	23.83	0.66	0.02	**2**
			0.88	41.5	37.4	0(a)	1.4		988	2695	737	1001	362	1358	8.93	0.04	119	0	**0**
M	cheese,natural,cheshire	1-2/3x1x1"	28	110	1.4	0.00	6.6	**15**	251	412	552	350	443	33	unk.	unk.	unk.	0.01	**1**
		piece	*3.52*	10.7	8.7	0(a)	1.4		233	636	174	236	85	320	unk.	unk.	29	0	**0**
N	cheese,natural,colby	1-2/3x1x1"	28	112	0.7	0.00	6.8	**15**	255	419	562	355	450	34	0.27	5.74	0.19	tr	**0**
		piece	*3.52*	10.8	9.1	0(a)	0.7		237	646	177	240	87	326	2.22	0.05	27	0	**0**
O	cheese,natural,edam	1-2/3x1x1"	28	101	0.4	0.00	7.1	**16**	274	371	755	407	514	unk.	0.19	4.99	0.12	0.01	**1**
		piece	*3.52*	11.8	7.9	0(a)	0.4		294	730	205	265	unk.	414	1.96	0.04	25	0	**0**
P	cheese,natural,feta,made from	1-3/4x1x1"	28	75	1.2	0.00	4.0	**9**	unk.	unk.	unk.	unk.	unk.	unk.	0.17	4.25	0.09	unk.	**unk.**
	sheep's milk	piece	*3.52*	15.7	6.0	0(a)	1.2		unk.	unk.	unk.	unk.	unk.	unk.	1.13	0.04	25	0	**0**
Q	cheese,natural,fontina	1-2/3x1x1"	28	110	0.4	0.00	7.3	**16**	unk.	unk.	unk.	unk.	unk.	unk.	0.47	5.45	0.24	0.01	**0**
		piece	*3.52*	10.8	8.8	0(a)	0.4		unk.	unk.	unk.	unk.	unk.	unk.	2.02	0.09	33	0	**0**
R	cheese,natural,gjetost,made from goat's	1 oz	28	132	12.1	0.00	2.7	**6**	94	147	231	153	217	16	0.27	5.44	0.14	unk.	**unk.**
	& cow's milk		*3.52*	3.8	8.4	0(a)	12.1		83	282	90	112	38	154	1.98	0.05	unk.	0	**0**
S	cheese,natural,gouda	1-2/3x1x1"	28	101	0.6	0.00	7.1	**16**	273	371	754	406	513	unk.	0.19	5.00	0.07	0.01	**1**
		piece	*3.52*	11.8	7.8	0(a)	0.6		293	728	204	264	unk.	413	1.81	0.04	32	0	**0**

Riboflavin mg / Niacin mg	% USRDA	Vitamin B_6 mg / Folacin mcg	% USRDA	Vitamin B_{12} mcg / Pantothenic acid mg	% USRDA	Biotin mg / Vitamin A IU	% USRDA	Preformed A RE / Beta carotene RE	Vitamin D IU / Vitamin E IU	% USRDA	Total tocopherol mg / Alpha tocopherol mg	Other tocopherol mg / Total ash g	Calcium mg / Phosphorus mg	% USRDA	Sodium mg / Sodium meq	Potassium mg / Potassium meq	Chlorine mg / Chlorine meq	Iron mg / Magnesium mg	% USRDA	Zinc mg / Copper mg	% USRDA	Iodine mcg / Selenium mcg	% USRDA	Manganese mcg / Chromium mcg	
0.11	**6**	unk.	**unk.**	0.62	**10**	unk.	**unk.**	unk.	unk.	**unk.**	unk.	unk.	195	**20**	419	77	unk.	0.2	**1**	1.0	**6**	unk.	**unk.**	unk.	A
tr	**0**	unk.	**unk.**	0.13	**1**	231	**5**	unk.	unk.	**unk.**	unk.	1.56	142	**14**	18.2	2.0	unk.	8	**2**	unk.	**unk.**	unk.		unk.	
0.22	**13**	0.09	**5**	0.75	**13**	0.005	**2**	unk.	12	**3**	0.4	tr(a)	102	**10**	268	179	50	1.2	**7**	0.9	**6**	73.9	**49**	78	B
0.7	**3**	19	**5**	0.66	**7**	177	**4**	unk.	0.4	**2**	0.1	1.36	201	**20**	11.7	4.6	1.4	57	**14**	0.09	**4**	9		12	
0.16	**9**	0.07	**4**	0.91	**15**	unk.	**unk.**	unk.	0	**0**	0.0	0(a)	581	**58**	194	103	unk.	0.3	**2**	0.1	**0**	unk.	**unk.**	tr	C
0.07	**1**	6	**1**	0.32	**3**	692	**14**	unk.	0.0	**0**	0(a)	2.57	351	**35**	8.4	2.6	unk.	6	**1**	0.07	**3**	tr(a)		unk.	
0.20	**12**	0.05	**3**	0.33	**5**	0.002	**1**	unk.	3	**1**	0.1	0(a)	215	**22**	317	68	35	1.2	**7**	1.2	**8**	29.5	**20**	119	D
1.0	**5**	14	**3**	0.37	**4**	463	**9**	unk.	0.2	**1**	0.0	1.64	192	**19**	13.8	1.7	1.0	37	**9**	0.04	**2**	15		17	
0.11	**6**	0.05	**2**	0.34	**6**	unk.	**unk.**	unk.	unk.	**unk.**	unk.	unk.	150	**15**	396	73	unk.	0.1	**1**	0.8	**5**	unk.	**unk.**	unk.	E
0.3	**1**	10	**3**	0.49	**5**	205	**4**	unk.	unk.	**unk.**	unk.	1.45	110	**11**	17.2	1.9	unk.	7	**2**	unk.	**unk.**	unk.		unk.	
0.03	**2**	0.01	**1**	0.10	**2**	unk.	**unk.**	unk.	unk.	**unk.**	unk.	unk.	44	**4**	117	22	unk.	tr	**0**	0.2	**2**	unk.	**unk.**	unk.	F
0.1	**0**	3	**1**	0.15	**2**	61	**1**	unk.	unk.	**unk.**	unk.	0.43	33	**3**	5.1	0.5	unk.	2	**1**	unk.	**unk.**	unk.		unk.	
0.10	**6**	0.02	**1**	0.35	**6**	unk.	**unk.**	unk.	unk.	**unk.**	unk.	unk.	191	**19**	159	39	unk.	0.1	**1**	0.7	**5**	unk.	**unk.**	unk.	G
tr	**0**	6	**1**	0.08	**1**	308	**6**	unk.	unk.	**unk.**	unk.	0.90	128	**13**	6.9	1.0	unk.	7	**2**	unk.	**unk.**	unk.		unk.	
0.15	**9**	0.07	**3**	0.47	**8**	unk.	**unk.**	unk.	unk.	**unk.**	unk.	unk.	52	**5**	179	43	unk.	0.1	**1**	unk.	**unk.**	unk.	**unk.**	unk.	H
0.1	**1**	18	**5**	0.20	**2**	189	**4**	unk.	unk.	**unk.**	unk.	0.77	53	**5**	7.8	1.1	unk.	unk.	**unk.**	unk.	**unk.**	unk.		unk.	
0.18	**11**	0.09	**4**	0.49	**8**	unk.	**unk.**	unk.	unk.	**unk.**	unk.	unk.	147	**15**	320	71	unk.	0.1	**1**	0.9	**6**	unk.	**unk.**	unk.	I
0.2	**1**	24	**6**	0.52	**5**	351	**7**	unk.	unk.	**unk.**	unk.	1.40	99	**10**	13.9	1.8	unk.	8	**2**	unk.	**unk.**	unk.		unk.	
0.13	**8**	unk.	**unk.**	0.08	**1**	unk.	**unk.**	unk.	unk.	**unk.**	unk.	unk.	191	**19**	196	unk.	unk.	unk.	**unk.**	unk.	**unk.**	unk.	**unk.**	unk.	J
0.1	**0**	unk.	**unk.**	0.05	**1**	299	**6**	unk.	unk.	**unk.**	unk.	0.93	139	**14**	8.5	unk.	unk.	6	**2**	unk.	**unk.**	unk.		unk.	
0.11	**6**	0.02	**1**	0.23	**4**	unk.	**unk.**	unk.	unk.	**unk.**	unk.	unk.	205	**21**	176	28	unk.	0.2	**1**	0.9	**6**	unk.	**unk.**	unk.	K
tr	**0**	5	**1**	0.12	**1**	301	**6**	unk.	unk.	**unk.**	unk.	1.12	145	**15**	7.7	0.7	unk.	8	**2**	unk.	**unk.**	unk.		unk.	
0.42	**25**	0.08	**4**	0.93	**15**	unk.	**unk.**	unk.	unk.	**unk.**	unk.	unk.	815	**82**	701	111	unk.	0.8	**4**	3.5	**23**	unk.	**unk.**	unk.	L
0.1	**1**	20	**5**	0.47	**5**	1197	**24**	unk.	unk.	**unk.**	unk.	4.44	579	**58**	30.5	2.8	unk.	32	**8**	unk.	**unk.**	unk.		unk.	
0.08	**5**	unk.	**unk.**	unk.	**unk.**	unk.	**unk.**	unk.	unk.	**unk.**	unk.	unk.	183	**18**	199	27	unk.	0.1	**0**	unk.	**unk.**	unk.	**unk.**	unk.	M
unk.	**unk.**	unk.	**unk.**	unk.	**unk.**	280	**6**	unk.	unk.	**unk.**	unk.	1.02	132	**13**	8.6	0.7	unk.	6	**2**	unk.	**unk.**	unk.		unk.	
0.11	**6**	0.02	**1**	0.23	**4**	unk.	**unk.**	unk.	unk.	**unk.**	unk.	unk.	195	**20**	172	36	unk.	0.2	**1**	0.9	**6**	unk.	**unk.**	unk.	N
tr	**0**	unk.	**unk.**	0.06	**1**	294	**6**	unk.	unk.	**unk.**	unk.	0.95	130	**13**	7.5	0.9	unk.	7	**2**	unk.	**unk.**	unk.		unk.	
0.11	**6**	0.02	**1**	0.43	**7**	unk.	**unk.**	unk.	unk.	**unk.**	unk.	unk.	208	**21**	274	53	unk.	0.1	**1**	1.1	**7**	unk.	**unk.**	unk.	O
tr	**0**	5	**1**	0.08	**1**	260	**5**	unk.	unk.	**unk.**	unk.	1.20	152	**15**	11.9	1.4	unk.	9	**2**	unk.	**unk.**	unk.		unk.	
unk.	**unk.**	unk.	**unk.**	unk.	**unk.**	unk.	**unk.**	unk.	unk.	**unk.**	unk.	unk.	140	**14**	317	18	unk.	0.2	**1**	0.8	**6**	unk.	**unk.**	unk.	P
unk.	**unk.**	unk.	**unk.**	unk.	**unk.**	unk.	**unk.**	unk.	unk.	**unk.**	unk.	1.48	96	**10**	13.8	0.4	unk.	5	**1**	unk.	**unk.**	unk.		unk.	
0.06	**3**	unk.	**unk.**	unk.	**unk.**	unk.	**unk.**	unk.	unk.	**unk.**	unk.	unk.	156	**16**	unk.	unk.	unk.	0.1	**0**	1.0	**7**	unk.	**unk.**	unk.	Q
tr	**0**	unk.	**unk.**	unk.	**unk.**	333	**7**	unk.	unk.	**unk.**	unk.	1.08	unk.	**unk.**	unk.	unk.	unk.	4	**1**	unk.	**unk.**	unk.		unk.	
unk.	**unk.**	unk.	**unk.**	unk.	**unk.**	unk.	**unk.**	unk.	unk.	**unk.**	unk.	unk.	114	**11**	170	unk.	unk.	unk.	**unk.**	unk.	**unk.**	unk.	**unk.**	unk.	R
0.2	**1**	1	**0**	unk.	**unk.**	unk.	**unk.**	unk.	unk.	**unk.**	unk.	1.35	126	**13**	7.4	unk.	unk.	unk.	**unk.**	unk.	**unk.**	unk.		unk.	
0.09	**6**	0.02	**1**	unk.	**unk.**	unk.	**unk.**	unk.	unk.	**unk.**	unk.	unk.	199	**20**	233	34	unk.	0.1	**0**	1.1	**7**	unk.	**unk.**	unk.	S
tr	**0**	6	**2**	0.10	**1**	183	**4**	unk.	unk.	**unk.**	unk.	1.12	155	**16**	10.1	0.9	unk.	8	**2**	unk.	**unk.**	unk.		unk.	

	FOOD	Portion	Weight in grams / Conversion for 100 g	Kilocalories / H_2O g	Total carbohydrate g / Total fats g	Crude fiber g / Dietary fiber g	Total protein g / Total sugar g	% USRDA	Arginine mg / Histidine mg	Isoleucine mg / Leucine mg	Lysine mg / Methionine mg	Phenylalanine mg / Threonine mg	Valine mg / Tryptophan mg	Cystine mg / Tyrosine mg	Polyunsat. fatty acids g / Monounsat. fatty acids g	Saturated fatty acids g / P/S ratio	Linoleic acid g / Cholesterol mg	Thiamin mg / Ascorbic acid mg	% USRDA
A	cheese,natural,gruyere	2x1x1" piece	28	117	0.1	0.00	8.5	**19**	276	458	770	492	637	86	0.49	5.37	0.37	0.02	**1**
			3.52	9.4	9.2	0(a)	0.1		317	881	233	309	120	504	2.44	0.09	31	0	**0**
B	cheese,natural,liederkranz	1-2/3x1x1"	28	90	0.6	0.00	5.3	**12**	unk.	unk.	unk.	unk.	unk.	unk.	unk.	unk.	unk.	tr	**0**
		piece	*3.52*	14.2	7.4	unk.	unk.		unk.	unk.	unk.	unk.	unk.	unk.	unk.	unk.	unk.	unk.	**unk.**
C	cheese,natural,limburger	1-3/5x1x1"	28	93	0.1	0.00	5.7	**13**	198	346	476	317	407	unk.	0.14	4.76	0.10	0.02	**2**
		piece	*3.52*	13.8	7.7	0(a)	0.1		164	594	176	210	82	340	2.04	0.03	26	0	**0**
D	cheese,natural,monterey	1-2/3x1x1"	28	106	0.2	0.00	6.9	**15**	263	431	579	366	464	35	unk.	unk.	unk.	unk.	**unk.**
		piece	*3.52*	11.6	8.6	0(a)	unk.		244	666	182	247	89	336	unk.	unk.	unk.	0	**0**
E	cheese,natural,mozzarella	3-1/2x3-3/8	28	80	0.6	0.00	5.5	**12**	237	264	560	288	345	33	0.22	3.73	0.11	tr	**0**
		x1/8" piece	*3.52*	15.4	6.1	0(a)	0.6		208	538	154	210	unk.	319	1.60	0.06	22	0	**0**
F	cheese,natural,mozzarella, low moisture	3-1/2x3-3/8	28	90	0.7	0.00	6.1	**14**	264	294	623	320	384	37	0.22	4.42	0.16	tr	**0**
		x1/8" piece	*3.52*	13.7	7.0	0(a)	0.7		231	598	171	234	unk.	355	1.68	0.05	25	0	**0**
G	cheese,natural,mozzarella, low moisture,part skim	3-1/2x3-3/8	28	80	0.9	0.00	7.8	**17**	335	374	792	407	488	47	0.14	3.09	0.10	0.01	**0**
		x1/8" piece	*3.52*	13.8	4.9	0(a)	09		293	761	218	297	unk.	451	1.18	0.05	15	0	**0**
H	cheese,natural,mozzarella,part skim	3-1/2x3-3/8	28	72	0.8	0.00	6.9	**15**	296	331	700	360	431	41	0.13	2.87	0.10	tr	**0**
		x1/8" piece	*3.52*	15.3	4.5	0(a)	0.8		259	672	192	262	unk.	398	1.10	0.05	16	0	**0**
I	cheese,natural,muenster	1-2/3x1x1"	28	105	0.3	0.00	6.6	**15**	250	325	607	352	421	37	0.19	5.43	0.12	tr	**0**
		piece	*3.52*	11.9	8.5	0(a)	0.3		235	642	162	252	93	319	2.08	0.03	27	0	**0**
J	cheese,natural,neufchatel	1-2/3x1x1"	28	74	0.8	0.00	2.8	**6**	107	149	253	157	166	25	0.18	4.20	0.13	tr	**0**
		piece	*3.52*	17.7	6.7	0(a)	0.8		101	274	68	120	25	135	1.61	0.04	22	0	**0**
K	cheese,natural,parmesan	1 oz	28	111	0.9	0.00	10.1	**23**	374	538	939	546	697	67	0.16	4.66	0.08	0.01	**1**
			3.52	8.3	7.3	0(a)	0.9		393	980	272	374	137	567	1.89	0.03	19	0	**0**
L	cheese,natural,parmesan,grated	1 Tbsp	5.0	23	0.2	0.00	2.1	**5**	77	110	192	112	143	14	0.03	0.95	0.02	tr	**0**
			20.00	0.9	1.5	0(a)	0.2		80	201	56	77	28	116	0.39	0.03	4	0	**0**
M	cheese,natural,port du salut	1 oz	28	100	0.2	0.00	6.8	**15**	235	411	564	376	485	unk.	0.21	4.74	0.11	unk.	**unk.**
			3.52	12.9	8.0	0(a)	0.2		195	705	208	249	97	403	2.29	0.04	35	0	**0**
N	cheese,natural,provolone	4-5/8" dia x	28	100	0.6	0.00	7.3	**16**	290	310	751	366	466	33	0.22	4.85	0.14	tr	**0**
		1/8" slice	*3.52*	11.6	7.6	0(a)	0.6		317	652	195	279	unk.	432	1.75	0.05	20	0	**0**
O	cheese,natural,ricotta,made w/part skim milk	1/2 C	123	170	6.3	0.00	14.0	**31**	786	733	1664	691	861	123	0.32	6.06	0.23	0.02	**2**
			0.81	91.5	9.7	0(a)	6.3		571	1519	349	643	unk.	733	2.36	0.05	38	0	**0**
P	cheese,natural,ricotta,made w/whole milk	1/2 C	123	214	3.7	0.00	13.8	**31**	777	724	1646	684	851	122	0.47	10.21	0.33	0.01	**1**
			0.81	88.2	16.0	0(a)	3.7		565	1502	346	636	unk.	724	3.53	0.05	63	0	**0**
Q	cheese,natural,romano,grated	1 Tbsp	5.0	19	0.2	0.00	1.6	**4**	unk.	unk.	unk.	unk.	unk.	unk.	unk.	unk.	unk.	unk.	**unk.**
			20.00	1.5	1.3	0(a)	0.2		unk.	unk.	unk.	unk.	unk.	unk.	unk.	unk.	5	0	**0**
R	cheese,natural,roquefort	1-3/4x1x1"	28	105	0.6	0.00	6.1	**14**	unk.	unk	unk.	unk.	unk.	unk.	0.37	5.47	0.18	0.01	**1**
		piece	*3.52*	11.2	8.7	0(a)	0.6		unk.	unk.	unk.	unk.	unk.	unk.	2.12	0.07	26	0	**0**
S	cheese,natural,Swiss	1 oz	28	107	1.0	0.00	8.1	**18**	263	437	734	472	607	82	0.28	5.05	0.18	0.01	**0**
			3.52	10.6	7.8	0(a)	1.0		302	840	223	295	114	481	1.71	0.05	26	0	**0**

Riboflavin mg / Niacin mg	% USRDA	Vitamin B_6 mg / Folacin mcg	% USRDA	Vitamin B_{12} mcg / Pantothenic acid mg	% USRDA	Biotin mg / Vitamin A IU	% USRDA	Preformed A RE / Beta carotene RE	Vitamin D IU / Vitamin E IU	% USRDA	Total tocopherol mg / Alpha tocopherol mg	Other tocopherol mg / Total ash g	Calcium mg / Phosphorus mg	% USRDA	Sodium mg / Sodium meq	Potassium mg / Potassium meq	Chlorine mg / Chlorine meq	Iron mg / Magnesium mg	% USRDA	Zinc mg / Copper mg	% USRDA	Iodine mcg / Selenium mcg	% USRDA	Manganese mcg / Chromium mcg	
0.08	**5**	0.02	**1**	0.45	**8**	unk.	**unk.**	unk.	unk.	**unk.**	unk.	unk.	287	**29**	95	23	unk.	unk.	**unk.**	unk.	**unk.**	unk.	**unk.**	unk.	A
tr	**0**	3	**1**	0.16	**2**	346	**7**	unk.	unk.	**unk.**	unk.	1.22	172	**17**	4.1	0.6	unk.	unk.	**unk.**	unk.	**unk.**	unk.		unk.	
0.16	**10**	unk.	**unk.**	unk.	**unk.**	unk.	**unk.**	unk.	unk.	**unk.**	unk.	unk.	174	**17**	unk.	unk.	unk.	unk.	**unk.**	unk.	**unk.**	unk.	**unk.**	unk.	B
0.2	**1**	unk.	**unk.**	unk.	**unk.**	344	**7**	unk.	unk.	**unk.**	unk.	0.97	unk.	**unk.**	unk.	unk.	unk.	unk.	**unk.**	unk.	**unk.**	unk.		unk.	
0.14	**8**	0.02	**1**	0.30	**5**	unk.	**unk.**	unk	unk.	**unk.**	unk.	unk.	141	**14**	227	36	unk.	tr	**0**	0.6	**4**	unk.	**unk.**	unk.	C
tr	**0**	16	**4**	0.33	**3**	364	**7**	unk.	unk.	**unk.**	unk.	1.08	112	**11**	9.9	0.9	unk.	6	**2**	unk.	**unk.**	unk.		unk.	
0.11	**7**	unk.	**unk.**	unk.	**unk.**	unk.	**unk.**	unk.	unk.	**unk.**	unk.	unk.	212	**21**	152	23	unk.	0.02	**1**	0.9	**6**	unk.	**unk.**	unk.	D
unk.	**unk.**	unk.	**unk.**	0.08	**1**	270	**5**	unk.	unk.	**unk.**	unk.	1.01	126	**13**	6.6	0.6	unk.	8	**2**	unk.	**unk.**	unk.		unk.	
0.07	**4**	0.02	**1**	0.18	**3**	unk.	**unk.**	unk.	unk.	**unk.**	unk.	unk.	147	**15**	106	19	unk.	0.1	**0**	0.6	**4**	unk.	**unk.**	unk.	E
tr	**0**	2	**1**	0.02	**0**	225	**5**	unk.	unk.	**unk.**	unk.	0.74	105	**11**	4.6	0.5	unk.	5	**1**	unk.	**unk.**	unk.		unk.	
0.08	**5**	0.02	**1**	0.20	**3**	unk.	**unk.**	unk.	unk.	**unk.**	unk.	unk.	163	**16**	118	21	unk.	0.1	**0**	0.7	**5**	unk.	**unk.**	unk.	F
tr	**0**	2	**1**	0.02	**0**	257	**5**	unk.	unk.	**unk.**	unk.	0.83	117	**12**	5.1	0.5	unk.	6	**2**	unk.	**unk.**	unk.		unk.	
0.10	**6**	0.02	**1**	0.26	**4**	unk.	**unk.**	unk.	unk.	**unk.**	unk.	unk.	208	**21**	150	27	unk.	0.1	**0**	0.9	**6**	unk.	**unk.**	unk.	G
tr	**0**	3	**1**	0.03	**0**	178	**4**	unk.	unk.	**unk.**	unk.	1.05	149	**15**	6.5	0.7	unk.	7	**2**	unk.	**unk.**	unk.		unk.	
0.09	**5**	0.02	**1**	0.23	**4**	unk.	**unk.**	unk.	unk.	**unk.**	unk.	unk.	183	**18**	132	24	unk.	0.1	**0**	0.8	**5**	unk.	**unk.**	unk.	H
tr	**0**	3	**1**	0.02	**0**	166	**3**	unk.	unk.	**unk.**	unk.	0.93	132	**13**	5.8	0.6	unk.	7	**2**	unk.	**unk.**	unk.		unk.	
0.09	**5**	0.02	**1**	0.42	**7**	unk.	**unk.**	unk.	unk.	**unk.**	unk.	unk.	204	**20**	178	38	unk.	0.1	**1**	0.8	**5**	unk.	**unk.**	unk.	I
tr	**0**	3	**1**	0.05	**1**	318	**6**	unk.	unk.	**unk.**	unk.	1.04	133	**13**	7.8	1.0	unk.	8	**2**	unk.	**unk.**	unk.		unk.	
0.05	**3**	0.01	**1**	0.07	**1**	unk.	**unk.**	unk.	unk.	**unk.**	unk.	unk.	21	**2**	113	32	unk.	0.1	**0**	0.1	**1**	unk.	**unk.**	unk.	J
tr	**0**	3	**1**	0.16	**2**	322	**6**	unk.	unk.	**unk.**	unk.	0.41	39	**4**	4.9	0.8	unk.	2	**1**	unk.	**unk.**	unk.		unk.	
0.09	**6**	0.03	**1**	unk.	**unk.**	unk.	**unk.**	unk.	unk.	**unk.**	unk.	unk.	336	**34**	455	26	unk.	0.2	**1**	0.8	**5**	unk.	**unk.**	unk.	K
0.1	**0**	2	**1**	0.13	**1**	171	**3**	unk.	unk.	**unk.**	unk.	1.72	197	**20**	19.8	0.7	unk.	12	**3**	unk.	**unk.**	unk.		unk.	
0.02	**1**	0.01	**0**	unk.	**unk.**	unk.	**unk.**	unk.	unk.	**unk.**	unk.	unk.	69	**7**	93	5	unk.	0.0	**0**	0.2	**1**	unk.	**unk.**	unk.	L
tr	**0**	tr	**0**	0.03	**0**	35	**1**	unk.	unk.	**unk.**	unk.	0.35	40	**4**	4.0	0.1	unk.	3	**1**	unk.	**unk.**	unk.		unk.	
0.07	**4**	0.02	**1**	0.43	**7**	unk.	**unk.**	unk.	unk.	**unk.**	unk.	unk.	185	**19**	152	unk.	unk.	unk.	**unk.**	unk.	**unk.**	unk.	**unk.**	unk.	M
tr	**0**	5	**1**	0.06	**1**	379	**8**	unk.	unk.	**unk.**	unk.	0.57	102	**10**	6.6	unk.	unk.	unk.	**unk.**	unk.	**unk.**	unk.		unk.	
0.09	**5**	0.02	**1**	0.41	**7**	unk.	**unk.**	unk.	unk.	**unk.**	unk.	unk.	215	**22**	249	39	unk.	0.1	**1**	0.9	**6**	unk.	**unk.**	unk.	N
tr	**0**	3	**1**	0.14	**1**	231	**5**	unk.	unk.	**unk.**	unk.	1.34	141	**14**	10.8	1.0	unk.	8	**2**	unk.	**unk.**	unk.		unk.	
0.22	**13**	0.02	**1**	0.36	**6**	unk.	**unk.**	unk.	unk.	**unk.**	unk.	unk.	335	**34**	154	154	unk.	0.5	**3**	1.6	**11**	unk.	**unk.**	unk.	O
0.1	**0**	unk.	**unk.**	unk.	**unk.**	531	**11**	unk.	unk.	**unk.**	unk.	1.41	225	**23**	6.7	3.9	unk.	18	**5**	unk.	**unk.**	unk.		unk.	
0.23	**14**	0.05	**3**	0.41	**7**	unk.	**unk.**	unk.	unk.	**unk.**	unk.	unk.	255	**26**	103	129	unk.	0.5	**3**	1.4	**10**	unk.	**unk.**	unk.	P
0.1	**1**	unk.	**unk.**	unk.	**unk.**	603	**12**	unk.	unk.	**unk.**	unk.	1.25	194	**19**	4.5	3.3	unk.	14	**3**	unk.	**unk.**	unk.		unk.	
0.02	**1**	unk.	**unk.**	unk.	**unk.**	unk.	**unk.**	unk.	unk.	**unk.**	unk.	unk.	53	**5**	60	unk.	unk.	unk.	**unk.**	unk.	**unk.**	unk.	**unk.**	unk.	Q
0.0	**0**	tr	**0**	0.02	**0**	29	**1**	unk.	unk.	**unk.**	unk.	0.34	38	**4**	2.6	unk.	unk.	unk.	**unk.**	unk.	**unk.**	unk.		unk.	
0.16	**10**	0.04	**2**	0.18	**3**	unk.	**unk.**	unk.	unk.	**unk.**	unk.	unk.	188	**19**	514	26	unk.	0.2	**1**	0.6	**4**	unk.	**unk.**	unk.	R
0.2	**1**	14	**4**	0.49	**5**	297	**6**	unk.	unk.	**unk.**	unk.	1.83	111	**11**	22.3	0.7	unk.	9	**2**	unk.	**unk.**	unk.		unk.	
0.10	**6**	0.02	**1**	0.47	**8**	unk.	**unk.**	unk.	unk.	**unk.**	unk.	unk.	273	**27**	74	32	unk.	0.0	**0**	1.1	**7**	unk.	**unk.**	unk.	S
tr	**0**	2	**0**	0.12	**1**	240	**5**	unk.	unk.	**unk.**	unk.	1.00	172	**17**	3.2	0.8	unk.	10	**3**	unk.	**unk.**	unk.		unk.	

	FOOD	Portion	Weight in grams / *Conversion for 100 g*	Kilocalories / H_2O g	Total carbohydrate g / Total fats g	Crude fiber g / Dietary fiber g	Total protein g / Total sugar g	**% USRDA**	Arginine mg / Histidine mg	Isoleucine mg / Leucine mg	Lysine mg / Methionine mg	Phenylalanine mg / Threonine mg	Valine mg / Tryptophan mg	Cystine mg / Tyrosine mg	Polyunsat. fatty acids g / Monounsat. fatty acids g	Saturated fatty acids g / P/S ratio	Linoleic acid g / Cholesterol mg	Thiamin mg / Ascorbic acid mg	**% USRDA**
A	cheese,natural,tilsit	1-2/3x1x1"	28	97	0.5	0.00	6.9	**15**	241	421	579	386	498	unk.	0.20	4.77	0.11	0.02	**1**
		piece	*3.52*	12.2	7.4	0(a)	0.5		200	724	214	255	100	414	1.72	0.04	29	0	**0**
B	CHEESE,PASTEURIZED	3-1/2x3-3/8x	28	106	0.5	0.00	6.3	**14**	263	291	624	319	377	40	0.28	5.59	0.17	0.01	**0**
	PROCESSED,American	1/8" slice	*3.52*	11.1	8.9	0(a)	0.5		256	556	163	204	92	344	2.13	0.05	27	0	**0**
C	cheese,pasteurized processed,	3-1/2x3-3/8x	28	52	0.7	0(a)	7.0	**16**	263	431	579	366	464	35	0.08	1.47	0.05	0.01	**1**
	American,low calorie	1/8" slice	*3.52*	16.8	2.4	0(a)	0.7		244	666	182	247	89	336	0.67	0.05	11	tr	**0**
D	cheese,pasteurized processed,pimento	3-1/2x3-3/8x	28	106	0.5	tr	6.3	**14**	263	291	624	319	376	40	0.28	5.58	0.17	0.01	**0**
		1/8" slice	*3.52*	11.1	8.9	0(a)	0.5		256	556	162	204	92	344	2.13	0.05	27	unk.	**unk.**
E	cheese,pasteurized processed,spread,	2-3/4x2-1/4x	27	78	2.4	0.00	4.4	**10**	147	225	407	251	369	unk.	0.17	3.60	0.11	0.01	**1**
	American	1/4" slice	*3.70*	12.9	5.7	0(a)	2.4		137	481	145	170	unk.	240	1.41	0.05	15	0	**0**
F	cheese,pasteurized processed,Swiss	3-1/2x3-3/8	28	95	0.6	0.00	7.0	**16**	294	325	697	357	421	45	0.18	4.56	0.10	tr	**0**
		x1/8" slice	*3.52*	12.0	7.1	0(a)	0.6		286	621	182	228	102	384	1.68	0.04	24	0	**0**
G	CHEESE,SUBSTITUTE,low sodium,	1 oz	28	89	0.6	0.00	6.8	**11**	unk.	unk.	unk.	unk.	unk.	unk.	5.11	1.14	unk.	unk.	**unk.**
	pasteurized processed-Cheezola		*3.52*	13.1	6.2	0(a)	0.6		unk.	unk.	unk.	unk.	unk.	unk.	unk.	4.50	1	unk.	**unk.**
H	cheese,substitute,pasteurized	1 oz	28	89	0.6	0.00	6.8	**11**	unk.	unk.	unk.	unk.	unk.	unk.	5.11	1.14	unk.	unk.	**unk.**
	processed filled-Cheezola		*3.52*	13.1	6.2	0(a)	0.6		unk.	unk.	unk.	unk.	unk.	unk.	unk.	4.50	1	unk.	**unk.**
I	cheese,substitute,skim milk	.1 oz	28	89	0.6	0.00	6.8	**11**	unk.	unk.	unk.	unk.	unk.	unk.	5.11	1.14	unk.	unk.	**unk.**
	-Count Down Fisher		*3.52*	13.1	6.2	0(a)	0.6		unk.	unk.	unk.	unk.	unk.	unk.	unk.	4.50	1	unk.	**unk.**
J	cheese,substitute,skim milk,smokey flavor	1 oz	28	34	2.8	0.00	5.1	**8**	unk.	unk.	unk.	unk.	unk.	unk.	unk.	unk.	unk.	unk.	**unk.**
	-Count Down Fisher		*3.52*	19.3	0.3	0(a)	2.8		unk.	unk.	unk.	unk.	unk.	unk.	unk.	unk.	1	unk.	**unk.**
K	CHERRIES,maraschino	1 average	10	12	2.9	0.03	0.0	**0**	unk.	unk.	unk.	unk.	unk.	unk.	unk.	tr(a)	unk.	unk.	**unk.**
			10.00	7.0	tr	unk.	2.9		unk.	unk.	unk.	unk.	unk.	unk.	unk.	unk.	0(a)	unk.	**unk.**
L	cherries,sour,canned,pitted,	1/2 C	129	114	29.2	0.13	1.0	**2**	unk.	unk.	unk.	unk.	unk.	unk.	unk.	tr(a)	unk.	0.04	**3**
	heavy syrup,solids & liquid		*0.78*	97.7	0.3	1.36	29.0		unk.	unk.	unk.	unk.	unk.	unk.	unk.	unk.	0	6	**11**
M	cherries,sour,canned,pitted,	1/2 C	122	52	13.1	0.12	1.0	**2**	unk.	unk.	unk.	unk.	unk.	unk.	unk.	tr(a)	unk.	0.04	**2**
	water pack,solids & liquid		*0.82*	107.4	0.2	1.26	12.9		unk.	unk.	unk.	unk.	unk.	unk.	unk.	unk.	0	6	**10**
N	cherries,sour,raw,edible portion	1/2 C	78	45	11.1	0.15	0.9	**1**	unk.	unk.	unk.	unk.	unk.	unk.	unk.	tr(a)	unk.	0.04	**3**
			1.29	64.9	0.2	1.09	9.8		unk.	unk.	unk.	unk.	unk.	unk.	unk.	unk.	0	8	**13**
O	cherries,sweet,canned,pitted,heavy	1/2 C	129	104	26.3	0.39	1.2	**2**	unk.	unk.	unk.	unk.	unk.	unk.	unk.	tr(a)	unk.	0.03	**2**
	syrup,solids & liquid		*0.78*	100.2	0.3	unk.	26.0		unk.	unk.	unk.	unk.	unk.	unk.	unk.	unk.	0	4	**6**
P	cherries,sweet,canned,unpitted,water	1/2 C	124	60	14.8	0.37	1.1	**2**	unk.	unk.	unk.	unk.	unk.	unk.	unk.	tr(a)	unk.	0.02	**2**
	pack,solids & liquid		*0.81*	107.4	0.2	unk.	14.4		unk.	unk.	unk.	unk.	unk.	unk.	unk.	unk.	0	4	**6**
Q	cherries,sweet,raw,unpitted	1/2 C	65	45	11.3	0.26	0.8	**1**	unk.	unk.	unk.	unk.	unk.	unk.	unk.	tr(a)	unk.	0.03	**2**
			1.54	52.3	0.2	1.07	9.0		unk.	unk.	unk.	unk.	unk.	unk.	unk.	unk.	0	6	**11**
R	CHERVIL,dried	1 tsp	0.6	1	0.3	0.07	0.1	**0**	unk.	unk.	unk.	unk.	unk.	unk.	unk.	unk.	unk.	unk.	**unk.**
			166.67	tr	tr	unk.	0.0		unk.	unk.	unk.	unk.	unk.	unk.	unk.	unk.	0	unk.	**unk.**
S	CHESTNUT,fresh,in shell	1 average	7.3	14	3.1	0.08	0.2	**0**	unk.	unk.	unk.	unk.	unk.	unk.	0.08	0.03	0.07	0.02	**1**
			13.70	3.8	0.1	0.42	0.5		unk.	unk.	unk.	unk.	unk.	unk.	0.07	2.43	0	tr	**1**

Riboflavin mg / Niacin mg	% USRDA	Vitamin B6 mg / Folacin mcg	% USRDA	Vitamin B12 mcg / Pantothenic acid mg	% USRDA	Biotin mg / Vitamin A IU	% USRDA	Preformed A RE / Beta carotene RE	Vitamin D IU / Vitamin E IU	% USRDA	Total tocopherol mg / Alpha tocopherol mg	Other tocopherol mg / Total ash g	Calcium mg / Phosphorus mg	% USRDA	Sodium mg / Sodium meq	Potassium mg / Potassium meq	Chlorine mg / Chlorine meq	Iron mg / Magnesium mg	% USRDA	Zinc mg / Copper mg	% USRDA	Iodine mcg / Selenium mcg	% USRDA	Manganese mcg / Chromium mcg	
0.10	**6**	unk.	**unk.**	0.60	**10**	unk.	**unk.**	unk.	unk.	**unk.**	unk.	unk.	199	**20**	214	18	unk.	0.1	**0**	1.0	**7**	unk.	**unk.**	unk.	A
0.1	**0**	unk.	**unk.**	0.10	**1**	297	**6**	unk.	unk.	**unk.**	unk.	1.38	142	**14**	9.3	0.5	unk.	4	**1**	unk.	**unk.**	unk.		unk.	
0.10	**6**	0.02	**1**	0.20	**3**	unk.	**unk.**	unk.	unk.	**unk.**	unk.	unk.	175	**18**	406	46	unk.	0.1	**1**	0.8	**6**	unk.	**unk.**	unk.	B
tr	**0**	2	**1**	0.14	**1**	344	**7**	unk.	unk.	**unk.**	unk.	1.66	212	**21**	17.7	1.2	unk.	6	**2**	unk.	**unk.**	unk.		unk.	
0.05	**3**	unk.	**unk.**	unk.	**unk.**	unk.	**unk.**	unk.	unk.	**unk.**	unk.	unk.	200	**20**	430	20	unk.	0.1	**1**	1.0	**7**	unk.	**unk.**	unk.	C
tr	**0**	unk.	**unk.**	unk.	**unk.**	45	**1**	unk.	unk.	**unk.**	unk.	1.56	278	**28**	18.7	0.5	unk.	7	**2**	unk.	**unk.**	unk.		unk.	
0.10	**6**	0.02	**1**	0.20	**3**	unk.	**unk.**	unk.	unk.	**unk.**	unk.	unk.	174	**17**	406	46	unk.	0.1	**1**	0.8	**6**	unk.	**unk.**	unk.	D
tr	**0**	2	**1**	0.14	**1**	358	**7**	unk.	unk.	**unk.**	unk.	1.66	211	**21**	17.6	1.2	unk.	6	**2**	unk.	**unk.**	unk.		unk.	
0.12	**7**	0.03	**2**	0.11	**2**	unk.	**unk.**	unk.	unk.	**unk.**	unk.	unk.	152	**15**	363	65	unk.	0.1	**1**	0.7	**5**	unk.	**unk.**	unk.	E
tr	**0**	2	**1**	0.19	**2**	213	**4**	unk.	unk.	**unk.**	unk.	1.61	192	**19**	15.8	1.7	unk	8	**2**	unk.	**unk.**	unk.		unk.	
0.08	**5**	0.01	**1**	0.35	**6**	unk.	**unk.**	unk.	unk.	**unk.**	unk.	unk.	219	**22**	389	61	unk.	0.2	**1**	1.0	**7**	unk.	**unk.**	unk.	F
tr	**0**	unk.	**unk.**	0.07	**1**	229	**5**	unk.	unk.	**unk.**	unk.	1.66	216	**22**	16.9	1.6	unk.	8	**2**	unk.	**unk.**	unk.		unk.	
unk.	**unk.**	unk.	**unk.**	unk.	**unk.**	unk.	**unk.**	unk.	unk.	**unk.**	unk.	unk.	177	**18**	156	239	unk.	unk.	**unk.**	unk.	**unk.**	unk.	**unk.**	unk.	G
unk.	**unk.**	unk.	**unk.**	unk.	**unk.**	unk.	**unk.**	unk.	unk.	**unk.**	unk.	1.70	114	**11**	6.8	6.1	unk.	unk.	**unk.**	unk.	**unk.**	unk.		unk.	
unk.	**unk.**	unk.	**unk.**	unk.	**unk.**	unk.	**unk.**	unk.	unk.	**unk.**	unk.	unk.	177	**18**	417	31	unk.	unk.	**unk.**	unk.	**unk.**	unk.	**unk.**	unk.	H
unk.	**unk.**	unk.	**unk.**	unk.	**unk.**	unk.	**unk.**	unk.	unk.	**unk.**	unk.	1.70	281	**28**	18.2	0.8	unk.	unk.	**unk.**	unk.	**unk.**	unk.		unk.	
unk.	**unk.**	unk.	**unk.**	unk.	**unk.**	unk.	**unk.**	unk.	unk.	**unk.**	unk.	unk.	177	**18**	156	239	unk.	unk.	**unk.**	unk.	**unk.**	unk.	**unk.**	unk.	I
unk.	**unk.**	unk.	**unk.**	unk.	**unk.**	unk.	**unk.**	unk.	unk.	**unk.**	unk.	1.70	114	**11**	6.8	6.1	unk.	unk.	**unk.**	unk.	**unk.**	unk.		unk.	
unk.	**unk.**	unk.	**unk.**	unk.	**unk.**	unk.	**unk.**	unk.	unk.	**unk.**	unk.	unk.	153	**15**	398	82	unk.	unk.	**unk.**	unk.	**unk.**	unk.	**unk.**	unk.	J
unk.	**unk.**	unk.	**unk.**	unk.	**unk.**	unk.	**unk.**	unk.	unk.	**unk.**	unk.	1.14	220	**22**	17.3	2.1	unk.	unk.	**unk.**	unk.	**unk.**	unk.		unk.	
unk.	**unk.**	unk.	**unk.**	0(a)	**0**	unk.	**unk.**	unk.	0(a)	**0**	unk.	unk.	unk.	**unk.**	unk.	unk.	unk.	unk.	**unk.**	unk.	**unk.**	unk.	**unk.**	unk.	K
unk.	**unk.**	unk.	**unk.**	unk.	**unk.**	unk.	**unk.**	unk.	unk.	**unk.**	unk.	0.02	unk.	**unk.**	unk.	unk.	unk.	unk.	**unk.**	tr(a)	**0**	tr(a)		unk.	
0.03	**2**	0.06	**3**	0.00	**0**	tr	**0**	0	0(a)	**0**	unk.	unk.	18	**2**	1	159	tr	0.4	**2**	0.1	**1**	unk.	**unk.**	unk.	L
0.3	**1**	4	**1**	0.13	**1**	835	**17**	84	unk.	**unk.**	unk.	0.39	15	**2**	0.1	4.1	0.0	10	**3**	0.06	**3**	tr(a)		unk.	
0.02	**1**	0.05	**3**	0.00	**0**	0.001	**0**	0	0(a)	**0**	0.2	unk.	18	**2**	2	159	tr	0.4	**2**	0.2	**1**	unk.	**unk.**	unk.	M
0.2	**1**	unk.	**unk.**	0.13	**1**	830	**17**	83	0.2	**1**	unk.	0.37	16	**2**	0.1	4.1	0.0	9	**2**	tr(a)	**0**	tr(a)		unk.	
0.05	**3**	0.05	**3**	0.00	**0**	tr	**0**	0	0(a)	**0**	0.1	unk.	5	**1**	2	148	2	0.3	**2**	0.1	**1**	unk.	**unk.**	77	N
0.3	**2**	5	**1**	0.11	**1**	775	**16**	77	0.1	**0**	unk.	0.39	9	**1**	0.1	3.8	0.1	9	**2**	0.07	**4**	0		0	
0.03	**2**	0.04	**2**	0.00	**0**	unk.	**unk.**	0	0(a)	**0**	unk.	unk.	19	**2**	1	162	unk.	0.4	**2**	unk.	**unk.**	unk.	**unk.**	37	O
0.3	**1**	unk.	**unk.**	unk.	**unk.**	77	**2**	8	unk.	**unk.**	unk.	0.51	17	**2**	0.1	4.1	unk.	12	**3**	0.08	**4**	tr(a)		19	
0.02	**2**	0.04	**2**	0.00	**0**	unk.	**unk.**	0	0(a)	**0**	unk.	unk.	19	**2**	1	161	unk.	0.4	**2**	unk.	**unk.**	unk.	**unk.**	36	P
0.2	**1**	4	**1**	0.15	**2**	74	**2**	7	unk.	**unk.**	unk.	0.50	16	**2**	0.0	4.1	unk.	11	**3**	0.08	**4**	tr(a)		19	
0.04	**2**	0.02	**1**	0.00	**0**	tr	**0**	0	0(a)	**0**	0.1	unk.	6	**1**	1	124	tr	0.1	**1**	0.1	**0**	unk.	**unk.**	42	Q
0.3	**1**	5	**1**	0.17	**2**	71	**1**	7	0.1	**0**	unk.	0.39	8	**1**	0.1	3.2	0.0	11	**3**	0.00	**0**	0		0	
unk.	**unk.**	0.01	**0**	0.00	**0**	unk.	**unk.**	0(a)	0(a)	**0**	unk.	unk.	8	**1**	0	28	unk.	0.2	**1**	0.1	**0**	unk.	**unk.**	unk.	R
unk.	**unk.**	unk.	**unk.**	unk.	**unk.**	unk.	**unk.**	unk.	unk.	**unk.**	unk.	0.10	3	**0**	0.0	0.7	unk.	1	**0**	unk.	**unk.**	unk.		unk.	
0.02	**1**	0.03	**2**	0.00	**0**	unk.	**unk.**	0	0(a)	**0**	unk.	unk.	2	**0**	tr	33	1	0.1	**1**	unk.	**unk.**	unk.	**unk.**	270	S
tr	**0**	unk.	**unk.**	0.04	**0**	6	**0**	1	unk.	**unk.**	tr	0.07	6	**1**	0.0	0.8	0.0	3	**1**	tr	**0**	unk.		unk.	

	FOOD	Portion	Weight in grams / Conversion for 100 g	Kilocalories / H_2O g	Total carbohydrate g / Total fats g	Crude fiber g / Dietary fiber g	Total protein g / Total sugar g	% USRDA	Arginine mg / Histidine mg	Isoleucine mg / Leucine mg	Lysine mg / Methionine mg	Phenylalanine mg / Threonine mg	Valine mg / Tryptophan mg	Cystine mg / Tyrosine mg	Polyunsat. fatty acids g / Monounsat. fatty acids g	Saturated fatty acids g / P/S ratio	Linoleic acid g / Cholesterol mg	Thiamin mg / Ascorbic acid mg	% USRDA
A	CHICK PEAS (garbanzos),dry,raw	1/2 C	42	133	20.8	unk.	8.4	**13**	793	376	578	484	376	121	unk.	unk.	unk.	0.21	**14**
			2.40	4.1	2.4	6.24	4.2		228	632	107	322	67	242	unk.	unk.	0(a)	1	**2**
B	CHICKEN FAT	1 Tbsp	14	126	0.0	tr(a)	0.0	**0**	unk.	0(a)	0(a)	0(a)	0(a)	0(a)	2.93	4.17	2.73	0.00	**0**
			7.14	unk.	14.0	0(a)	unk.		0(a)	0(a)	0(a)	0(a)	0(a)	0(a)	5.22	0.70	12	tr(a)	**0**
C	CHICKEN,BROILER/FRYER,BACK,meat	1 piece	240	794	24.6	0.00	52.7	**117**	3223	2621	4097	2107	2570	734	12.48	13.99	11.16	0.29	**19**
	w/skin,batter dipped,fried		*0.42*	106.8	52.6	0(a)	tr(a)		1524	3862	1375	2150	602	1706	19.25	0.89	211	0	**0**
D	chicken,broiler/fryer,back,meat	1 piece	144	477	9.4	0.00	40.0	**89**	2470	2002	3198	1580	1947	544	6.93	8.08	6.06	0.16	**11**
	w/skin,flour coated,fried		*0.69*	63.3	29.9	0(a)	tr(a)		1174	2926	1058	1644	452	1295	10.15	0.86	128	0	**0**
E	chicken,broiler/fryer,back,meat	1 piece	106	318	0.0	0.00	27.5	**61**	1761	1328	2185	1056	1319	375	4.90	6.17	4.28	0.06	**4**
	w/skin,roasted		*0.94*	56.7	22.2	0(a)	0(a)		782	1965	715	1123	299	862	7;25	0.79	93	0	**0**
F	chicken,broiler/fryer,back,meat	1 piece	122	315	0.0	0.00	27.1	**60**	1730	1308	2151	1039	1298	368	4.88	6.12	4.26	0.05	**3**
	w/skin,stewed		*0.82*	74.3	22.1	0(a)	0(a)		771	1935	704	1104	294	849	7.17	0.80	95	0	**0**
G	chicken,broiler/fryer,back,meat	1 piece	116	334	6.6	0.00	34.8	**77**	2080	1829	2900	1394	1724	455	4.22	4.78	3.57	0.13	**9**
	w/o skin,fried		*0.86*	55.5	17.8	0(a)	unk.		1072	2611	955	1459	408	1172	5.73	0.88	108	0	**0**
H	chicken,broiler/fryer,back,meat	1 piece	80	191	0.0	0.00	22.5	**50**	1360	1190	1916	894	1118	289	2.44	2.88	2.02	0.06	**4**
	w/o skin,roasted		*1.25*	47.0	10.5	0(a)	0(a)		700	1692	624	953	263	762	3.22	0.85	72	0	**0**
I	chicken,broiler/fryer,back,meat	1 piece	84	176	0.0	0.00	21.3	**47**	1282	1122	1806	843	1054	272	2.20	2.55	1.81	0.04	**3**
	w/o skin,stewed		*1.19*	54.0	9.4	0(a)	0(a)		660	1595	588	898	249	717	2.81	0.86	71	0	**0**
J	CHICKEN,BROILER/FRYER,BREAST,meat	1/2 breast	140	364	12.6	0.00	34.8	**77**	2121	1753	2758	1387	1702	475	4.31	4.93	3.89	0.17	**11**
	w/skin,batter dipped,fried		*0.71*	72.3	18.5	0(a)	tr(a)		1022	2562	920	1431	399	1137	6.96	0.88	119	0	**0**
K	chicken,broiler/fryer,breast,meat	1/2 breast	98	218	1.6	0.00	31.2	**69**	1915	1599	2581	1228	1531	410	1.92	2.40	1.66	0.08	**5**
	w/skin,flour dipped,fried		*1.02*	55.5	8.7	0(a)	tr(a)		940	2306	845	1300	357	1028	2.94	0.80	87	0	**0**
L	chicken,broiler/fryer,breast,meat	1/2 breast	98	193	0.0	0.00	29.2	**65**	1800	1494	2424	1146	1432	382	1.63	2.15	1.38	0.07	**5**
	w/skin,roasted		*1.02*	61.2	7.6	0(a)	0(a)		879	2154	791	1219	333	960	2.46	0.76	82	0	**0**
M	chicken,broiler/fryer,breast,meat	1/2 breast	110	202	0.0	0.00	30.1	**67**	1858	1541	2499	1181	1476	395	1.74	2.29	1.48	0.04	**3**
	w/skin,stewed		*0.91*	72.8	8.2	0(a)	0(a)		906	2221	815	1257	343	990	2.63	0.76	82	0	**0**
N	chicken,broiler/fryer,breast,meat	1/2 breast	86	161	0.4	0.00	28.8	**64**	1733	1518	2439	1142	1427	368	0.92	1.11	0.74	0.07	**5**
	w/o skin,fried		*1.16*	51.8	4.1	0(a)	unk.		892	2158	795	1214	335	970	1.31	0.83	78	0	**0**
O	chicken,broiler/fryer,breast,meat	1/2 breast	86	142	0.0	0.00	26.7	**59**	1609	1409	2266	1059	1324	341	0.66	0.87	0.51	0.06	**4**
	w/o skin,roasted		*1.16*	56.1	3.1	0(a)	0(a)		828	2002	739	1127	311	900	0.89	0.76	73	0	**0**
P	chicken,broiler/fryer,breast,meat	1/2 breast	95	143	0.0	0.00	27.5	**61**	1661	1453	2339	1092	1365	352	0.63	0.81	0.47	0.04	**3**
	w/o skin,stewed		*1.05*	64.9	2.9	0(a)	0(a)		855	2066	762	1163	322	929	0.83	0.78	73	0	**0**
Q	CHICKEN,BROILER/FRYER,DARK MEAT	2 pieces(2	85	253	8.0	0.00	18.6	**41**	1133	930	1456	743	907	257	3.77	4.21	3.37	0.10	**7**
	w/skin,batter dipped,fried	3/8x2x1/2")	*1.18*	41.5	15.8	0(a)	tr(a)		541	1364	488	761	212	604	5.81	0.89	76	0	**0**
R	chicken,broiler/fryer,dark meat	2 pieces(2	85	242	3.5	0.00	23.1	**51**	1429	1164	1872	910	1127	310	3.32	3.89	2.91	0.08	**6**
	w/skin,flour coated,fried	3/8x2x1/2")	*1.18*	43.2	14.4	0(a)	tr(a)		683	1695	616	955	262	751	4.87	0.85	78	0	**0**
S	chicken,broiler/fryer,dark meat	2 pieces(2	85	215	0.0	0.00	22.1	**49**	1389	1095	1789	856	1069	296	2.97	3.71	2.58	0.06	**4**
	w/skin,roasted	3/8x2x1/2")	*1.18*	49.8	13.4	0(a)	0(a)		645	1601	585	910	246	707	4.34	0.80	77	0	**0**

Riboflavin mg / Niacin mg	% USRDA	Vitamin B_6 mg / Folacin mcg	% USRDA	Vitamin B_{12} mcg / Pantothenic acid mg	% USRDA	Biotin mg / Vitamin A IU	% USRDA	Preformed A RE / Beta carotene RE	Vitamin D IU / Vitamin E IU	% USRDA	Total tocopherol mg / Alpha tocopherol mg	Other tocopherol mg / Total ash g	Calcium mg / Phosphorus mg	% USRDA	Sodium mg / Sodium meq	Potassium mg / Potassium meq	Chlorine mg / Chlorine meq	Iron mg / Magnesium mg	% USRDA	Zinc mg / Copper mg	% USRDA	Iodine mcg / Selenium mcg	% USRDA	Manganese mcg / Chromium mcg	
0.06	**4**	unk.	**unk.**	0.00	**0**	unk.	**unk.**	0(a)	0	**0**	unk.	unk.	58	**6**	17	333	25	2.7	**15**	unk.	**unk.**	unk.	**unk.**	unk.	A
0.6	**3**	75	**19**	unk.	**unk.**	unk.	**unk.**	13	unk.	**unk.**	unk.	unk.	125	**13**	0.7	8.5	0.7	67	**17**	0.32	**16**		**unk.**		
0.00	**0**	unk.	**unk.**	tr(a)	**0**	tr(a)	**0**	0	0(a)	**0**	0.4	unk.	0	**0**	unk.	tr(a)	tr(a)	0.0	**0**	tr(a)	**0**	unk.	**unk.**	tr(a)	B
0.0	**0**	0(a)	**0**	tr(a)	**0**	0	**0**	0	0.5	**2**	unk.	tr(a)	0	**0**	unk.	0.0	0.0	unk.	**unk.**	tr(a)	**0**	tr(a)		tr(a)	
0.50	**30**	0.55	**28**	0.62	**10**	unk.	**unk.**	unk.	unk.	**unk.**	unk.	unk.	62	**6**	761	432	unk.	3.6	**20**	4.7	**31**	unk.	**unk.**	149	C
14.0	**70**	22	**5**	2.16	**22**	286	**6**	unk.	unk.	**unk.**	unk.	3.29	329	**33**	33.1	11.0	unk.	46	**11**	0.19	**10**	unk.		unk.	
0.35	**20**	0.43	**22**	0.40	**7**	unk.	**unk.**	unk.	unk.	**unk.**	unk.	unk.	35	**4**	130	325	unk.	2.3	**13**	3.6	**24**	unk.	**unk.**	72	D
10.5	**53**	12	**3**	1.57	**16**	177	**4**	unk.	unk.	**unk.**	unk.	1.45	239	**24**	5.6	8.3	unk.	33	**8**	0.13	**7**	unk.		unk.	
0.21	**13**	0.29	**14**	0.29	**5**	unk.	**unk.**	unk.	unk.	**unk.**	unk.	unk.	22	**2**	92	223	unk.	1.5	**8**	2.4	**16**	unk.	**unk.**	23	E
7.1	**36**	6	**2**	1.07	**11**	369	**7**	unk.	unk.	**unk.**	unk.	0.98	163	**16**	4.0	5.7	unk.	21	**5**	0.08	**4**	unk.		unk.	
0.18	**11**	0.18	**9**	0.22	**4**	unk.	**unk.**	unk.	unk.	**unk.**	unk.	unk.	22	**2**	78	177	unk.	1.5	**8**	2.4	**16**	unk.	**unk.**	24	F
5.3	**27**	6	**2**	0.85	**9**	376	**8**	unk.	unk.	**unk.**	unk.	0.78	146	**15**	3.4	4.5	unk.	20	**5**	0.08	**4**	unk.		unk.	
0.29	**17**	0.41	**20**	0.36	**6**	unk.	**unk.**	unk.	unk.	**unk.**	unk.	unk.	30	**3**	115	291	unk.	1.9	**11**	3.2	**22**	unk.	**unk.**	55	G
8.9	**45**	10	**3**	1.39	**14**	114	**2**	unk.	unk.	**unk.**	unk.	1.33	204	**20**	5.0	7.4	unk.	29	**7**	0.11	**6**	unk.		unk.	
0.18	**10**	0.27	**14**	0.24	**4**	unk.	**unk.**	unk.	unk.	**unk.**	unk.	unk.	19	**2**	77	190	unk.	1.1	**6**	2.1	**14**	unk.	**unk.**	18	H
5.7	**28**	6	**1**	0.90	**9**	76	**2**	unk.	unk.	**unk.**	unk.	0.86	132	**13**	3.3	4.8	unk.	18	**4**	0.06	**3**	unk.		unk.	
0.14	**8**	0.17	**8**	0.18	**3**	unk.	**unk.**	unk.	unk.	**unk.**	unk.	unk.	18	**2**	56	133	unk.	1.1	**6**	2.0	**13**	unk.	**unk.**	17	I
3.8	**19**	6	**2**	0.70	**7**	76	**2**	unk.	unk.	**unk.**	unk.	0.61	109	**11**	2.4	3.4	unk.	14	**4**	0.06	**3**	unk.		unk.	
0.21	**12**	0.60	**30**	0.42	**7**	unk.	**unk.**	unk.	unk.	**unk.**	unk.	unk.	28	**3**	385	281	unk.	1.7	**10**	1.3	**9**	unk.	**unk.**	84	J
14.7	**74**	8	**2**	1.15	**12**	94	**2**	unk.	unk.	**unk.**	0.6	1.86	259	**26**	16.8	7.2	unk.	34	**8**	0.08	**4**	unk.		unk.	
0.13	**8**	0.57	**28**	0.33	**6**	unk.	**unk.**	unk.	unk.	**unk.**	unk.	unk.	16	**2**	74	254	unk.	1.2	**7**	1.1	**7**	unk.	**unk.**	25	K
13.5	**67**	4	**1**	0.98	**10**	49	**1**	unk.	unk.	**unk.**	0.4	1.04	228	**23**	3.2	6.5	unk.	29	**7**	0.06	**3**	unk.		unk.	
0.12	**7**	0.55	**27**	0.31	**5**	unk.	**unk.**	unk.	unk.	**unk.**	unk.	unk.	14	**1**	70	240	unk.	1.0	**6**	1.0	**7**	unk.	**unk.**	18	L
12.5	**62**	4	**1**	0.92	**9**	91	**2**	unk.	unk.	**unk.**	0.4	0.97	210	**21**	3.0	6.1	unk.	26	**7**	0.05	**2**	unk.		unk.	
0.13	**8**	0.32	**16**	0.23	**4**	unk.	**unk.**	unk.	unk.	**unk.**	unk.	unk.	14	**1**	68	196	unk.	1.0	**6**	1.1	**7**	unk.	**unk.**	20	M
8.6	**43**	3	**1**	0.60	**6**	90	**2**	unk.	unk.	**unk.**	0.5	0.92	172	**17**	3.0	5.0	unk.	24	**6**	0.05	**2**	unk.		unk.	
0.11	**7**	0.55	**28**	0.32	**5**	unk.	**unk.**	unk.	unk.	**unk.**	unk.	unk.	14	**1**	68	237	unk.	1.0	**5**	0.9	**6**	unk.	**unk.**	18	N
12.7	**64**	3	**1**	0.89	**9**	20	**0**	unk.	unk.	**unk.**	0.4	0.98	212	**21**	3.0	6.1	unk.	27	**7**	0.05	**2**	unk.		unk.	
0.09	**6**	0.52	**26**	0.29	**5**	unk.	**unk.**	unk.	unk.	**unk.**	unk.	unk.	13	**1**	64	220	unk.	0.9	**5**	0.9	**6**	unk.	**unk.**	15	O
11.8	**59**	3	**1**	0.83	**8**	18	**0**	unk.	unk.	**unk.**	0.4	0.91	196	**20**	2.8	5.6	unk.	25	**6**	0.04	**2**	unk.		unk.	
0.11	**7**	0.31	**16**	0.22	**4**	unk.	**unk.**	unk.	unk.	**unk.**	unk.	unk.	12	**1**	60	178	unk.	0.8	**5**	0.9	**6**	unk.	**unk.**	17	P
8.0	**40**	3	**1**	0.54	**5**	18	**0**	unk.	unk.	**unk.**	0.4	0.86	157	**16**	2.6	4.5	unk.	23	**6**	0.04	**2**	unk.		unk.	
0.19	**11**	0.00	**0**	0.23	**4**	unk.	**unk.**	unk.	unk.	**unk.**	unk.	unk.	18	**2**	251	157	unk.	1.2	**7**	1.8	**12**	unk.	**unk.**	49	Q
4.8	**24**	8	**2**	0.81	**8**	88	**2**	unk.	unk.	**unk.**	unk.	1.11	123	**12**	10.9	4.0	unk.	17	**4**	0.07	**3**	unk.		unk.	
0.20	**12**	0.27	**14**	0.25	**4**	unk.	**unk.**	unk.	unk.	**unk.**	unk.	unk.	14	**1**	76	195	unk.	1.3	**7**	2.2	**15**	unk.	**unk.**	33	R
5.8	**29**	7	**2**	0.98	**10**	88	**2**	unk.	unk.	**unk.**	unk.	0.82	150	**15**	3.3	5.0	unk.	20	**5**	0.07	**4**	unk.		unk.	
0.18	**11**	0.26	**13**	0.25	**4**	unk.	**unk.**	unk.	unk.	**unk.**	unk.	unk.	13	**1**	74	187	unk.	1.2	**6**	2.1	**14**	unk.	**unk.**	18	S
5.4	**27**	6	**2**	0.94	**9**	171	**3**	unk.	unk.	**unk.**	unk.	0.78	143	**14**	3.2	4.8	unk.	19	**5**	0.07	**3**	unk.		unk.	

	FOOD	Portion	Weight in grams *Conversion for 100 g*	Kilocalories H_2O g	Total carbohydrate g Total fats g	Crude fiber g Dietary fiber g	Total protein g Total sugar g	**% USRDA**	Arginine mg Histidine mg	Isoleucine mg Leucine mg	Lysine mg Methionine mg	Phenylalanine mg Threonine mg	Valine mg Tryptophan mg	Cystine mg Tyrosine mg	Polyunsat. fatty acids g Monounsat. fatty acids g	Saturated fatty acids g P/S ratio	Linoleic acid g Cholesterol mg	Thiamin mg Ascorbic acid mg	**% USRDA**
A	chicken,broiler/fryer,dark meat	2 pieces(2-	85	198	0.0	0.00	20.0	**44**	1255	992	1620	774	968	267	2.75	3.45	2.40	0.04	**3**
	w/skin,stewed	3/8x2x1/2")	*1.18*	53.5	12.5	0(a)	0(a)		584	1449	530	824	222	641	4.03	0.80	70	0	**0**
B	chicken,broiler/fryer,dark meat	1/2 C	70	167	1.8	0.00	20.3	**45**	1219	1070	1709	809	1007	262	1.94	2.18	2.33	0.06	**4**
	w/o skin,fried,chopped		*1.43*	39.0	8.1	0(a)	unk.		628	1523	559	855	238	685	2.60	0.89	67	0	**0**
C	chicken,broiler/fryer,dark meat	1/2 C	70	143	0.0	0.00	19.2	**43**	1156	1011	1627	760	950	245	1.58	1.86	1.31	0.05	**3**
	w/o skin,roasted,chopped		*1.43*	44.1	6.8	0(a)	0(a)		594	1437	530	809	224	647	2.08	0.85	65	0	**0**
D	chicken,broiler/fryer,dark meat	1/2 C	70	134	0.0	0.00	18.2	**40**	1096	960	1544	721	902	232	1.46	1.71	1.21	0.04	**3**
	w/o skin,stewed,chopped		*1.43*	46.1	6.3	0(a)	0(a)		564	1364	503	768	212	614	1.90	0.85	62	0	**0**
E	CHICKEN,BROILER/FRYER,DRUMSTICK,	1 piece	72	193	6.0	0.00	15.8	**35**	962	798	1253	631	774	216	2.73	2.98	2.45	0.08	**5**
	meat w/skin,batter dipped,fried		*1.39*	38.0	11.3	0(a)	tr(a)		465	1165	418	650	181	518	4.23	0.92	62	0	**0**
F	chicken,broiler/fryer,drumstick,meat	1 piece	49	120	0.8	0.00	13.2	**29**	815	672	1085	518	646	174	1.58	1.79	1.39	0.04	**3**
	w/skin,flour coated,fried		*2.04*	27.8	6.7	0(a)	tr(a)		395	972	355	549	150	432	2.33	0.88	44	0	**0**
G	chicken,broiler/fryer,drumstick,meat	1 piece	52	112	0.0	0.00	14.1	**31**	876	708	1152	548	684	186	1.30	1.59	1.11	0.04	**2**
	w/skin,roasted		*1.92*	32.6	5.8	0(a)	0(a)		417	1028	376	583	158	456	1.82	0.82	47	0	**0**
H	chicken,broiler/fryer,drumstick,meat	1 piece	57	116	0.0	0.00	14.4	**32**	897	730	1187	563	704	191	1.36	1.66	1.16	0.03	**2**
	w/skin,stewed		*1.75*	37.1	6.1	0(a)	0(a)		429	1057	388	599	163	470	1.91	0.82	47	0	**0**
I	chicken,broiler/fryer,drumstick,meat	1 piece	42	82	0.0	0.00	12.0	**27**	725	635	1021	477	596	154	0.83	0.89	0.69	0.03	**2**
	w/o skin,fried		*2.38*	26.1	3.4	0(a)	unk.		373	902	333	508	141	406	1.08	0.92	39	0	**0**
J	chicken,broiler/fryer,drumstick,meat	1 piece	44	76	0.0	0.00	12.4	**28**	751	657	1057	494	617	159	0.60	0.65	0.48	0.04	**2**
	w/o skin,roasted		*2.27*	29.4	2.5	0(a)	0(a)		386	934	345	526	145	420	0.69	0.93	41	0	**0**
K	chicken,broiler/fryer,drumstick,meat	1 piece	46	78	0.0	0.00	12.6	**28**	763	668	1075	502	627	162	0.63	0.69	0.50	0.02	**2**
	w/o skin,stewed		*2.17*	31.1	2.6	0(a)	0(a)		393	949	350	535	148	427	0.74	0.91	40	0	**0**
L	CHICKEN,BROILER/FRYER,LEG,meat	1 piece	158	431	13.8	0.00	34.4	**76**	2095	1730	2714	1375	1683	471	6.08	6.76	5.45	0.19	**13**
	w/skin,batter dipped,fried		*0.63*	82.3	25.5	unk.	tr(a)		1008	2533	907	1413	395	1123	9.40	0.90	142	0	**0**
M	chicken,broiler/fryer,leg,meat	1 piece	112	284	2.8	0.00	30.1	**67**	1858	1520	2452	1180	1467	400	3.73	4.37	3.26	0.10	**7**
	w/skin,flour coated,fried		*0.89*	61.9	16.2	unk.	tr(a)		893	2205	804	1245	340	980	5.45	0.85	105	0	**0**
N	chicken,broiler/fryer,leg,meat	1 piece	114	264	0.0	0.00	29.6	**66**	1847	1487	2421	1153	1440	393	3.42	4.24	2.95	0.08	**5**
	w/skin,roasted		*0.88*	69.4	15.3	0(a)	0(a)		876	2160	791	1225	332	959	4.92	0.81	105	0	**0**
O	chicken,broiler/fryer,leg,meat	1 piece	125	275	0.0	0.00	30.2	**67**	1884	1520	2474	1178	1471	401	3.59	4.46	3.11	0.06	**4**
	w/skin,stewed		*0.80*	80.0	16.1	0(a)	0(a)		894	2208	808	1253	339	980	5.20	0.80	105	0	**0**
P	chicken,broiler/fryer,leg,meat	1 piece	94	196	0.6	0.00	26.7	**59**	1607	1407	2261	1059	1323	342	2.09	2.34	1.75	0.08	**5**
	w/o skin,fried		*1.06*	57.0	8.8	0(a)	unk.		827	2001	737	1126	312	901	2.78	0.89	93	0	**0**
Q	chicken,broiler/fryer,leg,meat	1 piece	95	181	0.0	0.00	25.7	**57**	1548	1356	2181	1018	1273	329	1.87	2.18	1.54	0.08	**5**
	w/o skin,roasted		*1.05*	61.5	8.0	0(a)	0(a)		797	1927	711	1085	300	866	2.41	0.86	89	0	**0**
R	chicken,broiler/fryer,leg,meat	1 piece	101	187	0.0	0.00	26.5	**59**	1600	1401	2253	1052	1316	339	1.90	2.22	1.57	0.06	**4**
	w/o skin,stewed		*0.99*	67.1	8.1	0(a)	0(a)		823	1991	734	1120	310	896	2.46	0.85	90	0	**0**
S	CHICKEN,BROILER/FRYER,LIGHT MEAT	3x2-1/8x3/4"	85	235	8.1	0.00	20.0	**45**	1228	995	1565	797	975	277	3.06	3.50	2.75	0.09	**6**
	w/skin,batter dipped,fried		*1.18*	42.7	13.1	0(a)	tr(a)		580	1465	524	819	228	648	4.92	0.87	71	0	**0**

Riboflavin mg / Niacin mg	% USRDA	Vitamin B_6 mg / Folacin mcg	% USRDA	Vitamin B_{12} mcg / Pantothenic acid mg	% USRDA	Biotin mg / Vitamin A IU	% USRDA	Preformed A RE / Beta carotene RE	Vitamin D IU / Vitamin E IU	% USRDA	Total tocopherol mg / Alpha tocopherol mg	Other tocopherol mg / Total ash g	Calcium mg / Phosphorus mg	% USRDA	Sodium mg / Sodium meq	Potassium mg / Potassium meq	Chlorine mg / Chlorine meq	Iron mg / Magnesium mg	% USRDA	Zinc mg / Copper mg	% USRDA	Iodine mcg / Selenium mcg	% USRDA	Manganese mcg / Chromium mcg	
0.15	**9**	0.14	**7**	0.17	**3**	unk.	**unk.**	unk.	unk.	**unk.**	unk.	unk.	12	**1**	59	141	unk.	1.1	**6**	1.9	**13**	unk.	**unk.**	16	A
3.8	**19**	5	**1**	0.66	**7**	158	**3**	unk.	unk.	**unk.**	unk.	0.63	113	**11**	2.6	3.6	unk.	15	**4**	0.06	**3**	unk.		unk.	
0.17	**10**	0.26	**13**	0.23	**4**	unk.	**unk.**	unk.	unk.	**unk.**	unk.	unk.	13	**1**	68	177	unk.	1.0	**6**	2.0	**14**	unk.	**unk.**	23	B
4.9	**25**	6	**2**	0.88	**9**	55	**1**	unk.	unk.	**unk.**	unk.	0.76	131	**13**	2.9	4.5	unk.	17	**4**	0.06	**3**	unk.		unk.	
0.16	**10**	0.25	**13**	0.22	**4**	unk.	**unk.**	unk.	unk.	**unk.**	unk.	unk.	10	**1**	65	168	unk.	0.9	**5**	2.0	**13**	unk.	**unk.**	15	C
4.6	**23**	6	**1**	0.85	**9**	50	**1**	unk.	unk.	**unk.**	unk.	0.71	125	**13**	2.8	4.3	unk.	16	**4**	0.06	**3**	unk.		unk.	
0.14	**8**	0.15	**7**	0.15	**3**	unk.	**unk.**	unk.	unk.	**unk.**	unk.	unk.	10	**1**	52	127	unk.	1.0	**5**	1.9	**12**	unk.	**unk.**	14	D
3.2	**17**	5	**1**	0.62	**6**	48	**1**	unk.	unk.	**unk.**	unk.	0.59	100	**10**	2.3	3.2	unk.	14	**4**	0.05	**3**	unk.		unk.	
0.19	**11**	0.19	**10**	0.20	**3**	unk.	**unk.**	unk.	unk.	**unk.**	unk.	unk.	12	**1**	194	134	unk.	1.0	**5**	1.7	**11**	unk.	**unk.**	37	E
3.7	**18**	6	**2**	0.73	**7**	62	**1**	unk.	unk.	**unk.**	unk.	0.90	106	**11**	8.4	3.4	unk.	14	**4**	0.05	**3**	unk.		unk.	
0.11	**7**	0.17	**9**	0.16	**3**	unk.	**unk.**	unk.	unk.	**unk.**	unk.	unk.	6	**1**	44	112	unk.	0.7	**4**	1.4	**9**	unk.	**unk.**	14	F
3.0	**15**	4	**1**	0.60	**6**	41	**1**	unk.	unk.	**unk.**	unk.	0.47	86	**9**	1.9	2.9	unk.	11	**3**	0.04	**2**	unk.		unk.	
0.11	**7**	0.18	**9**	0.17	**3**	unk.	**unk.**	unk.	unk.	**unk.**	unk.	unk.	6	**1**	47	119	unk.	0.7	**4**	1.5	**10**	unk.	**unk.**	11	G
3.1	**16**	4	**1**	0.63	**6**	52	**1**	unk.	unk.	**unk.**	unk.	0.50	91	**9**	2.0	3.0	unk.	12	**3**	0.04	**2**	unk.		unk.	
0.11	**6**	0.11	**5**	0.13	**2**	unk.	**unk.**	unk.	unk.	**unk.**	unk.	unk.	6	**1**	43	105	unk.	0.8	**4**	1.5	**10**	unk.	**unk.**	11	H
2.4	**12**	4	**1**	0.49	**5**	52	**1**	unk.	unk.	**unk.**	unk.	0.47	80	**8**	1.9	2.7	unk.	11	**3**	0.04	**2**	unk.		unk.	
0.10	**6**	0.16	**8**	0.15	**2**	unk.	**unk.**	unk.	unk.	**unk.**	unk.	unk.	5	**1**	40	105	unk.	0.6	**3**	1.4	**9**	unk.	**unk.**	9	I
2.6	**13**	4	**1**	0.55	**6**	26	**1**	unk.	unk.	**unk.**	unk.	0.45	78	**8**	1.8	2.7	unk.	10	**3**	0.03	**2**	unk.		unk.	
0.10	**6**	0.17	**9**	0.15	**3**	unk.	**unk.**	unk.	unk.	**unk.**	unk.	unk.	5	**1**	42	108	unk.	0.6	**3**	1.4	**9**	unk.	**unk.**	9	J
2.7	**13**	4	**1**	0.57	**6**	26	**1**	unk.	unk.	**unk.**	unk.	0.46	81	**8**	1.8	2.8	unk.	11	**3**	0.03	**2**	unk.		unk.	
0.10	**6**	0.11	**5**	0.11	**2**	unk.	**unk.**	unk.	unk.	**unk.**	unk.	unk.	5	**1**	37	92	unk.	0.6	**4**	1.4	**9**	unk.	**unk.**	9	K
2.0	**10**	4	**1**	0.44	**4**	26	**1**	unk.	unk.	**unk.**	unk.	0.42	69	**7**	1.6	2.3	unk.	10	**2**	0.04	**2**	unk.		unk.	
0.35	**20**	0.43	**21**	0.44	**7**	unk.	**unk.**	unk.	unk.	**unk.**	unk.	unk.	28	**3**	441	299	unk.	2.2	**12**	3.4	**23**	unk.	**unk.**	85	L
8.6	**43**	14	**4**	1.57	**16**	144	**3**	unk.	unk.	**unk.**	unk.	1.99	240	**24**	19.2	7.6	unk.	32	**8**	0.13	**6**	unk.		unk.	
0.27	**16**	0.38	**19**	0.35	**6**	unk.	**unk.**	unk.	unk.	**unk.**	unk.	unk.	15	**2**	99	261	unk.	1.6	**9**	3.0	**20**	unk.	**unk.**	36	M
7.3	**37**	9	**2**	1.34	**13**	103	**2**	unk.	unk.	**unk.**	unk.	1.06	204	**20**	4.3	6.7	unk.	27	**7**	0.10	**5**	unk.		unk.	
0.24	**14**	0.38	**19**	0.35	**6**	unk.	**unk.**	unk.	unk.	**unk.**	unk.	unk.	14	**1**	99	256	unk.	1.5	**8**	3.0	**20**	unk.	**unk.**	24	N
7.1	**35**	8	**2**	1.32	**13**	154	**3**	unk.	unk.	**unk.**	unk.	1.05	198	**20**	4.3	6.6	unk.	26	**7**	0.09	**4**	unk.		unk.	
0.24	**14**	0.22	**11**	0.25	**4**	unk.	**unk.**	unk.	unk.	**unk.**	unk.	0.99	14	**1**	91	220	unk.	1.7	**9**	3.0	**20**	unk.	**unk.**	24	O
5.7	**29**	8	**2**	1.01	**10**	155	**3**	unk.	unk.	**unk.**	unk.	unk.	174	**17**	4.0	5.6	unk.	25	**6**	0.09	**4**	unk.		unk.	
0.23	**14**	0.37	**18**	0.32	**5**	unk.	**unk.**	unk.	unk.	**unk.**	unk.	unk.	12	**1**	90	239	unk.	1.3	**7**	2.8	**19**	unk.	**unk.**	23	P
6.3	**31**	8	**2**	1.22	**12**	62	**1**	unk.	unk.	**unk.**	unk.	0.99	181	**18**	3.9	6.1	unk.	23	**6**	0.08	**4**	unk.		unk.	
0.22	**13**	0.35	**18**	0.30	**5**	unk.	**unk.**	unk.	unk.	**unk.**	unk.	unk.	11	**1**	86	230	unk.	1.2	**7**	2.7	**18**	unk.	**unk.**	20	Q
6.0	**30**	8	**2**	1.18	**12**	60	**1**	unk.	unk.	**unk.**	unk.	0.95	174	**17**	3.8	5.9	unk.	23	**6**	0.08	**4**	unk.		unk.	
0.22	**13**	0.21	**11**	0.23	**4**	unk.	**unk.**	unk.	unk.	**unk.**	unk.	unk.	11	**1**	79	192	unk.	1.4	**8**	2.8	**19**	unk.	**unk.**	19	R
4.8	**24**	8	**2**	0.92	**9**	61	**1**	unk.	unk.	**unk.**	unk.	0.89	151	**15**	3.4	4.9	unk.	21	**5**	0.08	**4**	unk.		unk.	
0.13	**8**	0.33	**17**	0.24	**4**	unk.	**unk.**	unk.	unk.	**unk.**	unk.	unk.	17	**2**	244	157	unk.	1.1	**6**	0.9	**6**	unk.	**unk.**	47	S
7.8	**39**	5	**1**	0.67	**7**	67	**1**	unk.	unk.	**unk.**	0.4	1.0	143	**14**	10.6	4.0	unk.	19	**5**	0.05	**3**	unk.		unk.	

	FOOD	Portion	Weight in grams / *Conversion for 100 g*	Kilocalories / H_2O g	Total carbohydrate g / Total fats g	Crude fiber g / Dietary fiber g	Total protein g / Total sugar g	**% USRDA**	Arginine mg / Histidine mg	Isoleucine mg / Leucine mg	Lysine mg / Methionine mg	Phenylalanine mg / Threonine mg	Valine mg / Tryptophan mg	Cystine mg / Tyrosine mg	Polyunsat. fatty acids g / Monounsat. fatty acids g	Saturated fatty acids g / P/S ratio	Linoleic acid g / Cholesterol mg	Thiamin mg / Ascorbic acid mg	**% USRDA**
A	chicken,broiler/fryer,light meat	3x2-1/8x3/4"	85	209	1.5	0.00	25.9	**58**	1605	1306	2114	1013	1262	344	2.29	2.82	1.98	0.07	**5**
	w/skin,flour coated,fried		*1.18*	46.5	10.3	0(a)	tr(a)		768	1896	693	1073	292	842	3.51	0.81	74	0	**0**
B	chicken,broiler/fryer,light meat	3x2-1/8x3/4"	85	189	0.0	0.00	24.7	**55**	1539	1239	2018	960	1200	327	1.96	2.59	1.68	0.05	**3**
	w/skin,roasted		*1.18*	51.4	9.2	0(a)	0(a)		729	1801	660	1022	277	799	2.98	0.76	71	0	**0**
C	chicken,broiler/fryer,light meat	3x2-1/8x3/4"	85	171	0.0	0.00	22.2	**49**	1385	1119	1821	866	1082	295	1.80	2.38	1.56	0.03	**2**
	w/skin,stewed		*1.18*	55.4	8.5	0(a)	0(a)		658	1623	594	921	250	721	2.75	0.76	63	0	**0**
D	chicken,broiler/fryer,light meat	1/2 C	70	134	0.3	0.00	23.0	**51**	1385	1212	1949	912	1140	294	0.88	1.06	0.69	0.05	**3**
	w/o skin,fried,chopped		*1.43*	42.1	3.9	0(a)	unk.		713	1724	636	970	268	776	1.22	0.83	63	0	**0**
E	chicken,broiler/fryer,light meat	1/2 C	70	121	0.0	0.00	21.6	**48**	1305	1142	1838	858	1073	277	0.69	0.89	0.52	0.05	**3**
	w/o skin,roasted,chopped		*1.43*	45.3	3.2	0(a)	0(a)		671	1623	598	913	253	730	0.91	0.77	59	0	**0**
F	chicken,broiler/fryer,light meat	1/2 C	70	111	0.0	0.00	20.2	**45**	1219	1067	1718	802	1003	259	0.61	0.78	0.45	0.29	**20**
	w/o skin,stewed,chopped		*1.43*	47.6	2.8	0(a)	0(a)		627	1517	559	854	236	682	0.80	0.78	54	0	**0**
G	CHICKEN,BROILER/FRYER,MEAT	1/2 bird	466	1347	43.9	0.00	105.0	**233**	6421	5242	8225	4189	5126	1449	19.11	21.48	17.10	0.56	**37**
	w/skin,batter dipped,fried		*0.21*	230.2	80.9	0(a)	tr(a)		3052	7703	2754	4297	1198	3411	29.92	0.89	405	0	**0**
H	chicken,broiler/fryer,meat w/skin,	1/2 bird	314	845	9.9	0.00	89.7	**199**	5545	4518	7285	3520	4371	1199	10.71	12.75	9.33	0.28	**19**
	flour coated,fried		*0.32*	164.6	46.8	0(a)	tr(a)		2653	6569	2393	3708	1014	2914	15.92	0.84	283	0	**0**
I	chicken,broiler/fryer,meat w/skin,	1/2 bird	299	715	0.0	0.00	81.6	**181**	5116	4072	6647	3172	3962	1088	8.88	11.33	7.68	0.18	**12**
	roasted		*0.33*	177.8	40.7	0(a)	0(a)		2398	5938	2171	3373	912	2628	13.16	0.78	263	0	**0**
J	chicken,broiler/fryer,meat w/skin,	1/2 bird	334	731	0.0	0.00	82.4	**183**	5160	4118	6717	3203	4005	1099	9.15	11.69	7.95	0.17	**11**
	stewed		*0.30*	213.5	42.0	0(a)	0(a)		2425	6002	2194	3407	922	2659	13.59	0.78	261	0	**0**
K	chicken,broiler/fryer,meat w/o skin,	1/2 C	70	153	1.2	0.00	21.4	**48**	1287	1128	1808	852	1061	275	1.50	1.72	1.25	0.06	**4**
	fried,chopped		*1.43*	40.3	6.4	0(a)	unk.		663	1606	591	902	251	722	2.04	0.87	66	0	**0**
L	chicken,broiler/fryer,meat w/o skin,	1/2 C	70	133	0.0	0.00	20.3	**45**	1221	1070	1721	804	1004	259	1.18	1.43	0.96	0.05	**3**
	roasted,chopped		*1.43*	44.7	5.2	0(a)	0(a)		629	1520	561	855	237	684	1.55	0.83	62	0	**0**
M	chicken,broiler/fryer,meat w/o skin,	1/2 C	70	124	0.0	0.00	19.1	**43**	1152	1009	1623	758	947	244	1.08	1.29	0.87	0.03	**2**
	stewed,chopped		*1.43*	46.8	4.7	0(a)	0(a)		593	1434	528	807	223	645	1.40	0.84	58	0	**0**
N	CHICKEN,BROILER/FRYER,NECK,meat	1 piece	52	172	4.5	0.00	10.3	**23**	657	481	764	402	491	149	2.89	3.23	2.63	0.05	**4**
	w/skin,batter dipped,fried		*1.92*	24.4	12.2	unk.	tr(a)		280	730	257	411	112	317	4.61	0.89	47	0	**0**
O	chicken,broiler/fryer,neck,meat	1 piece	36	120	1.5	0.00	8.6	**19**	559	403	658	331	410	122	1.97	2.28	1.77	0.03	**2**
	w/skin,flour coated,fried		*2.78*	17.1	8.5	unk.	tr(a)		237	608	218	346	92	264	3.07	0.87	34	0	**0**
P	chicken,broiler/fryer,neck,meat	1 piece	38	94	0.0	0.00	7.4	**17**	500	331	557	278	347	107	1.50	1.90	1.32	0.02	**1**
	w/skin,stewed		*2.63*	23.5	6.9	0(a)	0(a)		196	510	183	295	76	218	2.25	0.79	27	0	**0**
Q	chicken,broiler/fryer,neck,meat	1 piece	22	50	0.4	0.00	5.9	**13**	356	312	499	235	293	76	0.67	0.66	0.58	0.02	**1**
	w/o skin,fried		*4.55*	12.9	2.6	unk.	unk.		183	444	163	249	69	200	0.95	1.01	23	0	**0**
R	chicken,broiler/fryer,neck,meat	1 piece	18	32	0.0	0.00	4.4	**10**	267	233	375	175	219	57	0.37	0.38	0.28	0.01	**1**
	w/o skin,stewed		*5.56*	12.1	1.5	0(a)	0(a)		137	332	122	187	52	149	0.38	0.97	14	0	**0**
S	CHICKEN,BROILER/FRYER,SKIN only,fried	1 oz	28	112	6.6	0.03	2.9	**7**	190	110	145	120	133	53	1.94	2.16	1.80	0.05	**3**
			3.52	10.3	8.2	0(a)	unk.		60	193	58	103	30	79	3.23	0.90	21	0	**0**

Riboflavin mg / Niacin mg	**% USRDA**	Vitamin B_6 mg / Folacin mcg	**% USRDA**	Vitamin B_{12} mcg / Pantothenic acid mg	**% USRDA**	Biotin mg / Vitamin A IU	**% USRDA**	Preformed A RE / Beta carotene RE	Vitamin D IU / Vitamin E IU	**% USRDA**	Total tocopherol mg / Alpha tocopherol mg	Other tocopherol mg / Total ash g	Calcium mg / Phosphorus mg	**% USRDA**	Sodium mg / Sodium meq	Potassium mg / Potassium meq	Chlorine mg / Chlorine meq	Iron mg / Magnesium mg	**% USRDA**	Zinc mg / Copper mg	**% USRDA**	Iodine mcg / Selenium mcg	**% USRDA**	Manganese mcg / Chromium mcg	
0.11	**7**	0.46	**23**	0.28	**5**	unk.	**unk.**	unk.	unk.	**unk.**	unk.	unk.	14	**1**	65	203	unk.	1.0	**6**	1.1	**7**	unk.	**unk.**	22	A
10.2	**51**	3	**1**	0.82	**8**	58	**1**	unk.	unk.	**unk.**	0.4	0.83	181	**18**	2.8	5.2	unk.	23	**6**	0.05	**3**	unk.		unk.	
0.10	**6**	0.44	**22**	0.27	**5**	unk.	**unk.**	unk.	unk.	**unk.**	unk.	unk.	13	**1**	64	193	unk.	1.0	**5**	1.0	**7**	unk.	**unk.**	15	B
9.5	**47**	3	**1**	0.79	**8**	93	**2**	unk.	unk.	**unk.**	0.4	0.79	170	**17**	2.8	4.9	unk.	21	**5**	0.05	**2**	unk.		unk.	
0.09	**6**	0.23	**12**	0.17	**3**	unk.	**unk.**	unk.	unk.	**unk.**	unk.	unk.	11	**1**	54	142	unk.	0.8	**5**	1.0	**7**	unk.	**unk.**	15	C
5.9	**30**	3	**1**	0.45	**5**	82	**2**	unk.	unk.	**unk.**	0.4	0.66	124	**12**	2.3	3.6	unk.	17	**4**	0.04	**2**	unk.		unk.	
0.09	**5**	0.44	**22**	0.25	**4**	unk.	**unk.**	unk.	unk.	**unk.**	unk.	unk.	11	**1**	57	184	unk.	0.8	**4**	0.9	**6**	unk.	**unk.**	14	D
9.4	**47**	3	**1**	0.72	**7**	21	**0**	unk.	unk.	**unk.**	0.3	0.76	162	**16**	2.5	4.7	unk.	20	**5**	0.04	**2**	unk.		unk.	
0.08	**5**	0.42	**21**	0.24	**4**	unk.	**unk.**	unk.	unk.	**unk.**	unk.	unk.	10	**1**	54	173	unk.	0.7	**4**	0.9	**6**	unk.	**unk.**	12	E
8.7	**44**	3	**1**	0.68	**7**	20	**0**	unk.	unk.	**unk.**	0.3	0.71	151	**15**	2.3	4.4	unk.	19	**5**	0.03	**2**	unk.		unk.	
0.13	**7**	0.23	**12**	0.16	**3**	unk.	**unk.**	unk.	unk.	**unk.**	unk.	unk.	9	**1**	45	126	unk.	0.7	**4**	0.8	**6**	unk.	**unk.**	13	F
5.5	**27**	2	**1**	0.40	**4**	19	**0**	unk.	unk.	**unk.**	0.3	0.61	111	**11**	2.0	3.2	unk.	15	**4**	0.03	**2**	unk.		unk.	
0.89	**52**	1.44	**72**	1.30	**22**	unk.	**unk.**	unk.	unk.	**unk.**	unk.	unk.	98	**10**	1361	862	unk.	6.4	**36**	7.8	**52**	unk.	**unk.**	266	G
32.8	**164**	37	**9**	4.14	**41**	433	**9**	unk.	unk.	**unk.**	2.7	6.06	722	**72**	59.2	22.0	unk.	98	**25**	0.34	**17**	unk.		unk.	
0.60	**35**	1.29	**64**	0.97	**16**	unk.	**unk.**	unk.	unk.	**unk.**	unk.	unk.	53	**5**	264	735	unk.	4.3	**24**	6.4	**43**	unk.	**unk.**	107	H
28.2	**141**	19	**5**	3.39	**34**	279	**6**	unk.	unk.	**unk.**	1.8	3.08	600	**60**	11.5	18.8	unk.	78	**20**	0.24	**12**	unk.		unk.	
0.51	**30**	0.01	**1**	0.90	**15**	0.001	**0**	unk.	unk.	**unk.**	2.6	unk.	45	**5**	245	667	unk.	3.8	**21**	5.8	**39**	unk.	**unk.**	60	I
25.4	**127**	15	**4**	3.08	**31**	481	**10**	unk.	3.2	**11**	unk.	2.75	544	**54**	10.7	17.0	unk.	69	**17**	0.20	**10**	unk.		unk.	
0.50	**30**	0.73	**37**	0.67	**11**	0.001	**0**	unk.	unk.	**unk.**	2.9	unk.	43	**4**	224	554	unk.	3.9	**22**	5.9	**39**	unk.	**unk.**	63	J
18.7	**93**	17	**4**	2.23	**22**	488	**10**	unk.	3.5	**12**	unk.	2.54	464	**46**	9.7	14.2	unk.	63	**16**	0.19	**10**	unk.		unk.	
0.14	**8**	0.34	**17**	0.24	**4**	unk.	**unk.**	unk.	unk.	**unk.**	unk.	unk.	12	**1**	64	180	unk.	0.9	**5**	unk.	**unk.**	unk.	**unk.**	20	K
6.8	**34**	5	**1**	0.82	**8**	41	**1**	unk.	unk.	**unk.**	0.4	0.76	144	**14**	2.8	4.6	unk.	19	**5**	0.05	**3**	unk.		unk.	
0.13	**7**	tr	**0**	0.23	**4**	unk.	**unk.**	unk.	unk.	**unk.**	0.6	0.5	10	**1**	60	170	unk.	0.8	**5**	1.5	**10**	unk.	**unk.**	13	L
6.4	**32**	4	**1**	0.77	**8**	37	**1**	unk.	0.7	**3**	0.1	0.71	137	**14**	2.6	4.3	unk.	17	**4**	0.05	**2**	unk.		unk.	
0.11	**7**	0.18	**9**	0.15	**3**	unk.	**unk.**	unk.	unk.	**unk.**	0.6	0.5	10	**1**	49	126	unk.	0.8	**5**	1.4	**9**	unk.	**unk.**	13	M
4.3	**21**	4	**1**	0.52	**5**	35	**1**	unk.	0.7	**3**	0.1	0.59	105	**11**	2.1	3.2	unk.	15	**4**	0.04	**2**	unk.		unk.	
0.12	**7**	tr	**0**	0.12	**2**	unk.	**unk.**	unk.	unk.	**unk.**	unk.	unk.	16	**2**	144	79	unk.	1.1	**6**	1.3	**9**	unk.	**unk.**	34	N
2.4	**12**	4	**1**	0.44	**4**	88	**2**	unk.	unk.	**unk.**	unk.	0.54	60	**6**	6.2	2.0	unk.	8	**2**	0.06	**3**	unk.		unk.	
0.09	**6**	0.09	**5**	0.09	**2**	unk.	**unk.**	unk.	unk.	**unk.**	unk.	unk.	11	**1**	30	65	unk.	0.9	**5**	1.1	**7**	unk.	**unk.**	19	O
1.9	**10**	2	**1**	0.35	**4**	68	**1**	unk.	unk.	**unk.**	unk.	0.23	48	**5**	1.3	1.7	unk	7	**2**	0.05	**2**	unk.		unk.	
0.09	**6**	0.04	**2**	0.05	**1**	unk.	**unk.**	unk.	unk.	**unk.**	unk.	unk.	10	**1**	20	41	unk.	0.9	**5**	1.0	**7**	unk.	**unk.**	17	P
1.3	**6**	1	**0**	0.20	**2**	61	**1**	unk.	unk.	**unk.**	unk.	0.23	46	**5**	0.9	1.0	unk.	5	**1**	0.04	**2**	unk.		unk.	
0.07	**4**	0.08	**4**	0.07	**1**	unk.	**unk.**	unk.	unk.	**unk.**	unk.	unk.	9	**1**	22	47	unk.	0.7	**4**	0.9	**6**	unk.	**unk.**	11	Q
1.1	**6**	2	**0**	0.26	**3**	36	**1**	unk.	unk.	**unk.**	unk.	0.17	30	**3**	0.9	1.2	unk.	4	**1**	0.04	**2**	unk.		unk.	
0.05	**3**	0.03	**1**	0.03	**1**	unk.	**unk.**	unk.	unk.	**unk.**	unk.	unk.	8	**1**	12	25	unk.	0.5	**3**	0.7	**5**	unk.	**unk.**	9	R
0.7	**4**	1	**0**	0.12	**1**	22	**0**	unk.	unk.	**unk.**	unk.	0.14	23	**2**	0.5	0.6	unk.	3	**1**	0.02	**1**	unk.		unk.	
0.05	**3**	0.02	**1**	0.05	**1**	unk.	**unk.**	unk.	unk.	**unk.**	unk.	unk.	7	**1**	165	21	unk.	0.4	**2**	0.2	**1**	unk.	**unk.**	unk.	S
0.9	**5**	3	**1**	0.13	**1**	39	**1**	unk.	unk.	**unk.**	unk.	0.45	23	**2**	7.2	0.5	unk.	3	**1**	0.02	**1**	unk.		unk.	

	FOOD	Portion	Weight in grams / *Conversion for 100 g*	Kilocalories / H_2O g	Total carbohydrate g / Total fats g	Crude fiber g / Dietary fiber g	Total protein g / Total sugar g	% USRDA	Arginine mg / Histidine mg	Isoleucine mg / Leucine mg	Lysine mg / Methionine mg	Phenylalanine mg / Threonine mg	Valine mg / Tryptophan mg	Cystine mg / Tyrosine mg	Polyunsat. fatty acids g / Monounsat. fatty acids g	Saturated fatty acids g / P/S ratio	Linoleic acid g / Cholesterol mg	Thiamin mg / Ascorbic acid mg	% USRDA
A	chicken,broiler/fryer,skin only,	1 oz	28	129	0.0	0.00	5.8	**13**	446	186	345	195	243	96	2.43	3.24	2.22	0.01	**1**
	roasted		*3.52*	11.4	11.6	0(a)	0(a)		111	340	116	206	46	131	3.95	0.75	24	0	**0**
B	chicken,broiler/fryer,skin only,	1 oz	28	103	0.0	0.00	4.3	**10**	333	139	258	146	182	72	1.98	2.64	1.81	0.01	**1**
	stewed		*3.52*	15.1	9.4	0(a)	0(a)		83	254	86	154	35	98	3.21	0.75	18	0	**0**
C	CHICKEN,BROILER/FRYER,THIGH,meat	1 piece	86	238	7.8	0.00	18.6	**41**	1133	933	1461	743	908	255	3.35	3.79	3.00	0.10	**7**
	w/skin,batter dipped,fried		*1.16*	44.3	14.2	unk.	tr(a)		544	1367	488	762	213	606	5.17	0.88	80	0	**0**
D	chicken,broiler/fryer,thigh,meat	1 piece	62	162	2.0	0.00	16.6	**37**	1027	835	1345	652	808	222	2.11	2.54	1.84	0.06	**4**
	w/skin,flour coated,fried		*1.61*	33.6	9.3	unk.	tr(a)		490	1214	442	685	187	539	3.08	0.83	60	0	**0**
E	chicken,broiler/fryer,thigh,meat	1 piece	62	153	0.0	0.00	15.5	**35**	971	779	1269	604	755	206	2.12	2.68	1.85	0.04	**3**
	w/skin,roasted		*1.61*	36.8	9.6	0(a)	0(a)		458	1133	415	643	174	502	3.15	0.79	58	0	**0**
F	chicken,broiler/fryer,thigh,meat	1 piece	68	158	0.0	0.00	15.8	**35**	990	792	1291	615	768	211	2.21	2.79	1.93	0.04	**3**
	w/skin,stewed		*1.47*	42.9	10.0	0(a)	0(a)		466	1153	422	654	177	511	3.28	0.79	57	0	**0**
G	chicken,broiler/fryer,thigh,meat	1 piece	52	113	0.6	0.00	14.6	**33**	882	773	1240	583	726	188	1.26	1.45	1.06	0.05	**3**
	w/o skin,fried		*1.92*	30.8	5.4	0(a)	unk.		454	1100	405	618	171	495	1.70	0.87	53	0	**0**
H	chicken,broiler/fryer,thigh,meat	1 piece	52	109	0.0	0.00	13.5	**30**	814	712	1146	535	669	173	1.29	1.58	1.09	0.04	**2**
	w/o skin,roasted		*1.92*	32.7	5.7	0(a)	0(a)		419	1012	373	570	158	456	1.80	0.82	49	0	**0**
I	chicken,broiler/fryer,thigh,meat	1 piece	55	107	0.0	0.00	13.8	**31**	829	726	1168	546	682	176	1.23	1.49	1.04	0.03	**2**
	w/o skin,stewed		*1.82*	36.1	5.4	0(a)	0(a)		427	1032	381	581	161	464	1.69	0.83	49	0	**0**
J	CHICKEN,BROILER/FRYER,WING,meat	1 piece	49	159	5.4	0.00	9.7	**22**	611	461	722	385	467	141	2.48	2.86	2.22	0.05	**4**
	w/skin,batter dipped,fried		*2.04*	22.6	10.7	unk.	tr(a)		268	696	245	390	107	304	3.97	0.87	39	0	**0**
K	chicken,broiler/fryer,wing,meat	1 piece	32	103	0.8	0.00	8.4	**19**	538	396	650	321	398	116	1.58	1.94	1.39	0.02	**1**
	w/skin,flour coated,fried		*3.13*	15.6	7.1	0(a)	tr(a)		233	592	214	338	90	258	2.46	0.82	26	0	**0**
L	chicken,broiler/fryer,wing,meat	1 piece	34	99	0.0	0.00	9.1	**20**	592	432	714	348	435	126	1.41	1.85	1.21	0.01	**1**
	w/skin,roasted		*2.94*	18.7	6.6	0(a)	0(a)		255	645	234	370	98	281	2.14	0.76	29	0	**0**
M	chicken,broiler/fryer,wing,meat	1 piece	40	100	0.0	0.00	9.1	**20**	588	434	716	348	435	125	1.43	1.88	1.23	0.02	**1**
	w/skin,stewed		*2.50*	24.9	6.7	0(a)	0(a)		256	646	235	370	98	282	2.18	0.76	28	0	**0**
N	chicken,broiler/fryer,wing,meat	1 piece	20	42	0.0	0.00	6.0	**13**	364	318	512	239	299	77	0.41	0.50	0.31	0.01	**1**
	w/o skin,fried		*5.00*	12.0	1.8	0(a)	0(a)		187	452	167	255	70	204	0.54	0.83	17	0	**0**
O	chicken,broiler/fryer,wing,meat	1 piece	21	43	0.0	0.00	6.4	**14**	386	338	543	254	317	82	0.37	0.47	0.27	0.01	**1**
	w/o skin,roasted		*4.76*	13.2	1.7	0(a)	0(a)		198	480	177	270	75	216	0.46	0.79	18	0	**0**
P	chicken,broiler/fryer,wing,meat	1 piece	24	43	0.0	0.00	6.5	**15**	393	344	554	259	324	84	0.38	0.48	0.27	0.01	**1**
	w/o skin,stewed		*4.17*	16.1	1.7	0(a)	0(a)		203	489	180	276	76	220	0.47	0.78	18	0	**0**
Q	CHICKEN,CAPONS,meat w/skin,roasted	3x2-1/8x3/4"	85	195	0.0	0.00	24.6	**55**	1535	1237	2014	959	1198	326	2.14	2.77	1.89	0.06	**4**
			1.18	49.9	9.9	0(a)	0(a)		728	1798	658	1029	276	797	3.32	0.77	73	0	**0**
R	CHICKEN,DEBONED,rolled,light meat	3 sl (2-7/8"	85	135	2.1	unk.	16.6	**37**	unk.	unk.	unk.	unk.	unk.	unk.	1.36	1.72	1.16	0.06	**4**
		dia x 1/4")	*1.18*	58.3	6.3	unk.	unk.		unk.	unk.	unk.	unk.	unk.	unk.	2.07	0.79	42	unk.	**unk.**
S	CHICKEN,FROZEN ENTREE,creamed	1/2 pkg	92	150	3.0	unk.	10.0	**15**	unk.	unk.	unk.	unk.	unk.	unk.	unk.	unk.	unk.	0.06	**4**
	-Stouffer		*1.08*	64.6	10.9	unk.	unk.		unk.	unk.	unk.	unk.	unk.	unk.	unk.	unk.	unk.	tr	**0**

Riboflavin mg / Niacin mg	% USRDA	Vitamin B_6 mg / Folacin mcg	% USRDA	Vitamin B_{12} mcg / Pantothenic acid mg	% USRDA	Biotin mg / Vitamin A IU	% USRDA	Preformed A RE / Beta carotene RE	Vitamin D IU / Vitamin E IU	% USRDA	Total tocopherol mg / Alpha tocopherol mg	Other tocopherol mg / Total ash g	Calcium mg / Phosphorus mg	% USRDA	Sodium mg / Sodium meq	Potassium mg / Potassium meq	Chlorine mg / Chlorine meq	Iron mg / Magnesium mg	% USRDA	Zinc mg / Copper mg	% USRDA	Iodine mcg / Selenium mcg	% USRDA	Manganese mcg / Chromium mcg	
0.04	2	0.03	1	0.06	1	unk.	unk.	unk.	unk.	unk.	unk.	unk.	4	0	18	39	unk.	0.4	2	0.3	2	unk.	unk.	6	A
1.6	8	1	0	0.20	2	74	2	unk.	unk.	unk.	unk.	0.14	36	4	0.8	1.0	unk.	4	1	0.02	1	unk.		unk.	
0.03	2	0.01	1	0.03	1	unk.	unk.	unk.	unk.	unk.	unk.	unk.	3	0	16	33	unk.	0.3	2	0.3	2	unk.	unk.	5	B
1.1	5	1	0	0.11	1	56	1	unk.	unk.	unk.	unk.	0.12	28	3	0.7	0.8	unk.	4	1	0.01	1	unk.		unk.	
0.20	12	0.22	11	0.24	4	unk.	unk.	unk.	unk.	unk.	unk.	unk.	15	2	248	165	unk.	1.2	7	1.8	12	unk.	unk.	49	C
4.9	25	8	2	0.84	8	82	2	unk.	unk.	unk.	unk.	1.09	133	13	10.8	4.2	unk.	18	5	0.07	4	unk.		unk.	
0.15	9	0.20	10	0.19	3	unk.	unk.	unk.	unk.	unk.	unk.	unk.	9	1	55	147	unk.	0.9	5	1.6	10	unk.	unk.	22	D
4.3	22	5	1	0.73	7	61	1	unk.	unk.	unk.	unk.	0.58	116	12	2.4	3.8	unk.	15	4	0.06	3	unk.		unk.	
0.13	8	0.19	10	0.18	3	unk.	unk.	unk.	unk.	unk.	unk.	unk.	7	1	52	138	unk.	0.8	5	1.5	10	unk.	unk.	13	E
3.9	20	4	1	0.69	7	102	2	unk.	unk.	unk.	unk.	0.55	108	11	2.3	3.5	unk.	14	3	0.05	2	unk.		unk.	
0.13	8	0.12	6	0.13	2	unk.	unk.	unk.	unk.	unk.	unk.	unk.	7	1	48	116	unk.	0.9	5	1.5	10	unk.	unk.	13	F
3.3	17	4	1	0.53	5	103	2	unk.	unk.	unk.	unk.	0.53	95	9	2.1	3.0	unk.	13	3	0.05	2	unk.		unk.	
0.14	8	0.20	10	0.17	3	unk.	unk.	unk.	unk.	unk.	unk.	unk.	7	1	49	135	unk.	0.8	4	1.5	10	unk.	unk.	14	G
3.7	19	5	1	0.67	7	36	1	unk.	unk.	unk.	unk.	0.54	104	10	2.1	3.4	unk.	14	3	0.05	2	unk.		unk.	
0.12	7	0.18	9	0.16	3	unk.	unk.	unk.	unk.	unk.	unk.	unk.	6	1	46	124	unk.	0.7	4	1.3	9	unk.	unk.	11	H
3.4	17	4	1	0.62	6	34	1	unk.	unk.	unk.	unk.	0.49	95	10	2.0	3.2	unk.	12	3	0.04	2	unk.		unk.	
0.12	7	0.12	6	0.12	2	unk.	unk.	unk.	unk.	unk.	unk.	unk.	6	1	41	101	unk.	0.8	4	1.4	10	unk.	unk.	11	I
2.9	14	4	1	0.48	5	34	1	unk.	unk.	unk.	unk.	0.47	82	8	1.8	2.6	unk.	12	3	0.04	2	unk.		unk.	
0.07	4	0.15	7	0.12	2	unk.	unk.	unk.	unk.	unk.	unk.	unk.	10	1	157	68	unk.	0.6	4	0.7	5	unk.	unk.	29	J
2.6	13	3	1	0.35	4	55	1	unk.	unk.	unk.	unk.	0.57	59	6	6.8	1.7	unk.	8	2	0.03	2	unk.		unk.	
0.04	3	0.13	7	0.09	2	unk.	unk.	unk.	unk.	unk.	unk.	unk.	5	1	25	57	unk.	0.4	2	0.6	4	unk.	unk.	9	K
2.1	11	1	0	0.28	3	40	1	unk.	unk.	unk.	unk.	0.23	48	5	1.1	1.4	unk.	6	2	0.02	1	unk.		unk.	
0.04	3	0.14	7	0.10	2	unk.	unk.	unk.	unk.	unk.	unk.	unk.	5	1	28	63	unk.	0.4	2	0.6	4	unk.	unk.	6	L
2.3	11	1	0	0.30	3	54	1	unk.	unk.	unk.	unk.	0.25	51	5	1.2	1.6	unk.	6	2	0.02	1	unk.		unk.	
0.04	2	0.09	4	0.07	1	unk.	unk.	unk.	unk.	unk.	unk.	unk.	5	1	27	56	unk.	0.5	3	0.7	4	unk.	unk.	7	M
1.8	9	1	0	0.20	2	53	1	unk.	unk.	unk.	unk.	0.25	48	5	1.2	1.4	unk.	6	2	0.02	1	unk.		unk.	
0.03	2	0.12	6	0.07	1	unk.	unk.	unk.	unk.	unk.	unk.	unk.	3	0	18	42	unk.	0.2	1	0.4	3	unk.	unk.	12	N
1.4	7	1	0	0.20	2	12	0	unk.	unk.	unk.	unk.	0.17	33	3	0.8	1.1	unk.	4	1	0.01	1	unk.		unk.	
0.03	2	0.12	6	0.07	1	unk.	unk.	unk.	unk.	unk.	unk.	unk.	3	0	19	44	unk.	0.2	1	0.4	3	unk.	unk.	3	O
1.5	8	1	0	0.21	2	13	0	unk.	unk.	unk.	unk.	0.19	35	4	0.8	1.1	unk.	4	1	0.01	1	unk.		unk.	
0.03	2	0.08	4	0.05	1	unk.	unk.	unk.	unk.	unk.	unk.	unk.	3	0	18	37	unk.	0.3	2	0.05	3	unk.	unk.	4	P
1.2	6	1	0	0.14	1	13	0	unk.	unk.	unk.	unk.	0.18	32	3	0.8	0.9	unk.	4	1	0.01	1	unk.		unk.	
0.14	9	0.37	18	0.28	5	unk.	unk.	unk.	unk.	unk.	unk.	unk.	12	1	42	217	unk.	1.3	7	1.5	10	unk.	unk.	18	Q
7.6	38	5	1	0.94	9	58	1	unk.	unk.	unk.	unk.	0.88	209	21	1.8	5.5	unk.	20	5	0.06	3	unk.		unk.	
0.11	7	unk.	unk.	unk.	unk.	unk.	unk.	unk.	unk.	unk.	unk.	unk.	37	4	496	194	unk.	0.8	5	0.6	4	unk.	unk.	unk.	R
4.5	23	unk.	unk.	unk.	unk.	unk.	unk.	unk.	unk.	unk.	unk.	1.74	133	13	21.6	5.0	unk.	16	4	0.03	2	unk.		unk.	
0.06	4	unk.	unk.	unk.	unk.	unk.	unk.	unk.	unk.	unk.	unk.	unk.	30	3	338	112	unk.	0.7	4	unk.	unk.	unk.	unk.	unk.	S
2.0	10	unk.	unk.	unk.	unk.	198	4	unk.	unk.	unk.	unk.	1.29	unk.	unk.	14.7	2.9	unk.	unk.	unk.	unk.	unk.	unk.		unk.	

	FOOD	Portion	Weight in grams / *Conversion for 100 g*	Kilocalories / H_2O g	Total carbohydrate g / Total fats g	Crude fiber g / Dietary fiber g	Total protein g / Total sugar g	% USRDA	Arginine mg / Histidine mg	Isoleucine mg / Leucine mg	Lysine mg / Methionine mg	Phenylalanine mg / Threonine mg	Valine mg / Tryptophan mg	Cystine mg / Tyrosine mg	Polyunsat. fatty acids g / Monounsat. fatty acids g	Saturated fatty acids g / P/S ratio	Linoleic acid g / Cholesterol mg	Thiamin mg / Ascorbic acid mg	% USRDA
A	chicken,frozen entree,escalloped & noodles-Stouffer	1/2 pkg	163	248	15.8	unk.	13.7	**21**	unk.	unk.	unk.	unk.	unk.	unk.	unk.	unk.	unk.	0.15	**10**
			0.61	114.3	14.9	unk.	unk.		unk.	unk.	unk.	unk.	unk.	unk.	unk.	unk.	unk.	tr	**0**
B	chicken,frozen entree,fricassee	3/4 C	180	290	5.8	unk.	27.5	**61**	unk.	unk.	unk.	unk.	unk.	unk.	unk.	5.40	3.60	0.04	**2**
			0.56	128.3	16.7	unk.	unk.		unk.	unk.	unk.	unk.	unk.	unk.	7.20	unk.	72	0	**0**
C	CHICKEN,HOME RECIPE,a la king,creamed	3/4 C	161	234	7.5	0.09	15.7	**34**	748	841	1256	635	838	200	0.33	7.53	1.45	0.05	**4**
			0.62	120.6	15.5	0.62	1.7		404	1140	394	631	183	535	4.52	0.04	76	13	**22**
D	chicken,home recipe,cacciatore	3/4 C	180	394	9.3	0.92	32.6	**71**	1289	1688	2515	1278	1587	410	9.09	5.67	7.39	0.16	**11**
			0.56	112.9	23.8	1.77	6.5		825	2326	786	1270	333	1059	6.57	1.60	99	40	**66**
E	chicken,home recipe,casserole	3/4 C	168	198	17.3	0.08	17.6	**38**	861	948	1318	727	949	216	0.86	1.56	1.60	0.10	**7**
			0.60	119.4	6.2	3.30	4.6		445	1358	426	698	189	630	2.74	0.55	39	6	**11**
F	chicken,home recipe,chicken & dumplings	3/4 C	159	256	12.4	0.04	23.3	**51**	1357	1115	1757	914	1128	305	0.21	4.12	1.98	0.11	**7**
			0.63	108.5	11.8	0.49	0.8		576	1564	542	892	261	753	3.99	0.05	109	4	**6**
G	chicken,home recipe,chicken & noodles	3/4 C	173	191	22.9	0.17	11.1	**22**	402	501	614	412	517	147	2.10	1.85	1.90	0.13	**9**
			0.58	132.0	5.9	unk.	0.5		245	685	215	390	102	294	1.50	1.14	43	0	**0**
H	chicken,home recipe,potpie	1 average	302	706	42.6	0.90	29.4	**61**	1182	1374	2006	1178	1364	393	0.72	16.44	4.11	0.34	**23**
			0.33	178.2	46.3	1.40	2.7		691	1952	628	1068	326	931	21.34	0.04	118	20	**34**
I	chicken,home recipe,teriyaki	3/4 C	146	399	7.3	0.01	30.3	**66**	1477	1624	2383	1232	1528	406	11.09	5.84	9.48	0.09	**6**
			0.68	78.9	26.8	unk.	4.6		793	2243	742	1206	318	1013	7.21	1.90	92	tr	**0**
J	CHICKEN,ROASTING,dark meat w/o skin, roasted,chopped	1/2 C	70	125	0.0	0.00	16.3	**36**	981	860	1382	646	807	209	1.40	1.70	1.18	0.04	**3**
			1.43	46.9	6.1	0(a)	0(a)		505	1221	451	687	190	549	1.93	0.82	52	0	**0**
K	chicken,roasting,light meat w/o skin, roasted,chopped	1/2 C	70	107	0.0	0.00	19.0	**42**	1146	1003	1613	754	942	243	0.65	0.76	0.48	0.04	**3**
			1.43	47.5	2.8	0(a)	0(a)		589	1425	526	802	222	641	0.88	0.86	52	0	**0**
L	chicken,roasting,meat w/o skin, roasted,chopped	1/2 C	70	117	0.0	0.00	17.5	**39**	1056	925	1487	695	868	224	1.06	1.27	0.87	0.04	**3**
			1.43	47.2	4.6	0(a)	0(a)		543	1314	485	739	204	591	1.46	0.83	52	0	**0**
M	CHICKEN,STEWING,dark meat w/o skin, stewed,chopped	1/2 C	70	181	0.0	0.00	19.7	**44**	1189	1040	1674	782	977	252	2.55	2.85	2.05	0.09	**6**
			1.43	38.6	10.7	0(a)	0(a)		612	1478	545	832	230	665	3.06	0.90	66	0	**0**
N	chicken,stewing,light meat w/o skin, stewed,chopped	1/2 C	70	149	0.0	0.00	23.1	**51**	1395	1221	1965	918	1147	296	1.32	1.39	0.88	0.06	**4**
			1.43	40.5	5.6	0(a)	0(a)		717	1735	640	977	270	780	1.55	0.95	49	0	**0**
O	chicken,stewing,meat w/o skin,stewed, chopped	1/2 C	70	166	0.0	0.00	21.3	**47**	1284	1124	1809	845	1056	272	1.98	2.17	1.50	0.08	**5**
			1.43	39.4	8.3	0(a)	0(a)		661	1598	589	899	249	719	2.36	0.91	58	0	**0**
P	CHICORY greens,raw	1/2 C	15	3	0.6	0.12	0.3	**0**	unk.	10	8	unk.	unk.	1	unk.	tr(a)	unk.	0.01	**1**
			6.90	13.5	tr	unk.	0.0		3	18	2	10	3	6	unk.	unk.	0	3	**5**
Q	chicory,witlof,bleached head (French/ Belgian endive),chopped,raw	1/2 C	45	7	1.4	unk.	0.4	**1**	unk.	18	19	19	20	3	unk.	tr(a)	unk.	unk.	**unk.**
			2.22	42.8	tr	unk.	0.1		8	31	5	18	7	33	unk.	unk.	0	unk.	**unk.**
R	CHILI con carne,w/beans,canned	1 C	230	306	28.1	1.38	17.3	**36**	unk.	unk.	unk.	unk.	unk.	unk.	unk.	6.90	tr	0.07	**5**
			0.43	166.5	14.0	unk.	unk.		unk.	unk.	unk.	unk.	unk.	unk.	6.90	unk.	unk.	unk.	**unk.**
S	chili w/beans,home recipe	1 C	250	339	29.8	1.51	24.4	**48**	1338	1273	1979	1069	1338	270	1.84	6.20	0.38	0.23	**15**
			0.40	188.3	14.2	2.63	8.1		725	1950	489	1066	275	839	5.93	0.30	61	64	**106**

Riboflavin mg / Niacin mg	% USRDA	Vitamin B_6 mg / Folacin mcg	% USRDA	Vitamin B_{12} mcg / Pantothenic acid mg	% USRDA	Biotin mg / Vitamin A IU	% USRDA	Preformed A RE / Beta carotene RE	Vitamin D IU / Vitamin E IU	% USRDA	Total tocopherol mg / Alpha tocopherol mg	Other tocopherol mg / Total ash g	Calcium mg / Phosphorus mg	% USRDA	Sodium mg / Sodium meq	Potassium mg / Potassium meq	Chlorine mg / Chlorine meq	Iron mg / Magnesium mg	% USRDA	Zinc mg / Copper mg	% USRDA	Iodine mcg / Selenium mcg	% USRDA	Manganese mcg / Chromium mcg	
0.16	**10**	unk.	**unk.**	unk.	**unk.**	unk.	**unk.**	unk.	unk.	**unk.**	unk.	unk.	60	**6**	715	154	unk.	1.4	**8**	unk.	**unk.**	unk.	**unk.**	unk.	A
2.0	**10**	unk.	**unk.**	unk.	**unk.**	tr	**0**	unk.	unk.	**unk.**	unk.	1.96	unk.	**unk.**	31.1	3.9	unk.	unk.	**unk.**	unk.	**unk.**	unk.		unk.	
0.13	**7**	0.45	**23**	0.47	**8**	unk.	**unk.**	unk.	unk.	**unk.**	0.4	tr(a)	11	**1**	277	252	unk.	1.6	**9**	4.3	**29**	unk.	**unk.**	unk.	B
4.3	**22**	7	**2**	1.62	**16**	126	**3**	unk.	0.4	**1**	tr(a)	1.62	203	**20**	12.1	6.4	unk.	unk.	**unk.**	0.32	**16**	tr(a)		tr(a)	
0.20	**12**	0.14	**7**	0.56	**9**	0.002	**1**	30	15	**4**	0.2	tr(a)	63	**6**	729	149	4	1.4	**8**	1.4	**9**	44.6	**30**	23	C
4.8	**24**	8	**2**	0.42	**4**	703	**14**	30	0.2	**1**	0.1	2.19	214	**21**	31.7	3.8	0.1	21	**5**	0.21	**10**	3		3	
0.22	**13**	0.47	**24**	0.44	**7**	0.002	**1**	100	0	**0**	6.4	tr	45	**5**	671	765	11	3.5	**20**	2.2	**15**	104.9	**0**	70	D
10.3	**51**	18	**5**	0.70	**7**	1586	**32**	115	7.6	**25**	0.1	3.86	313	**31**	29.2	19.6	0.3	20	**5**	0.41	**20**	tr		2	
0.20	**12**	0.18	**9**	0.25	**4**	0.001	**1**	10	0(a)	**0**	2.7	1.2	132	**13**	460	314	2	1.4	**8**	1.4	**10**	70.8	**47**	10	E
4.7	**23**	12	**3**	0.58	**6**	496	**10**	16	3.2	**11**	1.3	2.24	220	**22**	20.0	8.0	0.0	25	**6**	0.02	**1**	1		1	
0.17	**10**	0.18	**9**	0.74	**12**	0.001	**0**	43	6	**2**	0.2	tr(a)	61	**6**	1283	163	21	2.4	**13**	1.9	**13**	111.6	**74**	60	F
4.5	**22**	10	**3**	0.57	**6**	233	**5**	6	0.2	**1**	0.0	3.97	314	**31**	55.8	4.2	0.6	40	**10**	0.25	**12**	8		9	
0.13	**8**	0.11	**6**	0.08	**1**	0.002	**1**	16	1	**0**	0.4	tr(a)	28	**3**	518	151	unk.	1.9	**10**	1.0	**6**	unk.	**unk.**	unk.	G
3.1	**15**	4	**1**	0.26	**3**	106	**2**	2	0.4	**2**	tr(a)	2.36	133	**13**	22.5	3.9	unk.	24	**6**	0.05	**2**	tr(a)		tr(a)	
0.30	**18**	0.26	**13**	0.72	**12**	0.001	**0**	50	0	**0**	0.2	tr(a)	52	**5**	1927	278	34	3.7	**20**	2.4	**16**	190.0	**127**	192	H
11.8	**59**	20	**5**	0.79	**8**	4674	**94**	407	0.3	**1**	0.2	5.54	332	**33**	83.8	7.1	1.0	47	**12**	0.31	**15**	25		29	
0.21	**13**	0.32	**16**	0.41	**7**	0(a)	**0**	93	0	**0**	9.7	tr(a)	39	**4**	2190	511	3	3.4	**19**	1.8	**12**	tr(a)	**0**	13	I
8	**44**	14	**4**	0.48	**5**	443	**9**	13	11.6	**39**	0.1	7.44	285	**29**	95.3	13.1	0.1	3	**1**	0.27	**13**	tr(a)		4	
0.13	**8**	0.22	**11**	0.15	**2**	unk.	**unk.**	unk.	unk.	**unk.**	unk.	unk.	8	**1**	66	157	unk.	0.9	**5**	1.5	**10**	unk.	**unk.**	13	J
4.0	**20**	5	**1**	0.72	**7**	38	**1**	unk.	unk.	**unk.**	unk.	0.66	120	**12**	2.9	4.0	unk.	14	**4**	0.05	**2**	unk.		unk.	
0.06	**4**	0.38	**19**	0.22	**4**	unk.	**unk.**	unk.	unk.	**unk.**	unk.	unk.	2	**0**	36	165	unk.	0.8	**4**	0.5	**4**	unk.	**unk.**	10	K
7.3	**37**	2	**1**	0.63	**6**	17	**0**	unk.	unk.	**unk.**	0.3	0.66	152	**15**	1.5	4.2	unk.	16	**4**	0.03	**2**	unk.		unk.	
0.10	**6**	0.29	**14**	0.20	**3**	unk.	**unk.**	unk.	unk.	**unk.**	0.6	0.5	8	**1**	52	160	unk.	0.8	**5**	1.1	**7**	unk.	**unk.**	12	L
5.5	**28**	3	**1**	0.68	**7**	29	**1**	unk.	0.7	**3**	0.1	0.66	134	**13**	2.3	4.1	unk.	15	**4**	0.04	**2**	unk.		unk.	
0.24	**14**	0.17	**8**	0.17	**3**	unk.	**unk.**	unk.	unk.	**unk.**	unk.	unk.	8	**1**	66	143	unk.	1.1	**6**	2.2	**15**	unk.	**unk.**	16	M
3.2	**16**	6	**1**	0.71	**7**	101	**2**	unk.	unk.	**unk.**	unk.	1.04	131	**13**	2.9	3.6	unk.	15	**4**	0.10	**5**	unk.		unk.	
0.14	**8**	0.27	**14**	0.19	**3**	unk.	**unk.**	unk.	unk.	**unk.**	unk.	unk.	10	**1**	41	139	unk.	0.8	**5**	0.6	**4**	unk.	**unk.**	14	N
6.0	**30**	3	**1**	0.48	**5**	51	**1**	unk.	unk.	**unk.**	0.3	0.83	158	**16**	1.8	3.6	unk.	16	**4**	0.06	**3**	unk.		unk.	
0.20	**12**	0.22	**11**	0.18	**3**	unk.	**unk.**	unk.	unk.	**unk.**	0.6	0.5	9	**1**	55	141	unk.	1.0	**6**	1.4	**10**	unk.	**unk.**	15	O
4.5	**22**	4	**1**	0.60	**6**	78	**2**	unk.	0.7	**3**	0.1	0.94	143	**14**	2.4	3.6	unk.	15	**4**	0.08	**4**	unk.		unk.	
0.01	**1**	0.01	**0**	0.00	**0**	unk.	**unk.**	0	0(a)	**0**	unk.	unk.	12	**1**	3	61	10	0.1	**1**	unk.	**unk.**	unk.	**unk.**	unk.	P
0.1	**0**	8	**2**	unk.	**unk.**	580	**12**	58	unk.	**unk.**	unk.	0.19	6	**1**	0.1	1.6	0.3	2	**1**	tr(a)	**0**	tr(a)		unk.	
unk.	**unk.**	0.02	**1**	0.00	**0**	unk.	**unk.**	unk.	0(a)	**0**	unk.	unk.	8	**1**	3	82	unk.	0.2	**1**	unk.	**unk.**	unk.	**unk.**	unk.	Q
unk.	**unk.**	23	**6**	unk.	**unk.**	tr	**0**	unk.	unk.	**unk.**	unk.	0.27	9	**1**	0.1	2.1	unk.	6	**2**	tr(a)	**0**	tr(a)		unk.	
0.16	**10**	0.23	**12**	0.53	**9**	unk.	**unk.**	unk.	0	**0**	0.5	tr(a)	74	**7**	1221	536	unk.	3.9	**22**	4.1	**28**	unk.	**unk.**	unk.	R
3.0	**15**	21	**5**	0.23	**2**	138	**3**	unk.	0.6	**2**	0.5	4.14	290	**29**	53.1	13.7	unk.	unk.	**unk.**	0.76	**38**	tr(a)		tr(a)	
0.27	**16**	0.64	**32**	0.80	**13**	0.003	**1**	7	0	**0**	0.5	0.0	78	**8**	977	1090	8	6.0	**33**	3.3	**22**	167.9	**112**	69	S
5.7	**28**	20	**5**	0.55	**6**	2030	**41**	159	0.6	**2**	0.3	6.12	297	**30**	42.5	27.9	0.2	80	**20**	0.30	**15**	tr		2	

	FOOD	Portion	Weight in grams / Conversion for 100 g	Kilocalories / H_2O g	Total carbohydrate g / Total fats g	Crude fiber g / Dietary fiber g	Total protein g / Total sugar g	% USRDA	Arginine mg / Histidine mg	Isoleucine mg / Leucine mg	Lysine mg / Methionine mg	Phenylalanine mg / Threonine mg	Valine mg / Tryptophan mg	Cystine mg / Tyrosine mg	Polyunsat. fatty acids g / Monounsat. fatty acids g	Saturated fatty acids g / P/S ratio	Linoleic acid g / Cholesterol mg	Thiamin mg / Ascorbic acid mg	% USRDA
A	CHILI POWDER	1 tsp	2.5	8	1.4	0.56	0.3	**1**	unk.	unk.	unk.	unk.	unk.	unk.	unk.	unk.	unk.	0.01	**1**
			40.00	tr	0.4	unk.	0.0		unk.	unk.	unk.	unk.	unk.	unk.	unk.	unk.	0	2	**3**
B	CHIVES,raw,chopped	1 Tbsp	3.0	1	0.2	0.03	0.0	**0**	unk.	unk.	unk.	unk.	unk.	unk.	unk.	tr(a)	unk.	tr	**0**
			33.33	2.7	tr	unk.	0.1		unk.	unk.	unk.	unk.	unk.	unk.	unk.	unk.	unk.	2	**3**
C	CHOCOLATE bits (toll house)	1/4 C	43	215	24.2	0.42	1.8	**3**	unk.	unk.	unk.	unk.	unk.	unk.	unk.	8.50	0.42	tr	**0**
			2.35	0.5	15.2	unk.	21.7		unk.	unk.	unk.	unk.	unk.	unk.	5.52	unk.	tr(a)	0	**0**
D	chocolate,bitter or baking	1 square	28	143	8.2	0.71	3.0	**5**	unk.	unk.	unk.	unk.	unk.	unk.	unk.	8.43	0.31	0.01	**1**
			3.52	0.7	15.1	unk.	0.0		unk.	unk.	unk.	unk.	unk.	unk.	5.57	unk.	0	0	**0**
E	CHOP SUEY,canned,w/meat	1 C	232	144	9.7	1.86	10.2	**16**	unk.	unk.	unk.	unk.	unk.	unk.	unk.	2.32	tr	0.12	**8**
			0.43	198.4	7.4	unk.	unk.		unk.	unk.	unk.	unk.	unk.	unk.	2.32	unk.	unk.	5	**8**
F	chop suey,home recipe,w/pork	1 C	250	375	8.2	1.17	18.9	**40**	1076	1028	1418	720	1014	186	3.97	8.45	5.58	0.42	**28**
			0.40	189.0	28.6	2.25	unk.		545	1411	461	860	240	597	10.90	0.47	62	58	**97**
G	CHOW CHOW,home recipe	1 Tbsp	15	8	2.0	0.07	0.2	**0**	7	5	6	5	5	1	tr	tr	tr	0.01	**0**
			6.67	12.7	0.0	0.10	1.3		2	7	2	4	2	3	0.01	2.00	0	3	**5**
H	CHOW MEIN,canned,chicken	1 C	250	95	17.7	0.75	6.5	**0**	unk.	unk.	unk.	unk.	unk.	unk.	unk.	unk.	unk.	0.05	**3**
			0.40	222.0	0.2	unk.	unk.		unk.	unk.	unk.	unk.	unk.	unk.	unk.	unk.	8	13	**21**
I	chow mein,home recipe,chicken	1 C	250	255	10.0	0.75	31.0	**67**	unk.	unk.	unk.	unk.	unk.	unk.	unk.	2.50	2.50	0.07	**5**
			0.40	195.0	10.0	unk.	unk.		unk.	unk.	unk.	unk.	unk.	unk.	2.50	unk.	78	10	**17**
J	chow mein,home recipe,pork	1 C	250	425	21.2	1.30	31.6	**67**	1851	1599	2380	1209	1628	322	8.33	7.82	6.85	0.72	**48**
			0.40	180.6	24.0	3.12	4.6		924	2299	739	1429	401	1028	10.44	1.06	89	15	**25**
K	chow mein,home recipe,shrimp	1 C	250	221	21.4	1.38	13.2	**27**	856	660	961	456	670	129	5.30	1.15	5.20	0.11	**7**
			0.40	203.3	10.0	3.12	4.6		196	919	311	476	119	375	2.22	4.63	55	15	**25**
L	CHUTNEY,tomato	1 Tbsp	20	41	7.8	unk.	0.2	**0**	unk.	6	9	6	6	unk.	0(a)	0(a)	0(a)	0.00	**0**
			5.00	unk.	0.0	unk.	unk.		3	9	2	7	2	3	0(a)	0.00	0(a)	0	**0**
M	CINNAMON,ground	1 tsp	2.3	6	1.8	0.56	0.1	**0**	unk.	unk.	unk.	unk.	unk.	unk.	0.01	0.01	0.01	tr	**0**
			43.48	0.2	0.1	unk.	0.0		unk.	unk.	unk.	unk.	unk.	unk.	0.01	0.82	0	1	**2**
N	CLAM fritter	2-1/2" dia x 1-3/4"	40	124	12.4	unk.	4.6	**10**	378	215	364	unk.	unk.	72	unk.	unk.	unk.	0.01	**1**
			2.50	16.1	6.0	unk.	0.0		108	352	unk.	214	59	190	unk.	unk.	unk.	unk.	**unk.**
O	clam,raw,hard or soft shell	1 med	17	13	0.3	unk.	2.1	**5**	177	101	171	79	65	34	unk.	unk.	unk.	0.02	**1**
			5.88	13.9	0.3	0(a)	0.0		51	166	62	100	28	89	unk.	unk.	8	2	**3**
P	CLOVES,ground	1 tsp	2.2	7	1.3	0.00	0.1	**0**	unk.	unk.	unk.	unk.	unk.	unk.	unk.	0.10	unk.	tr	**0**
			45.45	0.2	0.4	unk.	0.0		unk.	unk.	unk.	unk.	unk.	unk.	unk.	unk.	0	2	**3**
Q	COBBLER,home recipe,peach	1/2 C	102	201	36.9	0.03	2.1	**4**	48	88	93	97	111	39	0.31	3.31	0.24	0.05	**3**
			0.98	56.7	5.7	1.14	28.4		53	158	52	80	24	76	1.47	0.09	35	3	**5**
R	COCOA BUTTER	1 C	218	1927	0.0	0.00	0.0	**0**	0(a)	0(a)	0(a)	0(a)	0(a)	0(a)	6.54	130.15	6.10	0(a)	**0**
			0.46	0.0	218.0	0(a)	0(a)		0(a)	0(a)	0(a)	0(a)	0(a)	0(a)	71.07	0.05	0(a)	0(a)	**0**
S	COCONUT milk	1/2 C	120	26	5.6	unk.	0.4	**1**	tr(a)	tr(a)	tr(a)	tr(a)	tr(a)	tr(a)	tr	0.22	tr	tr	**0**
			0.83	113.0	0.2	tr	5.6		tr(a)	tr(a)	tr(a)	tr(a)	tr(a)	tr(a)	0.01	0.00	0	2	**4**

Riboflavin mg / Niacin mg	% USRDA	Vitamin B_6 mg / Folacin mcg	% USRDA	Vitamin B_{12} mcg / Pantothenic acid mg	% USRDA	Biotin mg / Vitamin A IU	% USRDA	Preformed A RE / Beta carotene RE	Vitamin D IU / Vitamin E IU	% USRDA	Total tocopherol mg / Alpha tocopherol mg	Other tocopherol mg / Total ash g	Calcium mg / Phosphorus mg	% USRDA	Sodium mg / Sodium meq	Potassium mg / Potassium meq	Chlorine mg / Chlorine meq	Iron mg / Magnesium mg	% USRDA	Zinc mg / Copper mg	% USRDA	Iodine mcg / Selenium mcg	% USRDA	Manganese mcg / Chromium mcg	
0.02	1	unk.	unk.	0.00	0	unk.	unk.	unk.	0(a)	0	unk.	unk.	7	1	25	48	unk.	0.4	2	0.7	5	unk.	unk.	unk.	A
0.2	1	unk.	unk.	unk.	unk.	873	18	unk.	unk.	unk.	unk.	0.21	8	1	1.1	1.2	unk	4	1	unk.	unk.	unk.		unk.	
tr	0	0.01	0	0.00	0	unk.	unk.	0	0(a)	0	unk.	unk.	2	0	unk.	7	unk.	0.1	0	unk.	unk.	unk.	unk.	unk.	B
tr	0	unk.	unk.	unk.	unk.	174	4	17	unk.	unk.	unk.	0.02	1	0	unk.	0.2	unk.	1	0	tr(a)	0	tr(a)		unk.	
0.03	2	unk.	unk.	tr(a)	0	unk.	unk.	unk.	39	10	unk.	unk.	13	1	1	138	unk.	1.1	6	unk.	unk.	unk.	unk.	unk.	C
0.2	1	unk.	unk.	unk.	unk.	8	0	unk.	unk.	unk.	unk.	0.51	64	6	0.0	3.5	unk.	51	13	unk.	unk.	unk.		unk.	
0.07	4	0.01	1	0.00	0	0.009	3	0	0	0	1.5	unk.	22	2	1	236	20	1.9	11	unk.	unk.	unk.	unk.	unk.	D
0.4	2	28	7	0.05	1	17	0	2	1.8	6	unk.	0.88	109	11	0.0	6.0	0.6	83	21	unk.	unk.	unk.		unk.	
0.12	7	0.00	0	0.00	0	unk.	unk.	unk.	0	0	0.0	0.0	81	8	1278	320	unk.	4	25	0.0	0	unk.	unk.	0	E
1.6	8	0	0	0.00	0	70	1	unk.	0.0	0	0.0	6.26	269	27	55.6	8.2	unk.	unk.	unk.	0.00	0	0		0	
0.37	22	0.30	15	0.34	6	0.004	1	0	0	0	6.2	6.0	36	4	1378	584	30	3.2	18	2.5	17	142.5	95	61	F
9.1	45	24	6	0.78	8	215	4	21	0.1	1	0.1	5.38	231	23	59.9	14.9	0.8	33	8	0.26	13	tr		13	
0.01	0	0.01	1	0.00	0	tr	0	0	0	0	tr	tr	3	0	4	22	1	0.1	1	tr	0	0.9	1	19	G
0.0	0	2	1	0.02	0	27	1	3	tr	0	tr	0.08	4	0	0.2	0.5	0.0	1	0	0.01	1	4		1	
0.10	6	0.45	23	1.65	28	unk.	unk.	unk.	unk.	unk.	0.0	0(a)	45	5	725	418	unk.	1.3	7	1.3	8	unk.	unk.	unk.	H
1.0	5	13	3	1.25	13	150	3	unk.	0.0	0	0(a)	3.50	85	9	31.5	10.7	unk.	unk.	unk.	0.27	14	tr(a)		tr(a)	
0.22	13	0.45	23	1.65	28	unk.	unk.	unk.	unk.	unk.	3.0	tr(a)	58	6	718	473	unk.	2.5	14	4.7	32	unk.	unk.	unk.	I
4.2	21	13	3	1.25	13	275	6	unk.	3.6	12	tr(a)	4.00	293	29	31.2	12.1	unk.	unk.	unk.	0.47	24	tr(a)		tr(a)	
0.51	30	0.38	19	0.63	11	0.003	1	0	0	0	8.4	tr	75	8	1673	1050	122	5.2	29	4.9	33	tr(a)	0	55	J
10.0	50	27	7	1.05	11	214	4	21	10.1	34	0.5	7.06	383	38	72.8	26.9	3.4	17	4	0.37	19	1		26	
0.25	15	0.13	7	0.25	4	0.003	1	6	55	14	8.6	0.1	105	11	1658	701	122	2.9	16	1.1	8	tr(a)	0	55	K
5.4	27	28	7	0.73	7	236	5	22	10.3	34	0.5	7.17	200	20	72.1	17.9	3.4	44	11	0.22	11	1		26	
0.00	0	unk.	unk.	0(a)	0	unk.	unk.	0	0	0	unk.	unk.	5	1	26	56	unk.	0.2	1	unk.	unk.	unk.	unk.	unk.	L
0.0	0	1	0	unk.	unk.	0	0	0	unk.	unk.	unk.	unk.	7	1	1.1	1.4	unk.	4	1	tr(a)	0	unk.		unk.	
tr	0	unk.	unk.	0.00	0	unk.	unk.	0(a)	0(a)	0	unk.	unk.	28	3	1	11	unk.	0.9	5	0.0	0	unk.	unk.	unk.	M
tr	0	unk.	unk.	unk.	unk.	6	0	unk.	unk.	unk.	unk.	0.08	1	0	0.0	0.3	unk.	1	0	unk.	unk.	unk.		unk.	
0.05	3	unk.	unk.	unk.	unk.	unk.	unk.	unk.	unk.	unk.	unk.	unk.	30	3	unk.	59	unk.	1.4	8	unk.	unk.	unk.	unk.	unk.	N
0.4	2	unk.	unk.	unk.	unk.	unk.	unk.	unk.	unk.	unk.	unk.	0.96	78	8	unk.	1.5	unk.	unk.	unk.	unk.	unk.	unk.		unk.	
0.03	2	0.01	1	3.25	54	unk.	unk.	5	unk.	unk.	0.1	tr(a)	12	1	20	31	unk.	1.0	6	0.3	2	unk.	unk.	unk.	O
0.2	1	1	0	0.10	1	17	0	tr	0.1	0	tr(a)	0.36	28	3	0.9	0.8	unk.	unk.	unk.	0.00	0	tr(a)		tr(a)	
0.01	0	unk.	unk.	0.00	0	unk.	unk.	0(a)	0(a)	0	unk.	unk.	14	1	5	24	unk.	0.2	1	tr	0	unk.	unk.	unk.	P
tr	0	unk.	unk.	unk.	unk.	12	0	unk.	unk.	unk.	unk.	0.13	2	0	0.2	0.6	unk.	6	2	unk.	unk.	unk.		unk.	
0.06	4	0.03	2	0.06	1	0.001	0	0	4	1	0.1	0.0	45	5	107	219	21	1.0	5	17.3	115	15.1	10	80	Q
0.9	4	5	1	0.16	2	564	11	37	0.1	0	0.1	0.76	70	7	4.7	5.6	0.6	25	6	0.08	4	15		25	
0(a)	0	0(a)	0	0(a)	0	0(a)	0	unk.	unk.	unk.	43.4	39.5	unk.	unk.	unk.	unk.	unk.	unk.	unk.	unk.	unk.	unk.	unk.	unk.	R
0(a)	0	0(a)	0	0(a)	0	unk.	unk.	unk.	3.9	13	3.9	0.00	unk.	unk.	unk.	unk.	unk.	unk.	unk.	unk.	unk.	unk.		unk.	
tr	0	0.04	2	0.00	0	unk.	unk.	0	0	0	unk.	unk.	24	2	30	176	unk.	0.4	2	unk.	unk.	unk.	unk.	unk.	S
0.1	1	3	1	0.06	1	0	0	0	unk.	unk.	unk.	0.72	16	2	1.3	4.5	unk.	34	8	tr	0	unk.		unk.	

	FOOD	Portion	Weight in grams / *Conversion for 100 g*	Kilocalories / H_2O g	Total carbohydrate g / Total fats g	Crude fiber g / Dietary fiber g	Total protein g / Total sugar g	% USRDA	Arginine mg / Histidine mg	Isoleucine mg / Leucine mg	Lysine mg / Methionine mg	Phenylalanine mg / Threonine mg	Valine mg / Tryptophan mg	Cystine mg / Tyrosine mg	Polyunsat. fatty acids g / Monounsat. fatty acids g	Saturated fatty acids g / P/S ratio	Linoleic acid g / Cholesterol mg	Thiamin mg / Ascorbic acid mg	% USRDA
A	coconut,dry,sweetened,chopped	1/4 C	16	85	8.2	0.64	0.6	**1**	86	29	25	28	35	10	0.05	5.27	tr	0.01	**0**
			6.45	0.5	6.1	3.64	4.7		11	44	12	21	5	17	0.46	0.01	0	0	**0**
B	coconut,fresh,chopped	1/4 C	24	84	2.3	0.97	0.8	**1**	128	44	37	42	52	15	0.19	7.58	0.16	0.01	**1**
			4.12	12.4	8.6	3.31	1.6		17	65	17	31	8	25	0.49	0.03	0	1	**2**
C	coconut,fresh,piece	1 piece	45	156	4.2	1.80	1.5	**2**	237	81	68	78	95	28	0.36	14.04	0.30	0.02	**2**
		(1/4 of med)	*2.22*	22.9	15.9	unk.	2.9		31	121	32	58	15	45	0.91	0.03	0	1	**2**
D	COD,fillet,broiled w/butter or	1 med	65	110	0.0	0.00	18.5	**41**	1221	910	1574	656	951	217	unk.	unk.	unk.	0.05	**4**
	margarine		*1.54*	42.0	3.4	0(a)	0.0		545	1355	520	786	177	726	unk.	unk.	32	unk.	**unk.**
E	cod,fillet,cooked	1 med	65	84	0.0	0.00	18.5	**41**	1221	910	1574	656	951	217	0.18	0.07	unk.	0.05	**4**
			1.54	42.0	0.5	0(a)	0.0		545	1355	520	786	177	726	0.04	2.45	32	unk.	**unk.**
F	COFFEE SUBSTITUTE,Postum,instant,prep	6 oz cup	173	11	2.4	tr	0.2	**0**	unk.	unk.	unk.	unk.	unk.	unk.	tr	tr	tr	0.03	**2**
			0.58	169.6	tr	0(a)	unk.		unk.	unk.	unk.	unk.	unk.	unk.	tr	0.00	0	0	**0**
G	COFFEE WHITENER,non-dairy liquid	1 Tbsp	15	20	1.7	0.00	0.1	**0**	12	8	10	8	8	3	tr	0.29	tr	0.00	**0**
	w/(H) vegetable oil & soy protein		*6.67*	11.6	1.5	0(a)	1.7		4	13	2	6	2	6	1.13	0.02	0	0	**0**
H	coffee whitener,non-dairy liquid	1 Tbsp	15	20	1.7	0.00	0.1	**0**	6	9	12	8	11	1	tr	1.39	tr	0.00	**0**
	w/lauric acid & casein		*6.67*	11.6	1.5	0(a)	1.7		4	15	4	6	2	9	unk.	0.00	0	0	**0**
I	coffee whitener,non-dairy powder	1 tsp	2.0	11	1.1	0.00	0.1	**0**	4	6	8	5	7	tr	0.00	0.64	tr	0.00	**0**
			51.02	tr	0.7	0(a)	1.1		3	9	3	4	1	5	0.02	0.00	0	0	**0**
J	COFFEE,brewed	6 oz cup	173	5	0.8	0.00	0.3	**0**	unk.	tr(a)	tr(a)	tr(a)	tr(a)	tr(a)	tr(a)	tr(a)	tr(a)	0.00	**0**
			0.58	unk.	0.2	0(a)	unk.		tr(a)	tr(a)	tr(a)	tr(a)	tr(a)	tr(a)	tr(a)	0.00	0(a)	0	**0**
K	coffee,instant,decaffeinated,prep	6 oz cup	173	2	tr	tr	tr	**0**	unk.	tr(â	tr(a)	tr(a)	tr(a)	tr(a)	tr(a)	tr(a)	tr(a)	0.00	**0**
			0.58	169.2	tr	0(a)	tr		tr(a)	tr(a)	tr(a)	tr(a)	tr(a)	tr(a)	tr(a)	0.00	0(a)	0	**0**
L	coffee,instant,freeze-dry powder	1 tsp	1.5	2	0.5	tr	tr	**0**	unk.	unk.	unk.	unk.	unk.	unk.	unk.	unk.	unk.	0.00	**0**
			68.12	tr	tr	0(a)	0.5		unk.	unk.	unk.	unk.	unk.	unk.	unk.	unk.	unk.	0	**0**
M	coffee,instant,freeze-dry powder,	1 tsp	1.5	2	0.5	tr	tr	**0**	unk.	unk.	unk.	unk.	unk.	unk.	unk.	unk.	unk.	0.00	**0**
	decaffeinated		*68.49*	tr	tr	0(a)	0.5		unk.	unk.	unk.	unk.	unk.	unk.	unk.	unk.	unk.	0	**0**
N	coffee,instant,powder	1 tsp	1.1	1	0.4	tr	tr	**0**	0	unk.	unk.	unk.	unk.	unk.	unk.	unk.	unk.	0.00	**0**
			86.96	tr	tr	0(a)	0.4		unk.	unk.	unk.	unk.	unk.	unk.	unk.	unk.	unk.	0	**0**
O	coffee,instant,powder,decaffeinated	1 tsp	1.1	1	0.4	tr	tr	**0**	unk.	unk.	unk.	unk.	unk.	unk.	unk.	unk.	unk.	0.00	**0**
			86.96	tr	tr	0(a)	0.4		unk.	unk.	unk.	unk.	unk.	unk.	unk.	unk.	unk.	0	**0**
P	coffee,instant,prep	6 oz cup	173	2	tr	tr	tr	**0**	0	tr(a)	tr(a)	tr(a)	tr(a)	tr(a)	tr(a)	tr(a)	tr(a)	0.00	**0**
			0.58	169.2	tr	0(a)	tr		tr(a)	tr(a)	tr(a)	tr(a)	tr(a)	tr(a)	tr(a)	0.00	0(a)	0	**0**
Q	coffee,international,Suisse Mocha	2 tsp	11	55	7.3	unk.	0.6	**1**	unk.	unk.	unk.	unk.	unk.	unk.	unk.	unk.	unk.	unk.	**unk.**
	-General Foods		*8.85*	0.2	2.7	unk.	unk.		unk.	unk.	unk.	unk.	unk.	unk.	unk.	unk.	0(a)	unk.	**unk.**
R	COLESLAW w/salad dressing	1/2 C	60	59	4.3	0.42	0.7	**1**	unk.	28	34	15	22	unk.	0.96	0.60	2.40	0.03	**2**
			1.67	49.7	4.7	1.68	unk.		unk.	29	7	20	6	unk.	1.20	1.60	0(a)	17	**29**
S	COLLARDS,fresh,leaves & stems,cooked,	1/2 C	73	21	3.6	0.58	2.0	**3**	unk.	61	101	62	98	unk.	unk.	tr(a)	unk.	0.10	**7**
	drained solids		*1.38*	65.8	0.4	unk.	0.2		unk.	109	23	57	28	unk.	unk.	unk.	0	33	**56**

Riboflavin mg / Niacin mg	% USRDA	Vitamin B_6 mg / Folacin mcg	% USRDA	Vitamin B_{12} mcg / Pantothenic acid mg	% USRDA	Biotin mg / Vitamin A IU	% USRDA	Preformed A RE / Beta carotene RE	Vitamin D IU / Vitamin E IU	% USRDA	Total tocopherol mg / Alpha tocopherol mg	Other tocopherol mg / Total ash g	Calcium mg / Phosphorus mg	% USRDA	Sodium mg / Sodium meq	Potassium mg / Potassium meq	Chlorine mg / Chlorine meq	Iron mg / Magnesium mg	% USRDA	Zinc mg / Copper mg	% USRDA	Iodine mcg / Selenium mcg	% USRDA	Manganese mcg / Chromium mcg	
0.00	**0**	0.01	**0**	0(a)	**0**	unk.	**unk.**	0	0	**0**	0.4	unk.	2	**0**	3	55	35	0.3	**2**	unk.	**unk.**	unk.	**unk.**	tr	A
0.1	**0**	5	**1**	0.03	**0**	0	**0**	0	0.5	**2**	0.1	0.12	17	**2**	0.1	1.4	1.0	12	**3**	tr	**0**	unk.		unk.	
0.00	**0**	0.01	**1**	0.00	**0**	unk.	**unk.**	0	0	**0**	0.7	unk.	3	**0**	6	62	3	0.4	**2**	unk.	**unk.**	unk.	**unk.**	252	B
0.1	**1**	7	**2**	0.05	**1**	0	**0**	0	0.8	**3**	0.2	0.22	23	**2**	0.2	1.6	0.1	11	**3**	0.00	**0**	unk.		unk.	
0.01	**1**	0.02	**1**	0.00	**0**	unk.	**unk.**	0	0	**0**	1.2	unk.	6	**1**	10	115	6	0.8	**4**	unk.	**unk.**	unk.	**unk.**	467	C
0.2	**1**	12	**3**	0.09	**1**	0	**0**	0	1.5	**5**	0.3	0.40	43	**4**	0.4	2.9	0.2	21	**5**	0.01	**0**	unk.		unk.	
0.07	**4**	0.09	**4**	0.47	**8**	unk.	**unk.**	unk.	unk.	**unk.**	tr(a)	tr(a)	20	**2**	71	265	unk.	0.6	**4**	unk.	**unk.**	unk.	**unk.**	unk.	D
1.9	**10**	12	**3**	0.05	**1**	117	**2**	unk.	0.0	**0**	tr(a)	unk.	178	**18**	3.1	6.8	unk.	unk.	**unk.**	unk.	**unk.**	tr(a)		tr(a)	
0.07	**4**	0.09	**4**	0.47	**8**	unk.	**unk.**	0	unk.	**unk.**	tr(a)	tr(a)	20	**2**	71	265	unk.	0.6	**4**	unk.	**unk.**	unk.	**unk.**	unk.	E
1.9	**10**	12	**3**	0.05	**1**	0	**0**	0	0.0	**0**	tr(a)	unk.	178	**18**	3.1	6.8	unk.	unk.	**unk.**	unk.	**unk.**	tr(a)		tr(a)	
0.00	**0**	unk.	**unk.**	0.00	**0**	unk.	**unk.**	0	0	**0**	unk.	unk.	5	**1**	3	93	unk.	0.2	**1**	unk.	**unk.**	unk.	**unk.**	unk.	F
0.7	**4**	unk.	**unk.**	unk.	**unk.**	0	**0**	0	unk.	**unk.**	unk.	0.22	20	**2**	0.1	2.4	unk.	10	**2**	unk.	**unk.**	unk.		unk.	
0.00	**0**	0.00	**0**	0.00	**0**	unk.	**unk.**	unk.	unk.	**unk.**	unk.	unk.	1	**0**	12	28	unk.	0.0	**0**	0.0	**0**	unk.	**unk.**	unk.	G
0.0	**0**	0	**0**	0.00	**0**	13	**0**	unk.	unk.	**unk.**	unk.	0.06	10	**1**	0.5	0.7	unk.	tr	**0**	unk.	**unk.**	unk.		unk.	
0.00	**0**	0.00	**0**	0.00	**0**	unk.	**unk.**	unk.	unk.	**unk.**	unk.	unk.	1	**0**	12	28	unk.	0.0	**0**	0.0	**0**	unk.	**unk.**	unk.	H
0.0	**0**	0	**0**	0.00	**0**	13	**0**	unk.	unk.	**unk.**	unk.	0.06	10	**1**	0.5	0.7	unk.	tr	**0**	unk.	**unk.**	unk.		unk.	
tr	**0**	0.00	**0**	0.00	**0**	unk.	**unk.**	unk.	unk.	**unk.**	unk.	unk.	tr	**0**	4	16	unk.	tr	**0**	tr	**0**	unk.	**unk.**	unk.	I
0.0	**0**	0	**0**	0.00	**0**	4	**0**	unk.	unk.	**unk.**	unk.	0.05	8	**1**	0.1	0.4	unk.	tr	**0**	unk.	**unk.**	unk.		unk.	
0.00	**0**	tr	**0**	unk.	**unk.**	unk.	**unk.**	0	0	**0**	0.8	0.8	5	**1**	unk.	158	unk.	0.2	**1**	0.1	**0**	unk.	**unk.**	24	J
0.9	**4**	0	**0**	0.01	**0**	0	**0**	0	1.0	**3**	0.0	unk.	5	**1**	unk.	4.0	unk.	9	**2**	0.01	**0**	tr(a)		21	
tr	**0**	tr	**0**	0.00	**0**	unk.	**unk.**	0	0	**0**	0.8	0.8	3	**0**	2	62	unk.	0.2	**1**	0.1	**0**	unk.	**unk.**	24	K
0.5	**3**	0	**0**	0.01	**0**	0	**0**	0	1.0	**3**	0.0	0.17	7	**1**	0.1	1.6	unk.	9	**2**	0.01	**0**	tr(a)		21	
tr	**0**	0.00	**0**	0.00	**0**	unk.	**unk.**	unk.	0	**0**	0.0	0(a)	3	**0**	1	48	unk.	0.1	**1**	tr	**0**	unk.	**unk.**	unk.	L
0.4	**2**	0	**0**	0.01	**0**	0	**0**	unk.	0.0	**0**	0.0	0.14	6	**1**	0.0	1.2	unk.	7	**2**	0.00	**0**	tr(a)		tr(a)	
tr	**0**	0.00	**0**	0.00	**0**	unk.	**unk.**	0	0	**0**	0.0	0(a)	3	**0**	1	48	unk.	0.1	**1**	tr	**0**	unk.	**unk.**	unk.	M
0.4	**2**	0	**0**	0.01	**0**	0	**0**	0	0.0	**0**	0.0	0.14	6	**1**	0.0	1.2	unk.	7	**2**	0.00	**0**	tr(a)		tr(a)	
tr	**0**	0.00	**0**	0.00	**0**	unk.	**unk.**	unk.	0	**0**	0.0	0(a)	2	**0**	1	37	unk.	0.1	**0**	tr	**0**	unk.	**unk.**	unk.	N
0.4	**2**	0	**0**	0.00	**0**	0	**0**	unk.	0.0	**0**	0.0	0.11	4	**0**	0.0	1.0	unk.	6	**1**	0.00	**0**	tr(a)		tr(a)	
tr	**0**	0.00	**0**	0.00	**0**	unk.	**unk.**	0	0	**0**	0.0	0(a)	2	**0**	1	37	unk.	0.1	**0**	tr	**0**	unk.	**unk.**	unk.	O
0.4	**2**	0	**0**	0.00	**0**	0	**0**	0	0.0	**0**	0.0	0.11	4	**0**	0.0	1.0	unk.	6	**1**	0.00	**0**	tr(a)		tr(a)	
tr	**0**	unk.	**unk.**	0.00	**0**	unk.	**unk.**	0	unk.	**unk.**	0.8	0.8	3	**0**	2	62	unk.	0.2	**1**	0.1	**0**	unk.	**unk.**	24	P
0.5	**3**	unk.	**unk.**	0.01	**0**	0	**0**	0	1.0	**3**	0.0	0.17	7	**1**	0.1	1.6	unk.	12	**3**	0.01	**0**	unk.		21	
unk.	**unk.**	unk.	**unk.**	unk.	**unk.**	unk.	**unk.**	unk.	0(a)	**0**	unk.	unk.	4	**0**	40	120	unk.	unk.	**unk.**	unk.	**unk.**	unk.	**unk.**	unk.	Q
0.2	**1**	unk.	**unk.**	unk.	**unk.**	unk.	**unk.**	unk.	unk.	**unk.**	unk.	0.37	34	**3**	1.7	3.1	unk.	unk.	**unk.**	unk.	**unk.**	unk.		unk.	
0.03	**2**	unk.	**unk.**	0(a)	**0**	unk.	**unk.**	0	unk.	**unk.**	unk.	unk.	26	**3**	74	115	unk.	0.2	**1**	0.1	**1**	unk.	**unk.**	unk.	R
0.2	**1**	unk.	**unk.**	unk.	**unk.**	90	**2**	9	unk.	**unk.**	unk.	0.54	17	**2**	3.2	2.9	unk.	7	**2**	unk.	**unk.**	unk.		unk.	
0.14	**9**	0.21	**11**	0.00	**0**	0.001	**1**	0	0(a)	**0**	unk.	unk.	110	**11**	18	170	unk.	0.4	**2**	unk.	**unk.**	unk.	**unk.**	unk.	S
0.9	**4**	1	**0**	0.29	**3**	3915	**78**	391	unk.	**unk.**	unk.	0.72	28	**3**	0.8	4.3	unk.	22	**6**	0.07	**3**	tr(a)		unk.	

	FOOD	Portion	Weight in grams / *Conversion for 100 g*	Kilocalories / H_2O g	Total carbohydrate g / Total fats g	Crude fiber g / Dietary fiber g	Total protein g / Total sugar g	% USRDA	Arginine mg / Histidine mg	Isoleucine mg / Leucine mg	Lysine mg / Methionine mg	Phenylalanine mg / Threonine mg	Valine mg / Tryptophan mg	Cystine mg / Tyrosine mg	Polyunsat. fatty acids g / Monounsat. fatty acids g	Saturated fatty acids g / P/S ratio	Linoleic acid g / Cholesterol mg	Thiamin mg / Ascorbic acid mg	% USRDA
A	collards,frozen,chopped,cooked,	1/2 C	85	25	4.8	0.85	2.5	**4**	unk.	76	128	79	123	unk.	unk.	tr(a)	unk.	0.05	**3**
	drained solids		*1.18*	76.7	0.3	unk.	1.9		unk.	138	30	71	35	unk.	unk.	unk.	unk.	28	**47**
B	CONE,ice cream	1 cone	4.4	18	3.6	unk.	0.3	**1**	unk.	unk.	unk.	unk.	unk.	unk.	unk.	unk.	unk.	0.01	**1**
			22.73	0.2	0.2	unk.	unk.		unk.	unk.	unk.	unk.	unk.	unk.	unk.	unk.	unk.	unk.	**unk.**
C	cone,sugar,ice cream	1 cone	10	40	8.5	unk.	0.9	**1**	unk.	unk.	unk.	unk.	unk.	unk.	unk.	unk.	unk.	0.04	**3**
			10.00	0.2	0.3	unk.	unk.		unk.	unk.	unk.	unk.	unk.	unk.	unk.	unk.	unk.	unk.	**unk.**
D	COOKIE,almond windmill-Nabisco	1 cookie	14	66	10.3	unk.	0.9	**2**	unk.	unk.	unk.	unk.	unk.	unk.	unk.	unk.	unk.	0.05	**3**
			6.94	0.6	2.4	unk.	unk.		unk.	unk.	unk.	unk.	unk.	unk.	unk.	unk.	unk.	unk.	**unk.**
E	cookies,animal crackers	11 pieces	29	123	22.9	0.03	1.9	**3**	unk.	87	43	104	82	38	unk.	0.69	0.54	0.01	**1**
			3.50	0.9	2.7	unk.	unk.		38	146	25	54	23	65	1.32	unk.	unk.	0(a)	**0**
F	cookies,animal crackers-Fireside	12 pieces	29	128	21.7	unk.	2.0	**3**	unk.	unk.	unk.	unk.	unk.	unk.	unk.	unk.	unk.	0.11	**7**
			3.44	1.1	3.7	unk.	unk.		unk.	unk.	unk.	unk.	unk.	unk.	unk.	unk.	unk.	unk.	**unk.**
G	cookie,animal crackers,Barnum's	11 pieces	29	128	21.2	unk.	1.7	**3**	unk.	unk.	unk.	unk.	unk.	unk.	unk.	unk.	unk.	0.09	**6**
	Animals-Nabisco		*3.50*	1.2	4.0	unk.	unk.		unk.	unk.	unk.	unk.	unk.	unk.	unk.	unk.	unk.	unk.	**unk.**
H	cookie,apple crisp-Nabisco	1 cookie	10.3	50	6.93	unk.	**0.6**	**1**	unk.	unk.	unk.	unk.	unk.	unk.	unk.	unk.	unk.	0.04	**3**
			9.71	0.4	2.14	unk.	**unk.**		unk.	unk.	unk.	unk.	unk.	unk.	unk.	unk.	unk.	unk.	**unk.**
I	cookie,assorted packaged	1 cookie	8.7	42	6.2	0.01	**0.4**	**1**	unk.	20	10	24	19	9	unk.	0.35	unk.	tr	**0**
			11.49	0.2	1.8	unk.	**unk.**		9	34	6	13	5	15	unk.	unk.	9	tr	**0**
J	cookie,baronet creme sandwich,Famous	1 cookie	11	54	7.9	unk.	**0.5**	**1**	unk.	unk.	unk.	unk.	unk.	unk.	unk.	unk.	unk.	0.04	**3**
	Cookie Assortment-Nabisco		*9.01*	0.4	2.3	unk.	**unk.**		unk.	unk.	unk.	unk.	unk.	unk.	unk.	unk.	unk.	unk.	**unk.**
K	cookie,Biscos sugar wafer,Famous	1 cookie	9.5	49	6.8	unk.	**0.4**	**1**	unk.	unk.	unk.	unk.	unk.	unk.	unk.	unk.	unk.	tr	**0**
	Cookie Assortment-Nabisco		*10.53*	0.1	2.2	unk.	**unk.**		unk.	unk.	unk.	unk.	unk.	unk.	unk.	unk.	unk.	unk.	**unk.**
L	cookie,brown edge sandwich-Nabisco	1 cookie	16	82	9.8	unk.	**0.9**	**1**	unk.	unk.	unk.	unk.	unk.	unk.	unk.	unk.	unk.	0.04	**3**
			6.37	0.4	4.4	unk.	**unk.**		unk.	unk.	unk.	unk.	unk.	unk.	unk.	unk.	unk.	unk.	**unk.**
M	cookie,brown edge wafer-Nabisco	1 cookie	5.8	28	4.1	unk.	**0.3**	**0**	unk.	unk.	unk.	unk.	unk.	unk.	unk.	unk.	unk.	0.02	**1**
			17.24	0.2	1.1	unk.	**unk.**		unk.	unk.	unk.	unk.	unk.	unk.	unk.	unk.	unk.	unk.	**unk.**
N	cookie,brownie thin wafers-Nabisco	1 cookie	9.1	39	6.7	unk.	**1.0**	**2**	unk.	unk.	unk.	unk.	unk.	unk.	unk.	unk.	unk.	0.03	**2**
			10.99	0.4	0.9	unk.	**unk.**		unk.	unk.	unk.	unk.	unk.	unk.	unk.	unk.	unk.	unk.	**unk.**
O	cookie,brownie w/nuts,home recipe	1-3/4x1-3/4x	24	97	10.2	0.08	**1.0**	**2**	56	43	40	44	51	18	0.68	1.22	1.33	0.02	**1**
		7/8"	*4.13*	7.8	4.9	0.15	**5.7**		23	70	20	35	11	34	1.99	0.56	15	0	**0**
P	cookie,butter flavored-Nabisco	1 cookie	4.9	23	3.4	unk.	**0.3**	**1**	unk.	unk.	unk.	unk.	unk.	unk.	unk.	unk.	unk.	0.02	**1**
			20.41	0.2	0.9	unk.	**unk.**		unk.	unk.	unk.	unk.	unk.	unk.	unk.	unk.	unk.	unk.	**unk.**
Q	cookie,butter flavored,Famous	1 cookie	10	47	7.0	unk.	**0.6**	**1**	unk.	unk.	unk.	unk.	unk.	unk.	unk.	unk.	unk.	0.04	**3**
	Cookie Assortment-Nabisco		*10.00*	0.3	1.9	unk.	**unk.**		unk.	unk.	unk.	unk.	unk.	unk.	unk.	unk.	unk.	unk.	**unk.**
R	cookie,butter/Christmas/rolled sugar,	2-1/2" dia x	11	50	5.9	0.01	**0.6**	**1**	unk.	27	19	30	27	12	0.04	0.63	0.21	0.02	**1**
	home recipe	1/4"	*9.17*	1.6	2.6	unk.	**2.8**		13	45	10	19	7	21	1.66	0.07	5	0	**0**
S	cookie,butter,bulk pack-Nabisco	1 cookie	5.0	23	3.5	unk.	**0.3**	**1**	unk.	unk.	unk.	unk.	unk.	unk.	unk.	unk.	unk.	0.03	**2**
			20.00	0.2	0.9	unk.	**unk.**		unk.	unk.	unk.	unk.	unk.	unk.	unk.	unk.	unk.	unk.	**unk.**

Riboflavin mg / Niacin mg	% USRDA	Vitamin B_6 mg / Folacin mcg	% USRDA	Vitamin B_{12} mcg / Pantothenic acid mg	% USRDA	Biotin mg / Vitamin A IU	% USRDA	Preformed A RE / Beta carotene RE	Vitamin D IU / Vitamin E IU	% USRDA	Total tocopherol mg / Alpha tocopherol mg	Other tocopherol mg / Total ash g	Calcium mg / Phosphorus mg	% USRDA	Sodium mg / Sodium meq	Potassium mg / Potassium meq	Chlorine mg / Chlorine meq	Iron mg / Magnesium mg	% USRDA	Zinc mg / Copper mg	% USRDA	Iodine mcg / Selenium mcg	% USRDA	Manganese mcg / Chromium mcg	
0.12	**7**	0.25	**12**	0.00	**0**	0.002	**1**	0	0(a)	**0**	unk.	unk.	150	**15**	14	201	unk.	0.8	**5**	unk.	**unk.**	unk.	**unk.**	unk.	A
0.5	**3**	2	**0**	0.34	**3**	5780	**116**	578	unk.	**unk.**	unk.	0.76	43	**4**	0.6	5.1	unk.	26	**7**	0.08	**4**	tr(a)		unk.	
0.02	**1**	unk.	**unk.**	unk.	**unk.**	unk.	**unk.**	unk.	0(a)	**0**	unk.	unk.	1	**0**	6	5	unk.	0.1	**1**	tr	**0**	unk.	**unk.**	22	B
0.2	**1**	unk.	**unk.**	unk.	**unk.**	unk.	**unk.**	unk.	unk.	**unk.**	unk.	0.03	4	**0**	0.2	0.1	unk.	1	**0**	0.01	**0**	unk.		unk.	
0.04	**2**	unk.	**unk.**	unk.	**unk.**	unk.	**unk.**	unk.	0(a)	**0**	unk.	unk.	5	**1**	33	17	unk.	0.4	**2**	0.1	**1**	unk.	**unk.**	72	C
0.5	**2**	unk.	**unk.**	unk.	**unk.**	unk.	**unk.**	unk.	unk.	**unk.**	unk.	0.14	10	**1**	1.4	0.4	unk.	3	**1**	0.03	**2**	unk.		unk.	
0.04	**3**	unk.	**unk.**	unk.	**unk.**	unk.	**unk.**	unk.	0(a)	**0**	unk.	unk.	4	**0**	57	16	unk.	0.4	**2**	0.1	**0**	unk.	**unk.**	72	D
0.4	**2**	unk.	**unk.**	unk.	**unk.**	unk.	**unk.**	unk.	unk.	**unk.**	unk.	0.17	11	**1**	2.5	0.4	unk.	3	**1**	0.01	**1**	unk.		unk.	
0.03	**2**	unk.	**unk.**	0(a)	**0**	unk.	**unk.**	unk.	unk.	**unk.**	unk.	unk.	15	**2**	87	27	unk.	0.1	**1**	unk.	**unk.**	unk.	**unk.**	unk.	E
0.1	**0**	unk.	**unk.**	unk.	**unk.**	37	**1**	unk.	unk.	**unk.**	0.1	0.31	33	**3**	3.8	0.7	unk.	unk.	**unk.**	unk.	**unk.**	unk.		unk.	
0.07	**4**	unk.	**unk.**	unk.	**unk.**	unk.	**unk.**	unk.	0(a)	**0**	unk.	unk.	5	**1**	176	30	unk.	0.5	**3**	0.1	**1**	unk.	**unk.**	135	F
1.2	**6**	unk.	**unk.**	unk.	**unk.**	unk.	**unk.**	unk.	unk.	**unk.**	unk.	0.55	43	**4**	7.6	0.8	unk.	5	**1**	0.02	**1**	unk.		unk.	
0.10	**6**	unk.	**unk.**	unk.	**unk.**	unk.	**unk.**	unk.	0(a)	**0**	unk.	unk.	8	**1**	138	33	unk.	0.7	**4**	0.2	**1**	unk.	**unk.**	102	G
1.0	**5**	unk.	**unk.**	unk.	**unk.**	unk.	**unk.**	unk.	unk.	**unk.**	unk.	0.38	26	**3**	6.0	0.8	unk.	5	**1**	0.04	**2**	unk.		unk.	
0.04	**2**	unk.	**unk.**	unk.	**unk.**	unk.	**unk.**	unk.	0(a)	**0**	unk.	unk.	3	**0**	43	13.4	unk.	0.3	**1**	0.1	**1**	unk.	**unk.**	45	H
0.5	**2**	unk.	**unk.**	unk.	**unk.**	unk.	**unk.**	unk.	unk.	**unk.**	unk.	0.14	9	**1**	1.8	0.4	unk.	2	**1**	0.01	**1**	unk.		unk.	
tr	**0**	tr	**0**	tr(a)	**0**	tr	**0**	unk.	tr	**0**	tr	unk.	3	**0**	32	6	unk.	0.1	**0**	0.1	**1**	unk.	**unk.**	unk.	I
tr	**0**	1	**0**	0.03	**0**	7	**0**	unk.	0.1	**0**	unk.	0.10	14	**1**	1.4	0.1	unk.	1	**0**	0.01	**1**	unk.		unk.	
0.03	**2**	unk.	**unk.**	unk.	**unk.**	unk.	**unk.**	unk.	0(a)	**0**	unk.	unk.	3	**0**	29	11	unk.	0.2	**1**	tr	**0**	unk.	**unk.**	19	J
0.4	**2**	unk.	**unk.**	unk.	**unk.**	unk.	**unk.**	unk.	unk.	**unk.**	unk.	0.09	9	**1**	1.3	0.3	unk.	1	**0**	0.01	**0**	unk.		unk.	
0.02	**1**	unk.	**unk.**	unk.	**unk.**	unk.	**unk.**	unk.	0(a)	**0**	unk.	unk.	1	**0**	16	6	unk.	0.2	**1**	tr	**0**	unk.	**unk.**	25	K
0.3	**1**	unk.	**unk.**	unk.	**unk.**	unk.	**unk.**	unk.	unk.	**unk.**	unk.	0.04	5	**1**	0.7	0.1	unk.	1	**0**	0.01	**0**	unk.		unk.	
0.05	**3**	unk.	**unk.**	unk.	**unk.**	unk.	**unk.**	unk.	0(a)	**0**	unk.	unk.	9	**1**	39	29	unk.	0.5	**3**	0.1	**1**	unk.	**unk.**	55	L
0.5	**2**	unk.	**unk.**	unk.	**unk.**	unk.	**unk.**	unk.	unk.	**unk.**	unk.	0.17	17	**2**	1.7	0.8	unk.	6	**2**	0.04	**2**	unk.		unk.	
0.01	**1**	unk.	**unk.**	unk.	**unk.**	unk.	**unk.**	unk.	0(a)	**0**	unk.	unk.	2	**0**	17	6	unk.	0.1	**1**	tr	**0**	unk.	**unk.**	14	M
0.2	**1**	unk.	**unk.**	unk.	**unk.**	unk.	**unk.**	unk.	unk.	**unk.**	unk.	0.06	4	**0**	0.8	0.2	unk.	1	**0**	0.00	**0**	unk.		unk.	
0.03	**2**	unk.	**unk.**	unk.	**unk.**	unk.	**unk.**	unk.	0(a)	**0**	unk.	unk.	2	**0**	43	11	unk.	0.3	**2**	0.1	**0**	unk.	**unk.**	unk.	N
0.2	**1**	unk.	**unk.**	unk.	**unk.**	unk.	**unk.**	unk.	unk.	**unk.**	unk.	0.14	8	**1**	1.9	0.3	unk.	2	**1**	0.02	**1**	unk.		unk.	
0.02	**1**	0.02	**1**	0.03	**1**	0.002	**1**	0	1	**0**	1.9	0.9	6	**1**	37	28	9	0.3	**2**	0.1	**1**	2.2	**1**	44	O
0.1	**1**	4	**1**	0.06	**1**	118	**2**	tr	2.3	**8**	1.0	0.20	22	**2**	1.6	0.7	0.2	15	**4**	0.03	**1**	1		6	
0.02	**1**	unk.	**unk.**	unk.	**unk.**	unk.	**unk.**	unk.	0(a)	**0**	unk.	unk.	1	**0**	25	6	unk.	0.1	**1**	tr	**0**	unk.	**unk.**	16	P
0.1	**1**	unk.	**unk.**	unk.	**unk.**	unk.	**unk.**	unk.	unk.	**unk.**	unk.	0.08	6	**1**	1.1	0.2	unk.	1	**0**	0.01	**0**	unk.		unk.	
0.03	**2**	unk.	**unk.**	unk.	**unk.**	unk.	**unk.**	unk.	0(a)	**0**	unk.	unk.	3	**0**	52	13	unk.	0.2	**1**	0.1	**0**	unk.	**unk.**	40	Q
0.3	**2**	unk.	**unk.**	unk.	**unk.**	unk.	**unk.**	unk.	unk.	**unk.**	unk.	0.16	12	**1**	2.2	0.3	unk.	2	**1**	0.01	**1**	unk.		unk.	
0.01	**1**	tr	**0**	0.01	**0**	tr	**0**	0	0	**0**	tr	0.0	4	**0**	23	6	5	0.1	**1**	0.1	**0**	3.3	**2**	18	R
0.1	**1**	1	**0**	0.03	**0**	3	**0**	0	0.0	**0**	0.0	0.12	9	**1**	1.0	0.2	0.1	4	**1**	0.01	**0**	2		4	
0.01	**1**	unk.	**unk.**	unk.	**unk.**	unk.	**unk.**	unk.	0(a)	**0**	unk.	unk.	1	**0**	24	6	unk.	0.1	**1**	tr	**0**	unk.	**unk.**	23	S
0.2	**1**	unk.	**unk.**	unk.	**unk.**	unk.	**unk.**	unk.	unk.	**unk.**	unk.	0.08	13	**1**	1.0	0.2	unk.	1	**0**	0.03	**1**	unk.		unk.	

	FOOD	Portion	Weight in grams / Conversion for 100 g	Kilocalories / H_2O g	Total carbohydrate g / Total fats g	Crude fiber g / Dietary fiber g	Total protein g / Total sugar g	% USRDA	Arginine mg / Histidine mg	Isoleucine mg / Leucine mg	Lysine mg / Methionine mg	Phenylalanine mg / Threonine mg	Valine mg / Tryptophan mg	Cystine mg / Tyrosine mg	Polyunsat. fatty acids g / Monounsat. fatty acids g	Saturated fatty acids g / P/S ratio	Linoleic acid g / Cholesterol mg	Thiamin mg / Ascorbic acid mg	% USRDA
A	cookie,butter,large-Fireside	1 cookie	10	46	7.7	unk.	0.7	1	unk.	unk.	unk.	unk.	unk.	unk.	unk.	unk.	unk.	0.05	3
			9.71	0.4	1.4	unk.	unk.		unk.	unk.	unk.	unk.	unk.	unk.	unk.	unk.	unk.	unk.	unk.
B	cookie,butter,small-Fireside	1 cookie	4.8	21	3.6	unk.	0.3	1	unk.	unk.	unk.	unk.	unk.	unk.	unk.	unk.	unk.	0.02	1
			20.83	0.2	0.6	unk.	unk.		unk.	unk.	unk.	unk.	unk.	unk.	unk.	unk.	unk.	unk.	unk.
C	cookie,Cameo creme sandwich-Nabisco	1 cookie	14	68	10.5	unk.	0.7	1	unk.	unk.	unk.	unk.	unk.	unk.	unk.	unk.	unk.	0.04	3
			7.04	0.3	2.6	unk.	unk.		unk.	unk.	unk.	unk.	unk.	unk.	unk.	unk.	unk.	unk.	unk.
D	cookie,Cameo creme sandwich,Famous	1 cookie	15	72	11.2	unk.	0.7	1	unk.	unk.	unk.	unk.	unk.	unk.	unk.	unk.	unk.	0.05	3
	Cookie Assortment-Nabisco		*6.62*	0.3	2.7	unk.	unk.		unk.	unk.	unk.	unk.	unk.	unk.	unk.	unk.	unk.	unk.	unk.
E	cookie,Chipits chocolate chip,vending	1-1/8 oz pkg	32	157	21.0	unk.	1.8	3	unk.	unk.	unk.	unk.	unk.	unk.	unk.	unk.	unk.	0.12	8
	-Nabisco		*3.13*	1.5	7.3	unk.	unk.		unk.	unk.	unk.	unk.	unk.	unk.	unk.	unk.	unk.	unk.	unk.
F	cookie,Chips Ahoy chocolate chip	1 cookie	11	52	7.1	unk.	0.6	1	unk.	unk.	unk.	unk.	unk.	unk.	unk.	unk.	unk.	0.02	2
	-Nabisco		*9.43*	0.4	2.4	unk.	unk.		unk.	unk.	unk.	unk.	unk.	unk.	unk.	unk.	unk.	unk.	unk.
G	cookie,chocolate chip snaps-Nabisco	1 cookie	4.5	21	3.3	unk.	0.3	0	unk.	unk.	unk.	unk.	unk.	unk.	unk.	unk.	unk.	0.01	1
			22.22	0.2	0.7	unk.	unk.		unk.	unk.	unk.	unk.	unk.	unk.	unk.	unk.	unk.	unk.	unk.
H	cookie,chocolate chip,home recipe	2-1/2" dia x	12	59	6.4	0.06	0.6	1	28	23	17	25	26	10	0.42	1.17	0.51	0.01	1
		1/4"	*8.33*	1.1	3.7	unk.	4.5		12	38	9	18	6	18	1.76	0.36	5	0	0
I	cookie,chocolate chip,large-Fireside	1 cookie	10	49	7.1	unk.	0.5	1	unk.	unk.	unk.	unk.	unk.	unk.	unk.	unk.	unk.	0.04	3
			9.71	0.5	2.0	unk.	unk.		unk.	unk.	unk.	unk.	unk.	unk.	unk.	unk.	unk.	unk.	unk.
J	cookie,chocolate chip,small-Fireside	1 cookie	6.0	29	4.0	unk.	0.4	1	unk.	unk.	unk.	unk.	unk.	unk.	unk.	unk.	unk.	0.02	1
			16.67	0.3	1.3	unk.	unk.		unk.	unk.	unk.	unk.	unk.	unk.	unk.	unk.	unk.	unk.	unk.
K	cookie,chocolate chip,vending-Nabisco	2 oz bag	53	263	34.8	unk.	3.1	5	unk.	unk.	unk.	unk.	unk.	unk.	unk.	unk.	unk.	0.18	12
			1.89	2.2	12.3	unk.	unk.		unk.	unk.	unk.	unk.	unk.	unk.	unk.	unk.	unk.	unk.	unk.
L	cookie,chocolate chocolate chip	1 cookie	11	52	7.2	unk.	0.6	1	unk.	unk.	unk.	unk.	unk.	unk.	unk.	unk.	unk.	0.01	1
	-Nabisco		*9.43*	0.4	2.3	unk.	unk.		unk.	unk.	unk.	unk.	unk.	unk.	unk.	unk.	unk.	unk.	unk.
M	cookie,chocolate cremes-Fireside	1 cookie	10	48	7.6	unk.	0.5	1	unk.	unk.	unk.	unk.	unk.	unk.	unk.	unk.	unk.	0.05	3
			9.71	0.2	1.8	unk.	unk.		unk.	unk.	unk.	unk.	unk.	unk.	unk.	unk.	unk.	unk.	unk.
N	cookie,chocolate flavored chip,bulk	1 cookie	14	68	9.2	unk.	0.8	1	unk.	unk.	unk.	unk.	unk.	unk.	unk.	unk.	unk.	0.04	3
	pack-Nabisco		*7.14*	0.6	3.1	unk.	unk.		unk.	unk.	unk.	unk.	unk.	unk.	unk.	unk.	unk.	unk.	unk.
O	cookie,chocolate graham-Nabisco	1 cookie	11	54	7.1	unk.	0.7	1	unk.	unk.	unk.	unk.	unk.	unk.	unk.	unk.	unk.	0.01	1
			9.26	0.3	2.6	unk.	unk.		unk.	unk.	unk.	unk.	unk.	unk.	unk.	unk.	unk.	unk.	unk.
P	cookie,chocolate oatmeal	2" dia x 1/4"	14	66	7.5	0.09	0.9	1	unk.	37	29	40	40	16	0.85	1.03	0.81	0.03	2
			7.25	1.4	3.9	0.17	4.1		17	57	13	26	10	27	1.79	0.83	5	0	0
Q	cookie,chocolate Pinwheel-Nabisco	1 cookie	30	139	20.9	unk.	1.1	2	unk.	unk.	unk.	unk.	unk.	unk.	unk.	unk.	unk.	0.03	2
			3.30	2.4	5.6	unk.	unk.		unk.	unk.	unk.	unk.	unk.	unk.	unk.	unk.	unk.	unk.	unk.
R	cookie,chocolate snap-Nabisco	1 cookie	3.8	17	2.8	unk.	0.3	0	unk.	unk.	unk.	unk.	unk.	unk.	unk.	unk.	unk.	0.01	0
			26.32	0.2	0.5	unk.	unk.		unk.	unk.	unk.	unk.	unk.	unk.	unk.	unk.	unk.	unk.	unk.
S	cookie,chocolate wafer-Nabisco	1 cookie	6.3	28	4.7	unk.	0.6	1	unk.	unk.	unk.	unk.	unk.	unk.	unk.	unk.	unk.	0.02	1
			15.87	unk.	0.7	unk.	2.3		unk.	unk.	unk.	unk.	unk.	unk.	unk.	unk.	unk.	unk.	unk.

Riboflavin mg / Niacin mg	% USRDA	Vitamin B_6 mg / Folacin mcg	% USRDA	Vitamin B_{12} mcg / Pantothenic acid mg	% USRDA	Biotin mg / Vitamin A IU	% USRDA	Preformed A RE / Beta carotene RE	Vitamin D IU / Vitamin E IU	% USRDA	Total tocopherol mg / Alpha tocopherol mg	Other tocopherol mg / Total ash g	Calcium mg / Phosphorus mg	% USRDA	Sodium mg / Sodium meq	Potassium mg / Potassium meq	Chlorine mg / Chlorine meq	Iron mg / Magnesium mg	% USRDA	Zinc mg / Copper mg	% USRDA	Iodine mcg / Selenium mcg	% USRDA	Manganese mcg / Chromium mcg	
0.02 / 0.4	2 / 2	unk. / unk.	unk. / unk.	unk. / unk.	unk. / unk.	unk. / unk.	unk. / unk.	unk. / unk.	0(a) / unk.	0 / unk.	unk. / unk.	unk. / 0.15	3 / 14	0 / 1	48 / 2.1	13 / 0.3	unk. / unk.	0.2 / 2	1 / 1	0.1 / 0.01	0 / 0	unk. / unk.	unk.	44 / unk.	A
0.01 / 0.2	1 / 1	unk. / unk.	unk. / unk.	unk. / unk.	unk. / unk.	unk. / unk.	unk. / unk.	unk. / unk.	0(a) / unk.	0 / unk.	unk. / unk.	unk. / 0.07	1 / 7	0 / 1	22 / 1.0	6 / 0.2	unk. / unk.	0.1 / 0	1 / 0	tr / tr	0 / 0	unk. / unk.	unk.	21 / unk.	B
0.04 / 0.4	2 / 2	unk. / unk.	unk. / unk.	unk. / unk.	unk. / unk.	unk. / unk.	unk. / unk.	unk. / unk.	0(a) / unk.	0 / unk.	unk. / unk.	unk. / 0.15	6 / 14	1 / 1	43 / 1.9	21 / 0.5	unk. / unk.	0.3 / 2	1 / 1	0.1 / 0.02	0 / 1	unk. / unk.	unk.	42 / unk.	C
0.04 / 0.4	2 / 2	unk. / unk.	unk. / unk.	unk. / unk.	unk. / unk.	unk. / unk.	unk. / unk.	unk. / unk.	0(a) / unk.	0 / unk.	unk. / unk.	unk. / 0.17	7 / 15	1 / 2	46 / 2.0	17 / 0.4	unk. / unk.	0.3 / 2	2 / 1	0.1 / 0.02	0 / 1	unk. / unk.	unk.	42 / unk.	D
0.09 / 1.0	2 / 5	unk. / unk.	unk. / unk.	unk. / unk.	unk. / unk.	unk. / unk.	unk. / unk.	unk. / unk.	0(a) / unk.	0 / unk.	unk. / unk.	unk. / 0.37	7 / 33	1 / 3	111 / 4.8	32 / 0.8	unk. / unk.	0.7 / 6	4 / 2	0.2 / 0.07	1 / 4	unk. / unk.	unk.	96 / unk.	E
0.03 / 0.3	2 / 2	unk. / unk.	unk. / unk.	unk. / unk.	unk. / unk.	unk. / unk.	unk. / unk.	unk. / unk.	0(a) / unk.	0 / unk.	unk. / unk.	unk. / 0.12	2 / 12	0 / 1	32 / 1.4	14 / 0.3	unk. / unk.	0.2 / 3	1 / 1	0.1 / 0.02	0 / 1	unk. / unk.	unk.	46 / unk.	F
0.01 / 0.1	1 / 1	unk. / unk.	unk. / unk.	unk. / unk.	unk. / unk.	unk. / unk.	unk. / unk.	unk. / unk.	0(a) / unk.	0 / unk.	unk. / unk.	unk. / 0.05	1 / 4	0 / 0	16 / 0.7	5 / 0.1	unk. / unk.	0.1 / 1	1 / 0	tr / 0.01	0 / 1	unk. / unk.	unk.	17 / unk.	G
0.01 / 0.1	1 / 1	0.01 / 1	1 / 0	0.01 / 0.02	0 / 0	0.001 / 3	0 / 0	0 / 0	3 / tr	1 / 0	tr / tr	0.0 / 0.17	4 / 13	0 / 1	43 / 1.9	21 / 0.5	4 / 0.1	0.2 / 8	1 / 2	0.1 / 0.02	0 / 1	8.7 / 1	6	30 / 4	H
0.03 / 0.3	2 / 2	unk. / unk.	unk. / unk.	unk. / unk.	unk. / unk.	unk. / unk.	unk. / unk.	unk. / unk.	0(a) / unk.	0 / unk.	unk. / unk.	unk. / 0.11	2 / 9	0 / 1	38 / 1.7	13 / 0.3	unk. / unk.	0.3 / 2	2 / 1	0.1 / 0.02	0 / 1	unk. / unk.	unk.	46 / unk.	I
0.02 / 0.2	1 / 1	unk. / unk.	unk. / unk.	unk. / unk.	unk. / unk.	unk. / unk.	unk. / unk.	unk. / unk.	0(a) / unk.	0 / unk.	unk. / unk.	unk. / 0.07	2 / 5	0 / 1	22 / 0.9	8 / 0.2	unk. / unk.	0.1 / 1	1 / 0	tr / 0.01	0 / 1	unk. / unk.	unk.	35 / unk.	J
0.14 / 1.8	8 / 9	unk. / unk.	unk. / unk.	unk. / unk.	unk. / unk.	unk. / unk.	unk. / unk.	unk. / unk.	0(a) / unk.	0 / unk.	unk. / unk.	unk. / 0.60	9 / 58	1 / 6	175 / 7.6	56 / 1.4	unk. / unk.	1.1 / 52	6 / 13	0.3 / 0.12	2 / 6	unk. / unk.	unk.	191 / unk.	K
0.02 / 0.2	1 / 1	unk. / unk.	unk. / unk.	unk. / unk.	unk. / unk.	unk. / unk.	unk. / unk.	unk. / unk.	0(a) / unk.	0 / unk.	unk. / unk.	unk. / 0.12	2 / 11	0 / 1	31 / 1.3	19 / 0.5	unk. / unk.	0.3 / 5	2 / 1	0.1 / 0.03	1 / 1	unk. / unk.	unk.	53 / unk.	L
0.04 / 0.3	2 / 1	unk. / unk.	unk. / unk.	unk. / unk.	unk. / unk.	unk. / unk.	unk. / unk.	unk. / unk.	0(a) / unk.	0 / unk.	unk. / unk.	unk. / 0.19	2 / 8	0 / 1	72 / 3.1	10 / 0.3	unk. / unk.	0.2 / 2	1 / 1	0.1 / 0.01	0 / 1	unk. / unk.	unk.	43 / unk.	M
0.04 / 0.4	2 / 2	unk. / unk.	unk. / unk.	unk. / unk.	unk. / unk.	unk. / unk.	unk. / unk.	unk. / unk.	0(a) / unk.	0 / unk.	unk. / unk.	unk. / 0.18	3 / 11	0 / 1	53 / 2.3	15 / 0.4	unk. / unk.	0.3 / 3	2 / 1	0.1 / 0.01	1 / 1	unk. / unk.	unk.	73 / unk.	N
0.02 / 0.3	1 / 1	unk. / unk.	unk. / unk.	unk. / unk.	unk. / unk.	unk. / unk.	unk. / unk.	unk. / unk.	0(a) / unk.	0 / unk.	unk. / unk.	unk. / 0.13	4 / 16	0 / 2	26 / 1.1	31 / 0.8	unk. / unk.	0.5 / 9	3 / 2	0.2 / 0.07	1 / 3	unk. / unk.	unk.	82 / unk.	O
0.02 / 0.1	1 / 1	0.01 / 2	0 / 1	0.01 / 0.07	0 / 1	0.001 / 5	0 / 0	0 / tr	tr / 0.4	0 / 1	0.3 / tr	0.0 / 0.11	4 / 20	0 / 2	14 / 0.6	27 / 0.7	5 / 0.1	0.3 / 10	2 / 3	0.2 / 0.02	1 / 1	0.8 / 1	1	92 / 5	P
0.05 / 0.2	3 / 1	unk. / unk.	unk. / unk.	unk. / unk.	unk. / unk.	unk. / unk.	unk. / unk.	unk. / unk.	0(a) / unk.	0 / unk.	unk. / unk.	unk. / 0.26	6 / 22	1 / 2	41 / 1.8	44 / 1.1	unk. / unk.	0.6 / 11	4 / 3	0.2 / 0.09	1 / 5	unk. / unk.	unk.	95 / unk.	Q
0.01 / 0.1	1 / 1	unk. / unk.	unk. / unk.	unk. / unk.	unk. / unk.	unk. / unk.	unk. / unk.	unk. / unk.	0(a) / unk.	0 / unk.	unk. / unk.	unk. / 0.07	1 / 5	0 / 1	20 / 0.8	7 / 0.2	unk. / unk.	0.1 / 2	1 / 0	tr / 0.02	0 / 1	unk. / unk.	unk.	18 / unk.	R
0.02 / 0.1	1 / 1	unk. / unk.	unk. / unk.	tr(a) / unk.	0 / unk.	unk. / unk.	unk. / unk.	unk. / unk.	tr(a) / 0.1	0 / 0	0.1 / tr	unk. / 0.10	2 / 8	0 / 1	29 / 1.3	9 / 0.2	unk. / unk.	0.2 / unk.	1 / unk.	unk. / unk.	unk. / unk.	unk. / unk.	unk.	unk. / unk.	S

	FOOD	Portion	Weight in grams / *Conversion for 100 g*	Kilocalories / H_2O g	Total carbohydrate g / Total fats g	Crude fiber g / Dietary fiber g	Total protein g / Total sugar g	**% USRDA**	Arginine mg / Histidine mg	Isoleucine mg / Leucine mg	Lysine mg / Methionine mg	Phenylalanine mg / Threonine mg	Valine mg / Tryptophan mg	Cystine mg / Tyrosine mg	Polyunsat. fatty acids g / Monounsat. fatty acids g	Saturated fatty acids g / P/S ratio	Linoleic acid g / Cholesterol mg	Thiamin mg / Ascorbic acid mg	**% USRDA**
A	**cookie,chocolate wafer,Famous**	1 cookie	6.3	28	4.6	unk.	0.5	**1**	unk.	unk.	unk.	unk.	unk.	unk.	unk.	unk.	unk.	0.01	**1**
	Cookie Assortment-Nabisco		*15.87*	0.2	0.9	unk.	**unk.**		unk.	unk.	unk.	unk.	unk.	unk.	unk.	unk.	unk.	unk.	**unk.**
B	cookie,Cinnamon Treat-Nabisco	1 cookie	6.4	27	4.8	unk.	0.5	**1**	unk.	unk.	unk.	unk.	unk.	unk.	unk.	unk.	unk.	0.01	**0**
			15.63	0.3	0.6	unk.	**unk.**		unk.	unk.	unk.	unk.	unk.	unk.	unk.	unk.	unk.	unk.	**unk.**
C	cookie,coconut bar,Bakers Bonus	1 cookie	8.3	42	5.2	unk.	0.5	**1**	unk.	unk.	unk.	unk.	unk.	unk.	unk.	unk.	unk.	0.01	**1**
	-Nabisco		*12.05*	0.4	2.1	unk.	**unk.**		unk.	unk.	unk.	unk.	unk.	unk.	unk.	unk.	unk.	unk.	**unk.**
D	cookie,coconut bar	2-3/8x1-5/8x	9.0	44	5.8	0.05	0.6	**1**	unk.	unk.	unk.	unk.	unk.	unk.	unk.	0.85	0.36	tr	**0**
		3/8"	*11.11*	0.3	2.2	unk.	3.1		unk.	unk.	unk.	unk.	unk.	unk.	0.08	unk.	unk.	0	**0**
E	cookie,coconut chocolate chip-Nabisco	1 cookie	15	76	9.2	unk.	0.8	**1**	unk.	unk.	unk.	unk.	unk.	unk.	unk.	unk.	unk.	0.03	**2**
			6.80	0.5	4.1	unk.	**unk.**		unk.	unk.	unk.	unk.	unk.	unk.	unk.	unk.	unk.	unk.	**unk.**
F	cookie,coconut macaroon cake-Nabisco	1 cookie	20	94	11.5	2.46	0.8	**1**	unk.	unk.	unk.	unk.	unk.	unk.	unk.	unk.	unk.	tr	**0**
			5.05	2.2	5.0	unk.	**unk.**		unk.	unk.	unk.	unk.	unk.	unk.	unk.	unk.	unk.	unk.	**unk.**
G	cookie,country cremes,naturally flavored	1 cookie	10.33	49.16	7.52	unk.	0.5	**1**	unk.	unk.	unk.	unk.	unk.	unk.	unk.	unk.	unk.	0.03	**2**
	apple cinnamon sandwich-Nabisco		*9.68*	.26	1.88	unk.	**unk.**		unk.	unk.	unk.	unk.	unk.	unk.	unk.	unk.	unk.	unk.	**unk.**
H	cookie,creme sandwich,chocolate	1 cookie	15	69	10.5	unk.	0.8	**1**	unk.	unk.	unk.	unk.	unk.	unk.	unk.	unk.	unk.	0.05	**3**
	-Fireside		*6.80*	0.5	2.7	unk.	**unk.**		unk.	unk.	unk.	unk.	unk.	unk.	unk.	unk.	unk.	unk.	**unk.**
I	cookie,creme sandwich,crown,Mayfair	1 cookie	11	53	8.0	unk.	0.6	**1**	unk.	unk.	unk.	unk.	unk.	unk.	unk.	unk.	unk.	0.01	**1**
	assortment-Nabisco		*9.01*	0.2	2.1	unk.	**unk.**		unk.	unk.	unk.	unk.	unk.	unk.	unk.	unk.	unk.	unk.	**unk.**
J	cookie,creme sandwich,duplex,large	1 cookie	15	70	10.4	unk.	0.7	**1**	unk.	unk.	unk.	unk.	unk.	unk.	unk.	unk.	unk.	0.03	**2**
	-Fireside		*6.80*	0.5	2.9	unk.	**unk.**		unk.	unk.	unk.	unk.	unk.	unk.	unk.	unk.	unk.	unk.	**unk.**
K	cookie,creme sandwich,duplex,small	1 cookie	10	50	7.5	unk.	0.4	**1**	unk.	unk.	unk.	unk.	unk.	unk.	unk.	unk.	unk.	0.02	**2**
	-Fireside		*9.71*	0.2	2.0	unk.	**unk.**		unk.	unk.	unk.	unk.	unk.	unk.	unk.	unk.	unk.	unk.	**unk.**
L	cookie,creme sandwich,filigree,Mayfair	1 cookie	12	58	8.4	unk.	0.6	**1**	unk.	unk.	unk.	unk.	unk.	unk.	unk.	unk.	unk.	0.03	**2**
	assortment-Nabisco		*8.40*	0.3	2.4	unk.	**unk.**		unk.	unk.	unk.	unk.	unk.	unk.	unk.	unk.	unk.	unk.	**unk.**
M	cookie,creme sandwich,fudge	1 cookie	11	52	7.7	unk.	0.5	**1**	unk.	unk.	unk.	unk.	unk.	unk.	unk.	unk.	unk.	0.02	**1**
	-Nabisco		*9.26*	0.2	2.2	unk.	**unk.**		unk.	unk.	unk.	unk.	unk.	unk.	unk.	unk.	unk.	unk.	**unk.**
N	cookie,creme sandwich,Gaiety fudge	1 cookie	10	52	7.1	unk.	0.5	**1**	unk.	unk.	unk.	unk.	unk.	unk.	unk.	unk.	unk.	0.02	**1**
	artif. flavored-Nabisco		*9.71*	0.2	2.4	unk.	**unk.**		unk.	unk.	unk.	unk.	unk.	unk.	unk.	unk.	unk.	unk.	**unk.**
O	cookie,creme sandwich,lemon,Fireside	1 cookie	15	71	10.5	unk.	0.7	**1**	unk.	unk.	unk.	unk.	unk.	unk.	unk.	unk.	unk.	0.03	**2**
	-Nabisco		*6.83*	0.4	2.8	unk.	**unk.**		unk.	unk.	unk.	unk.	unk.	unk.	unk.	unk.	unk.	unk.	**unk.**
P	cookie,creme sandwich,lemon,vending	1-1/2 oz pkg	43	202	30.4	unk.	2.2	**3**	unk.	unk.	unk.	unk.	unk.	unk.	unk.	unk.	unk.	0.16	**11**
	-Nabisco		*2.35*	1.4	8.0	unk.	**unk.**		unk.	unk.	unk.	unk.	unk.	unk.	unk.	unk.	unk.	unk.	**unk.**
Q	cookie,creme sandwich,Mayfair	1 cookie	13	66	9.2	unk.	0.7	**1**	unk.	unk.	unk.	unk.	unk.	unk.	unk.	unk.	unk.	0.02	**1**
	assortment-Nabisco		*7.52*	0.4	2.9	unk.	**unk.**		unk.	unk.	unk.	unk.	unk.	unk.	unk.	unk.	unk.	unk.	**unk.**
R	cookie,creme sandwich,mixed,Cookie	1 cookie	10	51	7.3	unk.	0.5	**1**	unk.	unk.	unk.	unk.	unk.	unk.	unk.	unk.	unk.	0.02	**1**
	Break-Nabisco		*9.71*	0.1	2.2	unk.	**unk.**		unk.	unk.	unk.	unk.	unk.	unk.	unk.	unk.	unk.	unk.	**unk.**
S	cookie,creme sandwich,tea rose,Mayfair	1 cookie	11	53	7.5	unk.	0.5	**1**	unk.	unk.	unk.	unk.	unk.	unk.	unk.	unk.	unk.	0.02	**1**
	assortment-Nabisco		*9.26*	0.3	2.3	unk.	**unk.**		unk.	unk.	unk.	unk.	unk.	unk.	unk.	unk.	unk.	unk.	**unk.**

Riboflavin mg / Niacin mg	% USRDA	Vitamin B_6 mg / Folacin mcg	% USRDA	Vitamin B_{12} mcg / Pantothenic acid mg	% USRDA	Biotin mg / Vitamin A IU	% USRDA	Preformed A RE / Beta carotene RE	Vitamin D IU / Vitamin E IU	% USRDA	Total tocopherol mg / Alpha tocopherol mg	Other tocopherol mg / Total ash g	Calcium mg / Phosphorus mg	% USRDA	Sodium mg / Sodium meq	Potassium mg / Potassium meq	Chlorine mg / Chlorine meq	Iron mg / Magnesium mg	% USRDA	Zinc mg / Copper mg	% USRDA	Iodine mcg / Selenium mcg	% USRDA	Manganese mcg / Chromium mcg	
0.02	**1**	unk.	**unk.**	unk.	**unk.**	unk.	**unk.**	unk.	0(a)	**0**	unk.	unk.	2	**0**	49	17	unk.	0.2	**1**	0.1	**1**	unk.	**unk.**	48	A
0.2	**1**	unk.	**unk.**	unk.	**unk.**	unk.	**unk.**	unk.	unk.	**unk.**	unk.	0.15	9	**1**	2.1	0.4	unk.	4	**1**	0.03	**2**	unk.		unk.	
0.02	**1**	unk.	**unk.**	unk.	**unk.**	unk.	**unk.**	unk.	0(a)	**0**	unk.	unk.	2	**0**	48	10	unk.	0.2	**1**	0.1	**0**	unk.	**unk.**	59	B
0.3	**2**	unk.	**unk.**	unk.	**unk.**	unk.	**unk.**	unk.	unk.	**unk.**	unk.	0.11	6	**1**	2.1	0.3	unk.	2	**0**	0.02	**1**	unk.		unk.	
0.02	**1**	unk.	**unk.**	unk.	**unk.**	unk.	**unk.**	unk.	0(a)	**0**	unk.	unk.	2	**0**	37	19	unk.	0.2	**1**	0.1	**0**	unk.	**unk.**	52	C
0.2	**1**	unk.	**unk.**	unk.	**unk.**	unk.	**unk.**	unk.	unk.	**unk.**	unk.	0.14	9	**1**	1.6	0.5	unk.	3	**1**	0.02	**1**	unk.		unk.	
0.01	**0**	unk.	**unk.**	tr(a)	**0**	unk.	**unk.**	unk.	tr(a)	**0**	unk.	unk.	6	**1**	13	21	unk.	0.1	**1**	unk.	**unk.**	unk.	**unk.**	unk.	D
tr	**0**	unk.	**unk.**	unk.	**unk.**	14	**0**	unk.	unk.	**unk.**	unk.	0.14	11	**1**	0.6	0.5	unk.	1	**0**	unk.	**unk.**	unk.		unk.	
0.03	**2**	unk.	**unk.**	unk.	**unk.**	unk.	**unk.**	unk.	0(a)	**0**	unk.	unk.	8	**1**	48	23	unk.	0.4	**2**	0.1	**1**	unk.	**unk.**	77	E
0.3	**2**	unk.	**unk.**	unk.	**unk.**	unk.	**unk.**	unk.	unk.	**unk.**	unk.	0.21	22	**2**	2.1	0.6	unk.	5	**1**	0.03	**2**	unk.		unk.	
0.02	**1**	unk.	**unk.**	unk.	**unk.**	unk.	**unk.**	unk.	0(a)	**0**	unk.	unk.	3	**0**	48	54	unk.	0.4	**2**	0.1	**1**	unk.	**unk.**	139	F
0.0	**0**	unk.	**unk.**	unk.	**unk.**	unk.	**unk.**	unk.	unk.	**unk.**	unk.	0.24	25	**3**	2.1	1.4	unk.	9	**2**	0.07	**4**	unk.		unk.	
0.02	**1**	unk.	**unk.**	unk.	**unk.**	unk.	**unk.**	unk.	0(a)	**0**	unk.	unk.	4	**0**	28	11	unk.	0.2	**1**	0.1	**0**	unk.	**unk.**	53	G
0.3	**2**	unk.	**unk.**	unk.	**unk.**	unk.	**unk.**	unk.	unk.	**unk.**	unk.	0.14	8	**1**	1.2	0.3	unk.	2	**1**	0.02	**1**	unk.		unk.	
0.03	**2**	unk.	**unk.**	unk.	**unk.**	unk.	**unk.**	unk.	0(a)	**0**	unk.	unk.	4	**0**	54	40	unk.	0.5	**3**	0.1	**1**	unk.	**unk.**	unk.	H
0.5	**2**	unk.	**unk.**	unk.	**unk.**	unk.	**unk.**	unk.	unk.	**unk.**	unk.	0.24	15	**2**	2.4	1.0	unk.	6	**1**	0.06	**3**	unk.		unk.	
0.02	**1**	unk.	**unk.**	unk.	**unk.**	unk.	**unk.**	unk.	0(a)	**0**	unk.	unk.	5	**1**	64	22	unk.	0.5	**3**	0.1	**1**	unk.	**unk.**	75	I
0.2	**1**	unk.	**unk.**	unk.	**unk.**	unk.	**unk.**	unk.	unk.	**unk.**	unk.	0.19	14	**1**	2.8	0.6	unk.	5	**1**	0.04	**2**	unk.		unk.	
0.02	**2**	unk.	**unk.**	unk.	**unk.**	unk.	**unk.**	unk.	0(a)	**0**	unk.	unk.	3	**0**	54	27	unk.	0.4	**2**	0.1	**1**	unk.	**unk.**	unk.	J
0.4	**2**	unk.	**unk.**	unk.	**unk.**	unk.	**unk.**	unk.	unk.	**unk.**	unk.	0.19	12	**1**	2.3	0.7	unk.	4	**1**	0.02	**1**	unk.		unk.	
0.02	**1**	unk.	**unk.**	unk.	**unk.**	unk.	**unk.**	unk.	0(a)	**0**	unk.	unk.	1	**0**	54	6	unk.	0.2	**1**	tr	**0**	unk.	**unk.**	unk.	K
0.2	**1**	unk.	**unk.**	unk.	**unk.**	unk.	**unk.**	unk.	unk.	**unk.**	unk.	0.17	5	**1**	2.4	0.2	unk.	unk.	**unk.**	unk.	**unk.**	unk.		unk.	
0.04	**2**	unk.	**unk.**	unk.	**unk.**	unk.	**unk.**	unk.	0(a)	**0**	unk.	unk.	2	**0**	41	11	unk.	0.2	**1**	0.1	**0**	unk.	**unk.**	42	L
0.3	**2**	unk.	**unk.**	unk.	**unk.**	unk.	**unk.**	unk.	unk.	**unk.**	unk.	0.11	8	**1**	1.8	0.3	unk.	2	**0**	0.02	**1**	unk.		unk.	
0.02	**1**	unk.	**unk.**	unk.	**unk.**	unk.	**unk.**	unk.	0(a)	**0**	unk.	unk.	4	**0**	50	22	unk.	0.3	**2**	0.1	**1**	unk.	**unk.**	45	M
0.3	**1**	unk.	**unk.**	unk.	**unk.**	unk.	**unk.**	unk.	unk.	**unk.**	unk.	0.17	12	**1**	2.2	0.6	unk.	4	**1**	0.04	**2**	unk.		unk.	
0.02	**1**	unk.	**unk.**	unk.	**unk.**	unk.	**unk.**	unk.	0(a)	**0**	unk.	unk.	3	**0**	42	18	unk.	0.3	**2**	0.1	**1**	unk.	**unk.**	49	N
0.2	**1**	unk.	**unk.**	unk.	**unk.**	unk.	**unk.**	unk.	unk.	**unk.**	unk.	0.16	12	**1**	1.8	0.4	unk.	5	**1**	0.04	**2**	unk.		unk.	
0.04	**2**	unk.	**unk.**	unk.	**unk.**	unk.	**unk.**	unk.	0(a)	**0**	unk.	unk.	2	**0**	56	12	unk.	0.3	**2**	0.1	**1**	unk.	**unk.**	48	O
0.2	**1**	unk.	**unk.**	unk.	**unk.**	unk.	**unk.**	unk.	unk.	**unk.**	unk.	0.12	9	**1**	2.4	0.3	unk.	2	**1**	0.02	**1**	unk.		unk.	
0.09	**6**	unk.	**unk.**	unk.	**unk.**	unk.	**unk.**	unk.	0(a)	**0**	unk.	unk.	9	**1**	125	27	unk.	1.0	**5**	0.2	**1**	unk.	**unk.**	140	P
1.3	**7**	unk.	**unk.**	unk.	**unk.**	unk.	**unk.**	unk.	unk.	**unk.**	unk.	0.59	33	**3**	5.4	0.7	unk.	6	**2**	0.05	**2**	unk.		unk.	
0.04	**3**	unk.	**unk.**	unk.	**unk.**	unk.	**unk.**	unk.	0(a)	**0**	unk.	unk.	3	**0**	38	15	unk.	0.3	**2**	0.1	**1**	unk.	**unk.**	55	Q
0.4	**2**	unk.	**unk.**	unk.	**unk.**	unk.	**unk.**	unk.	unk.	**unk.**	unk.	0.11	10	**1**	1.6	0.4	unk.	3	**1**	0.03	**1**	unk.		unk.	
0.03	**2**	unk.	**unk.**	unk.	**unk.**	unk.	**unk.**	unk.	0(a)	**0**	unk.	unk.	2	**0**	38	9	unk.	0.3	**1**	0.0	**0**	unk.	**unk.**	34	R
0.3	**2**	unk.	**unk.**	unk.	**unk.**	unk.	**unk.**	unk.	unk.	**unk.**	unk.	0.11	7	**1**	1.7	0.2	unk.	2	**1**	0.02	**1**	unk.		unk.	
0.04	**2**	unk.	**unk.**	unk.	**unk.**	unk.	**unk.**	unk.	0(a)	**0**	unk.	unk.	2	**0**	44	17	unk.	0.3	**2**	0.1	**0**	unk.	**unk.**	45	S
0.4	**2**	unk.	**unk.**	unk.	**unk.**	unk.	**unk.**	unk.	unk.	**unk.**	unk.	0.12	8	**1**	1.9	0.4	unk.	2	**1**	0.02	**1**	unk.		unk.	

	FOOD	Portion	Weight in grams / *Conversion for 100 g*	Kilocalories / H_2O g	Total carbohydrate g / Total fats g	Crude fiber g / Dietary fiber g	Total protein g / Total sugar g	% USRDA	Arginine mg / Histidine mg	Isoleucine mg / Leucine mg	Lysine mg / Methionine mg	Phenylalanine mg / Threonine mg	Valine mg / Tryptophan mg	Cystine mg / Tyrosine mg	Polyunsat. fatty acids g / Monounsat. fatty acids g	Saturated fatty acids g / P/S ratio	Linoleic acid g / Cholesterol mg	Thiamin mg / Ascorbic acid mg	% USRDA
A	cookie,creme sandwich,vanilla	1 cookie	10	51	7.4	unk.	0.5	**1**	unk.	unk.	unk.	unk.	unk.	unk.	unk.	unk.	unk.	0.02	**1**
	artif flavored,Cookie Break-Nabisco		*9.71*	0.2	2.1	unk.	unk.		unk.	unk.	unk.	unk.	unk.	unk.	unk.	unk.	unk.	unk.	**unk.**
B	cookie,creme sandwich,vanilla	1 cookie	15	72	10.4	unk.	0.7	**1**	unk.	unk.	unk.	unk.	unk.	unk.	unk.	unk.	unk.	0.03	**2**
	-Fireside		*6.80*	0.3	3.1	unk.	unk.		unk.	unk.	unk.	unk.	unk.	unk.	unk.	unk.	unk.	unk.	**unk.**
C	cookie,creme sandwich,vanilla	1 cookie	11	52	7.7	unk.	0.5	**1**	unk.	unk.	unk.	unk.	unk.	unk.	unk.	unk.	unk.	0.04	**3**
	-Nabisco		*9.26*	0.3	2.2	unk.	unk.		unk.	unk.	unk.	unk.	unk.	unk.	unk.	unk.	unk.	unk.	**unk.**
D	cookie,creme wafer stick-Nabisco	1 cookie	9.1	48	6.2	unk.	0.3	**1**	unk.	unk.	unk.	unk.	unk.	unk.	unk.	unk.	unk.	tr	**0**
			10.99	0.1	2.4	unk.	unk.		unk.	unk.	unk.	unk.	unk.	unk.	unk.	unk.	unk.	unk.	**unk.**
E	cookie,cremes,assorted-Fireside	1 cookie	10	50	7.5	unk.	0.4	**1**	unk.	unk.	unk.	unk.	unk.	unk.	unk.	unk.	unk.	0.02	**2**
			9.71	0.2	2.0	unk.	unk.		unk.	unk.	unk.	unk.	unk.	unk.	unk.	unk.	unk.	unk.	**unk.**
F	cookie,deluxe graham-Fireside	1 cookie	12	63	8.2	unk.	0.6	**1**	unk.	unk.	unk.	unk.	unk.	unk.	unk.	unk.	unk.	0.02	**2**
			8.13	0.2	3.1	unk.	unk.		unk.	unk.	unk.	unk.	unk.	unk.	unk.	unk.	unk.	unk.	**unk.**
G	cookie,devil's food cake-Nabisco	1 cookie	20	72	15.3	unk.	1.0	**2**	unk.	unk.	unk.	unk.	unk.	unk.	unk.	unk.	unk.	0.04	**3**
			5.13	2.3	0.7	unk.	unk.		unk.	unk.	unk.	unk.	unk.	unk.	unk.	unk.	unk.	unk.	**unk.**
H	cookie,fancy dip graham-Nabisco	1 cookie	12	63	8.2	unk.	0.6	**1**	unk.	unk.	unk.	unk.	unk.	unk.	unk.	unk.	unk.	0.02	**2**
			8.13	0.2	3.1	unk.	unk.		unk.	unk.	unk.	unk.	unk.	unk.	unk.	unk.	unk.	unk.	**unk.**
I	cookie,fig bar	1-1/2x1-3/4x	14	49	10.3	0.24	0.5	**1**	unk.	27	13	29	24	11	unk.	0.21	0.15	0.03	**2**
		1/2"	*7.14*	2.3	0.7	unk.	5.5		11	42	7	15	7	18	0.38	unk.	8	tr	**0**
J	cookie,fig bar-Keebler	1 cookie	19	71	14.4	unk.	0.6	**1**	unk.	unk.	unk.	unk.	unk.	unk.	unk.	unk.	unk.	unk.	**unk.**
			5.18	2.8	1.2	unk.	7.6		unk.	unk.	unk.	unk.	unk.	unk.	unk.	unk.	12	unk.	**unk.**
K	cookie,fig bar,large-Fireside	1 cookie	23	82	16.7	unk.	0.8	**1**	unk.	unk.	unk.	unk.	unk.	unk.	unk.	unk.	unk.	0.07	**4**
			4.41	3.7	1.3	unk.	unk.		unk.	unk.	unk.	unk.	unk.	unk.	unk.	unk.	unk.	unk.	**unk.**
L	cookie,Fig Newtons cake-Nabisco	1 cookie	16	58	11.1	unk.	0.6	**1**	unk.	unk.	unk.	unk.	unk.	unk.	unk.	unk.	unk.	0.03	**2**
			6.37	2.6	1.2	unk.	unk.		unk.	unk.	unk.	unk.	unk.	unk.	unk.	unk.	unk.	unk.	**unk.**
M	cookie,Fireside assorted-Nabisco	1 cookie	5.7	28	3.8	unk.	0.3	**1**	unk.	unk.	unk.	unk.	unk.	unk.	unk.	unk.	unk.	0.01	**1**
			17.54	0.2	1.3	unk.	unk.		unk.	unk.	unk.	unk.	unk.	unk.	unk.	unk.	unk.	unk.	**unk.**
N	cookie,fortune,home recipe	1 cookie	14	66	7.1	0.05	0.8	**2**	55	41	37	45	49	20	0.30	1.99	0.24	0.01	**1**
			7.04	2.1	4.0	0.14	5.4		20	68	19	30	10	31	1.37	0.15	8	0	**0**
O	cookie,frosted spice-Fireside	1 cookie	10	47	7.6	unk.	0.5	**1**	unk.	unk.	unk.	unk.	unk.	unk.	unk.	unk.	unk.	0.03	**2**
			9.71	0.4	1.7	unk.	unk.		unk.	unk.	unk.	unk.	unk.	unk.	unk.	unk.	unk.	unk.	**unk.**
P	cookie,fudge drop,vending-Nabisco	2 oz bag	53	260	34.3	unk.	3.4	**5**	unk.	unk.	unk.	unk.	unk.	unk.	unk.	unk.	unk.	0.13	**9**
			1.89	2.4	12.2	unk.	unk.		unk.	unk.	unk.	unk.	unk.	unk.	unk.	unk.	unk.	unk.	**unk.**
Q	cookie,fudge sugar-Fireside	1 cookie	11	54	7.8	unk.	0.6	**1**	unk.	unk.	unk.	unk.	unk.	unk.	unk.	unk.	unk.	0.03	**2**
			8.85	0.4	2.2	unk.	unk.		unk.	unk.	unk.	unk.	unk.	unk.	unk.	unk.	unk.	unk.	**unk.**
R	cookie,fudgettes deluxe cookie	1 cookie	8.3	39	5.8	unk.	0.4	**1**	unk.	unk.	unk.	unk.	unk.	unk.	unk.	unk.	unk.	0.01	**1**
	assortment-Nabisco		*12.05*	0.4	1.6	unk.	unk.		unk.	unk.	unk.	unk.	unk.	unk.	unk.	unk.	unk.	unk.	**unk.**
S	cookie,ginger snap	2" dia x 1/4"	7.0	30	4.2	0.01	0.3	**1**	4	14	10	16	14	6	0.02	0.34	0.11	0.01	**1**
			14.29	0.8	1.4	unk.	2.4		7	24	5	10	4	11	0.90	0.07	2	0	**0**

Riboflavin mg / Niacin mg	% USRDA	Vitamin B_6 mg / Folacin mcg	% USRDA	Vitamin B_{12} mcg / Pantothenic acid mg	% USRDA	Biotin mg / Vitamin A IU	% USRDA	Preformed A RE / Beta carotene RE	Vitamin D IU / Vitamin E IU	% USRDA	Total tocopherol mg / Alpha tocopherol mg	Other tocopherol mg / Total ash g	Calcium mg / Phosphorus mg	% USRDA	Sodium mg / Sodium meq	Potassium mg / Potassium meq	Chlorine mg / Chlorine meq	Iron mg / Magnesium mg	% USRDA	Zinc mg / Copper mg	% USRDA	Iodine mcg / Selenium mcg	% USRDA	Manganese mcg / Chromium mcg	
0.03	2	unk.	unk.	unk.	unk.	unk.	unk.	unk.	0(a)	0	unk.	unk.	1	0	40	8	unk.	0.2	1	tr	0	unk.	unk.	30	A
0.3	1	unk.	unk.	unk.	unk.	unk.	unk.	unk.	unk.	unk.	unk.	0.11	6	1	1.8	0.2	unk.	1	0	0.01	1	unk.		unk.	
0.04	2	unk.	unk.	unk.	unk.	unk.	unk.	unk.	0(a)	0	unk.	unk.	2	0	56	12	unk.	0.3	2	0.1	0	unk.	unk.	unk.	B
0.4	2	unk.	unk.	unk.	unk.	unk.	unk.	unk.	unk.	unk.	unk.	0.16	9	1	2.4	0.3	unk.	2	1	0.01	1	unk.		unk.	
0.03	2	unk.	unk.	unk.	unk.	unk.	unk.	unk.	0(a)	0	unk.	unk.	7	1	46	14	unk.	0.2	1	0.0	0	unk.	unk.	33	C
0.4	2	unk.	unk.	unk.	unk.	unk.	unk.	unk.	unk.	unk.	unk.	0.15	14	1	2.0	0.3	unk.	10	3	0.01	1	unk.		unk.	
0.02	1	unk.	unk.	unk.	unk.	unk.	unk.	unk.	0(a)	0	unk.	unk.	4	0	14	22	unk.	0.1	1	0.1	0	unk.	unk.	22	D
0.1	1	unk.	unk.	unk.	unk.	unk.	unk.	unk.	unk.	unk.	unk.	0.08	9	1	0.6	0.6	unk.	3	1	0.02	1	unk.		unk.	
0.02	1	unk.	unk.	unk.	unk.	unk.	unk.	unk.	0(a)	0	unk.	unk.	1	0	54	6	unk.	0.2	1	tr	0	unk.	unk.	unk.	E
0.2	1	unk.	unk.	unk.	unk.	unk.	unk.	unk.	unk.	unk.	unk.	0.17	5	1	2.4	0.2	unk.	1	0	unk.	unk.	unk.		unk.	
0.05	3	unk.	unk.	unk.	unk.	unk.	unk.	unk.	0(a)	0	unk.	unk.	7	1	43	47	unk.	0.4	2	0.1	1	unk.	unk.	65	F
0.4	2	unk.	unk.	unk.	unk.	unk.	unk.	unk.	unk.	unk.	unk.	0.24	22	2	1.9	1.2	unk.	5	1	0.04	2	unk.		unk.	
0.04	2	unk.	unk.	unk.	unk.	unk.	unk.	unk.	0(a)	0	unk.	unk.	7	1	37	27	unk.	0.4	2	0.1	1	unk.	unk.	66	G
0.2	1	unk.	unk.	unk.	unk.	unk.	unk.	unk.	unk.	unk.	unk.	0.18	16	2	1.6	0.7	unk.	7	2	0.05	2	unk.		unk.	
0.05	3	unk.	unk.	unk.	unk.	unk.	unk.	unk.	0(a)	0	unk.	unk.	7	1	43	47	unk.	0.4	2	0.1	1	unk.	unk.	65	H
0.4	2	unk.	unk.	unk.	unk.	unk.	unk.	unk.	unk.	unk.	unk.	0.24	23	2	1.9	1.2	unk.	5	1	0.04	2	unk.		unk.	
0.03	2	0.01	1	tr(a)	0	0.001	0	unk.	tr	0	0.1	unk.	5	1	66	16	unk.	0.2	2	0.1	0	unk.	unk.	31	I
0.3	2	1	0	0.04	0	15	0	unk.	0.1	0	unk.	0.18	8	1	2.9	0.4	unk.	2	1	0.02	1	unk.		unk.	
unk.	unk.	unk.	unk.	tr(a)	0	unk.	unk.	unk.	tr(a)	0	unk.	unk.	13	1	84	49	unk.	0.3	2	unk.	unk.	unk.	unk.	unk.	J
unk.	unk.	unk.	unk.	unk.	unk.	unk.	unk.	unk.	unk.	unk.	unk.	0.37	13	1	3.7	1.2	unk.	unk.	unk.	unk.	unk.	unk.		unk.	
0.03	2	unk.	unk.	unk.	unk.	unk.	unk.	unk.	0(a)	0	unk.	unk.	10	1	64	37	unk.	0.5	3	0.1	1	unk.	unk.	82	K
0.6	3	unk.	unk.	unk.	unk.	unk.	unk.	unk.	unk.	unk.	unk.	0.22	17	2	2.8	0.9	unk.	7	2	0.03	2	unk.		unk.	
0.03	2	unk.	unk.	unk.	unk.	unk.	unk.	unk.	0(a)	0	unk.	unk.	10	1	64	34	unk.	0.4	2	0.1	0	unk.	unk.	49	L
0.2	1	unk.	unk.	unk.	unk.	unk.	unk.	unk.	unk.	unk.	unk.	0.26	10	1	2.8	0.9	unk.	4	1	0.02	1	unk.		unk.	
0.02	1	unk.	unk.	unk.	unk.	unk.	unk.	unk.	0(a)	0	unk.	unk.	1	0	22	6	unk.	0.1	1	tr	0	unk.	unk.	30	M
0.2	1	unk.	unk.	unk.	unk.	unk.	unk.	unk.	unk.	unk.	unk.	0.08	4	0	1.0	0.2	unk.	1	0	0.01	0	unk.		unk.	
0.03	2	tr	0	0.00	0	0.000	0	0	0	0	0.1	tr	5	1	46	20	7	0.1	1	0.1	1	2.6	2	49	N
0.1	1	2	1	0.02	0	116	2	0	0.1	1	0.3	0.18	11	1	2.0	0.5	0.2	6	2	0.03	1	1		4	
0.03	2	unk.	unk.	unk.	unk.	unk.	unk.	unk.	0(a)	0	unk.	unk.	6	1	44	13	unk.	0.2	1	0.1	0	unk.	unk.	unk.	O
0.3	1	unk.	unk.	unk.	unk.	unk.	unk.	unk.	unk.	unk.	unk.	0.16	11	1	1.9	0.3	unk.	unk.	unk.	unk.	unk.	unk.		unk.	
0.15	9	unk.	unk.	unk.	unk.	unk.	unk.	unk.	0(a)	0	unk.	unk.	20	2	197	78	unk.	1.5	8	0.4	3	unk.	unk.	249	P
1.4	7	unk.	unk.	unk.	unk.	unk.	unk.	unk.	unk.	unk.	unk.	0.72	63	6	8.6	2.0	unk.	18	5	0.14	7	unk.		unk.	
0.03	2	unk.	unk.	unk.	unk.	unk.	unk.	unk.	0(a)	0	unk.	unk.	6	1	47	16	unk.	0.2	1	0.1	1	unk.	unk.	unk.	Q
0.3	2	unk.	unk.	unk.	unk.	unk.	unk.	unk.	unk.	unk.	unk.	0.21	16	2	2.1	0.4	unk.	4	1	unk.	unk.	unk.		unk.	
0.01	1	unk.	unk.	unk.	unk.	unk.	unk.	unk.	0(a)	0	unk.	unk.	3	0	29	9	unk.	0.2	1	tr	0	unk.	unk.	30	R
0.1	1	unk.	unk.	unk	unk.	unk.	unk.	unk.	unk.	unk.	unk.	0.10	6	1	1.3	0.2	unk.	2	1	0.01	1	unk.		unk.	
0.01	1	tr	0	0.00	0	tr	0	0	tr	0	0.0	tr(a)	5	1	31	18	8	0.2	1	tr	0	2.8	2	10	S
0.1	1	1	0	0.02	0	1	0	0	unk.	unk.	tr(a)	0.13	4	0	1.4	0.4	0.2	4	1	0.01	0	1		3	

	FOOD	Portion	Weight in grams / *Conversion for 100 g*	Kilocalories / H_2O g	Total carbohydrate g / Total fats g	Crude fiber g / Dietary fiber g	Total protein g / Total sugar g	**% USRDA**	Arginine mg / Histidine mg	Isoleucine mg / Leucine mg	Lysine mg / Methionine mg	Phenylalanine mg / Threonine mg	Valine mg / Tryptophan mg	Cystine mg / Tyrosine mg	Polyunsat. fatty acids g / Monounsat. fatty acids g	Saturated fatty acids g / P/S ratio	Linoleic acid g / Cholesterol mg	Thiamin mg / Ascorbic acid mg	**% USRDA**
A	cookie,Heyday caramel peanut logs	1 cookie	24	123	12.7	unk.	2.5	**4**	unk.	unk.	unk.	unk.	unk.	unk.	unk.	unk.	unk.	0.02	**1**
	-Nabisco		*4.18*	1.4	6.9	unk.	unk.		unk.	unk.	unk.	unk.	unk.	unk.	unk.	unk.	unk.	unk.	**unk.**
B	cookie,home-style chocolate chip,bulk	1 cookie	16	72	10.7	unk.	0.8	**1**	unk.	unk.	unk.	unk.	unk.	unk.	unk.	unk.	unk.	0.01	**1**
	institutional-Nabisco		*6.41*	0.9	2.9	unk.	unk.		unk.	unk.	unk.	unk.	unk.	unk.	unk.	unk.	unk.	unk.	**unk.**
C	cookie,home-style fudge sugar,bulk	1 cookie	14	59	9.7	unk.	0.8	**1**	unk.	unk.	unk.	unk.	unk.	unk.	unk.	unk.	unk.	0.03	**2**
	institutional-Nabisco		*7.41*	0.9	1.9	unk.	unk.		unk.	unk.	unk.	unk.	unk.	unk.	unk.	unk.	unk.	unk.	**unk.**
D	cookie,home-style sugar,bulk	1 cookie	13	56	9.5	unk.	0.7	**1**	unk.	unk.	unk.	unk.	unk.	unk.	unk.	unk.	unk.	0.01	**1**
	institutional-Nabisco		*7.81*	0.8	1.7	unk.	unk.		unk.	unk.	unk.	unk.	unk.	unk.	unk.	unk.	unk.	unk.	**unk.**
E	cookie,iced oatmeal-Fireside	1 cookie	15	71	9.7	unk.	0.9	**1**	unk.	unk.	unk.	unk.	unk.	unk.	unk.	unk.	unk.	0.04	**2**
			6.80	0.8	3.2	unk.	unk.		unk.	unk.	unk.	unk.	unk.	unk.	unk.	unk.	unk.	unk.	**unk.**
F	cookie,iced oatmeal,deluxe cookie	1 cookie	10	44	7.3	unk.	0.6	**1**	unk.	unk.	unk.	unk.	unk.	unk.	unk.	unk.	unk.	0.02	**1**
	assortment-Nabisco		*10.00*	0.6	1.3	unk.	unk.		unk.	unk.	unk.	unk.	unk.	unk.	unk.	unk.	unk.	unk.	**unk.**
G	cookie,Ideal chocolate peanut bar	1 cookie	18	94	10.2	unk.	1.4	**2**	unk.	unk.	unk.	unk.	unk.	unk.	unk.	unk.	unk.	0.05	**4**
	-Nabisco		*5.71*	0.3	5.2	unk.	unk.		unk.	unk.	unk.	unk.	unk.	unk.	unk.	unk.	unk.	unk.	**unk.**
H	cookie,kettle cookies,famous	1 cookie	7.3	34	5.2	unk.	0.5	**1**	unk.	unk.	unk.	unk.	unk.	unk.	unk.	unk.	unk.	0.02	**2**
	assortment-Nabisco		*13.70*	0.3	1.2	unk.	unk.		unk.	unk.	unk.	unk.	unk.	unk.	unk.	unk.	unk.	unk.	**unk.**
I	cookie,lemon bar,home recipe	2x2x3/4"	29	110	14.7	0.02	1.3	**3**	41	65	63	69	72	28	0.29	3.11	0.22	0.03	**2**
			3.42	7.7	5.3	0.19	9.9		35	108	30	52	18	52	1.41	0.09	37	1	**2**
J	cookie,lemon creme,small-Fireside	1 cookie	10	49	7.5	unk.	0.5	**1**	unk.	unk.	unk.	unk.	unk.	unk.	unk.	unk.	unk.	0.05	**3**
			9.62	0.3	1.9	unk.	unk.		unk.	unk.	unk.	unk.	unk.	unk.	unk.	unk.	unk.	unk.	**unk.**
K	cookie,Lorna Doone shortbread-Nabisco	1 cookie	10	53	6.7	unk.	0.6	**1**	unk.	unk.	unk.	unk.	unk.	unk.	unk.	unk.	unk.	0.04	**2**
			9.62	0.4	2.6	unk.	unk.		unk.	unk.	unk.	unk.	unk.	unk.	unk.	unk.	unk.	unk.	**unk.**
L	cookie,macaroon	2-1/2" dia x	15	54	9.5	0.19	0.5	**1**	48	27	32	25	35	14	0.01	1.56	0.00	tr	**0**
		1/4"	*6.49*	3.5	1.8	unk.	8.5		11	43	18	22	7	20	0.14	0.01	0	0	**0**
M	cookie,macaroon-Fireside	1 cookie	11	54	7.8	unk.	0.6	**1**	unk.	unk.	unk.	unk.	unk.	unk.	unk.	unk.	unk.	0.04	**2**
			8.85	0.5	2.2	unk.	unk.		unk.	unk.	unk.	unk.	unk.	unk.	unk.	unk.	unk.	unk.	**unk.**
N	cookie,Mallomars chocolate	1 cookie	13	60	8.7	unk.	0.6	**1**	unk.	unk.	unk.	unk.	unk.	unk.	unk.	unk.	unk.	0.01	**1**
	cake,vending-Nabisco		*7.69*	1.0	2.5	unk.	unk.		unk.	unk.	unk.	unk.	unk.	unk.	unk.	unk.	unk.	unk.	**unk.**
O	cookie,malted milk peanut butter	1 cookie	7.4	36	4.5	unk.	0.8	**1**	unk.	unk.	unk.	unk.	unk.	unk.	unk.	unk.	unk.	0.04	**3**
	artif flavored-Nabisco		*13.51*	0.2	1.6	unk.	unk.		unk.	unk.	unk.	unk.	unk.	unk.	unk.	unk.	unk.	unk.	**unk.**
P	cookie,marshmellow puff-Nabisco	1 cookie	20	83	13.8	unk.	0.8	**1**	unk.	unk.	unk.	unk.	unk.	unk.	unk.	unk.	unk.	0.02	**1**
			5.05	2.3	2.8	unk.	unk.		unk.	unk.	unk.	unk.	unk.	unk.	unk.	unk.	unk.	unk.	**unk.**
Q	cookie,marshmellow sandwich-Nabisco	1 cookie	7.7	31	5.9	unk.	0.3	**1**	unk.	unk.	unk.	unk.	unk.	unk.	unk.	unk.	unk.	0.01	**1**
			12.99	0.8	0.7	unk.	unk.		unk.	unk.	unk.	unk.	unk.	unk.	unk.	unk.	unk.	unk.	**unk.**
R	cookie,marshmellow,coconut coated	2-1/8" dia x	18	74	13.0	0.05	0.7	**1**	unk.	unk.	unk.	unk.	unk.	unk.	unk.	1.44	0.18	tr	**0**
		1-1/8"	*5.56*	1.8	2.4	unk.	unk.		unk.	unk.	unk.	unk.	unk.	unk.	0.90	unk.	unk.	unk.	**unk.**
S	cookie,Melt-A-Way shortbread-Nabisco	1 cookie	13	68	7.9	unk.	0.7	**1**	unk.	unk.	unk.	unk.	unk.	unk.	unk.	unk.	unk.	0.02	**1**
			7.81	0.3	3.7	unk.	unk.		unk.	unk.	unk.	unk.	unk.	unk.	unk.	unk.	unk.	unk.	**unk.**

Riboflavin mg / Niacin mg	% USRDA	Vitamin B_6 mg / Folacin mcg	% USRDA	Vitamin B_{12} mcg / Pantothenic acid mg	% USRDA	Biotin mg / Vitamin A IU	% USRDA	Preformed A RE / Beta carotene RE	Vitamin D IU / Vitamin E IU	% USRDA	Total tocopherol mg / Alpha tocopherol mg	Other tocopherol mg / Total ash g	Calcium mg / Phosphorus mg	% USRDA	Sodium mg / Sodium meq	Potassium mg / Potassium meq	Chlorine mg / Chlorine meq	Iron mg / Magnesium mg	% USRDA	Zinc mg / Copper mg	% USRDA	Iodine mcg / Selenium mcg	% USRDA	Manganese mcg / Chromium mcg	
0.04	2	unk.	unk.	unk.	unk.	unk.	unk.	unk.	0(a)	0	unk.	unk.	13	1	44	69	unk.	0.3	2	0.3	2	unk.	unk.	167	A
1.1	5	unk.	unk.	unk.	unk.	unk.	unk.	unk.	unk.	unk.	unk.	0.35	39	4	1.9	1.8	unk.	14	3	0.07	3	unk.		unk.	
0.00	0	unk.	unk.	unk.	unk.	unk.	unk.	unk.	0(a)	0	unk.	unk.	8	1	96	22	unk.	0.1	1	0.1	1	unk.	unk.	62	B
0.1	1	unk.	unk.	unk.	unk.	unk.	unk.	unk.	unk.	unk.	unk.	0.31	43	4	4.2	0.6	unk.	5	1	0.05	2	unk.		unk.	
0.02	1	unk.	unk.	unk.	un	unk.	unk.	unk.	0(a)	0	unk.	unk.	5	1	54	15	unk.	0.3	2	0.1	1	unk.	unk.	78	C
0.4	2	unk.	unk.	unk.	unk.	unk.	unk.	unk.	unk.	unk.	unk.	0.17	12	1	2.4	0.4	unk.	3	1	0.02	1	unk.		unk.	
0.01	1	unk.	unk.	unk.	unk.	unk.	unk.	unk.	0(a)	0	unk.	unk.	2	0	51	9	unk.	0.1	1	tr	0	unk.	unk.	45	D
0.2	1	unk.	unk.	unk.	unk.	unk.	unk.	unk.	unk.	unk.	unk.	0.13	8	1	2.2	0.2	unk.	2	0	0.00	0	unk.		unk.	
0.02	1	unk.	unk.	unk.	unk.	unk.	unk.	unk.	0(a)	0	unk.	unk.	4	0	55	20	unk.	0.3	2	0.1	1	unk.	unk.	138	E
0.3	2	unk.	unk.	unk.	unk.	unk.	unk.	unk.	unk.	unk.	unk.	0.21	22	2	2.4	0.5	unk.	6	1	0.03	2	unk.		unk.	
0.02	1	unk.	unk.	unk.	unk.	unk.	unk.	unk.	0(a)	0	unk.	unk.	2	0	32	11	unk.	0.2	1	0.1	1	unk.	unk.	80	F
0.2	1	unk.	unk.	unk.	unk.	unk.	unk.	unk.	unk.	unk.	unk.	0.13	11	1	1.4	0.3	unk.	3	1	0.01	0	unk.		unk.	
0.03	2	unk.	unk.	unk.	unk.	unk.	unk.	unk.	0(a)	0	unk.	unk.	6	1	71	48	unk.	0.4	2	0.2	1	unk.	unk.	130	G
0.7	3	unk.	unk.	unk.	unk.	unk.	unk.	unk.	unk.	unk.	unk.	0.36	25	3	3.1	1.2	unk.	10	2	0.05	3	unk.		unk.	
0.02	1	unk.	unk.	unk.	unk.	unk.	unk.	unk.	0(a)	0	unk.	unk.	2	0	27	8	unk.	0.2	1	tr	0	unk.	unk.	28	H
0.3	1	unk.	unk.	unk.	unk.	unk.	unk.	unk.	unk.	unk.	unk.	0.09	10	1	1.2	0.2	unk.	1	0	0.01	1	unk.		unk.	
0.03	2	0.01	1	0.06	1	0.001	0	0	2	1	0.4	0.1	6	1	59	18	13	0.3	2	0.2	1	3.7	3	32	I
0.2	1	3	1	0.10	1	185	4	0	0.4	1	0.2	0.23	21	2	2.6	0.5	0.4	14	4	0.01	1	3		10	
0.03	2	unk.	unk.	unk.	unk.	unk.	unk.	unk.	0(a)	0	unk.	unk.	2	0	49	10	unk.	0.2	1	0.1	0	unk.	unk.	47	J
0.3	1	unk.	unk.	unk.	unk.	unk.	unk.	unk.	unk.	unk.	unk.	0.14	8	1	2.1	0.3	unk.	2	1	0.01	1	unk.		unk.	
0.03	2	unk.	unk.	unk.	unk.	unk.	unk.	unk.	unk.	unk.	unk.	unk.	4	0	51	11	unk.	0.2	1	0.1	0	unk.	unk.	41	K
0.4	2	unk.	unk.	unk.	unk.	unk.	unk.	unk.	unk.	unk.	unk.	0.16	11	1	2.2	0.3	unk.	2	0	0.01	1	unk.		unk.	
0.01	1	tr	0	0.00	0	tr	0	0	0	0	0.1	0.0	1	0	21	22	17	0.1	1	tr	0	3.6	2	3	L
tr	0	1	0	0.01	0	0	0	0	tr	0	tr	0.09	6	1	0.9	0.6	0.5	4	1	tr	0	0		4	
0.03	2	unk.	unk.	unk.	unk.	unk.	unk.	unk.	0(a)	0	unk.	unk.	2	0	44	14	unk.	0.2	1	0.1	1	unk.	unk.	46	M
0.3	2	unk.	unk.	unk.	unk.	unk.	unk.	unk.	unk.	unk.	unk.	0.13	12	1	1.9	0.3	unk.	2	1	0.01	1	unk.		unk.	
0.03	2	unk.	unk.	unk.	unk.	unk.	unk.	unk.	0(a)	0	unk.	unk.	4	0	20	26	unk.	0.3	2	0.1	1	unk.	unk.	50	N
0.2	1	unk.	unk.	unk.	unk.	unk.	unk.	unk.	unk.	unk.	unk.	0.13	14	1	0.8	0.7	unk.	7	2	0.06	3	unk.		unk.	
0.02	1	unk.	unk.	unk.	unk.	unk.	unk.	unk.	0(a)	0	unk.	unk.	6	1	76	16	unk.	0.2	1	0.1	1	unk.	unk.	60	O
0.4	2	unk.	unk.	unk.	unk.	unk.	unk.	unk.	unk.	unk.	unk.	0.24	17	2	3.3	0.4	unk.	4	1	0.02	1	unk.		unk.	
0.05	3	unk.	unk.	unk.	unk.	unk.	unk.	unk.	0(a)	0	unk.	unk.	10	1	37	34	unk.	0.2	1	0.1	1	unk.	unk.	25	P
0.1	1	unk.	unk.	unk.	unk.	unk.	unk.	unk.	unk.	unk.	unk.	0.20	15	2	1.6	0.9	unk.	4	1	0.02	1	unk.		unk.	
0.02	1	unk.	unk.	unk.	unk.	unk.	unk.	unk.	0(a)	0	unk.	unk.	2	0	21	6	unk.	0.1	1	tr	0	unk.	unk.	19	Q
0.1	1	unk.	unk.	unk.	unk.	unk.	unk.	unk.	unk.	unk.	unk.	0.07	6	1	0.9	0.2	unk.	1	0	0.01	0	unk.		unk.	
0.01	1	unk.	unk.	unk.	unk.	unk.	unk.	unk.	unk.	unk.	unk.	unk.	4	0	38	16	unk.	0.1	1	unk.	unk.	unk.	unk.	unk.	R
tr	0	unk.	unk.	unk.	unk.	47	1	unk.	unk.	unk.	unk.	0.13	10	1	1.6	0.4	unk.	3	1	unk.	unk.	unk.		unk.	
0.02	1	unk.	unk.	unk.	unk.	unk.	unk.	unk.	0(a)	0	unk.	unk.	3	0	60	12	unk.	0.1	1	0.1	0	unk.	unk.	26	S
0.3	1	unk.	unk.	unk.	unk.	unk.	unk.	unk.	unk.	unk.	unk.	0.20	15	2	2.6	0.3	unk.	2	0	0.01	1	unk.		unk.	

	FOOD	Portion	Weight in grams / *Conversion for 100 g*	Kilocalories / H_2O g	Total carbohydrate g / Total fats g	Crude fiber g / Dietary fiber g	Total protein g / Total sugar g	% USRDA	Arginine mg / Histidine mg	Isoleucine mg / Leucine mg	Lysine mg / Methionine mg	Phenylalanine mg / Threonine mg	Valine mg / Tryptophan mg	Cystine mg / Tyrosine mg	Polyunsat. fatty acids g / Monounsat. fatty acids g	Saturated fatty acids g / P/S ratio	Linoleic acid g / Cholesterol mg	Thiamin mg / Ascorbic acid mg	% USRDA
A	cookie,molasses,home recipe	3-5/8" dia x	33	138	21.7	0.07	1.9	3	29	92	73	100	95	36	0.29	3.04	0.22	0.06	4
		3/4"	*3.08*	12.0	5.1	unk.	10.4		44	152	33	65	24	70	1.31	0.10	25	tr	0
B	cookie,Mystic mint sandwich-Nabisco	1 cookie	17	88	11.1	unk.	0.6	1	unk.	unk.	unk.	unk.	unk.	unk.	unk.	unk.	unk.	0.02	1
			5.95	0.3	4.6	unk.	unk.		unk.	unk.	unk.	unk.	unk.	unk.	unk.	unk.	unk.	unk.	unk.
C	cookie,National arrowroot biscuit	1 cookie	4.8	22	3.4	unk.	0.4	1	unk.	unk.	unk.	unk.	unk.	unk.	unk.	unk.	unk.	0.02	1
	-Nabisco		*20.83*	0.2	0.7	unk.	unk.		unk.	unk.	unk.	unk.	unk.	unk.	unk.	unk.	unk.	unk.	unk.
D	cookie,nut fudge brownie,vending	1-1/4 oz pkg	35	158	22.9	unk.	1.4	2	unk.	unk.	unk.	unk.	unk.	unk.	unk.	unk.	unk.	0.04	3
	-Nabisco		*2.86*	3.7	6.8	unk.	unk.		unk.	unk.	unk.	unk.	unk.	unk.	unk.	unk.	unk.	unk.	unk.
E	cookie,Nutter Butter peanut sandwich	1 cookie	14	69	9.1	unk.	1.3	2	unk.	unk.	unk.	unk.	unk.	unk.	unk.	unk.	unk.	0.05	3
	-Nabisco		*7.04*	0.5	3.1	unk.	unk.		unk.	unk.	unk.	unk.	unk.	unk.	unk.	unk.	unk.	unk.	unk.
F	cookie,nutter nougat-Nabisco	1 cookie	10	52	6.6	unk.	0.7	1	unk.	unk.	unk.	unk.	unk.	unk.	unk.	unk.	unk.	0.03	2
			9.71	0.4	2.5	unk.	unk.		unk.	unk.	unk.	unk.	unk.	unk.	unk.	unk.	unk.	unk.	unk.
G	cookie,oatmeal w/raisin,home recipe	2-5/8" dia x	13	54	7.0	0.05	0.8	1	20	22	19	23	25	9	0.24	0.55	0.34	0.03	2
		1/4"	*7.69*	2.5	2.6	0.13	3.8		11	36	9	17	6	17	1.42	0.43	5	0	0
H	cookie,oatmeal-Fireside	1 cookie	11	55	7.5	unk.	0.7	1	unk.	unk.	unk.	unk.	unk.	unk.	unk.	unk.	unk.	0.04	3
			8.85	0.5	2.4	unk.	unk.		unk.	unk.	unk.	unk.	unk.	unk.	unk.	unk.	unk.	unk.	unk.
I	cookie,oatmeal,Bakers Bonus-Nabisco	1 cookie	17	80	11.8	unk.	1.0	2	unk.	unk.	unk.	unk.	unk.	unk.	unk.	unk.	unk.	0.04	3
			5.85	0.9	3.1	unk.	unk.		unk.	unk.	unk.	unk.	unk.	unk.	unk.	unk.	unk.	unk.	unk.
J	cookie,oatmeal,vending-Nabisco	2 oz bag	53	230	38.3	unk.	3.6	6	unk.	unk.	unk.	unk.	unk.	unk.	unk.	unk.	unk.	0.12	8
			1.89	3.4	7.0	unk.	unk.		unk.	unk.	unk.	unk.	unk.	unk.	unk.	unk.	unk.	unk.	unk.
K	cookie,old fashion ginger snaps	1 cookie	7.0	29	5.4	unk.	0.4	1	unk.	unk.	unk.	unk.	unk.	unk.	unk.	unk.	unk.	0.01	1
	-Nabisco		*14.29*	0.4	0.7	unk.	unk.		unk.	unk.	unk.	unk.	unk.	unk.	unk.	unk.	unk.	unk.	unk.
L	cookie,Oreo chocolate sandwich	1 cookie	13	68	9.0	unk.	0.5	1	unk.	unk.	unk.	unk.	unk.	unk.	unk.	unk.	unk.	0.01	1
	w/double stuff-Nabisco		*7.52*	0.2	3.4	unk.	unk.		unk.	unk.	unk.	unk.	unk.	unk.	unk.	unk.	unk.	unk.	unk.
M	cookie,Oreo creme sandwich-Nabisco	1 cookie	10	50	7.2	unk.	0.5	1	unk.	unk.	unk.	unk.	unk.	unk.	unk.	unk.	unk.	0.01	0
			9.71	0.2	2.2	unk.	unk.		unk.	unk.	unk.	unk.	unk.	unk.	unk.	unk.	unk.	unk.	unk.
N	cookie,peanut brittle cookies-Nabisco	1 cookie	16	76	9.5	unk.	1.7	3	unk.	unk.	unk.	unk.	unk.	unk.	unk.	unk.	unk.	0.05	3
			6.45	0.6	3.5	unk.	unk.		unk.	unk.	unk.	unk.	unk.	unk.	unk.	unk.	unk.	unk.	unk.
O	cookie,peanut butter creme-Fireside	1 cookie	10	50	7.2	unk.	0.7	1	unk.	unk.	unk.	unk.	unk.	unk.	unk.	unk.	unk.	0.03	2
			9.71	0.2	2.0	unk.	unk.		unk.	unk.	unk.	unk.	unk.	unk.	unk.	unk.	unk.	unk.	unk.
P	cookie,peanut butter fudge cookies	1 cookie	16	79	10.2	unk.	1.3	2	unk.	unk.	unk.	unk.	unk.	unk.	unk.	unk.	unk.	0.03	2
	-Nabisco		*6.29*	0.6	3.6	unk.	unk.		unk.	unk.	unk.	unk.	unk.	unk.	unk.	unk.	unk.	unk.	unk.
Q	cookie,peanut butter wafer,vending	1-1/8 oz pkg	32	170	19.0	unk.	2.8	4	unk.	unk.	unk.	unk.	unk.	unk.	unk.	unk.	unk.	0.02	1
	-Nabisco		*3.13*	0.6	9.2	unk.	unk.		unk.	unk.	unk.	unk.	unk.	unk.	unk.	unk.	unk.	unk.	unk.
R	cookie,peanut butter,home recipe	2" dia x 1/4"	13	61	7.0	0.06	1.1	2	92	53	44	62	60	21	0.33	0.63	0.91	0.02	1
			7.46	1.6	3.3	0.28	4.3		29	82	15	36	14	44	1.66	0.53	5	0	0
S	cookie,peanut sandwich type	1-3/4" dia x	12	58	8.2	0.10	1.2	2	unk.	unk.	unk.	unk.	unk.	unk.	unk.	0.49	0.12	0.01	1
	w/peanut filling	1/2"	*8.20*	0.3	2.3	unk.	unk.		unk.	unk.	unk.	unk.	unk.	unk.	1.22	unk.	unk.	unk.	unk.

Riboflavin mg / Niacin mg	% USRDA	Vitamin B_6 mg / Folacin mcg	% USRDA	Vitamin B_{12} mcg / Pantothenic acid mg	% USRDA	Biotin mg / Vitamin A IU	% USRDA	Preformed A RE / Beta carotene RE	Vitamin D IU / Vitamin E IU	% USRDA	Total tocopherol mg / Alpha tocopherol mg	Other tocopherol mg / Total ash g	Calcium mg / Phosphorus mg	% USRDA	Sodium mg / Sodium meq	Potassium mg / Potassium meq	Chlorine mg / Chlorine meq	Iron mg / Magnesium mg	% USRDA	Zinc mg / Copper mg	% USRDA	Iodine mcg / Selenium mcg	% USRDA	Manganese mcg / Chromium mcg	
0.06	**4**	0.03	**2**	0.04	**1**	0.001	**0**	0	1	**0**	0.1	0.0	37	**4**	143	113	53	1.0	**5**	0.2	**1**	8.8	**6**	53	A
0.5	**3**	4	**1**	0.14	**1**	182	**4**	0	0.1	**0**	0.1	1.01	30	**3**	6.2	2.9	1.5	18	**4**	0.04	**2**	7		10	
0.03	**2**	unk.	**unk.**	unk.	**unk.**	unk.	**unk.**	unk.	0(a)	**0**	unk.	unk.	6	**1**	54	42	unk.	0.4	**2**	0.1	**1**	unk.	**unk.**	58	B
0.2	**1**	unk.	**unk.**	unk.	**unk.**	unk.	**unk.**	unk.	unk.	**unk.**	unk.	0.22	15	**2**	2.4	1.1	unk.	7	**2**	0.05	**3**	unk.		unk.	
0.01	**1**	unk.	**unk.**	unk.	**unk.**	unk.	**unk.**	unk.	0(a)	**0**	unk.	unk.	4	**0**	16	5	unk.	0.1	**1**	tr	**0**	unk.	**unk.**	16	C
0.2	**1**	unk.	**unk.**	unk.	**unk.**	unk.	**unk.**	unk.	unk.	**unk.**	unk.	0.06	8	**1**	0.7	0.1	unk.	1	**0**	0.01	**0**	unk.		unk.	
0.07	**4**	unk.	**unk.**	unk.	**unk.**	unk.	**unk.**	unk.	0(a)	**0**	unk.	unk.	11	**1**	72	37	unk.	0.5	**3**	0.1	**1**	unk.	**unk.**	35	D
0.3	**2**	unk.	**unk.**	unk.	**unk.**	unk.	**unk.**	unk.	unk.	**unk.**	unk.	0.29	18	**2**	3.1	1.0	unk.	5	**1**	0.05	**3**	unk.		unk.	
0.04	**2**	unk.	**unk.**	unk.	**unk.**	unk.	**unk.**	unk.	0(a)	**0**	unk.	unk.	8	**1**	52	28	unk.	0.3	**2**	0.1	**1**	unk.	**unk.**	124	E
0.5	**3**	unk.	**unk.**	unk.	**unk.**	unk.	**unk.**	unk.	unk.	**unk.**	unk.	0.23	27	**3**	2.2	0.7	unk.	7	**2**	0.03	**2**	unk.		unk.	
0.02	**1**	unk.	**unk.**	unk.	**unk.**	unk.	**unk.**	unk.	0(a)	**0**	unk.	unk.	3	**0**	32	19	unk.	0.3	**2**	0.1	**1**	unk.	**unk.**	70	F
0.3	**2**	unk.	**unk.**	unk.	**unk.**	unk.	**unk.**	unk.	unk.	**unk.**	unk.	0.12	12	**1**	1.4	0.5	unk.	4	**1**	0.03	**2**	unk.		unk.	
0.01	**1**	0.01	**1**	0.01	**0**	0.001	**0**	0	tr	**0**	0.1	0(a)	8	**1**	41	33	7	0.3	**2**	0.1	**1**	6.4	**4**	118	G
0.1	**1**	4	**1**	0.05	**1**	3	**0**	0	0.1	**0**	0.1	0.23	20	**2**	1.8	0.8	0.2	10	**2**	0.02	**1**	1		3	
0.04	**2**	unk.	**unk.**	unk.	**unk.**	unk.	**unk.**	unk.	0(a)	**0**	unk.	unk.	4	**0**	44	17	unk.	0.3	**2**	0.1	**1**	unk.	**unk.**	108	H
0.3	**2**	unk.	**unk.**	unk.	**unk.**	unk.	**unk.**	unk.	unk.	**unk.**	unk.	0.15	18	**2**	1.9	0.4	unk.	5	**1**	0.03	**2**	unk.		unk.	
0.04	**2**	unk.	**unk.**	unk.	**unk.**	unk.	**unk.**	unk.	0(a)	**0**	unk.	unk.	6	**1**	69	23	unk.	0.4	**2**	0.1	**1**	unk.	**unk.**	140	I
0.4	**2**	unk.	**unk.**	unk.	**unk.**	unk.	**unk.**	unk.	unk.	**unk.**	unk.	0.24	24	**2**	3.0	0.6	unk.	5	**1**	0.02	**1**	unk.		unk.	
0.08	**5**	unk.	**unk.**	unk.	**unk.**	unk.	**unk.**	unk.	0(a)	**0**	unk.	unk.	10	**1**	250	62	unk.	1.0	**6**	0.4	**3**	unk.	**unk.**	535	J
1.2	**6**	unk.	**unk.**	unk.	**unk.**	unk.	**unk.**	unk.	unk.	**unk.**	unk.	0.72	60	**6**	10.9	1.6	unk.	16	**4**	0.15	**7**	unk.		unk.	
0.02	**1**	unk.	**unk.**	unk.	**unk.**	unk.	**unk.**	unk.	0(a)	**0**	unk.	unk.	5	**1**	48	25	unk.	0.4	**2**	tr	**0**	unk.	**unk.**	124	K
0.2	**1**	unk.	**unk.**	unk.	**unk.**	unk.	**unk.**	unk.	unk.	**unk.**	unk.	0.17	6	**1**	2.1	0.6	unk.	3	**1**	0.02	**1**	unk.		unk.	
0.02	**1**	unk.	**unk.**	unk.	**unk.**	unk.	**unk.**	unk.	0(a)	**0**	unk.	unk.	4	**0**	67	17	unk.	0.3	**2**	0.1	**1**	unk.	**unk.**	45	L
0.2	**1**	unk.	**unk.**	unk.	**unk.**	unk.	**unk.**	unk.	unk.	**unk.**	unk.	0.19	13	**1**	2.9	0.4	unk.	4	**1**	0.04	**2**	unk.		unk.	
0.02	**1**	unk.	**unk.**	unk.	**unk.**	unk.	**unk.**	unk.	unk.	**unk.**	unk.	unk.	3	**0**	69	19	unk.	0.3	**2**	0.1	**1**	unk.	**unk.**	50	M
0.2	**1**	unk.	**unk.**	unk.	**unk.**	unk.	**unk.**	unk.	unk.	**unk.**	unk.	0.20	11	**1**	3.0	0.5	unk.	5	**1**	0.04	**2**	unk.		unk.	
0.06	**4**	unk.	**unk.**	unk.	**unk.**	unk.	**unk.**	unk.	0(a)	**0**	unk.	unk.	12	**1**	50	35	unk.	0.3	**2**	0.2	**1**	unk.	**unk.**	107	N
0.9	**5**	unk.	**unk.**	unk.	**unk.**	unk.	**unk.**	unk.	unk.	**unk.**	unk.	0.24	34	**3**	2.2	0.9	unk.	8	**2**	0.04	**2**	unk.		unk.	
0.02	**1**	unk.	**unk.**	unk.	**unk.**	unk.	**unk.**	unk.	0(a)	**0**	unk.	unk.	2	**0**	54	13	unk.	0.2	**1**	0.1	**1**	unk.	**unk.**	unk.	O
0.4	**2**	unk.	**unk.**	unk.	**unk.**	unk.	**unk.**	unk.	unk.	**unk.**	unk.	0.18	11	**1**	2.3	0.3	unk.	3	**1**	unk.	**unk.**	unk.		unk.	
0.04	**3**	unk.	**unk.**	unk.	**unk.**	unk.	**unk.**	unk.	0(a)	**0**	unk.	unk.	8	**1**	55	41	unk.	0.4	**2**	0.1	**1**	unk.	**unk.**	90	P
0.5	**3**	unk.	**unk.**	unk.	**unk.**	unk.	**unk.**	unk.	unk.	**unk.**	unk.	0.23	21	**2**	2.4	1.1	unk.	7	**2**	0.05	**2**	unk.		unk.	
0.03	**2**	unk.	**unk.**	unk.	**unk.**	unk.	**unk.**	unk.	0(a)	**0**	unk.	unk.	8	**1**	90	68	unk.	0.4	**3**	0.3	**2**	unk.	**unk.**	150	Q
1.0	**5**	unk.	**unk.**	unk.	**unk.**	unk.	**unk.**	unk.	unk.	**unk.**	unk.	0.44	36	**4**	3.9	1.7	unk.	17	**4**	0.05	**3**	unk.		unk.	
0.01	**1**	0.01	**1**	0.01	**0**	0.001	**0**	0	tr	**0**	1.3	0.6	5	**1**	70	30	6	0.2	**1**	0.1	**1**	3.7	**3**	62	R
0.5	**3**	3	**1**	0.09	**1**	80	**2**	0	1.6	**5**	0.7	0.24	16	**2**	3.0	0.8	0.2	5	**1**	0.02	**1**	2		6	
0.01	**1**	unk.	**unk.**	unk.	**unk.**	unk.	**unk.**	unk.	unk.	**unk.**	unk.	unk.	5	**1**	21	21	unk.	0.1	**1**	unk.	**unk.**	unk.	**unk.**	unk.	S
0.3	**2**	unk.	**unk.**	unk.	**unk.**	24	**1**	unk.	unk.	**unk.**	unk.	0.20	14	**1**	0.9	0.5	unk.	2	**1**	unk.	**unk.**	unk.		unk.	

	FOOD	Portion	Weight in grams / *Conversion for 100 g*	Kilocalories / H_2O g	Total carbohydrate g / Total fats g	Crude fiber g / Dietary fiber g	Total protein g / Total sugar g	% **USRDA**	Arginine mg / Histidine mg	Isoleucine mg / Leucine mg	Lysine mg / Methionine mg	Phenylalanine mg / Threonine mg	Valine mg / Tryptophan mg	Cystine mg / Tyrosine mg	Polyunsat. fatty acids g / Monounsat. fatty acids g	Saturated fatty acids g / P/S ratio	Linoleic acid g / Cholesterol mg	Thiamin mg / Ascorbic acid mg	% **USRDA**
A	cookie,Pecan Sandie-Keebler	1 cookie	16	85	9.2	unk.	0.8	**1**	unk.	unk.	unk.	unk.	unk.	unk.	unk.	unk.	unk.	unk.	**unk.**
			6.37	0.5	5.1	unk.	3.5		unk.	unk.	unk.	unk.	unk.	unk.	unk.	unk.	0(a)	tr(a)	**0**
B	cookie,Piccolo crepe-Nabisco	1 cookie	4.1	19	3.1	unk.	0.2	**0**	unk.	unk.	unk.	unk.	unk.	unk.	unk.	unk.	unk.	tr	**0**
			24.39	0.0	0.7	unk.	unk.		unk.	unk.	unk.	unk.	unk.	unk.	unk.	unk.	unk.	unk.	**unk.**
C	cookie,pumpkin bars,home recipe	2x2x1-1/2"	35	151	17.9	0.27	1.8	**3**	unk.	80	56	85	85	32	1.70	4.15	1.37	0.06	**4**
			2.86	6.7	8.6	unk.	7.0		39	130	29	58	20	61	1.96	0.41	25	1	**2**
D	cookie,raisin fruit biscuit-Nabisco	1 cookie	15	58	12.0	unk.	0.8	**1**	unk.	unk.	unk.	unk.	unk.	unk.	unk.	unk.	unk.	0.05	**3**
			6.58	1.4	0.7	unk.	unk.		unk.	unk.	unk.	unk.	unk.	unk.	unk.	unk.	unk.	unk.	**unk.**
E	cookie,raisin sugar-Nabisco	1 cookie	11	52	7.0	unk.	0.6	**1**	unk.	unk.	unk.	unk.	unk.	unk.	unk.	unk.	unk.	0.04	**3**
			9.52	0.4	2.4	unk.	unk.		unk.	unk.	unk.	unk.	unk.	unk.	unk.	unk.	unk.	unk.	**unk.**
F	cookie,raspberry bar,home recipe	1 cookie	24	76	13.3	0.08	0.7	**1**	unk.	33	21	37	33	9	0.15	1.46	0.11	0.03	**2**
			4.17	7.3	2.4	unk.	9.2		10	56	10	22	9	17	0.62	0.10	6	1	**2**
G	cookie,sand tart,deluxe cookie assortment-Nabisco	1 cookie	7.8	37	5.5	unk.	0.4	**1**	unk.	unk.	unk.	unk.	unk.	unk.	unk.	unk.	unk.	0.01	**1**
			12.82	0.4	1.5	unk.	unk.		unk.	unk.	unk.	unk.	unk.	unk.	unk.	unk.	unk.	unk.	**unk.**
H	cookie,sandwich type (chocolate or vanilla)	1-3/4" dia x 3/8"	10	49	6.9	0.01	0.5	**1**	unk.	unk.	unk.	unk.	unk.	unk.	unk.	0.61	0.46	tr	**0**
			10.00	0.2	2.2	unk.	3.6		unk.	unk.	unk.	unk.	unk.	unk.	1.10	unk.	unk.	0	**0**
I	cookie,shortbread	1-5/8x1-5/8x 1/4"	7.5	37	4.9	0.01	0.5	**1**	unk.	unk.	unk.	unk.	unk.	unk.	unk.	0.43	0.38	tr	**0**
			13.33	0.2	1.7	unk.	unk.		unk.	unk.	unk.	unk.	unk.	unk.	0.85	unk.	unk.	0	**0**
J	cookie,shortbread biscuit,fancy, Mayfair assortment-Nabisco	1 cookie	4.7	22	3.3	unk.	0.3	**0**	unk.	unk.	unk.	unk.	unk.	unk.	unk.	unk.	unk.	0.01	**1**
			21.28	0.2	0.8	unk.	unk.		unk.	unk.	unk.	unk.	unk.	unk.	unk.	unk.	unk.	unk.	**unk.**
K	cookie,Social Tea biscuit-Nabisco	6 cookies	28	127	21.3	unk.	2.0	**3**	unk.	unk.	unk.	unk.	unk.	unk.	unk.	unk.	unk.	0.11	**8**
			3.55	0.8	3.8	unk.	unk.		unk.	unk.	unk.	unk.	unk.	unk.	unk.	unk.	unk.	unk.	**unk.**
L	cookie,spiced wafers-Nabisco	4 cookies	31	133	23.6	unk.	1.6	**2**	unk.	unk.	unk.	unk.	unk.	unk.	unk.	unk.	unk.	0.05	**4**
			3.25	1.5	3.5	unk.	unk.		unk.	unk.	unk.	unk.	unk.	unk.	unk.	unk.	unk.	unk.	**unk.**
M	cookie,striped shortbread-Nabisco	1 cookie	9.5	48	6.4	unk.	0.5	**1**	unk.	unk.	unk.	unk.	unk.	unk.	unk.	unk.	unk.	0.02	**1**
			10.53	0.1	2.3	unk.	unk.		unk.	unk.	unk.	unk.	unk.	unk.	unk.	unk.	unk.	unk.	**unk.**
N	cookie,sugar gem-Fireside	1 cookie	11	51	8.5	unk.	0.6	**1**	unk.	unk.	unk.	unk.	unk.	unk.	unk.	unk.	unk.	0.04	**3**
			8.85	0.4	1.6	unk.	unk.		unk.	unk.	unk.	unk.	unk.	unk.	unk.	unk.	unk.	unk.	**unk.**
O	cookie,sugar ring,Bakers Bonus -Nabisco	1 cookie	15	69	10.5	unk.	0.9	**1**	unk.	unk.	unk.	unk.	unk.	unk.	unk.	unk.	unk.	0.05	**3**
			6.85	0.4	2.6	unk.	unk.		unk.	unk.	unk.	unk.	unk.	unk.	unk.	unk.	unk.	unk.	**unk.**
P	cookie,sugar wafer	3-1/2x1x 1/8"	9.5	46	7.0	0.01	0.5	**1**	unk.	unk.	unk.	unk.	unk.	unk.	unk.	0.47	0.39	tr	**0**
			10.53	0.1	1.8	unk.	unk.		unk.	unk.	unk.	unk.	unk.	unk.	0.91	unk.	unk.	0	**0**
Q	cookie,sugar wafer,assorted-Fireside	1 cookie	12	62	8.1	unk.	0.2	**0**	unk.	unk.	unk.	unk.	unk.	unk.	unk.	unk.	unk.	tr	**0**
			8.33	0.5	3.2	unk.	unk.		unk.	unk.	unk.	unk.	unk.	unk.	unk.	unk.	unk.	unk.	**unk.**
R	cookie,sugar wafer,Biscos-Nabisco	1 cookie	3.6	18	2.6	unk.	0.1	**0**	unk.	unk.	unk.	unk.	unk.	unk.	unk.	unk.	unk.	tr	**0**
			27.78	tr	0.8	unk.	unk.		unk.	unk.	unk.	unk.	unk.	unk.	unk.	unk.	unk.	unk.	**unk.**
S	cookie,sugar wafer,triple decker peanut butter,Biscos-Nabisco	1 cookie	9.3	47	5.7	unk.	1.1	**2**	unk.	unk.	unk.	unk.	unk.	unk.	unk.	unk.	unk.	0.01	**0**
			10.75	0.1	2.3	unk.	unk.		unk.	unk.	unk.	unk.	unk.	unk.	unk.	unk.	unk.	unk.	**unk.**

Riboflavin mg / Niacin mg	% USRDA	Vitamin B_6 mg / Folacin mcg	% USRDA	Vitamin B_{12} mcg / Pantothenic acid mg	% USRDA	Biotin mg / Vitamin A IU	% USRDA	Preformed A RE / Beta carotene RE	Vitamin D IU / Vitamin E IU	% USRDA	Total tocopherol mg / Alpha tocopherol mg	Other tocopherol mg / Total ash g	Calcium mg / Phosphorus mg	% USRDA	Sodium mg / Sodium meq	Potassium mg / Potassium meq	Chlorine mg / Chlorine meq	Iron mg / Magnesium mg	% USRDA	Zinc mg / Copper mg	% USRDA	Iodine mcg / Selenium mcg	% USRDA	Manganese mcg / Chromium mcg	
unk.	**unk.**	unk.	**unk.**	0(a)	**0**	unk.	**unk.**	0(a)	0(a)	**0**	unk.	unk.	unk.	**unk.**	52	25	unk.	0.1	**0**	unk.	**unk.**	unk.	**unk.**	unk.	A
tr(a)	**0**	unk.	**unk.**	unk.	**unk.**	tr(a)	**0**	tr(a)	unk.	**unk.**	unk.	0.20	12	**1**	2.3	0.6	unk.	unk.	**unk.**	unk.	**unk.**	unk.		unk.	
0.01	**0**	unk.	**unk.**	unk.	**unk.**	unk.	**unk.**	unk.	0(a)	**0**	unk.	unk.	2	**0**	7	6	unk.	tr	**0**	tr	**0**	unk.	**unk.**	10	B
tr	**0**	unk.	**unk.**	unk.	**unk.**	unk.	**unk.**	unk.	unk.	**unk.**	unk.	0.03	4	**0**	0.3	0.1	unk.	1	**0**	0.01	**0**	unk.		unk.	
0.04	**2**	0.05	**3**	0.02	**0**	0.002	**1**	0	1	**0**	0.1	0.0	16	**2**	135	86	10	0.6	**3**	0.3	**2**	10.5	**7**	unk.	C
0.4	**2**	7	**2**	0.09	**1**	760	**15**	52	0.1	**0**	0.1	0.58	37	**4**	5.9	2.2	0.3	14	**3**	0.08	**4**	4		10	
0.03	**2**	unk.	**unk.**	unk.	**unk.**	unk.	**unk.**	unk.	0(a)	**0**	unk.	unk.	7	**1**	35	64	unk.	0.5	**3**	0.1	**0**	unk.	**unk.**	60	D
0.2	**1**	unk.	**unk.**	unk.	**unk.**	unk.	**unk.**	unk.	unk.	**unk.**	unk.	0.23	15	**2**	1.5	1.6	unk.	4	**1**	0.05	**2**	unk.		unk.	
0.03	**2**	unk.	**unk.**	unk.	**unk.**	unk.	**unk.**	unk.	0(a)	**0**	unk.	unk.	6	**1**	28	16	unk.	0.2	**1**	tr	**0**	unk.	**unk.**	33	E
0.3	**1**	unk.	**unk.**	unk.	**unk.**	unk.	**unk.**	unk.	unk.	**unk.**	unk.	0.13	11	**1**	1.2	0.4	unk.	2	**0**	0.01	**1**	unk.		unk.	
0.02	**1**	tr	**0**	0.00	**0**	0.000	**0**	0	0	**0**	0.3	0.1	8	**1**	87	34	7	0.4	**2**	0.1	**1**	7.0	**5**	20	F
0.2	**1**	1	**0**	0.03	**0**	87	**2**	tr	0.4	**1**	0.2	0.32	11	**1**	3.8	0.9	0.2	7	**2**	0.01	**0**	2		6	
0.01	**1**	unk.	**unk.**	unk.	**unk.**	unk.	**unk.**	unk.	0(a)	**0**	unk.	unk.	1	**0**	20	4	unk.	0.1	**1**	tr	**0**	unk.	**unk.**	26	G
0.1	**1**	unk.	**unk.**	unk.	**unk.**	unk.	**unk.**	unk.	unk.	**unk.**	unk.	0.05	4	**0**	0.9	0.1	unk.	1	**0**	tr	**0**	unk.		unk.	
tr	**0**	unk.	**unk.**	unk.	**unk.**	unk.	**unk.**	0	tr(a)	**0**	0.3	0.2	3	**0**	48	4	unk.	0.1	**0**	unk.	**unk.**	unk.	**unk.**	unk.	H
0.0	**0**	unk.	**unk.**	unk.	**unk.**	0	**0**	0	0.3	**1**	0.1	0.12	24	**2**	2.1	0.1	unk.	1	**0**	unk.	**unk.**	unk.		unk.	
tr	**0**	unk.	**unk.**	unk.	**unk.**	unk.	**unk.**	unk.	tr(a)	**0**	0.1	0.1	5	**1**	4	5	unk.	tr	**0**	tr	**0**	unk.	**unk.**	unk.	I
tr	**0**	1	**0**	unk.	**unk.**	6	**0**	unk.	0.1	**0**	tr	0.12	12	**1**	0.2	0.1	unk.	1	**0**	unk.	**unk.**	unk.		unk.	
0.02	**1**	unk.	**unk.**	unk.	**unk.**	unk.	**unk.**	unk.	0(a)	**0**	unk.	unk.	1	**0**	16	4	unk.	0.1	**1**	tr	**0**	unk.	**unk.**	22	J
0.2	**1**	unk.	**unk.**	unk.	**unk.**	unk.	**unk.**	unk.	unk.	**unk.**	unk.	0.04	3	**0**	0.7	0.1	unk.	1	**0**	0.01	**0**	unk.		unk.	
0.08	**5**	unk.	**unk.**	unk.	**unk.**	unk.	**unk.**	unk.	0(a)	**0**	unk.	unk.	6	**1**	109	30	unk.	0.7	**4**	0.2	**1**	unk.	**unk.**	129	K
1.0	**5**	unk.	**unk.**	unk.	**unk.**	unk.	**unk.**	unk.	unk.	**unk.**	unk.	0.35	27	**3**	4.7	0.8	unk.	5	**1**	0.06	**3**	unk.		unk.	
0.08	**5**	unk.	**unk.**	unk.	**unk.**	unk.	**unk.**	unk.	0(a)	**0**	unk.	unk.	10	**1**	198	39	unk.	0.8	**5**	0.2	**1**	unk.	**unk.**	160	L
0.9	**4**	unk.	**unk.**	unk.	**unk.**	unk.	**unk.**	unk.	unk.	**unk.**	unk.	0.61	17	**2**	8.6	1.0	unk.	5	**1**	0.03	**1**	unk.		unk.	
0.03	**2**	unk.	**unk.**	unk.	**unk.**	unk.	**unk.**	unk.	0(a)	**0**	unk.	unk.	3	**0**	28	24	unk.	0.2	**1**	0.1	**0**	unk.	**unk.**	34	M
0.3	**2**	unk.	**unk.**	unk.	**unk.**	unk.	**unk.**	unk.	unk.	**unk.**	unk.	0.14	10	**1**	1.2	0.6	unk.	3	**1**	0.02	**1**	unk.		unk.	
0.03	**2**	unk.	**unk.**	unk.	**unk.**	unk.	**unk.**	unk.	0(a)	**0**	unk.	unk.	4	**0**	45	9	unk.	0.2	**1**	0.1	**0**	unk.	**unk.**	unk.	N
0.4	**2**	unk.	**unk.**	unk.	**unk.**	unk.	**unk.**	unk.	unk.	**unk.**	unk.	0.16	12	**1**	2.0	0.2	unk.	2	**0**	unk.	**unk.**	unk.		unk.	
0.04	**2**	unk.	**unk.**	unk.	**unk.**	unk.	**unk.**	unk.	0(a)	**0**	unk.	unk.	4	**0**	52	13	unk.	0.3	**2**	0.1	**1**	unk.	**unk.**	61	O
0.5	**3**	unk.	**unk.**	unk.	**unk.**	unk.	**unk.**	unk.	unk.	**unk.**	unk.	0.17	18	**2**	2.3	0.3	unk.	2	**1**	0.02	**1**	unk.		unk.	
tr	**0**	unk.	**unk.**	unk.	**unk.**	unk.	**unk.**	unk.	tr(a)	**0**	0.1	0.1	3	**0**	18	6	unk.	tr	**0**	tr	**0**	unk.	**unk.**	25	P
0.0	**0**	unk.	**unk.**	unk.	**unk.**	13	**0**	unk.	0.2	**1**	0.1	0.09	8	**1**	0.8	0.1	unk.	1	**0**	0.01	**0**	unk.		unk.	
0.09	**5**	unk.	**unk.**	unk.	**unk.**	unk.	**unk.**	unk.	0(a)	**0**	unk.	unk.	1	**0**	3	4	unk.	0.1	**0**	tr	**0**	unk.	**unk.**	13	Q
0.0	**0**	unk.	**unk.**	unk.	**unk.**	unk.	**unk.**	unk.	unk.	**unk.**	unk.	0.03	3	**0**	0.1	0.1	unk.	1	**0**	0.01	**1**	unk.		unk.	
0.01	**0**	unk.	**unk.**	unk.	**unk.**	unk.	**unk.**	unk.	0(a)	**0**	unk.	unk.	0	**0**	6	2	unk.	0.1	**0**	tr	**0**	unk.	**unk.**	10	R
0.1	**1**	unk.	**unk.**	unk.	**unk.**	unk.	**unk.**	unk.	unk.	**unk.**	unk.	0.02	2	**0**	0.3	0.1	unk.	tr	**0**	tr	**0**	unk.		unk.	
0.03	**2**	unk.	**unk.**	unk.	**unk.**	unk.	**unk.**	unk.	0(a)	**0**	unk.	unk.	4	**0**	40	30	unk.	0.2	**1**	0.1	**1**	unk.	**unk.**	70	S
0.6	**3**	unk.	**unk.**	unk.	**unk.**	unk.	**unk.**	unk.	unk.	**unk.**	unk.	0.18	17	**2**	1.7	0.8	unk.	6	**2**	0.02	**1**	unk.		unk.	

	FOOD	Portion	Weight in grams / Conversion for 100 g	Kilocalories / H_2O g	Total carbohydrate g / Total fats g	Crude fiber g / Dietary fiber g	Total protein g / Total sugar g	% USRDA	Arginine mg / Histidine mg	Isoleucine mg / Leucine mg	Lysine mg / Methionine mg	Phenylalanine mg / Threonine mg	Valine mg / Tryptophan mg	Cystine mg / Tyrosine mg	Polyunsat. fatty acids g / Monounsat. fatty acids g	Saturated fatty acids g / P/S ratio	Linoleic acid g / Cholesterol mg	Thiamin mg / Ascorbic acid mg	% USRDA
A	cookie,sugar wafer,triple decker,	1 cookie	9.3	48	6.8	unk.	0.3	**0**	unk.	unk.	unk.	unk.	unk.	unk.	unk.	unk.	unk.	tr	**0**
	Biscos-Nabisco		*10.75*	0.1	2.2	unk.	unk.		unk.	unk.	unk.	unk.	unk.	unk.	unk.	unk.	unk.	unk.	**unk.**
B	cookie,sugar-Fireside	1 cookie	11	53	7.9	unk.	0.6	**1**	unk.	unk.	unk.	unk.	unk.	unk.	unk.	unk.	unk.	0.04	**3**
			8.85	0.6	2.1	unk.	unk.		unk.	unk.	unk.	unk.	unk.	unk.	unk.	unk.	unk.	unk.	**unk.**
C	cookie,sugar,home recipe	2-1/4x1/4"	8.0	36	4.4	0.01	0.4	**1**	12	21	16	24	22	9	0.09	1.09	0.07	0.01	**1**
			12.50	1.5	1.8	0.10	2.1		10	35	8	15	6	16	0.46	0.09	8	0	**0**
D	cookie,Swiss creme sandwich Oreo,	1 cookie	10	51	7.4	unk.	0.4	**1**	unk.	unk.	unk.	unk.	unk.	unk.	unk.	unk.	unk.	0.03	**2**
	Swiss assortment-Nabisco		*9.71*	0.2	2.1	unk.	unk.		unk.	unk.	unk.	unk.	unk.	unk.	unk.	unk.	unk.	unk.	**unk.**
E	cookie,tea time biscuit,Mayfair	1 cookie	5.2	24	3.7	unk.	0.4	**1**	unk.	unk.	unk.	unk.	unk.	unk.	unk.	unk.	unk.	0.02	**1**
	assortment-Nabisco		*19.23*	0.2	0.8	unk.	unk.		unk.	unk.	unk.	unk.	unk.	unk.	unk.	unk.	unk.	unk.	**unk.**
F	cookie,toll house,deluxe cookie	1 cookie	10	53	6.5	unk.	0.5	**1**	unk.	unk.	unk.	unk.	unk.	unk.	unk.	unk.	unk.	tr	**0**
	assortment-Nabisco		*9.71*	0.5	2.8	unk.	unk.		unk.	unk.	unk.	unk.	unk.	unk.	unk.	unk.	unk.	unk.	**unk.**
G	cookie,Twiddle Sticks crunchy creme	1 cookie	10	54	7.1	unk.	0.3	**1**	unk.	unk.	unk.	unk.	unk.	unk.	unk.	unk.	unk.	0.02	**1**
	wafers-Nabisco		*9.71*	0.1	2.7	unk.	unk.		unk.	unk.	unk.	unk.	unk.	unk.	unk.	unk.	unk.	unk.	**unk.**
H	cookie,vanilla creme-Fireside	1 cookie	10	48	7.6	unk.	0.6	**1**	unk.	unk.	unk.	unk.	unk.	unk.	unk.	unk.	unk.	0.05	**3**
			9.71	0.3	1.7	unk.	unk.		unk.	unk.	unk.	unk.	unk.	unk.	unk.	unk.	unk.	unk.	**unk.**
I	cookie,vanilla wafer	1-3/4" dia x	4.0	18	3.0	tr	0.2	**0**	unk.	unk.	unk.	unk.	unk.	unk.	unk.	0.16	0.04	tr	**0**
		1/4"	*25.00*	0.1	0.6	0.03	unk.		unk.	unk.	unk.	unk.	unk.	unk.	0.40	unk.	tr(a)	0	**0**
J	cookie,vanilla wafer-Fireside	1 cookie	2.4	11	1.9	unk.	0.1	**0**	unk.	unk.	unk.	unk.	unk.	unk.	unk.	unk.	unk.	0.01	**1**
			41.67	0.1	0.3	unk.	unk.		unk.	unk.	unk.	unk.	unk.	unk.	unk.	unk.	unk.	unk.	**unk.**
K	cookie,vanilla wafer,Nilla-Nabisco	1 cookie	4.0	18	2.9	unk.	0.2	**0**	unk.	unk.	unk.	unk.	unk.	unk.	unk.	unk.	unk.	0.01	**1**
			25.00	0.2	0.6	unk.	unk.		unk.	unk.	unk.	unk.	unk.	unk.	unk.	unk.	unk.	unk.	**unk.**
L	cookie,waffle creme,Bisco-Nabisco	1 cookie	8.4	43	6.1	unk.	0.3	**0**	unk.	unk.	unk.	unk.	unk.	unk.	unk.	unk.	unk.	tr	**0**
			11.90	0.1	1.9	unk.	unk.		unk.	unk.	unk.	unk.	unk.	unk.	unk.	unk.	unk.	unk.	**unk.**
M	CORIANDER leaf,dried	1 tsp	0.6	2	0.3	0.06	0.1	**0**	unk.	unk.	unk.	unk.	unk.	unk.	unk.	unk.	unk.	0.01	**1**
			166.67	tr	tr	unk.	0.0		unk.	unk.	unk.	unk.	unk.	unk.	unk.	unk.	0	3	**6**
N	coriander seed	1 tsp	1.7	5	0.9	0.50	0.2	**0**	unk.	unk.	unk.	unk.	unk.	unk.	0.03	0.02	0.03	tr	**0**
			58.82	0.2	0.3	unk.	0.0		unk.	unk.	unk.	unk.	unk.	unk.	0.23	1.77	0	unk.	**unk.**
O	CORN,SWEET,CANNED,cream style,	1/2 C	128	105	23.7	0.38	3.3	**6**	unk.	123	123	187	210	unk.	unk.	unk.	unk.	0.04	**3**
	low sodium,solids & liquid		*0.78*	98.9	1.4	unk.	4.7		unk.	366	63	137	20	unk.	unk.	unk.	0	6	**11**
P	corn,sweet,canned,cream style,	1/2 C	128	105	25.6	0.64	2.7	**5**	unk.	123	123	187	210	unk.	unk.	unk.	unk.	0.04	**3**
	solids & liquid		*0.78*	97.7	0.8	unk.	4.7		unk.	366	63	137	20	unk.	unk.	unk.	0	6	**11**
Q	corn,sweet,canned,whole kernel,	1/2 C	83	69	16.3	0.66	2.1	**3**	unk.	80	80	115	130	unk.	unk.	tr(a)	unk.	0.02	**2**
	drained solids		*1.21*	62.6	0.7	4.62	3.0		unk.	240	41	89	12	55	unk.	unk.	0	3	**6**
R	corn,sweet,canned,whole kernel,	1/2 C	83	63	14.8	0.58	2.1	**3**	unk.	76	76	115	130	unk.	unk.	tr(a)	unk.	0.02	**2**
	low sodium,drained solids		*1.21*	64.7	0.6	4.62	3.0		unk.	227	40	84	12	unk.	unk.	unk.	0	3	**6**
S	CORN,SWEET,FRESH,off cob,cooked,	1/2 C	83	68	15.5	0.58	2.6	**4**	unk.	unk.	unk.	unk.	unk.	unk.	unk.	tr(a)	unk.	0.09	**6**
	drained solids		*1.21*	63.1	0.8	4.62	3.0		unk.	unk.	unk.	unk.	unk.	unk.	unk.	unk.	0	6	**10**

Riboflavin mg / Niacin mg	% USRDA	Vitamin B_6 mg / Folacin mcg	% USRDA	Vitamin B_{12} mcg / Pantothenic acid mg	% USRDA	Biotin mg / Vitamin A IU	% USRDA	Preformed A RE / Beta carotene RE	Vitamin D IU / Vitamin E IU	% USRDA	Total tocopherol mg / Alpha tocopherol mg	Other tocopherol mg / Total ash g	Calcium mg / Phosphorus mg	% USRDA	Sodium mg / Sodium meq	Potassium mg / Potassium meq	Chlorine mg / Chlorine meq	Iron mg / Magnesium mg	% USRDA	Zinc mg / Copper mg	% USRDA	Iodine mcg / Selenium mcg	% USRDA	Manganese mcg / Chromium mcg	
0.01	1	unk.	unk.	unk.	unk.	unk.	unk.	unk.	0(a)	0	unk.	unk.	1	0	14	4	unk.	0.1	1	tr	0	unk.	unk.	17	A
0.2	1	unk.	unk.	unk.	unk.	unk.	unk.	unk.	unk.	unk.	unk.	0.05	4	0	0.6	0.1	unk.	1	0	0.01	0	unk.		unk.	
0.03	2	unk.	unk.	unk.	unk.	unk.	unk.	unk.	0(a)	0	unk.	unk.	2	0	45	10	unk.	0.2	1	0.1	0	unk.	unk.	41	B
0.4	2	unk.	unk.	unk.	unk.	unk.	unk.	unk.	unk.	unk.	unk.	0.11	8	1	2.0	0.3	unk.	2	0	0.01	1	unk.		unk.	
0.01	1	tr	0	0.01	0	tr	0	0	tr	0	tr	0.0	3	0	34	5	3	0.1	1	0.1	0	2.5	2	14	C
0.1	1	1	0	0.03	0	65	1	0	tr	0	tr	0.13	7	1	1.5	0.1	0.1	3	1	0.01	0	2		3	
0.03	2	unk.	unk.	unk.	unk.	unk.	unk.	unk.	0(a)	0	unk.	unk.	4	0	40	10	unk.	0.2	1	tr	0	unk.	unk.	25	D
0.3	1	unk.	unk.	unk.	unk.	unk.	unk.	unk.	unk.	unk.	unk.	0.13	10	1	1.7	0.3	unk.	1	0	0.01	0	unk.		unk.	
0.02	1	unk.	unk.	unk.	unk.	unk.	unk.	unk.	0(a)	0	unk.	unk.	1	0	22	6	unk.	0.1	1	tr	0	unk.	unk.	23	E
0.2	1	unk.	unk.	unk.	unk.	unk.	unk.	unk.	unk.	unk.	unk.	0.06	5	0	0.9	0.2	unk.	1	0	0.01	0	unk.		unk.	
0.01	1	unk.	unk.	unk.	unk.	unk.	unk.	unk.	0(a)	0	unk.	unk.	2	0	35	11	unk.	0.1	1	0.1	0	unk.	unk.	31	F
0.0	0	unk.	unk.	unk.	unk.	unk.	unk.	unk.	unk.	unk.	unk.	0.11	7	1	1.5	0.3	unk.	2	1	0.01	0	unk.		unk.	
0.03	2	unk.	unk.	unk.	unk.	unk.	unk.	unk.	0(a)	0	unk.	unk.	4	0	14	21	unk.	0.2	1	0.1	0	unk.	unk.	30	G
0.2	1	unk.	unk.	unk.	unk.	unk.	unk.	unk.	unk.	unk.	unk.	0.10	10	1	0.6	0.5	unk.	3	1	0.02	1	unk.		unk.	
0.03	2	unk.	unk.	unk.	unk.	unk.	unk.	unk.	0(a)	0	unk.	unk.	2	0	51	12	unk.	0.2	1	0.1	1	unk.	unk.	65	H
0.3	1	unk.	unk.	unk.	unk.	unk.	unk.	unk.	unk.	unk.	unk.	0.12	6	1	2.2	0.3	unk.	2	1	0.02	1	unk.		unk.	
tr	0	unk.	unk.	unk.	unk.	unk.	unk.	unk.	tr(a)	0	tr	unk.	2	0	10	3	unk.	tr	0	tr	0	unk.	unk.	unk.	I
tr	0	unk.	unk.	unk.	unk.	5	0	unk.	tr	0	tr	0.05	3	0	0.4	0.1	unk.	1	0	tr	0	unk.		unk.	
0.01	1	unk.	unk.	unk.	unk.	unk.	unk.	unk.	0(a)	0	unk.	unk.	1	0	8	2	unk.	tr	0	tr	0	unk.	unk.	10	J
0.1	0	unk.	unk.	unk.	unk.	unk.	unk.	unk.	unk.	unk.	unk.	0.02	4	0	0.3	0.0	unk.	tr	0	tr	0	unk.		unk.	
0.01	1	unk.	unk.	unk.	unk.	unk.	unk.	unk.	unk.	unk.	unk.	unk.	2	0	12	4	unk.	0.1	0	tr	0	unk.	unk.	10	K
0.1	1	unk.	unk.	unk.	unk.	unk.	unk.	unk.	unk.	unk.	unk.	0.04	4	0	0.5	0.1	unk.	1	0	tr	0	unk.		unk.	
0.01	1	unk.	unk.	unk.	unk.	unk.	unk.	unk.	0(a)	0	unk.	unk.	1	0	11	4	unk.	0.1	1	tr	0	unk.	unk.	19	L
0.2	1	unk.	unk.	unk.	unk.	unk.	unk.	unk.	unk.	unk.	unk.	0.03	3	0	0.5	0.1	unk.	1	0	0.01	0	unk.		unk.	
0.01	1	unk.	unk.	0.00	0	unk.	unk.	0(a)	0(a)	0	unk.	unk.	7	1	1	27	unk.	0.3	1	unk.	unk.	unk.	unk.	unk.	M
0.1	0	unk.	unk.	unk.	unk.	unk.	unk.	unk.	unk.	unk.	unk.	0.08	3	0	0.1	0.7	unk.	4	1	unk.	unk.	unk.		unk.	
0.00	0	unk.	unk.	0.00	0	unk.	unk.	0(a)	0(a)	0	unk.	unk.	12	1	1	22	unk.	0.3	2	0.1	1	unk.	unk.	unk.	N
tr	0	unk.	unk.	unk.	unk.	tr	0	unk.	unk.	unk.	unk.	0.10	7	1	0.0	0.5	unk.	6	1	unk.	unk.	unk.		unk.	
0.06	4	0.26	13	0.00	0	0.003	1	6	unk.	unk.	0.3	0.3	4	0	3	124	unk.	0.8	4	0.6	4	unk.	unk.	unk.	O
1.3	6	unk.	unk.	0.38	4	346	7	33	0.4	1	0.1	0.64	72	7	0.1	3.2	unk.	unk.	unk.	0.08	4	tr(a)		unk.	
0.06	4	0.26	13	0.00	0	0.003	1	7	unk.	unk.	0.3	0.3	4	0	302	124	unk.	0.8	4	0.6	4	unk.	unk.	unk.	P
1.3	6	11	3	0.38	4	423	9	40	0.4	1	0.1	1.28	72	7	13.1	3.2	unk.	24	6	0.08	4	tr(a)		unk.	
0.04	2	0.16	8	0.00	0	unk.	unk.	0	0(a)	0	0.2	0.1	4	0	195	80	unk.	0.4	2	0.3	2	unk.	unk.	unk.	Q
0.7	4	7	2	0.18	2	289	6	29	0.2	1	tr	0.74	40	4	8.5	2.0	unk.	16	4	0.05	3	tr(a)		unk.	
0.04	2	0.16	8	0.00	0	unk.	unk.	0	0(a)	0	0.2	0.1	4	0	2	80	unk.	0.4	2	0.3	2	unk.	unk.	0	R
0.7	4	7	2	0.18	2	289	6	29	0.2	1	tr	0.33	40	4	0.1	2.0	unk.	9	2	0.01	1	tr(a)		0	
0.08	5	0.24	12	0.00	0	0.002	1	0	0(a)	0	0.4	0.3	2	0	tr	136	unk.	0.5	3	0.3	2	unk.	unk.	unk.	S
1.1	5	2	0	0.33	3	330	7	33	0.5	2	0.0	0.41	73	7	0.0	3.5	unk.	26	6	0.07	4	tr(a)		unk.	

	FOOD	Portion	Weight in grams / *Conversion for 100 g*	Kilocalories / H_2O g	Total carbohydrate g / Total fats g	Crude fiber g / Dietary fiber g	Total protein g / Total sugar g	**% USRDA**	Arginine mg / Histidine mg	Isoleucine mg / Leucine mg	Lysine mg / Methionine mg	Phenylalanine mg / Threonine mg	Valine mg / Tryptophan mg	Cystine mg / Tyrosine mg	Polyunsat. fatty acids g / Monounsat. fatty acids g	Saturated fatty acids g / P/S ratio	Linoleic acid g / Cholesterol mg	Thiamin mg / Ascorbic acid mg	**% USRDA**
A	corn,sweet,fresh,on cob,cooked,	1 med	77	70	16.2	0.54	2.5	**4**	unk.	94	94	142	160	unk.	unk.	tr(a)	unk.	0.09	**6**
	edible portion		*1.30*	57.1	0.8	4.31	2.8		unk.	280	49	104	15	68	unk.	unk.	0	7	**12**
B	CORN,SWEET,FROZEN,off cob,cooked,	1/2 C	83	65	15.5	0.41	2.5	**4**	unk.	unk.	unk.	unk.	unk.	unk.	unk.	tr(a)	unk.	0.07	**5**
	drained solids		*1.21*	63.7	0.4	4.62	3.0		unk.	unk.	unk.	unk.	unk.	unk.	unk.	unk.	0	4	**7**
C	corn,sweet,frozen,souffle-Stouffer	1/2 pkg	170	232	28.3	unk.	6.0	**9**	unk.	unk.	unk.	unk.	unk.	unk.	unk.	unk.	unk.	0.14	**9**
			0.59	122.7	10.4	unk.	unk.		unk.	unk.	unk.	unk.	unk.	unk.	unk.	unk.	unk.	tr	**0**
D	CORN,SWEET,HOME RECIPE,corn &	3/4 C	187	215	20.5	0.45	9.4	**19**	304	477	686	474	594	75	0.48	6.72	0.33	0.13	**8**
	cheese chowder		*0.53*	164.7	11.7	3.36	8.3		236	899	228	385	118	421	2.86	0.07	66	7	**11**
E	corn,sweet,home recipe,fritter	1 average	30	62	8.7	0.11	1.7	**3**	25	78	71	90	93	23	1.08	0.48	1.06	0.05	**3**
			3.33	16.8	2.4	0.93	1.0		30	151	33	63	19	60	0.59	2.23	12	1	**2**
F	corn,sweet,home recipe,scalloped	1/2 C	151	258	43.1	0.70	7.4	**13**	121	410	360	471	481	119	0.58	1.48	0.60	0.07	**5**
			0.66	89.2	7.3	unk.	5.7		155	703	164	328	99	256	1.32	0.39	47	14	**23**
G	CORNBREAD/johnny cake	2x2x7/8"	40	107	18.2	0.12	3.5	**5**	unk.	160	104	160	176	unk.	unk.	0.80	0.40	0.08	**5**
			2.50	15.2	2.1	unk.	unk.		unk.	452	64	140	21	unk.	1.20	unk.	28	tr	**1**
H	cornbread/johnny cake,home recipe	2x2x7/8"	42	112	14.0	0.09	2.3	**4**	47	118	110	115	130	37	0.39	2.92	0.31	0.05	**3**
			2.38	19.5	5.1	0.19	2.4		57	236	50	94	26	113	1.34	0.13	29	tr	**0**
I	cornbread,cornstick,home recipe	1 average	38	101	13.4	0.06	2.3	**4**	60	115	114	114	129	40	0.24	1.21	0.42	0.05	**3**
			2.63	17.7	4.2	0.19	3.6		58	218	52	94	27	106	2.24	0.20	30	tr	**0**
J	CORNISH HEN,roasted,whole	1 bird	232	316	0.0	0.00	55.2	**123**	unk.	2951	4389	2209	2810	724	2.27	2.37	1.30	0.12	**8**
			0.43	164.7	8.8	0(a)	0.0		1450	4065	1385	2192	566	1847	1.79	0.96	unk.	unk.	**unk.**
K	CORNMEAL,white,bolted,self-rising	1 C	134	468	96.2	1.34	10.8	**17**	unk.	501	312	493	555	141	unk.	unk.	unk.	0.59	**39**
	-Aunt Jemima		*0.75*	14.1	4.4	unk.	tr(a)		224	1406	202	433	66	663	unk.	unk.	0(a)	0	**0**
L	cornmeal,white,degerminated,self-rising	1 C	141	485	104.9	0.85	11.4	**18**	unk.	527	330	519	582	148	unk.	unk.	unk.	0.62	**41**
	-Aunt Jemima		*0.71*	14.8	2.3	unk.	tr(a)		235	1480	213	455	69	698	unk.	unk.	0(a)	0	**0**
M	CORNSTARCH	1 Tbsp	8.0	29	7.0	0.00	0.0	**0**	tr(a)	tr(a)	tr(a)	tr(a)	tr(a)	tr(a)	tr(a)	tr(a)	tr(a)	0.00	**0**
			12.50	1.0	tr	unk.	0.0		tr(a)	tr(a)	tr(a)	tr(a)	tr(a)	tr(a)	tr(a)	0.00	0	0	**0**
N	CRAB leg,steamed in shell	1 average	43	40	0.2	0(a)	7.4	**16**	611	343	581	unk.	unk.	93	unk.	unk.	unk.	0.07	**5**
			2.35	33.4	0.8	0(a)	0.0		138	639	unk.	336	85	268	unk.	unk.	43	1	**2**
O	crab,blue hard shell,steamed whole	1 average	21	20	0.1	0(a)	3.7	**8**	305	172	291	unk.	unk.	46	unk.	unk.	unk.	0.03	**2**
			4.69	16.7	0.4	0(a)	0.0		69	320	unk.	168	42	134	unk.	unk.	21	tr	**1**
P	crab,deviled	1/2 C	120	182	7.7	0(a)	11.4	**25**	944	463	794	463	492	19	0.43	3.20	1.97	0.08	**6**
			0.83	87.3	11.5	0(a)	0.1		unk.	852	274	483	150	46	4.45	0.13	125	1	**2**
Q	crab,dungeness,steamed whole	1 average	68	64	0.3	0(a)	11.8	**26**	981	551	934	unk.	unk.	149	unk.	unk.	unk.	0.11	**7**
			1.46	53.7	1.3	0(a)	0.0		222	1027	unk.	540	136	430	unk.	unk.	68	1	**2**
R	crab,home recipe,stuffed	3/4 C	105	129	5.9	0.17	9.5	**21**	666	463	703	174	214	139	0.41	3.41	0.34	0.08	**6**
			0.95	78.8	7.4	unk.	1.8		213	810	unk.	432	119	370	1.85	0.12	123	9	**15**
S	crab,imperial	1/2 C	110	162	4.3	unk.	16.1	**36**	1331	748	1267	unk.	unk.	202	unk.	unk.	unk.	0.07	**4**
			0.91	79.1	8.4	unk.	unk.		301	1394	unk.	733	185	583	unk.	unk.	154	5	**9**

Riboflavin mg / Niacin mg	% USRDA	Vitamin B_6 mg / Folacin mcg	% USRDA	Vitamin B_{12} mcg / Pantothenic acid mg	% USRDA	Biotin mg / Vitamin A IU	% USRDA	Preformed A RE / Beta carotene RE	Vitamin D IU / Vitamin E IU	% USRDA	Total tocopherol mg / Alpha tocopherol mg	Other tocopherol mg / Total ash g	Calcium mg / Phosphorus mg	% USRDA	Sodium mg / Sodium meq	Potassium mg / Potassium meq	Chlorine mg / Chlorine meq	Iron mg / Magnesium mg	% USRDA	Zinc mg / Copper mg	% USRDA	Iodine mcg / Selenium mcg	% USRDA	Manganese mcg / Chromium mcg	
0.08	**5**	0.22	**11**	0.00	**0**	0.002	**1**	0	0(a)	**0**	0.4	0.3	2	**0**	tr	151	unk.	0.5	**3**	0.3	**2**	unk.	**unk.**	unk.	A
1.1	**5**	2	**0**	0.31	**3**	308	**6**	31	0.4	**1**	0.0	0.46	69	**7**	0.0	3.9	unk.	24	**6**	0.07	**4**	tr(a)		unk.	
0.05	**3**	0.24	**12**	0.00	**0**	0.002	**1**	0	0(a)	**0**	0.4	0.3	2	**0**	1	152	unk.	0.7	**4**	unk.	**unk.**	unk.	**unk.**	0	B
1.2	**6**	2	**0**	0.33	**3**	289	**6**	29	0.5	**2**	0.0	0.41	60	**6**	0.0	3.9	unk.	18	**5**	0.07	**4**	tr(a)		0	
0.58	**34**	unk.	**unk.**	unk.	**unk.**	unk.	**unk.**	unk.	unk.	**unk.**	unk.	unk.	29	**3**	760	283	unk.	1.1	**6**	unk.	**unk.**	unk.	**unk.**	unk.	C
1.5	**8**	unk.	**unk.**	unk.	**unk.**	712	**14**	unk.	unk.	**unk.**	unk.	2.39	unk.	**unk.**	33.1	7.2	unk.	unk.	**unk.**	unk.	**unk.**	unk.		unk.	
0.31	**18**	0.24	**12**	0.56	**9**	0.003	**1**	0	52	**13**	0.9	tr(a)	220	**22**	386	337	13	0.7	**4**	1.1	**8**	48.1	**32**	13	D
1.0	**5**	12	**3**	0.77	**8**	636	**13**	23	1.1	**4**	0.1	2.38	265	**27**	16.8	8.6	0.4	53	**13**	0.24	**12**	1		2	
0.05	**3**	0.05	**2**	0.05	**1**	0.001	**0**	0	4	**1**	1.7	tr(a)	21	**2**	126	47	9	0.3	**2**	0.2	**1**	20.0	**13**	31	E
0.4	**2**	3	**1**	0.14	**1**	66	**1**	5	2.0	**7**	tr	0.64	43	**4**	5.5	1.2	0.3	12	**3**	0.04	**2**	4		5	
0.14	**9**	0.20	**10**	0.24	**4**	0.003	**1**	5	27	**7**	1.1	tr(a)	64	**6**	246	162	15	1.2	**7**	0.8	**5**	9.0	**6**	3	F
1.3	**6**	12	**3**	0.52	**5**	620	**12**	48	1.3	**4**	tr	1.32	139	**14**	10.7	4.1	0.4	42	**11**	0.12	**6**	tr(a)		unk.	
0.12	**7**	unk.	**unk.**	unk.	**unk.**	unk.	**unk.**	unk.	unk.	**unk.**	unk.	unk.	44	**4**	276	75	unk.	0.7	**4**	unk.	**unk.**	unk.	**unk.**	unk.	G
0.06	**3**	unk.	**unk.**	unk.	**unk.**	136	**3**	unk.	unk.	**unk.**	unk.	1.08	62	**6**	12.0	1.9	unk.	unk.	**unk.**	unk.	**unk.**	unk.		unk.	
0.06	**4**	0.02	**1**	0.09	**2**	0.001	**0**	0	8	**2**	0.1	0.0	36	**4**	229	37	10	0.4	**2**	0.3	**2**	32.3	**22**	25	H
0.4	**2**	3	**1**	0.12	**1**	206	**4**	3	0.1	**0**	0.1	1.01	63	**6**	10.0	0.9	0.3	13	**3**	0.04	**2**	3		4	
0.06	**3**	0.02	**1**	0.11	**2**	0.001	**0**	0	8	**2**	0.1	0.0	37	**4**	195	36	14	0.4	**2**	0.2	**2**	28.9	**19**	29	I
0.3	**2**	4	**1**	0.15	**2**	52	**1**	2	0.1	**0**	tr	0.97	66	**7**	8.5	0.9	0.4	18	**4**	0.03	**2**	3		5	
0.44	**26**	unk.	**unk.**	unk.	**unk.**	unk.	**unk.**	44	unk.	**unk.**	unk.	unk.	21	**2**	153	636	unk.	3.9	**22**	unk.	**unk.**	unk.	**unk.**	unk.	J
20.4	**102**	unk.	**unk.**	unk.	**unk.**	209	**4**	6	unk.	**unk.**	unk.	2.55	466	**47**	6.7	16.3	unk.	unk.	**unk.**	unk.	**unk.**	unk.		unk.	
0.35	**21**	0.72	**36**	0.00	**0**	0.008	**3**	0	0	**0**	unk.	unk.	516	**52**	1656	326	unk.	3.8	**21**	2.7	**18**	unk.	**unk.**	0	K
4.7	**24**	76	**19**	1.34	**13**	0	**0**	0	2.7	**9**	2.7	7.10	978	**98**	72.0	8.3	unk.	115	**29**	0.20	**10**	unk.		unk.	
0.37	**22**	0.55	**28**	0.00	**0**	unk.	**unk.**	0	0	**0**	unk.	unk.	543	**54**	1753	226	unk.	4.0	**22**	unk.	**unk.**	unk.	**unk.**	unk.	L
5.0	**25**	34	**9**	1.02	**10**	0	**0**	0	unk.	**unk.**	unk.	6.77	888	**89**	76.2	5.8	unk.	69	**17**	unk.	**unk.**	unk.		unk.	
0.00	**0**	unk.	**unk.**	0(a)	**0**	unk.	**unk.**	0	0	**0**	unk.	unk.	0	**0**	tr	tr	unk.	0.0	**0**	0.0	**0**	unk.	**unk.**	unk.	M
0.0	**0**	unk.	**unk.**	unk.	**unk.**	0	**0**	0	unk.	**unk.**	unk.	0.01	0	**0**	0.0	0.0	unk.	tr	**0**	unk.	**unk.**	unk.		unk.	
0.03	**2**	0.13	**6**	4.26	**71**	unk.	**unk.**	250	unk.	**unk.**	tr(a)	tr(a)	18	**2**	unk.	unk.	unk.	0.3	**2**	1.8	**12**	unk.	**unk.**	unk.	N
1.2	**6**	9	**2**	0.26	**3**	924	**19**	9	0.0	**0**	tr(a)	0.77	75	**7**	unk.	unk.	unk.	unk.	**unk.**	unk.	**unk.**	tr(a)		tr(a)	
0.02	**1**	0.06	**3**	2.13	**36**	unk.	**unk.**	125	unk.	**unk.**	tr(a)	tr(a)	9	**1**	unk.	unk.	unk.	0.2	**1**	0.9	**6**	unk.	**unk.**	unk.	O
0.6	**3**	4	**1**	0.13	**1**	462	**9**	5	0.0	**0**	tr(a)	0.38	37	**4**	unk.	unk.	unk.	7	**2**	unk.	**unk.**	tr(a)		tr(a)	
0.16	**9**	0.18	**9**	6.00	**100**	0.002	**1**	65	28	**7**	tr(a)	tr(a)	95	**10**	711	183	59	0.9	**5**	2.4	**16**	unk.	**unk.**	unk.	P
1.2	**6**	12	**3**	0.36	**4**	572	**11**	10	0.0	**0**	tr(a)	2.00	168	**17**	30.9	4.7	1.7	28	**7**	unk.	**unk.**	tr(a)		tr(a)	
0.05	**3**	0.21	**10**	6.84	**114**	unk.	**unk.**	401	unk.	**unk.**	tr(a)	tr(a)	29	**3**	unk.	unk.	unk.	0.5	**3**	2.9	**20**	unk.	**unk.**	unk.	Q
1.9	**10**	14	**3**	0.41	**4**	1484	**30**	15	0.0	**0**	tr(a)	1.23	120	**12**	unk.	unk.	unk.	unk.	**unk.**	unk.	**unk.**	tr(a)		tr(a)	
0.12	**7**	0.16	**8**	3.75	**62**	0.003	**1**	205	17	**4**	0.3	tr	63	**6**	77	101	35	0.8	**4**	1.9	**13**	13.0	**9**	19	R
1.1	**6**	16	**4**	0.55	**6**	993	**20**	11	0.4	**1**	0.1	1.17	133	**13**	3.3	2.6	1.0	42	**11**	0.06	**3**	tr		3	
0.13	**8**	unk.	**unk.**	unk.	**unk.**	unk.	**unk.**	unk.	unk.	**unk.**	unk.	unk.	66	**7**	801	144	unk.	1.0	**6**	unk.	**unk.**	unk.	**unk.**	unk.	S
1.2	**6**	unk.	**unk.**	unk.	**unk.**	1199	**24**	unk.	unk.	**unk.**	unk.	2.20	183	**18**	34.8	3.7	unk.	unk.	**unk.**	unk.	**unk.**	unk.		unk.	

	FOOD	Portion	Weight in grams / Conversion for 100 g	Kilocalories / H_2O g	Total carbohydrate g / Total fats g	Crude fiber g / Dietary fiber g	Total protein g / Total sugar g	% USRDA	Arginine mg / Histidine mg	Isoleucine mg / Leucine mg	Lysine mg / Methionine mg	Phenylalanine mg / Threonine mg	Valine mg / Tryptophan mg	Cystine mg / Tyrosine mg	Polyunsat. fatty acids g / Monounsat. fatty acids g	Saturated fatty acids g / P/S ratio	Linoleic acid g / Cholesterol mg	Thiamin mg / Ascorbic acid mg	% USRDA
A	CRABMEAT,all varieties,steamed,	1/2 C	63	58	0.3	0(a)	10.8	24	896	504	853	unk.	unk.	136	unk.	unk.	unk.	0.10	7
	flaked,not packed		*1.60*	49.1	1.2	0(a)	0.0		203	938	unk.	493	124	393	unk.	unk.	63	1	2
B	crabmeat,all varieties,steamed,	1/2 C	105	98	0.5	0(a)	18.2	40	1506	846	1433	unk.	unk.	229	unk.	unk.	unk.	0.17	11
	pieces or flaked,packed		*0.95*	82.4	2.0	0(a)	0.0		340	1576	unk.	828	209	659	unk.	unk.	105	2	4
C	crabmeat,all varieties,steamed,	1/2 C	78	72	0.4	0(a)	13.4	30	1111	625	1058	unk.	unk.	169	unk.	unk.	unk.	0.12	8
	pieces,not packed		*1.29*	60.8	1.5	0(a)	0.0		251	1163	unk.	611	154	487	unk.	unk.	77	2	3
D	crabmeat,canned,drained solids,not	1/2 C	68	68	0.7	0(a)	11.7	26	973	547	926	unk.	unk.	148	unk.	unk.	unk.	0.05	4
	packed (white or king)		*1.48*	52.1	1.7	0(a)	0.0		220	1019	unk.	536	135	427	unk.	unk.	68	unk.	unk.
E	crabmeat,canned,drained solids,packed	1/2 C	80	81	0.9	0(a)	13.9	31	1154	648	1098	561	666	176	unk.	unk.	unk.	0.06	4
	(claw; white or king)		*1.25*	61.8	2.0	0(a)	0.0		261	1207	406	635	160	506	unk.	unk.	81	unk.	unk.
F	CRACKER MEAL	1 C	128	538	93.1	0.77	12.3	19	unk.	765	379	914	718	333	unk.	unk.	unk.	0.08	5
			0.78	unk.	12.3	unk.	1.9		333	947	219	479	205	568	unk.	unk.	0(a)	0	0
G	cracker meal-Nabisco	1 C	115	436	93.8	unk.	11.1	17	unk.	unk.	unk.	unk.	unk.	unk.	unk.	unk.	unk.	0.92	61
			0.87	7.7	1.8	unk.	unk.		unk.	unk.	unk.	unk.	unk.	unk.	unk.	unk.	unk.	unk.	unk.
H	CRACKERS,bacon 'n dip snack-Nabisco	17 pieces	28	146	16.2	unk.	2.3	4	unk.	unk.	unk.	unk.	unk.	unk.	unk.	unk.	unk.	0.16	11
			3.52	1.0	7.9	unk.	unk.		unk.	unk.	unk.	unk.	unk.	unk.	unk.	unk.	unk.	unk.	unk.
I	crackers,bacon flavored thins-Nabisco	14 pieces	29	150	16.8	unk.	2.3	4	unk.	unk.	unk.	unk.	unk.	unk.	unk.	unk.	unk.	0.18	12
			3.40	1.0	8.1	unk.	unk.		unk.	unk.	unk.	unk.	unk.	unk.	unk.	unk.	unk.	unk.	unk.
J	crackers,buttery flavored sesame	9 pieces	29	147	16.6	unk.	2.6	4	unk.	unk.	unk.	unk.	unk.	unk.	unk.	unk.	unk.	0.14	10
	snack-Nabisco		*3.47*	1.0	7.8	unk.	unk.		unk.	unk.	unk.	unk.	unk.	unk.	unk.	unk.	unk.	unk.	unk.
K	crackers,cheese artif flavored sandwich	4 pieces	24	113	13.4	unk.	2.4	4	unk.	unk.	unk.	unk.	unk.	unk.	unk.	unk.	unk.	0.12	8
	-Nabisco		*4.22*	1.3	5.6	unk.	unk.		unk.	unk.	unk.	unk.	unk.	unk.	unk.	unk.	unk.	unk.	unk.
L	crackers,cheese Nips naturally	26 pieces	28	136	17.2	unk.	2.6	4	unk.	unk.	unk.	unk.	unk.	unk.	unk.	unk.	unk.	0.21	14
	flavored-Nabisco		*3.56*	1.0	6.3	unk.	unk.		unk.	unk.	unk.	unk.	unk.	unk.	unk.	unk.	unk.	unk.	unk.
M	crackers,cheese peanut artif flavored	4 pieces	28	134	16.8	unk.	3.2	5	unk.	unk.	unk.	unk.	unk.	unk.	unk.	unk.	unk.	0.20	13
	sandwich-Nabisco		*3.52*	1.4	6.0	unk.	unk.		unk.	unk.	unk.	unk.	unk.	unk.	unk.	unk.	unk.	unk.	unk.
N	crackers,cheese swirls parmesan snack	13 pieces	27	134	15.5	unk.	2.7	4	unk.	unk.	unk.	unk.	unk.	unk.	unk.	unk.	unk.	0.18	12
	-Nabisco		*3.66*	1.1	6.8	unk.			unk.	unk.	unk.	unk.	unk.	unk.	unk.	unk.	unk.	unk.	unk.
O	crackers,cheese Tid-Bit artif flavored	32 pieces	29	148	15.6	unk.	2.4	4	unk.	unk.	unk.	unk.	unk.	unk.	unk.	unk.	unk.	0.15	10
	-Nabisco		*3.47*	1.2	8.5	unk.	unk.		unk.	unk.	unk.	unk.	unk.	unk.	unk.	unk.	unk.	unk.	unk.
P	crackers,Chicken In A Biskit flavored	14 pieces	28	149	15.6	unk.	2.0	3	unk.	unk.	unk.	unk.	unk.	unk.	unk.	unk.	unk.	0.11	8
	-Nabisco		*3.57*	1.0	8.7	unk.	unk.		unk.	unk.	unk.	unk.	unk.	unk.	unk.	unk.	unk.	unk.	unk.
Q	crackers,Chippers potato 'n cheese	10 pieces	28	145	16.7	unk.	2.2	3	unk.	unk.	unk.	unk.	unk.	unk.	unk.	unk.	unk.	0.16	11
	snack-Nabisco		*3.52*	0.7	7.7	unk.	unk.		unk.	unk.	unk.	unk.	unk.	unk.	unk.	unk.	unk.	unk.	unk.
R	crackers,country cheddar 'n sesame	16 pieces	29	147	15.6	unk.	3.0	5	unk.	unk.	unk.	unk.	unk.	unk.	unk.	unk.	unk.	0.20	13
	snack-Nabisco		*3.51*	0.9	8.1	unk.	unk.		unk.	unk.	unk.	unk.	unk.	unk.	unk.	unk.	unk.	unk.	unk.
S	crackers,Crown Pilot-Nabisco	2 pieces	34	145	25.5	unk.	2.8	4	unk.	unk.	unk.	unk.	unk.	unk.	unk.	unk.	unk.	0.23	15
			2.92	1.8	3.5	unk.	unk.		unk.	unk.	unk.	unk.	unk.	unk.	unk.	unk.	unk.	0	0

Riboflavin mg / Niacin mg	% USRDA	Vitamin B6 mg / Folacin mcg	% USRDA	Vitamin B12 mcg / Pantothenic acid mg	% USRDA	Biotin mg / Vitamin A IU	% USRDA	Preformed A RE / Beta carotene RE	Vitamin D IU / Vitamin E IU	% USRDA	Total tocopherol mg / Alpha tocopherol mg	Other tocopherol mg / Total ash g	Calcium mg / Phosphorus mg	% USRDA	Sodium mg / Sodium meq	Potassium mg / Potassium meq	Chlorine mg / Chlorine meq	Iron mg / Magnesium mg	% USRDA	Zinc mg / Copper mg	% USRDA	Iodine mcg / Selenium mcg	% USRDA	Manganese mcg / Chromium mcg	
0.05	**3**	0.19	**9**	6.25	**104**	unk.	**unk.**	367	unk.	**unk.**	tr(a)	tr(a)	27	**3**	132	113	unk.	0.5	**3**	2.7	**18**	unk.	**unk.**	unk.	A
1.7	**9**	13	**3**	0.37	**4**	1356	**27**	14	0.0	**0**	tr(a)	1.12	109	**11**	5.8	2.9	unk.	unk.	**unk.**	unk.	**unk.**	tr(a)		tr(a)	
0.08	**5**	0.31	**16**	10.50	**175**	unk.	**unk.**	616	unk.	**unk.**	tr(a)	tr(a)	45	**5**	221	189	unk.	0.8	**5**	4.5	**30**	unk.	**unk.**	unk.	B
2.9	**15**	21	**5**	0.63	**6**	2278	**46**	23	0.0	**0**	tr(a)	1.89	184	**18**	9.6	4.8	unk.	unk.	**unk.**	unk.	**unk.**	tr(a)		tr(a)	
0.06	**4**	0.23	**12**	7.75	**129**	unk.	**unk.**	455	unk.	**unk.**	tr(a)	tr(a)	33	**3**	164	140	unk.	0.6	**3**	3.3	**22**	unk.	**unk.**	unk.	C
2.2	**11**	15	**4**	0.46	**5**	1682	**34**	17	0.0	**0**	tr(a)	1.39	136	**14**	7.1	3.6	unk.	unk.	**unk.**	unk.	**unk.**	tr(a)		tr(a)	
0.05	**3**	0.20	**10**	6.75	**113**	unk.	**unk.**	unk.	unk.	**unk.**	tr(a)	tr(a)	30	**3**	675	74	unk.	0.5	**3**	2.9	**19**	unk.	**unk.**	unk.	D
1.3	**6**	13	**3**	0.40	**4**	741	**15**	unk.	0.0	**0**	tr(a)	1.21	123	**12**	29.4	1.9	unk.	unk.	**unk.**	unk.	**unk.**	tr(a)		tr(a)	
0.06	**4**	0.24	**12**	8.00	**133**	unk.	**unk.**	unk.	unk.	**unk.**	tr(a)	tr(a)	36	**4**	800	88	unk.	0.6	**4**	3.4	**23**	unk.	**unk.**	unk.	E
1.5	**8**	16	**4**	0.48	**5**	872	**17**	unk.	0.0	**0**	tr(a)	1.44	146	**15**	34.8	2.3	unk.	unk.	**unk.**	unk.	**unk.**	tr(a)		tr(a)	
0.06	**4**	unk.	**unk.**	0(a)	**0**	unk.	**unk.**	0(a)	0(a)	**0**	unk.	unk.	26	**3**	unk.	unk.	unk.	1.4	**8**	0.5	**3**	unk.	**unk.**	unk.	F
1.4	**7**	unk.	**unk.**	unk.	**unk.**	0	**0**	0	unk.	**unk.**	unk.	unk.	123	**12**	unk.	unk.	unk.	unk.	**unk.**	unk.	**unk.**	unk.		unk.	
0.59	**35**	unk.	**unk.**	unk.	**unk.**	unk.	**unk.**	unk.	0(a)	**0**	unk.	unk.	26	**3**	8	147	unk.	4.5	**25**	0.8	**5**	unk.	**unk.**	746	G
6.8	**34**	unk.	**unk.**	unk.	**unk.**	unk.	**unk.**	unk.	unk.	**unk.**	unk.	0.59	119	**12**	0.3	3.8	unk.	29	**7**	0.22	**11**	unk.		unk.	
0.12	**7**	unk.	**unk.**	unk.	**unk.**	unk.	**unk.**	unk.	0(a)	**0**	unk.	unk.	19	**2**	307	48	unk.	0.8	**5**	0.2	**2**	unk.	**unk.**	123	H
1.3	**6**	unk.	**unk.**	unk.	**unk.**	unk.	**unk.**	unk.	unk.	**unk.**	unk.	0.89	35	**4**	13.3	1.2	unk.	6	**2**	0.03	**2**	unk.		unk.	
0.11	**7**	unk.	**unk.**	unk.	**unk.**	unk.	**unk.**	unk.	0(a)	**0**	unk.	unk.	14	**1**	366	43	unk.	0.9	**5**	0.2	**1**	unk.	**unk.**	154	I
1.2	**6**	unk.	**unk.**	unk.	**unk.**	unk.	**unk.**	unk.	unk.	**unk.**	unk.	1.11	32	**3**	15.9	1.1	unk.	6	**2**	0.04	**2**	unk.		unk.	
0.10	**6**	unk.	**unk.**	unk.	**unk.**	unk.	**unk.**	unk.	0(a)	**0**	unk.	unk.	36	**4**	232	40	unk.	1.0	**5**	0.4	**3**	unk.	**unk.**	154	J
1.4	**7**	unk.	**unk.**	unk.	**unk.**	unk.	**unk.**	unk.	unk.	**unk.**	unk.	0.82	86	**9**	10.1	1.0	unk.	19	**5**	0.12	**6**	unk.		unk.	
0.12	**7**	unk.	**unk.**	unk.	**unk.**	unk.	**unk.**	unk.	0(a)	**0**	unk.	unk.	53	**5**	292	60	unk.	0.7	**4**	0.2	**2**	unk.	**unk.**	129	K
0.9	**5**	unk.	**unk.**	unk.	**unk.**	unk.	**unk.**	unk.	unk.	**unk.**	unk.	1.07	96	**10**	12.7	1.5	unk.	7	**2**	0.19	**9**	unk.		unk.	
0.14	**8**	unk.	**unk.**	unk.	**unk.**	unk.	**unk.**	unk.	0(a)	**0**	unk.	unk.	30	**3**	298	55	unk.	0.9	**5**	0.2	**1**	unk.	**unk.**	138	L
1.6	**8**	unk.	**unk.**	unk.	**unk.**	unk.	**unk.**	unk.	unk.	**unk.**	unk.	0.96	66	**7**	13.0	1.4	unk.	8	**2**	0.06	**3**	unk.		unk.	
0.10	**6**	unk.	**unk.**	unk.	**unk.**	unk.	**unk.**	unk.	0(a)	**0**	unk.	unk.	40	**4**	286	60	unk.	0.8	**5**	0.3	**2**	unk.	**unk.**	267	M
1.6	**8**	unk.	**unk.**	unk.	**unk.**	unk.	**unk.**	unk.	unk.	**unk.**	unk.	1.02	89	**9**	12.4	1.5	unk.	13	**3**	0.07	**4**	unk.		unk.	
0.12	**7**	unk.	**unk.**	unk.	**unk.**	unk.	**unk.**	unk.	0(a)	**0**	unk.	unk.	23	**2**	350	27	unk.	0.9	**5**	0.2	**1**	unk.	**unk.**	136	N
1.4	**7**	unk.	**unk.**	unk.	**unk.**	unk.	**unk.**	unk.	unk.	**unk.**	unk.	1.13	72	**7**	15.2	0.7	unk.	7	**2**	0.04	**2**	unk.		unk.	
0.12	**7**	unk.	**unk.**	unk.	**unk.**	unk.	**unk.**	unk.	0(a)	**0**	unk.	unk.	46	**5**	384	41	unk.	0.8	**4**	0.2	**2**	unk.	**unk.**	130	O
1.1	**6**	unk.	**unk.**	unk.	**unk.**	unk.	**unk.**	unk.	unk.	**unk.**	unk.	1.19	72	**7**	16.7	1.0	unk.	6	**2**	0.05	**3**	unk.		unk.	
0.10	**6**	unk.	**unk.**	unk.	**unk.**	unk.	**unk.**	unk.	0(a)	**0**	unk.	unk.	9	**1**	245	41	unk.	0.7	**4**	0.2	**1**	unk.	**unk.**	123	P
1.1	**5**	unk.	**unk.**	unk.	**unk.**	unk.	**unk.**	unk.	unk.	**unk.**	unk.	0.71	25	**3**	10.6	1.0	unk.	6	**2**	0.02	**1**	unk.		unk.	
0.10	**6**	unk.	**unk.**	unk.	**unk.**	unk.	**unk.**	unk.	0(a)	**0**	unk.	unk.	22	**2**	334	59	unk.	0.7	**4**	0.2	**1**	unk.	**unk.**	111	Q
1.2	**6**	unk.	**unk.**	unk.	**unk.**	unk.	**unk.**	unk.	unk.	**unk.**	unk.	1.02	45	**5**	14.5	1.5	unk.	8	**2**	0.04	**2**	unk.		unk.	
0.13	**7**	unk.	**unk.**	unk.	**unk.**	unk.	**unk.**	unk.	0(a)	**0**	unk.	unk.	26	**3**	305	43	unk.	0.8	**5**	0.3	**2**	unk.	**unk.**	125	R
1.4	**7**	unk.	**unk.**	unk.	**unk.**	unk.	**unk.**	unk.	unk.	**unk.**	unk.	0.99	54	**5**	13.3	1.1	unk.	11	**3**	0.07	**4**	unk.		unk.	
0.14	**8**	unk.	**unk.**	unk.	**unk.**	unk.	**unk.**	unk.	0(a)	**0**	unk.	unk.	8	**1**	177	45	unk.	1.3	**7**	0.2	**2**	unk.	**unk.**	212	S
1.9	**9**	unk.	**unk.**	unk.	**unk.**	unk.	**unk.**	unk.	unk.	**unk.**	unk.	0.55	30	**3**	7.7	1.1	unk.	7	**2**	0.04	**2**	unk.		unk.	

	FOOD	Portion	Weight in grams / *Conversion for 100 g*	Kilocalories / H_2O g	Total carbohydrate g / Total fats g	Crude fiber g / Dietary fiber g	Total protein g / Total sugar g	% USRDA	Arginine mg / Histidine mg	Isoleucine mg / Leucine mg	Lysine mg / Methionine mg	Phenylalanine mg / Threonine mg	Valine mg / Tryptophan mg	Cystine mg / Tyrosine mg	Polyunsat. fatty acids g / Monounsat. fatty acids g	Saturated fatty acids g / P/S ratio	Linoleic acid g / Cholesterol mg	Thiamin mg / Ascorbic acid mg	% USRDA
A	crackers,crumbs,graham-Keebler	1 C	128	563	97.3	unk.	9.0	**14**	unk.	438	191	467	404	169	unk.	unk.	unk.	0.12	**8**
			0.78	5.1	15.4	unk.	unk.		169	572	138	200	106	285	unk.	unk.	0(a)	0(a)	**0**
B	crackers,dairy wafers round-Nabisco	1 pkg of 2	23	93	17.2	unk.	1.4	**2**	unk.	unk.	unk.	unk.	unk.	unk.	unk.	unk.	unk.	0.09	**6**
			4.42	1.6	2.0	unk.	unk.		unk.	unk.	unk.	unk.	unk.	unk.	unk.	unk.	unk.	unk.	**unk.**
C	crackers,Dip In A Chip cheese 'n	15 pieces	28	148	16.0	unk.	2.0	**3**	unk.	unk.	unk.	unk.	unk.	unk.	unk.	unk.	unk.	0.13	**9**
	chive snack-Nabisco		*3.52*	1.0	8.4	unk.	unk.		unk.	unk.	unk.	unk.	unk.	unk.	unk.	unk.	unk.	unk.	**unk.**
D	crackers,Dixies drumstick snack	18 pieces	28	140	16.8	unk.	2.0	**3**	unk.	unk.	unk.	unk.	unk.	unk.	unk.	unk.	unk.	0.15	**10**
	-Nabisco		*3.53*	1.2	7.2	unk.	unk.		unk.	unk.	unk.	unk.	unk.	unk.	unk.	unk.	unk.	unk.	**unk.**
E	crackers,Doo Dads mixed snacks-Nabisco	57 pieces	28	136	17.7	unk.	2.8	**4**	unk.	unk.	unk.	unk.	unk.	unk.	unk.	unk.	unk.	0.10	**6**
			3.52	0.9	6.3	unk.	unk.		unk.	unk.	unk.	unk.	unk.	unk.	unk.	unk.	unk.	unk.	**unk.**
F	crackers,Escort-Nabisco	7 pieces	29	152	17.7	unk.	2.3	**4**	unk.	unk.	unk.	unk.	unk.	unk.	unk.	unk.	unk.	0.13	**9**
			3.40	0.8	7.9	unk.	unk.		unk.	unk.	unk.	unk.	unk.	unk.	unk.	unk.	unk.	unk.	**unk.**
G	crackers,French onion-Nabisco	12 pieces	29	145	17.7	unk.	2.0	**3**	unk.	unk.	unk.	unk.	unk.	unk.	unk.	unk.	unk.	0.14	**10**
			3.47	1.0	7.4	unk.	unk.		unk.	unk.	unk.	unk.	unk.	unk.	unk.	unk.	unk.	unk.	**unk.**
H	crackers,Gitana soda-Nabisco	8 pieces	28	115	20.5	unk.	2.7	**4**	unk.	unk.	unk.	unk.	unk.	unk.	unk.	unk.	unk.	0.08	**5**
			3.62	1.3	2.5	unk.	unk.		unk.	unk.	unk.	unk.	unk.	unk.	unk.	unk.	unk.	unk.	**unk.**
I	crackers,graham	4 pieces	28	108	20.5	0.31	2.2	**3**	unk.	116	50	123	88	45	unk.	0.64	0.63	0.01	**1**
			3.57	1.8	2.6	unk.	unk.		45	152	37	53	28	76	1.26	unk.	0(a)	0	**0**
J	crackers,graham-Nabisco	4 pieces	28	119	21.2	unk.	2.2	**3**	unk.	unk.	unk.	unk.	unk.	unk.	unk.	unk.	unk.	0.05	**3**
			3.57	1.3	2.8	unk.	unk.		unk.	unk.	unk.	unk.	unk.	unk.	unk.	unk.	unk.	unk.	**unk.**
K	crackers,graham,chocolate covered	2 pieces	26	123	17.7	0.21	1.3	**2**	unk.	69	38	73	52	27	unk.	1.82	0.39	0.02	**1**
			3.85	0.5	6.1	unk.	unk.		27	90	22	31	17	45	3.74	unk.	tr(a)	0	**0**
L	crackers,graham,Honey Maid-Nabisco	4 pieces	28	119	21.6	unk.	1.9	**3**	unk.	unk.	unk.	unk.	unk.	unk.	unk.	unk.	unk.	0.04	**3**
			3.57	1.2	2.7	unk.	unk.		unk.	unk.	unk.	unk.	unk.	unk.	unk.	unk.	unk.	unk.	**unk.**
M	crackers,graham,sugar honey-Fireside	3 pieces	28	119	21.6	unk.	2.0	**3**	unk.	unk.	unk.	unk.	unk.	unk.	unk.	unk.	unk.	0.08	**5**
			3.57	0.8	2.8	unk.	unk.		unk.	unk.	unk.	unk.	unk.	unk.	unk.	unk.	unk.	unk.	**unk.**
N	crackers,matzo,unsalted,Manischewitz	1 cracker	30	119	25.2	0.10	3.0	**5**	unk.	119	59	143	112	52	unk.	unk.	unk.	0.00	**0**
			3.33	1.2	0.3	unk.	unk.		52	200	34	75	32	89	unk.	unk.	0(a)	0	**0**
O	crackers,Meal Mates sesame bread	6 pieces	28	129	18.5	unk.	2.7	**4**	unk.	unk.	unk.	unk.	unk.	unk.	unk.	unk.	unk.	0.30	**20**
	wafers-Nabisco		*3.55*	1.0	4.9	unk.	unk.		unk.	unk.	unk.	unk.	unk.	unk.	unk.	unk.	unk.	unk.	**unk.**
P	crackers,melba snacks-Nabisco	1 pkg of 2	4.7	18	3.4	unk.	0.6	**1**	unk.	unk.	unk.	unk.	unk.	unk.	unk.	unk.	unk.	0.02	**1**
			21.28	0.2	0.2	unk.	unk.		unk.	unk.	unk.	unk.	unk.	unk.	unk.	unk.	unk.	unk.	**unk.**
Q	crackers,melba toast,vending	1 pkg of 2	8.1	31	5.9	unk.	1.2	**2**	unk.	unk.	unk.	unk.	unk.	unk.	unk.	unk.	unk.	0.04	**3**
	-Nabisco		*12.35*	0.4	0.3	unk.	unk.		unk.	unk.	unk.	unk.	unk.	unk.	unk.	unk.	0(a)	0	**0**
R	crackers,melba variety pack,garlic,	1 pkg of 2	5.3	21	3.8	unk.	0.7	**1**	unk.	unk.	unk.	unk.	unk.	unk.	unk.	unk.	unk.	0.02	**1**
	vending-Nabisco		*18.87*	0.2	0.3	unk.	unk.		unk.	unk.	unk.	unk.	unk.	unk.	unk.	unk.	unk.	unk.	**unk.**
S	crackers,melba variety pack,onion,	1 pkg of 2	4.9	20	3.4	unk.	0.8	**1**	unk.	unk.	unk.	unk.	unk.	unk.	unk.	unk.	unk.	0.03	**2**
	vending-Nabisco		*20.41*	0.2	0.3	unk.	unk.		unk.	unk.	unk.	unk.	unk.	unk.	unk.	unk.	unk.	unk.	**unk.**

Riboflavin mg / Niacin mg	% USRDA	Vitamin B6 mg / Folacin mcg	% USRDA	Vitamin B12 mcg / Pantothenic acid mg	% USRDA	Biotin mg / Vitamin A IU	% USRDA	Preformed A RE / Beta carotene RE	Vitamin D IU / Vitamin E IU	% USRDA	Total tocopherol mg / Alpha tocopherol mg	Other tocopherol mg / Total ash g	Calcium mg / Phosphorus mg	% USRDA	Sodium mg / Sodium meq	Potassium mg / Potassium meq	Chlorine mg / Chlorine meq	Iron mg / Magnesium mg	% USRDA	Zinc mg / Copper mg	% USRDA	Iodine mcg / Selenium mcg	% USRDA	Manganese mcg / Chromium mcg	
0.38	23	unk.	unk.	0(a)	0	unk.	unk.	0(a)	0(a)	0	unk.	unk.	31	3	781	260	unk.	4.4	24	1.4	9	unk.	unk.	unk.	A
4.9	24	unk.	unk.	unk.	unk.	0(a)	0	0(a)	unk.	unk.	unk.	2.94	unk.	unk.	34.0	6.6	unk.	19	5	unk.	unk.	unk.		unk.	
0.06	4	unk.	unk.	unk.	unk.	unk.	unk.	unk.	0(a)	0	unk.	unk.	4	0	101	33	unk.	0.7	4	0.1	1	unk.	unk.	unk.	B
0.7	4	unk.	unk.	unk.	unk.	unk.	unk.	unk.	unk.	unk.	unk.	0.32	21	2	4.4	0.9	unk.	6	1	0.05	3	unk.		unk.	
0.11	7	unk.	unk.	unk.	unk.	unk.	unk.	unk.	0(a)	0	unk.	unk.	32	3	245	81	unk.	0.8	5	0.2	2	unk.	unk.	131	C
1.2	6	unk.	unk.	unk.	unk.	unk.	unk.	unk.	unk.	unk.	unk.	0.95	43	4	10.6	2.1	unk.	8	2	0.05	2	unk.		unk.	
0.11	6	unk.	unk.	unk.	unk.	unk.	unk.	unk.	0(a)	0	unk.	unk.	11	1	315	40	unk.	0.8	4	0.2	1	unk.	unk.	120	D
1.1	5	unk.	unk.	unk.	unk.	unk.	unk.	unk.	unk.	unk.	unk.	0.96	36	4	13.7	1.0	unk.	5	1	0.03	2	unk.		unk.	
0.09	5	unk.	unk.	unk.	unk.	unk.	unk.	unk.	0(a)	0	unk.	unk.	21	2	375	71	unk.	0.8	5	0.4	3	unk.	unk.	301	E
1.6	8	unk.	unk.	unk.	unk.	unk.	unk.	unk.	unk.	unk.	unk.	1.21	68	7	16.3	1.8	unk.	18	4	1.05	52	unk.		unk.	
0.10	6	unk.	unk.	unk.	unk.	unk.	unk.	unk.	0(a)	0	unk.	unk.	8	1	220	35	unk.	0.8	4	0.2	1	unk.	unk.	154	F
1.3	6	unk.	unk.	unk.	unk.	unk.	unk.	unk.	unk.	unk.	unk.	0.69	28	3	9.5	0.9	unk.	6	1	0.03	2	unk.		unk.	
0.10	6	unk.	unk.	unk.	unk.	unk.	unk.	unk.	0(a)	0	unk.	unk.	44	4	239	47	unk.	0.9	5	0.2	1	unk.	unk.	169	G
1.2	6	unk.	unk.	unk.	unk.	unk.	unk.	unk.	unk.	unk.	unk.	0.80	70	7	10.4	1.2	unk.	8	2	0.06	3	unk.		unk.	
0.14	8	unk.	unk.	unk.	unk.	unk.	unk.	unk.	0(a)	0	unk.	unk.	6	1	227	38	unk.	1.1	6	0.2	1	unk.	unk.	196	H
1.6	8	unk.	unk.	unk.	unk.	unk.	unk.	unk.	unk.	unk.	unk.	0.62	30	3	9.9	1.0	unk.	6	2	0.04	2	unk.		unk.	
0.06	4	unk.	unk.	0(a)	0	unk.	unk.	0	0(a)	0	unk.	unk.	11	1	188	108	unk.	0.4	2	0.3	2	unk.	unk.	120	I
0.4	2	unk.	unk.	0.13	1	0	0	0	unk.	unk.	unk.	0.81	42	4	8.2	2.8	unk.	5	1	0.06	3	unk.		17	
0.11	7	unk.	unk.	unk.	unk.	unk.	unk.	unk.	0(a)	0	unk.	unk.	10	1	204	61	unk.	1.3	7	0.2	2	unk.	unk.	202	J
1.4	7	unk.	unk.	0.13	1	unk.	unk.	unk.	unk.	unk.	unk.	0.57	36	4	8.9	1.6	unk.	13	3	0.09	5	unk.		unk.	
0.07	4	unk.	unk.	0(a)	0	unk.	unk.	0(a)	0(a)	0	unk.	unk.	29	3	106	83	unk.	0.7	4	unk.	unk.	unk.	unk.	unk.	K
0.3	2	unk.	unk.	unk.	unk.	16	0	unk.	unk.	unk.	unk.	0.42	53	5	4.6	2.1	unk.	11	3	unk.	unk.	unk.		unk.	
0.10	6	unk.	unk.	unk.	unk.	unk.	unk.	unk.	0(a)	0	unk.	unk.	6	1	176	38	unk.	0.9	5	0.2	2	unk.	unk.	230	L
1.3	7	unk.	unk.	0.13	1	unk.	unk.	unk.	unk.	unk.	unk.	0.51	28	3	7.6	1.0	unk.	9	2	0.06	3	unk.		unk.	
0.13	8	unk.	unk.	unk.	unk.	unk.	unk.	unk.	0(a)	0	unk.	unk.	7	1	206	43	unk.	1.3	8	0.2	2	unk.	unk.	162	M
1.3	7	unk.	unk.	0.13	1	unk.	unk.	unk.	unk.	unk.	unk.	0.77	30	3	9.0	1.1	unk.	unk.	unk.	0.07	4	unk.		unk.	
0.00	0	unk.	unk.	0(a)	0	unk.	unk.	0(a)	0(a)	0	unk.	unk.	0	0	tr	unk.	unk.	0.0	0	unk.	unk.	unk.	unk.	unk.	N
0.0	0	unk.	unk.	unk.	unk.	0(a)	0	0(a)	unk.	unk.	unk.	unk.	0	0	0.0	unk.	unk.	unk.	unk.	unk.	unk.	unk.		unk.	
0.17	10	unk.	unk.	unk.	unk.	unk.	unk.	unk.	0(a)	0	unk.	unk.	58	6	295	60	unk.	0.9	5	0.4	3	unk.	unk.	299	O
1.5	8	unk.	unk.	unk.	unk.	unk.	unk.	unk.	unk.	unk	unk.	1.05	92	9	12.8	1.5	unk.	19	5	0.13	6	unk.		unk.	
0.01	1	unk.	unk.	unk.	unk.	unk.	unk.	unk.	0(a)	0	unk.	unk.	3	0	63	12	unk.	0.1	1	0.2	1	unk.	unk.	unk.	P
0.2	1	unk.	unk.	unk.	unk.	unk.	unk.	unk.	unk.	unk.	unk.	0.19	7	1	2.8	0.3	unk.	3	1	0.17	9	unk.		unk.	
0.02	1	unk.	unk.	0(a)	0	unk.	unk.	unk.	unk.	unk.	unk.	unk.	6	1	77	21	unk.	0.2	1	0.3	2	unk.	unk.	unk.	Q
0.3	2	unk.	unk.	unk.	unk.	unk.	unk.	unk.	unk.	unk.	unk.	0.24	16	2	3.3	0.5	unk.	6	1	0.21	10	unk.		unk.	
0.01	1	unk.	unk.	unk.	unk.	unk.	unk.	unk.	0(a)	0	unk.	unk.	4	0	72	14	unk.	0.2	1	0.3	2	unk.	unk.	unk.	R
0.2	1	unk.	unk.	unk.	unk.	unk.	unk.	unk.	unk.	unk.	unk.	0.21	9	1	3.1	0.4	unk.	3	1	0.15	7	unk.		unk.	
0.02	1	unk.	unk.	unk.	unk.	unk.	unk.	unk.	0(a)	0	unk.	unk.	6	1	59	15	unk.	0.2	1	0.2	1	unk.	unk.	unk.	S
0.3	1	unk.	unk.	unk.	unk.	unk.	unk.	unk.	unk.	unk.	unk.	0.22	10	1	2.6	0.4	unk.	4	1	0.15	8	unk.		unk.	

	FOOD	Portion	Weight in grams / Conversion for 100 g	Kilocalories / H_2O g	Total carbohydrate g / Total fats g	Crude fiber g / Dietary fiber g	Total protein g / Total sugar g	% **USRDA**	Arginine mg / Histidine mg	Isoleucine mg / Leucine mg	Lysine mg / Methionine mg	Phenylalanine mg / Threonine mg	Valine mg / Tryptophan mg	Cystine mg / Tyrosine mg	Polyunsat. fatty acids g / Monounsat. fatty acids g	Saturated fatty acids g / P/S ratio	Linoleic acid g / Cholesterol mg	Thiamin mg / Ascorbic acid mg	% **USRDA**
A	crackers,melba variety pack,plain,	1 pkg of 2	4.7	18	3.4	unk.	0.6	**1**	unk.	unk.	unk.	unk.	unk.	unk.	unk.	unk.	unk.	0.02	**1**
	vending-Nabisco		*21.28*	0.2	0.2	unk.	unk.		unk.	unk.	unk.	unk.	unk.	unk.	unk.	unk.	unk.	unk.	**unk.**
B	crackers,melba variety pack,sesame,	1 pkg of 2	5.5	24	3.4	unk.	0.9	**1**	unk.	unk.	unk.	unk.	unk.	unk.	unk.	unk.	unk.	0.03	**2**
	vending-Nabisco		*18.18*	0.2	0.7	unk.	unk.		unk.	unk.	unk.	unk.	unk.	unk.	unk.	unk.	unk.	unk.	**unk.**
C	crackers,oyster-Nabisco	37 pieces	30	124	21.5	tr(a)	2.9	**4**	unk.	121	60	145	114	53	unk.	unk.	unk.	0.12	**8**
			3.33	1.4	3.0	unk.	unk.		53	203	35	76	32	90	unk.	unk.	0(a)	0	**0**
D	crackers,Oysterettes soup & oyster	36 pieces	28	114	20.3	unk.	2.6	**4**	unk.	unk.	unk.	unk.	unk.	unk.	unk.	unk.	unk.	0.14	**10**
	-Nabisco		*3.58*	1.5	2.4	unk.	unk.		unk.	unk.	unk.	unk.	unk.	unk.	unk.	unk.	unk.	unk.	**unk.**
E	crackers,peanut butter/cheese	4 crackers	28	137	15.7	0.14	4.3	**7**	unk.	unk.	unk.	unk.	unk.	unk.	unk.	1.76	1.57	0.01	**1**
	sandwich		*3.57*	0.7	6.7	unk.	unk.		unk.	unk.	unk.	unk.	unk.	unk.	3.08	unk.	unk.	0	**0**
F	crackers,potato Chipsters	57 pieces	29	130	18.5	unk.	1.4	**2**	unk.	unk.	unk.	unk.	unk.	unk.	unk.	unk.	unk.	0.04	**3**
	light 'n crisp snacks-Nabisco		*3.51*	1.2	5.6	unk.	unk.		unk.	unk.	unk.	unk.	unk.	unk.	unk.	unk.	unk.	unk.	**unk.**
G	crackers,Premium,unsalted	10 pieces	28	122	19.9	unk.	2.7	**4**	unk.	unk.	unk.	unk.	unk.	unk.	unk.	unk.	unk.	0.19	**13**
	tops-Nabisco		*3.57*	1.2	3.5	unk.	unk.		unk.	unk.	unk.	unk.	unk.	unk.	unk.	unk.	unk.	unk.	**unk.**
H	crackers,Ritz-Nabisco	9 pieces	30	149	17.9	unk.	2.1	**3**	unk.	91	45	109	86	unk.	unk.	unk.	unk.	0.11	**8**
			3.37	1.2	7.7	unk.	unk.		40	153	26	57	24	68	unk.	unk.	0(a)	0	**0**
I	crackers,Royal Lunch milk-Nabisco	2 pieces	23	106	16.1	unk.	1.8	**3**	unk.	unk.	unk.	unk.	unk.	unk.	unk.	unk.	unk.	0.13	**9**
			4.27	1.2	3.8	unk.	unk.		unk.	unk.	unk.	unk.	unk.	unk.	unk.	unk.	unk.	unk.	**unk.**
J	crackers,rusk (Holland)	3 pieces	30	123	21.9	0.06	4.0	**6**	unk.	unk.	unk.	unk.	unk.	unk.	unk.	0.78	0.48	0.13	**9**
			3.33	1.6	2.2	unk.	unk.		unk.	unk.	unk.	unk.	unk.	unk.	1.17	unk.	unk.	tr	**0**
K	crackers,rusketts	9 pieces	27	113	19.2	0.05	3.7	**6**	unk.	unk.	unk.	unk.	unk.	unk.	unk.	0.70	0.43	0.02	**1**
			3.70	1.3	2.3	unk.	unk.		unk.	unk.	unk.	unk.	unk.	unk.	1.05	unk.	unk.	tr	**0**
L	crackers,rye wafer,seasoned-Nabisco	1 pkg of 2	11	45	8.7	unk.	1.3	**2**	unk.	unk.	unk.	unk.	unk.	unk.	unk.	unk.	unk.	0.03	**2**
			8.77	0.8	0.5	unk.	unk.		unk.	unk.	unk.	unk.	unk.	unk.	unk.	unk.	unk.	unk.	**unk.**
M	crackers,Rykrisp-Ralston	4 pieces	25	91	19.5	0.48	2.6	**4**	unk.	110	106	122	135	52	unk.	unk.	unk.	0.11	**7**
			3.97	1.3	0.2	3.28	unk.		59	174	46	96	29	84	unk.	unk.	0(a)	0	**0**
N	crackers,saltine	10 pieces	28	121	20.0	0.11	2.5	**4**	unk.	116	59	137	109	50	unk.	0.81	0.81	tr	**0**
			3.57	1.2	3.4	unk.	unk.		50	193	34	73	31	87	1.62	unk.	0(a)	0	**0**
O	crackers,saltine-Fireside	10 pieces	30	126	21.7	unk.	2.8	**4**	unk.	unk.	unk.	unk.	unk.	unk.	unk.	unk.	unk.	0.20	**13**
			3.29	1.5	3.1	unk.	unk.		unk.	unk.	unk.	unk.	unk.	unk.	unk.	unk.	unk.	unk.	**unk.**
P	crackers,saltine,Premium-Nabisco	10 pieces	28	117	19.7	unk.	2.6	**4**	unk.	unk.	unk.	unk.	unk.	unk.	unk.	unk.	unk.	0.17	**12**
			3.57	1.1	3.4	unk.	unk.		unk.	unk.	unk.	unk.	unk.	unk.	unk.	unk.	unk.	unk.	**unk.**
Q	crackers,saltine,Premium,crushed	1 C	84	350	59.0	unk.	7.9	**12**	unk.	unk.	unk.	unk.	unk.	unk.	unk.	unk.	unk.	0.52	**35**
	-Nabisco		*1.19*	3.3	10.2	unk.	unk.		unk.	unk.	unk.	unk.	unk.	unk.	unk.	unk.	unk.	unk.	**unk.**
R	crackers,Sea Rounds-Nabisco	2 pieces	21	89	15.1	unk.	2.0	**3**	unk.	unk.	unk.	unk.	unk.	unk.	unk.	unk.	unk.	0.26	**18**
			4.69	1.3	2.4	unk.	unk.		unk.	unk.	unk.	unk.	unk.	unk.	unk.	unk.	unk.	unk.	**unk.**
S	crackers,snack & party-Fireside	9 pieces	30	141	19.1	unk.	2.1	**3**	unk.	unk.	unk.	unk.	unk.	unk.	unk.	unk.	unk.	0.14	**9**
			3.38	1.4	6.3	unk.	unk.		unk.	unk.	unk.	unk.	unk.	unk.	unk.	unk.	unk.	unk.	**unk.**

	Riboflavin mg / Niacin mg	% USRDA	Vitamin B_6 mg / Folacin mcg	% USRDA	Vitamin B_{12} mcg / Pantothenic acid mg	% USRDA	Biotin mg / Vitamin A IU	% USRDA	Preformed A RE / Beta carotene RE	Vitamin D IU / Vitamin E IU	% USRDA	Total tocopherol mg / Alpha tocopherol mg	Other tocopherol mg / Total ash g	Calcium mg / Phosphorus mg	% USRDA	Sodium mg / Sodium meq	Potassium mg / Potassium meq	Chlorine mg / Chlorine meq	Iron mg / Magnesium mg	% USRDA	Zinc mg / Copper mg	% USRDA	Iodine mcg / Selenium mcg	% USRDA	Manganese mcg / Chromium mcg
A	0.01	**1**	unk.	**unk.**	unk.	**unk.**	unk.	**unk.**	unk.	0(a)	**0**	unk.	unk.	3	**0**	63	12	unk.	0.1	**1**	0.2	**1**	unk.	**unk.**	unk.
	0.2	**1**	unk.	**unk.**	unk.	**unk.**	unk.	**unk.**	unk.	unk.	**unk.**	unk.	0.19	7	**1**	2.8	0.3	unk.	3	**1**	0.17	**9**	unk.		unk.
B	0.02	**1**	unk.	**unk.**	unk.	**unk.**	unk.	**unk.**	unk.	0(a)	**0**	unk.	unk.	11	**1**	57	17	unk.	0.2	**1**	0.3	**2**	unk.	**unk.**	unk.
	0.3	**2**	unk.	**unk.**	unk.	**unk.**	unk.	**unk.**	unk.	unk.	**unk.**	unk.	0.21	15	**2**	2.5	0.4	unk.	6	**2**	0.23	**12**	unk.		unk.
C	0.16	**9**	unk.	**unk.**	0(a)	**0**	unk.	**unk.**	0	tr(a)	**0**	unk.	unk.	8	**1**	484	45	unk.	1.3	**7**	0.3	**2**	unk.	**unk.**	208
	1.8	**9**	unk.	**unk.**	unk.	**unk.**	0	**0**	0	unk.	**unk.**	unk.	1.27	34	**3**	21.0	1.2	unk.	8	**2**	0.07	**4**	unk.		unk.
D	0.11	**7**	unk.	**unk.**	unk.	**unk.**	unk.	**unk.**	unk.	0(a)	**0**	unk.	unk.	7	**1**	285	40	unk.	1.1	**6**	0.2	**2**	unk.	**unk.**	unk.
	1.5	**7**	unk.	**unk.**	unk.	**unk.**	unk.	**unk.**	unk.	unk.	**unk.**	unk.	1.06	32	**3**	12.4	1.0	unk.	7	**2**	tr	**0**	unk.		unk.
E	0.02	**1**	unk.	**unk.**	tr(a)	**0**	unk.	**unk.**	tr(a)	0(a)	**0**	unk.	unk.	16	**2**	278	63	unk.	0.2	**1**	unk.	**unk.**	unk.	**unk.**	unk.
	1.0	**5**	unk.	**unk.**	unk.	**unk.**	11	**0**	unk.	unk.	**unk.**	unk.	0.67	50	**5**	12.1	1.6	unk.	unk.	**unk.**	unk.	**unk.**	unk.		unk.
F	0.02	**1**	unk.	**unk.**	unk.	**unk.**	unk.	**unk.**	unk.	0(a)	**0**	unk.	unk.	12	**1**	568	182	unk.	0.7	**4**	0.1	**1**	unk.	**unk.**	69
	0.9	**4**	unk.	**unk.**	unk.	**unk.**	unk.	**unk.**	unk.	unk.	**unk.**	unk.	1.80	39	**4**	24.7	4.7	unk.	12	**3**	0.06	**3**	unk.		unk.
G	0.13	**7**	unk.	**unk.**	unk.	**unk.**	unk.	**unk.**	unk.	0(a)	**0**	unk.	unk.	45	**5**	215	39	unk.	1.4	**8**	0.2	**2**	unk.	**unk.**	204
	1.5	**8**	unk.	**unk.**	0.08	**1**	unk.	**unk.**	unk.	unk.	**unk.**	unk.	0.71	31	**3**	9.3	1.0	unk.	8	**2**	0.06	**3**	unk.		unk.
H	0.10	**6**	unk.	**unk.**	0(a)	**0**	unk.	**unk.**	tr(a)	0(a)	**0**	unk.	unk.	41	**4**	228	30	unk.	0.8	**5**	0.2	**1**	unk.	**unk.**	153
	1.2	**6**	unk.	**unk.**	unk.	**unk.**	unk.	**unk.**	unk.	0.2	**1**	0.2	0.78	75	**7**	9.9	0.8	unk.	6	**1**	0.05	**3**	unk.		unk.
I	0.10	**6**	unk.	**unk.**	unk.	**unk.**	unk.	**unk.**	unk.	0(a)	**0**	unk.	unk.	40	**4**	140	27	unk.	0.7	**4**	0.2	**1**	unk.	**unk.**	124
	1.1	**5**	unk.	**unk.**	unk.	**unk.**	unk.	**unk.**	unk.	unk.	**unk.**	unk.	0.51	70	**7**	6.1	0.7	unk.	5	**1**	0.05	**3**	unk.		unk.
J	0.13	**7**	0.03	**1**	tr(a)	**0**	unk.	**unk.**	0	tr(a)	**0**	unk.	unk.	8	**1**	76	73	unk.	0.8	**4**	0.3	**2**	unk.	**unk.**	121
	1.5	**8**	unk.	**unk.**	unk.	**unk.**	69	**1**	7	unk.	**unk.**	unk.	0.37	46	**5**	3.3	1.9	unk.	11	**3**	0.06	**3**	unk.		unk.
K	0.06	**4**	0.02	**1**	tr(a)	**0**	unk.	**unk.**	0	tr(a)	**0**	unk.	unk.	5	**1**	66	43	unk.	0.4	**2**	unk.	**unk.**	unk.	**unk.**	unk.
	0.3	**2**	unk.	**unk.**	unk.	**unk.**	62	**1**	6	0.3	**1**	0.3	0.46	32	**3**	2.9	1.1	unk.	unk.	**unk.**	unk.	**unk.**	unk.		unk.
L	0.01	**1**	unk.	**unk.**	unk.	**unk.**	unk.	**unk.**	unk.	0(a)	**0**	unk.	unk.	5	**1**	85	57	unk.	0.3	**2**	0.3	**2**	unk.	**unk.**	unk.
	0.1	**1**	unk.	**unk.**	unk.	**unk.**	unk.	**unk.**	unk.	unk.	**unk.**	unk.	unk.	39	**4**	3.7	1.5	unk.	14	**4**	0.06	**3**	unk.		unk.
M	0.06	**4**	0.07	**3**	0.05	**1**	unk.	**unk.**	0(a)	0(a)	**0**	unk.	unk.	10	**1**	200	129	unk.	1.2	**7**	0.7	**5**	unk.	**unk.**	unk.
	0.2	**1**	10	**3**	0.14	**1**	0(a)	**0**	0(a)	0.3	**1**	0.3	1.03	86	**9**	8.7	3.3	unk.	30	**8**	0.13	**6**	unk.		unk.
N	0.01	**1**	0.02	**1**	0.00	**0**	tr	**0**	0	0(a)	**0**	unk.	unk.	6	**1**	308	34	unk.	0.3	**2**	0.1	**1**	unk.	**unk.**	109
	0.3	**1**	3	**1**	0.08	**1**	0	**0**	0	0.2	**1**	0.2	0.90	25	**3**	13.4	0.9	unk.	5	**1**	0.03	**1**	unk.		17
O	0.15	**9**	unk.	**unk.**	unk.	**unk.**	unk.	**unk.**	unk.	0(a)	**0**	unk.	unk.	14	**1**	484	41	unk.	1.4	**8**	0.2	**1**	unk.	**unk.**	192
	1.7	**9**	unk.	**unk.**	unk.	**unk.**	unk.	**unk.**	unk.	unk.	**unk.**	unk.	1.28	29	**3**	21.1	1.0	unk.	8	**2**	0.05	**2**	unk.		unk.
P	0.14	**8**	unk.	**unk.**	unk.	**unk.**	unk.	**unk.**	unk.	0(a)	**0**	unk.	unk.	47	**5**	389	37	unk.	1.3	**7**	0.2	**1**	unk.	**unk.**	190
	1.5	**8**	unk.	**unk.**	0.08	**1**	unk.	**unk.**	unk.	unk.	**unk.**	unk.	1.21	27	**3**	16.9	0.9	unk.	7	**2**	0.05	**3**	unk.		unk.
Q	0.41	**24**	unk.	**unk.**	unk.	**unk.**	unk.	**unk.**	unk.	0(a)	**0**	unk.	unk.	141	**14**	1168	110	unk.	3.8	**21**	0.6	**4**	unk.	**unk.**	570
	4.6	**23**	unk.	**unk.**	0.25	**3**	unk.	**unk.**	unk.	unk.	**unk.**	unk.	3.64	82	**8**	50.8	2.8	unk.	21	**5**	0.16	**8**	unk.		unk.
R	0.21	**12**	unk.	**unk.**	unk.	**unk.**	unk.	**unk.**	unk.	0(a)	**0**	unk.	unk.	18	**2**	209	29	unk.	1.3	**7**	0.2	**1**	unk.	**unk.**	143
	1.7	**8**	unk.	**unk.**	unk.	**unk.**	unk.	**unk.**	unk.	unk.	**unk.**	unk.	0.65	44	**4**	9.1	0.7	unk.	5	**1**	0.03	**2**	unk.		unk.
S	0.11	**6**	unk.	**unk.**	unk.	**unk.**	unk.	**unk.**	unk.	0(a)	**0**	unk.	unk.	35	**4**	232	29	unk.	0.8	**5**	0.2	**1**	unk.	**unk.**	146
	1.3	**6**	unk.	**unk.**	unk.	**unk.**	unk.	**unk.**	unk.	unk.	**unk.**	unk.	0.75	66	**7**	10.1	0.8	unk.	6	**1**	0.05	**2**	unk.		unk.

	FOOD	Portion	Weight in grams / *Conversion for 100 g*	Kilocalories / H_2O g	Total carbohydrate g / Total fats g	Crude fiber g / Dietary fiber g	Total protein g / Total sugar g	% USRDA	Arginine mg / Histidine mg	Isoleucine mg / Leucine mg	Lysine mg / Methionine mg	Phenylalanine mg / Threonine mg	Valine mg / Tryptophan mg	Cystine mg / Tyrosine mg	Polyunsat. fatty acids g / Monounsat. fatty acids g	Saturated fatty acids g / P/S ratio	Linoleic acid g / Cholesterol mg	Thiamin mg / Ascorbic acid mg	% USRDA
A	crackers,Sociables-Nabisco	14 pieces	29	143	17.7	unk.	2.8	**4**	unk.	unk.	unk.	unk.	unk.	unk.	unk.	unk.	unk.	0.16	**11**
			3.40	1.0	6.6	unk.	unk.		unk.	unk.	unk.	unk.	unk.	unk.	unk.	unk.	unk.	unk.	**unk.**
B	crackers,Swiss cheese naturally flavored snack-Nabisco	15 pieces	29	145	16.8	unk.	2.6	**4**	unk.	unk.	unk.	unk.	unk.	unk.	unk.	unk.	unk.	0.19	**13**
			3.44	1.1	7.5	unk.	unk.		unk.	unk.	unk.	unk.	unk.	unk.	unk.	unk.	unk.	unk.	**unk.**
C	crackers,Town House-Keebler	9 pieces	30	157	17.7	unk.	2.0	**3**	unk.	91	45	109	86	40	unk.	unk.	unk.	0.10	**7**
			3.37	0.7	8.7	unk.	unk.		40	153	26	57	24	68	unk.	unk.	tr(a)	0(a)	**0**
D	crackers,Triscuit wafer-Nabisco	7 pieces	31	143	20.9	unk.	2.8	**4**	unk.	140	93	135	162	57	unk.	unk.	unk.	0.06	**4**
			3.25	0.9	5.3	unk.	unk.		66	192	114	114	24	66	unk.	unk.	0(a)	0	**0**
E	crackers,Twigs sesame/cheese snack sticks-Nabisco	10 pieces	28	138	15.7	unk.	2.9	**5**	unk.	unk.	unk.	unk.	unk.	unk.	unk.	unk.	unk.	0.15	**10**
			3.57	1.0	7.1	unk.	unk.		unk.	unk.	unk.	unk.	unk.	unk.	unk.	unk.	unk.	unk.	**unk.**
F	crackers,Uneeda biscuits,unsalted tops-Nabisco	6 pieces	31	131	21.5	unk.	2.9	**5**	unk.	unk.	unk.	unk.	unk.	unk.	unk.	unk.	unk.	0.13	**8**
			3.27	1.8	3.7	unk.	unk.		unk.	unk.	unk.	unk.	unk.	unk.	unk.	unk.	unk.	unk.	**unk.**
G	crackers,unsalted tops -Fireside	10 pieces	30	127	22.2	unk.	2.9	**4**	unk.	unk.	unk.	unk.	unk.	unk.	unk.	unk.	unk.	0.24	**16**
			3.29	1.5	3.0	unk.	unk.		unk.	unk.	unk.	unk.	unk.	unk.	unk.	unk.	unk.	unk.	**unk.**
H	crackers,vegetable thins-Nabisco	13 pieces	28	144	16.8	unk.	2.0	**3**	unk.	unk.	unk.	unk.	unk.	unk.	unk.	unk.	unk.	0.13	**9**
			3.52	0.9	7.7	unk.	unk.		unk.	unk.	unk.	unk.	unk.	unk.	unk.	unk.	unk.	unk.	**unk.**
I	crackers,Waldorf,low salt-Keebler	9 pieces	29	130	21.5	unk.	2.9	**4**	unk.	unk.	unk.	unk.	unk.	unk.	unk.	unk.	unk.	0.20	**13**
			3.47	0.9	3.7	unk.	unk.		unk.	unk.	unk.	unk.	unk.	unk.	unk.	unk.	0(a)	0(a)	**0**
J	crackers,Waverly wafers-Nabisco	8 pieces	30	141	20.5	unk.	2.3	**4**	unk.	unk.	unk.	unk.	unk.	unk.	unk.	unk.	unk.	0.15	**10**
			3.29	1.0	5.5	unk.	unk.		unk.	unk.	unk.	unk.	unk.	unk.	unk.	unk.	unk.	unk.	**unk.**
K	crackers,Wheat Thins-Nabisco	16 pieces	29	138	18.5	unk.	2.2	**4**	unk.	unk.	unk.	unk.	unk.	unk.	unk.	unk.	unk.	0.17	**11**
			3.47	1.0	6.2	unk.	unk.		unk.	unk.	unk.	unk.	unk.	unk.	unk.	unk.	0(a)	0	**0**
L	crackers,Wheatsworth stone ground wheat wafers-Nabisco	10 pieces	30	145	18.2	unk.	3.3	**5**	unk.	unk.	unk.	unk.	unk.	unk.	unk.	unk.	unk.	0.17	**12**
			3.36	0.7	6.6	unk.	unk.		unk.	unk.	unk.	unk.	unk.	unk.	unk.	unk.	unk.	unk.	**unk.**
M	crackers,whole rice wafer	12 pieces	29	108	23.6	unk.	2.4	**4**	unk.	unk.	unk.	unk.	unk.	unk.	unk.	unk.	unk.	unk.	**unk.**
			3.47	unk.	0.5	unk.	unk.		unk.	unk.	unk.	unk.	unk.	unk.	unk.	unk.	0(a)	tr(a)	**0**
N	crackers,zwieback	4 pieces	28	119	20.8	0.06	2.8	**11**	unk.	unk.	unk.	unk.	unk.	unk.	0.32	1.11	0.03	0.06	**8**
			3.57	1.3	2.7	unk.	unk.		unk.	unk.	unk.	unk.	unk.	unk.	0.91	0.29	6	1	**4**
O	CRANBERRIES,raw	1/2 C	76	35	8.2	1.06	0.3	**1**	unk.	tr(a)	tr(a)	tr(a)	tr(a)	tr(a)	unk.	unk.	unk.	0.02	**2**
			1.32	66.4	0.5	unk.	2.6		tr(a)	tr(a)	tr(a)	tr(a)	tr(a)	tr(a)	unk.	unk.	0	8	**14**
P	cranberry crunch,home recipe	1/2 C	61	179	31.1	0.42	0.9	**1**	unk.	40	31	41	46	16	0.36	3.65	0.27	0.03	**2**
			1.64	22.3	6.3	unk.	26.7		15	59	12	26	10	29	1.58	0.10	16	1	**2**
Q	cranberry sauce,canned,strnd,sweetened	1/2 C	139	202	51.9	0.28	0.1	**0**	unk.	unk.	unk.	unk.	unk.	unk.	unk.	tr(a)	unk.	0.01	**1**
			0.72	86.0	0.3	unk.	51.7		unk.	unk.	unk.	unk.	unk.	unk.	unk.	unk.	0	3	**5**
R	cranberry sauce,unstrnd,sweetened, home recipe	1/2 C	139	247	63.0	0.97	0.3	**0**	unk.	unk.	unk.	unk.	unk.	unk.	unk.	tr(a)	unk.	0.01	**1**
			0.72	74.7	0.4	unk.	62.0		unk.	unk.	unk.	unk.	unk.	unk.	unk.	unk.	0	3	**5**
S	cranberry-orange relish,uncooked	1/2 C	138	245	62.4	unk.	0.5	**1**	unk.	unk.	unk.	unk.	unk.	unk.	unk.	tr(a)	unk.	0.04	**3**
			0.73	73.7	0.5	unk.	62.4		unk.	unk.	unk.	unk.	unk.	unk.	unk.	unk.	0	25	**41**

Riboflavin mg / Niacin mg	% USRDA	Vitamin B_6 mg / Folacin mcg	% USRDA	Vitamin B_{12} mcg / Pantothenic acid mg	% USRDA	Biotin mg / Vitamin A IU	% USRDA	Preformed A RE / Beta carotene RE	Vitamin D IU / Vitamin E IU	% USRDA	Total tocopherol mg / Alpha tocopherol mg	Other tocopherol mg / Total ash g	Calcium mg / Phosphorus mg	% USRDA	Sodium mg / Sodium meq	Potassium mg / Potassium meq	Chlorine mg / Chlorine meq	Iron mg / Magnesium mg	% USRDA	Zinc mg / Copper mg	% USRDA	Iodine mcg / Selenium mcg	% USRDA	Manganese mcg / Chromium mcg	
0.10	**6**	unk.	**unk.**	unk.	**unk.**	unk.	**unk.**	unk.	0(a)	**0**	unk.	unk.	45	**5**	289	62	unk.	1.1	**6**	0.3	**2**	unk.	**unk.**	223	A
1.3	**6**	unk.	**unk.**	unk.	**unk.**	unk.	**unk.**	unk.	unk.	**unk.**	unk.	1.05	87	**9**	12.6	1.6	unk.	16	**4**	0.10	**5**	unk.		unk.	
0.13	**8**	unk.	**unk.**	unk.	**unk.**	unk.	**unk.**	unk.	0(a)	**0**	unk.	unk.	50	**5**	304	72	unk.	0.7	**4**	0.3	**2**	unk.	**unk.**	136	B
1.2	**6**	unk.	**unk.**	unk.	**unk.**	unk.	**unk.**	unk.	unk.	**unk.**	unk.	1.07	61	**6**	13.2	1.8	unk.	9	**2**	0.04	**2**	unk.		unk.	
0.11	**7**	unk.	**unk.**	0(a)	**0**	unk.	**unk.**	0(a)	0(a)	**0**	unk.	unk.	7	**1**	285	40	unk.	1.1	**6**	unk.	**unk.**	unk.	**unk.**	unk.	C
1.3	**6**	unk.	**unk.**	unk.	**unk.**	unk.	**unk.**	unk.	0.2	**1**	0.2	0.59	unk.	**unk.**	12.4	1.0	unk.	unk.	**unk.**	unk.	**unk.**	unk.		unk.	
0.04	**2**	unk.	**unk.**	0(a)	**0**	unk.	**unk.**	0(a)	0(a)	**0**	unk.	unk.	12	**1**	174	89	unk.	0.8	**5**	0.7	**5**	unk.	**unk.**	541	D
1.4	**7**	unk.	**unk.**	unk.	**unk.**	0(a)	**0**	0(a)	unk.	**unk.**	unk.	0.82	87	**9**	7.6	2.3	unk.	31	**8**	0.14	**7**	unk.		unk.	
0.11	**6**	unk.	**unk.**	unk.	**unk.**	unk.	**unk.**	unk.	0(a)	**0**	unk.	unk.	77	**8**	310	50	unk.	1.1	**6**	0.4	**2**	unk.	**unk.**	201	E
1.3	**7**	unk.	**unk.**	unk.	**unk.**	unk.	**unk.**	unk.	unk.	**unk.**	unk.	1.12	94	**9**	13.5	1.3	unk.	17	**4**	0.09	**4**	unk.		unk.	
0.16	**9**	unk.	**unk.**	unk.	**unk.**	unk.	**unk.**	unk.	0(a)	**0**	unk.	unk.	7	**1**	223	42	unk.	1.7	**10**	0.3	**2**	unk.	**unk.**	278	F
1.9	**9**	unk.	**unk.**	unk.	**unk.**	unk.	**unk.**	unk.	unk.	**unk.**	unk.	0.66	36	**4**	9.7	1.1	unk.	8	**2**	0.07	**4**	unk.		unk.	
0.16	**10**	unk.	**unk.**	unk.	**unk.**	unk.	**unk.**	unk.	0(a)	**0**	unk.	unk.	5	**1**	163	34	unk.	0.9	**5**	0.2	**2**	unk.	**unk.**	131	G
1.8	**9**	unk.	**unk.**	0.09	**1**	unk.	**unk.**	unk.	unk.	**unk.**	unk.	0.82	67	**7**	7.1	0.9	unk.	6	**1**	0.06	**3**	unk.		unk.	
0.10	**6**	unk.	**unk.**	unk.	**unk.**	unk.	**unk.**	unk.	0(a)	**0**	unk.	unk.	41	**4**	287	61	unk.	0.8	**4**	0.2	**1**	unk.	**unk.**	154	H
1.2	**6**	unk.	**unk.**	unk.	**unk.**	unk.	**unk.**	unk.	unk.	**unk.**	unk.	0.96	68	**7**	12.5	1.5	unk.	8	**2**	0.05	**2**	unk.		unk.	
0.12	**7**	unk.	**unk.**	0(a)	**0**	unk.	**unk.**	0(a)	0(a)	**0**	unk.	unk.	5	**1**	3	57	unk.	1.2	**6**	unk.	**unk.**	unk.	**unk.**	unk.	I
1.6	**8**	unk.	**unk.**	unk.	**unk.**	unk.	**unk.**	unk.	0.2	**1**	0.2	0.12	unk.	**unk.**	0.1	1.5	unk.	unk.	**unk.**	unk.	**unk.**	unk.		unk.	
0.12	**7**	unk.	**unk.**	unk.	**unk.**	unk.	**unk.**	unk.	0(a)	**0**	unk.	unk.	47	**5**	297	36	unk.	0.9	**5**	0.2	**1**	unk.	**unk.**	165	J
1.4	**7**	unk.	**unk.**	unk.	**unk.**	unk.	**unk.**	unk.	unk.	**unk.**	unk.	0.95	86	**9**	12.9	0.9	unk.	7	**2**	0.06	**3**	unk.		unk.	
0.11	**6**	unk.	**unk.**	0(a)	**0**	unk.	**unk.**	tr(a)	0(a)	**0**	unk.	unk.	7	**1**	213	48	unk.	1.1	**6**	0.3	**2**	unk.	**unk.**	462	K
1.7	**9**	unk.	**unk.**	unk.	**unk.**	unk.	**unk.**	unk.	unk.	**unk.**	unk.	0.73	46	**5**	9.3	1.2	unk.	15	**4**	0.06	**3**	unk.		unk.	
0.12	**7**	unk.	**unk.**	unk.	**unk.**	unk.	**unk.**	unk.	0(a)	**0**	unk.	unk.	25	**3**	308	75	unk.	1.1	**6**	0.7	**5**	unk.	**unk.**	650	L
1.5	**8**	unk.	**unk.**	unk.	**unk.**	unk.	**unk.**	unk.	unk.	**unk.**	unk.	1.08	93	**9**	13.4	1.9	unk.	22	**6**	0.09	**4**	unk.		unk.	
unk.	**unk.**	unk.	**unk.**	0(a)	**0**	unk.	**unk.**	0(a)	0(a)	**0**	unk.	unk.	unk.	**unk.**	1	7	unk.	unk.	**unk.**	unk.	**unk.**	unk.	**unk.**	unk.	M
unk.	**unk.**	unk.	**unk.**	unk.	**unk.**	unk.	**unk.**	unk.	unk.	**unk.**	unk.	unk.	unk.	**unk.**	0.0	0.2	unk.	unk.	**unk.**	unk.	**unk.**	unk.		unk.	
0.07	**8**	0.02	**3**	unk.	**unk.**	unk.	**unk.**	unk.	unk.	**unk.**	unk.	unk.	6	**1**	65	85	unk.	0.2	**2**	0.4	**5**	unk.	**unk.**	76	N
3.7	**41**	unk.	**unk.**	unk.	**unk.**	16	**1**	unk.	unk.	**unk.**	unk.	0.42	15	**2**	2.8	2.2	unk.	4	**2**	0.08	**8**	unk.		unk.	
0.02	**1**	0.03	**1**	0.00	**0**	unk.	**unk.**	unk.	0(a)	**0**	unk.	unk.	11	**1**	2	62	unk.	0.4	**2**	unk.	**unk.**	unk.	**unk.**	unk.	O
0.1	**0**	2	**0**	0.17	**2**	30	**1**	unk.	unk.	**unk.**	unk.	unk.	8	**1**	0.1	1.6	unk.	unk.	**unk.**	unk.	**unk.**	unk.		unk.	
0.01	**1**	0.01	**0**	0.00	**0**	0(a)	**0**	0	0	**0**	0.1	0.0	12	**1**	118	51	3	0.5	**3**	0.2	**1**	8.8	**6**	186	P
0.1	**1**	tr	**0**	0.08	**1**	228	**5**	1	0.1	**0**	0.1	0.47	27	**3**	5.1	1.3	0.1	10	**2**	tr	**0**	0		2	
0.01	**1**	0.03	**2**	0.00	**0**	unk.	**unk.**	0	0(a)	**0**	unk.	unk.	8	**1**	1	42	unk.	0.3	**2**	tr	**0**	unk.	**unk.**	unk.	Q
tr	**0**	unk.	**unk.**	unk.	**unk.**	28	**1**	3	unk.	**unk.**	unk.	0.14	6	**1**	0.1	1.1	unk.	3	**1**	tr(a)	**0**	tr(a)		unk.	
0.01	**1**	unk.	**unk.**	unk.	**unk.**	unk.	**unk.**	0	0(a)	**0**	unk.	unk.	10	**1**	1	53	unk.	0.3	**2**	tr	**0**	unk.	**unk.**	unk.	R
0.1	**1**	unk.	**unk.**	unk.	**unk.**	28	**1**	3	unk.	**unk.**	unk.	0.14	7	**1**	0.1	1.3	unk.	unk.	**unk.**	tr(a)	**0**	tr(a)		unk.	
0.03	**2**	unk.	**unk.**	0(a)	**0**	unk.	**unk.**	0	0(a)	**0**	unk.	unk.	26	**3**	1	99	unk.	0.5	**3**	unk.	**unk.**	unk.	**unk.**	unk.	S
0.1	**1**	unk.	**unk.**	unk.	**unk.**	96	**2**	10	unk.	**unk.**	unk.	0.27	11	**1**	0.1	2.5	unk.	unk.	**unk.**	tr(a)	**0**	tr(a)		unk.	

	FOOD	Portion	Weight in grams / Conversion for 100 g	Kilocalories / H_2O g	Total carbohydrate g / Total fats g	Crude fiber g / Dietary fiber g	Total protein g / Total sugar g	% USRDA	Arginine mg / Histidine mg	Isoleucine mg / Leucine mg	Lysine mg / Methionine mg	Phenylalanine mg / Threonine mg	Valine mg / Tryptophan mg	Cystine mg / Tyrosine mg	Polyunsat. fatty acids g / Monounsat. fatty acids g	Saturated fatty acids g / P/S ratio	Linoleic acid g / Cholesterol mg	Thiamin mg / Ascorbic acid mg	% USRDA
A	CREAM PUFF w/custard filling	1 average	130	303	26.6	0.00	8.4	17	unk.	unk.	unk.	unk.	unk.	unk.	unk.	5.59	2.86	0.05	4
			0.77	75.8	18.1	unk.	18.6		unk.	unk.	unk.	unk.	unk.	unk.	8.32	unk.	187	0	0
B	cream puff/eclair w/whip cream filling	1 average	165	465	23.6	0.00	6.1	12	unk.	unk.	unk.	unk.	unk.	unk.	unk.	unk.	unk.	0.10	7
			0.61	unk.	40.4	unk.	23.6		unk.	unk.	unk.	unk.	unk.	unk.	unk.	unk.	unk.	0	0
C	cream puff,shell only,home recipe	3-1/2" dia x 2"	66	156	8.8	0.03	3.7	7	159	181	206	186	215	81	3.26	2.65	3.19	0.07	5
			1.52	42.4	11.7	0.35	0.3		106	302	96	159	51	152	5.18	1.23	100	0	0
D	CREAM,half & half,fluid	1 Tbsp	15	20	0.7	0.00	0.4	1	16	27	36	22	30	4	0.07	1.08	0.04	0.00	0
			6.61	12.2	1.7	0(a)	0.7		12	44	11	20	6	22	0.44	0.06	6	tr	0
E	cream,heavy whipping,fluid	1 Tbsp	15	51	0.4	0.00	0.3	1	11	18	24	15	20	3	0.20	3.43	0.12	tr	0
			6.72	8.6	5.5	0(a)	0.4		8	30	8	14	4	15	1.39	0.06	20	tr	0
F	cream,heavy whipping,whipped	1/4 C	30	103	0.8	0.00	0.6	1	22	37	48	29	41	6	0.41	6.85	0.25	0.01	0
			3.36	17.2	11.0	0(a)	0.8		17	60	15	28	9	29	2.77	0.06	41	tr	0
G	cream,light (coffee or table),fluid	1 Tbsp	15	29	0.5	0.00	0.4	1	15	24	32	19	27	4	0.11	1.80	0.07	tr	0
			6.67	11.1	2.9	0(a)	0.5		11	40	10	18	6	19	0.73	0.06	10	tr	0
H	cream,light whipping,fluid	1 Tbsp	15	44	0.4	0.00	0.3	1	12	20	26	16	22	3	0.13	2.89	0.09	tr	0
			6.69	9.5	4.6	0(a)	0.4		9	32	8	15	5	16	1.14	0.05	17	tr	0
I	cream,medium fat (25% fat),fluid	1 Tbsp	15	36	0.5	0.00	0.4	1	13	22	29	18	25	3	0.14	2.32	0.08	tr	0
			6.69	10.2	3.7	0(a)	0.5		10	36	9	17	5	18	0.94	0.06	13	tr	0
J	cream,sour cream,cultured	2 Tbsp	29	62	1.2	0.00	0.9	1	unk.	unk.	unk.	unk.	unk.	unk.	0.22	3.75	0.14	0.01	1
			3.48	20.4	6.0	0(a)	1.2		unk.	unk.	unk.	unk.	unk.	unk.	1.52	0.06	13	tr	0
K	cream,sour cream,imitation,non-dairy cultured	2 Tbsp	29	60	1.9	0.00	0.7	1	unk.	unk.	unk.	unk.	unk.	unk.	0.02	5.11	0.02	0.00	0
			3.48	20.5	5.6	0(a)	1.9		unk.	unk.	unk.	unk.	unk.	unk.	0.17	0.00	0	0	0
L	cream,sour half & half,cultured	2 Tbsp	30	40	1.3	0.00	0.9	1	unk.	unk.	unk.	unk.	unk.	unk.	0.13	2.24	0.08	0.01	1
			3.33	24.0	3.6	0(a)	1.3		unk.	unk.	unk.	unk.	unk.	unk.	0.91	0.06	11	tr	0
M	cream,whipped cream topping, pressurized	1/4 C	15	39	1.9	0.00	0.5	1	17	29	38	23	32	4	0.12	2.07	0.07	tr	0
			6.67	9.2	3.3	0(a)	1.9		13	47	12	22	7	23	0.84	0.06	11	0	0
N	cream,whipped topping,reduced calorie, prep-D-Zerta	1/4 C	19	47	2.0	0.03	0.9	1	unk.	unk.	unk.	unk.	unk.	unk.	0.48	2.59	0.44	0.01	1
			5.26	12.0	4.0	unk.	1.9		unk.	unk.	unk.	unk.	unk.	unk.	0.63	0.18	tr	0	0
O	CREAMSICLE (orange)-Sealtest	1	66	103	17.6	unk.	1.2	3	unk.	unk.	unk.	unk.	unk.	unk.	unk.	unk.	unk.	0.02	1
			1.52	43.7	3.1	unk.	16.4		unk.	unk.	unk.	unk.	unk.	unk.	unk.	unk.	unk.	unk.	unk.
P	CROUTONS,croutettes-Kellogg	1 C	40	144	29.4	0.24	5.2	8	unk.	239	118	286	224	104	unk.	unk.	unk.	0.08	6
			2.50	3.2	0.5	unk.	0.6		104	403	68	150	64	178	unk.	unk.	0(a)	0	0
Q	CUCUMBER,raw,not pared,edible portion	1 average	170	25	5.8	1.02	1.5	2	115	54	75	41	58	unk.	0.00	0.00	unk.	0.05	3
			0.59	161.7	0.2	1.21	4.4		19	78	17	48	14	unk.	unk.	0.00	0(a)	19	31
R	cucumber,raw,pared & sliced	1/2 C	70	10	2.2	0.21	0.4	1	32	13	18	10	14	unk.	0.00	0.00	unk.	0.02	1
			1.43	67.0	0.1	0.25	1.8		6	18	4	11	3	unk.	unk.	0.00	0(a)	8	13
S	CUMIN seed	1 tsp	2.0	7	0.9	0.21	0.4	1	unk.	unk.	unk.	unk.	unk.	unk.	unk.	unk.	unk.	0.01	1
			50.00	0.2	0.4	unk.	0.0		unk.	unk.	unk.	unk.	unk.	unk.	unk.	unk.	unk.	tr	0

Riboflavin mg / Niacin mg	% USRDA	Vitamin B_6 mg / Folacin mcg	% USRDA	Vitamin B_{12} mcg / Pantothenic acid mg	% USRDA	Biotin mg / Vitamin A IU	% USRDA	Preformed A RE / Beta carotene RE	Vitamin D IU / Vitamin E IU	% USRDA	Total tocopherol mg / Alpha tocopherol mg	Other tocopherol mg / Total ash g	Calcium mg / Phosphorus mg	% USRDA	Sodium mg / Sodium meq	Potassium mg / Potassium meq	Chlorine mg / Chlorine meq	Iron mg / Magnesium mg	% USRDA	Zinc mg / Copper mg	% USRDA	Iodine mcg / Selenium mcg	% USRDA	Manganese mcg / Chromium mcg	
0.22	**13**	unk.	**unk.**	unk.	**unk.**	unk.	**unk.**	unk.	unk.	**unk.**	unk.	unk.	105	**11**	108	157	unk.	0.9	**5**	unk.	**unk.**	unk.	**unk.**	unk.	A
0.1	**1**	unk.	**unk.**	unk.	**unk.**	455	**9**	unk.	unk.	**unk.**	unk.	1.04	148	**15**	4.7	4.0	unk.	17	**4**	unk.	**unk.**	unk.		unk.	
0.18	**11**	unk.	**unk.**	unk.	**unk.**	unk.	**unk.**	unk.	unk.	**unk.**	unk.	unk.	75	**8**	unk.	unk.	unk.	1.1	**6**	unk.	**unk.**	unk.	**unk.**	unk.	B
0.5	**2**	unk.	**unk.**	unk.	**unk.**	1759	**35**	unk.	unk.	**unk.**	unk.	unk.	110	**11**	unk.	unk.	unk.	unk.	**unk.**	unk.	**unk.**	unk.		unk.	
0.09	**6**	0.04	**2**	0.23	**4**	0.004	**1**	0	7	**2**	6.8	3.2	16	**2**	195	42	43	0.8	**4**	0.4	**3**	28.7	**19**	54	C
0.4	**2**	8	**2**	0.33	**3**	420	**8**	0	8.1	**27**	3.4	0.44	60	**6**	8.5	1.1	1.2	52	**13**	0.03	**1**	6		8	
0.02	**1**	0.01	**0**	0.05	**1**	unk.	**unk.**	unk.	unk.	**unk.**	unk.	unk.	16	**2**	6	20	unk.	tr	**0**	0.1	**1**	unk.	**unk.**	unk.	D
tr	**0**	tr	**0**	0.04	**0**	66	**1**	unk.	unk.	**unk.**	unk.	0.10	14	**1**	0.3	0.5	unk.	2	**0**	unk.	**unk.**	unk.		unk.	
0.02	**1**	tr	**0**	0.03	**0**	unk.	**unk.**	unk.	unk.	**unk.**	unk.	unk.	10	**1**	6	11	unk.	0.0	**0**	tr	**0**	unk.	**unk.**	unk.	E
0.0	**0**	1	**0**	0.04	**0**	219	**4**	unk.	unk.	**unk.**	unk.	0.07	9	**1**	0.3	0.3	unk.	1	**0**	unk.	**unk.**	unk.		unk.	
0.03	**2**	0.01	**0**	0.05	**1**	unk.	**unk.**	unk.	unk.	**unk.**	unk.	unk.	19	**2**	11	22	unk.	tr	**0**	0.1	**1**	unk.	**unk.**	unk.	F
tr	**0**	1	**0**	0.08	**1**	437	**9**	unk.	unk.	**unk.**	unk.	0.13	18	**2**	0.5	0.6	unk.	2	**1**	unk.	**unk.**	unk.		unk.	
0.02	**1**	0.00	**0**	0.03	**1**	unk.	**unk.**	unk.	unk.	**unk.**	unk.	unk.	14	**1**	6	18	unk.	tr	**0**	tr	**0**	unk.	**unk.**	unk.	G
tr	**0**	tr	**0**	0.04	**0**	108	**2**	unk.	unk.	**unk.**	unk.	0.09	12	**1**	0.3	0.5	unk.	1	**0**	unk.	**unk.**	unk.		unk.	
0.02	**1**	tr	**0**	0.03	**1**	unk.	**unk.**	unk.	unk.	**unk.**	unk.	unk.	10	**1**	5	14	unk.	0.0	**0**	tr	**0**	unk.	**unk.**	unk.	H
tr	**0**	1	**0**	0.04	**0**	168	**3**	unk.	unk.	**unk.**	unk.	0.07	9	**1**	0.2	0.4	unk.	1	**0**	unk.	**unk.**	unk.		unk.	
0.02	**1**	0.00	**0**	0.03	**1**	unk.	**unk.**	unk.	unk.	**unk.**	unk.	unk.	13	**1**	6	17	unk.	tr	**0**	tr	**0**	unk.	**unk.**	unk.	I
tr	**0**	tr	**0**	0.04	**0**	141	**3**	unk.	unk.	**unk.**	unk.	0.08	11	**1**	0.2	0.4	unk.	1	**0**	unk.	**unk.**	unk.		unk.	
0.04	**2**	0.00	**0**	0.09	**1**	unk.	**unk.**	unk.	unk.	**unk.**	unk.	unk.	33	**3**	15	41	unk.	tr	**0**	0.1	**1**	unk.	**unk.**	unk.	J
tr	**0**	3	**1**	0.10	**1**	227	**5**	unk.	unk.	**unk.**	unk.	0.19	24	**2**	0.7	1.1	unk.	3	**1**	unk.	**unk.**	unk.		unk.	
0.00	**0**	0.00	**0**	0.00	**0**	unk.	**unk.**	unk.	unk.	**unk.**	unk.	unk.	1	**0**	29	46	unk.	unk.	**unk.**	unk.	**unk.**	unk.	**unk.**	unk.	K
0.0	**0**	0	**0**	0.00	**0**	0	**0**	unk.	unk.	**unk.**	unk.	0.09	13	**1**	1.3	1.2	unk.	unk.	**unk.**	unk.	**unk.**	unk.		unk.	
0.04	**3**	0.00	**0**	0.09	**2**	unk.	**unk.**	unk.	unk.	**unk.**	unk.	unk.	31	**3**	12	39	unk.	tr	**0**	0.1	**1**	unk.	**unk.**	unk.	L
tr	**0**	3	**1**	0.11	**1**	136	**3**	unk.	unk.	**unk.**	unk.	0.20	29	**3**	0.5	1.0	unk.	3	**1**	unk.	**unk.**	unk.		unk.	
0.01	**1**	0.01	**0**	0.04	**1**	unk.	**unk.**	unk.	unk.	**unk.**	unk.	unk.	15	**2**	19	22	unk.	tr	**0**	0.1	**0**	unk.	**unk.**	unk.	M
tr	**0**	unk.	**unk.**	0.05	**1**	137	**3**	unk.	unk.	**unk.**	unk.	0.11	13	**1**	0.8	0.6	unk.	2	**0**	unk.	**unk.**	unk.		unk.	
0.06	**3**	0.01	**1**	0.06	**1**	unk.	**unk.**	0	0	**0**	0(a)	0(a)	21	**2**	43	52	unk.	tr	**0**	0.0	**0**	unk.	**unk.**	unk.	N
tr	**0**	0	**0**	0.14	**1**	53	**1**	0	0.0	**0**	0(a)	0.02	30	**3**	1.8	1.3	unk.	4	**1**	tr	**0**	unk.		unk.	
0.08	**5**	0.02	**1**	0.20	**3**	unk.	**unk.**	unk.	0	**0**	unk.	unk.	46	**5**	27	82	unk.	0.0	**0**	unk.	**unk.**	unk.	**unk.**	unk.	O
0.3	**2**	unk.	**unk.**	unk.	**unk.**	125	**3**	unk.	unk.	**unk.**	0.1	0.40	37	**4**	1.2	2.1	unk.	5	**1**	unk.	**unk.**	unk.		unk.	
0.06	**4**	tr	**0**	0(a)	**0**	unk.	**unk.**	0(a)	0(a)	**0**	unk.	unk.	36	**4**	544	56	unk.	1.0	**6**	unk.	**unk.**	unk.	**unk.**	unk.	P
0.4	**2**	400	**100**	unk.	**unk.**	59	**1**	unk.	unk.	**unk.**	unk.	unk.	61	**6**	23.7	1.4	unk.	12	**3**	0.06	**3**	unk.		unk.	
0.07	**4**	0.07	**4**	0.00	**0**	0.005	**2**	0	0(a)	**0**	0.2	tr(a)	42	**4**	10	272	51	1.9	**10**	0.4	**3**	unk.	**unk.**	95	Q
0.3	**2**	25	**6**	0.42	**4**	425	**9**	42	0.2	**1**	tr(a)	0.85	46	**5**	0.4	7.0	1.4	20	**5**	0.02	**1**	tr(a)		tr	
0.03	**2**	0.03	**2**	unk.	**unk.**	0.002	**1**	0	0(a)	**0**	0.1	tr(a)	12	**1**	4	112	unk.	0.2	**1**	0.1	**1**	unk.	**unk.**	42	R
0.1	**1**	10	**3**	unk.	**unk.**	tr	**0**	tr	0.1	**0**	tr(a)	0.28	13	**1**	0.2	2.9	unk.	7	**2**	tr	**0**	tr(a)		tr	
0.01	**0**	unk.	**unk.**	0.00	**0**	unk.	**unk.**	0(a)	0(a)	**0**	unk.	unk.	19	**2**	3	36	unk.	1.3	**7**	0.1	**1**	unk.	**unk.**	unk.	S
0.1	**1**	unk.	**unk.**	unk.	**unk.**	25	**1**	unk.	unk.	**unk.**	unk.	0.15	10	**1**	0.1	0.9	unk.	7	**2**	unk.	**unk.**	unk.		unk.	

	FOOD	Portion	Weight in grams / *Conversion for 100 g*	Kilocalories / H_2O g	Total carbohydrate g / Total fats g	Crude fiber g / Dietary fiber g	Total protein g / Total sugar g	% USRDA	Arginine mg / Histidine mg	Isoleucine mg / Leucine mg	Lysine mg / Methionine mg	Phenylalanine mg / Threonine mg	Valine mg / Tryptophan mg	Cystine mg / Tyrosine mg	Polyunsat. fatty acids g / Monounsat. fatty acids g	Saturated fatty acids g / P/S ratio	Linoleic acid g / Cholesterol mg	Thiamin mg / Ascorbic acid mg	% USRDA
A	CUPCAKE,chocolate or devil's food,	1 average	33	103	14.0	0.07	1.6	**3**	42	67	70	67	75	24	0.11	1.70	0.33	0.03	**2**
	home recipe		*3.03*	12.1	4.9	0.16	8.8		35	112	30	53	18	54	2.59	0.06	21	tr	**0**
B	cupcake,chocolate or devil's food,	1 average	46	175	25.6	0.14	2.0	**3**	unk.	156	124	110	161	55	unk.	4.14	0.64	0.01	**1**
	w/chocolate icing,home recipe		*2.17*	10.9	8.1	0.41	19.8		152	244	55	97	37	115	2.39	unk.	32	tr	**0**
C	cupcake,white,home recipe	1 average	33	114	15.8	0.02	1.5	**3**	41	76	71	77	84	33	0.05	1.25	0.38	0.04	**2**
			3.03	10.4	5.0	0.22	9.5		33	125	36	55	20	59	3.12	0.04	1	tr	**0**
D	cupcake,white,uncooked white icing,	1 average	46	164	27.1	0.02	1.2	**2**	33	63	60	62	69	26	0.11	2.19	0.34	0.03	**2**
	home recipe		*2.17*	10.7	5.9	0.17	22.1		27	103	30	45	17	49	2.92	0.05	6	tr	**0**
E	cupcake,yellow,chocolate icing,	1 average	46	186	31.3	0.06	1.6	**3**	unk.	73	67	75	77	26	0.29	3.89	0.22	0.04	**2**
	home recipe		*2.17*	6.7	6.5	0.23	24.1		35	120	29	54	19	57	1.76	0.07	28	tr	**0**
F	cupcake,yellow,home recipe	1 average	33	127	19.6	0.03	1.7	**3**	unk.	86	77	90	91	32	0.12	1.07	1.17	0.05	**3**
			3.03	7.0	4.7	0.28	11.1		42	142	34	64	23	67	2.21	0.12	17	tr	**0**
G	CURRY powder	1 tsp	2.1	7	1.2	0.34	0.3	**0**	unk.	unk.	unk.	unk.	unk.	unk.	unk.	unk.	unk.	0.01	**0**
			47.62	0.2	0.3	unk.	0.0		unk.	unk.	unk.	unk.	unk.	unk.	unk.	unk.	0	tr	**0**
H	CUSTARD,baked	1/2 C	133	152	14.7	0.00	7.1	**16**	unk.	unk.	unk.	unk.	unk.	unk.	0.66	3.31	0.38	0.05	**4**
			0.75	102.3	7.3	unk.	14.7		unk.	unk.	unk.	unk.	unk.	unk.	1.99	0.20	139	1	**2**
I	custard,reduced-calorie,made w/skim	1/2 C	133	97	17.1	0(a)	6.8	**15**	unk.	unk.	unk.	unk.	unk.	unk.	unk.	unk.	unk.	0.05	**4**
	milk-Delmark		*0.75*	unk.	0.2	unk.	unk.		unk.	unk.	unk.	unk.	unk.	unk.	unk.	unk.	6	1	**2**
J	DAIRY QUEEN Buster Bar	1 serving	149	390	37.0	unk.	10.0	**15**	unk.	unk.	unk.	unk.	unk.	unk.	unk.	unk.	unk.	0.09	**6**
			0.67	unk.	22.0	unk.	unk.		unk.	unk.	unk.	unk.	unk.	unk.	unk.	unk.	unk.	tr	**0**
K	Dairy Queen Dilly Bar	1 serving	85	240	22.0	unk.	4.0	**6**	unk.	unk.	unk.	unk.	unk.	unk.	unk.	unk.	unk.	0.06	**4**
			1.18	unk.	15.0	unk.	unk.		unk.	unk.	unk.	unk.	unk.	unk.	unk.	unk.	unk.	tr	**0**
L	Dairy Queen Mr. Misty Freeze	1 serving	411	500	87.0	unk.	10.0	**15**	unk.	unk.	unk.	unk.	unk.	unk.	unk.	unk.	unk.	0.16	**11**
			0.24	unk.	12.0	unk.	unk.		unk.	unk.	unk.	unk.	unk.	unk.	unk.	unk.	unk.	tr	**0**
M	DANDELION greens,fresh,cooked,	1/2 C	53	17	3.4	0.68	1.0	**2**	unk.	unk.	unk.	unk.	unk.	unk.	unk.	tr(a)	unk.	0.07	**5**
	drained solids		*1.90*	47.1	0.3	unk.	1.3		unk.	unk.	unk.	unk.	unk.	unk.	unk.	unk.	unk.	9	**16**
N	dandelion greens,raw,edible portion	1/2 C	39	17	3.5	0.62	1.0	**2**	unk.	unk.	unk.	unk.	unk.	unk.	unk.	tr(a)	unk.	0.07	**5**
			2.60	33.0	0.3	unk.	0.9		unk.	unk.	unk.	unk.	unk.	unk.	unk.	unk.	0	13	**23**
O	DANISH pastry,plain,home recipe	4-1/4" dia x	65	294	21.1	0.09	4.0	**7**	unk.	191	155	207	199	74	1.06	13.19	0.74	0.13	**9**
		1"	*1.54*	18.1	21.8	0.77	2.0		94	319	71	140	51	141	5.59	0.08	96	0	**0**
P	DATES,moisturized or hydrated	1 average	9.6	26	7.0	0.22	0.2	**0**	7	7	6	6	9	unk.	unk.	unk.	unk.	0.01	**1**
			10.42	2.2	0.0	0.84	4.9		5	7	3	6	6	2	unk.	unk.	0	0	**0**
Q	DESSERT TOPPING,non-dairy frozen,	1/4 C	19	60	4.3	0.00	0.2	**0**	9	14	19	13	17	1	0.10	4.08	0.06	0.00	**0**
	semisolid		*5.33*	9.4	4.7	0(a)	4.3		7	23	7	10	3	14	0.26	0.02	0	0	**0**
R	dessert topping,non-dairy powdered	1 Tbsp	1.3	8	0.7	0.00	0.1	**0**	3	4	5	3	5	tr	0.01	0.48	0.01	0.00	**0**
			76.92	tr	0.5	0(a)	0.7		2	6	2	3	1	4	0.01	0.01	0	0	**0**
S	dessert topping,non-dairy powdered,	1/4 C	20	38	3.3	0.00	0.7	**1**	27	41	57	36	49	6	0.04	2.14	0.03	tr	**0**
	made w/whole milk		*5.00*	13.3	2.5	0(a)	3.3		20	71	19	32	10	37	0.15	0.02	2	tr	**0**

Riboflavin mg / Niacin mg	% USRDA	Vitamin B_6 mg / Folacin mcg	% USRDA	Vitamin B_{12} mcg / Pantothenic acid mg	% USRDA	Biotin mg / Vitamin A IU	% USRDA	Preformed A RE / Beta carotene RE	Vitamin D IU / Vitamin E IU	% USRDA	Total tocopherol mg / Alpha tocopherol mg	Other tocopherol mg / Total ash g	Calcium mg / Phosphorus mg	% USRDA	Sodium mg / Sodium meq	Potassium mg / Potassium meq	Chlorine mg / Chlorine meq	Iron mg / Magnesium mg	% USRDA	Zinc mg / Copper mg	% USRDA	Iodine mcg / Selenium mcg	% USRDA	Manganese mcg / Chromium mcg	
0.05	3	0.01	1	0.08	1	0.001	1	0	5	1	0.2	0.0	16	2	67	43	12	0.4	2	0.2	1	10.7	7	26	A
0.2	1	5	1	0.11	1	22	0	tr	0.2	1	0.0	0.27	31	3	2.9	1.1	0.3	19	5	0.03	1	3		8	
0.04	2	0.02	1	tr(a)	0	0.003	1	unk.	4	1	0.1	unk.	27	3	193	60	unk.	0.4	2	0.3	2	unk.	unk.	unk.	B
0.1	1	3	1	0.09	1	198	4	unk.	0.1	0	0.1	0.64	42	4	8.4	1.5	unk.	7	2	0.14	7	unk.		unk.	
0.04	3	0.01	0	0.03	1	tr	0	0	3	1	tr	0.0	18	2	119	27	14	0.2	1	0.1	1	18.7	13	33	C
0.3	1	2	1	0.06	1	10	0	0	tr	0	tr	0.51	24	2	5.2	0.7	0.4	4	1	0.02	1	4		10	
0.04	2	0.01	0	0.02	0	tr	0	0	2	1	tr	0.0	16	2	111	24	11	0.2	1	0.2	1	14.5	10	31	D
0.2	1	2	0	0.05	1	81	2	0	tr	0	tr	0.45	21	2	4.8	0.6	0.3	3	1	0.02	1	3		16	
0.05	3	0.01	1	0.06	1	0.001	0	0	4	1	0.2	0.0	20	2	137	39	11	0.4	2	0.3	2	16.6	11	41	E
0.3	2	4	1	0.10	1	209	4	tr	0.2	1	0.1	0.57	35	4	6.0	1.0	0.3	14	4	0.03	1	4		19	
0.05	3	0.01	1	0.07	1	0.001	0	0	5	1	2.6	1.2	22	2	156	30	12	0.4	2	0.2	2	20.8	14	43	F
0.3	2	3	1	0.12	1	176	4	0	3.2	11	1.4	0.61	34	3	6.8	0.8	0.3	12	3	0.03	2	5		12	
0.01	0	unk.	unk.	0.00	0	unk.	unk.	0(a)	0(a)	0	unk.	unk.	10	1	1	32	unk.	0.6	4	0.1	1	unk.	unk.	unk.	G
0.1	0	unk.	unk.	unk.	unk.	21	0	unk.	unk.	unk.	unk.	0.12	7	1	0.0	0.8	unk.	5	1	unk.	unk.	unk.		unk.	
0.25	15	unk.	unk.	unk.	unk.	unk.	unk.	97	unk.	unk.	unk.	unk.	148	15	105	193	unk.	0.5	3	unk.	unk.	unk.	unk.	unk.	H
0.1	1	unk.	unk.	unk.	unk.	464	9	14	unk.	unk.	unk.	1.06	155	16	4.5	4.9	unk.	unk.	unk.	unk.	unk.	unk.		unk.	
0.24	14	unk.	unk.	unk.	unk.	unk.	unk.	unk.	58	15	unk.	unk.	222	22	129	234	unk.	0.9	5	unk.	unk.	unk.	unk.	unk.	I
0.1	1	unk.	unk.	unk.	unk.	643	13	unk.	unk.	unk.	unk.	unk.	164	16	5.6	6.0	unk.	unk.	unk.	unk.	unk.	unk.		unk.	
0.34	20	0.12	6	0.89	15	unk.	unk.	unk.	unk.	unk.	unk.	unk.	200	20	150	unk.	unk.	0.7	4	1.2	8	unk.	unk.	unk.	J
1.6	8	unk.	unk.	unk.	unk.	300	6	unk.	unk.	unk.	unk.	unk.	60	6	6.5	unk.	unk.	60	15	0.16	8	unk.		unk.	
0.17	10	0.04	2	0.48	8	unk.	unk.	unk.	unk.	unk.	unk.	unk.	100	10	unk.	unk.	unk.	0.4	2	0.3	2	unk.	unk.	unk.	K
tr	0	unk.	unk.	unk.	unk.	100	2	unk.	unk.	unk.	unk.	unk.	100	10	unk.	unk.	unk.	18	5	0.08	4	unk.		unk.	
0.33	19	unk.	unk.	1.19	20	unk.	unk.	unk.	0(a)	0	unk.	unk.	300	30	200	unk.	unk.	tr	0	unk.	unk.	unk.	unk.	unk.	L
tr	0	unk.	unk.	unk.	unk.	200	4	unk.	unk.	unk.	unk.	unk.	unk.	unk.	8.7	unk.	unk.	unk.	unk.	unk.	unk.	unk.		unk.	
0.08	5	unk.	unk.	0(a)	0	unk.	unk.	0	unk.	unk.	unk.	unk.	73	7	23	122	unk.	0.9	5	unk.	unk.	unk.	unk.	unk.	M
unk.	unk.	unk.	unk.	unk.	unk.	6142	123	614	unk.	unk.	unk.	0.63	22	2	1.0	3.1	unk.	unk.	unk.	tr(a)	0	tr(a)		unk.	
0.10	6	unk.	unk.	0(a)	0	unk.	unk.	0	0(a)	0	unk.	unk.	72	7	29	153	38	1.2	7	unk.	unk.	unk.	unk.	115	N
0.3	2	unk.	unk.	unk.	unk.	5390	108	539	1.2	4	1.2	0.69	25	3	1.3	3.9	1.1	14	4	0.06	3	tr(a)		unk.	
0.12	7	0.04	2	0.09	2	0.002	1	0	4	1	0.3	0.0	18	2	341	52	29	1.1	6	0.4	3	37.4	25	106	O
1.1	5	30	8	0.30	3	793	16	0	0.4	1	0.3	1.08	55	6	14.8	1.3	0.8	23	6	0.05	3	14		16	
0.01	1	0.01	1	0.00	0	unk.	unk.	0	0(a)	0	unk.	unk.	6	1	tr	62	27	0.3	2	tr	0	unk.	unk.	14	P
0.2	1	2	1	0.07	1	5	0	0	unk.	unk.	unk.	0.18	6	1	0.0	1.6	0.8	6	1	0.02	1	tr(a)		unk.	
0.00	0	0.00	0	0.00	0	unk.	unk.	unk.	unk.	unk.	unk.	unk.	1	0	5	3	unk.	tr	0	tr	0	unk.	unk.	unk.	Q
0.0	0	0	0	0.00	0	161	3	unk.	unk.	unk.	unk.	0.03	2	0	0.2	0.1	unk.	tr	0	unk.	unk.	unk.		unk.	
0.00	0	0.00	0	0.00	0	unk.	unk.	unk.	unk.	unk.	unk.	unk.	tr	0	2	2	unk.	unk.	unk.	unk.	unk.	unk.	unk.	unk.	R
0.0	0	0	0	0.00	0	14	0	unk.	unk.	unk.	unk.	0.02	1	0	0.1	0.1	unk.	tr	0	unk.	unk.	unk.		unk.	
0.02	1	0.01	0	0.05	1	unk.	unk.	unk.	unk.	unk.	unk.	unk.	18	2	13	30	unk.	tr	0	0.1	0	unk.	unk.	unk.	S
tr	0	1	0	0.05	1	72	1	unk.	unk.	unk.	unk.	0.16	17	2	0.6	0.8	unk.	2	1	unk.	unk.	unk.		unk.	

	FOOD	Portion	Weight in grams / Conversion for 100 g	Kilocalories / H_2O g	Total carbohydrate g / Total fats g	Crude fiber g / Dietary fiber g	Total protein g / Total sugar g	% USRDA	Arginine mg / Histidine mg	Isoleucine mg / Leucine mg	Lysine mg / Methionine mg	Phenylalanine mg / Threonine mg	Valine mg / Tryptophan mg	Cystine mg / Tyrosine mg	Polyunsat. fatty acids g / Monounsat. fatty acids g	Saturated fatty acids g / P/S ratio	Linoleic acid g / Cholesterol mg	Thiamin mg / Ascorbic acid mg	% USRDA
A	dessert topping,non-dairy pressurized	1/4 C	18	46	2.8	0.00	0.2	**0**	7	10	14	9	12	1	0.04	3.31	0.04	0.00	**0**
			5.71	10.6	3.9	0(a)	2.8		5	17	5	7	2	10	0.34	0.01	0	0	**0**
B	dessert topping,whipped,reduced calorie,prep-D-Zerta	1/4 C	19	47	2.0	0.03	0.9	**1**	unk.	unk.	unk.	unk.	unk.	unk.	0.48	2.59	0.44	0.01	**1**
			5.26	12.0	4.0	0(a)	1.9		unk.	unk.	unk.	unk.	unk.	unk.	0.63	0.18	tr	0	**0**
C	DILL seed	1 tsp	2.2	7	1.2	0.46	0.3	**1**	28	17	23	15	25	unk.	0.02	0.02	0.02	0.01	**1**
			45.45	0.2	0.3	unk.	0.0		7	20	3	13	unk.	unk.	0.21	1.38	0	unk.	**unk.**
D	dill weed,dried	1 tsp	1.0	3	0.6	0.12	0.2	**0**	unk.	unk.	unk.	unk.	unk.	unk.	unk.	unk.	unk.	tr	**0**
			100.00	0.1	tr	unk.	0.0		unk.	unk.	unk.	unk.	unk.	unk.	unk.	unk.	0	unk.	**unk.**
E	DIP,French onion-Sealtest	1 Tbsp	14	23	1.1	unk.	0.5	**1**	unk.	unk.	unk.	unk.	unk.	unk.	unk.	unk.	unk.	tr	**0**
			7.04	10.1	1.8	unk.	0.6		unk.	unk.	unk.	unk.	unk.	unk.	unk.	unk.	unk.	0	**0**
F	DOUGHNUT,cake type,iced or sugared	3-1/4" dia x	45	184	26.4	tr	2.6	**5**	unk.	131	91	140	130	unk.	unk.	unk.	unk.	0.08	**5**
		1"	*2.22*	unk.	7.9	unk.	9.4		unk.	208	44	88	33	unk.	unk.	unk.	20	tr	**0**
G	doughnut,cake type,plain	3-1/4" dia x	42	164	21.6	0.04	1.9	**3**	unk.	101	69	108	100	59	unk.	1.68	0.42	0.07	**5**
		1"	*2.38*	10.0	7.8	unk.	8.4		38	160	34	67	25	84	5.04	unk.	19	tr	**0**
H	doughnut,raised,jelly-filled	1 average	65	226	30.0	tr	3.4	**6**	unk.	174	120	186	172	38	unk.	unk.	unk.	0.12	**8**
			1.54	unk.	8.8	unk.	18.9		49	276	58	116	44	21	unk.	unk.	21	tr	**0**
I	doughnut,raised,plain	1 average	42	174	15.8	0.08	2.6	**5**	unk.	136	94	145	134	29	unk.	2.52	0.84	0.07	**5**
			2.38	11.9	11.2	unk.	4.7		38	215	45	91	34	67	7.14	unk.	17	0	**0**
J	DRUMSTICK,ice cream-Sealtest	1 average	60	186	21.5	unk.	2.6	**0**	unk.	unk.	unk.	unk.	unk.	unk.	unk.	unk.	unk.	0.02	**1**
			1.67	25.4	9.9	unk.	unk.		unk.	unk.	unk.	unk.	unk.	unk.	unk.	unk.	unk.	0	**0**
K	DUCK,domesticated,meat w/skin, roasted	1/2 bird	382	1287	0.0	0.00	72.5	**161**	4905	3331	5677	2873	3583	1142	13.94	36.94	12.84	0.65	**43**
			0.26	198.0	108.3	0(a)	0(a)		1765	5596	1814	2953	886	2445	44.01	0.38	321	0	**0**
L	duck,domesticated,meat w/o skin, roasted,one-half	1/2 bird	227	456	0.0	0.00	53.3	**118**	3403	2738	4560	2234	2788	819	3.25	9.47	2.93	0.59	**39**
			0.44	145.8	25.4	0(a)	0(a)		1407	4501	1441	2277	742	2029	7.35	0.34	202	0	**0**
M	duck,wild,breast,meat w/o skin,raw	1/2 breast	83	102	0.0	0.00	16.5	**37**	1052	846	1410	691	862	253	0.48	1.10	0.42	0.35	**23**
			1.20	62.7	3.5	0(a)	0(a)		435	1392	446	704	229	627	0.85	0.44	unk.	5	**9**
N	DUMPLING,home recipe	2 med	29	70	11.7	0.04	2.6	**5**	61	128	115	136	137	51	0.21	0.53	0.19	0.07	**5**
			3.45	12.4	1.3	0.43	0.8		64	213	53	97	34	101	0.41	0.39	33	tr	**0**
O	dumpling,home recipe,apple	1 average	300	566	78.7	1.22	4.0	**7**	unk.	192	134	212	188	73	5.64	9.25	5.23	0.17	**12**
			0.33	176.6	27.7	3.34	52.2		85	321	59	128	50	142	10.73	0.61	19	5	**8**
P	dumpling,home recipe,pear	1 average	290	540	96.6	2.26	7.3	**12**	unk.	338	255	358	346	109	0.80	9.03	0.59	0.24	**16**
			0.34	171.0	15.4	4.75	52.8		146	554	110	231	88	274	3.66	0.09	41	8	**13**
Q	ECLAIR w/custard filling & chocolate icing	1 average	110	263	25.5	0.00	6.8	**14**	unk.	unk.	unk.	unk.	unk.	unk.	unk.	4.84	2.31	0.04	**3**
			0.91	61.8	15.0	unk.	15.7		unk.	unk.	unk.	unk.	unk.	unk.	6.82	unk.	unk.	0	**0**
R	EGG,CHICKEN,white,dry,desugared	1 C	86	294	tr(a)	0(a)	67.4	**150**	3999	3414	4610	4076	4438	2047	0(a)	0(a)	0(a)	0.16	**11**
			1.16	5.5	0(a)	0(a)	0(a)		1574	6321	2537	3053	1187	2752	0(a)	0.00	0(a)	0(a)	**0**
S	egg,chicken,white,fresh or frozen,raw	1 lg	33	14	tr(a)	0(a)	3.1	**7**	198	168	218	148	219	98	0(a)	0(a)	0(a)	tr	**0**
			3.03	29.5	0(a)	0(a)	0(a)		70	273	128	139	49	135	0(a)	0.00	0(a)	0(a)	**0**

Riboflavin mg / Niacin mg	% USRDA	Vitamin B_6 mg / Folacin mcg	% USRDA	Vitamin B_{12} mcg / Pantothenic acid mg	% USRDA	Biotin mg / Vitamin A IU	% USRDA	Preformed A RE / Beta carotene RE	Vitamin D IU / Vitamin E IU	% USRDA	Total tocopherol mg / Alpha tocopherol mg	Other tocopherol mg / Total ash g	Calcium mg / Phosphorus mg	% USRDA	Sodium mg / Sodium meq	Potassium mg / Potassium meq	Chlorine mg / Chlorine meq	Iron mg / Magnesium mg	% USRDA	Zinc mg / Copper mg	% USRDA	Iodine mcg / Selenium mcg	% USRDA	Manganese mcg / Chromium mcg	
0.00	**0**	0.00	**0**	0.00	**0**	unk.	**unk.**	unk.	unk.	**unk.**	unk.	unk.	1	**0**	11	3	unk.	0.0	**0**	0.0	**0**	unk.	**unk.**	unk.	A
0.0	**0**	0	**0**	0.00	**0**	83	**2**	unk.	unk.	**unk.**	unk.	0.05	3	**0**	0.5	0.1	unk.	tr	**0**	unk.	**unk.**	unk.		unk.	
0.06	**3**	0.01	**1**	0.06	**1**	unk.	**unk.**	0	0	**0**	0(a)	0(a)	21	**2**	43	52	unk.	tr	**0**	0.0	**0**	unk.	**unk.**	unk.	B
tr	**0**	0	**0**	0.14	**1**	53	**1**	0	0.0	**0**	0(a)	0.02	30	**3**	1.8	1.3	unk.	4	**1**	tr	**0**	unk.		unk.	
0.01	**0**	unk.	**unk.**	0.00	**0**	unk.	**unk.**	0(a)	0(a)	**0**	unk.	unk.	33	**3**	tr	26	unk.	0.4	**2**	0.1	**1**	unk.	**unk.**	unk.	C
0.1	**0**	unk.	**unk.**	unk.	**unk.**	1	**0**	unk.	unk.	**unk.**	unk.	0.15	6	**1**	0.0	0.7	unk.	6	**1**	unk.	**unk.**	unk.		unk.	
tr	**0**	0.01	**1**	0.00	**0**	unk.	**unk.**	0(a)	0(a)	**0**	unk.	unk.	18	**2**	2	33	unk.	0.5	**3**	tr	**0**	unk.	**unk.**	unk.	D
tr	**0**	unk.	**unk.**	unk.	**unk.**	unk.	**unk.**	unk.	unk.	**unk.**	unk.	0.13	5	**1**	0.1	0.8	unk.	5	**1**	unk.	**unk.**	unk.		unk.	
0.02	**1**	0.01	**1**	0.08	**1**	unk.	**unk.**	unk.	1	**0**	unk.	unk.	17	**2**	80	25	unk.	tr	**0**	unk.	**unk.**	unk.	**unk.**	unk.	E
0.1	**1**	unk.	**unk.**	unk.	**unk.**	74	**2**	unk.	unk.	**unk.**	tr	0.28	13	**1**	3.5	0.6	unk.	2	**0**	unk.	**unk.**	unk.		unk.	
0.07	**4**	0.02	**1**	tr(a)	**0**	0.001	**0**	unk.	tr(a)	**0**	unk.	unk.	15	**2**	unk.	unk.	unk.	0.7	**4**	unk.	**unk.**	unk.	**unk.**	unk.	F
0.6	**3**	4	**1**	0.17	**2**	34	**1**	unk.	unk.	**unk.**	unk.	unk.	32	**3**	unk.	unk.	unk.	10	**2**	unk.	**unk.**	unk.		unk.	
0.07	**4**	0.02	**1**	tr(a)	**0**	0.001	**0**	unk.	1	**0**	0.2	unk.	17	**2**	210	38	unk.	0.6	**3**	0.2	**1**	unk.	**unk.**	147	G
0.5	**3**	3	**1**	0.16	**2**	34	**1**	unk.	0.2	**1**	unk.	0.71	80	**8**	9.1	1.0	unk.	9	**2**	0.07	**4**	unk.		10	
0.10	**6**	0.02	**1**	tr(a)	**0**	0.002	**1**	27	1	**0**	unk.	unk.	28	**3**	unk.	unk.	unk.	0.8	**4**	0.5	**3**	unk.	**unk.**	unk.	H
0.9	**5**	12	**3**	0.25	**3**	121	**2**	3	unk.	**unk.**	unk.	unk.	42	**4**	unk.	unk.	unk.	16	**4**	0.07	**4**	unk.		unk.	
0.07	**4**	0.02	**1**	tr(a)	**0**	0.001	**0**	unk.	tr	**0**	1.1	unk.	16	**2**	98	34	unk.	0.6	**4**	0.3	**2**	unk.	**unk.**	unk.	I
0.5	**3**	9	**2**	0.16	**2**	25	**1**	unk.	1.3	**4**	unk.	0.42	32	**3**	4.3	0.9	unk.	8	**2**	0.05	**2**	unk.		unk.	
0.09	**5**	0.03	**2**	0.29	**5**	unk.	**unk.**	unk.	0	**0**	unk.	unk.	67	**7**	57	99	unk.	0.1	**1**	unk.	**unk.**	unk.	**unk.**	unk.	J
0.5	**3**	unk.	**unk.**	unk.	**unk.**	185	**4**	unk.	unk.	**unk.**	0.1	0.48	59	**6**	2.5	2.5	unk.	7	**2**	unk.	**unk.**	unk.		unk.	
1.03	**61**	0.69	**34**	1.15	**19**	unk.	**unk.**	unk.	unk.	**unk.**	unk.	unk.	42	**4**	225	779	unk.	10.3	**57**	7.1	**47**	unk.	**unk.**	unk.	K
18.5	**92**	23	**6**	4.19	**42**	802	**16**	unk.	unk.	**unk.**	unk.	3.13	596	**60**	9.8	19.9	unk.	61	**15**	0.87	**43**	unk.		unk.	
1.07	**63**	0.57	**28**	0.91	**15**	unk.	**unk.**	unk.	unk.	**unk.**	unk.	unk.	27	**3**	148	572	unk.	6.1	**34**	5.9	**39**	unk.	**unk.**	unk.	L
11.6	**58**	23	**6**	3.40	**34**	175	**4**	unk.	unk.	**unk.**	unk.	2.52	461	**46**	6.4	14.6	unk.	45	**11**	0.52	**26**	unk.		unk.	
0.26	**15**	0.52	**26**	0.63	**11**	unk.	**unk.**	unk.	unk.	**unk.**	unk.	unk.	2	**0**	47	222	unk.	3.7	**21**	0.6	**4**	unk.	**unk.**	unk.	M
2.9	**14**	unk.	**unk.**	0.64	**6**	44	**1**	unk.	unk.	**unk.**	unk.	1.10	154	**15**	2.1	5.7	unk.	18	**5**	0.27	**14**	unk.		unk.	
0.07	**4**	0.02	**1**	0.10	**2**	0.001	**0**	0	6	**2**	0.1	tr(a)	36	**4**	249	39	21	0.6	**3**	0.2	**2**	40.3	**27**	60	N
0.5	**3**	5	**1**	0.19	**2**	27	**1**	0	0.1	**0**	tr	1.10	64	**6**	10.8	1.0	0.6	21	**5**	0.04	**2**	8		9	
0.12	**7**	0.06	**3**	0.06	**1**	0.001	**0**	0	7	**2**	2.0	0.5	51	**5**	473	181	23	1.1	**6**	0.6	**4**	77.1	**51**	170	O
1.2	**6**	16	**4**	0.31	**3**	352	**7**	10	2.4	**8**	1.4	2.08	80	**8**	20.6	4.6	0.6	17	**4**	0.12	**6**	16		41	
0.27	**16**	0.07	**4**	0.15	**3**	0.001	**0**	0	17	**4**	1.8	0.8	157	**16**	922	442	67	3.2	**18**	0.6	**4**	128.9	**86**	238	P
1.9	**9**	33	**8**	0.44	**4**	571	**11**	3	2.1	**7**	1.0	4.75	178	**18**	40.1	11.3	1.9	56	**14**	0.31	**15**	24		51	
0.18	**10**	unk.	**unk.**	unk.	**unk.**	unk.	**unk.**	unk.	unk.	**unk.**	unk.	unk.	88	**9**	90	134	unk.	0.8	**4**	unk.	**unk.**	unk.	**unk.**	unk.	Q
0.1	**1**	unk.	**unk.**	unk.	**unk.**	374	**8**	unk.	unk.	**unk.**	unk.	0.88	123	**12**	3.9	3.4	unk.	unk.	**unk.**	unk.	**unk.**	unk.		unk.	
2.36	**139**	0(a)	**0**	0.18	**3**	0.033	**11**	0(a)	0(a)	**0**	0(a)	0(a)	46	**5**	1101	1109	1182	0.4	**2**	0.1	**1**	25.8	**17**	43	R
1.0	**5**	69	**17**	0.58	**6**	0(a)	**0**	0(a)	0.0	**0**	0(a)	5.16	98	**10**	47.9	28.4	33.3	76	**19**	0.15	**8**	tr(a)		unk.	
0.08	**5**	tr	**0**	0.01	**0**	0.002	**1**	0(a)	0(a)	**0**	0(a)	0(a)	3	**0**	61	49	58	0.0	**0**	0.0	**0**	2.3	**2**	2	S
tr	**0**	tr	**0**	tr	**0**	0(a)	**0**	0(a)	0.0	**0**	0(a)	0.23	5	**1**	2.6	1.2	1.6	4	**1**	0.01	**0**	tr(a)		unk.	

	FOOD	Portion	Weight in grams / Conversion for 100 g	Kilocalories / H_2O g	Total carbohydrate g / Total fats g	Crude fiber g / Dietary fiber g	Total protein g / Total sugar g	**% USRDA**	Arginine mg / Histidine mg	Isoleucine mg / Leucine mg	Lysine mg / Methionine mg	Phenylalanine mg / Threonine mg	Valine mg / Tryptophan mg	Cystine mg / Tyrosine mg	Polyunsat. fatty acids g / Monounsat. fatty acids g	Saturated fatty acids g / P/S ratio	Linoleic acid g / Cholesterol mg	Thiamin mg / Ascorbic acid mg	**% USRDA**
A	egg,chicken,whole,dry,plain	1 C	85	507	2.9	0.00	38.3	**85**	2493	1895	2592	2040	2303	952	4.25	15.30	3.91	0.36	**24**
			1.18	2.7	37.7	0(a)	0.0		926	3570	1190	1861	637	1598	14.42	0.28	1615	0	**0**
B	egg,chicken,whole,dry,stabilized	1 C	85	523	2.0	0.00	40.9	**91**	2621	2562	2764	2313	2948	976	4.85	11.22	4.15	0.27	**18**
			1.18	1.6	37.4	0(a)	0.0		989	3596	1323	2011	655	1703	13.68	0.43	1714	0	**0**
C	egg,chicken,whole,fresh or frozen,raw	1 lg	51	79	0.6	0.00	6.2	**14**	396	387	418	350	446	147	0.74	1.71	0.63	0.05	**3**
			1.96	38.0	5.7	0(a)	0.0		149	544	200	304	258	99	2.08	0.43	279	0	**0**
D	egg,chicken,whole,fried in butter	1 lg	46	83	0.5	0.00	5.4	**12**	344	336	363	304	387	128	0.69	2.41	0.58	0.03	**2**
			2.17	33.1	6.4	0(a)	0.0		130	472	173	264	86	224	2.18	0.29	246	0	**0**
E	egg,chicken,whole,hard-cooked	1 lg	50	79	0.6	0.00	6.1	**14**	389	380	410	343	437	145	0.72	1.67	0.62	0.03	**2**
			2.00	37.3	5.6	0(a)	0.0		147	533	196	298	97	253	2.04	0.43	274	0	**0**
F	egg,chicken,whole,hard-cooked,chopped	1 C	136	215	1.6	0.00	16.5	**37**	1057	1032	1115	933	1189	393	1.97	4.56	1.69	0.10	**6**
			0.74	101.4	15.2	0(a)	0.0		398	1450	533	811	264	687	5.55	0.43	745	0	**0**
G	egg,chicken,whole,poached	1 lg	50	79	0.6	0.00	6.0	**13**	387	378	408	342	435	144	0.72	1.66	0.61	0.03	**2**
			2.00	37.1	5.5	0(a)	0.0		146	531	196	297	97	252	2.03	0.43	273	0	**0**
H	egg,chicken,whole,scrambled w/butter & milk	1/2 C	110	163	2.4	0.00	10.3	**23**	626	639	705	571	733	229	1.23	4.85	1.02	0.07	**4**
			0.91	83.9	12.2	0(a)	0.0		250	971	323	498	162	433	4.04	0.25	427	tr	**0**
I	egg,chicken,yolk,dry	1 C	80	551	tr(a)	0(a)	25.8	**57**	1824	1296	1864	1200	1488	536	5.68	20.08	5.26	0.54	**36**
			1.25	2.2	48.6	0(a)	tr(a)		672	2400	704	1376	432	1208	18.89	0.28	10240	0(a)	**0**
J	egg,chicken,yolk,fresh,raw	1 lg	18	68	tr(a)	0(a)	2.9	**7**	205	145	217	111	170	47	0.81	2.11	0.74	0.04	**3**
			5.56	8.9	6.1	0(a)	tr(a)		71	249	68	145	43	130	2.38	0.38	247	0(a)	**0**
K	EGG,HOME RECIPE,creamed	1/2 C	158	231	9.0	0.02	9.7	**21**	501	593	669	524	669	184	1.19	8.96	0.92	0.08	**5**
			0.63	120.1	17.3	0.16	4.4		242	880	282	455	147	422	4.98	0.13	310	1	**2**
L	egg,home recipe,deviled	2 halves	61	145	0.9	0.01	6.2	**14**	395	379	410	343	437	144	0.72	2.95	4.28	0.04	**2**
			1.64	40.1	12.9	unk.	tr		146	533	196	298	97	252	3.59	0.25	280	0	**0**
M	egg,home recipe,foo young	1 average	144	150	4.5	0.64	11.0	**23**	611	527	818	473	679	176	2.99	2.79	2.76	0.12	**8**
			0.70	113.1	10.1	1.98	1.1		310	853	301	477	131	383	3.39	1.07	250	10	**17**
N	egg,home recipe,omelet,two eggs, butter & milk	1 serving	128	189	2.7	0.00	11.9	**27**	728	744	820	664	852	266	1.43	5.64	1.19	0.08	**5**
			0.78	97.7	14.2	0(a)	0.0		291	1130	376	580	188	504	4.70	0.25	497	tr	**0**
O	egg,home recipe,salad	1/2 C	124	307	1.8	0.03	12.5	**28**	792	759	820	686	874	289	1.45	6.23	9.46	0.07	**5**
			0.81	79.8	27.7	unk.	0.0		293	1066	392	596	194	505	7.56	0.23	562	0	**0**
P	EGG,OTHER,duck,whole,fresh,raw	1 med	70	129	1.0	0.00	9.0	**20**	535	419	666	588	619	199	0.85	2.58	0.39	0.10	**7**
			1.43	49.6	9.6	0(a)	0.0		224	768	403	515	182	429	4.26	0.33	619	0	**0**
Q	egg,other,goose,whole,fresh,raw	1 med	144	266	1.9	0.00	20.0	**44**	unk.	unk.	unk.	unk.	unk.	unk.	2.40	5.18	0.98	unk.	**unk.**
			0.69	101.4	19.1	0(a)	0.0		unk.	unk.	unk.	unk.	unk.	unk.	7.70	0.46	unk.	0	**0**
R	egg,other,quail,whole,fresh,raw	1 med	9.0	14	tr	0.00	1.2	**3**	unk.	unk.	unk.	unk.	unk.	unk.	0.12	0.32	0.08	0.01	**1**
			11.11	6.7	1.0	0(a)	0.0		unk.	unk.	unk.	unk.	unk.	unk.	0.35	0.37	76	0	**0**
S	egg,other,turkey,whole,fresh,raw	1 med	79	135	0.9	0.00	10.8	**24**	unk.	unk.	unk.	unk.	unk.	unk.	1.31	2.87	0.92	0.09	**6**
			1.27	57.3	9.4	0(a)	0.0		unk.	unk.	unk.	unk.	unk.	unk.	3.08	0.46	737	0	**0**

Riboflavin mg / Niacin mg	% USRDA	Vitamin B_6 mg / Folacin mcg	% USRDA	Vitamin B_{12} mcg / Pantothenic acid mg	% USRDA	Biotin mg / Vitamin A IU	% USRDA	Preformed A RE / Beta carotene RE	Vitamin D IU / Vitamin E IU	% USRDA	Total tocopherol mg / Alpha tocopherol mg	Other tocopherol mg / Total ash g	Calcium mg / Phosphorus mg	% USRDA	Sodium mg / Sodium meq	Potassium mg / Potassium meq	Chlorine mg / Chlorine meq	Iron mg / Magnesium mg	% USRDA	Zinc mg / Copper mg	% USRDA	Iodine mcg / Selenium mcg	% USRDA	Manganese mcg / Chromium mcg	
1.16	**68**	0.29	**14**	2.59	**43**	0.007	**2**	unk.	161	**40**	7.9	unk.	165	**17**	412	476	583	6.6	**37**	4.8	**32**	178.5	**119**	127	A
0.3	**2**	238	**60**	4.77	**48**	1070	**21**	unk.	9.5	**32**	unk.	3.40	710	**71**	17.9	12.2	16.4	36	**9**	0.23	**12**	tr(a)		unk.	
1.05	**62**	0.36	**18**	8.93	**149**	unk.	**unk.**	unk.	unk.	**unk.**	unk.	unk.	189	**19**	466	438	unk.	7.0	**39**	4.9	**32**	unk.	**unk.**	unk.	B
0.2	**1**	164	**41**	5.70	**57**	1742	**35**	unk.	unk.	**unk.**	unk.	3.09	608	**61**	20.3	11.2	unk.	42	**10**	unk.	**unk.**	unk.		unk.	
0.15	**9**	0.06	**3**	0.79	**13**	0.009	**3**	unk.	18	**5**	0.5	unk.	29	**3**	70	66	88	1.1	**6**	0.7	**5**	36.7	**25**	21	C
tr	**0**	33	**8**	0.88	**9**	265	**5**	unk.	0.6	**2**	unk.	0.48	92	**9**	3.1	1.7	2.5	6	**2**	0.03	**2**	tr(a)		3	
0.12	**7**	0.05	**3**	0.58	**10**	unk.	**unk.**	unk.	unk.	**unk.**	unk.	unk.	26	**3**	144	58	79	0.9	**5**	0.6	**4**	unk.	**unk.**	19	D
tr	**0**	22	**5**	0.76	**8**	286	**6**	unk.	unk.	**unk.**	unk.	0.63	80	**8**	6.3	1.5	2.2	6	**1**	0.03	**1**	unk.		3	
0.14	**8**	0.06	**3**	0.65	**11**	unk.	**unk.**	unk.	unk.	**unk.**	unk.	unk.	28	**3**	69	65	86	1.0	**6**	0.7	**5**	unk.	**unk.**	21	E
tr	**0**	25	**6**	0.86	**9**	260	**5**	unk.	unk.	**unk.**	unk.	0.47	90	**9**	3.0	1.7	2.4	6	**2**	0.03	**2**	unk.		3	
0.38	**22**	0.16	**8**	1.78	**30**	unk.	**unk.**	unk.	unk.	**unk.**	unk.	unk.	76	**8**	188	177	234	2.8	**16**	2.0	**13**	unk.	**unk.**	56	F
0.1	**0**	67	**17**	2.35	**24**	707	**14**	unk.	unk.	**unk.**	unk.	1.28	245	**25**	8.2	4.5	6.6	16	**4**	0.08	**4**	unk.		8	
0.13	**7**	0.05	**3**	0.61	**10**	unk.	**unk.**	unk.	unk.	**unk.**	unk.	unk.	29	**3**	147	65	86	1.0	**6**	0.7	**5**	unk.	**unk.**	21	G
tr	**0**	25	**6**	0.86	**9**	259	**5**	unk.	unk.	**unk.**	unk.	0.67	90	**9**	6.4	1.6	2.4	6	**2**	0.03	**2**	unk.		3	
0.26	**16**	0.10	**5**	1.09	**18**	unk.	**unk.**	unk.	unk.	**unk.**	unk.	unk.	81	**8**	266	146	unk.	1.6	**9**	1.2	**8**	unk.	**unk.**	unk.	H
0.1	**0**	38	**10**	1.41	**14**	535	**11**	unk.	unk.	**unk.**	unk.	1.31	816	**17**	11.6	3.7	unk.	13	**3**	unk.	**unk.**	unk.		unk.	
0.74	**43**	0.45	**22**	2.02	**34**	0.072	**24**	unk.	202	**51**	8.6	unk.	201	**20**	142	210	324	8.4	**47**	0.1	**1**	216.0	**144**	40	I
0.1	**1**	344	**86**	7.02	**70**	1401	**28**	unk.	10.3	**34**	unk.	2.80	872	**87**	6.2	5.4	9.1	21	**5**	0.14	**7**	tr(a)		unk.	
0.08	**5**	0.06	**3**	0.57	**10**	0.007	**2**	unk.	31	**8**	0.7	unk.	25	**3**	11	20	29	1.1	**6**	0.7	**5**	30.1	**20**	20	J
tr	**0**	12	**3**	0.88	**9**	143	**3**	unk.	0.8	**3**	unk.	0.30	109	**11**	0.5	0.5	0.8	2	**1**	0.02	**1**	tr(a)		unk.	
0.30	**18**	0.10	**5**	0.97	**16**	0.000	**0**	0	38	**9**	0.2	tr(a)	143	**14**	619	212	4	1.3	**7**	1.2	**8**	113.9	**76**	22	K
0.3	**1**	31	**8**	1.18	**12**	701	**14**	0	0.2	**1**	0.2	2.44	184	**18**	26.9	5.4	0.1	22	**5**	0.15	**7**	3		3	
0.15	**9**	0.06	**3**	0.66	**11**	tr(a)	**0**	0	0	**0**	4.6	2.4	31	**3**	180	70	tr(a)	1.1	**6**	0.7	**5**	11.7	**8**	2	L
tr	**0**	25	**6**	0.86	**9**	285	**6**	0	5.5	**18**	2.2	0.78	94	**9**	7.8	1.8	0.0	7	**2**	tr	**0**	1		0	
0.23	**14**	0.10	**5**	0.69	**12**	0.009	**3**	0	18	**5**	3.5	0.0	60	**6**	475	262	97	2.1	**12**	0.8	**6**	131.7	**88**	26	M
0.7	**4**	24	**6**	0.80	**8**	564	**11**	45	4.2	**14**	0.1	2.17	191	**19**	20.7	6.7	2.7	124	**31**	0.04	**2**	0		2	
0.31	**18**	0.12	**6**	1.27	**21**	unk.	**unk.**	unk.	unk.	**unk.**	unk.	unk.	95	**10**	310	170	unk.	1.9	**10**	1.4	**9**	unk.	**unk.**	unk.	N
0.1	**0**	45	**11**	1.64	**16**	622	**12**	unk.	unk.	**unk.**	unk.	1.52	193	**19**	13.5	4.3	unk.	15	**4**	unk.	**unk.**	unk.		unk.	
0.28	**17**	0.11	**6**	1.31	**22**	0(a)	**0**	0	0(a)	**0**	10.4	5.3	64	**6**	565	140	unk.	2.2	**12**	1.5	**10**	70.0	**47**	0	O
0.0	**0**	49	**12**	1.73	**17**	578	**12**	0	12.4	**42**	5.0	2.09	189	**19**	24.6	3.6	unk.	15	**4**	0.01	**1**	0		0	
0.28	**17**	0.17	**9**	3.77	**63**	unk.	**unk.**	unk.	unk.	**unk.**	unk.	unk.	45	**5**	102	155	unk.	2.7	**15**	1.0	**7**	unk.	**unk.**	unk.	P
0.1	**1**	56	**14**	unk.	**unk.**	930	**19**	unk.	unk.	**unk.**	unk.	0.80	154	**15**	4.4	4.0	unk.	11	**3**	unk.	**unk.**	unk.		unk.	
unk.	**unk.**	unk.	**unk.**	unk.	**unk.**	unk.	**unk.**	unk.	unk.	**unk.**	unk.	unk.	unk.	**unk.**	unk.	unk.	unk.	unk.	**unk.**	unk.	**unk.**	unk.	**unk.**	unk.	Q
unk.	**unk.**	unk.	**unk.**	unk.	**unk.**	unk.	**unk.**	unk.	unk.	**unk.**	unk.	1.56	unk.	**unk.**	unk.	unk.	unk.	unk.	**unk.**	unk.	**unk.**	unk.		unk.	
0.07	**4**	0.01	**1**	unk.	**unk.**	unk.	**unk.**	unk.	unk.	**unk.**	unk.	unk.	6	**1**	unk.	unk.	unk.	0.3	**2**	unk.	**unk.**	unk.	**unk.**	unk.	R
tr	**0**	unk.	**unk.**	unk.	**unk.**	27	**1**	unk.	unk.	**unk.**	unk.	0.10	20	**2**	unk.	unk.	unk.	unk.	**unk.**	unk.	**unk.**	unk.		unk.	
0.37	**22**	unk.	**unk.**	unk.	**unk.**	unk.	**unk.**	unk.	unk.	**unk.**	unk.	unk.	78	**8**	unk.	unk.	unk.	3.2	**18**	unk.	**unk.**	unk.	**unk.**	unk.	S
tr	**0**	unk.	**unk.**	unk.	**unk.**	unk.	**unk.**	unk.	unk.	**unk.**	unk.	0.62	134	**13**	unk.	unk.	unk.	unk.	**unk.**	unk.	**unk.**	unk.		unk.	

	FOOD	Portion	Weight in grams / Conversion for 100 g	Kilocalories / H_2O g	Total carbohydrate g / Total fats g	Crude fiber g / Dietary fiber g	Total protein g / Total sugar g	% USRDA	Arginine mg / Histidine mg	Isoleucine mg / Leucine mg	Lysine mg / Methionine mg	Phenylalanine mg / Threonine mg	Valine mg / Tryptophan mg	Cystine mg / Tyrosine mg	Polyunsat. fatty acids g / Monounsat. fatty acids g	Saturated fatty acids g / P/S ratio	Linoleic acid g / Cholesterol mg	Thiamin mg / Ascorbic acid mg	% USRDA
A	EGG,SUBSTITUTE,frozen,egg beaters -Fleischmann	1/4 C	60	40	3.0	unk.	7.0	**16**	unk.	unk.	unk.	unk.	unk.	unk.	unk.	unk.	unk.	0.07	**5**
			1.67	unk.	7.5	0(a)	unk.		unk.	unk.	unk.	unk.	unk.	unk.	unk.	unk.	0	0	**0**
B	egg,substitute,frozen,made w/egg white,corn oil	1/4 C	60	96	1.9	0.00	6.8	**15**	unk.	unk.	unk.	unk.	unk.	unk.	3.74	1.16	3.71	0.07	**5**
			1.67	43.9	6.7	0(a)	0.0		unk.	unk.	unk.	unk.	unk.	unk.	1.46	3.23	1	unk.	**unk.**
C	egg,substitute,powder	1 oz	28	126	6.2	0.00	15.8	**35**	unk.	unk.	unk.	unk.	unk.	unk.	0.48	1.07	0.40	0.06	**4**
			3.52	1.1	3.7	0(a)	unk.		unk.	unk.	unk.	unk.	unk.	unk.	1.37	0.45	162	tr	**0**
D	EGGPLANT,fresh,chopped,cooked, drained solids	1/2 C	100	19	4.1	0.90	1.0	**2**	unk.	56	30	43	65	6	unk.	tr(a)	unk.	0.05	**3**
			1.00	94.3	0.2	2.50	2.8		19	68	6	38	10	38	unk.	unk.	0(a)	3	**5**
E	eggplant,home recipe,fresh,fried	1/2 C	169	100	7.3	0.02	2.2	**4**	47	112	93	107	128	36	0.10	1.71	0.58	0.05	**4**
			0.53	52.5	7.1	1.79	2.4		53	169	38	86	27	85	4.36	0.06	30	2	**4**
F	eggplant,home recipe,Italian style	1/2 C	137	43	2.9	0.46	0.7	**1**	9	26	24	24	25	1	0.33	0.48	0.28	0.04	**3**
			0.73	60.9	3.5	0.76	1.9		7	30	7	23	6	9	2.41	0.70	0	31	**51**
G	ENDIVE, curly (escarole),raw	1/2 C	25	5	1.0	0.22	0.4	**1**	unk.	18	20	20	20	3	unk.	tr(a)	unk.	0.02	**1**
			4.00	23.3	tr	0.55	0.1		8	31	6	18	unk.	14	unk.	unk.	0	3	**4**
H	endive,French/Belgian,witlof,bleached head,chopped,raw	1/2 C	45	7	1.4	unk.	0.4	**1**	unk.	18	19	19	20	3	unk.	tr(a)	unk.	unk.	**unk.**
			2.22	42.8	tr	unk.	0.1		8	31	5	18	7	33	unk.	unk.	0	unk.	**unk.**
I	ENGLISH MUFFIN,toasted-Newly Wed	1 average	30	71	14.1	0.09	3.0	**5**	unk.	unk.	unk.	unk.	unk.	unk.	unk.	unk.	unk.	0.09	**6**
			3.33	unk.	0.3	unk.	unk.		unk.	unk.	unk.	unk.	unk.	unk.	unk.	unk.	unk.	0	**0**
J	ESCAROLE,curly endive,raw	1/2 C	25	5	1.0	0.22	0.4	**1**	unk.	18	20	20	20	3	unk.	tr(a)	unk.	0.02	**1**
			4.00	23.3	tr	0.55	0.1		8	31	6	18	unk.	14	unk.	unk.	0	3	**4**
K	FENNEL seed	1 tsp	1.9	7	1.0	0.30	0.3	**1**	13	13	14	12	17	4	0.03	0.01	0.03	0.01	**1**
			52.63	0.2	0.3	unk.	0.0		6	19	6	11	5	8	0.19	3.52	0	unk.	**unk.**
L	FENUGREEK seed	1 tsp	3.7	12	2.2	0.37	0.8	**1**	91	46	62	40	41	14	unk.	unk.	unk.	0.01	**1**
			27.03	0.3	0.2	unk.	0.0		25	65	13	33	14	28	unk.	unk.	0	tr	**0**
M	FIG,dry,uncooked	1 average	14	38	9.7	0.78	0.6	**1**	13	unk.	unk.	15	unk.	unk.	unk.	unk.	unk.	0.01	**1**
			7.14	3.2	0.2	2.38	9.0		unk.	unk.	unk.	unk.	unk.	25	unk.	unk.	0	0	**0**
N	fig,whole,canned,heavy syrup, solids & liquid	1/2 C	130	109	28.2	0.91	0.6	**1**	15	20	26	16	25	10	unk.	tr(a)	unk.	0.04	**3**
			0.77	100.0	0.3	2.72	27.3		9	28	5	21	5	28	unk.	unk.	0	1	**2**
O	fig,whole,canned,water pack, solids & liquid	1/2 C	124	60	15.4	0.87	0.6	**1**	14	19	25	15	24	10	unk.	tr(a)	unk.	0.04	**3**
			0.81	107.4	0.2	2.60	14.5		9	27	5	20	5	27	unk.	unk.	0	1	**2**
P	FILBERTS/hazelnuts,shelled	1/4 C	34	214	5.6	1.01	4.3	**7**	737	288	141	182	316	56	2.33	1.56	2.23	0.16	**10**
			2.96	2.0	21.1	unk.	0.0		97	317	47	140	71	147	16.83	1.50	0	1	**2**
Q	FISH cake-Mrs Paul's	3 oz	85	159	17.7	unk.	7.8	**17**	477	403	768	unk.	unk.	99	unk.	unk.	unk.	0.05	**3**
			1.18	51.2	6.1	unk.	unk.		298	648	unk.	386	94	309	unk.	unk.	unk.	0	**0**
R	fish stick,breaded,cooked,frozen	3 sticks	84	200	5.7	tr	18.1	**40**	1029	913	1483	678	962	38	5.92	1.62	5.61	0.07	**5**
			1.19	45.3	11.2	0(a)	0.0		43	1379	508	779	181	91	2.81	3.65	80	unk.	**unk.**
S	FLOUNDER,fillet,baked w/butter or margarine	1 fillet	56	113	0.0	0.00	16.8	**37**	1079	863	1712	622	890	139	unk.	unk.	unk.	0.04	**3**
			1.79	32.5	4.6	0(a)	0.0		481	1411	487	839	231	659	unk.	unk.	28	1	**2**

Row	Riboflavin mg / Niacin mg	% USRDA	Vitamin B6 mg / Folacin mcg	% USRDA	Vitamin B12 mcg / Pantothenic acid mg	% USRDA	Biotin mg / Vitamin A IU	% USRDA	Preformed A RE / Beta carotene RE	Vitamin D IU / Vitamin E IU	% USRDA	Total tocopherol mg / Alpha tocopherol mg	Other tocopherol mg / Total ash g	Calcium mg / Phosphorus mg	% USRDA	Sodium mg / Sodium meq	Potassium mg / Potassium meq	Chlorine mg / Chlorine meq	Iron mg / Magnesium mg	% USRDA	Zinc mg / Copper mg	% USRDA	Iodine mcg / Selenium mcg	% USRDA	Manganese mcg / Chromium mcg
A	0.26	**15**	unk.	**unk.**	0.90	**15**	unk.	**unk.**	unk.	24	**6**	0.6	unk.	20	**2**	130	unk.	unk.	1.1	**6**	0.6	**4**	unk.	**unk.**	unk.
	0.0	**0**	40	**10**	0.40	**4**	810	**16**	unk.	0.7	**2**	0.6	unk.	43	**4**	5.6	unk.	unk.	unk.	**unk.**	unk.	**unk.**	unk.		unk.
B	0.23	**13**	0.08	**4**	unk.	**unk.**	unk.	**unk.**	unk.	unk.	**unk.**	unk.	unk.	44	**4**	119	128	unk.	1.2	**7**	0.6	**4**	unk.	**unk.**	unk.
	unk.	**unk.**	unk.	**unk.**	1.00	**10**	810	**16**	unk.	unk.	**unk.**	unk.	0.78	43	**4**	5.2	3.3	unk.	unk.	**unk.**	unk.	**unk.**	unk.		unk.
C	0.50	**29**	unk.	**unk.**	unk.	**unk.**	unk.	**unk.**	unk.	unk.	**unk.**	unk.	unk.	93	**9**	227	211	unk.	0.9	**5**	unk.	**unk.**	unk.	**unk.**	unk.
	0.2	**1**	unk.	**unk.**	unk.	**unk.**	349	**7**	unk.	unk.	**unk.**	unk.	1.66	136	**14**	9.9	5.4	unk.	unk.	**unk.**	unk.	**unk.**	unk.		unk.
D	0.04	**2**	0.05	**2**	0.00	**0**	unk.	**unk.**	0	0(a)	**0**	unk.	unk.	11	**1**	1	150	unk.	0.6	**3**	unk.	**unk.**	unk.	**unk.**	unk.
	0.5	**3**	16	**4**	0.11	**1**	10	**0**	1	unk.	**unk.**	unk.	0.40	21	**2**	0.0	3.8	unk.	unk.	**unk.**	tr(a)	**0**	tr(a)		unk.
E	0.06	**4**	0.05	**2**	0.06	**1**	0.001	**0**	0	2	**1**	0.1	0(a)	20	**2**	183	125	11	0.8	**4**	0.1	**1**	37.8	**25**	3
	0.6	**3**	15	**4**	0.20	**2**	21	**0**	1	0.1	**0**	0(a)	0.85	39	**4**	7.9	3.2	0.3	17	**4**	0.02	**1**	0		0
F	0.04	**2**	0.08	**4**	0.00	**0**	0.001	**0**	0	0(a)	**0**	0.6	tr	11	**1**	38	117	7	0.4	**2**	0.1	**1**	8.7	**6**	28
	0.4	**2**	10	**2**	0.09	**1**	224	**5**	22	0.7	**2**	tr	0.36	16	**2**	1.6	3.0	0.2	8	**2**	0.03	**2**	0		4
G	0.03	**2**	unk.	**unk.**	0(a)	**0**	unk.	**unk.**	0	0(a)	**0**	unk.	unk.	20	**2**	4	74	18	0.4	**2**	unk.	**unk.**	unk.	**unk.**	unk.
	0.1	**1**	12	**3**	unk.	**unk.**	825	**17**	83	unk.	**unk.**	unk.	0.25	14	**1**	0.1	1.9	0.5	3	**1**	tr(a)	**0**	tr(a)		unk.
H	unk.	**unk.**	0.02	**1**	0.00	**0**	unk.	**unk.**	unk.	0(a)	**0**	unk.	unk.	8	**1**	3	82	unk.	0.2	**1**	unk.	**unk.**	unk.	**unk.**	unk.
	unk.	**unk.**	23	**6**	unk.	**unk.**	tr	**0**	unk.	unk.	**unk.**	unk.	0.27	9	**1**	0.1	2.1	unk.	6	**2**	tr(a)	**0**	tr(a)		unk.
I	0.04	**3**	unk.	**unk.**	unk.	**unk.**	unk.	**unk.**	unk.	unk.	**unk.**	unk.	unk.	48	**5**	201	48	unk.	0.7	**4**	unk.	**unk.**	unk.	**unk.**	unk.
	0.9	**5**	unk.	**unk.**	0.14	**1**	0	**0**	unk.	unk.	**unk.**	unk.	0.51	unk.	**unk.**	8.7	1.2	unk.	5	**1**	unk.	**unk.**	unk.		unk.
J	0.03	**2**	unk.	**unk.**	0(a)	**0**	unk.	**unk.**	0	0(a)	**0**	unk.	unk.	20	**2**	4	74	18	0.4	**2**	unk.	**unk.**	unk.	**unk.**	unk.
	0.1	**1**	12	**3**	unk.	**unk.**	825	**17**	83	unk.	**unk.**	unk.	0.25	14	**1**	0.1	1.9	0.5	3	**1**	tr(a)	**0**	tr(a)		unk.
K	0.01	**0**	unk.	**unk.**	0.00	**0**	unk.	**unk.**	0(a)	0(a)	**0**	unk.	unk.	23	**2**	2	32	unk.	0.4	**2**	0.1	**1**	unk.	**unk.**	unk.
	0.1	**1**	unk.	**unk.**	unk.	**unk.**	3	**0**	unk.	unk.	**unk.**	unk.	0.16	9	**1**	0.1	0.8	unk.	7	**2**	unk.	**unk.**	unk.		unk.
L	0.01	**1**	unk.	**unk.**	0.00	**0**	unk.	**unk.**	0(a)	0(a)	**0**	unk.	unk.	7	**1**	2	28	unk.	1.2	**7**	0.1	**1**	unk.	**unk.**	unk.
	0.1	**0**	2	**1**	unk.	**unk.**	unk.	**unk.**	unk.	unk.	**unk.**	unk.	0.13	11	**1**	0.1	0.7	unk.	7	**2**	unk.	**unk.**	unk.		unk.
M	0.01	**1**	0.02	**1**	0.00	**0**	unk.	**unk.**	0	0(a)	**0**	unk.	unk.	18	**2**	5	90	15	0.4	**2**	0.1	**1**	unk.	**unk.**	unk.
	0.1	**1**	4	**1**	0.06	**1**	11	**0**	1	unk.	**unk.**	unk.	0.32	11	**1**	0.2	2.3	0.4	10	**3**	0.06	**3**	tr(a)		unk.
N	0.04	**2**	unk.	**unk.**	0.00	**0**	unk.	**unk.**	0	0(a)	**0**	unk.	unk.	17	**2**	3	193	15	0.5	**3**	0.3	**2**	unk.	**unk.**	unk.
	0.3	**1**	1	**0**	0.09	**1**	39	**1**	4	unk.	**unk.**	unk.	0.39	17	**2**	0.1	4.9	0.4	26	**7**	tr(a)	**0**	tr(a)		unk.
O	0.04	**2**	unk.	**unk.**	0.00	**0**	unk.	**unk.**	0	0(a)	**0**	unk.	unk.	17	**2**	2	192	14	0.5	**3**	0.3	**2**	unk.	**unk.**	unk.
	0.2	**1**	unk.	**unk.**	0.09	**1**	37	**1**	4	unk.	**unk.**	unk.	0.37	17	**2**	0.1	4.9	0.4	25	**6**	tr(a)	**0**	tr(a)		unk.
P	0.18	**11**	0.18	**9**	0.00	**0**	unk.	**unk.**	0	0	**0**	9.5	unk.	71	**7**	1	238	23	1.1	**6**	1.0	**7**	unk.	**unk.**	1420
	0.3	**2**	24	**6**	0.39	**4**	36	**1**	4	11.4	**38**	7.1	0.84	114	**11**	0.0	6.1	0.6	59	**15**	0.43	**22**	1		unk.
Q	0.02	**1**	unk.	**unk.**	unk.	**unk.**	unk.	**unk.**	unk.	unk.	**unk.**	tr(a)	tr(a)	43	**4**	unk.	unk.	unk.	0.2	**1**	unk.	**unk.**	unk.	**unk.**	unk.
	2.7	**14**	tr(a)	**0**	tr(a)	**0**	42	**1**	unk.	0.0	**0**	tr(a)	2.24	unk.	**unk.**	unk.	unk.	unk.	unk.	**unk.**	unk.	**unk.**	tr(a)		tr(a)
R	0.09	**5**	0.01	**1**	0.20	**3**	0.002	**1**	25	5	**1**	8.4	unk.	24	**2**	75	246	15	0.9	**5**	0.2	**1**	unk.	**unk.**	1
	1.8	**9**	1	**0**	0.28	**3**	145	**3**	6	0.1	**1**	0.1	0.16	189	**19**	3.3	6.3	0.4	20	**5**	0.00	**0**	tr(a)		6
S	0.04	**3**	0.06	**3**	0.60	**10**	unk.	**unk.**	unk.	unk.	**unk.**	tr(a)	tr(a)	13	**1**	133	329	unk.	0.8	**4**	unk.	**unk.**	unk.	**unk.**	unk.
	1.4	**7**	tr(a)	**0**	0.24	**2**	unk.	**unk.**	unk.	0.0	**0**	tr(a)	1.23	193	**19**	5.8	8.4	unk.	unk.	**unk.**	unk.	**unk.**	tr(a)		tr(a)

	FOOD	Portion	Weight in grams / *Conversion for 100 g*	Kilocalories / H_2O g	Total carbohydrate g / Total fats g	Crude fiber g / Dietary fiber g	Total protein g / Total sugar g	**% USRDA**	Arginine mg / Histidine mg	Isoleucine mg / Leucine mg	Lysine mg / Methionine mg	Phenylalanine mg / Threonine mg	Valine mg / Tryptophan mg	Cystine mg / Tyrosine mg	Polyunsat. fatty acids g / Monounsat. fatty acids g	Saturated fatty acids g / P/S ratio	Linoleic acid g / Cholesterol mg	Thiamin mg / Ascorbic acid mg	**% USRDA**
A	flounder,fillet,cooked	1 fillet	56	78	0.0	0.00	16.8	**37**	1079	857	1478	622	890	139	0.33	0.16	0.01	0.04	**3**
			1.79	32.5	0.7	0(a)	0.0		481	1260	487	722	168	659	0.06	2.11	28	1	**2**
B	FLOUR,corn	1 C	117	431	89.9	0.82	9.1	**14**	unk.	422	263	414	466	118	1.59	0.35	1.57	0.23	**16**
			0.85	14.0	3.0	unk.	3.7		188	1183	170	364	55	558	0.75	4.53	0	0	**0**
C	flour,rye	1 C	115	385	87.3	unk.	9.4	**15**	46	354	338	451	483	193	unk.	unk.	unk.	0.46	**31**
			0.87	17.2	2.3	unk.	tr		225	628	145	338	113	193	unk.	unk.	0(a)	0	**0**
D	flour,soy,defatted	1 C	85	286	25.5	2.55	44.2	**68**	3488	2171	3101	2403	2326	775	unk.	unk.	unk.	0.68	**45**
			1.18	6.8	0.8	unk.	unk.		1240	3798	620	1860	620	1550	unk.	unk.	0(a)	0	**0**
E	flour,soy,full-fat	1 C	60	268	14.1	unk.	22.1	**34**	1741	1084	1548	1200	1161	387	unk.	unk.	unk.	0.45	**30**
			1.67	4.2	14.1	7.14	6.7		619	1896	310	929	310	774	unk.	unk.	0(a)	0	**0**
F	flour,soy,low fat	1 C	83	292	234.1	unk.	37.6	**58**	2966	1845	2636	2043	1977	659	unk.	unk.	unk.	0.75	**50**
			1.20	5.8	6.0	11.87	11.1		1054	3229	527	1582	527	1318	unk.	unk.	0(a)	0	**0**
G	flour,wheat,white,all purpose,enriched, sifted	1 C	115	419	87.5	0.34	12.1	**19**	unk.	555	275	664	521	241	0.70	0.26	0.67	0.74	**49**
			0.87	13.8	1.1	3.45	2.5		241	930	159	347	148	413	0.13	2.65	0	0	**0**
H	flour,wheat,white,self-rising-Aunt Jemima	1 C	125	479	104.1	0.25	13.1	**20**	unk.	604	299	721	566	263	0.76	0.29	0.72	0.80	**53**
			0.80	13.8	1.1	3.84	2.7		263	1011	173	378	161	449	0.14	2.65	0	0	**0**
I	flour,whole wheat	1 C	120	400	85.2	2.76	16.0	**25**	unk.	unk.	unk.	unk.	unk.	unk.	1.09	0.48	1.04	0.66	**44**
			0.83	14.4	2.4	11.62	3.1		unk.	unk.	unk.	unk.	unk.	unk.	0.34	2.27	unk.	0	**0**
J	FRANKFURTER/weiner,beef,cooked-Oscar Mayer	1 (10 per lb)	45	144	0.9	0(a)	4.9	**11**	313	238	440	218	248	63	0.81	5.89	0.63	0.02	**2**
			2.22	24.3	13.5	0(a)	unk.		169	401	134	227	unk.	178	5.67	0.14	27	11	**18**
K	frankfurter/weiner,cocktail (Vienna sausage)	2	32	77	0.1	0.00	4.5	**10**	unk.	245	407	184	254	63	1.06	3.01	0.51	0.03	**2**
			3.13	20.2	6.3	0(a)	tr		136	363	107	207	43	164	3.55	0.35	unk.	0	**0**
L	frankfurter/weiner,cooked	1 (10 per lb)	44	134	0.7	unk.	5.5	**12**	unk.	unk.	unk.	202	286	unk.	6.95	4.93	0.40	0.07	**4**
			2.27	25.2	12.0	0(a)	tr		unk.	unk.	132	unk.	unk.	unk.	5.41	1.41	27	0	**0**
M	frankfurter/weiner,cooked-Oscar Mayer	1 (8 per lb)	57	182	1.0	unk.	6.3	**14**	unk.	unk.	unk.	233	329	unk.	1.88	6.78	1.65	0.11	**8**
			1.75	30.8	17.1	unk.	0.0		unk.	unk.	152	unk.	unk.	unk.	7.30	0.28	30	15	**25**
N	frankfurter/weiner,turkey	1	45	102	0.7	unk.	6.4	**14**	unk.	unk.	unk.	unk.	unk.	unk.	unk.	unk.	unk.	0.02	**1**
			2.22	28.3	8.0	unk.	unk.		unk.	unk.	unk.	unk.	unk.	unk.	unk.	unk.	48	unk.	**unk.**
O	FRENCH TOAST-Aunt Jemima	1 slice	43	85	13.2	0.04	3.2	**5**	unk.	unk.	unk.	unk.	unk.	unk.	unk.	unk.	unk.	0.08	**5**
			2.35	23.2	2.2	unk.	unk.		unk.	unk.	unk.	unk.	unk.	unk.	unk.	unk.	unk.	0	**0**
P	French toast,cinnamon swirl -Aunt Jemima	1 slice	47	106	15.0	0.14	3.5	**5**	unk.	unk.	unk.	unk.	unk.	unk.	unk.	unk.	unk.	0.07	**5**
			2.14	23.9	3.6	unk.	unk.		unk.	unk.	unk.	unk.	unk.	unk.	unk.	unk.	unk.	0	**0**
Q	FROG LEG,w/seasoned flour,fried	1 med	24	70	2.0	0.00	4.3	**10**	unk.	unk.	unk.	unk.	unk.	unk.	unk.	unk.	unk.	0.03	**2**
			4.17	unk.	4.8	0(a)	0.0		unk.	unk.	unk.	unk.	unk.	unk.	unk.	unk.	12	0	**0**
R	FRUIT COCKTAIL,canned,heavy syrup, solids & liquid	1/2 C	128	97	25.1	0.51	0.5	**1**	unk.	unk.	unk.	unk.	unk.	unk.	unk.	tr(a)	unk.	0.03	**2**
			0.78	101.5	0.1	1.45	24.7		unk.	unk.	unk.	unk.	unk.	unk.	unk.	unk.	0	3	**4**
S	fruit cocktail,canned,light syrup, solids & liquid	1/2 C	123	73	19.2	0.49	0.5	**1**	unk.	unk.	unk.	unk.	unk.	unk.	unk.	tr(a)	unk.	0.02	**2**
			0.82	102.4	0.1	1.40	18.7		unk.	unk.	unk.	unk.	unk.	unk.	unk.	unk.	0	2	**4**

Riboflavin mg / Niacin mg	% USRDA	Vitamin B_6 mg / Folacin mcg	% USRDA	Vitamin B_{12} mcg / Pantothenic acid mg	% USRDA	Biotin mg / Vitamin A IU	% USRDA	Preformed A RE / Beta carotene RE	Vitamin D IU / Vitamin E IU	% USRDA	Total tocopherol mg / Alpha tocopherol mg	Other tocopherol mg / Total ash g	Calcium mg / Phosphorus mg	% USRDA	Sodium mg / Sodium meq	Potassium mg / Potassium meq	Chlorine mg / Chlorine meq	Iron mg / Magnesium mg	% USRDA	Zinc mg / Copper mg	% USRDA	Iodine mcg / Selenium mcg	% USRDA	Manganese mcg / Chromium mcg	
0.04	**3**	0.06	**3**	0.60	**10**	unk.	**unk.**	unk.	unk.	**unk.**	tr(a)	tr(a)	13	**1**	133	329	unk.	0.8	**4**	unk.	**unk.**	unk.	**unk.**	unk.	A
1.4	**7**	tr(a)	**0**	0.24	**2**	unk.	**unk.**	unk.	0.0	**0**	tr(a)	1.23	193	**19**	5.8	8.4	unk.	17	**4**	unk.	**unk.**	tr(a)		tr(a)	
0.07	**4**	unk.	**unk.**	0(a)	**0**	unk.	**unk.**	0	0(a)	**0**	unk.	unk.	7	**1**	1	unk.	unk.	2.1	**12**	0.9	**6**	unk.	**unk.**	unk.	B
1.6	**8**	unk.	**unk.**	unk.	**unk.**	398	**8**	40	unk.	**unk.**	unk.	0.94	192	**19**	0.0	unk.	unk.	unk.	**unk.**	unk.	**unk.**	unk.		unk.	
0.25	**15**	0.40	**20**	0.00	**0**	0.007	**2**	0	0	**0**	unk.	unk.	37	**4**	1	471	unk.	3.1	**17**	3.2	**22**	unk.	**unk.**	unk.	C
1.1	**6**	90	**22**	1.15	**12**	0	**0**	0	unk.	**unk.**	0.9	unk.	414	**41**	0.0	12.1	unk.	106	**26**	0.48	**24**	unk.		unk.	
0.20	**12**	0.51	**26**	0.00	**0**	unk.	**unk.**	unk.	unk.	**unk.**	unk.	unk.	212	**21**	13	2125	unk.	204	**33**	4.3	**29**	unk.	**unk.**	2975	D
2.3	**12**	255	**64**	2.12	**21**	unk.	**unk.**	unk.	unk.	**unk.**	unk.	5.10	595	**60**	0.5	54.3	unk.	5.9	**51**	1.36	**68**	unk.		unk.	
0.19	**11**	0.34	**17**	0.00	**0**	unk.	**unk.**	0	0	**0**	unk.	unk.	126	**13**	1	996	unk.	4.1	**23**	unk.	**unk.**	unk.	**unk.**	unk.	E
1.2	**6**	unk.	**unk.**	1.08	**11**	unk.	**unk.**	unk.	unk.	**unk.**	0.0	unk.	360	**36**	0.0	25.5	unk.	144	**36**	unk.	**unk.**	unk.		unk.	
0.30	**18**	0.56	**28**	0.00	**0**	unk.	**unk.**	0	0	**0**	unk.	unk.	199	**20**	1	1685	unk.	7.6	**42**	unk.	**unk.**	unk.	**unk.**	unk.	F
2.0	**10**	unk.	**unk.**	1.74	**17**	unk.	**unk.**	unk.	unk.	**unk.**	0.0	unk.	531	**53**	0.0	43.1	unk.	241	**60**	unk.	**unk.**	unk.		unk.	
0.46	**27**	0.07	**3**	0.00	**0**	0.001	**0**	0	0(a)	**0**	3.8	2.0	18	**2**	2	109	82	3.7	**21**	0.8	**5**	unk.	**unk.**	460	G
6.1	**31**	24	**6**	0.53	**5**	0	**0**	0	4.5	**15**	1.8	0.46	100	**10**	0.1	2.8	2.3	34	**9**	0.15	**8**	61		69	
0.50	**29**	0.07	**4**	0.00	**0**	0.001	**0**	0	0	**0**	unk.	unk.	264	**26**	1566	129	unk.	3.6	**20**	0.9	**6**	unk.	**unk.**	unk.	H
6.6	**33**	26	**7**	0.58	**6**	0	**0**	0	unk.	**unk.**	unk.	5.12	674	**67**	68.1	3.3	unk.	25	**6**	0.16	**8**	tr		75	
0.14	**9**	1.12	**56**	0(a)	**0**	0.011	**4**	0	0(a)	**0**	3.9	2.1	49	**5**	4	444	unk.	4.0	**22**	2.9	**19**	unk.	**unk.**	unk.	I
5.2	**26**	46	**11**	1.30	**13**	0	**0**	0	4.7	**16**	1.8	2.04	446	**45**	0.2	11.3	unk.	136	**34**	unk.	**unk.**	unk.		unk.	
0.04	**3**	0.05	**3**	0.73	**12**	tr(a)	**0**	0(a)	10	**3**	tr(a)	tr(a)	6	**1**	466	71	675	0.6	**4**	0.9	**6**	unk.	**unk.**	18	J
1.1	**6**	tr(a)	**0**	tr(a)	**0**	0	**0**	0(a)	0.0	**0**	tr(a)	1.30	39	**4**	20.3	1.8	19.0	4	**1**	0.03	**2**	tr(a)		2	
0.04	**2**	0.03	**1**	unk.	**unk.**	unk.	**unk.**	unk.	0	**0**	tr(a)	tr(a)	3	**0**	unk.	unk.	unk.	0.7	**4**	unk.	**unk.**	unk.	**unk.**	unk.	K
0.8	**4**	unk.	**unk.**	unk.	**unk.**	unk.	**unk.**	unk.	0.0	**0**	tr(a)	0.93	49	**5**	unk.	unk.	unk.	unk.	**unk.**	unk.	**unk.**	tr(a)		tr(a)	
0.09	**5**	0.05	**2**	0.57	**10**	unk.	**unk.**	unk.	0	**0**	unk.	unk.	2	**0**	477	95	unk.	0.7	**4**	0.9	**6**	unk.	**unk.**	18	L
1.1	**6**	2	**0**	0.18	**2**	unk.	**unk.**	unk.	unk.	**unk.**	unk.	0.66	45	**5**	20.8	2.4	unk.	unk.	**unk.**	0.04	**2**	tr(a)		2	
0.06	**4**	0.07	**3**	0.74	**12**	unk.	**unk.**	0	21	**5**	unk.	unk.	6	**1**	651	93	unk.	0.7	**4**	1.0	**7**	unk.	**unk.**	23	M
1.5	**7**	10	**2**	unk.	**unk.**	0	**0**	0	unk.	**unk.**	unk.	1.82	51	**5**	28.3	2.4	unk.	6	**2**	0.04	**2**	tr(a)		2	
0.08	**5**	unk.	**unk.**	unk.	**unk.**	unk.	**unk.**	unk.	unk.	**unk.**	unk.	unk.	48	**5**	642	81	unk.	0.8	**5**	unk.	**unk.**	unk.	**unk.**	unk.	N
1.9	**9**	unk.	**unk.**	unk.	**unk.**	unk.	**unk.**	unk.	unk.	**unk.**	unk.	1.59	60	**6**	27.9	2.1	unk.	unk.	**unk.**	unk.	**unk.**	unk.		unk.	
0.09	**5**	0.15	**8**	0.45	**8**	0.002	**1**	unk.	unk.	**unk.**	unk.	unk.	47	**5**	216	49	unk.	0.7	**4**	0.4	**3**	unk.	**unk.**	0	O
0.8	**4**	18	**5**	0.43	**4**	86	**2**	unk.	unk.	**unk.**	unk.	0.81	60	**6**	9.4	1.2	unk.	6	**2**	0.04	**2**	unk.		unk.	
0.08	**5**	0.15	**8**	0.45	**8**	0.002	**1**	unk.	unk.	**unk.**	unk.	unk.	51	**5**	198	69	unk.	0.7	**4**	0.5	**3**	unk.	**unk.**	0	P
0.8	**4**	27	**7**	0.47	**5**	101	**2**	unk.	0.9	**3**	0.9	0.75	69	**7**	8.6	1.8	unk.	9	**2**	0.04	**2**	unk.		unk.	
0.06	**3**	0.02	**1**	tr(a)	**0**	unk.	**unk.**	0	0	**0**	tr(a)	tr(a)	5	**1**	unk.	unk.	unk.	0.3	**2**	unk.	**unk.**	unk.	**unk.**	unk.	Q
0.3	**2**	tr(a)	**0**	0.04	**0**	0	**0**	0	0.0	**0**	tr(a)	unk.	38	**4**	unk.	unk.	unk.	unk.	**unk.**	unk.	**unk.**	tr(a)		tr(a)	
0.01	**1**	0.04	**2**	0.00	**0**	unk.	**unk.**	0	0(a)	**0**	unk.	unk.	11	**1**	6	205	unk.	0.5	**3**	1.7	**11**	unk.	**unk.**	unk.	R
0.5	**3**	unk.	**unk.**	0.05	**1**	178	**4**	18	unk.	**unk.**	unk.	0.25	15	**2**	0.3	5.3	unk.	9	**2**	tr(a)	**0**	tr(a)		unk.	
0.01	**1**	0.04	**2**	0.00	**0**	unk.	**unk.**	0	0(a)	**0**	unk.	unk.	11	**1**	6	201	unk.	0.5	**3**	unk.	**unk.**	unk.	**unk.**	unk.	S
0.6	**3**	unk.	**unk.**	0.05	**1**	171	**3**	17	unk.	**unk.**	unk.	0.24	15	**2**	0.3	5.1	unk.	9	**2**	tr(a)	**0**	tr(a)		unk.	

	FOOD	Portion	Weight in grams / *Conversion for 100 g*	Kilocalories / H_2O g	Total carbohydrate g / Total fats g	Crude fiber g / Dietary fiber g	Total protein g / Total sugar g	**% USRDA**	Arginine mg / Histidine mg	Isoleucine mg / Leucine mg	Lysine mg / Methionine mg	Phenylalanine mg / Threonine mg	Valine mg / Tryptophan mg	Cystine mg / Tyrosine mg	Polyunsat. fatty acids g / Monounsat. fatty acids g	Saturated fatty acids g / P/S ratio	Linoleic acid g / Cholesterol mg	Thiamin mg / Ascorbic acid mg	**% USRDA**
A	fruit cocktail,canned,water pack,	1/2 C	123	45	11.9	0.49	0.5	**1**	unk.	unk.	unk.	unk.	unk.	unk.	unk.	tr(a)	unk.	0.02	**2**
	solids & liquid		*0.82*	109.8	0.1	1.40	11.4		unk.	unk.	unk.	unk.	unk.	unk.	unk.	unk.	0	2	**4**
B	FRUIT SALAD,canned,heavy syrup,	1/2 C	128	96	24.7	0.51	0.4	**1**	unk.	unk.	unk.	unk.	unk.	unk.	unk.	tr(a)	unk.	0.01	**1**
	solids & liquid		*0.78*	102.0	0.1	1.45	24.2		unk.	unk.	unk.	unk.	unk.	unk.	unk.	unk.	0	3	**4**
C	fruit salad,canned,water pack,	1/2 C	123	43	11.1	0.61	0.5	**1**	unk.	unk.	unk.	unk.	unk.	unk.	unk.	tr(a)	unk.	0.01	**1**
	solids & liquid		*0.82*	110.4	0.1	1.40	10.5		unk.	unk.	unk.	unk.	unk.	unk.	unk.	unk.	0	4	**6**
D	FUDGESICLE-Sealtest	1 average	73	91	18.6	tr(a)	3.8	**0**	unk.	unk.	unk.	unk.	unk.	unk.	unk.	unk.	unk.	0.03	**2**
			1.37	49.2	0.2	tr(a)	14.5		unk.	unk.	unk.	unk.	unk.	unk.	unk.	unk.	unk.	0	**0**
E	GARLIC clove,raw	1 average	3.0	4	0.9	0.04	0.2	**0**	unk.	unk.	unk.	unk.	unk.	unk.	tr	tr	tr	0.01	**1**
			33.33	1.8	tr	unk.	unk.		unk.	unk.	unk.	unk.	unk.	unk.	0.00	3.00	0	tr	**1**
F	garlic powder	1 tsp	2.8	9	2.0	0.05	0.5	**1**	47	18	16	14	20	5	0.02	0.01	0.01	0.01	**1**
			35.71	0.2	tr	unk.	0.0		9	29	9	13	6	6	tr	2.62	0	unk.	**unk.**
G	GELATIN,low calorie-D-Zerta	1/2 C	121	8	0.0	0.00	1.6	**2**	unk.	27	81	39	46	1	tr	tr	tr	0.00	**0**
			0.83	118.8	tr	0.00	tr		15	57	15	36	tr	7	tr	0.00	0	0	**0**
H	gelatin,sweetened,dry powder	1 pkg	85	315	74.8	0.00	8.0	**12**	unk.	80	256	127	144	7	0(a)	0(a)	unk.	tr(a)	**0**
			1.18	1.4	0.0	0.00	74.8		72	176	48	120	1	37	unk.	0.00	0(a)	tr(a)	**0**
I	gelatin,sweetened,made w/water,plain	1/2 C	120	71	16.9	0.00	1.8	**3**	unk.	29	89	43	51	2	0(a)	0(a)	0(a)	0.00	**0**
			0.83	101.0	0.0	0.00	16.9		16	61	18	40	1	8	0(a)	0.00	0	0	**0**
J	gelatin,sweetened,made w/water,	1/2 C	120	80	19.7	0.24	1.6	**3**	unk.	25	77	37	44	1	tr(a)	tr(a)	tr(a)	0.02	**2**
	w/fruit added		*0.83*	98.2	0.1	unk.	19.7		14	53	14	35	tr	7	tr(a)	0.00	0	4	**6**
K	gelatin,unsweetened,dry powder	1 Tbsp	7.0	23	0.0	0.00	6.0	**9**	unk.	95	296	143	169	5	0(a)	0(a)	unk.	0.00	**0**
			14.29	0.9	tr	unk.	0.0		54	205	55	134	tr	28	unk.	0.00	0(a)	0	**0**
L	GINGER root,fresh	1/2 oz	14	7	1.4	0.16	0.2	**0**	unk.	unk.	unk.	unk.	unk.	unk.	unk.	unk.	unk.	tr	**0**
			6.99	12.4	0.1	unk.	0.0		unk.	unk.	unk.	unk.	unk.	unk.	unk.	unk.	0	1	**2**
M	ginger,ground	1 tsp	1.8	6	1.3	0.11	0.2	**0**	3	5	5	4	7	1	0.02	0.03	0.02	tr	**0**
			55.56	0.2	0.1	unk.	0.0		3	7	1	3	1	2	0.02	0.68	0	unk.	**unk.**
N	GOOSE FAT	1 C	205	1846	0.0	0.00	0.0	**0**	0(a)	0(a)	0(a)	0(a)	0(a)	0	22.55	56.78	20.09	0(a)	**0**
			0.49	0.4	204.6	0(a)	0(a)		0(a)	0(a)	0(a)	0(a)	0(a)	0(a)	109.67	0.40	205	0(a)	**0**
O	GOOSE,domesticated,meat w/skin,	3x2-1/8x3/4"	85	259	0.0	0.00	21.4	**48**	1331	1006	1690	897	1047	unk.	2.14	5.84	1.90	0.07	**5**
	roasted		*1.18*	44.2	18.6	0(a)	0(a)		595	1793	517	955	unk.	684	8.09	0.37	77	0	**0**
P	goose,domesticated,meat w/o skin,	3x2-1/8x3/4"	85	202	0.0	0.00	24.6	**55**	unk.	unk.	unk.	unk.	unk.	unk.	1.31	3.88	1.16	0.08	**5**
	roasted		*1.18*	48.6	10.8	0(a)	0(a)		unk.	unk.	unk.	unk.	unk.	unk.	3.30	0.34	82	0	**0**
Q	GRANOLA BAR,cinnamon-General Mills	1 bar	24	110	16.1	unk.	2.0	**3**	unk.	unk.	unk.	unk.	unk.	unk.	unk.	unk.	unk.	0.06	**4**
			4.22	unk.	4.0	unk.	unk.		unk.	unk.	unk.	unk.	unk.	unk.	unk.	unk.	0(a)	tr	**0**
R	granola bar,coconut-General Mills	1 bar	24	120	15.0	unk.	2.0	**3**	unk.	unk.	unk.	unk.	unk.	unk.	unk.	unk.	unk.	0.06	**4**
			4.22	unk.	6.0	1.92	unk.		unk.	unk.	unk.	unk.	unk.	unk.	unk.	unk.	0(a)	tr	**0**
S	granola bar,honey & oats-	1 bar	24	110	16.1	unk.	2.0	**3**	unk.	unk.	unk.	unk.	unk.	unk.	unk.	unk.	unk.	tr	**0**
	General Mills		*4.22*	unk.	4.0	unk.	unk.		unk.	unk.	unk.	unk.	unk.	unk.	unk.	unk.	0(a)	tr	**0**

Riboflavin mg / Niacin mg	% USRDA	Vitamin B_6 mg / Folacin mcg	% USRDA	Vitamin B_{12} mcg / Pantothenic acid mg	% USRDA	Biotin mg / Vitamin A IU	% USRDA	Preformed A RE / Beta carotene RE	Vitamin D IU / Vitamin E IU	% USRDA	Total tocopherol mg / Alpha tocopherol mg	Other tocopherol mg / Total ash g	Calcium mg / Phosphorus mg	% USRDA	Sodium mg / Sodium meq	Potassium mg / Potassium meq	Chlorine mg / Chlorine meq	Iron mg / Magnesium mg	% USRDA	Zinc mg / Copper mg	% USRDA	Iodine mcg / Selenium mcg	% USRDA	Manganese mcg / Chromium mcg	
0.01	1	0.04	2	0.00	0	unk.	unk.	0	0(a)	0	unk.	unk.	11	1	6	206	unk.	0.5	3	unk.	unk.	unk.	unk.	unk.	A
0.6	3	unk.	unk.	0.05	1	184	4	18	unk.	unk.	unk.	0.24	16	2	0.3	5.3	unk.	9	2	tr(a)	0	tr(a)		unk.	
0.04	2	0.04	2	0.00	0	tr	0	0	0(a)	0	unk.	unk.	10	1	1	171	3	0.4	2	unk.	unk.	unk.	unk.	unk.	B
0.8	4	unk.	unk.	0.07	1	574	12	57	unk.	unk.	unk.	0.25	14	1	0.1	4.4	0.1	7	2	tr(a)	0	tr(a)		unk.	
0.04	2	0.04	2	0.00	0	unk.	unk.	0	0(a)	0	unk.	unk.	10	1	1	170	unk.	0.4	2	unk.	unk.	unk.	unk.	unk.	C
0.7	4	unk.	unk.	unk.	unk.	576	12	58	unk.	unk.	unk.	0.37	14	1	0.0	4.3	unk.	unk.	unk.	tr(a)	0	tr(a)		unk.	
0.18	10	0.04	2	0.61	10	unk.	unk.	unk.	0	0	0.0	unk.	129	13	55	173	unk.	0.1	0	unk.	unk.	unk.	unk.	unk.	D
0.7	3	unk.	unk.	unk.	unk.	0	0	unk.	0.0	0	unk.	1.02	99	10	2.4	4.4	unk.	18	5	unk.	unk.	unk.		unk.	
tr	0	unk.	unk.	0(a)	0	unk.	unk.	0	0(a)	0	unk.	unk.	1	0	1	16	unk.	tr	0	unk.	unk.	unk.	unk.	3	E
tr	0	unk.	unk.	unk.	unk.	tr	0	tr	unk.	unk.	unk.	0.04	6	1	0.0	0.4	unk.	1	0	tr(a)	0	tr(a)		unk.	
tr	0	unk.	unk.	0.00	0	unk.	unk.	0(a)	0(a)	0	unk.	unk.	2	0	1	31	unk.	0.1	0	0.1	1	unk.	unk.	unk.	F
tr	0	unk.	unk.	unk.	unk.	tr	0	unk.	unk.	unk.	unk.	0.09	12	1	0.0	0.8	unk.	2	0	unk.	unk.	unk.		unk.	
0.00	0	0.00	0	0.00	0	unk.	unk.	0	0	0	0(a)	0(a)	tr	0	9	45	tr(a)	tr	0	tr	0	tr(a)	0	tr(a)	G
0.0	0	0	0	0.00	0	0	0	0	0.0	0	0(a)	0.07	tr	0	0.4	1.2	0.0	tr	0	tr	0	tr(a)		tr(a)	
tr(a)	0	0.93	47	tr	0	unk.	unk.	tr	unk.	unk.	0(a)	0(a)	tr	0	270	8	unk.	0.0	0	0.1	1	unk.	unk.	unk.	H
tr(a)	0	tr(a)	0	tr	0	tr	0	tr	0.0	0	0(a)	0.85	241	24	11.8	0.2	unk.	tr	0	0.10	5	unk.		unk.	
0.00	0	0.14	7	tr	0	unk.	unk.	0	0	0	0(a)	0(a)	0	0	61	2	tr(a)	0.0	0	tr	0	unk.	unk.	tr(a)	I
0.0	0	tr(a)	0	tr	0	0	0	0	0.0	0	0(a)	0.24	54	5	2.7	0.0	0.0	5	1	0.02	1	tr(a)		tr(a)	
0.02	1	0.10	5	tr	0	unk.	unk.	0(a)	0(a)	0	unk.	unk.	5	1	41	123	unk.	tr(a)	0	tr	0	unk.	unk.	tr(a)	J
0.2	1	tr(a)	0	unk.	unk.	67	1	7	unk.	unk.	unk.	0.48	55	6	1.8	3.1	unk.	3	1	unk.	unk.	tr(a)		unk.	
0.00	0	0.00	0	tr	0	unk.	unk.	0	0	0	0(a)	0(a)	0	0	8	tr	unk.	0.0	0	unk.	unk.	unk.	unk.	unk.	K
0.0	0	tr(a)	0	unk.	unk.	0	0	0	0.0	0	0(a)	0.09	0	0	0.3	0.0	unk.	2	1	0.03	1	1		unk.	
0.01	0	unk.	unk.	0.00	0	unk.	unk.	0(a)	0(a)	0	unk.	unk.	3	0	1	38	unk.	0.3	2	unk.	unk.	unk.	unk.	unk.	L
0.1	1	unk.	unk.	0.03	0	1	0	tr	unk.	unk.	unk.	0.16	5	1	0.0	1.0	unk.	unk.	unk.	unk.	unk.	unk.		unk.	
tr	0	unk.	unk.	0.00	0	unk.	unk.	0(a)	0(a)	0	unk.	unk.	2	0	1	24	unk.	0.2	1	0.1	1	unk.	unk.	unk.	M
0.1	1	unk.	unk.	unk.	unk.	3	0	unk.	unk.	unk.	unk.	0.09	3	0	0.0	0.6	unk.	3	1	unk.	unk.	unk.		unk.	
0(a)	0	0(a)	0	0(a)	0	0(a)	0	unk.	unk.	unk.	unk.	unk.	unk.	unk.	unk.	unk.	unk.	unk.	unk.	unk.	unk.	unk.	unk.	unk.	N
0(a)	0	0(a)	0	0(a)	0	unk.	unk.	unk.	unk.	unk.	unk.	0.00	unk.	unk.	unk.	unk.	unk.	unk.	unk.	unk.	unk.	unk.		unk.	
0.27	16	0.31	16	unk.	unk.	unk.	unk.	unk.	unk.	unk.	unk.	unk.	11	1	59	280	unk.	2.4	13	unk.	unk.	unk.	unk.	unk.	O
3.5	18	2	0	unk.	unk.	59	1	unk.	unk.	unk.	unk.	0.82	230	23	2.6	7.1	unk.	19	5	0.22	11	unk.		unk.	
0.33	20	0.40	20	unk.	unk.	unk.	unk.	unk.	unk.	unk.	unk.	unk.	12	1	65	330	unk.	2.4	14	unk.	unk.	unk.	unk.	unk.	P
3.5	17	unk.	unk.	unk.	unk.	unk.	unk.	unk.	unk.	unk.	unk.	0.97	263	26	2.8	8.4	unk.	21	5	0.23	12	unk.		unk.	
tr	0	unk.	unk.	unk.	unk.	tr(a)	0	unk.	unk.	unk.	tr(a)	tr(a)	tr	0	76	78	unk.	0.7	4	unk.	unk.	unk.	unk.	tr(a)	Q
tr	0	unk.	unk.	unk.	unk.	tr	0	unk.	0.0	0	tr(a)	unk.	68	7	3.3	2.0	unk.	unk.	unk.	unk.	unk.	tr(a)		unk.	
0.03	2	unk.	unk.	unk.	unk.	tr(a)	0	unk.	unk.	unk.	tr(a)	tr(a)	tr	0	62	78	unk.	0.7	4	unk.	unk.	unk.	unk.	tr(a)	R
unk.	unk.	unk.	unk.	unk.	unk.	unk.	unk.	unk.	0.0	0	tr(a)	unk.	unk.	unk.	2.7	2.0	unk.	unk.	unk.	unk.	unk.	tr(a)		unk.	
0.03	2	unk.	unk.	unk.	unk.	tr(a)	0	unk.	unk.	unk.	tr(a)	tr(a)	tr	0	68	76	unk.	0.7	4	unk.	unk.	unk.	unk.	tr(a)	S
tr	0	unk.	unk.	unk.	unk.	tr	0	unk.	0.0	0	tr(a)	unk.	unk.	unk.	2.9	1.9	unk.	unk.	unk.	tr(a)	0	tr(a)		unk.	

	FOOD	Portion	Weight in grams / *Conversion for 100 g*	Kilocalories / H_2O g	Total carbohydrate g / Total fats g	Crude fiber g / Dietary fiber g	Total protein g / Total sugar g	% **USRDA**	Arginine mg / Histidine mg	Isoleucine mg / Leucine mg	Lysine mg / Methionine mg	Phenylalanine mg / Threonine mg	Valine mg / Tryptophan mg	Cystine mg / Tyrosine mg	Polyunsat. fatty acids g / Monounsat. fatty acids g	Saturated fatty acids g / P/S ratio	Linoleic acid g / Cholesterol mg	Thiamin mg / Ascorbic acid mg	% **USRDA**
A	granola bar,peanut-General Mills	1 bar	24	120	15.0	unk.	3.0	**5**	unk.	unk.	unk.	unk.	unk.	unk.	unk.	unk.	unk.	0.06	**4**
			4.22	unk.	5.0	unk.	unk.		unk.	unk.	unk.	unk.	unk.	unk.	unk.	unk.	0(a)	unk.	**unk.**
B	GRAPEFRUIT,canned,sections,syrup pack,solids & liquid	1/2 C	127	89	22.6	0.25	0.8	**1**	unk.	unk.	9	unk.	tr(a)	unk.	unk.	unk.	unk.	0.04	**3**
			0.79	103.0	0.1	0.51	6.3		unk.	unk.	tr(a)	unk.	1	unk.	unk.	unk.	0	38	**64**
C	grapefruit,canned,sections,water pack,solids & liquid	1/2 C	122	37	9.3	0.24	0.7	**1**	unk.	unk.	9	unk.	unk.	unk.	unk.	unk.	unk.	0.04	**2**
			0.82	111.4	0.1	unk.	6.1		unk.	unk.	tr(a)	unk.	1	unk.	unk.	unk.	0	37	**61**
D	grapefruit,fresh,any,half, edible portion	1/2 average	98	40	10.3	0.19	0.5	**1**	unk.	unk.	6	unk.	unk.	unk.	unk.	tr(a)	unk.	0.04	**3**
			1.03	86.2	0.1	0.59	4.9		unk.	unk.	0	unk.	1	unk.	unk.	unk.	0	37	**62**
E	grapefruit,fresh,any,sections	1/2 C	100	41	10.6	0.20	0.5	**1**	unk.	unk.	6	unk.	unk.	unk.	unk.	tr(a)	unk.	0.04	**3**
			1.00	88.4	0.1	0.60	5.0		unk.	unk.	0	unk.	1	unk.	unk.	unk.	0	38	**63**
F	GRAPES,canned,Thompson seedless, heavy syrup,solids & liquid	1/2 C	128	99	25.6	0.26	0.6	**1**	47	5	14	13	17	10	unk.	tr(a)	unk.	0.05	**3**
			0.78	101.2	0.1	unk.	25.3		27	13	22	17	3	11	unk.	unk.	0	3	**4**
G	grapes,canned,Thompson seedless,water pack,solids & liquid	1/2 C	123	62	16.7	0.24	0.6	**1**	45	5	14	13	17	10	unk.	tr(a)	unk.	0.05	**3**
			0.82	104.7	0.1	unk.	15.6		25	13	21	17	3	11	unk.	unk.	0	2	**4**
H	grapes,fresh,adherent skin (red/dark blue)	1/2 C	80	54	13.8	0.40	0.5	**1**	35	4	11	10	13	8	unk.	tr(a)	unk.	0.04	**3**
			1.25	65.1	0.2	1.32	13.1		18	10	16	13	2	8	unk.	unk.	0	3	**5**
I	grapes,fresh,slip skin (Concord/Delware)	1/2 C	77	53	12.0	0.46	1.0	**2**	73	8	22	21	27	16	unk.	tr(a)	unk.	0.04	**3**
			1.31	62.4	0.8	unk.	7.4		37	21	33	27	5	18	unk.	unk.	0	3	**5**
J	grapes,fresh,Thompson,green,seedless	1/2 C	80	54	13.8	0.40	0.5	**1**	35	4	11	10	13	8	unk.	tr(a)	unk.	0.04	**3**
			1.25	65.1	0.2	0.72	13.1		18	10	16	13	2	8	unk.	unk.	0	3	**5**
K	GRAVY,CANNED,au jus	1/4 C	60	10	1.5	unk.	0.7	**1**	unk.	unk.	unk.	unk.	unk.	unk.	0.01	0.06	0.01	0.01	**1**
			1.68	unk.	0.1	unk.	unk.		unk.	unk.	unk.	unk.	unk.	unk.	0.04	0.10	tr	1	**2**
L	gravy,canned,beef	1/4 C	58	31	2.8	unk.	2.2	**3**	unk.	unk.	unk.	unk.	unk.	unk.	0.05	0.69	0.04	0.02	**1**
			1.72	50.9	1.4	unk.	unk.		unk.	unk.	unk.	unk.	unk.	unk.	0.49	0.08	2	0	**0**
M	gravy,canned,chicken	1/4 C	60	47	3.2	unk.	1.1	**2**	unk.	unk.	unk.	unk.	unk.	unk.	0.89	0.84	0.85	0.01	**1**
			1.68	50.9	3.4	unk.	unk.		unk.	unk.	unk.	unk.	unk.	unk.	1.39	1.06	1	0	**0**
N	gravy,canned,mushroom	1/4 C	60	30	3.3	unk.	0.8	**1**	unk.	unk.	unk.	unk.	unk.	unk.	0.61	0.24	0.57	0.02	**1**
			1.68	53.0	1.6	unk.	unk.		unk.	unk.	unk.	unk.	unk.	unk.	0.69	2.55	0	0	**0**
O	gravy,canned,turkey	1/4 C	60	30	3.0	unk.	1.5	**2**	unk.	unk.	unk.	unk.	unk.	unk.	0.29	0.37	0.27	0.01	**1**
			1.68	52.8	1.3	unk.	unk.		unk.	unk.	unk.	unk.	unk.	unk.	0.45	0.79	1	0	**0**
P	GRAVY,DRY,au jus	4/5 oz pkg	24	79	9.9	0.05	3.0	**5**	unk.	unk.	unk.	unk.	unk.	unk.	0.13	1.66	0.10	unk.	**unk.**
			4.20	0.6	3.3	unk.	unk.		unk.	unk.	unk.	unk.	unk.	unk.	1.20	0.08	4	unk.	**unk.**
Q	gravy,dry,au jus,prep w/water	1/4 C	62	5	0.6	unk.	0.2	**0**	unk.	unk.	unk.	unk.	unk.	unk.	0.01	0.10	0.01	unk.	**unk.**
			1.63	60.1	0.2	unk.	unk.		unk.	unk.	unk.	unk.	unk.	unk.	0.07	0.06	tr	unk.	**unk.**
R	gravy,dry,brown	7/8 oz pkg	25	85	14.8	0.11	2.7	**4**	unk.	unk.	unk.	unk.	unk.	unk.	0.08	0.98	0.06	0.04	**3**
			4.03	1.3	2.0	unk.	unk.		unk.	unk.	unk.	unk.	unk.	unk.	0.71	0.08	2	unk.	**unk.**
S	gravy,dry,brown,prep w/water	1/4 C	65	2	0.4	0.00	0.1	**0**	unk.	unk.	unk.	unk.	unk.	unk.	0.00	0.03	0.00	tr	**0**
			1.53	64.6	0.1	unk.	unk.		unk.	unk.	unk.	unk.	unk.	unk.	0.02	0.00	tr	unk.	**unk.**

Riboflavin mg Niacin mg	% USRDA	Vitamin B_6 mg Folacin mcg	% USRDA	Vitamin B_{12} mcg Pantothenic acid mg	% USRDA	Biotin mg Vitamin A IU	% USRDA	Preformed A RE Beta carotene RE	Vitamin D IU Vitamin E IU	% USRDA	Total tocopherol mg Alpha tocopherol mg	Other tocopherol mg Total ash g	Calcium mg Phosphorus mg	% USRDA	Sodium mg Sodium meq	Potassium mg Potassium meq	Chlorine mg Chlorine meq	Iron mg Magnesium mg	% USRDA	Zinc mg Copper mg	% USRDA	Iodine mcg Selenium mcg	% USRDA	Manganese mcg Chromium mcg	
0.03	2	unk.	unk.	unk.	unk.	tr(a)	0	unk.	unk.	unk.	tr(a)	tr(a)	unk.	unk.	82	98	unk.	0.7	4	unk.	unk.	unk.	unk.	tr(a)	A
unk.	unk.	unk.	unk.	unk.	unk.	unk.	unk.	unk.	unk.	unk.	tr(a)	unk.	unk.	unk.	3.6	2.5	unk.	unk.	unk.	tr(a)	0	tr(a)		unk.	
0.03	2	0.03	1	0.00	0	0.001	0	0	0(a)	0	tr	unk.	17	2	1	171	3	0.4	2	0.1	0	unk.	unk.	13	B
0.3	1	14	4	0.15	2	13	0	1	0.0	0	unk.	0.51	18	2	0.1	4.4	0.1	14	4	0.02	1	unk.		19	
0.02	1	0.02	1	0.00	0	unk.	unk.	0	0(a)	0	unk.	unk.	16	2	5	176	unk.	0.4	2	0.1	0	unk.	unk.	12	C
0.2	1	13	3	0.15	2	12	0	1	unk.	unk.	unk.	0.49	17	2	0.2	4.5	unk.	13	3	0.02	1	unk.		18	
0.02	1	0.03	2	0.00	0	0.003	1	0	0(a)	0	0.3	tr	16	2	1	132	3	0.4	2	0.0	0	unk.	unk.	10	D
0.2	1	11	3	0.28	3	78	2	8	0.3	1	0.2	0.39	16	2	0.0	3.4	0.1	12	3	0.04	2	0		15	
0.02	1	0.03	2	0.00	0	0.003	1	0	0(a)	0	0.3	tr	16	2	1	135	3	0.4	2	0.0	0	unk.	unk.	10	E
0.2	1	11	3	0.28	3	80	2	8	0.3	1	0.3	0.40	16	2	0.0	3.4	0.1	12	3	0.04	2	0		15	
0.01	1	unk.	unk.	0(a)	0	unk.	unk.	0	0(a)	0	unk.	unk.	10	1	5	134	unk.	0.4	2	unk.	unk.	unk.	unk.	unk.	F
0.3	1	unk.	unk.	unk.	unk.	90	2	9	unk.	unk.	unk.	0.38	17	2	0.2	3.4	unk.	unk.	unk.	tr(a)	0	tr(a)		unk.	
0.01	1	unk.	unk.	0(a)	0	unk.	unk.	0	unk.	unk.	unk.	unk.	10	1	5	135	unk.	0.4	2	unk.	unk.	unk.	unk.	unk.	G
0.2	1	unk.	unk.	unk.	unk.	86	2	9	unk.	unk.	unk.	0.37	16	2	0.2	3.4	unk.	unk.	unk.	unk.	unk.	tr(a)		unk.	
0.02	1	0.06	3	0.00	0	0.001	0	0	0(a)	0	unk.	unk.	10	1	2	138	tr	0.3	2	0.1	1	unk.	unk.	52	H
0.2	1	4	1	0.06	1	80	2	8	unk.	unk.	0.6	0.32	16	2	0.1	3.5	0.0	5	1	0.00	0	0		12	
0.02	1	0.06	3	0.00	0	0.001	0	0	0(a)	0	unk.	unk.	12	1	2	121	2	0.3	2	unk.	unk.	unk.	unk.	50	I
0.2	1	4	1	0.06	1	76	2	8	unk.	unk.	unk.	0.31	9	1	0.1	3.1	0.0	10	3	0.03	1	tr(a)		11	
0.02	1	0.06	3	0.00	0	0.001	0	0	0(a)	0	unk.	unk.	10	1	2	138	tr	0.3	2	tr	0	unk.	unk.	52	J
0.2	1	4	1	0.06	1	80	2	8	unk.	unk.	0.6	0.32	16	2	0.1	3.5	0.0	5	1	0.00	0	0		12	
0.04	2	unk.	unk.	unk.	unk.	unk.	unk.	0	0(a)	0	unk.	unk.	2	0	unk.	unk.	unk.	0.4	2	unk.	unk.	unk.	unk.	unk.	K
0.5	3	unk.	unk.	unk.	unk.	0	0	0	unk.	unk.	unk.	unk.	unk.	unk.	unk.	unk.	unk.	unk.	unk.	unk.	unk.	unk.		unk.	
0.02	1	0.01	0	0.06	1	unk.	unk.	0	0(a)	0	unk.	unk.	3	0	29	47	unk.	0.4	2	0.6	4	unk.	unk.	116	L
0.4	2	unk.	unk.	unk.	unk.	0	0	0	unk.	unk.	unk.	0.93	18	2	1.3	1.2	unk.	unk.	unk.	0.06	3	unk.		unk.	
0.02	1	0.01	0	unk.	unk.	unk.	unk.	unk.	0(a)	0	unk.	unk.	12	1	344	65	unk.	0.3	2	0.5	3	unk.	unk.	119	M
0.3	1	unk.	unk.	unk.	unk.	220	4	unk.	unk.	unk.	unk.	0.95	17	2	15.0	1.7	unk.	unk.	unk.	0.06	3	unk.		unk.	
0.04	2	0.01	1	0.00	0	unk.	unk.	0	0(a)	0	unk.	unk.	4	0	340	63	unk.	0.4	2	0.4	3	unk.	unk.	179	N
0.4	2	unk.	unk.	unk.	unk.	0	0	0	unk.	unk.	unk.	0.93	9	1	14.8	1.6	unk.	unk.	unk.	0.06	3	unk.		unk.	
0.05	3	unk.	unk.	unk.	unk.	unk.	unk.	0	0(a)	0	unk.	unk.	2	0	unk.	unk.	unk.	0.4	2	unk.	unk.	unk.	unk.	unk.	O
0.8	4	unk.	unk.	unk.	unk.	0	0	0	unk.	unk.	unk.	0.95	unk.	unk.	unk.	unk.	unk.	unk.	unk.	unk.	unk.	unk.		unk.	
unk.	unk.	unk.	unk.	unk.	unk.	unk.	unk.	unk.	0(a)	0	unk.	unk.	45	5	2392	unk.	unk.	unk.	unk.	0.2	1	unk.	unk.	unk.	P
unk.	unk.	unk.	unk.	unk.	unk.	unk.	unk.	unk.	unk.	unk.	unk.	7.05	unk.	unk.	104.0	unk.	unk.	13	3	0.03	1	unk.		unk.	
unk.	unk.	unk.	unk.	unk.	unk.	unk.	unk.	unk.	0(a)	0	unk.	unk.	2	0	145	unk.	unk.	unk.	unk.	tr	0	unk.	unk.	unk.	Q
unk.	unk.	unk.	unk.	unk.	unk.	unk.	unk.	unk.	unk.	unk.	unk.	0.42	unk.	unk.	6.3	unk.	unk.	unk.	unk.	tr	0	unk.		unk.	
0.10	6	unk.	unk.	unk.	unk.	unk.	unk.	unk.	0(a)	0	unk.	unk.	70	7	1213	65	unk.	0.2	1	0.3	2	unk.	unk.	102	R
0.9	5	unk.	unk.	unk.	unk.	unk.	unk.	unk.	unk.	unk.	unk.	4.04	50	5	52.8	1.7	unk.	8	2	0.05	2	unk.		unk.	
tr	0	unk.	unk.	unk.	unk.	unk.	unk.	unk.	0(a)	0	unk.	unk.	2	0	31	2	unk.	tr	0	tr	0	unk.	unk.	3	S
tr	0	unk.	unk.	unk.	unk.	unk.	unk.	unk.	unk.	unk.	unk.	0.10	1	0	1.4	0.0	unk.	0	0	tr	0	unk.		unk.	

	FOOD	Portion	Weight in grams / Conversion for 100 g	Kilocalories / H_2O g	Total carbohydrate g / Total fats g	Crude fiber g / Dietary fiber g	Total protein g / Total sugar g	% USRDA	Arginine mg / Histidine mg	Isoleucine mg / Leucine mg	Lysine mg / Methionine mg	Phenylalanine mg / Threonine mg	Valine mg / Tryptophan mg	Cystine mg / Tyrosine mg	Polyunsat. fatty acids g / Monounsat. fatty acids g	Saturated fatty acids g / P/S ratio	Linoleic acid g / Cholesterol mg	Thiamin mg / Ascorbic acid mg	% USRDA
A	gravy,dry,chicken	4/5 oz pkg	23	83	14.3	0.06	2.6	**4**	unk.	unk.	unk.	unk.	unk.	unk.	0.45	0.53	0.42	unk.	**unk.**
			4.33	0.9	1.9	unk.	unk.		unk.	unk.	unk.	unk.	unk.	unk.	0.86	0.86	2	unk.	**unk.**
B	gravy,dry,chicken,prep w/water	1/4 C	65	21	3.6	0.02	0.7	**1**	unk.	unk.	unk.	unk.	unk.	unk.	0.11	0.13	0.10	unk.	**unk.**
			1.54	59.3	0.5	unk.	unk.		unk.	unk.	unk.	unk.	unk.	unk.	0.19	0.85	1	unk.	**unk.**
C	gravy,dry,mushroom	3/4 oz pkg	21	70	13.8	unk.	2.1	**3**	unk.	unk.	unk.	unk.	unk.	unk.	0.03	0.50	0.02	unk.	**unk.**
			4.69	0.7	0.9	unk.	unk.		unk.	unk.	unk.	unk.	unk.	unk.	0.24	0.06	1	unk.	**unk.**
D	gravy,dry,mushroom,prep w/water	1/4 C	65	17	3.4	unk.	0.5	**1**	unk.	unk.	unk.	unk.	unk.	unk.	0.01	0.13	0.01	unk.	**unk.**
			1.55	59.4	0.2	unk.	unk.		unk.	unk.	unk.	unk.	unk.	unk.	0.06	0.05	tr	unk.	**unk.**
E	gravy,dry,onion	5/6 oz pkg	24	77	16.2	unk.	2.2	**3**	unk.	unk.	unk.	unk.	unk.	unk.	0.03	0.45	0.02	unk.	**unk.**
			4.17	1.0	0.7	unk.	unk.		unk.	unk.	unk.	unk.	unk.	unk.	0.17	0.08	0	unk.	**unk.**
F	gravy,dry,onion,prep w/water	1/4 C	65	20	4.2	unk.	0.6	**1**	unk.	unk.	unk.	unk.	unk.	unk.	0.01	0.12	0.01	unk.	**unk.**
			1.53	59.5	0.2	unk.	unk.		unk.	unk.	unk.	unk.	unk.	unk.	0.05	0.06	tr	unk.	**unk.**
G	gravy,dry,pork	3/4 oz pkg	21	76	13.4	unk.	1.9	**3**	unk.	unk.	unk.	unk.	unk.	unk.	0.22	0.75	0.20	unk.	**unk.**
			4.69	0.9	1.9	unk.	unk.		unk.	unk.	unk.	unk.	unk.	unk.	0.79	0.29	2	unk.	**unk.**
H	gravy,dry,pork,prep w/water	1/4 C	65	19	3.4	unk.	0.5	**1**	unk.	unk.	unk.	unk.	unk.	unk.	0.05	0.19	0.05	unk.	**unk.**
			1.55	59.4	0.5	unk.	unk.		unk.	unk.	unk.	unk.	unk.	unk.	0.20	0.28	1	unk.	**unk.**
I	gravy,dry,turkey	7/8 oz pkg	25	87	15.1	0.10	2.9	**5**	unk.	unk.	unk.	unk.	unk.	unk.	0.43	0.55	0.40	unk.	**unk.**
			4.03	1.0	1.9	unk.	unk.		unk.	unk.	unk.	unk.	unk.	unk.	0.67	0.79	2	unk.	**unk.**
J	gravy,dry,turkey,prep w/water	1/4 C	65	22	3.8	0.03	0.7	**1**	unk.	unk.	unk.	unk.	unk.	unk.	0.11	0.14	0.10	unk.	**unk.**
			1.53	59.5	0.5	unk.	unk.		unk.	unk.	unk.	unk.	unk.	unk.	0.17	0.81	1	unk.	**unk.**
K	gravy,dry,unspecified	7/8 oz pkg	25	85	14.4	unk.	3.2	**5**	unk.	unk.	unk.	unk.	unk.	unk.	0.40	0.71	0.38	unk.	**unk.**
			4.03	1.0	2.0	unk.	unk.		unk.	unk.	unk.	unk.	unk.	unk.	0.77	0.57	1	unk.	**unk.**
L	gravy,dry,unspecified,prep w/water	1/4 C	65	22	3.6	unk.	0.8	**1**	unk.	unk.	unk.	unk.	unk.	unk.	0.10	0.18	0.09	unk.	**unk.**
			1.53	59.4	0.5	unk.	tr(a)		unk.	unk.	unk.	unk.	unk.	unk.	0.19	0.56	tr	unk.	**unk.**
M	GUINEA FOWL,meat w/skin,raw	1/2 bird	345	545	0.0	0.00	80.7	**179**	unk.	unk.	unk.	unk.	unk.	unk.	unk.	unk.	unk.	unk.	**unk.**
			0.29	237.7	22.3	0(a)	0(a)		unk.	unk.	unk.	unk.	unk.	unk.	unk.	unk.	unk.	unk.	**unk.**
N	guinea fowl,meat w/o skin,raw	1/2 bird	264	290	0.0	0.00	54.5	**121**	unk.	unk.	unk.	unk.	unk.	unk.	unk.	unk.	unk.	unk.	**unk.**
			0.38	196.5	6.5	0(a)	0(a)		unk.	unk.	unk.	unk.	unk.	unk.	unk.	unk.	166	unk.	**unk.**
O	HADDOCK,fillet,breaded,fried	1 med	110	181	6.4	unk.	21.6	**48**	1421	1011	2159	797	1143	252	unk.	unk.	unk.	0.04	**3**
			0.91	72.9	7.0	unk.	0.0		634	1839	625	1114	241	845	unk.	unk.	66	2	**4**
P	haddock,fillet,cooked	1 med	79	71	0.0	0(a)	15.5	**34**	1021	790	1363	573	821	181	0.19	0.09	0.01	0.03	**2**
			1.27	52.4	0.6	0(a)	0.0		455	1178	449	666	155	607	0.05	2.18	47	2	**3**
Q	HALIBUT,fillet,broiled w/butter or margarine	1 med	125	214	0.0	0.00	31.5	**70**	1786	1508	2870	1166	1670	369	unk.	unk.	unk.	0.06	**4**
			0.80	83.2	8.8	0(a)	0.0		1114	2421	914	1443	354	1155	unk.	unk.	63	unk.	**unk.**
R	halibut,fillet,cooked	1 med	125	163	0.0	0.00	31.5	**70**	1786	1606	2773	1166	1670	369	0.57	0.25	tr	0.06	**4**
			0.80	83.2	1.4	0(a)	0.0		1114	2394	914	1354	315	1155	0.12	2.30	63	unk.	**unk.**
S	halibut,steak,broiled w/butter or margarine	1 med (4 per lb raw)	84	144	0.0	0.00	21.2	**47**	1200	1013	1929	784	1122	248	unk.	unk.	unk.	0.04	**3**
			1.19	55.9	5.9	0(a)	0.0		748	1627	614	969	238	776	unk.	unk.	42	unk.	**unk.**

Riboflavin mg Niacin mg	% USRDA	Vitamin B_6 mg Folacin mcg	% USRDA	Vitamin B_{12} mcg Pantothenic acid mg	% USRDA	Biotin mg Vitamin A IU	% USRDA	Preformed A RE Beta carotene RE	Vitamin D IU Vitamin E IU	% USRDA	Total tocopherol mg Alpha tocopherol mg	Other tocopherol mg Total ash g	Calcium mg Phosphorus mg	% USRDA	Sodium mg Sodium meq	Potassium mg Potassium meq	Chlorine mg Chlorine meq	Iron mg Magnesium mg	% USRDA	Zinc mg Copper mg	% USRDA	Iodine mcg Selenium mcg	% USRDA	Manganese mcg Chromium mcg	
0.15	9	unk.	unk.	unk.	unk.	unk.	unk.	unk.	0(a)	0	unk.	unk.	39	4	1133	unk.	unk.	unk.	unk.	0.3	2	unk.	unk.	unk.	A
unk.	unk.	unk.	unk.	unk.	unk.	unk.	unk.	unk.	unk.	unk.	unk.	3.35	unk.	unk.	49.3	unk.	unk.	unk.	unk.	0.03	1	unk.		unk.	
0.04	2	unk.	unk.	unk.	unk.	unk.	unk.	unk.	0(a)	0	unk.	unk.	10	1	283	unk.	unk.	unk.	unk.	0.1	1	unk.	unk.	unk.	B
unk.	unk.	unk.	unk.	unk.	unk.	unk.	unk.	unk.	unk.	unk.	unk.	0.84	unk.	unk.	12.3	unk.	unk.	unk.	unk.	0.01	0	unk.		unk.	
unk.	unk.	unk.	unk.	unk.	unk.	unk.	unk.	unk.	0(a)	0	unk.	unk.	49	5	1402	unk.	unk.	unk.	unk.	0.3	2	unk.	unk.	unk.	C
unk.	unk.	unk.	unk.	unk.	unk.	unk.	unk.	unk.	unk.	unk.	unk.	3.83	unk.	unk.	61.0	unk.	unk.	unk.	unk.	0.11	6	unk.		unk.	
unk.	unk.	unk.	unk.	unk.	unk.	unk.	unk.	unk.	0(a)	0	unk.	unk.	12	1	350	unk.	unk.	unk.	unk.	0.1	1	unk.	unk.	unk.	D
unk.	unk.	unk.	unk.	unk.	unk.	unk.	unk.	unk.	unk.	unk.	unk.	0.96	unk.	unk.	15.2	unk.	unk.	unk.	unk.	0.03	1	unk.		unk.	
unk.	unk.	unk.	unk.	unk.	unk.	unk.	unk.	unk.	0(a)	0	unk.	unk.	67	7	1005	unk.	unk.	unk.	unk.	0.2	1	unk.	unk.	unk.	E
unk.	unk.	unk.	unk.	unk.	unk.	unk.	unk.	unk.	unk.	unk.	unk.	3.84	unk.	unk.	43.7	unk.	unk.	unk.	unk.	0.04	2	unk.		unk.	
unk.	unk.	unk.	unk.	unk.	unk.	unk.	unk.	unk.	0(a)	0	unk.	unk.	18	2	259	unk.	unk.	unk.	unk.	0.1	0	unk.	unk.	unk.	F
unk.	unk.	unk.	unk.	unk.	unk.	unk.	unk.	unk.	unk.	unk.	unk.	0.99	unk.	unk.	11.3	unk.	unk.	unk.	unk.	0.01	1	unk.		unk.	
0.06	4	unk.	unk.	unk.	unk.	unk.	unk.	unk.	0(a)	0	unk.	unk.	32	3	1235	unk.	unk.	unk.	unk.	unk.	unk.	unk.	unk.	unk.	G
unk.	unk.	unk.	unk.	unk.	unk.	unk.	unk.	unk.	unk.	unk.	unk.	3.19	unk.	unk.	53.7	unk.	unk.	unk.	unk.	unk.	unk.	unk.		unk.	
0.01	1	unk.	unk.	unk.	unk.	unk.	unk.	unk.	0(a)	0	unk.	unk.	8	1	309	unk.	unk.	unk.	unk.	unk.	unk.	unk.	unk.	unk.	H
unk.	unk.	unk.	unk.	unk.	unk.	unk.	unk.	unk.	unk.	unk.	unk.	0.80	unk.	unk.	13.4	unk.	unk.	unk.	unk.	unk.	unk.	unk.		unk.	
0.11	7	unk.	unk.	unk.	unk.	unk.	unk.	unk.	0(a)	0	unk.	unk.	50	5	1500	unk.	unk.	unk.	unk.	unk.	unk.	unk.	unk.	unk.	I
unk.	unk.	unk.	unk.	unk.	unk.	unk.	unk.	unk.	unk.	unk.	unk.	3.93	unk.	unk.	65.3	unk.	unk.	unk.	unk.	unk.	unk.	unk.		unk.	
0.03	2	unk.	unk.	unk.	unk.	unk.	unk.	unk.	0(a)	0	unk.	unk.	12	1	375	unk.	unk.	unk.	unk.	unk.	unk.	unk.	unk.	unk.	J
unk.	unk.	unk.	unk.	unk.	unk.	unk.	unk.	unk.	unk.	unk.	unk.	0.98	unk.	unk.	16.3	unk.	unk.	unk.	unk.	unk.	unk.	unk.		unk.	
0.11	6	unk.	unk.	unk.	unk.	unk.	unk.	unk.	0(a)	0	unk.	unk.	37	4	1421	unk.	unk.	unk.	unk.	unk.	unk.	unk.	unk.	unk.	K
unk.	unk.	unk.	unk.	unk.	unk.	unk.	unk.	unk.	unk.	unk.	unk.	4.22	unk.	unk.	61.8	unk.	unk.	unk.	unk.	unk.	unk.	unk.		unk.	
0.03	2	unk.	unk.	unk.	unk.	unk.	unk.	unk.	0(a)	0	unk.	unk.	9	1	356	unk.	unk.	unk.	unk.	unk.	unk.	unk.	unk.	unk.	L
unk.	unk.	unk.	unk.	unk.	unk.	unk.	unk.	unk.	unk.	unk.	unk.	1.05	unk.	unk.	15.5	unk.	unk.	unk.	unk.	unk.	unk.	unk.		unk.	
unk.	unk.	unk.	unk.	unk.	unk.	unk.	unk.	unk.	unk.	unk.	unk.	unk.	unk.	unk.	unk.	unk.	unk.	unk.	unk.	unk.	unk.	unk.	unk.	unk.	M
unk.	unk.	unk.	unk.	unk.	unk.	unk.	unk.	unk.	unk.	unk.	unk.	4.31	unk.	unk.	unk.	unk.	unk.	unk.	unk.	unk.	unk.	unk.		unk.	
unk.	unk.	unk.	unk.	unk.	unk.	unk.	unk.	unk.	unk.	unk.	unk.	unk.	unk.	unk.	unk.	unk.	unk.	unk.	unk.	unk.	unk.	unk.	unk.	unk.	N
unk.	unk.	unk.	unk.	unk.	unk.	unk.	unk.	unk.	unk.	unk.	unk.	3.30	unk.	unk.	unk.	unk.	unk.	unk.	unk.	unk.	unk.	unk.		unk.	
0.08	5	0.12	6	1.29	21	unk.	unk.	unk.	unk.	unk.	1.3	0.7	44	4	195	383	unk.	1.3	7	0.3	2	unk.	unk.	unk.	O
3.5	18	11	3	0.07	1	unk.	unk.	unk.	1.6	5	0.7	2.09	272	27	8.5	9.8	unk.	40	10	0.16	8	tr(a)		tr(a)	
0.06	3	0.09	4	0.92	15	unk.	unk.	unk.	unk.	unk.	0.9	0.5	32	3	140	275	unk.	0.9	5	0.2	2	869.0	579	unk.	P
2.5	13	8	2	0.05	1	unk.	unk.	unk.	1.1	4	0.5	1.50	195	20	6.1	7.0	unk.	28	7	0.12	6	tr(a)		tr(a)	
0.09	5	0.32	16	1.12	19	unk.	unk.	unk.	unk.	unk.	0.7	tr(a)	20	2	105	656	unk.	1.0	6	1.3	8	unk.	unk.	unk.	Q
10.4	52	15	4	0.17	2	850	17	unk.	0.9	3	tr(a)	2.12	310	31	4.6	16.8	unk.	29	7	0.24	12	tr(a)		tr(a)	
0.09	5	0.32	16	1.12	19	unk.	unk.	189	unk.	unk.	0.7	tr(a)	20	2	105	656	unk.	1.0	6	1.3	8	unk.	unk.	unk.	R
10.4	52	15	4	0.17	2	699	14	7	0.9	3	tr(a)	2.12	310	31	4.6	16.8	unk.	29	7	0.24	12	tr(a)		tr(a)	
0.06	4	0.22	11	0.76	13	unk.	unk.	unk.	unk.	unk.	0.5	tr(a)	13	1	71	441	unk.	0.7	4	unk.	unk.	unk.	unk.	unk.	S
7.0	35	10	3	0.12	1	571	11	unk.	0.6	2	tr(a)	1.43	208	21	3.1	11.3	unk.	19	5	unk.	unk.	unk.		tr(a)	

	FOOD	Portion	Weight in grams / Conversion for 100 g	Kilocalories / H_2O g	Total carbohydrate g / Total fats g	Crude fiber g / Dietary fiber g	Total protein g / Total sugar g	% USRDA	Arginine mg / Histidine mg	Isoleucine mg / Leucine mg	Lysine mg / Methionine mg	Phenylalanine mg / Threonine mg	Valine mg / Tryptophan mg	Cystine mg / Tyrosine mg	Polyunsat. fatty acids g / Monounsat. fatty acids g	Saturated fatty acids g / P/S ratio	Linoleic acid g / Cholesterol mg	Thiamin mg / Ascorbic acid mg	% USRDA
A	HAM,HOME RECIPE,croquettes,fried	1 (3" x 1")	89	217	10.9	0.06	12.0	**26**	640	612	872	522	665	155	3.27	5.13	3.02	0.25	**17**
			1.12	50.7	13.7	0.12	1.8		369	931	296	555	162	443	4.20	0.64	77	tr	**1**
B	ham,home recipe,salad	1/2 C	121	287	4.6	0.16	16.2	**36**	unk.	862	1231	718	956	228	1.11	5.06	8.06	0.30	**20**
			0.83	74.7	22.5	0.28	0.6		487	1255	440	782	227	591	6.17	0.22	237	1	**2**
C	HAM,SMOKED MEAT,canned,	3-1/4x2-7/8x	85	102	0.3	0(a)	15.3	**24**	972	782	1428	599	792	171	0.34	1.53	0.34	0.70	**47**
	jubilee-Oscar Mayer	1/2"	*1.18*	62.0	4.2	0(a)	0.3		581	1333	413	750	208	548	1.78	0.22	35	24	**40**
D	ham,smoked meat,center slice,w/o bone,	2 sl 4-1/2x	85	159	0.0	0.00	21.5	**48**	1366	1099	1737	880	1198	241	1.16	2.53	0.62	0.49	**33**
	no visible fat	2-1/4x1/4"	*1.18*	52.6	7.5	0(a)	0.0		706	1621	560	1054	292	770	3.12	0.46	75	unk.	**unk.**
E	ham,smoked meat,center slice,w/o bone,	2 sl 4-1/2x	85	246	0.0	0.00	17.8	**40**	1129	908	1435	727	990	199	2.92	6.59	1.78	0.40	**27**
	w/fat	2-1/4x1/4"	*1.18*	45.6	18.8	0(a)	0.0		584	1339	462	870	242	636	8.12	0.44	76	unk.	**unk.**
F	ham,smoked meat,cooked,chopped,	1/2 C	70	131	0.0	0.00	17.7	**39**	1125	905	1430	724	987	198	0.95	2.09	0.51	0.41	**27**
	no visible fat		*1.43*	43.3	6.2	0(a)	0.0		582	1335	461	868	241	634	2.57	0.46	62	unk.	**unk.**
G	ham,smoked meat,cooked,ground,w/fat	1/2 C	55	159	0.0	0.00	11.5	**26**	730	587	928	470	641	129	1.89	4.26	1.15	0.26	**17**
			1.82	29.5	12.2	0(a)	0.0		378	866	299	563	157	411	5.25	0.44	49	unk.	**unk.**
H	ham,smoked meat,cubed,cooked,w/fat	four 1" cubes	84	243	0.0	0.00	17.6	**39**	1116	897	1418	718	979	197	2.88	6.51	1.76	0.39	**26**
			1.19	45.0	18.6	0(a)	0.0		577	1323	457	860	239	628	8.02	0.44	75	unk.	**unk.**
I	ham,smoked meat,w/bone,sliced,	2 med slices	86	161	0.0	0.00	21.8	**48**	1382	1112	1757	890	1213	243	1.17	2.56	0.63	0.50	**33**
	no visible fat		*1.16*	53.2	7.6	0(a)	0.0		715	1640	567	1066	296	779	3.16	0.46	76	unk.	**unk.**
J	ham,smoked meat,w/bone,sliced,w/fat	2 med slices	86	249	0.0	0.00	18.0	**40**	1142	918	1452	735	1002	201	2.95	6.66	1.80	0.40	**27**
			1.16	46.1	19.0	0(a)	0.0		591	1354	468	881	245	643	8.21	0.44	77	unk.	**unk.**
K	ham,smoked meat,w/o bone,cubed,	1/2 C	70	131	0.0	0.00	17.7	**39**	1125	905	1430	724	987	198	0.95	2.09	0.51	0.41	**27**
	no visible fat		*1.43*	43.3	6.2	0(a)	0.0		582	1335	461	868	241	634	2.57	0.46	62	unk.	**unk.**
L	ham,smoked meat,w/o bone,cubed,w/fat	1/2 C	70	202	0.0	0.00	14.6	**33**	930	748	1182	598	815	164	2.40	5.42	1.46	0.33	**22**
			1.43	37.5	15.5	0(a)	0.0		481	1102	381	717	199	524	6.68	0.44	62	unk.	**unk.**
M	HAMBURGER HELPER,beef noodle,dry mix	1 pkg	164	624	115.8	unk.	22.3	**34**	unk.	unk.	unk.	unk.	unk.	unk.	unk.	unk.	unk.	1.00	**67**
	-General Mills		*0.61*	unk.	8.9	unk.	unk.		unk.	unk.	unk.	unk.	unk.	unk.	unk.	unk.	unk.	tr	**0**
N	Hamburger Helper,beef noodle,prep	1 serving	174	326	26.0	unk.	20.8	**35**	1016	875	1462	688	928	201	1.83	6.19	0.39	0.14	**10**
			0.57	107.3	15.2	unk.	0.0		537	1370	415	738	196	568	5.94	0.29	61	0	**0**
O	Hamburger Helper,cheeseburger	1 pkg	226	900	140.0	unk.	30.0	**46**	unk.	unk.	unk.	unk.	unk.	unk.	unk.	unk.	unk.	1.13	**75**
	macaroni,dry mix-General Mills		*0.44*	unk.	25.0	unk.	unk.		unk.	unk.	unk.	unk.	unk.	unk.	unk.	unk.	unk.	tr	**0**
P	Hamburger Helper,cheeseburger	1 serving	174	366	28.0	unk.	21.8	**35**	1016	875	1462	688	928	201	1.82	6.19	0.39	0.15	**10**
	macaroni,prep		*0.57*	99.3	18.2	unk.	0.0		537	1370	415	738	196	568	5.93	0.29	61	0	**0**
Q	Hamburger Helper,chili tomato,dry mix	1 pkg	205	700	145.0	unk.	20.0	**31**	unk.	unk.	unk.	unk.	unk.	unk.	unk.	unk.	unk.	1.13	**75**
	-General Mills		*0.49*	unk.	5.0	unk.	unk.		unk.	unk.	unk.	unk.	unk.	unk.	unk.	unk.	unk.	tr	**0**
R	Hamburger Helper,chili tomato,prep	1 serving	174	336	29.0	unk.	19.8	**35**	1016	875	1462	688	928	201	1.83	6.19	0.39	0.15	**10**
			0.57	103.3	15.2	unk.	0.0		537	1370	415	738	196	568	5.94	0.30	61	0	**0**
S	Hamburger Helper,lasagne,dry mix	1 pkg	219	750	160.0	unk.	20.0	**31**	unk.	unk.	unk.	unk.	unk.	unk.	unk.	unk.	unk.	1.12	**75**
	-General Mills		*0.46*	unk.	5.0	0(a)	unk.		unk.	unk.	unk.	unk.	unk.	unk.	unk.	unk.	unk.	tr	**0**

Riboflavin mg / Niacin mg	% USRDA	Vitamin B_6 mg / Folacin mcg	% USRDA	Vitamin B_{12} mcg / Pantothenic acid mg	% USRDA	Biotin mg / Vitamin A IU	% USRDA	Preformed A RE / Beta carotene RE	Vitamin D IU / Vitamin E IU	% USRDA	Total tocopherol mg / Alpha tocopherol mg	Other tocopherol mg / Total ash g	Calcium mg / Phosphorus mg	% USRDA	Sodium mg / Sodium meq	Potassium mg / Potassium meq	Chlorine mg / Chlorine meq	Iron mg / Magnesium mg	% USRDA	Zinc mg / Copper mg	% USRDA	Iodine mcg / Selenium mcg	% USRDA	Manganese mcg / Chromium mcg	
0.17	10	0.11	6	0.34	6	0.001	0	0	12	3	3.9	unk.	50	5	475	180	14	1.8	10	1.7	11	6.2	4	21	A
2.1	10	11	3	0.35	4	206	4	0	4.7	16	unk.	2.08	125	13	20.7	4.6	0.4	30	8	0.08	4	2		3	
0.21	12	0.16	8	0.69	12	unk.	unk.	0	0(a)	0	9.2	4.7	33	3	671	232	unk.	2.4	13	2.4	16	unk.	unk.	unk.	B
2.2	11	22	6	0.76	8	253	5	3	5.0	17	4.5	2.77	163	16	29.2	5.9	unk.	14	4	0.01	0	tr(a)		tr(a)	
0.20	12	0.35	17	0.72	12	tr(a)	0	tr(a)	22	6	0.4	0.2	5	1	1036	301	1275	0.9	5	1.6	11	unk.	unk.	6	C
4.6	23	tr(a)	0	tr(a)	0	tr(a)	0	0(a)	0.5	2	0.2	3.14	208	21	45.1	7.7	36.0	15	4	0.09	5	tr(a)		9	
0.20	12	0.20	10	0.46	8	unk.	unk.	0	unk.	unk.	0.4	0.2	9	1	790	277	unk.	2.7	15	3.4	23	unk.	unk.	13	D
3.8	19	10	3	0.29	3	0	0	0	0.5	2	0.2	3.40	170	17	34.4	7.1	unk.	17	4	0.06	3	tr(a)		9	
0.15	9	0.16	8	0.38	6	unk.	unk.	0	0(a)	0	0.4	0.2	8	1	723	unk.	unk.	2.2	12	3.0	20	unk.	unk.	13	E
3.1	15	10	3	0.22	2	0	0	0	0.5	2	0.2	2.89	146	15	31.4	unk.	unk.	14	4	0.06	3	unk.		9	
0.16	10	0.17	8	0.38	6	unk.	unk.	0	tr	0	0.4	0.2	8	1	651	228	unk.	2.2	12	2.8	19	unk.	unk.	10	F
3.1	16	8	2	0.24	2	0	0	0	0.4	2	0.2	2.80	140	14	28.3	5.8	unk.	14	4	0.05	3	tr(a)		8	
0.10	6	0.11	5	0.25	4	unk.	unk.	0	tr	0	0.3	0.1	5	1	468	unk.	unk.	1.4	8	1.9	13	unk.	unk.	6	G
2.0	10	6	2	0.14	1	0	0	0	0.3	1	0.2	1.87	95	10	20.3	unk.	unk.	9	2	0.04	2	tr(a)		6	
0.15	9	0.16	8	0.38	6	unk.	unk.	0	0(a)	0	0.4	0.2	8	1	714	unk.	unk.	2.2	12	2.9	20	unk.	unk.	13	H
3.0	15	10	3	0.22	2	0	0	0	0.5	2	0.2	2.86	145	14	31.0	unk.	unk.	14	4	0.06	3	tr(a)		9	
0.20	12	0.21	10	0.46	8	unk.	unk.	0	0(a)	0	0.4	0.2	9	1	800	280	unk.	2.8	15	3.4	23	unk.	unk.	13	I
3.9	19	10	3	0.29	3	0	0	0	0.5	2	0.2	3.44	172	17	34.8	7.2	unk.	17	4	0.06	3	tr(a)		9	
0.15	9	0.17	8	0.39	6	unk.	unk.	0	0(a)	0	0.4	0.2	8	1	731	unk.	unk.	2.2	12	3.0	20	unk.	unk.	13	J
3.1	16	10	3	0.23	2	0	0	0	0.5	2	0.2	2.92	148	15	31.8	unk.	unk.	15	4	0.06	3	tr(a)		9	
0.16	10	0.17	8	0.38	6	unk.	unk.	0	0(a)	0	0.4	0.2	8	1	651	228	unk.	2.2	12	2.8	19	unk.	unk.	10	K
3.1	16	8	2	0.24	2	0	0	0	0.4	2	0.2	2.80	140	14	28.3	5.8	unk.	14	4	0.05	3	tr(a)		8	
0.13	7	0.13	7	0.31	5	unk.	unk.	0	0(a)	0	0.4	0.2	6	1	595	unk.	unk.	1.8	10	2.4	16	unk.	unk.	10	L
2.5	13	8	2	0.18	2	0	0	0	0.4	2	0.2	2.38	120	12	25.9	unk.	unk.	12	3	0.05	3	tr(a)		8	
0.75	44	unk.	unk.	unk.	unk.	unk.	unk.	unk.	unk.	unk.	unk.	unk.	89	9	4133	unk.	unk.	6.4	36	unk.	unk.	unk.	unk.	unk.	M
8.9	45	unk.	unk.	unk.	unk.	tr	0	unk.	unk.	unk.	unk.	unk.	unk.	unk.	179.8	unk.	unk.	unk.	unk.	unk.	unk.	unk.		unk.	
0.20	12	0.13	7	0.82	14	unk.	unk.	7	0	0	0.3	tr(a)	15	2	31	293	unk.	2.6	15	2.9	19	unk.	unk.	unk.	N
3.6	18	3	1	0.15	2	26	1	tr	0.3	1	0.3	0.85	127	13	1.3	7.5	unk.	14	3	0.05	3	tr(a)		tr(a)	
0.86	51	unk.	unk.	unk.	unk.	unk.	unk.	unk.	unk.	unk.	unk.	unk.	300	30	4893	unk.	unk.	3.6	20	unk.	unk.	unk.	unk.	unk.	O
10.0	50	unk.	unk.	unk.	unk.	1000	20	unk.	unk.	unk.	unk.	unk.	unk.	unk.	212.8	unk.	unk.	unk.	unk.	unk.	unk.	unk.		unk.	
0.22	13	0.13	7	0.82	14	unk.	unk.	7	0	0	0.3	tr(a)	25	3	31	293	unk.	2.4	13	2.9	19	unk.	unk.	unk.	P
3.6	18	3	1	0.15	2	116	2	tr	0.3	1	0.3	0.85	127	13	1.3	7.5	unk.	14	3	0.05	3	tr(a)		tr(a)	
0.68	40	unk.	unk.	unk.	unk.	unk.	unk.	unk.	unk.	unk.	unk.	unk.	100	10	5914	unk.	unk.	7.2	40	unk.	unk.	unk.	unk.	unk.	Q
10.0	50	unk.	unk.	unk.	unk.	2000	40	unk.	unk.	unk.	unk.	unk.	unk.	unk.	257.3	unk.	unk.	unk.	unk.	unk.	unk.	unk.		unk.	
0.22	13	0.13	7	0.82	14	unk.	unk.	7	0	0	0.3	tr(a)	15	2	31	293	unk.	2.7	15	2.9	19	unk.	unk.	unk.	R
3.6	18	3	1	0.15	2	190	4	tr	0.3	1	0.3	0.85	127	13	1.3	7.5	unk.	14	3	0.05	3	tr(a)		tr(a)	
0.50	30	unk.	unk.	unk.	unk.	unk.	unk.	unk.	unk.	unk.	unk.	unk.	100	10	unk.	unk.	unk.	5.4	30	unk.	unk.	unk.	unk.	unk.	S
10.0	50	unk.	unk.	unk.	unk.	2500	50	unk.	unk.	unk.	unk.	unk.	unk.	unk.	unk.	unk.	unk.	unk.	unk.	unk.	unk.	unk.		unk.	

	FOOD	Portion	Weight in grams / Conversion for 100 g	Kilocalories / H_2O g	Total carbohydrate g / Total fats g	Crude fiber g / Dietary fiber g	Total protein g / Total sugar g	% USRDA	Arginine mg / Histidine mg	Isoleucine mg / Leucine mg	Lysine mg / Methionine mg	Phenylalanine mg / Threonine mg	Valine mg / Tryptophan mg	Cystine mg / Tyrosine mg	Polyunsat. fatty acids g / Monounsat. fatty acids g	Saturated fatty acids g / P/S ratio	Linoleic acid g / Cholesterol mg	Thiamin mg / Ascorbic acid mg	% USRDA
A	Hamburger Helper,lasagne,prep	1 serving	174	336	32.0	unk.	19.8	**35**	1016	875	1462	688	928	201	1.82	6.19	0.39	0.15	**10**
			0.57	100.3	14.2	0(a)	0.0		537	1370	415	738	196	568	5.93	0.29	61	4	**7**
B	HAZELNUTS/filberts,shelled	1/4 C	34	214	5.6	1.01	4.3	**7**	737	288	141	182	316	56	2.33	1.56	2.23	0.16	**10**
			2.96	2.0	21.1	unk.	0.0		97	317	47	140	71	147	16.83	1.50	0	1	**2**
C	HERRING,canned,plain,solids & liquid	3-1/2x2x3/4"	75	156	0.0	0.00	14.9	**33**	946	788	1345	552	791	182	unk.	1.93	1.93	unk.	**unk.**
		piece	*1.33*	47.2	10.2	0(a)	0.0		458	1316	433	767	160	578	unk.	unk.	73	unk.	**unk.**
D	herring,canned,w/tomato sauce	1.7" long +	55	97	2.0	0.00	8.7	**19**	551	443	756	322	460	106	unk.	1.10	1.10	unk.	**unk.**
		1 Tbsp sauce	*1.82*	36.7	5.8	0(a)	0.0		267	652	252	373	87	336	unk.	unk.	53	unk.	**unk.**
E	herring,pickled,pieces	1-3/4x7/8x	15	33	0.0	0.00	3.1	**7**	194	156	266	113	162	37	unk.	0.43	0.43	unk.	**unk.**
		3/4"	*6.67*	8.9	2.3	0(a)	0.0		94	229	89	132	31	118	unk.	unk.	13	unk.	**unk.**
F	herring,pickled,whole	1.7" long	50	112	0.0	0.00	10.2	**23**	646	520	888	378	541	124	unk.	1.43	1.43	unk.	**unk.**
			2.00	29.7	7.5	0(a)	0.0		313	765	296	439	102	395	unk.	unk.	43	unk.	**unk.**
G	herring,smoked kippered,filleted	1 sm 2-3/8x	20	42	0.0	0.00	4.4	**10**	281	234	400	164	235	54	unk.	0.48	0.48	unk.	**unk.**
		1-3/8x1/4"	*5.00*	12.2	2.6	0(a)	0.0		136	391	129	228	48	172	unk.	unk.	17	unk.	**unk.**
H	herring,smoked kippered,filleted	1 med 4-3/8x	40	84	0.0	0.00	8.9	**20**	562	469	800	329	470	108	unk.	0.96	0.96	unk.	**unk.**
		1-3/8x1/4"	*2.50*	24.4	5.2	0(a)	0.0		273	783	258	456	95	344	unk.	unk.	34	unk.	**unk.**
I	HICKORY NUTS	1/4 C	16	108	2.0	0.30	2.1	**3**	unk.	unk.	unk.	unk.	unk.	unk.	2.03	0.96	1.92	0.08	**6**
			6.25	0.5	11.0	unk.	0.0		unk.	unk.	unk.	unk.	unk.	unk.	7.52	2.12	0	0	**0**
J	HONEY	1 Tbsp	21	64	17.3	0.02	0.1	**0**	tr(a)	tr(a)	tr(a)	tr(a)	tr(a)	tr(a)	0.00	0.00	0.00	tr	**0**
			4.76	3.6	0.0	tr(a)	17.3		tr(a)	tr(a)	tr(a)	tr(a)	tr(a)	tr(a)	0.00	0.00	0	tr	**0**
K	HONEYDEW MELON,raw,balls,	1/2 C	85	28	6.5	0.51	0.7	**1**	unk.	unk.	13	unk.	unk.	unk.	unk.	tr(a)	unk.	0.03	**2**
	edible portion		*1.18*	77.0	0.3	0.77	6.2		unk.	unk.	2	unk.	1	unk.	unk.	unk.	0	20	**33**
L	honeydew melon,raw,fourth,	1/4 of 7" dia	373	123	28.7	2.24	3.0	**5**	unk.	unk.	56	unk.	unk.	unk.	unk.	tr(a)	unk.	0.15	**10**
	edible portion		*0.27*	337.8	1.1	3.36	27.2		unk.	unk.	7	unk.	4	unk.	unk.	unk.	0	86	**143**
M	HORS D'OEUVRE,beef puff-Durkee	1 whole	14	47	3.1	unk.	2.2	**3**	unk.	unk.	unk.	unk.	unk.	unk.	unk.	unk.	unk.	unk.	**unk.**
			7.09	unk.	4.5	unk.	unk.		unk.	unk.	unk.	unk.	unk.	unk.	unk.	unk.	unk.	unk.	**unk.**
N	hors d'oeuvre,cheese puff-Durkee	1 whole	14	58	2.9	unk.	2.7	**4**	unk.	unk.	unk.	unk.	unk.	unk.	unk.	unk.	unk.	unk.	**unk.**
			7.19	unk.	5.7	unk.	unk.		unk.	unk.	unk.	unk.	unk.	unk.	unk.	unk.	unk.	unk.	**unk.**
O	hors d'oeuvre,franks-n-blanket-Durkee	1 whole	13	45	1.0	unk.	1.8	**3**	unk.	unk.	unk.	unk.	unk.	unk.	unk.	unk.	unk.	unk.	**unk.**
			7.87	unk.	3.8	unk.	unk.		unk.	unk.	unk.	unk.	unk.	unk.	unk.	unk.	unk.	unk.	**unk.**
P	hors d'oeuvre,shrimp puff-Durkee	1 whole	14	44	3.0	unk.	2.0	**3**	unk.	unk.	unk.	unk.	unk.	unk.	unk.	unk.	unk.	unk.	**unk.**
			7.09	unk.	4.3	unk.	unk.		unk.	unk.	unk.	unk.	unk.	unk.	unk.	unk.	unk.	unk.	**unk.**
Q	HORSERADISH,prep	1 Tbsp	15	6	1.4	0.00	0.2	**0**	unk.	unk.	unk.	unk.	unk.	unk.	tr(a)	tr(a)	tr(a)	unk.	**unk.**
			6.67	13.1	tr	0(a)	1.1		unk.	unk.	unk.	unk.	unk.	unk.	tr(a)	0.00	0(a)	unk.	**unk.**
R	HUSH PUPPIES,home recipe	1 average	56	147	17.6	0.26	3.6	**7**	100	176	181	165	205	53	2.83	1.38	2.78	0.11	**7**
			1.79	25.8	7.0	unk.	1.2		90	387	84	155	36	190	1.68	2.05	45	1	**2**
S	ICE CREAM bar,chocolate coated	1 average	47	149	12.1	tr(a)	1.6	**4**	unk.	unk.	unk.	unk.	unk.	unk.	unk.	unk.	unk.	0.01	**1**
			2.13	22.1	10.5	tr(a)	8.5		unk.	unk.	unk.	unk.	unk.	unk.	unk.	unk.	unk.	0	**0**

Riboflavin mg / Niacin mg	% USRDA	Vitamin B_6 mg / Folacin mcg	% USRDA	Vitamin B_{12} mcg / Pantothenic acid mg	% USRDA	Biotin mg / Vitamin A IU	% USRDA	Preformed A RE / Beta carotene RE	Vitamin D IU / Vitamin E IU	% USRDA	Total tocopherol mg / Alpha tocopherol mg	Other tocopherol mg / Total ash g	Calcium mg / Phosphorus mg	% USRDA	Sodium mg / Sodium meq	Potassium mg / Potassium meq	Chlorine mg / Chlorine meq	Iron mg / Magnesium mg	% USRDA	Zinc mg / Copper mg	% USRDA	Iodine mcg / Selenium mcg	% USRDA	Manganese mcg / Chromium mcg	
0.22	13	0.13	7	0.82	14	unk.	unk.	7	0	0	0.3	tr(a)	16	2	31	293	unk.	2.6	14	2.9	19	unk.	unk.	unk.	A
3.6	18	3	1	0.15	2	245	5	tr	0.3	1	0.3	0.85	127	13	1.3	7.5	unk.	14	3	0.05	3	tr(a)		tr(a)	
0.18	11	0.18	9	0.00	0	unk.	unk.	0	0	0	9.5	unk.	71	7	1	238	23	1.1	6	1.0	7	unk.	unk.	1420	B
0.3	2	24	6	0.39	4	36	1	4	11.4	38	7.1	0.84	114	11	0.0	6.1	0.6	59	15	0.43	22	1		unk.	
0.13	8	0.12	6	6.00	100	unk.	unk.	unk.	unk.	unk.	tr(a)	tr(a)	110	11	unk.	unk.	unk.	1.3	8	unk.	unk.	unk.	unk.	unk.	C
unk.	unk.	tr(a)	0	0.52	5	unk.	unk.	unk.	0.0	0	tr(a)	2.77	223	22	unk.	unk.	unk.	13	3	unk.	unk.	tr(a)		tr(a)	
0.06	4	unk.	unk.	unk.	unk.	unk.	unk.	unk.	unk.	unk.	unk.	unk.	unk.	unk.	unk.	unk.	unk.	unk.	unk.	unk.	unk.	unk.	unk.	unk.	D
1.9	10	unk.	unk.	unk.	unk.	unk.	unk.	unk.	unk.	unk.	unk.	1.81	134	13	unk.	unk.	unk.	unk.	unk.	unk.	unk.	unk.		unk.	
unk.	unk.	0.02	1	unk.	unk.	unk.	unk.	unk.	unk.	unk.	tr(a)	tr(a)	unk.	unk.	unk.	unk.	unk.	unk.	unk.	unk.	unk.	unk.	unk.	unk.	E
unk.	unk.	tr(a)	0	tr(a)	0	unk.	unk.	unk.	0.0	0	tr(a)	0.60	unk.	unk.	unk.	unk.	unk.	unk.	unk.	unk.	unk.	tr(a)		tr(a)	
unk.	unk.	0.06	3	unk.	unk.	unk.	unk.	unk.	unk.	unk.	tr(a)	unk.	unk.	unk.	unk.	unk.	unk.	unk.	unk.	unk.	unk.	unk.	unk.	unk.	F
unk.	unk.	tr(a)	0	tr(a)	0	unk.	unk.	unk.	0.0	0	tr(a)	2.00	unk.	unk.	unk.	unk.	unk.	unk.	unk.	unk.	unk.	tr(a)		tr(a)	
0.06	3	0.05	3	0.30	5	unk.	unk.	2	unk.	unk.	tr(a)	tr(a)	13	1	unk.	unk.	unk.	0.3	2	unk.	unk.	unk.	unk.	1	G
0.7	3	tr(a)	0	0.21	2	6	0	tr	0.0	0	tr(a)	0.80	51	5	unk.	unk.	unk.	unk.	unk.	unk.	unk.	tr(a)		tr(a)	
0.11	7	0.10	5	0.60	10	unk.	unk.	3	unk.	unk.	tr(a)	tr(a)	26	3	unk.	unk.	unk.	0.6	3	unk.	unk.	unk.	unk.	2	H
1.3	7	tr(a)	0	0.42	4	12	0	tr	0.0	0	tr(a)	1.60	102	10	unk.	unk.	unk.	unk.	unk.	unk.	unk.	tr(a)		tr(a)	
unk.	unk.	unk.	unk.	unk.	unk.	unk.	unk.	0(a)	0	0	unk.	unk.	tr	0	unk.	unk.	unk.	0.4	2	unk.	unk.	unk.	unk.	unk.	I
unk.	unk.	unk.	unk.	unk.	unk.	unk.	unk.	unk.	unk.	unk.	unk.	0.32	58	6	unk.	unk.	unk.	26	6	unk.	unk.	unk.		unk.	
0.01	1	tr	0	0.00	0	tr(a)	0	0	0	0	0.0	0.0	1	0	1	11	6	0.1	1	tr	0	unk.	unk.	6	J
0.1	0	1	0	0.04	0	0	0	0	0.0	0	0.0	0.04	1	0	0.0	0.3	0.2	1	0	0.04	2	unk.		unk.	
0.03	2	0.05	2	0.00	0	unk.	unk.	0	0(a)	0	0.1	unk.	3	0	10	213	35	0.1	1	0.1	1	unk.	unk.	15	K
0.5	3	unk.	unk.	0.18	2	34	1	3	0.1	0	unk.	0.51	9	1	0.4	5.5	1.0	6	1	0.03	2	0		0	
0.11	7	0.21	10	0.00	0	unk.	unk.	0	0(a)	0	0.4	unk.	11	1	45	936	153	0.4	2	0.4	3	unk.	unk.	67	L
2.2	11	unk.	unk.	0.77	8	149	3	15	0.4	2	unk.	2.24	37	4	1.9	23.9	4.3	25	6	0.15	8	0		0	
unk.	unk.	unk.	unk.	unk.	unk.	unk.	unk.	unk.	unk.	unk.	unk.	unk.	unk.	unk.	unk.	unk.	unk.	unk.	unk.	unk.	unk.	unk.	unk.	unk.	M
unk.	unk.	unk.	unk.	unk.	unk.	unk.	unk.	unk.	unk.	unk.	unk.	unk.	unk.	unk.	unk.	unk.	unk.	unk.	unk.	unk.	unk.	unk.		unk.	
unk.	unk.	unk.	unk.	unk.	unk.	unk.	unk.	unk.	unk.	unk.	unk.	unk.	unk.	unk.	unk.	unk.	unk.	unk.	unk.	unk.	unk.	unk.	unk.	unk.	N
unk.	unk.	unk.	unk.	unk.	unk.	unk.	unk.	unk.	unk.	unk.	unk.	unk.	unk.	unk.	unk.	unk.	unk.	unk.	unk.	unk.	unk.	unk.		unk.	
unk.	unk.	unk.	unk.	unk.	unk.	unk.	unk.	unk.	0(a)	0	unk.	unk.	unk.	unk.	unk.	unk.	unk.	unk.	unk.	unk.	unk.	unk.	unk.	unk.	O
unk.	unk.	unk.	unk.	unk.	unk.	unk.	unk.	unk.	unk.	unk.	unk.	unk.	unk.	unk.	unk.	unk.	unk.	unk.	unk.	unk.	unk.	unk.		unk.	
unk.	unk.	unk.	unk.	unk.	unk.	unk.	unk.	unk.	0(a)	0	unk.	unk.	unk.	unk.	unk.	unk.	unk.	unk.	unk.	unk.	unk.	unk.	unk.	unk.	P
unk.	unk.	unk.	unk.	unk.	unk.	unk.	unk.	unk.	unk.	unk.	unk.	unk.	unk.	unk.	unk.	unk.	unk.	unk.	unk.	unk.	unk.	unk.		unk.	
unk.	unk.	0.02	1	0(a)	0	unk.	unk.	0(a)	0(a)	0	unk.	unk.	9	1	14	43	unk.	0.1	1	0.2	1	unk.	unk.	unk.	Q
unk.	unk.	unk.	unk.	unk.	unk.	unk.	unk.	unk.	unk.	unk.	unk.	0.27	5	1	0.6	1.1	unk.	unk.	unk.	tr(a)	0	unk.		unk.	
0.12	7	0.15	7	0.16	3	0.003	1	0	11	3	4.1	tr	129	13	545	106	16	0.9	5	0.7	5	54.3	36	12	R
0.8	4	18	4	0.41	4	47	1	tr	1.1	4	1.1	2.29	223	22	23.7	2.7	0.4	44	11	0.09	4	tr		1	
0.07	4	0.02	1	0.26	4	unk.	unk.	unk.	2	1	unk.	unk.	55	6	24	84	unk.	0.0	0	unk.	unk.	unk.	unk.	unk.	S
0.4	2	unk.	unk.	unk.	unk.	145	3	unk.	unk.	unk.	0.1	0.42	42	4	1.0	2.1	unk.	9	2	unk.	unk.	unk.		unk.	

	FOOD	Portion	Weight in grams / Conversion for 100 g	Kilocalories / H_2O g	Total carbohydrate g / Total fats g	Crude fiber g / Dietary fiber g	Total protein g / Total sugar g	% USRDA	Arginine mg / Histidine mg	Isoleucine mg / Leucine mg	Lysine mg / Methionine mg	Phenylalanine mg / Threonine mg	Valine mg / Tryptophan mg	Cystine mg / Tyrosine mg	Polyunsat. fatty acids g / Monounsat. fatty acids g	Saturated fatty acids g / P/S ratio	Linoleic acid g / Cholesterol mg	Thiamin mg / Ascorbic acid mg	% USRDA
A	ice cream sandwich-Sealtest	1 average	62	173	26.1	unk.	3.1	0	unk.	unk.	unk.	unk.	unk.	unk.	0.44	2.91	0.22	0.02	1
			1.61	25.9	6.2	0(a)	5.9		unk.	unk.	unk.	unk.	unk.	unk.	1.48	0.15	unk.	unk.	unk.
B	ice cream,chocolate-Baskin-Robbins	2/3 C	89	186	32.0	unk.	3.1	7	unk.	unk.	unk.	unk.	unk.	unk.	unk.	unk.	unk.	0.04	2
			1.12	unk.	9.8	unk.	32.0		unk.	unk.	unk.	unk.	unk.	unk.	unk.	unk.	unk.	unk.	unk.
C	ice cream,french vanilla,soft serve	2/3 C	115	251	25.4	0.00	4.7	10	187	281	365	224	309	48	0.66	8.98	0.44	0.05	3
			0.87	68.7	15.0	0(a)	25.4		125	452	117	216	67	223	3.89	0.07	102	1	2
D	ice cream,strawberry-Baskin-Robbins	2/3 C	89	167	29.9	unk.	14.8	33	unk.	unk.	unk.	unk.	unk.	unk.	unk.	unk.	unk.	0.04	2
			1.12	unk.	7.9	tr(a)	29.9		unk.	unk.	unk.	unk.	unk.	unk.	unk.	unk.	unk.	unk.	unk.
E	ice cream,vanilla-Baskin-Robbins	2/3 C	89	185	18.4	unk.	12.9	29	unk.	unk.	unk.	unk.	unk.	unk.	unk.	unk.	unk.	0.04	2
			1.12	unk.	10.7	unk.	18.4		unk.	unk.	unk.	unk.	unk.	unk.	unk.	unk.	unk.	unk.	unk.
F	ice cream,vanilla,hardened,regular,	2/3 C	89	180	21.2	0.00	3.2	7	117	194	255	155	215	29	0.36	5.96	0.21	0.03	2
	10% fat		*1.12*	54.1	9.6	0(a)	21.2		87	315	81	145	45	155	2.41	0.06	40	0	0
G	ice cream,vanilla,hardened,rich,	2/3 C	99	234	21.4	0.00	2.8	6	100	167	219	134	185	26	0.58	9.86	0.36	0.03	2
	16% fat		*1.01*	58.3	15.8	0(a)	21.4		75	270	69	125	39	134	3.98	0.06	58	tr	1
H	ICE MILK,vanilla,hardened	2/3 C	87	122	19.2	0.00	3.4	8	124	207	271	165	230	31	0.14	2.33	0.09	0.04	3
			1.15	59.7	3.7	0(a)	14.7		93	336	86	155	49	165	0.94	0.06	12	1	2
I	ice milk,vanilla,soft serve	2/3 C	117	150	25.7	0.00	5.4	12	194	325	426	260	359	49	0.12	1.92	0.07	0.07	5
			0.85	81.5	3.1	0(a)	25.6		145	526	135	242	76	260	0.77	0.06	9	1	2
J	ICES,any except orange	2/3 C	123	96	40.1	tr(a)	0.5	1	unk.	unk.	unk.	unk.	unk.	unk.	tr(a)	tr(a)	tr(a)	tr	0
			0.81	82.3	tr(a)	tr(a)	40.1		unk.	unk.	unk.	unk.	unk.	unk.	tr(a)	0.00	0(a)	1	2
K	ices,orange-Baskin-Robbins	2/3 C	129	253	62.3	tr(a)	0.1	0	0(a)	tr(a)	tr(a)	tr(a)	tr(a)	tr(a)	tr(a)	tr(a)	tr(a)	0.00	0
			0.78	unk.	tr(a)	tr(a)	62.3		tr(a)	tr(a)	tr(a)	tr(a)	tr(a)	tr(a)	tr(a)	0.00	0(a)	unk.	unk.
L	ices,orange-Sealtest	2/3 C	123	175	43.5	tr(a)	0.1	0	0(a)	tr(a)	tr(a)	tr(a)	tr(a)	tr(a)	tr(a)	tr(a)	tr(a)	0.01	1
			0.81	77.5	tr	tr(a)	43.5		tr(a)	tr(a)	tr(a)	tr(a)	tr(a)	tr(a)	tr(a)	0.00	0(a)	0	0
M	ICING,home recipe,caramel	1 C	220	895	168.7	0.26	3.8	6	128	237	256	150	263	47	2.34	8.16	6.86	0.05	3
			0.45	20.8	26.1	0.29	167.9		104	366	92	168	54	180	11.23	0.29	0	tr	0
N	icing,home recipe,chocolate	1 C	260	1123	208.4	0.47	2.8	5	28	47	62	38	52	7	1.00	22.35	0.81	0.00	0
			0.38	16.1	36.9	unk.	202.4		21	76	20	35	11	38	10.48	0.04	75	tr	0
O	icing,home recipe,coconut fluff	1 C	144	533	50.7	1.20	7.0	14	336	403	474	359	450	112	3.12	12.87	6.00	0.16	11
			0.69	49.0	35.5	4.89	43.6		197	631	173	315	97	299	14.57	0.24	77	1	2
P	icing,home recipe,white,boiled	1 C	95	247	62.0	0.00	1.3	2	unk.	70	91	62	91	41	0.00	0.00	0.00	0.00	0
			1.05	30.1	0.0	0(a)	61.7		29	114	53	58	20	56	0.00	0.00	0	0	0
Q	icing,home recipe,white,uncooked	1 C	200	813	160.3	0.00	0.6	1	unk.	37	49	30	41	6	0.78	13.22	0.48	0.00	0
			0.50	16.7	21.2	unk.	159.7		17	60	15	28	9	30	5.34	0.06	60	tr	0
R	INSTANT BREAKFAST,powder,vanilla	1 pkg	35	130	24.0	unk.	7.0	11	unk.	363	493	347	422	86	unk.	unk.	unk.	0.30	20
	-Carnation		*2.84*	unk.	unk.	0(a)	unk.		210	685	184	293	100	319	unk.	unk.	2	27	45
S	JAM/jelly,artif sweetened-Smucker's	1 Tbsp	20	6	1.4	tr(a)	0.1	0	unk.	unk.	unk.	unk.	unk.	unk.	unk.	tr(a)	unk.	unk.	unk.
			5.00	unk.	0.1	0(a)	1.4		unk.	unk.	unk.	unk.	unk.	unk.	unk.	unk.	0	tr	0

	Riboflavin mg / Niacin mg	% USRDA	Vitamin B_6 mg / Folacin mcg	% USRDA	Vitamin B_{12} mcg / Pantothenic acid mg	% USRDA	Biotin mg / Vitamin A IU	% USRDA	Preformed A RE / Beta carotene RE	Vitamin D IU / Vitamin E IU	% USRDA	Total tocopherol mg / Alpha tocopherol mg	Other tocopherol mg / Total ash g	Calcium mg / Phosphorus mg	% USRDA	Sodium mg / Sodium meq	Potassium mg / Potassium meq	Chlorine mg / Chlorine meq	Iron mg / Magnesium mg	% USRDA	Zinc mg / Copper mg	% USRDA	Iodine mcg / Selenium mcg	% USRDA	Manganese mcg / Chromium mcg
A	0.10	6	0.02	1	0.29	5	unk.	unk.	unk.	2	1	unk.	unk.	73	7	92	102	unk.	0.1	1	unk.	unk.	unk.	unk.	unk.
	0.5	3	unk.	unk.	unk.	unk.	193	4	unk.	unk.	unk.	0.1	0.62	72	7	4.0	2.6	unk.	7	2	unk.	unk.	unk.		unk.
B	0.16	9	unk.	unk.	unk.	unk.	unk.	unk.	unk.	4	1	0.3	unk.	119	12	47	unk.	unk.	unk.	unk.	unk.	unk.	unk.	unk.	unk.
	0.1	0	unk.	unk.	unk.	unk.	344	7	unk.	0.4	1	0.3	unk.	107	11	2.0	unk.	unk.	20	5	unk.	unk.	unk.		unk.
C	0.29	17	0.06	3	0.66	11	0.003	1	unk.	5	1	0.1	unk.	156	16	102	224	unk.	0.3	2	1.3	9	unk.	unk.	unk.
	0.1	1	6	1	0.71	7	528	11	unk.	0.2	1	0.1	1.17	132	13	4.4	5.7	unk.	16	4	0.03	2	unk.		unk.
D	0.13	8	unk.	unk.	unk.	unk.	0.003	1	unk.	4	1	0.1	unk.	98	10	39	unk.	unk.	unk.	unk.	unk.	unk.	unk.	unk.	unk.
	0.1	0	unk.	unk.	unk.	unk.	336	7	unk.	0.1	0	0.1	unk.	83	8	1.7	unk.	unk.	unk.	unk.	0.03	1	unk.		unk.
E	0.19	11	unk.	unk.	unk.	unk.	unk.	unk.	unk.	4	1	0.1	unk.	109	11	53	unk.	unk.	unk.	unk.	unk.	unk.	unk.	unk.	unk.
	0.1	0	unk.	unk.	unk.	unk.	438	9	unk.	0.1	0	0.1	unk.	93	9	2.3	unk.	unk.	unk.	unk.	unk.	unk.	unk.		unk.
F	0.21	13	0.04	2	0.42	7	0.003	1	unk.	4	1	0.1	unk.	117	12	77	172	unk.	0.1	0	0.9	6	unk.	unk.	tr(a)
	0.1	0	2	0	0.44	4	363	7	unk.	0.1	0	0.1	0.86	90	9	3.4	4.4	unk.	12	3	0.03	1	unk.		9
G	0.19	11	0.04	2	0.36	6	0.003	1	unk.	4	1	0.1	unk.	101	10	72	148	unk.	0.1	0	0.8	5	unk.	unk.	unk.
	0.1	0	2	1	0.38	4	600	12	unk.	0.1	1	0.1	0.74	77	8	3.1	3.8	unk.	11	3	0.03	2	unk.		unk.
H	0.23	13	0.06	3	0.57	10	unk.	unk.	unk.	unk.	unk.	unk.	unk.	117	12	70	176	unk.	0.1	1	0.4	2	unk.	unk.	unk.
	0.1	0	2	0	0.44	4	142	3	unk.	unk.	unk.	unk.	0.90	86	9	3.0	4.5	unk.	12	3	unk.	unk.	unk.		unk.
I	0.35	21	0.09	4	0.91	15	unk.	unk.	unk.	unk.	unk.	unk.	unk.	184	18	109	276	unk.	0.2	1	0.6	4	unk.	unk.	unk.
	0.1	1	4	1	0.69	7	117	2	unk.	unk.	unk.	unk.	1.40	135	13	4.7	7.1	unk.	20	5	unk.	unk.	unk.		unk.
J	tr	0	tr(a)	0	0(a)	0	tr(a)	0	0	0(a)	0	unk.	unk.	tr	0	tr	4	unk.	tr	0	unk.	unk.	unk.	unk.	unk.
	tr	0	tr(a)	0	tr(a)	0	0	0	0	unk.	unk.	unk.	unk.	tr	0	0.0	0.1	unk.	unk.	unk.	unk.	unk.	unk.		unk.
K	tr	0	unk.	unk.	0(a)	0	unk.	unk.	0(a)	0	0	unk.	unk.	1	0	tr	unk.	unk.	unk.	unk.	unk.	unk.	unk.	unk.	unk.
	tr	0	tr(a)	0	unk.	unk.	24	1	unk.	unk.	unk.	unk.	unk.	3	0	0.0	unk.	unk.	unk.	unk.	unk.	unk.	unk.		unk.
L	tr	0	unk.	unk.	tr(a)	0	unk.	unk.	0	0	0	unk.	unk.	1	0	tr	18	unk.	unk.	unk.	unk.	unk.	unk.	unk.	unk.
	tr	0	unk.	unk.	unk.	unk.	17	0	2	unk.	unk.	unk.	unk.	3	0	0.0	0.5	unk.	unk.	unk.	unk.	unk.	unk.		unk.
M	0.12	7	0.05	3	0.00	0	0.006	2	0	28	7	10.3	4.9	115	12	453	169	2	1.3	7	1.1	8	26.7	18	162
	0.2	1	8	2	0.00	0	634	13	tr	5.5	18	5.5	1.91	112	11	19.7	4.3	0.1	10	3	0.13	7	0		72
N	0.08	5	0.01	1	0.05	1	0.006	2	0	0	0	1.3	0(a)	41	4	245	195	13	1.5	8	1.4	9	unk.	unk.	81
	0.3	2	20	5	0.09	1	1024	21	1	1.5	5	0.3	1.30	95	10	10.7	5.0	0.4	58	14	0.04	2	unk.		122
O	0.22	13	0.08	4	0.22	4	0.010	3	0	42	11	11.6	3.7	149	15	216	330	76	1.2	7	1.4	9	16.9	11	242
	0.3	2	18	5	0.78	8	637	13	2	4.5	15	4.5	1.63	196	20	9.4	8.4	2.2	75	19	0.23	11	tr		28
P	0.04	2	tr	0	0.00	0	0.001	0	0	0	0	0.0	0.0	3	0	76	22	26	0.2	1	0.4	3	13.5	9	25
	tr	0	tr	0	tr	0	0	0	0	0.0	0	0.0	0.23	3	0	3.3	0.6	0.7	2	1	0.02	1	tr		36
Q	0.02	1	0.01	0	0.04	1	0(a)	0	0	0	0	0.2	0(a)	21	2	193	30	unk.	0.2	1	1.1	8	unk.	unk.	64
	tr	0	1	0	0.04	0	799	16	0	0.3	1	0.2	0.56	18	2	8.4	0.8	unk.	2	1	0.03	2	unk.		96
R	0.09	5	0.40	20	0.30	5	unk.	unk.	unk.	unk.	unk.	unk.	unk.	50	5	120	360	unk.	4.5	25	3.0	20	7.4	5	unk.
	5.0	25	100	25	2.00	20	1000	20	unk.	unk.	unk.	unk.	unk.	50	5	5.2	9.2	unk.	80	20	0.50	25	unk.		unk.
S	unk.	unk.	0.01	0	0(a)	0	tr(a)	0	0	0(a)	0	unk.	unk.	1	0	tr	13	unk.	tr	0	unk.	unk.	unk.	unk.	unk.
	unk.	unk.	tr	0	0.01	0	8	0	1	unk.	unk.	unk.	unk.	1	0	0.0	0.3	unk.	1	0	unk.	unk.	unk.		unk.

	FOOD	Portion	Weight in grams / *Conversion for 100 g*	Kilocalories / H_2O g	Total carbohydrate g / Total fats g	Crude fiber g / Dietary fiber g	Total protein g / Total sugar g	% **USRDA**	Arginine mg / Histidine mg	Isoleucine mg / Leucine mg	Lysine mg / Methionine mg	Phenylalanine mg / Threonine mg	Valine mg / Tryptophan mg	Cystine mg / Tyrosine mg	Polyunsat. fatty acids g / Monounsat. fatty acids g	Saturated fatty acids g / P/S ratio	Linoleic acid g / Cholesterol mg	Thiamin mg / Ascorbic acid mg	% **USRDA**
A	jam / jelly,low calorie-Smucker's	1 Tbsp	20	24	6.2	tr(a)	0.0	**0**	tr(a)	tr(a)	tr(a)	tr(a)	tr(a)	tr(a)	tr(a)	tr(a)	tr(a)	tr	**0**
			5.00	unk.	tr	0(a)	6.2		tr(a)	tr(a)	tr(a)	tr(a)	tr(a)	tr(a)	tr(a)	0.00	0	tr	**0**
B	jam,assorted	1 Tbsp	20	54	14.0	0.20	0.1	**0**	unk.	unk.	unk.	unk.	unk.	unk.	tr(a)	tr(a)	tr(a)	tr	**0**
			5.00	5.8	tr	0(a)	14.0		unk.	unk.	unk.	unk.	unk.	unk.	tr(a)	0.00	0	tr	**1**
C	jam,red cherry / strawberry	1 Tbsp	20	54	14.0	0.20	0.1	**0**	unk.	unk.	unk.	unk.	unk.	unk.	tr(a)	tr(a)	tr(a)	tr	**0**
			5.00	5.8	tr	0(a)	14.0		unk.	unk.	unk.	unk.	unk.	unk.	tr(a)	0.00	0	3	**5**
D	JELLY,assorted	1 Tbsp	20	55	14.1	0.00	0.0	**0**	tr(a)	tr(a)	tr(a)	tr(a)	tr(a)	tr(a)	tr(a)	tr(a)	tr(a)	tr	**0**
			5.00	5.8	tr	0(a)	14.1		tr(a)	tr(a)	tr(a)	tr(a)	tr(a)	tr(a)	tr(a)	0.00	0	1	**2**
E	jelly,red cherry / strawberry	1 Tbsp	20	55	14.1	0.00	0.0	**0**	tr(a)	tr(a)	tr(a)	tr(a)	tr(a)	tr(a)	tr(a)	tr(a)	tr(a)	tr	**0**
			5.00	5.8	tr	0(a)	14.1		tr(a)	tr(a)	tr(a)	tr(a)	tr(a)	tr(a)	tr(a)	0.00	0	3	**5**
F	JUICE,CITRUS FRUIT,grapefruit &	4 oz glass	123	61	15.0	0.12	0.6	**1**	unk.	unk.	unk.	unk.	unk.	unk.	unk.	tr(a)	unk.	0.06	**4**
	orange,sweetened,canned		*0.81*	106.9	0.1	unk.	15.0		unk.	unk.	unk.	unk.	unk.	unk.	unk.	unk.	0	42	**70**
G	juice,citrus fruit,grapefruit &	4 oz glass	123	53	12.4	0.12	0.7	**1**	unk.	unk.	unk.	unk.	unk.	unk.	unk.	tr(a)	unk.	0.06	**4**
	orange,unsweetened,canned		*0.81*	109.1	0.2	unk.	12.4		unk.	unk.	unk.	unk.	unk.	unk.	unk.	unk.	0	42	**70**
H	juice,citrus fruit,grapefruit & orange,	4 oz glass	123	54	12.9	unk.	0.7	**1**	unk.	unk.	unk.	unk.	unk.	unk.	unk.	tr(a)	unk.	0.07	**5**
	unsweetened,frozen,diluted		*0.81*	108.7	0.1	unk.	12.9		unk.	unk.	unk.	unk.	unk.	unk.	unk.	unk.	0	50	**84**
I	juice,citrus fruit,grapefruit,fresh	4 oz glass	123	52	12.3	0.12	0.6	**1**	unk.	unk.	7	unk.	unk.	unk.	unk.	tr(a)	unk.	0.05	**3**
	(pink / red),Texas		*0.81*	109.7	0.1	unk.	12.2		unk.	unk.	0	unk.	1	7	unk.	unk.	0	47	**78**
J	juice,citrus fruit,grapefruit,fresh	4 oz glass	123	48	11.3	tr	0.6	**1**	unk.	unk.	7	14	unk.	unk.	tr(a)	tr(a)	tr(a)	0.05	**3**
	(pink / red / white),all varieties		*0.81*	110.7	0.1	unk.	10.0		unk.	unk.	unk.	unk.	1	7	tr(a)	0.00	0(a)	47	**78**
K	juice,citrus fruit,grapefruit,fresh	4 oz glass	123	52	12.3	unk.	0.6	**1**	unk.	unk.	7	14	0	unk.	unk.	tr(a)	unk.	0.05	**3**
	(white),Texas		*0.81*	109.7	0.1	unk.	12.3		unk.	unk.	0(a)	unk.	1	7	unk.	unk.	0	47	**78**
L	juice,citrus fruit,grapefruit,	4 oz glass	125	66	16.0	0.02	0.4	**1**	unk.	unk.	8	14	unk.	unk.	tr(a)	tr(a)	tr(a)	0.04	**3**
	sweetened,canned		*0.80*	107.7	0.1	unk.	12.1		unk.	unk.	0	unk.	1	8	tr(a)	0.00	0	39	**65**
M	juice,citrus fruit,grapefruit,	4 oz glass	123	58	14.0	tr	0.5	**1**	unk.	unk.	unk.	unk.	unk.	unk.	unk.	tr(a)	unk.	0.04	**3**
	sweetened,frozen,diluted		*0.81*	108.0	0.1	unk.	14.0		unk.	unk.	0(a)	unk.	unk.	7	unk.	unk.	0	41	**68**
N	juice,citrus fruit,grapefruit,	4 oz glass	123	50	12.1	0.02	0.4	**1**	unk.	unk.	7	14	unk.	unk.	unk.	tr(a)	tr(a)	0.04	**3**
	unsweetened,canned		*0.81*	109.7	0.1	unk.	9.7		unk.	unk.	0	unk.	1	7	tr(a)	unk.	0(a)	42	**70**
O	juice,citrus fruit,grapefruit,	4 oz glass	123	50	12.1	tr	0.6	**1**	unk.	unk.	unk.	unk.	0	unk.	unk.	tr(a)	unk.	0.05	**3**
	unsweetened,frozen,diluted		*0.81*	109.8	0.1	unk.	12.1		unk.	unk.	unk.	unk.	unk.	7	unk.	unk.	0	48	**80**
P	juice,citrus fruit,lemon,canned	1 Tbsp	15	4	1.2	tr	0.1	**0**	unk.	unk.	unk.	unk.	unk.	unk.	unk.	tr(a)	unk.	0.00	**0**
			6.56	14.0	tr	unk.	1.2		unk.	unk.	unk.	unk.	unk.	unk.	unk.	unk.	0	6	**11**
Q	juice,citrus fruit,lemon,fresh	1 Tbsp	15	4	1.2	tr	tr	**0**	unk.	unk.	unk.	unk.	unk.	unk.	unk.	unk.	unk.	0.00	**0**
			6.56	13.9	tr	unk.	0.2		unk.	unk.	tr(a)	unk.	unk.	unk.	unk.	unk.	0	7	**12**
R	juice,citrus fruit,lime,fresh	1 Tbsp	15	4	1.4	tr	0.0	**0**	unk.	unk.	unk.	unk.	unk.	unk.	unk.	tr(a)	unk.	tr	**0**
			6.50	13.9	tr	unk.	0.4		unk.	unk.	0(a)	unk.	unk.	unk.	unk.	unk.	0	5	**8**
S	juice,citrus fruit,orange,fresh,	4 oz glass	123	55	12.8	0.12	0.9	**1**	unk.	unk.	unk.	unk.	unk.	unk.	unk.	tr(a)	unk.	0.11	**7**
	all varieties		*0.81*	108.6	0.2	unk.	12.1		unk.	unk.	unk.	unk.	unk.	18	unk.	unk.	0	61	**103**

Riboflavin mg / Niacin mg	% USRDA	Vitamin B_6 mg / Folacin mcg	% USRDA	Vitamin B_{12} mcg / Pantothenic acid mg	% USRDA	Biotin mg / Vitamin A IU	% USRDA	Preformed A RE / Beta carotene RE	Vitamin D IU / Vitamin E IU	% USRDA	Total tocopherol mg / Alpha tocopherol mg	Other tocopherol mg / Total ash g	Calcium mg / Phosphorus mg	% USRDA	Sodium mg / Sodium meq	Potassium mg / Potassium meq	Chlorine mg / Chlorine meq	Iron mg / Magnesium mg	% USRDA	Zinc mg / Copper mg	% USRDA	Iodine mcg / Selenium mcg	% USRDA	Manganese mcg / Chromium mcg	
tr	**0**	tr	**0**	0.00	**0**	tr(a)	**0**	0	0(a)	**0**	0(a)	0(a)	1	**0**	23	27	unk.	tr	**0**	unk.	**unk.**	unk.	**unk.**	unk.	A
tr	**0**	tr(a)	**0**	0.01	**0**	7	**0**	1	0.0	**0**	0(a)	unk.	1	**0**	1.0	0.7	unk.	1	**0**	unk.	**unk.**	unk.		unk.	
0.01	**0**	0.01	**0**	0(a)	**0**	tr(a)	**0**	0	0	**0**	unk.	unk.	4	**0**	2	18	unk.	0.2	**1**	tr	**0**	unk.	**unk.**	tr(a)	B
tr	**0**	2	**0**	0.04	**0**	2	**0**	tr	unk.	**unk.**	unk.	0.06	2	**0**	0.1	0.4	unk.	tr(a)	**0**	unk.	**unk.**	unk.		unk.	
0.01	**0**	0.01	**0**	0(a)	**0**	tr(a)	**0**	0	0	**0**	unk.	unk.	4	**0**	2	18	unk.	0.2	**1**	0.0	**0**	unk.	**unk.**	tr(a)	C
tr	**0**	2	**0**	0.04	**0**	2	**0**	tr	unk.	**unk.**	unk.	0.06	2	**0**	0.1	0.4	unk.	tr(a)	**0**	unk.	**unk.**	unk.		unk.	
0.01	**0**	tr	**0**	0(a)	**0**	tr	**0**	0	0	**0**	0.0	0.0	4	**0**	3	15	unk.	0.3	**2**	0.1	**1**	unk.	**unk.**	2400	D
tr	**0**	tr	**0**	0.01	**0**	2	**0**	tr	0.0	**0**	0.0	0.04	1	**0**	0.1	0.4	unk.	1	**0**	0.07	**4**	unk.		unk.	
0.01	**0**	0.01	**0**	0(a)	**0**	tr	**0**	0	0	**0**	0.0	0.0	4	**0**	3	15	unk.	0.3	**2**	0.1	**1**	unk.	**unk.**	unk.	E
tr	**0**	tr	**0**	0.02	**0**	2	**0**	tr	0.0	**0**	0.0	0.04	1	**0**	0.1	0.4	unk.	1	**0**	0.02	**1**	unk.		unk.	
0.02	**1**	0.02	**1**	0.00	**0**	unk.	**unk.**	0	unk.	**unk.**	unk.	unk.	12	**1**	1	226	unk.	0.4	**2**	unk.	**unk.**	unk.	**unk.**	unk.	F
0.2	**1**	unk.	**unk.**	0.17	**2**	123	**3**	12	unk.	**unk.**	unk.	0.49	18	**2**	0.0	5.8	unk.	10	**3**	tr(a)	**0**	0(a)		unk.	
0.02	**1**	0.02	**1**	0.00	**0**	unk.	**unk.**	0	unk.	**unk.**	unk.	unk.	12	**1**	1	226	unk.	0.4	**2**	unk.	**unk.**	unk.	**unk.**	unk.	G
0.2	**1**	unk.	**unk.**	0.17	**2**	123	**3**	12	unk.	**unk.**	unk.	0.49	18	**2**	0.0	5.8	unk.	10	**3**	tr(a)	**0**	0(a)		unk.	
0.01	**1**	unk.	**unk.**	unk.	**unk.**	unk.	**unk.**	0	unk.	**unk.**	unk.	unk.	10	**1**	unk.	218	unk.	0.1	**1**	unk.	**unk.**	unk.	**unk.**	unk.	H
0.4	**2**	123	**31**	unk.	**unk.**	135	**3**	14	unk.	**unk.**	unk.	0.49	16	**2**	unk.	5.6	unk.	11	**3**	tr(a)	**0**	0(a)		unk.	
0.02	**1**	0.02	**1**	unk.	**unk.**	0.001	**0**	0	unk.	**unk.**	unk.	unk.	11	**1**	1	199	2	0.2	**1**	tr	**0**	unk.	**unk.**	10	I
0.2	**1**	26	**7**	0.20	**2**	541	**11**	54	unk.	**unk.**	unk.	0.25	18	**2**	0.0	5.1	0.1	10	**3**	0.01	**1**	tr(a)		unk.	
0.02	**1**	0.04	**2**	0.00	**0**	0.001	**0**	0	0(a)	**0**	unk.	unk.	11	**1**	1	199	2	0.2	**1**	tr	**0**	unk.	**unk.**	10	J
0.2	**1**	26	**7**	0.35	**4**	98	**2**	10	unk.	**unk.**	unk.	0.25	18	**2**	0.0	5.1	0.1	15	**4**	0.01	**1**	tr(a)		unk.	
0.02	**1**	0.02	**1**	unk.	**unk.**	0.001	**0**	0	0(a)	**0**	unk.	unk.	11	**1**	1	199	2	0.2	**1**	tr	**0**	unk.	**unk.**	10	K
0.2	**1**	2	**0**	0.20	**2**	12	**0**	1	unk.	**unk.**	unk.	0.25	18	**2**	0.0	5.1	0.1	15	**4**	0.01	**1**	tr(a)		unk.	
0.02	**2**	0.01	**1**	0.00	**0**	0.001	**0**	0	0(a)	**0**	0.2	0.2	10	**1**	1	203	unk.	0.5	**3**	tr	**0**	unk.	**unk.**	10	L
0.2	**1**	1	**0**	0.16	**2**	13	**0**	1	0.3	**1**	0.0	0.50	18	**2**	0.0	5.2	unk.	15	**4**	0.01	**1**	tr(a)		15	
0.01	**1**	0.02	**1**	0.00	**0**	0.001	**0**	0	0(a)	**0**	unk.	unk.	10	**1**	1	177	unk.	0.1	**1**	unk.	**unk.**	unk.	**unk.**	10	M
0.2	**1**	2	**0**	0.20	**2**	12	**0**	1	unk.	**unk.**	unk.	0.37	17	**2**	0.0	4.5	unk.	unk.	**unk.**	0.01	**1**	tr(a)		unk.	
0.02	**1**	0.01	**1**	0.00	**0**	0.001	**0**	0	0(a)	**0**	0.2	0.2	10	**1**	1	199	unk.	0.5	**3**	tr	**0**	unk.	**unk.**	10	N
0.2	**1**	1	**0**	0.16	**2**	12	**0**	1	0.3	**1**	0.0	0.49	17	**2**	0.0	5.1	unk.	15	**4**	0.01	**1**	tr(a)		15	
0.02	**1**	0.02	**1**	0.00	**0**	0.001	**0**	0	0(a)	**0**	unk.	unk.	12	**1**	1	209	unk.	0.1	**1**	unk.	**unk.**	unk.	**unk.**	10	O
0.2	**1**	26	**7**	0.20	**2**	12	**0**	1	unk.	**unk.**	unk.	0.37	21	**2**	0.0	5.3	unk.	11	**3**	0.01	**1**	0(a)		unk.	
tr	**0**	0.01	**0**	0.00	**0**	unk.	**unk.**	0	unk.	**unk.**	unk.	unk.	1	**0**	tr	22	unk.	tr	**0**	0.0	**0**	unk.	**unk.**	unk.	P
tr	**0**	0	**0**	unk.	**unk.**	3	**0**	tr	unk.	**unk.**	unk.	0.05	2	**0**	0.0	0.5	unk.	2	**0**	tr(a)	**0**	0(a)		unk.	
tr	**0**	0.01	**0**	0.00	**0**	unk.	**unk.**	0	0(a)	**0**	unk.	unk.	1	**0**	tr	22	1	tr	**0**	0.0	**0**	unk.	**unk.**	1	Q
tr	**0**	tr	**0**	0.02	**0**	3	**0**	tr	unk.	**unk.**	unk.	0.05	2	**0**	0.0	0.5	0.0	1	**0**	tr	**0**	0		0	
tr	**0**	0.01	**0**	0.00	**0**	unk.	**unk.**	0	0(a)	**0**	unk.	unk.	1	**0**	tr	16	unk.	tr	**0**	unk.	**unk.**	unk.	**unk.**	1	R
tr	**0**	unk.	**unk.**	0.02	**0**	2	**0**	tr	unk.	**unk.**	unk.	0.05	2	**0**	0.0	0.4	unk.	1	**0**	0.00	**0**	0		0	
0.04	**2**	0.05	**3**	0.00	**0**	tr	**0**	unk.	0(a)	**0**	0.2	0.2	14	**1**	1	246	5	0.2	**1**	tr	**0**	unk.	**unk.**	31	S
0.5	**3**	3	**1**	0.23	**2**	246	**5**	unk.	0.3	**1**	0.0	0.49	21	**2**	0.0	6.3	0.1	13	**3**	0.10	**5**	7		unk.	

	FOOD	Portion	Weight in grams / *Conversion for 100 g*	Kilocalories / H_2O g	Total carbohydrate g / Total fats g	Crude fiber g / Dietary fiber g	Total protein g / Total sugar g	% USRDA	Arginine mg / Histidine mg	Isoleucine mg / Leucine mg	Lysine mg / Methionine mg	Phenylalanine mg / Threonine mg	Valine mg / Tryptophan mg	Cystine mg / Tyrosine mg	Polyunsat. fatty acids g / Monounsat. fatty acids g	Saturated fatty acids g / P/S ratio	Linoleic acid g / Cholesterol mg	Thiamin mg / Ascorbic acid mg	% USRDA
A	juice,citrus fruit,orange,sweetened,	4 oz glass	125	65	15.2	0.12	0.9	**1**	unk.	unk.	unk.	11	unk.	unk.	unk.	tr(a)	unk.	0.09	**6**
	canned		*0.80*	108.1	0.2	unk.	16.0		unk.	unk.	unk.	unk.	unk.	19	unk.	unk.	0	50	**83**
B	juice,citrus fruit,orange,	4 oz glass	123	59	13.8	0.12	1.0	**2**	unk.	unk.	26	unk.	unk.	unk.	unk.	tr(a)	unk.	0.09	**6**
	unsweetened,canned		*0.81*	107.5	0.2	unk.	10.5		unk.	unk.	2	unk.	4	18	unk.	unk.	0	49	**82**
C	juice,citrus fruit,orange,	4 oz glass	123	55	13.2	tr	0.9	**1**	unk.	unk.	unk.	11	unk.	unk.	unk.	tr(a)	unk.	0.11	**7**
	unsweetened,frozen,diluted		*0.81*	108.4	0.1	unk.	9.6		unk.	unk.	unk.	unk.	unk.	18	unk.	unk.	unk.	55	**92**
D	juice,citrus fruit,tangelo,fresh	4 oz glass	123	50	11.9	tr	0.6	**1**	unk.	unk.	unk.	unk.	unk.	unk.	unk.	tr(a)	unk.	unk.	**unk.**
			0.81	110.0	0.1	unk.	11.9		unk.	unk.	unk.	unk.	unk.	unk.	unk.	unk.	0	33	**55**
E	juice,citrus fruit,tangerine,	4 oz glass	123	61	14.8	0.12	0.6	**1**	unk.	unk.	unk.	unk.	unk.	unk.	unk.	tr(a)	unk.	0.07	**5**
	sweetened,canned		*0.81*	107.0	0.2	unk.	14.6		unk.	unk.	unk.	unk.	unk.	unk.	unk.	unk.	0	27	**45**
F	juice,citrus fruit,tangerine,	4 oz glass	123	53	12.5	0.12	0.6	**1**	unk.	unk.	unk.	unk.	unk.	unk.	unk.	tr(a)	unk.	0.07	**5**
	unsweetened,canned		*0.81*	109.2	0.2	unk.	12.4		unk.	unk.	unk.	unk.	unk.	unk.	unk.	unk.	0	27	**45**
G	juice,citrus fruit,tangerine,	4 oz glass	123	57	13.3	0.12	0.6	**1**	unk.	unk.	unk.	unk.	unk.	unk.	unk.	tr(a)	unk.	0.07	**5**
	unsweetened,frozen,diluted		*0.81*	108.4	0.2	unk.	13.2		unk.	unk.	unk.	unk.	unk.	unk.	unk.	unk.	0	33	**55**
H	JUICE,FRUIT,apple-cherry	4 oz glass	124	66	16.1	unk.	0.3	**0**	unk.	unk.	unk.	unk.	unk.	unk.	unk.	tr(a)	unk.	unk.	**unk.**
			0.81	unk.	unk.	unk.	16.1		unk.	unk.	unk.	unk.	unk.	unk.	unk.	unk.	0	50	**83**
I	juice,fruit,apple,canned	4 oz glass	124	58	14.8	0.00	0.1	**0**	unk.	6	5	4	4	tr(a)	tr(a)	tr(a)	tr(a)	0.01	**1**
			0.81	108.9	tr	unk.	15.1		2	6	2	4	0	2	tr(a)	0.00	0	1	**2**
J	juice,fruit,apricot nector	4 oz glass	126	72	18.4	0.25	0.4	**1**	unk.	unk.	unk.	unk.	unk.	unk.	unk.	tr(a)	unk.	0.01	**1**
	w/vitamin C added		*0.79*	106.6	0.1	unk.	18.1		unk.	unk.	unk.	unk.	unk.	unk.	unk.	unk.	0	53	**88**
K	juice,fruit,cranberry juice cocktail	4 oz glass	127	83	21.0	unk.	0.1	**0**	unk.	unk.	unk.	unk.	unk.	unk.	unk.	tr(a)	unk.	0.01	**1**
			0.79	105.7	0.1	unk.	21.0		unk.	unk.	unk.	unk.	unk.	unk.	unk.	unk.	0	51	**85**
L	juice,fruit,cranberry juice cocktail,	4 oz glass	124	24	6.1	tr	0.0	**0**	unk.	tr(a)	unk.	unk.	unk.	tr(a)	unk.	unk.	unk.	unk.	**unk.**
	low calorie		*0.81*	117.7	0.1	unk.	6.1		unk.	unk.	unk.	tr(a)	tr(a)	tr(a)	unk.	unk.	0	40	**66**
M	juice,fruit,grape,canned	4 oz glass	127	84	21.1	tr	0.3	**0**	unk.	unk.	unk.	unk.	unk.	unk.	tr(a)	tr(a)	tr(a)	0.05	**3**
			0.79	105.3	tr	unk.	16.3		unk.	unk.	unk.	unk.	unk.	unk.	tr(a)	0.00	0	tr	**0**
N	juice,fruit,papaya,canned	4 oz glass	123	59	14.9	unk.	0.5	**1**	unk.	unk.	unk.	unk.	unk.	unk.	0(a)	0(a)	0(a)	0.02	**2**
			0.81	unk.	0.0	unk.	15.1		unk.	unk.	unk.	unk.	unk.	unk.	0(a)	0.00	0(a)	50	**84**
O	juice,fruit,peach nectar,canned	4 oz glass	125	60	15.5	0.00	0.3	**0**	unk.	tr(a)	tr(a)	tr(a)	tr(a)	tr(a)	tr(a)	tr(a)	tr(a)	0.01	**1**
			0.80	109.0	tr	unk.	15.4		tr(a)	tr(a)	tr(a)	tr(a)	tr(a)	tr(a)	tr(a)	0.00	0	14	**23**
P	juice,fruit,pear nectar,canned	4 oz glass	123	64	16.2	0.37	0.4	**1**	unk.	unk.	unk.	unk.	unk.	unk.	unk.	tr(a)	unk.	unk.	**unk.**
			0.81	106.0	0.2	unk.	15.9		unk.	unk.	unk.	unk.	unk.	unk.	unk.	unk.	0	14	**23**
Q	juice,fruit,pineapple,unsweetened,	4 oz glass	128	70	17.3	0.13	0.5	**1**	unk.	unk.	12	unk.	unk.	unk.	unk.	tr(a)	unk.	0.06	**4**
	canned		*0.78*	109.6	0.1	unk.	17.2		unk.	unk.	1	unk.	6	unk.	unk.	unk.	0	12	**19**
R	juice,fruit,prune,canned	4 oz glass	128	99	24.3	tr	0.5	**1**	unk.	unk.	uk.	unk.	unk.	unk.	unk.	tr(a)	unk.	0.01	**1**
			0.78	102.4	0.1	unk.	24.3		unk.	unk.	unk.	unk.	unk.	unk.	unk.	unk.	0	3	**4**
S	juice,fruit,raspberry	4 oz glass	123	50	13.2	tr	0.2	**0**	unk.	tr(a)	unk.	unk.	unk.	tr(a)	unk.	unk.	unk.	0.01	**1**
			0.81	unk.	0.0	unk.	13.2		unk.	unk.	unk.	tr(a)	tr(a)	tr(a)	unk.	unk.	unk.	18	**31**

Riboflavin mg / Niacin mg	% USRDA	Vitamin B_6 mg / Folacin mcg	% USRDA	Vitamin B_{12} mcg / Pantothenic acid mg	% USRDA	Biotin mg / Vitamin A IU	% USRDA	Preformed A RE / Beta carotene RE	Vitamin D IU / Vitamin E IU	% USRDA	Total tocopherol mg / Alpha tocopherol mg	Other tocopherol mg / Total ash g	Calcium mg / Phosphorus mg	% USRDA	Sodium mg / Sodium meq	Potassium mg / Potassium meq	Chlorine mg / Chlorine meq	Iron mg / Magnesium mg	% USRDA	Zinc mg / Copper mg	% USRDA	Iodine mcg / Selenium mcg	% USRDA	Manganese mcg / Chromium mcg	
0.02	**2**	0.04	**2**	0.00	**0**	0.001	**0**	0	0(a)	**0**	unk.	unk.	13	**1**	1	249	unk.	0.5	**3**	0.9	**6**	unk.	**unk.**	31	A
0.4	**2**	4	**1**	0.19	**2**	250	**5**	25	unk.	**unk.**	unk.	0.50	23	**2**	0.0	6.4	unk.	15	**4**	0.10	**5**	7		unk.	
0.02	**1**	0.04	**2**	0.00	**0**	0.001	**0**	0	0(a)	**0**	unk.	unk.	12	**1**	1	245	unk.	0.5	**3**	0.1	**1**	unk.	**unk.**	31	B
0.4	**2**	43	**11**	0.18	**2**	246	**5**	25	unk.	**unk.**	unk.	0.49	22	**2**	0.0	6.3	unk.	15	**4**	0.10	**5**	7		unk.	
0.01	**1**	0.03	**2**	0.00	**0**	tr	**0**	0	0(a)	**0**	unk.	unk.	11	**1**	1	229	unk.	0.1	**1**	tr	**0**	unk.	**unk.**	31	C
0.4	**2**	61	**15**	0.20	**2**	246	**5**	25	unk.	**unk.**	unk.	0.49	20	**2**	0.0	5.8	unk.	10	**3**	0.10	**5**	7		15	
unk.	**unk.**	unk.	**unk.**	0(a)	**0**	unk.	**unk.**	unk.	0(a)	**0**	unk.	unk.	unk.	**unk.**	unk.	unk.	unk.	unk.	**unk.**	unk.	**unk.**	unk.	**unk.**	unk.	D
unk.	**unk.**	unk.	**unk.**	unk.	**unk.**	unk.	**unk.**	unk.	unk.	**unk.**	unk.	0.37	unk.	**unk.**	unk.	unk.	unk.	unk.	**unk.**	unk.	**unk.**	tr(a)		unk.	
0.02	**1**	0.04	**2**	0.00	**0**	unk.	**unk.**	0	0(a)	**0**	unk.	unk.	22	**2**	1	219	unk.	0.2	**1**	unk.	**unk.**	unk.	**unk.**	unk.	E
0.1	**1**	unk.	**unk.**	unk.	**unk.**	517	**10**	52	unk.	**unk.**	unk.	0.37	17	**2**	0.0	5.6	unk.	tr(a)	**0**	unk.	**unk.**	tr(a)		unk.	
0.02	**1**	0.04	**2**	0.00	**0**	unk.	**unk.**	0	0(a)	**0**	unk.	unk.	22	**2**	1	219	unk.	0.2	**1**	unk.	**unk.**	unk.	**unk.**	unk.	F
0.1	**1**	2	**1**	unk.	**unk.**	517	**10**	52	unk.	**unk.**	unk.	0.37	17	**2**	0.0	5.6	unk.	tr(a)	**0**	tr(a)	**0**	tr(a)		unk.	
0.02	**1**	unk.	**unk.**	0(a)	**0**	unk.	**unk.**	0	0(a)	**0**	unk.	unk.	22	**2**	1	214	unk.	0.2	**1**	unk.	**unk.**	unk.	**unk.**	unk.	G
0.1	**1**	unk.	**unk.**	unk.	**unk.**	504	**10**	50	unk.	**unk.**	unk.	0.49	17	**2**	0.0	5.5	unk.	tr(a)	**0**	tr(a)	**0**	tr(a)		unk.	
0.01	**1**	unk.	**unk.**	unk.	**unk.**	unk.	**unk.**	0	0(a)	**0**	unk.	unk.	7	**1**	5	112	unk.	0.9	**5**	unk.	**unk.**	unk.	**unk.**	unk.	H
0.1	**1**	unk.	**unk.**	unk.	**unk.**	4	**0**	tr	unk.	**unk.**	unk.	unk.	10	**1**	0.2	2.8	unk.	unk.	**unk.**	tr(a)	**0**	tr(a)		unk.	
0.02	**2**	0.04	**2**	0.00	**0**	unk.	**unk.**	unk.	0(a)	**0**	unk.	unk.	7	**1**	1	125	unk.	0.7	**4**	tr	**0**	unk.	**unk.**	260	I
0.1	**1**	tr	**0**	unk.	**unk.**	unk.	**unk.**	unk.	unk.	**unk.**	unk.	0.25	11	**1**	0.0	3.2	unk.	5	**1**	0.03	**1**	tr(a)		7	
0.01	**1**	0.04	**2**	0.00	**0**	unk.	**unk.**	0	0(a)	**0**	unk.	unk.	11	**1**	tr	190	unk.	0.3	**1**	unk.	**unk.**	unk.	**unk.**	unk.	J
0.3	**1**	1	**0**	unk.	**unk.**	1197	**24**	120	unk.	**unk.**	unk.	0.50	15	**2**	0.0	4.9	unk.	6	**2**	tr(a)	**0**	tr(a)		unk.	
0.01	**1**	unk.	**unk.**	0(a)	**0**	unk.	**unk.**	0	0(a)	**0**	unk.	unk.	6	**1**	1	13	unk.	0.4	**2**	0.3	**2**	unk.	**unk.**	unk.	K
tr	**0**	unk.	**unk.**	0.07	**1**	tr	**0**	tr	unk.	**unk.**	unk.	0.13	4	**0**	0.1	0.3	unk.	unk.	**unk.**	tr(a)	**0**	tr(a)		unk.	
unk.	**unk.**	unk.	**unk.**	0(a)	**0**	unk.	**unk.**	unk.	0(a)	**0**	unk.	unk.	8	**1**	4	25	unk.	2.9	**16**	unk.	**unk.**	unk.	**unk.**	unk.	L
unk.	**unk.**	unk.	**unk.**	0.07	**1**	unk.	**unk.**	unk.	unk.	**unk.**	unk.	0.06	2	**0**	0.2	0.6	unk.	2	**1**	0.12	**6**	tr(a)		unk.	
0.03	**2**	0.10	**5**	0.00	**0**	unk.	**unk.**	0	0(a)	**0**	unk.	unk.	14	**1**	3	147	unk.	0.4	**2**	0.1	**0**	unk.	**unk.**	457	M
0.3	**1**	tr	**0**	0.10	**1**	0	**0**	0	unk.	**unk.**	unk.	0.38	15	**2**	0.1	3.8	unk.	15	**4**	0.11	**6**	5		tr(a)	
0.01	**1**	unk.	**unk.**	0(a)	**0**	unk.	**unk.**	0	0(a)	**0**	unk.	unk.	22	**2**	unk.	unk.	unk.	0.4	**2**	unk.	**unk.**	unk.	**unk.**	unk.	N
0.1	**1**	unk.	**unk.**	unk.	**unk.**	2460	**49**	246	unk.	**unk.**	unk.	unk.	12	**1**	unk.	unk.	unk.	unk.	**unk.**	tr(a)	**0**	tr(a)		unk.	
0.02	**2**	unk.	**unk.**	0(a)	**0**	unk.	**unk.**	0	0(a)	**0**	unk.	unk.	5	**1**	1	98	unk.	0.2	**1**	unk.	**unk.**	unk.	**unk.**	unk.	O
0.5	**3**	unk.	**unk.**	unk.	**unk.**	538	**11**	54	unk.	**unk.**	unk.	0.25	14	**1**	0.0	2.5	unk.	unk.	**unk.**	tr(a)	**0**	tr(a)		unk.	
0.02	**1**	unk.	**unk.**	0(a)	**0**	unk.	**unk.**	0	0(a)	**0**	unk.	unk.	4	**0**	1	48	unk.	0.1	**1**	unk.	**unk.**	unk.	**unk.**	unk.	P
unk.	**unk.**	unk.	**unk.**	unk.	**unk.**	tr	**0**	tr	unk.	**unk.**	unk.	0.12	6	**1**	0.0	1.2	unk.	unk.	**unk.**	tr(a)	**0**	tr(a)		unk.	
0.03	**2**	0.12	**6**	0.00	**0**	unk.	**unk.**	0	0(a)	**0**	unk.	unk.	19	**2**	1	191	unk.	0.4	**2**	0.1	**1**	unk.	**unk.**	unk.	Q
0.3	**1**	1	**0**	0.13	**1**	64	**1**	6	unk.	**unk.**	unk.	0.51	12	**1**	0.1	4.9	unk.	15	**4**	0.01	**0**	tr(a)		15	
0.01	**1**	0.08	**4**	0.00	**0**	unk.	**unk.**	unk.	0(a)	**0**	unk.	unk.	18	**2**	3	301	unk.	5.2	**29**	tr	**0**	unk.	**unk.**	26	R
0.5	**3**	tr	**0**	unk.	**unk.**	unk.	**unk.**	unk.	unk.	**unk.**	unk.	0.64	26	**3**	0.1	7.7	unk.	7	**2**	0.02	**1**	tr(a)		15	
0	**0**	unk.	**unk.**	0.00	**0**	unk.	**unk.**	0	unk.	**unk.**	unk.	unk.	30	**3**	unk.	unk.	unk.	1.0	**6**	unk.	**unk.**	unk.	**unk.**	unk.	S
0.0	**0**	unk.	**unk.**	unk.	**unk.**	123	**3**	12	unk.	**unk.**	unk.	unk.	14	**1**	unk.	unk.	unk.	unk.	**unk.**	tr(a)	**0**	tr(a)		unk.	

	FOOD	Portion	Weight in grams / *Conversion for 100 g*	Kilocalories / H_2O g	Total carbohydrate g / Total fats g	Crude fiber g / Dietary fiber g	Total protein g / Total sugar g	% USRDA	Arginine mg / Histidine mg	Isoleucine mg / Leucine mg	Lysine mg / Methionine mg	Phenylalanine mg / Threonine mg	Valine mg / Tryptophan mg	Cystine mg / Tyrosine mg	Polyunsat. fatty acids g / Monounsat. fatty acids g	Saturated fatty acids g / P/S ratio	Linoleic acid g / Cholesterol mg	Thiamin mg / Ascorbic acid mg	% USRDA
A	JUICE,FRUIT,tomato,canned	4 oz glass	122	23	5.2	0.24	1.1	2	unk.	32	46	30	30	unk.	unk.	tr(a)	unk.	0.06	4
			0.82	114.2	0.1	unk.	5.0		104	45	7	37	10	99	unk.	unk.	0	20	33
B	juice,fruit,tomato,low sodium,	4 oz glass	122	23	5.2	0.24	1.0	2	unk.	28	41	27	27	unk.	unk.	tr(a)	unk.	0.06	4
	canned		*0.82*	114.9	0.1	unk.	5.0		93	40	7	32	9	88	unk.	unk.	0	20	33
C	JUICE,VEGETABLE,vegetable cocktail,	4 oz glass	122	21	4.4	0.37	1.1	2	unk.	unk.	unk.	unk.	unk.	unk.	unk.	tr(a)	unk.	0.06	4
	canned		*0.82*	114.8	0.1	unk.	4.0		unk.	unk.	unk.	unk.	unk.	unk.	unk.	unk.	0	11	18
D	KALE,fresh,leaves & stems,cooked,	1/2 C	55	15	2.2	0.60	1.8	3	unk.	unk.	unk.	75	unk.	unk.	unk.	tr(a)	unk.	unk.	unk.
	drained solids		*1.82*	50.2	0.4	unk.	0.3		unk.	unk.	unk.	unk.	unk.	74	unk.	unk.	0	34	57
E	kale,fresh,leaves w/o stems,midribs,	1/2 C	55	21	3.4	unk.	2.5	4	unk.	unk.	unk.	unk.	unk.	unk.	unk.	tr(a)	unk.	0.05	4
	cooked,drained solids		*1.82*	48.3	0.4	unk.	0.3		unk.	unk.	unk.	unk.	unk.	unk.	unk.	unk.	0	51	85
F	kale,frozen,cooked,drained solids	1/2 C	65	20	3.5	0.58	1.9	3	unk.	unk.	unk.	89	unk.	unk.	unk.	tr(a)	unk.	0.04	3
			1.54	58.8	0.3	unk.	0.4		unk.	unk.	unk.	unk.	unk.	87	unk.	unk.	0	25	41
G	kale,raw,leaves & stems,chopped	1/2 C	85	32	5.1	1.10	3.6	6	unk.	unk.	unk.	152	unk.	unk.	unk.	tr(a)	unk.	unk.	unk.
			1.18	74.4	0.7	unk.	0.5		unk.	unk.	unk.	unk.	unk.	150	unk.	unk.	0	106	177
H	kale,raw,leaves w/o stems,midribs	1/2 C	28	15	2.5	unk.	1.6	3	unk.	unk.	unk.	unk.	unk.	unk.	unk.	tr(a)	unk.	0.04	3
			3.64	22.7	0.2	unk.	0.2		unk.	unk.	unk.	unk.	unk.	unk.	unk.	unk.	0	51	85
I	KENTUCKY FRIED CHICKEN,	1 piece	50	162	6.0	unk.	11.0	25	unk.	unk.	unk.	unk.	unk.	unk.	unk.	unk.	unk.	0.03	2
	extra crispy		*2.00*	21.5	10.4	unk.	unk.		unk.	unk.	unk.	unk.	unk.	unk.	unk.	unk.	unk.	1	1
J	Kentucky Fried Chicken,	1 piece	50	145	4.5	unk.	11.7	26	unk.	unk.	unk.	unk.	unk.	unk.	unk.	unk.	unk.	0.02	2
	original recipe		*2.00*	24.3	8.9	unk.	unk.		unk.	unk.	unk.	unk.	unk.	unk.	unk.	unk.	unk.	1	1
K	KOHLRABI,fresh,chopped,cooked,	1/2 C	83	20	4.4	0.82	1.4	2	unk.	unk.	unk.	35	unk.	unk.	unk.	tr(a)	unk.	0.05	3
	drained solids		*1.21*	76.1	0.1	unk.	2.6		unk.	unk.	unk.	unk.	unk.	unk.	unk.	unk.	0(a)	35	59
L	KUCHEN,home recipe	2-1/4x2-1/4x	95	315	44.9	0.07	4.6	9	90	180	155	197	195	79	0.76	7.88	0.57	0.11	8
		1-3/4".	*1.05*	30.7	13.4	0.67	28.1		93	301	77	138	49	144	3.53	0.10	84	tr	0
M	LADYFINGER	1 average	11	40	7.1	0.01	0.9	2	unk.	unk.	unk.	unk.	unk.	unk.	unk.	0.27	0.06	0.01	0
			9.09	2.1	0.9	unk.	5.0		unk.	unk.	unk.	unk.	unk.	unk.	0.33	unk.	39	0	0
N	LAMB,cooked,chopped,w/visible fat	1/2 C	70	195	0.0	0.00	17.7	39	1220	932	1457	731	887	227	0.99	6.68	0.55	0.10	7
			1.43	37.8	13.2	0(a)	0.0		486	1393	432	823	232	584	5.54	0.15	69	0	0
O	lamb,cooked,ground,w/visible fat	1/2 C	55	153	0.0	0.00	13.9	31	959	733	1145	575	697	178	0.78	5.25	0.43	0.08	6
			1.82	29.7	10.4	0(a)	0.0		382	1094	339	647	183	459	4.36	0.15	54	0	0
P	lamb,home recipe,curry	3/4 C	185	345	21.5	0.35	25.8	56	1641	1268	2021	1065	1311	329	4.86	3.03	4.62	0.24	16
			0.54	118.8	16.8	unk.	0.4		683	1982	612	1185	321	863	3.60	1.60	89	3	5
Q	lamb,home recipe,lamb & potato	3/4 C	173	277	16.0	1.25	16.1	34	1134	775	1196	639	778	192	1.11	7.51	0.59	0.23	15
	casserole		*0.58*	121.2	16.5	3.08	5.2		393	1160	341	691	194	512	6.13	0.15	58	21	34
R	lamb,leg,w/o bone,roasted,sliced,	2 sl 4-1/8x	85	158	0.0	0.00	24.4	5	1681	1216	1994	1007	1221	313	0.63	3.43	0.31	0.14	9
	no visible fat	2-1/4x1/4"	*1.18*	52.9	5.9	0(a)	0.0		669	1881	595	1147	309	805	2.89	0.18	85	0	0
S	lamb,leg,w/o bone,roasted,sliced,	2 sl 4-1/8x	85	237	0.0	0.00	21.5	48	1482	1132	1769	888	1077	275	1.21	8.11	0.67	0.13	9
	w/fat	2-1/4x1/4"	*1.18*	45.9	16.1	0(a)	0.0		590	1691	524	1000	282	710	6.73	0.15	83	0	0

Riboflavin mg / Niacin mg	% USRDA	Vitamin B_6 mg / Folacin mcg	% USRDA	Vitamin B_{12} mcg / Pantothenic acid mg	% USRDA	Biotin mg / Vitamin A IU	% USRDA	Preformed A RE / Beta carotene RE	Vitamin D IU / Vitamin E IU	% USRDA	Total tocopherol mg / Alpha tocopherol mg	Other tocopherol mg / Total ash g	Calcium mg / Phosphorus mg	% USRDA	Sodium mg / Sodium meq	Potassium mg / Potassium meq	Chlorine mg / Chlorine meq	Iron mg / Magnesium mg	% USRDA	Zinc mg / Copper mg	% USRDA	Iodine mcg / Selenium mcg	% USRDA	Manganese mcg / Chromium mcg	
0.04	2	0.23	12	0.00	0	unk.	unk.	0	0(a)	0	0.9	0.6	9	1	244	277	464	1.1	6	0.0	0	unk.	unk.	unk.	A
1.0	5	32	8	0.30	3	976	20	98	1.0	4	0.3	1.34	22	2	10.6	7.1	13.1	12	3	tr(a)	0	tr(a)		unk.	
0.04	2	0.23	12	0.00	0	unk.	unk.	0	0(a)	0	0.9	0.6	9	1	4	277	unk.	1.1	6	0.1	1	unk.	unk.	16	B
0.9	4	32	8	0.30	3	976	20	98	1.0	4	0.3	0.73	22	2	0.2	7.1	unk.	5	1	0.00	0	tr(a)		15	
0.04	2	unk.	unk.	0(a)	0	unk.	unk.	0	0(a)	0	unk.	unk.	15	2	244	270	unk.	0.6	3	unk.	unk.	unk.	unk.	unk.	C
1.0	5	unk.	unk.	0.27	3	854	17	85	unk.	unk.	unk.	1.59	27	3	10.6	6.9	unk.	unk.	unk.	tr(a)	0	tr(a)		unk.	
unk.	unk.	unk.	unk.	0(a)	0	unk.	unk.	0	0(a)	0	unk.	unk.	74	7	24	122	unk.	0.7	4	unk.	unk.	unk.	unk.	unk.	D
unk.	unk.	unk.	unk.	unk.	unk.	4070	81	407	unk.	unk.	unk.	0.49	25	3	1.0	3.1	unk.	16	4	tr(a)	0	tr(a)		unk.	
0.10	6	unk.	unk.	0(a)	0	unk.	unk.	0	0(a)	0	unk.	unk.	103	10	24	122	unk.	0.9	5	unk.	unk.	unk.	unk.	unk.	E
0.9	4	unk.	unk.	unk.	unk.	4565	91	456	unk.	unk.	unk.	0.49	32	3	1.0	3.1	unk.	17	4	tr(a)	0	tr(a)		unk.	
0.10	6	unk.	unk.	0(a)	0	unk.	unk.	0	0(a)	0	unk.	unk.	79	8	14	125	unk.	0.6	4	unk.	unk.	unk.	unk.	unk.	F
0.5	2	unk.	unk.	unk.	unk.	5330	107	533	unk.	unk.	unk.	0.39	31	3	0.6	3.2	unk.	20	5	tr(a)	0	tr(a)		unk.	
unk.	unk.	0.25	13	0.00	0	unk.	unk.	0	0(a)	0	unk.	unk.	152	15	64	321	unk.	1.9	10	unk.	unk.	unk.	unk.	unk.	G
unk.	unk.	51	13	0.85	9	7565	151	756	unk.	unk.	unk.	1.27	62	6	2.8	8.2	unk.	31	8	tr(a)	0	tr(a)		unk.	
0.07	4	0.08	4	0.00	0	unk.	unk.	0	0(a)	0	unk.	unk.	68	7	21	104	unk.	0.7	4	unk.	unk.	unk.	unk.	unk.	H
0.6	3	16	4	0.27	3	2750	55	275	unk.	unk.	unk.	0.41	26	3	0.9	2.7	unk.	10	3	tr(a)	0	tr(a)		unk.	
0.07	4	unk.	unk.	unk.	unk.	unk.	unk.	unk.	unk.	unk.	unk.	unk.	17	2	unk.	unk.	unk.	0.6	3	unk.	unk.	unk.	unk.	unk.	I
3.0	15	unk.	unk.	unk.	unk.	37	1	unk.	unk.	unk.	unk.	1.03	unk.	unk.	unk.	unk.	unk.	unk.	unk.	unk.	unk.	unk.		unk.	
0.09	6	unk.	unk.	unk.	unk.	unk.	unk.	unk.	unk.	unk.	unk.	unk.	20	2	unk.	unk.	unk.	0.7	4	unk.	unk.	unk.	unk.	unk.	J
3.8	19	unk.	unk.	unk.	unk.	50	1	unk.	unk.	unk.	unk.	1.33	unk.	unk.	unk.	unk.	unk.	unk.	unk.	unk.	unk.	unk.		unk.	
0.02	2	0.07	4	0.00	0	unk.	unk.	0	0(a)	0	unk.	unk.	27	3	5	214	unk.	0.2	1	unk.	unk.	unk.	unk.	unk.	K
0.2	1	unk.	unk.	0.07	1	16	0	2	unk.	unk.	unk.	0.58	34	3	0.2	5.5	unk.	unk.	unk.	tr(a)	0	tr(a)		unk.	
0.13	8	0.03	2	0.20	3	0.002	1	0	4	1	0.2	0.0	58	6	307	78	35	1.0	5	0.6	4	23.7	16	105	L
0.8	4	11	3	0.35	4	468	9	0	0.3	1	0.1	1.19	94	9	13.4	2.0	1.0	37	9	0.04	2	12		29	
0.02	1	unk.	unk.	unk.	unk.	unk.	unk.	unk.	unk.	unk.	unk.	unk.	5	1	8	8	unk.	0.2	1	unk.	unk.	unk.	unk.	unk.	M
tr	0	unk.	unk.	unk.	unk.	71	1	unk.	unk.	unk.	unk.	0.08	18	2	0.3	0.2	unk.	2	0	unk.	unk.	unk.		unk.	
0.19	11	0.12	6	1.36	23	unk.	unk.	0	0	0	0.2	0.1	8	1	49	203	unk.	1.2	7	2.6	17	unk.	unk.	unk.	N
3.8	19	2	1	0.19	2	0	0	0	0.3	1	0.1	1.19	146	15	2.1	5.2	unk.	15	4	0.57	29	tr(a)		tr(a)	
0.15	9	0.09	5	1.07	18	unk.	unk.	0	0	0	0.2	0.1	6	1	38	159	unk.	0.9	5	2.0	13	unk.	unk.	unk.	O
3.0	15	2	0	0.15	2	0	0	0	0.2	1	0.1	0.93	114	11	1.7	4.1	unk.	12	3	0.39	19	tr(a)		tr(a)	
0.26	16	0.16	8	1.75	29	0.002	1	0	0	0	5.8	4.7	24	2	258	317	10	2.9	16	4.3	29	tr(a)	0	10	P
7.8	39	16	4	0.34	3	145	3	13	6.9	23	1.1	2.40	222	22	11.2	8.1	0.3	26	7	0.79	40	tr		3	
0.21	12	0.30	15	1.19	20	0.003	1	0	0	0	0.7	0.5	40	4	65	561	26	1.9	10	2.5	17	unk.	unk.	23	Q
4.3	22	32	8	0.66	7	4688	94	468	0.8	3	0.2	1.86	172	17	2.8	14.3	0.7	39	10	0.50	25	tr(a)		1	
0.25	15	0.14	7	1.91	32	0.002	1	0	0	0	0.3	0.1	11	1	59	246	unk.	1.9	10	3.7	24	unk.	unk.	unk.	R
5.3	26	3	1	0.23	2	0	0	0	0.3	1	0.1	1.70	202	20	2.6	6.3	unk.	15	4	0.79	40	tr(a)		tr(a)	
0.23	14	0.14	7	1.91	32	unk.	unk.	0	0	0	0.3	0.1	9	1	59	246	unk.	1.4	8	3.1	21	unk.	unk.	unk.	S
4.7	23	3	1	0.23	2	0	0	0	0.3	1	0.1	1.44	177	18	2.6	6.3	unk.	18	5	0.70	35	tr(a)		tr(a)	

	FOOD	Portion	Weight in grams *Conversion for 100 g*	Kilocalories H_2O g	Total carbohydrate g Total fats g	Crude fiber g Dietary fiber g	Total protein g Total sugar g	% USRDA	Arginine mg Histidine mg	Isoleucine mg Leucine mg	Lysine mg Methionine mg	Phenylalanine mg Threonine mg	Valine mg Tryptophan mg	Cystine mg Tyrosine mg	Polyunsat. fatty acids g Monounsat. fatty acids g	Saturated fatty acids g P/S ratio	Linoleic acid g Cholesterol mg	Thiamin mg Ascorbic acid mg	% USRDA
A	lamb,loin chop,w/bone,broiled,	1 med	49	92	0.0	0.00	13.8	**31**	952	689	1129	571	692	177	0.23	1.26	0.11	0.07	**5**
	no visible fat	(4 per lb)	*2.04*	30.4	3.7	0(a)	0.0		379	1066	337	649	175	456	1.06	0.18	49	0(a)	**0**
B	lamb,loin chop,w/bone,broiled,w/fat	1 med	71	255	0.0	0.00	15.6	**35**	1076	779	1277	645	782	200	1.61	10.72	0.87	0.09	**6**
		(4 per lb)	*1.41*	33.4	20.9	0(a)	0.0		429	1205	381	734	198	515	8.80	0.15	70	0	**0**
C	lamb,patty,cooked,w/fat	3 oz cooked	82	229	0.0	0.00	20.8	**46**	1429	1092	1706	857	1039	266	1.16	7.82	0.65	0.12	**8**
			1.22	44.3	15.5	0(a)	0.0		569	1632	506	964	272	685	6.49	0.15	80	0	**0**
D	lamb,rib chop,w/bone,broiled,	1 med	43	91	0.0	0.00	11.7	**26**	806	584	956	483	599	150	0.23	3.04	0.11	0.06	**4**
	no visible fat	(4 per lb)	*2.33*	25.9	4.5	0(a)	0.0		321	902	285	550	148	386	1.07	0.08	43	0(a)	**0**
E	lamb,rib chop,w/bone,broiled,w/fat	1 med	67	273	0.0	0.00	13.5	**30**	928	705	1101	553	670	173	1.69	11.26	0.91	0.08	**5**
		(4 per lb)	*1.49*	28.7	23.9	0(a)	0.0		369	1053	326	622	176	445	9.25	0.15	66	0	**0**
F	lamb,shoulder shank,cooked,w/fat	2 servings	91	306	0.0	0.00	19.6	**44**	1353	979	1605	811	984	252	1.69	11.40	0.91	0.12	**8**
		per shank	*1.10*	44.9	24.6	0(a)	0.0		538	1514	479	923	249	648	9.32	0.15	89	0	**0**
G	lamb,shoulder,roasted,sliced,w/fat	3 sl 2-1/2x	85	287	0.0	0.00	18.4	**41**	1271	920	1508	762	924	236	1.59	10.71	0.86	0.11	**7**
		2-1/2x1/4"	*1.18*	42.2	23.1	0(a)	0.0		506	1422	450	867	234	609	8.75	0.15	83	0	**0**
H	lamb,shoulder,w/o bone,roasted,sliced,	3 sl 2-1/2x	85	174	0.0	0.00	22.8	**51**	1570	1136	1861	941	1141	292	0.36	1.98	0.18	0.13	**9**
	no visible fat	2-1/2x1/4"	*1.18*	52.2	8.5	0(a)	0.0		625	1757	556	1070	289	752	1.68	0.18	85	0	**0**
I	lamb,shoulder,w/o bone,roasted,sliced,	3 sl 2-1/2x	85	287	0.0	0.00	18.4	**41**	1271	920	1508	762	924	236	1.59	10.71	0.86	0.11	**7**
	w/fat	2-1/2x1/4"	*1.18*	42.2	23.1	0(a)	0.0		506	1422	450	867	234	609	8.75	0.15	83	0	**0**
J	LARD	1 C	205	1849	0.0	0.00	0.0	**0**	0(a)	0(a)	0(a)	0(a)	0(a)	0(a)	22.96	80.36	20.54	0.00	**0**
			0.49	0.0	205.0	0.00	0.0		0(a)	0(a)	0(a)	0(a)	0(a)	0(a)	84.46	0.29	195	0	**0**
K	LASAGNA,canned,w/cheese & meat	1 C	227	293	34.5	0.40	12.7	**20**	unk.	unk.	unk.	unk.	unk.	unk.	unk.	unk.	unk.	0.34	**23**
	sauce-Campbell		*0.44*	165.5	11.6	unk.	unk.		unk.	unk.	unk.	unk.	unk.	unk.	unk.	unk.	unk.	2	**3**
L	lasagna,frozen-Stouffer	1/2 pkg	150	191	18.0	unk.	13.9	**21**	unk.	unk.	unk.	unk.	unk.	unk.	unk.	unk.	unk.	0.12	**8**
			0.67	103.2	7.0	unk.	unk.		unk.	unk.	unk.	unk.	unk.	unk.	unk.	unk.	unk.	tr	**0**
M	lasagna,home recipe	2-1/2x2-1/2x	250	374	25.1	0.42	22.0	**46**	1031	919	1721	1070	1293	180	0.47	10.56	1.01	0.22	**15**
		1-3/4"	*0.40*	172.7	20.5	1.69	5.9		651	1734	542	761	51	927	6.02	0.04	107	13	**21**
N	LEMON peel,raw,grated	1 tsp	2.0	unk.	0.3	unk.	0.0	**0**	unk.	unk.	unk.	unk.	unk.	unk.	unk.	unk.	unk.	tr	**0**
			50.00	1.6	tr	unk.	0.0		unk.	unk.	unk.	unk.	unk.	unk.	unk.	unk.	0(a)	3	**4**
O	lemon,raw,edible portion	1 average	74	20	6.0	0.29	0.3	**14**	unk.	unk.	unk.	unk.	unk.	unk.	unk.	unk.	unk.	0.3	**2**
			1.36	66.4	0.2	2.58	2.4		unk.	unk.	unk.	unk.	unk.	unk.	unk.	unk.	0	39	**65**
P	LENTILS,whole seeds,cooked	1/2 C	100	106	19.3	1.20	7.8	**12**	674	413	476	359	421	64	unk.	unk.	unk.	0.07	**5**
			1.00	72.0	tr	unk.	unk.		unk.	554	55	273	70	207	unk.	unk.	unk.	0	**0**
Q	LETTUCE,bibb or Boston,raw	1/2 C	28	4	0.7	0.14	0.3	**1**	15	13	19	9	18	unk.	unk.	tr(a)	unk.	0.02	**1**
			3.52	26.2	0.1	0.41	0.2		5	21	1	14	3	26	unk.	unk.	0	2	**4**
R	lettuce,bibb or Boston,raw,	1 head	163	23	4.1	0.81	2.0	**3**	88	74	114	53	106	unk.	unk.	tr(a)	unk.	0.10	**7**
	edible portion		*0.61*	155.0	0.3	2.45	1.3		31	122	7	80	20	153	unk.	unk.	0	13	**22**
S	lettuce,iceberg,raw,chopped	1/2 C	28	4	0.8	0.14	0.3	**0**	11	9	9	10	13	unk.	unk.	tr(a)	unk.	0.02	**1**
			3.64	26.3	tr	0.41	0.2		11	16	4	10	2	7	unk.	unk.	0	2	**3**

Riboflavin mg / Niacin mg	% USRDA	Vitamin B_6 mg / Folacin mcg	% USRDA	Vitamin B_{12} mcg / Pantothenic acid mg	% USRDA	Biotin mg / Vitamin A IU	% USRDA	Preformed A RE / Beta carotene RE	Vitamin D IU / Vitamin E IU	% USRDA	Total tocopherol mg / Alpha tocopherol mg	Other tocopherol mg / Total ash g	Calcium mg / Phosphorus mg	% USRDA	Sodium mg / Sodium meq	Potassium mg / Potassium meq	Chlorine mg / Chlorine meq	Iron mg / Magnesium mg	% USRDA	Zinc mg / Copper mg	% USRDA	Iodine mcg / Selenium mcg	% USRDA	Manganese mcg / Chromium mcg	
0.14	**8**	0.08	**4**	0.95	**16**	unk.	**unk.**	0(a)	0(a)	**0**	0.2	0.1	6	**1**	34	142	unk.	1.0	**5**	2.1	**14**	unk.	**unk.**	unk.	A
3.0	**15**	1	**0**	0.13	**1**	0(a)	**0**	0(a)	0.2	**1**	0.1	1.08	107	**11**	1.5	3.6	unk.	unk.	**unk.**	0.45	**22**	tr(a)		tr(a)	
0.16	**10**	0.12	**6**	1.38	**23**	unk.	**unk.**	0	0	**0**	0.2	0.1	6	**1**	50	206	unk.	0.9	**5**	2.1	**14**	unk.	**unk.**	unk.	B
3.5	**18**	2	**1**	0.20	**2**	0	**0**	0	0.3	**1**	0.1	1.21	122	**12**	2.2	5.3	unk.	12	**3**	0.51	**25**	tr(a)		tr(a)	
0.22	**13**	0.14	**7**	1.59	**27**	unk.	**unk.**	0	0	**0**	0.3	0.1	9	**1**	57	238	unk.	1.4	**8**	3.0	**20**	unk.	**unk.**	unk.	C
4.5	**23**	2	**1**	0.23	**2**	0	**0**	0	0.3	**1**	0.1	1.39	171	**17**	2.5	6.1	unk.	17	**4**	0.58	**29**	tr(a)		tr(a)	
0.12	**7**	0.07	**4**	0.83	**14**	unk.	**unk.**	0(a)	0(a)	**0**	0.1	0.1	5	**1**	30	125	unk.	0.8	**5**	1.8	**12**	unk.	**unk.**	unk.	D
2.5	**13**	1	**0**	0.12	**1**	0(a)	**0**	0(a)	0.2	**1**	0.1	0.86	91	**9**	1.3	3.2	unk.	unk.	**unk.**	0.38	**19**	tr(a)		tr(a)	
0.14	**8**	0.11	**6**	1.30	**22**	unk.	**unk.**	0	0	**0**	0.2	0.1	6	**1**	47	194	unk.	0.7	**4**	1.9	**13**	unk.	**unk.**	unk.	E
3.1	**15**	2	**1**	0.18	**2**	0	**0**	0	0.3	**1**	0.1	0.94	105	**10**	2.0	5.0	unk.	unk.	**unk.**	0.48	**24**	tr(a)		tr(a)	
0.21	**12**	0.15	**8**	1.76	**29**	unk.	**unk.**	0	0	**0**	0.3	0.1	9	**1**	63	262	unk.	1.1	**6**	3.3	**22**	unk.	**unk.**	unk.	F
4.3	**21**	3	**1**	0.25	**3**	0	**0**	0	0.3	**1**	0.1	1.27	156	**16**	2.8	6.7	unk.	15	**4**	0.64	**32**	tr(a)		tr(a)	
0.20	**12**	0.14	**7**	1.68	**28**	unk.	**unk.**	0	0	**0**	0.3	0.1	8	**1**	59	246	unk.	1.0	**6**	2.8	**19**	unk.	**unk.**	unk.	G
4.0	**20**	3	**1**	0.23	**2**	0	**0**	0	0.3	**1**	0.1	1.19	146	**15**	2.6	6.3	unk.	14	**4**	0.60	**30**	tr(a)		tr(a)	
0.24	**14**	0.14	**7**	1.68	**28**	unk.	**unk.**	0	0	**0**	0.3	0.1	10	**1**	59	246	unk.	1.6	**9**	3.7	**24**	unk.	**unk.**	unk.	H
4.8	**24**	3	**1**	0.23	**2**	0	**0**	0	0.3	**1**	0.1	1.61	186	**19**	2.6	6.3	unk.	19	**5**	0.74	**37**	tr(a)		tr(a)	
0.20	**12**	0.14	**7**	1.68	**28**	unk.	**unk.**	0	0	**0**	0.3	0.1	8	**1**	59	246	unk.	1.0	**6**	2.8	**19**	unk.	**unk.**	unk.	I
4.0	**20**	3	**1**	0.23	**2**	0	**0**	0	0.3	**1**	0.1	1.19	146	**15**	2.6	6.3	unk.	14	**4**	0.60	**30**	tr(a)		tr(a)	
0.00	**0**	0.04	**2**	0.00	**0**	tr(a)	**0**	0	5740	**1435**	2.7	0.2	tr	**0**	0	0	0	0.0	**0**	0.2	**2**	tr(a)	**0**	205	J
0.0	**0**	0	**0**	tr(a)	**0**	0	**0**	0	3.2	**11**	2.5	0.00	0	**0**	0.0	0.0	0.0	0	**0**	0.53	**27**	tr(a)		14	
0.32	**19**	unk.	**unk.**	unk.	**unk.**	unk.	**unk.**	unk.	unk.	**unk.**	unk.	unk.	159	**16**	656	372	unk.	3.9	**22**	unk.	**unk.**	unk.	**unk.**	unk.	K
1.8	**9**	50	**13**	unk.	**unk.**	824	**17**	unk.	unk.	**unk.**	unk.	2.70	250	**25**	29.5	9.5	unk.	unk.	**unk.**	unk.	**unk.**	unk.		unk.	
0.13	**8**	unk.	**unk.**	unk.	**unk.**	unk.	**unk.**	unk.	unk.	**unk.**	unk.	unk.	184	**18**	598	289	unk.	1.3	**8**	unk.	**unk.**	unk.	**unk.**	unk.	L
1.5	**8**	unk.	**unk.**	unk.	**unk.**	501	**10**	unk.	unk.	**unk.**	unk.	2.99	unk.	**unk.**	26.0	7.4	unk.	unk.	**unk.**	unk.	**unk.**	unk.		unk.	
0.34	**20**	0.08	**4**	0.54	**9**	0.001	**0**	0	0	**0**	0.3	0.0	359	**36**	668	354	unk.	2.6	**15**	2.8	**19**	104.1	**69**	11	M
2.3	**12**	11	**3**	0.37	**4**	1315	**26**	74	0.4	**1**	0.0	5.26	351	**35**	29.1	9.0	unk.	43	**11**	0.07	**3**	0		0	
tr	**0**	unk.	**unk.**	0(a)	**0**	unk.	**unk.**	0	0(a)	**0**	unk.	unk.	3	**0**	tr	3	unk.	tr	**0**	unk.	**unk.**	unk.	**unk.**	unk.	N
tr	**0**	unk.	**unk.**	unk.	**unk.**	1	**0**	tr	unk.	**unk.**	unk.	0.01	0	**0**	0.0	0.1	unk.	unk.	**unk.**	unk.	**unk.**	unk.		unk.	
0.01	**1**	0.06	**3**	0.00	**0**	tr	**0**	0	0(a)	**0**	unk.	unk.	19	**2**	1	102	3	0.4	**3**	0.1	**0**	unk.	**unk.**	29	O
0.1	**0**	4	**1**	0.14	**1**	15	**0**	1	unk.	**unk.**	unk.	0.22	12	**1**	0.1	2.6	0.1	7	**2**	0.19	**10**	tr(a)		unk.	
0.06	**4**	0.14	**7**	0(a)	**0**	0.004	**1**	0	0(a)	**0**	unk.	unk.	25	**3**	unk.	249	unk.	2.1	**12**	1.0	**7**	unk.	**unk.**	unk.	P
0.6	**3**	30	**8**	0.42	**4**	20	**0**	2	unk.	**unk.**	unk.	0.90	119	**12**	unk.	6.4	unk.	unk.	**unk.**	unk.	**unk.**	unk.		unk.	
0.02	**1**	0.02	**1**	0.00	**0**	0.001	**0**	0	0(a)	**0**	0.0	tr	10	**1**	2	73	20	0.5	**3**	0.1	**1**	unk.	**unk.**	2	Q
0.1	**0**	10	**3**	0.05	**1**	267	**5**	27	0.1	**0**	tr	0.27	7	**1**	0.1	1.9	0.6	2	**1**	0.01	**1**	tr(a)		2	
0.10	**6**	0.09	**5**	0.00	**0**	0.005	**2**	0	0(a)	**0**	0.3	0.2	57	**6**	15	430	119	3.3	**18**	0.7	**4**	unk.	**unk.**	15	R
0.5	**2**	60	**15**	0.33	**3**	1582	**32**	158	0.3	**1**	0.1	1.63	42	**4**	0.6	11.0	3.4	13	**3**	0.06	**3**	tr(a)		11	
0.02	**1**	0.02	**1**	0.00	**0**	0.001	**0**	0	0(a)	**0**	0.0	tr	5	**1**	2	48	11	0.1	**1**	0.1	**1**	unk.	**unk.**	19	S
0.1	**0**	10	**3**	0.05	**1**	91	**2**	9	0.1	**0**	tr	0.16	6	**1**	0.1	1.2	0.3	3	**1**	0.01	**1**	tr(a)		2	

	FOOD	Portion	Weight in grams / *Conversion for 100 g*	Kilocalories / H_2O g	Total carbohydrate g / Total fats g	Crude fiber g / Dietary fiber g	Total protein g / Total sugar g	**% USRDA**	Arginine mg / Histidine mg	Isoleucine mg / Leucine mg	Lysine mg / Methionine mg	Phenylalanine mg / Threonine mg	Valine mg / Tryptophan mg	Cystine mg / Tyrosine mg	Polyunsat. fatty acids g / Monounsat. fatty acids g	Saturated fatty acids g / P/S ratio	Linoleic acid g / Cholesterol mg	Thiamin mg / Ascorbic acid mg	**% USRDA**
A	lettuce,iceberg,raw,head,	1 average	539	70	15.6	2.69	4.8	**8**	218	185	185	194	262	unk.	unk.	tr(a)	unk.	0.32	**22**
	edible portion		*0.19*	514.7	0.5	8.09	4.4		209	306	87	198	39	131	unk.	unk.	0	32	**54**
B	lettuce,iceberg,raw,leaf	1 average	10	1	0.3	0.05	0.1	**0**	4	3	3	4	5	unk.	unk.	tr(a)	unk.	0.01	**0**
			10.00	9.5	tr	0.15	0.1		4	6	2	4	1	2	unk.	unk.	0	1	**1**
C	lettuce,iceberg,raw,wedge	1/6 head	90	12	2.6	0.45	0.8	**1**	36	31	31	32	44	unk.	unk.	tr(a)	unk.	0.05	**4**
			1.11	85.9	0.1	1.35	0.7		35	51	14	33	6	22	unk.	unk.	0	5	**9**
D	lettuce,romaine,raw,chopped	1/2 C	28	5	1.0	0.19	0.4	**1**	16	14	14	10	19	unk.	unk.	tr(a)	unk.	0.01	**1**
			3.64	25.8	0.1	unk.	0.2		6	23	6	15	3	10	unk.	unk.	0	5	**8**
E	lettuce,romaine,raw,head,	1 average	454	82	15.9	3.18	5.9	**9**	unk.	unk.	318	unk.	unk.	unk.	unk.	tr(a)	unk.	0.23	**15**
	edible portion		*0.22*	426.8	1.4	unk.	3.7		unk.	unk.	18	unk.	54	unk.	unk.	unk.	0	82	**136**
F	lettuce,romaine,raw,leaf	1 average	10	2	0.3	0.07	0.1	**0**	6	5	5	3	7	unk.	unk.	tr(a)	unk.	0.00	**0**
			10.00	9.4	tr	unk.	0.1		2	8	2	5	1	4	unk.	unk.	0	2	**3**
G	LIME,raw,edible portion	1 average	67	19	6.4	0.33	0.5	**1**	unk.	unk.	10	unk.	unk.	unk.	unk.	unk.	unk.	0.02	**1**
			1.49	59.8	0.1	unk.	1.0		unk.	unk.	0	unk.	2	unk.	unk.	unk.	0	25	**41**
H	LOBSTER/shrimp paste	1 Tbsp	21	38	0.3	0(a)	4.4	**10**	unk.	unk.	unk.	unk.	unk.	unk.	unk.	unk.	unk.	unk.	**unk.**
			4.76	12.9	2.0	0(a)	0.0		unk.	unk.	unk.	unk.	unk.	unk.	unk.	unk.	unk.	unk.	**unk.**
I	lobster,home recipe,newburg	1 C	250	485	12.7	unk.	46.3	**103**	unk.	unk.	unk.	unk.	unk.	unk.	unk.	unk.	unk.	0.17	**12**
			0.40	160.0	26.5	unk.	0.0		unk.	unk.	unk.	unk.	unk.	unk.	unk.	unk.	455	unk.	**unk.**
J	lobster,home recipe,salad	1/2 C	103	113	2.4	unk.	10.3	**23**	unk.	437	1000	498	477	unk.	unk.	unk.	unk.	0.09	**6**
			0.98	82.3	6.6	unk.	unk.		unk.	600	334	469	97	unk.	unk.	unk.	unk.	18	**31**
K	lobster,meat,cooked,cubed	1/2 C	73	69	0.2	0.00	13.6	**30**	1124	538	1248	618	591	171	0.43	0.10	0.02	0.07	**5**
			1.38	55.7	0.9	0(a)	0.0		254	1130	420	579	119	492	0.07	4.21	62	0	**0**
L	lobster,tail,cooked	1 med	104	99	0.3	0.00	19.4	**43**	1612	772	1790	886	848	245	0.61	0.15	0.03	0.10	**7**
			0.96	79.9	1.2	0(a)	0.0		365	1620	603	830	171	706	0.09	4.21	88	0	**0**
M	lobster,whole,cooked	1 med	113	107	0.3	0.00	21.1	**47**	1751	838	1945	963	921	267	0.37	0.16	0.01	0.11	**8**
			0.88	86.8	1.1	0(a)	0.0		397	1761	655	902	185	767	0.14	2.36	96	0	**0**
N	LOGANBERRIES,canned,heavy syrup,	1/2 C	128	114	28.4	2.43	0.8	**1**	unk.	unk.	unk.	unk.	unk.	unk.	unk.	tr(a)	unk.	0.01	**1**
	solids & liquid		*0.78*	97.9	0.5	6.66	26.0		unk.	unk.	unk.	unk.	unk.	unk.	unk.	unk.	0	10	**17**
O	loganberries,canned,water pack,	1/2 C	122	49	11.5	2.44	0.8	**1**	unk.	unk.	unk.	unk.	unk.	unk.	unk.	tr(a)	unk.	0.01	**1**
	solids & liquid		*0.82*	108.8	0.5	5.37	9.0		unk.	unk.	unk.	unk.	unk.	unk.	unk.	unk.	0	10	**16**
P	LUNCHEON MEAT,bar-b-q loaf	2 slices	28	48	1.4	0(a)	4.5	**10**	unk.	unk.	unk.	unk.	unk.	unk.	0.26	0.45	0.23	0.11	**7**
	-Oscar Mayer		*3.52*	18.5	2.8	0(a)	unk.		unk.	unk.	unk.	unk.	unk.	unk.	1.14	0.56	11	5	**9**
Q	luncheon meat,beef,hard salami	2 thin slice	28	126	0.3	0.00	6.7	**15**	421	325	538	244	336	83	unk.	unk.	unk.	0.10	**7**
			3.57	8.3	10.7	0(a)	tr		180	480	141	274	57	217	unk.	unk.	17	0	**0**
R	luncheon meat,beef,phrasky	2 thin slice	28	87	0.4	0.00	4.9	**11**	309	236	435	216	245	62	2.97	2.44	0.17	0.07	**5**
			3.57	14.3	7.2	0(a)	tr		167	397	132	225	55	176	2.35	1.22	17	unk.	**unk.**
S	luncheon meat,beef,salami,cooked	4-1/2" dia x	28	87	0.4	0.00	4.9	**11**	309	236	435	216	245	62	2.97	2.44	0.17	0.07	**5**
		1/10"	*3.57*	14.3	7.2	0(a)	tr		167	397	132	225	55	176	2.35	1.22	17	unk.	**unk.**

Riboflavin mg / Niacin mg	% USRDA	Vitamin B_6 mg / Folacin mcg	% USRDA	Vitamin B_{12} mcg / Pantothenic acid mg	% USRDA	Biotin mg / Vitamin A IU	% USRDA	Preformed A RE / Beta carotene RE	Vitamin D IU / Vitamin E IU	% USRDA	Total tocopherol mg / Alpha tocopherol mg	Other tocopherol mg / Total ash g	Calcium mg / Phosphorus mg	% USRDA	Sodium mg / Sodium meq	Potassium mg / Potassium meq	Chlorine mg / Chlorine meq	Iron mg / Magnesium mg	% USRDA	Zinc mg / Copper mg	% USRDA	Iodine mcg / Selenium mcg	% USRDA	Manganese mcg / Chromium mcg	
0.32	19	0.30	15	0.00	0	0.017	6	0	0(a)	0	0.9	0.6	108	11	49	943	210	2.7	15	2.2	14	unk.	unk.	372	A
1.6	8	199	50	1.08	11	1780	36	178	1.1	4	0.3	3.23	119	12	2.1	24.1	5.9	59	15	0.20	10	tr(a)		38	
0.01	0	0.01	0	0.00	0	tr	0	0	0(a)	0	tr	tr	2	0	1	17	4	0.0	0	tr	0	unk.	unk.	7	B
tr	0	4	1	0.02	0	33	1	3	tr	0	tr	0.06	2	0	0.0	0.4	0.1	1	0	tr	0	tr(a)		1	
0.05	3	0.05	3	0.00	0	0.003	1	0	0(a)	0	0.2	0.1	18	2	8	157	35	0.4	3	0.4	2	unk.	unk.	62	C
0.3	1	33	8	0.18	2	297	6	30	0.2	1	0.1	0.54	20	2	0.3	4.0	1.0	10	3	0.03	2	tr(a)		6	
0.02	1	0.00	0	0.00	0	0.001	0	0	0(a)	0	0.0	tr	19	2	2	73	unk.	0.4	2	unk.	unk.	unk.	unk.	15	D
0.1	1	49	12	0.05	1	522	10	52	0.1	0	tr	0.25	7	1	0.1	1.9	unk.	unk.	unk.	0.01	1	tr(a)		8	
0.36	21	0.08	4	0.00	0	0.014	5	0	0(a)	0	0.8	0.5	309	31	41	1199	unk.	6.4	35	unk.	unk.	unk.	unk.	254	E
1.8	9	813	203	0.91	9	8626	173	863	0.9	3	0.3	4.09	114	11	1.8	30.6	unk.	unk.	unk.	0.17	8	tr(a)		136	
0.01	1	tr	0	0.00	0	tr	0	0	0(a)	0	tr	tr	7	1	1	26	unk.	0.1	1	unk.	unk.	unk.	unk.	6	F
tr	0	18	5	0.02	0	190	4	19	tr	0	tr	0.09	3	0	0.0	0.7	unk.	unk.	unk.	tr	0	tr(a)		3	
0.01	1	unk.	unk.	0.00	0	unk.	unk.	0	0(a)	0	unk.	unk.	22	2	1	68	unk.	0.4	2	unk.	unk.	unk.	unk.	unk.	G
0.1	1	3	1	0.15	2	7	0	1	unk.	unk.	unk.	0.20	12	1	0.1	1.8	unk.	unk.	unk.	tr(a)	0	unk.		unk.	
0.05	3	unk.	unk.	unk.	unk.	unk.	unk.	unk.	unk.	unk.	unk.	unk.	unk.	unk.	unk.	unk.	unk.	unk.	unk.	unk.	unk.	unk.	unk.	unk.	H
unk.	unk.	unk.	unk.	unk.	unk.	unk.	unk.	unk.	unk.	unk.	unk.	1.47	unk.	unk.	unk.	unk.	unk.	unk.	unk.	unk.	unk.	tr(a)		tr(a)	
0.27	16	unk.	unk.	unk.	unk.	unk.	unk.	unk.	unk.	unk.	unk.	unk.	218	22	573	428	unk.	2.2	13	unk.	unk.	unk.	unk.	unk.	I
unk.	unk.	unk.	unk.	unk.	unk.	unk.	unk.	unk.	unk.	unk.	unk.	4.50	480	48	24.9	10.9	unk.	unk.	unk.	unk.	unk.	unk.		unk.	
0.08	5	unk.	unk.	unk.	unk.	unk.	unk.	unk.	unk.	unk.	unk.	unk.	37	4	127	271	unk.	0.9	5	unk.	unk.	unk.	unk.	unk.	J
unk.	unk.	unk.	unk.	unk.	unk.	unk.	unk.	unk.	unk.	unk.	unk.	0.92	97	10	5.5	6.9	unk.	unk.	unk.	unk.	unk.	unk.		unk.	
0.05	3	unk.	unk.	0.33	5	unk.	unk.	unk.	unk.	unk.	tr(a)	tr(a)	47	5	152	130	unk.	0.6	3	1.6	11	unk.	unk.	unk.	K
1.6	8	12	3	0.54	5	unk.	unk.	unk.	0.0	0	tr(a)	1.96	139	14	6.6	3.3	unk.	unk.	unk.	unk.	unk.	tr(a)		tr(a)	
0.07	4	unk.	unk.	0.47	8	unk.	unk.	unk.	unk.	unk.	tr(a)	tr(a)	68	7	218	187	unk.	0.8	5	2.3	15	unk.	unk.	unk.	L
2.3	12	18	4	0.78	8	unk.	unk.	unk.	0.0	0	tr(a)	2.81	200	20	9.5	4.8	unk.	unk.	unk.	unk.	unk.	tr(a)		tr(a)	
0.08	5	unk.	unk.	0.51	9	unk.	unk.	unk.	unk.	unk.	tr(a)	tr(a)	73	7	237	203	unk.	0.9	5	2.5	17	unk.	unk.	unk.	M
2.5	13	19	5	0.85	9	unk.	unk.	unk.	0.0	0	tr(a)	3.05	217	22	10.3	5.2	unk.	unk.	unk.	unk.	unk.	tr(a)		tr(a)	
0.03	2	0.06	3	0(a)	0	unk.	unk.	0	0(a)	0	0.4	unk.	28	3	1	140	17	1.0	6	unk.	unk.	unk.	unk.	unk.	N
0.3	1	unk.	unk.	0.23	2	166	3	17	0.5	2	unk.	0.38	14	1	0.1	3.6	0.5	14	4	tr(a)	0	tr(a)		unk.	
0.02	1	0.05	2	0(a)	0	unk.	unk.	0	0(a)	0	0.2	unk.	29	3	1	140	14	1.0	5	unk.	unk.	unk.	unk.	unk.	O
0.2	1	unk.	unk.	0.24	2	171	3	17	0.2	1	unk.	0.37	13	1	0.0	3.6	0.4	13	3	tr(a)	0	tr(a)		unk.	
0.07	4	0.07	4	0.47	8	tr(a)	0	tr(a)	11	3	tr(a)	tr(a)	16	2	374	95	511	0.4	2	0.7	5	unk.	unk.	9	P
0.6	3	tr(a)	0	tr(a)	0	tr(a)	0	0	0.0	0	tr(a)	1.16	42	4	16.3	2.4	14.4	5	1	0.02	1	tr(a)		1	
0.07	4	0.03	2	0.39	7	unk.	unk.	unk.	0	0	0.2	0.2	4	0	510	106	unk.	1.0	6	0.6	4	unk.	unk.	12	Q
1.5	7	unk.	unk.	unk.	unk.	unk.	unk.	unk.	0.2	1	tr	1.99	79	8	22.2	2.7	unk.	4	1	0.02	1	tr(a)		1	
0.07	4	unk.	unk.	unk.	unk.	unk.	unk.	unk.	unk.	unk.	0.2	0.2	3	0	unk.	unk.	unk.	0.7	4	0.6	4	unk.	unk.	unk.	R
1.1	6	unk.	unk.	unk.	unk.	unk.	unk.	unk.	0.2	1	tr	1.26	56	6	unk.	unk.	unk.	4	1	unk.	unk.	tr(a)		tr(a)	
0.07	4	0.03	2	0.39	7	unk.	unk.	unk.	unk.	unk.	0.2	0.2	3	0	386	71	unk.	0.7	4	0.6	4	unk.	unk.	12	S
1.1	6	unk.	unk.	0.28	3	0	0	0	0.2	1	tr	1.26	56	6	16.8	1.8	unk.	4	1	0.02	1	tr(a)		1	

	FOOD	Portion	Weight in grams / Conversion for 100 g	Kilocalories / H_2O g	Total carbohydrate g / Total fats g	Crude fiber g / Dietary fiber g	Total protein g / Total sugar g	% USRDA	Arginine mg / Histidine mg	Isoleucine mg / Leucine mg	Lysine mg / Methionine mg	Phenylalanine mg / Threonine mg	Valine mg / Tryptophan mg	Cystine mg / Tyrosine mg	Polyunsat. fatty acids g / Monounsat. fatty acids g	Saturated fatty acids g / P/S ratio	Linoleic acid g / Cholesterol mg	Thiamin mg / Ascorbic acid mg	% USRDA
A	luncheon meat,blood sausage/pudding	5x4-5/8x	25	99	0.1	0.00	3.5	**8**	unk.	unk.	unk.	unk.	unk.	unk.	unk.	unk.	unk.	unk.	**unk.**
		1/16"	*4.00*	11.6	9.2	0(a)	0.0		unk.	unk.	unk.	unk.	unk.	unk.	unk.	unk.	unk.	0	**0**
B	luncheon meat,bologna	4-1/2" dia x	28	86	0.3	unk.	3.4	**8**	unk.	unk.	unk.	150	171	unk.	4.35	3.38	0.20	0.05	**3**
		1/8"	*3.52*	16.0	7.8	0(a)	tr		unk.	unk.	92	unk.	unk.	unk.	3.52	1.29	18	0	**0**
C	luncheon meat,bologna,all meat -Oscar Mayer	1 slice	28	89	0.8	unk.	3.1	**7**	unk.	unk.	unk.	138	156	unk.	0.71	3.49	0.60	0.07	**5**
			3.52	15.3	8.2	0(a)	0.8		unk.	unk.	85	unk.	unk.	unk.	3.72	0.20	15	5	**8**
D	luncheon meat,bologna,ground	2 Tbsp	28	86	0.3	unk.	3.4	**8**	unk.	unk.	unk.	153	211	unk.	4.35	3.38	0.20	0.05	**3**
			3.52	16.0	7.8	0(a)	tr		unk.	unk.	89	unk.	unk.	unk.	3.52	1.29	18	0	**0**
E	luncheon meat,bologna,pure beef -Oscar Mayer	1 slice	28	89	0.8	0(a)	3.1	**7**	197	150	278	137	156	40	0.37	3.72	0.28	0.01	**1**
			3.52	15.3	8.2	0(a)	0.8		106	253	84	143	unk.	112	3.72	0.10	16	5	**8**
F	luncheon meat,bologna,turkey	3-1/4" dia x	28	57	0.3	unk.	3.9	**9**	unk.	unk.	unk.	unk.	unk.	unk.	unk.	unk.	unk.	0.02	**1**
		1/8"	*3.52*	18.5	4.3	unk.	unk.		unk.	unk.	unk.	unk.	unk.	unk.	unk.	unk.	28	unk.	**unk.**
G	luncheon meat,braunschweiger (smoked liver sausage)	1 slice	28	89	0.6	unk.	4.3	**10**	273	211	336	196	268	52	unk.	2.80	0.56	0.05	**3**
			3.52	14.7	7.7	0(a)	0.0		128	361	90	187	48	132	3.36	unk.	17	0	**0**
H	luncheon meat,braunschweiger (smoked liver sausage)-Oscar Mayer	1 slice	28	99	0.7	0(a)	3.7	**8**	unk.	unk.	unk.	unk.	unk.	unk.	1.14	3.15	1.02	0.07	**4**
			3.52	13.9	9.1	0(a)	0.0		unk.	unk.	unk.	unk.	unk.	unk.	4.00	0.36	35	3	**4**
I	luncheon meat,chicken spread,canned	1 Tbsp	13	25	0.7	unk.	2.0	**5**	unk.	unk.	unk.	unk.	unk.	unk.	unk.	unk.	unk.	tr	**0**
			7.69	unk.	1.5	unk.	unk.		unk.	unk.	unk.	unk.	unk.	unk.	unk.	unk.	unk.	unk.	**unk.**
J	luncheon meat,cotto salami -Oscar Mayer	1-2/3 slices	28	64	0.5	0(a)	4.0	**9**	unk.	unk.	unk.	unk.	unk.	unk.	0.54	2.16	0.48	0.07	**5**
			3.52	17.9	5.1	0(a)	unk.		unk.	unk.	unk.	unk.	unk.	unk.	2.24	0.25	18	5	**8**
K	luncheon meat,ham & cheese loaf -Oscar Mayer	1 slice	28	77	0.6	0(a)	4.5	**10**	unk.	unk.	unk.	unk.	unk.	unk.	0.77	2.47	0.65	0.17	**11**
			3.52	15.9	6.2	0(a)	unk.		unk.	unk.	unk.	unk.	unk.	unk.	2.75	0.31	16	7	**11**
L	luncheon meat,ham,boiled	6-1/4x4-1/4x	28	66	0.0	0.00	5.3	**12**	338	272	430	202	275	59	0.74	1.66	0.44	0.12	**8**
		1/16"	*3.57*	16.5	4.8	0(a)	0.0		175	401	129	261	73	190	2.04	0.44	25	unk.	**unk.**
M	luncheon meat,ham,chipped/chopped	3-1/3x3-1/3x	23	68	0.3	0.00	3.4	**8**	219	170	288	131	178	55	0.64	1.31	0.40	0.07	**5**
		1/8"	*4.35*	12.6	5.7	0(a)	tr		110	265	83	140	33	132	1.77	0.49	20	unk.	**unk.**
N	luncheon meat,ham,chopped-Oscar Mayer	1 slice	28	64	0.9	0(a)	4.8	**11**	289	232	367	189	235	51	0.60	1.59	0.51	0.20	**13**
			3.52	17.0	4.5	0(a)	unk.		149	343	123	223	62	163	2.10	0.37	14	6	**10**
O	luncheon meat,ham,cooked-Oscar Mayer	1-1/2 slices	28	34	0.0	0(a)	5.4	**12**	361	290	459	237	294	63	0.11	0.43	0.11	0.27	**18**
			3.52	20.4	1.4	0(a)	0.0		187	428	153	278	77	203	0.57	0.27	17	7	**12**
P	luncheon meat,ham,deviled spread	2 Tbsp	26	91	0.0	0.00	3.6	**8**	unk.	unk.	unk.	unk.	unk.	unk.	unk.	3.12	0.78	0.04	**2**
			3.85	13.1	8.4	0(a)	0.0		unk.	unk.	unk.	unk.	unk.	unk.	3.64	unk.	unk.	unk.	**unk.**
Q	luncheon meat,ham,minced	3-1/3x3-1/3x	23	52	1.0	0.00	3.1	**7**	201	161	254	120	163	35	unk.	1.38	0.46	0.09	**6**
		1/8"	*4.35*	14.2	3.9	0(a)	tr		103	238	76	154	43	113	1.61	unk.	20	unk.	**unk.**
R	luncheon meat,ham,shaved	1/4 C	38	88	0.0	0.00	7.2	**16**	456	367	580	273	371	80	0.99	2.24	0.59	0.17	**11**
			2.65	22.3	6.4	0(a)	0.0		236	541	174	352	98	257	2.76	0.44	34	unk.	**unk.**
S	luncheon meat,hard salami-Oscar Mayer	1 slice	30	111	0.6	0(a)	6.3	**14**	unk.	unk.	unk.	unk.	unk.	unk.	0.96	3.56	0.87	0.17	**11**
			3.34	12.3	9.3	0(a)	0.2		unk.	unk.	unk.	unk.	unk.	unk.	4.40	0.27	25	7	**13**

Row	Riboflavin mg / Niacin mg	% USRDA	Vitamin B_6 mg / Folacin mcg	% USRDA	Vitamin B_{12} mcg / Pantothenic acid mg	% USRDA	Biotin mg / Vitamin A IU	% USRDA	Preformed A RE / Beta carotene RE	Vitamin D IU / Vitamin E IU	% USRDA	Total tocopherol mg / Alpha tocopherol mg	Other tocopherol mg / Total ash g	Calcium mg / Phosphorus mg	% USRDA	Sodium mg / Sodium meq	Potassium mg / Potassium meq	Chlorine mg / Chlorine meq	Iron mg / Magnesium mg	% USRDA	Zinc mg / Copper mg	% USRDA	Iodine mcg / Selenium mcg	% USRDA	Manganese mcg / Chromium mcg
A	unk.	**unk.**	0.01	**1**	unk.	**unk.**	unk.	**unk.**	unk.	0	**0**	unk.	unk.	2	**0**	unk.	unk.	unk.	0.5	**3**	unk.	**unk.**	unk.	**unk.**	unk.
	unk.	**unk.**	unk.	**unk.**	unk.	**unk.**	unk.	**unk.**	unk.	unk.	**unk.**	unk.	0.57	40	**4**	unk.	unk.	unk.	unk.	**unk.**	unk.	**unk.**	tr(a)		tr(a)
B	0.06	**4**	0.03	**1**	unk.	**unk.**	unk.	**unk.**	unk.	0	**0**	0.1	0.1	2	**0**	369	65	unk.	0.5	**3**	0.5	**3**	unk.	**unk.**	14
	0.7	**4**	1	**0**	0.14	**1**	unk.	**unk.**	unk.	0.2	**1**	tr	0.88	36	**4**	16.1	1.7	unk.	5	**1**	0.02	**1**	tr(a)		4
C	0.03	**2**	0.05	**3**	0.29	**5**	tr(a)	**0**	0	12	**3**	0.1	0.1	2	**0**	288	45	460	0.3	**2**	0.5	**3**	unk.	**unk.**	14
	0.7	**3**	5	**1**	tr(a)	**0**	0	**0**	0	0.2	**1**	tr	0.88	24	**2**	12.5	1.2	13.0	3	**1**	0.03	**2**	tr(a)		4
D	0.06	**4**	0.03	**1**	unk.	**unk.**	unk.	**unk.**	unk.	0	**0**	0.1	0.1	2	**0**	369	65	unk.	0.5	**3**	0.5	**3**	unk.	**unk.**	14
	0.7	**4**	1	**0**	0.14	**1**	unk.	**unk.**	unk.	0.2	**1**	tr	0.88	36	**4**	16.1	1.7	unk.	5	**1**	0.02	**1**	tr(a)		4
E	0.03	**2**	0.05	**3**	0.41	**7**	tr(a)	**0**	0	9	**2**	0.1	0.1	3	**0**	295	42	460	0.4	**2**	0.6	**4**	unk.	**unk.**	14
	0.7	**4**	4	**1**	tr(a)	**0**	0	**0**	0	0.2	**1**	tr	0.91	21	**2**	12.8	1.1	13.0	3	**1**	0.01	**0**	tr(a)		4
F	0.05	**3**	unk.	**unk.**	unk.	**unk.**	unk.	**unk.**	unk.	unk.	**unk.**	unk.	unk.	24	**2**	249	57	unk.	0.4	**2**	0.5	**3**	unk.	**unk.**	unk.
	1.0	**5**	unk.	**unk.**	unk.	**unk.**	unk.	**unk.**	unk.	unk.	**unk.**	unk.	0.94	37	**4**	10.8	1.4	unk.	4	**1**	0.01	**0**	unk.		unk.
G	0.40	**24**	unk.	**unk.**	unk.	**unk.**	unk.	**unk.**	494	4	**1**	unk.	unk.	3	**0**	330	51	unk.	1.7	**9**	0.8	**5**	unk.	**unk.**	12
	2.3	**12**	unk.	**unk.**	unk.	**unk.**	1828	**37**	18	unk.	**unk.**	unk.	0.81	69	**7**	14.4	1.3	unk.	4	**1**	0.02	**1**	tr(a)		1
H	0.45	**26**	0.09	**5**	5.55	**93**	tr(a)	**0**	1289	10	**3**	tr(a)	tr(a)	2	**0**	335	52	494	2.8	**16**	0.8	**6**	unk.	**unk.**	12
	2.4	**12**	15	**4**	tr	**0**	4292	**86**	0(a)	0.0	**0**	tr(a)	0.99	51	**5**	14.6	1.3	13.9	3	**1**	0.08	**4**	tr(a)		1
I	0.01	**1**	unk.	**unk.**	unk.	**unk.**	unk.	**unk.**	unk.	unk.	**unk.**	unk.	unk.	16	**2**	unk.	unk.	unk.	0.3	**2**	unk.	**unk.**	unk.	**unk.**	unk.
	0.4	**2**	unk.	**unk.**	unk.	**unk.**	unk.	**unk.**	unk.	unk.	**unk.**	unk.	unk.	unk.	**unk.**	unk.	unk.	unk.	unk.	**unk.**	unk.	**unk.**	unk.		unk.
J	0.11	**6**	0.06	**3**	0.95	**16**	tr(a)	**0**	tr(a)	10	**2**	0.2	tr(a)	3	**0**	303	55	494	0.7	**4**	0.6	**4**	unk.	**unk.**	12
	1.1	**6**	tr(a)	**0**	tr(a)	**0**	tr(a)	**0**	0(a)	0.2	**0**	tr	0.97	31	**3**	13.2	1.4	13.9	4	**1**	0.07	**3**	tr(a)		1
K	0.06	**3**	0.08	**4**	0.22	**4**	tr(a)	**0**	tr(a)	12	**3**	tr(a)	tr(a)	16	**2**	370	80	460	0.2	**1**	0.5	**3**	unk.	**unk.**	9
	1.0	**5**	tr(a)	**0**	tr(a)	**0**	tr(a)	**0**	0(a)	0.0	**0**	tr(a)	1.11	66	**7**	16.1	2.0	13.0	5	**1**	0.02	**1**	tr(a)		1
L	0.04	**3**	unk.	**unk.**	unk.	**unk.**	unk.	**unk.**	0	unk.	**unk.**	unk.	unk.	3	**0**	384	89	unk.	0.8	**4**	0.9	**6**	unk.	**unk.**	9
	0.7	**4**	unk.	**unk.**	unk.	**unk.**	0	**0**	0	unk.	**unk.**	unk.	1.37	47	**5**	16.7	2.3	unk.	3	**1**	0.02	**1**	tr(a)		1
M	0.05	**3**	unk.	**unk.**	unk.	**unk.**	unk.	**unk.**	0	unk.	**unk.**	unk.	unk.	2	**0**	284	51	unk.	0.5	**3**	0.8	**5**	unk.	**unk.**	10
	0.7	**3**	unk.	**unk.**	0.13	**1**	0	**0**	0	unk.	**unk.**	unk.	0.90	25	**3**	12.3	1.3	unk.	unk.	**unk.**	0.02	**1**	tr(a)		1
N	0.05	**3**	0.09	**5**	0.26	**4**	tr(a)	**0**	tr(a)	10	**2**	tr(a)	tr(a)	2	**0**	378	86	494	0.2	**1**	0.6	**4**	unk.	**unk.**	9
	1.1	**6**	tr(a)	**0**	tr(a)	**0**	tr(a)	**0**	0(a)	0.0	**0**	tr(a)	1.14	48	**5**	16.4	2.2	13.9	5	**1**	0.02	**1**	tr(a)		1
O	0.07	**4**	0.14	**7**	0.22	**4**	tr(a)	**0**	tr(a)	7	**2**	tr(a)	tr(a)	2	**0**	398	99	511	0.2	**1**	0.5	**4**	unk.	**unk.**	unk.
	1.3	**7**	tr(a)	**0**	tr(a)	**0**	tr(a)	**0**	0(a)	0.0	**0**	tr(a)	1.14	67	**7**	17.3	2.5	14.4	5	**1**	0.02	**1**	tr(a)		1
P	0.03	**2**	unk.	**unk.**	unk.	**unk.**	unk.	**unk.**	0	unk.	**unk.**	tr(a)	tr(a)	2	**0**	unk.	unk.	unk.	0.5	**3**	unk.	**unk.**	unk.	**unk.**	unk.
	0.4	**2**	unk.	**unk.**	unk.	**unk.**	0	**0**	0	0.0	**0**	tr(a)	0.86	24	**2**	unk.	unk.	unk.	unk.	**unk.**	unk.	**unk.**	tr(a)		tr(a)
Q	0.05	**3**	unk.	**unk.**	unk.	**unk.**	unk.	**unk.**	0	unk.	**unk.**	unk.	unk.	2	**0**	315	73	unk.	0.5	**3**	0.8	**5**	unk.	**unk.**	10
	0.8	**4**	unk.	**unk.**	unk.	**unk.**	0	**0**	0	unk.	**unk.**	unk.	0.76	21	**2**	13.7	1.9	unk.	unk.	**unk.**	0.02	**1**	tr(a)		1
R	0.06	**3**	unk.	**unk.**	unk.	**unk.**	unk.	**unk.**	0	unk.	**unk.**	unk.	unk.	4	**0**	518	121	unk.	1.1	**6**	1.3	**9**	unk.	**unk.**	12
	1.0	**5**	unk.	**unk.**	unk.	**unk.**	0	**0**	0	unk.	**unk.**	unk.	1.85	63	**6**	22.5	3.1	unk.	3	**1**	0.03	**2**	tr(a)		2
S	0.08	**5**	0.13	**7**	0.56	**9**	tr(a)	**0**	tr(a)	19	**5**	0.2	0.2	3	**0**	554	112	753	0.5	**3**	1.0	**7**	unk.	**unk.**	13
	1.5	**8**	tr(a)	**0**	tr(a)	**0**	tr(a)	**0**	0(a)	0.2	**1**	tr	1.52	47	**5**	24.1	2.8	21.3	5	**1**	0.02	**1**	tr(a)		1

	FOOD	Portion	Weight in grams / Conversion for 100 g	Kilocalories / H_2O g	Total carbohydrate g / Total fats g	Crude fiber g / Dietary fiber g	Total protein g / Total sugar g	% USRDA	Arginine mg / Histidine mg	Isoleucine mg / Leucine mg	Lysine mg / Methionine mg	Phenylalanine mg / Threonine mg	Valine mg / Tryptophan mg	Cystine mg / Tyrosine mg	Polyunsat. fatty acids g / Monounsat. fatty acids g	Saturated fatty acids g / P/S ratio	Linoleic acid g / Cholesterol mg	Thiamin mg / Ascorbic acid mg	% USRDA
A	luncheon meat,head cheese	4x4x3/32''	28	75	0.3	0.00	4.3	**10**	unk.	143	254	159	173	59	0.68	1.25	0.38	0.01	**1**
			3.57	16.5	6.2	0(a)	tr		78	265	70	117	22	159	1.88	0.54	unk.	0	**0**
B	luncheon meat,honey loaf-Oscar Mayer	1 slice	28	37	1.1	0(a)	4.8	**11**	unk.	unk.	unk.	unk.	unk.	unk.	0.14	0.43	0.11	0.20	**13**
			3.52	19.9	1.4	0(a)	1.1		unk.	unk.	unk.	unk.	unk.	unk.	0.54	0.33	11	5	**9**
C	luncheon meat,knockwurst (pork)	4'' link	68	189	1.5	0.00	9.6	**21**	609	490	775	404	501	107	unk.	unk.	unk.	0.12	**8**
			1.47	39.2	15.8	0(a)	tr		315	723	261	470	131	343	unk.	unk.	42	unk.	**unk.**
D	luncheon meat,liver cheese/pork fat wrap-Oscar Mayer	1-1/3 slices	28	84	0.4	0(a)	4.3	**10**	unk.	218	344	178	221	48	1.02	2.67	0.94	0.06	**4**
			3.52	15.3	7.4	0(a)	0.4		140	321	115	209	58	152	3.32	0.38	52	1	**1**
E	luncheon meat,liverwurst,fresh	2 Tbsp	28	86	0.5	0.00	4.5	**10**	unk.	229	364	213	290	57	unk.	unk.	unk.	0.06	**4**
			3.57	15.1	7.2	0(a)	tr		139	392	97	203	52	143	unk.	unk.	29	0	**0**
F	luncheon meat,liverwurst,fresh,sliced	1 oz slice	28	87	0.5	0.00	4.6	**10**	unk.	232	369	216	295	58	unk.	unk.	unk.	0.06	**4**
			3.52	15.3	7.3	0(a)	tr		141	398	99	206	53	145	unk.	unk.	30	0	**0**
G	luncheon meat,luncheon meat -Oscar Mayer	2 slices	28	98	0.4	0(a)	3.7	**8**	unk.	unk.	unk.	unk.	unk.	unk.	1.11	3.46	0.94	0.09	**6**
			3.52	14.2	9.1	0(a)	unk.		unk.	unk.	unk.	unk.	unk.	unk.	4.03	0.32	16	4	**7**
H	luncheon meat,mortadella	4-7/8'' dia x 3/32''	25	79	0.1	0.00	5.1	**11**	unk.	unk.	unk.	224	255	unk.	unk.	unk.	unk.	unk.	**unk.**
			4.00	12.2	6.3	0(a)	tr		unk.	unk.	138	unk.	unk.	unk.	unk.	unk.	unk.	unk.	**unk.**
I	luncheon meat,old fashioned loaf -Oscar Mayer	2 slices	28	65	2.3	0(a)	4.3	**10**	unk.	unk.	unk.	unk.	unk.	unk.	0.51	1.56	0.45	0.11	**7**
			3.52	16.5	4.3	0(a)	unk.		unk.	unk.	unk.	unk.	unk.	unk.	1.87	0.33	14	4	**6**
J	luncheon meat,olive loaf-Oscar Mayer	2 slices	28	65	2.8	0(a)	3.4	**8**	unk.	unk.	unk.	unk.	unk.	unk.	0.57	1.70	0.54	0.09	**6**
			3.52	16.5	4.5	0(a)	unk.		unk.	unk.	unk.	unk.	unk.	unk.	2.07	0.33	11	3	**4**
K	luncheon meat,pickle & pimento loaf -Oscar Mayer	2 slices	28	65	3.0	0(a)	3.7	**8**	unk.	unk.	unk.	unk.	unk.	unk.	0.57	1.70	0.51	0.10	**6**
			3.52	16.2	4.3	0(a)	unk.		unk.	unk.	unk.	unk.	unk.	unk.	1.87	0.33	11	3	**5**
L	luncheon meat,pork/ham	2 Tbsp	28	83	0.4	0.00	4.3	**10**	271	210	356	162	220	68	1.34	3.02	0.87	0.09	**6**
			3.52	15.6	7.1	0(a)	tr		136	327	103	173	41	163	3.75	0.44	25	unk.	**unk.**
M	luncheon meat,potted meat (beef/chicken/turkey)	2 Tbsp	26	64	0.0	0.00	4.5	**10**	unk.	167	276	167	245	unk.	unk.	unk.	unk.	0.01	**1**
			3.85	15.8	5.0	0(a)	0.0		84	313	94	172	39	unk.	unk.	unk.	unk.	unk.	**unk.**
N	luncheon meat,pressed,canned-Spam	1 oz	28	88	1.1	0.06	4.0	**9**	252	203	321	151	206	44	3.98	2.27	0.85	0.11	**7**
			3.52	14.9	7.5	0(a)	0.0		131	300	96	195	54	142	3.12	1.75	25	unk.	**unk.**
O	luncheon meat,salami for beer -Oscar Mayer	1-2/3 slices	28	67	0.5	0(a)	4.0	**9**	271	218	344	178	221	48	0.71	1.90	0.60	0.16	**10**
			3.52	17.6	5.4	0(a)	unk.		140	321	115	209	58	152	2.44	0.37	16	9	**14**
P	luncheon meat,sandwich spread -Oscar Mayer	1 Tbsp	14	34	1.6	0(a)	1.1	**3**	unk.	unk.	unk.	unk.	unk.	unk.	0.40	0.91	0.34	0.02	**2**
			7.04	8.5	2.6	0(a)	unk.		unk.	unk.	unk.	unk.	unk.	unk.	1.05	0.44	5	tr	**0**
Q	luncheon meat,sandwich spread,poultry	1 Tbsp	13	26	1.0	unk.	1.5	**3**	unk.	unk.	unk.	unk.	unk.	unk.	0.81	0.45	unk.	tr	**0**
			7.69	unk.	1.8	unk.	unk.		unk.	unk.	unk.	unk.	unk.	unk.	unk.	1.80	4	tr	**0**
R	luncheon meat,spiced luncheon meat (pork-ham type)	4x3x1/8'' slice	28	82	0.4	0.00	4.2	**9**	267	207	351	160	217	67	1.32	2.98	0.85	0.09	**6**
			3.57	15.4	7.0	0(a)	tr		134	322	101	171	40	161	3.69	0.44	25	unk.	**unk.**
S	luncheon meat,summer sausage (thuringer cervelat),beef	4-3/8'' dia x 1/10''	28	86	0.4	0.00	5.2	**12**	329	251	463	229	261	66	4.03	3.08	0.22	0.03	**2**
			3.57	13.6	6.9	0(a)	tr		178	422	141	239	58	187	3.14	1.31	17	unk.	**unk.**

Riboflavin mg / Niacin mg	% USRDA	Vitamin B_6 mg / Folacin mcg	% USRDA	Vitamin B_{12} mcg / Pantothenic acid mg	% USRDA	Biotin mg / Vitamin A IU	% USRDA	Preformed A RE / Beta carotene RE	Vitamin D IU / Vitamin E IU	% USRDA	Total tocopherol mg / Alpha tocopherol mg	Other tocopherol mg / Total ash g	Calcium mg / Phosphorus mg	% USRDA	Sodium mg / Sodium meq	Potassium mg / Potassium meq	Chlorine mg / Chlorine meq	Iron mg / Magnesium mg	% USRDA	Zinc mg / Copper mg	% USRDA	Iodine mcg / Selenium mcg	% USRDA	Manganese mcg / Chromium mcg	
0.03	2	unk.	unk.	unk.	unk.	unk.	unk.	0	0	0	unk.	unk.	3	0	unk.	unk.	unk.	0.6	4	unk.	unk.	unk.	unk.	unk.	A
0.3	1	1	0	unk.	unk.	0	0	0	unk.	unk.	unk.	0.76	48	5	unk.	unk.	unk.	unk.	unk.	unk.	unk.	tr(a)		tr(a)	
0.08	5	0.12	6	0.27	4	tr(a)	0	tr(a)	10	3	tr(a)	tr(a)	6	1	378	97	545	0.3	2	0.7	5	unk.	unk.	9	B
1.0	5	tr(a)	0	tr(a)	0	tr(a)	0	0(a)	0.0	0	tr(a)	1.14	42	4	16.4	2.5	15.4	5	1	0.02	1	tr(a)		1	
0.14	8	unk.	unk.	unk.	unk.	unk.	unk.	unk.	unk.	unk.	unk.	unk.	5	1	unk.	unk.	unk.	1.4	8	unk.	unk.	unk.	unk.	unk.	C
1.8	9	unk.	unk.	unk.	unk.	unk.	unk.	unk.	unk.	unk.	unk.	1.97	105	11	unk.	unk.	unk.	unk.	unk.	unk.	unk.	tr(a)		tr(a)	
0.62	37	0.13	6	7.05	118	tr(a)	0	1755	14	4	tr(a)	tr(a)	2	0	341	64	494	3.4	19	1.1	7	unk.	unk.	9	D
3.2	16	32	8	1.05	11	5845	117	0(a)	0.0	0	tr(a)	0.99	62	6	14.8	1.6	13.9	3	1	0.11	5	tr(a)		1	
0.36	21	0.05	3	3.89	65	unk.	unk.	487	4	1	0.2	0.1	3	0	unk.	unk.	unk.	1.5	8	0.8	5	unk.	unk.	unk.	E
1.6	8	1	0	0.78	8	1778	36	18	0.2	1	0.1	0.70	67	7	unk.	unk.	unk.	unk.	unk.	0.85	43	tr(a)		tr(a)	
0.37	22	0.05	3	3.95	66	unk.	unk.	487	4	1	0.2	0.1	3	0	unk.	unk.	unk.	1.5	9	0.8	5	unk.	unk.	unk.	F
1.6	8	1	0	0.79	8	1803	36	18	0.2	1	0.1	0.71	68	7	unk.	unk.	unk.	unk.	unk.	0.87	43	tr(a)		tr(a)	
0.04	3	0.05	3	0.36	6	tr(a)	0	tr(a)	17	4	tr(a)	tr(a)	3	0	363	58	528	0.3	2	0.5	3	unk.	unk.	9	G
0.8	4	tr(a)	0	tr(a)	0	tr(a)	0	0(a)	0.0	0	tr(a)	1.02	26	3	15.8	1.5	14.9	3	1	0.01	1	tr(a)		1	
unk.	unk.	unk.	unk.	unk.	unk.	unk.	unk.	unk.	unk.	unk.	unk.	unk.	3	0	unk.	unk.	unk.	0.8	4	unk.	unk.	unk.	unk.	unk.	H
unk.	unk.	unk.	unk.	unk.	unk.	unk.	unk.	unk.	unk.	unk.	unk.	1.27	60	6	unk.	unk.	unk.	unk.	unk.	unk.	unk.	tr(a)		tr(a)	
0.09	5	0.07	4	0.49	8	tr(a)	0	tr(a)	12	3	tr(a)	tr(a)	32	3	344	102	477	0.3	2	0.5	3	unk.	unk.	9	I
0.7	4	tr(a)	0	tr(a)	0	tr(a)	0	0(a)	0.0	0	tr(a)	1.08	48	5	14.9	2.6	13.5	6	2	0.02	1	tr(a)		1	
0.07	4	0.06	3	0.38	6	tr(a)	0	0(a)	11	3	tr(a)	tr(a)	32	3	406	82	579	0.2	1	0.4	3	unk.	unk.	unk.	J
0.5	3	tr(a)	0	tr(a)	0	52	1	5	0.0	0	tr(a)	1.22	37	4	17.6	2.1	16.3	5	1	0.01	1	tr(a)		1	
0.08	5	0.05	3	0.40	7	tr(a)	0	tr(a)	11	3	tr(a)	tr(a)	34	3	387	93	579	0.2	1	0.4	3	unk.	unk.	9	K
0.6	3	tr(a)	0	tr(a)	0	24	1	2	0.0	0	tr(a)	1.28	40	4	16.8	2.4	16.3	5	1	0.03	2	tr(a)		1	
0.06	4	unk.	unk.	unk.	unk.	unk.	unk.	0	unk.	unk.	unk.	unk.	3	0	350	63	unk.	0.6	4	1.0	6	unk.	unk.	9	L
0.9	4	unk.	unk.	0.16	2	0	0	0	unk.	unk.	unk.	1.11	31	3	15.2	1.6	unk.	unk.	unk.	0.02	1	tr(a)		1	
0.06	3	unk.	unk.	unk.	unk.	unk.	unk.	unk.	unk.	unk.	tr(a)	tr(a)	unk.	unk.	unk.	unk.	unk.	unk.	unk.	unk.	unk.	unk.	unk.	unk.	M
0.3	2	unk.	unk.	unk.	unk.	unk.	unk.	unk.	0.0	0	tr(a)	0.73	unk.	unk.	unk.	unk.	unk.	unk.	unk.	unk.	unk.	tr(a)		tr(a)	
0.04	2	0.06	3	0.21	4	unk.	unk.	unk.	unk.	unk.	unk.	unk.	3	0	341	59	unk.	0.5	3	1.0	6	unk.	unk.	8	N
0.6	3	1	0	unk.	unk.	unk.	unk.	unk.	unk.	unk.	unk.	0.82	36	4	14.8	1.5	unk.	unk.	unk.	0.02	1	tr(a)		1	
0.05	3	0.09	5	0.25	4	tr(a)	0	tr(a)	11	3	0.2	0.2	2	0	349	71	460	0.2	1	0.5	3	unk.	unk.	12	O
0.9	5	tr(a)	0	tr(a)	0	tr(a)	0	0(a)	0.2	1	tr	0.97	31	3	15.2	1.8	13.0	4	1	0.02	1	tr(a)		1	
0.02	1	0.02	1	0.17	3	tr(a)	0	0(a)	5	1	tr(a)	tr(a)	2	0	140	15	213	0.1	1	0.1	1	unk.	unk.	5	P
0.2	1	tr(a)	0	tr(a)	0	12	0	1	0.0	0	tr(a)	0.38	8	1	6.1	0.4	6.0	1	0	0.02	1	tr(a)		1	
0.01	1	0.01	1	0.05	1	unk.	unk.	unk.	unk.	unk.	unk.	unk.	1	0	49	24	unk.	0.1	0	unk.	unk.	unk.	unk.	unk.	Q
0.2	1	1	0	0.04	0	18	0	unk.	unk.	unk.	unk.	unk.	4	0	2.1	0.6	unk.	unk.	unk.	unk.	unk.	unk.		unk.	
0.06	4	unk.	unk.	unk.	unk.	unk.	unk.	0	unk.	unk.	unk.	unk.	3	0	346	62	unk.	0.6	3	0.9	6	unk.	unk.	9	R
0.8	4	1	0	0.15	2	0	0	0	unk.	unk.	unk.	1.09	30	3	15.0	1.6	unk.	3	1	0.02	1	tr(a)		1	
0.07	4	0.04	2	unk.	unk.	unk.	unk.	unk.	unk.	unk.	unk.	unk.	3	0	433	90	unk.	0.8	4	0.6	4	unk.	unk.	13	S
1.2	6	unk.	unk.	unk.	unk.	unk.	unk.	unk.	unk.	unk.	unk.	1.90	60	6	18.8	2.3	unk.	unk.	unk.	0.02	1	tr(a)		1	

	FOOD	Portion	Weight in grams / Conversion for 100 g	Kilocalories / H_2O g	Total carbohydrate g / Total fats g	Crude fiber g / Dietary fiber g	Total protein g / Total sugar g	% USRDA	Arginine mg / Histidine mg	Isoleucine mg / Leucine mg	Lysine mg / Methionine mg	Phenylalanine mg / Threonine mg	Valine mg / Tryptophan mg	Cystine mg / Tyrosine mg	Polyunsat. fatty acids g / Monounsat. fatty acids g	Saturated fatty acids g / P/S ratio	Linoleic acid g / Cholesterol mg	Thiamin mg / Ascorbic acid mg	% USRDA
A	luncheon meat,summer sausage (thuringer cervelat),beef-Oscar Mayer	1-2/3 slices	28	89	0.9	0(a)	4.3	**10**	unk.	unk.	unk.	unk.	unk.	unk.	0.37	3.55	0.31	0.04	**3**
			3.52	14.5	7.7	0(a)	0.6		unk.	unk.	unk.	unk.	unk.	unk.	3.32	0.10	20	6	**10**
B	luncheon meat,summer sausage (thuringer cervelat)-Oscar Mayer	1-2/3 slices	28	92	0.3	0(a)	4.3	**10**	unk.	unk.	unk.	unk.	unk.	unk.	0.77	3.46	0.65	0.06	**4**
			3.52	14.5	8.2	0(a)	0.3		unk.	unk.	unk.	unk.	unk.	unk.	3.46	0.22	22	7	**12**
C	luncheon meat,turkey ham	3 thin slices	85	109	0.3	0(a)	16.1	**36**	1125	838	1520	640	857	168	1.30	1.45	1.05	0.04	**3**
			1.17	60.8	4.3	0(a)	unk.		504	1285	467	717	183	637	0.81	0.89	unk.	unk.	**unk.**
D	luncheon meat,turkey pastrami	3-1/4" dia x 1/8"	28	40	0.5	unk.	5.2	**12**	366	261	472	203	270	59	0.45	0.51	0.37	0.02	**1**
			3.52	20.1	1.8	unk.	unk.		156	404	145	227	58	197	0.47	0.88	unk.	unk.	**unk.**
E	luncheon meat,turkey salami	3-1/4" dia x 1/8"	28	56	0.2	unk.	4.6	**10**	unk.	unk.	unk.	unk.	unk.	unk.	unk.	unk.	unk.	0.02	**1**
			3.52	18.7	3.9	unk.	unk.		unk.	unk.	unk.	unk.	unk.	unk.	unk.	unk.	23	unk.	**unk.**
F	MACADAMIA NUT	1 med	1.7	12	0.3	0.04	0.1	**0**	unk.	unk.	unk.	unk.	unk.	unk.	0.28	0.19	0.02	0.01	**0**
			58.82	0.1	1.2	unk.	0.1		unk.	unk.	unk.	unk.	unk.	unk.	0.73	1.49	0	0	**0**
G	MACARONI,canned,macaroni & cheese -Chef Boy-Ar-Dee	1 C	209	166	23.2	unk.	6.5	**10**	unk.	unk.	unk.	unk.	unk.	unk.	unk.	unk.	unk.	0.27	**18**
			0.48	171.4	5.2	unk.	unk.		unk.	unk.	unk.	unk.	unk.	unk.	unk.	unk.	unk.	0	**0**
H	macaroni,canned,macroni 'n beef in tomato sauce-Chef Boy-Ar-Dee	1 C	213	218	29.0	0.43	8.5	**13**	unk.	unk.	unk.	unk.	unk.	unk.	unk.	unk.	unk.	0.21	**14**
			0.47	165.5	6.6	unk.	unk.		unk.	unk.	unk.	unk.	unk.	unk.	unk.	unk.	unk.	4	**7**
I	macaroni,cooked,enriched, tender stage	1/2 C	70	230	16.1	0.07	2.4	**4**	unk.	120	77	125	137	45	unk.	tr(a)	unk.	0.10	**7**
			1.43	50.4	0.3	unk.	0.3		56	158	36	94	29	78	unk.	unk.	0	0	**0**
J	macaroni,frozen,macroni & beef -Stouffer	1/2 pkg	163	189	19.9	unk.	10.0	**22**	unk.	unk.	unk.	unk.	unk.	unk.	unk.	unk.	unk.	0.11	**8**
			0.61	124.1	8.0	unk.	unk.		unk.	unk.	unk.	unk.	unk.	unk.	unk.	unk.	unk.	tr	**0**
K	macaroni,frozen,macroni & cheese -Stouffer	1/2 pkg	170	259	23.9	unk.	11.9	**18**	unk.	unk.	unk.	unk.	unk.	unk.	unk.	unk.	unk.	0.14	**9**
			0.59	116.7	11.9	unk.	unk.		unk.	unk.	unk.	unk.	unk.	unk.	unk.	unk.	unk.	tr	**0**
L	macaroni,home recipe,macaroni & cheese,baked	3/4 C	150	323	30.1	0.15	12.6	**26**	unk.	1125	1095	660	1290	255	unk.	7.50	1.50	0.15	**10**
			0.67	87.3	16.6	unk.	3.9		555	1665	420	780	255	840	7.50	unk.	32	tr	**0**
M	macaroni,home recipe,macaroni & mushroom casserole	3/4 C	166	250	17.4	0.06	10.5	**22**	unk.	555	849	523	654	97	0.52	9.56	0.33	0.10	**7**
			0.60	103.7	15.5	unk.	3.6		358	912	257	373	145	509	3.72	0.05	47	1	**1**
N	macaroni,home recipe,salad	1/2 C	95	167	24.2	0.00	3.8	**6**	unk.	unk.	unk.	unk.	unk.	unk.	unk.	unk.	unk.	0.02	**1**
			1.05	unk.	5.9	unk.	unk.		unk.	unk.	unk.	unk.	unk.	unk.	unk.	unk.	0(a)	1	**2**
O	macaroni,prep,macaroni & cheese dinner-Kraft	3/4 C	156	289	31.8	0.16	9.2	**17**	unk.	unk.	unk.	unk.	unk.	unk.	2.03	4.21	unk.	0.27	**18**
			0.64	98.3	13.9	unk.	4.1		unk.	unk.	unk.	unk.	unk.	unk.	unk.	0.48	8	tr	**0**
P	MACE,ground	1 tsp	1.8	9	0.9	0.09	0.1	**0**	unk.	unk.	unk.	unk.	unk.	unk.	0.08	0.17	0.08	0.01	**0**
			55.56	0.1	0.6	unk.	0.0		unk.	unk.	unk.	unk.	unk.	unk.	0.19	0.46	0	unk.	**unk.**
Q	MANGO,raw,half,edible portion	1/2 average	101	66	16.9	0.90	0.7	**1**	unk.	unk.	93	unk.	unk.	unk.	unk.	tr(a)	unk.	0.05	**3**
			1.00	82.1	0.4	1.51	unk.		unk.	unk.	8	unk.	14	unk.	unk.	unk.	0	35	**59**
R	mango,raw,sliced,edible portion	1/2 C	83	54	13.9	0.74	0.6	**1**	unk.	unk.	77	unk.	unk.	unk.	unk.	tr(a)	unk.	0.04	**3**
			1.21	67.4	0.3	1.24	9.0		unk.	unk.	7	unk.	12	unk.	unk.	unk.	0	29	**48**
S	MANICOTTI,home recipe	3/4 C	181	273	27.8	0.55	13.5	**28**	568	679	938	577	732	193	0.25	6.03	0.77	0.18	**12**
			0.55	127.5	11.9	2.26	3.5		356	977	294	557	139	425	3.51	0.04	77	12	**21**

	Riboflavin mg / Niacin mg	% USRDA	Vitamin B_6 mg / Folacin mcg	% USRDA	Vitamin B_{12} mcg / Pantothenic acid mg	% USRDA	Biotin mg / Vitamin A IU	% USRDA	Preformed A RE / Beta carotene RE	Vitamin D IU / Vitamin E IU	% USRDA	Total tocopherol mg / Alpha tocopherol mg	Other tocopherol mg / Total ash g	Calcium mg / Phosphorus mg	% USRDA	Sodium mg / Sodium meq	Potassium mg / Potassium meq	Chlorine mg / Chlorine meq	Iron mg / Magnesium mg	% USRDA	Zinc mg / Copper mg	% USRDA	Iodine mcg / Selenium mcg	% USRDA	Manganese mcg / Chromium mcg
A	0.09	5	0.07	4	1.59	27	tr(a)	0	tr(a)	12	3	tr(a)	tr(a)	2	0	391	65	545	0.6	4	0.6	4	unk.	unk.	12
	1.2	6	tr(a)	0	tr(a)	0	tr(a)	0	0(a)	0.0	0	tr(a)	1.14	29	3	17.0	1.7	15.4	3	1	0.03	1	tr(a)		1
B	0.08	5	0.09	4	1.06	18	tr(a)	0	tr(a)	13	3	tr(a)	tr(a)	2	0	420	62	562	0.6	3	0.6	4	unk.	unk.	12
	1.2	6	tr(a)	0	tr(a)	0	tr(a)	0	0(a)	0.0	0	tr(a)	1.16	33	3	18.3	1.6	15.9	4	1	0.03	1	tr(a)		1
C	0.21	13	unk.	unk.	unk.	unk.	unk.	unk.	unk.	unk.	unk.	unk.	unk.	9	1	849	277	unk.	2.4	13	unk.	unk.	unk.	unk.	unk.
	3.0	15	unk.	unk.	unk.	unk.	unk.	unk.	unk.	unk.	unk.	unk.	3.60	163	16	36.9	7.1	unk.	unk.	unk.	unk.	unk.	unk.		unk.
D	0.07	4	unk.	unk.	unk.	unk.	unk.	unk.	unk.	unk.	unk.	unk.	unk.	3	0	297	74	unk.	0.5	3	0.6	4	unk.	unk.	unk.
	1.0	5	unk.	unk.	unk.	unk.	unk.	unk.	unk.	unk.	unk.	unk.	0.89	57	6	12.9	1.9	unk.	4	1	0.02	1	unk.		unk.
E	0.05	3	unk.	unk.	unk.	unk.	unk.	unk.	unk.	unk.	unk.	unk.	unk.	6	1	285	69	unk.	0.5	3	0.5	3	unk.	unk.	unk.
	1.0	5	unk.	unk.	unk.	unk.	unk.	unk.	unk.	unk.	unk.	unk.	0.97	30	3	12.4	1.8	unk.	4	1	0.02	1	unk.		unk.
F	tr	0	unk.	unk.	0(a)	0	unk.	unk.	0	0	0	unk.	unk.	1	0	unk.	0.1	unk.	tr	0	unk.	unk.	unk.	unk.	unk.
	tr	0	unk.	unk.	unk.	unk.	0	0	0	unk.	unk.	unk.	0.03	3	unk.	4	unk.	unk.	unk.	unk.	unk.	unk.	unk.		unk.
G	0.21	12	unk.	unk.	unk.	unk.	unk.	unk.	unk.	unk.	unk.	unk.	unk.	92	9	792	107	unk.	1.5	8	unk.	unk.	unk.	unk.	unk.
	2.3	12	unk.	unk.	unk.	unk.	558	11	unk.	unk.	unk.	unk.	2.60	unk.	unk.	34.4	2.7	unk.	unk.	unk.	unk.	unk.	unk.		unk.
H	0.19	11	unk.	unk.	unk.	unk.	unk.	unk.	unk.	0(a)	0	tr(a)	tr(a)	30	3	1022	398	unk.	2.1	12	unk.	unk.	unk.	unk.	unk.
	3.6	18	unk.	unk.	tr(a)	0	741	15	unk.	0.0	0	tr(a)	3.90	113	11	44.4	10.2	unk.	unk.	unk.	tr(a)	0	tr(a)		unk.
I	0.06	3	0.01	1	0.00	0	0.000	0	0	0(a)	0	unk.	unk.	6	1	1	43	unk.	0.6	4	0.3	2	unk.	unk.	tr
	0.8	4	3	1	0.07	1	0	0	0	unk.	unk.	unk.	0.84	35	4	0.0	1.1	unk.	29	7	0.03	1	unk.		unk.
J	0.16	10	unk.	unk.	unk.	unk.	unk.	unk.	unk.	unk.	unk.	unk.	unk.	39	4	805	318	unk.	1.8	10	unk.	unk.	unk.	unk.	unk.
	2.0	10	unk.	unk.	unk.	unk.	498	10	unk.	unk.	unk.	unk.	2.45	unk.	unk.	35.0	8.1	unk.	unk.	unk.	unk.	unk.	unk.		unk.
K	0.17	10	unk.	unk.	unk.	unk.	unk.	unk.	unk.	unk.	unk.	unk.	unk.	249	25	775	138	unk.	1.1	6	unk.	unk.	unk.	unk.	unk.
	0.8	4	unk.	unk.	unk.	unk.	196	4	unk.	unk.	unk.	unk.	2.56	unk.	unk.	33.7	3.5	unk.	unk.	unk.	unk.	unk.	unk.		unk.
L	0.30	18	0.06	3	0.52	9	0.003	1	136	30	8	unk.	unk.	272	27	815	180	unk.	1.3	8	0.4	3	unk.	unk.	unk.
	1.3	7	9	2	0.30	3	645	13	19	unk.	unk.	unk.	3.30	242	24	35.4	4.6	unk.	39	10	0.06	3	unk.		unk.
M	0.28	17	0.07	3	0.42	7	0.001	0	0	25	6	0.1	tr(a)	255	26	616	174	1	0.8	4	1.4	9	39.3	26	7
	1.0	5	8	2	0.39	4	583	12	0	0.1	1	0.1	3.19	312	31	26.8	4.4	0.0	26	7	0.12	6	1		1
N	0.01	1	unk.	unk.	0(a)	0	unk.	unk.	0	unk.	unk.	unk.	unk.	10	1	unk.	unk.	unk.	0.5	3	0.3	2	unk.	unk.	unk.
	0.3	2	unk.	unk.	unk.	unk.	20	0	2	unk.	unk.	unk.	unk.	51	5	unk.	unk.	unk.	21	5	unk.	unk.	unk.		unk.
O	0.20	12	0.22	11	0.42	7	unk.	unk.	unk.	23	6	2.0	unk.	114	11	524	181	unk.	1.6	9	0.9	6	unk.	unk.	unk.
	2.0	10	16	4	0.09	1	560	11	unk.	unk.	unk.	unk.	2.65	198	20	22.8	4.6	unk.	36	9	0.16	8	unk.		unk.
P	0.01	1	unk.	unk.	0.00	0	unk.	unk.	0(a)	0(a)	0	unk.	unk.	5	1	1	8	unk.	0.3	1	tr	0	unk.	unk.	unk.
	tr	0	unk.	unk.	unk.	unk.	14	0	unk.	unk.	unk.	unk.	0.04	2	0	0.1	0.2	unk.	3	1	unk.	unk.	unk.		unk.
Q	0.05	3	unk.	unk.	0.00	0	unk.	unk.	0	0(a)	0	unk.	unk.	10	1	7	190	unk.	0.4	2	unk.	unk.	unk.	unk.	26
	1.1	6	unk.	unk.	0.16	2	4824	97	482	unk.	unk.	unk.	0.40	13	1	0.3	4.9	unk.	9	2	0.12	6	0		0
R	0.04	2	unk.	unk.	0.00	0	unk.	unk.	0	0(a)	0	unk.	unk.	8	1	6	156	unk.	0.3	2	unk.	unk.	unk.	unk.	21
	0.9	5	unk.	unk.	0.13	1	3960	79	396	unk.	unk.	unk.	0.33	11	1	0.3	4.0	unk.	7	2	0.10	5	0		0
S	0.20	12	0.21	10	0.27	5	0.004	1	29	1	0	0.7	0.0	76	8	414	303	12	2.6	15	1.7	11	31.5	21	137
	3.1	15	17	4	0.55	6	2613	52	226	0.8	3	0.1	2.33	183	18	18.0	7.7	0.3	49	12	0.21	10	1		3

	FOOD	Portion	Weight in grams / Conversion for 100 g	Kilocalories / H_2O g	Total carbohydrate g / Total fats g	Crude fiber g / Dietary fiber g	Total protein g / Total sugar g	% USRDA	Arginine mg / Histidine mg	Isoleucine mg / Leucine mg	Lysine mg / Methionine mg	Phenylalanine mg / Threonine mg	Valine mg / Tryptophan mg	Cystine mg / Tyrosine mg	Polyunsat. fatty acids g / Monounsat. fatty acids g	Saturated fatty acids g / P/S ratio	Linoleic acid g / Cholesterol mg	Thiamin mg / Ascorbic acid mg	% USRDA
A	MARGARINE,IMITATION spread,	1 Tbsp	14	77	0.0	unk.	0.1	**0**	3	5	7	4	6	1	1.96	1.82	1.83	0.00	**0**
	unspecified ingredient oils		*7.04*	5.3	8.6	unk.	0(a)		2	8	2	4	1	4	4.47	1.08	0(a)	0	**0**
B	margarine,imitation,unspecified	1 Tbsp	14	50	0.1	unk.	0.1	**0**	3	4	6	4	5	1	1.99	1.11	1.90	0.00	**0**
	ingredient oils		*6.94*	8.4	5.6	unk.	unk.		2	7	2	3	1	4	2.25	1.79	0(a)	0	**0**
C	MARGARINE,LOW SODIUM (any)	1 Tbsp	14	101	0.1	0.00	0.1	**0**	3	4	6	3	5	1	3.55	2.13	3.51	tr	**0**
			7.04	2.6	11.4	0(a)	0.1		2	7	2	3	1	3	5.21	1.67	0	0	**0**
D	margarine,low sodium-Fleischmann	1 Tbsp	14	102	0.1	0(a)	0.1	**0**	4	7	10	6	8	1	3.42	1.99	3.37	tr	**0**
			7.04	2.2	11.4	0(a)	0.1		3	12	3	6	2	6	5.51	1.72	0	0	**0**
E	MARGARINE,REGULAR (any)	1 Tbsp	14	101	0.1	0.00	0.1	**0**	3	4	6	3	5	1	3.55	2.13	3.51	tr(a)	**0**
			7.04	2.2	11.4	0.00	0.1		2	7	2	3	1	3	5.21	1.67	0	0	**0**
F	margarine,regular soft	1 Tbsp	14	102	0.1	0.00	0.1	**0**	4	7	9	5	7	1	4.91	1.96	4.76	tr	**0**
			7.04	2.2	11.4	0.00	0.1		3	11	3	5	2	5	4.05	2.51	0	0	**0**
G	margarine,regular-Blue Bonnet	1 Tbsp	14	102	0.1	0(a)	0.1	**0**	4	7	10	6	8	1	2.87	2.22	2.70	tr	**0**
			7.04	2.2	11.4	0(a)	0.1		3	12	3	6	2	6	5.85	1.29	0	0	**0**
H	margarine,regular-Chiffon	1 Tbsp	9.4	68	0.1	0(a)	0.1	**0**	3	5	6	4	5	1	2.46	1.23	2.28	tr	**0**
			10.64	1.5	7.6	0(a)	0.1		2	8	2	4	1	4	3.53	2.00	0	0	**0**
I	margarine,regular-Fleischmann	1 Tbsp	14	102	0.1	0(a)	0.1	**0**	4	7	10	6	8	1	3.42	1.99	3.37	tr	**0**
			7.04	2.2	11.4	0(a)	0.1		3	12	3	6	2	6	5.51	1.72	0	0	**0**
J	margarine,regular-Promise	1 Tbsp	13	93	0.1	0(a)	0.1	**0**	4	7	9	5	7	1	4.76	1.55	4.74	tr	**0**
			7.69	2.0	10.5	0(a)	0.1		3	11	3	5	2	5	3.70	3.08	0(a)	0	**0**
K	margarine,regular,corn oil-Mazola	1 Tbsp	14	101	0.1	0.00	0.1	**0**	4	7	10	6	8	1	3.51	2.10	3.47	tr	**0**
			7.14	2.2	11.3	0(a)	0.1		3	12	3	5	2	6	5.15	1.67	0	0	**0**
L	margarine,regular,corn oil	1 Tbsp	14	102	0.1	0.00	0.1	**0**	unk.	3	5	3	4	1	3.41	2.10	unk.	0.00	**0**
	-Mrs. Filbert's		*7.04*	2.3	11.4	0(a)	0.1		2	6	1	3	1	3	5.86	1.62	0	0	**0**
M	margarine,regular,corn oil,unsalted	1 Tbsp	14	100	0.1	0.00	0.1	**0**	3	4	5	3	5	1	3.50	2.10	3.46	tr	**0**
	-Mazola		*7.14*	2.6	11.2	0(a)	0.1		2	7	2	3	1	3	5.14	1.67	0	0	**0**
N	margarine,regular,soft-Promise	1 Tbsp	14	102	0.1	0(a)	0.1	**0**	4	7	9	5	7	1	6.82	1.82	6.76	tr	**0**
			7.04	2.3	11.4	0(a)	0.1		3	11	3	5	2	5	2.29	3.75	0	0	**0**
O	margarine,regular,soft,corn oil	1 Tbsp	14	102	0.1	0.00	0.1	**0**	unk.	3	5	3	4	1	5.34	2.06	unk.	0.00	**0**
	-Mrs. Filbert's		*7.04*	2.3	11.4	0(a)	0.1		2	6	1	3	1	3	3.98	2.59	0	0	**0**
P	margarine,regular,whipped	1 Tbsp	9.4	67	0.0	0.00	0.1	**0**	3	4	6	3	5	1	3.25	1.30	3.15	tr	**0**
			10.64	1.5	7.6	0(a)	0.0		2	7	2	3	1	3	2.68	2.51	0	0	**0**
Q	MARJORAM,dried	1 tsp	0.9	2	0.5	0.16	0.1	**0**	unk.	unk.	unk.	unk.	unk.	unk.	unk.	unk.	unk.	tr	**0**
			111.11	0.1	0.1	unk.	0.0		unk.	unk.	unk.	unk.	unk.	unk.	unk.	unk.	0	0	**0**
R	MARMALADE,citrus,sweetened	1 Tbsp	20	51	14.0	0.08	0.1	**0**	unk.	unk.	unk.	unk.	unk.	unk.	tr(a)	tr(a)	tr(a)	tr	**0**
			5.00	5.8	tr	0(a)	14.0		unk.	unk.	unk.	unk.	unk.	unk.	tr(a)	0.00	0	1	**2**
S	MARSHMALLOW,miniature	1 average	0.7	2	0.6	0.00	0.6	**1**	tr	1	1	1	1	tr	0.30	0.36	tr(a)	0.07	**5**
			142.86	0.1	tr	unk.	9.1		tr	1	tr	1	tr	1	tr(a)	0.84	0	3	**5**

Riboflavin mg / Niacin mg	% USRDA	Vitamin B_6 mg / Folacin mcg	% USRDA	Vitamin B_{12} mcg / Pantothenic acid mg	% USRDA	Biotin mg / Vitamin A IU	% USRDA	Preformed A RE / Beta carotene RE	Vitamin D IU / Vitamin E IU	% USRDA	Total tocopherol mg / Alpha tocopherol mg	Other tocopherol mg / Total ash g	Calcium mg / Phosphorus mg	% USRDA	Sodium mg / Sodium meq	Potassium mg / Potassium meq	Chlorine mg / Chlorine meq	Iron mg / Magnesium mg	% USRDA	Zinc mg / Copper mg	% USRDA	Iodine mcg / Selenium mcg	% USRDA	Manganese mcg / Chromium mcg	
tr	**0**	tr	**0**	0.01	**0**	unk.	**unk.**	unk.	unk.	**unk.**	unk.	unk.	3	**0**	141	4	unk.	unk.	**unk.**	unk.	**unk.**	unk.	**unk.**	unk.	A
0.0	**0**	tr	**0**	0.01	**0**	470	**9**	unk.	unk.	**unk.**	unk.	0.24	2	**0**	6.1	0.1	unk.	tr	**0**	unk.	**unk.**	unk.		unk.	
tr	**0**	tr	**0**	0.01	**0**	unk.	**unk.**	unk.	unk.	**unk.**	unk.	unk.	3	**0**	138	4	unk.	unk.	**unk.**	unk.	**unk.**	unk.	**unk.**	unk.	B
0.0	**0**	tr	**0**	0.01	**0**	476	**10**	unk.	unk.	**unk.**	unk.	0.32	2	**0**	6.0	0.1	unk.	tr	**0**	unk.	**unk.**	unk.		unk.	
tr	**0**	tr	**0**	0.01	**0**	tr(a)	**0**	unk.	unk.	**unk.**	7.7	3.7	2	**0**	tr	4	unk.	0.0	**0**	tr	**0**	unk.	**unk.**	unk.	C
0.0	**0**	tr	**0**	0.01	**0**	470	**9**	unk.	9.3	**31**	4.0	0.03	2	**0**	0.0	0.1	unk.	tr	**0**	unk.	**unk.**	unk.		unk.	
tr	**0**	tr	**0**	0.01	**0**	tr(a)	**0**	unk.	63	**16**	7.4	5.7	4	**0**	4	6	unk.	unk.	**unk.**	tr	**0**	unk.	**unk.**	unk.	D
0.0	**0**	tr	**0**	0.01	**0**	470	**9**	unk.	8.8	**29**	1.6	0.28	3	**0**	0.1	0.1	unk.	tr	**0**	0.35	**18**	unk.		unk.	
tr	**0**	tr	**0**	0.01	**0**	tr(a)	**0**	unk.	unk.	**unk.**	7.7	3.7	2	**0**	140	4	unk.	0.0	**0**	tr	**0**	unk.	**unk.**	unk.	E
0.0	**0**	tr	**0**	tr(a)	**0**	469	**9**	unk.	9.3	**31**	4.0	0.07	2	**0**	6.1	0.1	unk.	tr	**0**	unk.	**unk.**	unk.		unk.	
0.05	**3**	tr	**0**	0.01	**0**	tr(a)	**0**	unk.	unk.	**unk.**	7.7	3.7	4	**0**	140	5	unk.	0.0	**0**	tr	**0**	unk.	**unk.**	unk.	F
0.0	**0**	tr	**0**	0.01	**0**	469	**9**	unk.	9.3	**31**	4.0	0.35	3	**0**	6.1	0.1	unk.	tr	**0**	unk.	**unk.**	unk.		unk.	
0.01	**0**	tr	**0**	0.01	**0**	tr(a)	**0**	unk.	unk.	**unk.**	8.4	6.6	4	**0**	134	6	unk.	tr(a)	**0**	tr	**0**	unk.	**unk.**	unk.	G
0.0	**0**	tr	**0**	0.01	**0**	470	**9**	unk.	10.1	**34**	0.4	0.28	3	**0**	5.8	0.1	unk.	tr	**0**	unk.	**unk.**	unk.		unk.	
tr	**0**	tr	**0**	0.01	**0**	tr(a)	**0**	unk.	unk.	**unk.**	unk.	unk.	3	**0**	89	4	unk.	tr(a)	**0**	tr	**0**	unk.	**unk.**	unk.	H
0.0	**0**	tr	**0**	0.01	**0**	311	**6**	unk.	unk.	**unk.**	unk.	0.19	2	**0**	3.9	0.1	unk.	tr	**0**	unk.	**unk.**	unk.		unk.	
0.01	**0**	tr	**0**	0.01	**0**	tr(a)	**0**	unk.	unk.	**unk.**	7.4	5.7	4	**0**	134	6	unk.	0.0	**0**	tr	**0**	unk.	**unk.**	unk.	I
0.0	**0**	tr	**0**	0.01	**0**	470	**9**	unk.	8.8	**29**	1.6	unk.	3	**0**	5.8	0.1	unk.	tr	**0**	unk.	**unk.**	unk.		5	
0.01	**0**	tr	**0**	0.01	**0**	unk.	**unk.**	unk.	0(a)	**0**	unk.	unk.	4	**0**	123	6	unk.	unk.	**unk.**	unk.	**unk.**	unk.	**unk.**	unk.	J
0.0	**0**	tr	**0**	0.01	**0**	430	**9**	unk.	unk.	**unk.**	unk.	0.26	3	**0**	5.3	0.1	unk.	tr	**0**	unk.	**unk.**	unk.		unk.	
0.01	**0**	tr	**0**	0.01	**0**	tr(a)	**0**	unk.	62	**15**	9.2	7.7	4	**0**	132	6	unk.	0(a)	**0**	unk.	**unk.**	unk.	**unk.**	unk.	K
0.0	**0**	tr	**0**	0.01	**0**	463	**9**	unk.	11.1	**37**	1.5	0.28	3	**0**	5.8	0.1	unk.	tr	**0**	unk.	**unk.**	unk.		unk.	
tr	**0**	tr(a)	**0**	tr(a)	**0**	tr(a)	**0**	unk.	unk.	**unk.**	9.9	unk.	2	**0**	112	3	unk.	0.0	**0**	unk.	**unk.**	unk.	**unk.**	unk.	L
0.0	**0**	0	**0**	tr(a)	**0**	501	**10**	unk.	11.9	**40**	0.7	0.28	2	**0**	4.9	0.1	unk.	unk.	**unk.**	unk.	**unk.**	unk.		unk.	
tr	**0**	tr	**0**	0.01	**0**	tr(a)	**0**	unk.	62	**15**	7.8	unk.	2	**0**	tr	3	unk.	0(a)	**0**	unk.	**unk.**	unk.	**unk.**	unk.	M
0.0	**0**	tr	**0**	0.01	**0**	463	**9**	unk.	9.4	**31**	unk.	0.03	2	**0**	0.0	0.1	unk.	tr	**0**	unk.	**unk.**	unk.		unk.	
tr	**0**	tr	**0**	0.01	**0**	tr(a)	**0**	unk.	61	**15**	unk.	unk.	4	**0**	153	5	unk.	tr(a)	**0**	tr	**0**	unk.	**unk.**	unk.	N
0.0	**0**	tr	**0**	0.01	**0**	470	**9**	unk.	unk.	**unk.**	unk.	0.28	3	**0**	6.7	0.1	unk.	tr	**0**	unk.	**unk.**	unk.		unk.	
tr	**0**	tr(a)	**0**	tr(a)	**0**	tr(a)	**0**	unk.	unk.	**unk.**	9.9	unk.	2	**0**	112	3	unk.	0.0	**0**	unk.	**unk.**	unk.	**unk.**	unk.	O
0.0	**0**	0	**0**	tr(a)	**0**	501	**10**	unk.	11.9	**40**	0.7	0.28	2	**0**	4.9	0.1	unk.	unk.	**unk.**	unk.	**unk.**	unk.		unk.	
tr	**0**	tr	**0**	0.01	**0**	tr(a)	**0**	unk.	unk.	**unk.**	5.1	2.4	2	**0**	101	4	unk.	0.0	**0**	tr	**0**	unk.	**unk.**	unk.	P
0.0	**0**	tr	**0**	0.01	**0**	311	**6**	unk.	6.1	**21**	2.7	0.19	2	**0**	4.4	0.1	unk.	tr	**0**	unk.	**unk.**	unk.		unk.	
tr	**0**	unk.	**unk.**	0.00	**0**	unk.	**unk.**	0(a)	0(a)	**0**	unk.	unk.	18	**2**	1	14	unk.	0.7	**4**	tr	**0**	unk.	**unk.**	unk.	Q
tr	**0**	unk.	**unk.**	unk.	**unk.**	73	**2**	unk.	unk.	**unk.**	unk.	0.11	3	**0**	0.0	0.3	unk.	3	**1**	unk.	**unk.**	unk.		unk.	
tr	**0**	tr	**0**	0(a)	**0**	tr(a)	**0**	0	0	**0**	unk.	unk.	7	**1**	3	7	unk.	0.1	**1**	0.0	**0**	unk.	**unk.**	unk.	R
tr	**0**	4	**1**	0.01	**0**	0	**0**	0	unk.	**unk.**	unk.	0.06	2	**0**	0.1	0.2	unk.	1	**0**	unk.	**unk.**	unk.		unk.	
0.08	**5**	0.09	**5**	0(a)	**0**	tr(a)	**0**	0	8	**2**	1.3	1.0	2	**0**	73	7	tr(a)	0.5	**3**	0.1	**0**	unk.	**unk.**	tr(a)	S
0.9	**5**	19	**5**	tr(a)	**0**	235	**5**	0	1.5	**5**	0.3	0.03	8	**1**	3.2	0.2	0.0	2	**1**	0.02	**1**	tr(a)		unk.	

	FOOD	Portion	Weight in grams / *Conversion for 100 g*	Kilocalories / H_2O g	Total carbohydrate g / Total fats g	Crude fiber g / Dietary fiber g	Total protein g / Total sugar g	% USRDA	Arginine mg / Histidine mg	Isoleucine mg / Leucine mg	Lysine mg / Methionine mg	Phenylalanine mg / Threonine mg	Valine mg / Tryptophan mg	Cystine mg / Tyrosine mg	Polyunsat. fatty acids g / Monounsat. fatty acids g	Saturated fatty acids g / P/S ratio	Linoleic acid g / Cholesterol mg	Thiamin mg / Ascorbic acid mg	% USRDA
A	marshmallow treats	2x2x1"	18	70	13.3	0.02	0.0	**0**	unk.	unk.	unk.	unk.	unk.	unk.	tr(a)	tr(a)	tr(a)	0.00	**0**
			5.52	1.2	1.7	unk.	0.6		unk.	unk.	unk.	unk.	unk.	unk.	tr(a)	0.00	0(a)	0	**0**
B	McDONALD'S apple pie	1 serving	84	277	28.6	0.18	1.8	**3**	unk.	unk.	unk.	unk.	unk.	unk.	unk.	unk.	unk.	0.02	**1**
			1.19	35.4	17.5	unk.	unk.		unk.	unk.	unk.	unk.	unk.	unk.	unk.	unk.	13	2	**4**
C	McDonald's Big Mac	1 serving	183	531	38.3	0.69	25.5	**39**	unk.	unk.	unk.	unk.	unk.	unk.	unk.	unk.	unk.	0.35	**23**
			0.55	84.7	30.4	unk.	unk.		unk.	unk.	unk.	unk.	unk.	unk.	unk.	unk.	74	2	**4**
D	McDonald's cheeseburger	1 serving	111	298	30.2	0.29	15.6	**24**	unk.	unk.	unk.	unk.	unk.	unk.	unk.	unk.	unk.	0.23	**16**
			0.90	50.0	12.6	unk.	unk.		unk.	unk.	unk.	unk.	unk.	unk.	unk.	unk.	40	2	**3**
E	McDonald's chocolate shake	1 serving	269	339	55.8	tr	10.3	**16**	unk.	unk.	unk.	unk.	unk.	unk.	unk.	unk.	unk.	0.11	**7**
			0.37	192.6	8.4	unk.	unk.		unk.	unk.	unk.	unk.	unk.	unk.	unk.	unk.	27	tr	**0**
F	McDonald's Egg McMuffin	1 serving	127	338	25.0	0.38	17.3	**27**	unk.	unk.	unk.	unk.	unk.	unk.	unk.	unk.	unk.	0.34	**23**
			0.79	62.7	19.2	unk.	unk.		unk.	unk.	unk.	unk.	unk.	unk.	unk.	unk.	184	2	**3**
G	McDonald's Filet-O-Fish	1 serving	136	416	35.2	0.72	15.5	**24**	unk.	unk.	unk.	unk.	unk.	unk.	unk.	unk.	unk.	0.28	**19**
			0.74	57.9	23.8	unk.	unk.		unk.	unk.	unk.	unk.	unk.	unk.	unk.	unk.	45	4	**7**
H	McDonald's french fries	1 regular pkg	69	210	25.9	0.60	3.0	**5**	unk.	unk.	unk.	unk.	unk.	unk.	unk.	unk.	unk.	0.15	**10**
			1.46	27.7	11.0	unk.	unk.		unk.	unk.	unk.	unk.	unk.	unk.	unk.	unk.	10	11	**18**
I	McDonald's hamburger	1 serving	97	251	29.3	0.20	12.7	**20**	unk.	unk.	unk.	unk.	unk.	unk.	unk.	unk.	unk.	0.22	**15**
			1.03	43.6	8.8	unk.	unk.		unk.	unk.	unk.	unk.	unk.	unk.	unk.	unk.	25	2	**3**
J	McDonald's McDonaldland cookies	1 box	63	294	45.0	0.20	4.0	**6**	unk.	1	unk.	unk.	unk.	tr	unk.	unk.	unk.	0.28	**19**
			1.59	1.9	11.0	unk.	unk.		unk.	unk.	unk.	tr	20	tr	unk.	unk.	9	1	**2**
K	McDonald's Quarter Pounder, cheeseburger	1 serving	186	500	32.8	0.77	29.9	**46**	unk.	unk.	unk.	unk.	unk.	unk.	unk.	unk.	unk.	0.34	**22**
			0.54	90.9	28.0	unk.	unk.		unk.	unk.	unk.	unk.	unk.	unk.	unk.	unk.	93	3	**5**
L	McDonald's Quarter Pounder, hamburger	1 serving	157	400	31.5	0.76	24.8	**38**	unk.	unk.	unk.	unk.	unk.	unk.	unk.	unk.	unk.	0.30	**20**
			0.64	77.6	20.1	unk.	unk.		unk.	unk.	unk.	unk.	unk.	unk.	unk.	unk.	66	2	**4**
M	McDonald's strawberry shake	1 serving	267	314	51.9	tr	9.1	**14**	unk.	unk.	unk.	unk.	unk.	unk.	unk.	unk.	unk.	0.11	**7**
			0.37	194.9	8.2	unk.	unk.		unk.	unk.	unk.	unk.	unk.	unk.	unk.	unk.	27	tr	**0**
N	McDonald's vanilla shake	1 serving	274	306	49.2	tr	9.5	**15**	unk.	unk.	unk.	unk.	unk.	unk.	unk.	unk.	unk.	0.11	**7**
			0.37	204.5	7.6	unk.	unk.		unk.	unk.	unk.	unk.	unk.	unk.	unk.	unk.	27	tr	**0**
O	MEAT SUBSTITUTE,breakfast link -Morning Star Farms	1 piece	20	49	2.2	unk.	4.2	**7**	unk.	unk.	unk.	unk.	unk.	unk.	0.96	0.64	unk.	0.10	**7**
			5.00	unk.	2.6	unk.	unk.		unk.	unk.	unk.	unk.	unk.	unk.	unk.	1.50	0	unk.	**unk.**
P	meat substitute,breakfast patty -Morning Star Farms	1 piece	33	91	3.6	unk.	6.8	**11**	unk.	unk.	unk.	unk.	unk.	unk.	2.18	1.32	unk.	0.12	**8**
			3.03	unk.	5.5	unk.	unk.		unk.	unk.	unk.	unk.	unk.	unk.	unk.	1.65	0	unk.	**unk.**
Q	meat substitute,breakfast strip -Morning Star Farms	3 pieces	14	85	2.1	unk.	4.0	**6**	unk.	unk.	unk.	unk.	unk.	unk.	3.88	1.07	unk.	0.11	**7**
			7.09	unk.	6.7	unk.	unk.		unk.	unk.	unk.	unk.	unk.	unk.	unk.	3.62	0	unk.	**unk.**
R	meat substitute,meatless hotdog -Loma Linda Foods	1 piece	37	101	3.6	1.05	9.1	**20**	unk.	unk.	unk.	unk.	unk.	unk.	unk.	unk.	unk.	0.37	**25**
			2.67	18.0	5.6	unk.	unk.		unk.	unk.	unk.	unk.	unk.	unk.	unk.	unk.	0(a)	0(a)	**0**
S	meat substitute,soya meat loaf, home recipe	3x2-3/4x3/4" slice	135	130	16.0	1.19	15.3	**25**	unk.	725	908	764	777	150	0.33	0.54	0.17	0.19	**12**
			0.74	100.2	2.4	unk.	4.4		393	1174	204	640	119	491	0.59	0.61	54	24	**39**

Riboflavin mg / Niacin mg	% USRDA	Vitamin B6 mg / Folacin mcg	% USRDA	Vitamin B12 mcg / Pantothenic acid mg	% USRDA	Biotin mg / Vitamin A IU	% USRDA	Preformed A RE / Beta carotene RE	Vitamin D IU / Vitamin E IU	% USRDA	Total tocopherol mg / Alpha tocopherol mg	Other tocopherol mg / Total ash g	Calcium mg / Phosphorus mg	% USRDA	Sodium mg / Sodium meq	Potassium mg / Potassium meq	Chlorine mg / Chlorine meq	Iron mg / Magnesium mg	% USRDA	Zinc mg / Copper mg	% USRDA	Iodine mcg / Selenium mcg	% USRDA	Manganese mcg / Chromium mcg	
tr	0	tr(a)	0	0(a)	0	tr(a)	0	0	0(a)	0	tr(a)	tr(a)	tr	0	tr	0	tr(a)	tr	0	0.0	0	unk.	unk.	tr(a)	A
tr	0	tr(a)	0	tr(a)	0	0	0	0	0.0	0	tr(a)	tr	0	0	0.0	0.0	0.0	0	0	tr	0	tr(a)		unk.	
0.03	2	0.07	4	0.01	0	unk.	unk.	unk.	5	1	0.2	unk.	11	1	382	36	unk.	0.6	3	0.1	1	6.7	5	tr	B
1.2	6	3	1	0.17	2	64	1	unk.	0.2	1	unk.	1.09	21	2	16.6	0.9	unk.	6	2	0.03	1	unk.		unk.	
7	22	0.22	11	1.85	31	unk.	unk.	unk.	36	9	1.3	unk.	172	17	943	379	unk.	4.2	23	3.8	26	205.4	137	tr	C
8.1	40	24	6	0.22	2	321	6	unk.	1.6	5	unk.	3.67	211	21	41.0	9.7	unk.	37	9	0.15	7	unk.		unk.	
0.29	17	0.10	5	0.94	16	unk.	unk.	unk.	14	3	0.2	unk.	154	15	705	237	unk.	2.8	16	1.9	13	124.2	83	tr	D
5.3	27	14	4	0.17	2	362	7	unk.	0.2	1	unk.	2.55	130	13	30.7	6.1	unk.	23	6	0.03	2	unk.		unk.	
0.83	49	0.11	6	0.78	13	unk.	unk.	unk.	329	82	unk.	unk.	314	31	306	610	unk.	0.9	5	1.2	8	196.3	131	tr	E
0.8	4	5	1	1.16	12	296	6	unk.	unk.	unk.	unk.	2.96	272	27	13.3	15.6	unk.	47	12	0.16	8	unk.		40	
0.57	34	0.13	7	0.68	11	unk.	unk.	unk.	38	10	0.7	unk.	179	18	877	213	unk.	3.1	17	1.6	11	40.5	27	tr	F
4.1	21	11	3	0.62	6	346	7	unk.	0.8	3	unk.	3.04	254	25	38.2	5.4	unk.	24	6	0.11	5	unk.		unk.	
0.28	17	0.08	4	0.80	13	unk.	unk.	unk.	38	10	1.3	unk.	109	11	734	303	unk.	1.9	10	0.7	5	187.1	125	tr	G
4.0	20	15	4	0.98	10	157	3	unk.	1.5	5	unk.	2.71	164	16	31.9	7.8	unk.	30	8	0.07	4	unk.		41	
.03	2	tr	0	tr	0	unk.	unk.	unk.	3	1	0.1	unk.	10	1	113	568	unk.	0.5	3	0.1	1	10.3	7	tr	H
2.9	14	5	1	0.60	6	tr	0	unk.	0.1	0	unk.	1.31	49	5	4.9	14.5	unk.	23	6	0.02	1	unk.		21	
0.22	13	0.11	5	1.01	17	unk.	unk.	unk.	11	3	0.3	unk.	62	6	514	229	unk.	2.9	16	1.8	12	47.4	32	21	I
5.0	25	13	3	0.18	2	226	5	unk.	0.3	1	unk.	1.94	86	9	22.4	5.8	unk.	20	5	0.08	4	unk.		unk.	
0.23	14	0.00	0	tr	0	unk.	unk.	unk.	10	3	unk.	unk.	10	1	330	58	unk.	1.4	8	0.2	1	unk.	unk.	unk.	J
0.8	4	unk.	unk.	unk.	unk.	48	1	unk.	unk.	unk.	unk.	unk.	51	5	14.3	1.5	unk.	10	3	0.03	2	unk.		unk.	
0.56	33	0.24	12	2.33	39	unk.	unk.	unk.	35	9	0.5	unk.	242	24	1166	455	unk.	4.4	25	4.6	31	150.8	101	tr	K
14.6	73	22	6	0.65	7	659	13	unk.	0.6	2	unk.	4.28	248	25	50.7	11.6	unk.	41	10	0.14	7	unk.		56	
0.39	23	0.24	12	2.20	37	unk.	unk.	unk.	22	6	0.3	unk.	76	8	680	423	unk.	4.9	27	4.2	28	156.8	105	tr	L
9.4	47	19	5	0.45	5	157	3	unk.	0.4	1	unk.	2.82	171	17	29.6	10.8	unk.	36	9	0.12	6	unk.		47	
0.61	36	0.10	5	0.77	13	unk.	unk.	unk.	285	71	unk.	unk.	307	31	233	495	unk.	0.2	1	1.0	7	191.5	128	0	M
0.5	2	5	1	1.57	16	293	6	unk.	unk.	unk.	unk.	2.40	271	27	10.1	12.7	unk.	32	8	0.08	4	unk.		40	
0.63	37	0.11	6	0.90	15	unk.	unk.	unk.	335	84	unk.	unk.	328	33	237	472	unk.	0.2	1	1.0	6	199.0	133	0	N
0.6	3	5	1	1.34	13	328	7	unk.	unk.	unk.	unk.	2.46	252	25	10.3	12.1	unk.	33	8	0.06	3	unk.		41	
0.08	5	0.01	0	0.20	3	unk.	unk.	unk.	unk.	unk.	unk.	unk.	7	1	216	52	unk.	0.9	5	unk.	unk.	unk.	unk.	unk.	O
2.0	10	unk.	unk.	tr	0	unk.	unk.	unk.	unk.	unk.	unk.	unk.	unk.	unk.	9.4	1.3	unk.	unk.	unk.	unk.	unk.	unk.		unk.	
0.13	8	0.10	5	1.20	20	unk.	unk.	unk.	unk.	unk.	unk.	unk.	10	1	411	87	unk.	1.4	8	unk.	unk.	unk.	unk.	unk.	P
3.5	18	unk.	unk.	unk.	unk.	unk.	unk.	unk.	unk.	unk.	unk.	unk.	unk.	unk.	17.9	2.2	unk.	unk.	unk.	unk.	unk.	unk.		unk.	
0.07	4	0.03	2	0.68	11	unk.	unk.	unk.	unk.	unk.	unk.	unk.	15	2	364	7	unk.	0.5	3	unk.	unk.	unk.	unk.	unk.	Q
1.2	6	unk.	unk.	unk.	unk.	unk.	unk.	unk.	unk.	unk.	unk.	unk.	unk.	unk.	15.8	0.2	unk.	unk.	unk.	unk.	unk.	unk.		unk.	
0.24	14	0.29	15	0.97	16	unk.	unk.	unk.	0(a)	0	unk.	unk.	17	2	400	unk.	unk.	1.2	7	unk.	unk.	unk.	unk.	unk.	R
4.2	21	unk.	unk.	1.35	14	unk.	unk.	unk.	unk.	unk.	unk.	1.05	unk.	unk.	17.4	unk.	unk.	unk.	unk.	unk.	unk.	unk.		unk.	
0.21	13	0.39	20	1.50	25	0.020	7	0	4	1	0.2	0.0	75	8	619	692	28	3.5	19	1.6	10	219.6	146	659	S
4.4	22	94	24	0.66	7	491	10	39	0.3	1	tr	3.05	188	19	26.9	17.7	0.8	96	24	0.33	17	unk.		1	

	FOOD	Portion	Weight in grams / Conversion for 100 g	Kilocalories / H_2O g	Total carbohydrate g / Total fats g	Crude fiber g / Dietary fiber g	Total protein g / Total sugar g	% USRDA	Arginine mg / Histidine mg	Isoleucine mg / Leucine mg	Lysine mg / Methionine mg	Phenylalanine mg / Threonine mg	Valine mg / Tryptophan mg	Cystine mg / Tyrosine mg	Polyunsat. fatty acids g / Monounsat. fatty acids g	Saturated fatty acids g / P/S ratio	Linoleic acid g / Cholesterol mg	Thiamin mg / Ascorbic acid mg	% USRDA
A	meat substitute,soya sloppy joe,	1/2 C	110	119	12.2	1.05	15.3	**24**	unk.	732	903	742	766	116	0.27	1.85	0.07	0.19	**13**
	home recipe		*0.91*	77.3	3.2	unk.	3.5		384	1176	178	622	101	466	0.78	0.15	8	3	**5**
B	meat substitute,Wham	2 slices	68	140	4.0	unk.	10.0	**22**	unk.	unk.	unk.	unk.	unk.	unk.	unk.	unk.	unk.	0.60	**40**
	-Worthington Foods		*1.47*	unk.	9.0	unk.	0.0		unk.	unk.	unk.	unk.	unk.	unk.	unk.	unk.	0(a)	unk.	**unk.**
C	MILK DRINK,chocolate,whole,3.3% fat	8 oz glass	250	208	25.8	0.15	7.9	**18**	288	480	628	383	530	73	0.30	5.25	0.20	0.07	**5**
			0.40	205.7	8.5	unk.	25.2		215	778	198	358	113	383	2.17	0.06	30	2	**4**
D	milk drink,chocolate,1% fat	8 oz glass	250	158	26.1	0.15	8.1	**18**	293	490	643	390	543	75	0.10	1.55	0.05	0.07	**5**
			0.40	211.3	2.5	unk.	25.2		220	793	203	365	115	390	0.67	0.06	8	2	**4**
E	milk drink,chocolate,2% fat	8 oz glass	250	180	26.0	0.15	8.0	**18**	290	485	638	388	538	75	0.17	3.10	0.12	0.07	**5**
			0.40	208.9	5.0	unk.	25.2		218	785	203	363	113	388	1.30	0.06	18	2	**4**
F	milk drink,eggnog	8 oz glass	254	343	34.4	0.00	9.7	**22**	378	584	757	462	643	97	0.86	11.28	0.58	0.08	**5**
			0.39	188.9	19.0	0(a)	35.1		241	937	221	444	137	462	4.98	0.08	150	4	**6**
G	milk drink,hot cocoa,homemade	8 oz, 1 C	250	218	25.8	0.20	9.1	**20**	330	550	723	440	610	85	0.32	5.60	0.20	0.10	**7**
	w/whole milk		*0.40*	204.0	9.0	unk.	15.7		248	893	228	410	128	440	2.35	0.06	33	2	**4**
H	milk drink,malted milk,chocolate	8 oz glass	265	233	29.2	0.08	9.4	**21**	unk.	162	209	135	188	29	0.42	5.51	unk.	0.13	**9**
			0.38	215.1	9.1	unk.	29.2		74	265	69	125	37	132	unk.	0.08	34	2	**4**
I	milk drink,malted milk,natural flavor	8 oz glass	265	236	26.6	0.13	10.8	**24**	unk.	567	713	480	631	130	0.56	5.96	unk.	0.19	**12**
			0.38	215.2	9.9	unk.	26.6		273	943	238	432	143	464	unk.	0.09	37	2	**4**
J	milk drink,milk shake,thick type,	8 oz glass	244	290	51.6	0.61	7.4	**17**	268	451	590	359	498	68	0.24	4.10	0.15	0.10	**7**
	chocolate		*0.41*	176.2	6.6	unk.	51.6		203	730	185	337	105	359	1.66	0.06	24	0	**0**
K	milk drink,milk shake,thick type,	8 oz glass	244	273	43.3	0.15	9.4	**21**	342	571	747	454	630	88	0.27	4.61	0.17	0.07	**5**
	vanilla		*0.41*	181.7	7.4	unk.	43.3		256	922	237	425	132	454	1.85	0.06	29	0	**0**
L	MILK FLAVORING,cocoa powder,high	1 Tbsp	7.0	21	3.4	0.30	1.2	**3**	64	unk.	unk.	unk.	unk.	unk.	unk.	0.91	tr	0.01	**1**
	fat,breakfast		*14.29*	0.2	1.7	unk.	0.0		unk.	unk.	unk.	unk.	unk.	unk.	0.63	unk.	0	0	**0**
M	milk flavoring,hot chocolate	1 Tbsp	7.0	27	5.2	0.06	0.7	**1**	unk.	unk.	unk.	unk.	unk.	unk.	unk.	0.42	tr	0.01	**0**
	sweetened powder		*14.29*	0.2	0.7	unk.	5.2		unk.	unk.	unk.	unk.	unk.	unk.	0.28	unk.	0	tr	**0**
N	milk flavoring,malt flavor-Ovaltine	1 Tbsp	7.0	27	5.6	unk.	0.5	**1**	unk.	unk.	unk.	unk.	unk.	unk.	unk.	unk.	unk.	0.25	**16**
			14.29	unk.	0.3	unk.	5.6		unk.	unk.	unk.	unk.	unk.	unk.	unk.	unk.	unk.	7	**12**
O	milk flavoring,malted milk powder,	1 Tbsp	7.0	28	5.9	0.03	0.4	**1**	unk.	13	13	17	16	5	0.04	0.15	unk.	0.01	**1**
	chocolate		*14.29*	0.1	0.3	unk.	5.9		8	24	6	12	3	12	unk.	0.30	tr	unk.	**unk.**
P	milk flavoring,malted milk powder,	1 Tbsp	7.0	29	5.1	0.04	0.9	**2**	unk.	27	25	31	31	18	0.09	0.30	unk.	0.04	**3**
	natural flavor		*14.29*	0.2	0.6	unk.	5.1		18	52	13	23	10	25	unk.	0.29	1	0	**0**
Q	milk flavoring,Quik chocolate flavor	1 Tbsp	11	35	9.5	unk.	0.5	**1**	unk.	unk.	unk.	unk.	unk.	unk.	unk.	unk.	unk.	0.00	**0**
	-Nestle		*9.52*	unk.	0.5	unk.	9.5		unk.	unk.	unk.	unk.	unk.	unk.	unk.	unk.	unk.	0	**0**
R	milk flavoring,Quik strawberry flavor	1 Tbsp	11	40	10.5	unk.	0.0	**0**	unk.	unk.	unk.	unk.	unk.	unk.	unk.	unk.	unk.	unk.	**unk.**
	-Nestle		*9.52*	unk.	0.0	unk.	10.5		unk.	unk.	unk.	unk.	unk.	unk.	unk.	unk.	unk.	unk.	**unk.**
S	MILK,COW,buttermilk,cultured	8 oz glass	245	98	11.7	0.00	8.1	**18**	309	500	679	426	595	76	0.07	1.35	0.05	0.07	**5**
			0.41	220.8	2.2	0(a)	11.7		233	806	198	387	88	341	0.54	0.05	10	2	**4**

Riboflavin mg Niacin mg	% USRDA	Vitamin B_6 mg Folacin mcg	% USRDA	Vitamin B_{12} mcg Pantothenic acid mg	% USRDA	Biotin mg Vitamin A IU	% USRDA	Preformed A RE Beta carotene RE	Vitamin D IU Vitamin E IU	% USRDA	Total tocopherol mg Alpha tocopherol mg	Other tocopherol mg Total ash g	Calcium mg Phosphorus mg	% USRDA	Sodium mg Sodium meq	Potassium mg Potassium meq	Chlorine mg Chlorine meq	Iron mg Magnesium mg	% USRDA	Zinc mg Copper mg	% USRDA	Iodine mcg Selenium mcg	% USRDA	Manganese mcg Chromium mcg	
0.21	**12**	0.41	**21**	1.62	**27**	0.022	**7**	0	0	**0**	0.1	0.0	71	**7**	340	714	17	3.0	**17**	1.7	**11**	142.1	**95**	740	A
4.9	**25**	106	**26**	0.71	**7**	232	**5**	12	0.1	**0**	0.1	2.29	178	**18**	14.8	18.3	0.5	76	**19**	0.37	**19**	tr(a)		2	
0.17	**10**	0.12	**6**	2.10	**35**	unk.	**unk.**	unk.	0(a)	**0**	tr(a)	tr(a)	20	**2**	952	unk.	unk.	1.1	**6**	unk.	**unk.**	unk.	**unk.**	unk.	B
5.0	**25**	unk.	**unk.**	tr(a)	**0**	unk.	**unk.**	unk.	0.0	**0**	tr(a)	unk.	unk.	**unk.**	41.4	unk.	unk.	unk.	**unk.**	unk.	**unk.**	tr(a)		unk.	
0.40	**24**	0.10	**5**	0.82	**14**	unk.	**unk.**	unk.	unk.	**unk.**	unk.	unk.	280	**28**	150	418	unk.	0.6	**3**	1.0	**7**	unk.	**unk.**	unk.	C
0.3	**2**	13	**3**	0.74	**7**	303	**6**	unk.	unk.	**unk.**	unk.	2.00	250	**25**	6.5	10.7	unk.	33	**8**	unk.	**unk.**	unk.		unk.	
0.40	**24**	0.10	**5**	0.85	**14**	unk.	**unk.**	unk.	unk.	**unk.**	unk.	unk.	288	**29**	153	425	unk.	0.6	**3**	1.0	**7**	unk.	**unk.**	unk.	D
0.3	**2**	13	**3**	0.75	**8**	500	**10**	unk.	unk.	**unk.**	unk.	2.05	258	**26**	6.6	10.9	unk.	33	**8**	unk.	**unk.**	unk.		unk.	
0.40	**24**	0.10	**5**	0.82	**14**	unk.	**unk.**	unk.	unk.	**unk.**	unk.	unk.	285	**29**	150	423	unk.	0.6	**3**	1.0	**7**	unk.	**unk.**	unk.	E
0.3	**2**	13	**3**	0.75	**8**	500	**10**	unk.	unk.	**unk.**	unk.	2.02	255	**26**	6.5	10.8	unk.	33	**8**	unk.	**unk.**	unk.		unk.	
0.48	**28**	0.13	**6**	1.14	**19**	unk.	**unk.**	unk.	unk.	**unk.**	unk.	unk.	330	**33**	137	419	unk.	0.5	**3**	1.2	**8**	unk.	**unk.**	unk.	F
0.3	**1**	3	**1**	1.06	**11**	894	**18**	unk.	unk.	**unk.**	unk.	2.03	277	**28**	6.0	10.7	unk.	46	**11**	unk.	**unk.**	unk.		unk.	
0.42	**25**	0.11	**5**	0.85	**14**	unk.	**unk.**	unk.	unk.	**unk.**	unk.	unk.	298	**30**	123	480	unk.	0.8	**4**	1.2	**8**	unk.	**unk.**	unk.	G
0.3	**2**	13	**3**	0.81	**8**	318	**6**	unk.	unk.	**unk.**	unk.	2.02	270	**27**	5.3	12.3	unk.	55	**14**	unk.	**unk.**	unk.		unk.	
0.42	**25**	0.13	**7**	0.90	**15**	unk.	**unk.**	unk.	unk.	**unk.**	unk.	unk.	305	**31**	170	498	unk.	0.5	**3**	1.1	**7**	unk.	**unk.**	unk.	H
0.7	**3**	16	**4**	0.77	**8**	326	**7**	unk.	unk.	**unk.**	unk.	2.20	265	**27**	7.4	12.7	unk.	48	**12**	unk.	**unk.**	unk.		unk.	
0.53	**31**	0.18	**9**	1.03	**17**	unk.	**unk.**	unk.	unk.	**unk.**	unk.	unk.	347	**35**	215	530	unk.	0.3	**2**	1.1	**8**	unk.	**unk.**	unk.	I
1.3	**6**	21	**5**	0.77	**8**	376	**8**	unk.	unk.	**unk.**	unk.	2.46	307	**31**	9.3	13.5	unk.	53	**13**	unk.	**unk.**	unk.		unk.	
0.54	**32**	0.06	**3**	0.76	**13**	unk.	**unk.**	unk.	unk.	**unk.**	unk.	unk.	322	**32**	271	547	unk.	0.8	**4**	1.2	**8**	unk.	**unk.**	unk.	J
0.3	**2**	12	**3**	0.89	**9**	210	**4**	unk.	unk.	**unk.**	unk.	2.20	307	**31**	11.8	14.0	unk.	39	**10**	unk.	**unk.**	unk.		unk.	
0.46	**27**	0.10	**5**	1.27	**21**	unk.	**unk.**	unk.	unk.	**unk.**	unk.	unk.	356	**36**	232	447	unk.	0.2	**1**	1.0	**6**	unk.	**unk.**	unk.	K
0.3	**2**	17	**4**	unk.	**unk.**	278	**6**	unk.	unk.	**unk.**	unk.	2.22	281	**28**	10.1	11.4	unk.	29	**7**	unk.	**unk.**	unk.		unk.	
0.03	**2**	unk.	**unk.**	unk.	**unk.**	unk.	**unk.**	0	unk.	**unk.**	unk.	unk.	9	**1**	tr	107	4	0.7	**4**	0.4	**3**	unk.	**unk.**	175	L
0.2	**1**	unk.	**unk.**	unk.	**unk.**	2	**0**	tr	unk.	**unk.**	unk.	0.35	45	**5**	0.0	2.7	0.1	31	**8**	0.35	**18**	unk.		4	
0.03	**2**	tr	**0**	0.00	**0**	0.002	**1**	0	unk.	**unk.**	tr	0(a)	19	**2**	27	42	unk.	0.1	**1**	0.2	**1**	unk.	**unk.**	unk.	M
tr	**0**	6	**1**	0.01	**0**	1	**0**	tr	tr	**0**	tr	0.18	20	**2**	1.2	1.1	unk.	8	**2**	0.26	**13**	unk.		unk.	
0.30	**17**	0.30	**15**	0.10	**2**	tr	**0**	unk.	99	**25**	unk.	unk.	20	**2**	15	42	unk.	0.7	**4**	unk.	**unk.**	2.6	**2**	unk.	N
2.5	**12**	unk.	**unk.**	unk.	**unk.**	986	**20**	unk.	unk.	**unk.**	unk.	unk.	20	**2**	0.6	1.1	unk.	7	**2**	unk.	**unk.**	unk.		unk.	
0.01	**1**	0.01	**1**	0.02	**0**	unk.	**unk.**	unk.	unk.	**unk.**	unk.	unk.	4	**0**	16	43	unk.	0.1	**1**	0.1	**0**	unk.	**unk.**	unk.	O
0.1	**1**	2	**0**	unk.	**unk.**	7	**0**	unk.	unk.	**unk.**	unk.	0.15	13	**1**	0.7	1.1	unk.	5	**1**	unk.	**unk.**	unk.		unk.	
0.05	**3**	0.03	**1**	0.05	**1**	unk.	**unk.**	unk.	unk.	**unk.**	unk.	unk.	19	**2**	32	53	unk.	0.1	**0**	0.1	**1**	unk.	**unk.**	unk.	P
0.4	**2**	3	**1**	unk.	**unk.**	23	**1**	unk.	unk.	**unk.**	unk.	0.24	26	**3**	1.4	1.4	unk.	7	**2**	unk.	**unk.**	unk.		unk.	
unk.	**unk.**	unk.	**unk.**	unk.	**unk.**	unk.	**unk.**	0	unk.	**unk.**	unk.	unk.	0	**0**	19	78	unk.	0.0	**0**	unk.	**unk.**	unk.	**unk.**	unk.	Q
0.0	**0**	unk.	**unk.**	unk.	**unk.**	0	**0**	0	unk.	**unk.**	unk.	unk.	unk.	**unk.**	0.8	2.0	unk.	7	**2**	unk.	**unk.**	unk.		unk.	
unk.	**unk.**	unk.	**unk.**	unk.	**unk.**	unk.	**unk.**	unk.	unk.	**unk.**	unk.	unk.	unk.	**unk.**	3	unk.	unk.	unk.	**unk.**	unk.	**unk.**	unk.	**unk.**	unk.	R
unk.	**unk.**	unk.	**unk.**	unk.	**unk.**	unk.	**unk.**	unk.	unk.	**unk.**	unk.	unk.	unk.	**unk.**	0.1	unk.	unk.	unk.	**unk.**	unk.	**unk.**	unk.		unk.	
0.37	**22**	0.08	**4**	0.51	**9**	unk.	**unk.**	unk.	unk.	**unk.**	unk.	unk.	284	**28**	257	370	unk.	0.1	**1**	1.0	**7**	unk.	**unk.**	unk.	S
0.1	**1**	unk.	**unk.**	0.67	**7**	81	**2**	unk.	unk.	**unk.**	unk.	2.18	218	**22**	11.2	9.5	unk.	27	**7**	unk.	**unk.**	unk.		unk.	

	FOOD	Portion	Weight in grams / Conversion for 100 g	Kilocalories / H_2O g	Total carbohydrate g / Total fats g	Crude fiber g / Dietary fiber g	Total protein g / Total sugar g	% USRDA	Arginine mg / Histidine mg	Isoleucine mg / Leucine mg	Lysine mg / Methionine mg	Phenylalanine mg / Threonine mg	Valine mg / Tryptophan mg	Cystine mg / Tyrosine mg	Polyunsat. fatty acids g / Monounsat. fatty acids g	Saturated fatty acids g / P/S ratio	Linoleic acid g / Cholesterol mg	Thiamin mg / Ascorbic acid mg	% USRDA
A	milk,cow,lowfat,1% fat,fortified	8 oz glass	244	102	11.7	0.00	8.0	**18**	290	486	637	388	537	73	0.10	1.61	0.05	0.07	**5**
	w/vitamin A		*0.41*	219.8	2.6	0(a)	11.7		217	786	203	364	112	388	0.66	0.06	10	2	**4**
B	milk,cow,lowfat,1% fat,fortified	8 oz glass	244	102	11.7	0.00	8.0	**18**	290	486	637	388	537	73	0.10	1.61	0.05	0.07	**5**
	w/vitamins A & D		*0.41*	219.8	2.6	0(a)	11.7		217	786	203	364	112	388	0.66	0.06	10	2	**4**
C	milk,cow,lowfat,1% fat,protein	8 oz glass	246	118	13.6	0.00	9.7	**22**	349	585	768	467	647	89	0.10	1.80	0.07	0.10	**7**
	fortified		*0.41*	218.3	2.9	0(a)	13.6		263	947	244	435	135	467	0.71	0.05	10	3	**5**
D	milk,cow,lowfat,1% fat,w/NFDM	8 oz glass	245	105	12.2	0.00	8.5	**19**	309	517	676	412	571	78	0.10	1.47	0.05	0.10	**7**
	solids added		*0.41*	220.0	2.4	0(a)	12.2		230	835	213	385	120	412	0.59	0.07	10	2	**4**
E	milk,cow,lowfat,2% fat,fortified	8 oz glass	244	122	11.7	0.00	8.1	**18**	295	490	644	393	544	76	0.17	2.93	0.10	0.07	**5**
	w/vitamin A		*0.41*	217.7	4.7	0(a)	11.7		220	795	205	366	115	393	1.17	0.06	20	2	**4**
F	milk,cow,lowfat,2% fat,fortified	8 oz glass	244	122	11.7	0.00	8.1	**18**	295	490	644	393	544	76	0.17	2.93	0.10	0.07	**5**
	w/vitamins A & D		*0.41*	217.7	4.7	0(a)	11.7		220	795	205	366	115	393	1.17	0.06	20	2	**4**
G	milk,cow,lowfat,2% fat,protein	8 oz glass	246	138	13.5	0.00	9.7	**22**	352	588	770	470	649	91	0.17	3.03	0.10	0.10	**7**
	fortified		*0.41*	215.8	4.9	0(a)	13.5		263	952	244	438	138	470	1.23	0.06	20	3	**5**
H	milk,cow,lowfat,2% fat,	8 oz glass	245	125	12.2	0.00	8.5	**19**	309	517	676	412	571	78	0.17	2.94	0.10	0.10	**7**
	w/NFDM solids added		*0.41*	217.7	4.7	0(a)	12.2		230	835	213	385	120	412	1.18	0.06	20	2	**4**
I	milk,cow,nonfat dry milk powder,	1/3 C	23	81	11.8	0.00	8.0	**18**	288	482	631	384	533	74	0.01	0.11	0.00	0.09	**6**
	instantized		*4.41*	0.9	0.2	0(a)	11.8		216	779	199	359	112	384	0.04	0.06	4	1	**2**
J	milk,cow,nonfat dry milk powder,	1/3 C	40	145	20.8	0.00	14.5	**32**	524	875	1147	698	940	134	0.01	0.20	0.01	0.16	**11**
	non-instantized		*2.50*	1.3	0.3	0(a)	20.8		392	1417	363	653	204	698	0.07	0.06	8	3	**5**
K	milk,cow,skim,evaporated,canned	1 C	256	200	29.1	0.00	19.3	**43**	699	1170	1533	932	1293	179	0.00	0.31	0.00	0.10	**7**
			0.39	203.3	0.5	0(a)	29.1		525	1894	484	873	274	932	0.13	0.00	10	3	**5**
L	milk,cow,skim,fortified w/vitamin A	8 oz glass	245	86	11.9	0.00	8.3	**19**	301	505	661	404	559	78	0.00	0.27	0.00	0.07	**5**
			0.41	222.5	0.4	0(a)	11.9		225	818	211	377	118	404	0.07	0.00	5	2	**4**
M	milk,cow,skim,fortified	8 oz glass	245	86	11.9	0.00	8.3	**19**	301	505	661	404	559	78	0.00	0.27	0.00	0.07	**5**
	w/vitamins A & D		*0.41*	222.5	0.4	0(a)	11.9		225	818	211	377	118	404	0.07	0.00	5	2	**4**
N	milk,cow,skim,protein fortified	8 oz glass	246	101	13.7	0.00	9.7	**22**	352	590	772	470	652	91	0.00	0.39	0.00	0.10	**7**
			0.41	219.8	0.6	0(a)	13.7		263	954	244	440	138	470	0.12	0.00	5	3	**5**
O	milk,cow,skim,w/NFDM solids added	8 oz glass	245	91	12.3	0.00	8.8	**19**	316	529	693	421	586	81	0.00	0.39	0.00	0.10	**7**
			0.41	221.4	0.6	0(a)	12.3		238	857	220	394	122	421	0.12	0.00	5	2	**4**
P	milk,cow,sweetened condensed,canned	1 C	306	982	169.8	0.00	24.2	**37**	875	1466	1919	1169	1619	223	1.04	16.80	0.67	0.28	**18**
			0.33	83.1	26.6	0(a)	169.8		655	2371	606	1092	343	1169	6.70	0.06	104	8	**13**
Q	milk,cow,whole,dry	1/3 C	43	212	16.4	0.00	11.2	**25**	407	679	891	542	752	104	0.28	7.14	0.20	0.12	**8**
			2.34	1.1	11.4	0(a)	16.4		305	1100	282	507	158	542	2.64	0.04	41	4	**6**
R	milk,cow,whole,evaporated,canned	1 C	244	327	24.5	0.00	16.6	**37**	603	1005	1318	803	1113	154	0.59	11.20	0.41	0.10	**7**
			0.41	180.7	18.4	0(a)	24.5		451	1627	417	749	234	803	5.12	0.05	71	5	**8**
S	milk,cow,whole,low sodium	8 oz glass	244	149	10.9	0.00	7.6	**17**	273	459	600	366	505	71	0.32	5.25	0.20	0.05	**3**
			0.41	215.2	8.4	0(a)	10.9		205	742	190	342	107	366	2.12	0.06	34	unk.	**unk.**

Riboflavin mg / Niacin mg	% USRDA	Vitamin B_6 mg / Folacin mcg	% USRDA	Vitamin B_{12} mcg / Pantothenic acid mg	% USRDA	Biotin mg / Vitamin A IU	% USRDA	Preformed A RE / Beta carotene RE	Vitamin D IU / Vitamin E IU	% USRDA	Total tocopherol mg / Alpha tocopherol mg	Other tocopherol mg / Total ash g	Calcium mg / Phosphorus mg	% USRDA	Sodium mg / Sodium meq	Potassium mg / Potassium meq	Chlorine mg / Chlorine meq	Iron mg / Magnesium mg	% USRDA	Zinc mg / Copper mg	% USRDA	Iodine mcg / Selenium mcg	% USRDA	Manganese mcg / Chromium mcg	
0.39	23	0.10	5	0.88	15	unk.	unk.	unk.	unk.	unk.	unk.	unk.	300	30	122	381	unk.	0.1	1	1.0	6	unk.	unk.	unk.	A
0.2	1	12	3	0.79	8	500	10	unk.	unk.	unk.	unk.	1.81	234	23	5.3	9.7	unk.	34	9	unk.	unk.	unk.		unk.	
0.39	23	0.10	5	0.88	15	unk.	unk.	unk.	100	25	unk.	unk.	300	30	122	381	unk.	0.1	1	1.0	6	unk.	unk.	unk.	B
0.2	1	12	3	0.79	8	500	10	unk.	unk.	unk.	unk.	1.81	234	23	5.3	9.7	unk.	34	9	unk.	unk.	unk.		unk.	
0.47	28	0.12	6	1.03	17	unk.	unk.	unk.	unk.	unk.	unk.	unk.	349	35	143	443	unk.	0.1	1	1.1	7	unk.	unk.	unk.	C
0.2	1	15	4	0.92	9	499	10	unk.	unk.	unk.	unk.	2.12	273	27	6.2	11.3	unk.	39	10	unk.	unk.	unk.		unk.	
0.42	25	0.11	6	0.93	16	unk.	unk.	unk.	unk.	unk.	unk.	unk.	314	31	127	397	unk.	0.1	1	1.0	7	unk.	unk.	unk.	D
0.2	1	12	3	0.82	8	500	10	unk.	unk.	unk.	unk.	1.89	245	25	5.5	10.1	unk.	34	9	unk.	unk.	unk.		unk.	
0.39	23	0.10	5	0.88	15	unk.	unk.	unk.	unk.	unk.	unk.	unk.	298	30	122	376	unk.	0.1	1	1.0	6	unk.	unk.	unk.	E
0.2	1	12	3	0.78	8	500	10	unk.	unk.	unk.	unk.	1.81	232	23	5.3	9.6	unk.	34	9	unk.	unk.	unk.		unk.	
0.39	23	0.10	5	0.88	15	unk.	unk.	unk.	100	25	unk.	unk.	298	30	122	376	unk.	0.1	1	1.0	6	unk.	unk.	unk.	F
0.2	1	12	3	0.78	8	500	10	unk.	unk.	unk.	unk.	1.81	232	23	5.3	9.6	unk.	unk.	unk.	unk.	unk.	unk.		unk.	
0.47	28	0.13	6	1.03	17	unk.	unk.	unk.	unk.	unk.	unk.	unk.	352	35	145	448	unk.	0.1	1	1.1	7	unk.	unk.	unk.	G
0.2	1	15	4	0.92	9	499	10	unk.	unk.	unk.	unk.	2.14	276	28	6.3	11.4	unk.	39	10	unk.	unk.	unk.		unk.	
0.42	25	0.11	6	0.93	16	0.007	2	unk.	100	25	0.2	tr(a)	314	31	127	397	unk.	0.1	1	1.0	7	19.6	13	unk.	H
0.2	1	12	3	0.82	8	500	10	unk.	0.3	1	0.0	1.89	245	25	5.5	10.1	unk.	34	9	0.05	2	unk.		unk.	
0.39	23	0.08	4	0.90	15	unk.	unk.	unk.	unk.	unk.	unk.	unk.	279	28	124	387	unk.	0.1	0	1.0	7	unk.	unk.	unk.	I
0.2	1	11	3	0.73	7	537	11	unk.	unk.	unk.	unk.	1.82	223	22	5.4	9.9	unk.	27	7	unk.	unk.	unk.		unk.	
0.62	37	0.14	7	1.61	27	unk.	unk.	unk.	unk.	unk.	unk.	unk.	503	50	214	718	unk.	0.1	1	1.6	11	unk.	unk.	unk.	J
0.4	2	20	5	1.43	14	14	0	unk.	unk.	unk.	unk.	3.17	387	39	9.3	18.3	unk.	44	11	unk.	unk.	unk.		unk.	
0.77	45	0.14	7	0.59	10	unk.	unk.	unk.	unk.	unk.	unk.	unk.	742	74	294	850	unk.	0.7	4	2.3	15	unk.	unk.	unk.	K
0.4	2	23	6	1.89	19	1004	20	unk.	unk.	unk.	unk.	3.84	499	50	12.8	21.7	unk.	69	17	unk.	unk.	unk.		unk.	
0.34	20	0.10	5	0.91	15	0.005	2	unk.	100	25	0.0	0(a)	301	30	127	407	unk.	0.1	1	1.0	7	17.1	11	unk.	L
0.2	1	12	3	0.81	8	500	10	unk.	0.0	0	0.0	1.86	247	25	5.5	10.4	unk.	27	7	0.05	2	unk.		unk.	
0.34	20	0.10	5	0.91	15	0.005	2	unk.	100	25	unk.	unk.	301	30	127	407	unk.	0.1	1	1.0	7	17.1	11	unk.	M
0.2	1	12	3	0.81	8	500	10	unk.	unk.	unk.	unk.	1.86	247	25	5.5	10.4	unk.	27	7	0.05	2	unk.		unk.	
0.47	28	0.12	6	1.03	17	unk.	unk.	unk.	unk.	unk.	unk.	unk.	352	35	145	448	unk.	0.1	1	1.1	7	unk.	unk.	unk.	N
0.2	1	15	4	0.92	9	499	10	unk.	unk.	unk.	unk.	2.14	276	28	6.3	11.4	unk.	39	10	unk.	unk.	unk.		unk.	
0.42	25	0.11	6	0.93	16	unk.	unk.	unk.	unk.	unk.	unk.	unk.	316	32	130	419	unk.	0.1	1	1.0	7	unk.	unk.	unk.	O
0.2	1	12	3	0.83	8	500	10	unk.	unk.	unk.	unk.	1.91	255	26	5.6	10.7	unk.	37	9	unk.	unk.	unk.		unk.	
1.25	74	0.16	8	1.35	22	unk.	unk.	unk.	unk.	unk.	unk.	unk.	869	87	389	1135	unk.	0.6	3	2.9	19	unk.	unk.	unk.	P
0.6	3	34	8	2.29	23	1004	20	unk.	unk.	unk.	unk.	5.60	774	77	16.9	29.0	unk.	80	20	unk.	unk.	unk.		unk.	
0.51	30	0.13	6	1.39	23	unk.	unk.	unk.	unk.	unk.	unk.	unk.	389	39	158	568	unk.	0.2	1	1.4	10	unk.	unk.	unk.	Q
0.3	1	16	4	0.97.	10	393	8	unk.	unk.	unk.	unk.	2.59	331	33	6.9	14.5	unk.	36	9	unk.	unk.	unk.		unk.	
0.76	45	0.12	6	0.39	7	0.020	7	unk.	193	48	0.5	tr(a)	637	64	259	739	unk.	0.5	3	1.9	13	39.0	26	unk.	R
0.5	2	20	5	1.56	16	593	12	unk.	0.6	2	0.5	3.78	493	49	11.3	18.9	unk.	59	15	0.22	11	unk.		unk.	
0.24	14	0.08	4	0.85	14	unk.	unk.	unk.	unk.	unk.	unk.	unk.	246	25	5	617	unk.	unk.	unk.	unk.	unk.	unk.	unk.	unk.	S
0.1	1	unk.	unk.	0.74	7	317	6	unk.	unk.	unk.	unk.	1.90	210	21	0.2	15.8	unk.	12	3	unk.	unk.	unk.		unk.	

	FOOD	Portion	Weight in grams / Conversion for 100 g	Kilocalories / H_2O g	Total carbohydrate g / Total fats g	Crude fiber g / Dietary fiber g	Total protein g / Total sugar g	% USRDA	Arginine mg / Histidine mg	Isoleucine mg / Leucine mg	Lysine mg / Methionine mg	Phenylalanine mg / Threonine mg	Valine mg / Tryptophan mg	Cystine mg / Tyrosine mg	Polyunsat. fatty acids g / Monounsat. fatty acids g	Saturated fatty acids g / P/S ratio	Linoleic acid g / Cholesterol mg	Thiamin mg / Ascorbic acid mg	% USRDA
A	milk,cow,whole,3.3% fat	8 oz glass	244	149	11.4	0.00	8.0	**18**	290	486	637	388	537	73	0.29	5.08	0.20	0.07	**5**
			0.41	214.7	8.1	0(a)	11.4		217	786	203	364	112	388	2.05	0.06	34	2	**4**
B	milk,cow,whole,3.7% fat	8 oz glass	244	156	11.3	0.00	8.0	**18**	290	483	634	386	537	73	0.34	5.56	0.20	0.07	**5**
			0.41	214.0	8.9	0(a)	11.3		217	783	200	361	112	386	2.24	0.06	34	4	**6**
C	MILK,GOAT,whole fluid	8 oz glass	244	168	10.9	0.00	8.7	**19**	290	505	708	378	586	112	0.37	6.51	0.27	0.10	**7**
			0.41	212.4	10.1	0(a)	10.9		217	766	195	398	107	437	2.39	0.06	27	3	**5**
D	MILK,HUMAN,whole mature fluid	8 oz glass	246	172	16.9	0.00	2.5	**6**	106	138	167	113	155	47	1.23	4.94	0.91	0.02	**2**
			0.41	215.2	10.8	0(a)	16.9		57	234	52	113	42	130	3.64	0.25	34	12	**21**
E	MISO	1 Tbsp	17	26	3.3	0.31	2.3	**5**	20	26	20	13	17	unk.	unk.	unk.	unk.	0.01	**0**
			5.88	8.3	0.8	unk.	unk.		3	27	5	10	2	10	unk.	unk.	unk.	unk.	**unk.**
F	MOLASSES	1 Tbsp	20	46	12.8	0.00	0.0	**0**	0(a)	0(a)	0(a)	0(a)	0(a)	0(a)	0.00	0.00	0.00	0.02	**1**
			5.00	4.8	0.0	tr	12.8		0(a)	0(a)	0(a)	0(a)	0(a)	0(a)	0.00	0.00	0	0	**0**
G	molasses,blackstrap	1 Tbsp	20	43	11.0	0.00	0.0	**0**	0(a)	0(a)	0(a)	0(a)	0(a)	0(a)	0.00	0.00	0.00	0.02	**2**
			5.00	4.8	0.0	0(a)	11.0		0(a)	0(a)	0(a)	0(a)	0(a)	0(a)	0.00	0.00	0	0	**0**
H	MONOSODIUM GLUTAMATE-Accent	1 tsp	5.0	14	0.0	0(a)	2.3	**4**	unk.	unk.	unk.	unk.	unk.	unk.	0(a)	0(a)	0(a)	unk.	**unk.**
			20.00	0.0	0.0	unk.	0.0		unk.	unk.	unk.	unk.	unk.	unk.	0(a)	0.00	0	unk.	**unk.**
I	MUFFIN,home recipe,apple	1 average	45	137	17.2	0.05	2.5	**5**	46	124	105	131	129	46	3.32	1.17	3.25	0.08	**5**
			2.22	18.3	6.5	0.63	5.5		59	206	47	89	32	97	1.64	2.83	21	0	**0**
J	muffin,home recipe,blueberry	1 average	55	147	18.5	0.52	2.8	**5**	49	131	111	140	137	49	3.54	1.25	3.46	0.08	**5**
			1.82	16.3	7.0	1.32	5.7		63	218	50	95	35	103	1.75	2.82	22	3	**4**
K	muffin,home recipe,bran	1 average	40	104	17.2	0.72	3.1	**6**	unk.	unk.	unk.	unk.	unk.	unk.	unk.	2.00	0.40	0.06	**4**
			2.50	14.0	3.9	3.20	unk.		unk.	unk.	unk.	unk.	unk.	unk.	1.20	unk.	21	tr	**0**
L	muffin,home recipe,corn	1 average	45	169	18.1	0.09	3.1	**6**	81	156	153	154	174	54	0.32	2.50	0.83	0.07	**5**
			2.22	13.9	9.4	unk.	4.9		79	294	70	126	37	143	5.47	0.13	40	tr	**0**
M	muffin,home recipe,orange	1 average	45	137	18.6	0.06	2.7	**5**	50	115	95	129	123	51	3.00	0.95	2.94	0.10	**7**
			2.22	15.7	5.8	0.70	5.9		59	192	48	87	32	93	1.53	3.16	33	8	**13**
N	muffin,home recipe,plain	1 average	45	158	19.1	0.05	3.0	**5**	55	148	125	157	154	56	3.98	1.41	3.90	0.09	**6**
			2.22	14.6	7.8	0.53	5.5		71	245	57	107	39	115	1.97	2.82	25	tr	**0**
O	muffin,home recipe,whole wheat	1 average	48	123	21.1	0.39	4.4	**8**	69	117	126	113	129	38	0.20	1.35	0.17	0.13	**9**
			2.08	18.2	2.8	1.78	4.3		58	193	51	91	30	94	0.70	0.15	31	tr	**0**
P	MUSHROOMS, button (*Aqaricus*	1/2 C	135	26	5.0	unk.	0.3	**1**	21	78	16	7	55	3	unk.	unk.	unk.	0.01	**1**
	campestris), canned,drained solids		*0.74*	unk.	0.2	3.47	tr		4	41	25	10	1	7	unk.	unk.	unk.	tr	**1**
Q	mushrooms, button (*Aqaricus campestris*),	1/2 C	35	10	1.5	0.28	1.6	**3**	94	328	71	33	234	13	unk.	unk.	unk.	0.03	**2**
	raw,sliced or chopped		*2.86*	31.6	0.1	0.90	unk.		20	174	104	43	4	31	unk.	unk.	unk.	unk.	**unk.**
R	mushrooms, button (*Aqariaus campestris*),	1 average	13	4	0.6	0.10	0.9	**2**	56	209	42	20	148	8	0.00	0.00	unk.	0.03	**2**
	raw,whole		*7.69*	11.8	tr	0.33	tr		12	111	66	26	3	18	unk.	0.00	0(a)	1	**2**
S	mushrooms (*Lactarius*),fresh,	1/2 C	135	150	5.4	1.35	3.3	**5**	194	225	99	19	131	131	0.00	0.00	unk.	0.11	**7**
	sauteed/fried		*0.74*	unk.	14.3	3.47	0.1		42	158	235	177	6	63	unk.	0.00	0(a)	tr	**0**

Riboflavin mg / Niacin mg	% USRDA	Vitamin B_6 mg / Folacin mcg	% USRDA	Vitamin B_{12} mcg / Pantothenic acid mg	% USRDA	Biotin mg / Vitamin A IU	% USRDA	Preformed A RE / Beta carotene RE	Vitamin D IU / Vitamin E IU	% USRDA	Total tocopherol mg / Alpha tocopherol mg	Other tocopherol mg / Total ash g	Calcium mg / Phosphorus mg	% USRDA	Sodium mg / Sodium meq	Potassium mg / Potassium meq	Chlorine mg / Chlorine meq	Iron mg / Magnesium mg	% USRDA	Zinc mg / Copper mg	% USRDA	Iodine mcg / Selenium mcg	% USRDA	Manganese mcg / Chromium mcg	
0.39	23	0.10	5	0.85	14	unk.	unk.	unk.	100	25	0.2	tr(a)	290	29	120	371	unk.	0.1	1	0.9	6	17.1	11	unk.	A
0.2	1	12	3	0.77	8	307	6	unk.	0.3	1	0.2	1.76	227	23	5.2	9.5	unk.	32	8	0.37	18	unk.		unk.	
0.39	23	0.10	5	0.85	14	unk.	unk.	unk.	unk.	unk.	0.2	tr(a)	290	29	120	368	unk.	0.1	1	0.9	6	17.1	11	unk.	B
0.2	1	12	3	0.76	8	337	7	unk.	0.3	1	0.2	1.76	227	23	5.2	9.4	unk.	32	8	0.37	18	unk.		unk.	
0.32	19	0.11	6	0.15	2	unk.	unk.	unk.	unk.	unk.	unk.	unk.	327	33	122	498	unk.	0.1	1	0.7	5	unk.	unk.	unk.	C
0.7	3	2	1	0.76	8	451	9	unk.	unk.	unk.	unk.	2.00	271	27	5.3	12.7	unk.	34	9	unk.	unk.	unk.		unk.	
0.07	4	0.03	1	0.10	2	unk.	unk.	unk.	unk.	unk.	unk.	unk.	79	8	42	125	unk.	0.1	0	0.4	3	unk.	unk.	unk.	D
0.4	2	12	3	0.55	6	593	12	unk.	unk.	unk.	unk.	0.49	34	3	1.8	3.2	unk.	7	2	unk.	unk.	unk.		unk.	
0.02	1	unk.	unk.	unk.	unk.	unk.	unk.	unk.	unk.	unk.	unk.	unk.	15	2	697	unk.	unk.	0.7	4	unk.	unk.	unk.	unk.	unk.	E
0.3	1	unk.	unk.	unk.	unk.	unk.	unk.	unk.	unk.	unk.	unk.	2.18	27	3	30.3	unk.	unk.	unk.	unk.	unk.	unk.	unk.		unk.	
0.02	1	0.04	2	0.00	0	0.002	1	0	0	0	tr(a)	tr(a)	58	6	7	213	100	1.2	7	unk.	unk.	unk.	unk.	unk.	F
0.2	1	2	1	0.07	1	0	0	0	0.0	0	tr(a)	1.70	14	1	0.3	5.4	2.8	16	4	0.04	2	unk.		unk.	
0.04	2	0.04	2	0.00	0	0.002	1	0	0	0	unk.	unk.	137	14	19	585	100	3.2	18	unk.	unk.	unk.	unk.	unk.	G
0.4	2	2	1	0.07	1	0	0	0	unk.	unk.	unk.	2.10	17	2	0.8	15.0	2.8	42	11	0.04	2	5		unk.	
unk.	unk.	unk.	unk.	0(a)	0	unk.	unk.	0(a)	0(a)	0	unk.	unk.	unk.	unk.	615	unk.	5	unk.	unk.	unk.	unk.	unk.	unk.	unk.	H
unk.	unk.	unk.	unk.	unk.	unk.	unk.	unk.	unk.	unk.	unk.	unk.	1.38	unk.	unk.	26.8	unk.	0.1	unk.	unk.	unk.	unk.	unk.		unk.	
0.08	5	0.02	1	0.09	2	0.001	0	0	7	2	4.8	unk.	34	3	86	177	17	0.5	3	0.2	2	3.7	3	71	I
0.5	3	6	1	0.17	2	33	1	1	5.8	19	0.1	0.65	55	6	3.8	4.5	0.5	16	4	0.06	3	8		12	
0.08	5	0.03	1	0.10	2	0.001	0	0	7	2	5.1	0.0	39	4	225	64	20	0.6	3	0.3	2	38.4	26	275	J
0.6	3	7	2	0.20	2	51	1	2	6.1	20	0.1	1.06	61	6	9.8	1.6	0.6	19	5	0.07	4	9		13	
0.10	6	unk.	unk.	unk.	unk.	unk.	unk.	15	unk.	unk.	unk.	unk.	57	6	179	172	unk.	1.5	8	unk.	unk.	unk.	unk.	unk.	K
1.6	8	unk.	unk.	unk.	unk.	92	2	4	unk.	unk.	unk.	1.72	162	16	7.8	4.4	unk.	unk.	unk.	unk.	unk.	unk.		unk.	
0.08	5	0.02	1	0.15	2	0.001	1	0	10	3	0.1	0.0	50	5	263	48	19	0.6	3	0.3	2	39.1	26	39	L
0.4	2	5	1	0.20	2	70	1	3	0.1	0	tr	1.31	89	9	11.5	1.2	0.5	24	6	0.05	2	5		7	
0.07	4	0.04	2	0.12	2	0.001	1	0	4	1	4.2	0.0	14	1	191	59	23	0.6	4	0.2	2	33.5	22	75	M
0.8	4	17	4	0.19	2	92	2	3	5.1	17	0.0	0.66	42	4	8.3	1.5	0.6	23	6	0.05	3	9		14	
0.09	5	0.02	1	0.11	2	0.001	0	0	8	2	5.7	0.0	42	4	254	50	21	0.6	4	0.3	2	43.2	29	83	N
0.6	3	6	2	0.20	2	31	1	0	6.8	23	0.1	1.13	66	7	11.0	1.3	0.6	19	5	0.07	3	10		14	
0.10	6	0.18	9	0.14	2	0.003	1	0	12	3	0.1	tr(a)	67	7	331	154	34	1.1	6	0.6	4	62.4	42	33	O
1.0	5	11	3	0.37	4	78	2	0	0.1	0	tr	1.91	126	13	14.4	3.9	1.0	40	10	0.06	3	4		5	
0.06	4	0.02	1	0.00	0	0.002	1	0	unk.	unk.	unk.	unk.	1	0	2	54	3	0.1	1	unk.	unk.	unk.	unk.	unk.	P
0.5	3	3	1	0.29	3	tr	0	tr	unk.	unk.	unk.	0.12	15	2	0.1	1.4	0.1	2	0	0.01	0	unk.		unk.	
0.30	18	0.08	4	0.00	0	0.010	3	0	0(a)	0	unk.	unk.	9	1	unk.	unk.	unk.	0.9	5	unk.	unk.	unk.	unk.	tr(a)	Q
2.4	12	5	1	tr	0	0	0	0	unk.	unk.	unk.	unk.	109	11	unk.	unk.	unk.	9	2	tr(a)	0	tr(a)		unk.	
0.16	10	0.04	2	0.00	0	0.006	2	0	0(a)	0	0.0	0(a)	2	0	5	145	9	0.3	2	0.5	3	unk.	unk.	unk.	R
1.5	7	8	2	0.77	8	tr	0	tr	0.0	0	0(a)	0.31	41	4	0.2	3.7	0.3	5	1	0.02	1	tr(a)		unk.	
0.53	31	0.08	4	0	0	0.009	3	0	0(a)	0	unk.	0.0	15	2	unk.	unk.	unk.	1.3	8	0.5	4	unk.	unk.	unk.	S
5.6	28	unk.	unk.	1.48	15	333	7	33	unk.	unk.	0.0	2.16	157	16	unk.	unk.	unk.	unk.	unk.	0.35	18	tr(a)		unk.	

	FOOD	Portion	Weight in grams / Conversion for 100 g	Kilocalories / H_2O g	Total carbohydrate g / Total fats g	Crude fiber g / Dietary fiber g	Total protein g / Total sugar g	% USRDA	Arginine mg / Histidine mg	Isoleucine mg / Leucine mg	Lysine mg / Methionine mg	Phenylalanine mg / Threonine mg	Valine mg / Tryptophan mg	Cystine mg / Tyrosine mg	Polyunsat. fatty acids g / Monounsat. fatty acids g	Saturated fatty acids g / P/S ratio	Linoleic acid g / Cholesterol mg	Thiamin mg / Ascorbic acid mg	% USRDA
A	mushrooms (*Lactarius*),fresh,	1 med	18	19	0.7	0.17	0.4	**1**	25	29	13	3	17	17	0.00	0.00	unk.	0.01	**1**
	sauteed/fried		*5.71*	unk.	1.9	0.45	unk.		5	20	30	23	1	8	unk.	0.00	0(a)	tr	**0**
B	MUSTARD GREENS,fresh,leaves	1/2 C	70	16	2.8	0.63	1.5	**2**	unk.	51	76	49	72	unk.	unk.	tr(a)	unk.	0.06	**4**
	w/o stems,cooked,drained solids		*1.43*	64.8	0.3	unk.	0.2		unk.	41	15	40	24	unk.	unk.	unk.	0	34	**56**
C	mustard greens,frozen,chopped,cooked,	1/2 C	75	15	2.3	0.75	1.6	**3**	unk.	unk.	unk.	unk.	unk.	unk.	unk.	tr(a)	unk.	0.02	**2**
	drained solids		*1.33*	70.3	0.3	unk.	0.2		unk.	unk.	unk.	unk.	unk.	unk.	unk.	unk.	0	15	**25**
D	MUSTARD seed,yellow	1 tsp	3.7	17	1.3	0.24	0.9	**1**	65	40	56	39	49	22	0.20	0.05	0.10	0.02	**1**
			27.03	0.3	1.1	0(a)	0.0		28	66	18	41	19	28	0.22	3.69	0	unk.	**unk.**
E	mustard,ground,yellow	1 tsp	1.5	9	0.3	0.03	0.5	**1**	unk.	unk.	unk.	unk.	unk.	unk.	unk.	unk.	unk.	0.01	**1**
			66.67	0.0	0.6	unk.	0.0		unk.	unk.	unk.	unk.	unk.	unk.	unk.	unk.	0	tr	**1**
F	mustard,low sodium,prep,yellow	1 Tbsp	15	10	1.0	unk.	0.7	**1**	unk.	unk.	unk.	unk.	unk.	unk.	unk.	unk.	unk.	unk.	**unk.**
			6.67	unk.	0.4	unk.	0.0		unk.	unk.	unk.	unk.	unk.	unk.	unk.	unk.	0(a)	unk.	**unk.**
G	mustard,regular,prep,yellow	1 Tbsp	16	12	1.0	0.16	0.7	**1**	unk.	unk.	unk.	unk.	unk.	unk.	unk.	unk.	0(a)	unk.	**unk.**
			6.41	12.5	0.7	unk.	0.0		unk.	unk.	unk.	unk.	unk.	unk.	unk.	unk.	unk.	unk.	**unk.**
H	NECTARINES,fresh,any,edible portion	1 average	150	96	25.6	0.60	0.9	**1**	unk.	unk.	unk.	unk.	unk.	unk.	unk.	tr(a)	unk.	unk.	**unk.**
			0.67	122.7	tr	3.30	11.4		unk.	unk.	unk.	unk.	unk.	unk.	unk.	unk.	0	20	**33**
I	NOODLE pudding,home recipe	1/2 C	92	132	11.4	0.10	5.7	**12**	167	279	328	262	308	65	0.38	4.12	0.28	0.07	**5**
			1.09	66.3	7.0	unk.	1.9		151	456	130	220	57	221	1.80	0.09	27	1	**2**
J	noodles romanoff,prep-Betty Crocker	1 serving	73	253	23.0	unk.	7.0	**8**	unk.	unk.	unk.	unk.	unk.	unk.	unk.	unk.	unk.	0.23	**15**
			1.37	unk.	12.0	unk.	unk.		unk.	unk.	unk.	unk.	unk.	unk.	unk.	unk.	unk.	tr	**0**
K	noodles stroganoff,prep-Betty Crocker	1 serving	91	324	25.7	unk.	5.9	**8**	unk.	unk.	unk.	unk.	unk.	unk.	unk.	unk.	unk.	0.22	**15**
			1.10	unk.	10.9	unk.	unk.		unk.	unk.	unk.	unk.	unk.	unk.	unk.	unk.	unk.	tr	**0**
L	noodles,chow mein,canned	1/2 C	23	228	13.0	tr(a)	3.0	**5**	unk.	137	67	163	128	59	unk.	unk.	unk.	unk.	**unk.**
			4.44	0.2	5.3	unk.	0.3		59	229	4	85	36	101	unk.	unk.	3	0(a)	**0**
M	noodles,egg,cooked (any shape)	1/2 C	80	100	18.6	0.08	3.3	**7**	unk.	162	110	158	194	64	unk.	0.50	tr	0.11	**8**
			1.25	56.3	1.2	unk.	0.5		78	218	56	138	37	82	0.50	unk.	25	0	**0**
N	NUTMEG,ground	1 tsp	2.3	12	1.1	0.09	0.1	**0**	unk.	unk.	unk.	unk.	unk.	unk.	0.01	0.60	0.01	0.01	**1**
			43.48	0.1	0.8	unk.	0.0		unk.	unk.	unk.	unk.	unk.	unk.	3.66	0.01	0	unk.	**unk.**
O	NUTS,mixed	1/4 C	35	219	6.3	unk.	5.8	**9**	unk.	428	277	331	557	unk.	unk.	9.17	9.20	0.21	**14**
			2.86	unk.	20.7	unk.	0.0		unk.	531	123	258	165	unk.	18.37	unk.	0	tr	**0**
P	OIL,almond	1 C	218	1927	0.0	0.00	0.0	**0**	0(a)	0(a)	0(a)	0(a)	0(a)	0(a)	37.93	17.88	37.93	0(a)	**0**
			0.46	0.0	218.0	0(a)	0(a)		0(a)	0(a)	0(a)	0(a)	0(a)	0(a)	151.29	2.12	0(a)	0(a)	**0**
Q	oil,apricot kernel	1 C	218	1927	0.0	0.00	0.0	**0**	0(a)	0(a)	0(a)	0(a)	0(a)	0(a)	63.87	13.73	63.87	0(a)	**0**
			0.46	0.0	218.0	0(a)	0(a)		0(a)	0(a)	0(a)	0(a)	0(a)	0(a)	127.53	4.65	0(a)	0(a)	**0**
R	oil,babassu	1 C	218	1927	0.0	0.00	0.0	**0**	0(a)	0(a)	0(a)	0(a)	0(a)	0(a)	3.49	177.02	3.49	0(a)	**0**
			0.46	0.0	218.0	0(a)	0(a)		0(a)	0(a)	0(a)	0(a)	0(a)	0(a)	24.85	0.02	0(a)	0(a)	**0**
S	oil,coconut	1 C	218	1927	0.0	0.00	0.0	**0**	0(a)	0(a)	0(a)	0(a)	0(a)	0(a)	3.92	188.57	3.92	0(a)	**0**
			0.46	0.0	218.0	0(a)	0(a)		0(a)	0(a)	0(a)	0(a)	0(a)	0(a)	12.64	0.02	0(a)	0(a)	**0**

Riboflavin mg / Niacin mg	% USRDA	Vitamin B_6 mg / Folacin mcg	% USRDA	Vitamin B_{12} mcg / Pantothenic acid mg	% USRDA	Biotin mg / Vitamin A IU	% USRDA	Preformed A RE / Beta carotene RE	Vitamin D IU / Vitamin E IU	% USRDA	Total tocopherol mg / Alpha tocopherol mg	Other tocopherol mg / Total ash g	Calcium mg / Phosphorus mg	% USRDA	Sodium mg / Sodium meq	Potassium mg / Potassium meq	Chlorine mg / Chlorine meq	Iron mg / Magnesium mg	% USRDA	Zinc mg / Copper mg	% USRDA	Iodine mcg / Selenium mcg	% USRDA	Manganese mcg / Chromium mcg	
0.07	**4**	0.01	**1**	0.00	**0**	0.001	**0**	0	0(a)	**0**	0.0	0.0	2	**0**	unk.	unk.	unk.	0.2	**1**	0.1	**1**	unk.	**unk.**	unk.	A
0.7	**4**	1	**0**	0.19	**2**	43	**1**	4	0.0	**0**	0.0	0.28	20	**2**	unk.	unk.	unk.	unk.	**unk.**	0.05	**2**	tr(a)		unk.	
0.10	**6**	0.09	**5**	0.00	**0**	0.001	**0**	0	0(a)	**0**	1.2	unk.	97	**10**	13	154	unk.	1.3	**7**	0.1	**1**	unk.	**unk.**	unk.	B
0.4	**2**	6	**1**	0.14	**1**	4060	**81**	406	1.4	**5**	unk.	0.56	22	**2**	0.5	3.9	unk.	12	**3**	0.06	**3**	tr(a)		unk.	
0.07	**4**	0.10	**5**	0.00	**0**	0.001	**0**	0	0(a)	**0**	1.3	unk.	78	**8**	8	118	unk.	1.1	**6**	0.1	**1**	unk.	**unk.**	unk.	C
0.3	**2**	6	**2**	0.15	**2**	4500	**90**	450	1.5	**5**	unk.	0.38	32	**3**	0.3	3.0	unk.	13	**3**	0.07	**3**	tr(a)		unk.	
0.01	**1**	unk.	**unk.**	0.00	**0**	unk.	**unk.**	0(a)	0(a)	**0**	unk.	unk.	19	**2**	tr	25	unk.	0.4	**2**	0.2	**1**	unk.	**unk.**	unk.	D
0.3	**2**	unk.	**unk.**	unk.	**unk.**	2	**0**	unk.	unk.	**unk.**	unk.	0.17	31	**3**	0.0	0.6	unk.	11	**3**	unk.	**unk.**	unk.		unk.	
0.01	**0**	unk.	**unk.**	0(a)	**0**	unk.	**unk.**	0(a)	0(a)	**0**	unk.	unk.	4	**0**	tr	10	unk.	0.1	**1**	unk.	**unk.**	unk.	**unk.**	unk.	E
0.1	**1**	unk.	**unk.**	unk.	**unk.**	3	**0**	tr	unk.	**unk.**	unk.	0.06	12	**1**	0.0	0.3	unk.	unk.	**unk.**	unk.	**unk.**	unk.		unk.	
unk.	**unk.**	unk.	**unk.**	0(a)	**0**	unk.	**unk.**	0(a)	0(a)	**0**	unk.	unk.	unk.	**unk.**	2	11	unk.	0.3	**2**	0.1	**1**	unk.	**unk.**	unk.	F
unk.	**unk.**	1	**0**	unk.	**unk.**	unk.	**unk.**	unk.	unk.	**unk.**	unk.	unk.	unk.	**unk.**	0.1	0.3	unk.	unk.	**unk.**	unk.	**unk.**	unk.		unk.	
unk.	**unk.**	unk.	**unk.**	0(a)	**0**	unk.	**unk.**	0(a)	0(a)	**0**	0.6	0.4	13	**1**	195	20	unk.	0.3	**2**	0.1	**1**	unk.	**unk.**	unk.	G
unk.	**unk.**	1	**0**	unk.	**unk.**	unk.	**unk.**	unk.	0.8	**3**	0.3	0.67	11	**1**	8.5	0.5	unk.	7	**2**	unk.	**unk.**	unk.		unk.	
unk.	**unk.**	0.03	**1**	0.00	**0**	unk.	**unk.**	0	0(a)	**0**	unk.	unk.	6	**1**	9	441	8	0.8	**4**	0.1	**1**	unk.	**unk.**	57	H
unk.	**unk.**	11	**3**	unk.	**unk.**	2475	**50**	248	unk.	**unk.**	unk.	0.75	36	**4**	0.4	11.3	0.2	9	**2**	0.21	**11**	unk.		unk.	
0.10	**6**	0.05	**3**	0.22	**4**	0.001	**0**	6	tr	**0**	0.2	0.0	51	**5**	222	81	1	0.7	**4**	0.5	**3**	23.3	**16**	3	I
0.5	**3**	8	**2**	0.21	**2**	260	**5**	1	0.2	**1**	tr	1.09	82	**8**	9.7	2.1	0.0	56	**14**	0.03	**1**	tr		tr	
0.17	**10**	unk.	**unk.**	unk.	**unk.**	unk.	**unk.**	unk.	unk.	**unk.**	unk.	unk.	151	**16**	1059	193	unk.	0.7	**4**	unk.	**unk.**	unk.	**unk.**	unk.	J
1.6	**8**	unk.	**unk.**	unk.	**unk.**	400	**7**	unk.	unk.	**unk.**	unk.	unk.	310	**31**	46.0	4.9	unk.	unk.	**unk.**	unk.	**unk.**	unk.		unk.	
0.14	**8**	unk.	**unk.**	unk.	**unk.**	unk.	**unk.**	unk.	unk.	**unk.**	unk.	unk.	114	**11**	1033	260	unk.	0.7	**4**	unk.	**unk.**	unk.	**unk.**	unk.	K
1.6	**8**	unk.	**unk.**	unk.	**unk.**	400	**9**	unk.	unk.	**unk.**	unk.	unk.	223	**22**	44.9	6.6	unk.	unk.	**unk.**	unk.	**unk.**	unk.		unk.	
unk.	**unk.**	unk.	**unk.**	unk.	**unk.**	unk.	**unk.**	0(a)	0(a)	**0**	unk.	unk.	unk.	**unk.**	unk.	unk.	unk.	unk.	**unk.**	unk.	**unk.**	unk.	**unk.**	unk.	L
unk.	**unk.**	unk.	**unk.**	unk.	**unk.**	unk.	**unk.**	unk.	unk.	**unk.**	unk.	0.94	unk.	**unk.**	unk.	unk.	unk.	unk.	**unk.**	unk.	**unk.**	unk.		unk.	
0.06	**4**	0.05	**3**	tr	**0**	0.002	**1**	12	1	**0**	0.3	unk.	8	**1**	2	35	unk.	1.2	**7**	0.6	**4**	unk.	**unk.**	unk.	M
1.0	**5**	3	**1**	0.16	**2**	56	**1**	2	unk.	**unk.**	unk.	0.56	47	**5**	0.1	0.9	unk.	101	**25**	0.04	**2**	unk.		unk.	
tr	**0**	unk.	**unk.**	0.00	**0**	unk.	**unk.**	0(a)	0(a)	**0**	unk.	unk.	4	**0**	tr	8	unk.	0.1	**0**	0.0	**0**	unk.	**unk.**	unk.	N
tr	**0**	unk.	**unk.**	unk.	**unk.**	2	**0**	unk.	unk.	**unk.**	unk.	0.05	5	**1**	0.0	0.2	unk.	4	**1**	unk.	**unk.**	unk.		unk.	
0.05	**3**	unk.	**unk.**	0(a)	**0**	unk.	**unk.**	0	0	**0**	unk.	unk.	33	**3**	5	196	unk.	1.2	**7**	unk.	**unk.**	unk.	**unk.**	unk.	O
1.4	**7**	unk.	**unk.**	unk.	**unk.**	7	**0**	1	unk.	**unk.**	unk.	unk.	156	**16**	0.2	5.0	unk.	unk.	**unk.**	unk.	**unk.**	unk.		unk.	
0(a)	**0**	0(a)	**0**	0(a)	**0**	0(a)	**0**	unk.	unk.	**unk.**	87.4	2.0	unk.	**unk.**	unk.	unk.	unk.	unk.	**unk.**	unk.	**unk.**	unk.	**unk.**	unk.	P
0(a)	**0**	0(a)	**0**	0(a)	**0**	unk.	**unk.**	unk.	85.5	**285**	85.5	0.00	unk.	**unk.**	unk.	unk.	unk.	unk.	**unk.**	unk.	**unk.**	unk.		unk.	
0(a)	**0**	0(a)	**0**	0(a)	**0**	0(a)	**0**	unk.	unk.	**unk.**	110.1	101.4	unk.	**unk.**	unk.	unk.	unk.	unk.	**unk.**	unk.	**unk.**	unk.	**unk.**	unk.	Q
0(a)	**0**	0(a)	**0**	0(a)	**0**	unk.	**unk.**	unk.	8.7	**29**	8.7	0.00	unk.	**unk.**	unk.	unk.	unk.	unk.	**unk.**	unk.	**unk.**	unk.		unk.	
0(a)	**0**	0(a)	**0**	0(a)	**0**	0(a)	**0**	unk.	unk.	**unk.**	unk.	unk.	unk.	**unk.**	unk.	unk.	unk.	unk.	**unk.**	unk.	**unk.**	unk.	**unk.**	unk.	R
0(a)	**0**	0(a)	**0**	0(a)	**0**	unk.	**unk.**	unk.	unk.	**unk.**	unk.	0.00	unk.	**unk.**	unk.	unk.	unk.	unk.	**unk.**	unk.	**unk.**	unk.		unk.	
0(a)	**0**	0(a)	**0**	0(a)	**0**	0(a)	**0**	unk.	unk.	**unk.**	7.8	7.0	unk.	**unk.**	unk.	unk.	unk.	0.1	**1**	unk.	**unk.**	unk.	**unk.**	unk.	S
0(a)	**0**	0(a)	**0**	0(a)	**0**	unk.	**unk.**	unk.	0.9	**3**	0.9	0.00	0	**0**	unk.	unk.	unk.	unk.	**unk.**	unk.	**unk.**	unk.		unk.	

	FOOD	Portion	Weight in grams / Conversion for 100 g	Kilocalories / H_2O g	Total carbohydrate g / Total fats g	Crude fiber g / Dietary fiber g	Total protein g / Total sugar g	% USRDA	Arginine mg / Histidine mg	Isoleucine mg / Leucine mg	Lysine mg / Methionine mg	Phenylalanine mg / Threonine mg	Valine mg / Tryptophan mg	Cystine mg / Tyrosine mg	Polyunsat. fatty acids g / Monounsat. fatty acids g	Saturated fatty acids g / P/S ratio	Linoleic acid g / Cholesterol mg	Thiamin mg / Ascorbic acid mg	% USRDA
A	oil,corn	1 C	218	1927	0.0	0.00	0.0	**0**	0	0(a)	0(a)	0(a)	0(a)	0(a)	127.97	27.69	126.44	0.00	**0**
			0.46	0.0	218.0	0.00	0.0		0(a)	0(a)	0(a)	0(a)	0(a)	0(a)	52.76	4.62	0	0	**0**
B	oil,cottonseed	1 C	218	1927	0.0	0.00	0.0	**0**	0	0(a)	0(a)	0(a)	0(a)	0(a)	113.14	56.46	112.27	0.00	**0**
			0.46	0.0	218.0	0.00	0.0		0(a)	0(a)	0(a)	0(a)	0(a)	0(a)	37.06	2.00	0	0	**0**
C	oil,cupu assu	1 C	218	1927	0.0	0.00	0.0	**0**	0(a)	0(a)	0(a)	0(a)	0(a)	0(a)	8.28	115.98	8.28	0(a)	**0**
			0.46	0.0	218.0	0(a)	0(a)		0(a)	0(a)	0(a)	0(a)	0(a)	0(a)	84.37	0.07	0(a)	0(a)	**0**
D	oil,grapeseed	1 C	218	1927	0.0	0.00	0.0	**0**	0(a)	0(a)	0(a)	0(a)	0(a)	0(a)	152.38	20.93	151.73	0(a)	**0**
			0.46	0.0	218.0	0(a)	0(a)		0(a)	0(a)	0(a)	0(a)	0(a)	0(a)	34.44	7.28	0(a)	0(a)	**0**
E	oil,hazelnut	1 C	218	1927	0.0	0.00	0.0	**0**	0(a)	0(a)	0(a)	0(a)	0(a)	0(a)	22.24	16.13	22.02	0(a)	**0**
			0.46	0.0	218.0	0(a)	0(a)		0(a)	0(a)	0(a)	0(a)	0(a)	0(a)	169.60	1.38	0(a)	0(a)	**0**
F	oil,linseed	1 C	182	1609	0.0	0.00	0.0	**0**	0(a)	0(a)	0(a)	0(a)	0(a)	0(a)	120.12	17.11	23.11	0(a)	**0**
			0.55	0.0	182.0	0(a)	0(a)		0(a)	0(a)	0(a)	0(a)	0(a)	0(a)	36.76	7.02	0(a)	0(a)	**0**
G	oil,nutmeg butter	1 C	218	1927	0.0	0.00	0.0	**0**	0(a)	0(a)	0(a)	0(a)	0(a)	0(a)	0.00	196.20	0(a)	0(a)	**0**
			0.46	0.0	218.0	0(a)	0(a)		0(a)	0(a)	0(a)	0(a)	0(a)	0(a)	10.46	0.00	0(a)	0(a)	**0**
H	oil,olive	1 C	218	1927	0.0	0.00	0.0	**0**	0	0(a)	0(a)	0(a)	0(a)	0(a)	18.31	29.43	17.22	0.00	**0**
			0.46	0.0	218.0	0.00	0.0		0(a)	0(a)	0(a)	0(a)	0(a)	0(a)	158.05	0.62	0	0	**0**
I	oil,palm	1 C	218	1927	0.0	0.00	0.0	**0**	0(a)	0(a)	0(a)	0(a)	0(a)	0(a)	20.27	107.47	19.84	0(a)	**0**
			0.46	0.0	218.0	0(a)	0(a)		0(a)	0(a)	0(a)	0(a)	0(a)	0(a)	79.79	0.19	0(a)	0(a)	**0**
J	oil,palm kernel	1 C	218	1927	0.0	0.00	0.0	**0**	0(a)	0(a)	0(a)	0(a)	0(a)	0(a)	3.49	177.45	3.49	0(a)	**0**
			0.46	0.0	218.0	0(a)	0(a)		0(a)	0(a)	0(a)	0(a)	0(a)	0(a)	24.85	0.02	0(a)	0(a)	**0**
K	oil,peanut	1 C	218	1927	0.0	0.00	0.0	**0**	0	0(a)	0(a)	0(a)	0(a)	0(a)	69.76	36.84	69.76	0.00	**0**
			0.46	0.0	218.0	0.00	0.0		0(a)	0(a)	0(a)	0(a)	0(a)	0(a)	97.66	1.89	0	0	**0**
L	oil,poppyseed	1 C	218	1927	0.0	0.00	0.0	**0**	0(a)	0(a)	0(a)	0(a)	0(a)	0(a)	136.03	29.43	136.03	0(a)	**0**
			0.46	0.0	218.0	0(a)	0(a)		0(a)	0(a)	0(a)	0(a)	0(a)	0(a)	42.95	4.62	0(a)	0(a)	**0**
M	oil,rapeseed	1 C	218	1927	0.0	0.00	0.0	**0**	0(a)	0(a)	0(a)	0(a)	0(a)	0(a)	49.05	10.90	27.90	0(a)	**0**
			0.46	0.0	218.0	0(a)	0(a)		0(a)	0(a)	0(a)	0(a)	0(a)	0(a)	24.42	4.50	0(a)	0(a)	**0**
N	oil,rice bran	1 C	218	1927	0.0	0.00	0.0	**0**	0(a)	0(a)	0(a)	0(a)	0(a)	0(a)	76.30	42.95	72.81	0(a)	**0**
			0.46	0.0	218.0	0(a)	0(a)		0(a)	0(a)	0(a)	0(a)	0(a)	0(a)	85.24	1.78	0(a)	0(a)	**0**
O	oil,safflower (linoleic variety)	1 C	218	1927	0.0	0.00	0.0	**0**	0(a)	0(a)	0(a)	0(a)	0(a)	0(a)	30.96	13.30	30.96	0(a)	**0**
			0.46	0.0	218.0	0(a)	0(a)		0(a)	0(a)	0(a)	0(a)	0(a)	0(a)	164.15	2.33	0(a)	0(a)	**0**
P	oil,safflower (oleic variety)	1 C	218	1927	0.0	0.00	0.0	**0**	0(a)	0(a)	0(a)	0(a)	0(a)	0(a)	162.41	19.84	161.54	0.00	**0**
			0.46	0.0	218.0	0.00	0.0		0(a)	0(a)	0(a)	0(a)	0(a)	0(a)	25.51	8.19	0	0	**0**
Q	oil,sesame	1 C	218	1927	0.0	0.00	0.0	**0**	0(a)	0(a)	0(a)	0(a)	0(a)	0(a)	90.91	30.96	90.03	0.00	**0**
			0.46	0.0	218.0	0.00	0.0		0(a)	0(a)	0(a)	0(a)	0(a)	0(a)	85.67	2.94	0	0	**0**
R	oil,sheanut	1 C	218	1927	0.0	0.00	0.0	**0**	0(a)	0(a)	0(a)	0(a)	0(a)	0(a)	11.34	101.59	10.68	0(a)	**0**
			0.46	0.0	218.0	0(a)	0(a)		0(a)	0(a)	0(a)	0(a)	0(a)	0(a)	94.83	0.11	0(a)	0(a)	**0**
S	oil,soybean	1 C	218	1927	0.0	0.00	0.0	**0**	0(a)	0(a)	0(a)	0(a)	0(a)	0(a)	126.22	31.39	111.18	0.00	**0**
			0.46	0.0	218.0	0.00	0.0		0(a)	0(a)	0(a)	0(a)	0(a)	0(a)	49.70	4.02	0	0	**0**

Riboflavin mg / Niacin mg	% USRDA	Vitamin B6 mg / Folacin mcg	% USRDA	Vitamin B_{12} mcg / Pantothenic acid mg	% USRDA	Biotin mg / Vitamin A IU	% USRDA	Preformed A RE / Beta carotene RE	Vitamin D IU / Vitamin E IU	% USRDA	Total tocopherol mg / Alpha tocopherol mg	Other tocopherol mg / Total ash g	Calcium mg / Phosphorus mg	% USRDA	Sodium mg / Sodium meq	Potassium mg / Potassium meq	Chlorine mg / Chlorine meq	Iron mg / Magnesium mg	% USRDA	Zinc mg / Copper mg	% USRDA	Iodine mcg / Selenium mcg	% USRDA	Manganese mcg / Chromium mcg	
0.00	0	0(a)	0	0(a)	0	0(a)	0	0(a)	unk.	unk.	159.6	135.2	0	0	0	0	tr(a)	0.0	0	0.4	3	tr(a)	0	218	A
0.0	0	0(a)	0	0(a)	0	0	0	0	24.4	81	24.4	0.00	0	0	0.0	0.0	0.0	1	0	0.48	24	tr(a)		26	
0.00	0	0(a)	0	0(a)	0	0(a)	0	0	unk.	unk.	142.1	65.2	0	0	0	0	tr(a)	0.0	0	0.4	3	tr(a)	0	545	B
0.0	0	0(a)	0	0(a)	0	0	0	0	77.0	257	77.0	0.00	0	0	0.0	0.0	0.0	tr	0	0.28	14	tr(a)		tr(a)	
0(a)	0	0(a)	0	0(a)	0	0(a)	0	unk.	unk.	unk.	unk.	unk.	unk.	unk.	unk.	unk.	unk.	unk.	unk.	unk.	unk.	unk.	unk.	unk.	C
0(a)	0	0(a)	0	0(a)	0	unk.	unk.	unk.	unk.	unk.	unk.	0.00	unk.	unk.	unk.	unk.	unk.	unk.	unk.	unk.	unk.	unk.		unk.	
0(a)	0	0(a)	0	0(a)	0	0(a)	0	unk.	unk.	unk.	unk.	unk.	unk.	unk.	unk.	unk.	unk.	unk.	unk.	unk.	unk.	unk.	unk.	unk.	D
0(a)	0	0(a)	0	0(a)	0	unk.	unk.	unk.	unk.	unk.	unk.	0.00	unk.	unk.	unk.	unk.	unk.	unk.	unk.	unk.	unk.	unk.		unk.	
0(a)	0	0(a)	0	0(a)	0	0(a)	0	unk.	unk.	unk.	unk.	unk.	unk.	unk.	unk.	unk.	unk.	unk.	unk.	unk.	unk.	unk.	unk.	unk.	E
0(a)	0	0(a)	0	0(a)	0	unk.	unk.	unk.	unk.	unk.	unk.	0.00	unk.	unk.	unk.	unk.	unk.	unk.	unk.	unk.	unk.	unk.		unk.	
0(a)	0	0(a)	0	0(a)	0	0(a)	0	unk.	unk.	unk.	unk.	unk.	tr	0	unk.	unk.	unk.	unk.	unk.	unk.	unk.	unk.	unk.	unk.	F
0(a)	0	0(a)	0	0(a)	0	unk.	unk.	unk.	unk.	unk.	unk.	0.00	0	0	unk.	unk.	unk.	tr	0	unk.	unk.	unk.		unk.	
0(a)	0	0(a)	0	0(a)	0	0(a)	0	unk.	unk.	unk.	unk.	unk.	unk.	unk.	unk.	unk.	unk.	unk.	unk.	unk.	unk.	unk.	unk.	unk.	G
0(a)	0	0(a)	0	0(a)	0	unk.	unk.	unk.	unk.	unk.	unk.	0.00	unk.	unk.	unk.	unk.	unk.	unk.	unk.	unk.	unk.	unk.		unk.	
0.00	0	0(a)	0	0(a)	0	0(a)	0	0	unk.	unk.	27.5	1.5	tr	0	tr	0	tr(a)	0.8	5	0.1	1	tr(a)	0	tr(a)	H
0.0	0	0(a)	0	0(a)	0	0	0	0	27.5	92	25.9	0.00	3	0	0.0	0.0	0.0	0	0	0.70	35	tr(a)		tr(a)	
0(a)	0	0(a)	0	0(a)	0	0(a)	0	unk.	unk.	unk.	83.7	42.1	unk.	unk.	unk.	unk.	unk.	tr	0	unk.	unk.	unk.	unk.	unk.	I
0(a)	0	0(a)	0	0(a)	0	unk.	unk.	unk.	41.6	139	41.6	0.00	0	0	unk.	unk.	unk.	unk.	unk.	unk.	unk.	unk.		unk.	
0(a)	0	0(a)	0	0(a)	0	0(a)	0	unk.	unk.	unk.	13.5	unk.	unk.	unk.	unk.	unk.	unk.	unk.	unk.	unk.	unk.	unk.	unk.	unk.	J
0(a)	0	0(a)	0	0(a)	0	unk.	unk.	unk.	unk.	unk.	unk.	0.00	unk.	unk.	unk.	unk.	unk.	unk.	unk.	unk.	unk.	unk.		unk.	
0.00	0	0(a)	0	0(a)	0	tr(a)	0	0	0(a)	0	54.5	29.2	tr	0	tr	0	tr(a)	0.1	0	tr	0	unk.	unk.	218	K
0.0	0	tr(a)	0	tr(a)	0	0	0	0	25.3	84	25.3	0.00	0	0	0.0	0.0	0.0	tr	0	0.18	9	tr(a)		tr(a)	
0(a)	0	0(a)	0	0(a)	0	0(a)	0	unk.	unk.	unk.	unk.	unk.	unk.	unk.	unk.	unk.	unk.	unk.	unk.	unk.	unk.	unk.	unk.	unk.	L
0(a)	0	0(a)	0	0(a)	0	unk.	unk.	unk.	unk.	unk.	unk.	0.00	unk.	unk.	unk.	unk.	unk.	unk.	unk.	unk.	unk.	unk.		unk.	
0(a)	0	0(a)	0	0(a)	0	0(a)	0	unk.	unk.	unk.	unk.	unk.	unk.	unk.	unk.	unk.	unk.	unk.	unk.	unk.	unk.	unk.	unk.	unk.	M
0(a)	0	0(a)	0	0(a)	0	unk.	unk.	unk.	unk.	unk.	unk.	0.00	unk.	unk.	unk.	unk.	unk.	unk.	unk.	unk.	unk.	unk.		unk.	
0(a)	0	0(a)	0	0(a)	0	0(a)	0	unk.	unk.	unk.	84.6	14.2	unk.	unk.	unk.	unk.	unk.	0.2	1	unk.	unk.	unk.	unk.	unk.	N
0(a)	0	0(a)	0	0(a)	0	unk.	unk.	unk.	70.4	235	70.4	0.00	unk.	unk.	unk.	unk.	unk.	unk.	unk.	unk.	unk.	unk.		unk.	
0(a)	0	0(a)	0	0(a)	0	0(a)	0	unk.	unk.	unk.	unk.	unk.	unk.	unk.	unk.	unk.	unk.	unk.	unk.	unk.	unk.	unk.	unk.	unk.	O
0(a)	0	0(a)	0	0(a)	0	unk.	unk.	unk.	unk.	unk.	unk.	0.00	unk.	unk.	unk.	unk.	unk.	unk.	unk.	unk.	unk.	unk.		unk.	
0.00	0	0(a)	0	0(a)	0	0(a)	0	0	0(a)	0	83.1	8.7	0	0	0	0	tr(a)	0.0	0	0.4	3	tr(a)	0	0	P
0.0	0	0(a)	0	0(a)	0	0	0	0	74.3	248	74.3	0.00	0	0	0.0	0.0	0.0	tr(a)	0	tr(a)	0	tr(a)		tr(a)	
0.00	0	0(a)	0	0(a)	0	0(a)	0	0	0(a)	0	63.4	60.4	0	0	0	0	tr(a)	0.0	0	0.4	3	tr(a)	0	tr(a)	Q
0.0	0	0(a)	0	0(a)	0	0	0	0	3.1	10	3.1	0.00	0	0	0.0	0.0	0.0	tr(a)	0	tr(a)	0	tr(a)		tr(a)	
0(a)	0	0(a)	0	0(a)	0	0(a)	0	unk.	unk.	unk.	unk.	unk.	unk.	unk.	unk.	unk.	unk.	unk.	unk.	unk.	unk.	unk.	unk.	unk.	R
0(a)	0	0(a)	0	0(a)	0	unk.	unk.	unk.	unk.	unk.	unk.	0.00	unk.	unk.	unk.	unk.	unk.	unk.	unk.	unk.	unk.	unk.		unk.	
0.00	0	0(a)	0	0(a)	0	tr(a)	0	0	0(a)	0	204.3	180.3	tr	0	0	0	tr(a)	tr	0	0.0	0	tr(a)	0	tr(a)	S
0.0	0	0(a)	0	tr(a)	0	0	0	0	24.0	80	24.0	0.00	1	0	0.0	0.0	0.0	tr	0	tr(a)	0	tr(a)		tr(a)	

	FOOD	Portion	Weight in grams / *Conversion for 100 g*	Kilocalories / H_2O g	Total carbohydrate g / Total fats g	Crude fiber g / Dietary fiber g	Total protein g / Total sugar g	**% USRDA**	Arginine mg / Histidine mg	Isoleucine mg / Leucine mg	Lysine mg / Methionine mg	Phenylalanine mg / Threonine mg	Valine mg / Tryptophan mg	Cystine mg / Tyrosine mg	Polyunsat. fatty acids g / Monounsat. fatty acids g	Saturated fatty acids g / P/S ratio	Linoleic acid g / Cholesterol mg	Thiamin mg / Ascorbic acid mg	**% USRDA**
A	oil,soybean lecithin	1 C	218	1927	0.0	0.00	0.0	**0**	0(a)	0(a)	0(a)	0(a)	0(a)	0(a)	98.32	33.35	86.98	0(a)	**0**
			0.46	0.0	218.0	0(a)	0(a)		0(a)	0(a)	0(a)	0(a)	0(a)	0(a)	22.89	2.95	0(a)	0(a)	**0**
B	oil,sunflower (H)	1 C	218	1927	0.0	0.00	0.0	**0**	0(a)	0(a)	0(a)	0(a)	0(a)	0(a)	79.35	28.34	76.95	0(a)	**0**
			0.46	0.0	218.0	0(a)	0(a)		0(a)	0(a)	0(a)	0(a)	0(a)	0(a)	100.28	2.80	0(a)	0(a)	**0**
C	oil,sunflower,linoleic less than 60%	1 C	218	1927	0.0	0.00	0.0	**0**	0(a)	0(a)	0(a)	0(a)	0(a)	0(a)	87.42	22.02	86.76	0(a)	**0**
			0.46	0.0	218.0	0(a)	0(a)		0(a)	0(a)	0(a)	0(a)	0(a)	0(a)	98.75	3.97	0(a)	0(a)	**0**
D	oil,sunflower,linoleic 60% & over	1 C	218	1927	0.0	0.00	0.0	**0**	0(a)	0(a)	0(a)	0(a)	0(a)	0(a)	143.23	22.45	143.23	0(a)	**0**
			0.46	0.0	218.0	0(a)	0(a)		0(a)	0(a)	0(a)	0(a)	0(a)	0(a)	42.51	6.38	0(a)	0(a)	**0**
E	oil,tea seed	1 C	218	1927	0.0	0.00	0.0	**0**	0(a)	0(a)	0(a)	0(a)	0(a)	0(a)	50.14	46.00	48.40	0(a)	**0**
			0.46	0.0	218.0	0(a)	0(a)		0(a)	0(a)	0(a)	0(a)	0(a)	0(a)	108.78	1.09	0(a)	0(a)	**0**
F	oil,tomato seed	1 C	218	1927	0.0	0.00	0.0	**0**	0(a)	0(a)	0(a)	0(a)	0(a)	0(a)	115.76	42.95	110.74	0(a)	**0**
			0.46	0.0	218.0	0(a)	0(a)		0(a)	0(a)	0(a)	0(a)	0(a)	0(a)	47.74	2.70	0(a)	0(a)	**0**
G	oil,ucuhuba butter	1 C	218	1927	0.0	0.00	0.0	**0**	0(a)	0(a)	0(a)	0(a)	0(a)	0(a)	6.32	185.74	6.32	0(a)	**0**
			0.46	0.0	218.0	0(a)	0(a)		0(a)	0(a)	0(a)	0(a)	0(a)	0(a)	14.61	0.03	0(a)	0(a)	**0**
H	oil,walnut	1 C	218	1927	0.0	0.00	0.0	**0**	0(a)	0(a)	0(a)	0(a)	0(a)	0(a)	137.99	19.84	115.32	0(a)	**0**
			0.46	0.0	218.0	0(a)	0(a)		0(a)	0(a)	0(a)	0(a)	0(a)	0(a)	48.40	6.96	0(a)	0(a)	**0**
I	oil,wheat germ	1 C	218	1927	0.0	0.00	0.0	**0**	0(a)	0(a)	0(a)	0(a)	0(a)	0(a)	134.51	40.98	119.46	0(a)	**0**
			0.46	0.0	218.0	0(a)	0(a)		0(a)	0(a)	0(a)	0(a)	0(a)	0(a)	31.83	3.28	0(a)	0(a)	**0**
J	OKRA,fresh,cooked,drained solids	1/2 C	80	23	4.8	0.80	0.6	**17**	unk.	61	67	58	80	10	unk.	tr(a)	unk.	0.10	**7**
			1.25	72.9	0.2	2.60	2.2		11	90	19	59	16	17	unk.	unk.	0	16	**27**
K	okra,fresh,fried,home recipe	1/2 C	116	183	14.0	0.84	4.7	**8**	98	212	218	217	250	67	6.17	2.00	6.18	0.14	**9**
			0.86	83.3	12.6	2.60	0.8		82	341	90	183	57	121	3.51	3.09	63	16	**27**
L	okra,frozen,cooked,drained solids	1/2 C	83	31	7.3	0.82	0.6	**18**	unk.	63	69	59	82	10	unk.	tr(a)	unk.	0.12	**8**
			1.21	72.8	0.1	2.68	2.3		14	92	20	61	16	17	unk.	unk.	0	10	**17**
M	OLIVE,green	1 med	3.9	5	0.1	0.05	0.0	**0**	unk.	unk.	unk.	unk.	unk.	unk.	unk.	0.05	0.03	tr	**0**
			25.64	3.0	0.5	0.20	0.0		unk.	unk.	unk.	unk.	unk.	unk.	0.37	unk.	0	unk.	**unk.**
N	olive,ripe	1 med	3.9	7	0.1	0.06	0.0	**0**	unk.	2	1	2	2	unk.	unk.	0.09	0.05	tr	**0**
			25.64	2.8	0.8	0.20	0.0		1	3	1	2	unk.	1	0.60	unk.	0	unk.	**unk.**
O	olive,ripe,salt-cured (Greek)	1 med	2.6	9	0.2	0.10	0.1	**0**	unk.	unk.	unk.	unk.	unk.	unk.	unk.	0.10	0.06	tr	**0**
			38.46	1.1	0.9	0.13	0.0		unk.	unk.	unk.	unk.	unk.	unk.	0.71	unk.	0	unk.	**unk.**
P	ONION POWDER	1 tsp	2.2	8	1.8	0.13	0.2	**0**	29	6	10	5	5	4	unk.	unk.	unk.	0.01	**1**
			45.45	0.1	tr	unk.	0.0		3	7	2	4	3	5	unk.	unk.	0	tr	**1**
Q	ONION,green,raw,bulb & top	1 average	21	8	1.7	0.26	0.3	**1**	41	unk.	unk.	unk.	unk.	unk.	unk.	unk.	unk.	0.01	**1**
			4.69	19.0	tr	0.53	0.5		unk.	unk.	unk.	unk.	unk.	unk.	unk.	unk.	unk.	7	**11**
R	onion,rings,canned	1/2 C	25	90	2.5	unk.	0.4	**1**	unk.	unk.	unk.	unk.	unk.	unk.	unk.	unk.	unk.	unk.	**unk.**
			4.00	unk.	8.3	unk.	unk.		unk.	unk.	unk.	unk.	unk.	unk.	unk.	unk.	unk.	unk.	**unk.**
S	onion,rings,fresh,French fried	1/2 C	60	175	18.0	unk.	2.7	**4**	unk.	unk.	unk.	unk.	unk.	unk.	unk.	unk.	unk.	0.02	**2**
			1.67	unk.	10.5	unk.	unk.		unk.	unk.	unk.	unk.	unk.	unk.	unk.	unk.	unk.	3	**5**

Riboflavin mg / Niacin mg	% USRDA	Vitamin B6 mg / Folacin mcg	% USRDA	Vitamin B_{12} mcg / Pantothenic acid mg	% USRDA	Biotin mg / Vitamin A IU	% USRDA	Preformed A RE / Beta carotene RE	Vitamin D IU / Vitamin E IU	% USRDA	Total tocopherol mg / Alpha tocopherol mg	Other tocopherol mg / Total ash g	Calcium mg / Phosphorus mg	% USRDA	Sodium mg / Sodium meq	Potassium mg / Potassium meq	Chlorine mg / Chlorine meq	Iron mg / Magnesium mg	% USRDA	Zinc mg / Copper mg	% USRDA	Iodine mcg / Selenium mcg	% USRDA	Manganese mcg / Chromium mcg	
0(a)	0	0(a)	0	0(a)	0	0(a)	0	unk.	unk.	unk.	unk.	unk.	unk.	unk.	unk.	unk.	unk.	unk.	unk.	unk.	unk.	unk.	unk.	unk.	A
0(a)	0	0(a)	0	0(a)	0	unk.	unk.	unk.	unk.	unk.	unk.	unk.	unk.	unk.	unk.	unk.	unk.	unk.	unk.	unk.	unk.	unk.		unk.	
0(a)	0	0(a)	0	0(a)	0	0(a)	0	unk.	unk.	unk.	unk.	unk.	unk.	unk.	unk.	unk.	unk.	unk.	unk.	unk.	unk.	unk.	unk.	unk.	B
0(a)	0	0(a)	0	0(a)	0	unk.	unk.	unk.	unk.	unk.	unk.	0.00	unk.	unk.	unk.	unk.	unk.	unk.	unk.	unk.	unk.	unk.		unk.	
0(a)	0	0(a)	0	0(a)	0	0(a)	0	unk.	unk.	unk.	unk.	unk.	tr	0	tr	unk.	unk.	0.1	0	unk.	unk.	0.4	0	unk.	C
0(a)	0	0(a)	0	0(a)	0	unk.	unk.	unk.	unk.	unk.	unk.	0.00	unk.	unk.	0.0	unk.	unk.	unk.	unk.	unk.	unk.	unk.		unk.	
0(a)	0	0(a)	0	0(a)	0	0(a)	0	unk.	unk.	unk.	104.2	6.3	unk.	unk.	unk.	unk.	unk.	unk.	unk.	unk.	unk.	unk.	unk.	unk.	D
0(a)	0	0(a)	0	0(a)	0	unk.	unk.	unk.	97.9	326	97.9	0.00	unk.	unk.	unk.	unk.	unk.	unk.	unk.	unk.	unk.	unk.		unk.	
0(a)	0	0(a)	0	0(a)	0	0(a)	0	unk.	unk.	unk.	unk.	unk.	unk.	unk.	unk.	unk.	unk.	unk.	unk.	unk.	unk.	unk.	unk.	unk.	E
0(a)	0	0(a)	0	0(a)	0	unk.	unk.	unk.	unk.	unk.	unk.	0.00	unk.	unk.	unk.	unk.	unk.	unk.	unk.	unk.	unk.	unk.		unk.	
0(a)	0	0(a)	0	0(a)	0	0(a)	0	unk.	unk.	unk.	129.3	121.0	unk.	unk.	unk.	unk.	unk.	unk.	unk.	unk.	unk.	unk.	unk.	unk.	F
0(a)	0	0(a)	0	0(a)	0	unk.	unk.	unk.	8.3	28	8.3	0.00	unk.	unk.	unk.	unk.	unk.	unk.	unk.	unk.	unk.	unk.		unk.	
0(a)	0	0(a)	0	0(a)	0	0(a)	0	unk.	unk.	unk.	unk.	unk.	unk.	unk.	unk.	unk.	unk.	unk.	unk.	unk.	unk.	unk.	unk.	unk.	G
0(a)	0	0(a)	0	0(a)	0	unk.	unk.	unk.	unk.	unk.	unk.	0.00	unk.	unk.	unk.	unk.	unk.	unk.	unk.	unk.	unk.	unk.		unk.	
0(a)	0	0(a)	0	0(a)	0	0(a)	0	unk.	unk.	unk.	70.0	69.1	unk.	unk.	unk.	unk.	unk.	unk.	unk.	unk.	unk.	unk.	unk.	unk.	H
0(a)	0	0(a)	0	0(a)	0	unk.	unk.	unk.	0.9	3	0.9	0.00	unk.	unk.	unk.	unk.	unk.	unk.	unk.	unk.	unk.	unk.		unk.	
0(a)	0	0(a)	0	0(a)	0	0(a)	0	unk.	unk.	unk.	555.0	229.3	unk.	unk.	unk.	unk.	unk.	unk.	unk.	8.3	55	unk.	unk.	unk.	I
0(a)	0	0(a)	0	0(a)	0	unk.	unk.	unk.	325.7	1086	325.7	0.00	unk.	unk.	unk.	unk.	unk.	unk.	unk.	unk.	unk.	unk.		unk.	
0.14	9	0.03	2	0.00	0	unk.	unk.	0	0(a)	0	unk.	unk.	74	7	2	139	unk.	0.4	2	unk.	unk.	unk.	unk.	unk.	J
0.7	4	unk.	unk.	0.10	1	392	8	39	unk.	unk.	unk.	0.48	33	3	0.1	3.6	unk.	unk.	unk.	tr(a)	0	tr(a)		unk.	
0.22	13	0.06	3	0.14	2	0.002	1	0	5	1	9.0	0(a)	98	10	383	176	22	1.1	6	0.3	2	79.2	53	15	K
1.2	6	7	2	0.35	4	420	8	39	10.8	36	0.1	1.66	82	8	16.6	4.5	0.6	36	9	0.07	4	0		1	
0.14	8	0.03	2	0.00	0	unk.	unk.	0	0(a)	0	unk.	unk.	78	8	2	135	unk.	0.4	2	unk.	unk.	unk.	unk.	unk.	L
0.8	4	unk.	unk.	0.11	1	396	8	40	unk.	unk.	unk.	0.49	36	4	0.1	3.5	unk.	unk.	unk.	tr(a)	0	tr(a)		unk.	
tr(a)	0	unk.	unk.	0.00	0	tr(a)	0	0	0	0	unk.	unk.	2	0	94	2	73	0.1	0	0.0	0	unk.	unk.	47	M
tr(a)	0	unk.	unk.	tr	0	12	0	1	unk.	unk.	unk.	0.25	1	0	4.1	0.0	2.1	1	0	0.06	3	unk.		unk.	
tr	0	tr	0	0.00	0	unk.	unk.	0	0	0	unk.	unk.	4	0	29	1	unk.	0.1	0	tr	0	unk.	unk.	unk.	N
unk.	unk.	unk.	unk.	tr	0	3	0	tr	unk.	unk.	unk.	0.10	1	0	1.3	0.0	unk.	unk.	unk.	unk.	unk.	unk.		unk.	
tr	0	0.00	0	0.00	0	unk.	unk.	0(a)	0(a)	0	unk.	unk.	unk.	unk.	85	unk.	unk.	unk.	unk.	tr	0	unk.	unk.	unk.	O
unk.	unk.	unk.	unk.	0.00	0	unk.	unk.	unk.	unk.	unk.	unk.	0.25	1	0	3.7	unk.	unk.	unk.	unk.	unk.	unk.	unk.		unk.	
tr	0	unk.	unk.	0.00	0	unk.	unk.	0(a)	0(a)	0	unk.	unk.	8	1	1	21	unk.	0.1	0	0.1	0	unk.	unk.	unk.	P
tr	0	unk.	unk.	unk.	unk.	tr	0	unk.	unk.	unk.	unk.	0.07	8	1	0.0	0.5	unk.	3	1	unk.	unk.	unk.		unk.	
0.01	1	unk.	unk.	0.00	0	unk.	unk.	0	0(a)	0	unk.	unk.	11	1	1	49	unk.	0.2	1	0.1	0	unk.	unk.	tr(a)	Q
0.1	0	8	2	0.03	0	426	9	43	unk.	unk.	unk.	0.15	8	1	0.0	1.3	unk.	unk.	unk.	tr(a)	0	tr(a)		unk.	
unk.	unk.	unk.	unk.	0(a)	0	unk.	unk.	0	0(a)	0	unk.	unk.	16	2	112	unk.	unk.	0.1	1	unk.	unk.	unk.	unk.	unk.	R
unk.	unk.	unk.	unk.	unk.	unk.	0	0	unk.	unk.	unk.	unk.	unk.	15	2	4.9	unk.	unk.	unk.	unk.	tr(a)	0	tr(a)		unk.	
0.03	2	unk.	unk.	0(a)	0	unk.	unk.	0	0(a)	0	3.8	3.4	24	2	437	87	unk.	0.3	2	0.2	1	unk.	unk.	unk.	S
0.2	1	unk.	unk.	unk.	unk.	26	1	3	4.5	15	0.4	unk.	42	4	19.0	2.2	unk.	unk.	unk.	tr(a)	0	tr(a)		unk.	

	FOOD	Portion	Weight in grams / *Conversion for 100 g*	Kilocalories / H_2O g	Total carbohydrate g / Total fats g	Crude fiber g / Dietary fiber g	Total protein g / Total sugar g	% USRDA	Arginine mg / Histidine mg	Isoleucine mg / Leucine mg	Lysine mg / Methionine mg	Phenylalanine mg / Threonine mg	Valine mg / Tryptophan mg	Cystine mg / Tyrosine mg	Polyunsat. fatty acids g / Monounsat. fatty acids g	Saturated fatty acids g / P/S ratio	Linoleic acid g / Cholesterol mg	Thiamin mg / Ascorbic acid mg	% USRDA
A	onion,white,cooked,	1/2 C	105	30	6.8	0.63	1.3	**2**	161	19	58	36	27	unk.	unk.	unk.	unk.	0.03	**2**
	drained solids,chopped		*0.95*	96.4	0.1	0.86	4.3		12	33	12	20	19	40	unk.	unk.	unk.	7	**12**
B	onion,white,cooked,	1 med	50	15	3.3	0.30	0.6	**1**	77	9	28	17	13	unk.	unk.	unk.	unk.	0.01	**1**
	drained solids,whole		*2.00*	45.9	0.0	0.41	2.0		6	16	6	10	9	19	unk.	unk.	unk.	4	**6**
C	onion,white,raw,chopped	1/2 C	85	32	7.4	0.51	1.3	**2**	163	18	54	32	26	unk.	unk.	unk.	unk.	0.03	**2**
			1.18	75.7	0.1	0.70	3.5		12	31	11	19	18	36	unk.	unk.	unk.	8	**14**
D	onion,white,raw,whole	1 med	100	38	8.7	0.60	1.5	**2**	192	22	unk.	unk.	unk.	unk.	unk.	unk.	unk.	0.03	**2**
			1.00	89.1	0.1	0.82	4.1		unk.	unk.	unk.	24	22	42	unk.	unk.	unk.	10	**17**
E	onion,yellow,cooked,	1/2 C	105	30	6.8	0.63	1.3	**2**	161	19	58	36	27	unk.	unk.	unk.	unk.	0.03	**2**
	drained solids,chopped		*0.95*	96.4	0.1	0.86	4.3		12	33	12	20	19	40	unk.	unk.	unk.	7	**12**
F	onion,yellow,cooked,	1 med	50	15	3.3	0.30	0.6	**1**	77	9	28	17	13	unk.	unk.	unk.	unk.	0.01	**1**
	drained solids,whole		*2.00*	45.9	0.1	0.41	2.0		6	16	6	10	9	19	unk.	unk.	unk.	4	**6**
G	onion,yellow,raw,chopped	1/2 C	85	32	7.4	0.51	1.3	**2**	163	18	54	32	26	unk.	unk.	unk.	unk.	0.03	**2**
			1.18	75.7	0.1	0.70	3.5		12	31	11	19	18	36	unk.	unk.	unk.	8	**14**
H	onion,yellow,raw,whole	1 med	100	38	8.7	0.60	1.5	**2**	192	21	64	38	31	unk.	unk.	unk.	unk.	0.03	**2**
			1.00	89.1	0.1	0.82	4.1		14	37	13	22	21	42	unk.	unk.	unk.	10	**17**
I	ORANGE peel,raw,grated	1 tsp	2.0	unk.	0.5	unk.	0.0	**0**	unk.	unk.	unk.	unk.	unk.	unk.	tr(a)	tr(a)	tr(a)	tr	**0**
			50.00	1.4	0.0	unk.	0.0		unk.	unk.	unk.	unk.	unk.	unk.	tr(a)	0.00	0(a)	3	**5**
J	orange,fresh,any,edible portion	1 average	131	64	16.0	0.65	1.3	**2**	84	38	69	48	50	17	unk.	tr(a)	unk.	0.13	**9**
			0.76	112.7	0.3	2.62	12.1		19	36	19	19	8	27	unk.	unk.	0	65	**109**
K	orange,fresh,any,sections	1/2 C	90	44	11.0	0.45	0.9	**1**	58	26	48	33	35	12	unk.	tr(a)	unk.	0.09	**6**
			1.11	77.4	0.2	1.80	8.3		13	24	13	13	6	19	unk.	unk.	0	45	**75**
L	OREGANO,dried	1 tsp	1.5	5	1.0	0.22	0.2	**0**	unk.	unk.	unk.	unk.	unk.	unk.	0.08	0.04	0.02	0.01	**0**
			66.67	0.1	0.2	unk.	0.0		unk.	unk.	unk.	unk.	unk.	unk.	0.01	1.97	0	unk.	**unk.**
M	ORGAN MEATS,chitterlings,cooked	1/2 C	63	211	unk.	0(a)	5.4	**12**	372	194	422	226	291	69	unk.	6.30	1.89	unk.	**unk.**
			1.59	unk.	16.2	unk.	unk.		106	288	122	251	59	144	unk.	unk.	unk.	unk.	**unk.**
N	organ meats,giblets,chicken,	from average	75	208	3.3	0.00	24.4	**54**	1619	1223	1761	1110	1303	328	2.53	2.85	2.02	0.07	**5**
	broiler/fryer,flour coated,fried	chicken	*1.33*	35.9	10.1	0(a)	tr(a)		569	1951	608	1100	280	800	3.09	0.89	335	7	**11**
O	organ meats,giblets,chicken,	1/2 C	73	114	0.7	0.00	18.7	**42**	1252	940	1367	848	1000	249	0.78	1.08	0.53	0.07	**4**
	broiler/fryer,stewed,chopped		*1.38*	49.0	3.5	0(a)	unk.		436	1498	468	850	214	615	0.74	0.72	285	6	**10**
P	organ meats,giblets,chicken,roasting,	1/2 C	73	120	0.6	0.00	19.4	**43**	1312	966	1410	866	1010	257	0.88	1.19	0.60	0.05	**3**
	stewed,chopped		*1.38*	48.1	3.8	0(a)	unk.		444	1524	489	882	214	631	0.82	0.74	259	5	**8**
Q	organ meats,giblets,chicken,stewing,	1/2 C	73	141	0.1	0.00	18.6	**42**	1250	934	1359	841	988	248	1.37	1.93	1.04	0.07	**4**
	stewed,chopped		*1.38*	46.4	6.7	0(a)	unk.		432	1483	468	846	211	610	1.93	0.71	257	4	**7**
R	organ meats,giblets,turkey,stewed,	1/2 C	73	121	1.5	0.00	19.3	**43**	1282	971	1414	875	1037	257	0.84	1.12	0.58	0.04	**2**
	chopped		*1.38*	47.4	3.7	0(a)	unk.		453	1551	480	873	223	636	0.67	0.75	303	1	**2**
S	organ meats,gizzard,chicken,stewed	1 average	20	31	0.2	0.00	5.4	**12**	390	256	375	226	243	71	0.21	0.21	0.15	0.01	**0**
			5.00	13.5	0.7	0(a)	unk.		109	381	142	250	49	165	0.16	1.02	39	tr	**1**

Riboflavin mg / Niacin mg	% USRDA	Vitamin B_6 mg / Folacin mcg	% USRDA	Vitamin B_{12} mcg / Pantothenic acid mg	% USRDA	Biotin mg / Vitamin A IU	% USRDA	Preformed A RE / Beta carotene RE	Vitamin D IU / Vitamin E IU	% USRDA	Total tocopherol mg / Alpha tocopherol mg	Other tocopherol mg / Total ash g	Calcium mg / Phosphorus mg	% USRDA	Sodium mg / Sodium meq	Potassium mg / Potassium meq	Chlorine mg / Chlorine meq	Iron mg / Magnesium mg	% USRDA	Zinc mg / Copper mg	% USRDA	Iodine mcg / Selenium mcg	% USRDA	Manganese mcg / Chromium mcg	
0.03	2	unk.	unk.	0(a)	0	unk.	unk.	0	0(a)	0	unk.	unk.	25	3	7	115	unk.	0.4	2	unk.	unk.	unk.	unk.	tr(a)	A
0.2	1	unk.	unk.	unk.	unk.	tr	0	tr	unk.	unk.	unk.	0.42	30	3	0.3	2.9	unk.	unk.	unk.	tr(a)	0	tr(a)		unk.	
0.01	1	unk.	unk.	0(a)	0	unk.	unk.	0	0(a)	0	unk.	unk.	12	1	4	55	unk.	0.2	1	unk.	unk.	unk.	unk.	tr(a)	B
0.1	1	unk.	unk.	unk.	unk.	tr	0	tr	unk.	unk.	unk.	0.20	15	1	0.1	1.4	unk.	unk.	unk.	tr(a)	0	tr(a)		unk.	
0.03	2	0.05	3	unk.	unk.	0.003	1	0	0(a)	0	unk.	unk.	23	2	8	133	unk.	0.4	2	0.3	2	unk.	unk.	tr(a)	C
0.2	1	21	5	0.11	1	tr	0	tr	unk.	unk.	unk.	0.51	31	3	0.4	3.4	unk.	10	3	tr(a)	0	tr(a)		unk.	
0.04	2	0.06	3	unk.	unk.	0.003	1	0	0(a)	0	unk.	unk.	27	3	10	157	unk.	0.5	3	unk.	unk.	unk.	unk.	unk.	D
0.2	1	25	6	0.13	1	tr	0	tr	unk.	unk.	unk.	0.60	36	4	0.4	4.0	unk.	12	3	unk.	unk.	unk.		unk.	
0.03	2	unk.	unk.	0(a)	0	unk.	unk.	0	0(a)	0	unk.	unk.	25	3	7	115	unk.	0.4	2	unk.	unk.	unk.	unk.	tr(a)	E
0.2	1	unk.	unk.	unk.	unk.	42	1	4	unk.	unk.	unk.	0.42	30	3	0.3	2.9	unk.	unk.	unk.	tr(a)	0	tr(a)		unk.	
0.01	1	unk.	unk.	0(a)	0	unk.	unk.	0	0(a)	0	unk.	unk.	12	1	4	55	unk.	0.2	1	unk.	unk.	unk.	unk.	tr(a)	F
0.1	1	unk.	unk.	unk.	unk.	20	0	2	unk.	unk.	unk.	0.20	15	1	0.1	1.4	unk.	unk.	unk.	tr(a)	0	tr(a)		unk.	
0.03	2	0.11	6	0.00	0	0.003	1	0	0(a)	0	0.3	0.1	23	2	8	133	20	0.4	2	0.3	2	unk.	unk.	66	G
0.2	1	21	5	0.11	1	34	1	3	0.3	1	0.2	0.51	31	3	0.4	3.4	0.6	10	3	0.08	4	3		1	
0.04	2	0.13	7	0.00	0	0.003	1	0	0(a)	0	0.3	0.1	27	3	10	157	24	0.5	3	0.3	2	unk.	unk.	80	H
0.2	1	25	6	0.13	1	40	1	4	0.4	1	0.2	0.60	36	4	0.4	4.0	0.7	12	3	0.10	5	tr		2	
tr	0	unk.	unk.	0(a)	0	unk.	unk.	0	0(a)	0	unk.	unk.	3	0	tr	4	unk.	tr	0	unk.	unk.	unk.	unk.	unk.	I
tr	0	unk.	unk.	unk.	unk.	8	0	1	unk.	unk.	unk.	0.02	0	0	0.0	0.1	unk.	unk.	unk.	unk.	unk.	unk.		unk.	
0.05	3	0.08	4	0.00	0	0.002	1	0	0(a)	0	0.3	tr	54	5	1	262	5	0.5	3	0.3	2	unk.	unk.	10	J
0.5	3	60	15	0.33	3	262	5	26	0.4	1	0.3	0.79	26	3	0.1	6.7	0.1	9	2	0.01	0	0		16	
0.04	2	0.05	3	0.00	0	0.002	1	0	0(a)	0	0.2	tr	37	4	1	180	4	0.4	2	0.2	1	unk.	unk.	7	K
0.4	2	41	10	0.22	2	180	4	18	0.3	1	0.2	0.54	18	2	0.0	4.6	0.1	6	2	tr	0	0		11	
unk.	unk.	unk.	unk.	0.00	0	unk.	unk.	0(a)	0(a)	0	unk.	unk.	24	2	tr	25	unk.	0.7	4	0.1	0	unk.	unk.	unk.	L
0.1	1	unk.	unk.	unk.	unk.	104	2	unk.	unk.	unk.	unk.	0.11	3	0	0.0	0.6	unk.	4	1	unk.	unk.	unk.		unk.	
unk.	unk.	unk.	unk.	unk.	unk.	unk.	unk.	unk.	unk.	unk.	unk.	unk.	unk.	unk.	unk.	unk.	unk.	unk.	unk.	unk.	unk.	unk.	unk.	unk.	M
unk.	unk.	unk.	unk.	unk.	unk.	unk.	unk.	unk.	unk.	unk.	unk.	unk.	unk.	unk.	unk.	unk.	unk.	unk.	unk.	unk.	unk.	unk.		unk.	
1.14	67	0.46	23	9.98	166	unk.	unk.	unk.	unk.	unk.	unk.	unk.	14	1	85	248	unk.	7.7	43	4.7	31	unk.	unk.	167	N
8.2	41	284	71	3.34	33	8947	179	unk.	unk.	unk.	unk.	1.33	215	21	3.7	6.3	unk.	19	5	0.32	16	unk.		unk.	
0.69	41	0.25	12	7.35	123	unk.	unk.	unk.	unk.	unk.	unk.	unk.	9	1	42	115	unk.	4.7	26	3.3	22	unk.	unk.	123	O
3.0	15	273	68	2.15	22	5387	108	unk.	unk.	unk.	unk.	0.62	166	17	1.8	2.9	unk.	14	4	0.18	9	unk.		unk.	
0.59	35	0.22	11	6.13	102	unk.	unk.	unk.	unk.	unk.	unk.	unk.	9	1	43	116	unk.	4.4	25	3.4	22	unk.	unk.	99	P
2.9	15	220	55	1.82	18	5899	118	unk.	unk.	unk.	unk.	0.60	154	15	1.9	3.0	unk.	14	4	0.17	9	unk.		unk.	
0.76	45	0.30	15	6.87	115	unk.	unk.	unk.	unk.	unk.	unk.	unk.	9	1	41	112	unk.	4.7	26	3.1	21	unk.	unk.	128	Q
3.6	18	266	67	2.15	22	6914	138	unk.	unk.	unk.	unk.	0.63	162	16	1.8	2.9	unk.	14	4	0.22	11	unk.		unk.	
0.65	38	0.24	12	17.42	290	unk.	unk.	unk.	unk.	unk.	unk.	unk.	9	1	43	145	unk.	4.9	27	2.7	18	unk.	unk.	127	R
3.3	16	250	63	2.51	25	4376	88	unk.	unk.	unk.	unk.	0.66	148	15	1.9	3.7	unk.	12	3	0.28	14	unk.		unk.	
0.05	3	0.02	1	0.39	7	unk.	unk.	unk.	unk.	unk.	unk.	unk.	2	0	13	36	unk.	0.8	5	0.9	6	unk.	unk.	12	S
0.8	4	11	3	0.14	1	38	1	unk.	unk.	unk.	unk.	0.15	31	3	0.6	0.9	unk.	4	1	0.02	1	unk.		unk.	

	FOOD	Portion	Weight in grams / Conversion for 100 g	Kilocalories / H_2O g	Total carbohydrate g / Total fats g	Crude fiber g / Dietary fiber g	Total protein g / Total sugar g	% USRDA	Arginine mg / Histidine mg	Isoleucine mg / Leucine mg	Lysine mg / Methionine mg	Phenylalanine mg / Threonine mg	Valine mg / Tryptophan mg	Cystine mg / Tyrosine mg	Polyunsat. fatty acids g / Monounsat. fatty acids g	Saturated fatty acids g / P/S ratio	Linoleic acid g / Cholesterol mg	Thiamin mg / Ascorbic acid mg	% USRDA
A	organ meats,gizzard,turkey,stewed,	1/2 C	73	118	0.4	0.00	21.3	**47**	1533	1007	1475	887	956	280	0.81	0.80	0.57	0.02	**1**
	chopped		*1.38*	47.4	2.8	0(a)	unk.		430	1499	560	983	191	649	0.43	1.01	168	1	**2**
B	organ meats,heart,beef,braised,	1/2 C	73	136	0.5	0.00	22.7	**50**	1433	1093	2017	998	1129	290	0.93	1.30	0.54	0.18	**12**
	chopped		*1.38*	44.4	4.1	0(a)	0.0		773	1840	613	1041	254	816	0.96	0.71	199	1	**1**
C	organ meats,heart,chicken,stewed	1 average	3.0	6	0.0	0.00	0.8	**2**	51	42	66	35	45	11	0.07	0.07	0.05	tr	**0**
			33.33	1.9	0.2	0(a)	unk.		21	69	19	36	10	28	0.05	1.02	7	tr	**0**
D	organ meats,heart,turkey,stewed,	1/2 C	73	128	1.5	0.00	19.4	**43**	1245	1040	1626	869	1099	264	1.28	1.27	0.88	0.05	**3**
	chopped		*1.38*	46.6	4.4	0(a)	unk.		509	1691	468	879	249	695	0.67	1.01	164	1	**2**
E	organ meats,kidney,beef,braised,	3-1/4x2-1/2x	30	76	0.2	0.00	9.9	**22**	625	283	421	273	339	126	unk.	unk.	unk.	0.20	**13**
	sliced	1/4"	*3.33*	15.9	3.6	0(a)	0.0		337	503	119	257	85	356	unk.	unk.	241	0	**0**
F	organ meats,liver pate,chicken,canned	1 Tbsp	13	26	0.9	unk.	1.8	**4**	unk.	unk.	unk.	unk.	unk.	unk.	unk.	unk.	unk.	0.01	**0**
			7.69	0.0	1.7	unk.	unk.		unk.	unk.	unk.	unk.	unk.	unk.	unk.	unk.	unk.	1	**2**
G	organ meats,liver pate,goose,smoked,	1 Tbsp	13	60	0.6	0.00	1.5	**3**	unk.	unk.	unk.	unk.	unk.	unk.	unk.	unk.	unk.	0.01	**1**
	canned		*7.69*	4.8	5.7	0(a)	unk.		unk.	unk.	unk.	unk.	unk.	unk.	unk.	unk.	19	unk.	**unk.**
H	organ meats,liver,beef,cooked	6-1/2x2-3/8x	85	150	4.5	0.00	22.4	**50**	1418	1059	1514	1020	1271	286	unk.	unk.	unk.	0.22	**15**
		3/8" piece	*1.18*	47.6	4.1	0(a)	0.0		764	1867	476	960	303	807	unk.	unk.	372	23	**38**
I	organ meats,liver,beef,fried in	6-1/2x2-3/8x	85	195	4.5	0.00	22.4	**50**	1418	1059	1514	1020	1271	286	1.70	2.46	0.93	0.22	**15**
	margarine	3/8" piece	*1.18*	unk.	9.0	0(a)	0.0		764	1867	476	960	303	807	3.48	0.69	372	23	**38**
J	organ meats,liver,calf,cooked	6-1/2x2-3/8x	85	166	3.4	0.00	25.1	**56**	1572	1001	1433	963	1202	320	unk.	unk.	unk.	0.20	**14**
		3/8" piece	*1.18*	43.7	5.0	0(a)	0.0		854	1766	451	909	1766	903	unk.	unk.	372	31	**52**
K	organ meats,liver,chicken,stewed	1 average	25	39	0.2	0.00	6.1	**14**	373	324	461	303	384	82	0.22	0.46	0.14	0.04	**3**
			4.00	17.1	1.4	0(a)	unk.		162	550	144	271	86	214	0.29	0.49	158	4	**7**
L	organ meats,liver,chicken,stewed,	1/2 C	70	110	0.6	0.00	17.0	**38**	1045	906	1290	848	1074	229	0.63	1.29	0.40	0.10	**7**
	chopped		*1.43*	47.8	3.8	0(a)	unk.		453	1539	404	758	240	600	0.82	0.49	442	11	**18**
M	organ meats,liver,duck,domesticated,	1 average	44	60	1.6	0.00	8.3	**18**	505	438	624	410	520	111	0.28	0.63	0.16	unk.	**unk.**
	raw		*2.27*	31.6	2.0	0(a)	unk.		219	744	195	367	116	290	0.29	0.44	227	unk.	**unk.**
N	organ meats,liver,goose,domesticated,	1 average	94	125	5.9	0.00	15.4	**34**	943	818	1165	766	970	207	0.24	1.49	0.17	0.53	**35**
	raw		*1.06*	67.5	4.0	0(a)	unk.		409	1388	365	684	216	541	0.70	0.16	unk.	unk.	**unk.**
O	organ meats,liver,home recipe,liver	3/4 C	109	208	22.6	0.11	13.1	**28**	537	703	802	619	796	176	0.25	3.17	0.25	0.18	**12**
	& mushroom casserole		*0.92*	72.2	6.9	0.43	3.0		394	1095	294	537	168	482	1.81	0.08	157	6	**11**
P	organ meats,liver,turkey,stewed,	1/2 C	70	118	2.4	0.00	16.8	**37**	1028	891	1270	835	1058	225	0.74	1.32	0.51	0.03	**2**
	chopped		*1.43*	45.9	4.2	0(a)	unk.		446	1514	398	746	237	591	0.85	0.56	438	1	**2**
Q	organ meats,sweet breads(thymus),	1 average	97	244	0.8	0.00	32.0	**71**	2021	914	1360	884	1096	408	unk.	unk.	unk.	0.65	**43**
	beef,cooked		*1.03*	51.4	11.6	0(a)	0.0		1090	1628	384	832	276	1152	unk.	unk.	780	0	**0**
R	organ meats,tongue,beef,cooked,sliced	4 slices	80	195	0.3	0.00	17.2	**38**	1086	634	1091	529	672	220	unk.	unk.	unk.	0.04	**3**
		3x2x1/5"	*1.25*	48.6	13.4	0(a)	0.0		586	1029	285	566	157	619	unk.	unk.	unk.	0	**0**
S	OYSTER,breaded,fried	1 med	11	27	2.1	tr	1.0	**2**	71	45	77	36	52	15	unk.	unk.	unk.	0.02	**1**
			8.93	6.1	1.6	unk.	0.0		23	74	28	45	13	40	unk.	unk.	unk.	unk.	**unk.**

Riboflavin mg / Niacin mg	% USRDA	Vitamin B_6 mg / Folacin mcg	% USRDA	Vitamin B_{12} mcg / Pantothenic acid mg	% USRDA	Biotin mg / Vitamin A IU	% USRDA	Preformed A RE / Beta carotene RE	Vitamin D IU / Vitamin E IU	% USRDA	Total tocopherol mg / Alpha tocopherol mg	Other tocopherol mg / Total ash g	Calcium mg / Phosphorus mg	% USRDA	Sodium mg / Sodium meq	Potassium mg / Potassium meq	Chlorine mg / Chlorine meq	Iron mg / Magnesium mg	% USRDA	Zinc mg / Copper mg	% USRDA	Iodine mcg / Selenium mcg	% USRDA	Manganese mcg / Chromium mcg	
0.24	14	0.09	4	1.38	23	unk.	unk.	unk.	unk.	unk.	unk.	unk.	11	1	39	153	unk.	3.9	22	3.0	20	unk.	unk.	71	A
2.2	11	38	9	0.61	6	134	3	unk.	unk.	unk.	unk.	0.51	93	9	1.7	3.9	unk.	14	3	0.13	6	unk.		unk.	
0.88	52	0.11	5	7.18	120	unk.	unk.	6	unk.	unk.	unk.	unk.	4	0	75	168	unk.	4.3	24	unk.	unk.	unk.	unk.	unk.	B
5.5	28	unk.	unk.	0.91	9	22	0	tr	unk.	unk.	unk.	0.80	131	13	3.3	4.3	unk.	unk.	unk.	unk.	unk.	tr(a)		tr(a)	
0.02	1	0.01	1	0.22	4	unk.	unk.	unk.	unk.	unk.	unk.	unk.	1	0	1	4	unk.	0.3	2	0.2	2	unk.	unk.	3	C
0.1	0	2	1	0.08	1	1	0	unk.	unk.	unk.	unk.	0.02	6	1	0.1	0.1	unk.	1	0	0.02	1	unk.		unk.	
0.64	38	0.23	12	5.18	86	unk.	unk.	unk.	unk.	unk.	unk.	unk.	9	1	40	133	unk.	5.0	28	3.8	26	unk.	unk.	67	D
2.4	12	57	14	1.97	20	20	0	unk.	unk.	unk.	unk.	0.63	149	15	1.7	3.4	unk.	16	4	0.45	23	unk.		unk.	
1.37	81	0.08	4	8.37	140	unk.	unk.	93	unk.	unk.	unk.	unk.	5	1	76	97	unk.	3.9	22	unk.	unk.	unk.	unk.	unk.	E
3.2	16	18	5	0.58	6	345	7	3	unk.	unk.	unk.	0.36	73	7	3.3	2.5	unk.	unk.	unk.	unk.	unk.	unk.		unk.	
0.18	11	unk.	unk.	unk.	unk.	unk.	unk.	unk.	unk.	unk.	unk.	unk.	1	0	unk.	unk.	unk.	1.2	7	unk.	unk.	unk.	unk.	unk.	F
1.0	5	unk.	unk.	unk.	unk.	94	2	unk.	unk.	unk.	unk.	unk.	unk.	unk.	unk.	unk.	unk.	unk.	unk.	unk.	unk.	unk.		unk.	
0.04	2	unk.	unk.	1.22	20	unk.	unk.	unk.	unk.	unk.	unk.	unk.	unk.	unk.	unk.	unk.	unk.	unk.	unk.	unk.	unk.	unk.	unk.	unk.	G
0.3	2	unk.	unk.	unk.	unk.	unk.	unk.	unk.	unk.	unk.	unk.	0.40	unk.	unk.	unk.	unk.	unk.	unk.	unk.	unk.	unk.	unk.		unk.	
3.56	210	0.42	21	61.20	1020	unk.	unk.	12268	16	4	1.4	0.8	9	1	156	323	unk.	7.5	42	4.3	29	unk.	unk.	unk.	H
14.0	70	123	31	3.10	31	45390	908	454	1.7	6	0.5	1.44	405	41	6.8	8.3	unk.	15	4	3.14	157	tr(a)		tr(a)	
3.56	210	0.42	21	61.20	1020	unk.	unk.	12268	43	11	unk.	unk.	9	1	156	323	unk.	7.5	42	4.3	29	unk.	unk.	unk.	I
14.0	70	123	31	3.10	31	45390	908	454	unk.	unk.	unk.	1.44	405	41	6.8	8.3	unk.	15	4	3.14	157	tr(a)		tr(a)	
3.54	209	0.42	21	61.20	1020	unk.	unk.	7512	12	3	1.4	0.8	11	1	100	385	unk.	12.1	67	5.2	35	unk.	unk.	unk.	J
14.0	70	123	31	3.10	31	27795	556	278	1.7	6	0.5	1.61	456	46	4.4	9.8	unk.	22	6	3.14	157	tr(a)		tr(a)	
0.44	26	0.14	7	4.85	81	unk.	unk.	unk.	unk.	unk.	unk.	unk.	4	0	13	35	unk.	2.1	12	1.1	7	unk.	unk.	74	K
1.1	6	193	48	1.35	14	4094	82	unk.	unk.	unk.	0.4	0.25	78	8	0.5	0.9	unk.	5	1	0.09	5	unk.		unk.	
1.22	72	0.41	20	13.57	226	unk.	unk.	unk.	unk.	unk.	unk.	unk.	10	1	36	98	unk.	5.9	33	3.0	20	unk.	unk.	208	L
3.1	16	539	135	3.79	38	11462	229	unk.	unk.	unk.	1.1	0.70	218	22	1.5	2.5	unk.	15	4	0.26	13	unk.		unk.	
unk.	unk.	unk.	unk.	23.76	396	unk.	unk.	unk.	unk.	unk.	unk.	unk.	5	1	unk.	unk.	unk.	13.4	75	unk.	unk.	unk.	unk.	unk.	M
unk.	unk.	unk.	unk.	unk.	unk.	17559	351	unk.	unk.	unk.	unk.	0.58	118	12	unk.	unk.	unk.	unk.	unk.	2.62	131	unk.		unk.	
0.84	49	0.71	36	unk.	unk.	unk.	unk.	unk.	unk.	unk.	unk.	unk.	40	4	132	216	unk.	unk.	unk.	unk.	unk.	unk.	unk.	unk.	N
6.1	31	unk.	unk.	unk.	unk.	29138	583	unk.	unk.	unk.	unk.	1.17	245	25	5.7	5.5	unk.	23	6	7.07	354	unk.		unk.	
1.12	66	0.19	10	15.62	260	0.004	1	3082	26	6	0.7	0.2	127	13	297	206	11	3.4	19	2.0	14	42.8	29	3	O
4.8	24	41	10	1.19	12	11649	233	115	0.8	3	0.2	1.99	254	25	12.9	5.3	0.3	45	11	0.92	46	unk.		unk.	
0.99	59	0.36	18	33.25	554	unk.	unk.	unk.	unk.	unk.	unk.	unk.	8	1	45	136	unk.	5.5	30	2.2	14	unk.	unk.	177	P
4.2	21	466	117	4.17	42	8807	176	unk.	unk.	unk.	unk.	0.78	190	19	1.9	3.5	unk.	10	3	0.39	20	unk.		unk.	
4.44	261	0.25	13	27.06	451	unk.	unk.	335	unk.	unk.	unk.	unk.	17	2	245	314	unk.	12.7	71	unk.	unk.	unk.	unk.	unk.	Q
10.4	52	59	15	1.87	19	1115	22	0	unk.	unk.	unk.	1.16	237	24	10.7	8.0	unk.	unk.	unk.	unk.	unk.	unk.		unk.	
0.23	14	0.06	3	unk.	unk.	unk.	unk.	unk.	0	0	unk.	unk.	6	1	49	131	unk.	1.8	10	unk.	unk.	unk.	unk.	unk.	R
2.8	14	unk.	unk.	0.80	8	unk.	unk.	unk.	unk.	unk.	unk.	0.48	94	9	2.1	3.4	unk.	13	3	unk.	unk.	unk.		unk.	
0.03	2	tr	0	2.02	34	unk.	unk.	unk.	unk.	unk.	0.1	tr(a)	17	2	23	23	unk.	0.9	5	8.4	56	unk.	unk.	33	S
0.4	2	tr	0	0.02	0	49	1	unk.	0.1	0	tr(a)	0.17	27	3	1.0	0.6	unk.	4	1	1.53	77	5		tr(a)	

	FOOD	Portion	Weight in grams / Conversion for 100 g	Kilocalories / H_2O g	Total carbohydrate g / Total fats g	Crude fiber g / Dietary fiber g	Total protein g / Total sugar g	% USRDA	Arginine mg / Histidine mg	Isoleucine mg / Leucine mg	Lysine mg / Methionine mg	Phenylalanine mg / Threonine mg	Valine mg / Tryptophan mg	Cystine mg / Tyrosine mg	Polyunsat. fatty acids g / Monounsat. fatty acids g	Saturated fatty acids g / P/S ratio	Linoleic acid g / Cholesterol mg	Thiamin mg / Ascorbic acid mg	% USRDA
A	oyster,canned,solids & liquid	1/3 C	80	61	3.9	0.08	6.8	**15**	511	321	542	256	360	107	unk.	unk.	unk.	0.02	**1**
			1.25	65.8	1.8	unk.	0.0		162	526	200	319	89	283	unk.	unk.	36	unk.	**unk.**
B	oyster,eastern,raw	1 average	15	10	0.5	0(a)	1.3	**3**	95	59	100	47	67	20	unk.	unk.	unk.	0.02	**1**
			6.67	12.7	0.3	0(a)	0.0		30	97	37	59	16	52	unk.	unk.	7	unk.	**unk.**
C	oyster,meat,eastern,raw	1/3 C	80	53	2.7	0(a)	6.7	**15**	506	317	535	253	356	106	unk.	unk.	unk.	0.11	**8**
			1.25	67.7	1.4	0(a)	0.0		160	519	198	315	87	279	unk.	unk.	40	unk.	**unk.**
D	PANCAKE,blueberry,baked-Aunt Jemima	4" dia x 1/2"	38	68	13.8	0.08	2.1	**3**	unk.	unk.	unk.	unk.	unk.	unk.	unk.	unk.	unk.	0.11	**7**
			2.65	20.4	0.5	unk.	unk.		unk.	unk.	unk.	unk.	unk.	unk.	unk.	unk.	unk.	unk.	**unk.**
E	pancake,blueberry,batter-Aunt Jemima	1 C	240	434	87.8	0.48	13.2	**20**	unk.	unk.	unk.	unk.	unk.	unk.	unk.	unk.	unk.	0.67	**45**
			0.42	129.6	3.4	unk.	0.0		unk.	unk.	unk.	unk.	unk.	unk.	unk.	unk.	unk.	unk.	**unk.**
F	pancake,buttermilk,batter-Aunt Jemima	1 C	240	449	90.2	0.24	15.1	**23**	unk.	unk.	unk.	unk.	unk.	unk.	unk.	unk.	unk.	0.72	**48**
			0.42	125.3	3.1	4.44	unk.		unk.	unk.	unk.	unk.	unk.	unk.	unk.	unk.	unk.	unk.	**unk.**
G	pancake,buttermilk,baked-Aunt Jemima	4" dia x 1/2"	38	71	14.2	0.04	2.4	**4**	unk.	unk.	unk.	unk.	unk.	unk.	unk.	unk.	unk.	0.11	**8**
			2.65	19.7	0.5	0.70	unk.		unk.	unk.	unk.	unk.	unk.	unk.	unk.	unk.	unk.	unk.	**unk.**
H	pancake,home recipe,buckwheat	6" dia x 1/2"	65	137	24.4	0.29	3.5	**5**	unk.	104	62	121	101	42	1.61	0.37	1.58	0.15	**10**
			1.53	33.6	2.9	1.54	3.2		49	175	31	69	27	69	0.65	4.39	0	0	**0**
I	pancake,home recipe,buckwheat	4" dia x 3/8"	29	68	12.2	0.14	1.8	**3**	unk.	52	31	61	51	20	0.80	0.18	0.79	0.08	**5**
			3.45	13.5	1.5	0.77	1.6		24	88	15	35	13	34	0.32	4.39	0	0	**0**
J	pancake,home recipe,potato	4" dia x 3/8"	35	78	4.3	0.13	2.0	**4**	118	98	131	96	125	50	2.76	1.10	2.72	0.03	**2**
			2.86	22.3	5.9	0.27	0.2		50	153	54	95	27	85	1.65	2.50	60	4	**7**
K	pancake,home recipe,zucchini	4" dia x 3/8"	40	69	8.8	0.26	2.2	**4**	73	94	99	99	110	36	1.11	0.57	1.16	0.04	**3**
			2.50	25.7	2.8	0.81	1.8		46	156	43	77	27	75	0.91	1.96	31	4	**7**
L	pancake,original,baked-Aunt Jemima	4" dia x 1/2"	38	70	14.1	0.08	2.2	**3**	unk.	unk.	unk.	unk.	unk.	unk.	unk.	unk.	unk.	0.12	**8**
			2.65	20.1	0.5	unk.	unk.		unk.	unk.	unk.	unk.	unk.	unk.	unk.	unk.	unk.	unk.	**unk.**
M	pancake,original,batter-Aunt Jemima	1 C	240	444	89.3	0.48	13.9	**21**	unk.	unk.	unk.	unk.	unk.	unk.	unk.	unk.	unk.	0.74	**50**
			0.42	127.4	3.4	unk.	unk.		unk.	unk.	unk.	unk.	unk.	unk.	unk.	unk.	unk.	unk.	**unk.**
N	pancake,plain or buttermilk	6" dia x 1/2"	73	164	23.7	0.07	5.3	**9**	unk.	292	248	285	299	139	0.80	2.19	0.73	0.11	**7**
			1.37	36.9	5.3	1.35	15.0		124	445	109	204	73	226	2.92	0.37	54	tr	**0**
O	pancake,plain or buttermilk	4" dia x 3/8"	27	61	8.7	0.03	1.9	**3**	unk.	108	92	105	111	51	0.30	0.81	0.27	0.04	**3**
			3.70	13.7	2.0	0.50	5.5		46	165	40	76	27	84	1.08	0.37	20	tr	**0**
P	PAPAYA,raw,edible portion	1/4 average	114	52	19.3	unk.	0.4	**1**	unk.	unk.	unk.	unk.	unk.	unk.	unk.	unk.	unk.	0.02	**2**
			0.88	98.5	0.1	0.57	19.3		unk.	unk.	unk.	unk.	unk.	unk.	unk.	unk.	0(a)	17	**28**
Q	PAPRIKA	1 tsp	2.3	7	1.3	0.48	0.3	**1**	unk.	unk.	unk.	unk.	unk.	unk.	0.19	0.05	0.17	0.01	**1**
			43.48	0.2	0.3	unk.	0.0		unk.	unk.	unk.	unk.	unk.	unk.	0.03	3.96	0	2	**3**
R	PARSLEY,dried	1 tsp	0.4	1	0.2	0.04	0.1	**0**	unk.	unk.	unk.	unk.	unk.	unk.	unk.	unk.	unk.	tr	**0**
			250.00	tr	tr	unk.	0.0		unk.	unk.	unk.	unk.	unk.	unk.	unk.	unk.	0	0	**0**
S	parsley,raw,chopped	1/2 C	30	13	2.5	0.45	1.1	**2**	unk.	unk.	69	unk.	unk.	unk.	unk.	unk.	unk.	0.04	**2**
			3.33	25.5	0.2	2.73	0.1		unk.	unk.	5	unk.	22	unk.	unk.	unk.	0	52	**86**

Riboflavin mg / Niacin mg	% USRDA	Vitamin B_6 mg / Folacin mcg	% USRDA	Vitamin B_{12} mcg / Pantothenic acid mg	% USRDA	Biotin mg / Vitamin A IU	% USRDA	Preformed A RE / Beta carotene RE	Vitamin D IU / Vitamin E IU	% USRDA	Total tocopherol mg / Alpha tocopherol mg	Other tocopherol mg / Total ash g	Calcium mg / Phosphorus mg	% USRDA	Sodium mg / Sodium meq	Potassium mg / Potassium meq	Chlorine mg / Chlorine meq	Iron mg / Magnesium mg	% USRDA	Zinc mg / Copper mg	% USRDA	Iodine mcg / Selenium mcg	% USRDA	Manganese mcg / Chromium mcg	
0.16	**9**	0.03	**2**	14.40	**240**	0.007	**2**	unk.	unk.	**unk.**	0.5	tr(a)	22	**2**	unk.	56	unk.	4.5	**25**	59.8	**398**	unk.	**unk.**	236	A
0.6	**3**	2	**1**	0.16	**2**	unk.	**unk.**	unk.	0.6	**2**	tr(a)	1.76	99	**10**	unk.	1.4	unk.	26	**7**	10.96	**548**	39		tr(a)	
0.03	**2**	0.01	**0**	2.70	**45**	unk.	**unk.**	13	unk.	**unk.**	0.1	tr(a)	14	**1**	11	18	94	0.8	**5**	11.2	**75**	unk.	**unk.**	44	B
0.4	**2**	1	**0**	0.04	**0**	46	**1**	0	0.1	**0**	tr(a)	0.27	21	**2**	0.5	0.5	2.7	5	**1**	0.56	**28**	7		unk.	
0.14	**9**	0.04	**2**	14.40	**240**	unk.	**unk.**	67	unk.	**unk.**	0.5	tr(a)	75	**8**	58	97	502	4.4	**24**	59.8	**398**	unk.	**unk.**	236	C
2.0	**10**	8	**2**	0.20	**2**	248	**5**	2	0.6	**2**	tr(a)	1.44	114	**11**	2.5	2.5	14.2	26	**6**	2.96	**148**	39		unk.	
0.08	**4**	unk.	**unk.**	tr(a)	**0**	unk.	**unk.**	unk.	tr(a)	**0**	unk.	unk.	23	**2**	unk.	unk.	unk.	0.5	**3**	unk.	**unk.**	unk.	**unk.**	unk.	D
0.7	**3**	unk.	**unk.**	unk.	**unk.**	18	**0**	unk.	unk.	**unk.**	unk.	0.87	110	**11**	unk.	unk.	unk.	unk.	**unk.**	unk.	**unk.**	unk.		unk.	
0.48	**28**	unk.	**unk.**	tr(a)	**0**	unk.	**unk.**	unk.	tr(a)	**0**	unk.	unk.	144	**14**	unk.	unk.	unk.	3.0	**17**	unk.	**unk.**	unk.	**unk.**	unk.	E
4.2	**21**	unk.	**unk.**	unk.	**unk.**	115	**2**	unk.	unk.	**unk.**	unk.	5.52	696	**70**	unk.	unk.	unk.	unk.	**unk.**	unk.	**unk.**	unk.		unk.	
0.60	**35**	unk.	**unk.**	tr(a)	**0**	unk.	**unk.**	unk.	tr(a)	**0**	unk.	unk.	192	**19**	unk.	unk.	unk.	3.6	**20**	unk.	**unk.**	unk.	**unk.**	unk.	F
4.7	**23**	unk.	**unk.**	unk.	**unk.**	118	**2**	unk.	unk.	**unk.**	unk.	6.00	720	**72**	unk.	unk.	unk.	unk.	**unk.**	unk.	**unk.**	unk.		unk.	
0.09	**6**	unk.	**unk.**	tr(a)	**0**	unk.	**unk.**	unk.	tr(a)	**0**	unk.	unk.	30	**3**	unk.	unk.	unk.	0.6	**3**	unk.	**unk.**	unk.	**unk.**	unk.	G
0.7	**4**	unk.	**unk.**	unk.	**unk.**	19	**0**	unk.	unk.	**unk.**	unk.	0.94	113	**11**	unk.	unk.	unk.	unk.	**unk.**	unk.	**unk.**	unk.		unk.	
0.09	**5**	0.11	**6**	0.00	**0**	0.001	**0**	0	0	**0**	2.2	0(a)	11	**1**	193	74	16	1.1	**6**	0.4	**3**	35.6	**24**	80	H
1.3	**6**	32	**8**	0.25	**3**	0	**0**	0	2.7	**9**	tr	0.68	62	**6**	8.4	1.9	0.4	20	**5**	0.04	**2**	10		13	
0.05	**3**	0.06	**3**	0.00	**0**	0.001	**0**	0	0	**0**	1.1	0(a)	5	**1**	97	37	8	0.5	**3**	0.2	**1**	17.8	**12**	40	I
0.6	**3**	16	**4**	0.13	**1**	0	**0**	0	1.3	**4**	tr	0.34	31	**3**	4.2	0.9	0.2	10	**2**	0.02	**1**	5		7	
0.05	**3**	0.07	**3**	0.13	**2**	0.003	**1**	0	4	**1**	3.9	0.0	10	**1**	238	88	32	0.4	**2**	0.3	**2**	65.6	**44**	13	J
0.3	**2**	7	**2**	0.27	**3**	27	**1**	0	4.7	**16**	0.6	0.86	41	**4**	10.3	2.3	0.9	36	**9**	0.06	**3**	0		1	
0.07	**4**	0.05	**2**	0.07	**1**	0.002	**1**	0	2	**1**	2.2	tr	25	**3**	160	79	15	0.7	**4**	0.2	**2**	22.1	**15**	14	K
0.5	**3**	11	**3**	0.16	**2**	155	**3**	13	2.6	**9**	0.3	0.71	38	**4**	7.0	2.0	0.4	24	**6**	0.06	**3**	0		tr	
0.08	**5**	unk.	**unk.**	tr(a)	**0**	unk.	**unk.**	unk.	tr(a)	**0**	unk.	unk.	23	**2**	unk.	unk.	unk.	0.5	**3**	0.2	**2**	unk.	**unk.**	unk.	L
0.7	**4**	unk.	**unk.**	unk.	**unk.**	18	**0**	unk.	unk.	**unk.**	unk.	0.87	110	**11**	unk.	unk.	unk.	unk.	**unk.**	unk.	**unk.**	unk.		unk.	
0.50	**30**	unk.	**unk.**	tr(a)	**0**	unk.	**unk.**	unk.	tr(a)	**0**	unk.	unk.	144	**14**	unk.	unk.	unk.	3.2	**18**	1.6	**11**	unk.	**unk.**	unk.	M
4.5	**23**	unk.	**unk.**	unk.	**unk.**	113	**2**	unk.	unk.	**unk.**	unk.	5.52	696	**70**	unk.	unk.	unk.	unk.	**unk.**	unk.	**unk.**	unk.		unk.	
0.18	**10**	0.03	**2**	tr(a)	**0**	0.004	**1**	unk.	5	**1**	0.7	unk.	157	**16**	412	112	unk.	0.9	**5**	0.6	**4**	unk.	**unk.**	unk.	N
0.6	**3**	6	**2**	0.51	**5**	182	**4**	unk.	0.8	**3**	unk.	1.82	190	**19**	17.9	2.9	unk.	18	**4**	0.04	**2**	unk.		unk.	
0.06	**4**	0.01	**1**	tr(a)	**0**	0.001	**0**	unk.	2	**1**	0.2	unk.	58	**6**	152	42	unk.	0.3	**2**	0.2	**2**	unk.	**unk.**	unk.	O
0.2	**1**	2	**1**	0.19	**2**	67	**1**	unk.	0.3	**1**	unk.	0.67	70	**7**	6.6	1.1	unk.	6	**2**	0.01	**1**	unk.		unk.	
0.02	**1**	unk.	**unk.**	0(a)	**0**	unk.	**unk.**	0	0(a)	**0**	unk.	unk.	26	**3**	9	125	45	0.5	**3**	0.3	**2**	unk.	**unk.**	unk.	P
0.2	**1**	unk.	**unk.**	0.23	**2**	unk.	**unk.**	95	unk.	**unk.**	unk.	unk.	7	**1**	0.4	3.2	1.3	9	**2**	0.11	**6**	unk.		unk.	
0.04	**2**	unk.	**unk.**	0.00	**0**	unk.	**unk.**	0(a)	0(a)	**0**	unk.	unk.	4	**0**	1	54	unk.	0.5	**3**	0.1	**1**	unk.	**unk.**	unk.	Q
0.4	**2**	unk.	**unk.**	unk.	**unk.**	1394	**28**	unk.	unk.	**unk.**	unk.	0.16	8	**1**	0.0	1.4	unk.	4	**1**	unk.	**unk.**	unk.		unk.	
0.00	**0**	tr	**0**	0.00	**0**	unk.	**unk.**	0(a)	0(a)	**0**	unk.	unk.	6	**1**	2	15	unk.	0.4	**2**	tr	**0**	unk.	**unk.**	unk.	R
tr	**0**	unk.	**unk.**	unk.	**unk.**	93	**2**	unk.	unk.	**unk.**	unk.	0.05	1	**0**	0.1	0.4	unk.	1	**0**	unk.	**unk.**	unk.		unk.	
0.08	**5**	0.05	**3**	0.00	**0**	unk.	**unk.**	0	0(a)	**0**	unk.	unk.	61	**6**	13	218	47	1.9	**10**	0.3	**2**	unk.	**unk.**	0	S
0.4	**2**	35	**9**	0.09	**1**	2550	**51**	255	0.5	**2**	0.5	0.66	19	**2**	0.6	5.6	1.3	12	**3**	0.01	**0**	tr(a)		unk.	

	FOOD	Portion	Weight in grams / Conversion for 100 g	Kilocalories / H_2O g	Total carbohydrate g / Total fats g	Crude fiber g / Dietary fiber g	Total protein g / Total sugar g	**% USRDA**	Arginine mg / Histidine mg	Isoleucine mg / Leucine mg	Lysine mg / Methionine mg	Phenylalanine mg / Threonine mg	Valine mg / Tryptophan mg	Cystine mg / Tyrosine mg	Polyunsat. fatty acids g / Monounsat. fatty acids g	Saturated fatty acids g / P/S ratio	Linoleic acid g / Cholesterol mg	Thiamin mg / Ascorbic acid mg	**% USRDA**
A	PARSNIPS,fresh,cooked & mashed	1/2 C	105	69	15.6	2.10	1.6	**2**	unk.	unk.	unk.	unk.	unk.	unk.	unk.	unk.	unk.	0.07	**5**
			0.95	86.3	0.5	unk.	4.4		unk.	unk.	unk.	unk.	unk.	unk.	unk.	unk.	unk.	10	**18**
B	PATE de foie gras,canned	1 Tbsp	13	60	0.6	0.00	1.5	**3**	unk.	unk.	unk.	unk.	unk.	unk.	unk.	unk.	unk.	0.01	**1**
			7.69	4.8	5.7	0(a)	0.0		unk.	unk.	unk.	unk.	unk.	unk.	unk.	unk.	unk.	0	**0**
C	PEACH,canned,halves,heavy syrup,	1 half	85	67	17.1	0.34	0.3	**1**	7	5	13	8	17	4	unk.	tr(a)	unk.	0.01	**1**
	solids & liquid		*1.17*	67.5	0.1	0.85	16.8		7	12	13	11	2	9	unk.	unk.	0	3	**4**
D	peach,canned,halves,light syrup,	1/2 C	122	71	18.4	0.49	0.5	**1**	10	8	18	11	24	5	unk.	tr(a)	unk.	0.01	**1**
	solids & liquid		*0.82*	102.6	0.1	1.71	17.9		10	17	19	16	2	12	unk.	unk.	0	4	**6**
E	peach,canned,halves,water pack,	1/2 C	122	38	9.9	0.49	0.5	**1**	10	8	18	11	24	5	unk.	tr(a)	unk.	0.01	**1**
	solids & liquid		*0.82*	111.1	0.1	1.71	9.0		10	17	19	16	2	12	unk.	unk.	0	4	**6**
F	peach,canned,sliced or halves,heavy	1/2 C	128	100	25.7	0.51	0.5	**1**	11	8	19	12	25	6	unk.	tr(a)	unk.	0.01	**1**
	syrup,solids & liquid		*0.78*	101.2	0.1	1.28	25.2		11	18	20	17	2	13	unk.	unk.	0	4	**6**
G	peach,fresh,any,edible portion	1 average	100	38	9.7	0.60	0.6	**1**	13	10	22	13	30	7	unk.	tr(a)	unk.	0.02	**1**
			1.00	89.1	0.1	1.40	7.6		13	21	23	20	3	15	unk.	unk.	0	7	**12**
H	peach,fresh,any,sliced,edible portion	1/2 C	65	25	6.3	0.39	0.4	**1**	8	6	14	9	19	4	unk.	tr(a)	unk.	0.01	**1**
			1.55	57.5	0.1	0.90	4.9		8	14	15	13	2	10	unk.	unk.	0	5	**8**
I	peach,frozen,sliced,sweetened	1/2 C	125	110	28.2	0.50	0.5	**1**	10	8	18	11	25	6	unk.	tr(a)	unk.	0.01	**1**
			0.80	95.6	0.1	2.75	28.0		10	18	19	17	2	13	unk.	unk.	0	50	**83**
J	peach,home recipe,mousse	1/2 C	108	244	17.0	0.17	1.6	**3**	43	74	109	64	90	12	0.72	12.18	0.44	0.01	**1**
			0.93	69.3	19.6	0.40	16.4		36	124	37	62	16	58	4.92	0.06	72	4	**6**
K	peach,home recipe,pickled	1 lg half	95	114	30.0	0.38	0.4	**1**	8	6	13	8	18	4	0.00	0.00	0.00	0.01	**1**
			1.05	62.3	0.1	0.84	27.6		8	13	14	12	2	9	0.00	0.00	0	4	**7**
L	PEANUT BUTTER	1 Tbsp	15	87	2.6	0.28	3.9	**6**	502	184	160	226	223	67	1.78	1.35	2.10	0.02	**1**
			6.67	0.3	7.4	1.14	0.1		109	272	39	120	49	161	3.75	1.32	0	0	**0**
M	peanut butter,salt free-Sexton	1 Tbsp	15	98	2.8	0.27	3.8	**6**	485	178	154	219	215	65	1.83	1.38	2.14	0.02	**1**
			6.67	0.3	7.6	1.14	0.1		105	263	38	116	48	155	3.84	1.33	0	0	**0**
N	PEANUTS,dry roasted-Planters	1/4 C	36	217	6.9	0.94	9.1	**14**	1163	432	375	531	523	160	4.68	2.40	2.98	0.11	**8**
			2.75	unk.	17.0	2.94	1.1		260	639	93	283	116	382	4.86	1.95	0	0	**0**
O	peanuts,roasted in shell,w/skin	1 average	1.8	7	0.2	0.03	0.3	**1**	40	15	13	18	18	6	unk.	0.08	0.11	tr	**0**
			55.56	tr	0.6	0.15	tr		9	22	3	10	4	13	0.17	unk.	0	0	**0**
P	peanuts,roasted in shell,shelled,	1/4 C	36	210	7.4	0.97	9.4	**15**	1209	449	390	552	544	167	4.86	2.52	3.24	0.12	**8**
	chopped		*2.78*	0.6	17.5	2.92	1.1		270	664	96	294	121	397	5.04	1.93	0	0	**0**
Q	peanuts,roasted,salted (Spanish)	1/4 C	36	211	6.8	0.86	9.4	**14**	1200	488	423	600	590	172	5.36	3.39	5.18	0.12	**8**
			2.78	0.6	17.9	2.92	1.1		279	721	104	319	131	411	8.24	1.58	0	0	**0**
R	peanuts,roasted,salted (Virginia)	1/4 C	36	211	6.8	0.86	9.4	**14**	1200	488	423	600	590	172	5.36	3.39	5.18	0.12	**8**
			2.78	0.6	17.9	2.92	1.1		270	721	104	319	131	411	8.24	1.58	0	0	**0**
S	PEAR,canned,halves,heavy syrup,	1/2 C	128	97	25.0	0.76	0.3	**0**	unk.	7	7	6	10	unk.	unk.	tr(a)	unk.	0.01	**1**
	solids & liquid		*0.78*	101.7	0.3	2.17	20.5		3	10	3	7	10	2	unk.	unk.	0	1	**2**

	Riboflavin mg / Niacin mg	% USRDA	Vitamin B_6 mg / Folacin mcg	% USRDA	Vitamin B_{12} mcg / Pantothenic acid mg	% USRDA	Biotin mg / Vitamin A IU	% USRDA	Preformed A RE / Beta carotene RE	Vitamin D IU / Vitamin E IU	% USRDA	Total tocopherol mg / Alpha tocopherol mg	Other tocopherol mg / Total ash g	Calcium mg / Phosphorus mg	% USRDA	Sodium mg / Sodium meq	Potassium mg / Potassium meq	Chlorine mg / Chlorine meq	Iron mg / Magnesium mg	% USRDA	Zinc mg / Copper mg	% USRDA	Iodine mcg / Selenium mcg	% USRDA	Manganese mcg / Chromium mcg
A	0.08	**5**	0.09	**5**	0.00	**0**	unk.	**unk.**	0	0(a)	**0**	unk.	unk.	47	**5**	8	398	unk.	0.6	**4**	unk.	**unk.**	unk.	**unk.**	tr(a)
	0.1	**1**	unk.	**unk.**	0.63	**6**	31	**1**	3	unk.	**unk.**	unk.	0.94	65	**7**	0.4	10.2	unk.	13	**3**	tr(a)	**0**	tr(a)		unk.
B	0.04	**2**	unk.	**unk.**	unk.	**unk.**	unk.	**unk.**	unk.	unk.	**unk.**	unk.	unk.	unk.	**unk.**	unk.	unk.	unk.	unk.	**unk.**	unk.	**unk.**	unk.	**unk.**	unk.
	0.3	**2**	unk.	**unk.**	unk.	**unk.**	unk.	**unk.**	unk.	unk.	**unk.**	unk.	0.39	unk.	**unk.**	unk.	unk.	unk.	unk.	**unk.**	unk.	**unk.**	unk.		unk.
C	0.02	**1**	0.02	**1**	0.00	**0**	tr	**0**	0	0(a)	**0**	unk.	unk.	3	**0**	2	111	tr	0.3	**1**	0.1	**1**	unk.	**unk.**	36
	0.5	**3**	1	**0**	0.04	**0**	367	**7**	37	unk.	**unk.**	unk.	0.26	10	**1**	0.1	2.8	0.0	5	**1**	0.06	**3**	0		13
D	0.04	**2**	0.02	**1**	0.00	**0**	tr	**0**	0	0(a)	**0**	unk.	unk.	5	**1**	2	162	tr	0.4	**2**	0.1	**1**	unk.	**unk.**	51
	0.7	**4**	4	**1**	0.02	**0**	537	**11**	54	unk.	**unk.**	unk.	0.37	16	**2**	0.1	4.1	0.0	7	**2**	0.08	**4**	0		unk.
E	0.04	**2**	0.02	**1**	0.00	**0**	tr	**0**	0	0(a)	**0**	unk.	unk.	5	**1**	2	167	tr	0.4	**2**	0.1	**1**	unk.	**unk.**	51
	0.7	**4**	1	**0**	0.06	**1**	549	**11**	55	unk.	**unk.**	unk.	0.37	16	**2**	0.1	4.3	0.0	5	**1**	0.08	**4**	0		18
F	0.03	**2**	0.02	**1**	0.00	**0**	tr	**0**	0	0(a)	**0**	unk.	unk.	5	**1**	3	166	tr	0.4	**2**	0.3	**2**	unk.	**unk.**	54
	0.8	**4**	2	**1**	0.06	**1**	550	**11**	55	unk.	**unk.**	unk.	0.38	15	**2**	0.1	4.3	0.0	8	**2**	0.08	**4**	15		19
G	0.05	**3**	0.02	**1**	0.00	**0**	0.002	**1**	0	0(a)	**0**	unk.	unk.	9	**1**	1	202	5	0.5	**3**	0.2	**1**	unk.	**unk.**	42
	1.0	**5**	4	**1**	0.17	**2**	1330	**27**	133	unk.	**unk.**	unk.	0.50	19	**2**	0.0	5.2	0.1	7	**2**	0.07	**3**	0		1
H	0.03	**2**	0.02	**1**	0.00	**0**	0.001	**0**	0	0(a)	**0**	unk.	unk.	6	**1**	1	130	3	0.3	**2**	0.1	**1**	unk.	**unk.**	27
	0.6	**3**	3	**1**	0.11	**1**	858	**17**	86	unk.	**unk.**	unk.	0.32	12	**1**	0.0	3.3	0.1	4	**1**	0.04	**2**	0		1
I	0.05	**3**	0.02	**1**	0.00	**0**	unk.	**unk.**	0	0(a)	**0**	unk.	unk.	5	**1**	3	155	unk.	0.6	**4**	0.2	**2**	unk.	**unk.**	53
	0.9	**4**	unk.	**unk.**	0.16	**2**	813	**16**	81	unk.	**unk.**	unk.	0.50	16	**2**	0.1	4.0	unk.	8	**2**	0.08	**4**	0		unk.
J	0.08	**4**	0.02	**1**	0.10	**2**	0.000	**0**	0	0	**0**	0(a)	0(a)	37	**4**	51	103	1	0.2	**1**	0.2	**2**	7.8	**5**	17
	0.3	**2**	3	**1**	0.18	**2**	1159	**23**	38	0.0	**0**	0(a)	0.47	39	**4**	2.2	2.6	0.0	6	**2**	0.02	**1**	tr		8
K	0.03	**2**	0.01	**1**	0.00	**0**	0.001	**0**	0	0	**0**	0.0	0.0	7	**1**	1	142	3	0.5	**3**	0.3	**2**	unk.	**unk.**	83
	0.6	**3**	2	**1**	0.10	**1**	798	**16**	80	0.0	**0**	0.0	0.36	13	**1**	0.0	3.6	0.1	4	**1**	0.06	**3**	17		14
L	0.02	**1**	0.05	**3**	0.00	**0**	0.006	**2**	0	0	**0**	unk.	unk.	9	**1**	91	100	unk.	0.3	**2**	0.4	**3**	unk.	**unk.**	285
	2.4	**12**	12	**3**	0.37	**4**	0	**0**	0	unk.	**unk.**	unk.	0.57	61	**6**	4.0	2.6	unk.	26	**7**	0.09	**5**	unk.		9
M	0.02	**1**	0.05	**2**	0(a)	**0**	0.006	**2**	0(a)	0(a)	**0**	unk.	unk.	9	**1**	tr	94	unk.	0.3	**2**	0.4	**3**	unk.	**unk.**	tr
	2.2	**11**	12	**3**	0.36	**4**	0(a)	**0**	0(a)	unk.	**unk.**	unk.	0.55	57	**6**	0.0	2.4	unk.	25	**6**	0.09	**4**	unk.		9
N	0.04	**3**	0.14	**7**	0.00	**0**	0.012	**4**	0(a)	0	**0**	4.2	1.5	25	**3**	435	256	14	0.8	**4**	1.1	**7**	unk.	**unk.**	1
	5.9	**30**	37	**9**	0.73	**7**	tr	**0**	tr(a)	5.1	**17**	2.8	0.94	141	**14**	18.9	6.5	0.4	61	**15**	0.00	**0**	unk.		unk.
O	tr	**0**	0.00	**0**	0.00	**0**	tr	**0**	0	0	**0**	unk.	unk.	1	**0**	tr	8	0	tr	**0**	tr	**0**	unk.	**unk.**	0
	0.2	**1**	1	**0**	0.03	**0**	tr	**0**	tr	unk.	**unk.**	unk.	0.03	5	**1**	0.0	0.2	0.0	2	**1**	tr	**0**	unk.		unk.
P	0.05	**3**	0.14	**7**	0.00	**0**	0.012	**4**	0	0	**0**	unk.	unk.	26	**3**	2	252	15	0.8	**4**	1.1	**7**	unk.	**unk.**	544
	6.2	**31**	38	**10**	0.76	**8**	tr	**0**	tr(a)	unk.	**unk.**	unk.	0.97	147	**15**	0.1	6.4	0.4	63	**16**	0.10	**5**	unk.		unk.
Q	0.05	**3**	0.14	**7**	0.00	**0**	0.012	**4**	0(a)	0	**0**	4.0	1.6	27	**3**	150	243	unk.	0.8	**4**	1.1	**7**	unk.	**unk.**	249
	6.2	**31**	38	**10**	0.76	**8**	tr	**0**	tr(a)	4.8	**16**	2.4	1.37	144	**14**	6.5	6.2	unk.	63	**16**	0.15	**8**	unk.		unk.
R	0.05	**3**	0.14	**7**	0.00	**0**	0.012	**4**	0(a)	0	**0**	4.0	1.6	27	**3**	150	243	unk.	0.8	**4**	1.1	**7**	unk.	**unk.**	249
	6.2	**31**	38	**10**	0.76	**8**	tr	**0**	tr(a)	4.8	**16**	2.4	1.37	144	**14**	6.5	6.2	unk.	63	**16**	0.15	**8**	unk.		unk.
S	0.03	**2**	0.02	**1**	0.00	**0**	tr	**0**	0	0(a)	**0**	tr	unk.	6	**1**	1	107	4	0.3	**1**	unk.	**unk.**	unk.	**unk.**	13
	0.1	**1**	6	**2**	0.03	**0**	tr	**0**	tr	0.0	**0**	unk.	0.25	9	**1**	0.1	2.7	0.1	6	**2**	0.05	**3**	tr(a)		19

	FOOD	Portion	Weight in grams / Conversion for 100 g	Kilocalories / H_2O g	Total carbohydrate g / Total fats g	Crude fiber g / Dietary fiber g	Total protein g / Total sugar g	% USRDA	Arginine mg / Histidine mg	Isoleucine mg / Leucine mg	Lysine mg / Methionine mg	Phenylalanine mg / Threonine mg	Valine mg / Tryptophan mg	Cystine mg / Tyrosine mg	Polyunsat. fatty acids g / Monounsat. fatty acids g	Saturated fatty acids g / P/S ratio	Linoleic acid g / Cholesterol mg	Thiamin mg / Ascorbic acid mg	% USRDA
A	pear,canned,halves,light syrup,	1/2 C	122	74	19.0	0.85	0.2	0	unk.	7	7	6	10	unk.	unk.	tr(a)	unk.	0.01	1
	solids & liquid		*0.82*	102.2	0.2	3.05	14.5		2	10	2	6	9	2	unk.	unk.	0	1	2
B	pear,canned,halves,water pack,	1/2 C	122	39	10.1	0.85	0.2	0	unk.	7	7	6	10	unk.	unk.	tr(a)	unk.	0.01	1
	solids & liquid		*0.82*	111.1	0.2	1.83	5.6		2	10	2	6	9	2	unk.	unk.	unk.	1	2
C	pear,fresh,any,w/skin,edible portion	1 average	164	100	25.1	2.30	1.1	2	unk.	32	32	28	46	unk.	unk.	tr(a)	unk.	0.03	2
			0.61	136.4	0.7	3.77	14.6		11	46	11	30	9	44	unk.	unk.	0	7	11
D	pear,fresh,any,w/skin,sliced or	1/2 C	83	50	12.6	1.15	0.6	1	unk.	16	16	14	23	unk.	unk.	tr(a)	unk.	0.02	1
	cubed		*1.21*	68.6	0.3	1.90	6.8		6	23	6	15	5	22	unk.	unk.	0	3	6
E	PEAS,green & carrots,frozen,drained	1/2 C	80	42	8.1	1.20	2.6	4	unk.	unk.	unk.	unk.	unk.	unk.	unk.	tr(a)	unk.	0.15	10
	solids		*1.25*	68.6	0.2	unk.	2.6		unk.	unk.	unk.	unk.	unk.	unk.	unk.	unk.	0	6	11
F	peas,green & pearl onions,frozen,	1/2 C	94	67	12.3	unk.	4.2	6	unk.	unk.	unk.	unk.	unk.	unk.	unk.	tr(a)	unk.	0.23	16
	cooked,drained solids-Birds Eye		*1.06*	unk.	0.3	0.85	6.3		unk.	unk.	unk.	unk.	unk.	unk.	unk.	unk.	0	16	27
G	peas,green in pod,fresh,prep	1/2 C	69	58	9.9	unk.	4.4	7	unk.	unk.	unk.	unk.	unk.	unk.	unk.	unk.	unk.	0.24	16
			1.45	unk.	0.3	unk.	unk.		unk.	unk.	unk.	unk.	unk.	unk.	unk.	unk.	0(a)	19	31
H	peas,green,immature,canned,	1/2 C	85	68	12.7	1.61	3.9	6	369	180	184	152	161	37	0.00	0.00	unk.	0.09	6
	drained solids		*1.18*	67.1	0.3	5.36	5.4		83	246	27	63	31	121	unk.	0.00	0	7	11
I	peas,green,immature,canned,low	1/2 C	85	61	11.0	1.70	3.7	6	353	172	176	146	153	36	0.00	0.00	unk.	0.09	6
	sodium,drained solids		*1.18*	69.5	0.3	5.36	5.4		79	235	26	59	30	116	unk.	0.00	0	7	11
J	peas,green,immature,fresh,cooked,	1/2 C	80	57	9.7	1.60	4.3	7	408	198	203	153	177	42	0.00	0.00	unk.	0.22	15
	drained solids		*1.25*	65.2	0.3	4.16	4.4		92	272	30	69	34	110	unk.	0.00	0	16	27
K	peas,green,immature,frozen,cooked,	1/2 C	80	54	9.4	1.52	4.1	6	385	188	192	159	167	39	0.00	0.00	unk.	0.22	14
	drained solids		*1.25*	65.7	0.2	4.16	4.4		86	257	29	66	33	126	unk.	0.00	0	10	17
L	peas,green,immature,raw	1/2 C	73	61	10.4	1.45	4.6	7	431	210	215	162	187	43	0.00	0.00	unk.	0.25	17
			1.38	56.5	0.3	3.77	4.0		97	288	32	73	36	117	unk.	0.00	0	20	33
M	PECANS,shelled,chopped	1/4 C	30	203	4.3	0.68	2.8	4	unk.	163	128	166	155	64	5.40	1.80	5.01	0.25	17
			3.39	1.0	21.0	unk.	0.3		81	228	45	115	41	93	12.66	3.00	0	1	1
N	pecans,shelled,halves	1 half	1.4	10	0.2	0.03	0.1	0	unk.	8	6	8	7	3	0.26	0.09	0.24	0.01	1
			71.43	0.0	1.0	unk.	tr		4	11	2	5	2	4	0.60	3.00	0	0	0
O	PEPPER,HOT,chili,green,canned	1 med	25	6	1.5	0.30	0.2	0	unk.	9	10	11	6	unk.	tr(a)	tr(a)	tr(a)	0.00	0
			4.00	23.1	tr	unk.	0.5		unk.	9	3	10	2	unk.	tr(a)	0.00	0	17	28
P	pepper,hot,ground,black	1 tsp	2.1	5	1.4	0.28	0.2	0	unk.	unk.	unk.	unk.	unk.	unk.	0.03	0.03	0.03	tr	0
			47.62	0.2	0.1	unk.	0.0		unk.	unk.	unk.	unk.	unk.	unk.	0.03	1.16	0	unk.	unk.
Q	pepper,hot,ground,red or cayenne	1 tsp	1.8	6	1.0	0.45	0.2	0	unk.	unk.	unk.	unk.	unk.	unk.	0.15	0.06	0.14	0.01	0
			55.56	0.1	0.3	unk.	0.0		unk.	unk.	unk.	unk.	unk.	unk.	0.05	2.57	0	1	2
R	pepper,hot,ground,white	1 tsp	2.4	7	1.6	0.10	0.3	0	unk.	unk.	unk.	unk.	unk.	unk.	unk.	unk.	unk.	0.00	0
			41.67	0.3	0.1	unk.	0.0		unk.	unk.	unk.	unk.	unk.	unk.	unk.	unk.	0	unk.	unk.
S	PEPPER,SWEET,green,canned,stuffed	3/4 C	454	403	40.4	0.70	22.7	35	unk.	unk.	unk.	unk.	unk.	unk.	unk.	unk.	unk.	0.32	21
	w/beef & sauce-Campbell		*0.22*	359.1	24.1	unk.	unk.		unk.	unk.	unk.	unk.	unk.	unk.	unk.	unk.	unk.	45	75

Riboflavin mg / Niacin mg	% USRDA	Vitamin B_6 mg / Folacin mcg	% USRDA	Vitamin B_{12} mcg / Pantothenic acid mg	% USRDA	Biotin mg / Vitamin A IU	% USRDA	Preformed A RE / Beta carotene RE	Vitamin D IU / Vitamin E IU	% USRDA	Total tocopherol mg / Alpha tocopherol mg	Other tocopherol mg / Total ash g	Calcium mg / Phosphorus mg	% USRDA	Sodium mg / Sodium meq	Potassium mg / Potassium meq	Chlorine mg / Chlorine meq	Iron mg / Magnesium mg	% USRDA	Zinc mg / Copper mg	% USRDA	Iodine mcg / Selenium mcg	% USRDA	Manganese mcg / Chromium mcg	
0.02	1	0.02	1	0(a)	0	0.001	0	0	0(a)	0	tr	unk.	6	1	1	104	2	0.2	1	0.1	1	unk.	unk.	unk.	A
0.1	1	6	2	0.03	0	tr	0	tr	0.0	0	unk.	0.24	9	1	0.0	2.6	0.1	6	2	0.05	3	tr(a)		unk.	
0.02	1	0.02	1	0.00	0	tr	0	0	0(a)	0	tr	unk.	6	1	1	107	2	0.2	1	0.1	1	unk.	unk.	12	B
0.1	1	17	4	0.03	0	tr	0	tr	0.0	0	unk.	0.24	9	1	0.0	2.8	0.0	6	2	0.05	3	tr(a)		18	
0.07	4	0.03	1	0.00	0	tr	0	0	0(a)	0	tr	unk.	13	1	3	213	7	0.5	3	0.2	1	unk.	unk.	46	C
0.2	1	23	6	0.11	1	33	1	3	0.0	0	unk.	0.66	18	2	0.1	5.4	0.2	11	3	0.19	10	0		2	
0.03	2	0.01	1	0.00	0	tr	0	0	0(a)	0	tr	unk.	7	1	2	107	3	0.2	1	0.1	1	unk.	unk.	23	D
0.1	0	12	3	0.06	1	17	0	2	0.0	0	unk.	0.33	9	1	0.1	2.7	0.1	6	1	0.10	5	0		1	
0.06	3	0.07	4	0(a)	0	unk.	unk.	0	0(a)	0	unk.	unk.	20	2	67	126	unk.	0.9	5	unk.	unk.	unk.	unk.	unk.	E
1.0	5	unk.	unk.	0.18	2	7440	149	744	unk.	unk.	unk.	0.48	46	5	2.9	3.2	unk.	15	4	tr(a)	0	tr(a)		unk.	
0.08	4	unk.	unk.	0(a)	0	unk.	unk.	0	0(a)	0	unk.	unk.	21	2	428	135	unk.	1.6	9	unk.	unk.	unk.	unk.	unk.	F
1.6	8	unk.	unk.	unk.	unk.	515	10	52	unk.	unk.	unk.	unk.	72	7	18.6	3.5	unk.	unk.	unk.	tr(a)	0	tr(a)		unk.	
0.10	6	unk.	unk.	unk.	unk.	unk.	unk.	unk.	unk.	unk.	unk.	unk.	18	2	1	219	unk.	1.3	7	unk.	unk.	unk.	unk.	unk.	G
2.0	10	unk.	unk.	unk.	unk.	441	9	unk.	unk.	unk.	unk.	unk.	80	8	0.0	5.6	unk.	unk.	unk.	unk.	unk.	unk.		unk.	
0.05	3	0.05	3	0.00	0	0.002	1	0	0(a)	0	tr	tr	21	2	201	82	297	1.4	8	0.7	5	unk.	unk.	unk.	H
0.8	4	10	3	0.13	1	586	12	59	tr	0	tr	0.85	57	6	8.7	2.1	8.4	14	4	0.14	7	tr(a)		unk.	
0.05	3	0.04	2	0.00	0	0.002	1	0	0(a)	0	unk.	unk.	21	2	3	82	unk.	1.4	8	0.7	5	unk.	unk.	93	I
0.8	4	10	3	0.17	2	586	12	59	unk.	unk.	unk.	0.34	57	6	0.1	2.1	unk.	17	4	0.14	7	tr(a)		tr	
0.09	5	0.08	4	0(a)	0	0.002	1	0	0(a)	0	1.4	0.9	18	2	1	157	6	1.4	8	0.6	4	unk.	unk.	unk.	J
1.8	9	unk.	unk.	0.26	3	432	9	43	1.7	6	0.4	0.48	79	8	0.0	4.0	0.2	17	4	unk.	unk.	tr(a)		unk.	
0.07	4	0.08	4	0(a)	0	0.002	1	0	0(a)	0	0.5	0.3	15	2	92	108	6	1.5	8	unk.	unk.	unk.	unk.	unk.	K
1.4	7	unk.	unk.	0.26	3	480	10	48	0.6	2	0.2	0.56	69	7	4.0	2.8	0.2	17	4	tr(a)	0	tr(a)		unk.	
0.10	6	0.12	6	0.00	0	0.001	1	0	0(a)	0	1.5	1.4	19	2	1	229	24	1.4	8	0.7	4	unk.	unk.	46	L
2.1	11	18	5	0.54	5	464	9	46	1.8	6	0.1	0.65	84	8	0.1	5.9	0.7	25	6	0.03	2	tr(a)		unk.	
0.04	2	0.05	3	0.00	0	0.008	3	0	0	0	5.9	unk.	22	2	3	178	15	0.7	4	1.2	8	unk.	unk.	442	M
0.3	1	8	2	0.50	5	38	1	4	0.4	1	0.4	0.47	85	9	0.1	4.5	0.4	32	8	0.32	16	1		18	
tr	0	tr	0	0.00	0	tr	0	0	0	0	0.3	unk.	1	0	tr	8	1	tr	0	0.1	0	unk.	unk.	21	N
tr	0	tr	0	0.02	0	2	0	tr	0.3	1	tr	0.02	4	0	0.0	0.2	0.0	2	0	0.02	1	0		1	
0.01	1	unk.	unk.	0.00	0	unk.	unk.	0	0(a)	0	unk.	unk.	2	0	unk.	unk.	unk.	0.1	1	unk.	unk.	unk.	unk.	unk.	O
0.2	1	13	3	0.17	2	153	3	15	unk.	unk.	unk.	0.10	4	0	unk.	unk.	unk.	unk.	unk.	unk.	unk.	unk.		unk.	
0.01	0	unk.	unk.	0.00	0	unk.	unk.	0(a)	0(a)	0	unk.	unk.	9	1	1	26	unk.	0.6	3	tr	0	unk.	unk.	unk.	P
tr	0	unk.	unk.	unk.	unk.	4	0	unk.	unk.	unk.	unk.	0.09	4	0	0.0	0.7	unk.	4	1	unk.	unk.	unk.		unk.	
0.02	1	unk.	unk.	0.00	0	unk.	unk.	0(a)	0(a)	0	unk.	unk.	3	0	1	36	unk.	0.1	1	tr	0	unk.	unk.	unk.	Q
0.2	1	unk.	unk.	unk.	unk.	749	15	unk.	unk.	unk.	unk.	0.11	5	1	0.0	0.9	unk.	3	1	unk.	unk.	unk.		unk.	
tr	0	unk.	unk.	0.00	0	unk.	unk.	0(a)	0(a)	0	unk.	unk.	6	1	tr	2	unk.	0.3	2	tr	0	unk.	unk.	unk.	R
tr	0	unk.	unk.	unk.	unk.	tr	0	unk.	unk.	unk.	unk.	0.04	4	0	0.0	0.0	unk.	2	1	unk.	unk.	unk.		unk.	
0.18	11	unk.	unk.	unk.	unk.	unk.	unk.	unk.	unk.	unk.	unk.	unk.	86	9	unk.	913	unk.	3.2	18	unk.	unk.	unk.	unk.	unk.	S
5.0	25	unk.	unk.	unk.	unk.	1566	31	unk.	unk.	unk.	unk.	6.80	96	10	unk.	23.4	unk.	unk.	unk.	unk.	unk.	unk.		unk.	

	FOOD	Portion	Weight in grams / *Conversion for 100 g*	Kilocalories / H₂O g	Total carbohydrate g / Total fats g	Crude fiber g / Dietary fiber g	Total protein g / Total sugar g	% USRDA	Arginine mg / Histidine mg	Isoleucine mg / Leucine mg	Lysine mg / Methionine mg	Phenylalanine mg / Threonine mg	Valine mg / Tryptophan mg	Cystine mg / Tyrosine mg	Polyunsat. fatty acids g / Monounsat. fatty acids g	Saturated fatty acids g / P/S ratio	Linoleic acid g / Cholesterol mg	Thiamin mg / Ascorbic acid mg	% USRDA
A	pepper,sweet,green,fresh,cooked,	1 average	160	29	6.1	2.24	1.6	**3**	unk.	61	67	74	43	unk.	0.00	tr(a)	unk.	0.10	**6**
	drained solids		*0.63*	151.5	0.3	unk.	3.0		unk.	61	21	67	11	unk.	unk.	0.00	0	154	**256**
B	pepper,sweet,green,fresh,strips,	1/2 C	68	12	2.6	0.94	0.7	**1**	unk.	26	28	31	18	unk.	0.00	0.00	unk.	0.04	**3**
	cooked,drained solids		*1.48*	63.9	0.1	unk.	1.3		unk.	26	9	28	5	unk.	unk.	0.00	0	65	**108**
C	pepper,sweet,green,home recipe,stuffed	2-3/4" long,	185	322	18.2	2.57	20.4	**43**	1038	941	1706	915	1018	209	1.44	7.80	0.40	0.20	**13**
		1-1/8C stuffing	*0.54*	124.2	18.5	5.09	5.9		641	1583	502	857	235	752	5.36	0.18	65	166	**277**
D	pepper,sweet,green,raw,chopped	1/2 C	75	17	3.6	1.05	0.9	**1**	unk.	35	38	41	25	unk.	0.00	0.00	unk.	0.06	**4**
			1.33	70.0	0.1	unk.	1.4		11	35	12	38	7	unk.	unk.	0.00	0	96	**160**
E	pepper,sweet,green,raw,edible portion	1 average	164	36	7.9	2.30	2.0	**3**	unk.	75	84	90	54	unk.	0.00	0.00	unk.	0.13	**9**
			0.61	153.2	0.3	unk.	3.1		23	75	26	82	15	unk.	unk.	0.00	0	210	**350**
F	PERCH,ocean,breaded,fried	1 med piece	70	159	4.8	0.00	13.3	**30**	751	624	1256	500	862	157	unk.	unk.	unk.	0.07	**5**
			1.43	41.3	9.3	0(a)	0.0		472	949	379	561	145	487	unk.	unk.	unk.	unk.	**unk.**
G	perch,ocean,cooked	1 med piece	65	60	0.0	0.00	12.3	**27**	697	579	1166	464	801	146	0.56	0.27	0.02	0.06	**4**
			1.54	38.3	0.8	0(a)	0.0		439	881	352	521	135	452	0.29	2.10	unk.	unk.	**unk.**
H	PERSIMMON,raw,edible portion	1 average	136	103	27.6	0.46	0.7	**1**	unk.	tr(a)	tr(a)	tr(a)	tr(a)	tr(a)	tr(a)	tr(a)	tr(a)	tr	**0**
			0.74	107.0	0.2	unk.	unk.		tr(a)	tr(a)	tr(a)	tr(a)	tr(a)	tr(a)	tr(a)	0.00	0(a)	10	**17**
I	PHEASANT,breast,meat w/o skin,raw	1/2 breast	182	242	0.0	0.00	44.3	**99**	2697	2492	4059	1731	2455	582	1.00	2.00	0.87	0.15	**10**
			0.55	131.8	5.9	0(a)	0(a)		1767	3755	1292	2220	617	1454	1.57	0.50	unk.	11	**18**
J	pheasant,leg,meat w/o skin,raw	1 piece	140	188	0.0	0.00	31.1	**69**	1890	1746	2845	1214	1721	407	1.02	2.04	0.90	0.10	**7**
			0.71	102.8	6.0	0(a)	0(a)		1890	2631	904	1557	433	1019	1.60	0.50	unk.	unk.	**unk.**
K	pheasant,meat w/skin,raw	1/2 bird	390	707	0.0	0.00	88.6	**197**	5512	4794	7867	3420	4802	1191	4.61	10.54	3.16	0.27	**18**
			0.26	264.6	36.3	0(a)	0(a)		3373	7300	2510	4326	1187	2826	12.02	0.44	unk.	21	**35**
L	PICKLE RELISH,sour	1 Tbsp	15	3	0.4	0.16	0.1	**0**	unk.	3	5	2	4	unk.	unk.	unk.	unk.	tr	**0**
			6.67	13.9	0.1	unk.	0.3		tr	4	1	3	1	unk.	unk.	unk.	0	unk.	**unk.**
M	pickle relish,sweet	1 Tbsp	15	21	5.1	0.12	0.1	**0**	unk.	2	3	2	3	unk.	unk.	unk.	unk.	0(a)	**0**
			6.67	9.4	0.1	unk.	0.6		tr	3	1	2	1	unk.	unk.	unk.	0	unk.	**unk.**
N	PICKLE,bread & butter	1 slice from	7.1	5	1.3	0.04	0.1	**0**	unk.	2	3	1	2	unk.	unk.	tr(a)	unk.	tr	**0**
		med pickle	*14.08*	5.6	tr	unk.	1.2		tr	3	1	2	tr	unk.	unk.	unk.	0	1	**2**
O	pickle,dill,sliced	1 slice from	6.7	1	0.1	0.03	0.0	**0**	unk.	1	2	1	2	unk.	unk.	tr(a)	unk.	tr	**0**
		med pickle	*14.93*	6.3	tr	0.07	0.1		unk.	2	0	1	tr	unk.	unk.	unk.	0	tr	**1**
P	pickle,dill,whole	1 med	100	11	2.2	0.50	0.7	**1**	unk.	22	31	16	24	unk.	unk.	tr(a)	unk.	tr	**0**
			1.00	93.3	0.2	1.10	1.8		1	30	7	19	5	unk.	unk.	unk.	0	6	**10**
Q	pickle,sweet (gherkin)	1 sm	15	22	5.5	0.07	0.1	**0**	unk.	3	5	2	4	unk.	unk.	tr(a)	unk.	0.00	**0**
			6.67	9.1	0.1	unk.	5.1		tr	4	1	3	1	unk.	unk.	unk.	0	1	**2**
R	pickle,sweet,sliced	1 slice from	3.0	4	1.1	0.01	0.0	**0**	unk.	1	1	0	1	unk.	unk.	tr(a)	unk.	0.00	**0**
		med pickle	*33.33*	1.8	tr	unk.	1.0		0	1	tr	1	tr	unk.	unk.	unk.	0	tr	**0**
S	PIE CRUST,graham cracker crumb	1/7 of pie	42	242	21.2	unk.	2.7	**4**	unk.	unk.	unk.	unk.	unk.	unk.	unk.	unk.	unk.	unk.	**unk.**
	-Nabisco	crust	*2.38*	0.9	16.3	unk.	unk.		unk.	unk.	unk.	unk.	unk.	unk.	unk.	unk.	unk.	unk.	**unk.**

Riboflavin mg / Niacin mg	% USRDA	Vitamin B_6 mg / Folacin mcg	% USRDA	Vitamin B_{12} mcg / Pantothenic acid mg	% USRDA	Biotin mg / Vitamin A IU	% USRDA	Preformed A RE / Beta carotene RE	Vitamin D IU / Vitamin E IU	% USRDA	Total tocopherol mg / Alpha tocopherol mg	Other tocopherol mg / Total ash g	Calcium mg / Phosphorus mg	% USRDA	Sodium mg / Sodium meq	Potassium mg / Potassium meq	Chlorine mg / Chlorine meq	Iron mg / Magnesium mg	% USRDA	Zinc mg / Copper mg	% USRDA	Iodine mcg / Selenium mcg	% USRDA	Manganese mcg / Chromium mcg	
0.11	**7**	0.22	**11**	unk.	**unk.**	unk.	**unk.**	0	0(a)	**0**	0.8	tr(a)	14	**1**	14	238	24	0.8	**4**	0.1	**1**	unk.	**unk.**	202	A
0.8	**4**	30	**8**	0.26	**3**	672	**13**	67	1.0	**3**	tr(a)	0.48	26	**3**	0.6	6.1	0.7	16	**4**	0.07	**3**	tr(a)		unk.	
0.05	**3**	0.09	**5**	unk.	**unk.**	unk.	**unk.**	0	0(a)	**0**	0.3	tr(a)	6	**1**	6	101	10	0.3	**2**	tr	**0**	unk.	**unk.**	85	B
0.3	**2**	13	**3**	0.11	**1**	283	**6**	28	0.4	**1**	tr(a)	0.20	11	**1**	0.3	2.6	0.3	7	**2**	0.03	**2**	tr(a)		unk.	
0.33	**20**	0.22	**11**	0.94	**16**	0.002	**1**	5	0	**0**	1.0	unk.	148	**15**	712	618	unk.	3.2	**18**	3.4	**23**	46.7	**31**	220	C
3.6	**18**	44	**11**	0.46	**5**	1465	**29**	122	1.2	**4**	unk.	3.35	263	**26**	31.0	15.8	unk.	24	**6**	0.19	**10**	0		0	
0.06	**4**	0.19	**10**	0.00	**0**	unk.	**unk.**	0	unk.	**unk.**	0.4	unk.	7	**1**	10	160	10	0.5	**3**	tr	**0**	unk.	**unk.**	95	D
0.4	**2**	14	**4**	0.17	**2**	315	**6**	32	0.4	**2**	unk.	0.30	17	**2**	0.4	4.1	0.3	14	**3**	0.03	**2**	tr(a)		14	
0.13	**8**	0.43	**21**	0.00	**0**	unk.	**unk.**	0	0(a)	**0**	0.8	tr(a)	15	**2**	21	349	21	1.1	**6**	0.1	**1**	unk.	**unk.**	207	E
0.8	**4**	31	**8**	0.38	**4**	689	**14**	69	1.0	**3**	tr(a)	0.66	36	**4**	0.9	8.9	0.6	30	**7**	0.07	**4**	tr(a)		31	
0.08	**5**	0.10	**5**	0.63	**11**	unk.	**unk.**	unk.	unk.	**unk.**	tr(a)	tr(a)	23	**2**	107	199	unk.	0.9	**5**	unk.	**unk.**	unk.	**unk.**	unk.	F
1.3	**6**	6	**2**	0.13	**1**	unk.	**unk.**	unk.	0.0	**0**	tr(a)	1.33	158	**16**	4.7	5.1	unk.	unk.	**unk.**	unk.	**unk.**	tr(a)		tr(a)	
0.07	**4**	0.09	**5**	0.58	**10**	unk.	**unk.**	unk.	unk.	**unk.**	tr(a)	tr(a)	21	**2**	99	185	unk.	0.8	**5**	unk.	**unk.**	unk.	**unk.**	unk.	G
1.2	**6**	6	**2**	0.12	**1**	unk.	**unk.**	unk.	0.0	**0**	tr(a)	1.23	147	**15**	4.3	4.7	unk.	unk.	**unk.**	unk.	**unk.**	tr(a)		tr(a)	
0.12	**7**	unk.	**unk.**	unk.	**unk.**	unk.	**unk.**	unk.	0(a)	**0**	tr(a)	tr(a)	8	**1**	30	880	unk.	0.2	**1**	unk.	**unk.**	unk.	**unk.**	unk.	H
0.2	**1**	unk.	**unk.**	unk.	**unk.**	unk.	**unk.**	unk.	0.0	**0**	tr(a)	0.46	26	**3**	1.3	22.5	unk.	unk.	**unk.**	unk.	**unk.**	unk.		unk.	
0.22	**13**	1.35	**67**	1.53	**26**	unk.	**unk.**	unk.	unk.	**unk.**	unk.	unk.	5	**1**	60	440	unk.	1.4	**8**	1.1	**8**	unk.	**unk.**	27	I
15.6	**78**	unk.	**unk.**	1.75	**18**	268	**5**	unk.	unk.	**unk.**	unk.	2.37	364	**36**	2.6	11.3	unk.	38	**10**	0.08	**4**	unk.		unk.	
0.29	**17**	unk.	**unk.**	unk.	**unk.**	unk.	**unk.**	unk.	unk.	**unk.**	unk.	unk.	41	**4**	63	414	unk.	2.5	**14**	2.1	**14**	unk.	**unk.**	28	J
5.2	**26**	unk.	**unk.**	unk.	**unk.**	273	**6**	unk.	unk.	**unk.**	unk.	2.02	392	**39**	2.7	10.6	unk.	28	**7**	0.15	**8**	unk.		unk.	
0.55	**32**	2.58	**129**	3.01	**50**	unk.	**unk.**	unk.	unk.	**unk.**	unk.	unk.	47	**5**	156	949	unk.	4.5	**25**	3.7	**25**	unk.	**unk.**	66	K
25.1	**126**	unk.	**unk.**	3.62	**36**	691	**14**	unk.	unk.	**unk.**	unk.	4.96	836	**84**	6.8	24.3	unk.	78	**20**	0.25	**13**	unk.		unk.	
unk.	**unk.**	tr	**0**	0.00	**0**	unk.	**unk.**	0(a)	0(a)	**0**	unk.	unk.	4	**0**	unk.	unk.	unk.	0.2	**1**	unk.	**unk.**	unk.	**unk.**	unk.	L
unk.	**unk.**	unk.	**unk.**	unk.	**unk.**	unk.	**unk.**	unk.	unk.	**unk.**	unk.	0.40	3	**0**	unk.	unk.	unk.	2	**0**	unk.	**unk.**	unk.		unk.	
tr	**0**	tr	**0**	0.00	**0**	unk.	**unk.**	0(a)	0(a)	**0**	unk.	unk.	3	**0**	107	unk.	unk.	0.1	**1**	tr	**0**	unk.	**unk.**	unk.	M
tr	**0**	unk.	**unk.**	unk.	**unk.**	unk.	**unk.**	unk.	unk.	**unk.**	unk.	0.28	2	**0**	4.6	unk.	unk.	tr	**0**	unk.	**unk.**	unk.		unk.	
tr	**0**	0.00	**0**	0.00	**0**	unk.	**unk.**	0	0(a)	**0**	unk.	unk.	2	**0**	48	unk.	unk.	0.1	**1**	unk.	**unk.**	unk.	**unk.**	unk.	N
tr	**0**	unk.	**unk.**	unk.	**unk.**	10	**0**	1	unk.	**unk.**	unk.	0.16	2	**0**	2.1	unk.	unk.	tr	**0**	unk.	**unk.**	unk.		unk.	
tr	**0**	0.00	**0**	0.00	**0**	unk.	**unk.**	0	0	**0**	unk.	unk.	2	**0**	96	13	unk.	0.1	**0**	tr	**0**	unk.	**unk.**	unk.	O
0.0	**0**	unk.	**unk.**	unk.	**unk.**	7	**0**	1	unk.	**unk.**	unk.	0.24	1	**0**	4.2	0.3	unk.	1	**0**	unk.	**unk.**	unk.		unk.	
0.02	**1**	0.01	**0**	0.00	**0**	unk.	**unk.**	0	0	**0**	unk.	unk.	26	**3**	1428	200	unk.	1.0	**6**	0.3	**2**	unk.	**unk.**	unk.	P
0.1	**0**	unk.	**unk.**	unk.	**unk.**	100	**2**	10	unk.	**unk.**	unk.	3.60	21	**2**	62.1	5.1	unk.	12	**3**	unk.	**unk.**	unk.		unk.	
tr	**0**	tr	**0**	0.00	**0**	unk.	**unk.**	0	0	**0**	unk.	unk.	2	**0**	unk.	unk.	unk.	0.2	**1**	tr	**0**	unk.	**unk.**	unk.	Q
tr	**0**	unk.	**unk.**	unk.	**unk.**	13	**0**	1	unk.	**unk.**	unk.	0.25	2	**0**	unk.	unk.	unk.	tr	**0**	unk.	**unk.**	unk.		unk.	
tr	**0**	0.00	**0**	0.00	**0**	unk.	**unk.**	0	0	**0**	unk.	unk.	tr	**0**	unk.	unk.	unk.	tr	**0**	0.0	**0**	unk.	**unk.**	unk.	R
tr	**0**	unk.	**unk.**	unk.	**unk.**	3	**0**	tr	unk.	**unk.**	unk.	0.05	1	**0**	unk.	unk.	unk.	0	**0**	unk.	**unk.**	unk.		unk.	
unk.	**unk.**	unk.	**unk.**	unk.	**unk.**	unk.	**unk.**	unk.	0(a)	**0**	unk.	unk.	unk.	**unk.**	unk.	unk.	unk.	unk.	**unk.**	unk.	**unk.**	unk.	**unk.**	unk.	S
unk.	**unk.**	unk.	**unk.**	0.19	**2**	unk.	**unk.**	unk.	unk.	**unk.**	unk.	0.92	unk.	**unk.**	unk.	unk.	unk.	unk.	**unk.**	unk.	**unk.**	unk.		unk.	

	FOOD	Portion	Weight in grams / *Conversion for 100 g*	Kilocalories / H_2O g	Total carbohydrate g / Total fats g	Crude fiber g / Dietary fiber g	Total protein g / Total sugar g	**% USRDA**	Arginine mg / Histidine mg	Isoleucine mg / Leucine mg	Lysine mg / Methionine mg	Phenylalanine mg / Threonine mg	Valine mg / Tryptophan mg	Cystine mg / Tyrosine mg	Polyunsat. fatty acids g / Monounsat. fatty acids g	Saturated fatty acids g / P/S ratio	Linoleic acid g / Cholesterol mg	Thiamin mg / Ascorbic acid mg	**% USRDA**
A	pie crust,patty shell-Pepperidge Farm	1 shell	47	249	18.1	0.05	3.2	**5**	unk.	unk.	unk.	unk.	unk.	unk.	unk.	unk.	unk.	0.20	**13**
			2.13	6.7	18.2	unk.	unk.		unk.	unk.	unk.	unk.	unk.	unk.	unk.	unk.	unk.	0	**0**
B	PIE,apple,home recipe	1/7 of 9" dia	135	402	56.2	0.12	3.2	**5**	unk.	151	78	174	139	61	0.27	5.37	1.32	0.15	**10**
			0.74	55.4	19.2	2.63	32.3		64	245	43	94	38	108	10.91	0.05	7	3	**5**
C	pie,banana cream,home recipe	1/7 of 9" dia	140	354	47.9	0.24	6.0	**12**	168	294	326	288	318	102	0.50	5.38	1.12	0.14	**9**
			0.71	69.1	16.0	1.54	30.9		129	518	132	225	83	243	8.18	0.09	83	4	**7**
D	pie,blackberry,home recipe	1/7 of 9" dia	135	372	57.3	3.08	4.2	**6**	unk.	153	76	182	143	66	0.24	3.84	1.09	0.16	**11**
			0.74	57.6	14.9	6.20	29.6		66	256	44	96	41	114	8.51	0.06	4	16	**26**
E	pie,blueberry,home recipe	1/7 of 9" dia	135	411	59.1	1.22	3.8	**6**	unk.	153	76	182	143	66	0.24	4.76	1.37	0.16	**11**
			0.74	52.5	18.6	unk.	32.3		66	256	44	96	41	114	11.11	0.05	4	11	**18**
F	pie,butterscotch,home recipe	1/7 of 9" dia	140	417	50.0	0.08	7.0	**14**	217	366	384	346	399	128	0.67	7.65	1.43	0.15	**10**
			0.71	60.7	21.3	0.73	30.3		162	605	160	276	96	290	10.60	0.09	108	1	**1**
G	pie,cherry,home recipe	1/7 of 9" dia	133	462	68.9	0.29	4.1	**6**	unk.	158	80	188	149	68	0.30	5.66	1.41	0.17	**11**
			0.75	39.3	19.8	1.77	43.2		69	265	45	99	42	118	11.48	0.05	8	5	**8**
H	pie,chess,home recipe	1/7 of 9" dia	148	682	75.8	0.31	5.0	**9**	136	183	224	156	199	52	2.63	19.69	3.84	0.17	**11**
			0.68	23.5	42.0	0.72	55.4		87	294	72	151	48	143	14.60	0.13	201	1	**2**
I	pie,chocolate chiffon,home recipe	1/7 of 9" dia	93	337	38.5	0.24	6.1	**11**	140	231	251	252	268	98	0.46	5.89	1.38	0.12	**8**
			1.08	28.7	18.8	0.61	20.3		127	391	111	196	61	182	10.47	0.08	94	0	**0**
J	pie,chocolate meringue,home recipe	1/7 of 9" dia	130	371	46.4	0.21	6.2	**12**	175	292	306	276	318	102	0.50	7.08	1.18	0.12	**8**
			0.77	57.8	18.9	0.63	29.0		129	482	128	220	77	232	9.29	0.07	83	0	**0**
K	pie,coconut custard,home recipe	1/7 of 9" dia	140	346	41.6	0.38	5.5	**11**	242	288	300	273	316	101	0.49	7.68	1.05	0.11	**7**
			0.71	74.0	17.9	2.80	25.0		127	474	126	217	75	226	7.90	0.06	78	tr	**1**
L	pie,custard,home recipe	1/7 of 9" dia	135	324	32.0	0.06	7.8	**16**	261	409	468	391	465	138	0.65	5.66	1.46	0.13	**9**
			0.74	75.8	18.4	0.59	16.0		213	674	192	331	107	333	10.03	0.12	136	1	**2**
M	pie,grasshopper,home recipe	1/7 of 9" dia	135	396	39.5	0.07	5.5	**12**	unk.	216	325	207	278	79	0.97	11.71	0.81	0.03	**2**
			0.74	62.4	20.3	unk.	39.1		135	364	127	209	56	180	5.86	0.08	179	tr	**0**
N	pie,lemon chiffon,home recipe	1/7 of 9" dia	105	381	41.9	0.06	6.4	**13**	189	288	330	303	338	119	0.76	7.50	1.52	0.13	**9**
			0.95	34.3	21.4	0.59	25.9		161	485	143	248	76	230	10.78	0.10	145	7	**11**
O	pie,lemon meringue,home recipe	1/7 of 9" dia	135	399	60.1	0.07	4.4	**8**	123	210	201	226	235	93	0.51	5.03	1.25	0.12	**8**
			0.74	53.7	16.2	0.60	41.5		114	351	98	170	58	171	8.83	0.10	91	4	**7**
P	pie,lime,key,home recipe	1/7 of 9" dia	130	393	55.2	0.04	8.8	**18**	304	491	594	431	545	118	2.70	6.62	2.46	0.13	**9**
			0.77	48.1	16.2	0.43	43.7		234	801	212	376	121	394	5.88	0.41	87	7	**11**
Q	pie,mince,home recipe	1/7 of 9" dia	135	441	49.5	0.06	4.2	**8**	76	156	161	174	151	55	0.44	3.61	0.92	0.12	**8**
			0.74	54.4	25.9	2.68	33.6		84	259	61	118	38	115	8.07	0.12	6	5	**8**
R	pie,peach,home recipe	1/7 of 9" dia	140	409	55.7	0.58	3.8	**6**	unk.	164	102	193	177	74	0.30	5.68	1.40	0.16	**11**
			0.71	59.7	19.8	2.07	30.7		81	281	69	118	44	126	11.59	0.05	8	6	**9**
S	pie,pecan,home recipe	1/7 of 9" dia	120	566	64.3	0.40	6.0	**11**	150	305	285	321	330	130	3.43	9.41	3.98	0.24	**16**
			0.83	18.5	33.2	0.58	46.4		163	487	129	240	82	229	17.31	0.36	114	tr	**1**

Riboflavin mg / Niacin mg	% USRDA	Vitamin B_6 mg / Folacin mcg	% USRDA	Vitamin B_{12} mcg / Pantothenic acid mg	% USRDA	Biotin mg / Vitamin A IU	% USRDA	Preformed A RE / Beta carotene RE	Vitamin D IU / Vitamin E IU	% USRDA	Total tocopherol mg / Alpha tocopherol mg	Other tocopherol mg / Total ash g	Calcium mg / Phosphorus mg	% USRDA	Sodium mg / Sodium meq	Potassium mg / Potassium meq	Chlorine mg / Chlorine meq	Iron mg / Magnesium mg	% USRDA	Zinc mg / Copper mg	% USRDA	Iodine mcg / Selenium mcg	% USRDA	Manganese mcg / Chromium mcg	
0.06	**4**	unk.	**unk.**	unk.	**unk.**	unk.	**unk.**	unk.	unk.	**unk.**	unk.	unk.	1	**0**	245	28	unk.	0.9	**5**	unk.	**unk.**	unk.	**unk.**	unk.	A
0.4	**2**	unk.	**unk.**	unk.	**unk.**	0	**0**	unk.	unk.	**unk.**	unk.	0.80	19	**2**	10.6	0.7	unk.	6	**1**	unk.	**unk.**	unk.		unk.	
0.09	**5**	0.04	**2**	0.00	**0**	0.001	**0**	0	0	**0**	1.7	tr	11	**1**	292	116	21	1.0	**6**	0.4	**3**	67.7	**45**	148	B
1.1	**6**	12	**3**	0.22	**2**	174	**4**	7	2.1	**7**	1.7	1.10	32	**3**	12.7	3.0	0.6	12	**3**	0.07	**4**	15		31	
0.20	**12**	0.23	**12**	0.33	**6**	0.004	**2**	0	29	**7**	0.4	0.1	72	**7**	232	258	85	1.0	**6**	0.7	**5**	54.6	**36**	146	C
1.0	**5**	20	**5**	0.59	**6**	257	**5**	7	0.5	**2**	0.2	1.34	104	**10**	10.1	6.6	2.4	23	**6**	0.15	**7**	11		31	
0.11	**7**	0.06	**3**	0.00	**0**	tr	**0**	0	0	**0**	tr	0.0	19	**2**	292	155	33	1.3	**7**	0.4	**3**	71.2	**48**	1395	D
1.4	**7**	17	**4**	0.32	**3**	199	**4**	14	tr	**0**	tr	1.23	43	**4**	12.7	4.0	0.9	25	**6**	0.13	**7**	17		33	
0.12	**7**	0.07	**3**	0.00	**0**	tr	**0**	0	0	**0**	4.6	0.0	20	**2**	292	92	28	1.7	**10**	0.4	**3**	71.2	**48**	1803	E
1.5	**7**	12	**3**	0.26	**3**	127	**3**	7	5.5	**18**	4.6	1.09	39	**4**	12.7	2.3	0.8	15	**4**	0.13	**6**	17		33	
0.23	**13**	0.06	**3**	0.41	**7**	0.003	**1**	0	36	**9**	0.4	0.0	89	**9**	305	145	51	1.2	**7**	0.8	**6**	68.3	**46**	124	F
1.0	**5**	13	**3**	0.63	**6**	290	**6**	0	0.4	**2**	0.1	1.37	122	**12**	13.3	3.7	1.4	19	**5**	0.14	**7**	14		32	
0.11	**7**	0.04	**2**	0.00	**0**	tr	**0**	0	0	**0**	tr	0.0	13	**1**	307	126	23	1.1	**6**	0.5	**3**	71.2	**48**	175	G
1.3	**7**	11	**3**	0.28	**3**	162	**3**	5	0.0	**0**	tr	1.20	37	**4**	13.4	3.2	0.6	19	**5**	0.05	**3**	17		41	
0.15	**9**	0.08	**4**	0.35	**6**	0.006	**2**	0	14	**4**	2.2	0.1	107	**11**	295	375	61	3.3	**18**	0.8	**5**	14.2	**10**	2164	H
0.7	**4**	11	**3**	0.67	**7**	1068	**21**	1	2.7	**9**	0.4	2.17	138	**14**	12.8	9.6	1.7	55	**14**	0.15	**7**	1		39	
0.14	**8**	0.04	**2**	0.20	**3**	0.006	**2**	0	7	**2**	0.6	0.0	22	**2**	232	106	53	1.5	**9**	0.6	**4**	66.3	**44**	102	I
0.9	**4**	17	**4**	0.38	**4**	47	**1**	tr	0.7	**2**	0.0	1.03	93	**9**	10.1	2.7	1.5	74	**19**	0.05	**3**	12		25	
0.20	**12**	0.05	**3**	0.33	**6**	0.005	**2**	0	29	**7**	0.6	0.0	75	**8**	232	163	45	1.3	**8**	0.7	**5**	54.6	**36**	101	J
0.9	**4**	16	**4**	0.51	**5**	192	**4**	tr	0.7	**2**	0.1	1.24	119	**12**	10.1	4.2	1.3	32	**8**	0.12	**6**	11		28	
0.17	**10**	0.05	**3**	0.31	**5**	0.003	**1**	0	27	**7**	0.5	0.0	68	**7**	219	136	55	1.0	**6**	0.6	**4**	51.2	**34**	93	K
0.8	**4**	12	**3**	0.49	**5**	177	**4**	0	0.6	**2**	0.1	1.06	100	**10**	9.5	3.5	1.6	20	**5**	0.11	**6**	10		24	
0.15	**15**	0.08	**4**	0.53	**9**	0.005	**2**	0	40	**10**	0.3	0.0	110	**11**	277	172	59	1.2	**7**	0.9	**6**	76.2	**51**	101	L
0.8	**4**	16	**4**	0.69	**7**	152	**3**	0	0.4	**1**	0.1	1.41	151	**15**	12.0	4.4	1.7	78	**20**	0.16	**8**	11		20	
0.13	**8**	0.20	**10**	0.34	**6**	0.005	**2**	0	16	**4**	0.3	0.0	43	**4**	141	90	44	0.8	**4**	0.6	**4**	30.1	**20**	17	M
0.1	**0**	9	**2**	0.45	**5**	641	**13**	0	0.3	**1**	0.0	0.77	137	**14**	6.1	2.3	1.2	71	**18**	0.04	**2**	10		10	
0.15	**9**	0.06	**3**	0.30	**5**	0.005	**2**	0	9	**2**	0.3	0.0	31	**3**	314	90	59	1.3	**7**	0.7	**5**	88.4	**59**	106	N
0.8	**4**	13	**3**	0.49	**5**	278	**6**	tr	0.3	**1**	0.0	1.15	92	**9**	13.7	2.3	1.7	71	**18**	0.05	**3**	12		27	
0.10	**6**	0.04	**2**	0.19	**3**	0.003	**1**	0	6	**2**	0.2	0.0	16	**2**	174	60	45	1.1	**6**	0.7	**4**	43.9	**29**	111	O
0.8	**4**	10	**2**	0.34	**3**	138	**3**	tr	0.2	**1**	tr	0.66	62	**6**	7.6	1.5	1.3	49	**12**	0.05	**2**	12		37	
0.37	**22**	0.07	**4**	0.45	**8**	0.002	**1**	0	5	**1**	0.6	0.3	217	**22**	313	314	32	0.9	**5**	1.0	**7**	61.7	**41**	65	P
0.7	**3**	15	**4**	0.80	**8**	266	**5**	tr	0.7	**2**	0.2	2.06	228	**23**	13.6	8.0	0.9	55	**14**	0.03	**2**	8		11	
0.09	**5**	0.09	**4**	0.07	**1**	0.002	**1**	1	0	**0**	0.3	tr	20	**2**	278	258	34	1.6	**9**	0.5	**3**	69.1	**46**	190	Q
1.0	**5**	12	**3**	0.18	**2**	62	**1**	6	0.4	**1**	0.3	1.35	48	**5**	12.1	6.6	1.0	17	**4**	0.10	**5**	10		19	
0.13	**7**	0.04	**2**	0.00	**0**	0.002	**1**	0	0	**0**	tr	0.0	15	**2**	307	195	26	1.3	**8**	0.5	**4**	71.2	**48**	169	R
1.9	**10**	10	**3**	0.28	**3**	1181	**24**	107	tr	**0**	tr	1.32	45	**5**	13.4	5.0	0.7	16	**4**	0.10	**5**	17		34	
0.13	**8**	0.07	**3**	0.20	**3**	0.008	**3**	0	7	**2**	3.2	0.0	47	**5**	269	140	85	3.1	**17**	1.7	**11**	48.8	**33**	322	S
0.9	**5**	14	**4**	0.62	**6**	348	**7**	2	3.9	**13**	0.3	1.33	116	**12**	11.7	3.6	2.4	70	**17**	0.24	**12**	17		31	

	FOOD	Portion	Weight in grams / *Conversion for 100 g*	Kilocalories / H_2O g	Total carbohydrate g / Total fats g	Crude fiber g / Dietary fiber g	Total protein g / Total sugar g	**% USRDA**	Arginine mg / Histidine mg	Isoleucine mg / Leucine mg	Lysine mg / Methionine mg	Phenylalanine mg / Threonine mg	Valine mg / Tryptophan mg	Cystine mg / Tyrosine mg	Polyunsat. fatty acids g / Monounsat. fatty acids g	Saturated fatty acids g / P/S ratio	Linoleic acid g / Cholesterol mg	Thiamin mg / Ascorbic acid mg	**% USRDA**
A	pie,pineapple chiffon,home recipe	1/7 of 9" dia	113	337	40.2	0.22	4.3	**8**	54	177	171	197	194	73	0.33	6.35	1.12	0.13	**9**
			0.88	49.6	18.3	0.99	23.0		78	298	77	131	46	132	9.54	0.05	20	7	**12**
B	pie,pumpkin,home recipe	1/7 of 9" dia	135	287	33.9	0.61	6.2	**12**	195	318	357	301	341	98	0.44	4.46	1.09	0.11	**8**
			0.74	78.9	14.5	0.67	19.5		160	521	137	250	83	256	8.07	0.10	84	3	**4**
C	pie,raisin,home recipe	1/7 of 9" dia	135	545	72.1	0.25	7.4	**14**	312	328	305	355	376	144	2.97	7.67	3.65	0.18	**12**
			0.74	29.2	26.9	2.60	49.4		179	545	153	266	89	265	13.57	0.39	132	3	**5**
D	pie,rhubarb,home recipe	1/7 of 9" dia	135	414	58.9	0.67	3.9	**6**	unk.	155	78	185	146	67	0.30	5.38	1.34	0.16	**11**
			0.74	52.4	18.8	2.59	34.8		68	260	45	97	41	116	10.92	0.05	7	6	**10**
E	pie,squash,home recipe	1/7 of 9" dia	145	311	35.7	0.88	6.4	**13**	194	280	373	236	325	61	0.48	7.25	1.86	0.10	**6**
			0.69	62.4	16.6	unk.	23.5		132	453	132	226	70	214	6.17	0.07	87	5	**9**
F	pie,strawberry,home recipe	1/7 of 9" dia	135	282	41.0	0.98	2.6	**4**	unk.	112	71	130	109	48	0.12	2.77	0.95	0.11	**7**
			0.74	78.2	12.4	2.65	20.8		54	195	29	79	33	92	7.74	0.04	0	41	**69**
G	pie,sweet potato,home recipe	1/7 of 9" dia	130	277	30.8	0.26	5.8	**9**	unk.	unk.	unk.	unk.	unk.	unk.	unk.	5.20	2.21	0.10	**7**
			0.77	77.1	14.7	unk.	unk.		unk.	unk.	unk.	unk.	unk.	unk.	6.37	unk.	unk.	5	**9**
H	PIEROGI (stuffed dumplings),home recipe	3/4 C	125	307	23.8	0.15	11.2	**23**	485	566	745	478	571	144	1.33	6.99	1.31	0.29	**19**
			0.80	68.1	18.5	unk.	3.5		323	869	241	488	149	409	8.36	0.19	49	1	**2**
I	PIMIENTOS,canned,solids & liquid	1 Tbsp	19	5	1.1	0.11	0.2	**0**	unk.	7	9	8	9	unk.	unk.	tr(a)	unk.	tr	**0**
			5.41	17.1	0.1	unk.	0.0		4	11	1	8	unk.	4	unk.	unk.	0	18	**29**
J	PINE NUTS/pignolias	1/4 C	18	97	2.0	0.00	5.4	**8**	unk.	unk.	unk.	unk.	unk.	unk.	3.99	1.08	3.88	0.11	**7**
			5.71	1.0	8.3	unk.	0.0		unk.	unk.	unk.	unk.	unk.	unk.	3.32	3.71	0	unk.	**unk.**
K	pine nuts/piñon	1/4 C	18	111	3.6	0.19	2.3	**4**	unk.	unk.	unk.	unk.	unk.	unk.	unk.	unk.	unk.	0.22	**15**
			5.71	0.5	10.6	unk.	0.0		unk.	unk.	unk.	unk.	unk.	unk.	unk.	unk.	0	tr	**0**
L	PINEAPPLE,canned,chunks,heavy syrup, solids & liquid	1/2 C	128	94	24.7	0.38	0.4	**1**	unk.	unk.	9	10	unk.	unk.	unk.	tr(a)	unk.	0.10	**7**
			0.78	101.9	0.1	1.15	21.5		unk.	unk.	1	unk.	5	10	unk.	unk.	0	9	**15**
M	pineapple,canned,chunks,juice pack, solids & liquid	1/2 C	123	71	18.6	0.37	0.5	**1**	unk.	unk.	11	13	unk.	unk.	unk.	tr(a)	unk.	0.12	**8**
			0.81	103.3	0.1	unk.	14.3		unk.	unk.	2	unk.	6	13	unk.	unk.	0	12	**21**
N	pineapple,canned,crushed,heavy syrup, solids & liquid	1/2 C	128	94	24.7	0.38	0.4	**1**	unk.	unk.	9	10	unk.	unk.	unk.	tr(a)	unk.	0.10	**7**
			0.78	101.9	0.1	1.15	21.5		unk.	unk.	1	unk.	5	10	unk.	unk.	0	9	**15**
O	pineapple,canned,sliced,heavy syrup, solids & liquid	1 slice	58	43	11.3	0.17	0.2	**0**	unk.	unk.	4	4	unk.	unk.	unk.	tr(a)	unk.	0.05	**3**
			1.72	46.3	0.1	0.52	9.8		unk.	unk.	1	unk.	2	4	unk.	unk.	0	4	**7**
P	pineapple,canned,sliced,juice pack, solids & liquid	1 slice	58	34	8.8	0.17	0.2	**0**	unk.	unk.	5	6	unk.	unk.	unk.	unk.	unk.	0.06	**4**
			1.72	48.7	0.1	unk.	6.7		unk.	unk.	1	unk.	3	6	unk.	unk.	0	6	**10**
Q	pineapple,canned,sliced,water pack, solids & liquid	1 slice	58	23	5.9	0.17	0.2	**0**	unk.	unk.	4	4	unk.	unk.	unk.	tr(a)	unk.	0.05	**3**
			1.72	51.7	0.1	unk.	4.5		unk.	unk.	1	unk.	2	4	unk.	unk.	0	4	**7**
R	pineapple,canned,tidbits,heavy syrup, solids & liquid	1/2 C	128	94	24.7	0.38	0.4	**1**	unk.	unk.	9	10	unk.	unk.	unk.	tr(a)	unk.	0.10	**7**
			0.78	101.9	0.1	1.15	21.5		unk.	unk.	1	unk.	5	10	unk.	unk.	0	9	**15**
S	pineapple,fresh,cubed,edible portion	1/2 C	78	40	10.6	0.31	0.3	**1**	unk.	unk.	7	8	unk.	unk.	unk.	tr(a)	unk.	0.07	**5**
			1.29	66.1	0.2	0.93	8.2		unk.	unk.	1	unk.	4	8	unk.	unk.	0	13	**22**

Riboflavin mg / Niacin mg	% USRDA	Vitamin B_6 mg / Folacin mcg	% USRDA	Vitamin B_{12} mcg / Pantothenic acid mg	% USRDA	Biotin mg / Vitamin A IU	% USRDA	Preformed A RE / Beta carotene RE	Vitamin D IU / Vitamin E IU	% USRDA	Total tocopherol mg / Alpha tocopherol mg	Other tocopherol mg / Total ash g	Calcium mg / Phosphorus mg	% USRDA	Sodium mg / Sodium meq	Potassium mg / Potassium meq	Chlorine mg / Chlorine meq	Iron mg / Magnesium mg	% USRDA	Zinc mg / Copper mg	% USRDA	Iodine mcg / Selenium mcg	% USRDA	Manganese mcg / Chromium mcg	
0.10	**6**	0.05	**3**	0.03	**0**	0.001	**0**	0	0	**0**	0.0	0.0	21	**2**	159	103	48	0.8	**5**	0.3	**2**	35.6	**24**	919	A
0.8	**4**	9	**2**	0.20	**2**	246	**5**	3	0.0	**0**	0.0	0.72	33	**3**	6.9	2.6	1.3	15	**4**	0.07	**3**	12		24	
0.21	**13**	0.07	**4**	0.30	**5**	0.004	**1**	0	31	**8**	0.2	0.0	106	**11**	272	227	39	1.2	**7**	0.8	**5**	71.8	**48**	121	B
0.9	**5**	18	**5**	0.64	**6**	2467	**49**	235	0.3	**1**	0.1	1.49	125	**13**	11.8	5.8	1.1	51	**13**	0.12	**6**	9		19	
0.17	**10**	0.15	**8**	0.27	**5**	0.008	**3**	0	9	**2**	0.3	0.0	45	**5**	345	276	89	2.5	**14**	0.9	**6**	86.1	**57**	357	C
1.2	**6**	20	**5**	0.50	**5**	204	**4**	1	0.4	**1**	0.1	1.71	131	**13**	15.0	7.1	2.5	86	**21**	0.18	**9**	15		39	
0.12	**7**	0.02	**1**	0.00	**0**	tr	**0**	0	0(a)	**0**	tr	0(a)	83	**8**	293	226	23	1.5	**8**	0.3	**2**	67.7	**45**	127	D
1.4	**7**	7	**2**	0.15	**2**	180	**4**	8	tr	**0**	tr	1.45	44	**4**	12.7	5.8	0.6	23	**6**	0.04	**2**	17		24	
0.27	**16**	0.09	**5**	0.21	**4**	0.006	**2**	0	38	**9**	0.4	0.1	150	**15**	315	333	25	1.3	**7**	0.8	**5**	46.6	**31**	67	E
0.7	**4**	7	**2**	0.57	**6**	2256	**45**	193	0.4	**2**	0.2	1.94	152	**15**	13.7	8.5	0.7	61	**15**	0.13	**6**	0		10	
0.11	**6**	0.05	**3**	0.00	**0**	0.003	**1**	0	0	**0**	0.2	0.1	19	**2**	130	135	22	1.3	**7**	0.3	**2**	33.3	**22**	319	F
1.1	**6**	16	**4**	0.33	**3**	42	**1**	4	0.2	**1**	0.1	0.77	33	**3**	5.7	3.5	0.6	15	**4**	0.05	**2**	11		23	
0.19	**12**	unk.	**unk.**	unk.	**unk.**	unk.	**unk.**	unk.	unk.	**unk.**	unk.	unk.	90	**9**	283	212	unk.	1.0	**6**	unk.	**unk.**	unk.	**unk.**	unk.	G
0.8	**4**	unk.	**unk.**	unk.	**unk.**	3120	**62**	unk.	unk.	**unk.**	unk.	1.56	109	**11**	12.3	5.4	unk.	unk.	**unk.**	unk.	**unk.**	unk.		unk.	
0.18	**10**	0.04	**2**	0.14	**2**	0.000	**0**	0	16	**4**	0.1	tr(a)	79	**8**	369	101	17	1.9	**10**	1.6	**11**	30.7	**21**	97	H
1.9	**10**	8	**2**	0.24	**2**	246	**5**	5	0.1	**0**	0.1	3.54	156	**16**	16.1	2.6	0.5	17	**4**	0.11	**6**	13		15	
0.01	**1**	unk.	**unk.**	0.00	**0**	unk.	**unk.**	0	0(a)	**0**	unk.	unk.	1	**0**	unk.	unk.	unk.	0.3	**2**	tr	**0**	unk.	**unk.**	unk.	I
0.1	**0**	unk.	**unk.**	0.03	**0**	425	**9**	43	unk.	**unk.**	unk.	0.07	3	**0**	unk.	unk.	unk.	unk.	**unk.**	unk.	**unk.**	unk.		unk.	
unk.	**unk.**	unk.	**unk.**	0(a)	**0**	unk.	**unk.**	0(a)	0(a)	**0**	unk.	unk.	unk.	**unk.**	unk.	unk.	unk.	unk.	**unk.**	unk.	**unk.**	unk.	**unk.**	unk.	J
unk.	**unk.**	unk.	**unk.**	unk.	**unk.**	unk.	**unk.**	unk.	unk.	**unk.**	unk.	0.75	unk.	**unk.**	unk.	unk.	unk.	unk.	**unk.**	unk.	**unk.**	unk.		unk.	
0.04	**2**	unk.	**unk.**	0(a)	**0**	unk.	**unk.**	0	0(a)	**0**	unk.	unk.	2	**0**	unk.	unk.	unk.	0.9	**5**	unk.	**unk.**	unk.	**unk.**	unk.	K
0.8	**4**	unk.	**unk.**	unk.	**unk.**	5	**0**	1	unk.	**unk.**	unk.	0.51	106	**11**	unk.	unk.	unk.	unk.	**unk.**	unk.	**unk.**	unk.		unk.	
0.03	**2**	0.09	**5**	0.00	**0**	tr	**0**	0	0(a)	**0**	unk.	unk.	14	**1**	1	122	5	0.4	**2**	0.2	**1**	unk.	**unk.**	1479	L
0.3	**1**	1	**0**	0.13	**1**	64	**1**	6	unk.	**unk.**	unk.	0.38	6	**1**	0.1	3.1	0.1	10	**3**	0.19	**10**	0(a)		19	
0.04	**2**	0.09	**5**	0.00	**0**	unk.	**unk.**	0	0(a)	**0**	unk.	unk.	20	**2**	1	181	unk.	0.5	**3**	0.2	**1**	unk.	**unk.**	1427	M
0.4	**2**	1	**0**	0.12	**1**	74	**2**	7	unk.	**unk.**	unk.	0.49	10	**1**	0.0	4.6	unk.	10	**3**	0.18	**9**	tr(a)		18	
0.03	**2**	0.09	**5**	0.00	**0**	tr	**0**	0	0(a)	**0**	unk.	unk.	14	**1**	1	122	5	0.4	**2**	0.2	**1**	unk.	**unk.**	1479	N
0.3	**1**	1	**0**	0.13	**1**	64	**1**	6	unk.	**unk.**	unk.	0.38	6	**1**	0.1	3.1	0.1	10	**3**	0.19	**10**	0(a)		19	
0.01	**1**	0.04	**2**	0.00	**0**	tr	**0**	0	0(a)	**0**	unk.	unk.	6	**1**	1	56	2	0.2	**1**	unk.	**unk.**	unk.	**unk.**	673	O
0.1	**1**	0	**0**	0.06	**1**	29	**1**	3	unk.	**unk.**	unk.	0.17	3	**0**	0.0	1.4	0.1	5	**1**	0.09	**4**	unk.		9	
0.02	**1**	0.04	**2**	0.00	**0**	unk.	**unk.**	0	0(a)	**0**	unk.	unk.	9	**1**	1	85	unk.	0.2	**1**	0.1	**1**	unk.	**unk.**	673	P
0.2	**1**	0	**0**	0.06	**1**	35	**1**	3	unk.	**unk.**	unk.	0.23	5	**1**	0.0	2.2	unk.	5	**1**	0.09	**4**	tr(a)		9	
0.01	**1**	0.04	**2**	0.00	**0**	unk.	**unk.**	0	0(a)	**0**	unk.	unk.	7	**1**	1	57	unk.	0.2	**1**	0.1	**1**	unk.	**unk.**	673	Q
0.1	**1**	0	**0**	0.06	**1**	29	**1**	3	unk.	**unk.**	unk.	0.17	3	**0**	0.0	1.5	unk.	5	**1**	0.09	**4**	0(a)		9	
0.03	**2**	0.09	**5**	0.00	**0**	tr	**0**	0	0(a)	**0**	unk.	unk.	14	**1**	1	122	5	0.4	**2**	0.2	**1**	unk.	**unk.**	1479	R
0.3	**1**	1	**0**	0.13	**1**	64	**1**	6	unk.	**unk.**	unk.	0.38	6	**1**	0.1	3.1	0.1	10	**3**	0.19	**10**	0(a)		19	
0.02	**1**	0.07	**3**	0.00	**0**	tr	**0**	0	0(a)	**0**	unk.	unk.	13	**1**	1	113	36	0.4	**2**	0.1	**1**	unk.	**unk.**	1649	S
0.2	**1**	9	**2**	0.12	**1**	54	**1**	5	unk.	**unk.**	unk.	0.31	6	**1**	0.0	2.9	1.0	10	**3**	0.05	**3**	0		0	

	FOOD	Portion	Weight in grams / Conversion for 100 g	Kilocalories / H_2O g	Total carbohydrate g / Total fats g	Crude fiber g / Dietary fiber g	Total protein g / Total sugar g	% USRDA	Arginine mg / Histidine mg	Isoleucine mg / Leucine mg	Lysine mg / Methionine mg	Phenylalanine mg / Threonine mg	Valine mg / Tryptophan mg	Cystine mg / Tyrosine mg	Polyunsat. fatty acids g / Monounsat. fatty acids g	Saturated fatty acids g / P/S ratio	Linoleic acid g / Cholesterol mg	Thiamin mg / Ascorbic acid mg	% USRDA
A	pineapple,fresh,sliced,edible portion	1 slice	84	44	11.5	0.34	0.3	**1**	unk.	unk.	8	9	unk.	unk.	unk.	tr(a)	unk.	0.08	**5**
			1.19	71.7	0.2	1.01	8.9		unk.	unk.	1	unk.	4	9	unk.	unk.	0	14	**24**
B	pineapple,frozen,chunks,sweetened	1/2 C	123	104	27.2	0.37	0.5	**1**	unk.	unk.	11	12	unk.	unk.	unk.	tr(a)	unk.	0.12	**8**
			0.82	94.4	0.1	unk.	24.1		unk.	unk.	2	unk.	6	12	unk.	unk.	0	10	**16**
C	PISTACHIO NUTS	1/4 C	31	186	5.9	0.59	5.9	**9**	unk.	276	338	341	421	121	2.28	2.32	2.11	0.21	**14**
			3.19	1.7	16.8	unk.	0.0		147	477	115	192	unk.	209	11.27	0.99	0	0	**0**
D	PIZZA,w/cheese & sausage,home recipe,	1/8 of 14"	107	266	19.9	0.22	10.3	**18**	454	389	773	381	465	80	0.25	5.12	0.92	0.21	**14**
	made w/enriched flour	dia	*0.93*	58.9	16.2	1.41	3.1		300	727	229	367	61	357	5.00	0.05	35	11	**18**
E	pizza,w/cheese,frozen-type,baked,made	1/7 of 10"	57	140	20.2	0.17	5.4	**8**	unk.	393	336	279	399	91	unk.	1.14	tr	0.10	**7**
	w/enriched flour	dia	*1.75*	25.8	4.0	0.68	unk.		188	621	137	228	91	296	1.71	unk.	23	3	**6**
F	PLANTAIN,fresh,deep yellow flesh	1 average	263	313	82.0	1.05	2.9	**4**	unk.	unk.	unk.	unk.	unk.	unk.	unk.	tr(a)	unk.	0.16	**11**
	variety		*0.38*	174.5	1.1	unk.	unk.		unk.	unk.	unk.	unk.	unk.	unk.	unk.	unk.	0	37	**61**
G	plantain,fresh,white flesh variety	1 average	263	313	82.0	1.05	2.9	**4**	unk.	unk.	unk.	unk.	unk.	unk.	unk.	tr(a)	unk.	0.16	**11**
			0.38	174.5	1.1	unk.	unk.		unk.	unk.	unk.	unk.	unk.	unk.	unk.	unk.	0	37	**61**
H	PLUMS,canned,greengage,water pack,	1/2 C	114	38	14.4	0.23	0.5	**1**	unk.	unk.	unk.	unk.	unk.	unk.	unk.	tr(a)	unk.	0.01	**1**
	solids & liquid		*0.88*	103.3	0.1	unk.	14.1		unk.	unk.	unk.	unk.	unk.	unk.	unk.	unk.	0	2	**4**
I	plums,canned,purple,heavy syrup,	1/2 C	129	107	27.9	0.39	0.5	**1**	unk.	unk.	unk.	unk.	unk.	unk.	unk.	tr(a)	unk.	0.03	**2**
	solids & liquid		*0.78*	99.8	0.1	unk.	23.6		unk.	unk.	unk.	unk.	unk.	unk.	unk.	unk.	unk.	3	**4**
J	plums,canned,purple,light syrup,	1/2 C	125	78	20.7	0.37	0.5	**1**	unk.	unk.	unk.	unk.	unk.	unk.	unk.	tr(a)	unk.	0.02	**2**
	solids & liquid		*0.80*	102.6	0.1	unk.	20.3		unk.	unk.	unk.	unk.	unk.	unk.	unk.	unk.	0	2	**4**
K	plums,canned,purple,water pack,	1/2 C	125	57	14.8	0.37	0.5	**1**	unk.	unk.	unk.	unk.	unk.	unk.	unk.	tr(a)	unk.	0.02	**2**
	solids & liquid		*0.80*	108.1	0.2	unk.	12.4		unk.	unk.	unk.	unk.	unk.	unk.	unk.	unk.	0	2	**4**
L	plums,fresh,prune-type,edible portion	1 average	28	21	5.6	0.11	0.2	**0**	unk.	unk.	unk.	unk.	unk.	unk.	unk.	tr(a)	unk.	0.01	**1**
			3.55	22.2	0.1	0.59	2.7		unk.	unk.	unk.	unk.	unk.	unk.	unk.	unk.	0	1	**2**
M	POMEGRANATE,raw,edible portion	1 average	275	440	114.7	unk.	3.6	**6**	unk.	tr(a)	tr(a)	tr(a)	tr(a)	tr(a)	tr(a)	tr(a)	tr(a)	0.19	**13**
			0.36	unk.	2.2	unk.	unk.		tr(a)	tr(a)	tr(a)	tr(a)	tr(a)	tr(a)	tr(a)	0.00	0(a)	28	**46**
N	POP-TARTS,frosted,any-Kellogg	1 whole	52	218	36.3	unk.	3.2	**5**	unk.	unk.	unk.	unk.	unk.	unk.	unk.	unk.	unk.	0.25	**17**
			1.92	4.9	6.8	unk.	unk.		unk.	unk.	unk.	unk.	unk.	unk.	unk.	unk.	unk.	7	**13**
O	POPOVER,home recipe	2-3/4" dia	40	98	11.3	0.04	3.7	**7**	123	189	204	187	212	68	1.22	1.33	1.17	0.08	**5**
		x 4"	*2.50*	20.3	4.1	0.39	1.6		98	313	87	151	50	153	1.19	0.91	61	tr	**0**
P	POPPY seed	1 tsp	2.9	15	0.7	0.18	0.5	**1**	58	26	32	26	37	13	0.89	0.14	0.88	0.02	**2**
			34.48	0.2	1.3	unk.	0.0		15	43	14	26	7	20	0.18	6.33	0	unk.	**unk.**
Q	POPSICLE	1 average	95	70	18.0	0(a)	0.0	**0**	0(a)	0(a)	0(a)	0(a)	0(a)	0(a)	0.00	0.00	0.00	0.00	**0**
			1.05	76.0	0.0	0(a)	18.0		0(a)	0(a)	0(a)	0(a)	0(a)	0(a)	0.00	0.00	0	0	**0**
R	popsicle,twin	1 whole	128	95	24.2	0(a)	0.0	**0**	0(a)	0(a)	0(a)	0(a)	0(a)	0(a)	0.00	0.00	0.00	0.00	**0**
			0.78	102.4	0.0	0(a)	24.2		0(a)	0(a)	0(a)	0(a)	0(a)	0(a)	0.00	0.00	0	0	**0**
S	PORK,LEG center slice (fresh ham),	3-1/4x2-7/8x	85	184	0.0	0.00	25.2	**56**	1604	1289	2038	1010	1334	282	unk.	2.72	1.36	0.54	**36**
	w/o bone,cooked,no visible fat	1/2"	*1.18*	50.1	8.5	0(a)	0.0		830	1903	640	1237	343	904	4.08	unk.	75	0	**0**

Riboflavin mg / Niacin mg	% USRDA	Vitamin B_6 mg / Folacin mcg	% USRDA	Vitamin B_{12} mcg / Pantothenic acid mg	% USRDA	Biotin mg / Vitamin A IU	% USRDA	Preformed A RE / Beta carotene RE	Vitamin D IU / Vitamin E IU	% USRDA	Total tocopherol mg / Alpha tocopherol mg	Other tocopherol mg / Total ash g	Calcium mg / Phosphorus mg	% USRDA	Sodium mg / Sodium meq	Potassium mg / Potassium meq	Chlorine mg / Chlorine meq	Iron mg / Magnesium mg	% USRDA	Zinc mg / Copper mg	% USRDA	Iodine mcg / Selenium mcg	% USRDA	Manganese mcg / Chromium mcg	
0.03	**2**	0.07	**4**	0.00	**0**	tr	**0**	0	0(a)	**0**	unk.	unk.	14	**1**	1	123	39	0.4	**2**	0.1	**1**	unk.	**unk.**	1788	A
0.2	**1**	9	**2**	0.13	**1**	59	**1**	6	unk.	**unk.**	unk.	0.34	7	**1**	0.0	3.1	1.1	11	**3**	0.06	**3**	0		0	
0.04	**2**	0.09	**5**	0.00	**0**	unk.	**unk.**	0	0(a)	**0**	unk.	unk.	11	**1**	2	122	unk.	0.5	**3**	unk.	**unk.**	unk.	**unk.**	2607	B
0.4	**2**	7	**2**	0.13	**1**	37	**1**	4	unk.	**unk.**	unk.	0.24	5	**1**	0.1	3.1	unk.	12	**3**	0.09	**4**	0(a)		unk.	
unk.	**unk.**	unk.	**unk.**	0(a)	**0**	unk.	**unk.**	0	0(a)	**0**	unk.	unk.	41	**4**	unk.	304	unk.	2.3	**13**	unk.	**unk.**	unk.	**unk.**	unk.	C
0.4	**2**	18	**5**	unk.	**unk.**	72	**1**	7	unk.	**unk.**	unk.	0.85	157	**16**	unk.	7.8	unk.	49	**12**	unk.	**unk.**	unk.		unk.	
0.22	**13**	0.13	**7**	0.09	**2**	tr	**0**	0	0	**0**	0.0	0.0	163	**16**	406	264	tr(a)	2.1	**12**	1.2	**8**	9.2	**6**	6	D
2.6	**13**	34	**9**	0.20	**2**	846	**17**	66	0.0	**0**	0.0	2.25	136	**14**	17.6	6.7	0.0	16	**4**	0.04	**2**	0		0	
0.14	**8**	0.03	**1**	0.11	**2**	0.001	**0**	unk.	5	**1**	0.2	unk.	89	**9**	369	65	unk.	1.0	**5**	0.7	**5**	unk.	**unk.**	unk.	E
1.1	**5**	21	**5**	0.17	**2**	251	**5**	unk.	0.2	**1**	unk.	1.54	89	**9**	16.0	1.7	unk.	5	**1**	0.19	**10**	unk.		unk.	
0.11	**6**	unk.	**unk.**	unk.	**unk.**	unk.	**unk.**	0	0(a)	**0**	unk.	unk.	18	**2**	13	1012	unk.	1.8	**10**	unk.	**unk.**	unk.	**unk.**	unk.	F
1.6	**8**	42	**11**	unk.	**unk.**	3154	**63**	315	unk.	**unk.**	unk.	2.37	79	**8**	0.6	25.9	unk.	unk.	**unk.**	tr(a)	**0**	tr(a)		unk.	
0.11	**6**	unk.	**unk.**	unk.	**unk.**	unk.	**unk.**	0	0(a)	**0**	unk.	unk.	18	**2**	13	1012	unk.	1.8	**10**	unk.	**unk.**	unk.	**unk.**	unk.	G
1.6	**8**	42	**11**	unk.	**unk.**	26	**1**	3	unk.	**unk.**	unk.	2.37	79	**8**	0.6	25.9	unk.	unk.	**unk.**	tr(a)	**0**	tr(a)		unk.	
0.02	**1**	0.03	**2**	0.00	**0**	unk.	**unk.**	0	0(a)	**0**	unk.	unk.	10	**1**	1	93	unk.	0.2	**1**	unk.	**unk.**	unk.	**unk.**	unk.	H
0.3	**2**	7	**2**	0.08	**1**	182	**4**	18	unk.	**unk.**	unk.	0.34	15	**2**	0.0	2.4	unk.	unk.	**unk.**	unk.	**unk.**	tr(a)		unk.	
0.03	**2**	0.03	**2**	0.00	**0**	unk.	**unk.**	0	0(a)	**0**	unk.	unk.	12	**1**	1	183	unk.	1.2	**6**	unk.	**unk.**	unk.	**unk.**	unk.	I
0.5	**3**	1	**0**	0.09	**1**	1561	**31**	156	unk.	**unk.**	unk.	0.64	13	**1**	0.1	4.7	unk.	6	**2**	tr(a)	**0**	tr(a)		unk.	
0.02	**2**	0.03	**2**	0.00	**0**	unk.	**unk.**	0	0(a)	**0**	unk.	unk.	11	**1**	1	181	unk.	1.1	**6**	unk.	**unk.**	unk.	**unk.**	unk.	J
0.5	**3**	unk.	**unk.**	0.09	**1**	1531	**31**	153	unk.	**unk.**	unk.	0.62	12	**1**	0.0	4.6	unk.	6	**2**	tr(a)	**0**	tr(a)		unk.	
0.02	**2**	0.03	**2**	0.00	**0**	unk.	**unk.**	0	0(a)	**0**	unk.	unk.	11	**1**	2	184	unk.	1.2	**7**	unk.	**unk.**	unk.	**unk.**	unk.	K
0.5	**3**	1	**0**	0.09	**1**	1556	**31**	156	unk.	**unk.**	unk.	0.87	12	**1**	0.1	4.7	unk.	unk.	**unk.**	tr(a)	**0**	tr(a)		unk.	
0.01	**1**	0.01	**1**	0.00	**0**	unk.	**unk.**	0	0(a)	**0**	unk.	unk.	3	**0**	tr	48	unk.	0.1	**1**	unk.	**unk.**	unk.	**unk.**	unk.	L
0.1	**1**	2	**0**	0.05	**1**	85	**2**	8	unk.	**unk.**	unk.	0.17	5	**1**	0.0	1.2	unk.	3	**1**	unk.	**unk.**	unk.		1	
0.19	**11**	unk.	**unk.**	0.00	**0**	unk.	**unk.**	tr(a)	0(a)	**0**	tr(a)	tr(a)	22	**2**	22	1810	unk.	0.2	**1**	unk.	**unk.**	unk.	**unk.**	unk.	M
1.9	**10**	unk.	**unk.**	1.64	**16**	tr	**0**	tr(a)	0.0	**0**	tr(a)	unk.	55	**6**	1.0	46.3	unk.	unk.	**unk.**	unk.	**unk.**	unk.		unk.	
0.26	**15**	0.47	**23**	unk.	**unk.**	unk.	**unk.**	unk.	unk.	**unk.**	unk.	unk.	25	**3**	127	96	unk.	2.5	**14**	unk.	**unk.**	unk.	**unk.**	unk.	N
2.5	**13**	unk.	**unk.**	unk.	**unk.**	unk.	**unk.**	unk.	unk.	**unk.**	unk.	unk.	51	**5**	5.5	2.5	unk.	9	**2**	unk.	**unk.**	unk.		unk.	
0.11	**7**	0.04	**2**	0.22	**4**	0.002	**1**	0	16	**4**	1.5	tr(a)	43	**4**	154	71	29	0.7	**4**	0.4	**3**	42.4	**28**	59	O
0.5	**3**	8	**2**	0.31	**3**	61	**1**	0	1.8	**6**	tr	0.69	66	**7**	6.7	1.8	0.8	36	**9**	0.07	**4**	7		8	
0.00	**0**	0.01	**1**	0.00	**0**	unk.	**unk.**	0(a)	0(a)	**0**	unk.	unk.	42	**4**	1	20	unk.	0.3	**2**	0.3	**2**	unk.	**unk.**	unk.	P
tr	**0**	unk.	**unk.**	unk.	**unk.**	tr	**0**	unk.	unk.	**unk.**	unk.	0.20	25	**3**	0.0	0.5	unk.	10	**2**	unk.	**unk.**	unk.		unk.	
0.00	**0**	tr(a)	**0**	0(a)	**0**	tr(a)	**0**	0	0(a)	**0**	unk.	unk.	0	**0**	7	unk.	unk.	tr	**0**	unk.	**unk.**	unk.	**unk.**	unk.	Q
0.0	**0**	tr(a)	**0**	tr(a)	**0**	0	**0**	0	unk.	**unk.**	unk.	unk.	tr(a)	**0**	0.3	unk.	unk.	unk.	**unk.**	unk.	**unk.**	unk.		unk.	
0.00	**0**	tr(a)	**0**	0(a)	**0**	tr(a)	**0**	0	0(a)	**0**	unk.	unk.	0	**0**	9	unk.	unk.	tr	**0**	unk.	**unk.**	unk.	**unk.**	unk.	R
0.0	**0**	tr(a)	**0**	tr(a)	**0**	0	**0**	0	unk.	**unk.**	unk.	unk.	tr(a)	**0**	0.4	unk.	unk.	unk.	**unk.**	unk.	**unk.**	unk.		unk.	
0.25	**15**	0.23	**12**	0.54	**9**	unk.	**unk.**	0	0	**0**	0.5	0.4	11	**1**	55	331	unk.	3.2	**18**	3.4	**23**	unk.	**unk.**	unk.	S
4.8	**24**	4	**1**	0.34	**3**	0	**0**	0	0.6	**2**	0.1	1.19	262	**26**	2.4	8.5	unk.	unk.	**unk.**	0.25	**13**	tr(a)		tr(a)	

	FOOD	Portion	Weight in grams / *Conversion for 100 g*	Kilocalories / H_2O g	Total carbohydrate g / Total fats g	Crude fiber g / Dietary fiber g	Total protein g / Total sugar g	% USRDA	Arginine mg / Histidine mg	Isoleucine mg / Leucine mg	Lysine mg / Methionine mg	Phenylalanine mg / Threonine mg	Valine mg / Tryptophan mg	Cystine mg / Tyrosine mg	Polyunsat. fatty acids g / Monounsat. fatty acids g	Saturated fatty acids g / P/S ratio	Linoleic acid g / Cholesterol mg	Thiamin mg / Ascorbic acid mg	% USRDA
A	pork,leg center slice (fresh ham),	3-1/4x2-7/8x	85	318	0.0	0.00	19.5	**43**	1242	999	1578	776	1025	218	unk.	9.18	2.04	0.43	**29**
	w/o bone,cooked,w/fat	1/2"	*1.18*	38.7	26.0	0(a)	0.0		643	1474	491	958	266	700	11.22	unk.	76	0	**0**
B	pork,leg roast (fresh ham),roasted,	1/2 C	70	152	0.0	0.00	20.8	**46**	1321	1062	1679	816	1084	232	unk.	2.24	1.12	0.45	**30**
	chopped,no visible fat		*1.43*	41.2	7.0	0(a)	0.0		683	1567	517	1018	283	744	3.36	unk.	62	0	**0**
C	pork,leg roast (fresh ham),roasted,	1/2 C	70	262	0.0	0.00	16.1	**36**	1023	822	1300	639	844	180	unk.	7.56	1.68	0.36	**24**
	chopped,w/fat		*1.43*	31.8	21.4	0(a)	0.0		529	1214	405	789	219	576	8.40	unk.	62	0	**0**
D	pork,leg roast (fresh ham),roasted,	2 slices	85	184	0.0	0.00	25.2	**56**	1604	1289	2038	991	1316	282	unk.	2.72	1.36	0.54	**36**
	sliced,no visible fat		*1.18*	50.1	8.5	0(a)	0.0		830	1903	628	1237	343	904	4.08	unk.	75	0	**0**
E	pork,leg roast (fresh ham),roasted,	2 slices	85	318	0.0	0.00	19.5	**43**	1242	999	1578	776	1025	218	unk.	9.18	2.04	0.43	**29**
	sliced,w/fat		*1.18*	38.7	26.0	0(a)	0.0		643	1474	491	958	266	700	10.20	unk.	76	0	**0**
F	PORK,LOIN CHOP,w/bone,broiled,	2 med	84	227	0.0	0.00	25.7	**40**	1633	1313	2076	1020	1347	287	unk.	4.62	1.15	0.95	**63**
	no visible fat		*1.19*	44.2	12.9	0(a)	0.0		844	1938	646	1259	350	920	5.43	unk.	74	0(a)	**0**
G	pork,loin chop,w/bone,broiled,w/fat	2 med	116	454	0.0	0.00	28.6	**44**	1820	1484	2371	1137	1502	320	5.68	13.22	3.25	1.11	**74**
			0.86	49.1	36.8	0(a)	0.0		942	2126	720	1340	375	1025	15.43	0.43	103	0(a)	**0**
H	pork,loin chop,w/o bone,broiled,	2-1/2x2-1/2x	85	229	0.0	0.00	26.0	**58**	1652	1329	2100	1032	1363	291	unk.	4.67	1.16	0.96	**64**
	no visible fat	3/4"	*1.18*	44.7	13.1	0(a)	0.0		854	1961	654	1274	354	931	5.49	unk.	75	0(a)	**0**
I	pork,loin chop,w/o bone,broiled,	2-1/2x2-1/2x	85	332	0.0	0.00	21.0	**47**	1334	1087	1737	833	1101	235	4.16	9.69	2.38	0.82	**54**
	w/fat	3/4"	*1.18*	36.0	26.9	0(a)	0.0		690	1558	528	982	275	751	11.30	0.43	76	0(a)	**0**
J	PORK,LOIN ROAST,w/o bone,roasted,	1/2 C	70	178	0.0	0.00	20.6	**46**	1308	1051	1662	816	1079	230	1.50	3.30	0.80	0.76	**50**
	chopped,no visible fat		*1.43*	38.5	9.9	0(a)	0.0		676	1551	517	1008	280	736	4.05	0.46	62	0	**0**
K	pork,loin roast,w/o bone,roasted,	1/2 C	70	253	0.0	0.00	17.1	**38**	1089	895	1431	686	906	192	3.07	6.89	1.84	0.64	**43**
	chopped,w/fat		*1.43*	32.1	19.9	0(a)	0.0		563	1283	435	808	226	614	8.48	0.45	62	0	**0**
L	pork,loin roast,w/o bone,roasted,	1/2 C	55	140	0.0	0.00	16.2	**36**	1027	826	1306	641	848	181	1.18	2.59	0.63	0.59	**40**
	ground,no visible fat		*1.82*	30.2	7.8	0(a)	0.0		531	1219	406	792	220	579	3.18	0.46	48	0	**0**
M	pork,loin roast,w/o bone,roasted,	1/2 C	55	199	0.0	0.00	13.5	**30**	856	703	1124	539	712	151	2.41	5.41	1.45	0.51	**34**
	ground,w/fat		*1.82*	25.2	15.7	0(a)	0.0		443	1008	342	635	178	482	6.67	0.45	49	0	**0**
N	pork,loin roast,w/o bone,roasted,sliced,	3 slices	85	216	0.0	0.00	25.0	**39**	1588	1277	2018	991	1316	280	1.83	4.00	0.98	0.92	**61**
	no visible fat		*1.18*	46.7	12.1	0(a)	0.0		821	1884	628	1224	340	894	4.92	0.46	75	0	**0**
O	pork,leg roast (fresh ham),roasted,	3 slices	85	308	0.0	0.00	20.8	**32**	1323	1087	1737	833	1101	233	3.72	8.36	2.24	0.78	**52**
	chopped,no visible fat		*1.18*	38.9	24.2	0(a)	0.0		684	1558	528	982	275	745	10.30	0.45	76	0	**0**
P	PORK,MEAT,cooked,cut up,w/fat	1/2 C	70	261	0.0	0.00	15.8	**35**	1005	808	1277	628	829	177	unk.	7.56	1.68	0.35	**23**
			1.43	31.6	21.4	0(a)	0.0		520	1193	398	775	216	566	9.24	unk.	62	0	**0**
Q	pork,meat,cooked,ground,w/fat	1/2 C	55	205	0.0	0.00	12.4	**28**	790	635	1004	493	652	139	unk.	5.94	1.32	0.27	**18**
			1.82	24.9	16.8	0(a)	0.0		409	937	312	609	169	445	7.26	unk.	49	0	**0**
R	pork,meat,home recipe,fritter	3/4 C	177	487	31.5	0.12	21.9	**46**	1100	1106	1491	974	1181	325	4.83	9.26	6.06	0.48	**32**
			0.56	48.7	29.6	1.15	2.7		659	1718	531	989	296	827	10.94	0.52	222	1	**2**
S	pork,meat,patty,cooked,w/fat	3" dia 5/8"	82	306	0.0	0.00	18.5	**41**	1178	947	1496	736	972	207	unk.	8.86	1.97	0.41	**27**
		thick	*1.22*	37.1	25.1	0(a)	0.0		609	1397	466	908	253	663	10.82	unk.	73	0	**0**

Riboflavin mg / Niacin mg	% USRDA	Vitamin B6 mg / Folacin mcg	% USRDA	Vitamin B12 mcg / Pantothenic acid mg	% USRDA	Biotin mg / Vitamin A IU	% USRDA	Preformed A RE / Beta carotene RE	Vitamin D IU / Vitamin E IU	% USRDA	Total tocopherol mg / Alpha tocopherol mg	Other tocopherol mg / Total ash g	Calcium mg / Phosphorus mg	% USRDA	Sodium mg / Sodium meq	Potassium mg / Potassium meq	Chlorine mg / Chlorine meq	Iron mg / Magnesium mg	% USRDA	Zinc mg / Copper mg	% USRDA	Iodine mcg / Selenium mcg	% USRDA	Manganese mcg / Chromium mcg	
0.20	**12**	0.18	**9**	0.42	**7**	unk.	**unk.**	0	0	**0**	0.5	0.4	8	**1**	55	331	unk.	2.5	**14**	2.6	**18**	unk.	**unk.**	unk.	A
3.9	**20**	4	**1**	0.25	**3**	0	**0**	0	0.6	**2**	0.1	0.76	201	**20**	2.4	8.5	unk.	unk.	**unk.**	0.25	**13**	tr(a)		tr(a)	
0.20	**12**	0.19	**9**	0.44	**7**	unk.	**unk.**	0	0	**0**	0.4	0.3	9	**1**	45	273	unk.	2.7	**15**	2.8	**19**	unk.	**unk.**	unk.	B
4.0	**20**	3	**1**	0.28	**3**	0	**0**	0	0.5	**2**	0.1	0.98	216	**22**	2.0	7.0	unk.	unk.	**unk.**	0.21	**11**	tr(a)		tr(a)	
0.16	**10**	0.15	**7**	0.35	**6**	unk.	**unk.**	0	0	**0**	0.4	0.3	7	**1**	45	273	unk.	2.1	**12**	2.2	**14**	unk.	**unk.**	unk.	C
3.2	**16**	3	**1**	0.21	**2**	0	**0**	0	0.5	**2**	0.1	0.63	165	**17**	2.0	7.0	unk.	unk.	**unk.**	0.21	**11**	tr(a)		tr(a)	
0.25	**15**	0.23	**12**	0.54	**9**	unk.	**unk.**	0	0	**0**	0.5	0.4	11	**1**	55	331	unk.	3.2	**18**	3.4	**23**	unk.	**unk.**	unk.	D
4.8	**24**	4	**1**	0.34	**3**	0	**0**	0	0.6	**2**	0.1	1.19	262	**26**	2.4	8.5	unk.	unk.	**unk.**	0.25	**13**	tr(a)		tr(a)	
0.20	**12**	0.18	**9**	0.42	**7**	unk.	**unk.**	0	0	**0**	0.5	0.4	8	**1**	55	331	unk.	2.5	**14**	2.6	**18**	unk.	**unk.**	unk.	E
3.9	**20**	4	**1**	0.25	**3**	0	**0**	0	0.6	**2**	0.1	0.76	201	**20**	2.4	8.5	unk.	unk.	**unk.**	0.25	**13**	tr(a)		tr(a)	
0.28	**16**	0.23	**11**	0.53	**9**	0.004	**1**	0	0	**0**	0.5	0.4	11	**1**	55	328	53	3.3	**18**	3.4	**22**	unk.	**unk.**	unk.	F
5.7	**29**	4	**1**	0.33	**3**	0	**0**	0	0.6	**2**	0.1	1.26	272	**27**	2.4	8.4	1.5	17	**4**	0.25	**13**	tr(a)		tr(a)	
0.32	**19**	0.24	**12**	0.58	**10**	unk.	**unk.**	0	0	**0**	0.7	0.5	14	**1**	75	452	73	3.9	**22**	3.8	**26**	unk.	**unk.**	unk.	G
6.7	**34**	6	**1**	0.35	**4**	0	**0**	0	0.8	**3**	0.2	1.51	311	**31**	3.3	11.6	2.1	23	**6**	0.35	**17**	tr(a)		tr(a)	
0.28	**17**	0.23	**12**	0.54	**9**	0.004	**1**	0	0	**0**	0.5	0.4	11	**1**	55	331	54	3.3	**18**	3.4	**23**	unk.	**unk.**	unk.	H
5.8	**29**	4	**1**	0.34	**3**	0	**0**	0	0.6	**2**	0.1	1.27	275	**28**	2.4	8.5	1.5	17	**4**	0.25	**13**	tr(a)		tr(a)	
0.24	**14**	0.18	**9**	0.42	**7**	unk.	**unk.**	0	0	**0**	0.5	0.4	10	**1**	55	331	54	2.9	**16**	2.8	**19**	unk.	**unk.**	unk.	I
4.9	**25**	4	**1**	0.25	**3**	0	**0**	0	0.6	**2**	0.1	1.10	228	**23**	2.4	8.5	1.5	17	**4**	0.25	**13**	tr(a)		tr(a)	
0.22	**13**	0.19	**9**	0.44	**7**	0.003	**1**	0	0	**0**	0.4	0.3	9	**1**	45	273	44	2.7	**15**	2.8	**19**	unk.	**unk.**	unk.	J
4.5	**23**	3	**1**	0.28	**3**	0	**0**	0	0.5	**2**	0.1	0.91	217	**22**	2.0	7.0	1.2	14	**4**	0.21	**11**	tr(a)		tr(a)	
0.18	**11**	0.15	**7**	0.35	**6**	unk.	**unk.**	0	0	**0**	0.4	0.3	8	**1**	35	246	44	2.2	**12**	2.3	**15**	unk.	**unk.**	unk.	K
3.9	**20**	3	**1**	0.21	**2**	0	**0**	0	0.5	**2**	0.1	0.84	179	**18**	1.5	6.3	1.2	14	**4**	0.21	**11**	tr(a)		tr(a)	
0.17	**10**	0.15	**7**	0.35	**6**	0.003	**1**	0	0	**0**	0.3	0.2	7	**1**	36	214	35	2.1	**12**	2.2	**15**	unk.	**unk.**	unk.	L
3.6	**18**	3	**1**	0.22	**2**	0	**0**	0	0.4	**1**	0.1	0.71	171	**17**	1.6	5.5	1.0	11	**3**	0.16	**8**	tr(a)		tr(a)	
0.14	**8**	0.12	**6**	0.27	**5**	unk.	**unk.**	0	0	**0**	0.3	0.2	6	**1**	27	194	35	1.8	**10**	1.8	**12**	unk.	**unk.**	unk.	M
3.1	**15**	3	**1**	0.16	**2**	0	**0**	0	0.4	**1**	0.1	0.66	141	**14**	1.2	4.9	1.0	11	**3**	0.16	**8**	tr(a)		tr(a)	
0.26	**16**	0.23	**12**	0.54	**9**	0.004	**1**	0	0	**0**	0.5	0.3	11	**1**	55	331	54	3.2	**18**	3.4	**23**	unk.	**unk.**	unk.	N
5.5	**28**	4	**1**	0.34	**3**	0	**0**	0	0.6	**2**	0.1	1.10	264	**26**	2.4	8.5	1.5	17	**4**	0.25	**13**	tr(a)		tr(a)	
0.22	**13**	0.18	**9**	0.42	**7**	unk.	**unk.**	0	0	**0**	0.5	0.4	9	**1**	42	299	54	2.7	**15**	2.8	**19**	unk.	**unk.**	unk.	O
4.8	**24**	4	**1**	0.25	**3**	0	**0**	0	0.6	**2**	0.1	1.02	218	**22**	1.8	7.6	1.5	17	**4**	0.25	**13**	tr(a)		tr(a)	
0.16	**10**	0.15	**7**	0.35	**6**	unk.	**unk.**	0	0	**0**	0.4	0.3	7	**1**	45	273	unk.	2.0	**11**	2.1	**14**	unk.	**unk.**	unk.	P
3.4	**17**	3	**1**	0.21	**2**	0	**0**	0	0.5	**2**	0.1	1.12	162	**16**	2.0	7.0	unk.	16	**4**	0.21	**11**	tr(a)		tr(a)	
0.13	**7**	0.12	**6**	0.27	**5**	unk.	**unk.**	0	0	**0**	0.3	0.2	5	**1**	36	214	unk.	1.6	**9**	1.7	**11**	unk.	**unk.**	unk.	Q
2.7	**14**	3	**1**	0.16	**2**	0	**0**	0	0.4	**1**	0.1	0.88	128	**13**	1.6	5.5	unk.	13	**3**	0.16	**8**	tr(a)		tr(a)	
0.39	**23**	0.20	**10**	0.78	**13**	0.007	**2**	0	29	**7**	6.3	tr(a)	105	**11**	416	363	86	3.5	**19**	2.6	**18**	74.0	**49**	174	R
4.1	**21**	23	**6**	0.94	**9**	127	**3**	0	7.6	**25**	1.0	2.87	316	**32**	18.1	9.3	2.4	111	**28**	0.31	**16**	20		24	
0.19	**11**	0.17	**9**	0.41	**7**	unk.	**unk.**	0	0	**0**	0.5	0.4	8	**1**	53	320	unk.	2.4	**13**	2.5	**17**	unk.	**unk.**	unk.	S
4.0	**20**	4	**1**	0.25	**3**	0	**0**	0	0.6	**2**	0.1	1.31	190	**19**	2.3	8.2	unk.	19	**5**	0.25	**12**	tr(a)		tr(a)	

	FOOD	Portion	Weight in grams / *Conversion for 100 g*	Kilocalories / H_2O g	Total carbohydrate g / Total fats g	Crude fiber g / Dietary fiber g	Total protein g / Total sugar g	**% USRDA**	Arginine mg / Histidine mg	Isoleucine mg / Leucine mg	Lysine mg / Methionine mg	Phenylalanine mg / Threonine mg	Valine mg / Tryptophan mg	Cystine mg / Tyrosine mg	Polyunsat. fatty acids g / Monounsat. fatty acids g	Saturated fatty acids g / P/S ratio	Linoleic acid g / Cholesterol mg	Thiamin mg / Ascorbic acid mg	**% USRDA**
A	PORK,PIGS FEET,boiled	1 average	87	140	0.5	0(a)	12.5	**28**	796	640	1012	497	657	140	unk.	unk.	unk.	unk.	**unk.**
			1.15	unk.	9.1	0(a)	0.0		412	944	163	613	171	449	unk.	unk.	77	unk.	**unk.**
B	pork,pigs feet,pickled	1 average	87	173	unk.	0.00	14.5	**32**	923	742	1174	606	752	163	unk.	4.35	0.87	1.11	**74**
			1.15	58.2	12.9	0(a)	0.0		478	1095	391	712	197	520	5.22	unk.	unk.	unk.	**unk.**
C	PORK,RIB CHOP,w/bone,broiled,	2 med	84	227	0.0	0.00	25.7	**50**	1633	1313	2076	1020	1347	287	unk.	4.62	1.15	0.95	**63**
	no visible fat		*1.19*	44.2	12.9	0(a)	0.0		844	1938	646	1259	350	920	5.43	unk.	74	0(a)	**0**
D	pork,rib chop,w/bone,broiled,	2 med	116	454	0.0	0.00	28.6	**44**	1820	1484	2371	1137	1502	320	5.68	13.22	3.25	1.11	**74**
	w/fat		*0.86*	49.1	36.8	0(a)	0.0		942	2126	720	1340	375	1025	15.43	0.43	103	0(a)	**0**
E	pork,rib chop,w/o bone,broiled,	2x2-1/8x1"	85	229	0.0	0.00	26.0	**58**	1652	1329	2100	1032	1363	291	unk.	4.67	1.16	0.96	**64**
	no visible fat	each	*1.18*	44.7	13.1	0(a)	0.0		854	1961	654	1274	354	931	5.49	unk.	75	0(a)	**0**
F	pork,rib chop,w/o bone,broiled,w/fat	2x2-1/8x1"	85	332	0.0	0.00	21.0	**47**	1334	1087	1737	833	1101	235	4.16	9.69	2.38	0.82	**54**
		each	*1.18*	36.0	26.9	0(a)	0.0		690	1558	528	982	275	751	11.30	0.43	76	0(a)	**0**
G	PORK,SALT,fried	1 slice	25	171	0.0	0.00	3.0	**5**	unk.	85	247	121	130	33	unk.	6.20	1.76	0.00	**0**
			4.00	unk.	17.5	0(a)	unk.		27	283	42	159	5	40	7.70	unk.	unk.	0	**0**
H	PORK,SHOULDER blade steak,w/bone,	1/3 lb raw	77	189	0.0	0.00	20.9	**46**	1327	1067	1687	829	1096	234	2.32	5.12	1.25	0.46	**30**
	cooked,no visible fat		*1.29*	44.5	11.1	0(a)	0.0		687	1575	526	1024	285	748	6.29	0.45	68	0	**0**
I	pork,shoulder blade steak,w/bone,	1/3 lb raw	97	341	0.0	0.00	21.8	**48**	1382	1118	1786	864	1141	243	4.90	11.29	2.92	0.48	**32**
	cooked,w/fat		*1.03*	46.5	27.6	0(a)	0.0		715	1601	547	1010	282	778	13.56	0.43	86	0	**0**
J	pork,shoulder blade steak,w/o bone,	3-1/4x2-7/8x	85	207	0.0	0.00	22.9	**51**	1458	1172	1853	910	1204	257	2.55	5.63	1.37	0.50	**33**
	cooked,no visible fat	1/2" each	*1.18*	48.9	12.2	0(a)	0.0		754	1730	577	1125	313	822	6.91	0.45	75	0	**0**
K	pork,shoulder blade steak,w/o bone,	3-1/4x2-7/8x	85	300	0.0	0.00	19.1	**43**	1215	983	1570	759	1003	213	4.31	9.93	2.57	0.42	**28**
	cooked,w/fat	1/2" each	*1.18*	40.9	24.2	0(a)	0.0		628	1408	481	887	248	684	11.92	0.43	76	0	**0**
L	pork,shoulder Boston butt,w/o bone,	3 slices	85	207	0.0	0.00	22.9	**51**	1458	1172	1853	910	1204	257	2.55	5.63	1.37	0.50	**33**
	roasted,sliced,no visible fat		*1.18*	48.9	12.2	0(a)	0.0		754	1730	577	1125	313	822	6.91	0.45	75	0(a)	**0**
M	pork,shoulder Boston butt,w/o bone,	3 slices	85	300	0.0	0.00	19.1	**43**	1215	977	1544	759	1003	213	4.31	9.93	2.57	0.42	**28**
	roasted,sliced,w/fat		*1.18*	40.9	24.2	0(a)	0.0		628	1442	481	937	260	684	11.92	0.43	76	0(a)	**0**
N	pork,shoulder picnic,smoked,w/o bone,	3 sl 2-1/2x	85	179	0.0	0.00	24.1	**54**	1533	1233	1950	988	1346	269	1.29	2.86	0.70	0.55	**37**
	no visible fat	2-1/2x1/4"	*1.18*	48.6	8.4	0(a)	0.0		793	1820	629	1182	329	864	3.51	0.45	75	unk.	**unk.**
O	pork,shoulder picnic,smoked,w/o bone,	3 sl 2-1/2x	85	275	0.0	0.00	19.0	**42**	1210	972	1538	779	1061	212	3.34	7.51	2.03	0.44	**30**
	w/fat	2-1/2x1/4"	*1.18*	41.5	21.4	0(a)	0.0		626	1436	496	932	259	682	9.26	0.44	76	unk.	**unk.**
P	pork,shoulder picnic,w/o bone,	3 slices	85	180	0.0	0.00	24.6	**55**	1566	1260	1991	978	1292	275	1.58	3.48	0.85	0.56	**37**
	roasted,sliced,no visible fat		*1.18*	51.3	8.3	0(a)	0.0		810	1858	620	1208	336	882	4.28	0.45	75	0(a)	**0**
Q	pork,shoulder picnic,w/o bone,	3 slices	85	318	0.0	0.00	19.7	**44**	1253	1007	1593	782	1034	220	3.29	7.41	1.99	0.46	**31**
	roasted,sliced,w/fat		*1.18*	38.8	25.9	0(a)	0.0		648	1487	496	966	269	706	9.13	0.44	76	0(a)	**0**
R	PORK,SPARE RIB,cooked,no visible fat	4" rib	16	66	0.0	0.00	3.5	**8**	223	179	283	139	184	39	unk.	unk.	unk.	0.07	**5**
			6.25	6.8	5.6	0(a)	0.0		115	264	88	172	48	125	unk.	unk.	14	0	**0**
S	pork,spare rib,cooked,no visible fat	5" rib	20	82	0.0	0.00	4.4	**10**	278	224	354	174	230	49	unk.	unk.	unk.	0.09	**6**
			5.00	8.5	7.0	0(a)	0.0		144	330	110	215	60	157	unk.	unk.	18	0	**0**

Riboflavin mg Niacin mg	% USRDA	Vitamin B_6 mg Folacin mcg	% USRDA	Vitamin B_{12} mcg Pantothenic acid mg	% USRDA	Biotin mg Vitamin A IU	% USRDA	Preformed A RE Beta carotene RE	Vitamin D IU Vitamin E IU	% USRDA	Total tocopherol mg Alpha tocopherol mg	Other tocopherol mg Total ash g	Calcium mg Phosphorus mg	% USRDA	Sodium mg Sodium meq	Potassium mg Potassium meq	Chlorine mg Chlorine meq	Iron mg Magnesium mg	% USRDA	Zinc mg Copper mg	% USRDA	Iodine mcg Selenium mcg	% USRDA	Manganese mcg Chromium mcg	
unk.	**unk.**	unk.	**unk.**	unk.	**unk.**	unk.	**unk.**	0	unk.	**unk.**	unk.	unk.	unk.	**unk.**	unk.	unk.	unk.	unk.	**unk.**	unk.	**unk.**	unk.	**unk.**	unk.	A
unk.	**unk.**	unk.	**unk.**	unk.	**unk.**	0	**0**	0	unk.	**unk.**	unk.	unk.	unk.	**unk.**	unk.	unk.	unk.	unk.	**unk.**	unk.	**unk.**	unk.		unk.	
unk.	**unk.**	unk.	**unk.**	unk.	**unk.**	unk.	**unk.**	unk.	unk.	**unk.**	unk.	unk.	10	**1**	61	248	unk.	2.7	**15**	unk.	**unk.**	unk.	**unk.**	unk.	B
4.7	**24**	unk.	**unk.**	unk.	**unk.**	unk.	**unk.**	unk.	unk.	**unk.**	unk.	1.48	171	**17**	2.6	6.4	unk.	unk.	**unk.**	unk.	**unk.**	unk.		unk.	
0.28	**16**	0.23	**11**	0.53	**9**	0.004	**1**	0	0	**0**	tr(a)	tr(a)	11	**1**	55	328	53	3.3	**18**	3.4	**22**	unk.	**unk.**	unk.	C
5.7	**29**	4	**1**	0.25	**3**	0	**0**	0	0.0	**0**	tr(a)	1.26	272	**27**	2.4	8.4	1.5	17	**4**	0.25	**13**	tr(a)		tr(a)	
0.32	**19**	0.24	**12**	0.58	**10**	unk.	**unk.**	0	0	**0**	0.7	0.5	14	**1**	75	452	73	3.9	**22**	3.8	**26**	unk.	**unk.**	unk.	D
6.7	**34**	6	**1**	0.35	**4**	0	**0**	0	0.8	**3**	0.2	1.51	311	**31**	3.3	11.6	2.1	23	**6**	0.35	**17**	tr(a)		tr(a)	
0.28	**17**	0.23	**12**	0.54	**9**	0.004	**1**	0	0	**0**	0.5	0.4	11	**1**	55	331	54	3.3	**18**	3.4	**23**	unk.	**unk.**	unk.	E
5.8	**29**	4	**1**	0.34	**3**	0	**0**	0	0.6	**2**	0.1	1.27	275	**28**	2.4	8.5	1.5	17	**4**	0.25	**13**	tr(a)		tr(a)	
0.24	**14**	0.18	**9**	0.42	**7**	unk.	**unk.**	0	0	**0**	0.5	0.4	10	**1**	55	331	54	2.9	**16**	2.8	**19**	unk.	**unk.**	unk.	F
4.9	**25**	5	**1**	0.25	**3**	0	**0**	0	0.6	**2**	0.1	1.10	228	**23**	2.4	8.5	1.5	17	**4**	0.25	**13**	tr(a)		tr(a)	
0.00	**0**	0.00	**0**	0.00	**0**	unk.	**unk.**	0	0	**0**	unk.	unk.	2	**0**	unk.	unk.	unk.	0.4	**2**	unk.	**unk.**	unk.	**unk.**	unk.	G
0.0	**0**	0	**0**	unk.	**unk.**	0	**0**	0	unk.	**unk.**	unk.	unk.	30	**3**	unk.	unk.	unk.	unk.	**unk.**	unk.	**unk.**	unk.		unk.	
0.21	**12**	0.21	**10**	0.49	**8**	unk.	**unk.**	0	0	**0**	0.5	0.3	9	**1**	50	302	unk.	2.6	**15**	3.5	**23**	unk.	**unk.**	unk.	H
4.0	**20**	4	**1**	0.31	**3**	0	**0**	0	0.6	**2**	0.1	0.93	214	**21**	2.2	7.7	unk.	unk.	**unk.**	0.23	**12**	tr(a)		tr(a)	
0.22	**13**	0.20	**10**	0.48	**8**	unk.	**unk.**	0	0	**0**	0.6	0.4	10	**1**	36	251	unk.	2.8	**16**	3.5	**24**	unk.	**unk.**	unk.	I
4.3	**21**	5	**1**	0.29	**3**	0	**0**	0	0.7	**2**	0.2	0.87	221	**22**	1.6	6.4	unk.	20	**5**	0.29	**15**	tr(a)		tr(a)	
0.23	**14**	0.23	**12**	0.54	**9**	unk.	**unk.**	0	0	**0**	0.5	0.4	10	**1**	55	331	unk.	2.9	**16**	3.8	**26**	unk.	**unk.**	unk.	J
4.4	**22**	4	**1**	0.34	**3**	0	**0**	0	0.6	**2**	0.1	1.02	235	**24**	2.4	8.5	unk.	unk.	**unk.**	0.25	**13**	tr(a)		tr(a)	
0.20	**12**	0.18	**9**	0.42	**7**	unk.	**unk.**	0	0	**0**	0.5	0.4	8	**1**	31	221	unk.	2.5	**14**	3.1	**21**	unk.	**unk.**	unk.	K
3.7	**19**	4	**1**	0.25	**3**	0	**0**	0	0.6	**2**	0.1	0.76	195	**20**	1.4	5.6	unk.	18	**5**	0.25	**13**	tr(a)		tr(a)	
0.23	**14**	0.18	**9**	0.54	**9**	unk.	**unk.**	0	0(a)	**0**	0.5	0.4	10	**1**	55	331	unk.	2.9	**16**	3.4	**23**	unk.	**unk.**	unk.	L
4.4	**22**	4	**1**	0.34	**3**	0	**0**	0	0.6	**2**	0.1	1.02	235	**24**	2.4	8.5	unk.	unk.	**unk.**	0.25	**13**	tr(a)		tr(a)	
0.20	**12**	0.18	**9**	0.42	**7**	unk.	**unk.**	0	0(a)	**0**	0.5	0.4	8	**1**	55	331	unk.	2.5	**14**	2.8	**19**	unk.	**unk.**	unk.	M
3.7	**19**	4	**1**	0.25	**3**	0	**0**	0	0.6	**2**	0.1	0.76	195	**20**	2.4	8.5	unk.	unk.	**unk.**	0.25	**13**	tr(a)		tr(a)	
0.22	**13**	0.20	**10**	0.46	**8**	unk.	**unk.**	0	0(a)	**0**	tr(a)	tr(a)	11	**1**	790	277	unk.	3.1	**18**	3.4	**23**	unk.	**unk.**	9	N
4.2	**21**	10	**3**	0.29	**3**	0	**0**	0	0.0	**0**	tr(a)	3.82	187	**19**	34.4	7.1	unk.	unk.	**unk.**	0.07	**4**	tr(a)		9	
0.17	**10**	0.16	**8**	0.38	**6**	unk.	**unk.**	0	0(a)	**0**	tr(a)	tr(a)	8	**1**	723	unk.	unk.	2.5	**14**	2.9	**19**	unk.	**unk.**	9	O
3.4	**17**	10	**3**	0.22	**2**	0	**0**	0	0.0	**0**	tr(a)	3.06	155	**16**	31.4	unk.	unk.	unk.	**unk.**	0.07	**4**	tr(a)		9	
0.25	**15**	0.23	**12**	0.54	**9**	unk.	**unk.**	0	0(a)	**0**	0.5	0.4	10	**1**	55	331	unk.	3.1	**17**	3.4	**23**	unk.	**unk.**	unk.	P
5.0	**25**	4	**1**	0.34	**3**	0	**0**	0	0.6	**2**	0.1	0.68	150	**15**	2.4	8.5	unk.	unk.	**unk.**	0.25	**13**	tr(a)		tr(a)	
0.21	**13**	0.18	**9**	0.42	**7**	unk.	**unk.**	0	0(a)	**0**	0.5	0.4	8	**1**	55	331	unk.	2.5	**14**	2.8	**19**	unk.	**unk.**	unk.	Q
4.1	**20**	4	**1**	0.25	**3**	0	**0**	0	0.6	**2**	0.1	0.51	118	**12**	2.4	8.5	unk.	unk.	**unk.**	0.25	**13**	tr(a)		tr(a)	
0.04	**2**	0.03	**2**	0.08	**1**	unk.	**unk.**	0	unk.	**unk.**	0.1	0.1	1	**0**	10	62	unk.	0.4	**3**	0.6	**4**	unk.	**unk.**	unk.	R
0.6	**3**	1	**0**	0.05	**1**	0	**0**	0	0.1	**0**	tr	0.11	21	**2**	0.4	1.6	unk.	unk.	**unk.**	0.05	**2**	tr(a)		tr(a)	
0.04	**3**	0.04	**2**	0.10	**2**	unk.	**unk.**	0	unk.	**unk.**	0.1	0.1	2	**0**	13	78	unk.	0.6	**3**	0.8	**5**	unk.	**unk.**	unk.	S
0.7	**4**	1	**0**	0.06	**1**	0	**0**	0	0.1	**1**	tr	0.14	26	**3**	0.6	2.0	unk.	unk.	**unk.**	0.06	**3**	tr(a)		tr(a)	

	FOOD	Portion	Weight in grams / Conversion for 100 g	Kilocalories / H_2O g	Total carbohydrate g / Total fats g	Crude fiber g / Dietary fiber g	Total protein g / Total sugar g	% USRDA	Arginine mg / Histidine mg	Isoleucine mg / Leucine mg	Lysine mg / Methionine mg	Phenylalanine mg / Threonine mg	Valine mg / Tryptophan mg	Cystine mg / Tyrosine mg	Polyunsat. fatty acids g / Monounsat. fatty acids g	Saturated fatty acids g / P/S ratio	Linoleic acid g / Cholesterol mg	Thiamin mg / Ascorbic acid mg	% USRDA
A	pork,spare rib,cooked,no visible fat	6" rib	25	103	0.0	0.00	5.5	**12**	348	280	442	217	287	61	unk.	unk.	unk.	0.11	**8**
			4.00	10.6	8.8	0(a)	0.0		180	413	138	268	75	196	unk.	unk.	22	0	**0**
B	pork,spare rib,cooked,w/fat	4" rib	16	70	0.0	0.00	3.3	**7**	211	170	269	132	175	37	0.96	2.15	0.56	0.07	**5**
			6.25	6.4	6.2	0(a)	0.0		109	251	84	163	45	119	2.65	0.45	14	0	**0**
C	pork,spare rib,cooked,w/fat	5" rib	20	88	0.0	0.00	4.2	**9**	264	213	336	165	218	46	1.21	2.69	0.70	0.09	**6**
			5.00	7.9	7.8	0(a)	0.0		137	314	105	204	57	149	3.31	0.45	18	0	**0**
D	pork,spare rib,cooked,w/fat	6" rib	25	110	0.0	0.00	5.2	**12**	330	266	420	206	273	58	1.51	3.36	0.87	0.11	**7**
			4.00	9.9	9.7	0(a)	0.0		171	392	131	255	71	186	4.14	0.45	22	0	**0**
E	PORK,TENDERLOIN,w/o bone,cooked	2-1/2x2-1/2x	85	216	0.0	0.00	25.0	**56**	1588	1277	2018	994	1316	280	1.83	4.00	0.98	0.92	**61**
		3/4 piece	*1.18*	46.7	12.1	0(a)	0.0		821	1884	628	1224	340	894	4.92	0.46	75	0	**0**
F	POTATO,au gratin (w/cheese)	1/2 C	123	178	16.7	0.37	6.5	**0**	unk.	unk.	unk.	unk.	unk.	unk.	1.10	3.80	tr	0.07	**5**
			0.82	87.1	9.7	unk.	1.1		unk.	unk.	unk.	unk.	unk.	unk.	3.67	0.29	18	12	**20**
G	potato,au gratin,frozen-Stouffer	1/2 pkg	163	201	19.4	unk.	4.6	**7**	unk.	unk.	unk.	unk.	unk.	unk.	unk.	unk.	unk.	0.05	**3**
			0.61	123.3	11.9	unk.	unk.		unk.	unk.	unk.	unk.	unk.	unk.	unk.	unk.	unk.	tr	**0**
H	potato,baked in skin	1 average	156	145	32.8	0.93	4.0	**6**	201	720	513	459	653	93	unk.	tr(a)	unk.	0.16	**10**
			0.64	116.8	0.2	3.11	1.4		157	756	109	350	140	131	unk.	unk.	0	31	**52**
I	potato,French fried from fresh	1/2 C	55	151	19.8	0.55	2.4	**4**	117	101	122	101	122	33	3.96	1.65	3.85	0.07	**5**
			1.82	24.6	7.3	1.49	0.5		unk.	115	27	95	23	37	1.65	2.40	tr	12	**19**
J	potato,French fried,frozen &	1/2 C	55	121	18.5	0.38	2.0	**3**	98	unk.	unk.	unk.	unk.	33	2.64	0.99	1.98	0.08	**5**
	oven-heated		*1.82*	29.1	4.6	1.49	0.5		unk.	unk.	unk.	unk.	unk.	37	0.99	2.67	0	12	**19**
K	potato,fresh,fried w/vegetable	1/2 C	85	228	27.7	0.85	3.4	**5**	169	150	180	150	180	51	1.87	2.55	0.85	0.10	**7**
	shortening		*1.18*	39.9	12.1	2.30	0.8		unk.	170	41	139	34	58	7.65	0.73	tr	16	**27**
L	potato,hashed brown in vegetable	1/2 C	78	177	22.6	0.62	2.4	**4**	119	105	127	105	127	46	0.54	2.32	0.77	0.06	**4**
	shortening		*1.29*	42.0	9.1	unk.	0.7		unk.	120	29	98	24	53	6.20	0.23	tr	7	**12**
M	potato,mashed,dry flakes made	1/2 C	105	98	15.2	0.31	2.0	**0**	unk.	unk.	unk.	unk.	unk.	unk.	0.94	0.90	0.69	0.04	**3**
	w/water,milk & margarine		*0.95*	83.3	3.4	0.95	0.9		unk.	unk.	unk.	unk.	unk.	unk.	1.52	1.05	unk.	5	**9**
N	potato,mashed,dry granules made	1/2 C	105	101	15.1	0.21	2.1	**0**	unk.	unk.	unk.	unk.	unk.	unk.	1.05	0.97	0.81	0.04	**3**
	w/water,milk,& margarine		*0.95*	82.5	3.8	0.95	0.9		unk.	unk.	unk.	unk.	unk.	unk.	1.73	1.09	unk.	3	**5**
O	potato,mashed,milk & margarine added	1/2 C	105	99	12.9	0.42	2.2	**4**	unk.	97	117	97	117	unk.	0.94	2.10	tr	0.08	**6**
			0.95	83.8	4.5	0.95	0.9		unk.	110	26	90	22	unk.	1.05	0.45	unk.	9	**16**
P	potato,mashed,milk added	1/2 C	105	68	13.6	0.42	2.2	**4**	unk.	unk.	unk.	unk.	unk.	unk.	unk.	unk.	unk.	0.08	**6**
			0.95	86.9	0.7	0.95	0.9		unk.	unk.	unk.	unk.	unk.	unk.	unk.	unk.	unk.	10	**18**
Q	potato,mashed,plain	1/2 C	104	67	15.0	0.52	2.0	**3**	98	87	105	87	105	62	unk.	tr(a)	unk.	0.09	**6**
			0.97	85.7	0.1	1.04	0.9		unk.	98	24	81	20	70	unk.	unk.	0	17	**28**
R	potato,pared,boiled,sliced	1/2 C	78	50	11.2	0.39	1.5	**2**	73	65	78	65	78	46	unk.	tr(a)	unk.	0.07	**5**
			1.29	64.2	0.1	0.78	0.7		unk.	74	18	60	15	53	unk.	unk.	0	12	**21**
S	potato,pared,boiled,whole,long type	1 average	206	134	29.9	1.03	3.9	**6**	194	173	208	173	208	124	unk.	tr(a)	unk.	0.19	**12**
			0.49	170.6	0.2	2.06	1.9		unk.	196	47	161	39	140	unk.	unk.	0	33	**55**

Riboflavin mg / Niacin mg	% USRDA	Vitamin B_6 mg / Folacin mcg	% USRDA	Vitamin B_{12} mcg / Pantothenic acid mg	% USRDA	Biotin mg / Vitamin A IU	% USRDA	Preformed A RE / Beta carotene RE	Vitamin D IU / Vitamin E IU	% USRDA	Total tocopherol mg / Alpha tocopherol mg	Other tocopherol mg / Total ash g	Calcium mg / Phosphorus mg	% USRDA	Sodium mg / Sodium meq	Potassium mg / Potassium meq	Chlorine mg / Chlorine meq	Iron mg / Magnesium mg	% USRDA	Zinc mg / Copper mg	% USRDA	Iodine mcg / Selenium mcg	% USRDA	Manganese mcg / Chromium mcg	
0.05	3	0.05	3	0.13	2	unk.	unk.	0	unk.	unk.	0.1	0.1	2	0	16	98	unk.	0.7	4	0.9	6	unk.	unk.	unk.	A
0.9	5	1	0	0.07	1	0	0	0	0.2	1	tr	0.17	32	3	0.7	2.5	unk.	unk.	unk.	0.07	4	tr(a)		tr(a)	
0.03	2	0.03	2	0.08	1	unk.	unk.	0	unk.	unk.	tr(a)	unk.	1	0	10	62	unk.	0.4	2	0.4	3	unk.	unk.	unk.	B
0.5	3	1	0	0.05	1	0	0	0	0.0	0	tr(a)	0.10	19	2	0.4	1.6	unk.	4	1	0.05	2	tr(a)		tr(a)	
0.04	3	0.04	2	0.10	2	unk.	unk.	0	unk.	unk.	0.1	0.1	2	0	13	78	unk.	0.5	3	0.5	4	unk.	unk.	unk.	C
0.7	3	1	0	0.06	1	0	0	0	0.1	1	tr	0.12	24	2	0.6	2.0	unk.	5	1	0.06	3	tr(a)		tr(a)	
0.05	3	0.05	3	0.13	2	unk.	unk.	0	unk.	unk.	0.1	0.1	2	0	16	98	unk.	0.6	4	0.7	5	unk.	unk.	unk.	D
0.8	4	1	0	0.07	1	0	0	0	0.2	1	tr	0.15	30	3	0.7	2.5	unk.	6	2	0.07	4	tr(a)		tr(a)	
0.26	16	0.23	12	0.54	9	0.004	1	0	0	0	0.5	0.4	11	1	55	331	54	3.2	18	3.4	23	unk.	unk.	unk.	E
5.5	28	4	1	0.34	3	0	0	0	0.6	2	0.1	1.10	264	26	2.4	8.5	1.5	17	4	0.25	13	tr(a)		tr(a)	
0.15	9	unk.	unk.	unk.	unk.	unk.	unk.	unk.	unk.	unk.	unk.	unk.	156	16	548	375	unk.	0.6	3	unk.	unk.	unk.	unk.	unk.	F
1.1	6	unk.	unk.	unk.	unk.	392	8	unk.	unk.	unk.	unk.	2.57	149	15	23.8	9.6	unk.	unk.	unk.	tr(a)	0	unk.		unk.	
0.15	9	unk.	unk.	unk.	unk.	unk.	unk.	unk.	unk.	unk.	unk.	unk.	tr	0	720	428	unk.	0.5	3	unk.	unk.	unk.	unk.	unk.	G
1.2	6	unk.	unk.	unk.	unk.	tr	0	unk.	unk.	unk.	unk.	2.61	unk.	unk.	31.3	10.9	unk.	unk.	unk.	unk.	unk.	unk.		unk.	
0.06	4	0.31	16	0.00	0	0.003	1	0	0(a)	0	0.1	0.0	14	1	6	782	118	1.1	6	0.3	2	unk.	unk.	unk.	H
2.6	13	19	5	0.62	6	tr	0	tr	0.1	0	0.0	1.71	101	10	0.3	20.0	3.3	34	9	0.23	12	tr(a)		unk.	
0.04	3	0.10	5	0.00	0	0.001	0	0	0(a)	0	unk.	unk.	8	1	3	469	unk.	0.7	4	0.2	1	unk.	unk.	unk.	I
1.7	9	12	3	0.30	3	tr	0	tr	unk.	unk.	unk.	0.99	61	6	0.1	12.0	unk.	9	2	0.15	7	tr(a)		unk.	
0.01	1	0.11	6	0.00	0	0.001	0	0	0(a)	0	unk.	unk.	5	1	2	359	unk.	1.0	6	0.2	1	unk.	unk.	unk.	J
1.4	7	7	2	0.22	2	tr	0	tr	unk.	unk.	0.1	0.77	47	5	0.1	9.2	unk.	12	3	0.08	4	tr(a)		unk.	
0.06	4	0.15	8	0.00	0	0.001	0	0	0(a)	0	unk.	unk.	13	1	190	659	unk.	0.9	5	0.2	2	unk.	unk.	unk.	K
2.4	12	19	5	0.46	5	tr	0	tr	unk.	unk.	unk.	1.95	86	9	8.2	16.8	unk.	14	4	0.23	12	tr(a)		unk.	
0.04	2	0.14	7	0.00	0	0.001	0	0	0(a)	0	unk.	unk.	9	1	223	368	unk.	0.7	4	0.2	1	unk.	unk.	unk.	L
1.6	8	13	3	0.42	4	tr	0	tr	unk.	unk.	unk.	1.47	61	6	9.7	9.4	unk.	13	3	0.21	11	tr(a)		unk.	
0.04	3	0.10	5	0.00	0	tr	0	unk.	unk.	unk.	unk.	unk.	33	3	243	300	75	0.3	2	0.4	3	unk.	unk.	24	M
0.9	5	tr	0	0.25	3	136	3	unk.	unk.	unk.	unk.	1.15	49	5	10.5	7.7	2.1	unk.	unk.	0.10	5	tr(a)		unk.	
0.05	3	0.10	5	0.00	0	tr	0	unk.	unk.	unk.	unk.	unk.	34	3	269	304	75	0.5	3	0.4	3	unk.	unk.	24	N
0.7	4	tr	0	0.25	3	115	2	unk.	unk.	unk.	unk.	1.47	55	6	11.7	7.8	2.1	unk.	unk.	0.10	5	tr(a)		unk.	
0.05	3	0.10	5	0.00	0	tr	0	unk.	unk.	unk.	unk.	unk.	25	3	348	262	75	0.4	2	0.4	3	unk.	unk.	24	O
1.0	5	tr	0	0.25	3	178	4	unk.	unk.	unk.	unk.	1.57	50	5	15.1	6.7	2.1	unk.	unk.	0.10	5	tr(a)		unk.	
0.05	3	0.10	5	0.00	0	unk.	unk.	4	unk.	unk.	unk.	unk.	25	3	316	274	unk.	0.4	2	unk.	unk.	unk.	unk.	24	P
1.0	5	10	3	0.25	3	21	0	1	unk.	unk.	unk.	1.47	51	5	13.8	7.0	unk.	unk.	unk.	0.10	5	tr(a)		unk.	
0.03	2	0.21	10	0.00	0	0.002	1	0	0(a)	0	0.1	tr	6	1	2	295	42	0.5	3	0.3	2	unk.	unk.	unk.	Q
1.2	6	10	3	0.41	4	tr	0	tr	0.1	0	tr	0.72	44	4	0.1	7.5	1.2	23	6	0.16	8	tr(a)		unk.	
0.02	1	0.15	8	0.00	0	0.002	1	0	0(a)	0	0.0	tr	5	1	2	221	32	0.4	2	0.2	2	unk.	unk.	unk.	R
0.9	5	9	2	0.31	3	tr	0	tr	0.1	0	tr	0.54	33	3	0.1	5.6	0.9	17	4	0.12	6	tr(a)		unk.	
0.06	4	0.41	21	0.00	0	0.004	1	0	0(a)	0	0.1	tr	12	1	4	587	84	1.0	6	0.6	4	unk.	unk.	unk.	S
2.5	12	25	6	0.82	8	tr	0	tr	0.1	1	0.1	1.44	87	9	0.2	15.0	2.4	45	11	0.31	15	tr(a)		unk.	

	FOOD	Portion	Weight in grams / *Conversion for 100 g*	Kilocalories / H_2O g	Total carbohydrate g / Total fats g	Crude fiber g / Dietary fiber g	Total protein g / Total sugar g	**% USRDA**	Arginine mg / Histidine mg	Isoleucine mg / Leucine mg	Lysine mg / Methionine mg	Phenylalanine mg / Threonine mg	Valine mg / Tryptophan mg	Cystine mg / Tyrosine mg	Polyunsat. fatty acids g / Monounsat. fatty acids g	Saturated fatty acids g / P/S ratio	Linoleic acid g / Cholesterol mg	Thiamin mg / Ascorbic acid mg	**% USRDA**
A	potato,pared,boiled,whole,round type	1 average	124	81	18.0	0.62	2.4	**4**	117	104	125	104	125	74	unk.	tr(a)	unk.	0.11	**7**
			0.81	102.7	0.1	1.24	1.1		unk.	118	29	97	24	84	unk.	unk.	0	20	**33**
B	potato,salad w/mayonnaise,French	1/2 C	100	145	13.4	0.40	3.0	**5**	unk.	unk.	unk.	unk.	unk.	unk.	unk.	2.00	4.00	0.07	**5**
	dressing,home recipe		*1.00*	72.4	9.2	unk.	unk.		unk.	unk.	unk.	unk.	unk.	unk.	2.00	unk.	65	11	**18**
C	potato,scalloped,frozen-Stouffer	1/2 pkg	170	187	20.8	unk.	4.4	**7**	unk.	unk.	unk.	unk.	unk.	unk.	unk.	unk.	unk.	0.05	**3**
			0.59	131.2	10.4	unk.	unk.		unk.	unk.	unk.	unk.	unk.	unk.	unk.	unk.	unk.	tr	**0**
D	potato,scalloped,home recipe	1/2 C	123	127	18.0	0.37	3.7	**7**	unk.	unk.	unk.	unk.	unk.	unk.	unk.	unk.	unk.	0.07	**5**
			0.82	94.0	4.8	unk.	1.1		unk.	unk.	unk.	unk.	unk.	unk.	unk.	unk.	7	13	**23**
E	POULTRY SEASONING	1 tsp	1.2	4	0.8	0.14	0.1	**0**	unk.	unk.	unk.	unk.	unk.	unk.	unk.	unk.	unk.	tr	**0**
			83.33	0.1	0.1	unk.	0.0		unk.	unk.	unk.	unk.	unk.	unk.	unk.	unk.	0	tr	**0**
F	PRUNE WHIP,baked	1/2 C	57	100	14.5	0.31	3.1	**6**	89	153	164	141	198	82	2.63	0.43	2.18	0.03	**2**
			1.76	34.7	4.1	2.62	13.5		69	248	99	124	41	121	0.61	6.08	0	1	**2**
G	PRUNES,dry,cooked w/added sugar,	1/2 C	119	205	53.7	0.71	0.9	**2**	unk.	unk.	unk.	unk.	unk.	unk.	unk.	tr(a)	unk.	0.04	**2**
	solids & liquid		*0.84*	63.3	0.2	9.16	53.0		unk.	unk.	unk.	unk.	unk.	unk.	unk.	unk.	0	1	**2**
H	prunes,dry,cooked w/o added sugar,	1/2 C	106	126	33.4	0.85	1.1	**2**	unk.	75	52	50	68	unk.	unk.	unk.	unk.	0.03	**2**
	solids & liquid		*0.94*	70.6	0.3	7.34	32.7		unk.	7	unk.	47	unk.	unk.	unk.	tr(a)	0	1	**2**
I	prunes,dry,uncooked	1 average	9.6	24	6.5	0.15	0.2	**0**	unk.	unk.	unk.	8	unk.	unk.	unk.	unk.	unk.	0.01	**1**
			10.42	2.7	0.1	1.55	6.3		unk.	unk.	unk.	unk.	unk.	unk.	unk.	unk.	0	tr	**1**
J	PUDDING,banana cream w/skim milk-Jello	1/2 C	145	123	26.5	tr	4.2	**9**	unk.	unk.	unk.	unk.	unk.	unk.	0.01	0.14	0.01	0.04	**3**
			0.69	113.0	0.2	unk.	21.2		unk.	unk.	unk.	unk.	unk.	unk.	0.06	0.07	2	1	**2**
K	pudding,banana cream w/whole milk	1/2 C	144	155	26.1	tr	4.0	**9**	unk.	unk.	unk.	unk.	unk.	unk.	0.15	2.53	0.09	0.05	**3**
	-Jello		*0.69*	109.0	4.1	unk.	20.9		unk.	unk.	unk.	unk.	unk.	unk.	1.18	0.06	17	1	**2**
L	pudding,bread (enriched) w/raisins,	1/2 C	110	180	30.6	0.06	5.2	**11**	192	269	333	245	308	85	0.32	2.20	0.32	0.07	**5**
	home recipe		*0.91*	68.7	4.5	unk.	25.6		152	441	126	217	68	217	1.46	0.14	77	1	**2**
M	pudding,butterscotch or vanilla	1/2 C	130	69	12.2	0.01	4.4	**10**	175	291	381	231	322	44	0.01	0.16	0.01	0.05	**4**
	w/skim milk-D-Zerta		*0.77*	111.8	0.2	unk.	7.1		131	472	121	217	68	231	0.05	0.08	2	1	**2**
N	pudding,butterscotch w/skim milk	1/2 C	147	169	29.7	0.02	4.0	**9**	unk.	unk.	unk.	unk.	unk.	unk.	0.15	2.54	0.09	0.05	**3**
			0.68	109.0	4.1	unk.	24.5		unk.	unk.	unk.	unk.	unk.	unk.	1.18	0.06	17	1	**2**
O	pudding,butterscotch w/whole milk	1/2 C	148	136	29.8	0.02	4.2	**9**	unk.	unk.	unk.	unk.	unk.	unk.	0.01	0.14	0.01	0.04	**3**
			0.68	112.0	0.2	unk.	24.7		unk.	unk.	unk.	unk.	unk.	unk.	0.06	0.07	2	1	**2**
P	pudding,butterscotch,canned	1/2 C	113	140	26.5	tr	1.3	**0**	unk.	unk.	unk.	unk.	unk.	unk.	unk.	unk.	unk.	0.02	**2**
			0.89	unk.	3.0	unk.	unk.		unk.	unk.	unk.	unk.	unk.	unk.	unk.	unk.	unk.	1	**2**
Q	pudding,butterscotch,instant	1/2 C	149	172	30.3	0.08	4.1	**9**	unk.	unk.	unk.	unk.	unk.	unk.	0.20	2.56	0.14	0.05	**3**
	w/whole milk-Jello		*0.67*	108.0	4.3	unk.	26.4		unk.	unk.	unk.	unk.	unk.	unk.	1.21	0.08	17	1	**2**
R	pudding,chocolate w/skim milk-Jello	1/2 C	147	134	28.9	0.17	4.9	**11**	151	251	329	200	278	38	0.01	0.32	tr	0.06	**4**
			0.68	111.2	0.5	unk.	22.0		113	406	104	187	57	200	0.18	0.05	2	1	**2**
S	pudding,chocolate w/whole milk-Jello	1/2 C	148	167	28.9	0.17	4.8	**11**	147	244	320	194	269	37	0.15	2.72	0.09	0.06	**4**
			0.68	108.5	4.4	unk.	21.9		110	394	102	182	56	194	1.18	0.05	17	1	**2**

Riboflavin mg / Niacin mg	% USRDA	Vitamin B_6 mg / Folacin mcg	% USRDA	Vitamin B_{12} mcg / Pantothenic acid mg	% USRDA	Biotin mg / Vitamin A IU	% USRDA	Preformed A RE / Beta carotene RE	Vitamin D IU / Vitamin E IU	% USRDA	Total tocopherol mg / Alpha tocopherol mg	Other tocopherol mg / Total ash g	Calcium mg / Phosphorus mg	% USRDA	Sodium mg / Sodium meq	Potassium mg / Potassium meq	Chlorine mg / Chlorine meq	Iron mg / Magnesium mg	% USRDA	Zinc mg / Copper mg	% USRDA	Iodine mcg / Selenium mcg	% USRDA	Manganese mcg / Chromium mcg	
0.04	**2**	0.25	**12**	0.00	**0**	0.002	**1**	0	0(a)	**0**	0.1	tr	7	**1**	2	353	51	0.6	**3**	0.4	**3**	unk.	**unk.**	unk.	A
1.5	**7**	15	**4**	0.50	**5**	tr	**0**	tr	0.1	**0**	0.0	0.87	52	**5**	0.1	9.0	1.4	27	**7**	0.19	**9**	tr(a)		unk.	
0.06	**4**	unk.	**unk.**	unk.	**unk.**	unk.	**unk.**	37	unk.	**unk.**	unk.	unk.	19	**2**	480	296	unk.	0.8	**4**	0.2	**2**	unk.	**unk.**	unk.	B
0.9	**5**	unk.	**unk.**	unk.	**unk.**	180	**4**	6	unk.	**unk.**	unk.	2.00	63	**6**	20.9	7.6	unk.	unk.	**unk.**	unk.	**unk.**	unk.		unk.	
0.15	**9**	unk.	**unk.**	unk.	**unk.**	unk.	**unk.**	unk.	unk.	**unk.**	unk.	unk.	119	**12**	671	373	unk.	0.5	**3**	unk.	**unk.**	unk.	**unk.**	unk.	C
0.6	**3**	unk.	**unk.**	unk.	**unk.**	tr	**0**	unk.	unk.	**unk.**	unk.	2.56	unk.	**unk.**	29.2	9.5	unk.	unk.	**unk.**	unk.	**unk.**	unk.		unk.	
0.11	**7**	unk.	**unk.**	unk.	**unk.**	unk.	**unk.**	unk.	unk.	**unk.**	unk.	unk.	66	**7**	435	401	unk.	0.5	**3**	unk.	**unk.**	unk.	**unk.**	unk.	D
1.2	**6**	unk.	**unk.**	unk.	**unk.**	196	**4**	unk.	unk.	**unk.**	unk.	2.08	91	**9**	18.9	10.2	unk.	unk.	**unk.**	tr(a)	**0**	unk.		unk.	
tr	**0**	unk.	**unk.**	0.00	**0**	unk.	**unk.**	0(a)	0(a)	**0**	unk.	unk.	12	**1**	tr	8	unk.	0.4	**2**	tr	**0**	unk.	**unk.**	unk.	E
tr	**0**	unk.	**unk.**	unk.	**unk.**	32	**1**	unk.	unk.	**unk.**	unk.	0.07	2	**0**	0.0	0.2	unk.	3	**1**	unk.	**unk.**	unk.		unk.	
0.07	**4**	0.05	**2**	0.01	**0**	0.003	**1**	0	0(a)	**0**	0(a)	0(a)	14	**1**	39	137	38	0.6	**4**	0.1	**1**	1.4	**1**	133	F
0.3	**1**	5	**1**	tr	**0**	181	**4**	18	0.0	**0**	0.1	0.47	36	**4**	1.7	3.5	1.1	19	**5**	0.09	**5**	tr(a)		8	
0.07	**4**	0.12	**6**	0(a)	**0**	tr	**0**	0	0(a)	**0**	unk.	unk.	23	**2**	4	312	1	1.8	**10**	0.4	**3**	unk.	**unk.**	unk.	G
0.7	**4**	tr	**0**	0.24	**2**	714	**14**	71	unk.	**unk.**	unk.	0.83	36	**4**	0.2	8.0	0.0	29	**7**	tr(a)	**0**	tr(a)		18	
0.07	**4**	0.13	**6**	0(a)	**0**	tr	**0**	0	0(a)	**0**	unk.	unk.	26	**3**	4	348	1	1.9	**11**	0.3	**2**	unk.	**unk.**	83	H
0.7	**4**	tr	**0**	0.27	**3**	797	**16**	80	unk.	**unk.**	unk.	0.96	39	**4**	0.2	8.9	0.0	21	**5**	0.27	**13**	tr(a)		16	
0.02	**1**	0.02	**1**	0.00	**0**	tr	**0**	0	0(a)	**0**	unk.	unk.	5	**1**	1	67	1	0.4	**2**	unk.	**unk.**	unk.	**unk.**	17	I
0.2	**1**	0	**0**	0.04	**0**	154	**3**	15	unk.	**unk.**	unk.	0.18	8	**1**	0.0	1.7	0.0	4	**1**	0.02	**1**	tr(a)		unk.	
0.17	**10**	0.05	**3**	0.46	**8**	unk.	**unk.**	unk.	50	**13**	unk.	unk.	155	**16**	255	205	unk.	0.1	**1**	0.5	**3**	unk.	**unk.**	unk.	J
0.1	**1**	6	**2**	0.40	**4**	251	**5**	unk.	unk.	**unk.**	unk.	1.46	125	**13**	11.1	5.3	unk.	15	**4**	0.03	**2**	unk.		unk.	
0.20	**12**	0.05	**3**	0.44	**7**	unk.	**unk.**	unk.	50	**13**	unk.	unk.	149	**15**	251	186	unk.	0.1	**1**	0.5	**3**	unk.	**unk.**	unk.	K
0.1	**1**	6	**2**	0.33	**3**	154	**3**	unk.	unk.	**unk.**	unk.	1.41	114	**11**	10.9	4.8	unk.	17	**4**	0.05	**3**	unk.		unk.	
0.18	**11**	0.08	**4**	0.40	**7**	0.003	**1**	0	35	**9**	0.2	tr	124	**12**	185	266	45	1.5	**8**	0.6	**4**	33.6	**22**	115	L
0.4	**2**	13	**3**	0.47	**5**	125	**3**	tr	0.3	**1**	0.1	1.46	123	**12**	8.0	6.8	1.3	57	**14**	0.15	**8**	unk.		9	
0.21	**12**	0.06	**3**	0.49	**8**	0.003	**1**	unk.	50	**13**	unk.	unk.	165	**17**	147	234	unk.	0.1	**0**	0.5	**4**	unk.	**unk.**	unk.	M
0.1	**1**	7	**2**	0.48	**5**	250	**5**	unk.	unk.	**unk.**	unk.	1.23	137	**14**	6.4	6.0	unk.	17	**4**	0.03	**1**	unk.		unk.	
0.20	**12**	0.05	**3**	0.44	**7**	unk.	**unk.**	unk.	50	**13**	unk.	unk.	149	**15**	244	187	unk.	0.1	**1**	0.5	**3**	unk.	**unk.**	unk.	N
0.1	**1**	6	**2**	0.38	**4**	154	**3**	unk.	unk.	**unk.**	unk.	1.39	114	**11**	10.6	4.8	unk.	17	**4**	0.05	**3**	unk.		unk.	
0.17	**10**	0.05	**3**	0.46	**8**	unk.	**unk.**	unk.	50	**13**	unk.	unk.	154	**15**	246	204	unk.	0.1	**1**	0.5	**3**	unk.	**unk.**	unk.	O
0.1	**1**	6	**2**	0.40	**4**	249	**5**	unk.	unk.	**unk.**	unk.	1.43	124	**12**	10.7	5.2	unk.	15	**4**	0.03	**2**	unk.		unk.	
0.12	**7**	unk.	**unk.**	unk.	**unk.**	unk.	**unk.**	unk.	unk.	**unk.**	unk.	unk.	71	**7**	220	82	unk.	0.1	**1**	unk.	**unk.**	unk.	**unk.**	unk.	P
0.1	**1**	unk.	**unk.**	unk.	**unk.**	14	**0**	unk.	unk.	**unk.**	unk.	0.56	82	**8**	9.6	2.1	unk.	11	**3**	unk.	**unk.**	unk.		unk.	
0.20	**12**	0.05	**3**	0.44	**7**	unk.	**unk.**	unk.	50	**13**	unk.	unk.	147	**15**	422	187	unk.	0.1	**1**	0.5	**3**	unk.	**unk.**	unk.	Q
0.1	**1**	6	**2**	0.39	**4**	154	**3**	unk.	unk.	**unk.**	unk.	1.60	327	**33**	18.3	4.8	unk.	17	**4**	0.05	**3**	unk.		unk.	
0.21	**12**	0.05	**3**	0.07	**1**	0.003	**1**	tr	50	**12**	unk.	unk.	172	**17**	193	233	unk.	0.3	**2**	0.7	**4**	unk.	**unk.**	unk.	R
0.2	**1**	7	**2**	0.44	**4**	249	**5**	tr	unk.	**unk.**	unk.	1.48	147	**15**	8.4	6.0	unk.	28	**7**	0.03	**2**	unk.		unk.	
0.22	**13**	0.06	**3**	0.07	**1**	0.007	**3**	unk.	50	**13**	unk.	unk.	168	**17**	192	218	unk.	0.3	**2**	0.7	**4**	unk.	**unk.**	unk.	S
0.2	**1**	6	**2**	0.43	**4**	155	**3**	unk.	unk.	**unk.**	unk.	1.45	139	**14**	8.3	5.6	unk.	31	**8**	0.05	**3**	unk.		unk.	

	FOOD	Portion	Weight in grams / Conversion for 100 g	Kilocalories / H_2O g	Total carbohydrate g / Total fats g	Crude fiber g / Dietary fiber g	Total protein g / Total sugar g	% USRDA	Arginine mg / Histidine mg	Isoleucine mg / Leucine mg	Lysine mg / Methionine mg	Phenylalanine mg / Threonine mg	Valine mg / Tryptophan mg	Cystine mg / Tyrosine mg	Polyunsat. fatty acids g / Monounsat. fatty acids g	Saturated fatty acids g / P/S ratio	Linoleic acid g / Cholesterol mg	Thiamin mg / Ascorbic acid mg	% USRDA
A	pudding,chocolate,canned	1/2 C	114	157	28.4	tr	2.3	**0**	unk.	unk.	unk.	unk.	unk.	unk.	unk.	unk.	unk.	0.03	**2**
			0.88	unk.	3.7	unk.	unk.		unk.	unk.	unk.	unk.	unk.	unk.	unk.	unk.	unk.	1	**2**
B	pudding,chocolate,instant w/whole milk-Jello	1/2 C	151	180	32.0	0.11	4.5	**10**	unk.	unk.	unk.	unk.	unk.	unk.	0.19	2.68	0.12	0.05	**3**
			0.62	108.0	4.5	unk.	27.0		unk.	unk.	unk.	unk.	unk.	unk.	1.28	0.07	17	1	**2**
C	pudding,chocolate,low calorie w/skim milk-D-Zerta	1/2 C	130	65	11.5	0.11	4.6	**11**	163	273	357	217	302	42	0.01	0.29	0.01	0.05	**4**
			0.77	112.2	0.4	unk.	6.6		122	442	113	203	64	217	0.12	0.05	2	1	**2**
D	pudding,chocolate,mix	1 pkg	113	404	101.9	0.75	3.2	**5**	unk.	unk.	unk.	unk.	unk.	unk.	tr	0.78	tr	0.01	**1**
			0.88	3.8	1.4	unk.	71.3		unk.	unk.	unk.	unk.	unk.	unk.	0.55	0.00	0	0	**0**
E	pudding,chocolate,mix,instant-Jello	1 pkg	117	725	105.6	0.44	1.9	**3**	unk.	unk.	unk.	unk.	unk.	unk.	0.16	0.58	0.11	0.01	**1**
			0.85	2.0	1.7	unk.	85.5		unk.	unk.	unk.	unk.	unk.	unk.	0.41	0.28	tr	tr	**0**
F	pudding,cool 'n creamy,any flavor -Birds Eye	1/2 C	149	173	30.4	0.00	4.1	**0**	unk.	unk.	unk.	unk.	unk.	unk.	0.21	2.56	0.13	0.04	**3**
			0.67	108.5	4.3	unk.	26.5		unk.	unk.	unk.	unk.	unk.	unk.	1.06	0.08	17	1	**2**
G	pudding,lemon-Jello	1/2 C	125	137	29.2	0.07	0.9	**2**	unk.	unk.	unk.	unk.	unk.	unk.	0.26	0.57	0.22	0.01	**1**
			0.80	92.2	1.9	unk.	22.6		unk.	unk.	unk.	unk.	unk.	unk.	0.71	0.46	91	0	**0**
H	pudding,lemon,canned	1/2 C	111	150	31.1	tr	0.0	**0**	unk.	unk.	unk.	unk.	unk.	unk.	unk.	unk.	unk.	0.02	**2**
			0.90	unk.	2.9	unk.	unk.		unk.	unk.	unk.	unk.	unk.	unk.	unk.	unk.	unk.	1	**2**
I	pudding,pumpkin,home recipe	1/2 C	129	170	25.9	0.69	5.5	**12**	245	295	396	253	330	77	0.42	2.65	0.37	0.04	**3**
			0.78	90.8	5.4	unk.	23.9		156	479	142	248	76	244	1.88	0.16	105	3	**5**
J	pudding,rice w/raisins,home recipe	1/2 C	156	246	41.7	0.06	7.1	**15**	293	364	469	331	450	108	0.56	2.86	0.50	0.11	**8**
			0.64	100.2	5.9	4.36	28.8		190	605	185	313	93	311	1.94	0.20	136	1	**2**
K	pudding,rice,canned	1/2 C	113	160	31.9	tr	2.0	**5**	unk.	unk.	unk.	unk.	unk.	unk.	unk.	unk.	unk.	0.03	**2**
			0.89	unk.	2.7	unk.	unk.		unk.	unk.	unk.	unk.	unk.	unk.	unk.	unk.	unk.	1	**2**
L	pudding,tapioca cream,home recipe	1/2 C	142	169	21.4	0.00	6.3	**14**	292	358	481	303	418	89	0.47	3.40	0.40	0.05	**3**
			0.70	106.9	6.4	unk.	17.6		184	584	175	296	89	295	1.96	0.14	111	1	**2**
M	pudding,tapioca,canned	1/2 C	116	150	27.4	tr	1.7	**0**	unk.	unk.	unk.	unk.	unk.	unk.	unk.	unk.	unk.	0.02	**2**
			0.86	unk.	3.6	unk.	unk.		unk.	unk.	unk.	unk.	unk.	unk.	unk.	unk.	unk.	1	**2**
N	pudding,tomato,home recipe	1/2 C	137	212	31.6	0.84	3.5	**5**	21	138	109	155	131	47	0.29	4.95	0.34	0.08	**6**
			0.73	90.8	8.6	1.97	17.8		52	216	40	105	36	89	2.43	0.06	21	45	**75**
O	pudding,vanilla w/skim milk-Jello	1/2 C	145	124	26.6	tr	4.2	**9**	156	262	344	209	290	41	0.01	0.14	tr	0.04	**3**
			0.69	112.6	0.2	unk.	21.6		117	425	109	196	61	209	0.04	0.10	2	1	**2**
P	pudding,vanilla w/whole milk-Jello	1/2 C	144	155	26.3	tr	4.0	**9**	143	252	331	202	323	37	0.16	2.53	0.09	0.04	**3**
			0.69	108.3	4.1	unk.	21.3		107	409	105	189	59	202	1.02	0.06	17	1	**2**
Q	pudding,vanilla,canned	1/2 C	111	140	25.9	tr	1.7	**0**	unk.	unk.	unk.	unk.	unk.	unk.	unk.	unk.	unk.	0.02	**2**
			0.90	unk.	3.3	unk.	unk.		unk.	unk.	unk.	unk.	unk.	unk.	unk.	unk.	unk.	1	**2**
R	PUMPKIN PIE SPICE	1 tsp	1.9	6	1.3	0.28	0.1	**0**	unk.	unk.	unk.	unk.	unk.	unk.	unk.	unk.	unk.	tr	**0**
			52.63	0.2	0.2	unk.	0.0		unk.	unk.	unk.	unk.	unk.	unk.	unk.	unk.	0	tr	**1**
S	PUMPKIN/squash seed kernels,dry, hulled	1/4 C	35	194	5.2	0.66	10.1	**16**	unk.	568	467	568	548	unk.	unk.	2.80	7.00	0.08	**6**
			2.86	1.5	16.3	unk.	0.0		233	802	193	304	183	unk.	5.95	unk.	0	unk.	**unk.**

Riboflavin mg / Niacin mg	% USRDA	Vitamin B_6 mg / Folacin mcg	% USRDA	Vitamin B_{12} mcg / Pantothenic acid mg	% USRDA	Biotin mg / Vitamin A IU	% USRDA	Preformed A RE / Beta carotene RE	Vitamin D IU / Vitamin E IU	% USRDA	Total tocopherol mg / Alpha tocopherol mg	Other tocopherol mg / Total ash g	Calcium mg / Phosphorus mg	% USRDA	Sodium mg / Sodium meq	Potassium mg / Potassium meq	Chlorine mg / Chlorine meq	Iron mg / Magnesium mg	% USRDA	Zinc mg / Copper mg	% USRDA	Iodine mcg / Selenium mcg	% USRDA	Manganese mcg / Chromium mcg	
0.15	**9**	unk.	**unk.**	unk.	**unk.**	unk.	**unk.**	unk.	unk.	**unk.**	unk.	unk.	58	**6**	233	199	unk.	1.0	**6**	unk.	**unk.**	unk.	**unk.**	unk.	A
0.1	**1**	unk.	**unk.**	unk.	**unk.**	23	**1**	unk.	unk.	**unk.**	unk.	1.48	238	**24**	10.1	5.1	unk.	21	**5**	unk.	**unk.**	unk.		unk.	
0.21	**12**	0.05	**3**	0.44	**7**	unk.	**unk.**	unk.	50	**13**	unk.	unk.	149	**15**	459	200	unk.	0.3	**2**	0.6	**4**	unk.	**unk.**	unk.	B
0.2	**1**	6	**2**	0.38	**4**	154	**3**	unk.	unk.	**unk.**	unk.	1.88	312	**31**	20.0	5.1	unk.	25	**6**	0.11	**6**	unk.		unk.	
0.18	**11**	0.05	**3**	0.06	**1**	0.003	**1**	unk.	tr	**0**	unk.	unk.	161	**16**	82	227	unk.	0.3	**2**	0.5	**3**	unk.	**unk.**	unk.	C
0.2	**1**	6	**2**	0.41	**4**	251	**5**	unk.	unk.	**unk.**	unk.	1.14	138	**14**	3.6	5.8	unk.	22	**5**	0.03	**1**	unk.		unk.	
0.12	**7**	0.02	**1**	0.08	**1**	tr(a)	**0**	0	0	**0**	unk.	unk.	96	**10**	579	143	unk.	1.1	**6**	0.8	**5**	unk.	**unk.**	unk.	D
0.3	**2**	1	**0**	0.19	**2**	4	**0**	tr	unk.	**unk.**	unk.	2.50	108	**11**	25.2	3.7	unk.	62	**16**	0.02	**1**	unk.		unk.	
0.04	**3**	tr	**0**	0.01	**0**	unk.	**unk.**	unk.	0	**0**	unk.	unk.	16	**2**	517	64	unk.	1.0	**6**	0.5	**3**	unk.	**unk.**	unk.	E
0.2	**1**	tr	**0**	0.01	**0**	2	**0**	unk.	unk.	**unk.**	unk.	4.01	113	**11**	22.2	1.6	unk.	37	**9**	0.25	**13**	unk.		unk.	
0.19	**11**	0.05	**3**	0.43	**7**	unk.	**unk.**	unk.	50	**13**	unk.	unk.	147	**15**	385	187	unk.	0.1	**0**	0.5	**3**	unk.	**unk.**	unk.	F
0.1	**1**	6	**2**	0.39	**4**	154	**3**	unk.	unk.	**unk.**	unk.	1.53	301	**30**	16.7	4.8	unk.	17	**4**	0.05	**3**	unk.		unk.	
0.02	**2**	0.02	**1**	0.21	**4**	unk.	**unk.**	unk.	10	**2**	unk.	unk.	9	**1**	77	6	unk.	0.3	**2**	0.2	**1**	unk.	**unk.**	unk.	G
0.0	**0**	9	**2**	0.25	**3**	104	**2**	unk.	unk.	**unk.**	unk.	0.30	29	**3**	3.3	0.1	unk.	1	**0**	0.02	**1**	unk.		unk.	
0.07	**4**	unk.	**unk.**	unk.	**unk.**	unk.	**unk.**	unk.	unk.	**unk.**	unk.	unk.	7	**1**	139	10	unk.	0.6	**3**	unk.	**unk.**	unk.	**unk.**	unk.	H
0.1	**1**	unk.	**unk.**	unk.	**unk.**	20	**0**	unk.	unk.	**unk.**	unk.	0.22	21	**2**	6.0	0.3	unk.	unk.	**unk.**	unk.	**unk.**	unk.		unk.	
0.20	**12**	0.08	**4**	0.37	**6**	0.005	**2**	0	39	**10**	0.3	0.0	129	**13**	205	265	33	0.8	**5**	0.8	**5**	54.8	**37**	66	I
0.4	**2**	18	**5**	0.70	**7**	3083	**62**	294	0.3	**1**	0.1	1.43	137	**14**	8.9	6.8	0.9	57	**14**	0.12	**6**	0		12	
0.26	**15**	0.12	**6**	0.56	**9**	0.006	**2**	0	43	**11**	0.5	tr	125	**13**	270	265	56	1.5	**8**	1.0	**7**	24.1	**16**	73	J
0.7	**3**	21	**5**	0.70	**7**	161	**3**	tr	0.6	**2**	0.2	1.63	163	**16**	11.8	6.8	1.6	80	**20**	0.17	**8**	tr(a)		10	
0.11	**7**	unk.	**unk.**	unk.	**unk.**	unk.	**unk.**	unk.	unk.	**unk.**	unk.	unk.	51	**5**	265	69	unk.	0.5	**3**	unk.	**unk.**	unk.	**unk.**	unk.	K
0.1	**1**	unk.	**unk.**	unk.	**unk.**	18	**0**	unk.	unk.	**unk.**	unk.	0.45	44	**4**	11.5	1.8	unk.	unk.	**unk.**	unk.	**unk.**	unk.		unk.	
0.25	**15**	0.08	**4**	0.63	**11**	0.003	**1**	0	57	**14**	0.3	0.0	157	**16**	154	212	33	0.5	**3**	0.8	**6**	39.8	**27**	13	L
0.1	**1**	12	**3**	0.65	**7**	196	**4**	0	0.4	**1**	0.1	1.23	159	**16**	6.7	5.4	0.9	62	**15**	0.20	**10**	0		7	
0.14	**8**	unk.	**unk.**	unk.	**unk.**	unk.	**unk.**	unk.	unk.	**unk.**	unk.	unk.	77	**8**	215	86	unk.	1.2	**7**	unk.	**unk.**	unk.	**unk.**	unk.	M
0.1	**1**	unk.	**unk.**	unk.	**unk.**	19	**0**	unk.	unk.	**unk.**	unk.	0.58	87	**9**	9.4	2.2	unk.	unk.	**unk.**	unk.	**unk.**	unk.		unk.	
0.11	**7**	0.08	**4**	0.00	**0**	tr(a)	**0**	0	0	**0**	0.4	0.0	60	**6**	676	341	19	2.5	**14**	0.1	**1**	79.5	**53**	47	N
1.4	**7**	11	**3**	0.12	**1**	1210	**24**	75	0.4	**1**	0.2	2.79	55	**6**	29.4	8.7	0.5	26	**7**	0.08	**4**	0		10	
0.17	**10**	0.05	**3**	0.46	**8**	0.003	**1**	unk.	50	**13**	unk.	unk.	154	**15**	201	205	unk.	0.1	**0**	0.5	**3**	unk.	**unk.**	unk.	O
0.1	**1**	6	**2**	0.40	**4**	251	**5**	unk.	unk.	**unk.**	unk.	1.32	124	**12**	8.8	5.2	unk.	14	**4**	0.03	**2**	unk.		unk.	
0.20	**12**	0.05	**3**	0.43	**7**	0.007	**2**	unk.	50	**13**	unk.	unk.	148	**15**	197	186	unk.	0.1	**0**	0.5	**3**	unk.	**unk.**	unk.	P
0.1	**1**	6	**2**	0.38	**4**	154	**3**	unk.	unk.	**unk.**	unk.	1.25	114	**11**	8.6	4.8	unk.	17	**4**	0.05	**3**	unk.		unk.	
0.12	**7**	unk.	**unk.**	unk.	**unk.**	unk.	**unk.**	unk.	unk.	**unk.**	unk.	unk.	69	**7**	233	84	unk.	0.6	**3**	unk.	**unk.**	unk.	**unk.**	unk.	Q
0.1	**1**	unk.	**unk.**	unk.	**unk.**	13	**0**	unk.	unk.	**unk.**	unk.	0.56	96	**10**	10.1	2.2	unk.	9	**2**	unk.	**unk.**	unk.		unk.	
tr	**0**	unk.	**unk.**	0.00	**0**	unk.	**unk.**	0(a)	0(a)	**0**	unk.	unk.	13	**1**	1	13	unk.	0.4	**2**	0.0	**0**	unk.	**unk.**	unk.	R
tr	**0**	unk.	**unk.**	unk.	**unk.**	5	**0**	unk.	unk.	**unk.**	unk.	0.07	2	**0**	0.0	0.3	unk.	3	**1**	unk.	**unk.**	unk.		unk.	
0.07	**4**	0.03	**2**	0.00	**0**	unk.	**unk.**	0	0(a)	**0**	unk.	unk.	18	**2**	unk.	unk.	unk.	3.9	**22**	unk.	**unk.**	unk.	**unk.**	unk.	S
0.8	**4**	unk.	**unk.**	unk.	**unk.**	24	**1**	2	unk.	**unk.**	unk.	1.71	400	**40**	unk.	unk.	unk.	unk.	**unk.**	unk.	**unk.**	unk.		unk.	

	FOOD	Portion	Weight in grams / Conversion for 100 g	Kilocalories / H_2O g	Total carbohydrate g / Total fats g	Crude fiber g / Dietary fiber g	Total protein g / Total sugar g	% USRDA	Arginine mg / Histidine mg	Isoleucine mg / Leucine mg	Lysine mg / Methionine mg	Phenylalanine mg / Threonine mg	Valine mg / Tryptophan mg	Cystine mg / Tyrosine mg	Polyunsat. fatty acids g / Monounsat. fatty acids g	Saturated fatty acids g / P/S ratio	Linoleic acid g / Cholesterol mg	Thiamin mg / Ascorbic acid mg	% USRDA
A	pumpkin/squash seeds	1/4 C	38	155	4.2	0.53	8.1	**13**	unk.	456	374	456	439	unk.	unk.	2.24	5.61	0.07	**5**
			2.64	1.7	13.1	unk.	0.0		187	642	155	244	146	unk.	4.78	unk.	0	tr(a)	**0**
B	pumpkin,canned	1/2 C	123	40	9.7	1.59	1.2	**2**	unk.	45	53	34	unk.	unk.	unk.	tr(a)	unk.	0.04	**2**
			0.82	110.5	0.4	unk.	5.7		20	64	unk.	33	13	34	unk.	unk.	0	6	**10**
C	QUAIL,breast,meat w/o skin,raw	1 breast	56	69	0.0	0.00	12.6	**28**	802	690	1107	549	686	221	0.43	0.49	0.35	0.13	**9**
			1.79	40.1	1.7	0(a)	0(a)		480	1085	401	633	198	587	0.38	0.89	unk.	3	**5**
D	quail,meat w/skin,raw	1 bird	109	209	0.0	0.00	21.4	**48**	1394	1104	1793	900	1126	371	3.25	3.68	2.51	0.26	**17**
			0.92	75.9	13.1	0(a)	0(a)		759	1758	644	1030	314	925	3.84	0.88	unk.	7	**11**
E	quail,meat w/o skin,raw	1 bird	92	123	0.0	0.00	20.0	**45**	1269	1092	1753	868	1086	350	1.08	1.21	0.86	0.26	**17**
			1.09	64.4	4.2	0(a)	0(a)		759	1717	634	1003	314	929	0.94	0.89	unk.	7	**11**
F	QUICHE LORRAINE,home recipe	1/6 of 9" dia	175	379	19.2	0.06	15.2	**32**	600	784	1085	788	969	224	3.77	11.04	3.69	0.16	**11**
		pie	*0.57*	110.8	26.8	0.58	4.9		468	1382	383	595	204	716	9.71	0.34	159	1	**2**
G	RADISHES,raw,common,sliced	1/2 C	58	10	2.1	0.40	0.6	**1**	unk.	unk.	20	unk.	17	unk.	unk.	unk.	unk.	0.02	**1**
			1.74	54.3	0.1	0.63	2.0		unk.	unk.	1	34	3	unk.	unk.	unk.	unk.	15	**25**
H	radishes,raw,common,whole	1 med	4.5	1	0.2	0.03	0.0	**0**	unk.	unk.	2	unk.	1	unk.	unk.	unk.	unk.	tr	**0**
			22.22	4.3	0.0	0.05	0.2		unk.	unk.	tr	3	tr	unk.	unk.	unk.	unk.	1	**2**
I	RAISINS,natural,uncooked	1/4 C	36	104	27.9	0.32	0.9	**1**	unk.	unk.	unk.	13	unk.	unk.	unk.	unk.	unk.	0.04	**3**
			2.78	6.5	0.1	2.45	26.9		unk.	unk.	unk.	unk.	unk.	unk.	unk.	unk.	0	tr	**1**
J	raisins,Thompson seedless,California	1/4 C	42	125	32.6	0.35	1.4	**2**	unk.	unk.	unk.	unk.	unk.	unk.	unk.	unk.	unk.	0.06	**4**
			2.36	7.2	0.1	2.88	unk.		unk.	unk.	unk.	unk.	unk.	unk.	unk.	unk.	0	tr	**1**
K	RASPBERRIES,canned,red,water pack, solids & liquid	1/2 C	122	43	10.7	3.16	0.8	**1**	unk.	unk.	unk.	unk.	unk.	unk.	unk.	tr(a)	unk.	0.01	**1**
			0.82	109.5	0.1	8.99	7.5		unk.	unk.	unk.	unk.	unk.	unk.	unk.	unk.	0	11	**18**
L	raspberries,fresh,red	1/2 C	62	35	8.4	1.84	0.7	**1**	unk.	unk.	unk.	unk.	unk.	unk.	unk.	tr(a)	unk.	0.02	**1**
			1.63	51.8	0.3	4.55	4.0		unk.	unk.	unk.	unk.	unk.	unk.	unk.	unk.	0	15	**26**
M	raspberries,frozen,red,sweetened	1/2 C	125	123	30.7	2.75	0.9	**1**	unk.	unk.	unk.	unk.	unk.	unk.	unk.	tr(a)	unk.	0.02	**2**
			0.80	92.9	0.2	8.75	21.9		unk.	unk.	unk.	unk.	unk.	unk.	unk.	unk.	0	26	**44**
N	RAVIOLI,cheese filled,canned -Chef Boy-Ar-Dee	1 C	250	129	21.1	unk.	5.9	**4**	unk.	unk.	unk.	unk.	unk.	unk.	unk.	unk.	unk.	unk.	**unk.**
			0.40	unk.	2.9	unk.	unk.		unk.	unk.	unk.	unk.	unk.	unk.	unk.	unk.	unk.	unk.	**unk.**
O	ravioli,in meat sauce,canned -Franco-American	1 C	213	223	35.8	0.22	9.2	**14**	unk.	unk.	unk.	unk.	unk.	unk.	unk.	unk.	unk.	0.17	**11**
			0.47	159.7	5.1	unk.	unk.		unk.	unk.	unk.	unk.	unk.	unk.	unk.	unk.	unk.	6	**10**
P	ravioli,meat filled,canned -Chef Boy-Ar-Dee	1 C	250	123	19.4	unk.	4.7	**3**	unk.	unk.	unk.	unk.	unk.	unk.	unk.	unk.	unk.	unk.	**unk.**
			0.40	unk.	2.9	unk.	unk.		unk.	unk.	unk.	unk.	unk.	unk.	unk.	unk.	unk.	unk.	**unk.**
Q	ravioli,mini,canned-Chef Boy-Ar-Dee	1 C	250	129	19.9	unk.	4.7	**7**	unk.	unk.	unk.	unk.	unk.	unk.	unk.	unk.	unk.	unk.	**unk.**
			0.40	unk.	3.5	unk.	unk.		unk.	unk.	unk.	unk.	unk.	unk.	unk.	unk.	unk.	unk.	**unk.**
R	RHUBARB,cooked w/sugar added	1/2 C	135	190	48.6	0.81	0.7	**1**	unk.	unk.	unk.	unk.	unk.	unk.	unk.	unk.	unk.	0.03	**2**
			0.74	84.8	0.1	2.97	47.8		unk.	unk.	unk.	unk.	unk.	unk.	unk.	unk.	0	8	**14**
S	RICE-A-RONI,beef flavored,prep -Golden Grain	1/2 C	116	140	27.2	tr(a)	3.1	**5**	unk.	unk.	unk.	unk.	unk.	unk.	unk.	unk.	unk.	0.37	**25**
			0.86	unk.	2.1	unk.	unk.		unk.	unk.	unk.	unk.	unk.	unk.	unk.	unk.	unk.	0(a)	**0**

Riboflavin mg / Niacin mg	% USRDA	Vitamin B_6 mg / Folacin mcg	% USRDA	Vitamin B_{12} mcg / Pantothenic acid mg	% USRDA	Biotin mg / Vitamin A IU	% USRDA	Preformed A RE / Beta carotene RE	Vitamin D IU / Vitamin E IU	% USRDA	Total tocopherol mg / Alpha tocopherol mg	Other tocopherol mg / Total ash g	Calcium mg / Phosphorus mg	% USRDA	Sodium mg / Sodium meq	Potassium mg / Potassium meq	Chlorine mg / Chlorine meq	Iron mg / Magnesium mg	% USRDA	Zinc mg / Copper mg	% USRDA	Iodine mcg / Selenium mcg	% USRDA	Manganese mcg / Chromium mcg	
0.05	**3**	0.03	**1**	0.00	**0**	unk.	**unk.**	0	0(a)	**0**	unk.	unk.	14	**1**	unk.	unk.	unk.	3.1	**18**	unk.	**unk.**	unk.	**unk.**	unk.	A
0.7	**3**	unk.	**unk.**	unk.	**unk.**	20	**0**	2	unk.	**unk.**	unk.	1.36	321	**32**	unk.	unk.	unk.	unk.	**unk.**	unk.	**unk.**	unk.		unk.	
0.06	**4**	0.07	**3**	0.00	**0**	unk.	**unk.**	0	0(a)	**0**	unk.	unk.	31	**3**	2	294	unk.	0.5	**3**	0.2	**2**	unk.	**unk.**	135	B
0.7	**4**	23	**6**	0.49	**5**	7840	**157**	784	unk.	**unk.**	unk.	0.73	32	**3**	0.1	7.5	unk.	unk.	**unk.**	0.06	**3**	tr(a)		tr	
0.13	**8**	0.30	**15**	0.26	**4**	unk.	**unk.**	unk.	unk.	**unk.**	unk.	unk.	6	**1**	31	146	unk.	1.3	**7**	1.5	**10**	unk.	**unk.**	unk.	C
4.6	**23**	unk.	**unk.**	0.44	**4**	21	**0**	unk.	unk.	**unk.**	unk.	0.71	128	**13**	1.3	3.7	unk.	unk.	**unk.**	0.24	**12**	unk.		unk.	
0.28	**17**	0.65	**33**	unk.	**unk.**	unk.	**unk.**	unk.	unk.	**unk.**	unk.	unk.	14	**1**	58	235	unk.	4.3	**24**	unk.	**unk.**	unk.	**unk.**	unk.	D
8.2	**41**	9	**2**	unk.	**unk.**	265	**5**	unk.	unk.	**unk.**	unk.	0.98	300	**30**	2.5	6.0	unk.	unk.	**unk.**	0.55	**28**	unk.		unk.	
0.27	**16**	unk.	**unk.**	unk.	**unk.**	unk.	**unk.**	unk.	unk.	**unk.**	unk.	unk.	12	**1**	47	218	unk.	4.1	**23**	unk.	**unk.**	unk.	**unk.**	unk.	E
7.5	**38**	unk.	**unk.**	unk.	**unk.**	52	**1**	unk.	unk.	**unk.**	unk.	1.21	282	**28**	2.0	5.6	unk.	unk.	**unk.**	0.55	**27**	unk.		unk.	
0.35	**21**	0.10	**5**	0.87	**15**	0.005	**2**	0	43	**11**	1.0	0.3	300	**30**	477	219	67	1.4	**8**	1.6	**10**	94.4	**63**	87	F
1.0	**5**	17	**4**	0.78	**8**	401	**8**	1	1.2	**4**	0.4	2.67	282	**28**	20.8	5.6	1.9	87	**22**	0.17	**8**	10		13	
0.02	**1**	0.04	**2**	0.00	**0**	unk.	**unk.**	0	0(a)	**0**	unk.	unk.	17	**2**	10	185	21	0.6	**3**	0.2	**1**	unk.	**unk.**	29	G
0.2	**1**	14	**3**	0.11	**1**	6	**0**	1	unk.	**unk.**	unk.	0.46	18	**2**	0.4	4.7	0.6	9	**2**	0.07	**4**	tr(a)		0	
tr	**0**	tr	**0**	0.00	**0**	unk.	**unk.**	0	0(a)	**0**	unk.	unk.	1	**0**	1	14	2	tr	**0**	tr	**0**	unk.	**unk.**	2	H
tr	**0**	1	**0**	0.01	**0**	tr	**0**	0	unk.	**unk.**	unk.	0.04	1	**0**	0.0	0.4	0.0	1	**0**	0.01	**0**	tr(a)		0	
0.03	**2**	0.09	**4**	0.00	**0**	0.002	**1**	0	0(a)	**0**	unk.	unk.	22	**2**	10	275	37	1.3	**7**	0.1	**0**	unk.	**unk.**	168	I
0.2	**1**	4	**1**	0.04	**0**	7	**0**	1	unk.	**unk.**	unk.	0.68	36	**4**	0.4	7.0	1.0	13	**3**	0.07	**4**	tr(a)		1	
0.01	**1**	0.10	**5**	unk.	**unk.**	0.002	**1**	0	unk.	**unk.**	unk.	unk.	23	**2**	7	287	unk.	0.9	**5**	0.1	**1**	unk.	**unk.**	81	J
0.2	**1**	2	**0**	0.02	**0**	7	**0**	1	unk.	**unk.**	unk.	0.67	43	**4**	0.3	7.3	unk.	15	**4**	0.16	**8**	tr(a)		unk.	
0.05	**3**	0.05	**2**	0.00	**0**	0.002	**1**	0	0(a)	**0**	0.4	4.9	18	**2**	1	139	28	0.7	**4**	unk.	**unk.**	unk.	**unk.**	1384	K
0.6	**3**	unk.	**unk.**	0.21	**2**	109	**2**	11	0.4	**2**	unk.	0.36	18	**2**	0.0	3.5	0.8	16	**4**	0.07	**4**	0(a)		0(a)	
0.06	**3**	0.04	**2**	0.00	**0**	0.001	**0**	0	0(a)	**0**	0.2	2.5	14	**1**	1	103	14	0.6	**3**	unk.	**unk.**	unk.	**unk.**	700	L
0.6	**3**	3	**1**	0.15	**2**	80	**2**	8	0.2	**1**	unk.	0.31	14	**1**	0.0	2.6	0.4	12	**3**	0.04	**2**	0		0	
0.07	**4**	0.05	**2**	0.00	**0**	unk.	**unk.**	0	0(a)	**0**	unk.	unk.	16	**2**	1	125	unk.	0.7	**4**	unk.	**unk.**	unk.	**unk.**	1424	M
0.7	**4**	unk.	**unk.**	0.26	**3**	88	**2**	9	unk.	**unk.**	unk.	0.25	21	**2**	0.0	3.2	unk.	14	**3**	0.07	**4**	0(a)		0(a)	
unk.	**unk.**	unk.	**unk.**	tr(a)	**0**	unk.	**unk.**	unk.	0(a)	**0**	unk.	unk.	unk.	**unk.**	1350	unk.	unk.	unk.	**unk.**	1.9	**13**	unk.	**unk.**	unk.	N
unk.	**unk.**	unk.	**unk.**	unk.	**unk.**	1250	**25**	unk.	unk.	**unk.**	unk.	unk.	unk.	**unk.**	58.7	unk.	unk.	unk.	**unk.**	unk.	**unk.**	unk.		unk.	
0.19	**11**	unk.	**unk.**	unk.	**unk.**	unk.	**unk.**	unk.	0(a)	**0**	tr(a)	tr(a)	36	**4**	907	328	unk.	2.3	**13**	unk.	**unk.**	unk.	**unk.**	unk.	O
3.0	**15**	unk.	**unk.**	tr(a)	**0**	1733	**35**	unk.	0.0	**0**	tr(a)	3.30	114	**11**	39.4	8.4	unk.	unk.	**unk.**	tr(a)	**0**	tr(a)		unk.	
unk.	**unk.**	unk.	**unk.**	tr(a)	**0**	unk.	**unk.**	unk.	0(a)	**0**	unk.	unk.	unk.	**unk.**	1250	unk.	unk.	unk.	**unk.**	1.2	**8**	unk.	**unk.**	unk.	P
unk.	**unk.**	unk.	**unk.**	unk.	**unk.**	1000	**20**	unk.	unk.	**unk.**	unk.	unk.	unk.	**unk.**	54.4	unk.	unk.	unk.	**unk.**	0.13	**7**	unk.		unk.	
unk.	**unk.**	unk.	**unk.**	unk.	**unk.**	unk.	**unk.**	unk.	0(a)	**0**	unk.	unk.	unk.	**unk.**	1188	unk.	unk.	unk.	**unk.**	unk.	**unk.**	unk.	**unk.**	unk.	Q
unk.	**unk.**	unk.	**unk.**	unk.	**unk.**	750	**15**	unk.	unk.	**unk.**	unk.	unk.	unk.	**unk.**	51.6	unk.	unk.	unk.	**unk.**	unk.	**unk.**	unk.		unk.	
0.05	**3**	0.07	**3**	0.00	**0**	unk.	**unk.**	0	0(a)	**0**	0.3	unk.	105	**11**	3	274	99	0.8	**5**	0.1	**1**	unk.	**unk.**	unk.	R
0.4	**2**	5	**1**	0.07	**1**	108	**2**	11	0.3	**1**	unk.	0.81	20	**2**	0.1	7.0	2.8	18	**4**	tr(a)	**0**	tr(a)		7	
0.14	**8**	unk.	**unk.**	0(a)	**0**	tr(a)	**0**	0(a)	tr(a)	**0**	unk.	unk.	unk.	**unk.**	unk.	unk.	unk.	1.1	**6**	unk.	**unk.**	unk.	**unk.**	unk.	S
1.6	**8**	13	**3**	unk.	**unk.**	unk.	**unk.**	unk.	unk.	**unk.**	unk.	unk.	unk.	**unk.**	unk.	unk.	unk.	unk.	**unk.**	unk.	**unk.**	unk.		unk.	

	FOOD	Portion	Weight in grams / Conversion for 100 g	Kilocalories / H_2O g	Total carbohydrate g / Total fats g	Crude fiber g / Dietary fiber g	Total protein g / Total sugar g	% USRDA	Arginine mg / Histidine mg	Isoleucine mg / Leucine mg	Lysine mg / Methionine mg	Phenylalanine mg / Threonine mg	Valine mg / Tryptophan mg	Cystine mg / Tyrosine mg	Polyunsat. fatty acids g / Monounsat. fatty acids g	Saturated fatty acids g / P/S ratio	Linoleic acid g / Cholesterol mg	Thiamin mg / Ascorbic acid mg	% USRDA
A	Rice-A-Roni,chicken flavored,prep	3/4 C	192	238	45.8	tr(a)	4.3	**7**	unk.	unk.	unk.	unk.	unk.	unk.	unk.	unk.	unk.	0.42	**28**
	-Golden Grain		*0.52*	unk.	3.1	unk.	unk.		unk.	unk.	unk.	unk.	unk.	unk.	unk.	unk.	unk.	0(a)	**0**
B	Rice-A-Roni,Spanish flavored,prep	3/4 C	172	193	38.9	unk.	4.6	**7**	unk.	unk.	unk.	unk.	unk.	unk.	unk.	unk.	unk.	0.30	**20**
	-Golden Grain		*0.58*	unk.	3.1	unk.	unk.		unk.	unk.	unk.	unk.	unk.	unk.	unk.	unk.	unk.	tr	**1**
C	RICE,beef flavour'd,cooked-Uncle Ben	1/2 C	120	103	21.6	0.10	2.8	**4**	unk.	unk.	unk.	unk.	unk.	unk.	unk.	unk.	unk.	0.11	**7**
			0.83	93.4	0.6	unk.	unk.		unk.	unk.	unk.	unk.	unk.	unk.	unk.	unk.	0(a)	2	**3**
D	rice,brown,cooked-Uncle Ben	1/2 C	89	100	21.4	0.22	2.2	**3**	unk.	103	85	109	153	38	unk.	unk.	unk.	0.08	**5**
			1.12	64.7	0.8	unk.	0.8		46	188	39	85	24	126	unk.	unk.	0(a)	0	**0**
E	rice,brown,long grain,cooked	1/2 C	88	104	22.3	0.26	2.2	**3**	unk.	104	87	111	155	38	0.24	0.18	0.24	0.08	**5**
			1.14	61.5	0.5	unk.	0.9		46	191	39	87	25	126	0.17	1.33	0	0	**0**
F	rice,chicken flavour'd,cooked	1/2 C	102	100	20.5	0.10	2.6	**4**	unk.	unk.	unk.	unk.	unk.	unk.	unk.	unk.	unk.	0.10	**7**
	-Uncle Ben		*0.98*	76.9	0.9	unk.	unk.		unk.	unk.	unk.	unk.	unk.	unk.	unk.	unk.	tr(a)	2	**3**
G	rice,curried,cooked-Uncle Ben	1/2 C	118	100	21.9	0.20	2.8	**4**	unk.	unk.	unk.	unk.	unk.	unk.	unk.	tr(a)	unk.	0.11	**7**
			0.85	91.1	0.2	unk.	unk.		unk.	unk.	unk.	unk.	unk.	unk.	unk.	unk.	0(a)	1	**2**
H	rice,home recipe,casserole	3/4 C	147	170	13.6	0.66	4.0	**8**	193	137	187	147	161	39	1.06	4.05	1.07	0.12	**8**
			0.68	116.4	11.2	3.64	4.3		85	242	48	115	37	110	4.74	0.26	7	18	**29**
I	rice,home recipe,loaf,sliced,fried	2/3 C	114	210	19.3	0.14	4.4	**9**	131	213	248	212	282	82	0.39	3.35	1.06	0.11	**8**
			0.88	76.0	12.9	4.09	0.5		115	364	112	194	57	191	7.54	0.12	101	1	**2**
J	rice,home recipe,pilaf	1/2 C	110	84	10.6	0.42	4.2	**8**	254	175	268	160	190	47	0.11	1.45	0.08	0.09	**6**
			0.91	91.1	2.9	3.12	2.4		71	315	96	170	44	145	0.59	0.08	22	15	**25**
K	rice,home recipe,Spanish tomato	3/4 C	185	363	18.7	0.52	11.2	**24**	unk.	525	792	434	543	117	3.84	10.34	2.97	0.30	**20**
			0.54	126.6	27.0	unk.	4.4		322	779	264	502	143	383	12.18	0.37	35	26	**43**
L	rice,Spanish,cooked-Uncle Ben	1/2 C	129	109	23.3	0.30	3.8	**6**	unk.	unk.	unk.	unk.	unk.	unk.	unk.	tr(a)	unk.	0.10	**7**
			0.78	98.9	0.2	unk.	unk.		unk.	unk.	unk.	unk.	unk.	unk.	unk.	unk.	0(a)	5	**8**
M	rice,white enriched,long grain,cooked	1/2 C	103	112	24.8	0.10	2.0	**3**	unk.	98	80	102	143	28	0.07	0.05	0.07	0.11	**8**
			0.98	74.4	0.1	7.09	0.2		35	176	37	80	23	93	0.05	1.40	0	0	**0**
N	rice,white,converted,cooked-Uncle Ben	1/2 C	92	91	20.7	0.07	1.7	**3**	unk.	81	64	86	120	23	unk.	tr(a)	unk.	0.11	**7**
			1.09	68.7	0.1	6.33	0.2		29	146	30	68	18	79	unk.	unk.	0(a)	0	**0**
O	rice,white,quick,cooked-Uncle Ben	1/2 C	86	79	18.0	0.07	1.4	**2**	unk.	67	56	72	100	20	tr(a)	tr(a)	tr(a)	0.09	**6**
			1.17	65.8	0.1	5.92	0.2		24	123	26	56	15	65	tr(a)	0.00	0(a)	0	**0**
P	RIGATONI w/sausage sauce,home recipe	3/4 C	185	260	28.4	0.39	10.4	**21**	396	265	181	409	527	85	0.00	3.92	0.91	0.27	**18**
			0.54	132.6	11.6	2.58	4.5		113	323	211	202	55	123	4.52	0.00	59	16	**27**
Q	ROLL,brown & serve,enriched	1 roll	26	85	14.2	0.05	2.3	**4**	unk.	105	70	114	102	unk.	unk.	0.26	0.26	0.07	**5**
			3.85	7.0	2.0	unk.	1.5		unk.	170	32	70	26	unk.	0.78	unk.	0	tr	**0**
R	roll,dinner,enriched	1 roll	28	83	14.8	0.06	2.3	**4**	unk.	113	75	122	110	unk.	0.15	0.28	0.28	0.08	**5**
			3.57	8.8	1.6	0.80	1.6		unk.	183	35	75	28	unk.	0.84	0.53	0	tr	**0**
S	roll,hard,enriched	1 lg	50	156	29.8	0.10	4.9	**8**	unk.	240	160	260	235	unk.	0.14	0.50	0.50	0.13	**9**
		rectangular	*2.00*	12.7	1.6	1.43	2.9		unk.	390	75	160	60	unk.	1.50	0.29	0	tr	**0**

Row	Riboflavin mg / Niacin mg	% USRDA	Vitamin B_6 mg / Folacin mcg	% USRDA	Vitamin B_{12} mcg / Pantothenic acid mg	% USRDA	Biotin mg / Vitamin A IU	% USRDA	Preformed A RE / Beta carotene RE	Vitamin D IU / Vitamin E IU	% USRDA	Total tocopherol mg / Alpha tocopherol mg	Other tocopherol mg / Total ash g	Calcium mg / Phosphorus mg	% USRDA	Sodium mg / Sodium meq	Potassium mg / Potassium meq	Chlorine mg / Chlorine meq	Iron mg / Magnesium mg	% USRDA	Zinc mg / Copper mg	% USRDA	Iodine mcg / Selenium mcg	% USRDA	Manganese mcg / Chromium mcg
A	0.13	8	unk.	unk.	tr(a)	0	tr(a)	0	0(a)	tr(a)	0	unk.	unk.	unk.	unk.	unk.	unk.	unk.	2.0	11	unk.	unk.	unk.	unk.	unk.
	1.7	8	21	5	unk.	unk.	unk.	unk.	unk.	unk.	unk.	unk.	unk.	unk.	unk.	unk.	unk.	unk.	unk.	unk.	unk.	unk.	unk.		unk.
B	0.10	6	unk.	unk.	0(a)	0	tr(a)	0	0(a)	tr(a)	0	unk.	unk.	unk.	unk.	unk.	unk.	unk.	1.6	9	unk.	unk.	unk.	unk.	unk.
	0.2	1	19	5	0.32	3	unk.	unk.	unk.	unk.	unk.	unk.	unk.	unk.	unk.	unk.	unk.	unk.	unk.	unk.	unk.	unk.	unk.		unk.
C	0.02	1	unk.	unk.	0(a)	0	tr(a)	0	0	0(a)	0	unk.	unk.	25	3	692	95	unk.	1.4	8	unk.	unk.	unk.	unk.	unk.
	1.3	7	13	3	unk.	unk.	12	0	1	unk.	unk.	unk.	1.90	60	6	30.1	2.4	unk.	11	3	unk.	unk.	unk.		unk.
D	0.02	1	0.16	8	0.00	0	0.004	1	0	0(a)	0	0.4	0.3	5	1	252	63	unk.	0.4	3	0.5	4	unk.	unk.	unk.
	1.3	6	14	4	0.32	3	0	0	0	0.5	2	0.1	0.38	65	7	10.9	1.6	unk.	26	7	0.10	5	unk.		unk.
E	0.02	1	0.16	8	0.00	0	0.003	1	0	0(a)	0	0.4	0.3	11	1	247	61	unk.	0.4	2	0.5	4	unk.	unk.	unk.
	1.2	6	14	4	0.32	3	0	0	0	0.5	2	0.1	0.96	64	6	10.7	1.6	unk.	25	6	0.10	5	unk.		unk.
F	0.01	1	unk.	unk.	0(a)	0	tr(a)	0	0	0(a)	0	unk.	unk.	24	2	416	71	unk.	1.2	7	unk.	unk.	unk.	unk.	unk.
	1.4	7	11	3	unk.	unk.	10	0	1	unk.	unk.	unk.	1.30	48	5	18.1	1.8	unk.	8	2	unk.	unk.	unk.		unk.
G	0.00	0	unk.	unk.	0(a)	0	tr(a)	0	0	0(a)	0	unk.	unk.	26	3	541	82	unk.	1.3	7	unk.	unk.	unk.	unk.	unk.
	1.2	6	13	3	unk.	unk.	7	0	1	unk.	unk.	unk.	1.50	48	5	23.5	2.1	unk.	9	2	unk.	unk.	unk.		unk.
H	0.06	4	0.10	5	0.00	0	0.003	1	0	0	0	0.1	tr	21	2	297	202	6	1.0	6	0.4	2	35.0	23	35
	1.4	7	11	3	0.22	2	2018	40	202	0.1	1	0.1	1.59	54	5	12.9	5.2	0.2	15	4	0.10	5	1		1
I	0.08	5	0.07	4	0.24	4	0.005	2	unk.	10	3	0.2	tr(a)	41	4	252	90	37	1.4	8	0.7	5	71.1	47	20
	1.1	6	15	4	0.43	4	259	5	unk.	0.3	1	tr	0.98	92	9	10.9	2.3	1.0	58	14	0.03	1	2		2
J	0.07	4	0.12	6	1.59	27	0.002	1	13	0	0	0.1	tr	21	2	362	206	tr	0.8	4	0.9	6	19.0	13	12
	1.1	5	24	6	0.26	3	726	15	60	0.1	0	0.1	1.22	59	6	15.7	5.3	0.0	3	1	0.08	4	0		9
K	0.16	10	0.18	9	0.48	8	0.002	1	0	15	4	0.1	tr	22	2	1339	369	873	1.5	9	1.4	9	71.2	48	18
	3.2	16	14	4	0.23	2	763	15	75	0.1	0	0.1	4.53	159	16	58.2	9.4	24.6	22	5	0.12	6	0		tr
L	0.01	1	unk.	unk.	0(a)	0	tr(a)	0	0	0(a)	0	unk.	unk.	26	3	888	168	unk.	1.5	8	unk.	unk.	unk.	unk.	unk.
	1.6	8	14	4	unk.	unk.	26	1	3	unk.	unk.	unk.	2.40	63	6	38.6	4.3	unk.	12	3	unk.	unk.	unk.		unk.
M	0.07	4	0.05	3	0.00	0	0.002	1	0	0(a)	0	0.3	0.1	10	1	383	29	unk.	0.9	5	0.4	3	unk.	unk.	tr
	1.0	5	16	4	0.16	2	0	0	0	0.2	1	0.2	1.13	29	3	16.7	0.7	unk.	8	2	tr	0	unk.		unk.
N	0.00	0	0.05	2	0.00	0	0.002	1	0	0(a)	0	unk.	unk.	20	2	3	43	unk.	0.7	4	0.5	3	unk.	unk.	tr
	1.2	6	10	3	0.15	2	0	0	0	unk.	unk.	unk.	0.15	40	4	0.1	1.1	unk.	7	2	tr	0	unk.		unk.
O	0.00	0	0.01	0	0.00	0	0.001	0	0	0(a)	0	unk.	unk.	2	0	7	8	unk.	0.7	4	0.3	2	unk.	unk.	unk.
	0.8	4	9	2	0.06	1	0	0	0	unk.	unk.	unk.	0.04	17	2	0.3	0.2	unk.	2	0	tr	0	unk.		unk.
P	0.22	13	0.22	11	0.00	0	0.005	2	14	1	0	0.4	0.0	44	4	106	286	2	3.0	17	1.7	11	unk.	unk.	23
	3.2	16	8	2	0.39	4	999	20	93	0.5	2	tr	2.81	159	16	4.6	7.3	0.0	138	34	0.14	7	tr		tr
Q	0.06	3	0.01	1	tr(a)	0	tr(a)	0	tr	tr(a)	0	0.1	unk.	13	1	146	26	unk.	0.5	3	0.2	1	unk.	unk.	unk.
	0.9	5	10	3	0.08	1	tr	0	0	0.1	0	unk.	0.47	23	2	6.4	0.7	unk.	10	2	unk.	unk.	unk.		unk.
R	0.05	3	0.01	1	tr(a)	0	unk.	unk.	tr	tr(a)	0	0.1	unk.	21	2	142	27	unk.	0.5	3	0.2	1	unk.	unk.	140
	0.8	4	10	3	0.09	1	tr	0	0	0.1	0	unk.	0.50	24	2	6.2	0.7	unk.	10	3	0.07	4	unk.		13
S	0.11	7	0.02	1	tr(a)	0	tr(a)	0	tr	tr(a)	0	unk.	unk.	24	2	313	49	unk.	1.1	6	0.6	4	unk.	unk.	unk.
	1.7	8	22	6	0.15	2	tr	0	0	unk.	unk.	unk.	1.05	46	5	13.6	1.2	unk.	11	3	unk.	unk.	unk.		unk.

	FOOD	Portion	Weight in grams / Conversion for 100 g	Kilocalories / H_2O g	Total carbohydrate g / Total fats g	Crude fiber g / Dietary fiber g	Total protein g / Total sugar g	% USRDA	Arginine mg / Histidine mg	Isoleucine mg / Leucine mg	Lysine mg / Methionine mg	Phenylalanine mg / Threonine mg	Valine mg / Tryptophan mg	Cystine mg / Tyrosine mg	Polyunsat. fatty acids g / Monounsat. fatty acids g	Saturated fatty acids g / P/S ratio	Linoleic acid g / Cholesterol mg	Thiamin mg / Ascorbic acid mg	% USRDA
A	roll,hard,enriched	1 med	25	78	14.9	0.05	2.4	**4**	unk.	120	80	130	117	unk.	0.07	0.25	0.25	0.06	**4**
			4.00	6.3	0.8	unk.	1.4		unk.	195	37	80	30	unk.	0.75	0.29	0	tr	**0**
B	roll,hoagie or submarine	1 lg	135	391	74.8	0.27	12.3	**19**	unk.	344	277	668	520	259	unk.	0.89	1.08	0.38	**25**
			0.74	41.3	4.0	4.05	7.8		246	938	155	351	148	420	1.81	unk.	0	tr	**0**
C	roll,sweet,any,home recipe	1 med	50	143	22.3	0.07	3.7	**6**	unk.	182	147	196	188	65	0.22	1.26	0.41	0.13	**8**
			2.00	19.8	4.2	0.57	4.2		87	304	64	129	47	132	2.28	0.17	18	tr	**0**
D	roll,sweet,cinnamon bun w/raisins	1 med	60	165	33.8	0.00	4.1	**7**	unk.	189	94	226	176	unk.	unk.	0.60	0.60	0.04	**2**
			1.67	19.2	1.7	unk.	3.7		unk.	316	53	119	49	unk.	1.80	unk.	unk.	0	**0**
E	roll,sweet,cinnamon bun,plain	1 med	50	158	25.6	0.10	3.1	**5**	unk.	142	71	170	134	unk.	unk.	unk.	unk.	0.08	**5**
			2.00	unk.	4.8	unk.	2.8		unk.	239	41	89	38	unk.	unk.	unk.	unk.	0	**0**
F	roll,sweet,hot cross bun,home recipe	1 med	50	172	23.9	0.09	3.9	**7**	unk.	183	160	197	196	74	3.45	1.27	3.37	0.11	**8**
			2.00	14.8	6.9	0.98	7.7		91	304	76	137	49	139	1.81	2.70	39	tr	**0**
G	roll,whole wheat	1 lg	35	90	18.3	0.56	3.5	**5**	unk.	unk.	unk.	unk.	unk.	unk.	unk.	0.35	0.35	0.12	**8**
			2.86	11.2	1.0	unk.	2.3		unk.	unk.	unk.	unk.	unk.	unk.	1.05	unk.	0	tr	**0**
H	ROLLERCOASTERS,canned-Chef Boy-Ar-Dee	1 C	213	213	27.7	0.43	8.5	**13**	unk.	unk.	unk.	unk.	unk.	unk.	unk.	unk.	unk.	0.19	**13**
			0.47	165.8	8.5	unk.	unk.		unk.	unk.	unk.	unk.	unk.	unk.	unk.	unk.	unk.	2	**4**
I	ROSEMARY,dried	1 tsp	1.1	4	0.7	0.19	0.0	**0**	unk.	unk.	unk.	unk.	unk.	unk.	unk.	unk.	unk.	0.01	**0**
			90.91	0.1	0.2	unk.	0.0		unk.	unk.	unk.	unk.	unk.	unk.	unk.	unk.	0	1	**2**
J	RUTABAGAS,fresh,cooked,sliced, drained solids	1/2 C	85	30	7.0	0.93	0.8	**1**	unk.	unk.	unk.	unk.	unk.	unk.	unk.	unk.	unk.	0.05	**3**
			1.18	76.7	0.1	unk.	3.2		unk.	unk.	unk.	unk.	unk.	unk.	unk.	unk.	unk.	22	**37**
K	SAFFRON	1 tsp	0.7	2	0.5	0.03	0.1	**0**	unk.	unk.	unk.	unk.	unk.	unk.	unk.	unk.	unk.	unk.	**unk.**
			142.86	0.1	tr	unk.	0.0		unk.	unk.	unk.	unk.	unk.	unk.	unk.	unk.	0	unk.	**unk.**
L	SAGE,ground	1 tsp	0.7	2	0.4	0.13	0.1	**0**	unk.	unk.	unk.	unk.	unk.	unk.	0.01	0.05	tr	0.01	**0**
			142.86	0.1	0.1	unk.	0.0		unk.	unk.	unk.	unk.	unk.	unk.	0.01	0.25	0	tr	**0**
M	SALAD DRESSING,blue/roquefort	2 Tbsp	30	151	2.2	0.03	1.4	**2**	unk.	61	66	77	103	5	8.34	2.97	7.05	tr(a)	**0**
			3.33	9.7	15.7	unk.	unk.		29	140	37	33	12	43	3.54	2.81	tr(a)	1	**2**
N	salad dressing,blue/roquefort, low calorie	2 Tbsp	32	24	1.3	0.03	1.0	**2**	unk.	unk.	unk.	unk.	unk.	unk.	unk.	0.96	tr	unk.	**unk.**
			3.13	26.8	1.9	unk.	unk.		unk.	unk.	unk.	unk.	unk.	unk.	0.64	unk.	unk.	1	**2**
O	salad dressing,cooked,home recipe	2 Tbsp	32	50	4.8	0.00	1.3	**2**	unk.	unk.	unk.	unk.	unk.	unk.	0.67	0.93	0.64	0.02	**1**
			3.13	22.1	3.0	0(a)	unk.		unk.	unk.	unk.	unk.	unk.	unk.	1.18	0.72	unk.	tr	**0**
P	salad dressing,farm style,mix,prep -Good Seasons	2 Tbsp	30	107	1.2	0.01	0.8	**1**	unk.	unk.	unk.	unk.	unk.	unk.	5.73	1.73	5.15	0.01	**0**
			3.33	16.1	11.2	unk.	0.8		unk.	unk.	unk.	unk.	unk.	unk.	3.16	3.32	9	tr	**0**
Q	salad dressing,French	2 Tbsp	30	129	5.2	0.08	0.2	**0**	unk.	unk.	unk.	unk.	unk.	unk.	6.50	2.84	6.08	unk.	**unk.**
			3.33	11.6	12.2	unk.	unk.		unk.	unk.	unk.	unk.	unk.	unk.	2.12	2.28	0(a)	unk.	**unk.**
R	salad dressing,French style,mix,prep -Good Seasons	2 Tbsp	38	193	5.9	0.05	0.8	**1**	unk.	unk.	unk.	unk.	unk.	unk.	6.93	2.83	6.42	0.01	**1**
			2.63	11.0	18.9	unk.	5.5		unk.	unk.	unk.	unk.	unk.	unk.	7.90	2.45	15	tr	**0**
S	salad dressing,French,home recipe	2 Tbsp	28	177	1.0	unk.	0.0	**0**	unk.	unk.	unk.	unk.	unk.	unk.	8.88	3.53	8.88	tr	**0**
			3.64	6.8	19.7	unk.	unk.		unk.	unk.	unk.	unk.	unk.	unk.	5.74	2.52	0(a)	tr	**0**

Riboflavin mg / Niacin mg	% USRDA	Vitamin B_6 mg / Folacin mcg	% USRDA	Vitamin B_{12} mcg / Pantothenic acid mg	% USRDA	Biotin mg / Vitamin A IU	% USRDA	Preformed A RE / Beta carotene RE	Vitamin D IU / Vitamin E IU	% USRDA	Total tocopherol mg / Alpha tocopherol mg	Other tocopherol mg / Total ash g	Calcium mg / Phosphorus mg	% USRDA	Sodium mg / Sodium meq	Potassium mg / Potassium meq	Chlorine mg / Chlorine meq	Iron mg / Magnesium mg	% USRDA	Zinc mg / Copper mg	% USRDA	Iodine mcg / Selenium mcg	% USRDA	Manganese mcg / Chromium mcg	
0.06	**3**	0.01	**0**	tr(a)	**0**	tr(a)	**0**	tr	tr(a)	**0**	unk.	unk.	12	**1**	156	24	unk.	0.6	**3**	0.3	**2**	unk.	**unk.**	unk.	A
0.7	**3**	11	**3**	0.08	**1**	tr	**0**	0	unk.	**unk.**	unk.	0.52	23	**2**	6.8	0.6	unk.	6	**1**	unk.	**unk.**	unk.		unk.	
0.30	**18**	0.07	**4**	0.00	**0**	tr(a)	**0**	tr	0(a)	**0**	unk.	unk.	58	**6**	783	121	unk.	3.0	**17**	0.8	**5**	unk.	**unk.**	unk.	B
3.4	**17**	53	**13**	0.51	**5**	tr	**0**	0	unk.	**unk.**	unk.	2.56	115	**12**	34.1	3.1	unk.	30	**7**	0.03	**2**	unk.		unk.	
0.13	**7**	0.04	**2**	0.10	**2**	0.001	**0**	0	0	**0**	0.0	0.0	29	**3**	147	65	22	0.9	**5**	0.3	**2**	37.7	**25**	99	C
1.0	**5**	30	**8**	0.26	**3**	30	**1**	0	0.1	**0**	tr	0.64	52	**5**	6.4	1.6	0.6	18	**4**	0.07	**4**	13		16	
0.06	**4**	unk.	**unk.**	tr(a)	**0**	unk.	**unk.**	0	tr(a)	**0**	unk.	unk.	45	**5**	230	147	unk.	0.8	**5**	unk.	**unk.**	unk.	**unk.**	unk.	D
0.4	**2**	unk.	**unk.**	unk.	**unk.**	tr	**0**	tr	unk.	**unk.**	unk.	1.08	55	**6**	10.0	3.8	unk.	12	**3**	unk.	**unk.**	unk.		unk.	
0.09	**5**	unk.	**unk.**	tr(a)	**0**	tr(a)	**0**	21	tr(a)	**0**	0.2	0.2	27	**3**	unk.	unk.	unk.	0.9	**5**	unk.	**unk.**	unk.	**unk.**	unk.	E
0.8	**4**	unk.	**unk.**	unk.	**unk.**	205	**4**	14	0.3	**1**	tr	unk.	39	**4**	unk.	unk.	unk.	11	**3**	unk.	**unk.**	unk.		unk.	
0.11	**7**	0.04	**2**	0.11	**2**	0.002	**1**	0	7	**2**	4.8	0.0	24	**2**	103	90	36	1.0	**6**	0.3	**2**	27.8	**19**	119	F
0.9	**5**	21	**5**	0.27	**3**	30	**1**	tr	0.1	**0**	0.1	0.57	55	**5**	4.5	2.3	1.0	28	**7**	0.07	**4**	11		16	
0.05	**3**	0.15	**7**	0(a)	**0**	tr(a)	**0**	tr	0(a)	**0**	unk.	unk.	37	**4**	197	102	unk.	0.8	**5**	0.6	**4**	unk.	**unk.**	unk.	G
1.0	**5**	12	**3**	0.21	**2**	tr	**0**	0	unk.	**unk.**	unk.	1.01	98	**10**	8.6	2.6	unk.	40	**10**	unk.	**unk.**	unk.		unk.	
0.17	**10**	unk.	**unk.**	unk.	**unk.**	unk.	**unk.**	unk.	0(a)	**0**	tr(a)	tr(a)	36	**4**	1224	375	unk.	3.0	**17**	unk.	**unk.**	unk.	**unk.**	unk.	H
3.4	**17**	unk.	**unk.**	tr(a)	**0**	551	**11**	unk.	0.0	**0**	tr(a)	3.83	113	**11**	53.2	9.6	unk.	unk.	**unk.**	tr(a)	**0**	tr(a)		unk.	
unk.	**unk.**	unk.	**unk.**	0.00	**0**	unk.	**unk.**	0(a)	0(a)	**0**	unk.	unk.	14	**1**	1	11	unk.	0.3	**2**	tr	**0**	unk.	**unk.**	unk.	I
tr	**0**	unk.	**unk.**	unk.	**unk.**	34	**1**	unk.	unk.	**unk.**	unk.	0.07	1	**0**	0.0	0.3	unk.	2	**1**	unk.	**unk.**	unk.		unk.	
0.05	**3**	0.08	**4**	0.00	**0**	unk.	**unk.**	0	0(a)	**0**	unk.	unk.	50	**5**	3	142	unk.	0.3	**1**	unk.	**unk.**	unk.	**unk.**	tr(a)	J
0.7	**3**	18	**5**	0.14	**1**	468	**9**	47	unk.	**unk.**	unk.	0.51	26	**3**	0.1	3.6	unk.	unk.	**unk.**	tr(a)	**0**	tr(a)		unk.	
unk.	**unk.**	unk.	**unk.**	0.00	**0**	unk.	**unk.**	0(a)	0(a)	**0**	unk.	unk.	1	**0**	1	12	unk.	0.1	**0**	unk.	**unk.**	unk.	**unk.**	unk.	K
unk.	**unk.**	unk.	**unk.**	unk.	**unk.**	unk.	**unk.**	unk.	unk.	**unk.**	unk.	0.04	2	**0**	0.0	0.3	unk.	unk.	**unk.**	unk.	**unk.**	unk.		unk.	
tr	**0**	unk.	**unk.**	0.00	**0**	unk.	**unk.**	0(a)	0(a)	**0**	unk.	unk.	12	**1**	tr	7	unk.	0.2	**1**	tr	**0**	unk.	**unk.**	unk.	L
tr	**0**	unk.	**unk.**	unk.	**unk.**	41	**1**	unk.	unk.	**unk.**	unk.	0.06	1	**0**	0.0	0.2	unk.	3	**1**	unk.	**unk.**	unk.		unk.	
0.03	**2**	unk.	**unk.**	unk.	**unk.**	unk.	**unk.**	unk.	unk.	**unk.**	unk.	unk.	24	**2**	328	11	unk.	0.1	**0**	0.1	**0**	unk.	**unk.**	unk.	M
tr	**0**	unk.	**unk.**	unk.	**unk.**	63	**1**	unk.	unk.	**unk.**	unk.	0.96	22	**2**	14.3	0.3	unk.	unk.	**unk.**	unk.	**unk.**	unk.		unk.	
0.02	**1**	tr	**unk.**	unk.	**unk.**	unk.	**unk.**	unk.	unk.	**unk.**	unk.	unk.	20	**2**	355	11	unk.	tr	**0**	0.1	**0**	unk.	**unk.**	unk.	N
tr	**0**	unk.	**unk.**	unk.	**unk.**	54	**1**	unk.	unk.	**unk.**	unk.	1.06	15	**2**	15.4	0.3	unk.	unk.	**unk.**	unk.	**unk.**	unk.		unk.	
0.05	**3**	unk.	**unk.**	unk.	**unk.**	unk.	**unk.**	unk.	unk.	**unk.**	unk.	unk.	27	**3**	235	39	unk.	0.2	**1**	unk.	**unk.**	unk.	**unk.**	unk.	O
0.1	**0**	unk.	**unk.**	unk.	**unk.**	131	**3**	unk.	unk.	**unk.**	unk.	0.77	28	**3**	10.2	1.0	unk.	unk.	**unk.**	unk.	**unk.**	unk.		unk.	
0.02	**1**	0.01	**0**	0.03	**1**	unk.	**unk.**	unk.	0	**0**	unk.	unk.	22	**2**	250	31	unk.	0.1	**1**	tr	**1**	unk.	**unk.**	unk.	P
tr	**0**	1	**0**	0.04	**0**	46	**1**	unk.	unk.	**unk.**	unk.	0.72	18	**2**	10.9	0.8	unk.	2	**1**	0.03	**2**	unk.		unk.	
unk.	**unk.**	unk.	**unk.**	0(a)	**0**	unk.	**unk.**	0(a)	unk.	**unk.**	unk.	unk.	3	**0**	410	24	unk.	0.2	**1**	tr	**0**	unk.	**unk.**	unk.	Q
unk.	**unk.**	unk.	**unk.**	unk.	**unk.**	unk.	**unk.**	unk.	unk.	**unk.**	unk.	1.26	4	**0**	17.8	0.6	unk.	2	**1**	unk.	**unk.**	unk.		unk.	
0.02	**1**	0.00	**0**	0.10	**2**	unk.	**unk.**	unk.	1	**0**	unk.	unk.	13	**1**	410	22	unk.	0.1	**1**	0.1	**1**	unk.	**unk.**	unk.	R
0.1	**0**	2	**0**	0.07	**1**	41	**1**	unk.	unk.	**unk.**	unk.	1.14	16	**2**	17.8	0.6	unk.	3	**1**	0.01	**0**	unk.		unk.	
0.01	**0**	unk.	**unk.**	0(a)	**0**	unk.	**unk.**	unk.	unk.	**unk.**	unk.	unk.	2	**0**	184	7	unk.	0.1	**0**	unk.	**unk.**	unk.	**unk.**	unk.	S
tr	**0**	unk.	**unk.**	unk.	**unk.**	144	**3**	unk.	unk.	**unk.**	unk.	0.50	1	**0**	8.0	0.2	unk.	unk.	**unk.**	unk.	**unk.**	unk.		unk.	

	FOOD	Portion	Weight in grams / Conversion for 100 g	Kilocalories / H_2O g	Total carbohydrate g / Total fats g	Crude fiber g / Dietary fiber g	Total protein g / Total sugar g	% USRDA	Arginine mg / Histidine mg	Isoleucine mg / Leucine mg	Lysine mg / Methionine mg	Phenylalanine mg / Threonine mg	Valine mg / Tryptophan mg	Cystine mg / Tyrosine mg	Polyunsat. fatty acids g / Monounsat. fatty acids g	Saturated fatty acids g / P/S ratio	Linoleic acid g / Cholesterol mg	Thiamin mg / Ascorbic acid mg	% USRDA
A	salad dressing,French,low calorie	2 Tbsp	32	43	6.9	0.10	0.1	**0**	unk.	unk.	unk.	unk.	unk.	unk.	1.09	0.26	0.96	unk.	**unk.**
			3.13	22.2	1.9	0(a)	unk.		unk.	unk.	unk.	unk.	unk.	unk.	0.42	4.25	2	unk.	**unk.**
B	salad dressing,Italian	2 Tbsp	30	140	2.1	tr	0.2	**1**	unk.	tr(a)	tr(a)	tr(a)	tr(a)	tr(a)	8.40	2.10	7.38	tr	**0**
			3.33	8.2	14.5	tr(a)	unk.		tr(a)	tr(a)	tr(a)	tr(a)	tr(a)	tr(a)	3.30	4.00	0	0	**0**
C	salad dressing,Italian,low calorie	2 Tbsp	30	32	0.8	0.09	0.0	**0**	unk.	tr(a)	tr(a)	tr(a)	tr(a)	tr(a)	1.80	0.39	1.56	tr	**0**
			3.33	24.6	2.9	unk.	unk.		tr(a)	tr(a)	tr(a)	tr(a)	tr(a)	tr(a)	0.60	4.62	2	tr(a)	**0**
D	salad dressing,Italian,mix,prep	2 Tbsp	30	157	1.3	0.06	0.1	**0**	unk.	unk.	unk.	unk.	unk.	unk.	6.37	2.52	5.91	tr	**0**
	-Good Seasons		*3.33*	10.3	16.9	unk.	0.9		unk.	unk.	unk.	unk.	unk.	unk.	7.19	2.52	0	tr	**0**
E	salad dressing,mayonaise,imitation	2 Tbsp	30	29	3.3	0.00	0.6	**1**	unk.	unk.	unk.	unk.	unk.	unk.	0.09	0.84	0.06	unk.	**unk.**
			3.33	23.9	1.5	0(a)	unk.		unk.	unk.	unk.	unk.	unk.	unk.	0.51	0.11	13	unk.	**unk.**
F	salad dressing,mayonnaise	2 Tbsp	27	196	0.7	0.00	0.3	**1**	20	18	20	16	20	6	11.32	3.23	10.17	0.00	**0**
			3.65	4.2	21.8	0(a)	unk.		7	26	10	15	5	13	6.16	3.50	16	unk.	**unk.**
G	salad dressing,mayonnaise type	2 Tbsp	29	114	4.2	tr(a)	0.3	**0**	unk.	unk.	unk.	unk.	unk.	unk.	5.26	1.43	4.67	tr	**0**
			3.42	11.9	9.8	0(a)	unk.		unk.	unk.	unk.	unk.	unk.	unk.	2.63	3.67	8	1	**2**
H	salad dressing,mayonnaise type,	2 Tbsp	31	42	1.5	0.16	0.3	**1**	unk.	unk.	unk.	unk.	unk.	unk.	unk.	0.62	1.87	tr	**0**
	low calorie		*3.21*	25.2	4.0	unk.	unk.		unk.	unk.	unk.	unk.	unk.	unk.	0.94	unk.	unk.	unk.	**unk.**
I	salad dressing,Russian	2 Tbsp	30	148	3.1	0.09	0.5	**1**	unk.	unk.	unk.	unk.	unk.	unk.	8.82	2.19	7.77	0.01	**1**
			3.33	10.3	15.2	unk.	unk.		unk.	unk.	unk.	unk.	unk.	unk.	3.48	4.03	0(a)	2	**3**
J	salad dressing,sesame seed,commercial	2 Tbsp	31	136	2.6	0.12	0.9	**2**	unk.	unk.	unk.	unk.	unk.	unk.	7.68	1.90	7.10	unk.	**unk.**
			3.27	12.0	13.8	unk.	unk.		unk.	unk.	unk.	unk.	unk.	unk.	3.64	4.05	0	unk.	**unk.**
K	salad dressing,soyumaise-Cellu	2 Tbsp	28	199	1.4	unk.	0.5	**1**	unk.	unk.	unk.	unk.	unk.	unk.	unk.	unk.	unk.	unk.	**unk.**
			3.57	unk.	21.7	unk.	unk.		unk.	unk.	unk.	unk.	unk.	unk.	unk.	unk.	unk.	unk.	**unk.**
L	salad dressing,thousand island	2 Tbsp	32	121	4.9	0.10	0.3	**1**	unk.	unk.	unk.	unk.	unk.	unk.	6.34	1.92	5.28	0.01	**0**
			3.13	10.2	11.4	unk.	unk.		unk.	unk.	unk.	unk.	unk.	unk.	2.46	3.30	unk.	1	**2**
M	salad dressing,thousand island,	2 Tbsp	30	48	4.7	0.09	0.2	**0**	unk.	unk.	unk.	unk.	unk.	unk.	1.86	0.48	1.59	0.01	**0**
	low calorie		*3.33*	20.5	3.2	unk.	unk.		unk.	unk.	unk.	unk.	unk.	unk.	0.72	3.88	4	1	**2**
N	salad dressing,vinegar & oil,	2 Tbsp	31	140	0.8	unk.	0.0	**0**	0(a)	0(a)	0(a)	0(a)	0(a)	0(a)	7.52	2.84	7.08	unk.	**unk.**
	home recipe		*3.19*	14.8	15.6	unk.	unk.		0(a)	0(a)	0(a)	0(a)	0(a)	0(a)	4.59	2.65	0(a)	unk.	**unk.**
O	SALAD,home recipe,chef	1-1/2 C	270	386	8.8	0.66	24.1	**53**	unk.	1317	1844	1106	1396	238	0.73	13.36	2.05	0.38	**25**
			0.37	203.7	28.3	unk.	2.6		751	1968	598	1006	315	962	9.31	0.05	244	15	**25**
P	salad,home recipe,Popeye	1/2 C	87	204	2.7	0.28	3.1	**6**	196	169	199	157	196	62	11.31	2.88	11.09	0.05	**4**
			1.15	59.1	20.7	2.77	0.6		75	253	77	143	49	120	5.24	3.93	75	23	**38**
Q	salad,home recipe,three bean	3/4 C	198	230	30.8	0.40	5.2	**8**	227	285	365	256	304	46	5.98	1.30	5.89	0.06	**4**
			0.51	148.0	10.7	1.24	15.5		130	420	58	221	54	178	2.53	4.59	0	3	**6**
R	salad,home recipe,tossed	1 C	161	32	6.9	0.97	1.7	**3**	61	19	50	33	23	6	0.00	0.00	unk.	0.08	**5**
			0.62	151.3	0.3	2.19	3.5		9	27	6	26	9	11	unk.	0.00	0	26	**43**
S	salad,home recipe,waldorf	1/2 C	44	79	5.6	0.15	0.9	**2**	84	41	31	39	51	15	2.70	1.67	2.27	0.02	**2**
			2.30	30.1	6.2	0.80	3.5		20	66	16	32	9	30	1.30	1.62	8	2	**3**

Row	Riboflavin mg / Niacin mg	% USRDA	Vitamin B_6 mg / Folacin mcg	% USRDA	Vitamin B_{12} mcg / Pantothenic acid mg	% USRDA	Biotin mg / Vitamin A IU	% USRDA	Preformed A RE / Beta carotene RE	Vitamin D IU / Vitamin E IU	% USRDA	Total tocopherol mg / Alpha tocopherol mg	Other tocopherol mg / Total ash g	Calcium mg / Phosphorus mg	% USRDA	Sodium mg / Sodium meq	Potassium mg / Potassium meq	Chlorine mg / Chlorine meq	Iron mg / Magnesium mg	% USRDA	Zinc mg / Copper mg	% USRDA	Iodine mcg / Selenium mcg	% USRDA	Manganese mcg / Chromium mcg
A	unk.	**unk.**	unk.	**unk.**	0(a)	**0**	unk.	**unk.**	0(a)	unk.	**unk.**	unk.	unk.	4	**0**	252	25	unk.	0.1	**1**	0.1	**0**	unk.	**unk.**	unk.
	unk.	**unk.**	unk.	**unk.**	unk.	**unk.**	unk.	**unk.**	unk.	unk.	**unk.**	unk.	0.93	5	**0**	10.9	0.6	unk.	3	**1**	unk.	**unk.**	unk.		unk.
B	tr	**0**	0.00	**0**	0.00	**0**	0.000	**0**	tr(a)	0	**0**	2.7	unk.	3	**0**	236	4	unk.	0.1	**0**	tr	**0**	unk.	**unk.**	unk.
	tr	**0**	0	**0**	0.00	**0**	tr	**0**	tr(a)	unk.	**unk.**	unk.	1.62	2	**0**	10.3	0.1	unk.	2	**1**	0.01	**1**	unk.		unk.
C	tr	**0**	tr(a)	**0**	0(a)	**0**	unk.	**unk.**	tr(a)	unk.	**unk.**	unk.	unk.	1	**0**	236	4	unk.	0.1	**0**	unk.	**unk.**	unk.	**unk.**	unk.
	tr	**0**	tr(a)	**0**	unk.	**unk.**	tr	**0**	tr(a)	unk.	**unk.**	unk.	0.99	2	**0**	10.3	0.1	unk.	unk.	**unk.**	unk.	**unk.**	unk.		unk.
D	tr	**1**	tr	**0**	0.00	**0**	unk.	**unk.**	unk.	0	**0**	unk.	unk.	2	**0**	356	9	unk.	0.0	**0**	tr	**0**	unk.	**unk.**	unk.
	tr	**0**	unk.	**unk.**	tr	**0**	1	**0**	unk.	unk.	**unk.**	unk.	0.76	3	**0**	15.5	0.2	unk.	1	**0**	tr	**0**	unk.		unk.
E	unk.	**unk.**	unk.	**unk**	unk.	**unk.**	unk.	**unk.**	unk.	unk.	**unk.**	unk.	unk.	unk.	**unk.**	151	unk.	unk.	unk.	**unk.**	unk.	**unk.**	unk.	**unk.**	unk.
	unk.	**unk.**	unk.	**unk.**	unk.	**unk.**	unk.	**unk.**	unk.	unk.	**unk.**	unk.	0.63	unk.	**unk.**	6.6	unk.	unk.	unk.	**unk.**	unk.	**unk.**	unk.		unk.
F	0.00	**0**	unk.	**unk.**	unk.	**unk.**	unk.	**unk.**	unk.	unk.	**unk.**	15.9	10.2	5	**1**	156	9	unk.	0.1	**1**	tr	**0**	unk.	**unk.**	unk.
	tr	**0**	unk.	**unk.**	unk.	**unk.**	77	**2**	unk.	5.7	**19**	5.7	0.41	8	**1**	6.8	0.2	unk.	unk.	**unk.**	0.01	**0**	unk.		unk.
G	0.01	**1**	0.00	**0**	tr(a)	**0**	0.000	**0**	unk.	2	**1**	1.5	unk.	4	**0**	208	3	unk.	0.1	**0**	0.1	**0**	unk.	**unk.**	unk.
	tr	**0**	0	**0**	0.03	**0**	64	**1**	unk.	unk.	**unk.**	unk.	0.50	8	**1**	9.0	0.1	unk.	1	**0**	0.07	**4**	unk.		unk.
H	0.01	**1**	unk.	**unk.**	0(a)	**0**	unk.	**unk.**	unk.	unk.	**unk.**	unk.	unk.	6	**1**	37	3	unk.	0.1	**0**	unk.	**unk.**	unk.	**unk.**	unk.
	unk.	**unk.**	unk.	**unk.**	unk.	**unk.**	69	**1**	unk.	unk.	**unk.**	unk.	0.22	9	**1**	1.6	0.1	unk.	unk.	**unk.**	unk.	**unk.**	unk.		unk.
I	0.01	**1**	unk.	**unk.**	0(a)	**0**	unk.	**unk.**	0(a)	0(a)	**0**	unk.	unk.	6	**1**	260	47	unk.	0.2	**1**	0.1	**1**	unk.	**unk.**	unk.
	0.2	**1**	unk.	**unk.**	unk.	**unk.**	207	**4**	unk.	unk.	**unk.**	unk.	0.81	11	**1**	11.3	1.2	unk.	unk.	**unk.**	unk.	**unk.**	unk.		unk.
J	unk.	**unk.**	unk.	**unk.**	unk.	**unk.**	unk.	**unk.**	unk.	unk.	**unk.**	unk.	unk.	unk.	**unk.**	306	unk.	unk.	unk.	**unk.**	unk.	**unk.**	unk.	**unk.**	unk.
	unk.	**unk.**	unk.	**unk.**	unk.	**unk.**	unk.	**unk.**	unk.	unk.	**unk.**	unk.	1.19	unk.	**unk.**	13.3	unk.	unk.	unk.	**unk.**	unk.	**unk.**	unk.		unk.
K	unk.	**unk.**	unk.	**unk.**	0(a)	**0**	unk.	**unk.**	unk.	tr(a)	**0**	11.8	9.4	4	**0**	4	52	unk.	unk.	**unk.**	unk.	**unk.**	unk.	**unk.**	unk.
	unk.	**unk.**	unk.	**unk.**	unk.	**unk.**	unk.	**unk.**	unk.	2.4	**8**	2.4	unk.	12	**1**	0.2	1.3	unk.	unk.	**unk.**	unk.	**unk.**	unk.		unk.
L	0.01	**1**	unk.	**unk.**	unk.	**unk.**	unk.	**unk.**	unk.	unk.	**unk.**	unk.	unk.	4	**0**	224	36	unk.	0.2	**1**	tr	**0**	unk.	**unk.**	unk.
	0.1	**0**	unk.	**unk.**	unk.	**unk.**	102	**2**	unk.	unk.	**unk.**	unk.	0.51	5	**1**	9.7	0.9	unk.	2	**0**	unk.	**unk.**	unk.		unk.
M	0.01	**1**	unk.	**unk.**	0(a)	**0**	unk.	**unk.**	unk.	unk.	**unk.**	unk.	unk.	3	**0**	300	34	unk.	0.2	**1**	0.1	**0**	unk.	**unk.**	unk.
	0.1	**0**	unk.	**unk.**	unk.	**unk.**	96	**2**	unk.	unk.	**unk.**	unk.	0.48	5	**1**	13.0	0.9	unk.	2	**0**	unk.	**unk.**	unk.		unk.
N	unk.	**unk.**	unk.	**unk.**	unk.	**unk.**	unk.	**unk.**	unk.	unk.	**unk.**	unk.	unk.	unk.	**unk.**	tr	2	unk.	unk.	**unk.**	unk.	**unk.**	unk.	**unk.**	unk.
	unk.	**unk.**	unk.	**unk.**	unk.	**unk.**	unk.	**unk.**	unk.	unk.	**unk.**	unk.	unk.	unk.	**unk.**	0.0	0.1	unk.	unk.	**unk.**	unk.	**unk.**	unk.		unk.
O	0.43	**25**	0.15	**7**	0.65	**11**	0.004	**2**	0	0(a)	**0**	0.5	0.3	317	**32**	279	330	56	3.0	**17**	4.5	**30**	unk.	**unk.**	71
	3.2	**16**	56	**14**	0.90	**9**	1197	**24**	67	0.6	**2**	0.2	4.97	338	**34**	12.1	8.4	1.6	31	**8**	0.04	**2**	tr(a)		7
P	0.13	**8**	0.14	**7**	0.18	**3**	0.003	**1**	0	0	**0**	19.2	tr(a)	49	**5**	105	230	29	1.7	**9**	0.6	**4**	14.0	**9**	361
	0.3	**1**	40	**10**	0.37	**4**	3635	**73**	356	0.2	**1**	0.2	0.94	49	**5**	4.6	5.9	0.8	41	**10**	0.13	**7**	0		3
Q	0.06	**4**	0.27	**14**	0.00	**0**	0.000	**0**	0	0	**0**	9.1	0.1	58	**6**	500	270	0	2.5	**14**	0.2	**2**	81.4	**54**	16
	0.7	**3**	5	**1**	0.11	**1**	220	**4**	22	0.2	**1**	0.2	2.76	101	**10**	21.8	6.9	0.0	34	**9**	0.14	**7**	0		9
R	0.08	**5**	0.10	**5**	0.00	**0**	0.005	**2**	0	0(a)	**0**	0.5	0.3	48	**5**	15	368	46	1.1	**6**	0.4	**3**	unk.	**unk.**	72
	0.7	**4**	82	**21**	0.34	**3**	2279	**46**	228	0.6	**2**	0.2	1.02	42	**4**	0.7	9.4	1.3	17	**4**	0.04	**2**	tr(a)		12
S	0.02	**1**	0.04	**2**	0.01	**0**	0.002	**1**	0	tr	**0**	0.4	0.0	12	**1**	49	78	15	0.2	**1**	0.1	**1**	unk.	**unk.**	unk.
	0.1	**0**	6	**1**	0.08	**1**	125	**3**	4	0.5	**2**	0.2	0.34	24	**2**	2.1	2.0	0.4	8	**2**	0.08	**4**	0		tr(a)

	FOOD	Portion	Weight in grams / Conversion for 100 g	Kilocalories / H_2O g	Total carbohydrate g / Total fats g	Crude fiber g / Dietary fiber g	Total protein g / Total sugar g	**% USRDA**	Arginine mg / Histidine mg	Isoleucine mg / Leucine mg	Lysine mg / Methionine mg	Phenylalanine mg / Threonine mg	Valine mg / Tryptophan mg	Cystine mg / Tyrosine mg	Polyunsat. fatty acids g / Monounsat. fatty acids g	Saturated fatty acids g / P/S ratio	Linoleic acid g / Cholesterol mg	Thiamin mg / Ascorbic acid mg	**% USRDA**
A	SALMON,broiled or baked w/butter or	1/2 C	115	209	0.0	0.00	31.0	**69**	1755	1552	2701	1149	1646	313	3.30	1.46	0.09	0.18	**12**
	margarine		*0.87*	72.9	8.5	0(a)	0.0		944	2329	900	1335	310	944	1.49	2.26	54	unk.	**unk.**
B	salmon,broiled or baked w/butter or	3 oz portion	85	155	0.0	0.00	22.9	**51**	1297	1147	1997	849	1216	231	2.44	1.08	0.07	0.14	**9**
	margarine		*1.18*	53.9	6.3	0(a)	0.0		698	1721	666	987	229	698	1.10	2.26	40	unk.	**unk.**
C	salmon,broiled or baked,steak	1 med	145	264	0.0	0.00	39.1	**87**	2213	1957	3406	1449	2075	394	4.16	1.84	0.12	0.23	**16**
	w/butter or margarine		*0.69*	91.9	10.7	0(a)	0.0		1190	2936	1135	1683	391	1190	1.88	2.26	68	unk.	**unk.**
D	salmon,canned,chum,low sodium,	1/2 C	110	153	0.0	0.00	23.6	**53**	1336	1070	2108	unk.	unk.	238	unk.	unk.	unk.	0.02	**2**
	solids & liquid		*0.91*	77.9	5.7	0(a)	0.0		718	1654	unk.	1033	262	718	unk.	unk.	38	unk.	**unk.**
E	salmon,canned,coho silver,	1/2 C	110	168	0.0	0.00	22.9	**51**	1292	1036	2039	847	1212	230	3.07	2.40	0.12	0.03	**2**
	solids & liquid		*0.91*	76.2	7.8	0(a)	0.0		695	1600	663	999	253	695	2.53	1.28	unk.	unk.	**unk.**
F	salmon,canned,pink humpback,	1/2 C	110	155	0.0	0.00	22.5	**50**	1276	1021	2010	unk.	unk.	227	unk.	1.69	0.12	0.03	**2**
	solids & liquid		*0.91*	77.9	6.5	0(a)	0.0		685	1577	unk.	984	250	685	1.50	unk.	unk.	unk.	**unk.**
G	salmon,canned,sockeye (red),	1/2 C	110	188	0.0	0.00	22.2	**49**	1255	1127	1948	825	1184	298	4.70	0.82	0.16	0.04	**3**
	solids & liquid		*0.91*	73.9	10.2	0(a)	0.0		675	1679	647	964	220	601	0.82	5.69	38	0	**0**
H	salmon,home recipe,cake	3" dia x 3/4"	97	241	6.2	0.05	18.4	**40**	996	847	1539	723	993	222	2.62	7.23	0.61	0.05	**4**
			1.03	54.9	15.4	unk.	0.5		554	1327	524	803	212	593	4.70	0.36	104	tr	**0**
I	salmon,home recipe,casserole	3/4 C	168	416	23.7	0.18	21.0	**44**	1022	966	1611	297	311	265	0.55	7.41	1.74	0.17	**11**
			0.60	93.8	26.0	1.03	1.9		603	1522	123	879	246	687	13.35	0.07	91	3	**5**
J	salmon,home recipe,rice loaf	3-3/4x2-1/2x	177	299	9.4	0.02	27.1	**59**	1522	797	1169	641	850	67	0.85	5.81	0.51	0.07	**5**
		1/2" slice	*0.56*	120.3	16.2	unk.	0.0		510	1173	423	651	173	124	3.33	0.15	133	2	**4**
K	salmon,smoked	1 piece	20	35	0.0	0.00	4.3	**10**	244	208	378	160	230	unk.	0.60	0.60	unk.	0.04	**3**
			5.00	11.8	1.9	0(a)	0.0		112	324	126	186	44	116	unk.	1.00	0	1	**2**
L	SALT SUBSTITUTE-Diamond	1 tsp	5.0	2	0.6	0.00	0.0	**0**	0(a)	0	0	0	0	0	0.00	0.00	0.00	0.00	**0**
			20.00	0.0	0.0	unk.	0.0		0	0	0	0	0	0	0.00	0.00	0	0	**0**
M	SALT,TABLE,iodized	1 tsp	5.7	0	0.0	0.00	0.0	**0**	0(a)	0(a)	0(a)	0(a)	0(a)	0(a)	0.00	0.00	0.00	0.00	**0**
			17.54	tr	0.0	unk.	0.0		0(a)	0(a)	0(a)	0(a)	0(a)	0(a)	0.00	0.00	0	0	**0**
N	SARDINES,Atlantic,canned w/oil,	1 med	12	24	0.0	0.00	2.9	**6**	182	152	259	106	134	35	unk.	unk.	unk.	tr	**0**
	drained solids		*8.33*	7.4	1.3	0(a)	0.0		88	254	73	148	31	111	unk.	unk.	17	0	**0**
O	sardines,Pacific,canned w/brine &	1 lg	20	39	0.3	0(a)	3.8	**8**	238	180	310	131	189	48	unk.	unk.	unk.	0.01	**0**
	mustard		*5.00*	12.8	2.4	0(a)	0.0		115	267	103	153	35	96	unk.	unk.	24	0(a)	**0**
P	sardines,Pacific,canned w/tomato	1 lg	20	39	0.3	unk.	3.7	**8**	236	180	310	131	189	48	unk.	unk.	unk.	tr	**0**
	sauce		*5.00*	12.9	2.4	unk.	0.0		115	267	103	153	35	96	unk.	unk.	24	unk.	**unk.**
Q	SAUCE,barbecue	1 Tbsp	16	12	2.0	0.09	0.3	**0**	unk.	unk.	unk.	unk.	unk.	unk.	0.11	0.04	0.10	0.00	**0**
			6.41	12.6	0.3	unk.	unk.		unk.	unk.	unk.	unk.	unk.	unk.	0.12	2.52	0	1	**2**
R	sauce,butterscotch,home recipe	2 Tbsp	44	151	25.3	0.00	0.4	**1**	24	19	27	15	22	6	0.28	3.46	0.19	tr	**0**
			2.27	9.1	5.9	0(a)	25.3		9	32	9	18	5	17	1.56	0.08	40	0	**0**
S	sauce,chili,any	1 Tbsp	15	16	3.7	0.10	0.4	**1**	unk.	unk.	unk.	unk.	unk.	unk.	unk.	tr(a)	unk.	0.01	**1**
			6.67	10.2	tr	unk.	unk.		unk.	unk.	unk.	unk.	unk.	unk.	unk.	unk.	0(a)	2	**4**

Riboflavin mg / Niacin mg	% USRDA	Vitamin B_6 mg / Folacin mcg	% USRDA	Vitamin B_{12} mcg / Pantothenic acid mg	% USRDA	Biotin mg / Vitamin A IU	% USRDA	Preformed A RE / Beta carotene RE	Vitamin D IU / Vitamin E IU	% USRDA	Total tocopherol mg / Alpha tocopherol mg	Other tocopherol mg / Total ash g	Calcium mg / Phosphorus mg	% USRDA	Sodium mg / Sodium meq	Potassium mg / Potassium meq	Chlorine mg / Chlorine meq	Iron mg / Magnesium mg	% USRDA	Zinc mg / Copper mg	% USRDA	Iodine mcg / Selenium mcg	% USRDA	Manganese mcg / Chromium mcg	
0.07	4	0.48	24	4.14	69	unk.	unk.	unk.	460	115	2.1	0.6	unk.	unk.	133	509	unk.	1.4	8	2.0	13	unk.	unk.	unk.	A
11.3	56	30	8	0.75	8	184	4	unk.	2.5	8	1.6	1.84	476	48	5.8	13.0	unk.	47	12	0.23	12	tr(a)		tr(a)	
0.05	3	0.36	18	3.06	51	unk.	unk.	unk.	340	85	1.5	0.5	unk.	unk.	99	377	unk.	1.0	6	1.4	10	unk.	unk.	unk.	B
8.3	42	22	6	0.55	6	136	3	unk.	1.8	6	1.1	1.36	352	35	4.3	9.6	unk.	35	9	0.17	9	tr(a)		tr(a)	
0.09	5	0.61	30	5.22	87	unk.	unk.	unk.	580	145	2.6	0.8	unk.	unk.	168	642	unk.	1.7	10	2.5	16	unk.	unk.	unk.	C
14.2	71	38	9	0.94	9	232	5	unk.	3.1	11	2.0	2.32	600	60	7.3	16.4	unk.	59	15	0.29	15	tr(a)		tr(a)	
0.18	10	0.33	17	7.58	126	0.016	6	18	407	102	unk.	unk.	274	27	58	370	unk.	0.8	4	1.0	7	unk.	unk.	unk.	D
7.8	39	1	0	0.60	6	66	1	1	unk.	unk.	unk.	2.86	387	39	2.5	9.4	unk.	33	8	0.08	4	tr(a)		tr(a)	
0.20	12	0.01	0	unk.	unk.	unk.	unk.	24	407	102	0.5	tr(a)	268	27	386	373	unk.	1.0	6	1.0	7	unk.	unk.	unk.	E
8.1	41	unk.	unk.	0.60	6	88	2	1	0.7	2	tr(a)	2.64	317	32	16.8	9.5	unk.	33	8	0.32	16	tr(a)		tr(a)	
0.20	12	0.33	17	7.58	126	unk.	unk.	21	407	102	0.5	tr(a)	216	22	426	397	unk.	0.9	5	1.0	7	unk.	unk.	unk.	F
8.8	44	22	6	0.60	6	77	2	1	0.7	2	tr(a)	2.53	315	32	18.5	10.2	unk.	33	8	0.32	16	tr(a)		tr(a)	
0.18	10	0.33	17	7.58	126	0.016	6	68	407	102	unk.	unk.	285	29	574	378	unk.	1.3	7	1.0	7	unk.	unk.	22	G
8.0	40	1	0	0.60	6	253	5	3	unk.	unk.	unk.	2.97	378	38	25.0	9.7	unk.	32	8	0.08	4	tr(a)		tr(a)	
0.22	13	0.03	2	0.18	3	0.003	1	16	278	69	0.7	0.0	203	20	602	290	29	1.4	8	1.0	6	58.9	39	7	H
5.8	29	7	2	0.69	7	446	9	1	0.8	3	0.2	2.84	267	27	26.2	7.4	0.8	67	17	0.25	13	0		0	
0.29	17	0.28	14	5.30	88	0.004	1	14	286	72	0.6	tr(a)	193	19	610	374	52	2.0	11	1.2	8	83.7	56	122	I
6.9	35	30	8	0.85	9	310	6	11	0.7	2	0.1	2.95	300	30	26.5	9.6	1.5	77	19	0.29	15	15		17	
0.32	19	0.33	17	7.40	123	0.016	5	139	22	6	0.1	0.0	295	30	573	468	31	1.8	10	0.3	2	unk.	unk.	1	J
8.6	43	1	0	0.01	0	688	14	23	0.2	1	0.1	3.40	393	39	24.9	12.0	0.9	42	11	tr	0	1		1	
0.02	1	0.14	7	1.40	23	0.002	1	unk.	80	20	0.3	tr(a)	3	0	1246	102	unk.	0.3	2	0.3	2	7.4	5	unk.	K
2.5	13	1	0	0.14	1	38	1	0(a)	0.3	1	tr(a)	1.88	49	5	54.2	2.6	unk.	7	2	0.26	13	tr(a)		tr(a)	
0.00	0	0.00	0	0.00	0	0.000	0	0	0	0	0.0	0.0	0	0	0	2206	0	0.0	0	0.0	0	0.0	0	0	L
0.0	0	0	0	0.00	0	0	0	0	0.0	0	0.0	0.00	0	0	0.0	56.4	0.0	0	0	0.00	0	0(a)		0(a)	
0.00	0	0(a)	0	0(a)	0	0(a)	0	0	0(a)	0	0(a)	0(a)	14	1	2209	tr	unk.	tr	0	tr	0	570.0	380	0	M
0.0	0	0(a)	0	0(a)	0	0	0	0	0.0	0	0(a)	5.69	11	1	96.1	0.0	unk.	10	2	0.02	1	tr		0	
0.02	1	0.02	1	1.20	20	0.003	1	7	60	15	0.1	tr(a)	52	5	99	71	unk.	0.3	2	0.3	2	unk.	unk.	0	N
0.6	3	2	1	0.00	0	26	1	tr	0.1	0	tr(a)	0.37	60	6	4.3	1.8	unk.	5	1	0.01	1	tr(a)		tr(a)	
0.04	2	0.06	3	2.00	33	0.005	2	2	unk.	unk.	0.1	tr(a)	61	6	152	52	unk.	1.0	6	0.6	4	unk.	unk.	unk.	O
1.1	5	3	1	0.12	1	6	0	tr	0.1	1	tr(a)	0.68	71	7	6.6	1.3	unk.	8	2	0.01	0	tr(a)		tr(a)	
0.05	3	0.04	2	2.00	33	unk.	unk.	unk.	unk.	unk.	0.1	tr(a)	90	9	80	64	unk.	0.8	5	0.6	4	unk.	unk.	unk.	P
1.1	5	3	1	0.08	1	6	0	unk.	0.1	1	tr(a)	0.62	96	10	3.5	1.6	unk.	unk.	unk.	0.01	0	tr(a)		tr(a)	
tr	0	0.01	1	0.00	0	unk.	unk.	0(a)	0(a)	0	unk.	unk.	3	0	127	27	unk.	0.1	1	unk.	unk.	unk.	unk.	unk.	Q
0.1	1	unk.	unk.	unk.	unk.	135	3	14	unk.	unk.	unk.	0.42	3	0	5.5	0.7	unk.	unk.	unk.	unk.	unk.	unk.		unk.	
0.01	1	0.01	0	0.06	1	0.001	0	0	3	1	0.1	0(a)	24	2	67	63	25	1.2	7	0.3	2	3.2	2	9	R
tr	0	1	0	0.09	1	212	4	0	0.2	1	0.1	0.50	18	2	2.9	1.6	0.7	3	1	0.02	1	2		10	
0.01	1	unk.	unk.	0(a)	0	unk.	unk.	0	0(a)	0	unk.	unk.	3	0	201	55	unk.	0.1	1	unk.	unk.	unk.	unk.	unk.	S
0.2	1	1	0	unk.	unk.	210	4	21	unk.	unk.	unk.	0.66	8	1	8.7	1.4	unk.	unk.	unk.	tr(a)	0	unk.		unk.	

	FOOD	Portion	Weight in grams / Conversion for 100 g	Kilocalories / H_2O g	Total carbohydrate g / Total fats g	Crude fiber g / Dietary fiber g	Total protein g / Total sugar g	% USRDA	Arginine mg / Histidine mg	Isoleucine mg / Leucine mg	Lysine mg / Methionine mg	Phenylalanine mg / Threonine mg	Valine mg / Tryptophan mg	Cystine mg / Tyrosine mg	Polyunsat. fatty acids g / Monounsat. fatty acids g	Saturated fatty acids g / P/S ratio	Linoleic acid g / Cholesterol mg	Thiamin mg / Ascorbic acid mg	% USRDA
A	sauce,chili,low sodium	1 Tbsp	15	16	3.7	0.10	0.4	**1**	unk.	unk.	unk.	unk.	unk.	unk.	unk.	tr(a)	unk.	0.01	**1**
			6.67	10.2	tr	unk.	unk.		unk.	unk.	unk.	unk.	unk.	unk.	unk.	unk.	0(a)	2	**4**
B	sauce,chili,red hot	1 Tbsp	15	3	0.6	0.26	0.1	**0**	unk.	unk.	unk.	unk.	unk.	unk.	unk.	unk.	unk.	tr	**0**
			6.54	14.4	0.1	unk.	0.0		unk.	unk.	unk.	unk.	unk.	unk.	unk.	unk.	0(a)	5	**8**
C	sauce,chocolate,home recipe	2 Tbsp	48	108	20.8	0.13	1.0	**2**	unk.	31	40	25	34	5	0.02	1.85	0.07	0.00	**0**
			2.08	17.8	3.2	unk.	19.2		14	50	13	23	7	25	1.15	0.01	2	tr	**0**
D	sauce,clam,home recipe	1/2 C	126	274	3.0	0.28	16.9	**37**	1360	812	1417	605	725	228	1.70	2.73	1.60	0.08	**5**
			0.79	81.7	21.6	1.02	tr		340	1255	479	731	192	630	14.68	0.62	87	25	**41**
E	sauce,custard,home recipe	2 Tbsp	40	49	4.7	0.00	1.8	**3**	tr(a)	tr(a)	tr(a)	79	tr(a)	21	0.24	1.19	0.21	0.02	**1**
			2.50	30.5	2.6	0(a)	4.6		tr(a)	tr(a)	tr(a)	tr(a)	tr(a)	83	0.86	0.20	66	tr	**1**
F	sauce,dry,bearnaise	7/8 oz pkg	25	90	14.8	0.05	3.5	**5**	unk.	unk.	unk.	unk.	unk.	unk.	0.84	0.33	0.78	unk.	**unk.**
			4.03	1.4	2.2	unk.	unk.		unk.	unk.	unk.	unk.	unk.	unk.	0.95	2.52	tr	unk.	**unk.**
G	sauce,dry,bearnaise,prep w/milk & butter	2 Tbsp	32	88	2.2	tr	1.0	**2**	unk.	unk.	unk.	unk.	unk.	unk.	0.38	5.23	0.26	unk.	**unk.**
			3.13	19.5	8.5	unk.	unk.		unk.	unk.	unk.	unk.	unk.	unk.	2.18	0.07	24	unk.	**unk.**
H	sauce,dry,cheese	1-1/4 oz pkg	35	158	11.8	0.04	8.0	**12**	unk.	unk.	unk.	unk.	unk.	unk.	1.29	4.24	1.15	0.06	**4**
			2.84	1.4	9.0	unk.	unk.		unk.	unk.	unk.	unk.	unk.	unk.	2.69	0.30	18	unk.	**unk.**
I	sauce,dry,cheese,prep w/milk	2 Tbsp	35	38	2.9	tr	2.0	**3**	unk.	unk.	unk.	unk.	unk.	unk.	0.20	1.17	0.17	0.02	**1**
			2.87	27.0	2.1	unk.	unk.		unk.	unk.	unk.	unk.	unk.	unk.	0.59	0.17	7	tr	**1**
J	sauce,dry,curry	1-1/4 oz pkg	35	151	17.9	0.46	3.3	**5**	unk.	unk.	unk.	unk.	unk.	unk.	3.08	1.22	2.85	unk.	**unk.**
			2.82	1.5	8.2	unk.	unk.		unk.	unk.	unk.	unk.	unk.	unk.	3.48	2.53	tr	unk.	**unk.**
K	sauce,dry,curry,prep w/milk	2 Tbsp	34	34	3.2	0.05	1.3	**2**	unk.	unk.	unk.	unk.	unk.	unk.	0.34	0.75	0.31	unk.	**unk.**
			2.94	27.0	1.8	unk.	unk.		unk.	unk.	unk.	unk.	unk.	unk.	0.61	0.45	4	unk.	**unk.**
L	sauce,dry,hollandaise,butterfat	1-1/5 oz pkg	34	187	10.8	0.03	3.7	**6**	unk.	unk.	unk.	unk.	unk.	unk.	0.73	9.12	0.50	unk.	**unk.**
			2.97	0.7	15.5	unk.	unk.		unk.	unk.	unk.	unk.	unk.	unk.	4.09	0.08	53	unk.	**unk.**
M	sauce,dry,hollandaise,butterfat,prep w/water	2 Tbsp	32	30	1.7	0.01	0.6	**1**	unk.	unk.	unk.	unk.	unk.	unk.	0.12	1.45	0.08	unk.	**unk.**
			3.09	27.2	2.5	unk.	unk.		unk.	unk.	unk.	unk.	unk.	unk.	0.65	0.08	6	unk.	**unk.**
N	sauce,dry,hollandaise,veg oil	7/8 oz pkg	25	93	15.5	0.05	3.4	**5**	unk.	unk.	unk.	unk.	unk.	unk.	0.68	0.48	0.65	unk.	**unk.**
			4.03	1.2	2.3	unk.	unk.		unk.	unk.	unk.	unk.	unk.	unk.	1.01	1.42	tr	unk.	**unk.**
O	sauce,dry,hollandaise,veg oil,prep w/milk & butter	2 Tbsp	32	88	2.2	tr	1.0	**2**	unk.	unk.	unk.	unk.	unk.	unk.	0.37	5.24	0.24	unk.	**unk.**
			3.13	19.5	8.5	unk.	unk.		unk.	unk.	unk.	unk.	unk.	unk.	2.19	0.07	24	unk.	**unk.**
P	sauce,dry,mushroom	1 oz pkg	28	99	15.5	0.28	4.1	**6**	unk.	unk.	unk.	unk.	unk.	unk.	1.01	0.40	0.94	unk.	**unk.**
			3.52	1.0	2.7	unk.	unk.		unk.	unk.	unk.	unk.	unk.	unk.	1.15	2.51	0	unk.	**unk.**
Q	sauce,dry,mushroom,prep w/milk	2 Tbsp	33	28	3.0	0.03	1.4	**2**	unk.	unk.	unk.	unk.	unk.	unk.	0.14	0.67	0.12	unk.	**unk.**
			3.00	26.9	1.3	unk.	unk.		unk.	unk.	unk.	unk.	unk.	unk.	0.37	0.20	4	unk.	**unk.**
R	sauce,dry,sour cream	1-1/4 oz pkg	35	180	17.0	unk.	5.5	**9**	unk.	unk.	unk.	unk.	unk.	unk.	1.24	5.51	1.07	unk.	**unk.**
			2.84	0.6	11.1	unk.	unk.		unk.	unk.	unk.	unk.	unk.	unk.	3.50	0.22	28	unk.	**unk.**
S	sauce,dry,sour cream,prep w/milk	2 Tbsp	39	64	5.7	unk.	2.4	**4**	unk.	unk.	unk.	unk.	unk.	unk.	0.35	2.01	0.29	unk.	**unk.**
			2.54	27.0	3.8	unk.	unk.		unk.	unk.	unk.	unk.	unk.	unk.	1.13	0.17	11	unk.	**unk.**

	Riboflavin mg / Niacin mg	% USRDA	Vitamin B_6 mg / Folacin mcg	% USRDA	Vitamin B_{12} mcg / Pantothenic acid mg	% USRDA	Biotin mg / Vitamin A IU	% USRDA	Preformed A RE / Beta carotene RE	Vitamin D IU / Vitamin E IU	% USRDA	Total tocopherol mg / Alpha tocopherol mg	Other tocopherol mg / Total ash g	Calcium mg / Phosphorus mg	% USRDA	Sodium mg / Sodium meq	Potassium mg / Potassium meq	Chlorine mg / Chlorine meq	Iron mg / Magnesium mg	% USRDA	Zinc mg / Copper mg	% USRDA	Iodine mcg / Selenium mcg	% USRDA	Manganese mcg / Chromium mcg	
	0.01	**1**	unk.	**unk.**	0(a)	**0**	unk.	**unk.**	0	0(a)	**0**	unk.	unk.	3	**0**	3	55	unk.	0.1	**1**	unk.	**unk.**	unk.	**unk.**	unk.	A
	0.2	**1**	1	**0**	unk.	**unk.**	210	**4**	21	unk.	**unk.**	unk.	0.66	8	**1**	0.1	1.4	unk.	unk.	**unk.**	tr(a)	**0**	unk.		unk.	
	0.01	**1**	unk.	**unk.**	0(a)	**0**	unk.	**unk.**	0	0(a)	**0**	unk.	unk.	1	**0**	unk.	unk.	unk.	0.1	**0**	unk.	**unk.**	unk.	**unk.**	unk.	B
	0.1	**1**	unk.	**unk.**	unk.	**unk.**	1467	**29**	147	unk.	**unk.**	unk.	0.08	2	**0**	unk.	unk.	unk.	unk.	**unk.**	tr(a)	**0**	unk.		unk.	
	0.03	**2**	0.01	**0**	0.01	**0**	0.002	**1**	0	6	**2**	0.3	0.0	28	**3**	15	65	11	0.8	**4**	0.2	**2**	1.2	**1**	5	C
	0.1	**1**	6	**1**	0.06	**1**	21	**0**	tr	0.3	**1**	tr	0.34	36	**4**	0.6	1.7	0.3	17	**4**	0.02	**1**	1		7	
	0.13	**7**	0.08	**4**	10.02	**167**	0(a)	**0**	20	62	**16**	2.9	0.1	106	**11**	124	236	18	5.2	**29**	1.7	**12**	tr(a)	**0**	2	D
	1.5	**8**	21	**5**	0.42	**4**	1032	**21**	96	3.5	**12**	2.6	2.83	203	**20**	5.4	6.0	0.5	36	**9**	0.23	**12**	tr(a)		tr(a)	
	0.07	**4**	0.03	**2**	0.25	**4**	0.002	**1**	0	21	**5**	0.2	0.0	44	**4**	36	54	7	0.3	**2**	0.3	**2**	14.3	**10**	6	E
	tr	**0**	5	**1**	0.32	**3**	76	**2**	0	0.2	**1**	tr	0.35	57	**6**	1.6	1.4	0.2	5	**1**	0.05	**3**	0		2	
	unk.	**unk.**	unk.	**unk.**	unk.	**unk.**	unk.	**unk.**	unk.	0(a)	**0**	unk.	unk.	unk.	**unk.**	841	unk.	unk.	unk.	**unk.**	unk.	**unk.**	unk.	**unk.**	unk.	F
	unk.	**unk.**	unk.	**unk.**	unk.	**unk.**	unk.	**unk.**	unk.	unk.	**unk.**	unk.	2.85	unk.	**unk.**	36.6	unk.	unk.	unk.	**unk.**	unk.	**unk.**	unk.		unk.	
	unk.	**unk.**	unk.	**unk.**	unk.	**unk.**	unk.	**unk.**	unk.	unk.	**unk.**	unk.	unk.	unk.	**unk.**	158	unk.	unk.	unk.	**unk.**	unk.	**unk.**	unk.	**unk.**	unk.	G
	unk.	**unk.**	unk.	**unk.**	unk.	**unk.**	unk.	**unk.**	unk.	unk.	**unk.**	unk.	0.58	unk.	**unk.**	6.9	unk.	unk.	unk.	**unk.**	unk.	**unk.**	unk.		unk.	
	0.17	**10**	unk.	**unk.**	unk.	**unk.**	unk.	**unk.**	unk.	unk.	**unk.**	unk.	unk.	280	**28**	1447	183	unk.	0.1	**1**	0.9	**6**	unk.	**unk.**	unk.	H
	0.1	**1**	unk.	**unk.**	unk.	**unk.**	unk.	**unk.**	unk.	unk.	**unk.**	unk.	4.97	210	**21**	62.9	4.7	unk.	15	**4**	unk.	**unk.**	unk.		unk.	
	0.07	**4**	unk.	**unk.**	unk.	**unk.**	unk.	**unk.**	unk.	unk.	**unk.**	unk.	unk.	71	**7**	196	69	unk.	tr	**0**	0.1	**1**	unk.	**unk.**	unk.	I
	tr	**0**	unk.	**unk.**	unk.	**unk.**	unk.	**unk.**	unk.	unk.	**unk.**	unk.	0.84	55	**6**	8.5	1.8	unk.	6	**2**	unk.	**unk.**	unk.		unk.	
	unk.	**unk.**	unk.	**unk.**	unk.	**unk.**	unk.	**unk.**	unk.	0(a)	**0**	unk.	unk.	unk.	**unk.**	1444	unk.	unk.	unk.	**unk.**	unk.	**unk.**	unk.	**unk.**	unk.	J
	unk.	**unk.**	unk.	**unk.**	unk.	**unk.**	unk.	**unk.**	unk.	unk.	**unk.**	unk.	4.50	unk.	**unk.**	62.8	unk.	unk.	unk.	**unk.**	unk.	**unk.**	unk.		unk.	
	unk.	**unk.**	unk.	**unk.**	unk.	**unk.**	unk.	**unk.**	unk.	0(a)	**0**	unk.	unk.	61	**6**	159	unk.	unk.	unk.	**unk.**	unk.	**unk.**	unk.	**unk.**	unk.	K
	unk.	**unk.**	unk.	**unk.**	unk.	**unk.**	unk.	**unk.**	0(a)	unk.	**unk.**	unk.	0.67	35	**4**	6.9	unk.	unk.	unk.	**unk.**	unk.	**unk.**	unk.		unk.	
	0.14	**8**	unk.	**unk.**	unk.	**unk.**	unk.	**unk.**	unk.	0(a)	**0**	unk.	unk.	97	**10**	1230	98	unk.	0.7	**4**	unk.	**unk.**	unk.	**unk.**	unk.	L
	tr	**0**	unk.	**unk.**	unk.	**unk.**	unk.	**unk.**	unk.	unk.	**unk.**	unk.	2.95	unk.	**unk.**	53.5	2.5	unk.	unk.	**unk.**	unk.	**unk.**	unk.		unk.	
	0.02	**1**	unk.	**unk.**	unk.	**unk.**	unk.	**unk.**	unk.	0(a)	**0**	unk.	unk.	16	**2**	196	16	unk.	0.1	**1**	unk.	**unk.**	unk.	**unk.**	unk.	M
	tr	**0**	unk.	**unk.**	unk.	**unk.**	unk.	**unk.**	unk.	unk.	**unk.**	unk.	0.47	unk.	**unk.**	8.5	0.4	unk.	unk.	**unk.**	unk.	**unk.**	unk.		unk.	
	unk.	**unk.**	unk.	**unk.**	unk.	**unk.**	unk.	**unk.**	unk.	0(a)	**0**	unk.	unk.	unk.	**unk.**	645	unk.	unk.	unk.	**unk.**	unk.	**unk.**	unk.	**unk.**	unk.	N
	unk.	**unk.**	unk.	**unk.**	unk.	**unk.**	unk.	**unk.**	unk.	unk.	**unk.**	unk.	2.36	unk.	**unk.**	28.0	unk.	unk.	unk.	**unk.**	unk.	**unk.**	unk.		unk.	
	unk.	**unk.**	unk.	**unk.**	unk.	**unk.**	unk.	**unk.**	unk.	unk.	**unk.**	unk.	unk.	unk.	**unk.**	142	unk.	unk.	unk.	**unk.**	unk.	**unk.**	unk.	**unk.**	unk.	O
	unk.	**unk.**	unk.	**unk.**	unk.	**unk.**	unk.	**unk.**	unk.	unk.	**unk.**	unk.	0.54	unk.	**unk.**	6.2	unk.	unk.	unk.	**unk.**	unk.	**unk.**	unk.		unk.	
	unk.	**unk.**	unk.	**unk.**	0.00	**0**	unk.	**unk.**	unk.	0(a)	**0**	unk.	unk.	unk.	**unk.**	1769	unk.	unk.	unk.	**unk.**	unk.	**unk.**	unk.	**unk.**	unk.	P
	unk.	**unk.**	unk.	**unk.**	unk.	**unk.**	unk.	**unk.**	unk.	unk.	**unk.**	unk.	5.08	unk.	**unk.**	77.0	unk.	unk.	unk.	**unk.**	unk.	**unk.**	unk.		unk.	
	unk.	**unk.**	unk.	**unk.**	unk.	**unk.**	unk.	**unk.**	unk.	unk.	**unk.**	unk.	unk.	unk.	**unk.**	191	unk.	unk.	unk.	**unk.**	unk.	**unk.**	unk.	**unk.**	unk.	Q
	unk.	**unk.**	unk.	**unk.**	unk.	**unk.**	unk.	**unk.**	unk.	unk.	**unk.**	unk.	0.73	unk.	**unk.**	8.3	unk.	unk.	unk.	**unk.**	unk.	**unk.**	unk.		unk.	
	0.15	**9**	unk.	**unk.**	unk.	**unk.**	unk.	**unk.**	unk.	0(a)	**0**	unk.	unk.	128	**13**	444	181	unk.	0.2	**1**	0.6	**4**	unk.	**unk.**	unk.	R
	0.2	**1**	unk.	**unk.**	unk.	**unk.**	unk.	**unk.**	unk.	unk.	**unk.**	unk.	1.06	unk.	**unk.**	19.3	4.6	unk.	un.	**unk.**	0.03	**1**	unk.		unk.	
	0.09	**5**	unk.	**unk.**	unk.	**unk.**	unk.	**unk.**	unk.	unk.	**unk.**	unk.	unk.	68	**7**	126	92	unk.	0.1	**0**	0.2	**1**	unk.	**unk.**	unk.	S
	0.1	**0**	unk.	**unk.**	unk.	**unk.**	unk.	**unk.**	unk.	unk.	**unk.**	unk.	0.48	unk.	**unk.**	5.5	2.3	unk.	un.	**unk.**	0.01	**1**	unk.		unk.	

	FOOD	Portion	Weight in grams / Conversion for 100 g	Kilocalories / H_2O g	Total carbohydrate g / Total fats g	Crude fiber g / Dietary fiber g	Total protein g / Total sugar g	% **USRDA**	Arginine mg / Histidine mg	Isoleucine mg / Leucine mg	Lysine mg / Methionine mg	Phenylalanine mg / Threonine mg	Valine mg / Tryptophan mg	Cystine mg / Tyrosine mg	Polyunsat. fatty acids g / Monounsat. fatty acids g	Saturated fatty acids g / P/S ratio	Linoleic acid g / Cholesterol mg	Thiamin mg / Ascorbic acid mg	% **USRDA**
A	sauce,dry,spaghetti	1/4 pkg	10	28	6.4	unk.	0.6	**1**	unk.	unk.	unk.	unk.	unk.	unk.	tr	0.06	tr	unk.	**unk.**
			10.00	0.4	0.1	unk.	unk.		unk.	unk.	unk.	unk.	unk.	unk.	0.02	0.05	0	unk.	**unk.**
B	sauce,dry,spaghetti w/mushrooms	1/4 pkg	10	30	4.9	unk.	1.0	**2**	unk.	unk.	unk.	unk.	unk.	unk.	0.03	0.57	0.02	unk.	**unk.**
			10.00	0.3	0.9	unk.	unk.		unk.	unk.	unk.	unk.	unk.	unk.	0.21	0.05	3	unk.	**unk.**
C	sauce,dry,stroganoff	1-3/5 oz pkg	46	161	26.5	0.60	5.6	**9**	unk.	unk.	unk.	unk.	unk.	unk.	0.12	2.85	0.08	0.84	**56**
			2.17	2.1	4.4	unk.	unk.		unk.	unk.	unk.	unk.	unk.	unk.	1.01	0.04	11	unk.	**unk.**
D	sauce,dry,stroganoff,prep	1 C	296	272	33.9	0.56	11.7	**18**	unk.	unk.	unk.	unk.	unk.	unk.	0.36	6.78	0.24	0.86	**57**
	w/milk & water		*0.34*	231.4	10.7	unk.	unk.		unk.	unk.	unk.	unk.	unk.	unk.	2.60	0.05	38	unk.	**unk.**
E	sauce,dry,sweet & sour	2 oz pkg	57	221	54.5	unk.	0.6	**1**	unk.	unk.	unk.	unk.	unk.	unk.	0.03	0.01	0.03	unk.	**unk.**
			1.76	0.3	0.1	unk.	unk.		unk.	unk.	unk.	unk.	unk.	unk.	0.01	6.00	0	unk.	**unk.**
F	sauce,dry,sweet & sour,prep w/water	2 Tbsp	39	37	9.1	unk.	0.1	**0**	tr(a)	tr(a)	tr(a)	tr(a)	tr(a)	tr(a)	tr	0.00	tr	unk.	**unk.**
			2.55	29.8	tr	unk.	unk.		tr(a)	tr(a)	tr(a)	tr(a)	tr(a)	tr(a)	tr	0.00	0	unk.	**unk.**
G	sauce,dry,teriyaki	1-3/5 oz pkg	46	130	27.6	unk.	4.1	**6**	unk.	unk.	unk.	unk.	unk.	unk.	0.53	0.13	0.47	unk.	**unk.**
			2.17	0.5	0.9	unk.	unk.		unk.	unk.	unk.	unk.	unk.	unk.	0.21	4.00	0	unk.	**unk.**
H	sauce,dry,teriyaki,prep w/water	1 Tbsp	18	8	1.7	unk.	0.3	**0**	unk.	unk.	unk.	unk.	unk.	unk.	0.03	0.01	0.03	unk.	**unk.**
			5.65	14.8	0.1	unk.	unk.		unk.	unk.	unk.	unk.	unk.	unk.	0.01	3.80	0	unk.	**unk.**
I	sauce,dry,white	1-3/4 oz pkg	50	230	25.1	unk.	5.4	**8**	unk.	unk.	unk.	unk.	unk.	unk.	3.46	3.31	3.24	unk.	**unk.**
			2.02	0.7	13.2	unk.	unk.		unk.	unk.	unk.	unk.	unk.	unk.	5.89	1.04	0	unk.	**unk.**
J	sauce,dry,white,prep w/milk	2 Tbsp	33	30	2.7	0.01	1.3	**2**	unk.	unk.	unk.	unk.	unk.	unk.	0.21	0.80	0.19	0.01	**1**
			3.03	26.9	1.7	unk.	unk.		unk.	unk.	unk.	unk.	unk.	unk.	0.55	0.26	4	unk.	**unk.**
K	sauce,grape,home recipe	1 Tbsp	11	26	6.5	0.03	0.1	**0**	5	1	2	1	2	1	0.00	0.00	0.00	tr	**0**
			8.93	4.6	0.1	0(a)	5.6		3	1	2	2	tr	1	0.00	0.00	0	tr	**0**
L	sauce,hard,home recipe	2 Tbsp	31	142	23.7	0.00	0.1	**0**	2	4	5	3	4	1	0.21	3.54	0.13	0.00	**0**
			3.23	1.3	5.7	0(a)	23.6		2	6	1	3	1	3	1.43	0.06	15	0	**0**
M	sauce,hollandaise,home recipe	2 Tbsp	24	134	0.1	0.00	1.5	**3**	unk.	76	112	58	88	23	0.81	8.17	0.61	0.02	**2**
			4.12	7.9	14.4	unk.	tr		37	130	35	74	22	68	4.02	0.10	148	1	**2**
N	sauce,jiffy hollandaise,home recipe	2 Tbsp	30	130	1.0	0.01	0.6	**1**	unk.	unk.	unk.	unk.	unk.	unk.	0.11	3.79	5.55	0.01	**0**
			3.33	13.9	14.0	unk.	0.6		unk.	unk.	tr(a)	unk.	unk.	unk.	3.08	0.03	16	1	**2**
O	sauce,lemon,home recipe	2 Tbsp	44	57	11.3	tr	0.6	**1**	45	29	42	22	33	9	0.16	0.42	0.15	0.01	**1**
			2.27	30.8	1.2	0.04	10.2		14	49	13	29	8	26	0.50	0.38	48	3	**5**
P	sauce,raisin,home recipe	2 Tbsp	36	51	13.3	0.00	0.2	**0**	tr(a)	tr(a)	tr(a)	2	tr(a)	0	0.00	0.00	0.00	0.01	**1**
			2.75	20.2	tr	0.27	12.2		tr(a)	tr(a)	tr(a)	tr(a)	tr(a)	2	0.00	0.00	0	5	**9**
Q	sauce,steak,A-1-Heublein	1 Tbsp	16	14	3.3	tr(a)	0.2	**0**	unk.	unk.	unk.	unk.	unk.	unk.	unk.	0(a)	unk.	unk.	**unk.**
			6.10	11.6	tr	tr(a)	unk.		unk.	unk.	unk.	unk.	unk.	unk.	unk.	unk.	0(a)	unk.	**unk.**
R	sauce,sweet & sour,home recipe	2 Tbsp	27	65	17.0	0.01	0.1	**0**	unk.	3	1	3	3	1	tr	0.00	tr	tr	**0**
			3.72	9.6	tr	0.02	16.0		1	5	1	2	1	2	0.00	0.00	0	0	**0**
S	sauce,tartar-Hellmann	1 Tbsp	14	71	0.2	0.01	0.1	**0**	unk.	unk.	unk.	unk.	unk.	unk.	4.20	1.29	unk.	tr(a)	**0**
			7.14	5.3	7.9	0(a)	unk.		unk.	unk.	unk.	unk.	unk.	unk.	unk.	3.26	5	unk.	**unk.**

Riboflavin mg / Niacin mg	% USRDA	Vitamin B6 mg / Folacin mcg	% USRDA	Vitamin B12 mcg / Pantothenic acid mg	% USRDA	Biotin mg / Vitamin A IU	% USRDA	Preformed A RE / Beta carotene RE	Vitamin D IU / Vitamin E IU	% USRDA	Total tocopherol mg / Alpha tocopherol mg	Other tocopherol mg / Total ash g	Calcium mg / Phosphorus mg	% USRDA	Sodium mg / Sodium meq	Potassium mg / Potassium meq	Chlorine mg / Chlorine meq	Iron mg / Magnesium mg	% USRDA	Zinc mg / Copper mg	% USRDA	Iodine mcg / Selenium mcg	% USRDA	Manganese mcg / Chromium mcg	
0.06	**3**	unk.	**unk.**	unk.	**unk.**	unk.	**unk.**	unk.	0(a)	**0**	unk.	unk.	17	**2**	848	84	unk.	0.3	**2**	tr	**0**	unk.	**unk.**	unk.	A
0.2	**1**	unk.	**unk.**	unk.	**unk.**	unk.	**unk.**	unk.	unk.	**unk.**	unk.	2.50	unk.	**unk.**	36.9	2.1	unk.	unk.	**unk.**	0.09	**5**	unk.		unk.	
0.05	**3**	unk.	**unk.**	unk.	**unk.**	unk.	**unk.**	unk.	0(a)	**0**	unk.	unk.	40	**4**	942	41	unk.	0.2	**1**	unk.	**unk.**	unk.	**unk.**	unk.	B
0.2	**1**	unk.	**unk.**	unk.	**unk.**	unk.	**unk.**	unk.	unk.	**unk.**	unk.	2.90	unk.	**unk.**	41.0	1.0	unk.	unk.	**unk.**	unk.	**unk.**	unk.		unk.	
0.48	**28**	unk.	**unk.**	unk.	**unk.**	unk.	**unk.**	unk.	unk.	**unk.**	unk.	unk.	307	**31**	1863	398	unk	1.3	**7**	1.1	**7**	unk.	**unk.**	unk.	C
0.6	**3**	unk.	**unk.**	unk.	**unk.**	unk.	**unk.**	unk.	unk.	**unk.**	unk.	7.43	126	**13**	81.0	10.2	unk.	unk.	**unk.**	0.07	**4**	unk.		unk.	
0.77	**45**	unk.	**unk.**	unk.	**unk.**	unk.	**unk.**	unk.	unk.	**unk.**	unk.	unk.	521	**52**	1829	672	unk.	1.3	**7**	1.1	**7**	unk.	**unk.**	unk.	D
0.8	**4**	unk.	**unk.**	unk.	**unk.**	unk.	**unk.**	unk.	unk.	**unk.**	unk.	8.35	302	**30**	79.6	17.2	unk.	unk.	**unk.**	0.09	**4**	unk.		unk.	
0.07	**4**	unk.	**unk.**	0.00	**0**	unk.	**unk.**	unk.	0(a)	**0**	unk.	unk.	31	**3**	584	49	unk.	1.2	**7**	0.1	**1**	unk.	**unk.**	unk.	E
unk.	**unk.**	unk.	**unk.**	unk.	**unk.**	unk.	**unk.**	unk.	unk.	**unk.**	unk.	1.25	unk.	**unk.**	25.4	1.3	unk.	unk.	**unk.**	0.02	**1**	unk.		unk.	
0.01	**1**	unk.	**unk.**	0.00	**0**	unk.	**unk.**	unk.	0(a)	**0**	unk.	unk.	5	**1**	98	8	unk.	0.2	**1**	tr	**0**	unk.	**unk.**	unk.	F
unk.	**unk.**	unk.	**unk.**	unk.	**unk.**	unk.	**unk.**	unk.	unk.	**unk.**	unk.	0.21	unk.	**unk.**	4.3	0.2	unk.	unk.	**unk.**	tr	**0**	unk.		unk.	
unk.	**unk.**	unk.	**unk.**	0.00	**0**	unk.	**unk.**	unk.	0(a)	**0**	unk.	unk.	112	**11**	4784	215	unk.	2.8	**16**	unk.	**unk.**	unk.	**unk.**	unk.	G
1.3	**7**	unk.	**unk.**	unk.	**unk.**	unk.	**unk.**	unk.	unk.	**unk.**	unk.	12.88	unk.	**unk.**	208.1	5.5	unk.	unk.	**unk.**	unk.	**unk.**	unk.		unk.	
unk.	**unk.**	unk.	**unk.**	0.00	**0**	unk.	**unk.**	unk.	0(a)	**0**	unk.	unk.	7	**1**	300	13	unk.	0.2	**1**	unk.	**unk.**	unk.	**unk.**	unk.	H
0.1	**0**	unk.	**unk.**	unk.	**unk.**	unk.	**unk.**	unk.	unk.	**unk.**	unk.	0.81	unk.	**unk.**	13.0	0.3	unk.	unk.	**unk.**	unk.	**unk.**	unk.		unk.	
unk.	**unk.**	unk.	**unk.**	unk.	**unk.**	unk.	**unk.**	unk.	0(a)	**0**	unk.	unk.	334	**33**	1691	184	unk.	unk.	**unk.**	unk.	**unk.**	unk.	**unk.**	unk.	I
unk.	**unk.**	unk.	**unk.**	unk.	**unk.**	unk.	**unk.**	unk.	unk.	**unk.**	unk.	5.11	unk.	**unk.**	73.6	4.7	unk.	unk.	**unk.**	unk.	**unk.**	unk.		unk.	
0.06	**3**	0.01	**0**	unk.	**unk.**	unk.	**unk.**	unk.	unk.	**unk.**	unk.	unk.	53	**5**	100	55	unk.	tr	**0**	0.1	**1**	unk.	**unk.**	unk.	J
0.1	**0**	unk.	**unk.**	unk.	**unk.**	unk.	**unk.**	unk.	unk.	**unk.**	unk.	0.48	32	**3**	4.3	1.4	unk.	33	**8**	0.01	**0**	unk.		unk.	
tr	**0**	tr	**0**	0.00	**0**	tr	**0**	0	0	**0**	0.0	0.0	1	**0**	9	9	tr	tr	**0**	tr	**0**	2.1	**1**	6	K
tr	**0**	tr	**0**	tr	**0**	5	**0**	1	0.0	**0**	0.0	0.04	1	**0**	0.4	0.2	0.0	tr	**0**	tr	**0**	0		4	
tr	**0**	0.00	**0**	0.00	**0**	0.000	**0**	0	0	**0**	0.1	0.0	2	**0**	58	3	0	tr	**0**	0.2	**1**	unk.	**unk.**	9	L
0.0	**0**	tr	**0**	0.00	**0**	214	**4**	0	0.1	**0**	0.1	0.15	2	**0**	2.5	0.1	0.0	tr	**0**	0.00	**0**	unk.		14	
0.04	**3**	0.03	**2**	0.27	**5**	0.003	**1**	0	14	**4**	0.5	0(a)	15	**2**	123	15	14	0.5	**3**	0.4	**3**	14.2	**10**	10	M
tr	**0**	6	**2**	0.42	**4**	503	**10**	0	0.6	**2**	0.1	0.44	55	**6**	5.3	0.4	0.4	1	**0**	0.01	**1**	0		0	
0.02	**1**	tr	**0**	0.04	**1**	unk.	**unk.**	0	0(a)	**0**	6.9	3.5	20	**2**	98	28	tr	0.1	**1**	0.1	**0**	unk.	**unk.**	tr	N
tr	**0**	2	**0**	0.05	**1**	152	**3**	0	8.3	**28**	3.3	0.35	17	**2**	4.2	0.7	0.0	2	**1**	0.00	**0**	0		0	
0.02	**1**	0.01	**1**	0.11	**2**	0.001	**1**	0	6	**2**	0.1	0.0	7	**1**	27	11	6	0.2	**1**	0.2	**1**	12.1	**8**	8	O
tr	**0**	2	**1**	0.17	**2**	29	**1**	tr	0.2	**1**	0.0	0.14	22	**2**	1.2	0.3	0.2	1	**0**	0.01	**0**	0		6	
0.01	**0**	0.01	**1**	0.00	**0**	tr	**0**	0	0	**0**	0.0	0.0	8	**1**	23	65	6	0.4	**2**	0.1	**0**	5.0	**3**	1079	P
0.1	**0**	6	**1**	0.02	**2**	23	**1**	2	0.0	**0**	0.0	0.22	7	**1**	1.0	1.6	0.2	4	**1**	0.05	**3**	1		3	
unk.	**unk.**	unk.	**unk.**	0(a)	**0**	unk.	**unk.**	0(a)	0(a)	**0**	0(a)	unk.	5	**1**	268	55	unk.	0.2	**1**	unk.	**unk.**	unk.	**unk.**	unk.	Q
unk.	**unk.**	unk.	**unk.**	unk.	**unk.**	unk.	**unk.**	unk.	0.0	**0**	unk.	unk.	unk.	**unk.**	11.7	1.4	unk.	unk.	**unk.**	unk.	**unk.**	unk.		unk.	
0.00	**0**	0.00	**0**	0.00	**0**	0.0000	**0**	0	0	**0**	0.0	0.0	1	**0**	91	11	tr	0.1	**1**	0.1	**1**	23.4	**16**	34	R
tr	**0**	tr	**0**	tr	**0**	0	**0**	0	0.0	**0**	0.0	0.26	2	**0**	4.0	0.3	0.0	1	**0**	0.01	**1**	9		10	
unk.	**unk.**	unk.	**unk.**	tr(a)	**0**	unk.	**unk.**	unk.	unk.	**unk.**	6.0	unk.	unk.	**unk.**	182	unk.	unk.	unk.	**unk.**	unk.	**unk.**	unk.	**unk.**	unk.	S
tr(a)	**0**	unk.	**unk.**	unk.	**unk.**	unk.	**unk.**	unk.	unk.	**unk.**	unk.	0.46	unk.	**unk.**	7.9	unk.	unk.	unk.	**unk.**	unk.	**unk.**	unk.		unk.	

	FOOD	Portion	Weight in grams / *Conversion for 100 g*	Kilocalories / H_2O g	Total carbohydrate g / Total fats g	Crude fiber g / Dietary fiber g	Total protein g / Total sugar g	**% USRDA**	Arginine mg / Histidine mg	Isoleucine mg / Leucine mg	Lysine mg / Methionine mg	Phenylalanine mg / Threonine mg	Valine mg / Tryptophan mg	Cystine mg / Tyrosine mg	Polyunsat. fatty acids g / Monounsat. fatty acids g	Saturated fatty acids g / P/S ratio	Linoleic acid g / Cholesterol mg	Thiamin mg / Ascorbic acid mg	**% USRDA**
A	sauce,tartar,low calorie	1 Tbsp	14	31	0.9	0.04	0.1	**0**	unk.	unk.	unk.	unk.	unk.	unk.	unk.	unk.	unk.	tr	**0**
			7.14	9.5	3.1	unk.	0.4		unk.	unk.	unk.	unk.	unk.	unk.	unk.	unk.	0(a)	tr	**0**
B	sauce,tartar,regular	1 Tbsp	14	74	0.6	0.04	0.2	**0**	unk.	unk.	unk.	unk.	unk.	unk.	unk.	0.98	4.06	tr	**0**
			7.14	4.8	8.1	unk.	0.4		unk.	unk.	unk.	unk.	unk.	unk.	0.98	unk.	0(a)	tr	**0**
C	sauce,teriyaki	1 Tbsp	18	15	2.9	unk.	1.1	**2**	unk.	unk.	unk.	unk.	unk.	unk.	0.00	0.00	0.00	0.01	**0**
			5.56	12.2	0.0	unk.	unk.		unk.	unk.	unk.	unk.	unk.	unk.	0.00	0.00	0	0	**0**
D	sauce,Tobasco-McIlhenny	1 Tbsp	15	1	0.2	tr(a)	tr(a)	**0**	tr(a)	tr(a)	tr(a)	tr(a)	tr(a)	tr(a)	unk.	tr(a)	unk.	tr	**0**
			6.67	14.4	0.1	unk.	0.0		tr(a)	tr(a)	tr(a)	tr(a)	tr(a)	tr(a)	unk.	unk.	0(a)	0(a)	**0**
E	sauce,tomato,home recipe	2 Tbsp	34	10	2.3	0.30	0.5	**1**	7	13	19	15	12	2	tr	0.00	tr	0.02	**1**
			2.93	30.6	0.1	0.78	1.3		3	17	4	15	4	5	0.00	0.00	0	18	**30**
F	sauce,Worchestershire-Lea & Perrins	1 Tbsp	15	59	2.5	tr(a)	0.0	**0**	tr(a)	tr(a)	tr(a)	tr(a)	tr(a)	tr(a)	unk.	0.00	0.00	0.01	**1**
			6.67	10.5	tr	unk.	0.4		tr(a)	tr(a)	tr(a)	tr(a)	tr(a)	tr(a)	0.00	unk.	0	27	**46**
G	SAUERBRATEN,home recipe	3x2x3/4''	100	190	3.2	0.13	24.0	**53**	1519	1145	2116	1047	1192	304	1.03	3.70	0.33	0.06	**4**
			1.00	62.1	8.7	0.15	1.2		810	1928	643	1091	268	858	3.14	0.28	75	2	**4**
H	SAUERKRAUT,canned,solids & liquid	1/2 C	18	21	4.7	0.82	1.2	**2**	unk.	unk.	unk.	unk.	unk.	unk.	unk.	unk.	unk.	0.04	**2**
			0.85	109.0	0.2	1.65	unk.		unk.	unk.	unk.	unk.	unk.	unk.	unk.	unk.	unk.	16	**27**
I	SAUSAGE,beef,patty	3-7/8'' dia x	27	86	0.6	unk.	3.0	**7**	188	182	302	137	188	38	0.32	3.43	0.24	0.01	**1**
		1/4''	*3.70*	14.6	8.1	0(a)	0.0		101	269	79	154	32	107	3.51	0.09	14	7	**12**
J	sausage,country-style,smoked links	1 average	43	148	0.0	0.00	6.5	**14**	unk.	unk.	unk.	215	330	unk.	unk.	4.73	1.29	0.09	**6**
			2.33	21.5	13.4	0(a)	0.0		unk.	unk.	136	unk.	unk.	unk.	5.59	unk.	38	unk.	**unk.**
K	sausage,home recipe, roll	1 average	100	365	21.1	0.29	13.6	**28**	633	678	969	492	689	140	3.22	6.54	2.44	0.63	**42**
			1.00	39.9	25.2	0.66	7.6		350	1013	262	645	175	403	8.02	0.49	47	2	**3**
L	sausage,home recipe,casserole	3/4 C	113	262	10.4	0.70	12.8	**27**	609	635	941	522	706	148	1.92	6.03	1.43	0.34	**23**
			0.88	73.9	18.9	unk.	5.6		383	1011	290	547	167	489	6.09	0.32	111	12	**20**
M	sausage,Italian,cooked	1 med (1'')	22	76	0.0	0.00	3.3	**7**	unk.	unk.	unk.	137	171	unk.	unk.	2.42	0.66	0.05	**3**
		piece	*4.55*	11.0	6.8	0(a)	0.0		unk.	unk.	89	unk.	unk.	unk.	2.86	unk.	20	unk.	**unk.**
N	sausage,Polish/kielbasa,cooked	5-3/8'' long x	76	231	0.9	0.00	11.9	**27**	unk.	1079	1305	608	821	unk.	unk.	unk.	unk.	0.26	**17**
		1'' dia	*1.32*	40.8	19.6	0(a)	tr		unk.	1201	385	676	112	unk.	unk.	unk.	68	0	**0**
O	sausage,pork,cooked	1/2 C	70	333	tr	0.00	12.7	**28**	805	647	1023	420	644	141	3.49	8.22	2.38	0.55	**37**
			1.43	24.4	30.9	0(a)	0.0		416	955	266	621	172	454	9.88	0.43	62	0	**0**
P	sausage,pork,link,cooked	2 links	26	124	tr	0.00	4.7	**11**	unk.	240	380	156	239	53	1.30	3.05	0.88	0.21	**14**
			3.85	9.0	11.5	0(a)	0.0		155	355	99	231	64	168	3.67	0.43	23	0	**0**
Q	sausage,pork,Little Friers,cooked	1 average	24	97	0.2	0(a)	3.8	**8**	301	242	383	198	242	53	1.28	3.13	1.09	0.12	**8**
	-Oscar Mayer		*4.22*	10.0	9.0	0(a)	unk.		156	357	128	232	65	170	4.05	0.41	21	tr	**0**
R	sausage,pork,patty,cooked	3-/8'' dia x	27	129	tr	0.00	4.9	**11**	310	250	395	162	248	55	1.35	3.17	0.92	0.21	**14**
		5/8''	*3.70*	9.4	11.9	0(a)	0.0		161	368	103	239	66	175	3.81	0.43	24	0	**0**
S	sausage,rice	1 oz	28	53	4.5	unk.	1.7	**4**	unk.	unk.	unk.	unk.	unk.	unk.	unk.	unk.	unk.	0.02	**1**
			3.52	unk.	3.6	unk.	tr		unk.	unk.	unk.	unk.	unk.	unk.	unk.	unk.	unk.	unk.	**unk.**

	Riboflavin mg / Niacin mg	% USRDA	Vitamin B_6 mg / Folacin mcg	% USRDA	Vitamin B_{12} mcg / Pantothenic acid mg	% USRDA	Biotin mg / Vitamin A IU	% USRDA	Preformed A RE / Beta carotene RE	Vitamin D IU / Vitamin E IU	% USRDA	Total tocopherol mg / Alpha tocopherol mg	Other tocopherol mg / Total ash g	Calcium mg / Phosphorus mg	% USRDA	Sodium mg / Sodium meq	Potassium mg / Potassium meq	Chlorine mg / Chlorine meq	Iron mg / Magnesium mg	% USRDA	Zinc mg / Copper mg	% USRDA	Iodine mcg / Selenium mcg	% USRDA	Manganese mcg / Chromium mcg
A	tr	0	unk.	unk.	0(a)	0	unk.	unk.	0(a)	0(a)	0	unk.	unk.	3	0	99	11	unk.	0.1	1	unk.	unk.	unk.	unk.	unk.
	tr	0	unk.	unk.	unk.	unk.	31	1	unk.	unk.	unk.	unk.	0.31	5	0	4.3	0.3	unk.	unk.	unk.	unk.	unk.	unk.		unk.
B	tr	0	unk.	unk.	0(a)	0	unk.	unk.	unk.	0(a)	0	unk.	unk.	3	0	99	11	unk.	0.1	1	unk.	unk.	unk.	unk.	unk.
	tr	0	unk.	unk.	unk.	unk.	31	1	unk.	unk.	unk.	unk.	0.31	5	0	4.3	0.3	unk.	tr	0	unk.	unk.	unk.		unk.
C	0.01	1	0.02	1	0.00	0	unk.	unk.	0	0(a)	0	0(a)	0(a)	4	0	690	40	unk.	0.3	2	tr	0	unk.	unk.	0
	0.2	1	4	1	0.04	0	0	0	0	0.0	0	0(a)	1.88	28	3	30.0	1.0	unk.	11	3	0.02	1	unk.		unk.
D	0.01	1	tr(a)	0	0(a)	0	tr(a)	0	0(a)	0(a)	0	0(a)	0(a)	unk.	unk.	67	9	unk.	unk.	unk.	tr(a)	0	unk.	unk.	tr(a)
	tr(a)	0	tr(a)	0	tr(a)	0	tr(a)	0	tr(a)	0.0	0	0(a)	0.27	unk.	unk.	2.9	0.2	unk.	tr(a)	0	tr(a)	0	unk.		unk.
E	0.02	1	0.03	1	0.00	0	0(a)	0	0	0(a)	0	0.1	0.0	9	1	150	102	3	0.5	3	tr	0	22.1	15	16
	0.3	2	3	1	0.01	0	357	7	30	0.1	0	tr	0.65	10	1	6.5	2.6	0.1	4	1	0.01	0	0		1
F	0.02	1	unk.	unk.	0(a)	0	unk.	unk.	0	0(a)	0	unk.	unk.	16	2	147	120	unk.	0.8	5	unk.	unk.	unk.	unk.	tr(a)
	0.1	0	unk.	unk.	unk.	unk.	16	0	2	unk.	unk.	unk.	0.57	9	1	6.4	3.1	unk.	2	1	tr(a)	0	unk.		unk.
G	0.19	11	0.23	11	1.32	22	tr	0	4	0	0	0.2	tr	20	2	381	346	2	3.4	19	4.8	32	85.0	57	55
	4.3	21	6	2	0.28	3	23	1	1	0.2	1	0.2	2.07	207	21	16.6	8.9	0.1	28	7	0.10	5	16		1
H	0.05	3	0.15	8	0.00	0	unk.	unk.	0	0(a)	0	unk.	unk.	42	4	878	164	unk.	0.6	3	1.0	6	unk.	unk.	unk.
	0.2	1	12	3	0.11	1	59	1	6	unk.	unk.	unk.	2.35	21	2	38.2	4.2	unk.	unk.	unk.	unk.	unk.	unk.		4
I	0.02	1	0.04	2	0.35	6	unk.	unk.	0	0	0	unk.	unk.	2	0	255	38	unk.	0.4	2	0.6	4	unk.	unk.	12
	0.6	3	4	1	unk.	unk.	0	0	0	unk.	unk.	unk.	0.76	21	2	11.1	1.0	unk.	2	1	0.02	1	tr(a)		1
J	0.08	5	unk.	unk.	unk.	unk.	unk.	unk.	unk.	unk.	unk.	0.1	0.1	4	0	unk.	unk.	unk.	1.0	6	1.3	9	unk.	unk.	unk.
	1.3	7	unk.	unk.	unk.	unk.	unk.	unk.	unk.	0.2	1	0.1	1.68	72	7	unk.	unk.	unk.	7	2	0.06	3	tr(a)		tr(a)
K	0.28	17	0.13	6	0.74	12	0.000	0	0	0	0	1.8	0.1	31	3	605	286	5	2.7	15	3.0	20	unk.	unk.	1183
	2.8	14	9	2	0.43	4	28	1	3	2.1	7	0.3	2.45	204	20	26.3	7.3	0.1	46	11	0.20	10	10		3
L	0.32	19	0.17	8	0.68	11	0.005	2	0	13	3	2.9	0.1	156	16	668	316	27	2.0	11	1.7	11	54.3	36	6
	2.4	12	20	5	0.54	5	577	12	47	3.5	12	0.1	2.73	194	19	29.0	8.1	0.8	67	17	0.20	10	0		0
M	0.04	3	0.03	2	unk.	unk.	unk.	unk.	unk.	0	0	0.1	tr	2	0	unk.	unk.	unk.	0.5	3	0.7	4	unk.	unk.	9
	0.7	3	unk.	unk.	unk.	unk.	unk.	unk.	unk.	0.1	0	tr	0.86	37	4	unk.	unk.	unk.	4	1	0.02	1	tr(a)		1
N	0.14	9	unk.	unk.	unk.	unk.	unk.	unk.	0	0	0	0.2	0.1	7	1	unk.	unk.	unk.	1.8	10	2.3	15	unk.	unk.	unk.
	2.4	12	unk.	unk.	unk.	unk.	0	0	0	0.3	1	0.1	2.74	134	13	unk.	unk.	unk.	unk.	unk.	unk.	unk.	tr(a)		tr(a)
O	0.24	14	0.13	7	0.98	16	unk.	unk.	0	0	0	0.2	0.1	5	1	671	188	unk.	1.7	9	2.1	14	unk.	unk.	unk.
	2.6	13	3	1	0.42	4	0	0	0	0.3	1	0.1	2.03	113	11	29.2	4.8	unk.	11	3	0.10	5	tr(a)		tr(a)
P	0.09	5	0.05	3	0.36	6	unk.	unk.	0	0	0	0.1	tr	2	0	249	70	unk.	0.6	4	0.8	5	unk.	unk.	unk.
	1.0	5	1	0	0.16	2	0	0	0	0.1	0	tr	0.75	42	4	10.8	1.8	unk.	4	1	0.04	2	tr(a)		tr(a)
Q	0.05	3	0.006	3	0.37	6	tr(a)	0	tr(a)	16	4	0.1	tr	9	1	264	64	398	0.3	2	0.6	4	unk.	unk.	10
	1.1	5	tr(a)	0	tr(a)	0	tr(a)	0	0(a)	0.1	0	tr	0.76	37	4	11.5	1.6	11.2	4	1	0.02	1	tr(a)		1
R	0.09	5	unk.	unk.	unk.	unk.	unk.	unk.	0	0	0	0.1	tr	2	0	259	73	unk.	0.6	4	0.8	5	unk.	unk.	unk.
	1.0	5	unk.	unk.	unk.	unk.	0	0	0	0.1	0	tr	0.78	44	4	11.3	1.9	unk.	4	1	unk.	unk.	unk.		unk.
S	tr	0	unk.	unk.	unk.	unk.	unk.	unk.	unk.	unk.	unk.	unk.	unk.	2	0	unk.	unk.	unk.	0.2	1	unk.	unk.	unk.	unk.	unk.
	0.2	1	unk.	unk.	unk.	unk.	unk.	unk.	unk.	unk.	unk.	unk.	unk.	5	1	unk.	unk.	unk.	unk.	unk.	unk.	unk.	tr(a)		tr(a)

	FOOD	Portion	Weight in grams / Conversion for 100 g	Kilocalories / H_2O g	Total carbohydrate g / Total fats g	Crude fiber g / Dietary fiber g	Total protein g / Total sugar g	% USRDA	Arginine mg / Histidine mg	Isoleucine mg / Leucine mg	Lysine mg / Methionine mg	Phenylalanine mg / Threonine mg	Valine mg / Tryptophan mg	Cystine mg / Tyrosine mg	Polyunsat. fatty acids g / Monounsat. fatty acids g	Saturated fatty acids g / P/S ratio	Linoleic acid g / Cholesterol mg	Thiamin mg / Ascorbic acid mg	% USRDA
A	sausage,rice/pork	1 oz	28	67	4.5	unk.	1.6	**3**	unk.	unk.	unk.	unk.	unk.	unk.	unk.	unk.	unk.	0.02	**1**
			3.52	unk.	5.7	unk.	tr		unk.	unk.	unk.	unk.	unk.	unk.	unk.	unk.	unk.	unk.	**unk.**
B	sausage,smokie links-Oscar Mayer	2 pieces	38	120	0.5	0(a)	4.9	**11**	unk.	unk.	unk.	unk.	unk.	unk.	1.24	4.21	1.13	0.11	**8**
			2.66	20.3	10.9	0(a)	0.0		unk.	unk.	unk.	unk.	unk.	unk.	4.89	0.29	23	7	**11**
C	SAVORY,ground	1 tsp	1.5	4	1.0	0.23	0.1	**0**	unk.	unk.	unk.	unk.	unk.	unk.	unk.	unk.	unk.	0.01	**0**
			66.67	0.1	0.1	unk.	0.0		unk.	unk.	unk.	unk.	unk.	unk.	unk.	unk.	0	unk.	**unk.**
D	SCALLOP,bay & sea,steamed	1 med	24	27	0(a)	0(a)	5.6	**12**	419	263	444	unk.	unk.	88	unk.	unk.	unk.	unk.	**unk.**
			4.17	17.5	0.3	0(a)	0(a)		132	430	unk.	261	72	232	unk.	unk.	13	unk.	**unk.**
E	scallop,frozen,breaded,fried,reheated	1 med	24	47	2.5	0(a)	4.3	**10**	325	204	344	unk.	unk.	68	unk.	unk.	unk.	unk.	**unk.**
			4.17	14.4	2.0	unk.	0.0		103	334	unk.	203	56	180	unk.	unk.	unk.	unk.	**unk.**
F	SCHOOL LUNCH,apple crisp	2/3 C	93	217	38.5	0.09	1.3	**2**	unk.	65	43	69	66	25	0.39	4.16	0.29	0.07	**4**
			1.08	39.3	7.2	1.47	29.9		26	100	19	41	15	46	1.76	0.09	18	2	**4**
G	school lunch,beef pin wheel w/gravy	1 w/3 Tbsp	121	293	21.9	0.10	16.9	**36**	859	897	1339	761	941	229	1.71	6.20	0.68	0.17	**11**
		gravy	*0.83*	64.9	14.5	0.82	1.6		519	1423	399	734	206	598	6.84	0.28	54	1	**2**
H	school lunch,berroy/burros	1 average	208	391	25.4	0.36	21.7	**46**	1165	1167	1760	969	1115	237	2.50	10.55	0.94	0.19	**13**
			0.48	129.4	22.1	1.41	1.5		615	1827	521	954	269	667	9.05	0.24	83	10	**17**
I	school lunch,cheese sticks	3/4 C	186	321	20.8	0.09	17.2	**36**	671	851	1338	859	1026	247	1.00	10.25	0.95	0.15	**10**
			0.54	113.1	18.7	1.02	6.4		619	1494	434	655	240	816	5.47	0.10	204	1	**2**
J	school lunch,chicken fried steak	3 oz	85	240	3.6	0.01	21.2	**47**	1303	1019	1848	935	1058	273	1.33	5.51	0.59	0.08	**5**
			1.18	43.9	15.1	0.13	0.3		713	1716	564	962	238	762	6.76	0.24	68	0	**0**
K	school lunch,Chinese pie	3/4 C	193	157	7.8	0.04	11.9	**26**	503	519	714	416	494	138	1.97	2.30	1.33	0.04	**3**
			0.52	158.9	8.4	0.01	0.4		252	732	220	379	104	333	2.62	0.86	32	3	**4**
L	school lunch,clam roll	1 average	215	493	28.0	0.09	16.3	**34**	1078	765	1130	643	491	228	18.92	5.44	16.98	0.17	**12**
			0.47	112.8	34.8	0.99	6.0		313	1248	441	705	209	553	8.64	3.48	76	15	**26**
M	school lunch,enchilada	1 average	249	447	35.6	2.12	26.2	**55**	802	807	1295	673	871	166	1.32	11.35	0.35	0.15	**10**
			0.40	140.7	22.9	unk.	1.1		497	1424	379	645	184	575	9.09	0.12	53	4	**7**
N	school lunch,flying saucer sandwich	1 average	148	341	26.4	0.15	18.5	**30**	417	831	1111	699	897	74	0.50	11.10	0.30	0.15	**10**
			0.68	108.3	17.9	unk.	3.0		458	1290	350	494	175	648	4.19	0.05	57	1	**1**
O	school lunch,goulash	3/4 C	190	171	8.3	0.83	15.2	**33**	898	733	1151	574	735	183	0.80	3.34	0.78	0.19	**13**
			0.53	139.2	8.7	1.22	0.9		455	1033	328	639	188	510	2.87	0.24	46	15	**25**
P	school lunch,hamburger gravy	1/2 C	123	263	5.1	0.02	17.8	**39**	1078	979	1598	783	1035	232	2.02	8.33	0.54	0.10	**7**
			0.81	80.4	18.2	0.20	0.2		595	1536	458	819	220	638	8.03	0.24	66	1	**2**
Q	school lunch,lobster roll	1 average	196	409	27.1	0.34	16.2	**34**	1079	649	1279	733	610	162	14.17	4.16	12.80	0.16	**11**
			0.51	106.6	26.5	1.67	5.1		242	1292	438	642	148	477	6.72	3.41	80	5	**9**
R	school lunch,Mexican noodles	3/4 C	170	256	34.9	0.48	10.3	**21**	unk.	554	550	510	633	136	0.15	4.45	0.10	0.20	**14**
			0.59	107.9	8.0	unk.	1.2		289	798	212	401	121	350	2.23	0.03	62	7	**12**
S	school lunch,mostaccioli	3/4 C	179	277	14.6	0.35	16.1	**34**	742	812	1388	738	924	174	1.18	8.48	0.52	0.14	**10**
			0.56	117.1	17.2	1.93	4.0		549	1335	386	652	207	660	6.52	0.14	54	17	**28**

Riboflavin mg / Niacin mg	% USRDA	Vitamin B_6 mg / Folacin mcg	% USRDA	Vitamin B_{12} mcg / Pantothenic acid mg	% USRDA	Biotin mg / Vitamin A IU	% USRDA	Preformed A RE / Beta carotene RE	Vitamin D IU / Vitamin E IU	% USRDA	Total tocopherol mg / Alpha tocopherol mg	Other tocopherol mg / Total ash g	Calcium mg / Phosphorus mg	% USRDA	Sodium mg / Sodium meq	Potassium mg / Potassium meq	Chlorine mg / Chlorine meq	Iron mg / Magnesium mg	% USRDA	Zinc mg / Copper mg	% USRDA	Iodine mcg / Selenium mcg	% USRDA	Manganese mcg / Chromium mcg	
tr	**0**	unk.	**unk.**	unk.	**unk.**	unk.	**unk.**	unk.	unk.	**unk.**	unk.	unk.	2	**0**	unk.	unk.	unk.	0.2	**1**	unk.	**unk.**	unk.	**unk.**	unk.	A
0.2	**1**	unk.	**unk.**	unk.	**unk.**	unk.	**unk.**	unk.	unk.	**unk.**	unk.	unk.	5	**1**	unk.	unk.	unk.	unk.	**unk.**	unk.	**unk.**	tr(a)		tr(a)	
0.05	**3**	0.05	**3**	0.41	**7**	tr(a)	**0**	tr(a)	17	**4**	0.1	0.1	3	**0**	347	62	519	0.4	**3**	0.8	**6**	unk.	**unk.**	16	B
1.0	**5**	tr(a)	**0**	tr(a)	**0**	tr(a)	**0**	0(a)	0.1	**1**	0.1	1.02	38	**4**	15.1	1.6	14.6	5	**1**	0.02	**1**	tr(a)		2	
unk.	**unk.**	unk.	**unk.**	0.00	**0**	unk.	**unk.**	0(a)	0(a)	**0**	unk.	unk.	32	**3**	tr	16	unk.	0.6	**3**	0.1	**0**	unk.	**unk.**	unk.	C
0.1	**0**	unk.	**unk.**	unk.	**unk.**	77	**2**	unk.	unk.	**unk.**	unk.	0.14	2	**0**	0.0	0.4	unk.	6	**1**	unk.	**unk.**	unk.		unk.	
unk.	**unk.**	unk.	**unk.**	0.26	**4**	unk.	**unk.**	unk.	unk.	**unk.**	unk.	unk.	28	**3**	64	114	unk.	0.7	**4**	unk.	**unk.**	unk.	**unk.**	unk.	D
unk.	**unk.**	4	**1**	0.02	**0**	unk.	**unk.**	unk.	unk.	**unk.**	unk.	unk.	81	**8**	2.8	2.9	unk.	unk.	**unk.**	unk.	**unk.**	tr(a)		tr(a)	
unk.	**unk.**	unk.	**unk.**	0.26	**4**	unk.	**unk.**	unk.	unk.	**unk.**	1.5	1.3	unk.	**unk.**	unk.	unk.	unk.	unk.	**unk.**	unk.	**unk.**	unk.	**unk.**	unk.	E
unk.	**unk.**	4	**1**	0.02	**0**	unk.	**unk.**	unk.	1.8	**6**	0.1	0.70	unk.	**unk.**	unk.	unk.	unk.	8	**2**	unk.	**unk.**	tr(a)		tr(a)	
0.04	**2**	0.02	**1**	0.00	**0**	0.001	**0**	0	0	**0**	0.4	tr	29	**3**	115	162	23	1.3	**7**	0.2	**2**	10.2	**7**	181	F
0.4	**2**	5	**1**	0.13	**1**	293	**6**	4	0.5	**2**	0.4	0.89	30	**3**	5.0	4.1	0.7	25	**6**	0.03	**2**	3		19	
0.22	**13**	0.13	**7**	0.75	**13**	tr	**0**	6	7	**2**	0.2	tr(a)	62	**6**	507	315	19	2.6	**15**	2.7	**18**	69.2	**46**	106	G
4.0	**20**	9	**2**	0.31	**3**	44	**1**	tr	0.3	**1**	0.2	2.70	191	**19**	22.0	8.1	0.5	23	**6**	0.11	**5**	14		16	
0.25	**15**	0.17	**9**	0.92	**15**	0.001	**0**	8	0	**0**	0.5	tr(a)	44	**4**	272	405	3	3.3	**19**	3.5	**23**	unk.	**unk.**	19	H
5.1	**25**	30	**7**	0.41	**4**	293	**6**	7	0.6	**2**	0.4	1.93	188	**19**	11.8	10.4	0.1	33	**8**	0.09	**5**	3		3	
0.43	**25**	0.12	**6**	0.89	**15**	0.006	**2**	0	47	**12**	0.5	tr	362	**36**	973	269	56	1.7	**10**	2.0	**14**	89.8	**60**	202	I
0.9	**5**	29	**7**	1.02	**10**	669	**13**	0	0.6	**2**	0.1	4.18	446	**45**	42.3	6.9	1.6	105	**26**	0.19	**9**	0		1	
0.18	**11**	0.19	**10**	1.18	**20**	0.000	**0**	6	2	**1**	0.1	tr(a)	16	**2**	171	277	3	2.6	**15**	3.5	**23**	32.7	**22**	18	J
4.2	**21**	4	**1**	0.26	**3**	27	**1**	tr	0.2	**1**	0.1	1.31	188	**19**	7.4	7.1	0.1	23	**6**	0.07	**4**	2		3	
0.08	**5**	0.07	**3**	0.12	**2**	tr	**0**	10	0(a)	**0**	0.1	0.0	13	**1**	455	107	3	1.4	**8**	0.9	**6**	unk.	**unk.**	11	K
1.8	**9**	4	**1**	0.17	**2**	110	**2**	8	0.1	**0**	0.0	2.16	116	**12**	19.8	2.7	0.1	8	**2**	0.01	**1**	1		1	
0.26	**15**	0.10	**5**	18.12	**302**	0.001	**0**	26	tr(a)	**0**	0.4	0.0	102	**10**	943	259	640	6.5	**36**	1.8	**12**	56.8	**38**	4	L
2.0	**10**	16	**4**	0.66	**7**	100	**2**	1	0.5	**2**	0.0	4.41	202	**20**	41.0	6.6	18.0	10	**3**	0.03	**2**	tr		tr	
0.30	**18**	0.27	**13**	0.95	**16**	0.001	**0**	5	0	**0**	0.5	0.0	219	**22**	1048	642	unk.	5.4	**30**	4.9	**33**	unk.	**unk.**	unk.	M
4.7	**24**	20	**5**	0.35	**4**	1622	**32**	1	0.6	**2**	0.4	4.81	429	**43**	45.6	16.4	unk.	55	**14**	0.56	**28**	tr(a)		0	
0.30	**17**	0.06	**3**	0.55	**9**	unk.	**unk.**	0	unk.	**unk.**	tr	tr(a)	481	**48**	653	unk.	0	1.5	**8**	1.7	**11**	2	**1**	0	N
1.6	**8**	11	**3**	0.35	**4**	570	**11**	0	0.1	**0**	tr	3.05	289	**29**	28.4	unk.	0.0	20	**5**	0.08	**4**	tr(a)		0	
0.19	**11**	0.24	**12**	0.51	**9**	0.001	**0**	0	98	**25**	tr	0.0	41	**4**	588	412	0	2.4	**13**	2.2	**15**	45.9	**31**	5	O
3.2	**16**	13	**3**	0.41	**4**	851	**17**	10	tr	**0**	0.1	2.03	137	**14**	25.6	10.5	0.0	20	**5**	0.13	**7**	0		3	
0.17	**10**	0.14	**7**	0.89	**15**	tr	**0**	8	0	**0**	0.3	tr(a)	11	**1**	218	329	5	2.5	**14**	3.1	**21**	45.1	**30**	26	P
4.0	**20**	4	**1**	0.20	**2**	29	**1**	tr	0.3	**1**	0.3	1.43	145	**14**	9.5	8.4	0.1	18	**4**	0.07	**3**	3		4	
0.12	**7**	0.04	**2**	0.31	**5**	0.000	**0**	0	tr(a)	**0**	0.3	tr	90	**9**	684	284	504	1.5	**9**	1.9	**13**	unk.	**unk.**	9	Q
2.3	**12**	31	**8**	0.74	**7**	211	**4**	21	0.3	**1**	0.1	3.72	182	**18**	29.8	7.3	14.2	10	**3**	0.03	**2**	tr		8	
0.19	**11**	0.10	**5**	0.14	**2**	0.003	**1**	21	1	**0**	0.6	unk.	141	**14**	126	107	unk.	2.5	**14**	1.6	**11**	unk.	**unk.**	tr(a)	R
1.9	**9**	13	**3**	0.41	**4**	718	**14**	8	0.7	**2**	unk.	1.82	177	**18**	5.5	2.8	unk.	41	**10**	0.08	**4**	unk.		unk.	
0.21	**13**	0.18	**9**	0.59	**10**	0.001	**0**	4	0	**0**	0.1	0.0	171	**17**	962	432	0	2.3	**13**	2.5	**17**	113.2	**76**	15	S
3.0	**15**	7	**2**	0.35	**4**	1130	**23**	84	0.2	**1**	0.1	4.42	286	**29**	41.8	11.0	0.0	33	**8**	0.11	**6**	tr		0	

	FOOD	Portion	Weight in grams / *Conversion for 100 g*	Kilocalories / H_2O g	Total carbohydrate g / Total fats g	Crude fiber g / Dietary fiber g	Total protein g / Total sugar g	% USRDA	Arginine mg / Histidine mg	Isoleucine mg / Leucine mg	Lysine mg / Methionine mg	Phenylalanine mg / Threonine mg	Valine mg / Tryptophan mg	Cystine mg / Tyrosine mg	Polyunsat. fatty acids g / Monounsat. fatty acids g	Saturated fatty acids g / P/S ratio	Linoleic acid g / Cholesterol mg	Thiamin mg / Ascorbic acid mg	% USRDA
A	school lunch,noodleburg	1 average	113	245	28.5	0.11	10.9	**22**	282	592	695	583	530	77	0.49	4.85	0.53	0.16	**11**
			0.88	62.3	9.5	1.20	1.3		253	999	260	431	130	380	3.06	0.10	28	0	**0**
B	school lunch,pepper steak	3/4 C	185	205	7.7	1.08	16.0	**34**	948	727	1300	672	746	182	0.89	4.13	0.52	0.11	**7**
			0.54	146.5	12.3	unk.	3.1		488	1196	393	709	173	524	5.88	0.22	46	82	**136**
C	school lunch,pizzaburger	1 average	141	384	22.4	0.21	21.7	**46**	932	1144	1798	961	1118	215	2.62	8.95	1.25	0.30	**20**
			0.71	66.6	22.6	2.27	2.7		634	1838	519	952	241	716	8.88	0.29	74	7	**12**
D	school lunch,steak fingers	4x1-1/2x	68	178	tr	0.01	14.9	**33**	940	816	1343	654	884	205	1.75	5.81	0.53	0.05	**4**
		3/4"	*1.47*	39.4	12.7	unk.	0.0		504	1284	400	700	187	547	5.63	0.30	114	0	**0**
E	school lunch,tacos	1 average	132	209	14.5	0.68	12.0	**25**	572	581	963	507	640	116	0.90	5.69	0.87	0.09	**6**
			0.76	81.2	11.8	0.83	1.7		369	1029	276	466	137	434	3.85	0.16	39	8	**14**
F	SCONES,home recipe	1 average	39	130	15.2	0.05	3.0	**6**	73	145	130	156	157	61	0.43	3.60	0.34	0.08	**6**
			2.56	14.0	6.3	0.50	2.6		75	242	62	111	39	116	1.72	0.12	56	tr	**0**
G	SCRAPPLE,fried	2-3/4x2-1/8x	25	54	3.6	0.02	2.2	**5**	unk.	unk.	unk.	unk.	unk.	unk.	unk.	unk.	unk.	0.05	**3**
		1/4"	*4.00*	15.3	3.4	unk.	unk.		unk.	unk.	unk.	unk.	unk.	unk.	unk.	unk.	unk.	0	**0**
H	SESAME SEED,decorticated	1 tsp	2.7	16	0.3	0.08	0.7	**1**	90	35	22	41	40	14	unk.	unk.	unk.	0.02	**1**
			37.04	0.1	1.5	unk.	0.0		18	58	24	32	13	30	unk.	unk.	0	unk.	**unk.**
I	SHAKE'N BAKE,oven fry seasoned	1 Tbsp	5.0	20	3.1	0.00	0.5	**1**	unk.	22	11	26	20	9	unk.	unk.	unk.	0.01	**0**
	coating mix-General Foods		*20.00*	0.2	0.8	unk.	unk.		9	36	6	13	6	16	unk.	unk.	0(a)	0	**0**
J	SHERBERT,orange	2/3 C	129	181	39.3	tr	1.4	**3**	53	88	115	70	97	13	0.09	1.59	0.05	0.01	**1**
			0.78	85.2	2.6	tr(a)	21.2		39	142	36	66	21	70	0.64	0.06	9	3	**4**
K	sherbert,orange-Baskin-Robbins	2/3 C	129	253	48.0	unk.	2.0	**4**	unk.	unk.	unk.	unk.	unk.	unk.	unk.	unk.	unk.	0.01	**1**
			0.78	unk.	2.0	unk.	48.0		unk.	unk.	unk.	unk.	unk.	unk.	unk.	unk.	unk.	unk.	**unk.**
L	SHORTENING SUBSTITUTE,	1 oz	28	62	0.0	0(a)	0.0	**0**	unk.	0(a)	0(a)	0(a)	0(a)	0(a)	3.55	0.80	unk.	0(a)	**0**
	no-stick-Mazola		*3.52*	17.0	7.1	0(a)	unk.		0(a)	0(a)	0(a)	0(a)	0(a)	0(a)	unk.	4.46	0	0(a)	**0**
M	SHORTENING,for cake mix,(H) soybean,	1 C	205	1812	0.0	0.00	0.0	**0**	0(a)	0(a)	0(a)	0(a)	0(a)	0(a)	28.90	55.76	26.85	0(a)	**0**
	(H) cottonseed		*0.49*	0.0	205.0	0(a)	0(a)		0(a)	0(a)	0(a)	0(a)	0(a)	0(a)	111.11	0.52	0(a)	0(a)	**0**
N	shortening,for confections,(H)	1 C	205	1812	0.0	0.00	0.0	**0**	0(a)	0(a)	0(a)	0(a)	0(a)	0(a)	2.05	187.16	2.05	0(a)	**0**
	coconut &/or (H) palm		*0.49*	0.0	205.0	0(a)	0(a)		0(a)	0(a)	0(a)	0(a)	0(a)	0(a)	4.51	0.01	0(a)	0(a)	**0**
O	shortening,for confections,	1 C	205	1812	0.0	0.00	0.0	**0**	0(a)	0(a)	0(a)	0(a)	0(a)	0(a)	1.02	134.27	1.02	0(a)	**0**
	fractionated palm		*0.49*	0.0	205.0	0(a)	0(a)		0(a)	0(a)	0(a)	0(a)	0(a)	0(a)	60.06	0.01	0(a)	0(a)	**0**
P	shortening,for frying,(H) soybean,	1 C	205	1812	0.0	0.00	0.0	**0**	0(a)	0(a)	0(a)	0(a)	0(a)	0(a)	45.10	31.57	43.46	0(a)	**0**
	(H) cottonseed		*0.49*	0.0	205.0	0(a)	0(a)		0(a)	0(a)	0(a)	0(a)	0(a)	0(a)	119.31	1.43	0(a)	0(a)	**0**
Q	shortening,for heavy duty frying,	1 C	205	1812	0.0	0.00	0.0	**0**	0(a)	0(a)	0(a)	0(a)	0(a)	0(a)	15.37	97.37	15.37	0(a)	**0**
	(H) palm		*0.49*	0.0	205.0	0(a)	0(a)		0(a)	0(a)	0(a)	0(a)	0(a)	0(a)	83.23	0.16	0(a)	0(a)	**0**
R	shortening,for heavy duty frying,	1 C	205	1845	0.0	0.00	0.0	**0**	0(a)	0(a)	0(a)	0(a)	0(a)	0(a)	17.01	92.04	17.01	0(a)	**0**
	beef tallow,cottonseed		*0.49*	0.0	205.0	0(a)	0(a)		0(a)	0(a)	0(a)	0(a)	0(a)	0(a)	70.11	0.18	unk.	0(a)	**0**
S	shortening,heavy frying,(H) soybean,	1 C	205	1812	0.0	0.00	0.0	**0**	0(a)	0(a)	0(a)	0(a)	0(a)	0(a)	68.67	37.72	63.75	0(a)	**0**
	high linoleic		*0.49*	0.0	205.0	0(a)	0(a)		0(a)	0(a)	0(a)	0(a)	0(a)	0(a)	89.58	1.82	0(a)	0(a)	**0**

Row	Riboflavin mg / Niacin mg	% USRDA	Vitamin B_6 mg / Folacin mcg	% USRDA	Vitamin B_{12} mcg / Pantothenic acid mg	% USRDA	Biotin mg / Vitamin A IU	% USRDA	Preformed A RE / Beta carotene RE	Vitamin D IU / Vitamin E IU	% USRDA	Total tocopherol mg / Alpha tocopherol mg	Other tocopherol mg / Total ash g	Calcium mg / Phosphorus mg	% USRDA	Sodium mg / Sodium meq	Potassium mg / Potassium meq	Chlorine mg / Chlorine meq	Iron mg / Magnesium mg	% USRDA	Zinc mg / Copper mg	% USRDA	Iodine mcg / Selenium mcg	% USRDA	Manganese mcg / Chromium mcg
A	0.17	**10**	0.06	**3**	0.23	**4**	0.001	**0**	4	tr	**0**	0.3	0(a)	116	**12**	475	87	unk.	1.3	**7**	0.7	**5**	unk.	**unk.**	unk.
	1.3	**6**	22	**5**	0.28	**3**	255	**5**	1	0.4	**1**	0.0	1.82	136	**14**	20.6	2.2	unk.	25	**6**	0.04	**2**	unk.		unk.
B	0.18	**11**	0.31	**15**	0.80	**13**	0.001	**0**	4	0	**0**	0.4	tr(a)	26	**3**	505	393	9	2.5	**14**	2.4	**16**	52.6	**35**	82
	3.2	**16**	25	**6**	0.35	**4**	270	**5**	26	0.5	**2**	0.1	2.39	157	**16**	21.9	10.0	0.3	30	**8**	0.07	**3**	1		12
C	0.27	**16**	0.19	**9**	0.97	**16**	0.000	**0**	6	0	**0**	0.4	tr	102	**10**	599	425	0	3.2	**18**	3.3	**22**	28.1	**19**	tr
	4.6	**23**	18	**4**	0.34	**3**	543	**11**	35	0.5	**2**	0.2	2.90	216	**22**	26.1	10.9	0.0	34	**9**	0.09	**5**	0		1
D	0.16	**9**	0.13	**6**	0.83	**14**	0.002	**1**	6	5	**1**	0.3	tr(a)	14	**1**	102	266	22	2.1	**11**	2.6	**17**	24.2	**16**	5
	3.0	**15**	6	**2**	0.31	**3**	50	**1**	tr	0.4	**1**	0.2	0.99	138	**14**	4.4	6.8	0.6	42	**11**	0.05	**3**	0		0.2
E	0.15	**9**	0.12	**6**	0.44	**7**	0.002	**1**	3	0	**0**	0.2	0.1	147	**15**	610	302	11	2.2	**12**	1.8	**12**	42.5	**28**	67
	2.1	**10**	14	**4**	0.23	**2**	743	**15**	33	0.3	**1**	0.2	2.89	213	**21**	26.5	7.7	0.3	41	**10**	0.07	**4**	0		1
F	0.08	**5**	0.02	**1**	0.11	**2**	0.002	**1**	0	3	**1**	0.7	0.3	29	**3**	218	38	27	0.7	**4**	0.3	**2**	29.5	**20**	71
	0.6	**3**	6	**2**	0.21	**2**	211	**4**	0	0.9	**3**	0.3	0.94	61	**6**	9.5	1.0	0.8	26	**7**	0.03	**1**	9		12
G	0.02	**1**	unk.	**unk.**	unk.	**unk.**	unk.	**unk.**	unk.	0	**0**	unk.	unk.	1	**0**	unk.	unk.	unk.	0.3	**2**	unk.	**unk.**	unk.	**unk.**	unk.
	0.4	**2**	unk.	**unk.**	unk.	**unk.**	unk.	**unk.**	unk.	unk.	**unk.**	unk.	0.42	16	**2**	unk.	unk.	unk.	unk.	**unk.**	unk.	**unk.**	tr(a)		tr(a)
H	tr	**0**	tr	**0**	0.00	**0**	unk.	**unk.**	0(a)	0(a)	**0**	unk.	unk.	4	**0**	1	11	unk.	0.2	**1**	0.3	**2**	unk.	**unk.**	unk.
	0.1	**1**	unk.	**unk.**	0.002	**0**	2	**0**	unk.	unk.	**unk.**	unk.	0.13	21	**2**	0.0	0.3	unk.	9	**2**	unk.	**unk.**	unk.		unk.
I	tr	**0**	0.00	**0**	0.00	**0**	unk.	**unk.**	0(a)	0(a)	**0**	unk.	unk.	4	**0**	170	3	unk.	0.1	**1**	unk.	**unk.**	unk.	**unk.**	unk.
	0.1	**0**	unk.	**unk.**	0.01	**0**	2	**0**	unk.	unk.	**unk.**	unk.	0.48	2	**0**	7.4	0.1	unk.	1	**0**	unk.	**unk.**	unk.		unk.
J	0.05	**3**	0.02	**1**	0.10	**2**	unk.	**unk.**	unk.	tr(a)	**0**	unk.	unk.	70	**7**	59	133	unk.	0.2	**1**	0.9	**6**	unk.	**unk.**	tr
	0.1	**0**	9	**2**	0.04	**0**	124	**3**	unk.	unk.	**unk.**	unk.	0.52	49	**5**	2.6	3.4	unk.	10	**3**	0.03	**1**	unk.		tr
K	0.03	**2**	unk.	**unk.**	unk.	**unk.**	unk.	**unk.**	unk.	tr	**0**	unk.	unk.	76	**8**	29	unk.	unk.	unk.	**unk.**	unk.	**unk.**	unk.	**unk.**	unk.
	0.1	**0**	unk.	**unk.**	unk.	**unk.**	82	**2**	unk.	unk.	**unk.**	unk.	unk.	58	**6**	1.3	unk.	unk.	unk.	**unk.**	unk.	**unk.**	unk.		unk.
L	0(a)	**0**	0(a)	**0**	0(a)	**0**	0(a)	**0**	0(a)	0(a)	**0**	0.0	0(a)	0(a)	**0**	0(a)	0(a)	0(a)	0(a)	**0**	0(a)	**0**	0(a)	**0**	0(a)
	0(a)	**0**	0(a)	**0**	0(a)	**0**	0(a)	**0**	0(a)	0.0	**0**	0(a)	0.00	0(a)	**0**	0.0	0.0	0.0	0(a)	**0**	0(a)	**0**	0(a)		0(a)
M	0(a)	**0**	0(a)	**0**	0(a)	**0**	0(a)	**0**	unk.	unk.	**unk.**	unk.	unk.	unk.	**unk.**	unk.	unk.	unk.	unk.	**unk.**	unk.	**unk.**	unk.	**unk.**	unk.
	0(a)	**0**	0(a)	**0**	0(a)	**0**	unk.	**unk.**	unk.	unk.	**unk.**	unk.	0.00	unk.	**unk.**	unk.	unk.	unk.	unk.	**unk.**	unk.	**unk.**	unk.		unk.
N	0(a)	**0**	0(a)	**0**	0(a)	**0**	0(a)	**0**	unk.	unk.	**unk.**	unk.	unk.	unk.	**unk.**	unk.	unk.	unk.	unk.	**unk.**	unk.	**unk.**	unk.	**unk.**	unk.
	0(a)	**0**	0(a)	**0**	0(a)	**0**	unk.	**unk.**	unk.	unk.	**unk.**	unk.	0.00	unk.	**unk.**	unk.	unk.	unk.	unk.	**unk.**	unk.	**unk.**	unk.		unk.
O	0(a)	**0**	0(a)	**0**	0(a)	**0**	0(a)	**0**	unk.	unk.	**unk.**	unk.	unk.	unk.	**unk.**	unk.	unk.	unk.	unk.	**unk.**	unk.	**unk.**	unk.	**unk.**	unk.
	0(a)	**0**	0(a)	**0**	0(a)	**0**	unk.	**unk.**	unk.	unk.	**unk.**	unk.	0.00	unk.	**unk.**	unk.	unk.	unk.	unk.	**unk.**	unk.	**unk.**	unk.		unk.
P	0(a)	**0**	0(a)	**0**	0(a)	**0**	0(a)	**0**	unk.	unk.	**unk.**	unk.	unk.	unk.	**unk.**	unk.	unk.	unk.	unk.	**unk.**	unk.	**unk.**	unk.	**unk.**	unk.
	0(a)	**0**	0(a)	**0**	0(a)	**0**	unk.	**unk.**	unk.	unk.	**unk.**	unk.	0.00	unk.	**unk.**	unk.	unk.	unk.	unk.	**unk.**	unk.	**unk.**	unk.		unk.
Q	0(a)	**0**	0(a)	**0**	0(a)	**0**	0(a)	**0**	unk.	unk.	**unk.**	unk.	unk.	unk.	**unk.**	unk.	unk.	unk.	unk.	**unk.**	unk.	**unk.**	unk.	**unk.**	unk.
	0(a)	**0**	0(a)	**0**	0(a)	**0**	unk.	**unk.**	unk.	unk.	**unk.**	unk.	0.00	unk.	**unk.**	unk.	unk.	unk.	unk.	**unk.**	unk.	**unk.**	unk.		unk.
R	0(a)	**0**	0(a)	**0**	0(a)	**0**	0(a)	**0**	unk.	unk.	**unk.**	unk.	unk.	unk.	**unk.**	unk.	unk.	unk.	unk.	**unk.**	unk.	**unk.**	unk.	**unk.**	unk.
	0(a)	**0**	0(a)	**0**	0(a)	**0**	unk.	**unk.**	unk.	unk.	**unk.**	unk.	0.00	unk.	**unk.**	unk.	unk.	unk.	unk.	**unk.**	unk.	**unk.**	unk.		unk.
S	0(a)	**0**	0(a)	**0**	0(a)	**0**	0(a)	**0**	unk.	unk.	**unk.**	unk.	unk.	unk.	**unk.**	unk.	unk.	unk.	unk.	**unk.**	unk.	**unk.**	unk.	**unk.**	unk.
	0(a)	**0**	0(a)	**0**	0(a)	**0**	unk.	**unk.**	unk.	unk.	**unk.**	unk.	0.00	unk.	**unk.**	unk.	unk.	unk.	unk.	**unk.**	unk.	**unk.**	unk.		unk.

	FOOD	Portion	Weight in grams / *Conversion for 100 g*	Kilocalories / H_2O g	Total carbohydrate g / Total fats g	Crude fiber g / Dietary fiber g	Total protein g / Total sugar g	**% USRDA**	Arginine mg / Histidine mg	Isoleucine mg / Leucine mg	Lysine mg / Methionine mg	Phenylalanine mg / Threonine mg	Valine mg / Tryptophan mg	Cystine mg / Tyrosine mg	Polyunsat. fatty acids g / Monounsat. fatty acids g	Saturated fatty acids g / P/S ratio	Linoleic acid g / Cholesterol mg	Thiamin mg / Ascorbic acid mg	**% USRDA**
A	shortening,heavy frying,(H) soybean,	1 C	205	1812	0.0	0.00	0.0	**0**	0(a)	0(a)	0(a)	0(a)	0(a)	0(a)	0.82	43.25	0.61	0(a)	**0**
	low linoleic		*0.49*	0.0	205.0	0(a)	0(a)		0(a)	0(a)	0(a)	0(a)	0(a)	0(a)	151.08	0.02	0(a)	0(a)	**0**
B	shortening,household,(H) soybean &	1 C	205	1812	0.0	0.00	0.0	**0**	0(a)	0(a)	0(a)	0(a)	0(a)	0(a)	29.11	62.73	27.67	0(a)	**0**
	palm		*0.49*	0.0	205.0	0(a)	0(a)		0(a)	0(a)	0(a)	0(a)	0(a)	0(a)	103.73	0.46	0(a)	0(a)	**0**
C	shortening,household,(H) soybean,(H)	1 C	205	1812	0.0	0.00	0.0	**0**	0(a)	0(a)	0(a)	0(a)	0(a)	0(a)	53.50	51.25	50.22	0(a)	**0**
	cottonseed		*0.49*	0.0	205.0	0(a)	0(a)		0(a)	0(a)	0(a)	0(a)	0(a)	0(a)	91.22	1.04	0(a)	0(a)	**0**
D	shortening,household,lard &	1 C	205	1845	0.0	0.00	0.0	**0**	0(a)	0(a)	0(a)	0(a)	0(a)	0(a)	22.34	82.61	19.88	0(a)	**0**
	vegetable oil		*0.49*	0.0	205.0	0(a)	0(a)		0(a)	0(a)	0(a)	0(a)	0(a)	0(a)	83.84	0.27	unk.	0(a)	**0**
E	shortening,industrial,(H) soybean &	1 C	205	1812	0.0	0.00	0.0	**0**	0(a)	0(a)	0(a)	0(a)	0(a)	0(a)	24.60	52.48	23.37	0(a)	**0**
	cottonseed		*0.49*	0.0	205.0	0(a)	0(a)		0(a)	0(a)	0(a)	0(a)	0(a)	0(a)	118.90	0.47	0(a)	0(a)	**0**
F	shortening,industrial,lard &	1 C	205	1845	0.0	0.00	0.0	**0**	0(a)	0(a)	0(a)	0(a)	0(a)	0(a)	39.36	73.18	37.10	0(a)	**0**
	vegetable oil		*0.49*	0.0	205.0	0(a)	0(a)		0(a)	0(a)	0(a)	0(a)	0(a)	0(a)	77.90	0.54	unk.	0(a)	**0**
G	shortening,vegetable,(H)	1 C	200	1768	0.0	0.00	0.0	**0**	unk.	0(a)	0(a)	0(a)	0(a)	0(a)	52.20	50.00	49.00	0.00	**0**
			0.50	0.0	200.0	0(a)	unk.		0(a)	0(a)	0(a)	0(a)	0(a)	0(a)	89.00	1.04	0	0	**0**
H	SHRIMP/lobster paste	1 Tbsp	21	38	0.3	0(a)	4.4	**10**	unk.	unk.	unk.	unk.	unk.	unk.	unk.	unk.	unk.	unk.	**unk.**
			4.76	12.9	2.0	0(a)	0.0		unk.	unk.	unk.	unk.	unk.	unk.	unk.	unk.	unk.	unk.	**unk.**
I	shrimp,breaded,French fried	1 lg	8.0	18	0.8	0(a)	1.6	**4**	135	55	82	37	36	unk.	unk.	0.56	unk.	tr	**0**
			12.50	4.6	0.9	unk.	0(a)		22	84	29	45	12	29	unk.	unk.	11	unk.	**unk.**
J	shrimp,canned,drained solids	1/2 C	55	64	0.4	0.11	13.3	**30**	1103	679	1171	492	706	168	unk.	unk.	unk.	0.01	**0**
			1.82	38.7	0.6	unk.	0.0		250	1011	386	573	133	483	unk.	unk.	82	0	**0**
K	shrimp,home recipe,jambalaya	3/4 C	170	188	25.5	0.17	11.1	**23**	684	509	797	409	562	132	0.46	1.55	0.41	0.19	**13**
			0.59	126.0	4.7	8.36	5.8		207	784	262	435	99	380	1.77	0.30	50	19	**32**
L	SMELT,Atlantic,jack & bay,cooked	4 smelts	80	78	0.0	0.00	14.9	**33**	841	745	1297	552	790	unk.	unk.	unk.	unk.	0.01	**1**
			1.25	63.2	1.7	0(a)	unk.		unk.	1118	432	641	149	unk.	unk.	unk.	unk.	unk.	**unk.**
M	SNACK FOOD,bacon rinds-Wonder	1 oz	28	140	0.5	0.20	19.4	**30**	1156	330	892	5	643	250	tr(a)	unk.	unk.	0.00	**0**
			3.52	0.8	7.6	unk.	0.0		281	810	182	411	61	273	unk.	unk.	unk.	0	**0**
N	snack food,Bugles-General Mills	47 pieces	28	161	14.9	unk.	1.6	**3**	unk.	unk.	unk.	unk.	unk.	unk.	unk.	unk.	unk.	0.02	**1**
			3.52	unk.	10.6	unk.	unk.		unk.	unk.	unk.	unk.	unk.	unk.	unk.	unk.	unk.	unk.	**unk.**
O	snack food,Chee Tos,cheese puffs,fried	24 pieces	29	163	14.9	0.23	2.2	**3**	57	73	51	88	99	unk.	1.58	3.46	1.44	0.01	**1**
	-Frito Lay		*3.47*	0.6	10.5	unk.	unk.		27	247	39	60	19	93	4.90	0.46	0	0	**0**
P	snack food,cheese twists-Nabisco	3/4 oz pkg	21	105	12.4	unk.	1.6	**2**	unk.	unk.	unk.	unk.	unk.	unk.	unk.	unk.	unk.	0.02	**2**
			4.76	1.0	5.5	unk.	unk.		unk.	unk.	unk.	unk.	unk.	unk.	unk.	unk.	unk.	unk.	**unk.**
Q	snack food,cheese twists-Wonder	1 C	12	65	6.2	unk.	1.0	**2**	unk.	unk.	unk.	unk.	unk.	unk.	unk.	unk.	unk.	0.01	**1**
			8.33	0.2	4.0	unk.	unk.		unk.	unk.	unk.	unk.	unk.	unk.	unk.	unk.	unk.	0	**0**
R	snack food,corn chips-Nabisco	1 oz pkg	28	157	16.0	unk.	1.7	**3**	unk.	unk.	unk.	unk.	unk.	unk.	unk.	unk.	unk.	0.01	**1**
			3.52	0.5	9.5	unk.	unk.		unk.	unk.	unk.	unk.	unk.	unk.	unk.	unk.	unk.	unk.	**unk.**
S	snack food,corn Diggers,popcorn	36 pieces	29	151	16.8	unk.	1.7	**3**	unk.	unk.	unk.	unk.	unk.	unk.	unk.	unk.	unk.	0.02	**2**
	tastin snack-Nabisco		*3.47*	1.0	8.5	unk.	unk.		unk.	unk.	unk.	unk.	unk.	unk.	unk.	unk.	unk.	unk.	**unk.**

Riboflavin mg / Niacin mg	% USRDA	Vitamin B_6 mg / Folacin mcg	% USRDA	Vitamin B_{12} mcg / Pantothenic acid mg	% USRDA	Biotin mg / Vitamin A IU	% USRDA	Preformed A RE / Beta carotene RE	Vitamin D IU / Vitamin E IU	% USRDA	Total tocopherol mg / Alpha tocopherol mg	Other tocopherol mg / Total ash g	Calcium mg / Phosphorus mg	% USRDA	Sodium mg / Sodium meq	Potassium mg / Potassium meq	Chlorine mg / Chlorine meq	Iron mg / Magnesium mg	% USRDA	Zinc mg / Copper mg	% USRDA	Iodine mcg / Selenium mcg	% USRDA	Manganese mcg / Chromium mcg	
0(a)	**0**	0(a)	**0**	0(a)	**0**	0(a)	**0**	unk.	unk.	**unk.**	unk.	unk.	unk.	**unk.**	unk.	unk.	unk.	unk.	**unk.**	unk.	**unk.**	unk.	**unk.**	unk.	A
0(a)	**0**	0(a)	**0**	0(a)	**0**	unk.	**unk.**	unk.	unk.	**unk.**	unk.	0.00	unk.	**unk.**	unk.	unk.	unk.	unk.	**unk.**	unk.	**unk.**	unk.		unk.	
0(a)	**0**	0(a)	**0**	0(a)	**0**	0(a)	**0**	unk.	unk.	**unk.**	unk.	unk.	unk.	**unk.**	unk.	unk.	unk.	unk.	**unk.**	unk.	**unk.**	unk.	**unk.**	unk.	B
0(a)	**0**	0(a)	**0**	0(a)	**0**	unk.	**unk.**	unk.	unk.	**unk.**	unk.	0.00	unk.	**unk.**	unk.	unk.	unk.	unk.	**unk.**	unk.	**unk.**	unk.		unk.	
0(a)	**0**	0(a)	**0**	0(a)	**0**	0(a)	**0**	unk.	unk.	**unk.**	unk.	unk.	unk.	**unk.**	unk.	unk.	unk.	unk.	**unk.**	unk.	**unk.**	unk.	**unk.**	unk.	C
0(a)	**0**	0(a)	**0**	0(a)	**0**	unk.	**unk.**	unk.	unk.	**unk.**	unk.	0.00	unk.	**unk.**	unk.	unk.	unk.	unk.	**unk.**	unk.	**unk.**	unk.		unk.	
0(a)	**0**	0(a)	**0**	0(a)	**0**	0(a)	**0**	unk.	unk.	**unk.**	unk.	unk.	unk.	**unk.**	unk.	unk.	unk.	unk.	**unk.**	unk.	**unk.**	unk.	**unk.**	unk.	D
0(a)	**0**	0(a)	**0**	0(a)	**0**	unk.	**unk.**	unk.	unk.	**unk.**	unk.	0.00	unk.	**unk.**	unk.	unk.	unk.	unk.	**unk.**	unk.	**unk.**	unk.		unk.	
0(a)	**0**	0(a)	**0**	0(a)	**0**	0(a)	**0**	unk.	unk.	**unk.**	unk.	unk.	unk.	**unk.**	unk.	unk.	unk.	unk.	**unk.**	unk.	**unk.**	unk.	**unk.**	unk.	E
0(a)	**0**	0(a)	**0**	0(a)	**0**	unk.	**unk.**	unk.	unk.	**unk.**	unk.	0.00	unk.	**unk.**	unk.	unk.	unk.	unk.	**unk.**	unk.	**unk.**	unk.		unk.	
0(a)	**0**	0(a)	**0**	0(a)	**0**	0(a)	**0**	unk.	unk.	**unk.**	unk.	unk.	unk.	**unk.**	unk.	unk.	unk.	unk.	**unk.**	unk.	**unk.**	unk.	**unk.**	unk.	F
0(a)	**0**	0(a)	**0**	0(a)	**0**	unk.	**unk.**	unk.	unk.	**unk.**	unk.	0.00	unk.	**unk.**	unk.	unk.	unk.	unk.	**unk.**	unk.	**unk.**	unk.		unk.	
0.00	**0**	0(a)	**0**	0(a)	**0**	0(a)	**0**	0	0(a)	**0**	unk.	unk.	0	**0**	0	0	tr(a)	0.0	**0**	tr(a)	**0**	tr(a)	**0**	tr(a)	G
0.0	**0**	0(a)	**0**	0(a)	**0**	0	**0**	0	unk.	**unk.**	unk.	0.00	0	**0**	0.0	0.0	0.0	tr(a)	**0**	tr(a)	**0**	tr(a)		tr(a)	
0.05	**3**	unk.	**unk.**	unk.	**unk.**	unk.	**unk.**	unk.	unk.	**unk.**	unk.	unk.	unk.	**unk.**	unk.	unk.	unk.	unk.	**unk.**	unk.	**unk.**	unk.	**unk.**	unk.	H
unk.	**unk.**	unk.	**unk.**	unk.	**unk.**	unk.	**unk.**	unk.	unk.	**unk.**	unk.	1.47	unk.	**unk.**	unk.	unk.	unk.	unk.	**unk.**	unk.	**unk.**	tr(a)		tr(a)	
0.01	**0**	0.00	**0**	0.06	**1**	unk.	**unk.**	unk.	12	**3**	0.5	0.3	6	**1**	15	18	unk.	0.2	**1**	0.1	**1**	unk.	**unk.**	unk.	I
0.2	**1**	1	**0**	0.01	**0**	unk.	**unk.**	unk.	0.6	**2**	0.2	0.16	15	**2**	0.6	0.5	unk.	4	**1**	0.03	**2**	tr(a)		tr(a)	
0.02	**1**	0.03	**2**	0.38	**6**	unk.	**unk.**	9	82	**21**	0.3	tr(a)	63	**6**	77	67	unk.	1.7	**10**	1.2	**8**	unk.	**unk.**	unk.	J
1.0	**5**	8	**2**	0.12	**1**	33	**1**	tr	0.3	**1**	tr(a)	1.98	145	**15**	3.3	1.7	unk.	41	**18**	0.22	**11**	tr(a)		tr(a)	
0.07	**4**	0.20	**10**	0.22	**4**	0.002	**1**	5	47	**12**	0.2	0.0	67	**7**	83	422	tr(a)	3.1	**17**	1.1	**7**	unk.	**unk.**	tr	K
2.9	**14**	15	**4**	0.36	**4**	1256	**25**	124	0.2	**1**	tr	2.41	151	**15**	3.6	10.8	0.0	38	**10**	0.13	**6**	tr(a)		tr(a)	
0.10	**6**	unk.	**unk.**	unk.	**unk.**	unk.	**unk.**	unk.	unk.	**unk.**	unk.	unk.	unk.	**unk.**	unk.	unk.	unk.	0.3	**2**	unk.	**unk.**	unk.	**unk.**	unk.	L
1.1	**6**	unk.	**unk.**	unk.	**unk.**	unk.	**unk.**	unk.	unk.	**unk.**	unk.	0.88	218	**22**	unk.	unk.	unk.	unk.	**unk.**	unk.	**unk.**	unk.		unk.	
0.00	**0**	0.00	**0**	0.00	**0**	unk.	**unk.**	20	0(a)	**0**	0.2	unk.	7	**1**	246	49	unk.	0.7	**4**	0.0	**0**	unk.	**unk.**	0	M
0.6	**3**	unk.	**unk.**	0.00	**0**	66	**1**	0	0.2	**1**	unk.	1.42	13	**1**	10.7	1.3	unk.	4	**1**	0.09	**5**	unk.		10	
0.01	**1**	unk.	**unk.**	0(a)	**0**	unk.	**unk.**	unk.	0(a)	**0**	unk.	unk.	2	**0**	276	14	unk.	0.2	**1**	unk.	**unk.**	unk.	**unk.**	unk.	N
0.3	**2**	unk.	**unk.**	unk.	**unk.**	unk.	**unk.**	unk.	unk.	**unk.**	unk.	unk.	13	**1**	12.0	0.4	unk.	unk.	**unk.**	unk.	**unk.**	unk.		unk.	
0.04	**2**	unk.	**unk.**	0(a)	**0**	unk.	**unk.**	tr(a)	tr(a)	**0**	2.4	unk.	18	**2**	301	23	unk.	0.2	**1**	0.0	**0**	unk.	**unk.**	245	O
0.2	**1**	unk.	**unk.**	unk.	**unk.**	67	**1**	unk.	2.9	**10**	unk.	0.98	29	**3**	13.1	0.6	unk.	9	**2**	0.03	**2**	unk.		40	
0.05	**3**	unk.	**unk.**	unk.	**unk.**	unk.	**unk.**	unk.	0(a)	**0**	unk.	unk.	17	**2**	216	59	unk.	0.1	**1**	0.1	**1**	unk.	**unk.**	27	P
0.2	**1**	unk.	**unk.**	unk.	**unk.**	unk.	**unk.**	unk.	unk.	**unk.**	unk.	0.65	31	**3**	9.4	1.5	unk.	6	**2**	0.03	**2**	unk.		unk.	
0.03	**2**	0.02	**1**	0.02	**0**	unk.	**unk.**	unk.	unk.	**unk.**	unk.	unk.	17	**2**	141	unk.	unk.	0.1	**1**	unk.	**unk.**	unk.	**unk.**	unk.	Q
0.1	**0**	unk.	**unk.**	0.08	**1**	46	**1**	unk.	unk.	**unk.**	unk.	unk.	22	**2**	6.1	unk.	unk.	4	**1**	unk.	**unk.**	unk.		unk.	
0.03	**2**	unk.	**unk.**	unk.	**unk.**	unk.	**unk.**	unk.	0(a)	**0**	unk.	unk.	33	**3**	129	39	unk.	0.4	**2**	0.4	**2**	unk.	**unk.**	156	R
0.2	**1**	unk.	**unk.**	unk.	**unk.**	unk.	**unk.**	unk.	unk.	**unk.**	unk.	0.66	51	**5**	5.6	1.0	unk.	21	**5**	0.04	**2**	unk.		unk.	
0.02	**1**	unk.	**unk.**	unk.	**unk.**	unk.	**unk.**	unk.	0(a)	**0**	unk.	unk.	3	**0**	300	25	unk.	0.3	**2**	0.1	**1**	unk.	**unk.**	69	S
0.2	**1**	unk.	**unk.**	unk.	**unk.**	unk.	**unk.**	unk.	unk.	**unk.**	unk.	0.80	18	**2**	13.0	0.6	unk.	6	**2**	0.02	**1**	unk.		unk.	

	FOOD	Portion	Weight in grams / Conversion for 100 g	Kilocalories / H_2O g	Total carbohydrate g / Total fats g	Crude fiber g / Dietary fiber g	Total protein g / Total sugar g	% USRDA	Arginine mg / Histidine mg	Isoleucine mg / Leucine mg	Lysine mg / Methionine mg	Phenylalanine mg / Threonine mg	Valine mg / Tryptophan mg	Cystine mg / Tyrosine mg	Polyunsat. fatty acids g / Monounsat. fatty acids g	Saturated fatty acids g / P/S ratio	Linoleic acid g / Cholesterol mg	Thiamin mg / Ascorbic acid mg	% USRDA
A	snack food,corn Korkers-Nabisco	19 pieces	29	158	15.8	unk.	1.8	3	unk.	unk.	unk.	unk.	unk.	unk.	unk.	unk.	unk.	0.01	**1**
			3.51	0.4	9.7	unk.	unk.		unk.	unk.	unk.	unk.	unk.	unk.	unk.	unk.	unk.	unk.	**unk.**
B	snack food,Doritos,tortilla chips	16 pieces	29	142	18.8	0.32	2.1	3	76	64	45	91	92	35	1.33	0.85	1.31	0.03	**2**
	-Frito Lay		*3.47*	0.0	6.8	unk.	unk.		35	249	30	64	30	55	3.27	1.57	0	0	**0**
C	snack food,egg roll,meat or shrimp	1 whole	28	61	8.6	unk.	2.3	4	unk.	unk.	unk.	unk.	unk.	unk.	unk.	unk.	unk.	unk.	**unk.**
	-La Choy		*3.57*	unk.	2.0	unk.	unk.		unk.	unk.	unk.	unk.	unk.	unk.	unk.	unk.	unk.	unk.	**unk.**
D	snack food,Flings,crispy corn curls	16 pieces	29	162	13.6	unk.	2.4	4	unk.	unk.	unk.	unk.	unk.	unk.	unk.	unk.	unk.	0.01	**1**
	-Nabisco		*3.47*	0.7	10.8	unk.	unk.		unk.	unk.	unk.	unk.	unk.	unk.	unk.	unk.	unk.	unk.	**unk.**
E	snack food,food sticks,chocolate	1 stick	10	41	7.1	unk.	0.9	1	unk.	unk.	unk.	unk.	unk.	unk.	unk.	unk.	unk.	0.03	**2**
	-Pillsbury		*10.00*	0.9	0.9	unk.	unk.		unk.	unk.	unk.	unk.	unk.	unk.	unk.	unk.	unk.	1	**2**
F	snack food,Fritos,corn chips	14 pieces	28	154	15.3	0.28	1.9	3	58	61	38	85	88	unk.	7.67	2.49	2.38	0.01	**1**
	-Frito-Lay		*3.57*	0.1	10.1	unk.	unk.		48	241	31	58	23	62	5.18	3.08	0	0	**0**
G	snack food,Funyuns,onion flavored	12 whole	28	140	19.1	0.32	2.3	4	unk.	76	30	111	99	0(a)	unk.	unk.	unk.	0.02	**1**
	-Frito-Lay	rings	*3.47*	0.5	6.1	unk.	0.5		42	402	46	65	24	84	unk.	unk.	unk.	0	**0**
H	snack food,Munchos-Frito-Lay	18 pieces	28	156	15.5	0.23	1.6	2	62	41	63	52	66	unk.	unk.	unk.	unk.	0.03	**2**
			3.47	0.7	10.1	unk.	unk.		22	149	19	41	17	46	unk.	unk.	0(a)	0	**0**
I	snack food,popcorn w/salt,popped	1 C	14	64	8.3	0.24	1.4	2	unk.	65	40	63	72	unk.	unk.	2.10	0.28	unk.	**unk.**
	in coconut oil		*7.14*	0.4	3.1	0.46	unk.		unk.	182	26	56	9	unk.	0.28	unk.	0	0	**0**
J	snack food,popcorn,plain	1 C	12	46	9.2	0.26	1.5	2	unk.	71	45	69	79	unk.	unk.	0.12	0.36	unk.	**unk.**
			8.33	0.5	0.6	0.40	unk.		unk.	201	29	62	9	unk.	0.12	unk.	0	0	**0**
K	snack food,potato chips	14 pieces	28	159	14.0	0.45	1.5	2	98	65	79	65	79	92	unk.	2.77	5.57	0.06	**4**
			3.57	0.5	11.1	unk.	2.2		22	74	18	61	15	44	2.32	unk.	0(a)	4	**8**
L	snack food,potato sticks	4/5 C	28	154	14.4	0.43	1.8	3	unk.	80	96	80	96	112	unk.	2.56	5.11	0.06	**4**
			3.52	0.4	10.3	unk.	unk.		26	91	22	74	18	54	2.27	unk.	0	11	**19**
M	snack food,potato sticks-Nabisco	1 oz bag	28	155	14.6	unk.	1.9	3	unk.	unk.	unk.	unk.	unk.	unk.	unk.	unk.	unk.	0.05	**3**
			3.52	0.9	9.9	unk.	unk.		unk.	unk.	unk.	unk.	unk.	unk.	unk.	unk.	unk.	unk.	**unk.**
N	snack food,pretzelettes,Mister Salty	1 oz bag	28	103	21.3	unk.	2.8	4	unk.	unk.	unk.	unk.	unk.	unk.	unk.	unk.	unk.	0.05	**4**
	-Nabisco		*3.52*	1.0	0.8	unk.	unk.		unk.	unk.	unk.	unk.	unk.	unk.	unk.	unk.	unk.	unk.	**unk.**
O	snack food,pretzels	1 med	6.0	23	4.6	0.02	0.6	1	14	22	10	31	29	unk.	unk.	0.06	unk.	tr	**0**
			16.67	0.3	0.3	unk.	unk.		19	52	10	17	10	16	unk.	unk.	0	0	**0**
P	snack food,pretzels,bite-size-Nabisco	1 oz bag	28	109	22.5	unk.	2.7	4	unk.	unk.	unk.	unk.	unk.	unk.	unk.	unk.	unk.	0.04	**3**
			3.52	1.0	0.9	unk.	unk.		unk.	unk.	unk.	unk.	unk.	unk.	unk.	unk.	unk.	unk.	**unk.**
Q	snack food,pretzels,Mister Salty	2 pieces	28	104	21.8	unk.	3.0	5	unk.	unk.	unk.	unk.	unk.	unk.	unk.	unk.	unk.	0.04	**3**
	Dutch-Nabisco		*3.57*	1.5	0.5	unk.	0(a)		unk.	unk.	unk.	unk.	unk.	unk.	unk.	unk.	unk.	unk.	**unk.**
R	snack food,pretzels,Mister Salty	19 pieces	28	107	21.7	unk.	2.5	4	unk.	unk.	unk.	unk.	unk.	unk.	unk.	unk.	unk.	0.03	**2**
	little shapes-Nabisco		*3.61*	1.1	1.1	unk.	unk.		unk.	unk.	unk.	unk.	unk.	unk.	unk.	unk.	unk.	unk.	**unk.**
S	snack food,pretzels,Mister Salty	94 pieces	28	104	21.5	unk.	2.9	4	unk.	unk.	unk.	unk.	unk.	unk.	unk.	unk.	unk.	0.10	**7**
	Veri-Thin sticks-Nabisco		*3.55*	1.1	0.8	unk.	unk.		unk.	unk.	unk.	unk.	unk.	unk.	unk.	unk.	unk.	unk.	**unk.**

Riboflavin mg / Niacin mg	% USRDA	Vitamin B_6 mg / Folacin mcg	% USRDA	Vitamin B_{12} mcg / Pantothenic acid mg	% USRDA	Biotin mg / Vitamin A IU	% USRDA	Preformed A RE / Beta carotene RE	Vitamin D IU / Vitamin E IU	% USRDA	Total tocopherol mg / Alpha tocopherol mg	Other tocopherol mg / Total ash g	Calcium mg / Phosphorus mg	% USRDA	Sodium mg / Sodium meq	Potassium mg / Potassium meq	Chlorine mg / Chlorine meq	Iron mg / Magnesium mg	% USRDA	Zinc mg / Copper mg	% USRDA	Iodine mcg / Selenium mcg	% USRDA	Manganese mcg / Chromium mcg	
0.01	1	unk.	unk.	unk.	unk.	unk.	unk.	unk.	0(a)	0	unk.	unk.	30	3	186	40	unk.	0.5	3	0.4	3	unk.	unk.	120	A
0.3	1	unk.	unk.	unk.	unk.	unk.	unk.	unk.	unk.	unk.	unk.	0.74	60	6	8.1	1.0	unk.	23	6	0.04	2	unk.		unk.	
0.03	2	0.12	6	0(a)	0	tr	0	unk.	unk.	unk.	unk.	unk.	24	2	186	50	unk.	0.6	3	0.1	1	unk.	unk.	611	B
0.3	2	4	1	0.07	1	53	1	unk.	unk.	unk.	unk.	0.63	58	6	8.1	1.3	unk.	20	5	0.18	9	unk.		39	
unk.	unk.	unk.	unk.	unk.	unk.	unk.	unk.	unk.	unk.	unk.	unk.	unk.	unk.	unk.	unk.	unk.	unk.	unk.	unk.	unk.	unk.	unk.	unk.	unk.	C
unk.	unk.	unk.	unk.	unk.	unk.	unk.	unk.	unk.	unk.	unk.	unk.	unk.	unk.	unk.	unk.	unk.	unk.	unk.	unk.	unk.	unk.	unk.		unk.	
unk.	4	unk.	unk.	unk.	unk.	unk.	unk.	unk.	0(a)	0	unk.	unk.	52	5	352	65	unk.	0.6	3	0.2	1	unk.	unk.	24	D
0.2	1	unk.	unk.	unk.	unk.	unk.	unk.	unk.	unk.	unk.	unk.	1.25	55	6	15.3	1.6	unk.	8	2	0.02	1	unk.		unk.	
0.06	3	0.05	2	0.12	2	unk.	unk.	unk.	10	3	unk.	unk.	14	1	28	18	unk.	0.4	2	0.1	1	5.0	3	unk.	E
0.5	3	2	1	unk.	unk.	102	2	unk.	0.8	3	0.8	0.15	17	2	1.2	0.5	unk.	10	3	0.03	2	unk.		unk.	
0.02	1	0.06	3	0(a)	0	0.001	0	0	0(a)	0	1.8	unk.	35	4	202	23	unk.	0.3	2	0.1	1	unk.	unk.	53	F
0.3	2	12	3	0.14	1	99	2	10	2.2	7	unk.	0.67	52	5	8.8	0.6	unk.	22	5	0.13	6	unk.		17	
0.03	2	unk.	unk.	unk.	unk.	unk.	unk.	unk.	unk.	unk.	unk.	unk.	11	1	228	64	unk.	0.3	2	0.0	0	unk.	unk.	153	G
0.3	2	unk.	unk.	unk.	unk.	47	1	unk.	2.4	8	2.4	0.75	16	2	9.9	1.6	unk.	7	2	0.05	3	unk.		12	
0.02	1	unk.	unk.	0(a)	0	unk.	unk.	0	0(a)	0	unk.	unk.	9	1	193	79	unk.	0.4	2	0.0	0	unk.	unk.	142	H
0.9	5	unk.	unk.	unk.	unk.	0	0	0	6.3	21	6.3	0.92	13	1	8.4	2.0	unk.	16	4	0.07	4	unk.		81	
0.01	1	0.03	1	0(a)	0	unk.	unk.	0(a)	0(a)	0	unk.	unk.	1	0	272	unk.	unk.	0.3	2	0.4	3	unk.	unk.	tr(a)	I
0.2	1	unk.	unk.	unk.	unk.	0(a)	0	0(a)	unk.	unk.	unk.	0.87	30	3	11.8	unk.	unk.	22	6	0.04	2	unk.		unk.	
0.01	1	0.02	1	0.00	0	unk.	unk.	0(a)	0(a)	0	unk.	unk.	1	0	tr	unk.	unk.	0.3	2	0.5	3	unk.	unk.	tr(a)	J
0.3	1	unk.	unk.	unk.	unk.	0(a)	0	0(a)	unk.	unk.	unk.	0.19	34	3	0.0	unk.	unk.	19	5	0.04	2	unk.		unk.	
0.02	1	0.05	3	0.00	0	unk.	unk.	0	0(a)	0	3.2	1.4	11	1	280	316	unk.	0.5	3	0.2	2	unk.	unk.	39	K
1.3	7	15	4	0.00	0	0	0	0	3.8	13	1.8	0.87	39	4	12.2	8.1	unk.	15	4	0.06	3	unk.		22	
0.02	1	0.05	3	tr	0	unk.	unk.	0(a)	0(a)	0	unk.	unk.	12	1	284	321	unk.	0.5	3	0.2	2	unk.	unk.	unk.	L
1.4	7	4	1	0.00	0	unk.	unk.	unk.	unk.	unk.	unk.	1.39	40	4	12.3	8.2	unk.	unk.	unk.	0.07	4	unk.		unk.	
0.03	2	unk.	unk.	unk.	unk.	unk.	unk.	unk.	0(a)	0	unk.	unk.	7	1	175	469	unk.	0.5	3	0.4	3	unk.	unk.	133	M
1.5	7	unk.	unk.	unk.	unk.	unk.	unk.	unk.	unk.	unk.	unk.	1.15	45	5	7.6	12.0	unk.	19	5	0.17	9	unk.		unk.	
0.02	1	unk.	unk.	unk.	unk.	unk.	unk.	unk.	0(a)	0	unk.	unk.	12	1	853	43	unk.	0.6	3	0.3	2	unk.	unk.	290	N
0.3	1	unk.	unk.	0.13	1	unk.	unk.	unk.	unk.	unk.	unk.	2.49	38	4	37.1	1.1	unk.	10	2	0.07	3	unk.		unk.	
tr	0	tr	0	tr	0	unk.	unk.	0	0(a)	0	0.0	tr	1	0	101	8	unk.	0.1	1	0.1	0	unk.	unk.	114	O
tr	0	1	0	0.03	0	0	0	0	0.1	0	tr	0.32	8	1	4.4	0.2	unk.	1	0	0.01	0	unk.		0	
0.02	1	unk.	unk.	unk.	unk.	unk.	unk.	unk.	0(a)	0	unk.	unk.	10	1	518	38	unk.	0.5	3	0.3	2	unk.	unk.	223	P
0.4	2	unk.	unk.	0.13	1	unk.	unk.	unk.	unk.	unk.	unk.	1.31	35	4	22.5	1.0	unk.	7	2	0.18	9	unk.		unk.	
0.03	2	unk.	unk.	unk.	unk.	unk.	unk.	unk.	0(a)	0	unk.	unk.	7	1	430	40	unk.	0.5	3	0.3	2	unk.	unk.	201	Q
0.4	2	unk.	unk.	unk.	unk.	unk.	unk.	unk.	unk.	unk.	unk.	1.21	34	3	18.7	1.0	unk.	8	2	0.07	3	unk.		unk.	
0.01	1	unk.	unk.	unk.	unk.	unk.	unk.	unk.	0(a)	0	unk.	unk.	8	1	476	26	unk.	0.5	3	0.2	2	unk.	unk.	219	R
0.3	2	unk.	unk.	0.13	1	unk.	unk.	unk.	unk.	unk.	unk.	1.15	34	3	20.7	0.7	unk.	7	2	0.07	4	unk.		unk.	
0.14	8	unk.	unk.	unk.	unk.	unk.	unk.	unk.	0(a)	0	unk.	unk.	9	1	719	38	unk.	1.1	6	0.3	2	unk.	unk.	221	S
1.6	8	unk.	unk.	0.13	1	unk.	unk.	unk.	unk.	unk.	unk.	2.03	34	3	31.3	1.0	unk.	7	2	0.05	3	unk.		unk.	

	FOOD	Portion	Weight in grams / Conversion for 100 g	Kilocalories / H_2O g	Total carbohydrate g / Total fats g	Crude fiber g / Dietary fiber g	Total protein g / Total sugar g	% USRDA	Arginine mg / Histidine mg	Isoleucine mg / Leucine mg	Lysine mg / Methionine mg	Phenylalanine mg / Threonine mg	Valine mg / Tryptophan mg	Cystine mg / Tyrosine mg	Polyunsat. fatty acids g / Monounsat. fatty acids g	Saturated fatty acids g / P/S ratio	Linoleic acid g / Cholesterol mg	Thiamin mg / Ascorbic acid mg	% USRDA
A	snack food,pretzels,Mister Salty	5 pieces	26	100	19.5	unk.	2.5	**4**	unk.	unk.	unk.	unk.	unk.	unk.	unk.	unk.	unk.	0.11	**7**
	Veri-Thin-Nabisco		*3.85*	1.2	1.3	unk.	unk.		unk.	unk.	unk.	unk.	unk.	unk.	unk.	unk.	unk.	unk.	**unk.**
B	snack food,pretzels,slim thin sticks	47 pieces	28	110	21.4	0.08	2.8	**4**	65	105	47	144	137	unk.	unk.	0.28	unk.	0.01	**0**
			3.55	1.3	1.3	unk.	unk.		91	246	45	78	47	76	unk.	unk.	0	0	**0**
C	snack food,pretzels,tiny twist	14 pieces	28	109	21.3	0.08	2.7	**4**	64	104	46	143	136	unk.	unk.	0.28	unk.	0.01	**0**
			3.57	1.3	1.3	unk.	unk.		66	244	45	77	47	75	unk.	unk.	0	0	**0**
D	snack food,tortilla chips-Nabisco	14 pieces	29	142	18.7	unk.	2.4	**4**	unk.	unk.	unk.	unk.	unk.	unk.	unk.	unk.	unk.	0.03	**2**
			3.47	0.5	6.5	unk.	unk.		unk.	unk.	unk.	unk.	unk.	unk.	unk.	unk.	unk.	unk.	**unk.**
E	snack food,tortilla chips,nacho	13 pieces	29	149	17.6	unk.	2.2	**3**	unk.	unk.	unk.	unk.	unk.	unk.	unk.	unk.	unk.	0.42	**28**
	cheese flavored-Nabisco		*3.47*	0.5	7.8	unk.	unk.		unk.	unk.	unk.	unk.	unk.	unk.	unk.	unk.	unk.	unk.	**unk.**
F	SNAIL,raw	1 average	5.0	4	0.1	0(a)	0.8	**2**	unk.	unk.	unk.	unk.	unk.	unk.	unk.	unk.	unk.	unk.	**unk.**
			20.00	4.0	0.1	0(a)	0.0		unk.	unk.	unk.	unk.	unk.	unk.	unk.	unk.	unk.	unk.	**unk.**
G	SOUFFLE,cheese,home recipe	1 C	95	308	6.5	0.02	14.0	**31**	629	682	1232	701	859	154	1.22	14.95	0.91	0.06	**4**
			1.05	49.2	25.2	0.15	2.7		515	1232	376	528	202	713	6.68	0.08	196	tr	**1**
H	souffle,pineapple,home recipe	1 C	98	244	31.0	0.45	4.5	**9**	199	220	286	214	277	101	0.69	7.33	0.55	0.08	**5**
			1.02	49.9	12.0	2.29	25.4		128	362	135	202	63	187	2.93	0.09	141	3	**5**
I	SOUP,CANNED,asparagus,cream of,	10-3/4 oz can	305	210	26.0	1.83	5.5	**9**	207	235	271	232	278	73	4.45	2.50	4.33	0.12	**8**
	condensed		*0.33*	256.4	9.9	unk.	unk.		119	396	101	189	70	183	2.10	1.78	12	7	**11**
J	soup,canned,asparagus,cream of,prep	1 C	248	161	16.4	0.74	6.3	**10**	231	340	432	290	384	67	2.23	3.32	2.13	0.10	**7**
	w/milk		*0.40*	213.3	8.2	unk.	unk.		159	558	144	260	84	270	1.88	0.67	22	4	**7**
K	soup,canned,asparagus,cream of,prep	1 C	244	85	10.7	0.73	2.3	**4**	85	98	112	95	115	29	1.85	1.05	1.78	0.05	**3**
	w/water		*0.41*	224.0	4.1	unk.	unk.		49	163	41	78	29	76	0.88	1.77	5	3	**5**
L	soup,canned,barley w/beef,prep	1 C	245	86	11.0	0.24	6.1	**9**	unk.	unk.	unk.	unk.	unk.	unk.	unk.	unk.	unk.	0.02	**2**
	w/water-Campbell		*0.41*	unk.	1.2	unk.	unk.		unk.	unk.	unk.	unk.	unk.	unk.	unk.	unk.	tr(a)	2	**4**
M	soup,canned,bean w/bacon,condensed	11-1/2 oz can	326	421	55.3	4.24	19.2	**30**	1004	936	1304	1069	1053	212	4.43	3.72	3.33	0.23	**15**
			0.31	228.8	14.4	unk.	tr(a)		499	1578	241	792	202	574	4.96	1.19	7	4	**7**
N	soup,canned,bean w/bacon,prep	1 C	253	172	22.8	1.52	7.9	**12**	415	385	536	440	435	89	1.82	1.52	1.37	0.10	**7**
	w/water		*0.40*	212.9	5.9	unk.	tr(a)		205	650	99	326	83	235	2.05	1.20	3	2	**3**
O	soup,canned,bean w/frankfurters,	11-1/4 oz can	319	453	53.4	4.15	24.2	**37**	1273	1183	1652	1353	1333	268	3.99	5.14	3.03	0.26	**17**
	condensed		*0.31*	216.0	16.9	unk.	tr(a)		632	1997	306	1002	255	724	6.16	0.78	29	2	**4**
P	soup,canned,bean w/frankfurters,prep	1 C	250	188	22.0	1.50	10.0	**15**	525	488	680	558	550	110	1.65	2.12	1.25	0.10	**7**
	w/water		*0.40*	207.6	7.0	unk.	tr(a)		260	823	125	413	105	298	2.52	0.78	13	1	**2**
Q	soup,canned,bean w/ham,chunky,	1 C	243	231	27.1	unk.	12.6	**19**	unk.	unk.	unk.	unk.	unk.	unk.	0.95	3.33	0.87	0.15	**10**
	ready-to-serve		*0.41*	191.1	8.5	unk.	tr(a)		unk.	unk.	unk.	unk.	unk.	unk.	3.50	0.28	22	4	**7**
R	soup canned,bean,black,condensed	11 oz can	312	284	48.1	unk.	15.1	**23**	883	768	1108	833	761	156	1.28	1.06	0.94	0.12	**8**
			0.32	235.0	4.1	unk.	tr(a)		440	1126	168	665	172	462	1.40	1.21	0	1	**2**
S	soup,canned,bean,black,prep w/water	1 C	247	116	19.8	1.31	5.6	**9**	331	287	415	311	284	59	0.47	0.40	0.35	0.07	**5**
			0.40	215.6	1.5	unk.	tr(a)		163	422	62	249	64	173	0.52	1.19	0	1	**2**

Riboflavin mg / Niacin mg	% USRDA	Vitamin B_6 mg / Folacin mcg	% USRDA	Vitamin B_{12} mcg / Pantothenic acid mg	% USRDA	Biotin mg / Vitamin A IU	% USRDA	Preformed A RE / Beta carotene RE	Vitamin D IU / Vitamin E IU	% USRDA	Total tocopherol mg / Alpha tocopherol mg	Other tocopherol mg / Total ash g	Calcium mg / Phosphorus mg	% USRDA	Sodium mg / Sodium meq	Potassium mg / Potassium meq	Chlorine mg / Chlorine meq	Iron mg / Magnesium mg	% USRDA	Zinc mg / Copper mg	% USRDA	Iodine mcg / Selenium mcg	% USRDA	Manganese mcg / Chromium mcg	
0.13	**8**	unk.	**unk.**	unk.	**unk.**	unk.	**unk.**	unk.	0(a)	**0**	unk.	unk.	8	**1**	566	33	unk.	1.0	**6**	0.2	**2**	unk.	**unk.**	180	A
1.5	**7**	unk.	**unk.**	0.12	**1**	unk.	**unk.**	unk.	unk.	**unk.**	unk.	1.52	30	**3**	24.6	0.8	unk.	7	**2**	0.05	**2**	unk.		unk.	
0.01	**1**	0.01	**0**	tr	**0**	unk.	**unk.**	0	0(a)	**0**	0.2	0.2	6	**1**	474	37	unk.	0.4	**2**	0.3	**2**	unk.	**unk.**	536	B
0.2	**1**	6	**1**	0.15	**2**	0	**0**	0	0.3	**1**	tr	1.49	37	**4**	20.6	0.9	unk.	11	**3**	0.04	**2**	unk.		0	
0.01	**1**	0.01	**0**	tr	**0**	unk.	**unk.**	0	0(a)	**0**	0.2	0.2	6	**1**	470	36	unk.	0.4	**2**	0.3	**2**	unk.	**unk.**	532	C
0.2	**1**	5	**1**	0.15	**2**	0	**0**	0	0.3	**1**	tr	1.48	37	**4**	20.5	0.9	unk.	7	**2**	0.04	**2**	unk.		0	
0.03	**2**	unk.	**unk.**	unk.	**unk.**	unk.	**unk.**	unk.	0(a)	**0**	unk.	unk.	43	**4**	196	47	unk.	0.5	**3**	0.6	**4**	unk.	**unk.**	133	D
0.4	**2**	unk.	**unk.**	unk.	**unk.**	unk.	**unk.**	unk.	unk.	**unk.**	unk.	0.71	71	**7**	8.5	1.2	unk.	31	**8**	0.05	**2**	unk.		unk.	
0.04	**2**	unk.	**unk.**	unk.	**unk.**	unk.	**unk.**	unk.	0(a)	**0**	unk.	unk.	47	**5**	206	72	unk.	0.5	**3**	0.5	**3**	unk.	**unk.**	113	E
0.4	**2**	unk.	**unk.**	unk.	**unk.**	unk.	**unk.**	unk.	unk.	**unk.**	unk.	0.82	78	**8**	9.0	1.8	unk.	26	**7**	0.04	**2**	unk.		unk.	
unk.	**unk.**	unk.	**unk.**	unk.	**unk.**	unk.	**unk.**	unk.	unk.	**unk.**	unk.	unk.	unk.	**unk.**	unk.	unk.	unk.	0.2	**1**	unk.	**unk.**	unk.	**unk.**	unk.	F
unk.	**unk.**	unk.	**unk.**	unk.	**unk.**	unk.	**unk.**	unk.	unk.	**unk.**	unk.	0.06	unk.	**unk.**	unk.	unk.	unk.	unk.	**unk.**	unk.	**unk.**	tr(a)		tr(a)	
0.30	**18**	0.09	**4**	0.71	**12**	0.005	**2**	0	27	**7**	0.6	0.1	316	**32**	899	174	50	0.9	**5**	1.8	**12**	71.3	**48**	31	G
0.3	**1**	14	**4**	0.72	**7**	898	**18**	0	0.8	**3**	0.3	3.60	407	**41**	39.1	4.4	1.4	80	**20**	0.09	**5**	3		5	
0.11	**6**	0.07	**4**	0.27	**5**	0.005	**2**	0	9	**2**	0.7	0.1	24	**2**	129	118	76	1.0	**5**	0.6	**4**	24.3	**16**	525	H
0.3	**1**	11	**3**	0.43	**4**	295	**6**	2	0.8	**3**	0.2	0.70	77	**8**	5.6	3.0	2.1	72	**18**	0.09	**5**	2		20	
0.18	**11**	0.03	**2**	unk.	**unk.**	unk.	**unk.**	unk.	0(a)	**0**	unk.	unk.	70	**7**	2385	421	unk.	2.0	**11**	2.1	**14**	unk.	**unk.**	915	I
1.9	**10**	unk.	**unk.**	unk.	**unk.**	1083	**22**	unk.	unk.	**unk.**	unk.	7.17	95	**9**	103.8	10.8	unk.	9	**2**	0.30	**15**	unk.		unk.	
0.27	**16**	0.06	**3**	unk.	**unk.**	unk.	**unk.**	unk.	unk.	**unk.**	unk.	unk.	174	**17**	1042	360	unk.	0.9	**5**	0.9	**6**	unk.	**unk.**	379	J
0.9	**5**	unk.	**unk.**	unk.	**unk.**	600	**12**	unk.	unk.	**unk.**	unk.	3.84	154	**15**	45.3	9.2	unk.	20	**5**	0.14	**7**	unk.		unk.	
0.07	**4**	0.01	**1**	unk.	**unk.**	unk.	**unk.**	unk.	0(a)	**0**	unk.	unk.	29	**3**	981	173	unk.	0.8	**5**	0.9	**6**	unk.	**unk.**	376	K
0.8	**4**	unk.	**unk.**	unk.	**unk.**	444	**9**	unk.	unk.	**unk.**	unk.	2.95	39	**4**	42.7	4.4	unk.	5	**1**	0.12	**6**	unk.		unk.	
0.05	**3**	0.09	**4**	tr(a)	**0**	unk.	**unk.**	unk.	0(a)	**0**	unk.	unk.	16	**2**	968	135	unk.	1.0	**5**	unk.	**unk.**	unk.	**unk.**	unk.	L
1.1	**6**	unk.	**unk.**	unk.	**unk.**	953	**19**	unk.	unk.	**unk.**	unk.	2.94	49	**5**	42.1	3.4	unk.	unk.	**unk.**	unk.	**unk.**	unk.		unk.	
0.10	**6**	0.10	**5**	unk.	**unk.**	unk.	**unk.**	unk.	0(a)	**0**	unk.	unk.	196	**20**	2311	978	unk.	5.0	**28**	2.5	**17**	unk.	**unk.**	1630	M
1.4	**7**	77	**19**	unk.	**unk.**	2158	**43**	unk.	unk.	**unk.**	unk.	8.38	320	**32**	100.5	25.0	unk.	108	**27**	0.98	**49**	unk.		unk.	
0.03	**2**	0.04	**2**	unk.	**unk.**	unk.	**unk.**	unk.	0(a)	**0**	unk.	unk.	81	**8**	951	402	unk.	2.0	**11**	1.0	**7**	unk.	**unk.**	670	N
0.6	**3**	32	**8**	unk.	**unk.**	888	**18**	unk.	unk.	**unk.**	unk.	3.44	132	**13**	41.4	10.3	unk.	46	**11**	0.40	**20**	unk.		unk.	
0.16	**9**	0.32	**16**	unk.	**unk.**	unk.	**unk.**	unk.	0(a)	**0**	unk.	unk.	211	**21**	2651	1158	unk.	5.7	**32**	2.9	**19**	unk.	**unk.**	1914	O
2.5	**12**	unk.	**unk.**	unk.	**unk.**	2112	**42**	unk.	unk.	**unk.**	unk.	8.42	402	**40**	115.3	29.6	unk.	118	**30**	0.96	**48**	unk.		unk.	
0.07	**4**	0.13	**7**	unk.	**unk.**	unk.	**unk.**	unk.	0(a)	**0**	unk.	unk.	88	**9**	1093	478	unk.	2.3	**13**	1.2	**8**	unk.	**unk.**	788	P
1.0	**5**	unk.	**unk.**	unk.	**unk.**	870	**17**	unk.	unk.	**unk.**	unk.	3.47	165	**17**	47.5	12.2	unk.	48	**12**	0.39	**20**	unk.		unk.	
0.15	**9**	unk.	**unk.**	unk.	**unk.**	unk.	**unk.**	unk.	0(a)	**0**	unk.	unk.	78	**8**	972	unk.	unk.	3.2	**18**	unk.	**unk.**	unk.	**unk.**	unk.	Q
1.7	**9**	unk.	**unk.**	unk.	**unk.**	3951	**79**	unk.	unk.	**unk.**	unk.	3.69	unk.	**unk.**	42.3	unk.	unk.	unk.	**unk.**	unk.	**unk.**	unk.		unk.	
0.12	**7**	0.22	**11**	0.00	**0**	unk.	**unk.**	unk.	0(a)	**0**	unk.	unk.	109	**11**	3026	780	unk.	4.7	**26**	3.4	**23**	unk.	**unk.**	1560	R
1.3	**6**	unk.	**unk.**	unk.	**unk.**	1388	**28**	unk.	unk.	**unk.**	unk.	9.67	234	**23**	131.6	19.9	unk.	103	**26**	0.94	**47**	unk.		unk.	
0.05	**3**	0.09	**5**	0.02	**0**	unk.	**unk.**	unk.	0(a)	**0**	unk.	unk.	44	**4**	1198	274	unk.	2.1	**12**	1.4	**9**	unk.	**unk.**	642	S
0.5	**3**	25	**6**	0.20	**2**	506	**10**	unk.	unk.	**unk.**	unk.	3.73	106	**11**	52.1	7.0	unk.	42	**11**	0.39	**19**	unk.		unk.	

	FOOD	Portion	Weight in grams Conversion for 100 g	Kilocalories H_2O g	Total carbohydrate g Total fats g	Crude fiber g Dietary fiber g	Total protein g Total sugar g	**% USRDA**	Arginine mg Histidine mg	Isoleucine mg Leucine mg	Lysine mg Methionine mg	Phenylalanine mg Threonine mg	Valine mg Tryptophan mg	Cystine mg Tyrosine mg	Polyunsat. fatty acids g Monounsat. fatty acids g	Saturated fatty acids g P/S ratio	Linoleic acid g Cholesterol mg	Thiamin mg Ascorbic acid mg	**% USRDA**
A	soup,canned,beef broth or bouillon,	1 C	240	17	0.1	tr	2.7	**4**	unk.	unk.	unk.	unk.	unk.	unk.	0.02	0.26	0.02	tr	**0**
	ready-to-serve		*0.42*	234.1	0.5	unk.	tr(a)		unk.	unk.	unk.	unk.	unk.	unk.	0.19	0.09	tr	0	**0**
B	soup,canned,beef mushroom,condensed	10-3/4 oz can	305	unk.	unk.	unk.	14.0	**22**	unk.	unk.	unk.	unk.	unk.	unk.	0.30	3.66	0.21	0.06	**4**
			0.33	unk.	7.3	unk.	unk.		unk.	unk.	unk.	unk.	unk.	unk.	2.62	0.08	15	0	**0**
C	soup,canned,beef mushroom,prep	1 C	244	unk.	unk.	unk.	5.8	**9**	unk.	unk.	unk.	unk.	unk.	unk.	0.12	1.49	0.10	0.05	**3**
	w/water		*0.41*	unk.	3.0	unk.	unk.		unk.	unk.	unk.	unk.	unk.	unk.	1.07	0.08	7	5	**8**
D	soup,canned,beef noodle,condensed	10-3/4 oz can	305	204	21.8	0.30	11.7	**18**	482	454	637	473	506	140	1.19	2.78	1.01	0.18	**12**
			0.33	257.5	7.5	unk.	unk.		271	766	220	375	113	305	2.65	0.43	12	1	**2**
E	soup,canned,beef noodle,prep w/water	1 C	244	83	9.0	tr	4.8	**7**	198	188	261	195	207	59	0.49	1.15	0.41	0.07	**5**
			0.41	224.5	3.1	unk.	unk.		112	315	90	154	46	124	1.10	0.43	5	tr	**0**
F	soup,canned,beef,chunky,	1 C	240	170	19.6	0.72	11.7	**18**	610	593	929	482	636	122	0.22	2.54	0.17	0.05	**3**
	ready-to-serve		*0.42*	200.0	5.1	unk.	unk.		276	898	247	466	113	353	1.85	0.08	14	7	**12**
G	soup,canned,celery,cream of,condensed	10-3/4 oz can	305	220	21.4	0.91	4.0	**6**	143	189	180	189	217	46	6.10	3.42	5.92	0.06	**4**
			0.33	259.1	13.6	unk.	unk.		95	302	73	143	46	131	2.90	1.79	34	1	**2**
H	soup,canned,celery,cream of,prep	1 C	248	164	14.5	0.37	5.7	**9**	206	322	394	273	360	55	2.65	3.94	2.53	0.07	**5**
	w/milk		*0.40*	214.4	9.7	unk.	unk.		149	518	131	241	74	248	2.23	0.67	32	1	**3**
I	soup,canned,celery,cream of,prep	1 C	244	90	8.8	0.37	1.7	**3**	59	78	73	78	90	20	2.51	1.42	2.44	0.02	**2**
	w/water		*0.41*	225.1	5.6	unk.	unk.		39	124	29	59	20	54	1.20	1.78	15	tr	**0**
J	soup,canned,cheese,condensed	11 oz can	312	378	25.6	unk.	13.2	**20**	393	783	924	677	924	109	0.72	16.19	0.44	0.03	**2**
			0.32	240.7	25.4	unk.	unk.		356	1245	284	462	175	605	6.05	0.04	72	0	**0**
K	soup,canned,cheese,prep w/milk	1 C	251	231	16.2	unk.	9.5	**15**	309	567	700	474	650	83	0.45	9.11	0.28	0.08	**5**
			0.40	206.9	14.6	unk.	unk.		256	906	218	371	128	444	3.49	0.05	48	1	**2**
L	soup,canned,cheese,prep w/water	1 C	247	156	10.5	unk.	5.4	**8**	163	321	380	279	380	44	0.30	6.67	0.17	0.02	**2**
			0.40	217.7	10.5	unk.	unk.		146	511	116	190	72	249	2.49	0.04	30	0	**0**
M	soup,canned,chicken & dumplings,	10-1/2 oz can	298	235	14.7	unk.	13.6	**21**	709	593	918	542	349	176	3.16	3.19	2.98	0.03	**2**
	condensed		*0.34*	249.8	13.4	unk.	unk.		334	992	262	471	131	349	5.51	0.99	80	0	**0**
N	soup,canned,chicken & dumplings,prep	1 C	241	96	6.0	unk.	5.6	**9**	292	243	378	224	277	72	1.30	1.30	1.23	0.02	**2**
	w/water		*0.41*	221.2	5.5	unk.	unk.		137	407	108	193	53	142	2.27	1.00	34	0	**0**
O	soup,canned,chicken gumbo,condensed	10-3/4 oz can	305	137	20.3	0.61	6.4	**10**	299	241	393	235	284	43	0.85	0.79	0.76	0.06	**4**
			0.33	268.5	3.5	unk.	unk.		140	409	110	201	55	168	1.40	1.08	9	12	**20**
P	soup,canned,chicken gumbo,prep	1 C	244	237	8.4	0.24	2.6	**4**	122	100	161	98	117	17	0.34	0.32	0.32	0.02	**2**
	w/water		*0.41*	229.0	1.4	unk.	unk.		59	168	46	83	22	68	0.59	1.08	5	5	**8**
Q	soup,canned,chicken mushroom,	10-3/4 oz can	305	unk.	unk.	unk.	10.7	**16**	unk.	unk.	unk.	unk.	unk.	unk.	5.64	5.86	5.34	0.06	**4**
	condensed		*0.33*	unk.	22.3	unk.	unk.		unk.	unk.	unk.	unk.	unk.	unk.	8.97	0.96	24	0	**0**
R	soup,canned,chicken mushroom,prep	1 C	244	unk.	unk.	unk.	4.4	**7**	unk.	unk.	unk.	unk.	unk.	unk.	2.32	2.39	2.20	0.02	**2**
	w/water		*0.41*	unk.	9.1	unk.	unk.		unk.	unk.	unk.	unk.	unk.	unk.	3.68	0.97	10	0	**0**
S	soup,canned,chicken noodle	1 C	248	99	8.4	0.55	8.1	**13**	unk.	unk.	unk.	unk.	unk.	unk.	0.74	1.07	0.69	0.12	**8**
	w/meatballs,ready-to-serve		*0.40*	225.0	3.6	unk.	unk.		unk.	unk.	unk.	unk.	unk.	unk.	1.34	0.70	10	8	**13**

Row	Riboflavin mg / Niacin mg	% USRDA	Vitamin B_6 mg / Folacin mcg	% USRDA	Vitamin B_{12} mcg / Pantothenic acid mg	% USRDA	Biotin mg / Vitamin A IU	% USRDA	Preformed A RE / Beta carotene RE	Vitamin D IU / Vitamin E IU	% USRDA	Total tocopherol mg / Alpha tocopherol mg	Other tocopherol mg / Total ash g	Calcium mg / Phosphorus mg	% USRDA	Sodium mg / Sodium meq	Potassium mg / Potassium meq	Chlorine mg / Chlorine meq	Iron mg / Magnesium mg	% USRDA	Zinc mg / Copper mg	% USRDA	Iodine mcg / Selenium mcg	% USRDA	Manganese mcg / Chromium mcg
A	0.05	3	unk.	unk.	unk.	unk.	unk.	unk.	0	0(a)	0	unk.	unk.	14	1	782	130	unk.	0.4	2	unk.	unk.	unk.	unk.	unk.
	1.9	9	unk.	unk.	unk.	unk.	0	0	0	unk.	unk.	unk.	2.52	31	3	34.0	3.3	unk.	unk.	unk.	unk.	unk.	unk.		unk.
B	0.18	11	unk.	unk.	unk.	unk.	unk.	unk.	0	0(a)	0	unk.	unk.	12	1	unk.	unk.	unk.	2.1	12	unk.	unk.	unk.	unk.	unk.
	2.7	14	unk.	unk.	unk.	unk.	0	0	0	unk.	unk.	unk.	unk.	unk.	unk.	unk.	unk.	unk.	unk.	unk.	unk.	unk.	unk.		unk.
C	0.05	3	unk.	unk.	unk.	unk.	unk.	unk.	0	0(a)	0	unk.	unk.	5	1	unk.	unk.	unk.	0.9	5	unk.	unk.	unk.	unk.	unk.
	1.0	5	unk.	unk.	unk.	unk.	0	0	0	unk.	unk.	unk.	unk.	unk.	unk.	unk.	unk.	unk.	unk.	unk.	unk.	unk.	unk.		unk.
D	0.15	9	0.09	5	0.49	8	unk.	unk.	unk.	0(a)	0	unk.	unk.	37	4	2315	241	unk.	2.7	15	3.8	25	unk.	unk.	665
	2.6	13	11	3	unk.	unk.	1531	31	unk.	unk.	unk.	unk.	6.47	113	11	100.7	6.2	unk.	15	4	0.34	17	unk.		unk.
E	0.05	3	0.04	2	0.20	3	unk.	unk.	unk.	0(a)	0	unk.	unk.	15	2	952	100	unk.	1.1	6	1.5	10	unk.	unk.	273
	1.1	5	4	1	unk.	unk.	630	13	unk.	unk.	unk.	unk.	2.66	46	5	41.4	2.6	unk.	5	1	0.14	7	unk.		unk.
F	0.14	9	0.13	7	0.62	10	unk.	unk.	unk.	0(a)	0	unk.	unk.	31	3	866	336	unk.	2.3	13	2.6	18	unk.	unk.	240
	2.7	14	13	3	unk.	unk.	2611	52	unk.	unk.	unk.	unk.	3.58	120	12	37.7	8.6	unk.	unk.	unk.	0.24	12	unk.		unk.
G	0.12	7	0.03	2	unk.	unk.	unk.	unk.	unk.	unk.	unk.	unk.	unk.	98	10	2309	299	unk.	1.5	9	0.4	2	unk.	unk.	610
	0.8	4	6	1	unk.	unk.	744	15	unk.	unk.	unk.	unk.	6.80	92	9	100.4	7.6	unk.	15	4	0.34	17	unk.		unk.
H	0.25	15	0.06	3	unk.	unk.	unk.	unk.	unk.	unk.	unk.	unk.	unk.	186	19	1009	310	unk.	0.7	4	0.2	1	unk.	unk.	253
	0.4	2	8	2	unk.	unk.	461	9	unk.	unk.	unk.	unk.	3.70	151	15	43.9	7.9	unk.	22	6	0.15	8	unk.		unk.
I	0.05	3	0.01	1	unk.	unk.	unk.	unk.	unk.	unk.	unk.	unk.	unk.	39	4	949	122	unk.	0.6	4	0.1	1	unk.	unk.	251
	0.3	2	2	1	unk.	unk.	307	6	unk.	unk.	unk.	unk.	2.81	37	4	41.3	3.1	unk.	7	2	0.14	7	unk.		unk.

Row	Riboflavin mg / Niacin mg	% USRDA	Vitamin B_6 mg / Folacin mcg	% USRDA	Vitamin B_{12} mcg / Pantothenic acid mg	% USRDA	Biotin mg / Vitamin A IU	% USRDA	Preformed A RE / Beta carotene RE	Vitamin D IU / Vitamin E IU	% USRDA	Total tocopherol mg / Alpha tocopherol mg	Other tocopherol mg / Total ash g	Calcium mg / Phosphorus mg	% USRDA	Sodium mg / Sodium meq	Potassium mg / Potassium meq	Chlorine mg / Chlorine meq	Iron mg / Magnesium mg	% USRDA	Zinc mg / Copper mg	% USRDA	Iodine mcg / Selenium mcg	% USRDA	Manganese mcg / Chromium mcg
J	0.34	20	0.06	3	0.00	0	unk.	unk.	unk.	unk.	unk.	unk.	unk.	346	35	2331	374	unk.	1.8	10	1.6	10	unk.	unk.	624
	1.0	5	unk.	unk.	unk.	unk.	2643	53	unk.	unk.	unk.	unk.	7.14	331	33	101.4	9.6	unk.	9	2	0.31	16	unk.		unk.
K	0.33	19	0.08	4	0.43	7	unk.	unk.	unk.	unk.	unk.	unk.	unk.	289	29	1019	341	unk.	0.8	5	0.7	5	unk.	unk.	259
	0.5	3	unk.	unk.	unk.	unk.	1242	25	unk.	unk.	unk.	unk.	3.82	251	25	44.3	8.7	unk.	20	5	0.14	7	unk.		unk.
L	0.15	9	0.02	1	0.00	0	unk.	unk.	unk.	unk.	unk.	unk.	unk.	141	14	958	153	unk.	0.7	4	0.6	4	unk.	unk.	257
	0.4	2	unk.	unk.	unk.	unk.	1087	22	unk.	unk.	unk.	unk.	2.94	136	14	41.7	3.9	unk.	5	1	0.13	6	unk.		unk.
M	0.18	11	0.09	5	0.39	7	unk.	unk.	unk.	0(a)	0	unk.	unk.	36	4	2095	283	unk.	1.5	8	0.9	6	unk.	unk.	894
	4.3	21	unk.	unk.	unk.	unk.	1261	25	unk.	unk.	unk.	unk.	6.44	149	15	91.1	7.2	unk.	9	2	0.30	15	unk.		unk.
N	0.07	4	0.04	2	0.17	3	unk.	unk.	unk.	0(a)	0	unk.	unk.	14	1	860	116	unk.	0.6	4	0.4	2	unk.	unk.	489
	1.8	9	unk.	unk.	unk.	unk.	518	10	unk.	unk.	unk.	unk.	2.65	60	6	37.4	3.0	unk.	5	1	0.12	6	unk.		unk.
O	0.09	5	0.15	8	unk.	unk.	unk.	unk.	unk.	0(a)	0	unk.	unk.	58	6	2321	183	unk.	2.2	12	0.9	6	unk.	unk.	610
	1.6	8	unk.	unk.	unk.	unk.	329	7	unk.	unk.	unk.	unk.	6.31	61	6	101.0	4.7	unk.	9	2	0.30	15	unk.		unk.
P	0.05	3	0.06	3	unk.	unk.	unk.	unk.	unk.	0(a)	0	unk.	unk.	24	2	954	76	unk.	0.9	5	0.4	2	unk.	unk.	251
	0.7	3	unk.	unk.	unk.	unk.	137	3	unk.	unk.	unk.	unk.	2.59	24	2	41.5	1.9	unk.	5	1	0.12	6	unk.		unk.
Q	0.27	16	unk.	unk.	unk.	unk.	unk.	unk.	unk.	0(a)	0	unk.	unk.	70	7	unk.	unk.	unk.	2.1	12	unk.	unk.	unk.	unk.	unk.
	4.0	20	unk.	unk.	unk.	unk.	2760	55	unk.	unk.	unk.	unk.	unk.	unk.	unk.	unk.	unk.	unk.	unk.	unk.	unk.	unk.	unk.		unk.
R	0.12	7	unk.	unk.	unk.	unk.	unk.	unk.	unk.	0(a)	0	unk.	unk.	29	3	unk.	unk.	unk.	0.9	5	unk.	unk.	unk.	unk.	unk.
	1.6	8	unk.	unk.	unk.	unk.	1135	23	unk.	unk.	unk.	unk.	unk.	unk.	unk.	unk.	unk.	unk.	unk.	unk.	unk.	unk.	unk.		unk.
S	0.12	7	unk.	unk.	unk.	unk.	unk.	unk.	unk.	0(a)	0	unk.	unk.	30	3	1039	unk.	unk.	1.7	10	unk.	unk.	unk.	unk.	unk.
	2.5	13	unk.	unk.	unk.	unk.	2326	47	unk.	unk.	unk.	unk.	2.98	unk.	unk.	45.2	unk.	unk.	unk.	unk.	unk.	unk.	unk.		unk.

	FOOD	Portion	Weight in grams Conversion for 100 g	Kilocalories H_2O g	Total carbohydrate g Total fats g	Crude fiber g Dietary fiber g	Total protein g Total sugar g	% USRDA	Arginine mg Histidine mg	Isoleucine mg Leucine mg	Lysine mg Methionine mg	Phenylalanine mg Threonine mg	Valine mg Tryptophan mg	Cystine mg Tyrosine mg	Polyunsat. fatty acids g Monounsat. fatty acids g	Saturated fatty acids g P/S ratio	Linoleic acid g Cholesterol mg	Thiamin mg Ascorbic acid mg	% USRDA
A	soup,canned,chicken noodle,chunky,	1 C	240	unk.	unk.	unk.	12.7	**20**	662	552	854	504	629	163	1.51	1.39	1.51	0.07	**5**
	ready-to-serve		*0.42*	unk.	6.0	unk.	unk.		312	924	245	437	unk.	326	2.40	1.09	19	0	**0**
B	soup,canned,chicken noodle,condensed	10-1/2 oz can	298	182	22.7	0.30	9.6	**15**	393	372	521	387	414	113	1.22	1.46	1.07	0.15	**10**
			0.34	253.5	5.5	unk.	unk.		223	626	179	307	92	250	2.29	0.84	15	0	**0**
C	soup,canned,chicken noodle,	1 C	241	75	9.4	0.24	4.0	**6**	166	159	219	164	176	46	0.55	0.65	0.51	0.05	**3**
	prep w/water		*0.41*	221.7	2.5	unk.	unk.		94	265	77	128	39	106	1.01	0.85	7	tr	**0**
D	soup,canned,chicken rice,chunky,	1 C	240	127	13.0	unk.	12.3	**19**	unk.	unk.	unk.	unk.	unk.	unk.	0.67	0.96	0.62	0.02	**2**
	ready-to-serve		*0.42*	208.3	3.2	unk.	unk.		unk.	unk.	unk.	unk.	unk.	unk.	1.20	0.70	12	4	**6**
E	soup,canned,chicken rice,condensed	10-1/2 oz can	298	146	17.4	0.30	8.6	**13**	569	432	608	367	459	122	1.01	1.10	0.95	0.03	**2**
			0.34	261.8	4.6	unk.	unk.		244	659	223	346	101	301	2.03	0.92	15	tr	**1**
F	soup,canned,chicken rice,prep w/water	1 C	241	60	7.2	tr	3.5	**6**	234	178	251	152	188	51	0.41	0.46	0.39	0.02	**2**
			0.41	226.1	1.9	unk.	unk.		101	270	92	142	41	123	0.82	0.89	7	tr	**0**
G	soup,canned,chicken vegetable,chunky,	1 C	240	166	18.9	unk.	12.3	**19**	unk.	unk.	unk.	unk.	unk.	unk.	1.01	1.44	0.94	0.05	**3**
	ready-to-serve		*0.42*	200.3	4.8	unk.	unk.		unk.	unk.	unk.	unk.	unk.	unk.	1.80	0.70	17	6	**9**
H	soup,canned,chicken vegetable,	10-1/2 oz can	298	182	20.9	0.30	8.8	**14**	411	331	539	322	387	60	1.46	2.06	1.34	0.12	**8**
	condensed		*0.34*	254.8	6.9	unk.	unk.		194	560	149	274	74	229	2.59	0.71	21	2	**4**
I	soup,canned,chicken vegetable,	1 C	241	75	8.6	0.12	3.6	**6**	169	135	222	133	159	24	0.60	0.84	0.55	0.05	**3**
	prep w/water		*0.41*	223.3	2.8	unk.	unk.		80	231	60	113	31	94	1.06	0.71	10	1	**2**
J	soup,canned,chicken broth,condensed	10-3/4 oz can	305	95	2.3	tr	13.5	**21**	unk.	unk.	unk.	unk.	unk.	unk.	0.67	0.95	0.61	0.03	**2**
			0.33	280.5	3.2	tr(a)	0(a)		unk.	unk.	unk.	unk.	unk.	unk.	1.19	0.71	3	0	**0**
K	soup, canned,chicken broth,	1 C	244	39	0.9	tr	4.9	**8**	unk.	unk.	unk.	unk.	unk.	unk.	0.29	0.41	0.27	tr	**0**
	prep w/water		*0.41*	234.1	1.4	unk.	0(a)		unk.	unk.	unk.	unk.	unk.	unk.	0.54	0.71	0	0	**0**
L	soup,canned,chicken,chunky,	1 C	251	178	17.3	0.25	12.7	**20**	660	552	853	505	627	163	1.38	1.98	1.31	0.08	**5**
	ready-to-serve		*0.40*	211.1	6.6	unk.	unk.		311	924	243	437	123	326	2.48	0.70	30	1	**2**
M	soup,canned,chicken,cream of,	10-3/4 oz can	305	284	22.5	0.30	8.3	**13**	406	415	522	372	421	122	3.63	5.06	3.42	0.06	**4**
	condensed		*0.33*	249.3	17.9	unk.	unk.		223	640	195	317	104	287	7.17	0.72	24	tr	**1**
N	soup,canned,chicken,cream of,	1 C	248	191	15.0	1.24	7.5	**12**	312	414	533	347	444	87	1.64	4.64	1.51	0.07	**5**
	prep w/milk		*0.40*	210.4	11.5	unk.	unk.		201	657	181	312	99	312	3.97	0.35	27	1	**2**
O	soup,canned,chicken,cream of,	1 C	244	117	9.3	0.12	3.4	**5**	166	171	215	154	173	51	1.49	2.07	1.42	0.02	**2**
	prep w/water		*0.41*	221.1	7.4	unk.	unk.		93	264	81	129	41	117	2.95	0.72	10	tr	**0**
P	soup,canned,chili beef,condensed	11-1/4 oz can	319	412	52.1	3.51	16.2	**25**	852	794	1107	906	893	179	0.64	7.97	0.51	0.16	**11**
			0.31	226.0	16.0	unk.	unk.		424	1340	204	670	172	485	5.77	0.08	32	10	**17**
Q	soup,canned,chili beef,prep w/water	1 C	250	170	21.4	1.45	6.7	**10**	350	328	455	373	368	73	0.27	3.35	0.20	0.05	**3**
			0.40	211.7	6.6	unk.	unk.		175	553	85	275	70	200	2.40	0.08	13	4	**7**
R	soup,canned,chunky chili beef,	1 C	276	255	32.8	unk.	18.8	**29**	unk.	unk.	unk.	unk.	unk.	unk.	unk.	unk.	unk.	0.11	**7**
	ready-to-serve-Campbell		*0.36*	214.7	5.5	unk.	unk.		unk.	unk.	unk.	unk.	unk.	unk.	unk.	unk.	unk.	0	**0**
S	soup,canned,clam chowder,Manhattan,	1 C	240	134	18.8	0.48	7.3	**11**	unk.	unk.	unk.	unk.	unk.	unk.	0.12	2.11	0.07	0.05	**3**
	chunky,ready-to-serve		*0.42*	206.5	3.4	unk.	unk.		unk.	unk.	unk.	unk.	unk.	unk.	0.84	0.06	14	12	**20**

Riboflavin mg / Niacin mg	% USRDA	Vitamin B6 mg / Folacin mcg	% USRDA	Vitamin B12 mcg / Pantothenic acid mg	% USRDA	Biotin mg / Vitamin A IU	% USRDA	Preformed A RE / Beta carotene RE	Vitamin D IU / Vitamin E IU	% USRDA	Total tocopherol mg / Alpha tocopherol mg	Other tocopherol mg / Total ash g	Calcium mg / Phosphorus mg	% USRDA	Sodium mg / Sodium meq	Potassium mg / Potassium meq	Chlorine mg / Chlorine meq	Iron mg / Magnesium mg	% USRDA	Zinc mg / Copper mg	% USRDA	Iodine mcg / Selenium mcg	% USRDA	Manganese mcg / Chromium mcg	
0.17	10	unk.	unk.	unk.	unk.	unk.	unk.	unk.	0(a)	0	unk.	unk.	24	2	unk.	unk.	unk.	1.4	8	unk.	unk.	unk.	unk.	unk.	A
4.3	22	unk.	unk.	unk.	unk.	1222	24	unk.	unk.	unk.	unk.	unk.	unk.	unk.	unk.	unk.	unk.	unk.	unk.	unk.	unk.	unk.		unk.	
0.15	9	0.06	3	unk.	unk.	unk.	unk.	unk.	0(a)	0	unk.	unk.	33	3	2256	134	unk.	1.8	10	0.7	5	unk.	unk.	700	B
3.7	18	5	1	unk.	unk.	1585	32	unk.	unk.	unk.	unk.	6.65	89	9	98.1	3.4	unk.	12	3	0.47	24	unk.		unk.	
0.07	4	0.03	1	unk.	unk.	unk.	unk.	unk.	0(a)	0	unk.	unk.	17	2	1106	55	unk.	0.8	4	0.4	3	unk.	unk.	289	C
1.4	7	2	1	unk.	unk.	711	14	unk.	unk.	unk.	unk.	2.89	36	4	48.1	1.4	unk.	5	1	0.20	10	unk.		unk.	
0.10	6	unk.	unk.	unk.	unk.	unk.	unk.	unk.	0(a)	0	unk.	unk.	34	3	888	unk.	unk.	1.9	10	unk.	unk.	unk.	unk.	unk.	D
4.1	21	4	1	unk.	unk.	5858	117	unk.	unk.	unk.	unk.	3.29	unk.	unk.	38.6	unk.	unk.	unk.	unk.	unk.	unk.	unk.		unk.	
0.06	4	0.06	3	unk.	unk.	unk.	unk.	unk.	0(a)	0	unk.	unk.	42	4	1982	244	unk.	1.8	10	0.6	4	unk.	unk.	894	E
2.7	14	3	1	unk.	unk.	1606	32	unk.	unk.	unk.	unk.	5.60	51	5	86.2	6.3	unk.	0	0	0.29	14	unk.		unk.	
0.02	1	0.02	1	unk.	unk.	unk.	unk.	unk.	0(a)	0	unk.	unk.	17	2	815	101	unk.	0.7	4	0.3	2	unk.	unk.	366	F
1.1	6	1	0	unk.	unk.	660	13	unk.	unk.	unk.	unk.	2.29	22	2	35.4	2.6	unk.	0	0	0.12	6	unk.		unk.	
0.17	10	unk.	unk.	unk.	unk.	unk.	unk.	unk.	0(a)	0	unk.	unk.	26	3	1068	unk.	unk.	1.5	8	unk.	unk.	unk.	unk.	unk.	G
3.3	16	unk.	unk.	unk.	unk.	5990	120	unk.	unk.	unk.	unk.	3.70	unk.	unk.	46.5	unk.	unk.	unk.	unk.	unk.	unk.	unk.		unk.	
0.15	9	0.12	6	unk.	unk.	unk.	unk.	unk.	0(a)	0	unk.	unk.	42	4	2298	375	unk.	2.1	12	0.9	6	unk.	unk.	894	H
3.0	15	unk.	unk.	unk.	unk.	6461	129	unk.	unk.	unk.	unk.	6.62	98	10	99.9	9.6	unk.	15	4	0.30	15	unk.		unk.	
0.05	3	0.05	2	unk.	unk.	unk.	unk.	unk.	0(a)	0	unk.	unk.	17	2	945	154	unk.	0.9	5	0.4	2	unk.	unk.	366	I
1.2	6	unk.	unk.	unk.	unk.	2656	53	unk.	unk.	unk.	unk.	2.72	41	4	41.1	3.9	unk.	7	2	0.12	6	unk.		unk.	
0.15	9	0.06	3	0.61	10	unk.	unk.	0	0(a)	0	unk.	unk.	18	2	1909	518	unk.	1.3	7	0.6	4	unk.	unk.	610	J
6.8	34	unk.	unk.	unk.	unk.	0	0	0	unk.	unk.	unk.	5.58	183	18	83.0	13.3	unk.	6	2	0.30	15	unk.		unk.	
0.07	4	0.02	1	0.24	4	unk.	unk.	0	0(a)	0	unk.	unk.	10	1	776	210	unk.	0.5	3	0.2	2	unk.	unk.	249	K
3.3	17	unk.	unk.	unk.	unk.	0	0	0	unk.	unk.	unk.	2.49	73	7	33.8	5.4	unk.	2	1	0.12	6	unk.		unk.	
0.18	10	0.05	3	0.25	4	unk.	unk.	unk.	0(a)	0	unk.	unk.	25	3	889	176	unk.	1.7	10	1.0	7	unk.	unk.	251	L
4.4	22	5	1	unk.	unk.	1300	26	unk.	unk.	unk.	unk.	3.29	113	11	38.6	4.5	unk.	unk.	unk.	0.25	13	unk.		unk.	
0.15	9	0.04	2	unk.	unk.	unk.	unk.	unk.	0(a)	0	unk.	unk.	82	8	2397	213	unk.	1.5	8	1.5	10	unk.	unk.	915	M
2.0	10	4	1	unk.	unk.	1360	27	unk.	unk.	unk.	unk.	6.92	92	9	104.3	5.5	unk.	6	2	0.30	15	unk.		unk.	
0.25	15	0.07	3	unk.	unk.	unk.	unk.	unk.	unk.	unk.	unk.	unk.	181	18	1047	273	unk.	0.7	4	0.7	5	unk.	unk.	379	N
0.9	5	8	2	unk.	unk.	714	14	unk.	unk.	unk.	unk.	3.74	151	15	45.5	7.0	unk.	17	4	0.14	7	unk.		unk.	
0.07	4	0.02	1	unk.	unk.	unk.	unk.	unk.	unk.	unk.	unk.	unk.	34	3	986	88	unk.	0.6	3	0.6	4	unk.	unk.	376	O
0.8	4	2	0	unk.	unk.	561	11	unk.	unk.	unk.	unk.	2.85	37	4	42.9	2.3	unk.	2	1	0.12	6	unk.		unk.	
0.19	11	0.38	19	0.77	13	unk.	unk.	unk.	0(a)	0	unk.	unk.	105	11	2514	1276	unk.	5.2	29	3.4	23	unk.	unk.	2552	P
2.6	13	unk.	unk.	unk.	unk.	3665	73	unk.	unk.	unk.	unk.	8.58	361	36	109.3	32.6	unk.	73	18	0.96	48	unk.		unk.	
0.07	4	0.16	8	0.32	5	unk.	unk.	unk.	0(a)	0	unk.	unk.	43	4	1035	525	unk.	2.1	12	1.4	9	unk.	unk.	1050	Q
1.1	5	unk.	unk.	unk.	unk.	1510	30	unk.	unk.	unk.	unk.	3.55	148	15	45.0	13.4	unk.	30	8	0.39	20	unk.		unk.	
0.14	8	unk.	unk.	unk.	unk.	unk.	unk.	unk.	unk.	unk.	unk.	unk.	58	6	1049	513	unk.	3.9	22	unk.	unk.	unk.	unk.	unk.	R
2.5	13	unk.	unk.	unk.	unk.	1212	24	unk.	unk.	unk.	unk.	4.10	153	15	45.6	13.1	unk.	unk.	unk.	unk.	unk.	unk.		unk.	
0.07	4	0.26	13	7.92	132	unk.	unk.	unk.	0(a)	0	unk.	unk.	67	7	1001	384	unk.	2.6	15	1.7	11	unk.	unk.	240	S
1.8	9	9	2	unk.	unk.	3293	66	unk.	unk.	unk.	unk.	4.06	84	8	43.5	9.8	unk.	unk.	unk.	0.24	12	unk.		unk.	

	FOOD	Portion	Weight in grams / *Conversion for 100 g*	Kilocalories / H_2O g	Total carbohydrate g / Total fats g	Crude fiber g / Dietary fiber g	Total protein g / Total sugar g	**% USRDA**	Arginine mg / Histidine mg	Isoleucine mg / Leucine mg	Lysine mg / Methionine mg	Phenylalanine mg / Threonine mg	Valine mg / Tryptophan mg	Cystine mg / Tyrosine mg	Polyunsat. fatty acids g / Monounsat. fatty acids g	Saturated fatty acids g / P/S ratio	Linoleic acid g / Cholesterol mg	Thiamin mg / Ascorbic acid mg	**% USRDA**
A	soup,canned,clam chowder,Manhattan,	10-3/4 oz can	305	186	29.7	1.10	5.3	**8**	unk.	unk.	unk.	unk.	unk.	unk.	3.08	1.01	2.87	0.06	**4**
	condensed		*0.33*	257.1	5.4	unk.	unk.		unk.	unk.	unk.	unk.	unk.	unk.	0.88	3.06	6	10	**16**
B	soup,canned,clam chowder,Manhattan,	1 C	244	78	12.2	0.49	4.2	**6**	unk.	unk.	unk.	unk.	unk.	unk.	1.32	0.44	1.22	0.07	**5**
	prep w/water		*0.41*	218.4	2.3	unk.	unk.		unk.	unk.	unk.	unk.	unk.	unk.	0.39	3.00	2	3	**5**
C	soup,canned,clam chowder,New England,	10-3/4 oz can	305	213	26.5	unk.	13.2	**20**	unk.	503	692	439	534	156	2.29	0.91	2.13	0.06	**4**
	condensed		*0.33*	251.7	6.1	unk.	unk.		378	787	250	409	146	348	2.59	2.50	12	6	**10**
D	soup,canned,clam chowder,New England,	1 C	248	164	16.6	unk.	9.5	**15**	407	451	605	374	489	102	1.09	2.95	0.97	0.07	**5**
	prep w/milk		*0.40*	211.4	6.6	unk.	unk.		265	719	203	350	117	337	2.08	0.37	22	3	**6**
E	soup,canned,clam chowder,New England,	1 C	244	95	12.4	0.27	4.8	**7**	229	183	251	159	195	56	1.10	0.41	1.00	0.02	**2**
	prep w/water		*0.41*	220.9	2.9	unk.	unk.		137	288	90	149	54	127	1.22	2.65	5	2	**3**
F	soup,canned,consomme w/gelatin,	10-1/2 oz can	298	72	4.3	unk.	13.0	**20**	unk.	unk.	unk.	unk.	unk.	unk.	0.00	0.00	0.00	0.06	**4**
	condensed		*0.34*	275.9	0.0	unk.	unk.		unk.	unk.	unk.	unk.	unk.	unk.	0.00	0.00	0	2	**4**
G	soup,canned,consomme w/gelatin,	1 C	241	29	1.8	unk.	5.3	**8**	unk.	unk.	unk.	unk.	unk.	unk.	0.00	0.00	0.00	0.02	**2**
	prep w/water		*0.41*	231.9	0.0	unk.	unk.		unk.	unk.	unk.	unk.	unk.	unk.	0.00	0.00	0	1	**2**
H	soup,canned,crab,ready-to-serve	1 C	244	76	10.3	0.54	5.5	**8**	unk.	unk.	unk.	unk.	unk.	unk.	0.39	0.39	0.37	0.20	**13**
			0.41	223.3	1.5	unk.	unk.		unk.	unk.	unk.	unk.	unk.	unk.	0.68	1.00	10	0	**0**
I	soup,canned,escarole,ready-to-serve	1 C	248	27	1.8	0.74	1.5	**2**	unk.	unk.	unk.	unk.	unk.	unk.	0.37	0.55	0.35	0.07	**5**
			0.40	240.3	1.8	unk.	unk.		unk.	unk.	unk.	unk.	unk.	unk.	0.67	0.68	2	4	**7**
J	soup,canned,gazpacho,ready-to-serve	1 C	244	56	0.8	0.78	8.7	**13**	unk.	unk.	unk.	unk.	unk.	unk.	1.32	0.29	1.29	0.05	**3**
			0.41	228.8	2.2	unk.	unk.		unk.	unk.	unk.	unk.	unk.	unk.	0.54	4.50	0	3	**5**
K	soup,canned,lentil w/ham,	1 C	248	139	20.2	1.41	9.3	**14**	unk.	unk.	unk.	unk.	unk.	unk.	0.32	1.12	0.30	0.17	**12**
	ready-to-serve		*0.40*	212.7	2.8	unk.	unk.		unk.	unk.	unk.	unk.	unk.	unk.	1.17	0.29	7	4	**7**
L	soup,canned,minestrone,chunky,	1 C	240	127	20.7	unk.	5.1	**8**	unk.	unk.	unk.	unk.	unk.	unk.	0.26	1.49	0.22	0.05	**3**
	ready-to-serve		*0.42*	208.1	2.8	unk.	unk.		unk.	unk.	unk.	unk.	unk.	unk.	0.84	0.18	5	5	**8**
M	soup,canned,minestrone,condensed	10-1/2 oz can	298	203	27.3	1.79	10.4	**16**	480	316	444	372	435	80	2.71	1.31	2.35	0.12	**8**
			0.34	247.2	6.1	unk.	unk.		179	572	107	253	74	206	1.61	2.07	3	3	**5**
N	soup,canned,minestrone,prep w/water	1 C	241	82	11.2	0.72	4.3	**7**	198	130	183	154	178	34	1.11	0.55	0.96	0.05	**3**
			0.41	220.1	2.5	1.73	unk.		72	236	43	104	31	84	0.65	2.00	2	1	**2**
O	soup,canned,mushroom w/beef stock,	10-3/4 oz can	305	207	22.6	unk.	7.7	**12**	unk.	unk.	unk.	unk.	unk.	unk.	1.89	3.78	1.65	0.09	**6**
	condensed		*0.33*	258.0	9.8	unk.	unk.		unk.	unk.	unk.	unk.	unk.	unk.	3.08	0.50	18	2	**4**
P	soup,canned,mushroom w/beef stock,	1 C	244	85	9.3	unk.	3.1	**5**	unk.	unk.	unk.	unk.	unk.	unk.	0.78	1.56	0.68	0.02	**2**
	prep w/water		*0.41*	224.7	4.0	unk.	unk.		unk.	unk.	unk.	unk.	unk.	unk.	1.27	0.50	7	1	**2**
Q	soup,canned,mushroom,cream of,	10-3/4 oz can	305	314	22.6	0.61	4.9	**8**	204	235	265	226	262	61	10.77	6.25	10.74	0.06	**4**
	condensed		*0.33*	247.7	23.1	unk.	unk.		204	384	95	189	70	183	4.15	1.72	3	3	**5**
R	soup,canned,mushroom,cream of,	1 C	248	203	15.0	0.25	6.0	**9**	231	340	429	288	377	62	4.61	5.13	4.51	0.07	**5**
	prep w/milk		*0.40*	209.7	13.6	unk.	unk.		156	553	141	260	84	270	2.73	0.90	20	2	**4**
S	soup,canned,mushroom,cream of,	1 C	244	129	9.3	0.46	2.3	**4**	95	112	127	107	122	27	4.22	2.44	4.17	0.05	**3**
	prep w/water		*0.41*	220.4	9.0	unk.	unk.		54	181	44	90	34	88	1.61	1.73	2	1	**2**

Riboflavin mg / Niacin mg	% USRDA	Vitamin B_6 mg / Folacin mcg	% USRDA	Vitamin B_{12} mcg / Pantothenic acid mg	% USRDA	Biotin mg / Vitamin A IU	% USRDA	Preformed A RE / Beta carotene RE	Vitamin D IU / Vitamin E IU	% USRDA	Total tocopherol mg / Alpha tocopherol mg	Other tocopherol mg / Total ash g	Calcium mg / Phosphorus mg	% USRDA	Sodium mg / Sodium meq	Potassium mg / Potassium meq	Chlorine mg / Chlorine meq	Iron mg / Magnesium mg	% USRDA	Zinc mg / Copper mg	% USRDA	Iodine mcg / Selenium mcg	% USRDA	Manganese mcg / Chromium mcg	
0.09	**5**	0.24	**12**	9.85	**164**	unk.	**unk.**	unk.	0(a)	**0**	unk.	unk.	58	**6**	2446	457	unk.	4.0	**22**	2.3	**15**	unk.	**unk.**	915	A
2.0	**10**	23	**6**	unk.	**unk.**	2339	**47**	unk.	unk.	**unk.**	unk.	7.50	101	**10**	106.4	11.7	unk.	24	**6**	0.30	**15**	unk.		unk.	
0.05	**3**	0.08	**4**	2.20	**37**	unk.	**unk.**	unk.	0(a)	**0**	unk.	unk.	34	**3**	1808	261	unk.	1.9	**11**	0.9	**6**	unk.	**unk.**	376	B
1.3	**7**	10	**2**	unk.	**unk.**	920	**18**	unk.	unk.	**unk.**	unk.	3.78	59	**6**	78.6	6.7	unk.	10	**2**	0.15	**7**	unk.		unk.	
0.09	**5**	0.18	**9**	23.85	**398**	unk.	**unk.**	unk.	0(a)	**0**	unk.	unk.	101	**10**	2266	278	unk.	3.4	**19**	18	**12**	unk.	**unk.**	610	C
2.3	**11**	9	**2**	unk.	**unk.**	24	**1**	unk.	unk.	**unk.**	unk.	7.47	104	**10**	98.6	7.1	unk.	1.8	**5**	0.30	**15**	unk.		unk.	
0.25	**15**	0.13	**6**	10.24	**171**	unk.	**unk.**	unk.	0(a)	**0**	unk.	unk.	186	**19**	992	300	unk.	1.5	**8**	0.8	**5**	unk.	**unk.**	253	D
1.0	**5**	10	**2**	unk.	**unk.**	164	**3**	unk.	unk.	**unk.**	unk.	3.97	156	**16**	43.1	7.7	unk.	22	**6**	0.14	**7**	unk.		unk.	
0.05	**3**	0.08	**4**	8.00	**133**	unk.	**unk.**	unk.	0(a)	**0**	unk.	unk.	44	**4**	915	146	unk.	1.5	**8**	0.8	**5**	unk.	**unk.**	251	E
1.0	**5**	4	**1**	0.32	**3**	7	**0**	unk.	unk.	**unk.**	unk.	2.98	54	**5**	39.8	3.7	unk.	7	**2**	0.12	**6**	unk.		unk.	
0.06	**4**	0.06	**3**	0.00	**0**	unk.	**unk.**	0	0(a)	**0**	0(a)	0(a)	21	**2**	1550	372	unk.	1.3	**7**	0.9	**6**	unk.	**unk.**	894	F
1.7	**9**	7	**2**	unk.	**unk.**	0	**0**	0	0.0	**0**	0(a)	4.74	78	**8**	67.4	9.5	unk.	0	**0**	0.60	**30**	unk.		unk.	
0.02	**1**	0.02	**1**	0.00	**0**	unk.	**unk.**	0	0(a)	**0**	0(a)	0(a)	10	**1**	636	154	unk.	0.5	**3**	0.4	**2**	unk.	**unk.**	366	G
0.7	**4**	3	**1**	unk.	**unk.**	0	**0**	0	0.0	**0**	0(a)	1.95	31	**3**	27.7	3.9	unk.	0	**0**	0.25	**12**	0		unk.	
0.07	**4**	0.12	**6**	0.20	**3**	unk.	**unk.**	unk.	0(a)	**0**	unk.	unk.	66	**7**	1235	327	unk.	1.2	**7**	unk.	**unk.**	unk.	**unk.**	unk.	H
1.3	**7**	unk.	**unk.**	0.29	**3**	505	**10**	unk.	unk.	**unk.**	unk.	3.44	88	**9**	53.7	8.4	unk.	unk.	**unk.**	unk.	**unk.**	unk.		unk.	
0.05	**3**	unk.	**unk.**	unk.	**unk.**	unk.	**unk.**	unk.	0(a)	**0**	unk.	unk.	32	**3**	3864	unk.	unk.	0.7	**4**	unk.	**unk.**	unk.	**unk.**	unk.	I
2.3	**12**	unk.	**unk.**	unk.	**unk.**	2170	**43**	unk.	unk.	**unk.**	unk.	2.58	unk.	**unk.**	168.1	unk.	unk.	unk.	**unk.**	unk.	**unk.**	unk.		unk.	
0.02	**1**	0.15	**7**	0.00	**0**	unk.	**unk.**	unk.	0(a)	**0**	unk.	unk.	24	**2**	1183	224	unk.	1.0	**5**	unk.	**unk.**	unk.	**unk.**	unk.	J
0.9	**5**	unk.	**unk.**	0.17	**2**	200	**4**	unk.	unk.	**unk.**	unk.	3.49	37	**4**	51.5	5.7	unk.	unk.	**unk.**	unk.	**unk.**	unk.		unk.	
0.12	**7**	0.22	**11**	0.30	**5**	unk.	**unk.**	unk.	0(a)	**0**	unk.	unk.	42	**4**	1319	357	unk.	2.7	**15**	unk.	**unk.**	unk.	**unk.**	unk.	K
1.4	**7**	50	**12**	0.35	**4**	360	**7**	unk.	unk.	**unk.**	unk.	3.08	184	**18**	57.4	9.1	unk.	unk.	**unk.**	unk.	**unk.**	unk.		unk.	
0.12	**7**	unk.	**unk.**	unk.	**unk.**	unk.	**unk.**	unk.	0(a)	**0**	unk.	unk.	60	**6**	864	unk.	unk.	1.8	**10**	unk.	**unk.**	unk.	**unk.**	unk.	L
1.2	**6**	unk.	**unk.**	unk.	**unk.**	4351	**87**	unk.	unk.	**unk.**	unk.	3.24	unk.	**unk.**	37.6	unk.	unk.	unk.	**unk.**	unk.	**unk.**	unk.		unk.	
0.12	**7**	0.24	**12**	0.00	**0**	unk.	**unk.**	unk.	0(a)	**0**	unk.	unk.	83	**8**	2217	760	unk.	2.2	**12**	1.8	**12**	unk.	**unk.**	894	M
2.3	**12**	39	**10**	unk.	**unk.**	5686	**114**	unk.	unk.	**unk.**	unk.	7.00	137	**14**	96.4	19.4	unk.	18	**5**	0.30	**15**	unk.		unk.	
0.05	**3**	0.10	**5**	0.00	**0**	unk.	**unk.**	unk.	0(a)	**0**	unk.	unk.	34	**3**	911	313	unk.	0.9	**5**	0.7	**5**	unk.	**unk.**	366	N
0.9	**5**	16	**4**	unk.	**unk.**	2338	**47**	unk.	unk.	**unk.**	unk.	2.87	55	**6**	39.6	8.0	unk.	7	**2**	0.12	**6**	unk.		unk.	
0.24	**14**	0.09	**5**	0.00	**0**	unk.	**unk.**	unk.	0(a)	**0**	unk.	unk.	24	**2**	2358	384	unk.	2.0	**11**	3.4	**22**	unk.	**unk.**	915	O
2.9	**15**	22	**6**	unk.	**unk.**	3050	**61**	unk.	unk.	**unk.**	unk.	6.92	88	**9**	102.5	9.8	unk.	21	**5**	0.61	**31**	unk.		unk.	
0.10	**6**	0.04	**2**	0.00	**0**	unk.	**unk.**	unk.	0(a)	**0**	unk.	unk.	10	**1**	969	159	unk.	0.8	**5**	1.4	**9**	unk.	**unk.**	376	P
1.2	**6**	9	**2**	unk.	**unk.**	1254	**25**	unk.	unk.	**unk.**	unk.	2.85	37	**4**	42.1	4.1	unk.	10	**2**	0.25	**13**	unk.		unk.	
0.21	**13**	0.03	**2**	unk.	**unk.**	unk.	**unk.**	0	0(a)	**0**	unk.	unk.	79	**8**	2470	204	unk.	1.3	**7**	1.4	**10**	unk.	**unk.**	610	Q
2.0	**10**	unk.	**unk.**	unk.	**unk.**	0	**0**	0	unk.	**unk.**	unk.	6.71	104	**10**	107.5	5.2	unk.	12	**3**	0.30	**15**	unk.		unk.	
0.27	**16**	0.06	**3**	unk.	**unk.**	unk.	**unk.**	unk.	0(a)	**0**	unk.	unk.	179	**18**	1076	270	unk.	0.6	**3**	0.6	**4**	unk.	**unk.**	253	R
0.9	**5**	unk.	**unk.**	unk.	**unk.**	154	**3**	unk.	unk.	**unk.**	unk.	3.65	156	**16**	46.8	6.9	unk.	20	**5**	0.14	**7**	unk.		unk.	
0.10	**6**	0.01	**1**	0.05	**1**	unk.	**unk.**	0	0(a)	**0**	unk.	unk.	46	**5**	1032	100	unk.	0.5	**3**	0.6	**4**	unk.	**unk.**	251	S
0.7	**4**	unk.	**unk.**	0.29	**3**	0	**0**	0	unk.	**unk.**	unk.	2.83	49	**5**	44.9	2.6	unk.	5	**1**	0.12	**6**	unk.		unk.	

	FOOD	Portion	Weight in grams / Conversion for 100 g	Kilocalories / H_2O g	Total carbohydrate g / Total fats g	Crude fiber g / Dietary fiber g	Total protein g / Total sugar g	% USRDA	Arginine mg / Histidine mg	Isoleucine mg / Leucine mg	Lysine mg / Methionine mg	Phenylalanine mg / Threonine mg	Valine mg / Tryptophan mg	Cystine mg / Tyrosine mg	Polyunsat. fatty acids g / Monounsat. fatty acids g	Saturated fatty acids g / P/S ratio	Linoleic acid g / Cholesterol mg	Thiamin mg / Ascorbic acid mg	% USRDA
A	soup,canned,onion,condensed	10-1/2 oz can	298	137	19.9	1.19	9.1	**14**	unk.	unk.	unk.	unk.	unk.	unk.	1.58	0.63	1.49	0.09	**6**
			0.34	257.3	4.2	unk.	unk.		unk.	unk.	unk.	unk.	unk.	unk.	1.79	2.52	0	3	**5**
B	soup,canned,onion,prep w/water	1 C	241	58	8.2	0.48	3.8	**6**	unk.	unk.	unk.	unk.	unk.	unk.	0.65	0.27	0.60	0.02	**2**
			0.41	224.3	1.7	unk.	unk.		unk.	unk.	unk.	unk.	unk.	unk.	0.75	2.45	0	1	**2**
C	soup,canned,oyster stew,condensed	10-1/2 oz can	298	143	9.9	unk.	5.1	**8**	unk.	unk.	unk.	unk.	unk.	unk.	0.39	6.08	0.30	0.06	**4**
			0.34	267.8	9.3	unk.	unk.		unk.	unk.	unk.	unk.	unk.	unk.	1.97	0.06	33	8	**13**
D	soup,canned,oyster stew,prep w/milk	1 C	245	135	9.8	unk.	6.1	**10**	unk.	unk.	unk.	unk.	unk.	unk.	0.32	5.05	0.22	0.07	**5**
			0.41	217.9	7.9	unk.	unk.		unk.	unk.	unk.	unk.	unk.	unk.	1.84	0.06	32	4	**7**
E	soup,canned,oyster stew,prep w/water	1 C	241	58	4.1	unk.	2.1	**3**	unk.	unk.	unk.	unk.	unk.	unk.	0.17	2.51	0.12	0.02	**2**
			0.41	228.6	3.8	unk.	unk.		unk.	unk.	unk.	unk.	unk.	unk.	0.80	0.07	14	3	**5**
F	soup,canned,pea,green,condensed	11-1/4 oz can	319	399	64.4	1.59	20.9	**32**	1716	721	1238	916	1072	195	0.93	3.41	0.83	0.26	**17**
			0.31	218.6	7.1	unk.	unk.		415	1509	252	737	175	609	2.23	0.27	0	4	**7**
G	soup,canned,pea,green,prep w/milk	1 C	254	239	32.2	0.66	12.6	**19**	853	541	831	571	711	117	0.51	4.01	0.43	0.15	**10**
			0.39	197.9	7.0	unk.	unk.		279	1016	206	485	130	447	1.96	0.13	18	3	**5**
H	soup,canned,pea,green,prep w/water	1 C	250	165	26.5	0.65	8.6	**13**	708	298	510	378	443	80	0.37	1.40	0.35	0.10	**7**
			0.40	208.7	2.9	unk.	unk.		170	623	105	303	73	250	0.92	0.27	0	2	**3**
I	soup,canned,pea,split w/ham,chunky, ready to serve	1 C	240	185	26.8	unk.	11.1	**17**	758	468	749	490	526	144	0.58	1.58	0.53	0.12	**8**
			0.42	194.3	4.0	unk.	unk.		233	766	149	391	110	343	1.51	0.36	7	7	**12**
J	soup,canned,pea,split w/ham, condensed	11-1/2 oz can	326	460	67.8	1.63	25.0	**39**	1708	1056	1689	1105	1190	323	1.53	4.27	1.40	0.36	**24**
			0.31	214.1	10.7	unk.	unk.		525	1725	336	883	248	776	4.07	0.36	20	4	**6**
K	soup,canned,pea,split w/ham, prep w/water	1 C	253	190	28.0	0.68	10.3	**16**	703	435	696	455	491	134	0.63	1.77	0.58	0.15	**10**
			0.40	206.9	4.4	unk.	unk.		215	711	139	364	101	319	1.67	0.36	8	2	**3**
L	soup,canned,pepperpot,condensed	10-1/2 oz can	298	250	22.8	1.19	15.5	**24**	1201	569	757	569	724	149	0.86	5.01	0.74	0.12	**8**
			0.34	240.2	11.3	unk.	unk.		221	977	221	474	101	378	4.35	0.17	24	3	**6**
M	soup,canned,pepperpot,prep w/water	1 C	241	104	9.4	0.48	6.4	**10**	494	234	311	234	299	60	0.36	2.05	0.31	0.05	**3**
			0.41	217.3	4.6	unk.	unk.		92	402	92	195	41	157	1.78	0.18	10	1	**2**
N	soup,canned,potato,cream of,condensed	10-3/4 oz can	305	180	27.9	unk.	4.2	**7**	183	186	201	201	226	67	1.01	2.96	0.91	0.09	**6**
			0.33	260.4	5.7	unk.	unk.		95	287	73	149	61	149	1.19	0.34	15	0	**0**
O	soup,canned,potato,cream of, prep w/milk	1 C	248	149	17.2	unk.	5.8	**9**	221	320	402	278	362	64	0.57	3.77	0.47	0.07	**5**
			0.40	214.9	6.4	unk.	unk.		149	513	131	243	82	255	1.51	0.15	22	1	**2**
P	soup,canned,potato,cream of, prep w/water	1 C	244	73	11.5	unk.	1.8	**3**	76	76	83	83	93	27	0.41	1.22	0.39	0.02	**2**
			0.41	225.7	2.4	unk.	unk.		39	117	29	61	24	61	0.49	0.34	5	0	**0**
Q	soup,canned,Scotch broth,condensed	10-1/2 oz can	298	197	23.0	unk.	12.1	**19**	560	453	739	441	530	83	1.34	2.71	1.10	0.06	**4**
			0.34	249.7	6.4	unk.	unk.		560	769	203	375	104	313	1.73	0.49	12	2	**4**
R	soup,canned,Scotch broth, prep w/water	1 C	241	80	9.5	0.00	5.0	**8**	231	0	304	181	217	0	0.55	1.11	0.46	0.02	**2**
			0.41	221.1	2.6	unk.	unk.		108	316	84	0	0	0	0.70	0.50	5	1	**2**
S	soup,canned,shrimp,cream of,condensed	10-3/4 oz can	305	220	19.9	unk.	6.8	**10**	unk.	unk.	unk.	unk.	unk.	unk.	0.46	7.87	0.30	0.06	**4**
			0.33	258.8	12.6	unk.	unk.		unk.	unk.	unk.	unk.	unk.	unk.	3.17	0.06	40	0	**0**

Riboflavin mg / Niacin mg	% USRDA	Vitamin B_6 mg / Folacin mcg	% USRDA	Vitamin B_{12} mcg / Pantothenic acid mg	% USRDA	Biotin mg / Vitamin A IU	% USRDA	Preformed A RE / Beta carotene RE	Vitamin D IU / Vitamin E IU	% USRDA	Total tocopherol mg / Alpha tocopherol mg	Other tocopherol mg / Total ash g	Calcium mg / Phosphorus mg	% USRDA	Sodium mg / Sodium meq	Potassium mg / Potassium meq	Chlorine mg / Chlorine meq	Iron mg / Magnesium mg	% USRDA	Zinc mg / Copper mg	% USRDA	Iodine mcg / Selenium mcg	% USRDA	Manganese mcg / Chromium mcg	
0.06	4	0.12	6	0.00	0	unk.	unk.	0	0(a)	0	unk.	unk.	66	7	2563	167	unk.	1.6	9	1.5	10	unk.	unk.	596	A
1.5	7	37	9	unk.	unk.	0	0	0	unk.	unk.	unk.	7.42	27	3	111.5	4.3	unk.	6	2	0.30	15	unk.		unk.	
0.02	1	0.05	2	0.00	0	unk.	unk.	0	0(a)	0	unk.	unk.	27	3	1053	67	unk.	0.7	4	0.6	4	unk.	unk.	246	B
0.6	3	15	4	unk.	unk.	0	0	0	unk.	unk.	unk.	3.06	12	1	45.8	1.7	unk.	2	1	0.12	6	unk.		unk.	
0.09	5	0.03	2	5.33	89	unk.	unk.	unk.	0(a)	0	unk.	unk.	54	5	2384	119	unk.	2.4	13	25.0	167	unk.	unk.	894	C
0.6	3	unk.	unk.	unk.	unk.	173	4	unk.	unk.	unk.	unk.	5.81	116	12	103.7	3.0	unk.	12	3	3.87	194	unk.		unk.	
0.24	14	0.06	3	2.62	44	unk.	unk.	unk.	0(a)	0	unk.	unk.	167	17	1041	235	unk.	1.1	6	10.3	69	unk.	unk.	370	D
0.3	2	unk.	unk.	unk.	unk.	225	5	unk.	unk.	unk.	unk.	3.28	162	16	45.3	6.0	unk.	20	5	1.60	80	unk.		unk.	
0.05	3	0.01	1	2.19	37	unk.	unk.	unk.	0(a)	0	unk.	unk.	22	2	981	48	unk.	1.0	6	10.3	69	unk.	unk.	366	E
0.2	1	unk.	unk.	unk.	unk.	70	1	unk.	unk.	unk.	unk.	2.39	48	5	42.7	1.2	unk.	5	1	1.59	80	unk.		unk.	
0.16	9	0.13	6	0.00	0	unk.	unk.	unk.	0(a)	0	unk.	unk.	67	7	2399	463	unk.	4.7	26	4.1	28	unk.	unk.	1595	F
3.0	15	4	1	unk.	unk.	488	10	unk.	unk.	unk.	unk.	8.01	303	30	104.3	11.8	unk.	96	24	0.92	46	unk.		unk.	
0.28	16	0.10	5	0.43	7	unk.	unk.	unk.	0(a)	0	unk.	unk.	173	17	1046	376	unk.	2.0	11	1.8	12	unk.	unk.	660	G
1.3	7	8	2	unk.	unk.	356	7	unk.	unk.	unk.	unk.	4.19	239	24	45.5	9.6	unk.	56	14	0.39	20	unk.		unk.	
0.07	4	0.05	3	0.00	0	unk.	unk.	unk.	0(a)	0	unk.	unk.	28	3	988	190	unk.	1.9	11	1.7	11	unk.	unk.	658	H
1.3	6	2	0	unk.	unk.	203	4	unk.	unk.	unk.	unk.	3.30	125	13	42.9	4.9	unk.	40	10	0.38	19	unk.		unk.	
0.10	6	unk.	unk.	unk.	unk.	unk.	unk.	unk.	0(a)	0	unk.	unk.	34	3	965	unk.	unk.	2.1	12	unk.	unk.	unk.	unk.	unk.	I
2.5	13	5	1	unk.	unk.	4872	97	unk.	unk.	unk.	unk.	3.84	unk.	unk.	42.0	unk.	unk.	unk.	unk.	unk.	unk.	unk.		unk.	
0.20	12	0.16	8	unk.	unk.	unk.	unk.	unk.	0(a)	0	unk.	unk.	52	5	2445	968	unk.	5.5	31	3.2	22	unk.	unk.	1630	J
3.6	18	6	2	unk.	unk.	1079	22	unk.	unk.	unk.	unk.	8.38	518	52	106.3	24.8	unk.	117	29	0.90	45	unk.		unk.	
0.08	5	0.07	3	unk.	unk.	unk.	unk.	unk.	0(a)	0	unk.	unk.	23	2	1007	400	unk.	2.3	13	1.3	9	unk.	unk.	670	K
1.5	7	3	1	unk.	unk.	445	9	unk.	unk.	unk.	unk.	3.44	213	21	43.8	10.2	unk.	48	12	0.37	19	unk.		unk.	
0.12	7	0.15	7	0.42	7	unk.	unk.	unk.	0(a)	0	unk.	unk.	57	6	2360	370	unk.	2.2	12	3.0	20	unk.	unk.	1490	L
3.0	15	unk.	unk.	unk.	unk.	2104	42	unk.	unk.	unk.	unk.	8.19	101	10	102.7	9.4	unk.	12	3	0.30	15	unk.		unk.	
0.05	3	0.06	3	0.17	3	unk.	unk.	unk.	0(a)	0	unk.	unk.	24	2	971	152	unk.	0.9	5	1.2	8	unk.	unk.	612	M
1.2	6	unk.	unk.	unk.	unk.	865	17	unk.	unk.	unk.	unk.	3.37	41	4	42.3	3.9	unk.	5	1	0.12	6	unk.		unk.	
0.09	5	0.09	5	unk.	unk.	unk.	unk.	unk.	0(a)	0	unk.	unk.	49	5	2431	332	unk.	1.2	6	1.5	10	unk.	unk.	915	N
1.3	7	7	2	unk.	unk.	701	14	unk.	unk.	unk.	unk.	6.74	113	11	105.7	8.5	unk.	3	1	0.61	31	unk.		unk.	
0.25	15	0.09	5	unk.	unk.	unk.	unk.	unk.	0(a)	0	unk.	unk.	166	17	1061	322	unk.	0.5	3	0.7	5	unk.	unk.	379	O
0.6	3	9	2	unk.	unk.	444	9	unk.	unk.	unk.	unk.	3.65	161	16	46.2	8.3	unk.	17	4	0.26	13	unk.		unk.	
0.05	3	0.04	2	unk.	unk.	unk.	unk.	unk.	0(a)	0	unk.	unk.	20	2	1000	137	unk.	0.5	3	0.6	4	unk.	unk.	376	P
0.5	3	3	1	unk.	unk.	288	6	unk.	unk.	unk.	unk.	2.78	46	5	43.5	3.5	unk.	2	1	0.25	13	unk.		unk.	
0.12	7	0.18	9	0.66	11	unk.	unk.	unk.	0(a)	0	unk.	unk.	36	4	2461	387	unk.	2.0	11	3.9	26	unk.	unk.	894	Q
2.8	14	unk.	unk.	unk.	unk.	5301	106	unk.	unk.	unk.	unk.	6.85	134	13	107.1	9.9	unk.	9	2	0.60	30	unk.		unk.	
0.05	3	0.07	4	0.27	4	unk.	unk.	unk.	0(a)	0	unk.	unk.	14	1	1012	159	unk.	0.8	5	1.6	11	unk.	unk.	366	R
1.2	6	unk.	unk.	unk.	unk.	2179	44	unk.	unk.	unk.	unk.	2.82	55	6	44.0	4.1	unk.	5	1	0.25	12	unk.		unk.	
0.06	4	unk.	unk.	unk.	unk.	unk.	unk.	unk.	0(a)	0	unk.	unk.	43	4	2373	unk.	unk.	1.3	7	1.8	12	unk.	unk.	unk.	S
1.0	5	unk.	unk.	unk.	unk.	384	8	unk.	unk.	unk.	unk.	6.89	unk.	unk.	103.2	unk.	unk.	unk.	unk.	unk.	unk.	unk.		unk.	

	FOOD	Portion	Weight in grams / Conversion for 100 g	Kilocalories / H_2O g	Total carbohydrate g / Total fats g	Crude fiber g / Dietary fiber g	Total protein g / Total sugar g	% USRDA	Arginine mg / Histidine mg	Isoleucine mg / Leucine mg	Lysine mg / Methionine mg	Phenylalanine mg / Threonine mg	Valine mg / Tryptophan mg	Cystine mg / Tyrosine mg	Polyunsat. fatty acids g / Monounsat. fatty acids g	Saturated fatty acids g / P/S ratio	Linoleic acid g / Cholesterol mg	Thiamin mg / Ascorbic acid mg	% USRDA
A	soup,canned,shrimp,cream of,	1 C	248	164	13.9	unk.	6.8	**11**	unk.	unk.	unk.	unk.	unk.	unk.	0.35	5.78	0.22	0.07	**5**
	prep w/milk		*0.40*	214.3	9.3	unk.	unk.		unk.	unk.	unk.	unk.	unk.	unk.	2.33	0.06	35	1	**2**
B	soup,canned,shrimp,cream of,	1 C	244	90	8.2	unk.	2.8	**4**	unk.	unk.	unk.	unk.	unk.	unk.	0.20	3.25	0.12	0.02	**2**
	prep w/water		*0.41*	225.0	5.2	unk.	unk.		unk.	unk.	unk.	unk.	unk.	unk.	1.32	0.06	17	0	**0**
C	soup,canned,stockpot,condensed	11 oz can	312	243	27.9	1.25	11.8	**18**	552	446	727	434	521	81	4.27	2.09	3.90	0.09	**6**
			0.32	255.5	9.5	unk.	unk.		262	755	200	368	103	309	2.34	2.04	9	5	**8**
D	soup,canned,stockpot,prep w/water	1 C	247	99	11.5	0.49	4.9	**8**	227	183	299	178	215	35	1.75	0.86	1.61	0.05	**3**
			0.40	223.7	3.9	unk.	unk.		109	311	82	151	42	126	0.96	2.03	5	2	**3**
E	soup,canned,tomato beef w/noodle,	10-3/4 oz can	305	342	51.5	unk.	10.8	**17**	442	418	589	436	467	128	1.65	3.84	1.40	0.21	**14**
	condensed		*0.33*	225.9	10.4	unk.	unk.		250	705	201	348	104	281	3.69	0.43	9	0	**0**
F	soup,canned,tomato beef w/noodle,	1 C	244	139	21.2	unk.	4.5	**7**	183	171	242	181	193	54	0.68	1.59	0.59	0.07	**5**
	prep w/water		*0.41*	211.5	4.3	unk.	unk.		102	290	83	144	41	115	1.51	0.43	5	0	**0**
G	soup,canned,tomato bisque,condensed	11 oz can	312	300	57.6	unk.	5.5	**8**	159	187	218	187	212	59	2.71	1.31	2.34	0.19	**13**
			0.32	235.0	6.1	unk.	unk.		106	306	72	162	56	144	1.59	2.07	12	14	**24**
H	soup,canned,tomato bisque,prep w/milk	1 C	251	198	29.4	unk.	6.3	**10**	211	321	409	271	356	60	1.23	3.14	1.05	0.13	**8**
			0.40	204.6	6.6	unk.	unk.		153	520	131	248	80	254	1.71	0.39	23	7	**12**
I	soup,canned,tomato bisque,	1 C	247	123	23.7	unk.	2.3	**4**	67	77	89	77	86	25	1.11	0.54	0.96	0.07	**5**
	prep w/water		*0.40*	215.3	2.5	unk.	unk.		44	126	30	67	22	59	0.67	2.05	5	6	**10**
J	soup,canned,tomato rice,condensed	11 oz can	312	290	53.3	1.56	5.1	**8**	unk.	unk.	unk.	unk.	unk.	unk.	3.28	1.25	2.71	0.16	**10**
			0.32	240.6	6.6	unk.	unk.		unk.	unk.	unk.	unk.	unk.	unk.	1.44	2.62	3	36	**60**
K	soup,canned,tomato rice,prep w/water	1 C	247	119	21.9	0.64	2.1	**3**	unk.	unk.	unk.	unk.	unk.	unk.	1.36	0.52	1.11	0.07	**5**
			0.40	217.6	2.7	unk.	unk.		unk.	unk.	unk.	unk.	unk.	unk.	0.59	2.62	2	15	**25**
L	soup,canned,tomato,condensed	10-3/4 oz can	305	207	40.3	1.22	5.0	**8**	146	143	122	171	162	67	2.32	0.88	1.92	0.21	**14**
			0.33	247.8	4.7	unk.	unk.		88	241	55	125	49	104	1.01	2.62	0	162	**269**
M	soup,canned,tomato,prep w/milk	1 C	248	161	22.3	0.50	6.1	**9**	206	303	370	265	335	64	1.12	2.90	0.89	0.12	**8**
			0.40	209.8	6.0	unk.	unk.		146	494	124	233	77	238	1.44	0.38	17	68	**113**
N	soup,canned,tomato,prep w/water	1 C	244	85	16.6	0.49	2.0	**3**	61	59	51	71	66	27	0.95	0.37	0.78	0.10	**7**
			0.41	220.5	1.9	unk.	unk.		37	100	22	51	20	41	0.41	2.60	0	66	**111**
O	soup,canned,turkey noodle,condensed	10-3/4 oz can	305	168	21.0	0.30	9.5	**15**	387	369	515	381	409	113	1.19	1.34	1.13	0.18	**12**
			0.33	263.4	4.8	unk.	tr(a)		220	619	177	305	91	247	1.68	0.89	12	tr	**1**
P	soup,canned,turkey noodle,	1 C	244	68	8.6	0.24	3.9	**6**	159	151	212	156	168	46	0.49	0.56	0.46	0.07	**5**
	prep w/water		*0.41*	226.9	2.0	unk.	tr(a)		90	254	73	124	37	102	0.68	0.87	5	tr	**0**
Q	soup,canned,turkey vegetable,	10-1/2 oz can	298	179	21.0	unk.	7.5	**12**	352	283	462	274	331	51	1.61	2.18	1.49	0.06	**4**
	condensed		*0.34*	256.3	7.4	unk.	unk.		167	480	128	235	66	197	2.71	0.74	3	0	**0**
R	soup,canned,turkey vegetable,	1 C	241	72	8.6	unk.	3.1	**5**	145	116	190	113	135	22	0.67	0.89	0.63	0.02	**2**
	prep w/water		*0.41*	223.9	3.0	unk.	unk.		67	198	53	96	27	82	1.11	0.76	2	0	**0**
S	soup,canned,turkey,chunky,	1 C	236	135	14.1	0.94	10.2	**16**	531	514	809	418	552	106	1.09	1.23	1.04	0.05	**3**
	ready-to-serve		*0.42*	203.8	4.4	unk.	unk.		238	781	215	404	99	307	1.53	0.88	9	6	**11**

	Riboflavin mg / Niacin mg	% USRDA	Vitamin B6 mg / Folacin mcg	% USRDA	Vitamin B12 mcg / Pantothenic acid mg	% USRDA	Biotin mg / Vitamin A IU	% USRDA	Preformed A RE / Beta carotene RE	Vitamin D IU / Vitamin E IU	% USRDA	Total tocopherol mg / Alpha tocopherol mg	Other tocopherol mg / Total ash g	Calcium mg / Phosphorus mg	% USRDA	Sodium mg / Sodium meq	Potassium mg / Potassium meq	Chlorine mg / Chlorine meq	Iron mg / Magnesium mg	% USRDA	Zinc mg / Copper mg	% USRDA	Iodine mcg / Selenium mcg	% USRDA	Manganese mcg / Chromium mcg
A	0.22	**13**	unk.	**unk.**	unk.	**unk.**	unk.	**unk.**	unk.	0(a)	**0**	unk.	unk.	164	**16**	1037	unk.	unk.	0.6	**3**	0.8	**5**	unk.	**unk.**	unk.
	0.5	**3**	unk.	**unk.**	unk.	**unk.**	312	**6**	unk.	unk.	**unk.**	unk.	3.72	unk.	**unk.**	45.1	unk.	unk.	unk.	**unk.**	unk.	**unk.**	unk.		unk.
B	0.02	**1**	unk.	**unk.**	unk.	**unk.**	unk.	**unk.**	unk.	0(a)	**0**	unk.	unk.	17	**2**	976	unk.	unk.	0.5	**3**	0.8	**5**	unk.	**unk.**	unk.
	0.4	**2**	unk.	**unk.**	unk.	**unk.**	159	**3**	unk.	unk.	**unk.**	unk.	2.83	unk.	**unk.**	42.4	unk.	unk.	unk.	**unk.**	unk.	**unk.**	unk.		unk.
C	0.12	**7**	0.22	**11**	0.00	**0**	unk.	**unk.**	unk.	0(a)	**0**	unk.	unk.	53	**5**	2546	577	unk.	2.1	**12**	2.8	**19**	unk.	**unk.**	624
	3.0	**15**	unk.	**unk.**	unk.	**unk.**	9672	**193**	unk.	unk.	**unk.**	unk.	7.30	131	**13**	110.7	14.8	unk.	9	**2**	0.31	**16**	unk.		unk.
D	0.05	**3**	0.09	**4**	0.00	**0**	unk.	**unk.**	unk.	0(a)	**0**	unk.	unk.	22	**2**	1047	237	unk.	0.9	**5**	1.2	**8**	unk.	**unk.**	257
	1.2	**6**	unk.	**unk.**	unk.	**unk.**	3979	**80**	unk.	unk.	**unk.**	unk.	3.01	54	**5**	45.5	6.1	unk.	5	**1**	0.13	**6**	unk.		unk.
E	0.21	**13**	0.21	**11**	0.46	**8**	unk.	**unk.**	unk.	0(a)	**0**	unk.	unk.	43	**4**	2230	537	unk.	2.7	**15**	1.8	**12**	unk.	**unk.**	610
	4.5	**23**	unk.	**unk.**	unk.	**unk.**	1296	**26**	unk.	unk.	**unk.**	unk.	6.34	137	**14**	97.0	13.7	unk.	18	**5**	0.30	**15**	unk.		unk.
F	0.10	**6**	0.09	**4**	0.20	**3**	unk.	**unk.**	unk.	0(a)	**0**	unk.	unk.	17	**2**	917	220	unk.	1.1	**6**	0.8	**5**	unk.	**unk.**	251
	1.9	**9**	unk.	**unk.**	unk.	**unk.**	534	**11**	unk.	unk.	**unk.**	unk.	2.61	56	**6**	39.9	5.6	unk.	7	**2**	0.12	**6**	unk.		unk.
G	0.19	**11**	0.22	**11**	0.00	**0**	unk.	**unk.**	unk.	0(a)	**0**	unk.	unk.	97	**10**	2546	1014	unk.	2.0	**11**	1.4	**10**	unk.	**unk.**	624
	2.8	**14**	unk.	**unk.**	unk.	**unk.**	1753	**35**	unk.	unk.	**unk.**	unk.	7.77	147	**15**	110.7	25.9	unk.	22	**6**	0.31	**16**	unk.		unk.
H	0.28	**16**	0.14	**7**	0.43	**7**	unk.	**unk.**	unk.	0(a)	**0**	unk.	unk.	186	**19**	1109	605	unk.	0.9	**5**	0.6	**4**	unk.	**unk.**	259
	1.3	**6**	unk.	**unk.**	unk.	**unk.**	878	**18**	unk.	unk.	**unk.**	unk.	4.09	173	**17**	48.3	15.5	unk.	25	**6**	0.14	**7**	unk.		unk.
I	0.07	**4**	0.09	**4**	0.00	**0**	unk.	**unk.**	unk.	0(a)	**0**	unk.	unk.	40	**4**	1047	417	unk.	0.8	**5**	0.6	**4**	unk.	**unk.**	257
	1.2	**6**	unk.	**unk.**	unk.	**unk.**	721	**14**	unk.	unk.	**unk.**	unk.	3.19	59	**6**	45.5	10.7	unk.	10	**3**	0.13	**6**	unk.		unk.
J	0.12	**7**	0.19	**9**	0.00	**0**	unk.	**unk.**	unk.	0(a)	**0**	unk.	unk.	56	**6**	1981	802	unk.	1.9	**11**	1.2	**8**	unk.	**unk.**	936
	2.6	**13**	unk.	**unk.**	unk.	**unk.**	1835	**37**	unk.	unk.	**unk.**	unk.	6.33	81	**8**	86.2	20.5	unk.	12	**3**	0.31	**16**	unk.		unk.
K	0.05	**3**	0.08	**4**	0.00	**0**	unk.	**unk.**	unk.	0(a)	**0**	unk.	unk.	22	**2**	815	331	unk.	0.8	**4**	0.5	**4**	unk.	**unk.**	385
	1.1	**5**	unk.	**unk.**	unk.	**unk.**	756	**15**	unk.	unk.	**unk.**	unk.	2.59	35	**4**	35.4	8.5	unk.	5	**1**	0.13	**6**	unk.		unk.
L	0.12	**7**	0.27	**14**	0.00	**0**	unk.	**unk.**	unk.	0(a)	**0**	unk.	unk.	34	**3**	2120	640	unk.	4.3	**24**	0.6	**4**	unk.	**unk.**	610
	3.4	**17**	36	**9**	unk.	**unk.**	1693	**34**	unk.	unk.	**unk.**	unk.	7.20	82	**8**	92.2	16.4	unk.	18	**5**	0.61	**31**	unk.		unk.
M	0.25	**15**	0.16	**8**	0.45	**7**	unk.	**unk.**	unk.	0(a)	**0**	unk.	unk.	159	**16**	932	449	unk.	1.8	**10**	0.3	**2**	unk.	**unk.**	253
	1.5	**8**	21	**5**	unk.	**unk.**	848	**17**	unk.	unk.	**unk.**	unk.	3.84	149	**15**	40.6	11.5	unk.	22	**6**	0.26	**13**	unk.		unk.
N	0.05	**3**	0.11	**6**	0.00	**0**	unk.	**unk.**	unk.	0(a)	**0**	unk.	unk.	12	**1**	871	264	unk.	1.8	**10**	0.2	**2**	unk.	**unk.**	251
	1.4	**7**	15	**4**	unk.	**unk.**	688	**14**	unk.	unk.	**unk.**	unk.	2.95	34	**3**	37.9	6.7	unk.	7	**2**	0.25	**13**	unk.		unk.
O	0.15	**9**	0.09	**5**	unk.	**unk.**	unk.	**unk.**	unk.	0(a)	**0**	unk.	unk.	27	**3**	1982	183	unk.	2.3	**13**	1.4	**9**	unk.	**unk.**	610
	3.4	**17**	unk.	**unk.**	unk.	**unk.**	711	**14**	unk.	unk.	**unk.**	unk.	6.28	116	**12**	86.2	4.7	unk.	12	**3**	0.30	**15**	unk.		unk.
P	0.07	**4**	0.04	**2**	unk.	**unk.**	unk.	**unk.**	unk.	0(a)	**0**	unk.	unk.	12	**1**	815	76	unk.	1.0	**5**	0.6	**4**	unk.	**unk.**	251
	1.4	**7**	unk.	**unk.**	unk.	**unk.**	293	**6**	unk.	unk.	**unk.**	unk.	2.59	49	**5**	35.4	1.9	unk.	5	**1**	unk.	**6**	unk.		unk.
Q	0.09	**5**	0.12	**6**	0.42	**7**	unk.	**unk.**	unk.	0(a)	**0**	unk.	unk.	42	**4**	2202	426	unk.	1.8	**10**	1.5	**10**	unk.	**unk.**	596
	2.4	**12**	unk.	**unk.**	unk.	**unk.**	5948	**119**	unk.	unk.	**unk.**	unk.	5.81	98	**10**	95.8	10.9	unk.	6	**2**	0.30	**15**	unk.		unk.
R	0.05	**3**	0.05	**2**	0.17	**3**	unk.	**unk.**	unk.	0(a)	**0**	unk.	unk.	17	**2**	906	176	unk.	0.8	**4**	0.6	**4**	unk.	**unk.**	246
	1.0	**5**	unk.	**unk.**	unk.	**unk.**	2444	**49**	unk.	unk.	**unk.**	unk.	2.39	41	**4**	39.4	4.5	unk.	5	**1**	0.12	**6**	unk.		unk.
S	0.12	**7**	0.31	**15**	2.12	**35**	unk.	**unk.**	unk.	0(a)	**0**	unk.	unk.	50	**5**	923	361	unk.	1.9	**11**	2.1	**14**	unk.	**unk.**	236
	3.6	**18**	11	**3**	unk.	**unk.**	7156	**143**	unk.	unk.	**unk.**	unk.	3.47	104	**10**	40.1	9.2	unk.	unk.	**unk.**	0.24	**12**	unk.		unk.

	FOOD	Portion	Weight in grams / Conversion for 100 g	Kilocalories / H_2O g	Total carbohydrate g / Total fats g	Crude fiber g / Dietary fiber g	Total protein g / Total sugar g	**% USRDA**	Arginine mg / Histidine mg	Isoleucine mg / Leucine mg	Lysine mg / Methionine mg	Phenylalanine mg / Threonine mg	Valine mg / Tryptophan mg	Cystine mg / Tyrosine mg	Polyunsat. fatty acids g / Monounsat. fatty acids g	Saturated fatty acids g / P/S ratio	Linoleic acid g / Cholesterol mg	Thiamin mg / Ascorbic acid mg	**% USRDA**
A	soup,canned,vegetable meatball, prep w/water-Campbell	1 C	227 / *0.44*	80 / unk.	10.4 / 1.6	unk. / unk.	5.6 / unk.	**9**	unk. / unk.	unk. / unk.	unk. / unk.	unk. / unk.	unk. / unk.	unk. / unk.	unk. / unk.	unk. / unk.	unk. / unk.	0.02 / 2	**2 / 3**
B	soup,canned,vegetable w/beef broth, condensed	10-1/2 oz can	298 / *0.34*	197 / 248.1	31.9 / 4.6	1.49 / unk.	7.2 / unk.	**11**	334 / 122	221 / 399	307 / 74	259 / 176	304 / 51	57 / 143	1.91 / 1.25	1.07 / 1.78	1.64 / 3	0.12 / 6	**8 / 9**
C	soup,canned,vegetable w/beef broth, prep w/water	1 C	241 / *0.41*	82 / 220.5	13.1 / 1.9	0.72 / unk.	3.0 / unk.	**5**	137 / 51	92 / 164	125 / 31	106 / 72	125 / 22	24 / 58	0.80 / 0.53	0.43 / 1.83	0.67 / 2	0.05 / 2	**3 / 4**
D	soup,canned,vegetable w/beef, condensed	10-3/4 oz can	305 / *0.33*	192 / 255.3	24.7 / 4.6	0.76 / unk.	13.6 / unk.	**21**	634 / 299	512 / 869	836 / 229	497 / 424	601 / 116	95 / 354	0.27 / 1.71	2.07 / 0.13	0.24 / 12	0.09 / 6	**6 / 10**
E	soup,canned,vegetable w/beef, prep w/water	1 C	244 / *0.41*	78 / 223.5	10.2 / 1.9	0.32 / unk.	5.6 / unk.	**9**	261 / 122	210 / 359	344 / 95	205 / 173	246 / 49	39 / 146	0.12 / 0.71	0.85 / 0.14	0.10 / 5	0.05 / 2	**3 / 4**
F	soup,canned,vegetable,chunky, ready-to-serve	1 C	240 / *0.42*	122 / 210.2	19.0 / 3.7	1.20 / unk.	3.5 / unk.	**5**	190 / 82	161 / 271	190 / 26	161 / 108	190 / 26	26 / 82	1.39 / 1.58	0.55 / 2.52	1.30 / 0	0.07 / 6	**5 / 10**
G	soup,canned,vegetable,vegetarian, condensed	10-1/2 oz can	298 / *0.34*	176 / 253.0	29.1 / 4.7	1.19 / unk.	5.1 / unk.	**8**	241 / 119	241 / 361	241 / 60	241 / 182	241 / 36	60 / 119	1.76 / 2.00	0.72 / 2.46	1.64 / 0	0.12 / 4	**8 / 6**
H	soup,canned,vegetable,vegetarian, prep w/water	1 C	241 / *0.41*	72 / 222.5	12.0 / 1.9	0.48 / unk.	2.1 / unk.	**3**	99 / 48	99 / 147	99 / 24	99 / 75	99 / 14	24 / 48	0.72 / 0.82	0.29 / 2.50	0.67 / 0	0.05 / 1	**3 / 2**
I	SOUP,DRY,asparagus,cream of	2-1/4 oz pkg	64 / *1.57*	234 / 3.1	35.6 / 6.9	0.51 / unk.	8.8 / unk.	**14**	unk. / unk.	unk. / unk.	unk. / unk.	unk. / unk.	unk. / unk.	unk. / unk.	2.59 / 2.93	1.03 / 2.52	2.41 / 1	unk. / unk.	**unk. / unk.**
J	soup,dry,asparagus,cream of, prep w/water	1 C	251 / *0.40*	58 / 235.7	9.0 / 1.7	0.13 / unk.	2.2 / unk.	**3**	unk. / unk.	unk. / unk.	unk. / unk.	unk. / unk.	unk. / unk.	unk. / unk.	0.65 / 0.73	0.25 / 2.60	0.60 / tr	unk. / unk.	**unk. / unk.**
K	soup,dry,bean w/bacon	1 oz pkg	28 / *3.52*	105 / 1.1	16.4 / 2.2	1.53 / unk.	5.5 / unk.	**9**	unk. / unk.	unk. / unk.	unk. / unk.	unk. / unk.	unk. / unk.	unk. / unk.	0.16 / 0.83	0.96 / 0.17	0.14 / 3	unk. / unk.	**unk. / unk.**
L	soup,dry,bean w/bacon,prep w/water	1 C	265 / *0.38*	106 / 237.8	16.4 / 2.1	1.54 / unk.	5.5 / unk.	**8**	unk. / unk.	unk. / unk.	unk. / unk.	unk. / unk.	unk. / unk.	unk. / unk.	0.16 / 0.82	0.95 / 0.17	0.13 / 3	unk. / unk.	**unk. / unk.**
M	soup,dry,beef broth or bouillion	1/5 oz pkg	6.0 / *16.67*	14 / 0.2	1.4 / 0.5	0.01 / 0.00	1.0 / unk.	**2**	unk. / unk.	unk. / unk.	unk. / unk.	unk. / unk.	unk. / unk.	unk. / unk.	0.02 / 0.19	0.27 / 0.08	0.02 / 1	tr / unk.	**0 / unk.**
N	soup,dry,beef broth or bouillion, prep w/water	1 C	244 / *0.41*	20 / 236.3	1.9 / 0.7	0.02 / unk.	1.3 / unk.	**2**	unk. / unk.	unk. / unk.	unk. / unk.	unk. / unk.	unk. / unk.	unk. / unk.	0.02 / 0.24	0.34 / 0.07	0.02 / 0	tr / unk.	**0 / unk.**
O	soup,dry,beef broth,cubed	1 cube	3.6 / *27.78*	6 / 0.1	0.6 / 0.1	unk. / 0.00	0.6 / unk.	**1**	unk. / unk.	unk. / unk.	unk. / unk.	unk. / unk.	unk. / unk.	unk. / unk.	0.01 / 0.05	0.07 / 0.08	tr / tr	0.01 / unk.	**1 / unk.**
P	soup,dry,beef broth,cubed, prep w/water	1 C	241 / *0.41*	7 / 236.3	0.8 / 0.2	unk. / unk.	0.8 / unk.	**1**	unk. / unk.	unk. / unk.	unk. / unk.	unk. / unk.	unk. / unk.	unk. / unk.	tr / 0.07	0.10 / 0.00	tr / tr	tr / unk.	**0 / unk.**
Q	soup,dry,beef noodle	1/3 oz pkg	9.2 / *10.87*	30 / 0.5	4.5 / 0.6	0.05 / unk.	1.6 / unk.	**3**	unk. / unk.	unk. / unk.	unk. / unk.	unk. / unk.	unk. / unk.	unk. / unk.	0.12 / 0.23	0.19 / 0.64	0.11 / 1	0.09 / tr	**6 / 1**
R	soup,dry,beef noodle,prep w/water	1 C	251 / *0.40*	40 / 239.3	6.0 / 0.8	0.05 / unk.	2.2 / unk.	**3**	unk. / unk.	unk. / unk.	unk. / unk.	unk. / unk.	unk. / unk.	unk. / unk.	0.18 / 0.33	0.25 / 0.70	0.15 / 3	0.13 / 1	**8 / 2**
S	soup,dry,cauliflower	2/3 oz pkg	19 / *5.29*	68 / 0.8	10.7 / 1.7	0.19 / unk.	2.9 / unk.	**5**	unk. / unk.	unk. / unk.	unk. / unk.	unk. / unk.	unk. / unk.	unk. / unk.	0.65 / 0.73	0.26 / 2.51	0.60 / tr	unk. / unk.	**unk. / unk.**

Row	Riboflavin mg / Niacin mg	% USRDA	Vitamin B_6 mg / Folacin mcg	% USRDA	Vitamin B_{12} mcg / Pantothenic acid mg	% USRDA	Biotin mg / Vitamin A IU	% USRDA	Preformed A RE / Beta carotene RE	Vitamin D IU / Vitamin E IU	% USRDA	Total tocopherol mg / Alpha tocopherol mg	Other tocopherol mg / Total ash g	Calcium mg / Phosphorus mg	% USRDA	Sodium mg / Sodium meq	Potassium mg / Potassium meq	Chlorine mg / Chlorine meq	Iron mg / Magnesium mg	% USRDA	Zinc mg / Copper mg	% USRDA	Iodine mcg / Selenium mcg	% USRDA	Manganese mcg / Chromium mcg
A	0.02	**1**	unk.	**unk.**	unk.	**unk.**	unk.	**unk.**	0	0(a)	**0**	0.0	tr(a)	0	**0**	896	unk.	unk.	0.9	**5**	unk.	**unk.**	unk.	**unk.**	unk.
	1.0	**5**	tr(a)	**0**	tr(a)	**0**	1000	**20**	0	0.0	**0**	tr(a)	unk.	unk.	**unk.**	39.0	unk.	unk.	unk.	**unk.**	tr(a)	**0**	tr(a)		unk.
B	0.12	**7**	0.14	**7**	0.00	**0**	unk.	**unk.**	unk.	0(a)	**0**	unk.	unk.	42	**4**	1970	468	unk.	2.4	**13**	1.9	**13**	unk.	**unk.**	819
	2.4	**12**	unk.	**unk.**	unk.	**unk.**	5087	**102**	unk.	unk.	**unk.**	unk.	6.14	95	**10**	85.7	12.0	unk.	15	**4**	0.37	**19**	unk.		unk.
C	0.05	**3**	0.06	**3**	0.00	**0**	unk.	**unk.**	unk.	0(a)	**0**	unk.	unk.	17	**2**	810	193	unk.	1.0	**5**	0.8	**5**	unk.	**unk.**	337
	1.0	**5**	unk.	**unk.**	unk.	**unk.**	2089	**42**	unk.	unk.	**unk.**	unk.	2.51	39	**4**	35.2	4.9	unk.	7	**2**	0.15	**8**	unk.		unk.
D	0.12	**7**	0.18	**9**	0.76	**13**	unk.	**unk.**	unk.	0(a)	**0**	unk.	unk.	40	**4**	2327	421	unk.	2.7	**15**	3.8	**25**	unk.	**unk.**	762
	2.5	**13**	26	**6**	unk.	**unk.**	4599	**92**	unk.	unk.	**unk.**	unk.	6.86	98	**10**	101.2	10.8	unk.	15	**4**	0.44	**22**	unk.		unk.
E	0.05	**3**	0.08	**4**	0.32	**5**	unk.	**unk.**	unk.	0(a)	**0**	unk.	unk.	17	**2**	956	173	unk.	1.1	**6**	1.5	**10**	unk.	**unk.**	315
	1.0	**5**	10	**3**	unk.	**unk.**	1891	**38**	unk.	unk.	**unk.**	unk.	2.83	42	**4**	41.6	4.4	unk.	5	**1**	0.18	**9**	unk.		unk.
F	0.07	**4**	0.19	**10**	0.00	**0**	unk.	**unk.**	unk.	0(a)	**0**	unk.	unk.	55	**6**	1010	396	unk.	1.6	**9**	3.1	**21**	unk.	**unk.**	480
	1.2	**6**	17	**4**	unk.	**unk.**	5878	**118**	unk.	unk.	**unk.**	unk.	3.60	72	**7**	43.9	10.1	unk.	unk.	**unk.**	0.24	**12**	unk.		unk.
G	0.12	**7**	0.13	**7**	0.00	**0**	unk.	**unk.**	unk.	0(a)	**0**	unk.	unk.	51	**5**	2003	510	unk.	2.6	**15**	1.1	**8**	unk.	**unk.**	1117
	2.2	**11**	26	**6**	unk.	**unk.**	7310	**146**	unk.	unk.	**unk.**	unk.	6.02	83	**8**	87.1	13.0	unk.	18	**5**	0.30	**15**	unk.		unk.
H	0.05	**3**	0.06	**3**	0.00	**0**	unk.	**unk.**	unk.	0(a)	**0**	unk.	unk.	22	**2**	822	210	unk.	1.1	**6**	0.5	**3**	unk.	**unk.**	460
	0.9	**5**	11	**3**	unk.	**unk.**	3005	**60**	unk.	unk.	**unk.**	unk.	2.48	34	**3**	35.8	5.4	unk.	7	**2**	0.12	**6**	unk.		unk.
I	unk.	**unk.**	unk.	**unk.**	unk.	**unk.**	unk.	**unk.**	unk.	0(a)	**0**	unk.	unk.	unk.	**unk.**	3177	unk.	unk.	unk.	**unk.**	unk.	**unk.**	unk.	**unk.**	unk.
	unk.	**unk.**	unk.	**unk.**	unk.	**unk.**	unk.	**unk.**	unk.	unk.	**unk.**	unk.	9.51	unk.	**unk.**	138.2	unk.	unk.	unk.	**unk.**	unk.	**unk.**	unk.		unk.
J	unk.	**unk.**	unk.	**unk.**	unk.	**unk.**	unk.	**unk.**	unk.	0(a)	**0**	unk.	unk.	unk.	**unk.**	801	unk.	unk.	unk.	**unk.**	unk.	**unk.**	unk.	**unk.**	unk.
	unk.	**unk.**	unk.	**unk.**	unk.	**unk.**	unk.	**unk.**	unk.	unk.	**unk.**	unk.	2.41	unk.	**unk.**	34.8	unk.	unk.	unk.	**unk.**	unk.	**unk.**	unk.		unk.
K	unk.	**unk.**	unk.	**unk.**	unk.	**unk.**	unk.	**unk.**	unk.	0(a)	**0**	unk.	unk.	unk.	**unk.**	930	327	unk.	unk.	**unk.**	unk.	**unk.**	unk.	**unk.**	unk.
	unk.	**unk.**	unk.	**unk.**	unk.	**unk.**	unk.	**unk.**	unk.	unk.	**unk.**	unk.	3.21	unk.	**unk.**	40.5	8.3	unk.	unk.	**unk.**	unk.	**unk.**	unk.		unk.
L	unk.	**unk.**	unk.	**unk.**	unk.	**unk.**	unk.	**unk.**	unk.	0(a)	**0**	unk.	unk.	unk.	**unk.**	927	326	unk.	unk.	**unk.**	unk.	**unk.**	unk.	**unk.**	unk.
	unk.	**unk.**	unk.	**unk.**	unk.	**unk.**	unk.	**unk.**	unk.	unk.	**unk.**	unk.	3.21	unk.	**unk.**	40.3	8.3	unk.	unk.	**unk.**	unk.	**unk.**	unk.		unk.
M	0.01	**1**	unk.	**unk.**	unk.	**unk.**	unk.	**unk.**	unk.	0(a)	**0**	unk.	unk.	4	**0**	1019	27	unk.	unk.	**unk.**	unk.	**unk.**	unk.	**unk.**	28
	0.3	**1**	unk.	**unk.**	unk.	**unk.**	3	**0**	unk.	unk.	**unk.**	unk.	2.89	19	**2**	44.3	0.7	unk.	3	**1**	unk.	**unk.**	unk.		unk.
N	0.02	**1**	unk.	**unk.**	unk.	**unk.**	unk.	**unk.**	unk.	0(a)	**0**	unk.	unk.	5	**1**	1357	37	unk.	unk.	**unk.**	unk.	**unk.**	unk.	**unk.**	37
	0.4	**2**	unk.	**unk.**	unk.	**unk.**	5	**0**	unk.	unk.	**unk.**	unk.	3.86	24	**2**	59.0	0.9	unk.	5	**1**	unk.	**unk.**	unk.		unk.
O	0.01	**1**	unk.	**unk.**	unk.	**unk.**	unk.	**unk.**	unk.	0(a)	**0**	unk.	unk.	unk.	**unk.**	864	15	unk.	unk.	**unk.**	tr	**0**	unk.	**unk.**	14
	0.1	**1**	unk.	**unk.**	unk.	**unk.**	unk.	**unk.**	unk.	unk.	**unk.**	unk.	2.13	8	**1**	37.6	0.4	unk.	2	**0**	unk.	**unk.**	unk.		unk.
P	0.02	**1**	unk.	**unk.**	unk.	**unk.**	unk.	**unk.**	unk.	0(a)	**0**	unk.	unk.	unk.	**unk.**	1157	19	unk.	0.1	**1**	tr	**0**	unk.	**unk.**	19
	0.2	**1**	unk.	**unk.**	unk.	**unk.**	unk.	**unk.**	unk.	unk.	**unk.**	unk.	2.92	12	**1**	50.3	0.5	unk.	2	**1**	unk.	**unk.**	unk.		unk.
Q	0.05	**3**	0.03	**1**	unk.	**unk.**	unk.	**unk.**	unk.	0(a)	**0**	unk.	unk.	4	**0**	774	60	unk.	0.2	**1**	0.1	**1**	unk.	**unk.**	unk.
	0.5	**3**	1	**0**	unk.	**unk.**	6	**0**	unk.	unk.	**unk.**	unk.	2.03	29	**3**	33.6	1.5	unk.	7	**2**	unk.	**unk.**	unk.		unk.
R	0.05	**3**	0.04	**2**	unk.	**unk.**	unk.	**unk.**	unk.	0(a)	**0**	unk.	unk.	5	**1**	1042	80	unk.	0.3	**2**	0.1	**1**	unk.	**unk.**	unk.
	0.7	**4**	2	**0**	unk.	**unk.**	8	**0**	unk.	unk.	**unk.**	unk.	2.74	40	**4**	45.3	2.0	unk.	10	**3**	unk.	**unk.**	unk.		unk.
S	unk.	**unk.**	unk.	**unk.**	unk.	**unk.**	unk.	**unk.**	unk.	0(a)	**0**	unk.	unk.	unk.	**unk.**	841	unk.	unk.	unk.	**unk.**	unk.	**unk.**	unk.	**unk.**	unk.
	unk.	**unk.**	unk.	**unk.**	unk.	**unk.**	unk.	**unk.**	unk.	unk.	**unk.**	unk.	2.76	unk.	**unk.**	36.6	unk.	unk.	unk.	**unk.**	unk.	**unk.**	unk.		unk.

	FOOD	Portion	Weight in grams / Conversion for 100 g	Kilocalories / H_2O g	Total carbohydrate g / Total fats g	Crude fiber g / Dietary fiber g	Total protein g / Total sugar g	% USRDA	Arginine mg / Histidine mg	Isoleucine mg / Leucine mg	Lysine mg / Methionine mg	Phenylalanine mg / Threonine mg	Valine mg / Tryptophan mg	Cystine mg / Tyrosine mg	Polyunsat. fatty acids g / Monounsat. fatty acids g	Saturated fatty acids g / P/S ratio	Linoleic acid g / Cholesterol mg	Thiamin mg / Ascorbic acid mg	% USRDA
A	soup,dry,cauliflower,prep w/water	1 C	256	69	10.7	0.18	2.9	**5**	unk.	unk.	unk.	unk.	unk.	unk.	0.64	0.26	0.59	unk.	**unk.**
			0.39	237.9	1.7	unk.	unk.		unk.	unk.	unk.	unk.	unk.	unk.	0.72	2.50	tr	unk.	**unk.**
B	soup,dry,celery,cream of	3/5 oz pkg	17	62	9.7	0.19	2.6	**4**	unk.	unk.	unk.	unk.	unk.	unk.	0.61	0.24	0.57	unk.	**unk.**
			5.75	0.8	1.6	unk.	unk.		unk.	unk.	unk.	unk.	unk.	unk.	0.69	2.52	1	unk.	**unk.**
C	soup,dry,celery,cream of,prep w/water	1 C	254	63	9.8	0.20	2.6	**4**	unk.	unk.	unk.	unk.	unk.	unk.	0.61	0.25	0.56	unk.	**unk.**
			0.39	237.4	1.6	unk.	unk.		unk.	unk.	unk.	unk.	unk.	unk.	0.69	2.40	tr	unk.	**unk.**
D	soup,dry,chicken noodle	2.6 oz pkg	74	257	36.0	0.31	14.3	**22**	unk.	unk.	unk.	unk.	unk.	unk.	1.67	1.28	1.55	0.34	**23**
			1.34	2.8	5.7	unk.	unk.		unk.	unk.	unk.	unk.	unk.	unk.	2.28	1.30	10	1	**2**
E	soup,dry,chicken noodle,prep w/water	1 C	252	53	7.4	0.08	2.9	**5**	unk.	unk.	unk.	unk.	unk.	unk.	0.35	0.25	0.30	0.08	**5**
			0.40	237.3	1.2	unk.	unk.		unk.	unk.	unk.	unk.	unk.	unk.	0.45	1.40	3	tr	**0**
F	soup,dry,chicken rice	3/5 oz pkg	16	60	9.2	0.03	2.4	**4**	unk.	unk.	unk.	unk.	unk.	unk.	0.42	0.32	0.39	0.01	**0**
			6.17	0.6	1.4	unk.	unk.		unk.	unk.	unk.	unk.	unk.	unk.	0.58	1.31	3	unk.	**unk.**
G	soup,dry,chicken rice,prep w/water	1 C	253	61	9.3	tr	2.4	**4**	unk.	unk.	unk.	unk.	unk.	unk.	0.43	0.33	0.40	tr	**0**
			0.40	237.4	1.4	unk.	unk.		unk.	unk.	unk.	unk.	unk.	unk.	0.58	1.31	3	tr	**0**
H	soup,dry,chicken vegetable	3/8 oz pkg	11	37	5.8	unk.	2.0	**3**	unk.	unk.	unk.	unk.	unk.	unk.	0.11	0.14	0.11	0.05	**4**
			9.43	0.5	0.6	unk.	unk.		unk.	unk.	unk.	unk.	unk.	unk.	0.21	0.84	2	1	**2**
I	soup,dry,chicken vegetable,	1 C	251	50	7.8	unk.	2.7	**4**	unk.	unk.	unk.	unk.	unk.	unk.	0.15	0.18	0.15	0.08	**5**
	prep w/water		*0.40*	237.5	0.8	unk.	unk.		unk.	unk.	unk.	unk.	unk.	unk.	0.28	0.86	3	1	**2**
J	soup,dry,chicken broth or bouillion	1/5 oz pkg	6.0	16	1.1	0.01	1.0	**2**	unk.	unk.	unk.	unk.	unk.	unk.	0.27	0.21	0.26	0.01	**0**
			16.67	0.1	0.8	unk.	unk.		unk.	unk.	unk.	unk.	unk.	unk.	0.28	1.31	1	tr	**0**
K	soup,dry,chicken broth or bouillion,	1 C	244	22	1.4	0.02	1.3	**2**	unk.	unk.	unk.	unk.	unk.	unk.	0.37	0.27	0.34	tr	**0**
	prep w/water		*0.41*	236.2	1.1	unk.	unk.		unk.	unk.	unk.	unk.	unk.	unk.	0.37	1.36	tr	0	**0**
L	soup,dry,chicken broth,cubed	1 cube	4.8	10	1.1	unk.	0.7	**1**	unk.	unk.	unk.	unk.	unk.	unk.	0.08	0.06	0.08	0.01	**1**
			20.83	0.1	0.2	unk.	unk.		unk.	unk.	unk.	unk.	unk.	unk.	0.08	1.31	1	unk.	**unk.**
M	soup,dry,chicken broth,cubed,	1 C	243	12	1.5	unk.	0.9	**2**	unk.	unk.	unk.	unk.	unk.	unk.	0.10	0.07	0.10	0.02	**2**
	prep w/water		*0.41*	236.8	0.3	unk.	unk.		unk.	unk.	unk.	unk.	unk.	unk.	0.10	1.33	tr	unk.	**unk.**
N	soup,dry,chicken,cream of	2/3 oz pkg	18	80	9.9	0.87	1.3	**2**	unk.	unk.	unk.	unk.	unk.	unk.	0.30	2.52	0.30	unk.	**unk.**
			5.46	0.7	4.0	unk.	unk.		unk.	unk.	unk.	unk.	unk.	unk.	0.79	0.12	2	unk.	**unk.**
O	soup,dry,chicken,cream of,	1 C	261	107	13.3	1.17	1.8	**3**	unk.	unk.	unk.	unk.	unk.	unk.	0.42	3.39	0.42	unk.	**unk.**
	prep w/water		*0.38*	237.4	5.3	unk.	unk.		unk.	unk.	unk.	unk.	unk.	unk.	1.07	0.12	3	unk.	**unk.**
P	soup,dry,consomme,w/gelatin	2 oz pkg	57	78	9.2	0.06	9.8	**15**	unk.	unk.	unk.	unk.	unk.	unk.	0.02	0.01	0.02	unk.	**unk.**
			1.76	0.3	0.1	unk.	unk.		unk.	unk.	unk.	unk.	unk.	unk.	0.02	1.50	0	unk.	**unk.**
Q	soup,dry,consomme,w/gelatin,	1 C	249	17	2.1	0.02	2.2	**3**	unk.	unk.	unk.	unk.	unk.	unk.	tr	tr	tr	unk.	**unk.**
	prep w/water		*0.40*	236.5	tr	unk.	unk.		unk.	unk.	unk.	unk.	unk.	unk.	tr	0.00	0	unk.	**unk.**
R	soup,dry,leek	2-3/4 oz pkg	78	294	47.4	1.09	8.8	**14**	unk.	unk.	unk.	unk.	unk.	unk.	0.34	4.24	0.27	unk.	**unk.**
			1.28	1.9	8.5	unk.	unk.		unk.	unk.	unk.	unk.	unk.	unk.	3.06	0.08	9	unk.	**unk.**
S	soup,dry,leek,prep w/water	1 C	254	71	11.4	0.25	2.1	**3**	unk.	unk.	unk.	unk.	unk.	unk.	0.08	1.02	0.05	unk.	**unk.**
			0.39	235.7	2.1	unk.	unk.		unk.	unk.	unk.	unk.	unk.	unk.	0.74	0.07	3	unk.	**unk.**

Riboflavin mg / Niacin mg	% USRDA	Vitamin B6 mg / Folacin mcg	% USRDA	Vitamin B12 mcg / Pantothenic acid mg	% USRDA	Biotin mg / Vitamin A IU	% USRDA	Preformed A RE / Beta carotene RE	Vitamin D IU / Vitamin E IU	% USRDA	Total tocopherol mg / Alpha tocopherol mg	Other tocopherol mg / Total ash g	Calcium mg / Phosphorus mg	% USRDA	Sodium mg / Sodium meq	Potassium mg / Potassium meq	Chlorine mg / Chlorine meq	Iron mg / Magnesium mg	% USRDA	Zinc mg / Copper mg	% USRDA	Iodine mcg / Selenium mcg	% USRDA	Manganese mcg / Chromium mcg	
unk.	**unk.**	unk.	**unk.**	unk.	**unk.**	unk.	**unk.**	unk.	0(a)	**0**	unk.	unk.	unk.	**unk.**	842	unk.	unk.	unk.	**unk.**	unk.	**unk.**	unk.	**unk.**	unk.	A
unk.	**unk.**	unk.	**unk.**	unk.	**unk.**	unk.	**unk.**	unk.	unk.	**unk.**	unk.	2.76	unk.	**unk.**	36.6	unk.	unk.	unk.	**unk.**	unk.	**unk.**	unk.		unk.	
unk.	**unk.**	unk.	**unk.**	unk.	**unk.**	unk.	**unk.**	unk.	0(a)	**0**	unk.	unk.	unk.	**unk.**	837	unk.	unk.	unk.	**unk.**	unk.	**unk.**	unk.	**unk.**	unk.	B
unk.	**unk.**	unk.	**unk.**	unk.	**unk.**	unk.	**unk.**	unk.	unk.	**unk.**	unk.	2.64	unk.	**unk.**	36.4	unk.	unk.	unk.	**unk.**	unk.	**unk.**	unk.		unk.	
unk.	**unk.**	unk.	**unk.**	unk.	**unk.**	unk.	**unk.**	unk.	0(a)	**0**	unk.	unk.	unk.	**unk.**	838	unk.	unk.	unk.	**unk.**	unk.	**unk.**	unk.	**unk.**	unk.	C
unk.	**unk.**	unk.	**unk.**	unk.	**unk.**	unk.	**unk.**	unk.	unk.	**unk.**	unk.	2.64	unk.	**unk.**	36.5	unk.	unk.	unk.	**unk.**	unk.	**unk.**	unk.		unk.	
0.28	**17**	0.40	**20**	unk.	**unk.**	unk.	**unk.**	unk.	0(a)	**0**	unk.	unk.	154	**15**	6243	153	unk.	2.4	**13**	1.0	**6**	unk.	**unk.**	391	D
4.3	**22**	7	**2**	unk.	**unk.**	307	**6**	unk.	unk.	**unk.**	unk.	15.62	158	**16**	271.5	3.9	unk.	33	**8**	0.17	**8**	unk.		unk.	
0.05	**3**	0.01	**0**	unk.	**unk.**	unk.	**unk.**	unk.	0(a)	**0**	unk.	unk.	33	**3**	1283	30	unk.	0.5	**3**	0.2	**1**	unk.	**unk.**	81	E
0.9	**4**	2	**0**	unk.	**unk.**	63	**1**	unk.	unk.	**unk.**	unk.	3.20	33	**3**	55.8	0.8	unk.	8	**2**	0.04	**2**	unk.		unk.	
tr	**0**	unk.	**unk.**	unk.	**unk.**	unk.	**unk.**	unk.	0(a)	**0**	unk.	unk.	7	**1**	980	12	unk.	0.1	**1**	unk.	**unk.**	unk.	**unk.**	unk.	F
0.4	**2**	unk.	**unk.**	unk.	**unk.**	unk.	**unk.**	unk.	unk.	**unk.**	unk.	2.44	11	**1**	42.6	0.3	unk.	unk.	**unk.**	unk.	**unk.**	unk.		unk.	
tr	**0**	unk.	**unk.**	unk.	**unk.**	unk.	**unk.**	unk.	0(a)	**0**	unk.	unk.	8	**1**	982	10	unk.	tr	**0**	unk.	**unk.**	unk.	**unk.**	unk.	G
0.4	**2**	unk.	**unk.**	unk.	**unk.**	tr	**0**	unk.	unk.	**unk.**	unk.	2.45	10	**1**	42.7	0.3	unk.	unk.	**unk.**	unk.	**unk.**	unk.		unk.	
0.04	**2**	0.07	**3**	unk.	**unk.**	unk.	**unk.**	unk.	0(a)	**0**	unk.	unk.	unk.	**unk.**	604	51	unk.	0.4	**2**	0.2	**1**	unk.	**unk.**	unk.	H
0.5	**3**	unk.	**unk.**	unk.	**unk.**	10	**0**	unk.	unk.	**unk.**	unk.	1.67	24	**2**	26.3	1.3	unk.	16	**4**	0.02	**1**	unk.		unk.	
0.05	**3**	0.09	**4**	unk.	**unk.**	unk.	**unk.**	unk.	0(a)	**0**	unk.	unk.	unk.	**unk.**	808	68	unk.	0.6	**3**	0.2	**1**	unk.	**unk.**	unk.	I
0.7	**3**	unk.	**unk.**	unk.	**unk.**	15	**0**	unk.	unk.	**unk.**	unk.	2.23	33	**3**	35.2	1.7	unk.	23	**6**	0.03	**1**	unk.		unk.	
0.03	**2**	unk.	**unk.**	unk.	**unk.**	unk.	**unk.**	unk.	0(a)	**0**	unk.	unk.	11	**1**	1115	19	unk.	0.1	**0**	tr	**0**	unk.	**unk.**	unk.	J
0.1	**1**	unk.	**unk.**	unk.	**unk.**	30	**1**	unk.	unk.	**unk.**	unk.	2.95	10	**1**	48.5	0.5	unk.	3	**1**	unk.	**unk.**	unk.		unk.	
0.02	**1**	unk.	**unk.**	unk.	**unk.**	unk.	**unk.**	unk.	0(a)	**0**	unk.	unk.	15	**2**	1484	24	unk.	0.1	**0**	tr	**0**	unk.	**unk.**	unk.	K
0.2	**1**	unk.	**unk.**	unk.	**unk.**	39	**1**	unk.	unk.	**unk.**	unk.	3.93	12	**1**	64.5	0.6	unk.	5	**1**	unk.	**unk.**	unk.		unk.	
0.02	**1**	unk.	**unk.**	unk.	**unk.**	unk.	**unk.**	unk.	0(a)	**0**	unk.	unk.	unk.	**unk.**	1152	18	unk.	0.1	**1**	tr	**0**	unk.	**unk.**	18	L
0.2	**1**	unk.	**unk.**	unk.	**unk.**	unk.	**unk.**	unk.	unk.	**unk.**	unk.	2.63	9	**1**	50.1	0.5	unk.	3	**1**	unk.	**unk.**	unk.		unk.	
0.02	**1**	unk.	**unk.**	unk.	**unk.**	unk.	**unk.**	unk.	0(a)	**0**	unk.	unk.	unk.	**unk.**	792	24	unk.	0.1	**1**	tr	**0**	unk.	**unk.**	24	M
0.2	**1**	unk.	**unk.**	unk.	**unk.**	unk.	**unk.**	unk.	unk.	**unk.**	unk.	3.50	12	**1**	34.5	0.6	unk.	2	**1**	unk.	**unk.**	unk.		unk.	
0.15	**9**	unk.	**unk.**	unk.	**unk.**	unk.	**unk.**	unk.	0(a)	**0**	unk.	unk.	57	**6**	882	160	unk.	unk.	**unk.**	unk.	**unk.**	unk.	**unk.**	unk.	N
unk.	**unk.**	unk.	**unk.**	unk.	**unk.**	unk.	**unk.**	unk.	unk.	**unk.**	unk.	2.38	71	**7**	38.4	4.1	unk.	unk.	**unk.**	unk.	**unk.**	unk.		unk.	
0.21	**12**	unk.	**unk.**	unk.	**unk.**	unk.	**unk.**	unk.	0(a)	**0**	unk.	unk.	76	**8**	1185	214	unk.	unk.	**unk.**	unk.	**unk.**	unk.	**unk.**	unk.	O
unk.	**unk.**	unk.	**unk.**	unk.	**unk.**	unk.	**unk.**	unk.	unk.	**unk.**	unk.	3.18	97	**10**	51.5	5.5	unk.	unk.	**unk.**	unk.	**unk.**	unk.		unk.	
unk.	**unk.**	unk.	**unk.**	unk.	**unk.**	unk.	**unk.**	unk.	0(a)	**0**	unk.	unk.	unk.	**unk.**	14855	unk.	unk.	unk.	**unk.**	unk.	**unk.**	unk.	**unk.**	unk.	P
unk.	**unk.**	unk.	**unk.**	unk.	**unk.**	unk.	**unk.**	unk.	unk.	**unk.**	unk.	37.31	unk.	**unk.**	646.2	unk.	unk.	unk.	**unk.**	unk.	**unk.**	unk.		unk.	
unk.	**unk.**	unk.	**unk.**	unk.	**unk.**	unk.	**unk.**	unk.	0(a)	**0**	unk.	unk.	unk.	**unk.**	3299	unk.	unk.	unk.	**unk.**	unk.	**unk.**	unk.	**unk.**	unk.	Q
unk.	**unk.**	unk.	**unk.**	unk.	**unk.**	unk.	**unk.**	unk.	unk.	**unk.**	unk.	8.29	unk.	**unk.**	143.5	unk.	unk.	unk.	**unk.**	unk.	**unk.**	unk.		unk.	
unk.	**unk.**	unk.	**unk.**	unk.	**unk.**	unk.	**unk.**	unk.	0(a)	**0**	unk.	unk.	unk.	**unk.**	4009	unk.	unk.	unk.	**unk.**	1.0	**7**	unk.	**unk.**	unk.	R
unk.	**unk.**	unk.	**unk.**	unk.	**unk.**	unk.	**unk.**	unk.	unk.	**unk.**	unk.	11.39	unk.	**unk.**	174.4	unk.	unk.	unk.	**unk.**	0.15	**7**	unk.		unk.	
unk.	**unk.**	unk.	**unk.**	unk.	**unk.**	unk.	**unk.**	unk.	0(a)	**0**	unk.	unk.	unk.	**unk.**	965	unk.	unk.	unk.	**unk.**	0.2	**2**	unk.	**unk.**	unk.	S
unk.	**unk.**	unk.	**unk.**	unk.	**unk.**	unk.	**unk.**	unk.	unk.	**unk.**	unk.	2.74	unk.	**unk.**	42.0	unk.	unk.	unk.	**unk.**	0.04	**2**	unk.		unk.	

	FOOD	Portion	Weight in grams / Conversion for 100 g	Kilocalories / H_2O g	Total carbohydrate g / Total fats g	Crude fiber g / Dietary fiber g	Total protein g / Total sugar g	% USRDA	Arginine mg / Histidine mg	Isoleucine mg / Leucine mg	Lysine mg / Methionine mg	Phenylalanine mg / Threonine mg	Valine mg / Tryptophan mg	Cystine mg / Tyrosine mg	Polyunsat. fatty acids g / Monounsat. fatty acids g	Saturated fatty acids g / P/S ratio	Linoleic acid g / Cholesterol mg	Thiamin mg / Ascorbic acid mg	% USRDA
A	soup,dry,minestrone	2-3/4 oz pkg	78	279	41.8	1.48	15.6	**24**	unk.	unk.	unk.	unk.	unk.	unk.	0.35	2.87	0.30	unk.	**unk.**
			1.28	3.7	6.1	unk.	unk.		unk.	unk.	unk.	unk.	unk.	unk.	2.27	0.12	6	unk.	**unk.**
B	soup,dry,minestrone,prep w/water	1 C	254	79	11.9	0.43	4.4	**7**	unk.	unk.	unk.	unk.	unk.	unk.	0.10	0.81	0.08	unk.	**unk.**
			0.39	232.8	1.7	unk.	unk.		unk.	unk.	unk.	unk.	unk.	unk.	0.63	0.12	3	unk.	**unk.**
C	soup,dry,mushroom	3/5 oz pkg	17	74	8.5	0.06	1.7	**3**	unk.	unk.	unk.	unk.	unk.	unk.	1.19	0.63	1.19	0.22	**14**
			5.99	0.5	3.7	unk.	unk.		unk.	unk.	unk.	unk.	unk.	unk.	1.67	1.89	tr	unk.	**unk.**
D	soup,dry,mushroom,prep w/water	1 C	253	96	11.1	0.08	2.2	**3**	unk.	unk.	unk.	unk.	unk.	unk.	1.54	0.81	1.54	0.28	**19**
			0.40	231.9	4.9	unk.	unk.		unk.	unk.	unk.	unk.	unk.	unk.	2.18	1.91	0	unk.	**unk.**
E	soup,dry,onion	1/4 oz pkg	7.1	21	3.8	0.17	0.8	**1**	unk.	unk.	unk.	unk.	unk.	unk.	0.05	0.10	0.05	0.02	**1**
			14.08	0.3	0.4	unk.	unk.		unk.	unk.	unk.	unk.	unk.	unk.	0.25	0.50	tr	tr	**0**
F	soup,dry,onion,prep w/water	1 C	246	27	5.1	0.22	1.1	**2**	unk.	unk.	unk.	unk.	unk.	unk.	0.07	0.12	0.07	0.02	**2**
			0.41	236.9	0.6	unk.	unk.		unk.	unk.	unk.	unk.	unk.	unk.	0.32	0.60	0	tr	**0**
G	soup,dry,oxtail	2-3/4 oz pkg	74	280	35.6	0.52	11.2	**17**	unk.	unk.	unk.	unk.	unk.	unk.	0.40	5.04	0.31	unk.	**unk.**
			1.34	3.6	10.1	unk.	unk.		unk.	unk.	unk.	unk.	unk.	unk.	3.65	0.08	11	unk.	**unk.**
H	soup,dry,oxtail,prep w/water	1 C	253	71	9.0	0.13	2.8	**4**	unk.	unk.	unk.	unk.	unk.	unk.	0.10	1.26	0.08	unk.	**unk.**
			0.40	235.2	2.6	unk.	unk.		unk.	unk.	unk.	unk.	unk.	unk.	0.91	0.08	3	unk.	**unk.**
I	soup,dry,pea,green or split	1 oz pkg	28	100	17.0	0.52	5.7	**9**	unk.	unk.	unk.	unk.	unk.	unk.	0.29	0.43	0.24	0.17	**11**
			3.52	1.2	1.2	unk.	unk.		unk.	unk.	unk.	unk.	unk.	unk.	0.72	0.67	tr	tr	**1**
J	soup,dry,pea,green or split, prep w/water	1 C	271	133	22.7	0.68	7.7	**12**	unk.	unk.	unk.	unk.	unk.	unk.	0.30	0.43	0.24	0.22	**15**
			0.37	235.2	1.6	unk.	unk.		unk.	unk.	unk.	unk.	unk.	unk.	0.73	0.69	3	tr	**0**
K	soup,dry,tomato	3/4 oz pkg	21	77	14.5	0.32	1.8	**3**	unk.	unk.	unk.	unk.	unk.	unk.	0.17	0.81	0.17	0.05	**3**
			4.69	0.8	1.8	unk.	unk.		unk.	unk.	unk.	unk.	unk.	unk.	0.64	0.22	1	3	**6**
L	soup,dry,tomato vegetable	1-1/3 oz pkg	39	125	23.1	1.19	4.5	**7**	unk.	unk.	unk.	unk.	unk.	unk.	0.19	0.89	0.19	0.13	**9**
			2.60	1.4	2.0	unk.	unk.		unk.	unk.	unk.	unk.	unk.	unk.	0.70	0.22	1	14	**23**
M	soup,dry,tomato vegetable, prep w/water	1 C	253	56	10.2	0.53	2.0	**3**	unk.	unk.	unk.	unk.	unk.	unk.	0.08	0.40	0.08	0.05	**3**
			0.40	236.6	0.9	unk.	unk.		unk.	unk.	unk.	unk.	unk.	unk.	0.30	0.19	tr	6	**10**
N	soup,dry,tomato,prep w/water	1 C	265	427	19.4	0.42	2.5	**4**	unk.	unk.	unk.	unk.	unk.	unk.	0.24	1.09	0.24	0.05	**4**
			0.38	237.7	2.4	unk.	unk.		unk.	unk.	unk.	unk.	unk.	unk.	0.85	0.22	tr	5	**8**
O	soup,dry,vegetable beef	2-3/5 oz pkg	74	256	38.5	0.74	14.1	**22**	unk.	unk.	unk.	unk.	unk.	unk.	0.22	2.67	0.16	0.14	**9**
			1.34	3.5	5.4	unk.	unk.		unk.	unk.	unk.	unk.	unk.	unk.	1.93	0.08	6	unk.	**unk.**
P	soup,dry,vegetable beef,prep w/water	1 C	253	53	8.0	0.15	2.9	**5**	unk.	unk.	unk.	unk.	unk.	unk.	0.05	0.56	0.03	0.03	**2**
			0.40	238.2	1.1	unk.	unk.		unk.	unk.	unk.	unk.	unk.	unk.	0.05	0.09	tr	unk.	**unk.**
Q	soup,dry,vegetable,cream of	3/5 oz pkg	18	79	9.2	0.11	1.4	**2**	unk.	unk.	unk.	unk.	unk.	unk.	1.11	1.07	1.04	0.92	**61**
			5.65	0.5	4.3	unk.	unk.		unk.	unk.	unk.	unk.	unk.	unk.	1.90	1.04	tr	3	**5**
R	soup,dry,vegetable,cream of, prep w/water	1 C	260	107	12.3	0.13	1.9	**3**	unk.	unk.	unk.	unk.	unk.	unk.	1.48	1.43	1.40	1.22	**82**
			0.38	237.1	5.7	unk.	unk.		unk.	unk.	unk.	unk.	unk.	unk.	2.55	1.04	0	4	**7**
S	SOUP,HOME RECIPE,corn chowder	1 C	264	233	27.2	0.60	9.4	**19**	274	490	603	472	603	80	0.50	6.15	0.37	0.16	**11**
			0.38	214.0	11.1	4.48	10.9		187	921	223	411	112	389	2.75	0.08	75	9	**15**

Riboflavin mg / Niacin mg	% USRDA	Vitamin B_6 mg / Folacin mcg	% USRDA	Vitamin B_{12} mcg / Pantothenic acid mg	% USRDA	Biotin mg / Vitamin A IU	% USRDA	Preformed A RE / Beta carotene RE	Vitamin D IU / Vitamin E IU	% USRDA	Total tocopherol mg / Alpha tocopherol mg	Other tocopherol mg / Total ash g	Calcium mg / Phosphorus mg	% USRDA	Sodium mg / Sodium meq	Potassium mg / Potassium meq	Chlorine mg / Chlorine meq	Iron mg / Magnesium mg	% USRDA	Zinc mg / Copper mg	% USRDA	Iodine mcg / Selenium mcg	% USRDA	Manganese mcg / Chromium mcg	
unk.	**unk.**	unk.	**unk.**	unk.	**unk.**	unk.	**unk.**	unk.	0(a)	**0**	unk.	unk.	unk.	**unk.**	3604	unk.	unk.	unk.	**unk.**	unk.	**unk.**	unk.	**unk.**	unk.	A
unk.	**unk.**	unk.	**unk.**	unk.	**unk.**	unk.	**unk.**	unk.	unk.	**unk.**	unk.	10.76	unk.	**unk.**	156.8	unk.	unk.	unk.	**unk.**	unk.	**unk.**	unk.		unk.	
unk.	**unk.**	unk.	**unk.**	unk.	**unk.**	unk.	**unk.**	unk.	0(a)	**0**	unk.	unk.	unk.	**unk.**	1026	unk.	unk.	unk.	**unk.**	unk.	**unk.**	unk.	**unk.**	unk.	B
unk.	**unk.**	unk.	**unk.**	unk.	**unk.**	unk.	**unk.**	unk.	unk.	**unk.**	unk.	3.07	unk.	**unk.**	44.6	unk.	unk.	unk.	**unk.**	unk.	**unk.**	unk.		unk.	
0.09	**5**	unk.	**unk.**	unk.	**unk.**	unk.	**unk.**	unk.	0(a)	**0**	unk.	unk.	51	**5**	782	153	unk.	unk.	**unk.**	0.1	**0**	unk.	**unk.**	unk.	C
0.4	**2**	unk.	**unk.**	0.00	**0**	5	**0**	unk.	unk.	**unk.**	unk.	2.21	59	**6**	34.0	3.9	unk.	unk.	**unk.**	0.02	**1**	unk.		unk.	
0.10	**6**	unk.	**unk.**	unk.	**unk.**	unk.	**unk.**	unk.	0(a)	**0**	unk.	unk.	unk.	**unk.**	1020	200	unk.	unk.	**unk.**	0.1	**1**	unk.	**unk.**	unk.	D
0.5	**3**	unk.	**unk.**	0.00	**0**	8	**0**	unk.	unk.	**unk.**	unk.	2.88	76	**8**	44.3	5.1	unk.	unk.	**unk.**	0.03	**2**	unk.		unk.	
0.04	**3**	unk.	**unk.**	unk.	**unk.**	unk.	**unk.**	unk.	0(a)	**0**	unk.	unk.	10	**1**	636	47	unk.	0.1	**1**	tr	**0**	unk.	**unk.**	45	E
0.4	**2**	1	**0**	unk.	**unk.**	1	**0**	unk.	unk.	**unk.**	unk.	1.79	23	**2**	27.7	1.2	unk.	5	**1**	0.01	**1**	unk.		unk.	
0.05	**3**	unk.	**unk.**	unk.	**unk.**	unk.	**unk.**	unk.	0(a)	**0**	unk.	unk.	12	**1**	849	64	unk.	0.1	**1**	0.0	**0**	unk.	**unk.**	59	F
0.5	**3**	1	**0**	unk.	**unk.**	2	**0**	unk.	unk.	**unk.**	unk.	2.39	30	**3**	36.9	1.6	unk.	5	**1**	0.01	**1**	unk.		unk.	
unk.	**unk.**	unk.	**unk.**	unk.	**unk.**	unk.	**unk.**	unk.	0(a)	**0**	unk.	unk.	unk.	**unk.**	4806	unk.	unk.	unk.	**unk.**	unk.	**unk.**	unk.	**unk.**	unk.	G
unk.	**unk.**	unk.	**unk.**	unk.	**unk.**	unk.	**unk.**	unk.	unk.	**unk.**	unk.	13.91	unk.	**unk.**	209.1	unk.	unk.	unk.	**unk.**	unk.	**unk.**	unk.		unk.	
unk.	**unk.**	unk.	**unk.**	unk.	**unk.**	unk.	**unk.**	unk.	0(a)	**0**	unk.	unk.	unk.	**unk.**	1209	unk.	unk.	unk.	**unk.**	unk.	**unk.**	unk.	**unk.**	unk.	H
unk.	**unk.**	unk.	**unk.**	unk.	**unk.**	unk.	**unk.**	unk.	unk.	**unk.**	unk.	3.49	unk.	**unk.**	52.6	unk.	unk.	unk.	**unk.**	unk.	**unk.**	unk.		unk.	
0.11	**7**	0.04	**2**	unk.	**unk.**	unk.	**unk.**	unk.	0(a)	**0**	unk.	unk.	17	**2**	914	178	unk.	0.8	**4**	0.4	**3**	unk.	**unk.**	199	I
1.0	**5**	11	**3**	0.20	**2**	36	**1**	unk.	unk.	**unk.**	unk.	2.89	100	**10**	39.7	4.5	unk.	35	**9**	0.14	**7**	unk.		unk.	
0.16	**10**	0.05	**2**	unk.	**unk.**	unk.	**unk.**	unk.	0(a)	**0**	unk.	unk.	22	**2**	1219	238	unk.	1.0	**6**	0.6	**4**	unk.	**unk.**	268	J
1.4	**7**	42	**11**	0.26	**3**	49	**1**	unk.	unk.	**unk.**	unk.	3.85	133	**13**	53.0	6.1	unk.	46	**12**	0.19	**10**	unk.		unk.	
0.04	**2**	0.07	**4**	unk.	**unk.**	unk.	**unk.**	unk.	0(a)	**0**	unk.	unk.	40	**4**	707	221	unk.	0.3	**2**	0.2	**1**	unk.	**unk.**	unk.	K
0.6	**3**	1	**0**	unk.	**unk.**	624	**13**	unk.	unk.	**unk.**	unk.	2.30	50	**5**	30.8	5.7	unk.	11	**3**	0.07	**4**	unk.		unk.	
0.10	**6**	unk.	**unk.**	unk.	**unk.**	unk.	**unk.**	unk.	0(a)	**0**	unk.	unk.	18	**2**	2588	233	unk.	1.4	**8**	0.4	**3**	unk.	**unk.**	unk.	L
1.8	**9**	unk.	**unk.**	0.33	**3**	429	**9**	unk.	unk.	**unk.**	unk.	7.52	67	**7**	112.6	6.0	unk.	44	**11**	0.01	**0**	unk.		unk.	
0.05	**3**	unk.	**unk.**	unk.	**unk.**	unk.	**unk.**	unk.	0(a)	**0**	unk.	unk.	8	**1**	1146	104	unk.	0.6	**4**	0.2	**1**	unk.	**unk.**	unk.	M
0.8	**4**	unk.	**unk.**	1.44	**14**	190	**4**	unk.	unk.	**unk.**	unk.	3.34	30	**3**	49.8	2.6	unk.	20	**5**	1.67	**84**	unk.		unk.	
0.05	**3**	0.10	**5**	unk.	**unk.**	unk.	**unk.**	unk.	0(a)	**0**	unk.	unk.	53	**5**	943	294	unk.	0.4	**2**	0.2	**1**	unk.	**unk.**	unk.	N
0.8	**4**	7	**2**	unk.	**unk.**	832	**17**	unk.	unk.	**unk.**	unk.	3.07	66	**7**	41.0	7.5	unk.	13	**3**	0.09	**5**	unk.		unk.	
0.17	**10**	0.26	**13**	unk.	**unk.**	unk.	**unk.**	unk.	0(a)	**0**	unk.	unk.	unk.	**unk.**	4806	unk.	unk.	4.1	**23**	1.3	**9**	unk.	**unk.**	unk.	O
2.2	**11**	unk.	**unk.**	unk.	**unk.**	1144	**23**	unk.	unk.	**unk.**	unk.	13.02	176	**18**	209.1	unk.	unk.	unk.	**unk.**	0.15	**7**	unk.		unk.	
0.03	**2**	0.05	**3**	unk.	**unk.**	unk.	**unk.**	unk.	0(a)	**0**	unk.	unk.	unk.	**unk.**	1002	unk.	unk.	0.9	**5**	0.3	**2**	unk.	**unk.**	unk.	P
0.5	**2**	unk.	**unk.**	unk.	**unk.**	238	**5**	unk.	unk.	**unk.**	unk.	2.71	35	**4**	43.6	unk.	unk.	unk.	**unk.**	0.03	**2**	unk.		unk.	
0.08	**5**	unk.	**unk.**	unk.	**unk.**	unk.	**unk.**	unk.	0(a)	**0**	unk.	unk.	unk.	**unk.**	877	72	unk.	unk.	**unk.**	unk.	**unk.**	unk.	**unk.**	unk.	Q
0.4	**2**	unk.	**unk.**	unk.	**unk.**	27	**1**	unk.	unk.	**unk.**	unk.	2.28	40	**4**	38.2	1.8	unk.	unk.	**unk.**	unk.	**unk.**	unk.		unk.	
0.10	**6**	unk.	**unk.**	unk.	**unk.**	unk.	**unk.**	unk.	0(a)	**0**	unk.	unk.	unk.	**unk.**	1170	96	unk.	unk.	**unk.**	unk.	**unk.**	unk.	**unk.**	unk.	R
0.5	**3**	unk.	**unk.**	unk.	**unk.**	36	**1**	unk.	unk.	**unk.**	unk.	3.04	55	**6**	50.9	2.5	unk.	unk.	**unk.**	unk.	**unk.**	unk.		unk.	
0.37	**22**	0.31	**16**	0.66	**11**	0.003	**1**	0	70	**17**	1.2	tr(a)	206	**21**	312	426	17	0.9	**5**	1.1	**7**	64.2	**43**	18	S
1.3	**7**	15	**4**	0.96	**10**	676	**14**	31	1.5	**5**	0.2	2.32	248	**25**	13.6	10.9	0.5	68	**17**	0.33	**16**	2		2	

	FOOD	Portion	Weight in grams / *Conversion for 100 g*	Kilocalories / H_2O g	Total carbohydrate g / Total fats g	Crude fiber g / Dietary fiber g	Total protein g / Total sugar g	**% USRDA**	Arginine mg / Histidine mg	Isoleucine mg / Leucine mg	Lysine mg / Methionine mg	Phenylalanine mg / Threonine mg	Valine mg / Tryptophan mg	Cystine mg / Tyrosine mg	Polyunsat. fatty acids g / Monounsat. fatty acids g	Saturated fatty acids g / P/S ratio	Linoleic acid g / Cholesterol mg	Thiamin mg / Ascorbic acid mg	**% USRDA**
A	soup,home recipe,Greek	3/4 C	168	63	6.7	0.02	4.2	**9**	130	122	161	119	163	52	0.28	0.77	0.27	0.05	**3**
			0.60	153.0	2.1	1.57	0.1		74	207	74	119	34	110	0.84	0.37	83	4	**6**
B	soup,home recipe,lentil	1 C	259	175	29.9	2.08	11.5	**18**	901	550	654	482	559	93	0.00	0.49	0.16	0.16	**10**
			0.39	210.3	1.8	2.57	4.5		39	751	83	383	104	293	0.72	0.00	0	23	**38**
C	soup,home recipe,mock turtle	1 C	235	256	11.3	0.85	20.4	**46**	1245	987	1677	872	1024	270	1.10	3.45	0.34	0.10	**7**
			0.43	187.9	14.3	2.92	5.3		646	1568	508	897	236	712	3.37	0.32	164	20	**33**
D	soup,home recipe,potato	1 C	155	201	18.8	0.56	6.6	**13**	297	354	470	299	401	82	0.44	7.10	0.26	0.10	**7**
			0.65	204.7	11.5	1.07	9.9		136	549	140	276	86	290	2.86	0.06	37	16	**27**
E	soup,home recipe,seafood chowder	1 C	155	170	6.5	0.24	21.7	**46**	1529	1092	1878	848	1133	222	2.44	1.40	2.21	0.07	**5**
			0.65	116.1	5.9	0.58	0.0		541	1681	636	923	213	816	0.90	1.74	68	13	**22**
F	soup,home recipe,vegetable	1 C	221	70	9.6	0.39	7.2	**15**	55	61	112	75	95	22	0.00	0.00	0.00	0.05	**4**
			0.45	162.6	0.7	2.24	3.0		unk.	107	22	106	14	27	0.00	0.00	0	19	**31**
G	soup,home recipe,vegetable beef	1 C	250	320	10.1	0.71	13.9	**30**	770	652	1150	604	676	160	3.01	10.16	0.52	0.08	**5**
			0.40	210.3	24.6	1.44	3.6		436	1083	341	618	148	471	9.97	0.30	54	17	**28**
H	soup,home recipe,wonton	1 C	250	205	26.1	0.22	16.3	**35**	632	770	968	636	803	231	0.40	1.25	0.27	0.17	**12**
			0.40	201.9	3.4	0.59	0.8		384	1069	341	609	161	468	1.20	0.32	89	5	**9**
I	SOY protein,texturized-ADM	1 oz	28	80	8.9	0.85	14.8	**33**	unk.	700	899	739	743	116	0.17	0.06	unk.	0.18	**12**
			3.52	1.7	0.3	unk.	2.8		383	1157	169	616	97	460	0.06	3.00	0	0	**0**
J	soy sauce	1 Tbsp	18	12	1.5	unk.	1.6	**2**	unk.	unk.	unk.	unk.	unk.	unk.	0.00	0.00	0.00	0.01	**1**
			5.56	12.2	0.0	unk.	unk.		unk.	unk.	unk.	unk.	unk.	unk.	0.00	0.00	0	0	**0**
K	SOYBEAN,home recipe,casserole	3/4 C	175	127	9.3	1.31	7.7	**13**	unk.	410	537	385	434	111	0.16	3.36	1.44	0.14	**9**
			0.57	150.0	7.4	0.83	3.7		72	625	130	278	106	279	1.64	0.05	12	15	**26**
L	soybeans,mature seeds,cooked	1/2 C	90	117	9.7	1.44	9.9	**15**	unk.	584	683	535	574	193	unk.	0.90	2.70	0.19	**13**
			1.11	63.9	5.1	unk.	unk.		unk.	841	148	381	148	345	0.90	unk.	unk.	0	**0**
M	SPAGHETTI,canned,spaghetti & meatballs	1 C	250	123	17.0	unk.	5.3	**12**	unk.	unk.	unk.	unk.	unk.	unk.	unk.	unk.	unk.	unk.	**unk.**
	-Chef Boy-Ar-Dee		*0.40*	unk.	3.5	unk.	unk.		unk.	unk.	unk.	unk.	unk.	unk.	unk.	unk.	unk.	unk.	**unk.**
N	spaghetti,canned,spaghetti'n beef	1 C	213	212	26.0	0.64	8.7	**13**	unk.	unk.	unk.	unk.	unk.	unk.	unk.	unk.	unk.	0.19	**13**
	-Franco-American		*0.47*	166.4	8.1	unk.	unk.		unk.	unk.	unk.	unk.	unk.	unk.	unk.	unk.	unk.	4	**7**
O	spaghetti,canned,tomato sauce &	1 C	250	88	18.4	unk.	2.3	**4**	unk.	unk.	unk.	unk.	unk.	unk.	unk.	unk.	unk.	unk.	**unk.**
	cheese-Chef Boy-Ar-Dee		*0.40*	unk.	0.6	unk.	unk.		unk.	unk.	unk.	unk.	unk.	unk.	unk.	unk.	unk.	unk.	**unk.**
P	spaghetti,canned,tomato sauce	1 C	250	190	38.5	0.50	5.5	**10**	unk.	275	175	275	325	100	unk.	1.00	1.00	0.35	**23**
	w/cheese		*0.40*	200.2	1.5	unk.	unk.		125	375	75	225	75	175	1.00	unk.	15	10	**17**
Q	spaghetti,canned,w/ground beef	1 C	250	123	16.4	unk.	4.1	**9**	unk.	unk.	unk.	unk.	unk.	unk.	unk.	unk.	unk.	unk.	**unk.**
	-Chef Boy-Ar-Dee		*0.40*	unk.	5.3	unk.	unk.		unk.	unk.	unk.	unk.	unk.	unk.	unk.	unk.	unk.	unk.	**unk.**
R	spaghetti,canned,w/meatballs &	1 C	250	258	28.5	0.25	12.3	**27**	unk.	unk.	unk.	unk.	unk.	unk.	unk.	2.15	4.00	0.15	**10**
	tomato sauce		*0.40*	195.0	10.2	unk.	unk.		unk.	unk.	unk.	unk.	unk.	unk.	3.25	unk.	23	5	**8**
S	spaghetti,enriched,all varieties,	1/2 C	70	78	16.1	0.07	2.4	**4**	unk.	119	76	124	136	45	unk.	tr(a)	unk.	0.10	**7**
	cooked		*1.43*	50.4	0.3	unk.	0.3		56	157	36	93	29	78	unk.	unk.	0	0	**0**

Riboflavin mg / Niacin mg	% USRDA	Vitamin B_6 mg / Folacin mcg	% USRDA	Vitamin B_{12} mcg / Pantothenic acid mg	% USRDA	Biotin mg / Vitamin A IU	% USRDA	Preformed A RE / Beta carotene RE	Vitamin D IU / Vitamin E IU	% USRDA	Total tocopherol mg / Alpha tocopherol mg	Other tocopherol mg / Total ash g	Calcium mg / Phosphorus mg	% USRDA	Sodium mg / Sodium meq	Potassium mg / Potassium meq	Chlorine mg / Chlorine meq	Iron mg / Magnesium mg	% USRDA	Zinc mg / Copper mg	% USRDA	Iodine mcg / Selenium mcg	% USRDA	Manganese mcg / Chromium mcg	
0.05	3	0.04	2	0.18	3	0.004	1	0	6	2	0.2	unk.	22	2	386	45	30	1.2	7	0.4	3	12.3	8	8	A
0.3	2	8	2	0.28	3	39	1	tr	0.2	1	unk.	1.42	87	9	16.8	1.1	0.8	43	11	0.01	1	0		0	
0.13	8	0.27	14	0.00	0	0.007	2	0	0	0	0.1	0.0	61	6	1242	614	22	3.4	19	1.5	10	276.2	184	47	B
1.7	8	47	12	0.80	8	1681	34	167	0.1	0	0.1	5.09	184	18	54.0	15.7	0.6	22	6	0.14	7	5		4	
0.20	12	0.27	13	0.90	15	0.002	1	4	0	0	0.4	tr	54	5	1011	646	77	2.8	16	2.8	18	212.5	142	51	C
2.9	15	23	6	0.76	8	3372	67	331	0.5	2	0.4	3.79	197	20	44.0	16.5	2.2	31	8	0.15	8	2		16	
0.26	15	0.23	12	0.52	9	0.002	1	0	0(a)	0	0.4	0.1	194	19	315	487	24	0.6	3	0.9	6	10.3	6.86	26	D
1.0	5	23	6	0.80	8	471	9	4	0.5	2	0.4	2.35	179	18	13.7	12.5	0.7	38	10	0.35	17	0		4	
0.14	8	0.07	4	0.65	11	0.002	1	3	28	7	0.1	tr(a)	67	7	1043	333	unk.	1.9	11	1.2	8	95.0	63	126	E
2.7	13	8	2	0.40	4	295	6	28	0.1	0	tr(a)	4.52	269	27	45.4	8.5	unk.	31	8	0.14	7	0		0	
0.05	3	0.12	6	0.00	0	0.001	0	0	0(a)	0	0.1	0.0	35	4	498	279	32	1.2	7	0.2	2	35.0	23	17	F
0.9	5	22	5	0.24	2	2230	45	223	0.1	0	0.1	1.87	69	7	21.6	7.1	0.9	14	4	0.10	5	9		tr	
0.13	8	0.27	14	1.13	19	0.001	0	12	0	0	0.4	0.2	26	3	746	504	148	2.6	14	2.5	17	141.9	95	15	G
2.7	14	22	6	0.42	4	1434	29	138	0.5	2	0.2	2.85	104	10	32.5	12.9	4.2	31	8	0.11	6	tr		2	
0.15	9	0.22	11	0.20	3	0.004	1	19	3	1	1.0	tr(a)	32	3	322	226	16	2.9	16	1.2	8	4.8	3	59	H
4.8	24	15	4	0.45	5	807	16	73	1.2	4	tr(a)	2.25	192	19	14.0	5.8	0.5	53	13	0.08	4	tr(a)		tr	
0.17	10	0.40	20	1.62	27	0.021	7	0	0	0	0.0	0.0	62	6	199	625	5	2.8	16	1.6	10	142.0	95	738	I
4.5	23	101	25	0.57	6	0	0	0	0.0	0	0.0	1.70	162	16	8.6	16.0	0.1	74	19	0.37	19	tr(a)		unk.	
0.02	1	0.03	2	0.00	0	unk.	unk.	0	0(a)	0	0(a)	0(a)	3	0	1029	64	unk.	0.5	3	tr	0	unk.	unk.	0	J
0.6	3	2	1	0.06	1	0	0	0	0.0	0	0(a)	2.78	38	4	44.8	1.6	unk.	8	2	0.02	1	unk.		unk.	
0.12	7	0.17	8	0.00	0	0.011	4	0	0(a)	0	1.7	0.2	112	11	114	434	21	1.8	10	0.3	2	unk.	unk.	13	K
1.0	5	40	10	0.37	4	752	15	59	2.0	7	0.2	1.51	142	14	4.9	11.1	0.6	18	5	0.05	3	tr		0	
0.08	5	0.17	8	0(a)	0	0.016	5	0	0(a)	0	unk.	unk.	66	7	2	486	unk.	2.4	14	unk.	unk.	unk.	unk.	unk.	L
0.5	3	58	15	0.44	4	27	1	3	unk.	unk.	unk.	1.35	161	16	0.1	12.4	unk.	unk.	unk.	unk.	unk.	unk.		unk.	
unk.	unk.	unk.	unk.	unk.	unk.	tr(a)	0	tr(a)	tr(a)	0	unk.	unk.	unk.	unk.	1138	unk.	unk.	unk.	unk.	2.2	15	unk.	unk.	unk.	M
unk.	unk.	unk.	unk.	unk.	unk.	500	10	unk.	unk.	unk.	unk.	unk.	unk.	unk.	49.5	unk.	unk.	unk.	unk.	unk.	unk.	unk.		unk.	
0.19	11	unk.	unk.	unk.	unk.	unk.	unk.	unk.	0(a)	0	unk.	unk.	32	3	956	392	unk.	2.3	13	unk.	unk.	unk.	unk.	unk.	N
3.2	16	unk.	unk.	unk.	unk.	850	17	unk.	unk.	unk.	unk.	3.90	113	11	41.6	10.0	unk.	unk.	unk.	unk.	unk.	unk.		unk.	
unk.	unk.	unk.	unk.	tr(a)	0	tr(a)	0	tr(a)	tr(a)	0	unk.	unk.	unk.	unk.	1300	unk.	unk.	unk.	unk.	unk.	unk.	unk.	unk.	unk.	O
unk.	unk.	unk.	unk.	unk.	unk.	250	5	unk.	unk.	unk.	unk.	unk.	unk.	unk.	56.5	unk.	unk.	unk.	unk.	unk.	unk.	unk.		unk.	
0.27	16	0.12	6	0.63	10	0.000	0	29	3	1	1.0	unk.	40	4	955	303	unk.	2.7	15	0.2	2	unk.	unk.	unk.	P
4.5	23	3	1	0.75	8	925	19	83	1.2	4	unk.	4.25	88	9	41.5	7.7	unk.	28	7	0.30	15	unk.		unk.	
unk.	unk.	unk.	unk.	unk.	unk.	tr(a)	0	tr(a)	tr(a)	0	unk.	unk.	unk.	unk.	1238	unk.	unk.	unk.	unk.	unk.	unk.	unk.	unk.	unk.	Q
unk.	unk.	unk.	unk.	unk.	unk.	750	15	unk.	unk.	unk.	unk.	unk.	unk.	unk.	53.8	unk.	unk.	unk.	unk.	unk.	unk.	unk.		unk.	
0.17	10	unk.	unk.	unk.	unk.	tr(a)	0	unk.	tr(a)	0	unk.	unk.	53	5	1220	245	unk.	3.2	18	unk.	unk.	unk.	unk.	unk.	R
2.2	11	unk.	unk.	unk.	unk.	1000	20	unk.	unk.	unk.	1.1	4.00	113	11	53.1	6.3	unk.	unk.	unk.	unk.	unk.	unk.		unk.	
0.06	3	0.01	1	0.00	0	0.000	0	0	0	0	tr	unk.	6	1	1	43	unk.	0.6	4	0.4	2	unk.	unk.	unk.	S
0.8	4	3	1	0.07	1	0	0	0	tr	0	tr	0.84	35	4	0.0	1.1	unk.	13	3	0.01	1	unk.		unk.	

	FOOD	Portion	Weight in grams / Conversion for 100 g	Kilocalories / H_2O g	Total carbohydrate g / Total fats g	Crude fiber g / Dietary fiber g	Total protein g / Total sugar g	% USRDA	Arginine mg / Histidine mg	Isoleucine mg / Leucine mg	Lysine mg / Methionine mg	Phenylalanine mg / Threonine mg	Valine mg / Tryptophan mg	Cystine mg / Tyrosine mg	Polyunsat. fatty acids g / Monounsat. fatty acids g	Saturated fatty acids g / P/S ratio	Linoleic acid g / Cholesterol mg	Thiamin mg / Ascorbic acid mg	% USRDA
A	spaghetti,home recipe,w/meatballs &	1 C	253	407	38.4	0.77	21.0	**42**	unk.	959	1480	920	1063	278	4.16	5.77	3.39	0.32	**22**
	cheese		*0.40*	172.5	19.1	unk.	12.9		649	1566	468	850	232	763	5.61	0.72	104	45	**75**
B	spaghetti,home recipe,w/meatballs &	1 C	248	332	38.7	0.74	18.6	**41**	unk.	868	1290	769	868	248	unk.	2.48	tr	0.25	**17**
	tomato sauce		*0.40*	173.6	11.7	unk.	unk.		546	1314	298	744	198	546	7.44	unk.	74	22	**37**
C	SPINACH,canned,drained solids	1/2 C	103	25	3.7	0.92	2.8	**4**	177	129	170	125	150	45	0.00	0.00	unk.	0.02	**1**
			0.98	93.7	0.6	6.46	1.2		71	211	47	120	44	100	unk.	0.00	0	14	**24**
D	spinach,canned,low sodium,	1/2 C	103	27	4.1	1.02	3.3	**5**	210	129	170	117	150	53	0.00	0.00	unk.	0.02	**1**
	drained solids		*0.98*	93.6	0.5	6.46	1.2		83	211	47	120	44	unk.	unk.	0.00	0	14	**24**
E	spinach,fresh,cooked,drained solids	1/2 C	90	21	3.2	0.54	2.7	**4**	173	127	167	122	148	44	unk.	tr(a)	unk.	0.06	**4**
			1.11	82.8	0.3	5.67	1.1		68	208	46	119	43	98	unk.	unk.	0	25	**42**
F	spinach,frozen,chopped,cooked,	1/2 C	103	24	3.8	0.82	3.1	**5**	197	145	191	132	169	50	0.00	0.00	unk.	0.07	**5**
	drained solids		*0.98*	94.2	0.3	6.46	1.2		78	237	52	135	49	unk.	unk.	0.00	0	19	**33**
G	spinach,frozen,souffle-Stouffer	1/2 pkg	170	201	17.9	unk.	7.5	**12**	unk.	unk.	unk.	unk.	unk.	unk.	unk.	unk.	unk.	0.19	**13**
			0.59	127.8	10.4	unk.	unk.		unk.	unk.	unk.	unk.	unk.	unk.	unk.	unk.	unk.	tr	**0**
H	spinach,pie,home recipe	1/8 of 9'' pie	183	329	28.8	0.92	6.5	**10**	224	269	274	297	285	103	7.17	4.81	6.86	0.22	**15**
			0.55	123.6	21.8	6.14	2.8		131	447	90	210	92	211	8.58	1.49	0	62	**104**
I	spinach,raw,chopped	1/2 C	28	7	1.2	0.16	0.9	**1**	56	41	54	40	48	14	unk.	tr(a)	unk.	0.03	**2**
			3.64	24.9	0.1	1.73	0.3		22	68	15	39	14	32	unk.	unk.	0	14	**23**
J	SQUAB (pigeon),breast,meat w/o skin,	1 bird	101	135	0.0	0.00	22.0	**49**	unk.	unk.	unk.	unk.	unk.	unk.	0.97	1.19	0.63	unk.	**unk.**
	raw		*0.99*	73.5	4.6	0(a)	0(a)		unk.	unk.	unk.	unk.	unk.	unk.	1.44	0.81	91	unk.	**unk.**
K	squab (pigeon),meat w/skin	1 bird	199	585	0.0	0.00	36.8	**82**	unk.	unk.	unk.	unk.	unk.	unk.	6.11	16.78	5.31	unk.	**unk.**
			0.50	112.6	47.4	0(a)	0(a)		unk.	unk.	unk.	unk.	unk.	unk.	12.46	0.36	unk.	unk.	**unk.**
L	squab (pigeon),meat w/o skin,raw	1 bird	168	239	0.0	0.00	29.4	**65**	unk.	unk.	unk.	unk.	unk.	unk.	2.69	3.29	1.73	unk.	**unk.**
			0.60	122.3	12.6	0(a)	0(a)		unk.	unk.	unk.	unk.	unk.	unk.	3.98	0.82	unk.	unk.	**unk.**
M	SQUASH/pumpkin seed kernels, dry,	1/4 C	35	194	5.2	0.66	10.1	**16**	unk.	568	467	568	548	unk.	unk.	2.80	7.00	0.08	**6**
	hulled		*2.86*	1.5	16.3	unk.	0.0		233	802	193	304	183	unk.	5.95	unk.	0	unk.	**unk.**
N	squash/pumpkin seeds	1/4 C	38	155	4.2	0.53	8.1	**13**	unk.	456	374	456	439	unk.	unk.	2.24	5.61	0.07	**5**
			2.64	1.7	13.1	unk.	0.0		187	642	155	244	146	unk.	4.78	unk.	0	tr(a)	**0**
O	squash,summer,fresh,sliced,cooked,	1/2 C	90	13	2.8	0.54	0.8	**1**	unk.	26	31	22	30	unk.	unk.	tr(a)	unk.	0.04	**3**
	drained solids		*1.11*	85.9	0.1	unk.	3.4		unk.	37	11	19	6	unk.	unk.	unk.	0	9	**15**
P	squash,winter,fresh,baked & mashed	1/2 C	103	65	15.8	1.84	1.8	**3**	unk.	59	70	502	68	unk.	unk.	tr(a)	unk.	0.05	**3**
			0.98	83.4	0.4	unk.	3.8		unk.	83	24	42	14	unk.	unk.	unk.	0	13	**22**
Q	squash,winter,fresh,boiled & mashed	1/2 C	123	47	11.3	1.71	1.3	**2**	unk.	43	51	37	50	unk.	unk.	tr(a)	unk.	0.05	**3**
			0.82	108.8	0.4	unk.	4.6		unk.	61	17	31	11	unk.	unk.	unk.	0	10	**16**
R	STEW,canned,beef w/vegetables,	1 C	235	240	16.9	unk.	16.4	**37**	unk.	unk.	unk.	unk.	unk.	unk.	unk.	unk.	unk.	unk.	**unk.**
	unsalted		*0.43*	unk.	11.7	unk.	unk.		unk.	unk.	unk.	unk.	unk.	unk.	unk.	unk.	unk.	unk.	**unk.**
S	stew,canned,beef w/vegetables	1 C	235	186	16.7	0.70	13.6	**21**	unk.	663	1079	552	724	unk.	unk.	unk.	unk.	0.07	**5**
			0.43	193.9	7.3	unk.	unk.		unk.	1003	296	566	157	unk.	unk.	unk.	33	7	**12**

Riboflavin mg / Niacin mg	% USRDA	Vitamin B_6 mg / Folacin mcg	% USRDA	Vitamin B_{12} mcg / Pantothenic acid mg	% USRDA	Biotin mg / Vitamin A IU	% USRDA	Preformed A RE / Beta carotene RE	Vitamin D IU / Vitamin E IU	% USRDA	Total tocopherol mg / Alpha tocopherol mg	Other tocopherol mg / Total ash g	Calcium mg / Phosphorus mg	% USRDA	Sodium mg / Sodium meq	Potassium mg / Potassium meq	Chlorine mg / Chlorine meq	Iron mg / Magnesium mg	% USRDA	Zinc mg / Copper mg	% USRDA	Iodine mcg / Selenium mcg	% USRDA	Manganese mcg / Chromium mcg	
0.32	19	0.43	21	0.77	13	0.005	2	4	5	1	5.0	0.0	164	16	696	913	23	5.0	28	3.0	20	80.5	54	120	A
4.9	24	22	6	0.95	10	2724	55	261	6.0	20	0.1	4.97	254	25	30.3	23.3	0.7	84	21	0.24	12	0		1	
0.30	18	0.37	19	0.55	9	0.002	1	unk.	tr(a)	0	0.7	unk.	124	12	1009	665	unk.	3.7	21	3.5	23	unk.	unk.	unk.	B
4.0	20	15	4	0.50	5	1587	32	unk.	0.9	3	unk.	5.46	236	24	43.9	17.0	unk.	42	11	0.42	21	unk.		unk.	
0.12	7	0.07	4	0.00	0	0.002	1	0	0(a)	0	0.1	tr	121	12	242	256	unk.	2.7	15	0.8	6	unk.	unk.	796	C
0.3	2	30	7	0.10	1	8200	164	820	0.1	0	tr	1.74	27	3	10.5	6.5	unk.	64	16	0.10	5	tr(a)		unk.	
0.12	7	0.07	4	0.00	0	0.002	1	0	0(a)	0	0.1	tr	121	12	33	256	unk.	2.7	15	0.8	6	unk.	unk.	796	D
0.3	2	30	7	0.10	1	8200	164	820	0.1	0	tr	1.02	27	3	1.4	6.5	unk.	64	16	0.10	5	tr(a)		unk.	
0.13	7	0.17	9	0.00	0	0.002	1	0	0(a)	0	1.0	tr(a)	84	8	45	292	50	2.0	11	0.6	4	unk.	unk.	699	E
0.4	2	26	7	0.09	1	7290	146	729	1.2	4	tr(a)	0.99	34	3	2.0	7.5	1.4	53	13	0.18	9	tr(a)		unk.	
0.15	9	0.20	10	0.00	0	0.002	1	0	0(a)	0	1.1	tr(a)	116	12	53	341	unk.	2.3	13	0.7	5	unk.	unk.	796	F
0.4	2	30	7	0.11	1	8097	162	810	1.4	5	tr(a)	1.13	45	5	2.3	8.7	unk.	67	17	0.20	10	tr(a)		unk.	
0.20	12	unk.	unk.	unk.	unk.	unk.	unk.	unk.	unk.	unk.	unk.	unk.	119	12	895	373	unk.	2.1	12	unk.	unk.	unk.	unk.	unk.	G
0.6	3	unk.	unk.	unk.	unk.	2541	51	unk.	unk.	unk.	unk.	3.07	unk.	unk.	38.9	9.5	unk.	unk.	unk.	unk.	unk.	unk.		unk.	
0.27	16	0.30	15	0.00	0	0.007	2	0	0(a)	0	9.0	tr	112	11	293	535	97	4.2	23	1.0	7	58.2	39	786	H
1.7	8	87	22	0.45	5	7586	152	758	10.8	36	0.9	2.32	85	9	12.7	13.7	2.7	90	23	0.24	12	16		20	
0.05	3	0.08	4	0.00	0	0.002	1	0	0(a)	0	0.5	tr	26	3	20	129	18	0.9	5	0.2	2	unk.	unk.	214	I
0.2	1	44	11	0.08	1	2227	45	223	0.6	2	0.5	0.41	14	1	0.8	3.3	0.5	24	6	0.05	3	tr(a)		1	
unk.	unk.	unk.	unk.	unk.	unk.	unk.	unk.	unk.	unk.	unk.	unk.	unk.	unk.	unk.	unk.	unk.	unk.	unk.	unk.	unk.	unk.	unk.	unk.	unk.	J
7.4	37	unk.	unk.	unk.	unk.	unk.	unk.	unk.	unk.	unk.	unk.	1.29	unk.	unk.	unk.	unk.	unk.	unk.	unk.	unk.	unk.	unk.		unk.	
unk.	unk.	unk.	unk.	unk.	unk.	unk.	unk.	unk.	unk.	unk.	unk.	unk.	unk.	unk.	unk.	unk.	unk.	unk.	unk.	unk.	unk.	unk.	unk.	unk.	K
unk.	unk.	unk.	unk.	unk.	unk.	unk.	unk.	unk.	unk.	unk.	unk.	2.79	unk.	unk.	unk.	unk.	unk.	unk.	unk.	unk.	unk.	unk.		unk.	
unk.	unk.	unk.	unk.	unk.	unk.	unk.	unk.	unk.	unk.	unk.	unk.	unk.	unk.	unk.	unk.	unk.	unk.	unk.	unk.	unk.	unk.	unk.	unk.	unk.	L
11.5	58	unk.	unk.	unk.	unk.	unk.	unk.	unk.	unk.	unk.	unk.	1.97	unk.	unk.	unk.	unk.	unk.	unk.	unk.	unk.	unk.	unk.		unk.	
0.07	4	0.03	2	0.00	0	unk.	unk.	0	0(a)	0	unk.	unk.	18	2	unk.	unk.	unk.	3.9	22	unk.	unk.	unk.	unk.	unk.	M
0.8	4	unk.	unk.	unk.	unk.	24	1	2	unk.	unk.	unk.	1.71	400	40	unk.	unk.	unk.	unk.	unk.	unk.	unk.	unk.		unk.	
0.05	3	0.03	1	0.00	0	unk.	unk.	0	0(a)	0	unk.	unk.	14	1	unk.	unk.	unk.	3.1	18	unk.	unk.	unk.	unk.	unk.	N
0.7	3	unk.	unk.	unk.	unk.	20	0	2	unk.	unk.	unk.	1.36	321	32	unk.	unk.	unk.	unk.	unk.	unk.	unk.	unk.		unk.	
0.07	4	0.05	3	0.00	0	0.002	1	0	0(a)	0	2.2	unk.	22	2	1	127	unk.	0.4	2	0.2	1	unk.	unk.	unk.	O
0.7	4	9	2	unk.	unk.	351	7	35	2.6	9	unk.	0.36	23	2	0.0	3.3	unk.	13	3	0.07	4	tr(a)		unk.	
0.13	8	0.09	5	0.00	0	unk.	unk.	0	0(a)	0	unk.	unk.	29	3	1	473	unk.	0.8	5	0.3	2	unk.	unk.	102	P
0.7	4	unk.	unk.	0.20	2	4305	86	430	unk.	unk.	unk.	1.02	49	5	0.0	12.1	unk.	27	7	0.14	7	tr(a)		unk.	
0.12	7	0.11	6	0.00	0	unk.	unk.	0	0(a)	0	unk.	unk.	24	2	1	316	unk.	0.6	3	0.4	2	unk.	unk.	122	Q
0.5	2	unk.	unk.	0.24	2	4287	86	429	unk.	unk.	unk.	0.73	39	4	0.0	8.1	unk.	32	8	0.17	9	tr(a)		unk.	
unk.	unk.	unk.	unk.	unk.	unk.	unk.	unk.	unk.	unk.	unk.	unk.	unk.	19	2	70	442	unk.	unk.	unk.	unk.	unk.	unk.	unk.	unk.	R
unk.	unk.	unk.	unk.	unk.	unk.	unk.	unk.	unk.	unk.	unk.	unk.	unk.	153	15	3.1	11.3	unk.	unk.	unk.	unk.	unk.	unk.		unk.	
0.12	7	unk.	unk.	1.53	26	unk.	unk.	unk.	0	0	unk.	unk.	28	3	966	409	unk.	2.1	12	unk.	unk.	unk.	unk.	unk.	S
2.3	12	unk.	unk.	unk.	unk.	2279	46	unk.	unk.	unk.	unk.	3.52	106	11	42.0	10.5	unk.	unk.	unk.	unk.	unk.	unk.		unk.	

	FOOD	Portion	Weight in grams / *Conversion for 100 g*	Kilocalories / H_2O g	Total carbohydrate g / Total fats g	Crude fiber g / Dietary fiber g	Total protein g / Total sugar g	**% USRDA**	Arginine mg / Histidine mg	Isoleucine mg / Leucine mg	Lysine mg / Methionine mg	Phenylalanine mg / Threonine mg	Valine mg / Tryptophan mg	Cystine mg / Tyrosine mg	Polyunsat. fatty acids g / Monounsat. fatty acids g	Saturated fatty acids g / P/S ratio	Linoleic acid g / Cholesterol mg	Thiamin mg / Ascorbic acid mg	**% USRDA**
A	stew,canned,chicken,low sodium	3/4 C	170	131	10.2	unk.	11.0	**25**	unk.	unk.	unk.	unk.	unk.	unk.	unk.	unk.	unk.	unk.	**unk.**
			0.59	unk.	5.1	0(a)	unk.		unk.	unk.	unk.	unk.	unk.	unk.	unk.	unk.	unk.	unk.	**unk.**
B	stew,home recipe,beef w/vegetables	1 C	235	209	14.6	0.94	15.0	**33**	unk.	unk.	unk.	unk.	unk.	unk.	unk.	4.70	0.23	0.14	**9**
			0.43	193.6	10.1	unk.	unk.		unk.	unk.	unk.	unk.	unk.	unk.	4.70	unk.	61	16	**27**
C	stew,home recipe,crab	1 C	233	208	9.6	0.01	11.2	**25**	628	612	887	315	435	119	0.49	8.36	0.30	0.07	**5**
			0.43	195.9	14.1	0.02	9.2		264	1044	164	509	145	485	3.38	0.06	73	2	**3**
D	stew,home recipe,frankfurter/weiner	3/4 C	176	291	15.6	0.75	10.3	**22**	90	78	85	388	533	51	11.96	7.89	1.79	0.18	**12**
			0.57	129.5	20.7	1.79	3.2		10	96	224	71	19	57	8.87	1.52	42	14	**24**
E	stew,home recipe,lamb	3/4 C	170	124	10.6	0.75	10.1	**21**	669	454	720	383	467	125	1.11	0.91	1.04	0.12	**8**
			0.59	141.0	4.8	1.57	4.5		229	692	210	448	121	307	1.01	1.22	29	20	**34**
F	stew,home recipe,oyster	1 C	242	278	14.7	0.01	14.9	**33**	843	788	1166	635	876	193	0.59	9.80	0.37	0.22	**14**
			0.41	192.9	17.5	0.09	9.2		374	1284	405	682	199	660	3.94	0.06	100	4	**6**
G	stew,home recipe,shrimp	1 C	228	207	9.4	0.06	11.8	**26**	680	665	985	512	717	127	0.50	8.36	0.32	0.07	**5**
			0.44	190.5	13.7	0.02	9.2		276	1041	319	524	145	508	3.37	0.06	79	2	**3**
H	stew,home recipe,turkey	3/4 C	170	336	14.8	0.21	45.4	**100**	2790	2411	3490	1833	2279	609	1.77	1.89	1.11	0.15	**10**
			0.59	98.9	9.2	0.45	0.9		1182	3318	1116	1802	467	1496	1.43	0.93	138	2	**3**
I	stew,home recipe,veal w/carrots &	1 C	238	242	7.2	0.71	17.6	**38**	unk.	unk.	unk.	unk.	unk.	unk.	unk.	unk.	unk.	0.05	**3**
	onions		*0.42*	unk.	15.6	unk.	unk.		unk.	unk.	unk.	unk.	unk.	unk.	unk.	unk.	unk.	0	**0**
J	STRAWBERRIES,fresh,unsweetened,	1/2 C	120	44	10.1	1.56	1.3	**21**	57	30	53	38	38	11	unk.	tr(a)	unk.	0.04	**2**
	crushed		*0.83*	107.9	0.6	2.76	7.1		25	68	2	40	15	44	unk.	unk.	0	71	**118**
K	strawberries,fresh,unsweetened,whole	1/2 C	75	28	6.3	0.97	0.8	**13**	35	18	33	24	24	7	unk.	tr(a)	unk.	0.02	**2**
			1.34	67.0	0.4	1.71	4.4		16	42	1	25	9	28	unk.	unk.	0	44	**73**
L	strawberries,frozen,sweetened,whole	1/2 C	128	117	30.0	0.76	0.5	**1**	22	11	20	15	15	4	unk.	tr(a)	unk.	0.03	**2**
			0.78	96.5	0.3	unk.	25.4		10	26	1	16	6	17	unk.	unk.	0	70	**117**
M	strawberries,home recipe,strawberry	1/2 C	45	140	15.6	0.22	1.4	**3**	unk.	71	63	71	80	28	1.13	4.47	0.90	0.03	**2**
	dessert		*2.22*	22.5	8.4	unk.	11.3		33	117	30	53	19	56	1.93	0.25	21	8	**13**
N	STRUDEL,home recipe	2-3/4x2-3/4x	118	273	50.2	0.42	3.2	**5**	47	135	107	143	137	51	0.41	4.38	0.33	0.10	**7**
		1-1/8''	*0.85*	55.9	7.9	2.56	26.7		63	219	50	96	34	101	1.96	0.09	39	4	**6**
O	strudel,pineapple-cheese	1 piece	67	209	21.2	0.07	4.4	**7**	unk.	unk.	unk.	unk.	unk.	unk.	unk.	unk.	unk.	0.13	**8**
	-Pepperidge Farm		*1.50*	unk.	12.1	unk.	unk.		unk.	unk.	unk.	unk.	unk.	unk.	unk.	unk.	unk.	unk.	**unk.**
P	STUFFING/dressing,home recipe,bread,	1/2 C	70	251	24.9	0.28	4.5	**7**	unk.	220	136	250	209	104	unk.	7.70	0.70	0.06	**4**
	dry,w/water & fat		*1.43*	23.2	15.3	unk.	2.9		173	361	68	141	52	131	5.60	unk.	tr(a)	tr	**0**
Q	stuffing/dressing,home recipe,bread,	1/2 C	69	107	8.5	0.16	3.1	**6**	unk.	146	158	151	168	75	0.40	3.56	0.44	0.04	**3**
	moist,w/water,egg & fat		*1.44*	49.9	6.7	unk.	1.1		103	243	71	122	41	114	2.05	0.11	75	3	**6**
R	stuffing,bread,Stuff'n Such,prep	1/2 C	83	118	24.2	0.20	4.2	**7**	unk.	unk.	unk.	unk.	unk.	unk.	unk.	unk.	unk.	0.06	**4**
	-Uncle Ben		*1.20*	52.4	0.6	unk.	unk.		unk.	unk.	unk.	unk.	unk.	unk.	unk.	unk.	0(a)	1	**2**
S	stuffing,home recipe,sausage	1/2 C	95	292	39.8	0.53	7.6	**13**	167	439	361	452	421	162	0.71	1.64	0.49	0.15	**10**
			1.05	52.1	11.2	0.52	2.5		217	575	141	319	117	320	1.98	0.43	12	2	**3**

Riboflavin mg / Niacin mg	% USRDA	Vitamin B₆ mg / Folacin mcg	% USRDA	Vitamin B₁₂ mcg / Pantothenic acid mg	% USRDA	Biotin mg / Vitamin A IU	% USRDA	Preformed A RE / Beta carotene RE	Vitamin D IU / Vitamin E IU	% USRDA	Total tocopherol mg / Alpha tocopherol mg	Other tocopherol mg / Total ash g	Calcium mg / Phosphorus mg	% USRDA	Sodium mg / Sodium meq	Potassium mg / Potassium meq	Chlorine mg / Chlorine meq	Iron mg / Magnesium mg	% USRDA	Zinc mg / Copper mg	% USRDA	Iodine mcg / Selenium mcg	% USRDA	Manganese mcg / Chromium mcg	
unk.	unk.	unk.	unk.	unk.	unk.	unk.	unk.	unk.	unk.	unk.	unk.	unk.	10	1	42	207	unk.	unk.	unk.	unk.	unk.	unk.	unk.	unk.	A
unk.	unk.	unk.	unk.	unk.	unk.	unk.	unk.	unk.	unk.	unk.	unk.	unk.	85	9	1.8	5.3	unk.	unk.	unk.	unk.	unk.	unk.		unk.	
0.16	10	unk.	unk.	unk.	unk.	unk.	unk.	unk.	unk.	unk.	unk.	unk.	28	3	87	587	unk.	2.8	16	unk.	unk.	unk.	unk.	unk.	B
4.5	22	unk.	unk.	unk.	unk.	2303	46	unk.	unk.	unk.	unk.	1.64	176	18	3.8	15.0	unk.	40	10	unk.	unk.	unk.		unk.	
0.33	19	0.17	8	3.38	56	tr	0	0	80	20	0.3	0.0	247	25	436	332	1	0.3	2	1.9	13	13.7	9	2	C
0.7	3	16	4	0.78	8	507	10	tr	0.4	1	0.3	2.07	233	23	19.0	8.5	0.0	26	7	0.29	15	tr		0	
0.18	11	0.26	13	0.88	15	0.002	1	0	0	0	2.0	tr	27	3	1255	467	16	1.7	10	1.7	12	127.5	85	16	D
2.7	14	24	6	0.64	6	4064	81	401	2.4	8	0.2	3.08	116	12	54.6	11.9	0.5	24	6	0.16	8	1		2	
0.15	9	0.18	9	0.59	10	0.003	1	0	0	0	1.6	0.0	54	5	140	364	13	1.4	8	1.5	10	tr(a)	0	8	E
2.4	12	37	9	0.31	3	4265	85	427	2.0	7	0.2	1.65	109	11	6.1	9.3	0.4	25	6	0.29	15	tr(a)		1	
0.49	29	0.13	7	17.96	299	0.000	0	80	80	20	1.0	0.1	331	33	928	427	605	5.5	31	72.5	483	183.7	122	295	F
2.6	13	20	5	0.87	9	892	18	3	1.3	4	0.4	5.11	328	33	40.4	10.9	17.1	60	15	3.86	193	49		2	
0.32	19	0.10	5	0.84	14	tr	0	4	113	28	0.4	0.0	260	26	197	329	1	0.8	4	1.2	8	13.7	9	2	G
0.5	2	14	4	0.66	7	521	10	tr	0.5	2	0.3	2.39	242	24	8.6	8.4	0.0	42	11	0.38	19	tr		0	
0.29	17	0.60	30	0.57	10	0.001	0	8	1	0	0.7	0.0	33	3	138	634	15	3.3	18	4.9	32	unk.	unk.	9	H
11.2	56	14	3	0.76	8	949	19	92	0.8	3	0.1	2.17	311	31	6.0	16.2	0.4	53	13	0.28	14	1		3	
0.17	10	unk.	unk.	unk.	unk.	unk.	unk.	0	0	0	unk.	unk.	32	3	unk.	unk.	unk.	3.0	17	unk.	unk.	unk.	unk.	unk.	I
3.0	15	unk.	unk.	unk.	unk.	3253	65	325	unk.	unk.	unk.	unk.	192	19	unk.	unk.	unk.	unk.	unk.	unk.	unk.	unk.		unk.	
0.08	5	0.07	3	0.00	0	0.005	2	0	0(a)	0	0.3	0.2	25	3	1	197	13	1.2	7	0.1	1	unk.	unk.	394	J
0.7	4	19	5	0.41	4	72	1	7	0.4	1	0.2	0.60	25	3	0.0	5.0	0.4	14	4	0.03	1	0		0	
0.05	3	0.04	2	0.00	0	0.003	1	0	0(a)	0	0.2	0.1	16	2	1	122	8	0.7	4	0.1	0	unk.	unk.	244	K
0.4	2	12	3	0.25	3	45	1	4	0.3	1	0.1	0.37	16	2	0.0	3.1	0.2	9	2	0.02	1	0		0	
0.08	5	0.05	3	0.00	0	0.005	2	0	0(a)	0	unk.	unk.	17	2	1	133	unk.	0.8	4	0.1	1	unk.	unk.	unk.	L
0.6	3	11	3	0.17	2	38	1	4	unk.	unk.	unk.	0.25	20	2	0.1	3.4	unk.	11	3	tr(a)	0	tr(a)		unk.	
0.04	3	0.03	1	0.02	0	0.001	1	0	0	0	0.1	tr	16	2	49	59	12	0.4	2	0.2	1	0.2	0	108	M
0.3	1	5	1	0.09	1	265	5	1	0.1	0	0.1	0.32	23	2	2.1	1.5	0.3	9	2	0.04	2	3		10	
0.08	5	0.06	3	0.07	1	0.002	1	0	4	1	0.6	tr	40	4	142	185	18	1.4	8	0.4	3	15.2	10	2099	N
0.7	4	11	3	0.23	2	322	6	6	0.7	2	0.6	0.90	50	5	6.2	4.7	0.5	23	6	0.11	5	8		20	
0.09	5	unk.	unk.	unk.	unk.	unk.	unk.	unk.	unk.	unk.	unk.	unk.	21	2	208	43	unk.	0.9	5	unk.	unk.	unk.	unk.	unk.	O
1.1	6	unk.	unk.	unk.	unk.	tr	0	unk.	unk.	unk.	unk.	0.70	42	4	9.0	1.1	unk.	5	1	unk.	unk.	unk.		unk.	
0.08	5	0.02	1	tr(a)	0	0.001	0	unk.	8	2	unk.	unk.	46	5	627	63	unk.	1.1	6	0.3	2	unk.	unk.	unk.	P
1.0	5	20	5	0.21	2	455	9	unk.	unk.	unk.	unk.	2.03	68	7	27.3	1.6	unk.	21	5	0.12	6	unk.		unk.	
0.08	5	0.03	2	0.14	2	0.003	1	0	5	1	0.3	tr	31	3	319	80	162	0.9	5	0.3	2	52.2	35	7	Q
0.5	2	12	3	0.27	3	431	9	15	0.3	1	0.1	1.11	52	5	13.9	2.0	4.6	36	9	0.03	2	tr		0	
0.02	2	unk.	unk.	0(a)	0	unk.	unk.	0	0(a)	0	unk.	unk.	14	1	647	64	unk.	1.1	6	unk.	unk.	unk.	unk.	unk.	R
0.7	4	unk.	unk.	unk.	unk.	10	0	1	unk.	unk.	unk.	1.70	39	4	28.1	1.6	unk.	30	8	unk.	unk.	unk.		unk.	
0.07	4	0.04	2	0.20	3	tr	0	0	0	0	0.2	tr	17	2	258	96	17	1.0	5	0.6	4	28.0	19	7	S
1.1	6	3	1	0.15	2	78	2	4	0.2	1	0.2	0.84	77	8	11.2	2.4	0.5	4	1	0.03	1	0		4	

	FOOD	Portion	Weight in grams / Conversion for 100 g	Kilocalories / H_2O g	Total carbohydrate g / Total fats g	Crude fiber g / Dietary fiber g	Total protein g / Total sugar g	% USRDA	Arginine mg / Histidine mg	Isoleucine mg / Leucine mg	Lysine mg / Methionine mg	Phenylalanine mg / Threonine mg	Valine mg / Tryptophan mg	Cystine mg / Tyrosine mg	Polyunsat. fatty acids g / Monounsat. fatty acids g	Saturated fatty acids g / P/S ratio	Linoleic acid g / Cholesterol mg	Thiamin mg / Ascorbic acid mg	% USRDA
A	SUCCOTASH (corn/lima beans),frozen,	1/2 C	85	79	17.4	0.76	3.6	**6**	unk.	unk.	unk.	unk.	unk.	unk.	unk.	unk.	unk.	0.08	**5**
	cooked,drained solids		*1.18*	63.0	0.3	unk.	2.9		unk.	unk.	unk.	unk.	unk.	unk.	unk.	unk.	0	5	**9**
B	SUGAR SUBSTITUTE-Adolph's	1 tsp	5.0	0	0.0	0.00	0.0	**0**	0	0	0	0	0	0	0.00	0.00	0.00	0.00	**0**
			20.00	0.3	0.0	0(a)	0.0		0	0	0	0	0	0	0.00	0.00	0	0	**0**
C	sugar substitute-Diamond	1 tsp	4.0	tr	3.6	0.00	0.0	**0**	0	0	0	0	0	0	0.00	0.00	0.00	0.00	**0**
			25.00	0.1	0.0	0(a)	0.0		0	0	0	0	0	0	0.00	0.00	0	0	**0**
D	sugar substitute-Sugar Twin	1 tsp	0.1	tr	0.1	unk.	tr	**0**	0(a)	0(a)	0(a)	0(a)	0(a)	0(a)	0.00	0.00	0.00	0(a)	**0**
			1000.00	unk.	0.0	0(a)	0.0		0(a)	0(a)	0(a)	0(a)	0(a)	0(a)	0.00	0.00	0	0(a)	**0**
E	sugar substitute-Sweet 'n Low	1 tsp	4.0	12	3.2	0(a)	0.0	**0**	0(a)	0(a)	0(a)	0(a)	0(a)	0(a)	0.00	0.00	0.00	0(a)	**0**
			25.00	unk.	0.0	0(a)	0.0		0(a)	0(a)	0(a)	0(a)	0(a)	0(a)	0.00	0.00	0	0(a)	**0**
F	sugar substitute,liquid-Sucaryl	1 tsp	5.0	0	0.0	0(a)	0.0	**0**	0	0	0	0	0	0	0.00	0.00	0.00	0.00	**0**
			20.00	4.9	0.0	0(a)	0.0		0	0	0	0	0	0	0.00	0.00	0	0	**0**
G	sugar substitute,powdered sweetener	1 tsp	4.0	14	unk.	unk.	unk.	**unk.**	0(a)	0(a)	0(a)	0(a)	0(a)	0(a)	0(a)	0(a)	0(a)	0(a)	**0**
	-Superose		*25.00*	unk.	0(a)	0(a)	unk.		0(a)	0(a)	0(a)	0(a)	0(a)	0(a)	0(a)	0.00	0(a)	0(a)	**0**
H	SUGAR,brown	1 C	224	836	216.0	0.00	0.0	**0**	0(a)	0(a)	0(a)	0(a)	0(a)	0(a)	0.00	0.00	0.00	0.02	**2**
			0.45	4.7	0.0	0(a)	216.0		0(a)	0(a)	0(a)	0(a)	0(a)	0(a)	0.00	0.00	0	0	**0**
I	sugar,powdered	1 C	128	493	127.4	0.00	0.0	**0**	0(a)	0(a)	0(a)	0(a)	0(a)	0(a)	0.00	0.00	0.00	0.00	**0**
			0.78	0.6	0.0	0(a)	127.4		0(a)	0(a)	0(a)	0(a)	0(a)	0(a)	0.00	0.00	0	0	**0**
J	sugar,white,cubed	1 cube	7.0	27	7.0	0.00	0.0	**0**	0	0	0	0	0	0	0.00	0.00	0.00	0.00	**0**
			14.29	tr	0.0	0(a)	7.0		0	0	0	0	0	0	0.00	0.00	0	0	**0**
K	sugar,white,granulated	1 C	200	770	199.0	0.00	0.0	**0**	0	0	0	0	0	0	0.00	0.00	0.00	0.00	**0**
			0.50	1.0	0.0	0(a)	199.0		0	0	0	0	0	0	0.00	0.00	0	0	**0**
L	SUNDAE TOPPING,cherry-Smucker's	1 Tbsp	20	53	13.6	0.01	0.1	**0**	unk.	tr(a)	tr(a)	tr(a)	tr(a)	tr(a)	tr(a)	tr(a)	tr(a)	tr	**0**
			5.00	unk.	tr	0(a)	13.6		tr(a)	tr(a)	tr(a)	tr(a)	tr(a)	tr(a)	tr(a)	0.00	0	tr	**0**
M	sundae topping,marshmallow-Smucker's	1 Tbsp	18	68	17.0	tr	0.2	**0**	unk.	unk.	unk.	unk.	unk.	unk.	tr(a)	tr(a)	tr(a)	tr	**0**
			5.65	0.4	tr	0(a)	17.0		unk.	unk.	unk.	unk.	unk.	unk.	tr(a)	0.00	0	tr	**0**
N	sundae topping,pineapple-Smucker's	1 Tbsp	20	54	13.8	tr(a)	0.0	**0**	tr(a)	tr(a)	tr(a)	tr(a)	tr(a)	tr(a)	tr(a)	tr(a)	tr(a)	tr	**0**
			5.00	unk.	tr	0(a)	13.8		tr(a)	tr(a)	tr(a)	tr(a)	tr(a)	tr(a)	tr(a)	0.00	0	tr	**1**
O	sundae topping,strawberry-Smucker's	1 Tbsp	20	44	11.4	unk.	tr	**0**	tr(a)	tr(a)	tr(a)	tr(a)	tr(a)	tr(a)	tr(a)	tr(a)	tr(a)	tr	**0**
			5.00	unk.	tr	0(a)	11.4		tr(a)	tr(a)	tr(a)	tr(a)	tr(a)	tr(a)	tr(a)	0.00	0	tr	**0**
P	SUNFLOWER seed kernels,hulled	1/4 C	38	213	7.6	1.45	8.8	**14**	unk.	486	331	465	516	177	unk.	2.29	11.43	0.75	**50**
			2.62	1.8	18.0	unk.	0.0		223	661	169	347	131	247	3.43	unk.	0	unk.	**unk.**
Q	SWEET POTATO,candied	1/2 C	110	185	37.6	0.66	1.4	**2**	71	68	67	79	108	16	unk.	2.20	0.11	0.07	**4**
			0.91	66.0	3.6	unk.	26.6		18	81	25	67	24	34	1.10	unk.	0	11	**18**
R	sweet potato,canned,low sodium,	1/2 C	109	118	27.1	1.09	2.2	**3**	108	105	102	120	163	24	unk.	unk.	unk.	0.05	**4**
	vacuum pack		*0.92*	78.4	0.2	2.62	18.1		28	124	39	102	37	unk.	unk.	unk.	0	15	**25**
S	sweet potato,canned,pieces,	1/2 C	100	108	24.9	1.00	2.0	**3**	99	96	94	100	150	22	unk.	unk.	unk.	0.05	**3**
	vacuum pack		*1.00*	71.9	0.2	2.40	16.6		26	114	36	94	34	58	unk.	unk.	0	14	**23**

Riboflavin mg / Niacin mg	% USRDA	Vitamin B_6 mg / Folacin mcg	% USRDA	Vitamin B_{12} mcg / Pantothenic acid mg	% USRDA	Biotin mg / Vitamin A IU	% USRDA	Preformed A RE / Beta carotene RE	Vitamin D IU / Vitamin E IU	% USRDA	Total tocopherol mg / Alpha tocopherol mg	Other tocopherol mg / Total ash g	Calcium mg / Phosphorus mg	% USRDA	Sodium mg / Sodium meq	Potassium mg / Potassium meq	Chlorine mg / Chlorine meq	Iron mg / Magnesium mg	% USRDA	Zinc mg / Copper mg	% USRDA	Iodine mcg / Selenium mcg	% USRDA	Manganese mcg / Chromium mcg	
0.04	3	unk.	unk.	0(a)	0	unk.	unk.	0	0(a)	0	unk.	unk.	11	1	32	209	unk.	0.8	5	unk.	unk.	unk.	unk.	unk.	A
1.1	6	unk.	unk.	unk.	unk.	255	5	25	unk.	unk.	unk.	0.68	72	7	1.4	5.3	unk.	unk.	unk.	tr(a)	0	tr(a)		unk.	
0.00	0	0.00	0	0.00	0	0.000	0	0	0	0	0.0	0.0	13	1	0	0	0	0.0	0	0.0	0	0.0	0	0	B
0.0	0	0	0	0.00	0	0	0	0	0.0	0	0.0	0.09	0	0	0.0	0.0	0.0	0	0	0.00	0	0		0	
0.00	0	0.00	0	0.00	0	0.000	0	0	0	0	0.0	0.0	34	3	0	0	0	0.0	0	0(a)	0	0.0	0	0	C
0.0	0	0	0	0.00	0	0	0	0	0.0	0	0.0	0.03	0	0	0.0	0.0	0.0	0	0	0.00	0	0(a)		0(a)	
0(a)	0	0(a)	0	0(a)	0	0(a)	0	0(a)	0(a)	0	0(a)	0(a)	0	0	tr	0	0(a)	0(a)	0	0(a)	0	0(a)	0	0(a)	D
0(a)	0	0(a)	0	0(a)	0	0(a)	0	0(a)	0.0	0	0(a)	unk.	0(a)	0	0.0	0.0	0.0	0(a)	0	0(a)	0	0(a)		0(a)	
0(a)	0	0(a)	0	0(a)	0	0(a)	0	0(a)	0(a)	0	0(a)	0(a)	unk.	unk.	unk.	0(a)	0(a)	0(a)	0	0(a)	0	0(a)	0	0(a)	E
0(a)	0	0(a)	0	0(a)	0	0(a)	0	0(a)	0.0	0	0(a)	unk.	0(a)	0	unk.	0.0	0.0	0(a)	0	0(a)	0	0(a)		0(a)	
0.00	0	0.00	0	0.00	0	0.000	0	0	0	0	0.0	0.0	unk.	unk.	8	0(a)	0(a)	0(a)	0	0(a)	0	0(a)	0	0(a)	F
0.0	0	0	0	0.00	0	0	0	0	0.0	0	0.0	unk.	0(a)	0	0.3	0.0	0.0	0(a)	0	0(a)	0	0(a)		0(a)	
0(a)	0	0(a)	0	0(a)	0	0(a)	0	0(a)	0(a)	0	0(a)	0(a)	unk.	unk.	4	0(a)	0(a)	0(a)	0	0(a)	0	0(a)	0	0(a)	G
0(a)	0	0(a)	0	0(a)	0	0(a)	0	0(a)	0.0	0	0(a)	unk.	0(a)	0	0.2	0.0	0.0	0(a)	0	0(a)	0	0(a)		0(a)	
0.07	4	0(a)	0	0(a)	0	0(a)	0	0	0	0	0(a)	0(a)	190	19	67	771	170	7.6	42	unk.	unk.	unk.	unk.	90	H
0.4	2	0(a)	0	0(a)	0	0	0	0	0.0	0	0(a)	3.36	43	4	2.9	19.7	4.8	139	35	0.04	2	unk.		134	
0.00	0	0(a)	0	0(a)	0	0(a)	0	0	0	0	0(a)	0(a)	0	0	1	4	unk.	0.1	1	0.8	5	unk.	unk.	51	I
0.0	0	0(a)	0	0(a)	0	0	0	0	0.0	0	0(a)	0(a)	0	0	0.1	0.1	unk.	0	0	0.03	1	unk.		77	
0.00	0	0.00	0	0.00	0	0.000	0	0	0	0	0.0	unk.	0	0	tr	tr	0	tr	0	tr	0	0.6	0	3	J
0.0	0	0	0	0.00	0	0	0	0	0.0	0	unk.	0.00	0	0	0.0	0.0	0.0	0	0	tr	0	unk.		4	
0.00	0	0.00	0	0.00	0	0.000	0	0	0	0	0.0	0.0	0	0	2	6	0	0.2	1	1.2	8	unk.	unk.	80	K
0.0	0	0	0	0.00	0	0	0	0	0.0	0	0.0	0.00	0	0	0.1	0.1	0.0	tr	0	0.04	2	unk.		120	
0.01	0	0.00	0	0.00	0	unk.	unk.	0	0(a)	0	unk.	unk.	6	1	7	152	unk.	0.5	3	0.0	0	unk.	unk.	unk.	L
tr	0	1	0	0.01	0	78	2	8	unk.	unk.	unk.	unk.	3	0	0.3	3.9	unk.	1	0	unk.	unk.	unk.		unk.	
0.01	0	tr	0	tr	0	tr(a)	0	0(a)	0(a)	0	tr(a)	tr(a)	7	1	12	7	0	tr	0	0.0	0	tr(a)	0	tr(a)	M
0.1	0	tr(a)	0	tr(a)	0	tr	0	tr(a)	0.0	0	tr(a)	0.12	3	0	0.5	0.2	0.0	tr	0	tr(a)	0	tr(a)		tr(a)	
tr	0	0.01	0	0.00	0	tr	0	0	0(a)	0	tr	unk.	6	1	7	11	unk.	0.4	2	0.0	0	unk.	unk.	unk.	N
tr	0	1	0	0.01	0	4	0	tr	tr	0	unk.	unk.	2	0	0.3	0.3	unk.	1	0	0.01	0	unk.		unk.	
tr	0	0.00	0	0.00	0	unk.	unk.	0	0(a)	0	unk.	unk.	5	1	7	2	unk.	0.4	2	0.0	0	unk.	unk.	unk.	O
tr	0	tr	0	tr	0	tr	0	0	unk.	unk.	unk.	unk.	2	0	0.3	0.0	unk.	tr	0	unk.	unk.	unk.		unk.	
0.09	5	0.48	24	0.00	0	unk.	unk.	0	0(a)	0	5.0	0.0	46	5	11	351	unk.	2.7	15	unk.	unk.	unk.	unk.	unk.	P
2.1	10	unk.	unk.	0.53	5	19	0	2	5.9	20	5.0	1.52	319	32	0.5	9.0	unk.	14	4	unk.	unk.	unk.		unk.	
0.04	3	0.08	4	0.00	0	0.002	1	0	0(a)	0	unk.	tr(a)	41	4	46	209	unk.	1.0	6	0.5	4	unk.	unk.	682	Q
0.4	2	21	5	unk.	unk.	770	15	693	unk.	unk.	6.6	1.32	47	5	2.0	5.3	unk.	20	5	0.07	3	tr(a)		tr	
0.04	3	0.07	4	0.00	0	unk.	unk.	0	0(a)	0	unk.	tr(a)	27	3	13	218	unk.	0.9	5	0.5	4	unk.	unk.	676	R
0.7	3	20	5	0.47	5	8502	170	850	unk.	unk.	6.5	1.09	45	5	0.6	5.6	unk.	20	5	0.07	3	tr(a)		tr	
0.04	2	0.07	3	0.00	0	unk.	unk.	0	0(a)	0	unk.	tr(a)	25	3	48	200	unk.	0.8	4	0.2	1	unk.	unk.	620	S
0.6	3	18	5	0.43	4	7800	156	780	unk.	unk.	6.0	1.00	41	4	2.1	5.1	unk.	18	5	0.06	3	tr(a)		tr	

	FOOD	Portion	Weight in grams / Conversion for 100 g	Kilocalories / H_2O g	Total carbohydrate g / Total fats g	Crude fiber g / Dietary fiber g	Total protein g / Total sugar g	% USRDA	Arginine mg / Histidine mg	Isoleucine mg / Leucine mg	Lysine mg / Methionine mg	Phenylalanine mg / Threonine mg	Valine mg / Tryptophan mg	Cystine mg / Tyrosine mg	Polyunsat. fatty acids g / Monounsat. fatty acids g	Saturated fatty acids g / P/S ratio	Linoleic acid g / Cholesterol mg	Thiamin mg / Ascorbic acid mg	% USRDA
A	sweet potato,fresh,baked in skin,	1 med	146	206	47.4	1.31	3.1	5	152	147	145	152	229	unk.	unk.	tr(a)	unk.	0.13	9
	edible portion		*0.68*	93.0	0.7	3.50	31.7		unk.	175	55	145	53	89	unk.	unk.	0	32	54
B	sweet potato,fresh,boiled in skin &	1/2 C	128	145	33.5	0.89	2.2	3	108	unk.	unk.	108	unk.	unk.	unk.	unk.	unk.	0.11	8
	mashed		*0.78*	90.0	0.5	3.06	22.3		unk.	unk.	unk.	unk.	unk.	64	unk.	unk.	0	22	36
C	sweet potato,fresh,boiled in skin,	1 med	151	172	39.7	1.06	2.6	4	127	95	86	128	115	29	unk.	tr(a)	unk.	0.14	9
	edible portion		*0.66*	106.6	0.6	3.62	26.4		33	140	41	99	45	75	unk.	unk.	0	26	43
D	SWISS CHARD,fresh,cooked,	1/2 C	88	16	2.9	0.61	1.6	2	unk.	unk.	unk.	unk.	unk.	unk.	unk.	tr(a)	unk.	0.03	2
	drained solids		*1.14*	82.0	0.2	unk.	0.3		unk.	unk.	unk.	unk.	unk.	unk.	unk.	unk.	0(a)	14	23
E	Swiss chard,raw,chopped	1/2 C	28	7	1.3	0.22	0.7	1	unk.	unk.	unk.	unk.	unk.	unk.	unk.	tr(a)	unk.	0.02	1
			3.64	25.1	0.1	unk.	0.1		unk.	unk.	unk.	unk.	unk.	unk.	unk.	unk.	0(a)	9	15
F	SWORDFISH,steak,broiled w/butter or	1/4 lb raw	84	146	0.0	0.00	23.5	52	1197	1200	2070	870	1247	256	unk.	unk.	unk.	0.03	2
	margarine		*1.19*	54.3	5.0	0(a)	0.0		1174	1764	682	1011	235	843	unk.	unk.	unk.	unk.	unk.
G	SYRUP,cane/maple,all table varieties	1 Tbsp	19	48	12.3	0.00	0.0	0	unk.	0(a)	0(a)	0(a)	0(a)	0(a)	0.00	0.00	0.00	0.00	0
			5.26	6.3	0.0	0(a)	12.3		0(a)	0(a)	0(a)	0(a)	0(a)	0(a)	0.00	0.00	0	0	0
H	syrup,chocolate fudge,home recipe	1 Tbsp	22	62	12.0	0.07	0.6	1	11	18	23	14	20	3	0.01	1.07	0.04	tr	0
			4.55	7.5	1.9	tr(a)	11.1		8	29	7	13	4	14	0.66	0.01	1	tr	0
I	syrup,chocolate,home recipe	1 Tbsp	20	36	6.6	0.06	0.3	0	tr(a)	tr(a)	tr(a)	tr(a)	tr(a)	tr(a)	0.00	0.76	0.03	tr	0
			5.00	11.0	1.4	tr(a)	5.7		tr(a)	tr(a)	tr(a)	tr(a)	tr(a)	tr(a)	0.50	0.00	0	0	0
J	syrup,corn,light/dark	1 Tbsp	20	58	15.0	0.00	0.0	0	unk.	0(a)	0(a)	0(a)	0(a)	0(a)	0.00	0.00	0.00	0.00	0
			5.00	4.8	0.0	0(a)	15.0		0(a)	0(a)	0(a)	0(a)	0(a)	0(a)	0.00	0.00	0	0	0
K	syrup,fruit flavored,any-Smucker's	1 Tbsp	20	50	12.8	unk.	0.0	0	unk.	tr(a)	tr(a)	tr(a)	tr(a)	tr(a)	tr(a)	tr(a)	tr(a)	tr	0
			5.00	unk.	tr	0(a)	12.8		tr(a)	tr(a)	tr(a)	tr(a)	tr(a)	tr(a)	tr(a)	0.00	0	tr	0
L	syrup,maple,pure	1 Tbsp	19	48	12.3	0.00	0.0	0	unk.	0(a)	0(a)	0(a)	0(a)	0(a)	0.00	0.00	0.00	0.00	0
			5.26	6.3	0.0	0(a)	12.3		0(a)	0(a)	0(a)	0(a)	0(a)	0(a)	0.00	0.00	0	0	0
M	syrup,pancake,low calorie	1 Tbsp	20	12	3.0	0(a)	0.0	0	unk.	0(a)	0(a)	0(a)	0(a)	0(a)	0(a)	0(a)	0(a)	0(a)	0
	-Featherweight		*5.00*	unk.	0.0	0(a)	3.0		0(a)	0(a)	0(a)	0(a)	0(a)	0(a)	0(a)	0.00	0(a)	0(a)	0
N	T.V. DINNER,beef-Swanson	1 dinner	312	286	21.5	1.59	27.8	43	unk.	unk.	unk.	unk.	unk.	unk.	unk.	unk.	unk.	0.09	6
			0.32	248.7	10.0	unk.	unk.		unk.	unk.	unk.	unk.	unk.	unk.	unk.	unk.	unk.	12	20
O	t.v. dinner,chicken,fried-Swanson	1 dinner	308	583	45.6	1.36	32.0	49	unk.	unk.	unk.	unk.	unk.	unk.	unk.	unk.	49.64	0.31	21
			0.32	195.0	30.5	unk.	unk.		unk.	unk.	unk.	unk.	unk.	unk.	146.54	unk.	184	3	5
P	t.v. dinner,frank & beans-Swanson	1 dinner	354	433	57.7	unk.	18.4	28	unk.	unk.	unk.	unk.	unk.	unk.	unk.	unk.	unk.	0.07	5
			0.28	252.8	19.5	unk.	unk.		unk.	unk.	unk.	unk.	unk.	unk.	unk.	unk.	unk.	0	0
Q	t.v. dinner,ham-Swanson	1 dinner	291	358	47.1	1.69	18.3	28	unk.	unk.	unk.	unk.	unk.	unk.	unk.	unk.	unk.	0.47	31
			0.34	213.3	7.6	unk.	unk.		unk.	unk.	unk.	unk.	unk.	unk.	unk.	unk.	unk.	23	38
R	t.v. dinner,loin of pork-Swanson	1 dinner	283	373	23.5	0.57	28.9	45	unk.	unk.	unk.	unk.	unk.	unk.	unk.	unk.	unk.	0.71	47
			0.35	208.3	18.1	unk.	unk.		unk.	unk.	unk.	unk.	unk.	unk.	unk.	unk.	unk.	3	5
S	t.v. dinner,meat loaf-Swanson	1 dinner	305	528	48.0	1.14	19.0	29	unk.	unk.	unk.	unk.	unk.	unk.	unk.	unk.	unk.	0.49	33
			0.33	189.0	29.0	unk.	unk.		unk.	unk.	unk.	unk.	unk.	unk.	unk.	unk.	unk.	5	8

Riboflavin *mg* / Niacin *mg*	% USRDA	Vitamin B_6 *mg* / Folacin *mcg*	% USRDA	Vitamin B_{12} *mcg* / Pantothenic acid *mg*	% USRDA	Biotin *mg* / Vitamin A *IU*	% USRDA	Preformed A *RE* / Beta carotene *RE*	Vitamin D *IU* / Vitamin E *IU*	% USRDA	Total tocopherol *mg* / Alpha tocopherol *mg*	Other tocopherol *mg* / Total ash *g*	Calcium *mg* / Phosphorus *mg*	% USRDA	Sodium *mg* / Sodium *meq*	Potassium *mg* / Potassium *meq*	Chlorine *mg* / Chlorine *meq*	Iron *mg* / Magnesium *mg*	% USRDA	Zinc *mg* / Copper *mg*	% USRDA	Iodine *mcg* / Selenium *mcg*	% USRDA	Manganese *mcg* / Chromium *mcg*	
0.10	**6**	0.25	**12**	0.00	**0**	0.003	**1**	0	0(a)	**0**	2.9	unk.	58	**6**	18	438	unk.	1.3	**7**	1.0	**7**	unk.	**unk.**	unk.	A
1.0	**5**	26	**7**	1.02	**10**	11826	**237**	1183	3.5	**12**	unk.	1.75	85	**9**	0.8	11.2	unk.	18	**4**	0.25	**12**	tr(a)		unk.	
0.08	**5**	0.22	**11**	0.00	**0**	0.003	**1**	0	0(a)	**0**	unk.	unk.	41	**4**	13	310	unk.	0.9	**5**	0.9	**6**	unk.	**unk.**	unk.	B
0.8	**4**	23	**6**	0.89	**9**	10072	**201**	1007	unk.	**unk.**	7.6	1.27	60	**6**	0.5	7.9	unk.	15	**4**	0.22	**11**	tr(a)		unk.	
0.09	**5**	0.26	**13**	0.00	**0**	0.003	**1**	0	0(a)	**0**	unk.	unk.	48	**5**	15	367	unk.	1.1	**6**	1.1	**7**	unk.	**unk.**	unk.	C
0.9	**5**	27	**7**	1.06	**11**	11929	**239**	1193	unk.	**unk.**	9.1	1.51	71	**7**	0.7	9.4	unk.	18	**5**	0.26	**13**	tr(a)		tr	
0.10	**6**	unk.	**unk.**	0(a)	**0**	unk.	**unk.**	0	0(a)	**0**	unk.	unk.	64	**6**	75	281	unk.	1.6	**9**	unk.	**unk.**	unk.	**unk.**	unk.	D
0.3	**2**	unk.	**unk.**	unk.	**unk.**	4725	**95**	473	unk.	**unk.**	unk.	0.88	21	**2**	3.3	7.2	unk.	unk.	**unk.**	tr(a)	**0**	tr(a)		unk.	
0.05	**3**	unk.	**unk.**	0(a)	**0**	unk.	**unk.**	0	0(a)	**0**	unk.	unk.	24	**2**	40	151	unk.	0.9	**5**	unk.	**unk.**	unk.	**unk.**	unk.	E
0.1	**1**	unk.	**unk.**	unk.	**unk.**	1787	**36**	179	unk.	**unk.**	unk.	0.44	11	**1**	1.8	3.9	unk.	unk.	**unk.**	tr(a)	**0**	tr(a)		unk.	
0.04	**3**	unk.	**unk.**	0.76	**13**	unk.	**unk.**	unk.	unk.	**unk.**	tr(a)	tr(a)	23	**2**	unk.	unk.	unk.	1.1	**6**	unk.	**unk.**	unk.	**unk.**	unk.	F
9.2	**46**	unk.	**unk.**	0.08	**1**	1722	**34**	unk.	0.0	**0**	tr(a)	1.43	231	**23**	unk.	unk.	unk.	unk.	**unk.**	unk.	**unk.**	tr(a)		tr(a)	
0.00	**0**	0(a)	**0**	0(a)	**0**	tr(a)	**0**	0	0(a)	**0**	0(a)	0(a)	3	**0**	tr	5	unk.	0(a)	**0**	tr	**0**	unk.	**unk.**	unk.	G
0.0	**0**	0(a)	**0**	tr(a)	**0**	0	**0**	0	0.0	**0**	0(a)	0.02	0	**0**	0.0	0.1	unk.	1	**0**	unk.	**unk.**	unk.		unk.	
0.02	**1**	tr	**0**	0.01	**0**	0.001	**0**	0	3	**1**	0.2	0.0	16	**2**	9	38	6	0.4	**2**	0.1	**1**	0.7	**1**	3	H
0.1	**0**	3	**1**	0.03	**0**	12	**0**	tr	0.2	**1**	tr	0.20	21	**2**	0.4	1.0	0.2	10	**2**	0.01	**1**	1		4	
0.01	**0**	tr	**0**	0.00	**0**	0.001	**0**	0	0	**0**	0.1	0.0	2	**0**	tr	21	2	0.2	**1**	tr	**0**	unk.	**unk.**	2	I
tr	**0**	3	**1**	0.00	**0**	2	**0**	tr	0.2	**1**	0.0	0.08	10	**1**	0.0	0.5	0.0	7	**2**	tr	**0**	unk.		3	
0.00	**0**	0.00	**0**	0.00	**0**	0.000	**0**	0	0	**0**	0.0	0(a)	9	**1**	14	1	15	0.8	**5**	0.3	**2**	unk.	**unk.**	unk.	J
0.0	**0**	0	**0**	0.00	**0**	0	**0**	0	0.0	**0**	0(a)	0.14	3	**0**	0.6	0.0	0.4	tr	**0**	0.02	**1**	3		unk.	
tr	**0**	tr	**0**	0.00	**0**	unk.	**unk.**	0	0(a)	**0**	unk.	unk.	2	**0**	6	5	unk.	0.2	**1**	unk.	**unk.**	unk.	**unk.**	unk.	K
tr	**0**	tr	**0**	0.01	**0**	4	**0**	tr	unk.	**unk.**	unk.	tr(a)	1	**0**	0.2	0.1	unk.	1	**0**	tr(a)	**0**	unk.		unk.	
0.00	**0**	0(a)	**0**	0(a)	**0**	0(a)	**0**	0	0(a)	**0**	0(a)	0(a)	20	**2**	2	33	unk.	0.2	**1**	unk.	**unk.**	unk.	**unk.**	unk.	L
0.0	**0**	0(a)	**0**	0(a)	**0**	0	**0**	0	0.0	**0**	0(a)	0.13	2	**0**	0.1	0.9	unk.	4	**1**	unk.	**unk.**	unk.		unk.	
0(a)	**0**	0(a)	**0**	0(a)	**0**	0(a)	**0**	0(a)	0(a)	**0**	0(a)	0(a)	unk.	**unk.**	unk.	unk.	unk.	unk.	**unk.**	unk.	**unk.**	unk.	**unk.**	unk.	M
0(a)	**0**	0(a)	**0**	0(a)	**0**	0(a)	**0**	0(a)	0.0	**0**	0(a)	unk.	unk.	**unk.**	unk.	unk.	unk.	unk.	**unk.**	unk.	**unk.**	unk.		unk.	
0.31	**18**	unk.	**unk.**	unk.	**unk.**	unk.	**unk.**	unk.	unk.	**unk.**	unk.	unk.	106	**11**	982	468	unk.	3.1	**17**	unk.	**unk.**	unk.	**unk.**	unk.	N
5.0	**25**	unk.	**unk.**	unk.	**unk.**	911	**18**	unk.	unk.	**unk.**	unk.	4.10	296	**30**	42.7	12.0	unk.	unk.	**unk.**	unk.	**unk.**	unk.		unk.	
0.31	**18**	unk.	**unk.**	unk.	**unk.**	unk.	**unk.**	unk.	unk.	**unk.**	unk.	unk.	71	**7**	1525	644	unk.	4.3	**24**	unk.	**unk.**	unk.	**unk.**	unk.	O
14.8	**74**	61	**15**	unk.	**unk.**	2519	**50**	unk.	unk.	**unk.**	unk.	4.90	357	**36**	66.3	16.5	unk.	unk.	**unk.**	unk.	**unk.**	14		unk.	
0.21	**12**	unk.	**unk.**	unk.	**unk.**	unk.	**unk.**	unk.	0(a)	**0**	unk.	unk.	159	**16**	1522	467	unk.	4.6	**26**	unk.	**unk.**	unk.	**unk.**	unk.	P
1.8	**9**	unk.	**unk.**	unk.	**unk.**	255	**5**	unk.	unk.	**unk.**	unk.	5.70	303	**30**	66.2	12.0	unk.	unk.	**unk.**	unk.	**unk.**	unk.		unk.	
0.26	**15**	unk.	**unk.**	unk.	**unk.**	unk.	**unk.**	3497	unk.	**unk.**	unk.	unk.	41	**4**	1106	512	unk.	3.5	**19**	unk.	**unk.**	unk.	**unk.**	unk.	Q
5.2	**26**	44	**11**	unk.	**unk.**	10816	**216**	unk.	unk.	**unk.**	unk.	4.70	276	**28**	48.1	13.1	unk.	unk.	**unk.**	unk.	**unk.**	unk.		unk.	
0.20	**12**	unk.	**unk.**	unk.	**unk.**	unk.	**unk.**	21	unk.	**unk.**	unk.	unk.	34	**3**	708	461	unk.	2.3	**13**	unk.	**unk.**	unk.	**unk.**	unk.	R
7.6	**38**	20	**5**	unk.	**unk.**	0	**0**	unk.	unk.	**unk.**	unk.	4.20	213	**21**	30.8	11.8	unk.	unk.	**unk.**	unk.	**unk.**	unk.		unk.	
0.30	**18**	unk.	**unk.**	unk.	**unk.**	unk.	**unk.**	unk.	unk.	**unk.**	unk.	unk.	103	**10**	1312	unk.	unk.	4.0	**22**	unk.	**unk.**	unk.	**unk.**	unk.	S
6.9	**35**	unk.	**unk.**	unk.	**unk.**	144	**3**	unk.	unk.	**unk.**	unk.	unk.	99	**10**	57.0	unk.	unk.	unk.	**unk.**	unk.	**unk.**	unk.		unk.	

	FOOD	Portion	Weight in grams / *Conversion for 100 g*	Kilocalories / H_2O g	Total carbohydrate g / Total fats g	Crude fiber g / Dietary fiber g	Total protein g / Total sugar g	**% USRDA**	Arginine mg / Histidine mg	Isoleucine mg / Leucine mg	Lysine mg / Methionine mg	Phenylalanine mg / Threonine mg	Valine mg / Tryptophan mg	Cystine mg / Tyrosine mg	Polyunsat. fatty acids g / Monounsat. fatty acids g	Saturated fatty acids g / P/S ratio	Linoleic acid g / Cholesterol mg	Thiamin mg / Ascorbic acid mg	**% USRDA**
A	t.v. dinner,Mexican style-Swanson	1 dinner	454	695	75.0	unk.	22.0	**34**	unk.	unk.	unk.	unk.	unk.	unk.	unk.	unk.	unk.	0.64	**43**
			0.22	304.6	35.0	unk.	unk.		unk.	unk.	unk.	unk.	unk.	unk.	unk.	unk.	unk.	0	**0**
B	t.v. dinner,perch-Swanson	1 dinner	197	265	23.9	unk.	15.6	**24**	unk.	unk.	unk.	unk.	unk.	unk.	unk.	unk.	unk.	0.16	**11**
			0.51	unk.	12.0	unk.	unk.		unk.	unk.	unk.	unk.	unk.	unk.	unk.	unk.	unk.	4	**7**
C	TACO BELL bean burrito	1 serving	166	343	48.0	unk.	11.0	**17**	unk.	unk.	unk.	unk.	unk.	unk.	unk.	unk.	unk.	0.37	**24**
			0.60	unk.	12.0	unk.	unk.		unk.	unk.	unk.	unk.	unk.	unk.	unk.	unk.	unk.	15	**25**
D	Taco Bell beef burrito	1 serving	184	466	37.0	unk.	30.0	**46**	unk.	unk.	unk.	unk.	unk.	unk.	unk.	unk.	unk.	0.29	**20**
			0.54	unk.	21.0	unk.	unk.		unk.	unk.	unk.	0	unk.	unk.	unk.	unk.	unk.	15	**25**
E	Taco Bell beefy tostada	1 serving	184	291	21.0	unk.	19.0	**29**	unk.	unk.	unk.	unk.	unk.	unk.	unk.	unk.	unk.	0.17	**11**
			0.54	unk.	15.0	unk.	unk.		unk.	unk.	unk.	unk.	unk.	unk.	unk.	unk.	unk.	13	**21**
F	Taco Bell Bellbeefer	1 serving	123	221	23.0	unk.	19.0	**29**	unk.	unk.	unk.	unk.	unk.	unk.	unk.	unk.	unk.	0.15	**10**
			0.81	unk.	15.0	unk.	unk.		unk.	unk.	unk.	unk.	unk.	unk.	unk.	unk.	unk.	10	**17**
G	Taco Bell Bellbeefer w/cheese	1 serving	137	278	23.0	unk.	19.0	**29**	unk.	unk.	unk.	unk.	unk.	unk.	unk.	unk.	unk.	0.16	**11**
			0.73	unk.	12.0	unk.	unk.		unk.	unk.	unk.	unk.	unk.	unk.	unk.	unk.	unk.	10	**17**
H	Taco Bell Burrito Supreme	1 serving	225	457	43.0	unk.	21.0	**32**	unk.	unk.	unk.	unk.	unk.	unk.	unk.	unk.	unk.	0.34	**23**
			0.44	unk.	22.0	unk.	unk.		unk.	unk.	unk.	unk.	unk.	unk.	unk.	unk.	unk.	16	**27**
I	Taco Bell combination burrito	1 serving	175	404	43.0	unk.	21.0	**32**	unk.	unk.	unk.	unk.	unk.	unk.	unk.	unk.	unk.	0.33	**22**
			0.57	unk.	16.0	unk.	unk.		unk.	unk.	unk.	unk.	unk.	unk.	unk.	unk.	unk.	15	**25**
J	Taco Bell Enchirito	1 serving	207	454	42.0	unk.	25.0	**39**	unk.	unk.	unk.	unk.	unk.	unk.	unk.	unk.	unk.	0.31	**21**
			0.48	unk.	21.0	unk.	unk.		unk.	unk.	unk.	unk.	unk.	unk.	unk.	unk.	unk.	10	**16**
K	Taco Bell pintos'n cheese	1 serving	158	168	21.0	unk.	11.0	**17**	unk.	unk.	unk.	unk.	unk.	unk.	unk.	unk.	unk.	0.25	**17**
			0.63	unk.	5.0	unk.	unk.		unk.	unk.	unk.	unk.	unk.	unk.	unk.	unk.	unk.	9	**16**
L	Taco Bell taco	1 serving	83	186	14.0	unk.	15.0	**23**	unk.	unk.	unk.	unk.	unk.	unk.	unk.	unk.	unk.	0.09	**6**
			1.20	unk.	8.0	unk.	unk.		unk.	unk.	unk.	unk.	unk.	unk.	unk.	unk.	unk.	tr	**0**
M	Taco Bell tostada	1 serving	138	179	unk.	unk.	9.0	**14**	unk.	unk.	unk.	unk.	unk.	unk.	unk.	unk.	unk.	0.18	**12**
			0.72	unk.	6.0	unk.	unk.		unk.	unk.	unk.	unk.	unk.	unk.	unk.	unk.	unk.	10	**16**
N	TACO salad,home recipe	1 C	247	292	8.2	0.79	20.0	**43**	1054	1045	1636	882	1117	194	3.06	9.69	0.91	0.12	**8**
			0.41	194.2	20.2	2.91	2.9		628	1654	478	807	232	725	8.06	0.32	71	14	**23**
O	tacos verde blanco y rojo,home recipe	1 average	160	296	17.1	1.39	10.5	**22**	469	451	711	357	445	109	4.64	5.44	5.51	0.12	**8**
			0.63	108.5	21.8	0.31	2.3		213	700	216	361	104	274	6.99	0.85	38	28	**47**
P	TALLOW,mutton	1 C	205	1849	0.0	0.00	0.0	**0**	0(a)	0(a)	0(a)	0(a)	0(a)	0(a)	15.99	96.96	11.27	0(a)	**0**
			0.49	0.0	205.0	0(a)	0(a)		0(a)	0(a)	0(a)	0(a)	0(a)	0(a)	77.08	0.16	209	0(a)	**0**
Q	TAMALE	1 average	110	155	15.8	tr(a)	5.0	**8**	unk.	165	264	143	165	44	unk.	3.30	unk.	0.00	**0**
			0.91	unk.	7.9	unk.	unk.		110	253	77	132	77	110	unk.	unk.	10	0	**0**
R	TANGERINE,fresh,any,edible portion	1 average	86	40	10.0	0.43	0.7	**1**	unk.	unk.	24	unk.	unk.	unk.	unk.	tr(a)	unk.	0.05	**3**
			1.16	74.8	0.2	1.60	5.4		unk.	unk.	3	unk.	4	unk.	unk.	unk.	0	27	**44**
S	TARRAGON,ground	1 tsp	1.6	5	0.8	0.12	0.4	**1**	unk.	unk.	unk.	unk.	unk.	unk.	unk.	unk.	unk.	tr	**0**
			62.50	0.1	0.1	unk.	0.0		unk.	unk.	unk.	unk.	unk.	unk.	unk.	unk.	0	unk.	**unk.**

Riboflavin mg / Niacin mg	% USRDA	Vitamin B_6 mg / Folacin mcg	% USRDA	Vitamin B_{12} mcg / Pantothenic acid mg	% USRDA	Biotin mg / Vitamin A IU	% USRDA	Preformed A RE / Beta carotene RE	Vitamin D IU / Vitamin E IU	% USRDA	Total tocopherol mg / Alpha tocopherol mg	Other tocopherol mg / Total ash g	Calcium mg / Phosphorus mg	% USRDA	Sodium mg / Sodium meq	Potassium mg / Potassium meq	Chlorine mg / Chlorine meq	Iron mg / Magnesium mg	% USRDA	Zinc mg / Copper mg	% USRDA	Iodine mcg / Selenium mcg	% USRDA	Manganese mcg / Chromium mcg	
0.27	16	unk.	unk.	unk.	unk.	unk.	unk.	unk.	0(a)	0	unk.	unk.	259	26	1793	unk.	unk.	4.1	23	unk.	unk.	unk.	unk.	unk.	A
4.5	23	unk.	unk.	unk.	unk.	3060	61	unk.	unk.	unk.	unk.	unk.	unk.	unk.	78.0	unk.	unk.	unk.	unk.	unk.	unk.	unk.		unk.	
0.16	9	unk.	unk.	unk.	unk.	unk.	unk.	unk.	unk.	unk.	unk.	unk.	unk.	unk.	651	424	unk.	1.8	10	unk.	unk.	unk.	unk.	unk.	B
2.8	14	unk.	unk.	unk.	unk.	282	6	unk.	unk.	unk.	unk.	unk.	197	20	28.3	10.8	unk.	unk.	unk.	unk.	unk.	unk.		unk.	
0.22	13	unk.	unk.	0(a)	0	unk.	unk.	unk.	0(a)	0	unk.	unk.	98	10	272	235	unk.	2.8	16	unk.	unk.	unk.	unk.	unk.	C
2.2	11	unk.	unk.	unk.	unk.	1657	33	unk.	unk.	unk.	unk.	unk.	173	17	11.8	6.0	unk.	unk.	unk.	unk.	unk.	unk.		unk.	
0.39	23	unk.	unk.	unk.	unk.	unk.	unk.	unk.	0(a)	0	unk.	unk.	83	8	327	320	unk.	4.6	26	unk.	unk.	unk.	unk.	unk.	D
7.0	35	unk.	unk.	unk.	unk.	1675	34	unk.	unk.	unk.	unk.	unk.	288	29	14.2	8.2	unk.	unk.	unk.	unk.	unk.	unk.		unk.	
0.28	16	unk.	unk.	unk.	unk.	unk.	unk.	unk.	0(a)	0	unk.	unk.	208	21	138	277	unk.	3.4	19	unk.	unk.	unk.	unk.	unk.	E
3.3	17	unk.	unk.	unk.	unk.	3450	69	unk.	unk.	unk.	unk.	unk.	265	27	6.0	7.1	unk.	unk.	unk.	unk.	unk.	unk.		unk.	
0.20	12	unk.	unk.	unk.	unk.	unk.	unk.	unk.	0(a)	0	unk.	unk.	40	4	231	183	unk.	2.6	14	unk.	unk.	unk.	unk.	unk.	F
3.7	19	unk.	unk.	unk.	unk.	2961	59	unk.	unk.	unk.	unk.	unk.	140	14	10.0	4.7	unk.	unk.	unk.	unk.	unk.	unk.		unk.	
0.27	16	unk.	unk.	unk.	unk.	unk.	unk.	unk.	0(a)	0	unk.	unk.	147	15	330	195	unk.	2.7	15	unk.	unk.	unk.	unk.	unk.	G
3.7	19	unk.	unk.	unk.	unk.	3146	63	unk.	unk.	unk.	unk.	unk.	208	21	14.3	5.0	unk.	unk.	unk.	unk.	unk.	unk.		unk.	
0.36	21	unk.	unk.	unk.	unk.	unk.	unk.	unk.	0(a)	0	unk.	unk.	121	12	367	350	unk.	3.8	21	unk.	unk.	unk.	unk.	unk.	H
4.7	24	unk.	unk.	unk.	unk.	3463	69	unk.	unk.	unk.	unk.	unk.	245	25	16.0	8.9	unk.	unk.	unk.	unk.	unk.	unk.		unk.	
0.31	19	unk.	unk.	unk.	unk.	unk.	unk.	unk.	0(a)	0	unk.	unk.	91	9	300	278	unk.	3.7	21	unk.	unk.	unk.	unk.	unk.	I
4.6	23	unk.	unk.	unk.	unk.	1666	33	unk.	unk.	unk.	unk.	unk.	230	23	13.0	7.1	unk.	unk.	unk.	unk.	unk.	unk.		unk.	
0.37	22	unk.	unk.	unk.	unk.	unk.	unk.	unk.	0(a)	0	unk.	unk.	259	26	1175	491	unk.	3.8		unk.	unk.	unk.	unk.	unk.	J
4.7	24	unk.	unk.	unk.	unk.	1178	24	unk.	unk.	unk.	unk.	unk.	338	34	51.1	12.6	unk.	unk.		unk.	unk.	unk.		unk.	
0.16	9	unk.	unk.	unk.	unk.	unk.	unk.	unk.	0(a)	0	unk.	unk.	150	15	102	307	unk.	2.3	13	unk.	unk.	unk.	unk.	unk.	K
0.9	5	unk.	unk.	unk.	unk.	3123	63	unk.	unk.	unk.	unk.	unk.	210	21	4.4	7.8	unk.	unk.	unk.	unk.	unk.	unk.		unk.	
0.16	9	unk.	unk.	unk.	unk.	unk.	unk.	unk.	0(a)	0	unk.	unk.	120	12	79	143	unk.	2.5	14	unk.	unk.	unk.	unk.	unk.	L
2.9	15	unk.	unk.	unk.	unk.	120	2	unk.	unk.	unk.	unk.	unk.	175	18	3.4	3.7	unk.	unk.	unk.	unk.	unk.	unk.		unk.	
0.15	9	unk.	unk.	unk.	unk.	unk.	unk.	unk.	0(a)	0	unk.	unk.	191	19	101	172	unk.	2.3	13	unk.	unk.	unk.	unk.	unk.	M
0.8	4	unk.	unk.	unk.	unk.	3152	63	unk.	unk.	unk.	unk.	unk.	186	19	4.4	4.4	unk.	unk.	unk.	unk.	unk.	unk.		unk.	
0.30	17	0.23	11	0.84	14	0.005	2	6	0	0	1.0	0.2	178	18	723	520	53	2.7	15	3.5	23	unk.	unk.	77	N
8.7	44	46	12	0.53	5	820	16	60	1.2	4	0.4	3.85	244	24	31.4	13.3	1.5	36	9	0.11	6	tr		10	
0.19	11	0.32	16	0.27	5	0.003	1	17	1	0	4.2	0.0	78	8	710	489	10	2.0	11	1.0	6	97.5	65	92	O
2.8	14	33	8	0.76	8	743	15	48	5.0	17	0.5	3.04	159	16	30.9	12.5	0.3	56	14	0.36	18	0		1	
0(a)	0	0(a)	0	0(a)	0	0(a)	0	unk.	unk.	unk.	unk.	unk.	unk.	unk.	unk.	unk.	unk.	unk.	unk.	unk.	unk.	unk.	unk.	unk.	P
0(a)	0	0(a)	0	0(a)	0	unk.	unk.	unk.	unk.	unk.	unk.	0.00	unk.	unk.	unk.	unk.	unk.	unk.	unk.	unk.	unk.	unk.		unk.	
0.00	0	0.22	11	tr(a)	0	0.011	4	0	0	0	0.0	0.0	22	2	738	0	unk.	1.3	7	1.0	7	unk.	unk.	unk.	Q
0.0	0	1	0	0.44	4	0	0	0	0.0	0	0.0	tr(a)	43	4	32.1	0.0	unk.	10	3	0.05	3	unk.		unk.	
0.02	1	0.06	3	0.00	0	unk.	unk.	0	0(a)	0	unk.	unk.	15	2	2	108	2	0.3	2	unk.	unk.	unk.	unk.	28	R
0.1	0	18	5	0.17	2	362	7	36	unk.	unk.	unk.	0.34	9	1	0.1	2.8	0.0	11	3	0.02	1	0		0	
0.02	1	unk.	unk.	0.00	0	unk.	unk.	0(a)	0(a)	0	unk.	unk.	18	2	1	48	unk.	0.5	3	0.1	0	unk.	unk.	unk.	S
0.1	1	unk.	unk.	unk.	unk.	67	1	unk.	unk.	unk.	unk.	0.19	5	1	0.0	1.2	unk.	6	1	unk.	unk.	unk.		unk.	

	FOOD	Portion	Weight in grams / *Conversion for 100 g*	Kilocalories / H_2O g	Total carbohydrate g / Total fats g	Crude fiber g / Dietary fiber g	Total protein g / Total sugar g	**% USRDA**	Arginine mg / Histidine mg	Isoleucine mg / Leucine mg	Lysine mg / Methionine mg	Phenylalanine mg / Threonine mg	Valine mg / Tryptophan mg	Cystine mg / Tyrosine mg	Polyunsat. fatty acids g / Monounsat. fatty acids g	Saturated fatty acids g / P/S ratio	Linoleic acid g / Cholesterol mg	Thiamin mg / Ascorbic acid mg	**% USRDA**
A	TEA,brewed	6 oz cup	180	2	0(a)	0(a)	tr	**0**	unk.	tr(a)	tr(a)	tr(a)	tr(a)	tr(a)	tr(a)	tr(a)	tr(a)	0.00	**0**
			0.56	unk.	tr	0(a)	0(a)		tr(a)	tr(a)	tr(a)	tr(a)	tr(a)	tr(a)	tr(a)	0.00	0(a)	1	**2**
B	tea,instant,dry powder	1 tsp	0.5	1	0.4	0.00	unk.	**unk.**	unk.	unk.	unk.	unk.	unk.	unk.	unk.	unk.	unk.	unk.	**unk.**
			200.00	tr	tr	0(a)	0.1		unk.	unk.	unk.	unk.	unk.	unk.	unk.	unk.	unk.	unk.	**unk.**
C	tea,instant,prep	6 oz cup	173	3	0.7	unk.	unk.	**unk.**	unk.	unk.	unk.	unk.	unk.	unk.	unk.	unk.	unk.	unk.	**unk.**
			0.58	171.5	tr	0(a)	0.1		unk.	unk.	unk.	unk.	unk.	unk.	unk.	unk.	unk.	unk.	**unk.**
D	THYME,ground	1 tsp	1.4	4	0.9	0.26	0.1	**0**	unk.	7	3	7	7	4	0.02	0.04	0.01	0.01	**1**
			71.43	0.1	0.1	unk.	0.0		unk.	6	4	4	3	7	0.01	0.44	0	unk.	**unk.**
E	TOFU (soybean curd)	1/2 C	125	90	2.9	0.00	9.8	**22**	unk.	unk.	unk.	unk.	unk.	unk.	unk.	unk.	unk.	0.02	**2**
			0.80	106.1	5.4	unk.	unk.		unk.	unk.	unk.	unk.	unk.	unk.	unk.	unk.	unk.	unk.	**unk.**
F	TOMATO,CANNED,low sodium, solids & liquid	1/2 C	121	24	5.1	0.48	1.2	**2**	25	35	51	30	34	8	0.00	0.00	unk.	0.06	**4**
			0.83	113.4	0.2	2.41	4.0		unk.	49	8	40	11	25	unk.	0.00	0	20	**34**
G	tomato,canned,paste,no salt added	1/2 C	131	107	24.4	1.18	4.4	**7**	unk.	92	128	94	102	29	0.00	0.00	unk.	0.26	**18**
			0.76	98.2	0.5	10.09	18.7		76	124	24	102	39	59	unk.	0.00	0	64	**107**
H	tomato,canned,puree	1/2 C	125	49	11.1	0.50	2.1	**3**	unk.	61	88	60	60	14	0.00	0.00	unk.	0.11	**8**
			0.80	108.3	0.2	4.98	8.6		26	87	15	70	19	27	unk.	0.00	0	41	**69**
I	tomato,canned,puree,low sodium	1/2 C	125	49	11.1	0.50	2.1	**3**	unk.	61	88	60	60	14	0.00	0.00	unk.	0.11	**8**
			0.80	109.6	0.2	4.98	8.6		26	87	15	70	19	27	unk.	0.00	unk.	41	**69**
J	tomato,canned,solids & liquid	1/2 C	121	25	5.2	0.48	1.2	**2**	25	35	51	30	34	8	0.00	0.00	unk.	0.06	**4**
			0.83	112.9	0.2	2.41	4.0		18	49	8	40	11	25	unk.	0.00	0	20	**34**
K	TOMATO,FRESH,cooked,solids & liquid	1/2 C	121	31	6.6	0.72	1.6	**2**	33	35	51	40	34	8	0.00	0.00	unk.	0.08	**6**
			0.83	111.3	0.2	2.41	4.0		18	49	8	40	11	34	unk.	0.00	0	29	**48**
L	tomato,fresh,green,raw,edible portion	1 med	123	30	6.3	0.61	1.5	**2**	31	36	52	34	34	9	unk.	tr(a)	unk.	0.07	**5**
			0.81	114.4	0.2	2.46	4.2		18	50	9	41	11	17	unk.	unk.	0	25	**41**
M	tomato,fresh,green,raw,slice/wedge	1/6 tomato	21	5	1.1	0.10	0.3	**0**	5	6	9	6	6	1	unk.	tr(a)	unk.	0.01	**1**
			4.88	19.1	tr	0.41	0.7		3	8	1	7	2	3	unk.	unk.	0	4	**7**
N	tomato,fresh,raw,edible portion	1 med	123	27	5.8	0.61	1.3	**2**	28	36	52	34	34	9	unk.	unk.	unk.	0.07	**5**
			0.81	115.0	0.2	2.46	4.2		18	50	9	41	11	30	unk.	unk.	0	28	**47**
O	tomato,fresh,raw,slice/wedge	1/6 tomato	21	5	1.0	0.10	0.2	**0**	5	6	9	6	6	1	0.00	0.00	unk.	0.01	**1**
			4.88	19.2	tr	0.41	0.7		3	8	1	7	2	5	unk.	0.00	0	5	**8**
P	TOMATO,HOME RECIPE,aspic	1/2 C	140	44	9.3	0.36	2.6	**5**	18	54	115	64	71	2	0.00	0.00	0.00	0.06	**4**
			0.71	127.2	0.1	0.34	8.3		105	91	20	66	10	97	0.00	0.00	0	22	**36**
Q	tomato,home recipe,preserves	1 Tbsp	15	25	6.5	0.04	0.1	**0**	2	2	3	2	2	1	0.00	0.00	0.00	tr	**0**
			6.67	8.3	tr	0.19	6.3		1	3	1	3	1	2	0.00	0.00	0	2	**4**
R	tomato,home recipe,scalloped	1/2 C	147	88	13.0	0.73	2.5	**4**	49	85	81	84	86	38	0.12	1.84	0.15	0.09	**6**
			0.68	125.9	3.4	3.29	6.2		47	126	25	73	25	60	0.82	0.06	8	20	**34**
S	TORTE,chocolate,home recipe	1/16 of	92	317	28.4	0.13	4.0	**6**	246	165	181	147	203	65	3.22	10.64	2.54	0.06	**4**
		8-1/2" dia	*1.09*	37.0	21.7	0.32	23.2		76	267	89	130	42	130	4.73	0.30	61	tr	**1**

Riboflavin mg / Niacin mg	% USRDA	Vitamin B₆ mg / Folacin mcg	% USRDA	Vitamin B₁₂ mcg / Pantothenic acid mg	% USRDA	Biotin mg / Vitamin A IU	% USRDA	Preformed A RE / Beta carotene RE	Vitamin D IU / Vitamin E IU	% USRDA	Total tocopherol mg / Alpha tocopherol mg	Other tocopherol mg / Total ash g	Calcium mg / Phosphorus mg	% USRDA	Sodium mg / Sodium meq	Potassium mg / Potassium meq	Chlorine mg / Chlorine meq	Iron mg / Magnesium mg	% USRDA	Zinc mg / Copper mg	% USRDA	Iodine mcg / Selenium mcg	% USRDA	Manganese mcg / Chromium mcg	
0.04	**2**	unk.	**unk.**	0(a)	**0**	unk.	**unk.**	0	0(a)	**0**	unk.	unk.	5	**1**	unk.	unk.	unk.	0.2	**1**	unk.	**unk.**	unk.	**unk.**	unk.	A
0.1	**1**	unk.	**unk.**	unk.	**unk.**	0	**0**	0	unk.	**unk.**	unk.	tr(a)	4	**0**	unk.	unk.	unk.	unk.	**unk.**	unk.	**unk.**	unk.		unk.	
0.00	**0**	unk.	**unk.**	unk.	**unk.**	unk.	**unk.**	unk.	unk.	**unk.**	unk.	unk.	tr	**0**	unk.	23	unk.	tr	**0**	unk.	**unk.**	unk.	**unk.**	unk.	B
tr	**0**	unk.	**unk.**	unk.	**unk.**	unk.	**unk.**	unk.	unk.	**unk.**	unk.	0.03	unk.	**unk.**	unk.	0.6	unk.	unk.	**unk.**	unk.	**unk.**	unk.		unk.	
0.02	**1**	0.00	**0**	0.00	**0**	unk.	**unk.**	unk.	0	**0**	0.0	0(a)	0	**0**	2	43	unk.	0.0	**0**	tr	**0**	unk.	**unk.**	379	C
tr	**0**	0	**0**	0.00	**0**	unk.	**unk.**	unk.	0.0	**0**	0.0	unk.	0	**0**	0.1	1.1	unk.	7	**2**	0.01	**1**	unk.		10	
0.01	**0**	unk.	**unk.**	0.00	**0**	unk.	**unk.**	0(a)	0(a)	**0**	unk.	unk.	26	**3**	1	11	unk.	1.7	**10**	0.1	**1**	unk.	**unk.**	unk.	D
0.1	**0**	unk.	**unk.**	unk.	**unk.**	53	**1**	unk.	unk.	**unk.**	unk.	0.16	3	**0**	0.0	0.3	unk.	3	**1**	unk.	**unk.**	unk.		unk.	
0.02	**2**	unk.	**unk.**	unk.	**unk.**	unk.	**unk.**	unk.	unk.	**unk.**	unk.	unk.	8	**1**	unk.	unk.	unk.	2.1	**12**	unk.	**unk.**	unk.	**unk.**	unk.	E
0.6	**3**	unk.	**unk.**	unk.	**unk.**	unk.	**unk.**	unk.	unk.	**unk.**	unk.	0.87	131	**13**	unk.	unk.	unk.	unk.	**unk.**	unk.	**unk.**	unk.		unk.	
0.04	**2**	0.11	**5**	0.00	**0**	0.002	**1**	0	0(a)	**0**	0.0	0(a)	7	**1**	4	261	unk.	0.6	**3**	0.2	**2**	unk.	**unk.**	36	F
0.8	**4**	4	**1**	0.28	**3**	1084	**22**	108	0.0	**0**	0(a)	0.60	23	**2**	0.2	6.7	unk.	14	**4**	0.16	**8**	tr(a)		unk.	
0.16	**9**	0.50	**25**	0.00	**0**	unk.	**unk.**	0	0(a)	**0**	0.0	0.0	35	**4**	50	1163	unk.	4.6	**26**	unk.	**unk.**	unk.	**unk.**	unk.	G
4.1	**20**	unk.	**unk.**	0.58	**6**	4323	**87**	432	0.0	**0**	0.0	3.41	92	**9**	2.2	29.8	unk.	26	**7**	tr(a)	**0**	tr(a)		unk.	
0.06	**4**	0.19	**10**	0.00	**0**	unk.	**unk.**	0	0(a)	**0**	0.0	0.0	16	**2**	497	530	unk.	2.1	**12**	unk.	**unk.**	unk.	**unk.**	unk.	H
1.7	**9**	unk.	**unk.**	unk.	**unk.**	1992	**40**	199	0.0	**0**	0.0	2.74	42	**4**	21.6	13.6	unk.	25	**6**	tr(a)	**0**	tr(a)		unk.	
0.06	**4**	0.19	**10**	0.00	**0**	unk.	**unk.**	0	0(a)	**0**	0.0	0.0	16	**2**	7	530	unk.	2.1	**12**	unk.	**unk.**	unk.	**unk.**	unk.	I
1.7	**9**	unk.	**unk.**	unk.	**unk.**	1992	**40**	199	0.0	**0**	0.0	1.49	42	**4**	0.3	13.6	unk.	25	**6**	tr(a)	**0**	tr(a)		unk.	
0.04	**2**	0.11	**5**	0.00	**0**	0.002	**1**	0	0(a)	**0**	0.0	0(a)	7	**1**	157	261	unk.	0.6	**3**	0.2	**2**	unk.	**unk.**	36	J
0.8	**4**	4	**1**	0.28	**3**	1084	**22**	108	0.0	**0**	0(a)	0.96	23	**2**	6.8	6.7	unk.	14	**4**	0.16	**8**	tr(a)		unk.	
0.06	**4**	0.11	**5**	0.00	**0**	0.002	**1**	0	0(a)	**0**	0.0	0(a)	18	**2**	5	346	unk.	0.7	**4**	0.2	**2**	unk.	**unk.**	24	K
1.0	**5**	31	**8**	0.24	**2**	1205	**24**	120	0.0	**0**	0(a)	0.72	39	**4**	0.2	8.8	unk.	unk.	**unk.**	0.16	**8**	tr(a)		17	
0.05	**3**	unk.	**unk.**	unk.	**unk.**	unk.	**unk.**	0	0(a)	**0**	unk.	unk.	16	**2**	4	300	unk.	0.6	**3**	unk.	**unk.**	unk.	**unk.**	unk.	L
0.6	**3**	unk.	**unk.**	unk.	**unk.**	333	**7**	33	unk.	**unk.**	unk.	0.61	33	**3**	0.2	7.7	unk.	17	**4**	tr(a)	**0**	tr(a)		unk.	
0.01	**1**	unk.	**unk.**	unk.	**unk.**	unk.	**unk.**	0	0(a)	**0**	unk.	unk.	3	**0**	1	50	unk.	0.1	**1**	unk.	**unk.**	unk.	**unk.**	unk.	M
0.1	**1**	unk.	**unk.**	unk.	**unk.**	56	**1**	6	unk.	**unk.**	unk.	0.10	6	**1**	0.0	1.3	unk.	3	**1**	tr(a)	**0**	tr(a)		unk.	
0.05	**3**	0.12	**6**	0.00	**0**	0.005	**2**	0	0(a)	**0**	1.0	0.6	16	**2**	4	300	63	0.6	**3**	0.2	**2**	unk.	**unk.**	25	N
0.9	**4**	10	**3**	0.41	**4**	1107	**22**	111	1.3	**4**	0.5	0.61	33	**3**	0.2	7.7	1.8	22	**5**	0.01	**1**	tr(a)		1	
0.01	**1**	0.02	**1**	0.00	**0**	0.001	**0**	0	0(a)	**0**	0.2	0.1	3	**0**	1	50	10	0.1	**1**	tr	**0**	unk.	**unk.**	4	O
0.1	**1**	2	**0**	0.07	**1**	184	**4**	18	0.2	**1**	0.1	0.10	6	**1**	0.0	1.3	0.3	4	**1**	tr	**0**	tr(a)		tr	
0.04	**3**	0.23	**11**	0.00	**0**	tr	**0**	0	0	**0**	0.9	0.5	21	**2**	486	330	437	1.2	**7**	0.1	**1**	63.3	**42**	9	P
0.9	**5**	31	**8**	0.35	**4**	908	**18**	91	0.3	**1**	0.3	2.11	29	**3**	21.1	8.4	12.3	17	**4**	0.02	**1**	1		7	
tr	**0**	0.01	**0**	0.00	**0**	tr	**0**	0	0	**0**	0.1	tr	1	**0**	tr	21	4	0.1	**0**	0.1	**0**	unk.	**unk.**	4	Q
0.1	**0**	1	**0**	0.03	**0**	69	**1**	7	0.1	**0**	tr	0.04	2	**0**	0.0	0.5	0.1	1	**0**	0.00	**0**	tr(a)		4	
0.06	**4**	0.13	**6**	0.00	**0**	0.002	**1**	0	0	**0**	0.3	tr	27	**3**	523	324	22	0.9	**5**	0.5	**3**	71.2	**48**	42	R
1.1	**6**	13	**3**	0.39	**4**	1093	**22**	98	0.4	**1**	0.1	2.10	52	**5**	22.8	8.3	0.6	24	**6**	0.19	**9**	1		12	
0.12	**7**	0.06	**3**	0.08	**1**	0.003	**1**	0	0	**0**	0.2	0.0	40	**4**	90	96	24	0.5	**3**	0.3	**2**	0.9	**1**	138	S
0.3	**2**	7	**2**	0.11	**1**	654	**13**	tr	0	**tr**	0.1	0.58	67	**7**	3.9	2.4	0.7	14	**3**	0.09	**5**	tr(a)		14	

	FOOD	Portion	Weight in grams *Conversion for 100 g*	Kilocalories H_2O g	Total carbohydrate g Total fats g	Crude fiber g Dietary fiber g	Total protein g Total sugar g	% USRDA	Arginine mg Histidine mg	Isoleucine mg Leucine mg	Lysine mg Methionine mg	Phenylalanine mg Threonine mg	Valine mg Tryptophan mg	Cystine mg Tyrosine mg	Polyunsat. fatty acids g Monounsat. fatty acids g	Saturated fatty acids g P/S ratio	Linoleic acid g Cholesterol mg	Thiamin mg Ascorbic acid mg	% USRDA
A	TORTILLA,baked or steamed	1 average	21 *4.88*	43 unk.	9.2 0.4	0.20 unk.	1.0 unk.	**2**	unk. 14	60 165	26 19	44 41	53 6	10 18	unk. unk.	0.20 unk.	unk. 0	0.03 0	**2** **0**
B	tortilla,home recipe,casserole	3/4 C	119 *0.84*	230 68.6	15.4 17.9	0.90 0.42	3.6 2.2	**7**	162 76	197 384	187 72	160 139	189 35	24 106	4.24 3.96	6.89 0.61	4.08 33	0.07 54	**5** **90**
C	tortilla,mix,corn,masa harina, uncooked-Quaker	1 C	114 *0.88*	421 10.0	84.5 4.5	2.04 unk.	10.7 3.6	**16**	unk. 235	635 1729	389 410	463 433	560 42	unk. unk.	unk. unk.	unk. unk.	unk. 0(a)	1.60 0	**107** **0**
D	tortilla,mix,wheat,masa trigo, uncooked-Quaker	1 C	114 *0.88*	458 11.5	76.0 12.4	0.23 unk.	10.7 unk.	**16**	unk. 254	551 822	317 176	651 349	519 unk.	285 300	unk. unk.	unk. unk.	unk. 0(a)	1.19 0	**80** **0**
E	TROUT,rainbow,fresh,whole	1 sm (8 oz raw)	114 *0.88*	222 0.0	0.0 13.0	0(a) 0(a)	24.5 0.0	**55**	unk. unk.	1642 1000	1040 844	219 635	1927 725	unk. unk.	4.33 unk.	unk. unk.	unk. 63	0.09 unk.	**6** **unk.**
F	TUNA HELPER,creamy noodles 'n tuna, dry mix-General Mills	1 pkg	248 *0.40*	1100 unk.	155.0 40.0	unk. unk.	25.0 unk.	**0**	unk. unk.	unk. unk.	unk. unk.	unk. unk.	unk. unk.	unk. unk.	unk. unk.	unk. unk.	unk. unk.	1.49 tr(a)	**99** **0**
G	Tuna Helper,creamy noodles 'n tuna, prep	1 serving	81 *1.23*	282 19.0	30.0 11.6	0.00 unk.	14.0 0.0	**20**	460 452	461 678	796 262	335 389	479 90	99 324	0.63 0.63	0.94 0.67	0.63 20	0.31 tr(a)	**21** **0**
H	Tuna Helper,creamy rice 'n tuna, dry mix-General Mills	1 pkg	255 *0.39*	950 unk.	165.0 25.0	unk. unk.	15.0 unk.	**0**	unk. unk.	unk. unk.	unk. unk.	unk. unk.	unk. unk.	unk. unk.	unk. unk.	unk. unk.	unk. unk.	1.12 24	**75** **40**
I	Tuna Helper,creamy rice 'n tuna,prep	1 serving	82 *1.21*	252 19.0	33.0 7.6	0.00 unk.	12.0 0.0	**20**	460 452	461 678	796 262	335 389	479 90	99 324	0.63 0.63	0.94 0.67	0.63 20	0.24 5	**16** **8**
J	TUNA,CANNED,in oil,drained solids	1/2 C	80 *1.25*	158 48.5	0.0 6.6	0.00 0(a)	23.0 0.0	**51**	1174 1150	1175 1728	2027 668	853 990	1221 230	251 826	1.60 1.60	2.40 0.67	1.60 52	0.04 unk.	**3** **unk.**
K	tuna,canned,in oil,drained solids	7 oz can	169 *0.59*	333 102.4	0.0 13.9	0.00 0(a)	48.7 0.0	**108**	2479 2430	2483 3650	4282 1411	1802 2092	2579 487	531 1746	3.38 3.38	5.07 0.67	3.38 110	0.08 unk.	**6** **unk.**
L	tuna,canned,in water,low sodium, solids & liquid	1/2 C	85 *1.18*	108 59.5	0.0 0.7	0.00 0(a)	23.8 0.0	**53**	1211 1188	1214 1785	2094 690	881 1023	1261 238	259 853	unk. unk.	unk. unk.	unk. 54	unk. unk.	**unk.** **unk.**
M	tuna,canned,in water,solids & liquid	1/2 C	85 *1.18*	108 59.5	0.0 0.7	0.00 0(a)	23.8 0.0	**53**	1211 1188	1214 1785	2094 690	881 1023	1261 238	259 853	unk. unk.	unk. unk.	unk. 54	unk. unk.	**unk.** **unk.**
N	TUNA,HOME RECIPE,casserole	3/4 C	173 *0.58*	299 136.5	23.0 17.8	0.38 0.27	12.0 1.9	**25**	406 442	560 864	845 277	498 456	643 131	136 426	1.17 4.16	5.68 0.21	5.78 56	0.13 9	**9** **16**
O	tuna,home recipe,pattie	3-1/2" dia x 5/8"	87 *1.15*	228 48.6	21.9 7.7	0.09 unk.	16.6 2.7	**35**	667 692	837 1277	1218 426	693 674	867 180	223 604	1.00 2.57	3.20 0.31	1.24 69	0.10 tr	**7** **1**
P	tuna,home recipe,salad w/celery, mayonnaise,pickle,onion & egg	1/2 C	103 *0.98*	174 71.5	3.6 10.8	unk. 0.31	14.9 unk.	**33**	unk. unk.	unk. unk.	unk. unk.	unk. unk.	unk. unk.	unk. unk.	unk. 3.07	3.07 unk.	3.07 unk.	0.04 1	**3** **2**
Q	tuna,home recipe,stuffed green pepper	1 average	281 *0.36*	261 228.8	10.9 13.1	2.43 1.87	25.4 3.3	**55**	1182 1174	1255 1808	2116 695	946 1076	1281 245	251 828	5.99 2.92	3.57 1.68	4.67 59	0.17 212	**11** **354**
R	tuna noodle casserole,frozen-Stouffer	1/2 pkg	163 *0.61*	199 123.3	18.0 9.0	unk. unk.	10.0 unk.	**15**	unk. unk.	unk. unk.	unk. unk.	unk. unk.	unk. unk.	unk. unk.	unk. unk.	unk. unk.	unk. unk.	0.15 tr	**10** **0**
S	TURKEY FAT	1 C	205 *0.49*	1846 0.4	0.0 204.6	0.00 0(a)	0.0 0(a)	**0**	0(a) 0(a)	0(a) 0(a)	0(a) 0(a)	0(a) 0(a)	0(a) 0(a)	0(a) 0(a)	47.35 73.59	60.27 0.79	43.46 209	0(a) 0(a)	**0** **0**

Riboflavin mg / Niacin mg	% USRDA	Vitamin B_6 mg / Folacin mcg	% USRDA	Vitamin B_{12} mcg / Pantothenic acid mg	% USRDA	Biotin mg / Vitamin A IU	% USRDA	Preformed A RE / Beta carotene RE	Vitamin D IU / Vitamin E IU	% USRDA	Total tocopherol mg / Alpha tocopherol mg	Other tocopherol mg / Total ash g	Calcium mg / Phosphorus mg	% USRDA	Sodium mg / Sodium meq	Potassium mg / Potassium meq	Chlorine mg / Chlorine meq	Iron mg / Magnesium mg	% USRDA	Zinc mg / Copper mg	% USRDA	Iodine mcg / Selenium mcg	% USRDA	Manganese mcg / Chromium mcg	
0.01	1	0.01	1	0.00	0	tr	0	0	0	0	tr	unk.	41	4	23	3	unk.	0.6	3	tr	0	unk.	unk.	unk.	A
0.2	1	tr	0	0.01	0	4	0	tr	tr	0	unk.	0.29	29	3	1.0	0.1	unk.	21	5	0.04	2	unk.		unk.	
0.10	6	0.14	7	0.06	1	0.001	0	0	0	0	6.1	tr(a)	110	11	343	142	5	1.2	6	0.3	2	71.2	48	59	B
0.5	3	12	3	0.20	2	552	11	18	7.4	25	0.1	1.58	87	9	14.9	3.6	0.1	40	10	0.08	4	0		9	
0.90	53	0.58	29	0.00	0	0.009	3	0(a)	0	0	unk.	unk.	239	24	8	370	unk.	9.0	50	2.3	15	unk.	unk.	0	C
10.8	54	51	13	0.00	0	unk.	unk.	unk.	unk.	unk.	unk.	1.82	264	26	0.3	9.5	unk.	125	31	0.25	13	unk.		unk.	
0.73	43	0.05	2	0.00	0	0.003	1	0	0	0	unk.	unk.	204	20	905	107	unk.	8.0	45	0.0	0	unk.	unk.	0	D
9.6	48	89	22	1.14	11	0	0	0	unk.	unk.	unk.	2.84	239	24	39.4	2.7	unk.	24	6	0.11	6	unk.		unk.	
0.23	13	unk.	unk.	unk.	unk.	unk.	unk.	unk.	unk.	unk.	unk.	unk.	unk.	unk.	unk.	unk.	unk.	unk.	unk.	unk.	unk.	unk.	unk.	unk.	E
9.6	48	unk.	unk.	unk.	unk.	unk.	unk.	unk.	unk.	unk.	unk.	1.48	unk.	unk.	unk.	unk.	unk.	unk.	unk.	unk.	unk.	unk.		unk.	
0.84	50	unk.	unk.	0(a)	0	unk.	unk.	unk.	unk.	unk.	unk.	unk.	100	10	3608	756	unk.	7.2	40	unk.	unk.	unk.	unk.	unk.	F
15.0	75	unk.	unk.	unk.	unk.	1500	30	unk.	unk.	unk.	unk.	unk.	570	57	157.0	19.3	unk.	unk.	unk.	unk.	unk.	unk.		unk.	
0.20	12	0.13	7	0.69	12	0.001	0	7	78	20	0.2	tr(a)	35	4	794	151	unk.	2.0	11	0.3	2	unk.	unk.	unk.	G
6.7	34	5	1	0.10	1	323	7	tr	0.2	1	tr(a)	0.63	188	19	34.5	3.9	unk.	10	3	0.04	2	tr(a)		tr(a)	
0.18	11	unk.	unk.	0(a)	0	unk.	unk.	unk.	unk.	unk.	unk.	unk.	166	17	4807	790	unk.	5.4	30	unk.	unk.	unk.	unk.	unk.	H
10.0	50	unk.	unk.	unk.	unk.	tr	0	unk.	unk.	unk.	unk.	unk.	459	46	209.1	20.2	unk.	unk.	unk.	unk.	unk.	unk.		unk.	
0.07	4	0.13	7	0.69	12	0.001	0	7	79	20	0.2	tr(a)	36	4	961	158	unk.	1.7	9	0.3	2	unk.	unk.	unk.	I
5.7	29	5	1	0.10	1	25	1	tr	0.2	1	tr(a)	0.63	165	17	41.8	4.0	unk.	10	3	0.04	2	tr(a)		tr(a)	
0.10	6	0.34	17	1.76	29	0.002	1	17	200	50	0.4	tr(a)	6	1	unk.	unk.	unk.	1.5	8	0.9	6	unk.	unk.	unk.	J
9.5	48	12	3	0.26	3	64	1	1	0.5	2	tr(a)	1.60	187	19	unk.	unk.	unk.	27	7	0.10	5	tr(a)		tr(a)	
0.20	12	0.72	36	3.72	62	0.005	2	37	422	106	0.8	tr(a)	14	1	unk.	unk.	unk.	3.2	18	1.9	12	unk.	unk.	unk.	K
20.1	101	25	6	0.54	5	135	3	1	1.0	3	tr(a)	3.38	396	40	unk.	unk.	unk.	56	14	0.20	10	tr(a)		tr(a)	
0.08	5	0.36	18	1.87	31	0.003	1	unk.	unk.	unk.	0.4	tr(a)	14	1	35	237	unk.	1.4	8	0.9	6	unk.	unk.	unk.	L
11.3	57	6	2	0.27	3	unk.	unk.	unk.	0.5	2	tr(a)	1.02	162	16	1.5	6.1	unk.	28	7	0.10	5	tr(a)		tr(a)	
0.08	5	0.36	18	1.87	31	0.003	1	unk.	212	53	0.4	tr(a)	14	1	744	234	unk.	1.4	8	0.9	6	unk.	unk.	unk.	M
11.3	57	13	3	0.27	3	unk.	unk.	unk.	0.5	2	tr(a)	1.02	162	16	32.3	6.0	unk.	unk.	unk.	0.10	5	tr(a)		tr(a)	
0.17	10	0.15	8	0.50	8	0.002	1	15	48	12	6.4	3.0	122	12	660	162	19	1.9	11	1.2	8	30.9	21	7	N
3.1	16	10	2	0.38	4	554	11	23	7.6	25	2.9	3.20	209	21	28.7	4.1	0.5	31	8	0.09	5	tr		4	
0.18	11	0.20	10	1.02	17	0.003	1	9	110	27	0.4	tr(a)	65	7	401	83	13	2.0	11	0.7	5	46.5	31	3	O
5.8	29	17	4	0.45	5	132	3	tr	0.4	1	tr	2.28	169	17	17.4	2.1	0.3	44	11	0.18	9	0		0	
0.11	7	unk.	unk.	unk.	unk.	unk.	unk.	unk.	unk.	unk.	unk.	unk.	20	2	unk.	unk.	unk.	1.3	7	0.3	2	unk.	unk.	unk.	P
5.1	26	unk.	unk.	unk.	unk.	297	6	unk.	unk.	unk.	unk.	1.64	146	15	unk.	unk.	unk.	unk.	unk.	unk.	unk.	unk.		unk.	
0.22	13	0.78	39	1.77	30	0.002	1	17	201	50	2.1	0.0	32	3	135	427	52	2.8	15	1.0	7	unk.	unk.	211	Q
10.4	52	45	11	0.74	7	839	17	75	2.5	8	0.1	2.73	233	23	5.9	10.9	1.5	58	15	0.20	10	tr(a)		38	
0.16	10	unk.	unk.	unk.	unk.	unk.	unk.	unk.	unk.	unk.	unk.	unk.	100	10	666	209	unk.	1.1	6	unk.	unk.	unk.	unk.	unk.	R
3.0	15	unk.	unk.	unk.	unk.	98	2	unk.	unk.	unk.	unk.	1.96	unk.	unk.	29.0	5.3	unk.	unk.	unk.	unk.	unk.	unk.		unk.	
0(a)	0	0(a)	0	0(a)	0	0(a)	0	unk.	unk.	unk.	unk.	unk.	unk.	unk.	unk.	unk.	unk.	unk.	unk.	unk.	unk.	unk.	unk.	unk.	S
0(a)	0	0(a)	0	0(a)	0	unk.	unk.	unk.	unk.	unk.	unk.	0.00	unk.	unk.	unk.	unk.	unk.	unk.	unk.	unk.	unk.	unk.		unk.	

	FOOD	Portion	Weight in grams / Conversion for 100 g	Kilocalories / H_2O g	Total carbohydrate g / Total fats g	Crude fiber g / Dietary fiber g	Total protein g / Total sugar g	% USRDA	Arginine mg / Histidine mg	Isoleucine mg / Leucine mg	Lysine mg / Methionine mg	Phenylalanine mg / Threonine mg	Valine mg / Tryptophan mg	Cystine mg / Tyrosine mg	Polyunsat. fatty acids g / Monounsat. fatty acids g	Saturated fatty acids g / P/S ratio	Linoleic acid g / Cholesterol mg	Thiamin mg / Ascorbic acid mg	% USRDA
A	TURKEY,frozen,tetrazzini-Stouffer	1/2 pkg	170	239	16.9	unk.	11.9	**27**	unk.	unk.	unk.	unk.	unk.	unk.	unk.	unk.	unk.	0.56	**38**
			0.59	122.7	14.0	unk.	unk.		unk.	unk.	unk.	unk.	unk.	unk.	unk.	unk.	unk.	tr	**0**
B	TURKEY,FRYER/ROASTER,back,meat w/skin,roasted	1 piece	260	530	0.0	0.00	68.0	**151**	4846	3258	5923	2616	3458	796	6.86	7.77	5.90	0.10	**7**
			0.38	164.5	26.6	0(a)	0(a)		1953	5125	1843	2902	725	2452	7.44	0.88	281	0	**0**
C	turkey,fryer/roaster,back,meat w/o skin,roasted	1 piece	192	326	0.0	0.00	53.8	**120**	3752	2797	5069	2135	2857	559	3.24	3.63	2.63	0.10	**6**
			0.52	126.7	10.8	0(a)	0(a)		1678	4285	1557	2392	611	2125	2.04	0.89	182	0	**0**
D	turkey,fryer/roaster,breast,meat w/skin,roasted	1/2	344	526	0.0	0.00	100.0	**222**	7031	5043	9150	3922	5225	1090	2.61	2.99	2.20	0.14	**9**
			0.29	232.6	11.0	0(a)	0(a)		3024	7802	2824	4379	1108	3818	3.34	0.87	310	0	**0**
E	turkey,fryer/roaster,breast,meat w/o skin,roasted	3x2-1/8x3/4"	85	115	0.0	0.00	25.5	**57**	1782	1329	2408	1014	1357	266	0.17	0.20	0.11	0.03	**2**
			1.18	58.1	0.6	0(a)	0(a)		797	2036	740	1136	291	1010	0.09	0.83	71	0	**0**
F	turkey,fryer/roaster,dark meat w/skin,roasted	3x2-1/8x3/4"	85	155	0.0	0.00	23.5	**52**	1661	1173	2129	919	1221	261	1.60	1.80	1.35	0.04	**3**
			1.18	55.1	6.0	0(a)	0(a)		704	1822	659	1024	258	887	1.57	0.89	99	0	**0**
G	turkey,fryer/roaster,dark meat w/o skin,roasted,chopped	1/2 C	70	113	0.0	0.00	20.2	**45**	1408	1050	1903	801	1072	210	0.90	1.01	0.73	0.03	**2**
			1.43	46.5	3.0	0(a)	0(a)		630	1609	584	898	230	798	0.57	0.89	78	0	**0**
H	turkey,fryer/roaster,leg,meat w/skin,roasted	1 piece	245	416	0.0	0.00	69.8	**155**	4902	3545	6431	2744	3660	752	3.63	4.09	3.04	0.12	**8**
			0.41	160.8	13.3	0(a)	0(a)		2127	5473	1982	3067	779	2688	3.23	0.89	171	0	**0**
I	turkey,fryer/roaster,leg,meat w/o skin,roasted	1 piece	224	356	0.0	0.00	65.4	**145**	4561	3400	6162	2594	3472	681	2.53	2.84	2.06	0.11	**8**
			0.45	149.0	8.4	0(a)	0(a)		2041	5210	1893	2908	744	2583	1.59	0.89	267	0	**0**
J	turkey,fryer/roaster,light meat w/skin,roasted	3x2-1/8x3/4"	85	139	0.0	0.00	24.4	**54**	1725	1217	2209	954	1268	271	0.93	1.06	0.78	0.03	**2**
			1.18	56.6	3.9	0(a)	0(a)		730	1890	683	1063	269	921	1.16	0.87	81	0	**0**
K	turkey,fryer/roaster,light meat w/o skin,chopped	1/2 C	70	98	0.0	0.00	21.1	**47**	1474	1099	1991	839	1123	220	0.22	0.27	0.15	0.03	**2**
			1.43	48.0	0.8	0(a)	0(a)		659	1683	612	940	240	835	0.12	0.82	60	0	**0**
L	turkey,fryer/roaster,meat w/o skin, roasted,chopped	1/2 C	70	105	0.0	0.00	20.7	**46**	1443	1076	1949	821	1099	215	0.54	0.61	0.42	0.03	**2**
			1.43	47.3	1.8	0(a)	0(a)		645	1648	599	920	235	818	0.33	0.89	69	0	**0**
M	turkey,fryer/roaster,skin only, roasted	1 oz	28	85	0.0	0.00	5.9	**13**	459	191	355	201	250	99	1.51	1.72	1.37	0.01	**1**
			3.52	15.8	6.6	0(a)	0(a)		114	349	119	212	47	135	2.27	0.88	41	0	**0**
N	turkey,fryer/roaster,wing,meat w/skin,roasted	1 piece	90	186	0.0	0.00	24.9	**55**	1781	1173	2134	951	1255	297	2.11	2.44	1.77	0.03	**2**
			1.11	56.1	8.9	0(a)	0(a)		703	1855	666	1053	262	881	2.63	0.87	103	0	**0**
O	turkey,fryer/roaster,wing,meat w/o skin,roasted	1 piece	60	98	0.0	0.00	18.5	**41**	1291	962	1744	734	983	193	0.55	0.66	0.37	0.02	**1**
			1.67	39.4	2.1	0(a)	0(a)		578	1475	536	823	210	731	0.30	0.84	61	0	**0**
P	TURKEY,HAM,cured,thigh meat	3 thin slices	85	109	0.3	0(a)	16.1	**36**	1125	838	1520	640	857	168	1.30	1.45	1.05	0.04	**3**
			1.17	60.8	4.3	0(a)	unk.		504	1285	467	717	183	637	0.81	0.89	unk.	unk.	**unk.**
Q	TURKEY,HOME RECIPE,loaf	3x2-3/4x3/4" slice	135	263	14.5	0.10	18.2	**38**	1027	920	1427	752	910	257	0.75	5.40	2.61	0.06	**4**
			0.74	86.1	14.2	unk.	2.5		461	1384	469	744	195	622	5.58	0.14	138	1	**1**
R	turkey,home recipe,potpie	1 average	302	710	42.6	0.90	28.6	**59**	1326	1383	2082	1146	1303	383	0.72	16.44	5.16	0.30	**20**
			0.33	178.0	47.0	1.70	3.2		669	2058	671	1081	286	903	22.36	0.04	122	16	**27**
S	turkey,home recipe,sticks,breaded, fried	1 average	64	179	10.9	unk.	9.1	**20**	unk.	unk.	unk.	unk.	unk.	unk.	unk.	unk.	unk.	0.06	**4**
			1.56	31.6	10.8	unk.	unk.		unk.	unk.	unk.	unk.	unk.	unk.	unk.	unk.	unk.	unk.	**unk.**

Riboflavin mg / Niacin mg	% USRDA	Vitamin B_6 mg / Folacin mcg	% USRDA	Vitamin B_{12} mcg / Pantothenic acid mg	% USRDA	Biotin mg / Vitamin A IU	% USRDA	Preformed A RE / Beta carotene RE	Vitamin D IU / Vitamin E IU	% USRDA	Total tocopherol mg / Alpha tocopherol mg	Other tocopherol mg / Total ash g	Calcium mg / Phosphorus mg	% USRDA	Sodium mg / Sodium meq	Potassium mg / Potassium meq	Chlorine mg / Chlorine meq	Iron mg / Magnesium mg	% USRDA	Zinc mg / Copper mg	% USRDA	Iodine mcg / Selenium mcg	% USRDA	Manganese mcg / Chromium mcg	
0.17	**10**	unk.	**unk.**	unk.	**unk.**	unk.	**unk.**	unk.	unk.	**unk.**	unk.	unk.	80	**8**	617	199	unk.	1.4	**8**	unk.	**unk.**	unk.	**unk.**	unk.	A
2.0	**10**	unk.	**unk.**	unk.	**unk.**	tr	**0**	unk.	unk.	**unk.**	unk.	2.04	unk.	**unk.**	26.8	5.1	unk.	unk.	**unk.**	unk.	**unk.**	unk.		unk.	
0.49	**29**	0.81	**40**	0.94	**16**	unk.	**unk.**	0	unk.	**unk.**	unk.	unk.	94	**9**	182	541	unk.	4.8	**27**	8.7	**58**	unk.	**unk.**	60	B
9.2	**46**	23	**6**	2.93	**29**	0	**0**	0	unk.	**unk.**	unk.	2.21	442	**44**	7.9	13.8	unk.	52	**13**	0.43	**22**	unk.		unk.	
0.40	**24**	0.77	**38**	0.79	**13**	unk.	**unk.**	0	unk.	**unk.**	unk.	unk.	69	**7**	140	417	unk.	3.6	**20**	7.3	**49**	unk.	**unk.**	48	C
7.4	**37**	19	**5**	2.73	**27**	0	**0**	0	unk.	**unk.**	unk.	1.73	340	**34**	6.1	10.7	unk.	42	**11**	0.34	**17**	unk.		unk.	
0.48	**28**	1.75	**88**	1.27	**21**	unk.	**unk.**	0	unk.	**unk.**	unk.	unk.	52	**5**	182	960	unk.	5.4	**30**	6.1	**41**	unk.	**unk.**	83	D
23.9	**120**	21	**5**	2.27	**23**	0	**0**	0	unk.	**unk.**	unk.	3.78	743	**74**	7.9	24.5	unk.	96	**24**	0.26	**13**	unk.		unk.	
0.11	**7**	0.48	**24**	0.33	**6**	unk.	**unk.**	0	unk.	**unk.**	unk.	unk.	10	**1**	44	248	unk.	1.3	**7**	1.5	**10**	unk.	**unk.**	20	E
6.4	**32**	5	**1**	0.60	**6**	0	**0**	0	unk.	**unk.**	unk.	0.97	190	**19**	1.9	6.3	unk.	25	**6**	0.06	**3**	unk.		unk.	
0.20	**12**	0.28	**14**	0.31	**5**	unk.	**unk.**	0(a)	unk.	**unk.**	unk.	unk.	23	**2**	65	201	unk.	2.0	**11**	3.3	**22**	unk.	**unk.**	20	F
2.8	**14**	8	**2**	1.02	**10**	0	**0**	0(a)	unk.	**unk.**	unk.	0.80	162	**16**	2.8	5.1	unk.	20	**5**	0.18	**9**	unk.		unk.	
0.17	**10**	0.27	**13**	0.27	**5**	unk.	**unk.**	0	unk.	**unk.**	unk.	unk.	18	**2**	55	172	unk.	1.7	**9**	2.9	**19**	unk.	**unk.**	17	G
2.4	**12**	7	**2**	0.95	**10**	0	**0**	0	unk.	**unk.**	unk.	0.69	137	**14**	2.4	4.4	unk.	17	**4**	0.16	**8**	unk.		unk.	
0.64	**38**	0.83	**42**	0.91	**15**	unk.	**unk.**	0	unk.	**unk.**	unk.	unk.	56	**6**	196	617	unk.	6.3	**35**	10.0	**67**	unk.	**unk.**	61	H
8.0	**40**	22	**6**	3.03	**30**	0	**0**	0	unk.	**unk.**	unk.	2.43	490	**49**	8.5	15.8	unk.	59	**15**	0.57	**29**	unk.		unk.	
0.60	**36**	0.08	**4**	0.85	**14**	unk.	**unk.**	0	unk.	**unk.**	unk.	unk.	49	**5**	181	578	unk.	6.0	**33**	9.6	**64**	unk.	**unk.**	tr	I
7.4	**37**	22	**6**	2.97	**30**	0	**0**	0	unk.	**unk.**	unk.	2.28	457	**46**	7.9	14.8	unk.	56	**14**	tr	**0**	unk.		unk.	
0.12	**7**	0.42	**21**	0.31	**5**	unk.	**unk.**	0(a)	unk	**unk.**	unk.	unk.	15	**2**	48	223	unk.	1.4	**8**	1.8	**12**	unk.	**unk.**	20	J
5.3	**27**	5	**1**	0.55	**6**	0	**0**	0(a)	unk.	**unk.**	unk.	0.88	174	**17**	2.1	5.7	unk.	22	**6**	0.08	**4**	unk.		unk.	
0.10	**6**	0.40	**20**	0.27	**5**	unk.	**unk.**	0	unk.	**unk.**	unk.	unk.	10	**1**	39	194	unk.	1.1	**6**	1.5	**10**	unk.	**unk.**	17	K
4.9	**24**	4	**1**	0.50	**5**	0	**0**	0	unk.	**unk.**	unk.	0.76	151	**15**	1.7	5.0	unk.	20	**5**	0.06	**3**	unk.		unk.	
0.13	**8**	0.34	**17**	0.27	**5**	unk.	**unk.**	0(a)	unk.	**unk.**	unk.	unk.	14	**1**	47	184	unk.	1.4	**8**	2.1	**14**	unk.	**unk.**	17	L
3.7	**19**	6	**1**	0.71	**7**	0	**0**	0(a)	unk.	**unk.**	unk.	0.73	145	**15**	2.0	4.7	unk.	18	**5**	0.10	**5**	unk.		unk.	
0.05	**3**	0.02	**1**	0.06	**1**	unk.	**unk.**	0(a)	unk.	**unk.**	unk.	unk.	10	**1**	17	51	unk.	0.5	**3**	0.6	**4**	unk.	**unk.**	6	M
0.7	**4**	1	**0**	0.08	**1**	0	**0**	0(a)	unk.	**unk.**	unk.	0.21	43	**4**	0.8	1.3	unk.	5	**1**	0.04	**2**	unk.		unk.	
0.14	**9**	0.38	**19**	0.31	**5**	unk.	**unk.**	0	unk.	**unk.**	unk.	unk.	26	**3**	66	176	unk.	1.6	**9**	2.9	**20**	unk.	**unk.**	22	N
3.3	**16**	5	**1**	0.54	**5**	0	**0**	0	unk.	**unk.**	unk.	0.73	149	**15**	2.9	4.5	unk.	18	**5**	0.14	**7**	unk.		unk.	
0.10	**6**	0.35	**18**	0.25	**4**	unk.	**unk.**	0	unk.	**unk.**	unk.	unk.	16	**2**	47	122	unk.	1.1	**6**	2.3	**15**	unk.	**unk.**	16	O
2.5	**12**	4	**1**	0.45	**5**	0	**0**	0	unk.	**unk.**	unk.	0.51	104	**10**	2.0	3.1	unk.	13	**3**	0.10	**5**	unk.		unk.	
0.21	**13**	unk.	**unk.**	unk.	**unk.**	unk.	**unk.**	unk.	unk.	**unk.**	unk.	unk.	9	**1**	849	277	unk.	2.4	**13**	unk.	**unk.**	unk.	**unk.**	unk.	P
3.0	**15**	unk.	**unk.**	unk.	**unk.**	unk.	**unk.**	unk.	unk.	**unk.**	unk.	3.60	163	**16**	36.9	7.1	unk.	unk.	**unk.**	unk.	**unk.**	unk.		unk.	
0.23	**14**	0.29	**15**	0.52	**9**	0.003	**1**	17	17	**4**	0.2	0.0	74	**7**	549	111	27	1.9	**11**	2.4	**16**	55.0	**37**	9	Q
3.6	**18**	14	**4**	0.65	**7**	312	**6**	3	0.3	**1**	0.1	2.94	94	**9**	23.9	2.8	0.8	46	**12**	0.22	**11**	tr		tr	
0.34	**20**	0.48	**24**	0.41	**7**	0.001	**0**	28	0	**0**	0.2	tr(a)	41	**4**	1371	136	34	3.5	**20**	3.8	**25**	190.0	**127**	192	R
12.1	**61**	20	**5**	0.79	**8**	4571	**91**	404	0.2	**1**	0.2	5.84	79	**8**	59.6	3.5	1.0	27	**7**	0.25	**13**	25		29	
0.12	**7**	unk.	**unk.**	unk.	**unk.**	unk.	**unk.**	unk.	unk.	**unk.**	unk.	unk.	9	**1**	536	166	unk.	1.4	**8**	unk.	**unk.**	unk.	**unk.**	unk.	S
1.3	**7**	unk.	**unk.**	unk.	**unk.**	unk.	**unk.**	unk.	unk.	**unk.**	unk.	1.60	150	**15**	23.3	4.3	unk.	unk.	**unk.**	unk.	**unk.**	unk.		unk.	

	FOOD	Portion	Weight in grams / Conversion for 100 g	Kilocalories / H_2O g	Total carbohydrate g / Total fats g	Crude fiber g / Dietary fiber g	Total protein g / Total sugar g	% USRDA	Arginine mg / Histidine mg	Isoleucine mg / Leucine mg	Lysine mg / Methionine mg	Phenylalanine mg / Threonine mg	Valine mg / Tryptophan mg	Cystine mg / Tyrosine mg	Polyunsat. fatty acids g / Monounsat. fatty acids g	Saturated fatty acids g / P/S ratio	Linoleic acid g / Cholesterol mg	Thiamin mg / Ascorbic acid mg	% USRDA
A	TURKEY,MEAT,breast,w/skin,	2 sl(3-3/8x	85	107	0.0	0.00	18.8	**42**	1321	949	1719	737	982	208	0.71	0.83	0.58	0.04	**3**
	prebasted product,cooked	2-3/4x1/4")	*1.18*	60.3	2.9	0(a)	0(a)		568	1466	530	822	209	718	0.78	0.86	36	0	**0**
B	turkey,meat,canned,boned,w/broth	1/2 C	71	116	0.0	0.00	16.8	**37**	1181	840	1522	655	872	189	1.24	1.42	1.04	0.01	**1**
			1.41	46.9	4.9	0(a)	0(a)		503	1301	469	731	185	635	1.30	0.88	unk.	1	**2**
C	turkey,meat,deboned roll,light	3 sl(2-3/8"	85	125	0.5	0(a)	15.9	**35**	unk.	unk.	unk.	unk.	unk.	unk.	1.48	1.72	1.21	0.08	**5**
		dia x 1/4")	*1.17*	61.0	6.2	0(a)	unk.		unk.	unk.	unk.	unk.	unk.	unk.	1.72	0.86	37	unk.	**unk.**
D	turkey,meat,deboned roll,light & dark	3 sl(2-7/8"	85	127	1.8	0(a)	15.5	**34**	unk.	unk.	unk.	unk.	unk.	unk.	1.52	1.74	1.26	0.08	**5**
		dia x 1/4")	*1.17*	59.8	6.0	0(a)	unk.		unk.	unk.	unk.	unk.	unk.	unk.	1.59	0.87	47	unk.	**unk.**
E	turkey,meat,patties,breaded,fried	3-3/8" dia x	64	181	10.0	unk.	9.0	**20**	unk.	unk.	unk.	unk.	unk.	unk.	unk.	unk.	unk.	0.06	**4**
		3/8"	*1.56*	31.8	11.5	unk.	unk.		unk.	unk.	unk.	unk.	unk.	unk.	unk.	unk.	unk.	unk.	**unk.**
F	turkey,meat,roast,w/o bone,light &	3x2-1/8x3/4"	85	132	2.6	0(a)	18.1	**40**	unk.	unk.	unk.	unk.	unk.	unk.	unk.	unk.	unk.	0.04	**3**
	dark,cooked		*1.18*	57.7	4.9	0(a)	unk.		unk.	unk.	unk.	unk.	unk.	unk.	unk.	unk.	45	unk.	**unk.**
G	turkey,meat,thigh,w/skin,prebasted	1 piece	350	550	0.0	0.00	65.8	**146**	4610	3339	6045	2580	3444	718	8.22	9.27	6.89	0.28	**19**
	product,cooked		*0.29*	247.1	29.9	0(a)	0(a)		1999	5149	1862	2884	735	2527	7.21	0.89	217	0	**0**
H	TURKEY ROLL,breast meat	3x2-1/8x3/4"	85	93	1.3	0.00	19.1	**43**	1334	994	1802	759	1016	199	0.24	0.41	0.22	0.03	**2**
			1.18	61.1	1.3	0(a)	unk.		597	1524	554	851	218	756	0.32	0.58	35	0	**0**
I	TURKEY,YOUNG HEN,back,meat w/skin,	1 piece	434	1102	0.0	0.00	114.6	**255**	8129	5607	0190	4444	5894	1306	17.40	19.70	15.02	0.22	**15**
	roasted		*0.23*	247.0	67.9	0(a)	0(a)		3363	8762	3160	4943	1241	4231	19.18	0.88	369	0	**0**
J	turkey,young hen,breast,meat w/skin,	3x2-1/8x3/4"	85	165	0.0	0.00	24.5	**54**	1720	1240	2250	961	1282	265	1.64	1.91	1.29	0.04	**3**
	roasted		*1.18*	53.3	6.7	0(a)	0(a)		745	1916	694	1074	273	939	1.73	0.86	61	0	**0**
K	turkey,young hen,dark meat w/skin,	3 sl(3x2-1/8	85	197	0.0	0.00	23.3	**52**	1638	1169	2121	911	1213	255	2.91	3.28	2.46	0.05	**3**
	roasted	x1/4")	*1.18*	50.7	10.9	0(a)	0(a)		701	1810	655	1017	258	885	2.80	0.89	71	0	**0**
L	turkey,young hen,dark meat w/o skin,	1/2 C	70	134	0.0	0.00	19.9	**44**	1388	1035	1875	790	1057	207	1.63	1.83	1.32	0.04	**3**
	roasted,chopped		*1.43*	43.9	5.5	0(a)	0(a)		621	1585	576	885	226	786	1.02	0.89	56	0	**0**
M	turkey,young hen,leg,meat w/skin,	1 piece	448	954	0.0	0.00	124.2	**276**	8714	6335	1491	4892	6527	1335	13.04	14.69	10.89	0.27	**18**
	roasted		*0.22*	272.1	47.0	0(a)	0(a)		3799	9766	3539	5470	1389	4803	11.11	0.89	367	0	**0**
N	turkey,young hen,light meat w/skin,	3 sl(3x2-1/8	85	176	0.0	0.00	24.3	**54**	1714	1225	2222	954	1270	266	1.94	2.25	1.56	0.04	**3**
	roasted	x1/4")	*1.18*	52.8	8.0	0(a)	0(a)		734	1896	686	1064	269	927	2.17	0.86	63	0	**0**
O	turkey,young hen,light meat w/o skin,	1/2 C	70	113	0.0	0.00	20.9	**47**	1459	1088	1972	830	1112	218	0.70	0.83	0.46	0.04	**3**
	roasted,chopped		*1.43*	46.0	2.6	0(a)	0(a)		653	1667	606	931	238	827	0.38	0.84	48	0	**0**
P	turkey,young hen,meat w/skin,roasted	3x2-1/8x3/4"	85	185	0.0	0.00	23.9	**53**	1680	1200	2178	935	1245	261	2.36	2.70	1.95	0.04	**3**
			1.18	51.9	9.2	0(a)	0(a)		720	1859	672	1044	264	909	2.44	0.87	66	0	**0**
Q	turkey,young hen,meat w/o skin,	1/2 C	70	122	0.0	0.00	20.5	**46**	1428	1065	1930	813	1088	213	1.11	1.27	0.84	0.04	**3**
	roasted,chopped		*1.43*	45.1	3.9	0(a)	0(a)		639	1632	593	911	232	809	0.66	0.87	51	0	**0**
R	turkey,young hen,skin only,roasted	1 oz	28	137	0.0	0.00	5.4	**12**	417	174	323	182	227	90	2.89	3.29	2.61	0.01	**0**
			3.52	10.1	12.6	0(a)	0(a)		104	318	108	193	43	123	4.33	0.88	30	0	**0**
S	turkey,young hen,wing,meat w/skin,	1 piece	174	414	0.0	0.00	47.5	**106**	3372	2312	4204	1837	2436	543	5.55	6.40	4.70	0.09	**6**
	roasted		*0.57*	101.8	23.4	0(a)	0(a)		1387	3619	1305	2044	513	1743	7.05	0.87	134	0	**0**

Riboflavin mg / Niacin mg	% USRDA	Vitamin B_6 mg / Folacin mcg	% USRDA	Vitamin B_{12} mcg / Pantothenic acid mg	% USRDA	Biotin mg / Vitamin A IU	% USRDA	Preformed A RE / Beta carotene RE	Vitamin D IU / Vitamin E IU	% USRDA	Total tocopherol mg / Alpha tocopherol mg	Other tocopherol mg / Total ash g	Calcium mg / Phosphorus mg	% USRDA	Sodium mg / Sodium meq	Potassium mg / Potassium meq	Chlorine mg / Chlorine meq	Iron mg / Magnesium mg	% USRDA	Zinc mg / Copper mg	% USRDA	Iodine mcg / Selenium mcg	% USRDA	Manganese mcg / Chromium mcg	
0.11	**7**	0.27	**14**	0.27	**5**	unk.	**unk.**	0	unk.	**unk.**	unk.	unk.	8	**1**	337	211	unk.	0.6	**3**	1.3	**9**	unk.	**unk.**	unk.	A
7.7	**39**	unk.	**unk.**	unk.	**unk.**	0	**0**	0	unk.	**unk.**	unk.	1.75	182	**18**	14.7	5.4	unk.	18	**5**	0.03	**2**	unk.		unk.	
0.12	**7**	unk.	**unk.**	unk.	**unk.**	unk.	**unk.**	0	unk.	**unk.**	unk.	unk.	9	**1**	332	unk.	unk.	1.3	**7**	unk.	**unk.**	unk.	**unk.**	unk.	B
4.7	**24**	unk.	**unk.**	unk.	**unk.**	0	**0**	0	unk.	**unk.**	unk.	1.36	unk.	**unk.**	14.4	unk.	unk.	unk.	**unk.**	unk.	**unk.**	unk.		unk.	
0.20	**12**	unk.	**unk.**	unk.	**unk.**	unk.	**unk.**	unk.	unk.	**unk.**	unk.	unk.	34	**3**	417	214	unk.	1.1	**6**	1.3	**9**	unk.	**unk.**	unk.	C
6.0	**30**	unk.	**unk.**	unk.	**unk.**	unk.	**unk.**	unk.	unk.	**unk.**	unk.	1.70	156	**16**	18.1	5.5	unk.	14	**3**	0.04	**2**	unk.		unk.	
0.24	**14**	unk.	**unk.**	unk.	**unk.**	unk.	**unk.**	unk.	unk.	**unk.**	unk.	unk.	27	**3**	499	230	unk.	1.2	**6**	1.7	**11**	unk.	**unk.**	unk.	D
4.1	**20**	unk.	**unk.**	unk.	**unk.**	unk.	**unk.**	unk.	unk.	**unk.**	unk.	2.22	143	**14**	21.7	5.9	unk.	15	**4**	0.06	**3**	unk.		unk.	
0.12	**7**	unk.	**unk.**	unk.	**unk.**	unk.	**unk.**	unk.	unk.	**unk.**	unk.	unk.	9	**1**	512	176	unk.	1.4	**8**	unk.	**unk.**	unk.	**unk.**	unk.	E
1.5	**7**	unk.	**unk.**	unk.	**unk.**	unk.	**unk.**	unk.	unk.	**unk.**	unk.	1.66	173	**17**	22.3	4.5	unk.	unk.	**unk.**	unk.	**unk.**	unk.		unk.	
0.14	**8**	0.23	**12**	1.29	**22**	unk.	**unk.**	unk.	unk.	**unk.**	unk.	unk.	4	**0**	578	253	unk.	1.4	**8**	2.2	**14**	unk.	**unk.**	unk.	F
5.3	**27**	unk.	**unk.**	0.69	**7**	unk.	**unk.**	unk.	unk.	**unk.**	unk.	1.69	207	**21**	25.1	6.5	unk.	19	**5**	0.05	**3**	unk.		unk.	
0.91	**54**	unk.	**unk.**	unk.	**unk.**	unk.	**unk.**	0	unk.	**unk.**	unk.	unk.	28	**3**	1530	844	unk.	5.3	**29**	14.4	**96**	unk.	**unk.**	unk.	G
unk.	**unk.**	unk.	**unk.**	unk.	**unk.**	0	**0**	0	unk.	**unk.**	unk.	7.21	599	**60**	66.5	21.6	unk.	60	**15**	0.49	**24**	unk.		unk.	
0.09	**6**	0.31	**15**	1.72	**29**	unk.	**unk.**	0	unk.	**unk.**	unk.	unk.	6	**1**	1216	236	unk.	0.3	**2**	1.0	**6**	unk.	**unk.**	unk.	H
7.1	**35**	unk.	**unk.**	0.50	**5**	0	**0**	0	unk.	**unk.**	unk.	3.55	195	**20**	52.9	6.0	unk.	17	**4**	0.05	**2**	unk.		unk.	
0.91	**54**	1.26	**63**	1.43	**24**	unk.	**unk.**	0	unk.	**unk.**	unk.	unk.	135	**14**	299	1141	unk.	9.6	**54**	16.9	**113**	unk.	**unk.**	100	I
15.4	**77**	35	**9**	4.49	**45**	0	**0**	0	unk.	**unk.**	unk.	4.08	820	**82**	13.0	29.2	unk.	100	**25**	0.58	**29**	unk.		unk.	
0.11	**7**	0.40	**20**	0.30	**5**	unk.	**unk.**	0	unk.	**unk.**	unk.	unk.	19	**2**	49	246	unk.	1.2	**6**	1.7	**11**	unk.	**unk.**	18	J
5.7	**28**	5	**1**	0.53	**5**	0	**0**	0	unk.	**unk.**	unk.	0.90	179	**18**	2.1	6.3	unk.	22	**6**	0.04	**2**	unk.		unk.	
0.19	**11**	0.26	**13**	0.29	**5**	unk.	**unk.**	0	unk.	**unk.**	unk.	unk.	26	**3**	61	235	unk.	1.9	**11**	3.5	**23**	unk.	**unk.**	20	K
3.1	**16**	7	**2**	0.95	**10**	0	**0**	0	unk.	**unk.**	unk.	0.83	167	**17**	2.7	6.0	unk.	20	**5**	0.12	**6**	unk.		unk.	
0.17	**10**	0.24	**12**	0.25	**4**	unk.	**unk.**	0	unk.	**unk.**	unk.	unk.	21	**2**	52	204	unk.	1.6	**9**	3.1	**21**	unk.	**unk.**	17	L
2.6	**13**	6	**2**	0.87	**9**	0	**0**	0	unk.	**unk.**	unk.	0.71	143	**14**	2.3	5.2	unk.	17	**4**	0.10	**5**	unk.		unk.	
1.03	**61**	1.43	**72**	1.57	**26**	unk.	**unk.**	0	unk.	**unk.**	unk.	unk.	134	**13**	327	1263	unk.	10.3	**57**	18.9	**126**	unk.	**unk.**	108	M
16.5	**83**	40	**10**	5.22	**52**	0	**0**	0	unk.	**unk.**	unk.	4.44	887	**89**	14.2	32.3	unk.	108	**27**	0.64	**32**	unk.		unk.	
0.11	**7**	0.39	**20**	0.29	**5**	unk.	**unk.**	0	unk.	**unk.**	unk.	unk.	20	**2**	49	243	unk.	1.2	**7**	unk.	**unk.**	unk.	**unk.**	18	N
5.6	**28**	5	**1**	0.52	**5**	0	**0**	0	unk.	**unk.**	unk.	0.89	177	**18**	2.1	6.2	unk.	22	**6**	0.04	**2**	unk.		unk.	
0.09	**5**	0.04	**2**	0.25	**4**	unk.	**unk.**	0	unk.	**unk.**	unk.	unk.	15	**2**	42	213	unk.	0.9	**5**	1.4	**9**	unk.	**unk.**	14	O
5.0	**25**	4	**1**	0.46	**5**	0	**0**	0	unk.	**unk.**	unk.	0.77	153	**15**	1.8	5.4	unk.	20	**5**	0.03	**1**	unk.		unk.	
0.14	**9**	0.34	**17**	0.29	**5**	unk.	**unk.**	0	unk.	**unk.**	unk.	unk.	22	**2**	54	240	unk.	1.5	**8**	2.5	**16**	unk.	**unk.**	19	P
4.5	**23**	6	**2**	0.71	**7**	0	**0**	0	unk.	**unk.**	unk.	0.87	173	**17**	2.4	6.1	unk.	21	**5**	0.07	**4**	unk.		unk.	
0.13	**7**	0.31	**15**	0.25	**4**	unk.	**unk.**	0	unk.	**unk.**	unk.	unk.	17	**2**	47	209	unk.	1.2	**7**	2.1	**14**	unk.	**unk.**	15	Q
4.0	**20**	5	**1**	0.64	**6**	0	**0**	0	unk.	**unk.**	unk.	0.75	148	**15**	2.0	5.3	unk.	18	**5**	0.06	**3**	unk.		unk.	
0.04	**2**	0.02	**1**	0.07	**1**	unk.	**unk.**	0	unk.	**unk.**	unk.	unk.	9	**1**	12	44	unk.	0.5	**3**	0.6	**4**	unk.	**unk.**	6	R
0.8	**4**	1	**0**	0.08	**1**	0	**0**	0	unk.	**unk.**	unk.	0.18	38	**4**	0.5	1.1	unk.	4	**1**	0.02	**1**	unk.		unk.	
0.23	**13**	0.71	**36**	0.57	**10**	unk.	**unk.**	0	unk.	**unk.**	unk.	unk.	42	**4**	97	466	unk.	2.5	**14**	3.5	**23**	unk.	**unk.**	37	S
10.6	**53**	9	**2**	0.99	**10**	0	**0**	0	unk.	**unk.**	unk.	1.72	345	**34**	4.2	11.9	unk.	43	**11**	0.09	**4**	unk.		unk.	

	FOOD	Portion	Weight in grams / *Conversion for 100 g*	Kilocalories / H_2O g	Total carbohydrate g / Total fats g	Crude fiber g / Dietary fiber g	Total protein g / Total sugar g	% USRDA	Arginine mg / Histidine mg	Isoleucine mg / Leucine mg	Lysine mg / Methionine mg	Phenylalanine mg / Threonine mg	Valine mg / Tryptophan mg	Cystine mg / Tyrosine mg	Polyunsat. fatty acids g / Monounsat. fatty acids g	Saturated fatty acids g / P/S ratio	Linoleic acid g / Cholesterol mg	Thiamin mg / Ascorbic acid mg	% USRDA
A	TURKEY,YOUNG TOM,dark meat w/skin,	3 sl(8x2-1/8	85	184	0.0	0.00	23.4	52	1652	1176	2133	917	1221	258	2.47	2.79	2.09	0.05	3
	roasted	x1/4")	*1.18*	51.3	9.2	0(a)	0(a)		705	1822	660	1023	259	890	2.35	0.89	77	0	0
B	turkey,young tom,dark meat w/o skin,	1/2 C	70	129	0.0	0.00	20.1	45	1401	1044	1892	797	1067	209	1.46	1.64	1.18	0.05	3
	roasted,chopped		*1.43*	44.1	4.9	0(a)	0(a)		626	1599	582	893	228	793	0.92	0.89	62	0	0
C	turkey,young tom,leg,meat w/skin,	1 piece	805	1658	0.0	0.00	224.8	500	15794	11407	20705	8839	11793	2423	21.41	24.07	17.87	0.48	32
	roasted		*0.12*	492.5	77.5	0(a)	0(a)		6851	17613	6392	9877	2512	8646	18.60	0.89	724	0	0
D	turkey,young tom,light meat w/skin,	3 sl(3x2-1/8	85	162	0.0	0.00	24.2	54	1707	1210	2197	946	1259	267	1.58	1.83	1.29	0.05	3
	roasted	x1/4")	*1.18*	53.7	6.5	0(a)	0(a)		726	1878	679	1056	267	915	1.83	0.87	64	0	0
E	turkey,young tom,light meat w/o skin,	1/2 C	70	108	0.0	0.00	20.9	47	1459	1088	1971	830	1111	218	0.55	0.65	0.36	0.05	3
	roasted,chopped		*1.43*	46.6	2.0	0(a)	0(a)		652	1667	605	930	237	827	0.29	0.84	48	0	0
F	turkey,young tom,meat w/skin,roasted	3 sl(3x2-1/8	85	172	0.0	0.00	23.9	53	1683	1196	2170	964	1243	263	1.96	2.24	1.63	0.05	3
		x1/4")	*1.18*	52.6	7.7	0(a)	0(a)		717	1854	671	1042	263	904	2.06	0.88	70	0	0
G	turkey,young tom,meat w/o skin,	3 sl(3x2-1/8	85	143	0.0	0.00	25.0	56	1741	1298	2352	990	1325	259	1.15	1.31	0.88	0.06	4
	roasted	x1/4")	*1.18*	55.3	4.0	0(a)	0(a)		779	1988	722	1110	283	986	0.69	0.88	65	0	0
H	turkey,young tom,skin only,roasted	1 oz	28	120	0.0	0.00	5.7	13	441	184	341	193	241	95	2.42	2.76	2.19	0.01	0
			3.52	11.8	10.6	0(a)	0(a)		110	336	114	204	46	130	3.63	0.88	33	0	0
I	turkey,young tom,thigh,meat w/o skin,	1 piece	200	406	0.0	0.00	60.0	133	unk.	3208	4770	2400	3054	786	3.52	3.14	2.22	0.08	5
	roasted		*0.50*	121.0	16.6	0(a)	0.0		1576	4416	1506	2382	616	2008	2.26	1.12	202	0	0
J	turkey,young tom,wing,meat w/skin,	1 piece	237	524	0.0	0.00	65.1	145	4626	3147	5719	2512	3323	751	6.45	7.42	5.47	0.14	10
	roasted		*0.42*	142.9	27.3	0(a)	0(a)		1887	4932	1777	2789	699	2370	8.32	0.87	192	0	0
K	TURMERIC,ground	1 tsp	2.3	8	1.5	0.15	0.2	0	unk.	unk.	unk.	unk.	unk.	unk.	unk.	unk.	unk.	tr	0
			43.48	0.3	0.2	unk.	0.0		unk.	unk.	unk.	unk.	unk.	unk.	unk.	unk.	0	1	2
L	TURNIP GREENS,fresh,cooked,	1/2 C	73	14	2.6	0.51	1.6	3	61	59	72	72	81	12	unk.	tr(a)	unk.	0.11	7
	drained solids		*1.38*	67.6	0.1	unk.	0.2		29	113	29	69	25	51	unk.	unk.	0	50	83
M	turnip greens,frozen,chopped,cooked,	1/2 C	83	19	3.2	0.82	2.1	3	79	70	105	unk.	unk.	16	unk.	unk.	unk.	0.04	3
	drained solids		*1.21*	76.5	0.2	unk.	0.3		38	139	unk.	84	28	57	unk.	unk.	0	16	26
N	TURNIP,fresh,cooked,drained solids,	1/2 C	78	18	3.8	0.70	0.6	1	10	11	33	13	16	1	unk.	unk.	unk.	0.03	2
	sliced		*1.29*	72.5	0.2	3.18	2.7		5	26	7	18	8	16	unk.	unk.	unk.	17	28
O	turnip,raw,cubed or sliced	1/2 C	65	19	4.3	0.58	0.6	1	10	8	14	8	11	1	unk.	unk.	unk.	0.03	2
			1.54	59.5	0.1	2.67	2.5		5	12	2	8	8	5	unk.	unk.	unk.	23	39
P	VANILLA,pure,double strength	1 tsp	4.7	14	1.4	0(a)	0.0	0	0(a)	0(a)	0(a)	0(a)	0(a)	0(a)	0.00	0.00	0.00	0.00	0
	-Foote & Jenks		*21.28*	1.9	0.0	0(a)	0.0		0(a)	0(a)	0(a)	0(a)	0(a)	0(a)	0.00	0.00	0	0	0
Q	VEAL,canned,parmigiana-Campbell	1 C	213	295	17.0	0.60	25.0	39	unk.	unk.	unk.	unk.	unk.	unk.	unk.	unk.	unk.	0.30	20
			0.47	unk.	14.0	unk.	unk.		unk.	unk.	unk.	unk.	unk.	unk.	unk.	unk.	unk.	unk.	unk.
R	veal, cutlet, w/o bone, braised or	4-1/8x2-1/2x	85	184	0.0	0.00	23.0	35	1506	1488	2354	1144	1457	294	0.90	3.98	0.40	0.06	4
	broiled, w/fat	1/2"	*1.18*	51.3	9.4	0(a)	0.0		785	2065	644	1222	370	829	3.37	0.23	86	unk.	unk.
S	veal,home recipe,loaf	3x2-3/4x3/4"	135	270	7.1	0.11	25.3	55	1582	1232	2104	1193	1525	328	2.75	6.30	2.00	0.22	15
			0.74	83.7	17.1	unk.	0.9		834	2000	683	1166	303	911	6.48	0.44	135	5	8

Riboflavin mg / Niacin mg	% USRDA	Vitamin B_6 mg / Folacin mcg	% USRDA	Vitamin B_{12} mcg / Pantothenic acid mg	% USRDA	Biotin mg / Vitamin A IU	% USRDA	Preformed A RE / Beta carotene RE	Vitamin D IU / Vitamin E IU	% USRDA	Total tocopherol mg / Alpha tocopherol mg	Other tocopherol mg / Total ash g	Calcium mg / Phosphorus mg	% USRDA	Sodium mg / Sodium meq	Potassium mg / Potassium meq	Chlorine mg / Chlorine meq	Iron mg / Magnesium mg	% USRDA	Zinc mg / Copper mg	% USRDA	Iodine mcg / Selenium mcg	% USRDA	Manganese mcg / Chromium mcg	
0.21	**13**	0.28	**14**	0.31	**5**	unk.	**unk.**	0	unk.	**unk.**	unk.	unk.	30	**3**	68	235	unk.	1.9	**11**	3.6	**24**	unk.	**unk.**	19	A
2.9	**15**	8	**2**	1.02	**10**	0	**0**	0	unk.	**unk.**	unk.	0.82	167	**17**	3.0	6.0	unk.	20	**5**	0.13	**7**	unk.		unk.	
0.18	**11**	0.26	**13**	0.27	**4**	unk.	**unk.**	0	unk.	**unk.**	unk.	unk.	24	**2**	57	205	unk.	1.6	**9**	3.2	**21**	unk.	**unk.**	15	B
2.5	**13**	7	**2**	0.93	**9**	0	**0**	0	unk.	**unk.**	unk.	0.71	144	**14**	2.5	5.3	unk.	16	**4**	0.11	**6**	unk.		unk.	
2.01	**118**	2.74	**137**	2.98	**50**	unk.	**unk.**	0	unk.	**unk.**	unk.	unk.	282	**28**	644	2262	unk.	18.4	**102**	34.9	**233**	unk.	**unk.**	177	C
27.9	**140**	72	**18**	9.96	**100**	0	**0**	0	unk.	**unk.**	unk.	7.89	1610	**161**	28.0	57.8	unk.	185	**46**	1.26	**63**	unk.		unk.	
0.11	**7**	0.41	**20**	0.31	**5**	unk.	**unk.**	0	unk.	**unk.**	unk.	unk.	18	**2**	57	244	unk.	1.2	**7**	1.8	**12**	unk.	**unk.**	16	D
5.1	**26**	5	**1**	0.54	**5**	0	**0**	0	unk.	**unk.**	unk.	0.85	178	**2.5**	2.5	6.2	unk.	23	**6**	0.04	**2**	unk.		unk.	
0.09	**5**	0.38	**19**	0.27	**4**	unk.	**unk.**	0	unk.	**unk.**	unk.	unk.	13	**1**	48	216	unk.	1.0	**5**	1.5	**10**	unk.	**unk.**	13	E
4.6	**23**	4	**1**	0.49	**5**	0	**0**	0	unk.	**unk.**	unk.	0.73	154	**15**	2.1	5.5	unk.	20	**5**	0.03	**2**	unk.		unk.	
0.15	**9**	0.36	**18**	0.31	**5**	unk.	**unk.**	0	unk.	**unk.**	unk.	unk.	23	**2**	61	240	unk.	1.5	**8**	2.6	**17**	unk.	**unk.**	17	F
4.2	**21**	6	**2**	0.75	**8**	0	**0**	0	unk.	**unk.**	unk.	0.84	173	**17**	2.7	6.1	unk.	21	**5**	0.08	**4**	unk.		unk.	
0.16	**10**	0.40	**20**	0.32	**5**	unk.	**unk.**	0	unk.	**unk.**	unk.	unk.	21	**2**	63	256	unk.	1.5	**8**	2.7	**18**	unk.	**unk.**	17	G
4.5	**22**	7	**2**	0.82	**8**	0	**0**	0	unk.	**unk.**	unk.	0.88	182	**18**	2.7	6.5	unk.	22	**6**	0.08	**4**	unk.		unk.	
0.05	**3**	0.02	**1**	0.07	**1**	unk.	**unk.**	0	unk.	**unk.**	unk.	unk.	11	**1**	17	46	unk.	0.5	**3**	0.6	**4**	unk.	**unk.**	6	H
0.7	**4**	1	**0**	0.09	**1**	0	**0**	0	unk.	**unk.**	unk.	0.19	40	**4**	0.7	1.2	unk.	5	**1**	0.02	**1**	unk.		unk.	
0.46	**27**	0.80	**40**	0.84	**14**	unk.	**unk.**	unk.	unk.	**unk.**	0.6	tr(a)	60	**6**	198	796	unk.	4.6	**26**	8.8	**59**	unk.	**unk.**	unk.	I
8.4	**42**	14	**4**	1.13	**11**	unk.	**unk.**	unk.	0.7	**2**	tr(a)	2.40	800	**80**	8.6	20.4	unk.	56	**14**	0.36	**18**	tr(a)		tr(a)	
0.33	**20**	1.02	**51**	0.83	**14**	unk.	**unk.**	0	unk.	**unk.**	unk.	unk.	55	**6**	156	642	unk.	3.5	**19**	5.0	**33**	unk.	**unk.**	47	J
13.2	**66**	14	**4**	1.42	**14**	0	**0**	0	unk.	**unk.**	unk.	2.28	474	**47**	6.8	16.4	unk.	59	**15**	0.13	**6**	unk.		unk.	
0.01	**0**	unk.	**unk.**	0.00	**0**	unk.	**unk.**	0(a)	0(a)	**0**	unk.	unk.	4	**0**	1	58	unk.	1.0	**5**	0.1	**1**	unk.	**unk.**	unk.	K
0.1	**1**	unk.	**unk.**	unk.	**unk.**	tr	**0**	unk.	unk.	**unk.**	unk.	0.14	6	**1**	0.0	1.5	unk.	4	**1**	unk.	**unk.**	unk.		unk.	
0.17	**10**	0.06	**3**	0(a)	**0**	unk.	**unk.**	0	0(a)	**0**	unk.	unk.	133	**13**	12	108	unk.	0.8	**4**	unk.	**unk.**	unk.	**unk.**	unk.	L
0.4	**2**	69	**17**	0.01	**0**	4567	**91**	457	unk.	**unk.**	1.6	0.58	27	**3**	0.5	2.8	unk.	17	**4**	tr(a)	**0**	tr(a)		unk.	
0.07	**4**	0.07	**4**	0(a)	**0**	unk.	**unk.**	0	0(a)	**0**	unk.	unk.	97	**10**	14	123	unk.	1.3	**7**	unk.	**unk.**	unk.	**unk.**	unk.	M
0.3	**2**	78	**20**	0.09	**1**	5692	**114**	569	unk.	**unk.**	1.8	0.49	32	**3**	0.6	3.1	unk.	19	**5**	tr(a)	**0**	tr(a)		unk.	
0.04	**2**	unk.	**unk.**	0(a)	**0**	unk.	**unk.**	0	0(a)	**0**	unk.	unk.	27	**3**	26	146	unk.	0.3	**2**	unk.	**unk.**	unk.	**unk.**	tr(a)	N
0.2	**1**	unk.	**unk.**	unk.	**unk.**	tr	**0**	tr	unk.	**unk.**	unk.	0.39	19	**2**	1.1	3.7	unk.	unk.	**unk.**	unk.	**unk.**	tr(a)		unk.	
0.05	**3**	0.06	**3**	0.00	**0**	unk.	**unk.**	0	0(a)	**0**	unk.	unk.	25	**3**	32	174	27	0.3	**2**	unk.	**unk.**	unk.	**unk.**	26	O
0.4	**2**	13	**3**	0.13	**1**	tr	**0**	tr	unk.	**unk.**	unk.	0.45	20	**2**	1.4	4.5	0.8	6	**2**	0.12	**6**	18		unk.	
0.00	**0**	unk.	**unk.**	0(a)	**0**	unk.	**unk.**	0	0(a)	**0**	unk.	unk.	0	**0**	0	0	unk.	0.0	**0**	unk.	**unk.**	unk.	**unk.**	unk.	P
0.0	**0**	unk.	**unk.**	unk.	**unk.**	0	**0**	0	unk.	**unk.**	unk.	unk.	unk.	**unk.**	0.0	0.0	unk.	unk.	**unk.**	unk.	**unk.**	unk.		unk.	
0.36	**21**	unk.	**unk.**	unk.	**unk.**	unk.	**unk.**	unk.	unk.	**unk.**	unk.	unk.	99	**10**	1828	469	unk.	2.4	**13**	unk.	**unk.**	unk.	**unk.**	unk.	Q
6.9	**35**	unk.	**unk.**	unk.	**unk.**	619	**12**	unk.	unk.	**unk.**	unk.	0.79	290	**29**	79.5	12.0	unk.	unk.	**unk.**	unk.	**unk.**	unk.		unk.	
0.21	**13**	0.17	**9**	1.22	**20**	unk.	**unk.**	unk.	unk.	**unk.**	0.2	tr(a)	9	**1**	68	425	unk.	2.7	**15**	3.2	**21**	unk.	**unk.**	unk.	R
4.6	**23**	3	**1**	0.38	**4**	unk.	**unk.**	unk.	0.2	**1**	0.2	1.19	196	**20**	3.0	10.9	unk.	15	**4**	0.07	**3**	tr(a)		tr(a)	
0.27	**16**	0.21	**11**	1.11	**19**	0.002	**1**	0	3	**1**	0.1	0.0	31	**3**	398	349	17	3.4	**19**	4.1	**27**	33.1	**22**	9	S
4.9	**24**	13	**3**	0.54	**5**	70	**1**	2	0.1	**1**	0.0	3.03	237	**24**	17.3	8.9	0.5	44	**11**	0.08	**4**	0		1	

	FOOD	Portion	Weight in grams / Conversion for 100 g	Kilocalories / H_2O g	Total carbohydrate g / Total fats g	Crude fiber g / Dietary fiber g	Total protein g / Total sugar g	% USRDA	Arginine mg / Histidine mg	Isoleucine mg / Leucine mg	Lysine mg / Methionine mg	Phenylalanine mg / Threonine mg	Valine mg / Tryptophan mg	Cystine mg / Tyrosine mg	Polyunsat. fatty acids g / Monounsat. fatty acids g	Saturated fatty acids g / P/S ratio	Linoleic acid g / Cholesterol mg	Thiamin mg / Ascorbic acid mg	% USRDA
A	veal,home recipe,parmigiana	5x4x3/4''	120	279	6.2	0.22	22.3	**49**	1299	1062	1947	1094	1384	264	1.16	9.65	0.65	0.11	**8**
			0.83	71.3	18.2	1.87	4.1		752	1832	607	980	228	871	5.40	0.12	136	15	**26**
B	veal,loin chop,w/bone,cooked, no visible fat	1 sm (4.75 oz raw)	69	143	0.0	0.00	23.6	**52**	1490	1244	1967	956	1217	301	0.28	0.50	0.09	0.14	**9**
			1.45	unk.	4.6	0(a)	0.0		803	1726	539	1022	310	848	0.41	0.55	68	0	**0**
C	veal,loin chop,w/bone,cooked, no visible fat	1 med (6.5 oz raw)	94	194	0.0	0.00	32.1	**71**	2030	1695	2680	1303	1658	410	0.38	0.69	0.12	0.20	**13**
			1.06	unk.	6.3	0(a)	0.0		1094	2351	734	1392	422	1156	0.56	0.55	93	0	**0**
D	veal,loin chop,w/bone,cooked,w/fat	1 sm (4.75 oz raw)	81	190	0.0	0.00	21.4	**48**	1351	963	1530	744	947	273	1.01	4.63	0.45	0.06	**4**
			1.23	47.7	10.9	0(a)	0.0		728	1342	419	795	241	769	3.93	0.22	82	0(a)	**0**
E	veal,loin chop,w/bone,cooked,w/fat	1 med (6.5 oz raw)	110	257	0.0	0.00	29.0	**65**	1835	1308	2078	1011	1286	371	1.37	6.28	0.62	0.08	**5**
			0.91	64.8	14.7	0(a)	0.0		989	1823	569	1079	327	1045	5.33	0.22	111	0(a)	**0**
F	veal,loin roast,w/o bone,roasted, sliced,no visible fat	2-1/2x2-1/2x 3/4''	85	130	0.0	0.00	24.6	**55**	1557	1312	2076	1008	1283	314	0.34	0.62	0.11	0.13	**9**
			1.18	unk.	2.7	0(a)	0.0		840	1821	568	1078	326	887	0.51	0.55	84	0	**0**
G	veal,loin roast,w/o bone,roasted, sliced,w/fat	2-1/2x2-1/2x 3/4''	85	199	0.0	0.00	22.4	**50**	1418	1011	1606	781	994	286	1.06	4.85	0.48	0.06	**4**
			1.18	50.1	11.4	0(a)	0.0		764	1408	439	834	252	807	4.12	0.22	86	unk.	**unk.**
H	veal,rib chop,w/bone,cooked,w/fat	1 med (6 oz raw)	100	269	0.0	0.00	27.2	**60**	1780	1776	2809	1366	1738	347	1.63	7.12	0.72	0.13	**9**
			1.00	54.6	16.9	0(a)	0.0		927	2464	769	1459	441	979	6.04	0.23	99	0(a)	**0**
I	veal,round,cooked,chopped,w/fat	1/2 C	70	151	0.0	0.00	19.0	**42**	1240	1046	1653	804	1023	242	0.74	3.28	0.33	0.05	**3**
			1.43	42.3	7.8	0(a)	0.0		646	1450	453	859	260	682	2.78	0.23	71	0(a)	**0**
J	veal,round,ground,cooked,w/fat	1/2 C	55	119	0.0	0.00	14.9	**33**	975	822	1299	631	804	190	0.58	2.57	0.26	0.04	**3**
			1.82	33.2	6.1	0(a)	0.0		508	1140	356	675	204	536	2.18	0.23	56	0(a)	**0**
K	veal,round,patty,cooked,w/fat	3'' dia x 5/8''	85	184	0.0	0.00	23.0	**35**	1506	1270	2008	976	1243	294	0.90	3.98	0.40	0.06	**4**
			1.18	51.3	9.4	0(a)	0.0		785	1761	550	1043	315	829	3.37	0.23	86	0(a)	**0**
L	veal,shoulder arm roast,w/o bone, braised,w/fat	2-1/2x2-1/2x 3/4''	85	200	0.0	0.00	23.7	**53**	1498	1328	2100	1022	1300	303	1.04	4.61	0.46	0.08	**5**
			1.18	49.7	10.9	0(a)	0.0		808	1844	569	1091	331	853	3.91	0.23	86	unk.	**unk.**
M	veal,shoulder arm roast,w/o bone, cooked,sliced,no visible fat	2-1/2x2-1/2x 3/4''	85	170	0.0	0.00	30.3	**67**	1912	1598	2527	1230	1564	387	0.34	0.62	0.11	0.10	**7**
			1.18	unk.	4.5	0(a)	0.0		1031	2219	694	1312	399	1089	0.51	0.55	84	0	**0**
N	veal,shoulder arm roast,w/o bone, cooked,sliced,w/fat	2-1/2x2-1/2x 3/4''	85	200	0.0	0.00	23.7	**37**	1498	1328	2100	1022	1300	303	1.04	4.61	0.46	0.08	**5**
			1.18	49.7	10.9	0(a)	0.0		808	1844	569	1091	331	853	3.91	0.23	86	unk.	**unk.**
O	veal,shoulder arm steak,w/bone, cooked,no visible fat	1/2 lb raw	98	196	0.0	0.00	34.9	**78**	2204	1842	2914	1418	1803	446	0.39	0.72	0.13	0.12	**8**
			1.02	unk.	5.2	0(a)	0.0		1189	2559	800	1513	460	1255	0.59	0.55	97	0	**0**
P	VEGETABLES,chow mein,frozen,cooked -Campbell	1/2 C	114	60	8.0	0.45	1.5	**2**	unk.	unk.	unk.	unk.	unk.	unk.	unk.	unk.	unk.	0.05	**3**
			0.88	101.5	2.5	unk.	unk.		unk.	unk.	unk.	unk.	unk.	unk.	unk.	unk.	unk.	unk.	**unk.**
Q	vegetables,Italian style-Birds Eye	1/2 C	94	55	9.3	0.65	2.1	**3**	unk.	unk.	unk.	unk.	unk.	unk.	0.06	0.08	0.05	0.04	**3**
			1.07	79.5	1.0	unk.	3.3		unk.	unk.	unk.	unk.	unk.	unk.	0.45	0.67	0	24	**39**
R	vegetables, Japanese style-Birds Eye	1/2 C	94	39	8.6	0.66	1.9	**3**	unk.	unk.	unk.	unk.	unk.	unk.	tr	tr	tr	0.04	**3**
			1.07	81.8	0.1	unk.	6.4		unk.	unk.	unk.	unk.	unk.	unk.	tr	0.00	0	27	**45**
S	vegetables,mixed,frozen,cooked, drained solids	1/2 C	91	58	12.2	1.09	2.9	**5**	unk.	unk.	unk.	unk.	unk.	unk.	unk.	tr(a)	unk.	0.11	**7**
			1.10	75.2	0.3	unk.	2.3		unk.	unk.	unk.	unk.	unk.	unk.	unk.	unk.	0	7	**12**

Riboflavin mg / Niacin mg	% USRDA	Vitamin B_6 mg / Folacin mcg	% USRDA	Vitamin B_{12} mcg / Pantothenic acid mg	% USRDA	Biotin mg / Vitamin A IU	% USRDA	Preformed A RE / Beta carotene RE	Vitamin D IU / Vitamin E IU	% USRDA	Total tocopherol mg / Alpha tocopherol mg	Other tocopherol mg / Total ash g	Calcium mg / Phosphorus mg	% USRDA	Sodium mg / Sodium meq	Potassium mg / Potassium meq	Chlorine mg / Chlorine meq	Iron mg / Magnesium mg	% USRDA	Zinc mg / Copper mg	% USRDA	Iodine mcg / Selenium mcg	% USRDA	Manganese mcg / Chromium mcg	
0.27	**16**	0.24	**12**	1.11	**19**	0.002	**1**	0	5	**1**	0.2	0.0	137	**14**	545	531	16	3.1	**17**	3.0	**20**	33.1	**22**	4	A
4.2	**21**	11	**3**	0.43	**4**	1204	**24**	75	0.2	**1**	0.1	3.02	255	**26**	23.7	13.6	0.5	48	**12**	0.06	**3**	0		tr	
0.21	**13**	0.17	**8**	1.09	**18**	unk.	**unk.**	0	0	**0**	0.1	tr(a)	6	**1**	45	328	unk.	3.0	**17**	2.8	**19**	unk.	**unk.**	unk.	B
4.9	**24**	2	**1**	0.37	**4**	0	**0**	0	0.2	**1**	0.1	unk.	180	**18**	1.9	8.4	unk.	14	**4**	0.06	**3**	tr(a)		tr(a)	
0.30	**18**	0.23	**11**	1.49	**25**	unk.	**unk.**	0	0	**0**	0.2	tr(a)	8	**1**	61	446	unk.	4.1	**23**	3.9	**26**	unk.	**unk.**	unk.	C
6.7	**33**	3	**1**	0.50	**5**	0	**0**	0	0.2	**1**	0.2	unk.	245	**25**	2.7	11.4	unk.	20	**5**	0.08	**4**	tr(a)		tr(a)	
0.20	**12**	0.17	**8**	1.17	**19**	unk.	**unk.**	unk.	unk.	**unk.**	0.2	tr(a)	9	**1**	65	405	unk.	2.6	**14**	2.6	**18**	unk.	**unk.**	unk.	D
4.4	**22**	2	**1**	0.36	**4**	unk.	**unk.**	unk.	0.2	**1**	0.2	1.05	182	**18**	2.8	10.4	unk.	15	**4**	0.06	**3**	tr(a)		tr(a)	
0.27	**16**	0.22	**11**	1.58	**26**	unk.	**unk.**	unk.	unk.	**unk.**	0.2	tr(a)	12	**1**	88	550	unk.	3.5	**20**	3.6	**24**	unk.	**unk.**	unk.	E
5.9	**30**	3	**1**	0.49	**5**	unk.	**unk.**	unk.	0.3	**1**	0.2	1.43	248	**25**	3.8	14.1	unk.	20	**5**	0.09	**4**	tr(a)		tr(a)	
0.19	**11**	0.14	**7**	1.22	**20**	unk.	**unk.**	0	0	**0**	0.2	tr(a)	7	**1**	54	493	unk.	3.2	**18**	3.5	**23**	unk.	**unk.**	unk.	F
6.3	**32**	3	**1**	0.38	**4**	0	**0**	0	0.2	**1**	0.2	unk.	219	**22**	2.4	12.6	unk.	19	**5**	0.07	**3**	tr(a)		tr(a)	
0.21	**13**	0.17	**9**	1.22	**20**	unk.	**unk.**	unk.	unk.	**unk.**	0.2	tr(a)	9	**1**	68	425	unk.	2.7	**15**	2.8	**19**	unk.	**unk.**	unk.	G
4.6	**23**	3	**1**	0.38	**4**	unk.	**unk.**	unk.	0.2	**1**	0.2	1.10	191	**19**	3.0	10.9	unk.	15	**4**	0.07	**3**	tr(a)		tr(a)	
0.31	**18**	0.20	**10**	1.44	**24**	unk.	**unk.**	unk.	unk.	**unk.**	0.2	tr(a)	12	**1**	80	500	unk.	3.4	**19**	4.1	**27**	unk.	**unk.**	unk.	H
7.8	**39**	3	**1**	0.45	**5**	unk.	**unk.**	unk.	0.2	**1**	0.2	1.30	248	**25**	3.5	12.8	unk.	20	**5**	0.08	**4**	tr(a)		tr(a)	
0.17	**10**	0.14	**7**	1.01	**17**	unk.	**unk.**	unk.	unk.	**unk.**	0.1	tr(a)	8	**1**	56	350	unk.	2.2	**12**	2.6	**17**	unk.	**unk.**	unk.	I
3.8	**19**	2	**1**	0.31	**3**	unk.	**unk.**	unk.	0.2	**1**	0.1	0.98	162	**16**	2.4	8.9	unk.	13	**3**	0.06	**3**	tr(a)		tr(a)	
0.14	**8**	0.11	**6**	0.79	**13**	unk.	**unk.**	unk.	unk.	**unk.**	0.1	tr(a)	6	**1**	44	275	unk.	1.8	**10**	2.0	**13**	unk.	**unk.**	unk.	J
3.0	**15**	2	**0**	0.25	**3**	unk.	**unk.**	unk.	0.1	**0**	0.1	0.77	127	**13**	1.9	7.0	unk.	10	**3**	0.04	**2**	tr(a)		tr(a)	
0.21	**13**	0.17	**9**	1.22	**20**	unk.	**unk.**	unk.	unk.	**unk.**	0.2	tr(a)	9	**1**	68	425	unk.	2.7	**15**	3.0	**20**	unk.	**unk.**	unk.	K
4.6	**23**	3	**1**	0.38	**4**	unk.	**unk.**	unk.	0.2	**1**	0.2	1.19	196	**20**	3.0	10.9	unk.	0(a)	**0**	0.07	**3**	tr(a)		tr(a)	
0.25	**15**	0.17	**9**	1.22	**20**	unk.	**unk.**	unk.	unk.	**unk.**	0.2	tr(a)	10	**1**	68	425	unk.	3.0	**17**	3.0	**20**	unk.	**unk.**	unk.	L
5.4	**27**	3	**1**	0.38	**4**	unk.	**unk.**	unk.	0.2	**1**	0.2	0.68	128	**13**	3.0	10.9	unk.	unk.	**unk.**	0.07	**3**	tr(a)		tr(a)	
0.25	**15**	0.20	**10**	1.34	**22**	unk.	**unk.**	0	0	**0**	0.2	tr(a)	8	**1**	43	427	unk.	3.7	**21**	3.5	**23**	unk.	**unk.**	unk.	M
5.1	**26**	3	**1**	0.45	**5**	0	**0**	0	0.2	**1**	0.2	unk.	238	**24**	1.9	10.9	unk.	19	**5**	0.07	**3**	tr(a)		tr(a)	
0.25	**15**	0.17	**9**	1.22	**20**	unk.	**unk.**	unk.	unk.	**unk.**	0.2	tr(a)	10	**1**	68	425	unk.	3.0	**17**	3.0	**20**	unk.	**unk.**	unk.	N
5.4	**27**	3	**1**	0.38	**4**	unk.	**unk.**	unk.	0.2	**1**	0.2	0.68	128	**13**	3.0	10.9	unk.	unk.	**unk.**	0.07	**3**	tr(a)		tr(a)	
0.29	**17**	0.24	**12**	1.55	**26**	unk.	**unk.**	0	0	**0**	0.2	tr(a)	9	**1**	50	492	unk.	4.3	**24**	4.0	**27**	unk.	**unk.**	unk.	O
5.9	**29**	3	**1**	0.52	**5**	0	**0**	0	0.2	**1**	0.2	unk.	274	**27**	2.2	12.6	unk.	22	**5**	0.08	**4**	tr(a)		tr(a)	
0.05	**3**	unk.	**unk.**	0(a)	**0**	unk.	**unk.**	0	0(a)	**0**	unk.	unk.	26	**3**	630	87	unk.	0.9	**5**	unk.	**unk.**	unk.	**unk.**	unk.	P
0.2	**1**	unk.	**unk.**	unk.	**unk.**	0	**0**	0	unk.	**unk.**	unk.	0.45	31	**3**	27.4	2.2	unk.	unk.	**unk.**	unk.	**unk.**	unk.		unk.	
0.07	**4**	0.09	**4**	0.00	**0**	unk.	**unk.**	unk.	0	**0**	unk.	unk.	39	**4**	471	141	unk.	0.9	**5**	0.2	**2**	unk.	**unk.**	unk.	Q
0.4	**2**	15	**4**	tr	**1**	729	**15**	unk.	unk.	**unk.**	unk.	1.59	36	**4**	20.5	3.6	unk.	17	**4**	0.08	**4**	unk.		unk.	
0.07	**4**	0.08	**4**	0.00	**0**	unk.	**unk.**	unk.	0	**0**	unk.	unk.	29	**3**	370	132	unk.	0.6	**3**	0.2	**2**	unk.	**unk.**	unk.	R
0.3	**1**	38	**10**	0.11	**1**	608	**12**	unk.	unk.	**unk.**	unk.	1.22	39	**4**	16.1	3.4	unk.	16	**4**	0.05	**2**	tr(a)		unk.	
0.06	**4**	0.09	**5**	0(a)	**0**	unk.	**unk.**	0	0(a)	**0**	unk.	unk.	23	**2**	48	174	unk.	1.2	**7**	0.3	**2**	unk.	**unk.**	unk.	S
1.0	**5**	unk.	**unk.**	0.24	**2**	4505	**90**	450	unk.	**unk.**	unk.	0.45	57	**6**	2.1	4.4	unk.	22	**6**	tr(a)	**0**	tr(a)		unk.	

	FOOD	Portion	Weight in grams / *Conversion for 100 g*	Kilocalories / H_2O g	Total carbohydrate g / Total fats g	Crude fiber g / Dietary fiber g	Total protein g / Total sugar g	**% USRDA**	Arginine mg / Histidine mg	Isoleucine mg / Leucine mg	Lysine mg / Methionine mg	Phenylalanine mg / Threonine mg	Valine mg / Tryptophan mg	Cystine mg / Tyrosine mg	Polyunsat. fatty acids g / Monounsat. fatty acids g	Saturated fatty acids g / P/S ratio	Linoleic acid g / Cholesterol mg	Thiamin mg / Ascorbic acid mg	**% USRDA**
A	VINEGAR,cider	1 Tbsp	15	2	0.9	0.00	tr	**0**	tr(a)	tr(a)	tr(a)	tr(a)	tr(a)	tr(a)	0.00	0.00	0.00	tr(a)	**0**
			6.67	14.1	0.0	0(a)	0.1		tr(a)	tr(a)	tr(a)	tr(a)	tr(a)	tr(a)	0.00	0.00	0	unk.	**unk.**
B	vinegar,distilled	1 Tbsp	15	2	0.7	0(a)	0.0	**0**	0(a)	0(a)	0(a)	0(a)	0(a)	0(a)	0.00	0.00	0.00	tr(a)	**0**
			6.67	14.2	0.0	0(a)	0.1		0(a)	0(a)	0(a)	0(a)	0(a)	0(a)	0.00	0.00	0	unk.	**unk.**
C	WAFFLE,blueberry-Aunt Jemima	1	24	59	9.3	0.05	1.4	**2**	unk.	unk.	unk.	unk.	unk.	unk.	unk.	unk.	unk.	0.08	**5**
			4.22	10.5	1.7	unk.	4.0		unk.	unk.	unk.	unk.	unk.	unk.	unk.	unk.	unk.	unk.	**unk.**
D	waffle,buttermilk-Aunt Jemima	1	24	57	8.9	0.05	1.4	**2**	unk.	unk.	unk.	unk.	unk.	unk.	unk.	unk.	unk.	0.08	**5**
			4.22	11.0	1.7	0.44	1.8		unk.	unk.	unk.	unk.	unk.	unk.	unk.	unk.	unk.	unk.	**unk.**
E	waffle,home recipe	7" dia	79	282	22.1	0.08	5.9	**11**	162	304	308	303	331	104	9.37	3.68	9.17	0.14	**9**
			1.27	31.4	18.9	0.76	3.0		151	503	129	232	79	243	4.86	2.54	70	0	**0**
F	waffle,home recipe	9" square	200	602	47.1	0.16	12.7	**24**	345	649	657	647	707	222	19.98	7.86	19.57	0.30	**20**
			0.50	80.0	40.3	1.61	6.5		321	1073	274	495	169	518	10.37	2.54	150	1	**2**
G	waffle,jumbo-Aunt Jemima	1	36	84	13.2	0.07	2.2	**3**	unk.	unk.	unk.	unk.	unk.	unk.	unk.	unk.	unk.	0.11	**8**
			2.82	16.5	2.5	unk.	2.6		unk.	unk.	unk.	unk.	unk.	unk.	unk.	unk.	unk.	unk.	**unk.**
H	waffle,original-Aunt Jemima	1	24	56	8.8	0.05	1.5	**2**	unk.	unk.	unk.	unk.	unk.	unk.	unk.	unk.	unk.	0.08	**5**
			4.22	11.0	1.7	0.44	1.8		unk.	unk.	unk.	unk.	unk.	unk.	unk.	unk.	unk.	unk.	**unk.**
I	WALNUTS,black,halves	1/4 C	31	197	4.6	0.53	6.4	**10**	956	unk.	unk.	unk.	unk.	unk.	12.80	1.59	11.52	0.07	**5**
			3.19	1.0	18.6	1.63	1.0		unk.	unk.	unk.	unk.	unk.	unk.	3.35	8.05	0	0	**0**
J	walnuts,English,halves	1/4 C	25	163	3.9	0.52	3.8	**6**	559	192	110	192	244	80	10.50	1.73	8.72	0.08	**6**
			4.00	0.9	16.0	1.30	0.8		101	307	77	147	44	146	2.42	6.05	0	1	**2**
K	WATERCHESTNUTS,raw, edible portion	1 average	15	12	2.8	0.12	0.2	**0**	unk.	unk.	unk.	unk.	unk.	unk.	unk.	unk.	unk.	0.02	**1**
			6.85	11.4	tr	unk.	unk.		unk.	unk.	unk.	unk.	unk.	unk.	unk.	unk.	unk.	1	**2**
L	WATERCRESS (garden cress),raw, edible portion	1/2 C	18	6	1.0	0.19	0.4	**1**	unk.	24	28	19	26	unk.	unk.	tr(a)	unk.	0.01	**1**
			5.71	15.6	0.1	0.60	0.1		9	40	3	26	9	unk.	unk.	unk.	0	12	**20**
M	WATERMELON,fresh,balls or cubes, edible portion	1/2 C	80	26	5.1	0.24	0.5	**1**	unk.	unk.	unk.	unk.	unk.	unk.	unk.	tr(a)	unk.	0.06	**6**
			1.25	74.1	0.3	unk.	6.2		unk.	unk.	unk.	unk.	unk.	46	unk.	unk.	0	8	**13**
N	watermelon,fresh,sliced	1 slice	426	139.5	31.3	1.28	2.7	**5**	unk.	unk.	unk.	unk.	unk.	unk.	unk.	tr(a)	unk.	0.35	**30**
			0.23	399.0	1.31	unk.	32.8		unk.	unk.	unk.	unk.	unk.	245	unk.	unk.	0	42	**70**
O	WHEAT GERM	1 oz (1/4 C)	28	103	13.2	0.71	7.5	**12**	unk.	359	467	275	413	unk.	1.88	0.53	1.66	0.57	**38**
			3.52	3.3	3.1	unk.	unk.		unk.	521	122	405	76	unk.	0.44	3.53	0	0	**0**
P	WHITEFISH,lake,cooked	3 oz serving	85	106	0.0	0(a)	11.6	**26**	658	659	1137	478	685	117	1.44	0.73	0.22	0.09	**6**
			1.18	53.7	4.4	0(a)	0.0		354	969	375	556	129	354	1.10	1.98	unk.	tr	**0**
Q	whitefish,lake,smoked	3 pieces	60	93	0.0	0.00	12.5	**28**	709	640	1103	464	665	126	1.02	0.52	0.16	unk.	**unk.**
			1.67	40.9	3.1	0(a)	0.0		381	941	364	539	125	381	0.78	1.98	unk.	unk.	**unk.**
R	whitefish,lake,stuffed,baked	3 oz serving	85	183	4.9	unk.	12.9	**29**	730	659	1137	478	685	130	unk.	unk.	unk.	0.09	**6**
			1.18	53.7	11.9	unk.	unk.		393	969	375	556	129	393	unk.	unk.	unk.	tr	**0**
S	WONTON,home recipe,fried	1/2 C	30	111	8.4	0.06	2.0	**4**	57	96	88	104	106	42	4.16	1.17	4.10	0.05	**4**
			3.29	11.8	7.7	0.59	0.3		51	161	42	76	27	77	1.97	3.55	31	1	**2**

Riboflavin mg / Niacin mg	% USRDA	Vitamin B6 mg / Folacin mcg	% USRDA	Vitamin B12 mcg / Pantothenic acid mg	% USRDA	Biotin mg / Vitamin A IU	% USRDA	Preformed A RE / Beta carotene RE	Vitamin D IU / Vitamin E IU	% USRDA	Total tocopherol mg / Alpha tocopherol mg	Other tocopherol mg / Total ash g	Calcium mg / Phosphorus mg	% USRDA	Sodium mg / Sodium meq	Potassium mg / Potassium meq	Chlorine mg / Chlorine meq	Iron mg / Magnesium mg	% USRDA	Zinc mg / Copper mg	% USRDA	Iodine mcg / Selenium mcg	% USRDA	Manganese mcg / Chromium mcg	
tr(a)	0	0.00	0	0.00	0	tr(a)	0	0	0	0	tr(a)	tr(a)	1	0	tr	15	tr(a)	0.1	1	tr	0	unk.	unk.	38	A
tr(a)	0	tr(a)	0	tr(a)	0	0	0	0	0.0	0	tr(a)	0.04	1	0	0.0	0.4	0.0	tr	0	0.01	1	13		tr(a)	
tr(a)	0	0.00	0	0.00	0	tr(a)	0	0	0	0	tr(a)	tr(a)	tr(a)	0	tr	2	tr(a)	0(a)	0	tr	0	tr(a)	0	tr(a)	B
tr(a)	0	0	0	tr(a)	0	0	0	0	0.0	0	tr(a)	tr	tr(a)	0	0.0	0.1	0.0	tr	0	tr(a)	0	tr(a)		tr(a)	
0.07	4	0.06	3	0.12	2	0.000	0	unk.	tr(a)	0	unk.	unk.	57	6	334	29	unk.	0.7	4	0.2	2	unk.	unk.	0	C
0.8	4	12	3	0.24	2	unk.	unk.	unk.	unk.	unk.	unk.	0.69	85	9	14.5	0.7	unk.	5	1	0.02	1	unk.		unk.	
0.07	4	0.06	3	0.12	2	0.000	0	unk.	tr(a)	0	unk.	unk.	59	6	179	26	unk.	0.7	4	0.2	2	unk.	unk.	0	D
0.8	4	10	3	0.24	2	unk.	unk.	unk.	unk.	unk.	unk.	0.66	101	10	7.8	0.7	unk.	5	1	0.02	1	unk.		unk.	
0.19	12	0.06	3	0.32	5	0.003	1	0	27	7	13.5	tr(a)	102	10	328	125	40	1.0	6	0.6	4	47.9	32	121	E
0.9	5	12	3	0.46	5	95	2	0	0.2	1	0.2	1.81	142	14	14.3	3.2	1.1	46	11	0.16	8	13		17	
0.42	25	0.12	6	0.69	12	0.006	2	0	57	14	28.8	tr(a)	218	22	700	266	85	2.2	12	1.3	9	102.3	68	259	F
2.0	10	25	6	0.98	10	204	4	0	0.4	1	0.4	3.86	303	30	30.4	6.8	2.4	97	24	0.33	17	28		36	
0.10	6	0.09	4	0.18	3	0.001	0	unk.	tr(a)	0	unk.	unk.	89	9	262	43	unk.	1.1	6	0.4	2	unk.	unk.	0	G
1.2	6	19	5	0.00	0	unk.	unk.	unk.	unk.	unk.	unk.	0.99	131	13	11.4	1.1	unk.	7	2	0.03	1	unk.		unk.	
0.07	4	0.06	3	0.12	2	0.001	0	unk.	tr(a)	0	unk.	unk.	59	6	175	29	unk.	0.7	4	0.2	2	unk.	unk.	0	H
0.8	4	13	3	0.00	0	unk.	unk.	unk.	unk.	unk.	unk.	0.66	87	9	7.6	0.7	unk.	5	1	0.02	1	unk.		unk.	
0.03	2	unk.	unk.	0(a)	0	unk.	unk.	0	0	0	unk.	unk.	tr	0	1	144	unk.	1.9	10	0.7	5	unk.	unk.	unk.	I
0.2	1	unk.	unk.	unk.	unk.	94	2	9	unk.	unk.	unk.	0.72	178	18	0.0	3.7	unk.	59	15	unk.	unk.	unk.		unk.	
0.03	2	0.18	9	0.00	0	0.009	3	0	0(a)	0	unk.	unk.	25	3	1	113	9	0.8	4	0.6	4	unk.	unk.	525	J
0.2	1	19	5	0.00	0	8	0	1	0.4	1	0.4	0.47	95	10	0.0	2.9	0.3	36	9	0.35	18	unk.		15	
0.03	2	unk.	unk.	0(a)	0	unk.	unk.	0	0(a)	0	unk.	unk.	1	0	3	73	unk.	0.1	1	unk.	unk.	unk.	unk.	tr(a)	K
0.1	1	unk.	unk.	unk.	unk.	0	0	0	unk.	unk.	unk.	0.16	10	1	0.1	1.9	unk.	2	0	tr(a)	0	tr(a)		unk.	
0.05	3	0.04	2	0.00	0	unk.	unk.	0	0(a)	0	unk.	unk.	14	1	2	106	unk.	0.2	1	unk.	unk.	unk.	unk.	unk.	L
0.2	1	unk.	unk.	0.05	1	1627	33	163	unk.	unk.	unk.	0.31	13	1	0.1	2.7	unk.	3	1	tr(a)	0	tr(a)		unk.	
0.02	1	0.12	6	0.00	0	0.003	1	0	0(a)	0	unk.	unk.	6	1	2	93	unk.	0.1	1	0.1	1	unk.	unk.	30	M
0.2	1	2	1	0.17	2	293	5	29	unk.	unk.	unk.	0.21	7	1	0.1	2.4	unk.	8	2	0.03	3	tr(a)		12	
0.09	8	0.63	15	0.00	0	0.015	5	0	0(a)	0	unk.	unk.	84	3	9	506	26	0.8	4	0.3	2	unk.	unk.	161	N
0.9	4	10	3	0.92	10	1596	50	160	unk.	unk.	unk.	1.13	38	4	0.4	13.0	0.7	43	11	0.14	6	tr(a)		64	
0.19	11	unk.	unk.	unk.	unk.	unk.	unk.	0	unk.	unk.	4.5	unk.	20	2	1	234	unk.	2.7	15	4.4	29	unk.	unk.	3888	O
1.2	6	unk.	unk.	unk.	unk.	0	0	0	unk.	18	unk.	1.22	316	32	0.0	6.0	unk.	96	24	0.21	11	31		2	
0.09	6	unk.	unk.	unk.	unk.	unk.	unk.	623	unk.	unk.	unk.	unk.	unk.	unk.	44	247	unk.	0.4	2	unk.	unk.	unk.	unk.	unk.	P
2.0	10	unk.	unk.	unk.	unk.	2305	46	23	unk.	unk.	unk.	1.53	209	21	1.9	6.3	unk.	unk.	unk.	unk.	unk.	tr(a)		tr(a)	
unk.	unk.	unk.	unk.	unk.	unk.	unk.	unk.	unk.	unk.	unk.	unk.	unk.	13	1	unk.	unk.	unk.	unk.	unk.	unk.	unk.	unk.	unk.	unk.	Q
unk.	unk.	unk.	unk.	unk.	unk.	unk.	unk.	unk.	unk.	unk.	unk.	2.22	164	16	unk.	unk.	unk.	unk.	unk.	unk.	unk.	tr(a)		tr(a)	
0.09	6	unk.	unk.	unk.	unk.	unk.	unk.	unk.	unk.	unk.	unk.	unk.	unk.	unk.	44	247	unk.	0.4	2	unk.	unk.	unk.	unk.	unk.	R
2.0	10	unk.	unk.	unk.	unk.	1700	34	unk.	unk.	unk.	unk.	1.53	209	21	1.9	6.3	unk.	unk.	unk.	unk.	unk.	tr(a)		tr(a)	
0.05	3	0.02	1	0.07	1	0.001	1	0	2	1	5.4	4.4	10	1	147	33	21	0.5	3	0.2	1	39.6	26	85	S
0.4	2	5	1	0.14	1	356	7	34	6.5	22	0.9	0.50	27	3	6.4	0.8	0.6	22	5	0.04	2	6		8	

	FOOD	Portion	Weight in grams / Conversion for 100 g	Kilocalories / H_2O g	Total carbohydrate g / Total fats g	Crude fiber g / Dietary fiber g	Total protein g / Total sugar g	% USRDA	Arginine mg / Histidine mg	Isoleucine mg / Leucine mg	Lysine mg / Methionine mg	Phenylalanine mg / Threonine mg	Valine mg / Tryptophan mg	Cystine mg / Tyrosine mg	Polyunsat. fatty acids g / Monounsat. fatty acids g	Saturated fatty acids g / P/S ratio	Linoleic acid g / Cholesterol mg	Thiamin mg / Ascorbic acid mg	% USRDA
A	YAM,cooked in skin	1/2 C	100	105	24.1	0.90	2.4	**4**	184	85	127	111	104	31	unk.	tr(a)	unk.	0.09	**6**
			1.00	unk.	0.2	unk.	0.2		46	146	38	81	41	77	unk.	unk.	0	9	**15**
B	YEAST,brewer's	1 Tbsp	8.0	23	3.1	0.00	3.1	**5**	unk.	192	264	152	218	44	unk.	unk.	unk.	1.25	**83**
			12.50	0.4	0.1	0(a)	0.0		100	258	67	188	57	152	unk.	unk.	0(a)	tr	**0**
C	yeast,compressed,fortified cake	1 cake	12	10	1.3	0(a)	1.4	**2**	unk.	79	110	73	101	14	unk.	unk.	unk.	0.17	**11**
			8.33	8.5	0.0	0(a)	0.0		64	138	30	79	15	70	unk.	unk.	0	unk.	**unk.**
D	yeast,dry (active)	1 Tbsp	8.0	23	3.1	0(a)	2.9	**5**	unk.	182	253	167	232	29	unk.	unk.	unk.	0.19	**12**
			12.50	0.4	0.1	0(a)	0.0		131	318	68	182	35	1	unk.	unk.	0	unk.	**unk.**
E	YOGURT,coffee & vanilla varities,low fat	1 C	227	193	31.3	0.00	11.2	**25**	336	611	1003	611	926	unk.	0.09	1.84	0.05	0.09	**6**
			0.44	179.3	2.8	0(a)	31.3		277	1128	329	459	64	565	0.66	0.05	11	2	**3**
F	yogurt,fruit varieties,low fat	1 C	227	232	43.2	0.27	9.9	**22**	300	540	890	540	822	unk.	0.07	1.59	0.05	0.07	**5**
			0.44	169.1	2.5	unk.	43.2		245	999	293	406	57	502	0.57	0.04	9	1	**2**
G	yogurt,plain,low fat	1 C	227	143	16.0	0.00	11.9	**27**	359	649	1069	649	985	unk.	0.09	2.27	0.07	0.09	**6**
			0.44	193.1	3.5	0(a)	16.0		295	1201	352	490	68	6.02	0.79	0.04	14	2	**3**
H	yogurt,plain,skim milk	1 C	227	127	17.4	0.00	13.0	**29**	390	711	1167	711	1076	unk.	0.00	0.25	0.00	0.09	**6**
			0.44	193.5	0.4	0(a)	17.4		322	1310	384	533	73	656	0.09	0.00	5	2	**3**
I	yogurt,plain,whole milk	1 C	227	138	10.6	0.00	7.9	**18**	236	429	706	429	651	unk.	0.20	4.77	0.14	0.05	**3**
			0.44	199.5	7.4	0(a)	10.6		195	794	232	322	45	397	1.68	0.04	30	1	**2**
J	ZUCCHINI,home recipe, casserole	1 C	226	295	23.3	0.67	8.4	**12**	182	251	290	265	282	62	1.58	9.75	1.33	0.11	**8**
			0.44	173.8	19.1	unk.	6.0		129	429	100	169	63	200	5.72	0.16	32	9	**15**
K	zucchini,home recipe,croquettes	1/2 C	105	118	11.4	0.36	4.1	**8**	136	211	235	204	239	70	0.42	3.07	0.33	0.07	**5**
			0.96	81.1	6.1	0.22	3.3		104	335	97	169	55	161	1.63	0.14	75	6	**10**
L	zucchini,home recipe,sticks	1/2 C	145	81	4.3	0.81	2.3	**4**	40	98	147	91	120	7	0.23	4.05	0.13	0.07	**5**
			0.69	130.5	6.6	unk.	5.1		42	162	46	69	24	61	1.64	0.06	18	13	**23**

Riboflavin mg / Niacin mg	% USRDA	Vitamin B_6 mg / Folacin mcg	% USRDA	Vitamin B_{12} mcg / Pantothenic acid mg	% USRDA	Biotin mg / Vitamin A IU	% USRDA	Preformed A RE / Beta carotene RE	Vitamin D IU / Vitamin E IU	% USRDA	Total tocopherol mg / Alpha tocopherol mg	Other tocopherol mg / Total ash g	Calcium mg / Phosphorus mg	% USRDA	Sodium mg / Sodium meq	Potassium mg / Potassium meq	Chlorine mg / Chlorine meq	Iron mg / Magnesium mg	% USRDA	Zinc mg / Copper mg	% USRDA	Iodine mcg / Selenium mcg	% USRDA	Manganese mcg / Chromium mcg	
0.04	2	unk.	unk.	0.00	0	unk.	unk.	0	0(a)	0	unk.	unk.	4	0	unk.	unk.	unk.	0.6	3	unk.	unk.	unk.	unk.	unk.	A
0.6	3	unk.	unk.	0.19	2	3000	60	300	unk.	unk.	unk.	unk.	50	5	unk.	unk.	unk.	unk.	unk.	tr(a)	0	tr(a)		unk.	
0.34	20	0.20	10	0.00	0	unk.	unk.	0(a)	0(a)	0	unk.	unk.	17	2	10	152	unk.	1.4	8	unk.	unk.	unk.	unk.	unk.	B
3.0	15	313	78	tr	0	tr	0	unk.	unk.	unk.	unk.	0.57	140	14	0.4	3.9	unk.	unk.	unk.	unk.	unk.	unk.		unk.	
0.20	12	0.07	4	0.00	0	unk.	unk.	0(a)	0(a)	0	unk.	unk.	2	0	2	73	unk.	0.6	3	unk.	unk.	unk.	unk.	unk.	C
17.3	86	unk.	unk.	0.42	4	unk.	unk.	unk.	unk.	unk.	unk.	0.29	47	5	0.1	1.9	unk.	7	2	0.07	4	unk.		unk.	
0.43	26	0.16	8	0.00	0	unk.	unk.	0(a)	0(a)	0	unk.	unk.	4	0	4	160	unk.	1.3	7	unk.	unk.	unk.	unk.	unk.	D
2.9	15	393	98	0.88	9	unk.	unk.	unk.	unk.	unk.	unk.	0.66	103	10	0.2	4.1	unk.	5	1	0.14	7	unk.		unk.	
0.45	27	0.10	5	1.18	20	unk.	unk.	unk.	unk.	unk.	unk.	unk.	388	39	150	497	unk.	0.2	1	1.9	13	unk.	unk.	unk.	E
0.2	1	23	6	1.25	13	123	3	unk.	unk.	unk.	unk.	2.32	306	31	6.5	12.7	unk.	36	9	unk.	unk.	unk.		unk.	
0.39	23	0.09	5	1.04	17	unk.	unk.	unk.	unk.	unk.	unk.	unk.	345	35	132	440	unk.	0.2	1	1.7	11	unk.	unk.	unk.	F
0.2	1	20	5	1.11	11	104	2	unk.	unk.	unk.	unk.	2.32	270	27	5.7	11.3	unk.	34	9	unk.	unk.	unk.		unk.	
0.48	28	0.11	6	1.27	21	unk.	unk.	unk.	unk.	unk.	unk.	unk.	415	42	159	531	unk.	0.2	1	2.0	14	unk.	unk.	unk.	G
0.2	1	25	6	1.34	13	150	3	unk.	unk.	unk.	unk.	2.47	327	33	6.9	13.6	unk.	39	10	unk.	unk.	unk.		unk.	
0.52	31	0.12	6	1.38	23	unk.	unk.	unk.	unk.	unk.	unk.	unk.	452	45	173	579	unk.	0.2	1	2.2	15	unk.	unk.	unk.	H
0.3	1	27	7	1.46	15	16	0	unk.	unk.	unk.	unk.	2.68	354	35	7.5	14.8	unk.	43	11	unk.	unk.	unk.		unk.	
0.32	19	0.07	4	0.84	14	unk.	unk.	unk.	unk.	unk.	unk.	unk.	275	28	104	352	unk.	0.1	1	1.3	9	unk.	unk.	unk.	I
0.2	1	16	4	0.88	9	279	6	unk.	unk.	unk.	unk.	1.63	216	22	4.5	9.0	unk.	27	7	unk.	unk.	unk.		unk.	
0.23	13	0.07	4	0.09	2	0.002	1	0	tr(a)	0	2.3	0(a)	171	17	945	255	unk.	1.5	8	0.5	3	unk.	unk.	unk.	J
1.9	10	18	4	0.24	2	1189	24	35	2.7	9	0.1	3.35	138	14	41.1	6.5	unk.	26	6	0.13	7	tr(a)		unk.	
0.15	9	0.06	3	0.24	4	0.003	1	0	5	1	1.3	tr(a)	61	6	340	145	27	0.8	5	0.5	3	2214.2	1476	34	K
0.7	4	11	3	0.30	3	447	9	18	1.5	5	0.1	1.36	84	8	14.8	3.7	0.8	45	11	0.10	5	4		4	
0.12	7	0.08	4	0.00	0	0.003	1	0	0(a)	0	3.3	0(a)	72	7	378	195	unk.	0.6	3	0.4	2	70.0	47	0	L
1.1	5	14	4	0.01	0	761	15	53	4.0	13	0.1	1.55	57	6	16.4	5.0	unk.	23	6	0.11	6	0		0	

Appendix Table A Caffeine Content of Selected Foods

FOOD	Portion	Wt.	Caffeine/ serving g
BEVERAGE, nonalcoholic, carbonated, Coca-Cola	12 oz. bottle	370	33.0
Beverage, nonalcoholic, carbonated, Dr. Pepper	12 oz. bottle	369	37.8
Beverage, nonalcoholic, carbonated, Dr. Pepper, sugar-free	12 oz. bottle	355	37.8
Beverage, nonalcoholic, carbonated, Mello Yellow	12 oz. bottle	370	51.0
Beverage, nonalcoholic, carbonated, Mountain Dew	12 oz. bottle	370	54.0
Beverage, nonalcoholic, carbonated, Mr. Pibb	12 oz. bottle	370	57.0
Beverage, nonalcoholic, carbonated, Mr. Pibb, sugar-free	12 oz. bottle	355	57.0
Beverage, nonalcoholic, carbonated, Pepsi Cola	12 oz. bottle	370	38.4
Beverage, nonalcoholic, carbonated, Pepsi, diet	12 oz. bottle	355	36.0
Beverage, nonalcoholic, carbonated, Pepsi Light	12 oz. bottle	365	36.0
Beverage, nonalcoholic, carbonated, Tab	12 oz. bottle	356	31.8
CAKE, Boston cream, home recipe	1/8th of 9" diam. cake	139	7.3
Cake, chocolate or devils food	1/6th of 9" diam.	92	13.8
Cake, roll, chocolate	1/4th of roll	85	8.5
CANDY, chocolate	1 oz. bar	28	7.7
Candy, chocolate-covered	1 oz. bar	28	2.8
Candy, chocolate-covered peanut butter	1 oz. pkg.	28	3.4
Candy, chocolate w/nuts	1 oz. bar	28	7.1
COFFEE, brewed	6 oz. cup	173	88.0
Coffee, instant, decaffeinated, prepared	6 oz. cup	173	2.1
Coffee, instant, freeze-dried powder	1 teaspoon	1	51.4
Coffee, instant, freeze-dried powder, decaffeinated	1 teaspoon	1	1.5
Coffee, instant, powder	1 teaspoon	1	40.2
Coffee, instant, powder, decaffeinated	1 teaspoon	1	1.2
Coffee, instant, prepared	6 oz. cup	173	70.7
Coffee, International, Suisse Moche, General Foods	2 teaspoons	11	38.8
COOKIE, brownie w/nuts, home recipe	1¾" X 1¾" X ⅞"	24	5.3
Cookie, chocolate chip, vending	1⅛ oz. pkg.	32	1.6
Cookie, chocolate chip, home recipe	2½" diam. X ¼"	12	1.4
Cookie, chocolate cremes	one	10	1.2
Cookie, chocolate grahams	one	11	1.1
Cookie, chocolate oatmeal	2" diam. X ¼"	14	1.7
Cookie, chocolate snaps	one	4	0.5
Cookie, chocolate wafers	one	6	0.5
Cookie, devils food cake	one	20	2.3
Cookie, nut fudge brownie, vending	1¼ oz. pkg.	35	7.7
DOUGHNUT, cake-type, chocolate	3¼" diam. X 1"	42	5.0
ICE CREAM, chocolate	⅔ C	89	4.5
Ice cream, neopolitan	⅔ C	89	1.8
Ice cream or ice product, drumstick, Sealtest	one	60	3.6
Ice cream or ice product, fudgesickle, Sealtest	one	73	2.9
Ice cream or ice product, ice cream bar, chocolate coated	one	47	3.3
Ice cream or ice product, ice cream sandwich, Sealtest	one	62	0.6
ICING, chocolate, home recipe	1 C	260	59.8
MILK DRINK, chocolate, 1% fat	8 oz. glass	250	7.5
Milk drink, chocolate, 2% fat	8 oz. glass	250	7.5
PIE, chocolate meringue, home recipe	1/7th of 9" diam.	130	7.9
Pie, chocolate chiffon, home recipe	1/7th of 9" diam.	93	9.7
PUDDING, chocolate, instant, w/whole milk	½ C	148	5.5
Pudding, chocolate, low calorie, w/skim milk, D-Zerta	½ C	130	5.1
Pudding, chocolate, w/skim milk, Jello	½ C	147	7.5
Pudding, chocolate, w/whole milk, Jello	½ C	148	7.6

FOOD	Portion	Wt.	Caffeine/ serving g
Pudding, chocolate tapioca, w/whole milk	½ C	148	7.4
Pudding, mix, chocolate	1 pkg.	103	30.4
Pudding, mix, chocolate, instant	1 pkg.	117	22.5
SAUCE, chocolate	2 tablespoons	48	10.6
SYRUP, chocolate fudge, home recipe	1 tablespoon	22	5.6
Syrup, chocolate, home recipe	1 tablespoon	20	4.3
TEA, brewed	6 oz. cup	180	41.4
Tea, instant, prepared	6 oz. cup	173	29.3
TORTE, chocolate	1/16th of 8½" diam.	92	13.8

Weight is given in grams.

Appendix Table B Alcohol Content in Foods

ITEM	Portion	Wt.	Alcohol/ serving g
ALE, mild	12 oz. bottle/can	360	13.9
ALMOND extract	1 teaspoon	5	2.2
BEER, 4.5% alcohol by volume	12 oz. bottle/can	360	13.0
Beer, light, 4.2% alcohol by volume	12 oz. bottle/can	360	12.0
BRANDY, California	1 oz. glass	30	10.5
COCKTAIL, Daiquiri	2½ oz. glass	71	10.7
Cocktail, highball	8 oz. glass	240	24.0
Cocktail, Manhattan	2½ oz. glass	71	13.6
Cocktail, martini	2½ oz. glass	71	13.1
Cocktail, mint julep	10 oz. glass	300	29.2
Cocktail, old-fashioned	2½ oz. glass	71	17.0
Cocktail, Tom Collins	10 oz. glass	300	21.5
Cocktail, sloe gin fizz	2½ oz. glass	71	16.9
EGGNOG w/alcohol	4 oz. glass	123	15.0
GIN, 80-proof	1½ oz. jigger	42	14.0
Gin, 86-proof	1½ oz. jigger	42	15.1
Gin, 90-proof	1½ oz. jigger	42	15.9
Gin, 94-proof	1½ oz. jigger	42	16.7
Gin, 100-proof	1½ oz. jigger	42	17.8
LIQUEUR, anisette	⅔ oz. glass	20	7.0
Liqueur, apricot brandy	⅔ oz. glass	20	6.0
Liqueur, benedictine	⅔ oz. glass	20	6.6
Liqueur, creme de menthe	⅔ oz. glass	20	7.0
Liqueur, Curacao	⅔ oz. glass	20	6.0
RUM, 80-proof	1½ oz. jigger	42	14.0
Rum, 86-proof	1½ oz. jigger	42	15.1
Rum, 90-proof	1½ oz. glass	42	15.9
Rum, 94-proof	1½ oz. jigger	42	16.7
Rum, 100-proof	1½ oz. glass	42	17.8
SHERRY (domestic)	2 oz. glass	60	9.0
VANILLA, pure, double strength, Foote and Jenks	1 teaspoon	5	1.4
VERMOUTH, French, dry	3½ oz. glass	100	15.0
Vermouth, Italian, sweet	3½ oz. glass	100	18.0
VODKA, 80-proof	1½ oz. jigger	42	14.0
Vodka, 86-proof	1½ oz. jigger	42	15.1
Vodka, 90-proof	1½ oz. jigger	42	15.9
Vodka, 94-proof	1½ oz. jigger	42	16.7
Vodka, 100-proof	1½ oz. jigger	42	17.8
WHISKEY, 80-proof	1½ oz. jigger	42	14.0
Whiskey, 86-proof	1½ oz. jigger	42	15.1
Whiskey, 90-proof	1½ oz. jigger	42	15.9
Whiskey, 94-proof	1½ oz. jigger	42	16.7
Whiskey, 100-proof	1½ oz. jigger	42	17.8
WINE, dessert, 18.8% alcohol by volume	3½ oz. glass	105	16.1
Wine, table, 12.2% alcohol by volume	3½ oz. glass	104	10.3

Weight is given in grams.

TABLE 9
List of Abbreviations

alc	alcoholic
approx	approximate
artif	artificial(artificially)
C	8-ounce measuring cup
carb	carbonated
dia	diameter
ea	each
EDTA	a common food preservative (Ethylenediaminetetraacetate)
FAO	Food and Agriculture Organization
FDA	Food and Drug Administration
fl oz	fluid ounce
g, G	gram
gal	gallon
(H)	hydrogenated
IU	International Unit
kcal	kilocalorie
kg	kilogram
kJ	kilojoule
l	liter
lb	pound
lg	large
mcg	microgram
med	medium
meq	milliequivalent
mg	milligram
ml	milliliter
MUFA	monounsaturated fatty acid
NFDM	non-fat dry milk
nonalc	nonalcoholic
noncarb	noncarbonated
oz	ounce
PER	protein efficiency ratio
p	pint
pkg	package
prep	prepared
P/S	polyunsaturated/saturated fatty acid ratio
PUFA	polyunsaturated fatty acid
qt	quart
RDA	recommended daily dietary allowance
RE	retinol equivalent
reg	regular
RTE	ready to eat
sl	slice
sm	small
strnd	strained
TBSP	tablespoon (U.S.)
TM	trademark
tr	trace
tr(a)	trace assumed
tsp	teaspoon (U.S.)
unk	unknown
unstrnd	unstrained
USDA	United States Department of Agriculture
U.S.RDA	United States Recommended Daily Allowance
WHO	World Health Organization
w/	with
w/o	without
"	inches
0(a)	zero amount assumed

TABLE 10
Weights and Measures

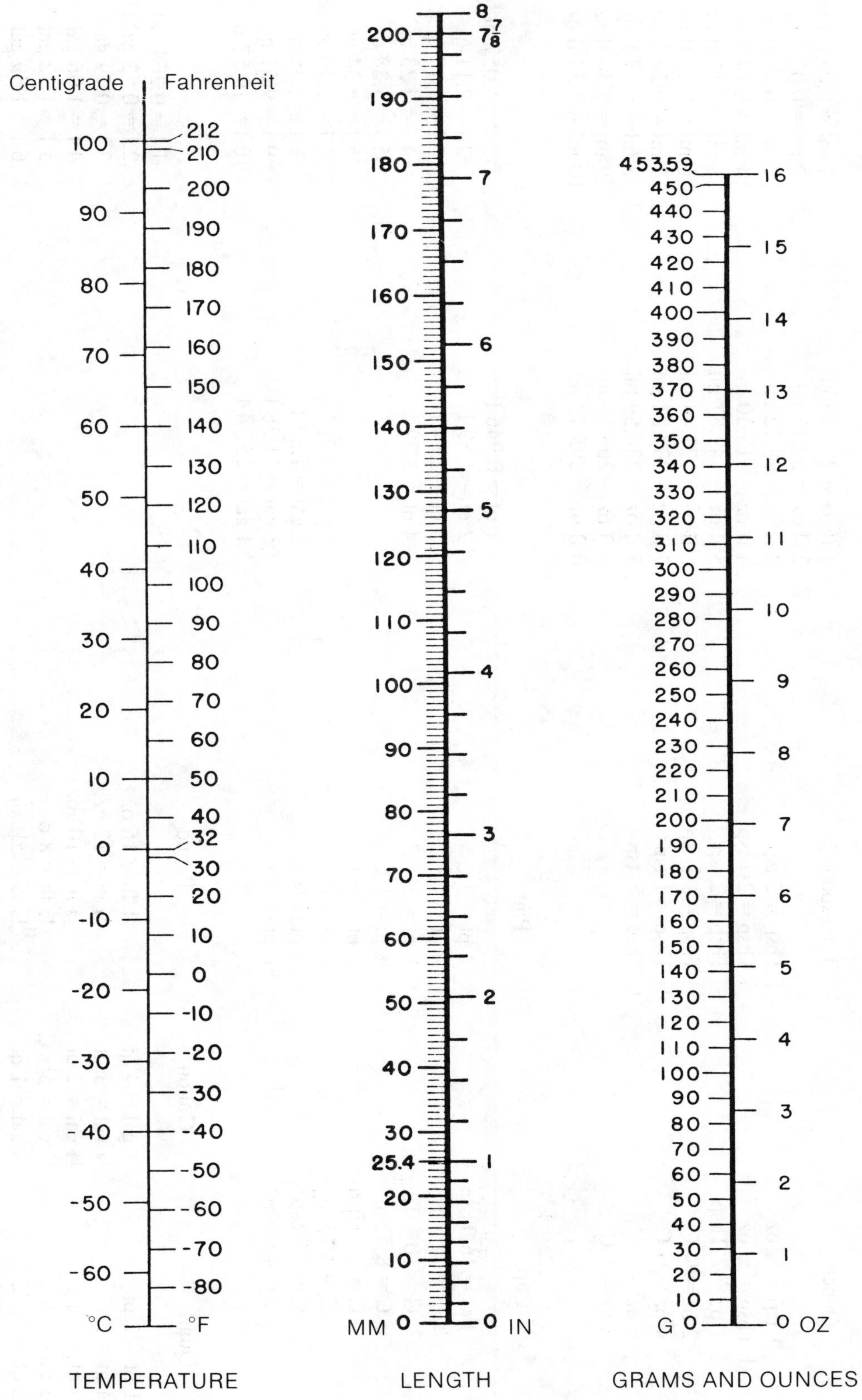

TABLE 10, Cont'd.
Weights and Measures

U.S Kitchen Measures

Ounce	Tablespoon
1 tsp = ⅙ oz	1 Tbsp = 3 tsp
1 Tbsp = ½ oz	¾ Tbsp = 2¼ tsp
1 oz = 2 Tbsp	⅔ Tbsp = 2 tsp
2 oz = ¼ C	½ Tbsp = 1½ tsp
4 oz = ½ C	⅓ Tbsp = 1 tsp
6 oz = ¾ C	¼ Tbsp = ¾ tsp
8 oz = 1 C	

Cup	Pint
1 C = 16 Tbsp	1 pt = 2 C
¾ C = 12 Tbsp	¾ pt = 1½ C
⅔ C = 10⅔ Tbsp	⅔ pt = 1⅓ C
½ C = 8 Tbsp	½ pt = 1 C
⅓ C = 5⅓ Tbsp	⅓ pt = ⅔ C
¼ C = 4 Tbsp	¼ pt = ½ C
⅛ C = 2 Tbsp	⅛ pt = ¼ C
1/16 C = 1 Tbsp	1/16 pt = ⅛ C

Quart	Gallon	Pound
1 qt = 2 pt	1 gal = 4 qt	1 lb = 16 oz
¾ qt = 3 C	¾ gal = 3 qt	¾ lb = 12 oz
⅔ qt = 2⅔ C	½ gal = 2 qt	⅔ lb = 10⅔oz
½ qt = 1 pt	⅓ gal = 5⅓ C	½ lb = 8 oz
⅓ qt = 1⅓ C	¼ gal = 1 qt	⅓ lb = 5⅓ oz
¼ qt = 1 C	⅛ gal = 1 pt	¼ lb = 4 oz
⅛ qt = ½ C	1/16 gal = 1 C	⅛ lb = 2 oz
1/16 qt = ¼ C		1/16 lb = 1 oz

Comparison of U.S. and Metric Units of Liquid Measure

1 fl oz = 29.573 ml	1 ml = 0.034 fl oz
2 fl oz = 59.15 ml	2 ml = 0.07 fl oz
3 fl oz = 88.72 ml	3 ml = 0.10 fl oz
4 fl oz = 118.30 ml	4 ml = 0.14 fl oz
5 fl oz = 147.87 ml	5 ml = 0.17 fl oz
6 fl oz = 177.44 ml	6 ml = 0.20 fl oz
7 fl oz = 207.02 ml	7 ml = 0.24 fl oz
8 fl oz = 236.59 ml	8 ml = 0.27 fl oz
9 fl oz = 266.16 ml	9 ml = 0.30 fl oz
10 fl oz = 295.73 ml	10 ml = 0.34 fl oz

1 qt = 0.946 l	1 l = 1.057 qt
2 qt = 1.89 l	2 l = 2.11 qt
3 qt = 2.84 l	3 l = 3.17 qt
4 qt = 3.79 l	4 l = 4.23 qt
	5 l = 5.28 qt
	6 l = 6.34 qt
1 gal = 3.785 l	7 l = 7.40 qt
2 gal = 7.57 l	8 l = 8.45 qt
3 gal = 11.36 l	9 l = 9.51 qt
4 gal = 15.14 l	10 l = 10.57 qt

1 l = 0.264 gal
2 l = 0.53 gal
3 l = 0.79 gal
4 l = 1.06 gal
5 l = 1.32 gal
6 l = 1.59 gal
7 l = 1.85 gal
8 l = 2.11 gal
9 l = 2.38 gal
10 l = 2.64 gal

TABLE 10, cont'd.
Weights and Measures

Comparison of Avoirdupois and Metric Units of Weight

1 oz = 0.06 lb = 28.35 g	1 lb = 0.454 kg
2 oz = 0.12 lb = 56.70 g	2 lb = 0.91 kg
3 oz = 0.19 lb = 85.05 g	3 lb = 1.36 kg
4 oz = 0.25 lb = 113.40 g	4 lb = 1.81 kg
5 oz = 0.31 lb = 141.75 g	5 lb = 2.27 kg
6 oz = 0.38 lb = 170.10 g	6 lb = 2.72 kg
7 oz = 0.44 lb = 198.45 g	7 lb = 3.18 kg
8 oz = 0.50 lb = 226.80 g	8 lb = 3.63 kg
9 oz = 0.56 lb = 255.15 g	9 lb = 4.08 kg
10 oz = 0.62 lb = 283.50 g	10 lb = 4.54 kg
11 oz = 0.69 lb = 311.85 g	
12 oz = 0.75 lb = 340.20 g	
13 oz = 0.81 lb = 368.55 g	
14 oz = 0.88 lb = 396.90 g	
15 oz = 0.94 lb = 425.25 g	
16 oz = 1.00 lb = 453.59 g	

1 g = 0.035 oz	1 kg = 2.205 lb
2 g = 0.07 oz	2 kg = 4.41 lb
3 g = 0.11 oz	3 kg = 6.61 lb
4 g = 0.14 oz	4 kg = 8.82 lb
5 g = 0.18 oz	5 kg = 11.02 lb
6 g = 0.21 oz	6 kg = 13.23 lb
7 g = 0.25 oz	7 kg = 15.43 lb
8 g = 0.28 oz	8 kg = 17.64 lb
9 g = 0.32 oz	9 kg = 19.84 lb
10 g = 0.35 oz	10 kg = 22.05 lb

Comparison of U.S. and British (Imperial) Kitchen Measures

	U.S.	British
1 fluid ounce	29.57 ml	28.4 l
1 teaspoon	4.7 ml	about 4 ml
1 dessertspoon	——	about 10 ml
1 tablespoon	14.0 ml	about 18 ml
1 standard cup	236.5 ml	about 285 ml
1 pint	473.0 ml	568.3 ml
1 quart	0.946 l	about 1.137 l
1 gallon	3.785 l	4.546 l
1 milliliter	0.0338 fl oz	0.0352 fl oz
1 liter	2.11 pt	1.76 pt

1 ml = 1 g water
1 pint = 1 lb water